PIPELINE RULES OF THUMB HANDBOOK

B 31.8 - Gas
B 31.4 - Liquid
API RP 520

SIXTH EDITION

PIPELINE RULES OF THUMB HANDBOOK

Quick and accurate solutions to your everyday pipeline problems

E.W. McAllister, Editor

AMSTERDAM • BOSTON • HEIDELBERG • LONDON
NEW YORK • OXFORD • PARIS • SAN DIEGO
SAN FRANCISCO • SINGAPORE • SYDNEY • TOKYO
Gulf Professional Publishing is an imprint of Elsevier

GPP

Gulf Professional Publishing is an imprint of Elsevier.
30 Corporate Drive, Suite 400, Burlington, MA 01803, USA
Linacre House, Jordan Hill, Oxford OX2 8DP, UK

Library of Congress Cataloging-in-Publication Data
Pipeline rules of thumb handbook: quick and accurate solutions to your everyday pipeline problems / E.W. McAllister—6th ed.
p. cm.
Includes index.
ISBN- 13: 978-0-7506-7852-0 ISBN-10: 0-7506-7852-6
1. Pipelines—Handbooks, manuals, etc. I. McAllister, E. W.

TJ930.P535 2005
665.5′44—dc22 2004059768

British Library Cataloguing-in-Publication Data
A catalogue record for this book is available from the British Library.

ISBN- 13: 978-0-7506-7852-0
ISBN-10: 0-7506-7852-6

For information on all Gulf Professional Publishing publications
Visit our Web site at www.books.elsevier.com

07 08 09 10 9 8 7 6 5 4

Printed in the United States of America

Contents

2: Construction, 35

3: Pipe Design, 85

8: Corrosion/Coatings, 203

9: Gas—General, 257

10: Gas—Compression, 289

11: Gas—Hydraulics, 313

12: Liquids—General, 341

13: Liquids—Hydraulics, 359

16: Instrumentation, 525

17: Leak Detection, 547

18: Tanks, 557

19: Maintenance, 569

20: Economics, 577

21: Rehabilitation–Risk Evaluation, 627

22: Conversion Factors, 645

1: General Information

Basic Formulas

1. Rate of Return Formulas:

$$S = P(1+i)^n$$

a. Single payment compound amount, SPCA. The $(1+i)^n$ factor is referred to as the compound amount of $1.00.

b. Single payment present worth, SPPW:

$$P = S\left[\frac{1}{(1+i)^n}\right]$$

The factor $[1/(1+i)^n]$ is referred to as the present worth of $1.00.

c. Uniform series compound amount, USCA:

$$S = R\left[\frac{(1+i)^n - 1}{i}\right]$$

$$\text{The factor} = \left[\frac{(1+i)^n - 1}{i}\right]$$

is referred to as the compound amount of $1.00 per period.

d. Sinking fund deposit, SFD:

$$R = S\left[\frac{i}{(1+i)^n - 1}\right]$$

$$\text{The factor} = \left[\frac{i}{(1+i)^n - 1}\right]$$

is referred to as the uniform series, which amounts to $1.00.

e. Capital recovery, CR:

$$R = S\left[\frac{i}{(1+i)^n - 1}\right] = P\left[\frac{i(1+i)^n}{(1+i)^n - 1}\right]$$

$$\text{The factor} = \left[\frac{i(1+i)^n}{(1+i)^n - 1}\right]$$

is referred to as the uniform series that $1.00 will purchase.

f. Uniform series present worth, USPW:

$$P = R\left[\frac{(1+i)^n - 1}{i(1+i)^n}\right]$$

The factor $[((1+i)^n - 1)/i(1+i)^n]$ is referred to as the present worth of $1.00 per period.

where:

P = a present sum of money
S = a sum of money at a specified future date
R = a uniform series of equal end-of-period payments
n = designates the number of interest periods
i = the interest rate earned at the end of each period

Mathematics—areas

Form	Method of Finding Area
Triangle	Base x 1/2 perpendicular height $\sqrt{s(s-a)(s-b)(s-c)}$ s = 1/2 sum of the three sides a,b,c
Trapezium	Sum of area of the two triangles
Trapezoid	1/2 sum of parallel sides x perpendicular height
Parallelogram	Base x perpendicular height
Reg. polygon	1/2 sum of sides x inside radius
Circle	πr^2 = 0.78540 x diam^2. = 0.07958 x circumference^2
Sector of a circle	$\frac{\pi r^2 A^\circ}{360}$ = 0.0087266r^2A° = arc x 1/2 radius
Segment of a circle	$r^2\left(\frac{\pi A^\circ}{180} - \sin A^\circ\right)$
Circle of same area as a square	Diameter = side x 1.12838
Square of same area as a circle	Side = diameter x 0.88623
Ellipse	Long diameter x short diameter x 0.78540
Parabola	Base x 2/3 perpendicular height
Irregular plane surface	The larger the value of n, the greater the accuracy of approximation (Simpson's Rule). $A = \frac{h}{3}\left[(y_o + y_n) + 4(y_1 + y_3 + \ldots + y_{n-1}) + 2(y_2 + y_4 + \ldots\ y_{n-2})\right]$

Mathematics—surfaces and volumes

Method of Finding Surfaces and Volumes of Solids
S = lateral or convex surface V = Volume

Parallelopiped

S = perimeter P perp. to sides x lat. length, l
V = area of base x perpendicular height, h
V = area of section A perp. to sides x lat. length, l

Prism-right, oblique, regular or irregular

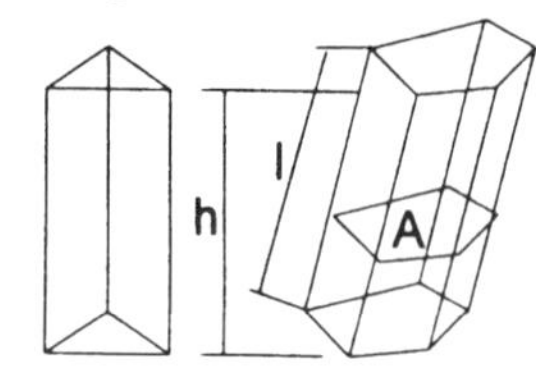

S = perimeter P perp. to sides x lat length l
V = area of base x perp. height, h
V = area of section A perp. to sides x lat. length, l

Cylinder-right, oblique, circular or elliptic, etc.

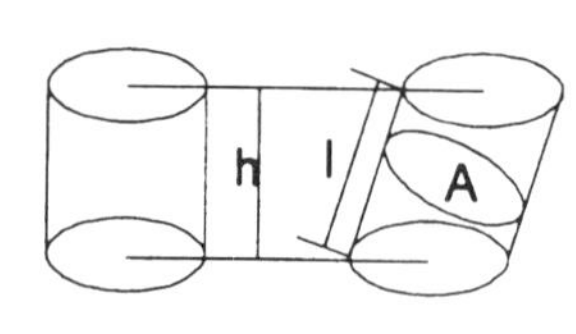

S = perimeter of base x perp. height
V = area of base x perp. height
V = area of section A perp. to sides x lat. length, l

Frustum of any prism or cylinder

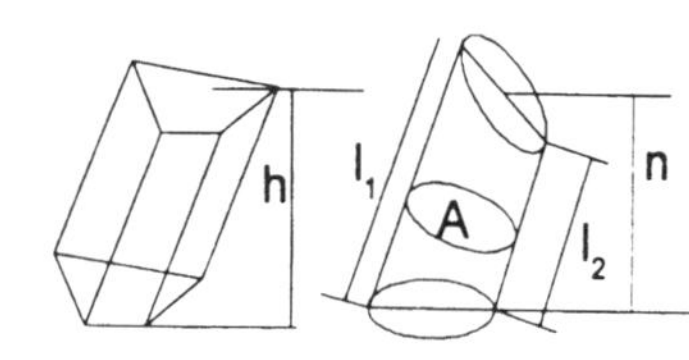

V = area of base x perpendicular distance h, from base to center of gravity of opposite face. For cylinder, $1/2A(l_1+l_2)$

Pyramid or Cone, right or regular

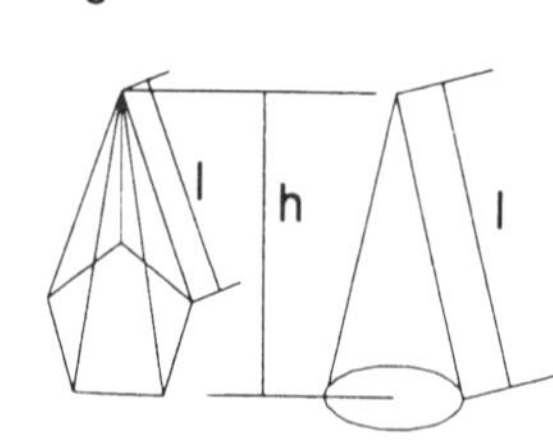

S = perimeter of base x 1/2 slant height, l

V = area of base x 1/3 perp. height, h

Pyramid or Cone-right, oblique, regular or irregular

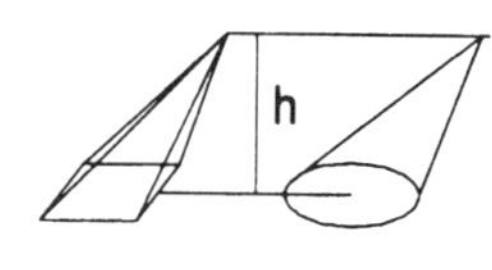

V = area of base x 1/3 perp. height, h
V = 1/3 vol. of prism or cylinder of same base and perp height
V = 1/2 vol. of hemisphere of same base and perp. height

Frustum of pyramid or cone, right and regular, parallel ends

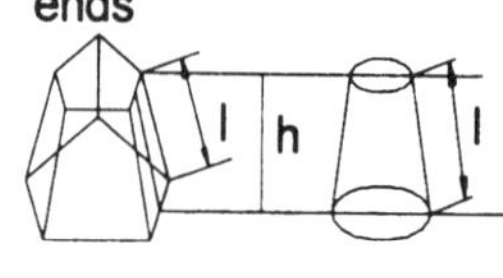

S = (sum of perimeter of base and top) x 1/2 slant height, l

$$V = \frac{h}{3}\left(A_1 + A_2 + \sqrt{A_1 x A_2}\right)$$

Where A_1 and A_2 are the areas of the bases

Frustum of any pyramid or cone, parallel ends

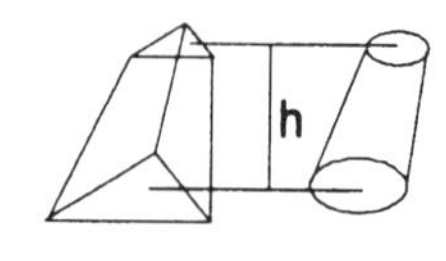

$$V = \frac{h}{3}\left(A_1 + A_2 + \sqrt{A_1 x A_2}\right)$$

Where A_1 and A_2 are the areas of the bases

Sphere

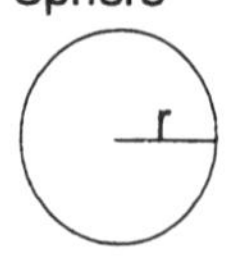

$S = 4\pi r^2$
$V = 4/3\pi r^3$

Spherical Sector

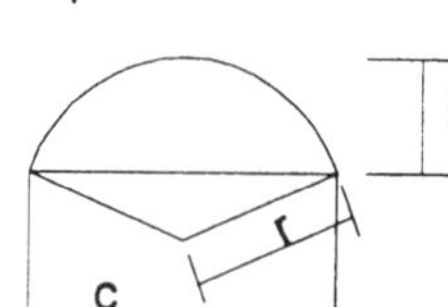

$V = 2/3\pi r^2 b$
$S = 1/2\pi r(4b+c)$

Spherical Zone

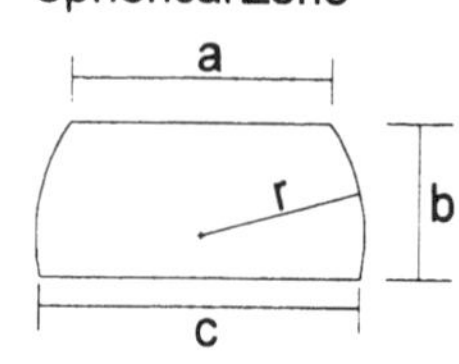

$S = 2\pi rb$
$V = 1/24\pi b(3a^2+3c^2+4b^2)$

CABLES AND ROPES

Estimating strength of cable

Rule.

1. Change line diameter to eighths
2. Square the numerator
3. Divide by the denominator
4. Read the answer in tons

Example. Estimate the strength of ½-in. steel cable:

$$\text{Diameter} = \frac{1}{2} = \frac{4}{8}$$

$$\frac{4^2}{8} = \frac{16}{8} = 2$$

The approximate strengh of ½-in. steel cable is 2 tons.

Find the working strength of Manila rope

The working strength of Manila rope is approximately $900 \times (\text{diameter})^2$:

$$W = 900\,d^2$$

where d is expressed in inches.

W is given in pounds.

Example. What is the working strength of a ¾-in. Manila rope?

The maximum recommended pull is:

$$W = 900 \times \frac{3 \times 3}{4 \times 4} = 506\,\text{lb.}$$

Example. Find the maximum working pull for a 1½-inch Manila rope.

$$W = 900 \times \frac{3}{2} \times \frac{3}{2} = 2{,}025\ \text{lb.}$$

For rope diameters greater than 2 in., a factor lower than 900 should be used. In working with heavier rigging it is advisable to refer to accepted handbooks to find safe working strengths.

How large should drums and sheaves be for various types of wire rope?

The diameter of sheaves or drums should preferably fall within the table* given below for most efficient utilization of the wire rope.

Type of Wire Rope	6 × 19	6 × 37	8 × 19	5 × 28	6 × 25	18 × 7	6 × 7
For best wear	45	27	31	36	45	51	72
Good practice	30	18	21	24	30	34	42
Critical	16	14	14	16	16	18	28

Example. What size should the hoisting drum on a dragline be, if the wire rope is 6 × 19 construction, ¾ in. in diameter?

From the table, good practice calls for 30 diameters, which in this instance would be 22½ in. Loads, speeds, bends, and service conditions will also affect the life of wire rope, so it is better to stay somewhere between the "good practice" and "best wear" factors in the table.

**Construction Methods and Machinery*, by F. H. Kellogg, Prentice-Hall, Inc., 1954.

Find advantages of block and tackle, taking into account pull out friction

The efficiency of various sheaves differs. For one with roller bearings the efficiency has been estimated at 96.2%. For plain bearing sheaves a commonly used figure is 91.7%. The following formula will give close results:

$$MA = \frac{W}{w} = E\frac{1 - E^n}{1 - E}$$

where: MA = Mechanical Advantage
W = Total weight to be lifted by the assembly
w = Maximum line pull at the hoist
n = Number of working parts in the tackle
E = Efficiency of individual sheaves

It is assumed that the line leaving the upper block goes directly to the hoist without additional direction change (requiring a snatch block).

Example. Find the Mechanical Advantage of a four-part block and tackle using upper and lower blocks having journal bearings, which have an efficiency of 91.7%.

$$MA = .917\frac{1 - .917^4}{1 - .917} = .917$$

$$\left(\frac{1 - .707^4}{1 - .917}\right) = .917\left(\frac{.293}{.083}\right)$$

MA = 3.25

If the load weighed 3,250 lb., what pull would be required on the lead line?

$$\frac{W}{w} = MA$$

$$\frac{3{,}250}{w} = 3.25$$

w = 1,000 lb.

Safe loads for wire rope

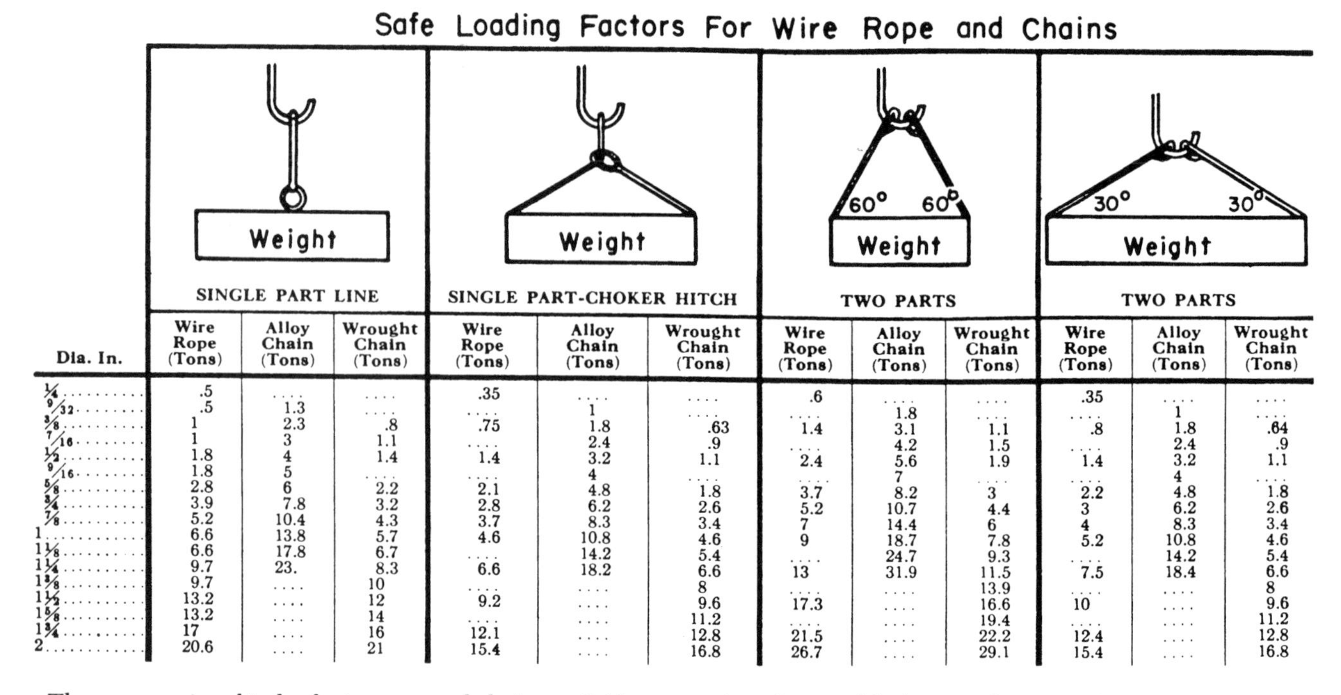

Safe Loading Factors For Wire Rope and Chains

Dia. In.	SINGLE PART LINE			SINGLE PART-CHOKER HITCH			TWO PARTS (60°)			TWO PARTS (30°)		
	Wire Rope (Tons)	Alloy Chain (Tons)	Wrought Chain (Tons)	Wire Rope (Tons)	Alloy Chain (Tons)	Wrought Chain (Tons)	Wire Rope (Tons)	Alloy Chain (Tons)	Wrought Chain (Tons)	Wire Rope (Tons)	Alloy Chain (Tons)	Wrought Chain (Tons)
1/4	.5	....	....	.35	....	....	.6	....	....	.35	....	....
9/32	.5	1.3	....	....	1	....	....	1.8	....	....	1	....
3/8	1	2.3	.8	.75	1.8	.63	1.4	3.1	1.1	.8	1.8	.64
7/16	1	3	1.1	....	2.4	.9	....	4.2	1.5	....	2.4	.9
1/2	1.8	4	1.4	1.4	3.2	1.1	2.4	5.6	1.9	1.4	3.2	1.1
9/16	1.8	5	....	....	4	....	....	7	....	....	4	....
5/8	2.8	6	2.2	2.1	4.8	1.8	3.7	8.2	3	2.2	4.8	1.8
3/4	3.9	7.8	3.2	2.8	6.2	2.6	5.2	10.7	4.4	3	6.2	2.6
7/8	5.2	10.4	4.3	3.7	8.3	3.4	7	14.4	6	4	8.3	3.4
1	6.6	13.8	5.7	4.6	10.8	4.6	9	18.7	7.8	5.2	10.8	4.6
1 1/8	6.6	17.8	6.7	....	14.2	5.4	....	24.7	9.3	....	14.2	5.4
1 1/4	9.7	23.	8.3	6.6	18.2	6.6	13	31.9	11.5	7.5	18.4	6.6
1 3/8	9.7	....	10	....	....	8	....	....	13.9	....	....	8
1 1/2	13.2	....	12	9.2	....	9.6	17.3	....	16.6	10	....	9.6
1 5/8	13.2	....	14	....	....	11.2	....	....	19.4	....	....	11.2
1 3/4	17	....	16	12.1	....	12.8	21.5	....	22.2	12.4	....	12.8
2	20.6	....	21	15.4	....	16.8	26.7	....	29.1	15.4	....	16.8

There are various kinds of wire rope and chain available, and where it is possible strength factors should be reviewed before using any particular kind. Tables here may be used on construction jobs. They allow for a safety factor of four as based on mild plow steels, etc. Where the wire rope is made of improved plow steel, etc., it will have a greater safety factor.

Stress in guy wires

Guys are wire ropes or strands used to hold a vertical structure in position against an overturning force. The most common types of guyed structures are stacks, derricks, masts for draglines, reversible tramways and radio transmission towers.

As a general rule, stresses in guys from temperature changes are neglected, but in structures such as radio masts this is an important feature and must be subject to special analysis.

The number of guys used for any particular installation is contingent on several variable factors such as type of structure, space available, contour of the ground, etc., and is not a part of this discussion.

It is desirable to space guys uniformly whenever possible. This equalizes the pull, P, on each guy insofar as possible, particularly against forces that change in direction, as when a derrick boom swings in its circle.

It is also desirable to equalize the erection tensions on the guys. When no external force is acting on the structure, the tension in each guy should be the same. A "Tension Indicator" is sometimes used to determine the tension in guys. If this instrument is not available, the tension can be very closely approximated by measuring the deflection at the center of the span from the chord drawn from the guy anchorage to the point of support on the structure. A good average figure to use for erection tension of guys is 20% of the maximum working tension of the guy.

This discussion outlines the method for determining the stresses in guys. One of the first considerations is the location of the guy anchorages. The anchorages should be so located that the angle α, between the horizontal plane and the guy line, is the same for all guys (to equalize erection tensions). Angle α, in good practice, seldom exceeds 45 degrees with 30 degrees being commonly used. The tension in the guys decreases as angle α becomes less. The direct load on the structure is also less with a smaller value of α.

To find the maximum extra tension, T, that will be applied to any single guy by the force, F; first, determine the pull, P, which is the amount required along the guys, in the same vertical plane as the force to resist the horizontal component

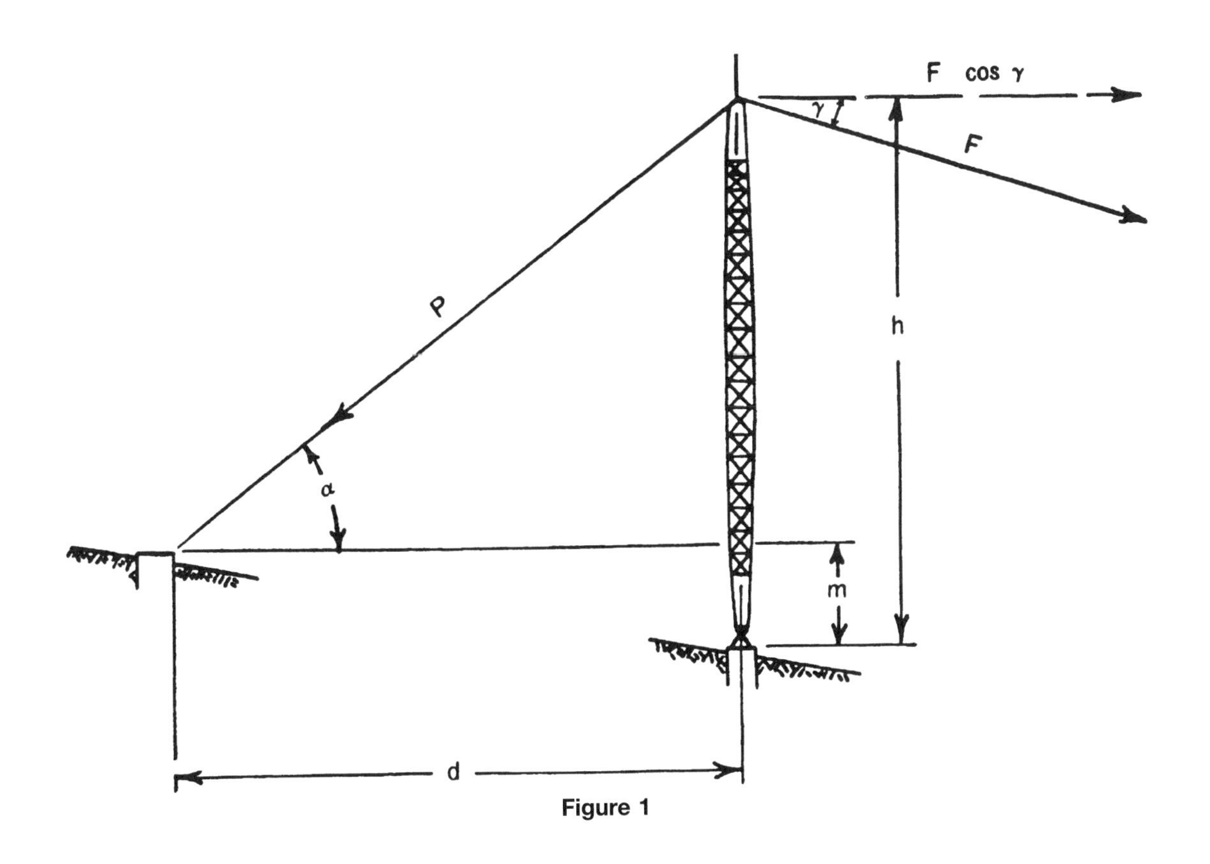

Figure 1

of the force. This pull is entirely independent of the number of guys. Assume that the following are known:

F = The total resultant external force acting on the structure
G = The angle between the horizontal plane and the force F
h = The height of the structure
d = The horizontal distance from structure to guy anchorage
m = The vertical height of anchorage above or below the base of the structure
The horizontal component of the force, F, = F cos γ.
α = The angle whose tangent is $(h \pm m) \div d$.
m is plus if the anchorage is below the base of the structure and subtracted if it is above.

$P = F \cos \gamma \div \cos \alpha$

As cos α is always less than 1, P is always greater than F cos γ, the horizontal component of force F.

It must be remembered that P represents the total pull acting along the guys at an angle, α, with the horizontal, and in the same vertical plane as the force, F.

If only one guy were used, P would represent the extra tension, T. In practice, however, a number of guys are always used and, therefore, the pull on any one guy will not be equal to P. The following table gives factors for any number of guys from 3 to 15, equally spaced about a central structure. To find the maximum extra tension, T, that will be applied to any single guy by the force, F, capable of rotating 360 degrees around a vertical axis, it is only necessary to multiply the value of P, as determined above, by the factor for the number of guys used. It must be clearly understood in using this table that the guys are uniformly spaced and under equal tension when no load is acting on the structure.

Table 1

No. of Guys	Factors*	No. of Guys	Factors*
3	1.15	10	0.45
4	1.00	11	0.40
5	0.90	12	0.37
6	0.75	13	0.35
7	0.65	14	0.32
8	0.55	15	0.30
9	0.50		

*These factors are for average conditions. If the guys are erected under accurately measured tensions of not less than 20% of the working load, the factors for five or more guys may be reduced by 10%. If the erecting tensions are low or not accurately equalized, the factors for five or more guys should be increased 10%.

Example. A derrick mast 90 ft high is supported by nine equally spaced guys anchored at a horizontal distance of 170 ft from the mast and the elevations of the guy anchorages are 10 ft below the base of the mast. The load on the structure is equivalent to a force of 10,000 lb., acting on an angle of 10 degrees below the horizontal. What is the maximum pull on any single cable?

From Figure 1—

h = 90 ft
d = 170 ft
m = 10 ft
g = 10°00′
F = 10,000 lb.

$$\tan \alpha = \frac{90 + 10}{170} = 100 = 0.588$$

$$\alpha = 30°28'$$

$$P = \frac{F \cos \gamma}{\cos \alpha} = \frac{10{,}000 \times 0.985}{0.862} = 11{,}427 \text{ lb.}$$

From Table 1, $T = 11{,}427 \times 0.50 = 5{,}714$ lb.

If erection tension is 10% of total working tension, 5,714 is 90% of total working tension. Therefore, working tension = $(5{,}714 \times 100)/90 = 6{,}349$ lb.

Strength and weight of popular wire rope

The following tables give the breaking strength for wire rope of popular construction made of improved plow steel.

	6 × 19	
SIZE	Breaking Strength	Weight
1/4	5,480	0.10
5/16	8,520	0.16
3/8	12,200	0.23
7/16	16,540	0.31
1/2	21,400	0.40
9/16	27,000	0.51
5/8	33,400	0.63
1 3/4	47,600	0.90
7/8	64,400	1.23
1	83,600	1.60
1 1/8	105,200	2.03
1 1/4	129,200	2.50
1 3/8	155,400	3.03
1 1/2	184,000	3.60
1 5/8	214,000	4.23
1 3/4	248,000	4.90
1 7/8	282,000	5.63
2	320,000	6.40

Conversion factors for wire rope of other construction

To apply the above table to wire rope of other construction, multiply by the following factors:

Wire Rope Construction	6 × 19	6 × 29	6 × 37	18 × 7
Strength Factors	1.00	0.96	0.95	0.92
Weight Factors	1.00	0.97	0.97	1.08

Example. Find the breaking strength of 6 × 29 improved plow steel wire rope 2 in. in diameter.

Strength = 320,000 × 0.96 = 307,000 lb.

The weight can be found the same way.

Measuring the diameter of wire rope

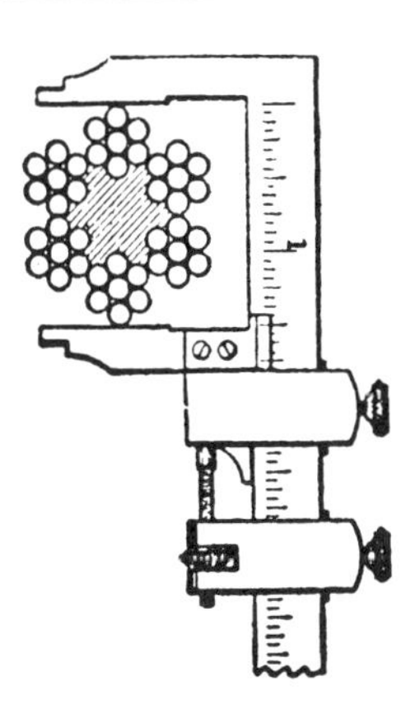

Incorrect way

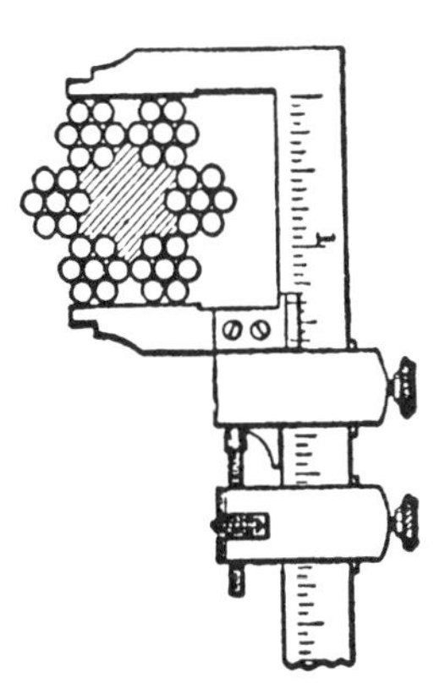

Wire rope: field troubles and their causes

All wire rope will eventually deteriorate in operation or have to be removed simply by virtue of the loads and reversals of load applied in normal service. There are, however, many conditions of service or inadvertent abuse that will materially shorten the normal life of a wire rope of proper construction although it is properly applied. The following field troubles and their causes give some of the field conditions and practices that result in the premature replacement of wire rope. It should be borne in mind that in all cases the contributory cause of removal may be one or more of these practices or conditions.

	Wire-Rope Trouble	Cause
a.	Rope broken (all strands).	Overload resulting from severe impact, kinking, damage, localized wear, weakening of one or more strands, or rust-bound condition and loss of elasticity.
b.	One or more whole strands parted.	Overloading, kinking, divider interference, localized wear, or rust-bound condition. Fatigue, excessive speed, slipping, or running too loosely. Concentration of vibration at dead sheave or dead-end anchor.
c.	Excessive corrosion.	Lack of lubrication. Exposure to salt spray, corrosive gases, alkaline water, acid water, mud, or dirt. Period of inactivity without adequate protection.
d.	Rope damage in hauling to the well or location.	Rolling reel over obstructions or dropping from car, truck, or platform. The use of chains for lashing, or the use of lever against rope instead of flange. Nailing through rope to flange.
e.	Damage by improper socketing.	Improper seizing, which allows slack from one or more strands to work back into rope; improper method of socketing or poor workmanship in socketing, frequently shown by rope being untwisted at socket, loose or drawn.
f.	Kinks, dog legs, and other distorted places.	Kinking the rope and pulling out the loops such as in improper coiling or unreeling. Improper winding on the drum. Improper tie-down. Open-drum reels having longitudinal spokes too widely spaced. Divider interference. The addition of improperly spaced cleats to increase the drum diameter. Stressing while rope is over small sheave or obstacle.
g.	Damage by hooking back slack too tightly to girt.	Operation of walking beam causing a bending action on wires at clamp and resulting in fatigue and cracking of wires, frequently before rope goes down into hole.
h.	Damage or failure on a fishing job.	Rope improperly used on a fishing job, resulting in damage or failure as a result of the nature of the work.
i.	Lengthening of lay and reduction of diameter.	Frequently produced by some type of overloading, such as an overload resulting in a collapse of the fiber core in swabbing lines. This may also occur in cable-tool lines as a result of concentrated pulsating or surging forces, which may contribute to fiber-core collapse.
j.	Premature breakage of wires.	Caused by frictional heat developed by pressure and slippage, regardless of drilling depth.
k.	Excessive wear in spots.	Kinks or bends in rope due to improper handling during installation or service. Divider interference; also, wear against casing or hard shells or abrasive formations in a crooked hole. Too infrequent cut-offs on working end.
l.	Spliced rope.	A splice is never as good as a continuous piece of rope, and slack is liable to work back and cause irregular wear.
m.	Abrasion and broken wires in a straight line. Drawn or loosened strands. Rapid fatigue breaks.	Injury due to slipping rope through clamps.
n.	Reduction in tensile strength or damage to rope.	Excessive heat due to careless exposure to fire or torch.
o.	Distortion of wire rope.	Damage due to improperly attached clamps or wire-rope clips.
p.	High strands.	Slipping through clamps, improper seizing, improper socketing or splicing, kinks, dog legs, and core popping.
q.	Wear by abrasion.	Lack of lubrication. Slipping clamp unduly. Sandy or gritty working conditions. Rubbing against stationary object or abrasive surface. Faulty alignment. Undersized grooves and sheaves.
r.	Fatigue breaks in wire.	Excessive vibration due to poor drilling conditions, i.e., high speed, rope slipping, concentration of vibration at dead sheave or

Wire-Rope Trouble	Cause
	dead-end anchor, undersized grooves and sheaves, and improper selection of rope construction. Prolonged bending action over spudder sheaves, such as that due to hard drilling.
s. Spiraling or curling.	Allowing rope to drag or rub over pipe, sill, or any object during installation or operation. It is recommended that a block with sheave diameter 16 times the nominal wire-rope diameter, or larger, be used during installation of the line.
t. Excessive flattening or crushing.	Heavy overload, loose winding on drum, or cross winding. Too infrequent cut-offs on working end of cable-tool lines. Improper cutoff and moving program for cable-tool lines.
u. Bird-caging or core-popping.	Sudden unloading of line such as hitting fluid with excessive speed. Improper drilling motion or jar action. Use of sheaves of too small diameter or passing line around sharp bend.
v. Whipping off of rope.	Running too loose.
w. Cutting in on drum.	Loose winding on drum. Improper cutoff and moving program for rotary drilling lines. Improper or worn drum grooving or line turn-back late.

Capacity of drums

The capacity of wire line drums may be figured from the following formula:

$$M = (A + B) \times A \times C \times K$$

where: M = rope capacity of drum, inches
A = depth of flange, inches
B = diameter of drum, inches
C = width of the drum between flanges, inches
K = constant depending on rope size shown below

Rope Size	K	Rope Size	K
3/8	1.86	1	.262
7/16	1.37	1 1/8	.207
1/2	1.05	1 1/4	.167
9/16	.828	1 3/8	.138
5/8	.672	1 1/2	.116
3/4	.465	1 5/8	.099
7/8	.342	1 3/4	.085

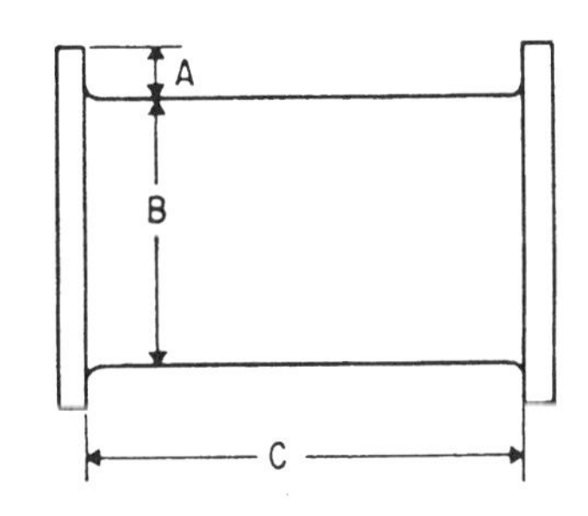

Figure 1. Capacity of wire line drums.

BELTS AND SHAFTS

Determine length of a V-belt

Rule: To find the nominal length of a V-belt, lay the belt on a table and place within it a pair of circular objects of the same diameter (flanges, tin cans, or whatever may be at hand). Pull them apart until the belt is fully extended without stretching. Then measure the shortest distance between the two circles, in inches.

The belt size is then twice this figure, plus 5.14 times the diameter of the circles. This value, for any standard belt, should be a whole number of inches, which is the belt size.

V-belts are made in four standard sections, classified as A, B, C, and D; the widths (at the widest part) are ½, ⅝, ⅞ and 1⅛ in., respectively. The complete designation of the belt is the letter showing the width, followed by the length in inches; thus, an A26 belt is ½-in. wide and 26 in. long on the inside edge. The *pitch length* of the belt is measured along a median section and corresponds to the length that runs on the pulley diameter, which determines the actual speed ratio—about half of the depth of the groove. Pitch lengths for A, B, C, and D belts are greater than their nominal lengths by 1.3, 1.8, 2.9, and 3.3 in., respectively.

Calculate stress in shaft key

The shear and compressive stresses in a key are calculated using the following equations:

$$Ss = \frac{2T}{d \times W \times L} \qquad Sc = \frac{2T}{d \times h_1 \times L}$$

Ss = Shear stress in psi
Sc = Compressive stress in psi
T = Shaft torque pounds-inches or
d = shaft diameter-inches
(For taper shafts, use average diameter)

w = width key-inches
L = effective length of key-inches
h_1 = height of key in the shaft or hub that bears against the keyway—inches
$h_1 = h_2$ for square keys. For designs where unequal portions of the key are in the hub or shaft, h_1 is the minimum portion.

Key material is usually AISI 1018 or AISI 1045 with the following allowable stresses:

Material	Heat Treatment	Allowable Stresses—psi	
		Shear	Compressive
AISI 1018	None	7,500	15,000
AISI 1045	225–300 Bhn	15,000	30,000

Example. Determine key stresses for these conditions: 300 hP @ 600 RPM; 3″ dia. shaft, ¾ × ¾ key, 4″ key engagement length.

$$T = \text{Torque} = \frac{HP \times 63{,}000}{RPM} = \frac{300 \times 63{,}000}{600} = 31.500 \text{ in.-lbs}$$

$$Ss = \frac{2T}{d \times W \times L} = \frac{2 \times 31{,}500}{3 \times 3/4 \times 4} = 7{,}000 \text{ psi}$$

$$Sc = \frac{2T}{d \times h_1 \times L} = \frac{2 \times 31{,}500}{3 \times 3/8 \times 4} = 14{,}000 \text{ psi}$$

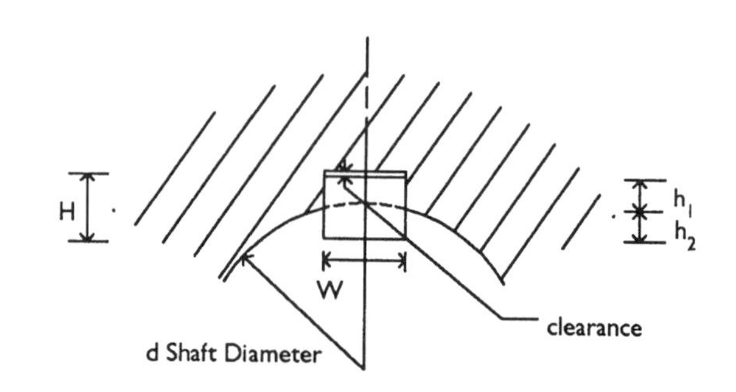

The AISI 1018 key should be used since it is within allowable stresses. *Note:* If shaft had been 2 ¾″ dia. (4″ long hub), the key would be ⅝ × ⅝ and Ss = 9,200 psi, Sc = 18,400 psi, and a heat treated key—AISI 1045 would have been required.

Reprinted with permission: The Falk Corporation

Calculate V-belt length using simple equation

Rule.

$$L = 1.57(D + d) + 2C$$

where: L = Belt length, inches
D = Diameter of larger sheave, inches
d = Diameter of smaller sheave, inches
C = Distance between sheave centers, inches

Estimate the horsepower that can be transmitted by a shaft

1. Where there are no stresses due to bending, weight of the shaft, pulleys, gears, or sprockets, use:

$$HP = \frac{D^3N}{50}$$

where:

D = diameter of shaft, inches
N = revolutions per minute

2. For heavy duty service use:

$$HP = \frac{D^3N}{125}$$

Example. What horsepower can be transmitted to an atmospheric cooling coil by a two-inch shaft turning at 1,800 revolutions per minute?

$$HP = \frac{(2)^3(1{,}800)}{125} = 115.2 \text{ horsepower}$$

MISCELLANEOUS

How to estimate length of material contained in roll

Where material of uniform thickness, like belting, is in a roll, the total length may be obtained by the following rule:

Measure the diameter of the hole in the center, and of the outside of the roll, both measurements in inches; count the number of turns; multiply the sum of the two measured diameters by the number of turns, and multiply this product by 0.13; the result is the total length of the material in feet.

Example. A roll of belting contains 24 turns. The diameter of the hole is 2 in., and of the outside of the roll is 13 in.

$(2 + 13) \times 24 \times 0.13 = 46.8$

The roll contains 46.8 feet of belting.

Note: The rule can even be applied to materials as thin as pipeline felt; counting the turns is not as difficult as might appear without a trial.

Convenient antifreeze chart for winterizing cooling systems

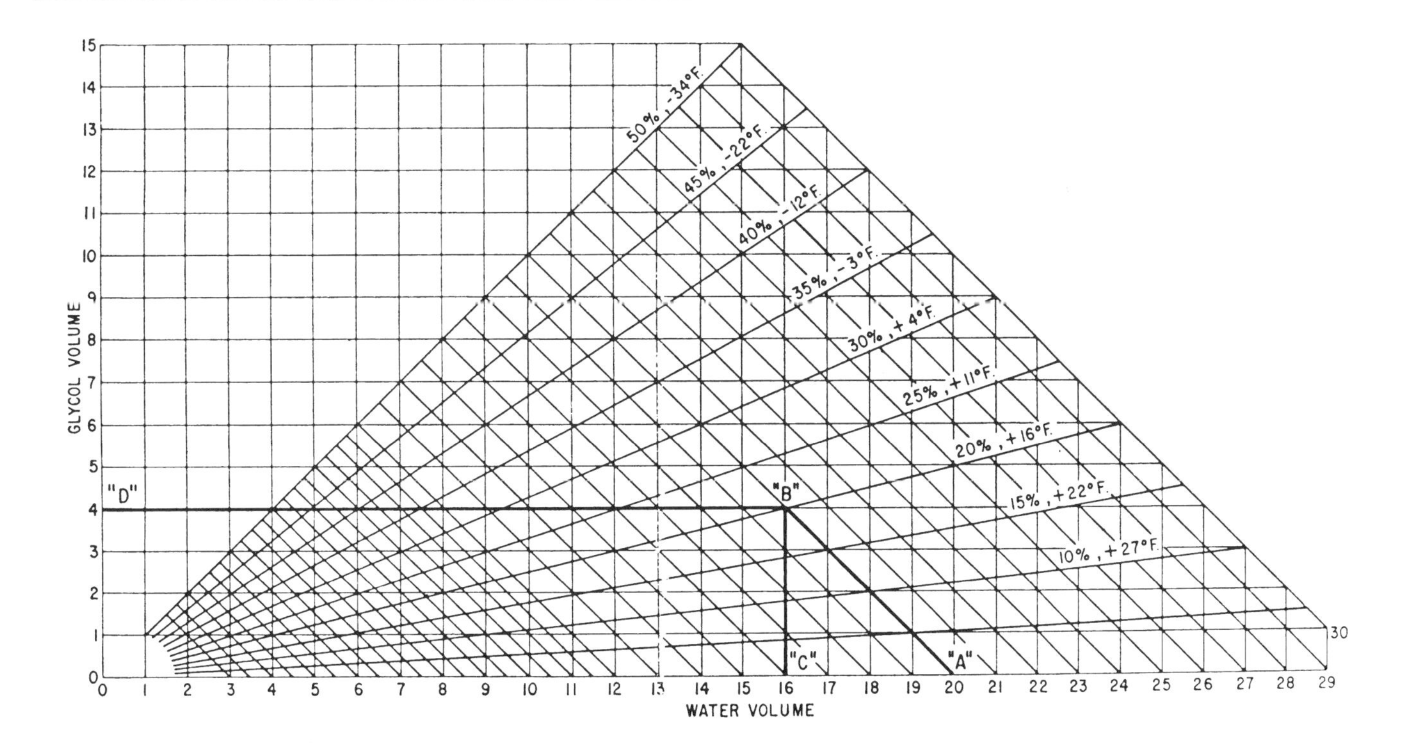

This plot of water volume versus glycol volume at various conditions of temperature and percent of glycol in the system makes winterizing field engines relatively easy.

Example. Determine the amount of glycol to be used in a 200 gallon system for protection to 16°F.

To use the chart, first find the total system capacity, 200 gallons, at Point A on the water volume axis. Point A also can represent 2, 20, 2,000, etc. Proceed along the 45° line from Point A to Point B at the intersection with the 16°F line. Then, move from Point B to Point C and read 160 on the water volume axis. This is the amount of water the system should contain. Move from Point B to Point D on the glycol axis and read 40. This is the amount of glycol that must be added to the system.

How to determine glycol requirements to bring a system to a desired temperature protection level

Solve the equation below to find the number of gallons of aqueous solution that must be removed and replaced with glycol.

$$d = \frac{D}{B}$$

where:

d = gallons of aqueous solution to be removed from the system and replaced with an equal number of gallons of glycol

D = total gallons of glycol required for protection at desired temperature less gallons of glycol in original system

B = 1.0 – fraction of glycol in system

Example. Find the number of gallons of existing aqueous solution that must be removed and replaced with glycol in a 1,200 gallon system to give protection to −10°F. Glycol in the system determined by a hydrometer test is 28% (0.28 fraction). The published figure for −10°F protection is 519 gallons of glycol in the 1,200 gallon system.

Then: Gallons of glycol in system

$= 1{,}200 \times 0.28 = 336$ gallons

$D = 519 - 336 = 183$ gallons difference

and $B = 1.0 - 0.28 = 0.72$

Thus: $d = \dfrac{183}{0.72} = 254$ gallons

Therefore, 254 gallons of existing aqueous solution need to be removed and replaced with 254 gallons of glycol for the desired protection to −10°F.

Weight in pounds of round steel shafting

Diameter of Shaft, Inches	LENGTH IN FEET																	Weight Per Inch
	1	2	3	4	5	6	7	8	9	10	12	14	16	18	20	22	24	
15/16	2	5	7	9	12	14	16	19	21	23	28	33	37	42	47	51	56	.195
1 3/16	4	8	11	15	19	23	26	30	34	38	45	53	60	68	76	83	91	.314
1 7/16	6	11	17	22	28	33	39	44	50	55	66	77	88	99	110	121	133	.460
1 11/16	8	15	23	30	38	46	53	61	68	76	91	107	122	137	152	167	183	.634
1 15/16	10	20	30	40	50	60	70	80	90	100	120	140	161	181	201	221	241	.835
2 3/8	13	26	38	51	64	77	90	102	115	128	153	179	205	230	256	281	307	1.06
2 7/16	16	32	48	63	79	95	111	127	143	159	190	222	254	286	317	349	381	1.32
2 11/16	19	39	58	77	97	116	135	154	174	193	232	270	309	348	386	425	463	1.61
2 15/16	23	46	69	92	115	138	161	184	208	231	277	323	369	415	461	507	553	1.92
3 3/16	27	54	81	109	136	163	190	217	244	272	326	380	435	489	543	598	652	2.26
3 7/16	32	63	95	126	158	189	221	253	284	316	379	442	505	568	631	695	758	2.63
3 11/16	36	73	109	146	182	218	255	291	328	364	437	510	582	655	728	801	874	3.03
3 15/16	41	83	124	166	207	248	290	331	373	414	497	580	662	745	828	911	994	3.45
4 7/16	53	105	158	210	263	315	368	421	473	526	631	736	841	946	1052	1157	1262	4.38
4 15/16	65	130	195	260	326	391	456	521	586	651	781	911	1041	1172	1302	1432	1562	5.43
5	67	134	200	267	334	401	467	534	601	668	801	935	1068	1202	1335	1469	1602	5.56
5 7/16	79	158	237	316	395	474	553	632	711	790	947	1105	1263	1421	1579	1737	1894	6.58
5 1/2	81	162	242	323	404	485	565	646	727	808	969	1131	1292	1454	1615	1777	1938	6.73
5 15/16	94	188	282	377	471	565	659	753	847	941	1130	1318	1506	1695	1883	2071	2259	7.85
6	96	192	288	385	481	577	673	769	865	961	1154	1346	1538	1731	1923	2115	2307	8.01
6 1/2	113	226	338	451	564	677	790	903	1015	1128	1354	1579	1805	2031	2256	2482	2708	9.40
7	131	262	393	524	654	785	916	1047	1178	1309	1570	1832	2094	2355	2677	2878	3140	10.90
7 1/2	150	300	451	601	751	901	1051	1202	1352	1502	1803	2103	2403	2704	3004	3305	3605	12.52
8	171	342	513	684	855	1024	1196	1367	1538	1709	2051	2303	2734	3076	3418	3760	4102	14.24

Properties of shafting

Diameter of Shaft, Inches	Area of Shaft	Weight per Inch	Weight per Foot	SECTION MODULUS		MOMENT OF INERTIA		Radius of Gyration
				Bending	Torsion	Bending	Torsion	
15/16	0.6903	0.195	2.34	0.0809	0.1618	0.0379	0.0758	0.2343
1 3/16	1.1075	0.314	3.76	0.1644	0.3288	0.0976	0.1952	0.2969
1 7/16	1.6230	0.460	5.52	0.2916	0.5832	0.2096	0.4192	0.3594
1 11/16	2.2365	0.634	7.61	0.4718	0.9435	0.3981	0.7961	0.4219
1 15/16	2.9483	0.835	10.03	0.7140	1.4281	0.6971	1.3835	0.4844
2 3/16	3.7583	1.06	12.78	1.0276	2.0553	1.1240	2.2480	0.5469
2 7/16	4.6664	1.32	15.86	1.4218	2.8436	1.7328	3.4656	0.6094
2 11/16	5.6727	1.61	19.29	1.9057	3.8113	2.5607	5.1215	0.6719
2 15/16	6.7771	1.92	23.04	2.4885	4.9770	3.6549	7.3099	0.7344
3 3/16	7.9798	2.26	27.12	3.1794	6.3589	5.0672	10.1345	0.7969
3 7/16	9.2806	2.63	31.56	3.9878	7.9755	6.8539	13.7079	0.8594
3 11/16	10.680	3.03	36.31	4.9226	9.8452	9.0761	18.1521	0.9219
3 15/16	12.177	3.45	41.40	5.9932	11.9865	11.7992	23.5984	0.9844
4 7/16	15.466	4.38	52.58	8.5786	17.1571	19.0337	38.0674	1.1094
4 15/16	19.147	5.43	65.10	11.8174	23.6348	29.1742	58.3483	1.2344
5	19.635	5.56	66.76	12.2718	24.5437	30.6796	61.3592	1.2500
5 7/16	23.221	6.58	78.96	15.7833	31.5666	42.9108	85.8217	1.3594
5 1/2	23.758	6.73	80.78	16.3338	32.6677	44.9180	89.8361	1.3750
5 15/16	27.688	7.85	94.14	20.5499	41.0999	61.0077	122.0153	1.4844
6	28.274	8.01	96.12	21.2058	42.4115	63.6172	127.2345	1.5000
6 1/2	33.183	9.40	112.82	26.9612	53.9225	87.6240	175.2481	1.6250
7	38.485	10.90	130.84	33.6739	67.3479	117.8588	235.7176	1.7500
7 1/2	44.179	12.52	150.21	41.4175	82.8350	155.3155	310.6311	1.8750
8	50.265	14.24	170.90	50.2655	100.5310	201.0619	402.1239	2.0000

Steel shaft sizes calculated for strength

Base Fiber Stress = 60,000 lb. sq. in.

$$\text{Base Formula}: d = \sqrt{\frac{C.(H.P.)}{N}}$$

Diameter of Shaft, Inches		$\frac{H.P.}{N}$		
Fractional	Decimal	C = 50 Very Little Bending Fac. Saf. 9.4	C = 75 Average Conditions of Bending Fac. Saf. 14.1	C = 100 Very Severe Conditions of Bending Fac. Saf. 18.8
1/16	0.0625	0.000004883	0.000003256	0.000002492
1/8	0.125	0.00003905	0.00002604	0.00001953
3/16	0.1875	0.0001318	0.00008788	0.00006592
1/4	0.25	0.0003124	0.0002083	0.0001563
5/16	0.3125	0.0006105	0.0004070	0.0003053
3/8	0.375	0.001054	0.0007030	0.0005272
7/16	0.4375	0.001675	0.001116	0.0008375
1/2	0.5	0.002500	0.001666	0.001250
9/16	0.5625	0.003558	0.002372	0.001779
5/8	0.625	0.004883	0.003256	0.002442

Diameter of Shaft, Inches		$\frac{H.P.}{N}$		
Fractional	Decimal	C = 50 Very Little Bending Fac. Saf. 9.4	C = 75 Average Conditions of Bending Fac. Saf. 14.1	C = 100 Very Severe Conditions of Bending Fac. Saf. 18.8
11/16	0.6875	0.006500	0.004333	0.003250
3/4	0.75	0.008439	0.005626	0.004220
13/16	0.8125	0.01074	0.007155	0.005367
7/8	0.875	0.01340	0.008932	0.006699
15/16	0.9375	0.01647	0.01098	0.008235
1	1.0	0.02000	0.01333	0.01000
1 1/16	1.0625	0.02396	0.01598	0.01199
1 1/8	1.125	0.02846	0.01898	0.01423
1 3/16	1.1875	0.03353	0.02236	0.01677
1 1/4	1.25	0.03905	0.02604	0.01953

(table continued on next page)

Diameter of Shaft, Inches		H.P./N		
Fractional	Decimal	C = 50 Very Little Bending Fac. Saf. 9.4	C = 75 Average Conditions of Bending Fac. Saf. 14.1	C = 100 Very Severe Conditions of Bending Fac. Saf. 18.8
15/16	1.3125	0.04519	0.03019	0.02259
1 3/8	1.375	0.5199	0.03465	0.02599
1 7/16	1.4375	0.05940	0.03960	0.02970
1 1/2	1.5	0.06750	0.04500	0.03375
1 9/16	1.5625	0.07628	0.05086	0.03815
1 5/8	1.625	0.08578	0.05719	0.04289
1 11/16	1.6875	0.09524	0.06409	0.04807
1 3/4	1.75	0.1072	0.07144	0.05358
1 13/16	1.8125	0.1191	0.07939	0.05955
1 7/8	1.875	0.1318	0.08788	0.06592
1 15/16	1.9375	0.1456	0.09707	0.07281
2	2.0	0.1600	0.1066	0.07998
2 1/16	2.0625	0.1755	0.1169	0.08774
2 1/8	2.125	0.1918	0.1278	0.09592
2 3/16	2.1875	0.2095	0.1397	0.1048
2 1/4	2.25	0.2278	0.1519	0.1140
2 5/16	2.3125	0.2474	0.1649	0.1237
2 3/8	2.375	0.2678	0.1785	0.1339
2 7/16	2.4375	0.2897	0.1932	0.1449
2 1/2	2.5	0.3124	0.2083	0.1563
2 9/16	2.5625	0.3363	0.2243	0.1682
2 5/8	2.625	0.3616	0.2411	0.1808
2 11/16	2.6875	0.3884	0.2589	0.1942
2 3/4	2.75	0.4159	0.2772	0.2079
2 13/16	2.8125	0.4449	0.2966	0.2225
2 7/8	2.875	0.4754	0.3170	0.2378
2 15/16	2.9375	0.5074	0.3383	0.2537
3	3.0	0.5399	0.3683	0.2700
3 1/16	3.0625	0.5742	0.3836	0.2871
3 1/8	3.125	0.6105	0.4070	0.3053

Diameter of Shaft, Inches		H.P./N		
Fractional	Decimal	C = 50 Very Little Bending Fac. Saf. 9.4	C = 75 Average Conditions of Bending Fac. Saf. 14.1	C = 100 Very Severe Conditions of Bending Fac. Saf. 18.8
3 3/16	3.1875	0.6479	0.4319	0.3240
3 1/4	3.25	0.6866	0.4577	0.3434
3 5/16	3.3125	0.7266	0.4844	0.3634
3 3/8	3.375	0.7685	0.5123	0.3842
3 7/16	3.4375	0.8127	0.5418	0.4061
3 1/2	3.5	0.8567	0.5718	0.4288
3 9/16	3.5625	0.9038	0.6026	0.4520
3 5/8	3.625	0.9526	0.6351	0.4763
3 11/16	3.6875	1.003	0.6688	0.5017
3 3/4	3.75	1.054	0.7030	0.5272
3 13/16	3.8125	1.108	0.7382	0.5538
3 7/8	3.875	1.163	0.7759	0.5820
3 15/16	3.9375	1.222	0.8143	0.6108
4	4.0	1.280	0.8535	0.6401
4 1/16	4.0625	1.340	0.8933	0.6701
4 1/8	4.125	1.404	0.9356	0.7018
4 3/16	4.1875	1.469	0.9792	0.7345
4 1/4	4.25	1.536	1.023	0.7678
4 5/16	4.3125	1.603	1.069	0.8019
4 3/8	4.375	1.675	1.116	0.8375
4 7/16	4.4375	1.748	1.165	0.8742
4 1/2	4.5	1.823	1.214	0.9111
4 9/16	4.5625	1.900	1.266	0.9497
4 5/8	4.625	1.978	1.319	0.9893
4 11/16	4.6875	2.059	1.373	1.030
4 3/4	4.75	2.143	1.429	1.072
4 13/16	4.8125	2.228	1.486	1.114
4 7/8	4.875	2.316	1.544	1.158
4 15/16	4.9375	2.407	1.604	1.203
5	5.0	2.500	1.666	1.250

Haven-Swett "Treatise on Leather Belting," American Leather Belting Association.

Tap drills and clearance drills for machine screws

Screw Size	Coarse Thread		Fine Thread		Clearance Drill
	TPI	Drill	TPI	Drill	
4	40	43	48	42	32
5	40	38	44	37	30
6	32	36	40	33	27
8	32	29	36	29	18
10	24	25	32	21	9
12	24	16	28	14	2
14	20	10	24	7	D
1/4	20	7	28	3	F
5/16	18	F	24	1	P
3/8	16	5/16	24	Q	W
7/16	14	U	20	25/64	29/64
1/2	13	27/64	20	29/64	33/64
9/16	12	31/64	18	33/64	37/64
5/8	11	17/32	18	37/64	41/64
3/4	10	21/32	16	11/16	49/64

Common nails

Size	Length	Gauge No.	Dia. Head	Approx. No./lb
2d	1.00	15	11/64	845
3d	1.25	14	13/64	540
4d	1.50	12.5	1/4	290
5d	1.75	12.5	1/4	250
6d	2.00	11.5	17/64	265
7d	2.25	11.5	17/64	250
8d	2.50	10.25	9/32	200
9d	2.75	10.25	9/32	90
10d	3.00	9	5/16	65
12d	3.25	9	5/16	60
16d	3.50	8	11/32	45
20d	4.00	6	13/32	30
30d	4.50	5	7/16	20
40d	5.00	4	15/32	17
50d	5.50	3	1/2	13
60d	6.00	2	17/32	10

Drill sizes for pipe taps

Tap Size (in.)	TPI	Drill Dia. (in.)
1/8	27	11/32
1/4	18	7/16
3/8	18	37/64
1/2	14	23/32
3/4	14	59/64
1	11 1/2	1 5/32
1 1/4	11 1/2	1 1/2
1 1/2	11 1/2	1 49/64
2	11 1/2	2 3/16
2 1/2	8	2 9/16
3	8	3 3/16
3 1/2	8	3 11/16
4	8	4 3/16
4 1/2	8	4 3/4
5	8	5 5/16
6	8	6 5/16

Carbon steel—color and approximate temperature

Color	Temperature (°F)
Black red	990
Dark blood red	1050
Dark cherry red	1175
Medium cherry red	1250
Light cherry, scaling	1550
Salmon, free scaling	1650
Light salmon	1725
Yellow	1825
Light yellow	1975
White	2220

Bolting dimensions for raised face weld neck flanges

Nom. Pipe Size	150# ANSI			300# ANSI			400# ANSI		
	Quan	Stud Dia.	Length	Quan	Stud Dia.	Length	Quan	Stud Dia.	Length
2	4	⅝	3¼	8	⅝	3½	Note1		
4	4	⅝	3¾	8	¾	4¼	8	⅞	5½
6	8	¾	4	8	¾	4½	12	⅞	6
8	8	¾	4¼	12	¾	5	12	1	6¾
10	12	⅞	4¾	12	⅞	5½	16	1⅛	7½
12	12	⅞	4¾	16	1	6¼	16	1¼	8
14	12	1	5¼	16	1⅛	6¾	20	1¼	8¼
16	16	1	5½	20	1⅛	7	20	1⅜	8¾
18	16	1⅛	6	20	1¼	7½	24	1⅜	9
20	20	1⅛	6¼	24	1¼	7¾	24	1½	9¾
24	20	1¼	7	24	1¼	8¼	24	1 1¾	10¾
26	24	1¼	8½	24	1¼	9¼	28	1¾	11¾
28	28	1¼	8¾	28	1⅝	10¼	28	1¾	11¾
30	28	1¼	9	28	1⅝	10¾	28	1⅞	12 ½
32	28	1½	10	28	1¾	11½	28	2	13¼
34	32	1½	10¼	28	1⅞	12¼	28	2	14
36	32	1½	10¾	32	2	13	32	2	14¼

Nom. Pipe Size	600# ANSI			900# ANSI			1500# ANSI		
	Quan	Stud Dia.	Length	Quan	Stud Dia.	Length	Quan	Stud Dia.	Length
2	8	5/8	4¼	Note 2			8	⅞	6¼
3	8	¾	5	8	⅞	5¾	8	1⅛	7
4	8	⅞	5¾	8	1⅛	6¾	8	3¾	7¾
6	12	1	6¾	12	1⅛	7¾	12	1⅜	10
8	12	1⅛	7¾	12	1⅜	8¾	12	1⅝	11½
10	16	1¼	8½	16	1⅜	8¾	12	1⅞	13½
12	20	1¼	8¾	20	1⅜	10	16	2	15
14	20	1⅜	9¼	20	1½	10¾	16	2¼	16¼
16	20	1½	10	20	1⅝	11¼	16	2½	17¾
18	20	1⅝	10¾	20	1⅞	12¾	16	2¾	19½
20	24	1⅝	11½	20	2	13½	16	3	21¼
24	24	1⅞	13	20	2½	17¼	16	3½	24¼
26	28	1⅞	13½	20	2¾	17¾			
28	28	2	14	20	3	18½			
30	28	2	14¼	20	3	19			
32	28	2¼	15	20	3¼	20¼			
34	28	2¼	15¼	20	3½	20¼			
36	28	2½	16	20	3½	12¾			

Notes:

1. For 3½″ and smaller, use ANSI 600 Dim.
2. For 2½″ and smaller, use ANSI 1500 Dim.

Steel fitting dimensions

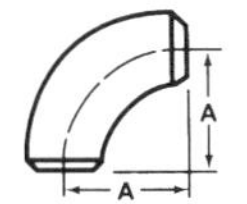

90° Long Radius Elbow

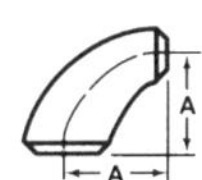

90° Reducing Long Radius Elbow

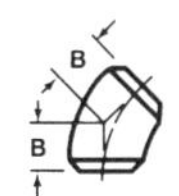

45° Long Radius Elbow

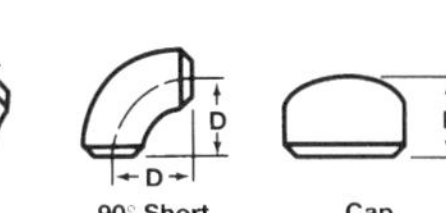

90° Short Radius Elbow

Cap

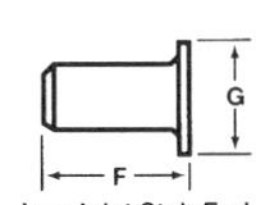

Lap Joint Stub End

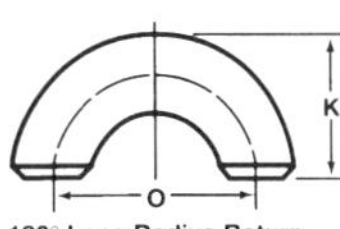

180° Long Radius Return

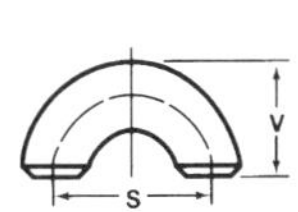

180° Short Radius Return

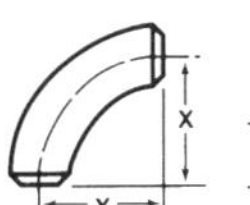

90° 3R Elbow

45° 3R Elbow

Nom. Pipe Size	Pipe O.D. (In.)	Wall thickness: Light Wall	SCH 20	SCH 30	STD	SCH 40	SCH 60	X STG	SCH 80	SCH 100	SCH 120	SCH 140	SCH 160	XX STG	A	B	D	E	F ANSI	F MSS	G	O	K	S	V	X	Y	Nom. Pipe Size
1/2	.84				.109	.109		.147	.147				.188	.294	1.50	.62		1.00	3.00	2.00	1.38	3.00	1.88					1/2
3/4	1.05				.113	.113		.154	.154				.219	.308	1.12	.44		1.00	3.00	2.00	1.69	2.25	1.69					3/4
1	1.32				.133	.133		.179	.179				.250	.358	1.50	.88	1.00	1.50	4.00	2.00	2.00	3.00	2.19	2.00	1.62			1
1 1/4	1.66				.140	.140		.191	.191				.250	.382	1.88	1.00	1.25	1.50	4.00	2.00	2.50	3.75	2.75	2.50	2.06			1 1/4
1 1/2	1.90				.145	.145		.200	.200				.281	.400	2.25	1.12	1.50	1.50	4.00	2.00	2.88	4.50	3.25	3.00	2.44			1 1/2
2	2.38				.154	.154		.218	.218				.344	.436	3.00	1.38	2.00	1.50	6.00	2.50	3.62	6.00	4.19	4.00	3.19	6.00	2.50	2
2 1/2	2.88				.203	.203		.276	.276				.375	.552	3.75	1.75	2.50	1.50	6.00	2.50	4.12	7.50	5.19	5.00	3.94			2 1/2
3	3.50	.188			.216	.216		.300	.300				.438	.600	4.50	2.00	3.00	2.00	6.00	2.50	5.00	9.00	6.25	6.00	4.75	9.00	3.75	3
3 1/2	4.00				.266	.266		.318	.318						5.25	2.25	3.50	2.50	6.00	3.00	5.50	10.50	7.25	7.00	5.50			3 1/2
4	4.50	.188			.237	.237		.337	.337		.438		.531	.674	6.00	2.50	4.00	2.50	6.00	3.00	6.19	12.00	8.25	8.00	6.25	12.00	5.00	4
5	5.56				.258	.258		.375	.375		.500		.625	.750	7.50	3.12	5.00	3.00	8.00	3.00	7.31	15.00	10.31	10.00	7.75			5
6	6.62	.219			.280	.280		.432	.432		.562		.719	.864	9.00	3.75	6.00	3.50	8.00	3.50	8.50	18.00	12.31	12.00	9.31	18.00	7.50	6
8	8.62	.219	.250	.277	.322	.322	.406	.500	.500	.594	.719	.812	.906	.875	12.00	5.00	8.00	4.00	8.00	4.00	10.62	24.00	16.31	16.00	12.31	24.00	10.00	8
10	10.75	.219	.250	.307	.365	.365	.500	.500	.594	.719	.844	1.000	1.125	1.000	15.00	6.25	10.00	5.00	10.00	5.00	12.75	30.00	20.38	20.00	15.38	30.00	12.50	10
12	12.75	.250	.250	.330	.375	.406	.562	.500	.688	.844	1.000	1.125	1.312	1.000	18.00	7.50	12.00	6.00	10.00	6.00	15.00	36.00	24.38	24.00	18.38	36.00	15.00	12
14	14.00	.250	.312	.375	.375	.438	.594	.500	.750	.938	1.094	1.250	1.406		21.00	8.75	14.00	6.50	12.00	6.00	16.25	42.00	28.00	28.00	21.00	42.00	17.50	14
16	16.00	.250	.312	.375	.375	.500	.656	.500	.844	1.031	1.219	1.438	1.594		24.00	10.00	16.00	7.00	12.00	6.00	18.50	48.00	32.00	32.00	24.00	48.00	19.88	16
18	18.00	.250	.312	.438	.375	.562	.750	.500	.938	1.156	1.375	1.562	1.781		27.00	11.25	18.00	8.00	12.00	6.00	21.00	54.00	36.00	36.00	27.00	54.00	22.38	18
20	20.00	.250	.375	.500	.375	.594	.812	.500	1.031	1.281	1.500	1.750	1.969		30.00	12.50	20.00	9.00	12.00	6.00	23.00	60.00	40.00	40.00	30.00	60.00	24.88	20
24	24.00	.250	.375	.562	.375	.688	.969	.500	1.219	1.531	1.812	2.062	2.344		36.00	15.00	24.00	10.50	12.00	6.00	27.25	72.00	48.00	48.00	36.00	72.00	29.81	24
30	30.00	.312	.500	.625	.375			.500							45.00	18.50	30.00	10.50				90.00	60.00	60.00	45.00	90.00	37.25	30
36	36.00	.312	.500	.625	.375	.750		.500							54.00	22.25	36.00	10.50						72.00	54.00	108.00	44.75	36
42	42.00				.375			.500							63.00	26.00	42.00	12.00								126.00	52.19	42
48	48.00				.375			.500							72.00	29.88	48.00	13.50								144.00	59.69	48

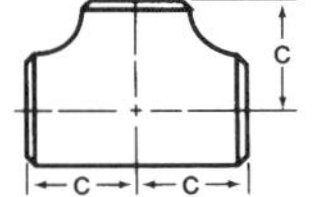

Straight Tee

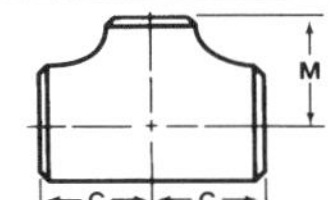

Reducing Tee

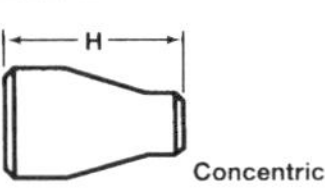

Concentric Reducer

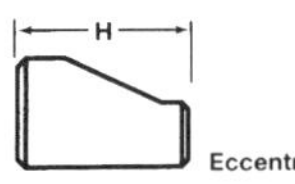

Eccentric Reducer

Nom. Pipe Size Run	Outlet	C	M	H
3/4	3/4	1.12		
	1/2	1.12	1.12	1.50
1	1	1.50		
	3/4	1.50	1.50	2.00
	1/2	1.50	1.50	2.00
1 1/4	1 1/4	1.88		
	1	1.88	1.88	2.00
	3/4	1.88	1.88	2.00
	1/2	1.88	1.88	2.00
1 1/2	1 1/2	2.25		
	1 1/4	2.25	2.25	2.50
	1	2.25	2.25	2.50
	3/4	2.25	2.25	2.50
	1/2	2.25	2.25	2.50
2	2	2.50		
	1 1/2	2.50	2.38	3.00
	1 1/4	2.50	2.25	3.00
	1	2.50	2.00	3.00
	3/4	2.50	1.75	3.00
2 1/2	2 1/2	3.00		
	2	3.00	2.75	3.50
	1 1/2	3.00	2.62	3.50
	1 1/4	3.00	2.50	3.50
	1	3.00	2.25	3.50
3	3	3.38		
	2 1/2	3.38	3.25	3.50
	2	3.38	3.00	3.50
	1 1/2	3.38	2.88	3.50
	1 1/4	3.38	2.75	3.50

Nom. Pipe Size Run	Outlet	C	M	H
3 1/2	3 1/2	3.75		
	3	3.75	3.62	4.00
	2 1/2	3.75	3.50	4.00
	2	3.75	3.25	4.00
	1 1/2	3.75	3.12	4.00
4	4	4.12		
	3 1/2	4.12	4.00	4.00
	3	4.12	3.88	4.00
	2 1/2	4.12	3.75	4.00
	2	4.12	3.50	4.00
	1 1/2	4.12	3.38	4.00
5	5	4.88		
	4	4.88	4.62	5.00
	3 1/2	4.88	4.50	5.00
	3	4.88	4.38	5.00
	2 1/2	4.88	4.25	5.00
	2	4.88	4.12	5.00
6	6	5.62		
	5	5.62	5.38	5.50
	4	5.62	5.12	5.50
	3 1/2	5.62	5.00	5.50
	3	5.62	4.88	5.50
	2 1/2	5.62	4.75	5.50
8	8	7.00		
	6	7.00	6.62	6.00
	5	7.00	6.38	6.00
	4	7.00	6.12	6.00
	3 1/2	7.00	6.00	6.00
10	10	8.50		
	8	8.50	8.00	7.00
	6	8.50	7.62	7.00
	5	8.50	7.50	7.00
	4	8.50	7.25	7.00

Nom. Pipe Size Run	Outlet	C	M	H
12	12	10.00		
	10	10.00	9.50	8.00
	8	10.00	9.00	8.00
	6	10.00	8.62	8.00
	5	10.00	8.50	8.00
14	14	11.00	11.00	
	12	11.00	10.62	13.00
	10	11.00	10.12	13.00
	8	11.00	9.75	13.00
	6	11.00	9.38	13.00
16	16	12.00		
	14	12.00	12.00	14.00
	12	12.00	11.62	14.00
	10	12.00	11.12	14.00
	8	12.00	10.75	14.00
	6	12.00	10.38	
18	18	13.50		
	16	13.50	13.00	15.00
	14	13.50	13.00	15.00
	12	13.50	12.62	15.00
	10	13.50	12.12	15.00
	8	13.50	11.75	
20	20	15.00		
	18	15.00	14.50	20.00
	16	15.00	14.00	20.00
	14	15.00	14.00	20.00
	12	15.00	13.62	20.00
	10	15.00	13.12	
	8	15.00	12.75	

Nom. Pipe Size Run	Outlet	C	M	H
24	24	17.00		
	22	17.00	17.00	20.00
	20	17.00	17.00	20.00
	18	17.00	16.50	20.00
	16	17.00	16.00	20.00
	14	17.00	16.00	
	12	17.00	15.62	
	10	17.00	15.12	
30	30	22.00		
	24	22.00	21.00	24.00
	20	22.00	20.00	24.00
	18	22.00	19.50	
	16	22.00	19.00	
	14	22.00	19.00	
36	36	26.50		
	30	26.50	25.00	24.00
	24	26.50	24.00	24.00
	20	26.50	23.00	
	18	26.50	22.50	
	16	26.50	22.00	
42	42	30.00		
	36	30.00	28.00	24.00
	30	30.00	28.00	24.00
	24	30.00	26.00	
	20	30.00	26.00	
48	48	35.00		
	42	35.00	32.00	28.00
	36	35.00	31.00	
	30	35.00	30.00	
	24	35.00	29.00	

NOTES:

Fittings are to ASTM A234 WPB with dimensions to ANSI B16.9 except that short radius elbows and returns are to ANSI B16.28 and 3R elbows are to MSS-SP75.

Other types, sizes and thicknesses of fittings on application.

ANSI forged steel flanges

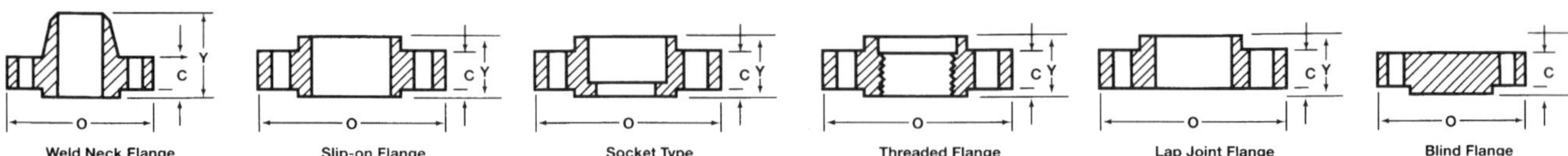

CLASS 150 FLANGES

Nom. Pipe Size	O	C†	Y† Weld Neck	Y† Slip-on Thrd. Socket	Y† Lap Joint	Bolt Circle	No. and Size of Holes
1/2	3.50	.44	1.88	.62	.62	2.38	4-0.62
3/4	3.88	.50	2.06	.62	.62	2.75	4-0.62
1	4.25	.56	2.19	.69	.69	3.12	4-0.62
1 1/4	4.62	.62	2.25	.81	.81	3.50	4-0.62
1 1/2	5.00	.69	2.44	.88	.88	3.88	4-0.62
2	6.00	.75	2.50	1.00	1.00	4.75	4-0.75
2 1/2	7.00	.88	2.75	1.12	1.12	5.50	4-0.75
3	7.50	.94	2.75	1.19	1.19	6.00	4-0.75
3 1/2	8.50	.94	2.81	1.25	1.25	7.00	8-0.75
4	9.00	.94	3.00	1.31	1.31	7.50	8-0.75
5	10.00	.94	3.50	1.44	1.44	8.50	8-0.88
6	11.00	1.00	3.50	1.56	1.56	9.50	8-0.88
8	13.50	1.12	4.00	1.75	1.75	11.75	8-0.88
10	16.00	1.19	4.00	1.94	1.94	14.25	12-1.00
12	19.00	1.25	4.50	2.19	2.19	17.00	12-1.00
14	21.00	1.38	5.00	2.25	USE SLIP-ON DIMS.	18.75	12-1.12
16	23.50	1.44	5.00	2.50		21.25	16-1.12
18	25.00	1.56	5.50	2.69		22.75	16-1.25
20	27.50	1.69	5.69	2.88		25.00	20-1.25
24	32.00	1.88	6.00	3.25		29.50	20-1.38
30	38.75	2.94	5.38			36.00	28-1.38
36	46.00	3.56	6.19			42.75	32-1.62
42	53.00	3.81	6.75			49.50	36-1.62
48	59.50	4.25	7.56			56.00	44-1.62
54	66.25	4.75	8.50			62.75	44-1.88
60	73.00	5.19	9.44			69.25	52-1.88

CLASS 300 FLANGES

Nom. Pipe Size	O	C†	Y† Weld Neck	Y† Slip-on Thrd. Socket	Y† Lap Joint	Bolt Circle	No. and Size of Holes
1/2	3.75	.56	2.06	.88	.88	2.62	4-0.62
3/4	4.62	.62	2.25	1.00	1.00	3.25	4-0.75
1	4.88	.69	2.44	1.06	1.06	3.50	4-0.75
1 1/4	5.25	.75	2.56	1.06	1.06	3.88	4-0.75
1 1/2	6.12	.81	2.69	1.19	1.19	4.50	4-0.88
2	6.50	.88	2.75	1.31	1.31	5.00	8-0.75
2 1/2	7.50	1.00	3.00	1.50	1.50	5.88	8-0.88
3	8.25	1.12	3.12	1.69	1.69	6.62	8-0.88
3 1/2	9.00	1.19	3.19	1.75	1.75	7.25	8-0.88
4	10.00	1.25	3.38	1.88	1.88	7.88	8-0.88
5	11.00	1.38	3.88	2.00	2.00	9.25	8-0.88
6	12.50	1.44	3.88	2.06	2.06	10.62	12-0.88
8	15.00	1.62	4.38	2.44	2.44	13.00	12-1.00
10	17.50	1.88	4.62	2.62		15.25	16-1.12
12	20.50	2.00	5.12	2.88		17.75	16-1.25
14	23.00	2.12	5.62	3.00	USE SLIP-ON DIMS.	20.25	20-1.25
16	25.50	2.25	5.75	3.25		22.50	20-1.38
18	28.00	2.38	6.25	3.50		24.75	24-1.38
20	30.50	2.50	6.38	3.75		27.00	24-1.38
24	36.00	2.75	6.62	4:19		32.00	24-1.62
30	43.00	3.62	8.25			39.25	28-1.88
36	50.00	4.12	9.50			46.00	32-2.12
42	50.75	4.69	7.88			47.50	32-1.75
48	57.75	5.25	8.81			54.00	32-2.00
54	65.25	6.00	9.94			61.00	28-2.38
60	71.25	6.44	10.75			67.00	32-2.38

CLASS 400 FLANGES

Nom. Pipe Size	O	C†	Y† Weld Neck	Y† Slip-on Thrd. Socket	Y† Lap Joint	Bolt Circle	No. and Size of Holes
1/2	3.75	.56	2.06	.88	.88	2.62	4-0.62
3/4	4.62	.62	2.25	1.00	1.00	3.25	4-0.75
1	4.88	.69	2.44	1.06	1.06	3.50	4-0.75
1 1/4	5.25	.81	2.62	1.12	1.12	3.88	4-0.75
1 1/2	6.12	.88	2.75	1.25	1.25	4.50	4-0.88
2	6.50	1.00	2.88	1.44	1.44	5.00	8-0.75
2 1/2	7.50	1.12	3.12	1.62	1.62	5.88	8-0.88
3	8.25	1.25	3.25	1.81	1.81	6.62	8-0.88
3 1/2	9.00	1.38	3.38	1.94	1.94	7.25	8-1.00
4	10.00	1.38	3.50	2.00	2.00	7.88	8-1.00
5	11.00	1.50	4.00	2.12	2.12	9.25	8-1.00
6	12.50	1.62	4.06	2.25	2.25	10.62	12-1.00
8	15.00	1.88	4.62	2.69	2.69	13.00	12-1.12
10	17.50	2.12	4.88	2.88		15.25	16-1.25
12	20.50	2.25	5.38	3.12		17.75	16-1.38
14	23.00	2.38	5.88	3.31	USE SLIP-ON DIMS.	20.25	20-1.38
16	25.50	2.50	6.00	3.69		22.50	20-1.50
18	28.00	2.62	6.50	3.88		24.75	24-1.50
20	30.50	2.75	6.62	4.00		27.00	24-1.62
24	36.00	3.00	6.88	4.50		32.00	24-1.88
30	43.00	4.00	8.62			39.25	28-2.12
36	50.00	4.50	9.88			46.00	32-2.12
42	52.00	5.25	8.81			48.25	32-2.00
48	59.50	6.00	10.12			55.25	28-2.38
54	67.00	6.69	11.38			62.25	28-2.62
60	74.25	7.31	12.56			69.00	32-2.88

CLASS 600 FLANGES

Nom. Pipe Size	O	C†	Y† Weld Neck	Y† Slip-on Thrd. Socket	Y† Lap Joint	Bolt Circle	No. and Size of Holes
1/2	3.75	.56	2.06	.88	.88	2.62	4-0.62
3/4	4.62	.62	2.25	1.00	1.00	3.25	4-0.75
1	4.88	.69	2.44	1.06	1.06	3.50	4-0.75
1 1/4	5.25	.81	2.62	1.12	1.12	3.88	4-0.75
1 1/2	6.12	.88	2.75	1.25	1.25	4.50	4-0.88
2	6.50	1.00	2.88	1.44	1.44	5.00	8-0.75
2 1/2	7.50	1.12	3.12	1.62	1.62	5.88	8-0.88
3	8.25	1.25	3.25	1.81	1.81	6.62	8-0.88
3 1/2	9.00	1.38	3.38	1.94	1.94	7.25	8-1.00
4	10.75	1.50	4.00	2.12	2.12	8.50	8-1.00
5	13.00	1.75	4.50	2.38	2.38	10.50	8-1.12
6	14.00	1.88	4.62	2.62	2.62	11.50	12-1.12
8	16.50	2.19	5.25	3.00	3.00	13.75	12-1.25
10	20.00	2.50	6.00	3.38	USE SLIP-ON DIMS.	17.00	16-1.38
12	22.00	2.62	6.12	3.62		19.25	20-1.38
14	23.75	2.75	6.50	3.69		20.75	20-1.50
16	27.00	3.00	7.00	4.19		23.75	20-1.62
18	29.25	3.25	7.25	4.62		25.75	20-1.75
20	32.00	3.50	7.50	5.00		28.50	24-1.75
24	37.00	4.00	8.00	5.50		33.00	24-2.00
30	44.50	4.50	9.75			40.25	28-2.12
36	51.75	4.88	11.12			47.00	28-2.62
42	55.25	6.62	11.00			50.50	28-2.62
48	62.75	7.44	12.44			57.50	32-2.88
54	70.00	8.25	13.75			64.25	32-3.12
60	78.50	9.19	15.31			71.75	28-3.62

CLASS 900 FLANGES

Nom. Pipe Size	O	C†	Y† Weld Neck	Y† Slip-on Thrd. Socket	Y† Lap Joint	Bolt Circle	No. and Size of Holes
1/2	4.75	.88	2.38	1.25	1.25	3.25	4-0.88
3/4	5.12	1.00	2.75	1.38	1.38	3.50	4-0.88
1	5.88	1.12	2.88	1.62	1.62	4.00	4-1.00
1 1/4	6.25	1.12	2.88	1.62	1.62	4.38	4-1.00
1 1/2	7.00	1.25	3.25	1.75	1.75	4.88	4-1.12
2	8.50	1.50	4.00	2.25	2.25	6.50	8-1.00
2 1/2	9.62	1.62	4.12	2.50	2.50	7.50	8-1.12
3	9.50	1.50	4.00	2.12	2.12	7.50	8-1.00
3 1/2							
4	11.50	1.75	4.50	2.75	2.75	9.25	8-1.25
5	13.75	2.00	5.00	3.12	3.12	11.00	8-1.38
6	15.00	2.19	5.50	3.38	3.38	12.50	12-1.25
8	18.50	2.50	6.38	4.00	USE SLIP-ON DIMS.	15.50	12-1.50
10	21.50	2.75	7.25	4.25		18.50	16-1.50
12	24.00	3.12	7.88	4.62		21.00	20-1.50
14	25.25	3.38	8.38	5.12		22.00	20-1.62
16	27.75	3.50	8.50	5.25		24.25	20-1.75
18	31.00	4.00	9.00	6.00		27.00	20-2.00
20	33.75	4.25	9.75	6.25		29.50	20-2.12
24	41.00	5.50	11.50	8.00		35.50	20-2.62
30	48.50	5.88	12.25			42.75	20-3.12
36	57.50	6.75	14.25			50.75	20-3.62
42	61.50	8.12	14.62			54.75	24-3.62
48	70.25	9.19	16.50			62.50	24-4.12
54							
60							

CLASS 1500 FLANGES

Nom. Pipe Size	O	C†	Y† Weld Neck	Y† Slip-on Thrd. Socket	Y† Lap Joint	Bolt Circle	No. and Size of Holes
1/2	4.75	.88	2.38	1.25	1.25	3.25	4-0.88
3/4	5.12	1.00	2.75	1.38	1.38	3.50	4-0.88
1	5.88	1.12	2.88	1.62	1.62	4.00	4-1.00
1 1/4	6.25	1.12	2.88	1.62	1.62	4.38	4-1.00
1 1/2	7.00	1.25	3.25	1.75	1.75	4.88	4-1.12
2	8.50	1.50	4.00	2.25	2.25	6.50	8-1.00
2 1/2	9.62	1.62	4.12	2.50	2.50	7.50	8-1.12
3	10.50	1.88	4.62	2.88	2.88	8.00	8-1.25
3 1/2							
4	12.25	2.12	4.88	3.56	3.56	9.50	8-1.38
5	14.75	2.88	6.12	4.12	4.12	11.50	8-1.62
6	15.50	3.25	6.75	4.69	4.69	12.50	12-1.50
8	19.00	3.62	8.38	5.62	5.62	15.50	12-1.75
10	23.00	4.25	10.00	6.25		19.00	12-2.00
12	26.50	4.88	11.12	7.12		22.50	16-2.12
14	29.50	5.25	11.75			25.00	16-2.38
16	32.50	5.75	12.25			27.75	16-2.62
18	36.00	6.38	12.88			30.50	16-2.88
20	38.75	7.00	14.00			32.75	16-3.12
24	46.00	8.00	16.00			39.00	16-3.62

CLASS 2500 FLANGES

Nom. Pipe Size	O	C†	Y† Weld Neck	Y† Slip-on Thrd. Socket	Y† Lap Joint	Bolt Circle	No. and Size of Holes
1/2	5.25	1.19	2.88	1.56	1.56	3.50	4-0.88
3/4	5.50	1.25	3.12	1.69	1.69	3.75	4-0.88
1	6.25	1.38	3.50	1.88	1.88	4.25	4-1.00
1 1/4	7.25	1.50	3.75	2.06	2.06	5.12	4-1.12
1 1/2	8.00	1.75	4.38	2.38	2.38	5.75	4-1.25
2	9.25	2.00	5.00	2.75	2.75	6.75	8-1.12
2 1/2	10.50	2.25	5.62	3.12	3.12	7.75	8-1.25
3	12.00	2.62	6.62	3.62	3.62	9.00	8-1.38
4	14.00	3.00	7.50	4.25	4.25	10.75	8-1.62
5	16.50	3.62	9.00	5.12	5.12	12.75	8-1.88
6	19.00	4.25	10.75	6.00	6.00	14.50	8-2.12
8	21.75	5.00	12.50	7.00	7.00	17.25	12-2.12
10	26.50	6.50	16.50	9.00	9.00	21.25	12-2.62
12	30.00	7.25	18.25	10.00	10.00	24.38	12-2.88

NOTES:

Flange material is ASTM A105 unless specified otherwise.

Flange dimensions are to ANSI B16.5 for sizes 24" and smaller. Sizes over 24" are to MSS-SP44.

Always specify bore or matching pipe wall when ordering.

Other types, sizes and facings on application.

† Includes .06 raised face in Class 150 and Class 300 standards. Does **NOT** include .25 raised face in Class 400 and heavier standards.

Trench shoring—minimum requirements

Trench jacks may be used in lieu of, or in combination with, cross braces. Shoring is not required in solid rock, hard shale, or hard slag. Where desirable, steel sheet piling and bracing of equal strength may be substituted for wood.

		Uprights		Stringers		Size & Spacing of Members						
						Cross Braces					Maximum Spacing	
						Width of Trench						
Depth of Trench Feet	Kind or Condition of Earth	Minimum Dimensions	Maximum Spacing	Minimum Dimensions	Maximum Spacing	Up to 3 Feet	3 to 6 Feet	6 to 9 Feet	9 to 12 Feet	12 to 15 Feet	Vertical	Horizontal
Feet		Inches	Feet	Inches	Feet	Inches	Inches	Inches	Inches	Inches	Feet	Feet
5 to 10	Hard, compact	3 × 4 or 2 × 6	6	—	—	2 × 6	4 × 4	4 × 6	6 × 6	6 × 8	4	6
	Likely to crack	3 × 4 or 2 × 6	3	4 × 6	4	2 × 6	4 × 4	4 × 6	6 × 6	6 × 8	4	6
	Soft, sandy, or filled	3 × 4 or 2 × 6	Close Sheeting	4 × 6	4	4 × 4	4 × 6	6 × 6	6 × 8	8 × 8	4	6
	Hydrostatic pressure	3 × 4 or 2 × 6	Close Sheeting	4 × 6	4	4 × 4	4 × 6	6 × 6	6 × 8	8 × 8	4	6
10 to 15	Hard	3 × 4 or 2 × 6	4	4 × 6	4	4 × 4	4 × 6	6 × 6	6 × 8	8 × 8	4	6
	Likely to crack	3 × 4 or 2 × 6	2	4 × 4	4	4 × 6	6 × 6	6 × 6	6 × 8	8 × 8	4	6
	Soft, sandy, or filled	3 × 4 or 2 × 6	Close Sheeting	4 × 6	4	4 × 6	6 × 6	6 × 8	8 × 8	8 × 10	4	6
	Hydrostatic pressure	3 × 6	Close Sheeting	8 × 10	4	4 × 6	6 × 6	6 × 8	8 × 8	8 × 10	4	6
15 to 20	All kinds or conditions	3 × 6	Close Sheeting	4 × 12	4	4 × 12	6 × 8	8 × 8	8 × 8	10 × 10	4	6
Over 20	All kinds or conditions	3 × 6	Close Sheeting	6 × 8	4	4 × 12	6 × 8	8 × 10	10 × 10	10 × 12	4	6

Reuniting separated mercury in thermometers

The largest single cause for failure of precision thermometers is due to separated mercury columns. This can occur in transit or in use. The mercury may be reunited by cooling the thermometer in a solution of solid CO_2 (Dry-Ice) and alcohol so that the mercury column retreats slowly into the bulb. Do not cool the stem or mercury column. Keep the bulb in the solution until the main column, as well as the separated portion, retreats into the bulb. Remove and swing the thermometer in a short arc, forcing all of the mercury into the bulb.

Most mercury thermometers can be reunited using this method regardless of range (with the exception of deep immersion thermometers) provided only the bulb is immersed in the CO_2.

Caution: Do not touch the bulb until it has warmed sufficiently for the mercury to emerge from the bulb into the capillary.

Never subject the stem or mercury column to the CO_2 solution as it will freeze the mercury column in the capillary and may cause the bulb to fracture.

Typical wire resistance

(Stranded Copper Conductors at 59°F)

Wire Size AWG	Resistance (ohms/ft)
0000	0.00005
000	0.00006
00	0.00008
0	0.00010
1	0.00012
2	0.00016
4	0.00025
6	0.00039
8	0.00063
10	0.00098
12	0.00160

How to cut odd-angle long radius elbows

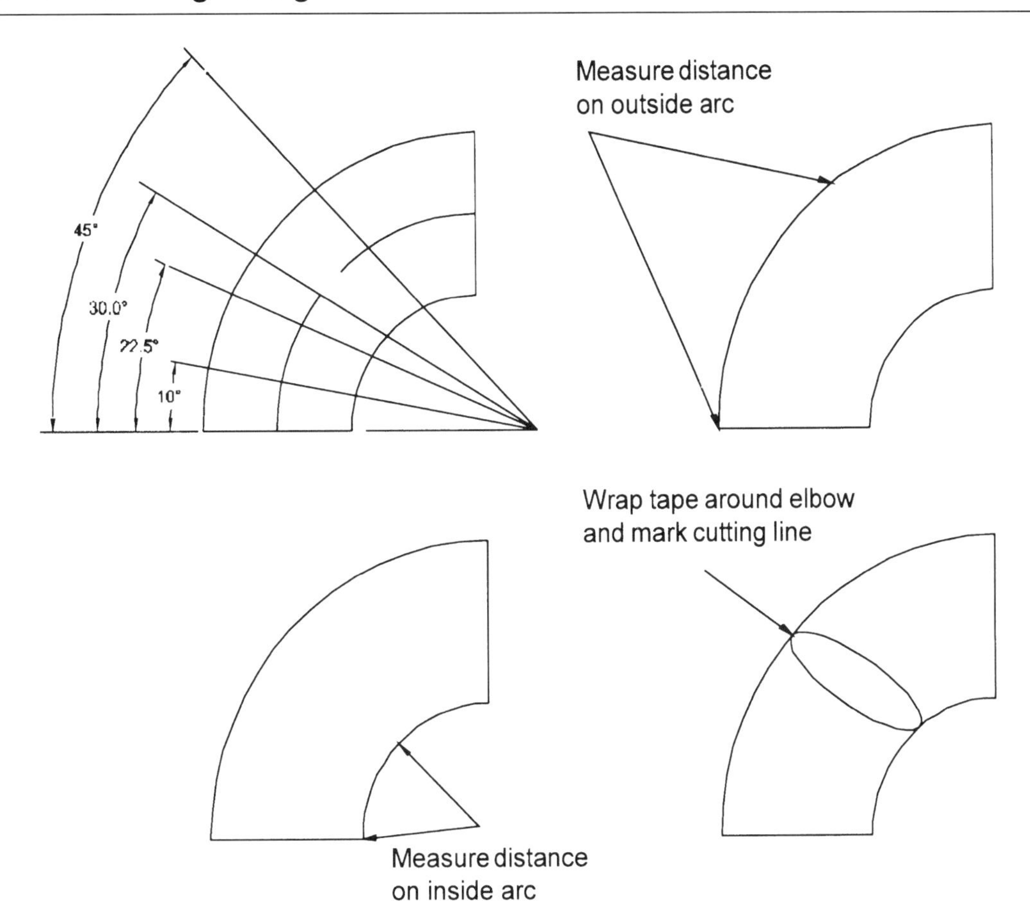

Nominal Pipe Size	Outside Arc Length*							Inside Arc Length*						
	1°	5°	10°	15°	22.5°	30°	45°	1°	5°	10°	15°	22.5°	30°	45°
2"	5/64	3/8	23/32	1-3/32	1-21/32	2-3/16	3-9/32	1/32	5/32	5/16	15/32	23/32	15/16	1-7/16
2 1/2"	3/32	7/16	29/32	1-11/32	2-1/32	2-23/32	4-1/16	3/64	3/16	13/32	19/32	29/32	1-7/32	1-13/16
3"	7/64	9/16	1-1/8	1-5/8	2-15/32	3-9/32	4-29/32	3/64	1/4	1/2	23/32	1-3/32	1-7/16	2-5/32
3 1/2"	1/8	5/8	1-9/32	1-29/32	2-27/32	3-13/16	5-11/16	1/16	9/32	9/16	27/32	1-9/32	1-11/16	2-9/16
4"	9/64	23/32	1-7/16	2-5/32	3-1/4	4-5/16	6-15/32	1/16	5/16	21/32	3-1/32	1-15/32	1 31/32	2-15/16
5"	3/16	29/32	1-25/32	2-11/16	4-1/32	5-3/8	8-1/16	5/64	13/32	13/16	1-1/4	1-27/32	2-15/32	3-23/32
6"	7/32	1-1/16	2-5/32	3-7/32	4-27/32	6-7/16	9-21/32	3/32	1/2	1	1-1/2	2-7/32	2-31/32	4-15/32
8"	9/32	1-7/16	2-27/32	4-9/32	6-13/32	8-17/32	12-13/16	1/8	11/16	1-11/32	2	3-1/32	4-1/32	6-1/32
10"	11/32	1-25/32	3-9/16	5-11/32	8	10-21/32	16	5/32	27/32	1-11/16	2-17/32	3-25/32	5-1/32	7-9/16
12"	7/16	2-1/8	4-1/4	6-3/8	9-9/16	12-3/4	19-5/32	7/32	1	2-1/32	3-1/16	4-9/16	6-3/32	9-1/8
14"	1/2	2-7/16	4-7/8	7-5/16	11	14-21/32	22	1/4	1-7/32	2-7/16	3-21/32	5-1/2	7-11/32	11
16"	9/16	2-13/16	5-19/32	8-3/8	12-9/16	16-3/4	25-1/8	9/32	1-13/32	2-13/16	4-3/16	6-9/32	8-3/8	12-5/8
18"	5/8	3 -1/8	6-9/32	9-7/16	14-1/8	18-27/32	28-9/32	5/16	1-9/16	3-1/8	4-23/32	7-1/16	9-7/16	14-1/8
20"	11/16	3-1/2	7	10-15/32	15-23/32	20-15/16	31-13/32	11/32	1-3/4	3-1/2	5-1/4	7-27/32	10-15/32	15-11/16
22"	3/4	3-27/32	7-11/16	11-17/32	17-9/32	23-1/32	34-9/16	3/8	1-29/32	3-27/32	5-3/4	8-5/8	11-17/32	17-9/32
24"	27/32	4-3/16	8-3/8	12-9/16	18-27/32	25-1/8	37-11/16	13/32	2-3/32	4-3/16	6-9/32	9-7/16	12-9/16	18-27/32
26"	29/32	4-17/32	9-3/32	13-5/8	20-13/32	27-7/32	40-27/32	15/32	2-9/32	4-17/32	6-13/16	10-7/32	13-5/8	20-13/32
30"	1-1/32	5-1/4	10-15/32	15-3/4	23-9/16	31-13/32	47-1/8	17/32	2-5/8	5-1/4	7-27/32	11-25/32	15-23/32	23-9/16
36"	1-7/32	6-9/32	12-17/32	18-7/8	28-7/32	37-11/16	56-17/32	5/8	2-13/16	6-1/4	9-7/16	14-1/8	18-27/32	28-1/4

*Rounded to nearest practical fraction.

How to read land descriptions

A land description is a description of a tract of land, in legally acceptable terms, that defines exactly where the tract of land is located and how many acres it contains.

Table 1
Land Measurements

Linear Measure

1 inch	0.833 feet	16 ½ feet	1 rod
7.92 inches	1 link	5 ½ yards	1 rod
12 inches	1 foot	4 rods	1 link
1 vara	33 inches	66 feet	1 chain
2 ¾ feet	1 vara	80 chains	1 mile
3 feet	1 yard	320 rods	1 mile
25 links	16 ½ feet	8000 links	1 mile
25 links	1 rod	5280 feet	1 mile
100 links	1 chain	1760 yards	1 mile

Square Measure

144 sq. in.	1 sq. ft.	43,560 sq. ft.	1 acre
9 sq. ft.	1 sq. yd.	640 acres	1 sq. mile
30 ¼ sq.yds	1 sq. rod	1 sq. mile	1 section
10 sq. rods	1 sq. chain	36 sq. miles	1 township
1 sq. rod	272 ¼ sq. ft.	6 miles sq.	1 township
1 sq. chain	4356 sq. ft.	208 ft. 8 insq.	1 acre
10 sq. chains	1 acre	80 rods sq.	40 acres
160 sq. rods	1 acre	160 rods sq.	160 acres
4840 sq. yds	1 acre		

In non-rectangular land descriptions, distance is usually described in terms of either feet or rods (this is especially true in surveying today), while square measure is in terms of acres. Such descriptions are called Metes and Bounds descriptions and will be explained in detail later. In rectangular land descriptions, square measure is again in terms of acres, and the location of the land is in such terms as N ½ (north one half), SE ¼ (south east one fourth or quarter), etc., as shown in Figures 2, 3, 4, and 5.

Meandered water & government lots

A meandered lake or stream is water, next to which the adjoining land owner pays taxes on the land only. Such land is divided into divisions of land called government lots. The location, acreage, and lot number of each such tract of land was determined, surveyed, and platted by the original government surveyors.

The original survey of your county (complete maps of each township, meandered lakes, government lots, etc.) is in your courthouse and is the basis for all land descriptions in your county. See Figure 1.

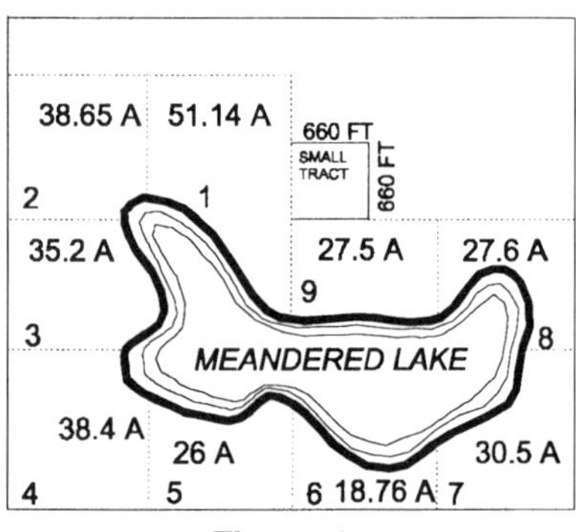

Figure 1

The government lot number given to a piece of land is the legal description of that tract of land.

How can you tell whether water is meandered or privately owned? If you find government lots adjoining a body of water or stream, those waters are meandered. If there are no government lots surrounding water, that water is privately owned; the owner is paying taxes on the land under the water and controls hunting, fishing, trapping rights, etc., on that water within the regulations of state and federal laws. Note that where such water is deemed navigable, other rulings may sometimes pertain.

As a generality, meandered water is public water that the public may use for recreational purposes, fishing, hunting, trapping, etc., provided that the public can reach the waters without trespassing. There still is much litigation concerning this that will have to be settled in court.

Reading land descriptions

Descriptions of land always read first from either the north or the south. In Figures 2, 3, 4, and 5, notice that they all start with N (north) or S (south), such as NW, SE, etc. They are never WN (west north), ES (east south), etc.

It is simple for anyone to understand a description. The secret is to read or analyze the description from the rear, or backwards.

Example. Under Figure 4, the first description reads E ½, SE ¼, SW ¼, SW ¼. The last part of the description reads SW ¼, which means that the tract of land we are looking for is somewhere in that quarter (as shown in Figure 2). Next back, we find SW ¼, which means the tract we are after is somewhere in the SW ¼ SW ¼ (as shown in Figure 3). Next back, we find the SE ¼, which means that the tract is in the SE ¼ SW ¼ SW ¼ (as shown in Figure 5). Next back and the last part to look up is the E ½ of the above, which is the location of the tract described by the whole description (as shown in Figure 4).

Sample sections showing rectangular land descriptions, acreages, and distances

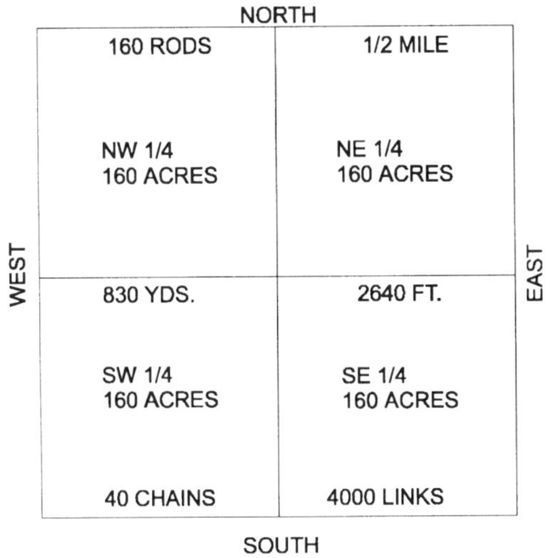

NORTH
160 RODS
1/2 MILE
NW 1/4
160 ACRES
NE 1/4
160 ACRES
WEST
EAST
830 YDS.
2640 FT.
SW 1/4
160 ACRES
SE 1/4
160 ACRES
40 CHAINS
4000 LINKS
SOUTH

Figure 2

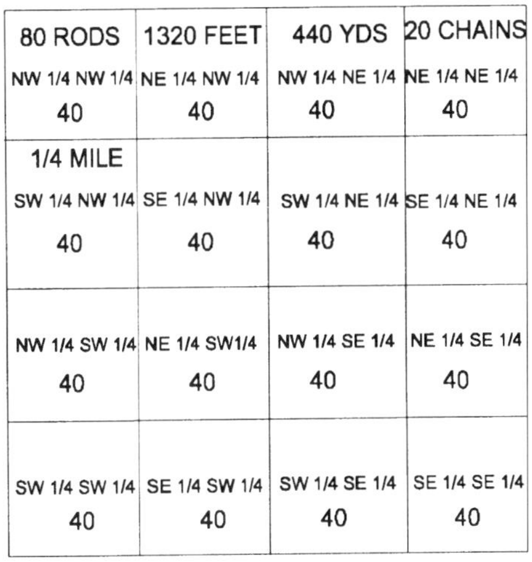

80 RODS
1320 FEET
440 YDS
20 CHAINS
NW 1/4 NW 1/4
40
NE 1/4 NW 1/4
40
NW 1/4 NE 1/4
40
NE 1/4 NE 1/4
40
1/4 MILE
SW 1/4 NW 1/4
40
SE 1/4 NW 1/4
40
SW 1/4 NE 1/4
40
SE 1/4 NE 1/4
40
NW 1/4 SW 1/4
40
NE 1/4 SW1/4
40
NW 1/4 SE 1/4
40
NE 1/4 SE 1/4
40
SW 1/4 SW 1/4
40
SE 1/4 SW 1/4
40
SW 1/4 SE 1/4
40
SE 1/4 SE 1/4
40

Figure 3

1 MILE - 320 RODS - 5280 FEET - 80 CHAINS
N 1/2
320
N 1/2 N 1/2 SW 1/4
40
S 1/2 N 1/2 SW 1/4
40
W 1/2 NW 1/4 SE 1/4
20
E 1/2 NW 1/4 SE 1/4
20
N 1/2 NE 1/4 SE 1/4
20
S 1/2 NE 1/4 SE 1/4
20
W 1/2 SW 1/4 SW 1/4
20
NE 1/4 SW 1/4 SW 1/4
10
SE 1/4 SW 1/4
40
S 1/2 SE 1/4
80
E 1/2 SE 1/4 SW 1/4 SW 1/4 - 5 A
W 1/2 SE 1/4 SW 1/4 SW 1/4 - 5 A

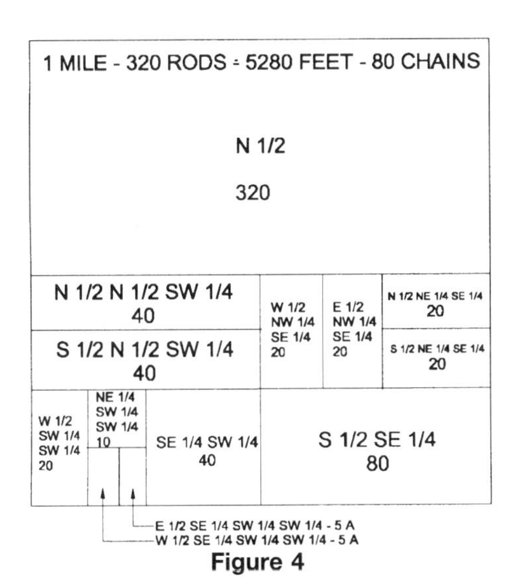

Figure 4

W 1/2
NW 1/4
80
E 1/2
NW 1/4
80
E 1/2
320
N 1/2 SW 1/4
80
N1/2SW1/4SW1/4
20
NW 1/4 SE 1/4 SW 1/4
10
NE 1/4 SE 1/4 SW 1/4
10
SE 1/4 SW 1/4 SW 1/4
10
SW 1/4 SE 1/4 SW 1/4
10
SE 1/4 SE 1/4 SW 1/4
10
S1/2SW1/4SW1/4SW1/4 - 5 A
N1/2SW1/4SW1/4SW1/4 - 5 A

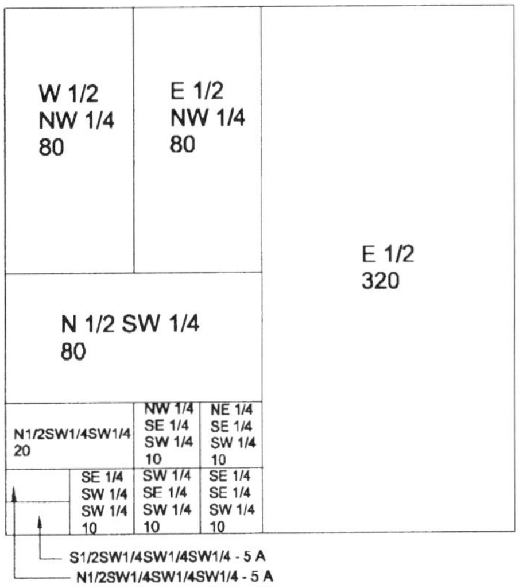

Figure 5

Metes and bounds descriptions

A metes and bounds description is a description of land obtained by starting at a given point, running so many feet in a certain direction, so many feet in another direction, etc., back to the point of beginning.

Example. In Figure 1, notice the small tract of land outlined. A typical metes and bounds description of this tract of land would be as follows: "Begin at the center of the section, thence north 660 ft, thence east 660 ft, thence south 660 ft, thence west 660 ft, back to the point of beginning and containing 10 acres, being a part of Section No. 2."

IMPORTANT: To locate a tract of land from a metes and bounds description, start from the point of beginning and follow it out (do not read backwards as in the case of a rectangular description).

The small tract of land just located by the above metes and bounds description could also be described as the SW $\frac{1}{4}$ SW $\frac{1}{4}$ NE $\frac{1}{4}$ of the section. In most cases, the same tract of land may be described in different ways. The rectangular system of describing and locating land as shown in Figures 2, 3, 4, and 5 is the simplest and is almost always used when possible.

In land descriptions, degree readings are not a measure of distance. They are combined with either north or south to show the direction a line runs from a given point.

Dimensions of hex nuts and hex jam nuts

Nominal size	F			G		H			J		
	Width across flats			Width across corners		Thickness hex nuts			Thickness hex jam nuts		
	Basic	Max	Min	Max	Min	Basic	Max	Min	Basic	Max	Min
1/4	7/16	0.438	0.428	0.505	0.488	7/32	0.263	0.212	5/32	0.163	0.150
5/16	1/2	0.500	0.489	0.557	0.577	17/64	0.273	0.258	3/16	0.195	0.180
3/8	9/16	0.562	0.551	0.650	0.628	21/64	0.337	0.320	7/23	0.227	0.210
7/16	11/16	0.688	0.675	0.794	0.768	3/8	0.385	0.365	1/4	0.260	0.240
1/2	3/4	0.750	0.736	0.855	0.840	7/16	0.448	0.427	5/16	0.323	0.302
9/16	7/8	0.875	0.861	1.010	0.982	31/64	0.496	0.473	5/16	0.324	0.301
5/8	15/16	0.938	0.922	1.083	1.051	35/64	0.559	0.535	3/8	0.387	0.363
3/4	1 1/8	1.125	1.088	1.299	1.240	41/64	0.665	0.617	27/64	0.446	0.398
7/8	1 5/16	1.312	1.269	1.516	1.447	3/4	0.776	0.724	31/64	0.510	0.458
1	1 1/2	1.500	1.450	1.732	1.653	55/64	0.887	0.831	35/64	0.575	0.519
1 1/8	1 11/16	1.688	1.631	1.949	1.859	31/32	0.999	0.939	39/64	0.639	0.579
1 1/4	1 7/8	1.875	1.812	2.165	2.066	1 1/16	1.094	1.030	32/32	0.751	0.687
1 3/8	2 1/16	2.062	1.994	2.382	2.273	1 11/64	1.206	1.138	25/32	0.815	0.747
1 1/2	2 1/4	2.250	2.175	2.598	2.480	1 9/32	1.317	1.245	27/32	0.880	0.808

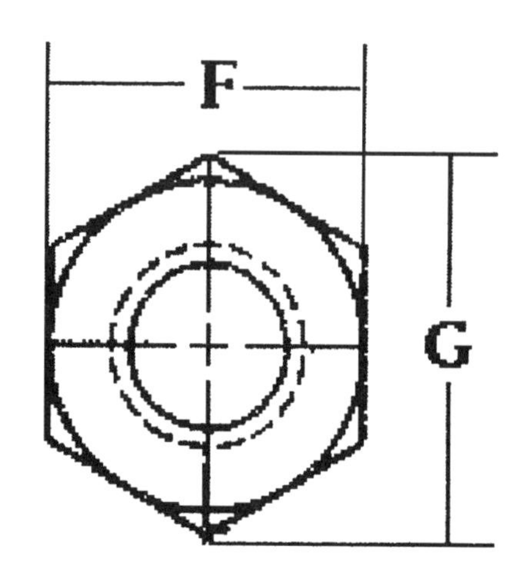

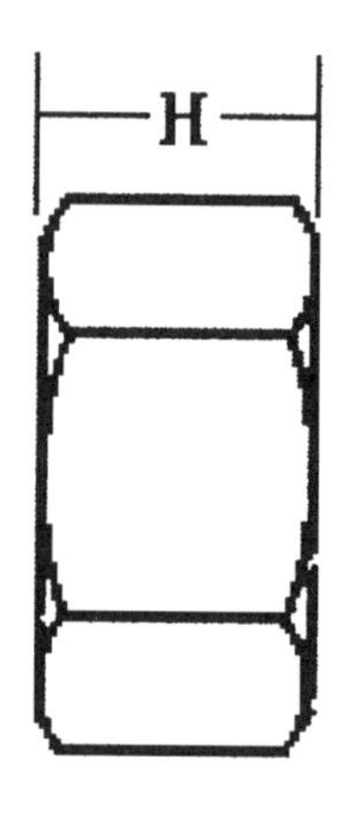

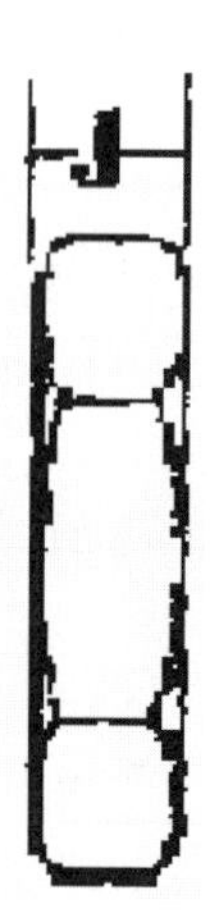

Reprinted with permission—Barnhill Bolt Co., Inc. Albuquerque, NM.

Color codes for locating underground utilities

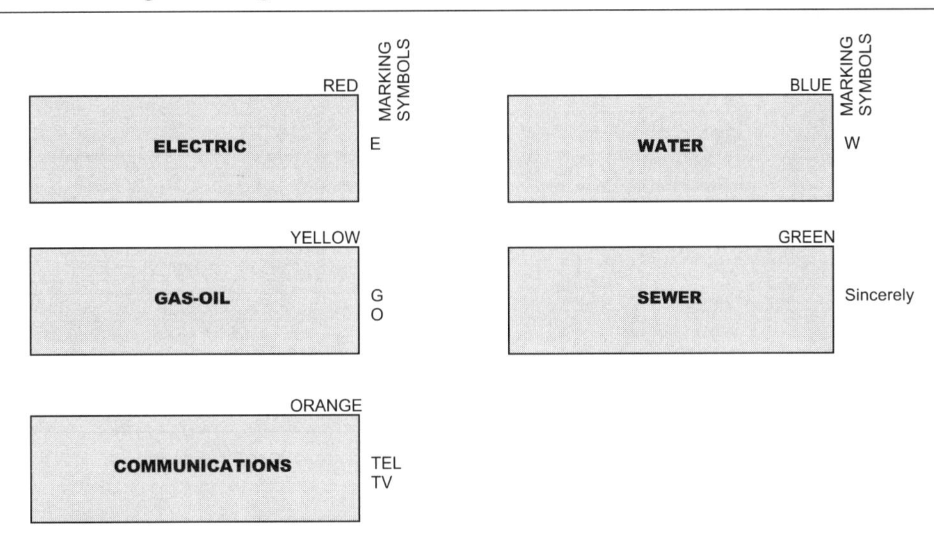

Approximate angle of repose for sloping sides of excavations

This information may be used by anyone desiring to do so. Where excavation activities are regulated by a regulatory body, the user should consult and use the regulations promulgated by the regulatory body.

Note: Clays, SIMS, loams, or non-hermorporus soils require shoring and bracing. The presence of groundwater requires special treatment.

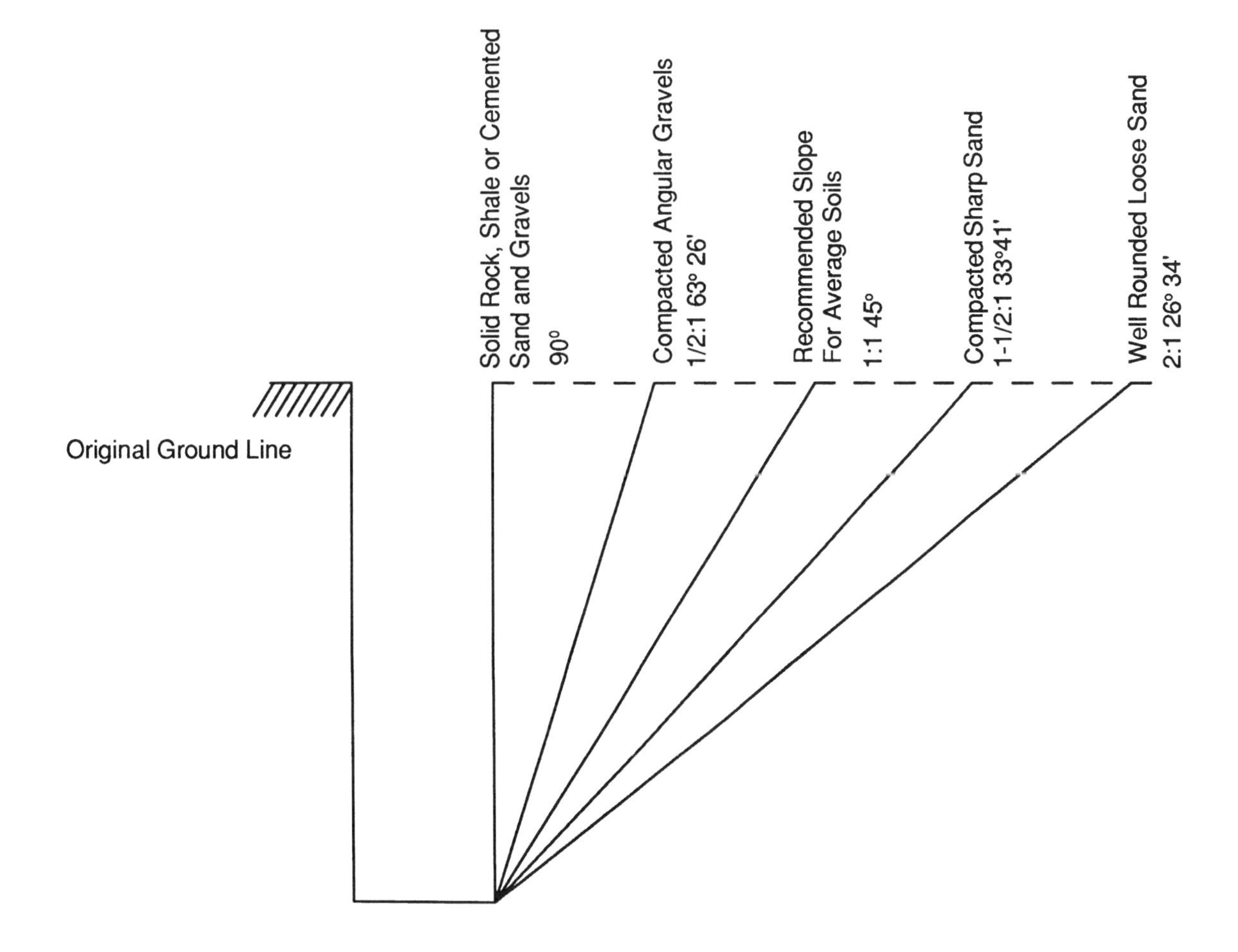

How to read descriptions that show directions in terms of degrees

In Figure 6, the north-south line and the east-west line divide the circle into four equal parts with each part containing 90 degrees. Several different directions are shown in this diagram with the number of degrees each varies east or west from the north and south starting points (remember that all descriptions read from the north or south). Northwest is a direction that is halfway between north and west. In terms of degrees, the direction north-west would read north 45 degrees west.

Notice the small tract of land in Figure 6. The following metes and bounds description will locate this tract: "Begin at the beginning point, thence N 20 degrees west—200 ft, thence N 75 degrees east—1,320 ft, thence S 30 degrees east—240 ft, thence S 45 degrees west—420 ft, thence west—900 ft back to the point of beginning, containing so many acres, etc."

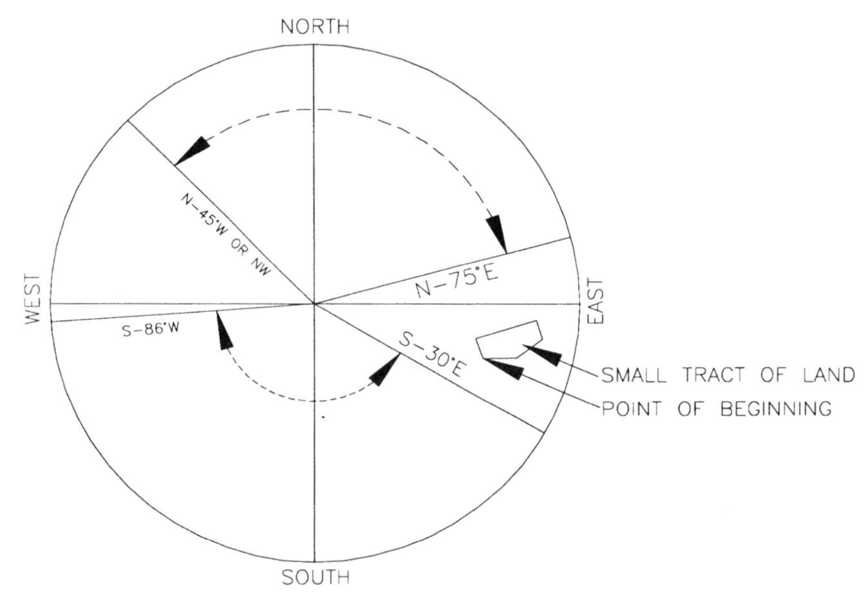

Figure 6. Land description in terms of degrees.

Size an air receiver for engine starting

Assume that the engine starter requires 16 standard cubic ft per second, that 10 seconds is required for starting, and that you want to store enough air in the receiver for three starts. Assume the initial pressure in the receiver is 215 psia and the final pressure is 115 psia.
where:

V = Receiver volume, cu. ft
R = 53.5 for air
T = Temperature (assumed constant at 520°R)
M = Weight of air required to start the engine based on density of air = 0.0763 pounds per cu. ft
P = Pressure, lb./sq. ft

Example.

M = 16 cu.ft./sec. × 10 sec./start × 3 starts × 0.0763 lb. per cu. ft
= 36.624 lb.

$$V = \frac{53.5 \times 520 \times 36.624}{144 \times (215 - 115)}$$

$$V = 70.76 \text{ cu.ft}$$

Assume receiver is 30″ or 2.5 ft inside diameter. Find the height of the receiver.

$$V = \frac{\Pi \times ID^2}{4} \times H$$

where:

V = Volume to be contained in receiver, cu. ft
ID = Inside diameter of the receiver, ft
H = Height of receiver, ft
H = 14.4 ft

Use two receivers, each 7.5 ft high.

Wind chill chart

Wind Speed		Cooling Power of Wind Expressed as "Equivalent Chill Temperature," °F																				
Knots	MPH																					
Calm	Calm	40	35	30	25	20	15	10	5	0	–5	–10	–15	–20	–25	–30	–35	–40	–45	–50	–55	–60
		Equivalent Chill Temperature																				
3–6	5	35	30	25	20	15	10	5	0	–5	–10	–15	–20	–25	–30	–35	–40	–45	–50	–55	–65	–70
7–10	10	30	20	15	10	5	0	–10	–15	–20	–25	–35	–40	–45	–50	–60	–60	–70	–75	–80	–90	–95
11–15	15	30	15	10	0	–5	–10	–20	–25	–30	–40	–45	–50	–60	–65	–70	–80	–85	–90	–100	–105	–110
16–19	20	25	10	5	0	–10	–15	–25	–30	–35	–45	–50	–60	–65	–75	–80	–85	–95	–100	–110	–115	–120
20–32	25	20	10	0	–5	–20	–20	–30	–35	–45	–50	–60	–65	–75	–80	–90	–95	–105	–110	–120	–125	–135
24–28	30	15	5	0	–10	–25	–25	–30	–40	–50	–55	–65	–70	–80	–85	–95	–100	–110	–115	–125	–130	–140
29–32	35	10	5	–5	–10	–30	–30	–35	–40	–50	–60	–65	–75	–80	–90	–100	–105	–115	–120	–130	–135	–145
33–36	40	10	0	–5	–10	–20	–30	–35	–45	–55	–60	–70	–75	–85	–95	–100	–110	–115	–125	–130	–140	–150
Winds above 40°F have little additional effect		Little Danger					Increasing Danger (Flesh may freeze within 1 minute)						Great Danger (Flesh may freeze within 30 seconds)									

PIPELINE PIGGING

Sizing plates

Often the debris-removal operation, after completion of construction, is combined with gauging to detect dents and buckles. This operation will prove that the pipeline has a circular hole from end to end. Typically an aluminum disc with a diameter of 95% of the nominal inside diameter of the pipe is attached to the front of a pig and is inspected for marks at the end of the run. The pig could also be equipped with a transmitter to facilitate tracking the location of the pig. If the pig hangs, this will facilitate easy location of the pig to locate the dent/buckle.

When constructing offshore pipelines, the most likely place for a buckle to occur during the lay operation is in the sag bend just before the pipe touches the bottom. A gauging pig can be placed inside the pipe and pulled along the pipe. If the lay barge moves forward and the pig encounters a buckle or dent, the pull line will become taut. This indicates that it will be necessary to pick up and replace the dented section of pipe.

Caliper pigging

Caliper pigs are used to measure pipe internal geometry. Typically they have an array of levers mounted in one of the cups. The levers are connected to a recording device in the body. As the pig travels through the pipeline, the deflections of the levers are recorded. The results can show up details such as girth-weld penetration, pipe ovality, and dents. The body is normally compact, about 60% of the internal diameter, which, combined with flexible cups, allows the pig to pass constrictions up to 15% of bore.

Caliper pigs can be used to gauge the pipeline. The ability to pass constrictions such as a dent or buckle means that the pig can be used to prove that the line is clear with minimum risk of jamming. This is particularly useful on subsea pipelines and long landlines where it would be difficult and expensive to locate a stuck pig. The results of a caliper pig run also form a baseline record for comparison with future similar surveys, as discussed further below.

Cleaning after construction

After construction, the pipeline bore typically contains dirt, rust, and millscale; for several reasons it is normal to clean these off. The most obvious of these reasons is to prevent contamination of the product. Gas feeding into the domestic grid, for example, must not be contaminated with particulate matter, since it could block the jets in the burners downstream. A similar argument applies to most product lines, in that the fluid is devalued by contamination.

A second reason for cleaning the pipeline after construction is to allow effective use of corrosion inhibitors during commissioning and operation. If product fluid contains corrosive components such as hydrogen sulfide carbon dioxide, or the pipeline has to be left full of water for some time before it can be commissioned, one way of protecting against corrosive attack is by introducing inhibitors into the pipeline. These are, however, less effective where the steel surface is already corroded or covered with millscale, because the inhibitors do not come into intimate contact with the surface they intended to protect.

Thirdly, the flow efficiency is improved by having a clean line and keeping it clean. This applies particularly to longer pipelines where the effect is more noticeable.

Therefore, most pipelines will be required to be clean commissioning. Increasingly, operators are specifying that the pipe should be sand blasted, coated with inhibitor, and have its ends capped after traction in order to minimize the postconstruction cleaning operation. A typical cleaning operation would consist of sending through a train of pigs driven by water. The pigs would have wire brushes and would permit some bypass flow of the water so that the rust and millscale dislodged by the brushing would be flushed out in front of the pigs and kept in suspension by the turbulent flow. The pipeline would then be flushed and swept out by batching pigs until the particulate matter in the flow has reduced to acceptable levels.

Following brushing, the longer the pipeline the longer it will take to flush and sweep out the particles to an acceptable level. Gel slugs are used to pick up the debris into suspension, cleaning the pipeline more efficiently. Gels are specially formulated viscous liquids that will wet the pipe surface and pick up and hold particles in suspension. A slug of gel would be contained between two batching pigs and would be followed by a slug of solvent to remove any traces of gel left behind.

Flooding for hydrotest

In order to demonstrate the strength and integrity of the pipeline, it is filled with water and pressure tested. The air must be removed so that the line can be pressurized efficiently because, if pockets of air remain, these will be compressed and absorb energy. It will also take longer to bring the line up to pressure and will be more hazardous in the event of a rupture during the test. It is therefore necessary to ensure that the line is properly flooded and all of the air is displaced.

A batching pig driven ahead of the water forms an efficient interface. Without a pig, in downhill portions of the line, the water will run down underneath the air trapping pockets at the high points. Even with a pig, in mountainous terrain with steep downhill slopes, the weight of water behind the pig can cause it to accelerate away, leaving a low pressure zone at the hill crest. This would cause dissolved air to come out of the solution and form an air lock. A pig with a high pressure drop across it would be required to prevent this.

Alternatives to using a pig include flushing out the air or installing vents at high point. For a long or large-diameter pipeline achieving sufficient flushing velocity becomes impractical. Installing vents reduces the pipeline integrity and should be avoided. So for flooding a pipeline, pigging is normally the best solution.

Dewatering and drying

After the hydrotest has been completed, the water is generally displaced by the product or by nitrogen. The same arguments apply to dewatering as to flooding. A pig is used to provide an interface between the hydrotest water and the displacing medium so that the water is swept out of all low points. A bidirectional batching pig may be used during hydrotest and left in the line during hydrotest and then reversed to dewater the line.

In some cases, it may be necessary to dry the pipeline. This is particularily so for gas pipelines where traces of water may combine with the gas to form hydrates. Drying is also required for chemical pipelines such as ethylene and propylene pipelines, since water will contaminate the material and make it unuseable. After dewatering the pipe walls will be damp and some water may remain trapped in valves and dead legs. This problem is solved by designing dead legs to be self draining and installing drains on valves.

Refer to Chapter 5, Pipeline Drying, for a detailed discussing of pipeline drying methods.

Estimate volume of onshore oil spill

An irregular shape can be converted into a series of rectangles that approximate the area of the irregular shape. There will be about the same amount of spill area outside the rectangle as there is dry area inside the rectangle. This can be done by stretching a steel tape along the ground outside the spill area. The area can then be quickly estimated by multiplying the length of the sides. In Figure 1, the following area is determined:

Area A = 70′ × 20′ = 1400 sq. ft
Area B = 60′ × 10′ = 600 sq. ft
Area C = 35′ × 20′ = 700 sq. ft
Total = 2700 sq. ft

The more rectangles used, the more accurate the estimate becomes.

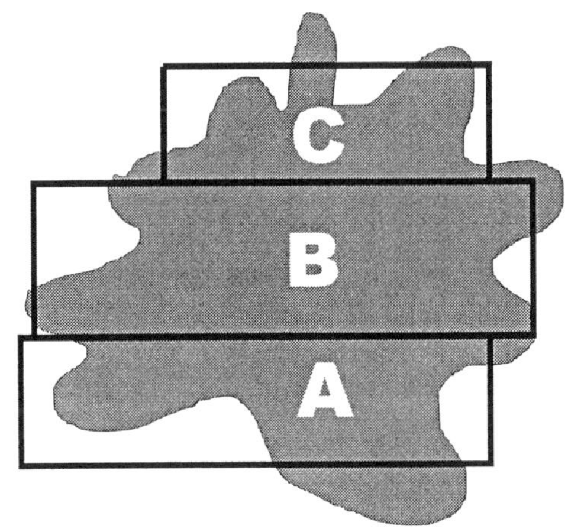

Figure 1

The next task is to estimate the average depth of oil in each of the areas. The oil will vary from very shallow at the edge to whatever depth the terrain is at the lowest point. This can be determined by "gauging" with a stick if it is shallow or accessible. If the pool is wider, you can heave a large stone into the pool to confirm depth. A good estimate can usually be made by observing the slope of the ground around the pool and assuming that the slope continues under the surface of the oil.

If you estimate that the deepest point in Area "A" is 20″ and Area A has three boundaries of "shore," divide the depth figure by 3 to obtain average depth. If it has two "shore" boundaries, like Area "B," divide the depth by 2 to obtain average area depth.

The irregular-shaped area with unseen bottom has now been reduced to a familiar shape. The volume of free oil in Area "A" is:

Area "A" 70′ × 20′ = 1400 sq. ft
Average depth = 20″/3 = 7″ or 0.6 ft
Area "A" volume = 1400 sq. ft × 0.6 ft = 840 cu. ft
The total volume will be the sum of volumes for Areas "A," "B," and "C."

Next, convert 840 cu. ft to barrels. Each cubic foot is equivalent to 0.178 bbls.

Area "A" volume = 840 cu. ft and therefore 840 × 0.178 = 150 bbls.

Determining how much additional oil has penetrated into the soil can be accurately measured by taking a core sample of the oil-covered soil; however, the following rule should suffice for estimates of oil spilled.

For penetration allowance in normal sand or soil, add 5% to the total volume for every foot of average depth.

In the case of Area "A," the average depth was 0.6 ft; therefore, 0.6 × 5% = 3% to be added. 150 bbls × 1.03 = 154.5 bbls total volume spilled in Area "A."

- Do not add a penetration allowance to areas with slopes that allow a reasonable flow rate.
- Add an allowance for slow-flowing areas.
- Reduce allowance by half if the area is wet from rain.

If more precise determination is required, drive a clear plastic tube, about 2″ or larger in diameter to a depth of 6″ in the uncontaminated soil adjacent to the spill. Twist and remove with soil core. Seal the bottom of the tube with plastic and tape. Pour free oil into the tube to the depth of the oil in the pool, mark the level, and let it set for 1 hour. Measure how much the oil level has dropped. Observe how deep the oil has penetrated. Retain the model to observe increased penetration with time.

Walk-around method

If the pool of oil is roughly circular, you can estimate its area by pacing around the pool and counting your paces. Walk as close to the pool edge as possible. Try to make your paces 3 ft, or 1 yard long. If you counted 700 paces, the circumference is 700 × 3 or 2100 ft. The next step is to guess how much smaller the actual pool is, compared to the circle you walked. If you were pretty close, deduct 10%.

2100 × 0.9 = 1890 ft adjusted circumference.

The diameter of a circle is related to the circumference by the following equation:

$C = \Pi D$

where:

$\Pi = 3.14186$
D = diameter
C = circumference

$D = 1890/\Pi = 602$ ft
The radius of the pool is D/2 or 301 ft

The area of the pool $= \Pi r^2$
A = 301 × 3.14186 × 301 × 301
A = 284,487 sq. ft

Now you can estimate the average depth by guessing the maximum depth. Assume the depth from the exposed slope to be 12″ at the deepest part, divide by four (four sloping sides) to estimate an average depth of 3″ or 0.25 ft.

The volume is:

V = 284487 × 0.25 = 71122 cu. ft

Volume of oil = 71122 × 0.178 = 12660 bbls.

The average depth was 3″ and, therefore, we need to add about 1% for penetration or 1.01 × 12660 bbls = 12,786 bbls.

Average diameters

You can also estimate the area of an oval-shaped pool by pacing off (3 ft per step) the width of the short diameter and the long diameter and averaging the diameters.

Pace off the short diameter, but stop short to allow for the irregular shape. Repeat the procedure for the long diameter. Add the diameters together and divide by 2 to get the average diameter.

Example. Short diameter = 75 paces = 75 × 3 = 225 ft
Long diameter = 120 paces = 120 × 3 = 360 ft
Average diameter = (225 + 360)/2 = 292 ft
Radius = 292/2 = 146 ft

$A = r^2 = 3.14186 \times 146 \times 146 = 66971$ sq. ft
Average depth = 3″ or 0.25 ft
Volume = 66971 × 0.25 = 16743 cu. ft
Volume = 2980 bbls.

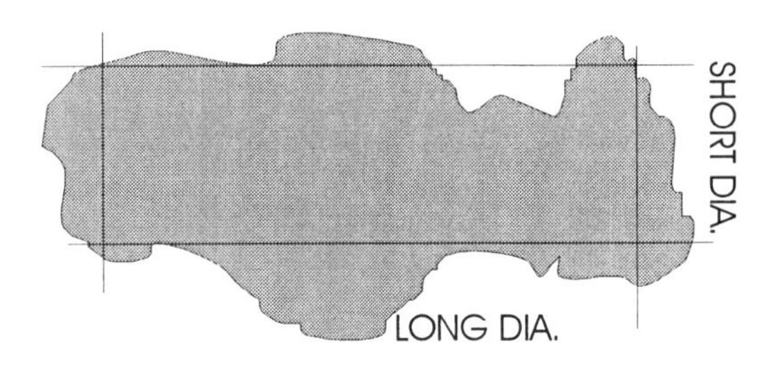

Figure 2. Average Diameters.

Estimating spill volume on water

When conditions permit, direct measurements of spill parameters are preferred over visual estimates.

A rough estimate of spill volume can be generated from observations of the oil slick's size and thickness. Figure 3 and Table 1 relate the appearance, thickness, and the light conditions. For example, slick thickness greater than 0.08 in. cannot be determined by appearance alone.

Since oil slick spreading is influenced by the spill volume as well as physical forces, stopping the spill at its source is critical in controlling the spread of a slick on water. The more conservative the first estimate of the spill volume, the better the chances that response forces will arrive at the spill site prepared with adequate and appropriate equipment. It is preferable to overrespond early rather than underrespond and risk unpreparedness. To underrespond will impede the effectiveness of spill control and cleanup efforts. A slow or poorly prepared initial response can incur more operational costs and increase the risk of damage to marine and shoreline resources and environments. Therefore, properly planning the initial response is critical in a spill situation.

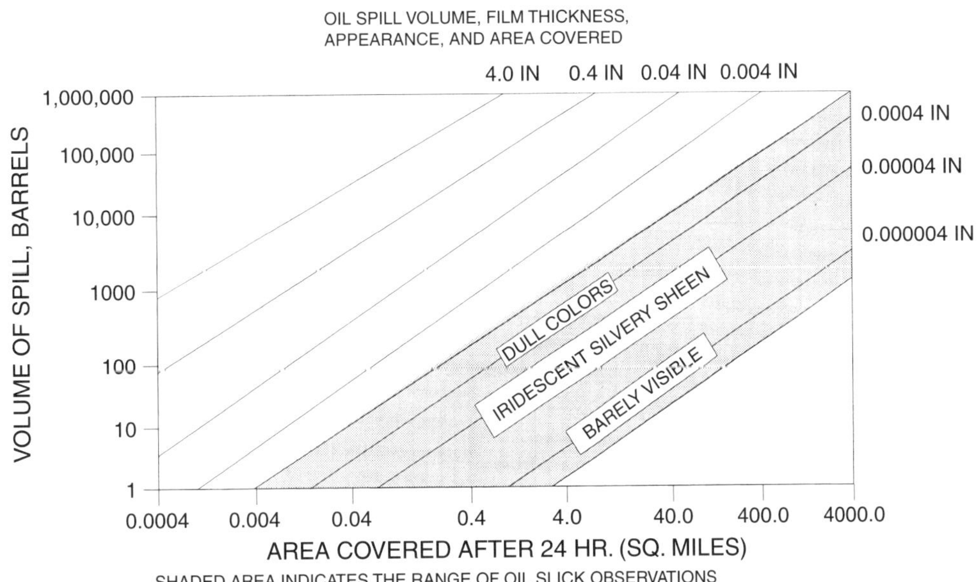

SHADED AREA INDICATES THE RANGE OF OIL SLICK OBSERVATIONS FOR WHICH THICKNESS AND AREA COVERED CAN BE DETERMINED BY APPEARANCE. ANY VALUE BELOW THE SHADED AREA WOULD NOT BE VISIBLE, AND ANY VALUE ABOVE THE SHADED AREA WOULD BE DARK BROWN OR BLACK.

Figure 3

Table 1
Estimating Spill Volume by Color and Coverage Area

		Silvery Sheen	Trace of Color (Yellow, Bronze Violet)	Bright Bands of Color (Purple, Blue to Green)	Colors Turning Dull (Brick Red, Turquoise, Pale Yellow)
Width × Length	Sq. Ft	Gals	Gals	Gals	Gals
100 × 500	50,000	0.1	0.2	0.4	1.2
100 × 1,000	100,000	0.2	0.4	0.7	2.5
100 × 2,000	200,000	0.4	0.7	1.5	4.9
200 × 2,000	400,000	0.7	1.5	3.0	9.9
500 × 1,000	500,000	0.9	1.9	3.7	12.3
200 × 5,000	1,000,000	1.9	3.7	7.5	24.7
500 × 5,000	2,500,000	4.7	9.4	18.7	61.7
500 × 10,000	5,000,000	9.4	18.7	37.4	123.4

Fluid Power Formulas

Heat equivalent of fluid power: BTU/H = PSI × GPM × 1½.

GPM = gallons per minute
PSI = pounds per square inch

Hydraulic cylinder piston travel speed: S = CIM ÷ A.

S = piston travel speed, inches per minute
CIM = oil flow into cylinder, cubic inches per minute

Force for piercing or shearing sheet metal: F = P × T × PSI.

F = force required, pounds
P = perimeter around area to be sheared, inches
T = sheet thickness, inches
PSI = shear strength rating of the material, pounds per square inch

Side load on pump or motor shaft: F = (HP × 63,024) ÷ (RPM × R).

F = side load, pounds, against shaft
R = pitch radius of sheave on pump shaft, inches
HP = driving power applied to shaft

Effective force of a cylinder working at an angle to direction of the load travel: F = T × sin A.

T = total cylinder force, pounds
F = effective force, pounds
A = least angle between cylinder axis and load direction, degrees

Relationship between displacement and torque of a hydraulic motor: T = D × PSI ÷ 24Π.

T = torque in foot pounds
D = displacement in cubic inches per revolution
PSI = pressure difference across the motor

Thrust or force of a cylinder: T = A × PSI.

T = thrust, pounds
A = effective area of cylinder, square inches
PSI = pounds per square inch gauge pressure

Recommended maximum oil velocity in hydraulic lines.

Pump suction lines: 2 to 4 feet per second (fps)
Pressure lines to 500 psi: 10 to 15 fps
Pressure lines to 3000 psi: 15 to 20 fps
Pressure lines over 3000 psi: 25 fps

2: Construction

PROJECT SCOPING DATA

Project scoping data worksheet for major facilities

Project Name

____

Location ______________________

Latitude ______________________

Longitude ______________________

☐ Offshore ☐ Onshore ☐ Arctic

☐ Marsh Site elevation, ft ______________

Project description ______________________

____

____

____

____

1.0 Type of contract required

1.1 Describe the type of contract that the project will be based on.
1.2 What is the completion date?
1.3 What type permits will be required and who will obtain them?
1.4 Will an environmental impact statement be required?
1.5 Describe currency requirements if in a foreign country.
1.6 If the project is located in a foreign country, does the country have regulations/codes that will need to be complied with?

2.0 General engineering

2.1 Does the client have engineering standards and specifications, or will it be necessary to develop these, or can the Engineering Contractor's specifications and standards be used?
2.2 Will in-country engineering concerns, such as design institutes, be required to provide engineering services?
2.3 If in-country engineering facilities are required, how much assistance will be required?
2.4 Describe any existing above or underground obstructions at the job site.
2.5 Obtain copies of any applicable local codes dealing with air and/or water pollution, sanitary systems, electrical systems, and structures. Obtain copies of any regulations dealing with waste disposal.
2.6 Is soil bearing data available?
2.7 If the project involves a pipeline that transports heat-sensitive fluid, are any soil thermal conductivity data available?
2.8 Describe type of soil—sandy, rocky, caliche, etc.
2.9 Does the site drain naturally? If fill material is required, what is the availability? Is sand-asphalt mix available?

3.0 Drafting

3.1 Will the drawings need to be sealed by a professional engineer?
3.2 Determine drawing forms to be used and type of drawings required—single line or double line.

4.0 Structural and architectural

4.1 Will all exposed steel need to be galvanized or have other protective coating?
4.2 Is sand, rock, cement, and ready-mixed concrete available locally? Price?
4.3 Requirements for building design:
    Steel
    Concrete/brick
    Cooling and heating requirements
    Other requirements

5.0 Piping

5.1 How will storm water, waste water, and sewage be disposed of? Will an oil-water separator be required?
5.2 Will connections to any existing pipelines be required?
5.3 Are there any facilities outside the job site that must be given consideration in the development of a plot plan?
5.4 What is the pressure rating for any existing manifolds, suction lines, or outgoing pipelines?
5.5 Are there any special requirements for isolating the facility?
5.6 Will scraper traps be required? Incoming, outgoing, or both?
5.7 How many relief headers are required?
5.8 If system relief valves are required, where will they relieve to?
5.9 Will a fire water system be required? Is a source of water available? If a source is available, what is the capacity of the source?
5.10 Is the water fresh, brackish, or seawater?

6.0 Electrical

6.1 Is power available from the local power grid? If yes, specify voltage, frequency, phase, and system capacity.
6.2 If power generation will be required in connection with the project, will it be a requirement to connect to a local grid?
6.3 If local power is available, define the approximate location where the power will enter the job site.
6.4 Will the local power entity furnish and install any substation equipment that will be required?
6.5 If power is to be supplied from a local power grid, attempt to determine the short-circuit capacity of the system.
6.6 Will across-the-line starting of large electric motors be permitted?
6.7 If across-the-line starting is not permitted, what is the maximum horsepower that will be permitted for across-the-line starting?

7.0 Instrumentation equipment

7.1 What type of control panel will be required?
7.2 Define electrical classification that will be required for a control house if a control house is required.
7.3 Are local communications facilities available? Describe.

8.0 Equipment

8.1 Have any long-delivery items been placed on order? List items and delivery schedule.
8.2 Who will supply general equipment specifications?
8.3 Will in-country purchasing be required?
8.4 List design criteria for equipment sizing.
8.5 Define spare equipment requirements. If equipment is purchased outside the country where the project is located, consideration should be given to transit time from country of origin to the job site.
8.6 Define the design life for the facility.
8.7 Will oil storage tanks be required? Will secondary seals be required? Will tank mixers be required?

9.0 Material purchasing

9.1 If a preferred vendor list is available, obtain a copy of the list.
9.2 If such list is not available, will it be a requirement to develop a list for the client's approval?
9.3 Determine the shipping address.
9.4 If material is shipped to a foreign location, who will handle receipt of the material and clearance with local customs?
9.5 Is the location near an established seaport? If not, determine the nearest seaport and logistics for moving material from the seaport to the job site. This should include any weight limitations, load width restrictions, and requirements for moving heavy equipment at night during periods of light road traffic.
9.6 Where will material be stored pending installation?
9.7 What procedure will be used for turning material over to the contractor?

10.0 Special considerations

10.1 Is heavy equipment available locally and does it have the capacity to lift the heaviest piece of material?
10.2 Is an adequate supply of sufficiently skilled local labor available?
10.3 If the project is in a foreign location, is there a limit on the use of expatriates?
10.4 What is the procedure for obtaining a work permit?
10.5 What is the availability of contractors that may be working in the area?
10.6 Will a construction camp be required?
10.7 If a construction camp is required, what is the availability of a local catering service?

11.0 Environmental information

11.1 Ambient Temp. (°F) Max. _______ Min._______
11.2 Relative humidity
Wet bulb (°F) Max. _____ Min. ____
Dry bulb (°F) Max. _____ Min. ____
11.3 Water temp. surface (°F) Max. _____ Min. ____
11.4 Water temp. bottom (°F) Max. _____ Min. ____
11.5 Potential for icing Y or N
11.6 Rainfall (inches/hour) Max. _____ Min. ____
11.7 Prevailing wind direction
11.8 Velocity Max. _____ Min. ____
11.9 Weather window (time of year) _________

12.0 Fluid characteristics

12.1 **Crude oil**
12.2 Sediment and water, % ________________
12.3 Free water __________________________
12.4 Salt content (#/1,000 bbls) ______________
12.5 API gravity ___________ @ ___________ °F and __________@ _____ °F (2 required)
12.6 Viscosity ___________ cP @ ___________ °F and __________cP @ _____ °F (2 required)
12.7 Oil pour point _______ °F
12.8 Reid vapor pressure ________________

12.9 Produced water: oil & grease (ppm) _____
12.10 Total suspended solids (mg/kg oil) ______
12.11 **Natural Gas**
12.12 MOL % inerts ____________________
12.13 MOL % CO_2 ____________________
12.14 ppm H_2S or grains/100 SCF _________
12.15 Water content _________ #/MMSCF or dew point _________ °F
12.16 Max. heating value (BTU/SCF) _______
12.17 Min heating value (BTU/SCF) ________
12.18 NGL content ______ gal/MCF or dew point _________ °F

13.0 Offshore Environmental Information

Parameter	1 Year	5 Year	25 Year	100 Year
Waves Direction Significant wave height, ft Maximum wave height, ft Crest elevation, ft Significant max. wave period, sec Peak period, sec				
Tides Astronomical tide, ft Storm surge, ft				
Wind @ 30 ft elevation above MLW 1-minute average, kt 1-hour average, kt 2-second gust, kt				
Current Surface speed, kt 3 ft from bottom, kt				
Wave force coefficients C_D with marine growth C_D with no marine growth C_M				
Marine growth to 150 ft, inches				
Sea temperature, °F Maximum surface Minimum surface Maximum bottom Minimum bottom				
Air temperature, °F Maximum Minimum				
Relative humidity Wet bulb, °F Dry bulb, °F				
Notes/Comments				

RIGHT-OF-WAY

How to determine the crop acreage included in a right-of-way strip

Multiply the width of the strip in feet by the length in rods; divide this by 2,640 to obtain the acreage. If the ends of the strip are not parallel, use the length of the center line of the right-of-way.

Example. A right-of-way 35 ft wide crosses a cultivated field for a length of 14 rods; how many acres of crop were destroyed?

35 × 14/2, 640 = 0.18 acres, or almost ⅕ acre.

Example. A right-of-way 50 ft wide crosses a field for 330 rods. How many acres of crop were destroyed?

50 × 330/2, 640 = 6.25 acres.

The rule is exact, not an approximation.

Clearing and grading right-of-way: labor/equipment considerations

To estimate labor crew and equipment spread for clearing and grading operations, the following items, as they may apply to a given project, should be given consideration:

1. Removal of trees, brush, and stumps.
2. Grubbing and removal of stumps that are in the way of the ditch.
3. Disposal of all debris, including method of disposal and length of haul.
4. Clearing area spoil a sufficient distance from the ditch line so that the spoil-bank from the ditching operations will not fall in any foreign material that might become mixed with the excavated spoil.
5. Cutting of merchantable timber into standard lengths and stacked along the right-of-way for disposition by others if specifically required by the right-of-way agreement.
6. Providing temporary walks, passageways, fences, or other structures so as not to interfere with traffic.
7. Providing sufficient and proper lighting where required.
8. Providing guards where required.
9. Preserving all trees, shrubs, hedges and lawns where required.
10. Grading irregularities where required.
11. Preserving topsoil for replacement, through all cultivated or improved fields and pastures, to its original position.
12. Proper grading of the terrain so as to allow passage of loaded trucks and equipment hauling materials and so ditching operations can be properly performed.
13. Protecting and preserving existing drainage facilities.
14. Protecting any existing structures or pipelines.
15. Protecting any telephone or utility lines and keeping them in service.
16. Cutting through fences and hedges where required and replacing these when necessary.
17. Installing gates and fences where required.

Estimating manhours for removing trees

NET MANHOURS—EACH

Average Tree Diameter in Inches	Softwood Trees		Hardwood Trees	
	Open Area	Congested Area	Open Area	Congested Area
Cross-cut Saws				
4	1.49	1.86	1.88	2.35
6	2.26	2.83	2.82	3.53
8	3.24	4.02	4.00	4.96
10	4.10	5.08	5.00	6.20
12	4.98	6.18	6.00	7.44
14	6.39	7.86	7.70	9.47
16	7.39	9.09	8.80	10.82
18	8.32	10.23	9.90	12.18
20	10.58	12.91	12.60	15.37
24	12.71	15.51	15.12	18.45
30	17.09	20.85	20.10	24.52
36	20.50	25.01	24.12	29.43

NET MANHOURS—EACH

Average Tree Diameter in Inches	Softwood Trees		Hardwood Trees	
	Open Area	Congested Area	Open Area	Congested Area
Chain Saws				
4	0.37	0.46	0.47	0.59
6	0.57	0.71	0.71	0.89
8	0.81	1.00	1.00	1.24
10	1.03	1.28	1.25	1.55
12	1.25	1.55	1.50	1.86
14	1.60	1.97	1.93	2.37
16	1.85	2.28	2.20	2.71
18	2.08	2.56	2.48	3.05
20	2.65	3.23	3.15	3.84
24	3.18	3.88	3.78	4.61
30	4.28	5.22	5.03	6.14
36	5.13	6.26	6.03	7.36

Manhours include ax trimming, cutting down with cross-cut saws or chain saws, and cutting into 4-ft lengths for the tree diameter sizes as listed previously. Manhours are average for various heights of trees.

Manhours do not include hauling, piling and burning of trees or branches, or the removal of stumps.

Estimating manhours for removing tree stumps

NET MANHOURS—EACH

Item	Laborer	Oper. Engr.	Powder Man	Total
Grub & Removal by Hand				
8″ to 12″ diameter	6.00	—	—	6.00
14″ to 18″ diameter	7.50	—	—	7.50
20″ to 24″ diameter	9.00	—	—	9.00
26″ to 36″ diameter	11.20	—	—	11.20

NET MANHOURS—EACH

Item	Laborer	Oper. Engr.	Powder Man	Total
Blast & Pull with Tractor				
8″ to 12″ diameter	0.83	0.11	1.50	2.44
14″ to 18″ diameter	1.05	0.23	2.33	3.61
20″ to 24″ diameter	1.50	0.30	3.38	5.18
26″ to 36″ diameter	2.11	0.42	4.76	7.29

Manhours include excavating and removing by hand or blasting and removing with cables and tractors. Manhours do not include burning or removal from premises.

Clearing and grading right-of-way

Equipment Spread

Equipment Description	NUMBER OF UNITS FOR 50 Linear Ft Width				80 Linear Ft Width				100 Linear Ft Width			
	L	M	MH	H	L	M	MH	H	L	M	MH	H
D8 Tractor W/Dozer	1	1	1	1	1	1	2	2	1	2	3	3
D7 Tractor W/Dozer	0	0	1	1	1	1	1	1	1	1	2	2
Truck—$2^1/_2$ Ton Dump	2	2	2	2	2	2	2	2	2	2	3	3
Truck—Pick-up	1	1	2	2	1	2	3	3	2	3	4	4
Ripper or Brushrake	1	1	1	1	1	1	1	1	1	2	2	2

Above equipment spread should be ample for clearing and grading 1 mile of right-of-way per 10-hour day for the width and conditions outlined. Haul trucks are based on round trip haul of 2 miles. If brush and trees are to be burned on site, omit above dump trucks. Small tools such as saws, axes, etc., must be added as required for the individual job.

Code description

L = Light—light brush and grass, no trees.
M = Medium—considerable brush of larger size.
MH = Medium Heavy—large brush and small trees.
H = Heavy—much small brush, many small trees, and occasional large trees.

Labor Crew

Personnel Description	NUMBER OF MEN FOR 50 Linear Ft Width				80 Linear Ft Width				100 Linear Ft Width			
	L	M	MH	H	L	M	MH	H	L	M	MH	H
Foreman	1	1	2	2	1	2	3	3	2	3	4	4
Operator	1	1	2	2	1	2	3	3	2	4	5	5
Mechanic	1	1	1	1	1	1	1	1	1	1	1	1
Swamper	1	1	2	2	1	2	3	3	2	4	5	5
Truck Driver	2	2	2	2	2	2	2	2	2	2	3	3
Laborer	10	15	20	30	15	25	30	40	20	35	40	50
Total Crew	16	21	29	39	21	34	42	52	29	49	58	68

Above total crew should be ample for clearing and grading 1 mile of right-of-way per 10-hour day for the width and conditions outlined. Crew spread includes cutting, stacking or piling, loading, and hauling a round-trip distance of 2 miles. If burning is necessary or permitted, substitute fire tenders for dump truck drivers. See Clearing and Grading equipment spread for number of dump trucks.

Code description

L = Light—light brush and grass, no trees.
M = Medium—considerable brush of larger size.
MH = Medium Heavy—large brush and small trees.
H = Heavy—much small brush, many small trees, and occasional large trees.

Source

Page, J. S., *Cost Estimating Man-Hour Manual for Pipelines and Marine Structures*, Gulf Publishing Co., Houston, Texas, 1977.

DITCHING

How many cubic yards of excavation in a mile of ditch?

Multiply the width in inches by the depth in inches by 1.36; the answer is cubic yards per mile.

Example. How many cubic yards of excavation in a mile of 12-in. ditch 30-in. deep?

$12 \times 30 \times 1.36 = 490$ cubic yards per mile.

The rule is correct within about ⅕ of 1%, actually, the errors in depth and width are much greater than this. To get the cubic yards per 1,000 ft, as in computing rock ditching, use 0.257 instead of 1.36.

Shrinkage and expansion of excavated and compacted soil

Ever notice how the spoil from a ditch occupies a greater volume than the ditch itself? There's a reason. *Excavate sand and it expands about 10%. Ordinary soil expands about 25 percent and clay expands about 40%.*

Here's a summary of the bulk you can expect from excavated soil:

Type soil	Undisturbed	Excavated	Compacted
Sand	1.00	1.11	0.95
Ordinary earth	1.00	1.25	0.90
Clay	1.00	1.43	0.90

Example. Find the volume of loose spoil from a pipeline ditch 42 inches wide, 60 inches deep. The excavation is through clay.

Volume of undisturbed clay soil:

$3.5 \times 5.0 = 17.5$ cu ft per lineal foot of ditch.

Volume of the spoil will be 143% of the undisturbed volume.

Volume of spoil $= 1.43 \times 17.5 = 25+$ cu ft per lineal ft.

Through mechanical compaction this volume can be reduced to 15.75 cu ft.

Ditching and trenching: labor/equipment considerations

In determining the labor crew and equipment spread for ditching and trenching operations, the following should be given consideration should they apply to the particular project:

1. Ditching or trenching for buried pipelines should be in accordance with the following table of minimum width and coverage for all soil formations.
2. In rock, cut ditches at least 6 in. wider.

3. If dirt-filled benches are used, ditch should be excavated deeper to obtain proper coverage.
4. Trench should be excavated to greater depth when required for proper installation of the pipe where the topography of the country warrants same.
5. Repair any damage to and maintain existing natural or other drainage facilities.
6. Do not open ditch too far in advance of pipelay crew.
7. Obtain permits for blasting.
8. When blasting, use extreme caution and protection.
9. Clean up blasted rock to prevent damage to coated pipe.

Nominal Pipe Size Inches	Minimum Width Inches	Normal Minimum Coverage Inches
4	22	30
6	26	30
8	26	30
10	26	30
12	30	30
14	32	30
16	36	30
18	38	30
20	40	30
24	44	30
30	50	30
36	52	36
42	58	36

CONCRETE WORK

How to approximate sacks of cement needed to fill a form

To obtain a close estimate of the number of sacks of cement that will be required, first determine the volume (cubic feet) to be filled in the form. Divide the volume by 4.86 to approximate the number of sacks of cement needed.

Example. How many 94-lb. sacks of cement will be required to fill a form for a concrete base 10 ft by 10 ft if it is to be 6 in. thick?

$$\frac{10 \times 10 \times 0.5}{4.86} = 10.3\text{, or 11 sacks needed.}$$

What you should know about mixing and finishing concrete

To determine proper mix, divide the constant 44 by the sum of the parts of cement, the parts of sand, and parts of gravel to determine the number of bags of Portland cement required. Multiply the number of bags of cement as determined above by parts of sand and the constant 0.035 to calculate the number of cubic yards of sand needed. To determine cubic yards of gravel needed, multiply bags of cement by parts of gravel × 0.035.

Example. Calculate the quantities of cement, sand, and gravel required for 1 cubic yard of 1 : 2 : 4 concrete.

$$C = \frac{44}{1+2+4}$$
$$= 6.28 \text{ bags of cement}$$

$$S = 6.28 \times 2 \times 0.035$$
$$= 0.44 \text{ cubic yard of sand}$$

$$G = 6.28 \times 4 \times 0.035$$
$$= 0.88 \text{ cubic yard of gravel}$$

To increase or decrease slump of concrete, add or subtract 1 gallon of water per cubic yard of mix and subtract or add 20 lb. of aggregate to maintain yield.

To adjust from no air to air-entrained concrete and maintain strength, reduce water $^1/_4$ gallon per sack of cement and reduce sand 10 lb. per sack of cement for each 1% of entrained air.

PIPE LAYING

How to determine the degrees of bend in a pipe that must fit a ditch calling for a bend in both horizontal and vertical planes

Rule. To find the number of degrees in the combination bend, square the side bend and the sag or overbend; add them together and extract the square root. The answer will be the number of degrees necessary to make the pipe fit the ditch.

Example. Determine the bend to make in a pipe whose ditch has a 3° overbend and a 4° side bend. Let X = the unknown angle:

$X^2 = 3^2 + 4^2$

$X^2 = 9 + 16$

$X^2 = 25$

$X = 5°$

Example. Determine the bend to make in a pipe whose ditch has a 9° sag and 12° sidebend. Let X = the unknown angle:

$X^2 = 9^2 + 12^2$

$X^2 = 225$

$X = 15°$

How to bend pipe to fit ditch—sags, overbends, and combination bends

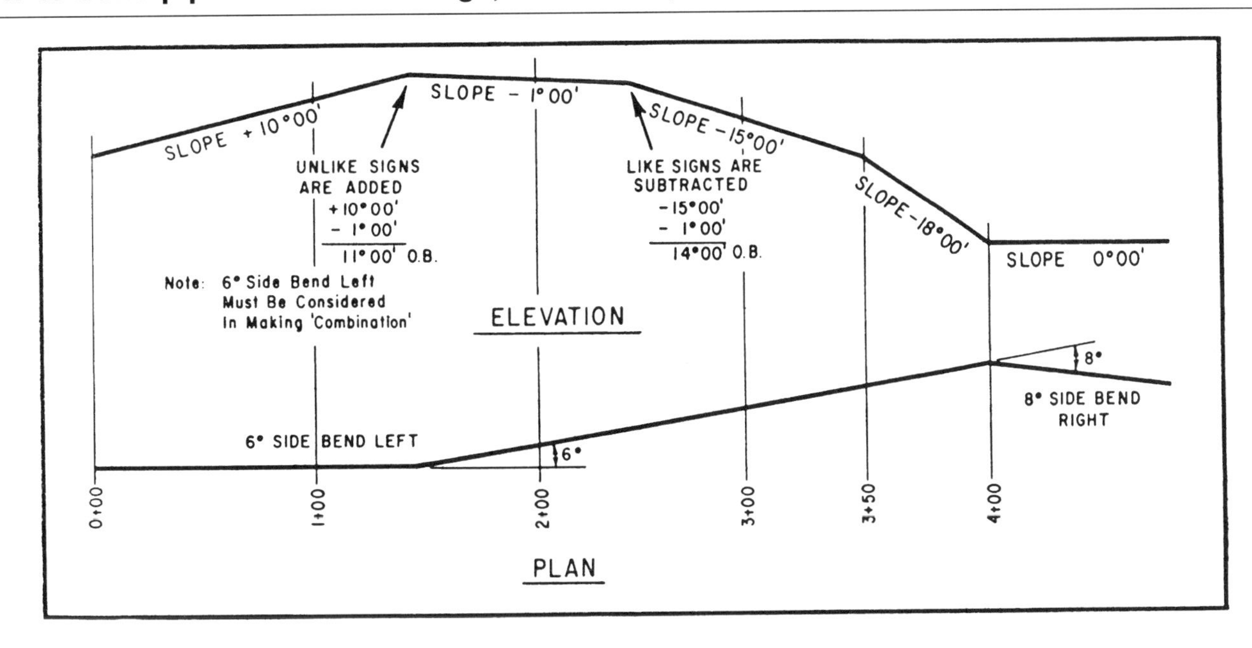

To make straight sags or overbends fit the ditch (see drawing), add angles of unlike signs and subtract those of like signs.

Example. (Sta. 2 + 40)

Slope−15°00′
Slope− 1°00′

(sub.) 14°00′ overbend

In cases involving either a sag or overbend, in addition to a side bend, the rule becomes: *Make the combination bend equal to the largest angle plus 1/3 of the smallest.*

Example. (Sta. 1 + 40)

Slope + 10°00′
Slope − 1°00′

(add) 11°00′ overbend (1)
6°00′ side bend left (2)

Combination bend = 11° + (× 6°) = 13°00′ overbend left.

Note: This rule gives an error of approximately 1° for a maximum bend of 18.5°.

Pipe bending computations made with hand-held calculator

Maximum code radii for pipe cold bends requires printer for calculations

Frank E. Hangs, Sovereign Engineering, Inc., Houston

Cold bending pipe is subject to provisions of Liquid Petroleum Transportation Piping Systems, B31.4, and Gas Transmission and Distribution Piping Systems, B31.8. These codes stipulate a minimum bending radius for each size pipe in addition to requirements for thinning, flattening, etc.

The following program (Table 1), written for the Hewlett Packard 41C/CV calculator, addresses the geometry of fabricating offset bends, sloping scraper traps, connections, and direction changes common to all pipelines. Minimum code radii are calculated for each size of oil or gas line. The vertical distance between below and above ground center lines is matched to a design distance with given tangents (5 ft is a convenient tangent for pipe bending machines). The bend angle, overall horizontal distance, and total length of pipe are calculated for each configuration.

A printer is a must. The prompting feature of this calculator asks for specific input data. The Results Recap routine prints out calculated data and inputs. Each item is identified for permanent record. Additional printouts of Results Recap may be made by XEQ "DATA."

Fig. 1 and Example 1 are a typical offset bend.
Fig. 2 and Example 2 are a conventional direction change.
Fig. 3 and Example 1 show receiver and launcher scraper trap connections, where the traps are not level.
Fig. 4 and Example 3 are a crossing under a foreign line, where offset bends are used.

The oil and gas codes differ in bending requirements. These are defined by the radius of bend/diameter ratio. The oil code, B31.4, specifies the R/D ratios as follows: for 12 in. and less, use 18; 14 in., 21; 16 in., 24; 18 in., 27; 20 in. and larger, use 30.

The gas code, B31.8, specifies for all pipe 12 in. and larger, the ratio is a constant of 38.1987. (The code states "longitudinal axis of the pipe shall not be deflected more than 1½ degrees in any length along the pipe axis equal to the diameter of the pipe.")

The program prompts: Gas pipeline greater than 12 in. Yes? or No? (Gas PL ≥ >12Y?N?). Is line gas or oil? When this information and other data are keyed in, the proper radius is determined and the results computed and printed out.

If the scraper trap routine is desired, it should be run immediately after "Bend" while pertinent data are in storage—XEQ 15 for each receiver and each launcher.

Here again the program asks if this is a receiver (R trap) Y? or N? Note that trap angle is requested for R & L traps (they could differ), and the length of the traps may not be equal.

Table 1
Examples of computations using the cold bend program

```
EXAMPLE 1

                 XEQ "BEND"

PIPE BENDS

∡1 TRIAL=?
            20.0000    RUN
D=? IN.
            12.7500    RUN
GAS PL=>12? Y?, N?
N                      RUN
V=?
             8.0000    RUN
T=?
             5.0000    RUN
∡1 INCR=?
             0.5000    RUN

RESULTS RECAP

∡1=25.5000
R/D=18.0000
R=19.1250
H=35.4929
L=37.0235
V CALC=8.0312
V GIVEN=8.0000

D IN=12.7500
D FT=1.0625
T=5.0000
∡1 INCR=0.5000

                 XEQ 15
SCRAPERTRAPS

TRAP ∡=?
             5.0000    RUN
R TRAP? Y? N?
Y                      RUN


VR=7.5227
HR=37.1407
LR=38.6925
TRAP ∡=5.0000
LRT=?
            10.0000    RUN

VRT=6.6511
HRT=47.1027
                 XEQ 15
SCRAPERTRAPS

TRAP ∡=?
             5.0000    RUN
R TRAP? Y? N?
N                      RUN

VL=8.3942
HL=33.8070
LL=35.3545
TRAP ∡=5.0000
```

```
EXAMPLE 1 CONT.

LLT=?
             5.0000    RUN

VLT=8.8300
HLT=38.7880

EXAMPLE 2

                 XEQ 17
 DIRECTION
 CHANGE
∡ TURN=?
            30.0000    RUN
R=?
            80.0000    RUN
T=?
             5.0000    RUN

RESULTS

V=13.2180
H=49.3301
L=51.8879


∡ TURN=30.0000
R=80.0000

EXAMPLE 3

                 XEQ "BEND"

PIPE BENDS

∡1 TRIAL=?
             8.0000    RUN
D=? IN.
            14.0000    RUN
GAS PL=>12? Y?, N?
N                      RUN
V=?
             2.5130    RUN
T=?
             5.0000    RUN
∡1 INCR=?
             0.5000    RUN

RESULTS RECAP

∡1=10.5000
R/D=21.0000
R=24.5000
H=28.7621
L=28.9797
V CALC=2.6429
V GIVEN=2.5130

D IN=14.0000
D FT=1.1667
T=5.0000
∡1 INCR=0.5000
```

Table 1
Examples of computations using the cold bend program (continued)

```
01◆LBL "BEND"
02 ADV
03 SF 12
04 "PIPE "
05 "⊢BENDS"
06 PRA
07 ADV
08 CF 12
09 "∡1 TRIAL=?"
10 PROMPT
11 STO 05
12 "D=? IN."
13 PROMPT
14 STO 06
15 12
16 /
17 STO 07
18 "GAS PL=>12?"
19 "⊢ Y?, N?"
20 AON
21 PROMPT
22 ASTO Y
23 AOFF
24 "Y"
25 ASTO X
26 X=Y?
27 GTO 12
28 12
29 RCL 06
30 X<Y?
31 GTO 06
32 14
33 RCL 06
34 X=Y?
35 GTO 07
36 16
37 RCL 06
38 X=Y?
39 GTO 08
40 18
41 RCL 06
42 X=Y?
43 GTO 09
44 19.99
45 RCL 06
46 X>Y?
47 GTO 10

48◆LBL 06
49 18
50 XEQ 13

51◆LBL 07
52 21
53 XEQ 13
54◆LBL 08
55 24
56 XEQ 13

57◆LBL 09
58 27
59 XEQ 13

60◆LBL 10
61 30
62 XEQ 13
63◆LBL 12
64 38.1972
65 XEQ 13

66◆LBL 13
67 STO 03
68 RCL 07
69 *
70 STO 08
71 GTO 05

72◆LBL 05
73 "V=?"
74 PROMPT
75 STO 09
76 "T=?"
77 PROMPT
78 STO 10
79 "∡1 INCR=?"
80 PROMPT
81 STO 11

82◆LBL 01
83 RCL 08
84 2
85 *
86 RCL 05
87 COS
88 CHS
89 1
90 +
91 *
92 RCL 10
93 2
94 *
95 RCL 05
96 SIN
97 *
98 +
99 STO 14
100 FS? 01
101 GTO 03
102 RCL 09
103 X<>Y
104 X<=Y?
105 GTO 02
106 GTO 03

107◆LBL 02
108 RCL 11
109 ST+ 05
110 GTO 01

111◆LBL 03
112 RCL 10
113 2
114 *
115 RCL 08
116 2
117 *
118 RCL 05
119 SIN
120 *
121 +
122 RCL 10
123 2
124 *
125 RCL 05
126 COS
127 *
128 +
129 STO 12
130 RCL 10
131 4
132 *
133 RCL 08
134 2
135 *
136 RCL 05
137 57.2958
138 /
139 *
140 +
141 STO 13

142◆LBL "DATA"
143 ADV
144 CLA
145 "RESULTS "
146 "⊢RECAP"
147 AVIEW
148 CLA
149 ADV
150 "∡1="
151 ARCL 05
152 AVIEW
153 CLA
154 "R/D="
155 ARCL 03
156 AVIEW
157 CLA
158 "R="
159 ARCL 08
160 AVIEW
161 CLA
162 "H="
163 ARCL 12
164 AVIEW
165 CLA
166 "L="
167 ARCL 13
168 AVIEW
169 CLA
170 "V CALC="
171 ARCL 14
172 AVIEW
173 CLA
174 "V GIVEN="
175 ARCL 09
176 AVIEW
177 CLA
178 ADV
179 "D IN="
180 ARCL 06
181 AVIEW
182 CLA
183 "D FT="
184 ARCL 07
185 AVIEW
186 CLA
187 "T="
188 ARCL 10
189 AVIEW
190 CLA
191 "∡1 INCR="
192 ARCL 11
193 AVIEW
194 CLA
195 BEEP
196 ADV
197 ADV
198 STOP

199◆LBL 15
200 SF 12
201 "SCRAPER"
202 "⊢TRAPS"
203 PRA
204 CF 12
205 ADV
206 "TRAP ∡=?"
207 PROMPT
208 STO 04
209 COS
210 CHS
211 1
212 +
213 RCL 08
214 *
215 STO 15
216 RCL 10
217 RCL 04
218 SIN
219 *
220 STO 16
221 RCL 08
222 RCL 04
223 SIN
224 *
225 STO 17
226 RCL 10
227 RCL 04
228 COS
229 *
230 STO 18
231 RCL 04
232 57.2958
233 /
234 RCL 08
235 *
236 STO 19
237 "R TRAP?"
238 "⊢ Y? N?"
239 AON
240 PROMPT
241 ASTO Y
242 AOFF
243 "Y"
244 ASTO X
245 X=Y?
246 GTO 16
247 RCL 14
248 RCL 15
249 -
250 RCL 16
251 +
252 STO 20
253 RCL 12
254 RCL 10
255 -
256 RCL 17
257 -
258 RCL 18
259 +
260 STO 21
261 RCL 13
262 RCL 19
263 -
264 STO 22
265 ADV
266 "VL="
267 ARCL 20
268 AVIEW
269 CLA
270 "HL="
271 ARCL 21
272 AVIEW
273 CLA
274 "LL="
275 ARCL 22
276 AVIEW
277 CLA
278 "TRAP ∡="
279 ARCL 04
280 AVIEW
281 CLA
282 "LLT=?"
283 PROMPT
284 STO 02
285 RCL 04
286 SIN
287 *
288 RCL 20
289 +
290 STO 26
291 RCL 02
292 RCL 04
293 COS
294 *
295 RCL 21
296 +
297 STO 27
298 ADV
299 "VLT="
300 ARCL 26
301 AVIEW
302 CLA
303 "HLT="
304 ARCL 27
305 AVIEW
306 CLA
307 ADV
308 ADV
309 STOP

310◆LBL 16
311 RCL 14
312 RCL 15
313 -
314 RCL 16
315 -
316 STO 23
317 RCL 12
318 RCL 10
319 -
320 RCL 17
321 +
322 RCL 18
323 +
324 STO 24
325 RCL 13
326 RCL 19
327 +
328 STO 25
329 ADV
330 ADV
331 "VR="
332 ARCL 23
333 AVIEW
334 CLA
335 "HR="
336 ARCL 24
337 AVIEW
338 CLA
339 "LR="
340 ARCL 25
341 AVIEW
342 CLA
343 "TRAP ∡="
344 ARCL 04
345 AVIEW
346 CLA
347 "LRT=?"
348 PROMPT
349 STO 01
350 RCL 04
351 SIN
352 *
353 CHS
354 RCL 23
355 +
356 STO 28
357 RCL 24
358 RCL 01
359 RCL 04
360 COS
361 *
362 +
363 STO 29
364 ADV
365 "VRT="
366 ARCL 28
367 AVIEW
368 CLA
369 "HRT="
370 ARCL 29
371 AVIEW
372 CLA
373 ADV
374 STOP

375◆LBL 17
376 SF 12
377 " DIRECTION "
378 "⊢ CHANGE"
379 PRA
380 CF 12
381 "∡ TURN=?"
382 PROMPT
383 STO 30
384 "R=?"
385 PROMPT
386 STO 31
387 "T=?"
388 PROMPT
389 STO 32
390 RCL 30
391 COS
392 CHS
393 1
394 +
395 RCL 31
396 *
397 RCL 30
398 SIN
399 RCL 32
400 *
401 +
402 STO 33
403 RCL 30
404 SIN
405 RCL 31
406 *
407 RCL 32
408 +
409 RCL 30
410 COS
411 RCL 32
412 *
413 +
414 STO 34
415 RCL 30
416 57.2958
417 /
418 RCL 31
419 *
420 RCL 32
421 2
422 *
423 +
424 STO 35
425 ADV
426 "RESULTS"
427 XEQ "PRA"
428 ADV
429 "V="
430 ARCL 33
431 AVIEW
432 CLA
433 "H="
434 ARCL 34
435 AVIEW
436 CLA
437 "L="
438 ARCL 35
439 AVIEW
440 CLA
441 ADV
442 ADV
443 "∡ TURN="
444 ARCL 30
445 AVIEW
446 CLA
447 "R="
448 ARCL 31
449 AVIEW
450 CLA
451 ADV
452 ADV
453 STOP
454 .END.
```

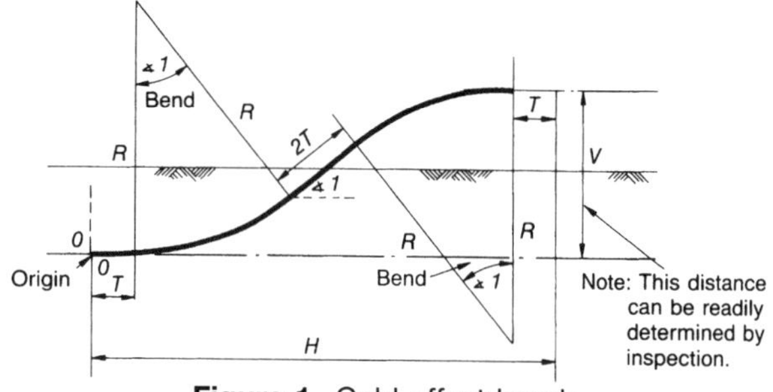

Figure 1. Cold offset bends.

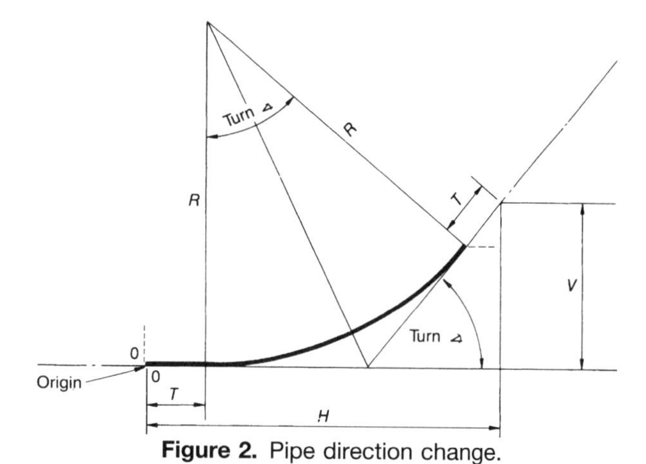

Figure 2. Pipe direction change.

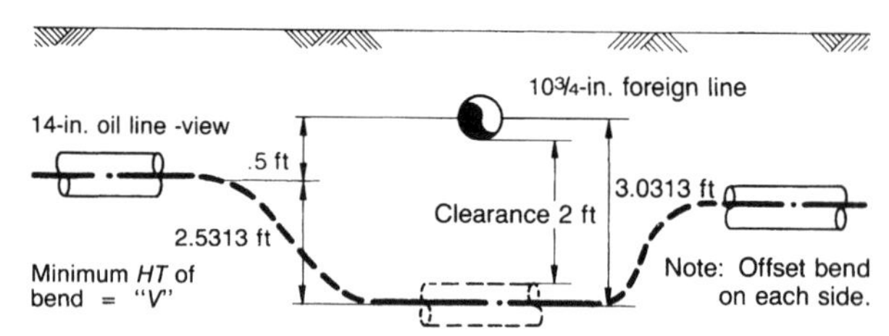

Figure 4. Crossing under foreign lines with offset bends.

Legend & registers

Reg. no.		
00	Not used	
01	LRT	Length of receiver trap (ft)
02	LLT	Length of launcher trap (ft)
03	R/D	Radius of bend/pipe diameter (ft/ft)
04	Trap ∡	Scraper trap angle in degrees
05	∡1	Bend angle in degrees
06	D–in.	Pipe diameter (in.)
07	D–ft	Pipe diameter (ft)
08	R	Bend radius (ft)
09	V	Vertical distance desired (ft between center lines)
10	T	Tangent (ft)
11	∡ Incr.	Bend angle increment in degrees
12	H	Horizontal distance, overall (ft)
13	L	Length of pipe (ft)
14	V Calc.	Vertical distance calculated (ft)
15	R(1–cos Trap ∡)	
16	T sin Trap ∡	
17	R sin Trap ∡	
18	T cos Trap ∡	
19	$R\frac{\text{Trap}\,\measuredangle}{57.2958}$	
20	VL	Vertical distance to launcher trap connection (ft)
21	HL	Horizontal distance to launcher trap connection (ft)
22	LL	Length of pipe to launcher trap connection (ft)
23	VR	Vertical distance to receiver trap connection (ft)
24	HR	Horizontal distance to receiver trap connection (ft)
25	LR	Length of pipe to receiver trap connection (ft)
26	VLT	Vertical distance to end of launcher trap (ft)
27	HLT	Horizontal distance to end of launcher trap (ft)
28	VRT	Vertical distance to end of receiver trap (ft)
29	HRT	Horizontal distance to end of receiver trap (ft)
30	∡	Change direction desired in degrees
31	Radius	Change direction (ft)
32	Tangent	Change direction (ft)
33	V	Vertical distance calculated, change direction
34	H	Horizontal distance calculated, change direction
35	L	Length of pipe calculated, change direction

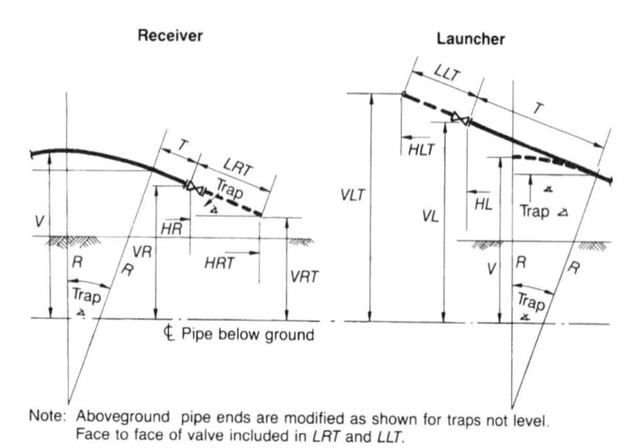

Note: Aboveground pipe ends are modified as shown for traps not level. Face to face of valve included in *LRT* and *LLT*.

Figure 3. Scraper traps.

The direction change, Example 2 (XEQ 17), can be run at any time since it prompts for input data. If this is a code bend, then R must be determined for pipe size and service of line by "bend" (use results of previous run for same diameter or key in data as shown on Example 1 to obtain R). Use this minimum radius or any larger radius. Points on a circle of given radius may be calculated for any angle. Subdivide angle, and for each subdivision key in R and let T = 0, thus determining points on curve.

Formulas for reverse bends

$$V = 2R(1 - \cos \measuredangle 1) + 2T \sin \measuredangle 1$$

$$H = 2R \sin \measuredangle 1 + 2T \cos \measuredangle 1 + 2T$$

$$L = 2R \frac{\measuredangle 1}{57.2958} + 4T$$

Formulas for scraper trap connections

Receiver.

$$VR = V - R(1 - \cos \text{ Trap } \measuredangle)T \sin \text{ Trap } \measuredangle$$

$$HR = H - T + R \sin \text{ Trap } \measuredangle + T \cos \text{ Trap } \measuredangle$$

$$LR = L + R \frac{\text{Trap } \measuredangle}{57.2958}$$

Launcher.

$$VL = V - R(1 - \cos \text{ Trap } \measuredangle) + T \sin \text{ Trap } \measuredangle$$

$$HL = H - T + R \sin \text{ Trap } \measuredangle + T \cos \text{ Trap } \measuredangle$$

$$LL = L - R \frac{\text{Trap } \measuredangle}{57.2958}$$

Receiver trap.

$$VRT = VR - LRT \sin \text{ Trap } \measuredangle$$

$$HRT = HR + LRT \cos \text{ Trap } \measuredangle$$

Launcher trap.

$$VLT = VL - LLT \sin \text{ Trap } \measuredangle$$

$$HLT = HL + LLT \cos \text{ Trap } \measuredangle$$

Formulas for direction change

$$V = R(1 - \cos \text{ Turn } \measuredangle) + T \sin \text{ Turn } \measuredangle$$

$$H = R(\sin \text{ Turn } \measuredangle) + T(1 + \cos \text{ Turn } \measuredangle)$$

$$L = 2T + R \frac{\text{Turn } \measuredangle}{57.2958}$$

User instructions

"Bend" program. Put in calculator, Size 35, XEQ "Bend." Key in prompted data and R/S each time. Key in a trial angle. Use a larger angle for smaller pipe as 15° for 12-in., 10° for 30-in. Results Recap prints out the calculated values and inputs. Each quantity is identified. Additional printouts may be made by XEQ "Data."

If one inadvertently puts in a larger trial angle than needed for a solution (i.e., calculated V approximates given V), only one calculation is made and printed out for this angle. If calculated V is too large, store smaller trial ∡ in 05, XEQ 01, to get proper result.

Remember to clear Flag 01 before resuming normal "Bend" calculations.

Parts of "Bend" routine may be used for small pipe in noncode work to determine bend angle for offsets. Do not XEQ "Bend" for noncode work as this determines R specified by code.

Procedure: XEQ clearing GTO "Bend."

Now select a radius (try: R = 18D: 3½ in. OD: R = 18 3.5/12 = 5.25). This can be changed if necessary. The V distance, offset, is determined by design configuration. Store R in 08, V in 09. Assume trial angle, say 20° for small pipe, store in 05. Let T = 0.5 ft, store in 10. SF 01. XEQ 01. (Only one calculation is performed.)

Inspect results. Do V given and V calc appear reasonably close? If not, take another "Fix" by changing R or angle (store new values, XEQ 01). This routine can be continued to a satisfactory solution. A point will be reached where R seems reasonable; choose an angle less than apparent solution, store in 05. CF 01, store 0.5 in 11, XEQ 01, and zero in. This returns to the iterative process for a more precise solution. T and R can be changed to suit.

"Bend" routine may be used to calculate bends for crossing under foreign lines. (See Example 3.)

The direction change routine (XEQ 17) can be used for code and noncode work. For code: determine minimum bend radius from "Bend" for pipe size and for gas or oil line. Use R from previous example or XEQ "Bend," and key in data as in Example 1 for desired diameter.

For noncode work: user may employ a radius that is suitable in his judgment. Do not use "Bend" program as this calculates radius in accordance with codes. Key in radius and desired tangent. Caution: Be aware that short radii for large pipes, say 8 in. and larger, soon get into hot bend category.

Example 1. Offset bend

$12\frac{3}{4}$-in. oil line. Let V = 8 ft, T = 5 ft. Smaller diameters have larger bend ∡. Try 20°, XEQ "Bend," key in data and R/S. Flag 01 Clear. Use ∡ Incr. = 0.5°. *Note:* R/D = 18 (meets B31.4). V Calc. is close to 8 ft. Now try scraper trap routine XEQ 15. Let trap ∡ = 5°, T = 5 ft, receiver Y. Length of trap = 10 ft. For launcher XEQ 15, same ∡, T, receiver N. Length of trap = 5 ft.

Example 2. Direction change

24-in. gas line. Note from "Bend" run Min. R = 76.3944 ft. R/D = 38.1972, meets B31.8. Say 30° turn, let R = 80 ft, T = 5 ft XEQ 17. Key in data R/S.

Example 3. Crossing

A 14-in. oil line crosses under $10\,^3/_4$ in. foreign line. Take 2 ft, 0 in. min. clearance. Center distance = 3.0313 ft. Min. height of bend = 2.513 ft = "V". Take 8° as trial ∡, T = 5 ft, ∡ Incr. = 0.5°. Note Calc. V is close to given V. This is a possible solution. Note, however, this is an alternative solution. Builders will take advantage of sag bends where possible, depending on site conditions, space available, pipe size, etc.

Calculate maximum bend on cold pipe

How much of a bend can cold straight pipe take without overstressing? In other words, in laying straight pipe crosscountry, what is the smallest radius of curvature possible within the allowable stress?

The following is an analytical solution to these questions.

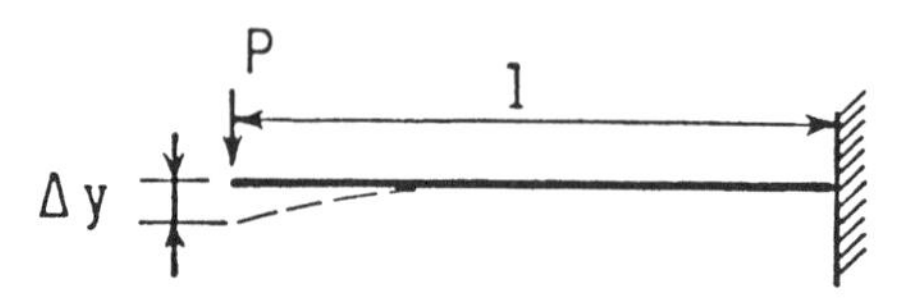

The deflection equation for a length of straight pipe used as a cantilever beam with a concentrated end load is:

$$\Delta y = \frac{Pl^3}{3EI} \quad (1)$$

The maximum stress in the beam is:

$$s = \frac{M}{Z} \quad (2)$$

The maximum bending moment is:

$$M = Pl \quad (3)$$

If the allowable stress is SA, rewriting Equation 2 gives:

$$S_A = \frac{M}{Z} \quad (4)$$

$$S_A = \frac{Pl}{Z} \quad (5)$$

$$Pl = ZS_A \quad (6)$$

Rewriting Equation 1:

$$\Delta y = \frac{(Pl)l^2}{3EI} = \frac{ZS_A l^2}{3EI} \quad (7)$$

Taking the first and second derivatives of Equation 7:

$$\frac{d\Delta y}{dl} = \frac{2lZS_A}{3EI} \quad (8)$$

$$\frac{d^2\Delta y}{dl^2} = \frac{2ZS_A}{3EI} \quad (9)$$

Radius of curvative is given by the equation:

$$R = \frac{[1 - (d\Delta y/dl)^2]^{3/2}}{d^2\Delta y/dl^2} \quad (10)$$

Substituting in Equation 10:

$$R = \frac{[1 + (2lZS_A/3EI)^2]^{3/2}}{2ZS_A/3EI} \quad (11)$$

$$\lim_{l \to 0} R = \frac{1^{8/2}}{2ZS_A/3EI} = \frac{3EI}{2ZS_A} \quad (12)$$

Notation

Δy	deflection, inches
l	length, inches
P	load, pounds
E	modulus of elasticity, pounds per square inch
I	cross-sectional moment of inertia, $inch^4$
S	stress, pounds per square inch
S_A	allowable stress, pounds per square inch from Code for Pressure Piping
M	bending moment, inch-pounds
Z	section modulus, $inch^3$
R	radius of curvature, inches

Example. What is the smallest radius of curvature that can be used without overstressing straight 24-in. Schedule 20 seamless steel pipe ASTM 53? The pipe is at 70°F and is to be used for oil outside of refinery limits.

$$R = \frac{3EI}{2ZS_A}$$

$$E = 29 \times 10^6 \text{ lb/sq. in.}$$

$I = 1{,}943\ in.^4$

$Z = 162\ in.^3$

$S_A = 25{,}500$ psi (from Code for Pressure Piping)

$$R = \frac{3 \times 29 \times 10^6 \times 1{,}943}{2 \times 162 \times 25{,}500} = 20{,}460 \text{ in. or } 1{,}705 \text{ ft}$$

The section modulus = Z = 2I/D, and may be simplified to Z = I/r and substituted for Z in Equation 12.

I = Moment of inertia
r = Distance from center pipe
D = Outside diameter of pipe

Equation 12 then becomes:

$$R = 3Er/2Sa \quad (13)$$

$$R = (21.75 \times 10^6/Sa) \times D \quad (14)$$

Sa, the allowable stress, is determined by the design factor permitted by the design code used. Equation 14 may be further simplified by using the following table.

Minimum Radius of Curvature, R				
Yield Stress (psi)	**F = 0.72**	**F = 0.6**	**F = 0.5**	**F = 0.4**
35,000	863 D	1,036 D	1,243 D	1,553 D
42,000	719 D	863 D	1,036 D	1,295 D
46,000	657 D	788 D	946 D	1,182 D
52,000	581 D	697 D	836 D	1,046 D
56,000	539 D	647 D	777 D	971 D
60,000	503 D	604 D	725 D	906 D
65,000	465 D	558 D	669 D	836 D
70,000	431 D	518 D	621 D	776 D
80,000	377 D	453 D	544 D	680 D

Example. What is the minimum radius of curvature that can be used without overstressing 42″ API 5LX X-65 line pipe at 72°F used in type C construction (gas line)? The design factor from ANSI B31.8 is 0.5; therefore, use the column headed F = 0.5.

R = 669 × 42″ = 28,098″ or 2,342 ft

For liquid lines designed in accordance with ANSI B31.4, use the column headed F = 0.72.

Determine length of a pipe bend

For smooth bends (other than wrinkle bends) use the following formula:

$$L = R \times D \times .017 + T$$

where:

L = length of bend
R = radius of bend (center line of pipe)
D = number of degrees in the bend
T = total length of the two tangents at either end

Example. Find the length of a piece of pipe required to make a 90° bend with a radius of 60 in. There should be a 4-in. tangent at either end.

L = ?
R = 60
D = 90
T = 8
L = 60 × 90 × .01745 + 8
= 94.23 + 8
– 102.23 in.

Length of pipe in arc subtended by any angle

Application

To find the length of pipe subtended by an angle of 42°–35 ft with a radius of 24 in.

Assuming R = 1 in.

angle of 42° subtends an arc = 0.7330″

$$\text{angle of } 35' \text{ subtends an arc} = \frac{0.01018''}{0.7432''}$$

DEGREES						MINUTES			
Degrees	Length Of Arc	Degrees	Length Of Arc	Degrees	Length Of Arc	Minutes	Length Of Arc	Minutes	Length Of Arc
1	.0175	31	.5411	61	1.0647	1	.000291	31	.009018
2	.0349	32	.5585	62	1.0821	2	.000582	32	.009308
3	.0524	33	.5760	63	1.0996	3	.000873	33	.009599
4	.0698	34	.5934	64	1.1170	4	.001164	34	.009890
5	.0873	35	.6109	65	1.1345	5	.001454	35	.010181
6	.1047	36	.6283	66	1.1519	6	.001745	36	.010472
7	.1222	37	.6458	67	1.1694	7	.002036	37	.010763
8	.1396	38	.6632	68	1.1868	8	.002327	38	.011054
9	.1571	39	.6807	69	1.2043	9	.002618	39	.011345
10	.1745	40	.6981	70	1.2217	10	.002909	40	.011636
11	.1920	41	.7156	71	1.2392	11	.003200	41	.011926
12	.2094	42	.7330	72	1.2566	12	.003491	42	.012217
13	.2269	43	.7505	73	1.2741	13	.003782	43	.012508
14	.2443	44	.7679	74	1.2915	14	.004072	44	.012799
15	.2618	45	.7854	75	1.3090	15	.004363	45	.013090
16	.2793	46	.8029	76	1.3265	16	.004654	46	.013381
17	.2967	47	.8203	77	1.3439	17	.004945	47	.013672
18	.3142	48	.8378	78	1.3614	18	.005236	48	.013963
19	.3316	49	.8552	79	1.3788	19	.005527	49	.014254
20	.3491	50	.8727	80	1.3963	20	.005818	50	.014544
21	.3665	51	.8901	81	1.4137	21	.006109	51	.014835
22	.3840	52	.9076	82	1.4312	22	.006400	52	.015126
23	.4014	53	.9250	83	1.4486	23	.006690	53	.015417
24	.4189	54	.9425	84	1.4661	24	.006981	54	.015708
25	.4363	55	.9599	85	1.4835	25	.007272	55	.015999
26	.4538	56	.9774	86	1.5010	26	.007563	56	.016290
27	.4712	57	.9948	87	1.5184	27	.007854	57	.016581
28	.4887	58	1.0123	88	1.5359	28	.008145	58	.016872
29	.5061	59	1.0297	89	1.5533	29	.008436	59	.017162
30	.5236	60	1.0472	90	1.5708	30	.008727	60	.017453

The arc subtended for any angle is directly proportional to radius:

$$\frac{1 \text{ in.}}{24 \text{ in.}} = \frac{0.7432 \text{ in.}}{\text{Arc}}$$

$$\text{Arc} = 0.7432 \times 24 = 17^{13}/_{16}''$$

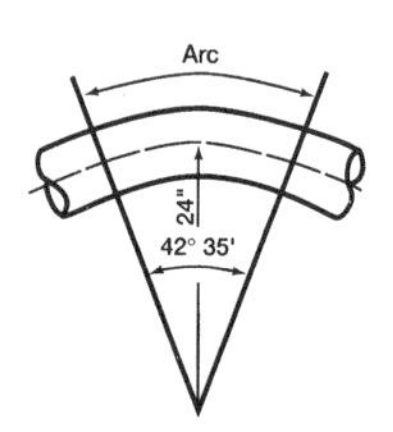

Average pipelay table—underground

Linear feet of pipelay per 10-hour day in trench in level, rock-free terrain

Nominal Pipe Size	PIPE WALL THICKNESS IN INCHES					
	0.000 through 0.250	0.251 through 0.375	0.376 through 0.500	0.501 through 0.625	0.626 through 0.750	0.756 through 1.000
4	10,800	10,500	10,300	—	—	—
6	9,600	9,400	9,210	—	—	—
8	9,120	8,900	8,720	8,550	—	—
10	8,900	8,760	8,580	8,410	—	—
12	8,600	8,400	8,230	8,070	7,910	—
14	8,400	8,230	8,160	8,000	7,840	—
16	8,000	7,840	7,680	7,530	7,380	7,230
18	7,800	7,640	7,500	7,350	7,200	7,050
20	7,400	7,250	7,100	6,960	6,820	6,680
24	—	6,600	6,470	6,340	6,210	6,080
30	—	6,000	5,880	5,760	5,640	5,530
36	—	—	5,400	5,300	5,200	5,090
42	—	—	4,800	4,700	4,600	4,500

Productivity will vary where different types of terrain or rock are encountered.
Above footage is based on installing double random joints of pipe.

Average pipelay table—on supports

Linear feet of pipelay per 10-hour day on waist-high supports on level ground						
Nominal Pipe Size	**PIPE WALL THICKNESS IN INCHES**					
	0.000 through 0.250	**0.251 through 0.375**	**0.376 through 0.500**	**0.501 through 0.625**	**0.626 through 0.750**	**0.756 through 1.000**
4	11,320	11,100	10,880	—	—	—
6	10,080	9,880	9,680	—	—	—
8	9,560	9,370	9,190	9,010	—	—
10	9,390	9,200	9,020	8,840	—	—
12	8,800	8,620	8,500	8,330	8,160	—
14	8,700	8,570	8,400	8,250	8,090	—
16	8,040	7,880	7,730	7,580	7,430	7,280
18	7,800	7,650	7,500	7,350	7,200	7,060
20	7,550	7,400	7,250	7,100	6,960	6,820
24	—	6,720	6,590	6,460	6,330	6,200
30	—	6,120	6,000	5,880	5,770	5,660
36	—	—	5,440	5,350	5,240	5,140
42	—	—	4,840	4,740	4,650	4,560

Productivity will vary with contour of terrain and accessibility.
Above footage is based on installing double random joints of pipe.

Allowable pipe span between supports

When it is desired to run pipelines on supports above the ground, the engineer must often calculate the allowable pipe span between supports from a structural viewpoint. The formula for the span length can be obtained by equating the maximum bending moment caused by the loading to the resisting moment of the pipe. This formula is:

$$L\left(\frac{SI}{wc}\right)^{0.5}$$

where:

L = span length, feet
c = radius to outside of pipe, inches
I = moment of inertia of pipe, inches^4
S = maximum allowable fiber stress of material, pounds per inch^2
w = uniform loading per unit length of pipe, pounds per foot

The above formula applies to pipes above 2 in. in diameter. Reference 1 gives calculated pipe span data, as well as estimated pipe loading for different pipe diameters, taking into account the fluid load, dead load, and the horizontal component of the wind loading.

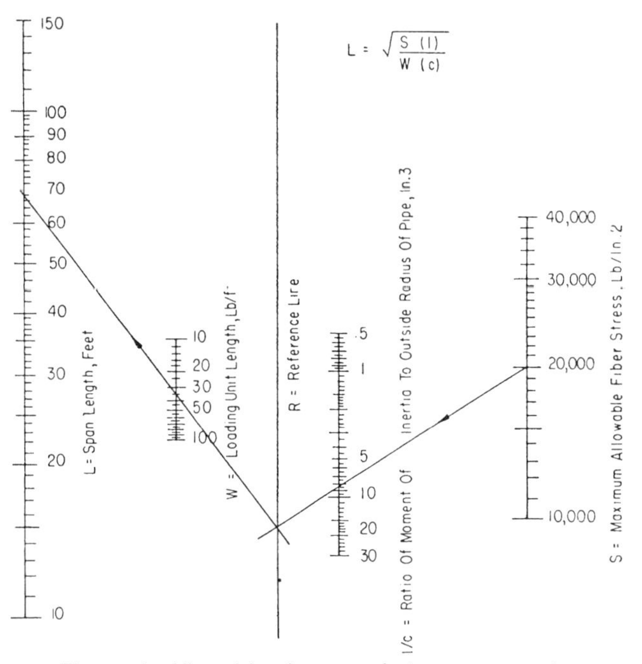

Figure 1. Allowable pipe scan between supports.

The design of the pipe span can be quickened up with Figure 1.

Example. Find the allowable span length for a 6-in. continuous steel pipe strung between supports. The estimated loading (sum of pipe, fluid, dead, and wind loads) is 36 lb/ft. The allowable fiber stress for steel can be taken at 20,000 lb/in.^{2}

From any suitable handbook, c for a 6-in. pipe is 6.625/2 = 3.3175 in., I is 28.14 in.^{4} Now calculate I/c = 28.14/3.3175 = 8.5 in.^{3} Connect S = 20,000 lb/in.^{2} on the S scale with 8.5 on the I/c scale to the intersection with the reference line. With the reference line as a pivot, connect the pivot point with W = 36. At the intersection with the L scale, obtain a pipe span of 68 ft.

Check the amount of deflection to make sure that it does not exceed your piping practices or applicable design code.

Source

Kuong, J. F., *Petroleum Refiner 39*, No. 10, 196 (1960).

Reference

1. Sweeney, R. J., *Chem. Eng.*, *63*, No. 3 199 (1956).

How engineers make pipe fit the ditch

By R. J. Brown

The five-men crews of engineers that keep track of the pipe and calculate side bends, overbends, sags, and combination bends can make money—or lose money—for any contractor crossing rough terrain. If they function smoothly, the bending crew and pipe gang have easy going and the lowering-in can be a breeze. But when they make mistakes, the welded-up pipe may not fit the ditch and this spells trouble for the contractor.

Bending crew

As a rule there are two rod men, two chairmen, and one instrument man in the measurement crew. Their duties are to measure each joint of pipe, number it, and stake its position in the line. In addition, they calculate degrees of bends, etc. Here is how they perform these duties:

Measuring and numbering joints

(See Figure 1) First, the crew measures all joints of pipe and numbers them from 0 to 100. (When 100 is reached, the numbering starts over.) Next, they stake the position of each joint of pipe in the line. They use stakes that are numbered according to the pipe, with the number of the joint behind on front of the stake and the joint ahead on back of the stake.

Stakes set back

Because the setup gang will move the pipe back half a joint's length and the pipe gang back another half, the engineers set the stakes from one to two joints' length behind the corresponding numbered pipe.

Notes on Figure 1

Note that joint 25 was measured 39.9 ft, and the gap between stakes 24 and 25 and 25 and 26 was set at 29.9 ft to allow a 10-ft lap for tie-in. This leaves a pup long enough to be carried ahead and worked into the line. If the pups are distributed over sections of bends, there is a possibility of saving a cut in a full joint. In event the stakes catch up with the pipe, one joint is carried ahead of the stringing.

Calculating bends

The method used to determine the size of overbend and sags is illustrated in Figure 2. Note that at 0 + 00 the instrument is set up on the ditch bottom at a height of 5 ft. The rod is set on the ditch bottom ahead of the radius of overbend. (Care should be taken to set neither transit nor rod on the arc of the change in slope.) The transit is moved behind the radius of overbend at 1 + 00, and the height of the instrument is measured to be 4.90 ft.

The rod is set just ahead of the change in radius of overbend and is read at 4.90 ft, establishing a slope of −1°. The magnitude of bend can now be determined.

Add unlike angles

If the slope is positive and the tangent of the slope ahead is decreasing, an overbend is established. If the slope is

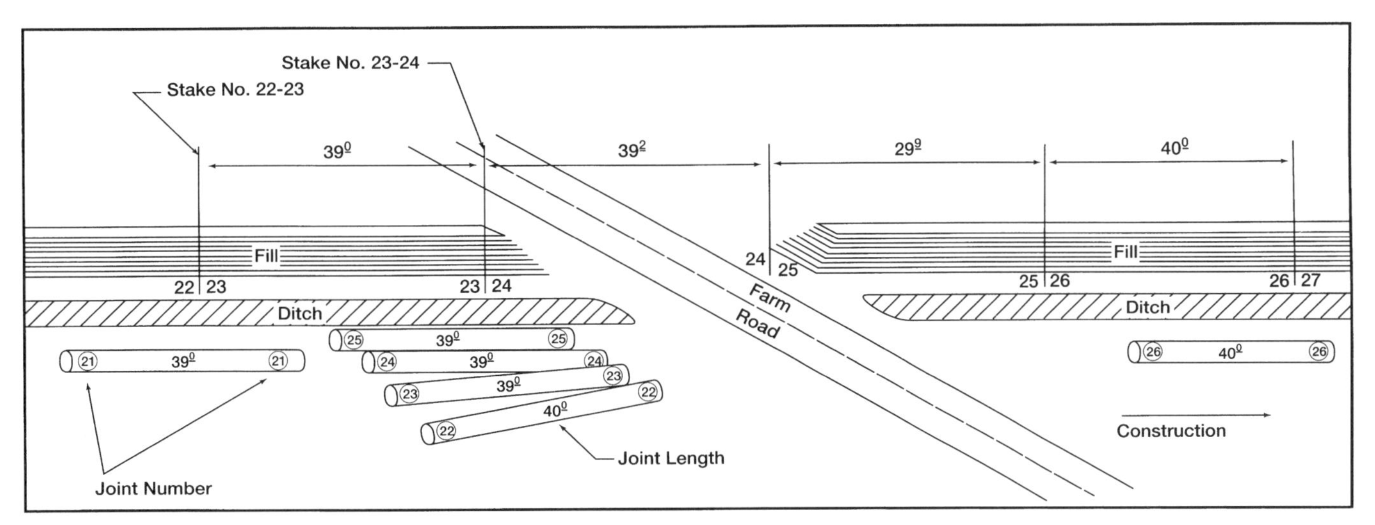

Figure 1

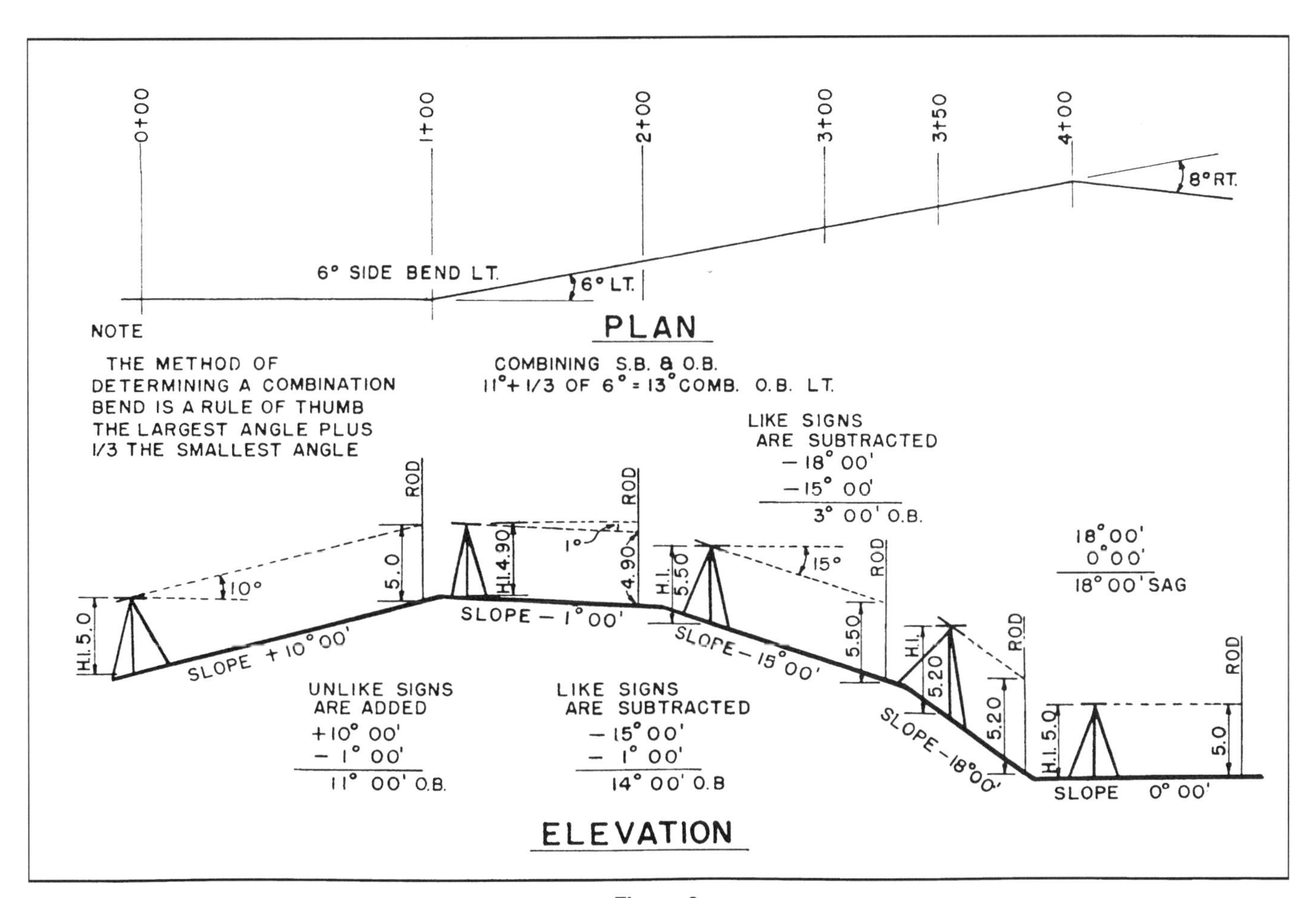

Figure 2

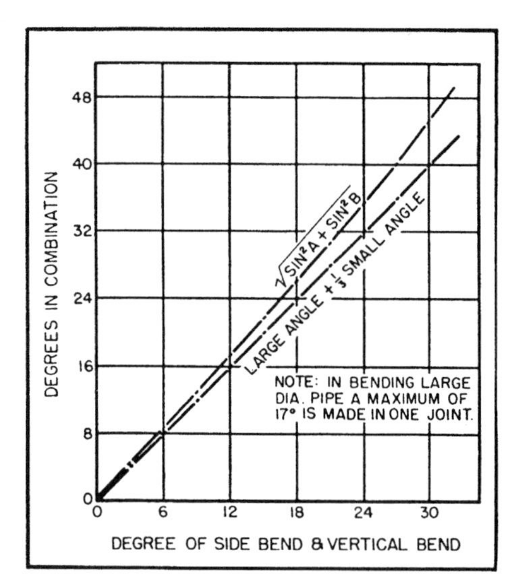

Figure 3

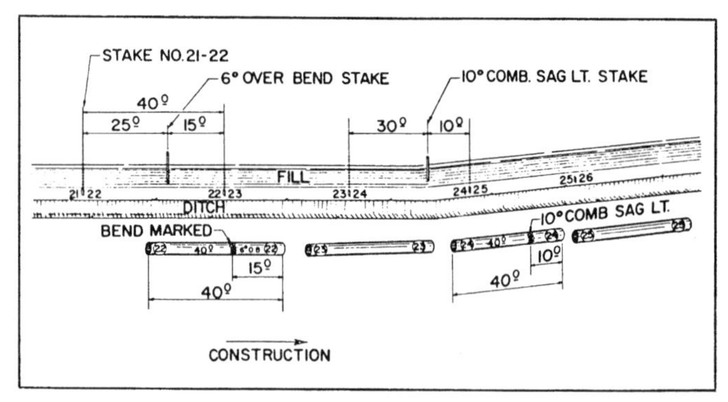

Figure 4

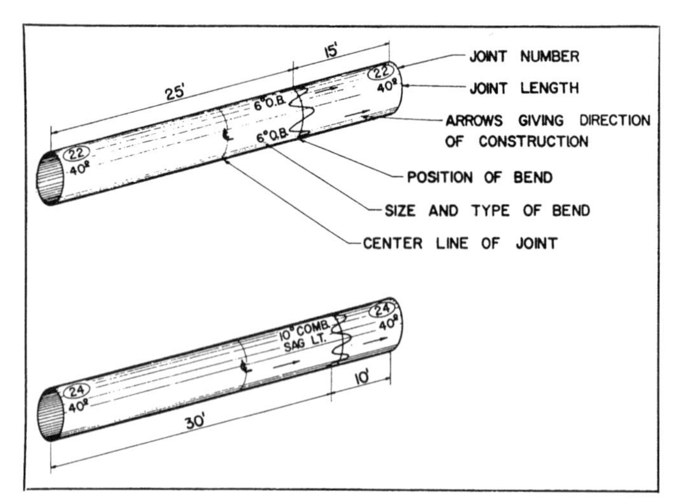

Figure 5

positive and the tangent ahead is greater positive slope, a sag is noted. The reverse holds true for change in negative slope: if the tangent ahead is increasing, the transition is an overbend; if it is decreasing, the transition is a sag. The unlike algebraic signs are added and like signs are subtracted. Figure 2, station 2+00, shows a positive 10° slope approaching the overbend and a negative 1° slope leaving it. Since the signs are unlike, the sum of the two angles is 11°, and since the slope is positive and decreasing, the notation overbend is used.

Side bend angles

Side bends are determined by "eyeballing" a P.I. in the ditch and setting the instrument up over it. A back sight is taken on the ditch or preceding P.I., whichever is more easily read. The scope is flipped and a foresight is taken on the ditch ahead. Referring to Figure 2 at station 1+00, the deflection angle is 6° left; station 4+00 shows a deflection of 8° right.

Combination bends

At points where side bends and vertical bends occur, the angles are combined to make one angle rolled. The method of calculating is a rule of thumb: *the larger angle plus one-third the smaller angle*.

This rule is accurate enough to give satisfactory results in actual practice. The actual derivation of this rule comes from an approximation of the following:

$$\sin(\text{combination angle}) = \sqrt{\sin^2 A + \sin^2 B}$$

Angles A and B are the horizontal and vertical angles. Figure 3 shows a plot of the ⅓ rule and of the equation. From the plot it can be seen that for the maximum bend of 18.5, an error of slightly over 1° is introduced. The maximum error evidenced on the graph is due to the side and vertical bends being equal.

Marking pipe

The final step of the measurement crew is marking the joints to be bent. After the bend is established, its position between the pipe stakes is then transferred to the pipe joint number matching the pipe stakes number. The bend size, the centerline of bend, and arrows indicating direction of construction are marked on the joint. (See Figure 5.) If a bend falls too close to the end of the pipe, the joint would have to be moved, so as to leave enough room for the bend. A pup could be inserted immediately behind the joint in question to give better bend positioning.

PIPE LOWERING

How to lower an existing pipeline that is still in service

Lowering a loaded pipeline is a low-cost alternative for constructing new facilities

Marshall D. Cromwell, Senior Project Engineer, PT Caltex Pacific Indonesia, Jakarta, Indonesia

Lowering an existing line is a dirty little job, but it can have big cost benefits. The line can be lowered while remaining in service with no lost production, and the cost of lowering an existing pipeline section is relatively cheap.

No expenses are incurred for new pipe, valves, stopples, and fittings. Construction is much faster than cutting and relocating a pipeline section.

Construction of new highways, buildings, airport runways, and other facilities is often planned at locations where aboveground pipelines are present. Relocating such lines can be extremely expensive in terms of downtime and new pipeline materials.

Long-delivery items, such as hot tap equipment, valves, and fittings can add to the pipeline construction time, delaying an already tight schedule for the new facility. The alternative is to lower the existing lines, with adequate protection, so as to eliminate the obstruction.

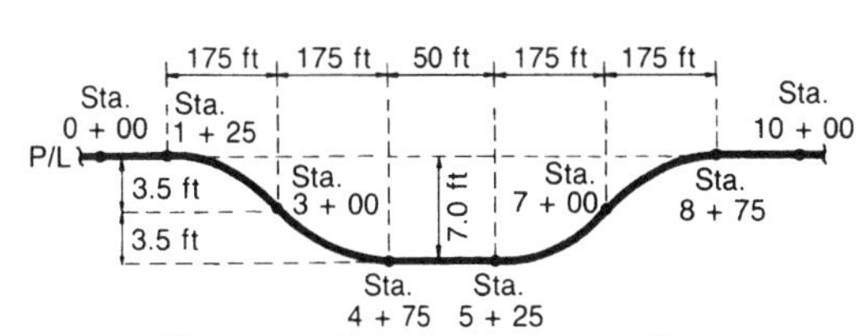

Figure 1. Road crossing profiles.

The following information describes how to calculate an optimum profile for smooth transitions, while keeping the pipe stresses within allowable limits, and how to physically lower the pipeline in the field to match this profile.

Precise engineering

There is more to lowering an existing line than just digging a hole and letting gravity take over. Engineering and design of a new profile must be precise so that the line is not dropped too suddenly, causing a buckle or rupture (the larger the pipeline, the more critical the profile).

For instance, in Saudi Arabia, an extensive highway modernization program resulted in scores of large-diameter

Table 1
Profile elevations

Station	X, ft	Deflection, ft	Top of Pipe Existing Elev.	Top of Pipe New Elevation
1+25	0	0.00	From field survey	Existing elevation minus deflection
1+50	25	0.07		
1+75	50	0.29		
2+00	75	0.65		
2+25	100	1.15		
2+50	125	1.80		
2+75	150	2.59	(Another column can be added for trench-bottom elevations if desired.)	
3+00	175	3.50		
3+25	200	4.41		
3+50	225	5.20		
3+75	250	5.85		
4+00	275	6.35		
4+25	300	6.71	Note: Depending on configuration of original pipeline, left profile does not have to be identical to right profile. Transition lengths, radii, and deflections may be different.	
4+50	325	6.93		
4+75	350	7.00		
5+00	375	7.00		
5+25	400	7.00		
5+50	425	6.93		
5+75	450	6.71		
etc.	etc.	etc.		

After elevations are received from the field, end-points (Sta 1+25 and 8+75) should be checked by use of equations 7 and 4, if pipeline is at uphill slope.

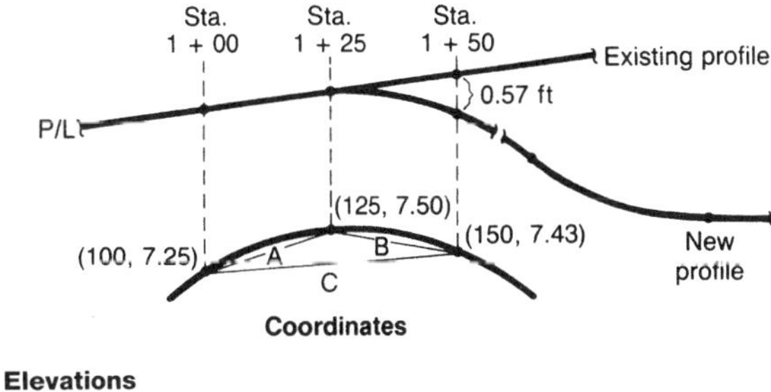

Elevations

Station	Feet
1 + 00	7.25
1 + 25	7.50
1 + 50	8.00
1 + 50	7.43 New

$$A = \sqrt{(125-100)^2 + (7.50-7.25)^2} = 25.00125 \text{ ft}$$

$$B = \sqrt{(150-125)^2 + (7.50-7.43)^2} = 25.00010 \text{ ft}$$

$$C = \sqrt{(150-100)^2 + (7.43-7.25)^2} = 50.00032 \text{ ft}$$

$$R = 1{,}947.5 \text{ ft} = 23{,}370\text{-in.}$$

$$S = 22{,}340 \text{ psi}$$

No good–must come up with a new profile, with wider transition length, and check again by trial-and-error.

Figure 2. Elevation coordinates.

1. $R = \dfrac{x^2 + y^2}{2y}$

2. $x = \sqrt{2Ry - y^2}$

3. $y = R - \sqrt{R^2 - x^2}$

4. $R = \dfrac{EC^1}{S}$ or $S = \dfrac{EC^1}{R}$

5. $y = \dfrac{EC^1}{S} - \sqrt{\left(\dfrac{EC^1}{S}\right)^2 - x^2}$

6. $x = \sqrt{2y\dfrac{EC^1}{S} - y^2}$

7. $R = \dfrac{A\ \text{SIN}\left[90 - \text{COS}^{-1}\dfrac{B^2 + C^2 - A^2}{2BC}\right]}{\text{SIN}\left[2\ \text{COS}^{-1}\dfrac{B^2 + C^2 - A^2}{2BC}\right]}$

(R = Feet here)

Note that formulas are independent of w.t., I, and cross-sectional area (cancelled out).

R = Radius of bend (in.)
x = Horizontal distance (in.)
y = Deflection or drop from the horizontal (in.)
S = Bending stress, psi
C^1 = One half diameter of pipeline (in.)
E = Young's modulus of elasticity (29×10^6 psi for steel)
A, B & C = Dimensions of triangle (ft.)

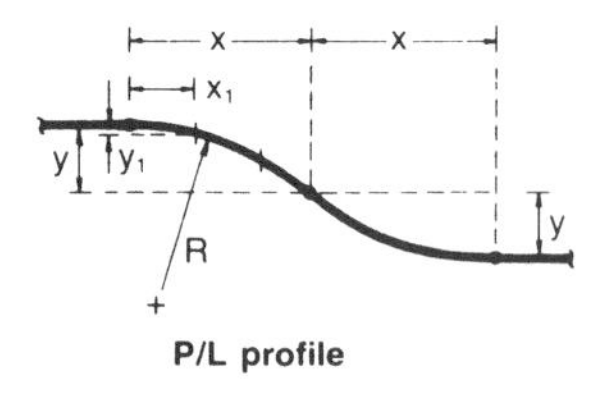

P/L profile

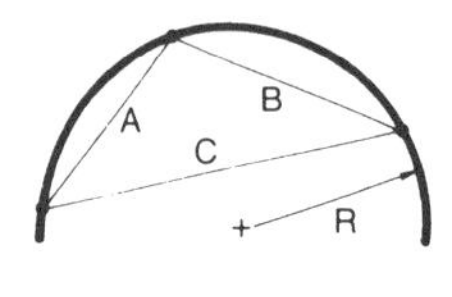

3 points on a circle

• = Inflection points
x_1 = 25′, x_2 = 50 , etc.

Figure 3. Formulas for changes in slope.

Excavation begins. Ten-foot soil plugs remained at 50-ft centers to act as temporary pipe supports for the 36-in. line.

pipelines being lowered. A few line sections ruptured during the pioneer stages due to the fact that each situation was not properly evaluated in terms of a critical engineering problem.

The key to the problem's solution is to provide a smooth profile with constant radius bends and to include a detailed list of instructions with the design package. Careful monitoring by survey equipment (levels) will assure smooth transitions in the field.

The following approach has been used effectively on numerous occasions in Saudi Arabia and Indonesia and can be recommended. Design steps to be included are:

- Determine how much soil cover there should be over the top of the pipeline to assure protection from construction equipment and wheel loads (see API-RP-1102). Four feet of cover is adequate for pipelines 12 in. and smaller. Larger pipelines may require additional protection by way of more cover, pipe sleeves, or concrete slabs.
- Determine allowable bending stress on pipeline, based on grade of pipe. Normally, 75% of allowable stress is acceptable for cross-country pipelines (ANSI B31.4, Section 419.6). However, 10,000-psi stress is very conservative and often preferred. If the pipeline is completely restrained (anchors on each end and no expansion loops), thermal stresses must also be considered.
- Once the total required drop of pipeline is established, calculate the minimum bending radius based on allowable bending stress. Then calculate the overall transition length from one end to the other.
- Obtain elevations on top of the pipeline every 25 ft (Table 1), marking station numbers and elevations with paint. If the line is on supports, also include grade elevations at each station.
- Starting at each end of transition, calculate the drop at 25-ft stations. Draw up a profile and fit it on the existing profile. Add stations and new elevations (top of pipe and trench bottom) to the profile drawing.
- Check the transition points at the beginning and end of new profiles. If the pipeline was going uphill, and suddenly it needs to go downhill, the end points may be critical in terms of stress.

Since any three points not on the same line form an arc of constant radius, the stress may be approximated by Equations 7 and 4 (Figure 3). The end point should become the center of three points for calculation. If overstressed, widen the transition length and check again.

Example 1. A new village was being built in the jungle, but access was obstructed by an aboveground, 36-in. pipeline. The village head requested that the line be buried or relocated so that bicycles, becaks, and water buffalo would have easier access.

It was determined that 4 ft of cover was required to protect the line from excessive wheel loads (people-packed buses and trucks). This means the line must be lowered 7 ft (Figure 1). The pipe is Grade B, allowable bending stress is 10,000 psi, and crossing width is 50 ft. To calculate minimum bending radius, total transition length, and deflections at 25-ft intervals, use:

$$R = EC'/S = (29{,}000{,}000)(18)/10{,}000 = 52{,}200\text{ in.}$$

$$= 4{,}350\text{ ft (min.)}$$

$$\text{Total transition length} = 4(X) + 50\text{ ft}$$

$$Y = 7.0\text{ ft}/2 = 3.5\text{ ft}$$

$$X = \sqrt{2RY - Y^2}$$

$$X = \sqrt{(2)(4{,}350)(3.5) - (3.5)^2} = 175\text{ ft}$$

$$\text{Total transition length} = 750\text{ ft}$$

It is suggested that field personnel start survey (elevations) 500 ft left of crossing centerline and stop at station 10 + 00.

Deflections at 25-ft intervals are calculated from Eq. 3 (Figure 3), and a table is prepared accordingly (Table 1). Note that calculations for Y are required for 175 ft only. Since the curve is then inverted (Sta 3 + 00 to Sta 4 + 75), results are subtracted from 7 ft, working backwards, to arrive at true deflection or drop. For example, deflection at Sta 4 + 50 = 7.0 − 0.07 = 6.93, etc.

Example 2. Given following elevations, is pipeline being dropped too suddenly, and will it become overstressed between stations 1 + 00 and 1 + 50 (Figure 2)?

Construction

The design package to the field should include drawing of new profiles, stations, tables of new pipeline elevations, trench elevations, and a very detailed construction sequence. The following may be used as a guide for construction:

- Starting at one end of transition, dig 40 ft of trench under the line. Leave 10 ft of soil for pipe support. Dig another 40 ft, repeat until opposite end is reached. When complete, 10-ft soil plugs are supporting the line at 50-ft centers.
- Clean the line between soil supports, take ultrasonic test (UT) readings, and install sleeve where required. Add primer and tape wrap between supports, maintaining station numbers and elevation points on the line.
- Add 12 in. of sand or soil/sand pre-mix to the bottom of the trench and compact to proper trench bottom elevation (profile elevation minus pipeline diameter).
- Install 6-in. × 6-in. wooden blocks every 50 ft (between soil plugs) to support the line.
- Remove soil plugs and complete line cleaning, UT readings, sleeve installations, tape wrapping, and trench preparation.
- Check elevations of new trench bottom and adjust as required. Check tape wrap with a holiday detector if desired.
- Starting at one end, use six sidebooms to raise approximately 250 ft of line off its supports. Remove 6 in. (or less) of blocks on outside supports and 12 in. of blocks on all inner supports. Lower line onto supports.

Hand cleaning. The 30-in. and 36-in. crude lines were cleaned by hand to expose corroded areas and prepare for tape wrapping.

Lowering lines. The two lines, full of crude oil, are gradually lowered with the use of six sidebooms.

Gradual lowering. A full sweep of the total transition length is made by lowering each line 6 in. or less at a time. The procedure is repeated until the lines finally reach desired profile elevations.

- Move convoy of sidebooms 250 ft, lift the line and remove 6-in. block at beginning and end supports and 12 in. of blocks at inner supports.
- Repeat until total transition length is lowered 12 in., except for several end supports, which are lowered less than 6 in. to match required deflections. Then come back and do it again, repeating several times until the pipeline is completely lowered. Note that each repeat becomes shorter. Check final pipeline elevations if desired.
- Check tape wrap for possible damage, add 12 in. of sand or pre-mix around pipe, backfill, and compact.

Surprisingly, this operation goes very quickly but must be well coordinated. One person should signal the sideboom operators to lower the pipeline concurrently, so as to eliminate excessive bends and sags. The pipeline is heavy when full of liquid, and if not handled carefully, it could easily rupture. Pressure should be reduced, and extreme caution exercised during the actual lowering process.

Strain calculations

A new profile means the pipeline has been stretched a certain amount, assuming the line was horizontal originally. (If the pipeline was convex, lowering it will put the steel into compression.) Fortunately, steel is very forgiving either way, and the added strain is generally not a problem. It can be checked, however, by use of Young's Modulus of Elasticity (E = stress over strain).

The new pipe length can be calculated from the formula:

Arc Length $= \pi R\theta/180$

where $\theta = \sin^{-1} X/R$

A comparison of the new pipe length with the original pipe length will result in a figure for strain as a result of the new profile. Knowing the strain, additional longitudinal stress can be easily calculated.

In Example 1, $\theta = \sin^{-1} 175\text{ ft}/4{,}350\text{ ft} = 2.3056°$
Arc Length (AL) $= \pi(4{,}350\text{ ft})(2.3056°)/180 = 175.047\text{ ft}$
Length of new line $= 4(\text{AL}) + 50 = 750.19\text{ ft}$

So the pipeline was actually stretched 0.19 ft or 2.28 in.

Strain $= \Delta L/L = (750.19 - 750.00)750.00 = 0.000253$
E = Stress/Strain, so longitudinal stress = E × Strain or
LS $= 29 \times 10^6 \times 0.000253 = 7{,}347$ psi.

This is well within the 35,000-psi yield strength of Grade B pipe.

Source

Pipe Line Industry, July 1986.

WELDING

When should steel be preheated before welding?

From the chemistry of the steel determine the carbon equivalent:

$$\text{Carbon Equivalent} = C + \frac{Mn}{4}$$

If it exceeds 0.58 the steel may be crack sensitive and should be preheated before welding in ambient temperatures below 40°F.

Example. If steel pipe having a carbon content of 0.25 and manganese content of 0.70 is to be welded in springtime temperatures, ranging from 40°F to 80°F, is preheat necessary?

$$\text{Carbon Equivalent} = .25 + \frac{.70}{4} = .25 + .175 = 0.425$$

It is not necessary to preheat this particular steel before welding it.

But for another example:

Carbon = 0.20

Manganese = 1.60

$$\text{Carbon Equivalent} = .20 + \frac{1.60}{4} = 0.60$$

This steel should be preheated, particularly for early morning welding.

Why does preheating prevent cracking? It slows the cooling rate and reduces the amount of austenite retained as the weld cools. This prevents microcracking. Other alloying elements and pipe wall thickness may also influence when joints of high strength pipe should be preheated.

Welding and brazing temperatures

Carbon Steel Welding	2700–2790°F
Stainless Steel Welding	2490–2730°F
Cast Iron Welding	1920–2500°F
Copper Welding and Brazing	1980°F
Brazing Copper-Silicon with Phosphor-Bronze	1850–1900°F
Brazing Naval Bronze with Manganese Bronze	1600–1700°F
Silver Solder	1175–1600°F
Low Temperature Brazing	1175–1530°F
Soft Solder	200–730°F
Wrought Iron	2700–2750°F

Reprinted with permission—Tube Turns, Inc.

Mechanical properties of pipe welding rods

When welded and tested in accordance with appropriate AWS or MIL specifications[(1)]

Electrode	AWS Class	Tensile psi	Yield psi	%Elong. in 2″	Tensile psi	Yield psi	%Elong. in 2″
Fleetweld 5P+	E6010	62–86,000[1]	50–75,000[1]	22–28[1]	67–78,000	51–67,000	30–34
Fleetweld 5P	E6010	62–75,000[1]	50–64,000[1]	22–30[1]	60–69,000	46–56,000	28–36
Shield-Arc 85	E7010-A1	70–78,000[1]	57–71,000[1]	22–26[1]	70–83,000	57–72,000	22–28
Shield-Arc 85P	E7010-A1	70–78,000[1]	57–63,000[1]	22–27[1]	70–77,000	57–68,000	22–25
Shield-Arc HYP	E7010-G	70–82,000[1]	60–71,000[1]	22–28[1]	72–81,000	60–72,000	25–29
Shield-Arc 70+	E8010-G	80–86,000[1]	67–74,000[1]	19–29[1]	76–80,000	68–72,000	23–29
Jetweld LH-70	E7018[4]	72–86,000	60–76,000	22–30	65–74,000	55–60,000	24–34
Jetweld LH-75MR	E8018-B2	72–87,000	60–74,000	22–30	70–82,000	56–72,000	29–32
Jetweld LH-90MR	E8018-C1	97–107,000	84–97,000	17–24	80–105,000[2]	67–93,000[2]	19–23[2]
Jet-LH8018-C1MR	E8018-C3	80–95,000	67–81,000	19–25	80–86,000	67–75,000	19–31
JEt-LH8018-C3MR	E8018-C3	80–90,000	68–80,000	24–30	75–84,000	66–73,000	24–32
Jetweld LH-100M1MR	MIL10018-M1	95–101,000	82–91,000	20–27	93–96,000[5]	80–90,000[5]	20–28[5]
Jetweld LH-110MMR	E11018-M	110–123,000	98–109,000	20–24	110–120,000[3]	95–107,000[3]	20–25[3]

1 Aged 48 hours @ 220°F
2 Stress relieved @ 1275°F
3 Stress relieved @ 1025°F
4 Imprinted identification 7018-1 to indicate
meeting minimum CVN impact requirement of 20 ft-lbs at −50°F
5 Stress relieved @ 1125°F

[(1)]Mechanical properties obtained with each type electrode also depend upon chemistry of the pipe, welding procedures, and the rate of cooling. Because these factors vary between actual applications and testing conditions, the mechanical properties may also vary.

Reprinted with permission—Lincoln Electric Co.

Lens shade selector

Operation	Shade No.
Soldering	2
Torch Brazing	3 or 4
Oxygen Cutting	
up to 1 in.	3 or 4
1 to 6 in.	4 or 5
6 in. and over	5 or 6
Gas Welding	
up to $\frac{1}{8}$ in.	4 or 5
$\frac{1}{8}$ to $\frac{1}{2}$ in.	5 or 6
$\frac{1}{2}$ in. and over	6 or 8
Shielded Metal-Arc Welding	
$\frac{1}{16}$, $\frac{3}{32}$, $\frac{1}{8}$, $\frac{5}{32}$ in. electrodes	10
Nonferrous	
Gas Tungsten-Arc Welding	
Gas Metal-Arc Welding	
$\frac{1}{16}$, $\frac{3}{32}$, $\frac{1}{8}$, $\frac{5}{32}$ in. electrodes	11
Ferrous	
Gas Tungsten-Arc Welding	
Gas Metal-Arc Welding	
$\frac{1}{16}$, $\frac{3}{32}$, $\frac{1}{8}$, $\frac{5}{32}$ in. electrodes	12
Shielded Metal-Arc Welding	
$\frac{3}{16}$, $\frac{7}{32}$, $\frac{1}{4}$ in. electrodes	12
$\frac{5}{16}$, $\frac{3}{8}$ in. electrodes	14
Atomic Hydrogen Welding	10 to 14
Carbon-Arc Welding	14

Reprinted with permission—Tube Turns, Inc.

PIPELINE WELDING

Meeting Today's Quality Requirements For Manual Vertical Down Techniques

Pipelines are inspected more critically than ever before, and today's radiographic equipment and techniques produce clearer radiographs with greater sensitivity than in the past. Although codes have not changed drastically, interpretation standards have been upgraded. The combination of more rigorous inspection, better testing methods, and high acceptability standards often approaches an attitude requiring zero defects.

This poses some serious problems because the job of welding cross-country pipelines under typical conditions has always been an extreme challenge requiring specialized and highly developed skills. Now that the demands are greater, even the best welding operators are having trouble. Rejectable defects usually require cutting out the entire weld. This is expensive and can cost competent pipeline welders their jobs.

The purpose of this bulletin is to discuss some of the more common reasons given for rejecting a pipeweld and to suggest what may be done to correct some of these conditions.

Change in attitude

Any discussion on this subject should begin by accepting the fact that the standards are higher than they were a few years ago. Therefore, it will take a change in the attitude of the welders and everyone else involved if this new level of quality and workmanship is to be met. In turn, a commitment to new methods, equipment, and theory will be required by all concerned.

Importance of joint preparation

It must be recognized that all joint preparation details (Figure 6) are critical and any variation could directly contribute to rejected welds.

Without good cooperation from those who prepare the pipe edges and the line-up crew, the welder has very little chance of meeting today's rigid inspection requirements. Too often the attitude exists that variations in fit-up and joint preparation are permissible and that the welder can compensate for them. This attitude cannot be tolerated. It puts too much responsibility on the welder and inevitably leads to rejects.

Recommended procedures for properly cleaning pipe

For lowest cost and highest quality, careful attention to the cleaning of pipe joint surfaces is critically important.

Courtesy of Pipeline & Gas Industry.

Today's pipe welders face pipe that may be covered with a variety of coatings; these include primers, epoxy, tar, paper, varnish, rust, scale, or moisture.

While joint cleanliness is important in all welding, it is especially so in the root pass of pipe. Even a thin film of contaminants, which may be difficult to see with the naked eye, may be present. Failure to recognize and properly clean joints can result in a hollow bead or other welding defects.

Follow these instructions to minimize costly defects:

1. Remove all moisture and condensations of any type prior to welding. The joint must be completely dry.
2. Clean BOTH ends of the pipe INSIDE AND OUT to remove traces of paint, rust, scale, oil, epoxy, or organic materials. The area to be cleaned should extend at least 1 inch (25 mm) from the end of the bevel (per ANSI/AWS D10.11-80 Page 1) on both the INSIDE and OUTSIDE surfaces.
3. A recommended method for cleaning in the field as described above is the use of a heavy duty straight shaft grinder with a rubber expanding wheel and a carbide coated sleeve. The small shaft and reduced overall weight allows easy access to the inside and outside surfaces of the pipe.

Internal undercut

See Figures 1 through 5 for x-rays of the various defects that can cause rejects. Of these, one of the most common and troublesome is internal undercut (undercut on the inside of the pipe). This is understandable because it occurs on the "blind side" and the operator cannot see it happening. Consequently, he cannot immediately do anything to correct it. To make matters worse, he seldom gets a chance to see the x-ray negative or the actual weld itself. All of this makes it difficult for him to correct the situation.

Undercut may be the direct result of poor joint preparation or fit-up. If the joint does not conform to the details specified in the established procedures (see Figure 6), every reasonable effort should be made to correct it before starting to weld.

Internal undercut will tend to occur if:

1. The root face (Land) is too small.
2. The root opening is too large.
3. If a severe high-low condition exists.
4. The current is too high.

When any undesirable variation occurs in the joint preparation, the normal reaction is to compensate for it by

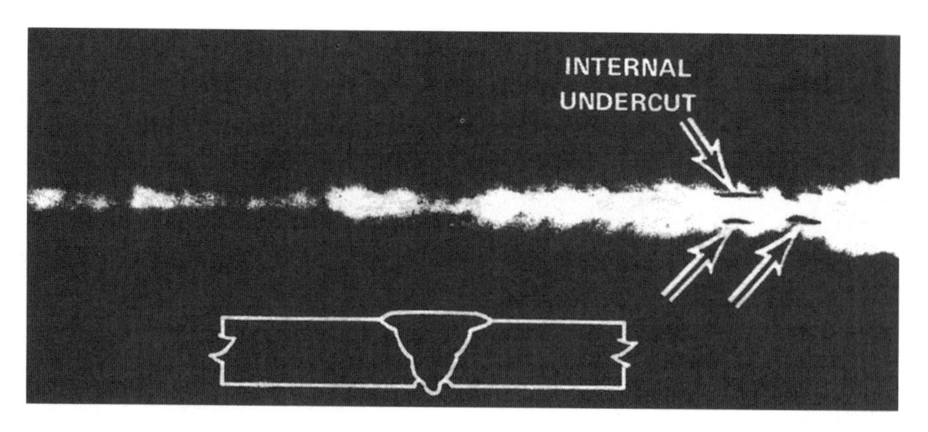

Figure 1. Radiograph of internal undercut. This defect may be intermittent or continuous (Figure 2) and either on one side or both sides of the weld centerline.

WHEN RADIOGRAPHS ARE INDICATIVE OF POOR WORKMANSHIP, BEST RESULTS CAN BE OBTAINED BY SHOWING THE RADIOGRAPHS TO THE WELDER SO THAT HE OR SHE CAN UNDERSTAND WHAT HE OR SHE IS DOING WRONG AND CORRECT IT.

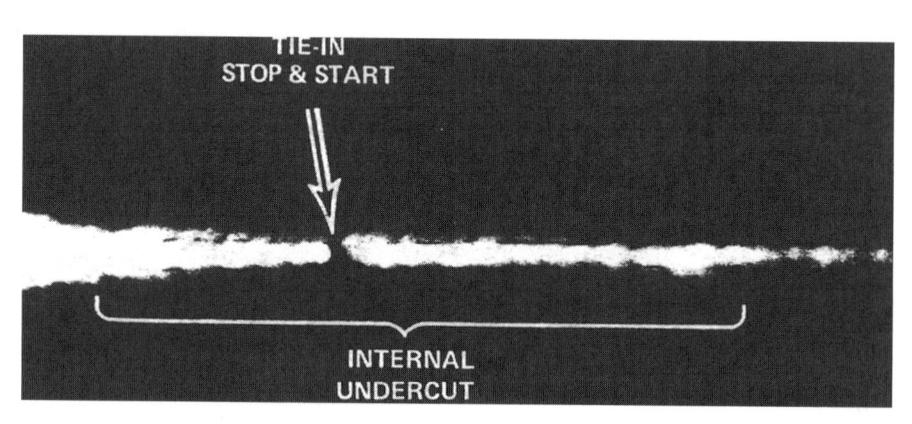

Figure 2. Radiograph of internal undercut and the lack of penetration that tends to occur at a stop and start.

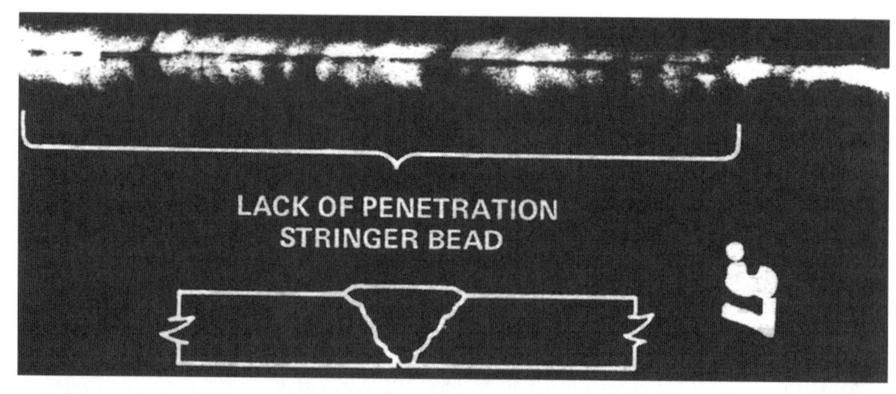

Figure 3. Radiograph of lack of penetration on the stringer bead appears as a single, straight, dark line.

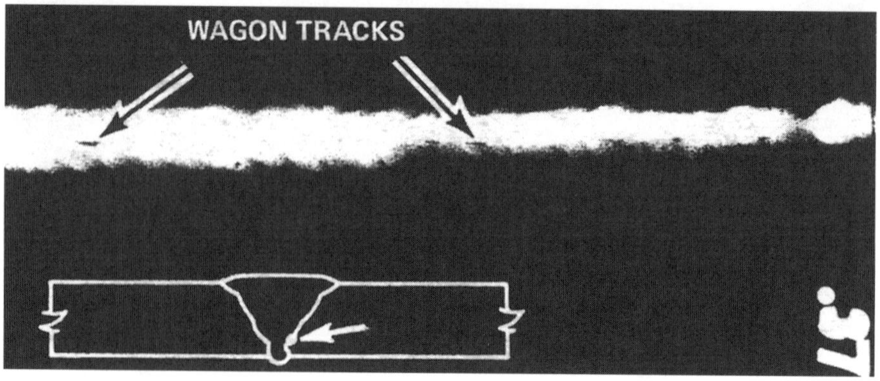

Figure 4. Radiograph of unfilled wagon tracks. Distinguishing between this defect and internal undercut required skill and experience.

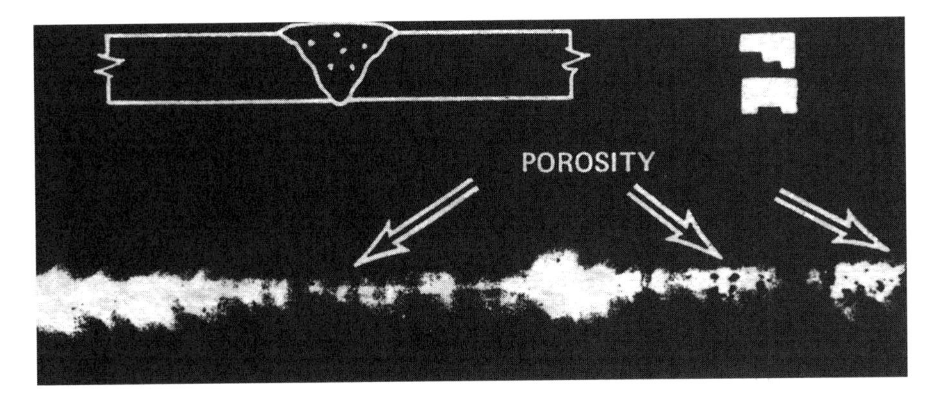

Figure 5. Radiograph of porosity.

juggling the root spacing. Within limits, this can be fairly effective. For example:

Condition	Change in Root Spacing
Land too small.	Decrease root spacing.
Land too large.	Increase root spacing.
High-low condition.	Decrease root spacing.
Bevel too small.	Increase root spacing.

A skillful line-up crew can be helpful in juggling the root spacing to bring about the most favorable results, but this has limitations. And it is at best a poor substitute for a uniform and consistent joint preparation.

Internal chamfer

It is important to remove any burr or overhang on the inside edge of the pipe after the root face has been machined

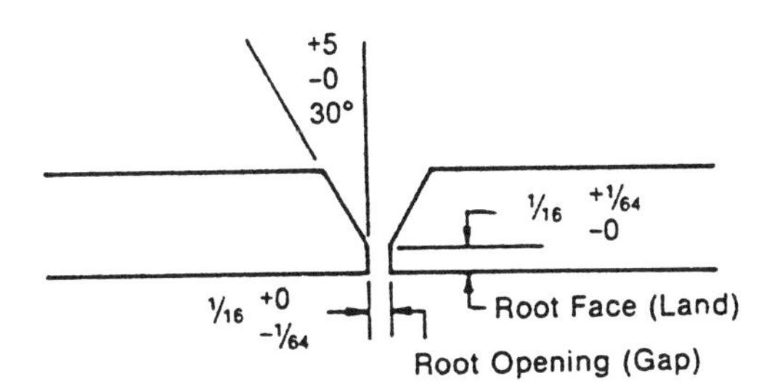

Pass	Electrode(1) in. (mm)	Current(2) Amps DC(+)(5)	Alternate Stringer Bead Procedure Amps DC(−)(5)
Stringer Bead	1/8 (3.2)(3) or 5/32 (4.0)(3)	90 — 120 140 — 165(4)	85—115 115—140
Hot Pass	5/32 (4.0)	160 — 185	—
Fills and Stripper	3/16 (4.8)	160 — 190	—
Cap Pass	3/16 (4.8)	150 — 180	—

(1) Fleetweld 5P+, Shield-Arc® HYP or 70+ suggested depending upon the pipe steel requirements and contractor specifications.

(2) The ideal current should be selected from within the range shown. In general, **the best quality will be obtained by operating at the lower end of the range.** Hot pass currents up to 200 amps are often used but cause early coating breakdown and larger stub losses.

(3) Use 1/8" (3.2 mm) diameter electrode for stringer beads when the wall thickness is .250" (6.35 mm) or less and when the gap is too small to permit use of the 5/32" (4.0 mm) size [1/8" (3.2 mm) diameter electrode should also be used on other passes when the wall thickness is .250" (6.35 mm) or less; however, 5/32" (4.0 mm) diameter can also be used on fill and cap passes on .250" (6.35 mm) wall].

(4) Weld stringer bead at 23-25 arc volts and 10-16 in/min arc speed. Measure arc voltage between the pipe and electrode holder.

(5) **Polarity — DC (−) (negative polarity)** should be used for stringer bead welding when burn-through, internal undercut and hollow bead defects are a problem. These problems generally occur on thin wall pipe steels containing over .1% silicon. Lower currents are used with DC(−) which helps to reduce these problems. Travel speeds with DC(−) will be equal to travel speeds with DC(+).

Hot pass and all other passes should be run DC(+) (positive polarity).

Welding DC(−) (negative polarity) on the stringer bead will not be harmful to either mechanical or metallurgical properties.

General Guidelines for Power Source Adjustment (SA- and SAE Machines)
In general, the "Current Range" or "Current Control" switch (whether a tap or continuous control), should always be set as low as possible to get the current wanted, and the "Fine Current Adjustment" or "Job Selector" control (which controls open circuit voltage) should be set as high as possible. It is usually better to set this control (for OCV) at the mid-point or higher for best arc stability and fewest popouts.

When using a Lincoln SA-250 engine driven power source, it will normally be necessary to set the "Current Range Selector" tap switch one position lower than used on a Lincoln SA-200 power source to get the same operating characteristics.

Figure 6. Recommended joint preparation and typical procedures—operate within these tolerances to help ensure good-quality welds.

(or ground). However, this clean-up operation should never produce an internal chamfer. A condition such as this is almost certain to result in "internal undercut" (Figure 7). See Recommended Procedures for Cleaning Pipe on the previous page.

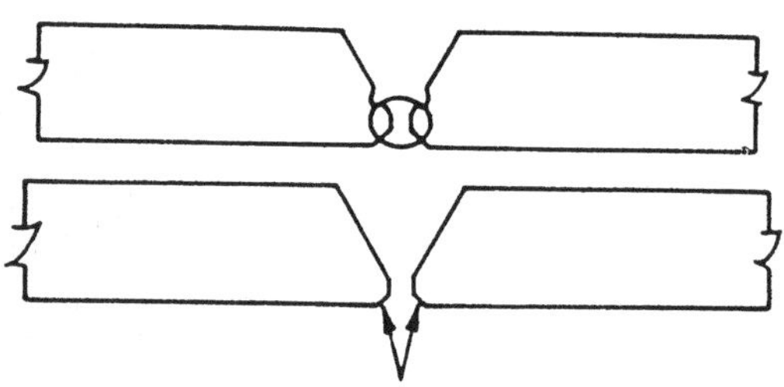

Figure 7. An internal chamfer will tend to leave an unfilled area on one or both sides of the stringer bead. On the x-ray this will be interpreted as internal undercut.

Undercut vs. welding current

Excessive current produces internal undercut. Further, the incidence of internal undercut is more prevalent in the 4-6-8 o'clock portions of the pipe. In this bottom area it is possible to use a considerably higher current than is necessary without getting burn-through, and this results in overheating and undercutting. Knowing this, it would appear to be a simple matter to specify an ideal current that would not give undercut. However, this is extremely difficult because of the many variables of the many interrelated factors.

A recommended current range is given in Figure 6, but it is necessarily rather broad to cover all reasonable variations. From a practical standpoint, the correct current should be determined by starting within the recommended range and then "fine tuning" to get precisely what "feels good" and produces the required results. This final current value will vary slightly from one operator to another and with different pipe materials, diameters, wall thickness, etc.

Because of inaccuracies in ammeters, the effect of arc voltage on current, the inability to accurately read a bobbing ammeter, etc., it is impractical to hold to an arbitrary current value. For the accomplished stringer bead welder, the selection of the ideal current is not too difficult—but he will be doing it primarily by "feel and touch."

Keyhole size vs. current and undercut

To get a good inside bead it is highly desirable to maintain a small, visible keyhole at all times. If the keyhole is allowed to get too big, it will result in internal undercut and/or windows (Figure 8).

Assuming that the joint preparation is correct, the keyhole size is a function of current, electrode angle, and pressure. The current should be "fine tuned" to produce a small keyhole when the electrode is dragged lightly using the normal electrode angle. If minor variations occur in the keyhole size (for any reason), the electrode angle and pressure can be manipulated to maintain the proper keyhole size (Figure 9).

In general, a keyhole that is about $^{1}/_{8}''$ (3.2 mm) in length is ideal; if it becomes $^{5}/_{32}''$ (4.0 mm) or longer, undercut and windows are imminent.

Frequently, a good inside bead is obtained without having a visible keyhole. For example, at a tight spot, the keyhole may disappear and the arc goes "inside the pipe." With the proper manipulative skill, this condition is tolerable and a good inside bead will be obtained. However, when it is done in this manner, the welder is largely dependent on the sound of the arc inside the pipe. A small, visible keyhole is easier to work with and is a much more controllable condition.

Significance of arc voltage

It is recommended that meters be used since a fairly good correlation can be established between the arc voltage and the keyhole size. For example, under controlled conditions, an ideal keyhole size is consistently accompanied by an arc voltage of no more than 25 volts. When keyhole size is

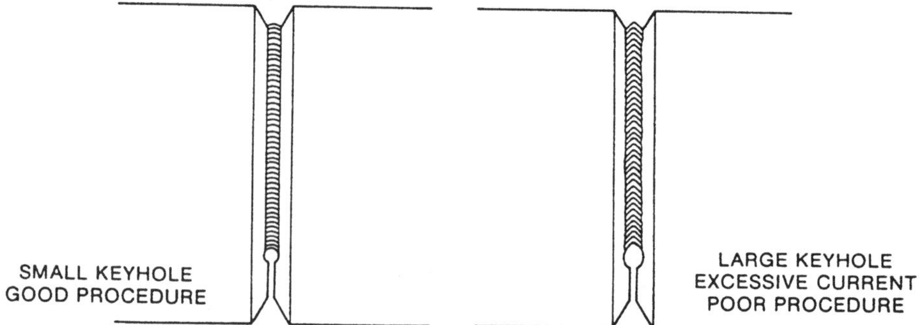

Figure 8. Maintaining a small keyhole will help assure a good inside bead. The larger keyhole at the right is almost certain to give internal undercut and/or windows.

ONE OF THE PRIME OBJECTIVES OF THE STRINGER BEAD CREW SHOULD BE TO PROVIDE A GOOD, STOUT STRINGER WITH AS MUCH THROAT AS POSSIBLE.

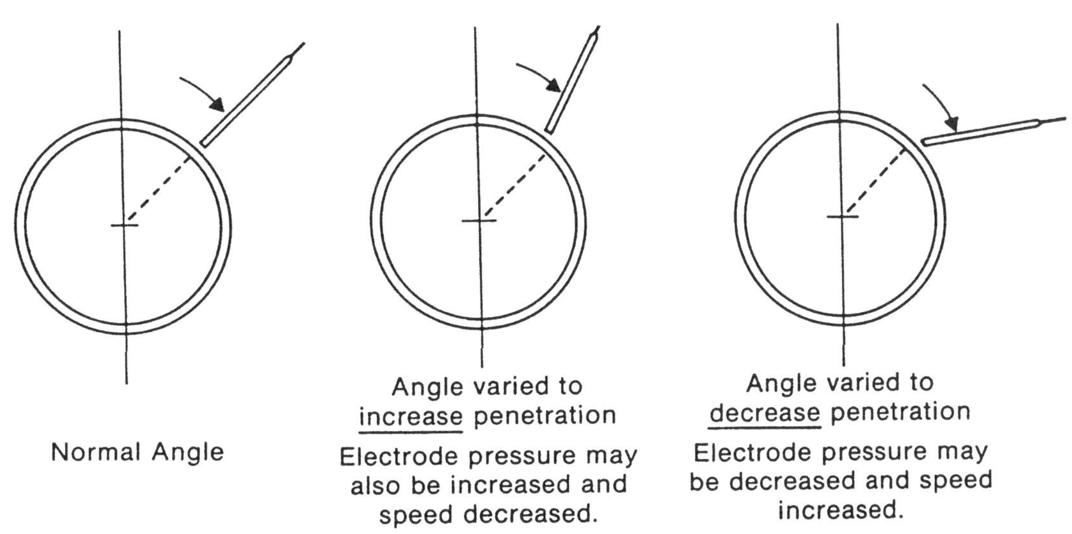

Figure 9. The penetration and keyhole size can be effectively controlled by varying the electrode angle as shown—if the joint preparation is uniform.

increased by increasing current, increasing the gap, changing the electrode angle, etc., the arc voltage increases to 26–28 volts. This is reasonable because the arc voltage reflects arc length. At the higher arc voltages, internal undercut and wagon tracks may occur.

The same arc voltage numbers quoted above may not occur in all instances. However, it is readily apparent that there is a dependable correlation between keyhole size and arc voltage. A maximum tolerable arc voltage can be determined by experimentation.

It should be further noted that this arc voltage is determined while welding and it should not be confused with the open circuit voltage. Also, it can be affected by the welding technique; changes in electrode angles and drag techniques should not be allowed to cloud the issue.

Varying current/varying gap

Greater root penetration naturally occurs at the top and bottom portions of the pipe, and least penetration tends to occur at the sides. This being the case, it would be desirable to alter the root spacing accordingly, but this obviously is not practical. It should also be noted that a condition which permits maximum spacing on top and bottom and minimum spacing on the sides is not tolerable.

Assuming a uniform gap all the way around, the ideal condition would be to vary the current as the weld progresses from 12 to 6 o'clock to compensate for the changes in penetration due to position. Instead, penetration changes are usually controlled by the manipulative skills of the welder (see Figure 9). In an extreme case, a second pair of hands may be required to adjust the current up or down on a signal from the welder. This requires a good communication system and a well-coordinated effort to avoid overshooting or undershooting the current. In some cases, this becomes a survival technique to make up for other undesirable variations.

Undercut—real or imaginary?

If visual inspection were practical for determining internal undercut, it would be easy to accurately determine its presence and severity. However, this is not the case. Radiography is the judge of internal conditions, and this has led to many questionable interpretations. If internal undercut is present, it will show up on the film as a very narrow, dark line immediately adjacent to the stringer bead—on one or both sides. The darkness of the line will vary with the depth of the undercut (Figures 1 and 2).

Proper identification of undercut is sometimes difficult because its location is directly in line with the "wagon track" area (Figure 4). Distinguishing one from the other may be difficult.

Correct interpretation of internal undercut may be further complicated by the fact that a "stout" inside bead will stand out clearly (lighter on the film) against the adjacent area, which is thinner and therefore darker. The thinner, darker area will be further darkened by the presence of any fairly deep, widely spaced surface ripples in the cap pass. The net effect is to produce a dark shading in the area where internal undercut might occur.

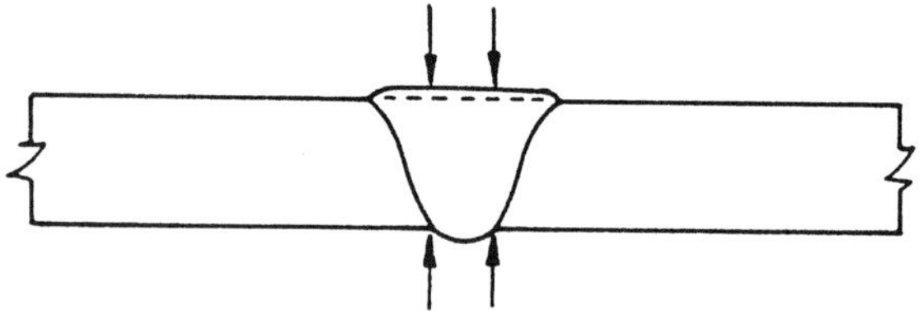

Figure 10. Any deep surface ripples located directly above a point adjacent to the inside bead will contribute to a dark shading of the film at this area. This has been misinterpreted by some as internal undercut.

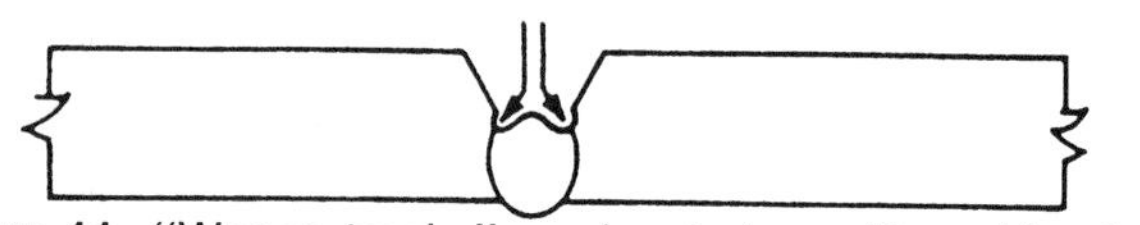

Figure 11. "Wagon tracks" are located on either side of the stringer bead.

Although this darker area is considerably wider than any undercut, it has been mistaken for undercut and resulted in cutting out good welds. Normally, a "stout" inside bead is considered good, but in this instance, the stouter the bead, the darker the adjacent area would appear (Figure 10).

Wagon tracks

It is not possible to completely eliminate the sidewall undercut produced by the stringer bead. This condition is generally referred to as "wagon tracks" (Figures 4 and 11).

It is the responsibility of the stringer bead crew to minimize wagon tracks and the responsibility of the hot pass crew to burn them out. This is normally done with 5/32″ (4.0 mm) electrode at about 175–180 amps. The secret in melting out wagon tracks lies primarily in the skill of the hot pass crew and the penetration factor of the electrode. The majority of the hot pass men use a rapid stitching technique that exposes the wagon track momentarily and permits them to direct the arc into the undercut area and melt them out.

Condition	Results
Bevel too small (Figure 12).	Increases depth of W.T.
Root spacing (gap) too small.	Increases depth of W.T.
Current and/or speed too high.	Increases depth of W.T.

The depth of the wagon tracks may be affected by the following:

In extremes, it may be necessary to use a grinder to open up the sidewalls to minimize deep "wagon tracks" or, if they do occur, to grind the stringer to eliminate the high peaked center (Figure 13). In all cases a $\frac{5}{32}$″ (4.0 mm) thickness disc grinder should be used to grind root beads on all pipe from 12″ (304.8 mm) up irrespective of wall thickness [a $\frac{1}{8}$″ (3.2 mm) disc grinder will grind the center and roll the sides over on the wagon track unless side pressure is applied]. This will make it easier to melt them out.

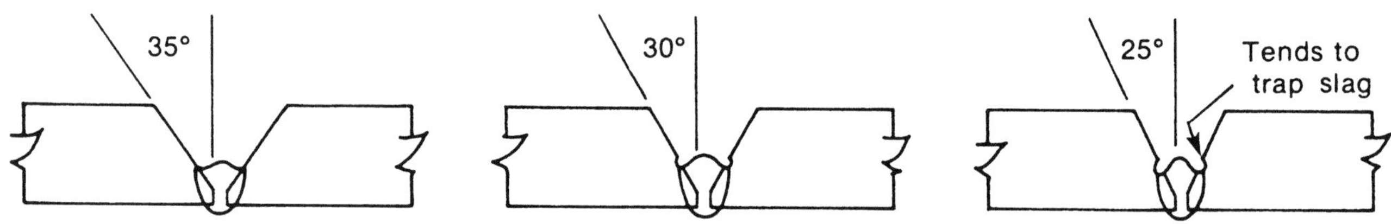

Figure 12. The depth of the wagon tracks will vary inversely with the bevel angle.

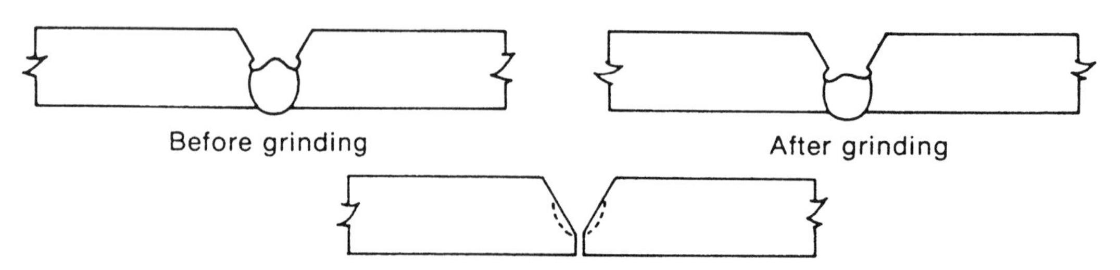

A narrow bevel or tight gap may require grinding as per dotted line.

Figure 13. A disc grinder, if used with restraint, can be helpful in correcting the condition shown above.

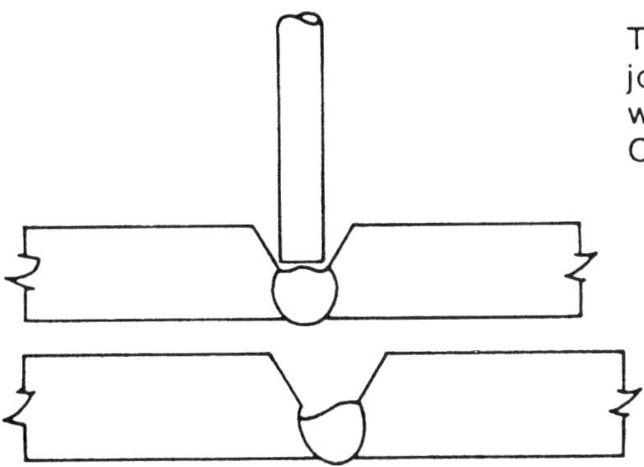

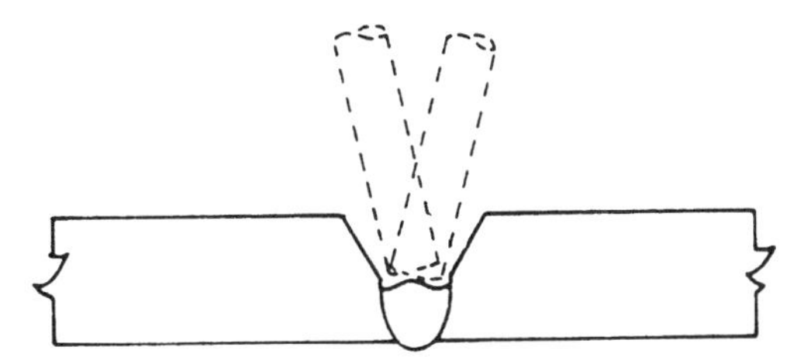

Figure 14. The correct angle can only be determined by observing the shape and location of the bead. If the bead drifts to one side for any reason, it will leave a deep undercut on the opposite sidewall. Tilting the electrode just a few degrees toward the undercut side will straighten the bead up. This change must take place rapidly if the correct bead placement is to be maintained.

If the stringer bead tends to wander to one side or the other, it will leave a deep undercut on the shallow side. This condition should be corrected immediately by changing the electrode angle (Figure 14).

The hollow bead

The stringer bead defect shown in this sketch is known by several names including the hollow bead, vermicular porosity, and wormhole porosity. Its length varies from a fraction of an inch to several inches. Radiography exposes the presence of the problem clearly.

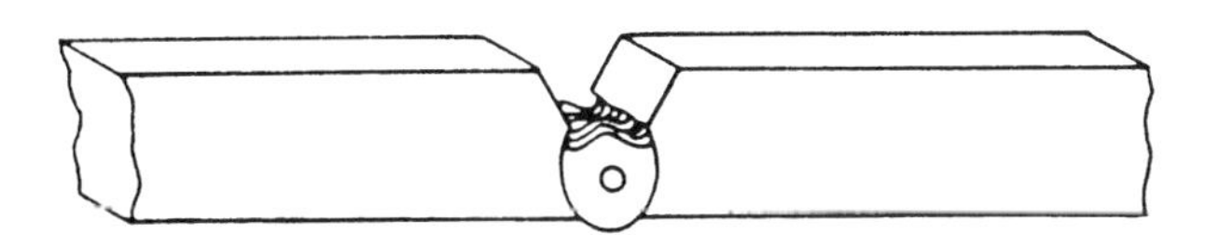

Vermicular porosity occurs most readily when welding high silicon pipe—generally above .10% silicon. It is aggravated by excessive travel speeds and high currents. Welding the stringer bead with $\frac{5}{32}$″ (4.0 mm) electrode at 130–165 amps DC(+) and 12 to 14 in/min (0.3 to 0.4 m/min) travel speed minimizes its occurrence. DC(−) (negative polarity) should be used for stringer bead welding when burn-through, internal undercut and hollow bead defects are a problem. These problems generally occur on thin wall pipe and on pipe steels containing over .1% silicon. Lower currents can be used with DC(−), which helps to reduce these problems. Travel speed with DC(−) will be equal to travel speeds with DC(+).

Hot pass and all other passes should be run DC(+) (positive polarity).

Welding DC(−) (negative polarity) on the stringer bead will not be harmful to either mechanical or metallurgical properties.

Hollow bead or wormhole porosity may also be caused by poor joint preparation and/or cleanliness. See Recommended Procedures for Properly Cleaning Pipe.

Filler, stripper, and cap

In general, there is little fault found with the fillers, strippers, and caps. Most of this welding today is being done with $\frac{3}{16}$″ (4.8 mm) Shield-Arc® HYP or 70+ at 160–180 amps, and the results have been excellent.

A reasonably competent crew of firing line welders armed with FW5P+, HYP, or 70+ can do a very fine job of finishing the weld if the stringer and hot pass crew have been doing their work properly.

The size and consistency of the final weld will have its influence on the x-ray radiograph. Thus, the firing line welders should be encouraged to produce a cap pass that is as uniform as possible with neatly stitched close ripples and as much reinforcement as required.

Welding cracks

Since the advent of the higher tensile pipe steels (5LX 52, 60, 65, 70 etc.), it has been necessary to exercise better procedural control to eliminate the possibility of weld and heat-affected zone (HAZ) cracks. To do this effectively, all of the following factors must be controlled.

1. Joint preparation and root spacing must be as specified in the approved procedure. (See Figure 6.)
2. Hi-Low conditions must be held to a minimum.
3. Weld 5LX70 with Shield-Arc® 70+. Weld 5LX65 and lower with Shield-Arc HYP or Fleetweld 5P+. Fleetweld 5P+ is recommended for all stringer beads when lower hardness is a consideration. Lower stringer bead hardness will result in improved resistance to HAZ cracking.
4. The need for preheat varies considerably between applications. Cracking tendencies increase with higher carbon and alloy content, thicker pipe wall and diameter (faster quench rate), and lower ambient temperature. Preheat cold pipe to at least 70°F (21°C). Preheat to as much as 300°F (149°C) may be required to slow the quench rate and prevent cracking when welding high-strength, large-diameter, or heavy-wall pipe. Specific preheat requirements must be determined for each situation based on these considerations.
5. Ideally the line-up clamp should not be removed nor should any movement of the pipe take place until the stringer bead is 100% completed. Any movement of the pipe could cause a partially completed stringer bead to be overstressed and to crack. The tendency for such cracking varies with the chemistry of the pipe, its diameter and wall thickness, and its temperature. Under favorable conditions it may be possible to remove the line-up clamp with as little as 60% of the stringer bead completed, but this should be done only when it has been clearly demonstrated that this practice does not cause cracks to occur.
6. After removal of the line-up clamp, the pipe must be gently and carefully lowered on skids.
7. Use only enough current on the stringer to get a good inside bead and travel slowly to get the maximum weld cross section.
8. Restrict lack of penetration on the inside bead at tie-ins to ¼″ (6.4 mm) or less. Use a disc grinder to improve this situation on starts and stops only.
9. Remove slag from each bead by power wire brushing. Grinding should not be necessary except possibly to clean up a lumpy start or humped up center or perhaps to improve a crater condition. *Note:* Grinding of the stringer bead should be done with a 5⁄32″ (4.0 mm) disc. Excessive grinding can be detrimental.
10. Weld stringer beads with two or more persons welding on opposite sides to equalize stress. Use three welders on 20–30″ (508–762 mm) pipe, and four welders on larger pipe.

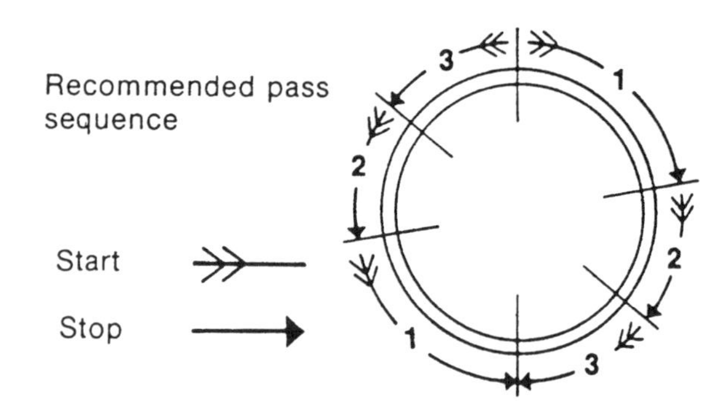

11. Start the hot pass immediately after completion of the stringer—always within 5 minutes. At least two hot pass welders should be used on each joint, and to put this pass in as soon as possible after the stringer bead it may require a total of four hot pass welders leap frogging each other to keep up.
12 Minimize the wagon tracks—this area is highly vulnerable to cracking until the hot pass is completed (Figure 15).

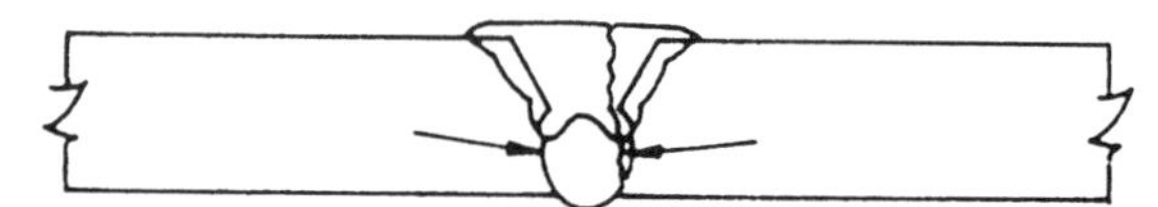

Figure 15. Cracks tend to occur at the area indicated on one side or the other of the stringer bead. This could eventually propagate up through the weld. A properly controlled procedure and good workmanship can eliminate this condition.

The Lincoln Electric Company strongly recommends for weldments intended for sour gas, sour crude, or other critical service applications that the customer verifies that both the level of hardness and variation in hardness of the weld, HAZ, and base plate are within acceptable limits.

How many welds will the average welder make per hour?

Where stringer beads have been run, the average welder can complete about 140 inches of weld per hour on ordinary ¼-inch wall line pipe. To find the average number of welds per hour, divide the circumference of the pipe into 140.

$N = \frac{140}{C}$ where circumference(C) = pipe diameter in inches times 3.14

Example. How many welds per hour will a welder complete on ¼-inch wall 10¾-inch line pipe, if the stringer beads have been run?

$$N = \frac{140}{(10.75)(3.14)}$$

$$N = \frac{140}{33.76}$$

N = 4.15 welds per hour or about 42 welds per 10-hour day.

On large diameter pipe where wall thickness has been increased, welders average about 100 inches of completed weld per hour.

Example. How many welds per hour will a welder complete on 30-inch line pipe if the stringer beads have been run?

$$N = \frac{100}{C}$$

$N = \frac{100}{(30)(3.14)} = \frac{100}{94.2} = 1.1$ welds per hour or 11 welds per 10-hour day

How much welding rod is required for a mile of schedule 40 pipeline?

Rule. For 40-foot joints of schedule 40 pipe, multiply the nominal pipe diameter by 22½; the answer is pounds of electrode per mile of pipeline.

Example. 6-in. pipe:

6 × 22½ = 135 lb. per mile

Example. 10-in. pipe, 30-ft joints:

10 × 22½ = 225 lb. per mile

For 40-ft joints:

$225 \times \frac{40}{30} = 300$ lb. per mile for 30-ft joints.

The rule is based on standard weight (schedule 40) pipe and the usual number of passes, which varies from two for small pipe up to four for 16 in.; larger pipe is usually thinner wall. The rule includes an allowance for wastage.

How many pounds of electrodes are required per weld on line pipe?

Divide the nominal pipe size by 2 and multiply the result by one-fourth pound.

$$\text{Pounds of electrode} = \frac{N}{2} \times 0.25$$

Example. How many pounds of electrode will be used per weld on 10-in. line pipe?

$$\text{Pounds of electrode} = \frac{10}{2} \times .25 = 1.25 \text{ lb. per weld.}$$

Example. How many pounds of electrode will be used per weld on 24-in. line pipe?

Solution. Pounds of electrode = 12 × 0.25 or 3 lb. per weld.

Welding criteria permit safe and effective pipeline repair

Battelle Laboratories studies show weld sleeves and deposited weld metal can be applied without removing defective lines from service

J. F. Kiefner, Senior Research Engineer, Battelle Houston Operations

Research conducted by the American Gas Association (AGA)[1–5] and others[6–8] showed that pipeline defects can be repaired without removing the lines from service. Two repair methods evaluated were full-encirclement sleeves and direct deposition of weld metal on defects.* To ensure minimum risk when using these methods, criteria for repairing pipelines in service have been formulated (Table 1).

The criteria in the table are intended as a repair guide for lines that operate at 40% of their specified minimum yield strength (SMYS) or more. The criteria may by unnecessarily restrictive for lines that operate at or below 40% of SMYS.

Note that defects in pressurized pipelines, especially pneumatically pressurized lines, can cause sudden catastrophic ruptures. When the procedures are carefully planned, risks are minimized. Nevertheless, repair errors could reduce the safety margin of these criteria. Consequently, users of Table 1 are urged to exercise caution when performing repairs.

Choice of repair method

The repair method options included in Table 1 are:

- Type A sleeve
- Type B sleeve
- Deposited weld metal
- Removal
- Removal by hot-tapping

Type A sleeve repair consists of placing a reinforcing band concentric with the pipe while leaving the band ends unsealed. The sleeve strengthens a defective area by restraining bulging that would otherwise occur when the weakened pipe shell is pressurized. Such a sleeve may also carry a small portion of the hoop stress, but in no case can it become the sole pressure-containing element since its ends are not sealed. As a result, it cannot be used to repair a leaking defect. If a gap exists over the defect, filling of the gap is required. Two classes of Type A sleeve exist: no filler and with filler.

Type B sleeve repair requires a pressure-tight reinforcing concentric band. Its ends are sealed to the carrier pipe, and it can contain pressure. It can be used to repair leaking defects. In addition, a Type B can also be used as a Type A sleeve. Three classes of Type B sleeves exist: no filler, with filler, and pressurized.

Deposited weld metal completely fills and eliminates clearly visible defects such as corrosion. Access to an entire defect is required so that weld metal can penetrate and bond to sound metal. It may be possible to enlarge pits, laps, or undercut by grinding a groove wide enough to permit access. Deposited weld metal can be selected when requirements to prevent burn-through are met. These requirements are discussed later.

Removal of a damaged pipe section is another way of eliminating a defect. The line must be taken out of service, purged, removed, and replaced with a sound tie-in piece. Removal is often a poor economic choice in terms of wasted product and interrupted service. Thus, removal is recommended in Table 1 with the understanding that it may not be practical in certain instances.

Removal by hot-tapping takes advantage of a widely accepted method for making branch connections on live pipelines. The coupon removed from the line must contain the entire defect. Limitations of this technique will be discussed later.

Types of defects

Classes of defects listed in Table 1 are:

- Manufacturing defects—cracks, undercut, lack of fusion, and lack of penetration in seam welds; laps, pits, cracks, and rolled-in slugs in the pipe body; hard spots.
- Environmentally caused defects—selective corrosion and hydrogen stress cracking in electric resistance welded or flash welded seams; general pipe corrosion; pitting corrosion; stress corrosion cracks; hydrogen stress cracking in hard spots.
- Defects caused by outside forces including dents and gouges.
- Construction defects.

Submerged arc-welded (SAW) seam defects, both straight seam and spiral, include undercut, incomplete fusion, incomplete penetration, and cracks. Repair methods recommended for these defects are shown in Table 1. No filler is required with sleeves, because after the weld reinforcement is ground flush, no gap between the pipe and sleeve should exist. Removal by hot-tapping is not recommended because the tap would involve cutting through the seam, a practice considered unacceptable by many companies. Repair by deposited weld metal is applicable only for undercuts and is subject to special requirements given in footnotes of Table 1.

Type B pressurized sleeves used on nonleaking defects can be pressurized only by using a tapping nipple. Since other means of sleeve repair are entirely adequate when used as directed, the intentional pressurization of a sleeve when no leak or near leak is present is not necessary.

Inside or interior defects require special consideration since they are not readily visible. Removal may be the best alternative unless one can be reasonably certain of the extent of the defects. Hot-tapping is not recommended because of the uncertainty of the extent of an inside or interior defect.

Electric resistance welded (ERW) or flash welded (FW) seam defects include: upturned fiber imperfections; incomplete fusion; penetrators; cold welds; cracks. These can only be repaired with pressurized sleeves and removal since the welds are susceptible to brittle fracture or low-resistance ductile fracture initiation. The value of restraint of bulging from either type of nonpressurized sleeve is uncertain. Hence they are not recommended for these defects. Hot-tapping also is not recommended because of the involvement of the seam weld. It is not recommended that weld metal be deposited because of possible low ductility. Required grinding before such welding would involve creating or enlarging a defect in a potentially low-toughness material. Only removal or use of a Type B pressurized sleeve that stress-relieves the defect is recommended.

Other seam defects include lap welds and furnace butt welds. These must be dealt with on a case-by-case basis. Such welds usually appear in older or smaller and lower pressure pipelines where fracture resistance requirements are less stringent. On the other hand, some of these older materials can be quite susceptible to low-toughness fracture behavior. It is probably best to treat these materials with the same caution as ERW and FW seams as they are being operated at stress levels exceeding 40% of SMYS.

Laps, pits, seams, cracks, and rolled-in plugs should be repaired subject to these limitations: special requirements of I.D. and interior defects and leaks must be observed; hot-tapping is acceptable on nonleaking, O.D. defects as long as the entire defect is removed with the coupon; deposited weld metal may be used to repair pits or laps if they can be entirely exposed by grinding.

Hard spots created in the plate by accidental quenching on the run-out table may become flat spots in the pipe since they do not yield when the plate is formed to pipe. Such hard spots do not fail spontaneously unless they contain quench cracks or unless they undergo hydrogen stress cracking. Such spots should be repaired if they are not cracked. Type A or B sleeves without filler would not be acceptable since they would not restrain flat spots from bulging.

Hard spots can be remedied by several methods if they occur as outside nonleaking defects. Sleeve repair methods with filler not only provide strengthening but shield the hard spot from the environment, which may cause it to become cracked.* Hard spots at the I.D. surface should not be repaired with Type A sleeves since they may be susceptible to cracking in certain types of product service. Such a sleeve would not prevent a leak if the resulting crack grew through the wall.

Selective corrosion and hydrogen stress cracking in ERW and FW seams are subject to low toughness fracture behavior. It is recommended that they be removed or repaired with pressurized Type B sleeves to relieve stress.

General or widespread corrosion is that which covers too wide an area to be repairable by means of hot-tapping or deposited weld metal. Any other repair means are acceptable if they cover the critically affected area. When internal corrosion is present, its extent must be reasonably well known and further corrosion must be prevented. Removal may sometimes be the only suitable choice.

Pitting corrosion may involve isolated pits or groups of pits. Generally, isolated single pits will not require repair if further corrosion can be halted. ASME guidelines can be used to determine whether or not a corroded region requires repair. Groups of interacting pits can substantially lower remaining strength. If strength has been lowered, the pipe should be repaired. Any of the methods of Table 1 are suitable if the noted precautions are observed.

Stress-corrosion cracks occur in clusters and often cover large areas. It is difficult to repair these cracks by hot-tapping or deposited weld metal. Repair methods for stress-corrosion cracks are confined to those which can strengthen the entire affected area.

Hydrogen stress cracking appears in hard spots attacked by hydrogen emitted from bacteria external to the pipe and from cathodic protection. Internally the hard spots can be attacked by certain types of products—especially those containing hydrogen sulfide. Because flatness often occurs near hard spots, nonfilled sleeves are not recommended. Deposited weld metal is not recommended because grinding of a hydrogen stress crack in a hard—and usually brittle—spot while the pipe is under pressure is not safe. Interior hydrogen stress cracking should not be repaired by a Type A sleeve since such a sleeve cannot prevent leaks if the crack grows through the wall.

Dents and gouges and a combination gouge-in-dent result from external encroachment by mechanical excavating equipment, other kinds of equipment, or rocks. Plain

* One method of protecting hard spots from hydrogen stress cracking that so far, has proven adequate, is using a concentric band of sheet metal spaced away from the pipe by rubber seals. The annular space is filled with coal tar to exclude ground water, and the metal band (which is coated) shields the hard spot from hydrogen generating cathodic protection current.

Table 1
Criteria for selecting an appropriate repair method (X indicates an acceptable repair)

Repair Method	Type of Defect								
	Manufacturing Defect				Environmentally Caused Defect in ERW or FW, Seam Weld in Body, SAW Seam or Girth Weld				
	In Seam Weld		In the Body of the Pipe						
	SAW Undercut, Incomplete Fusion, Incomplete Penetration, Crack	ERW and FW Upturned Fiber Imperfection, Incomplete Fusion, Penetrator, Cold Weld Crack	Lap, Pit, Seam, Crack Rolled-In Slug	Hard Spot Exceeding $R_c>35$ and 2 in. in Extent or More	Selective Corrosion, Hydrogen Stress Cracking in Weld Zone	General Corrosion	Pitting Corrosion	Stress-Corrosion Cracking	Hydrogen Stress Cracking in Hard Spot
	Nonleaking O.D. defect[a]								
Type A Sleeve, No Filler	X	...	X	...	...	X	X	X	...
Type A Sleeve, with Filler	...	...	...	X[g]	...	X	X	...	X
Type B Sleeve, No Filler	X	...	X	...	...	X	X	X	...
Type B Sleeve, with Filler	...	...	...	X	...	X	X	...	X
Type B Sleeve, Pressurized	X[b]	X[b]	X[b]	X[b]	X[b]	X[b]	X[b]	X[b]	X[b]
Deposited Weld Metal(h)	X[c]	...	X[c]	...	...	...	X[c]	...	...
Removal	X	X	X	X	X	X	X	X	X
Removal by Hot Tap	...	...	X[d]	X[d]	...	...	X[d]	...	X[d]
	Nonleaking interior or I.D. defect[a]								
Type A Sleeve, No Filler	X.	...	X	...	...	X	X.	...	...
Type B Sleeve, No Filler	X	...	X	X	...	X	X	...	X
Type B Sleeve, Pressurized	X[b]	X[b]	X[b]	X[b]	X[b]	X[b]	X[b]	...	X[b]
Removal	X	X	X	X	X	X	X	...	X
	Leaking defects								
Type B Sleeve Pressurized	X	X	X	X	X	X	X	X	X
Removal	X	X	X	X	X	X	X	X	X

[a] Corrosion or other defect that is known to exceed 80% of the wall thickness in depth should be treated as though it were a leak.
[b] Pressurization to be accomplished upon completion of repair by drilling hole through carrier pipe through a small tapping nipple.
[c] Only for undercut, laps, and pits where the defect can be safely enlarged to a weldable groove and subject to remaining wall thickness rules.
[d] Only if branch coupon entirely contains and removes defect.
[e] Only with humped sleeve or if flat sleeve is used. Girth weld must be ground flush. Do not use if extreme high-low condition exists.

Defect Caused by Outside Force					Construction Defect
In Seam or Other Weld		In Body of Pipe			In Girth Weld
Plain Dent or Gouge-in-Dent, SAW Seam or Girth Weld	Plain Dent, or Gouge-in-Dent ERW or FW Seam	Plain Dent Greater than 2% or Pipe Diameter	Gouge (Also Including Gouges in SAW Seams or Girth Welds)	Gouge-in-Dent	Undercut, Incomplete, Fusion, Incomplete Penetration, Crack
...	...	...	...	...	...
X	...	X	...	X	...
...	...	...	X	...	X[e]
X	...	X	...	X	...
X[b]	X[b]	X[b]	X[b]	X[b]	X[b,e]
...	...	...	X[b]	...	X[c]
X	X	X	X	X	X
...	...	...	X[d]	...	...
...	...	...	...	...	...
...	...	...	...	...	X[e]
...	...	...	...	...	X[b,e]
...	...	...	...	...	X
X	X	X	X	X	X[e]
X	X	X	X	X	X

One method of protecting hard spots from hydrogen stress cracking, which, so far, has proven adequate is that of using a concentric band of sheet metal spaced away from the pipe by rubber seals. The annular space is filled with coal tar to exclude ground water, and the metal band (which is itself coated) shields the hard spot from hydrogen-generating cathodic protection current. The role of this band is merely to shield the pipe from the cathodic protection and not to strengthen the pipe.
Use of low hydrogen electrodes is recommended.

dents are usually innocuous unless they are quite large*—2% or more of the diameter—or unless they involve a seam or girth weld. When the dent is large or welds are included in the dent or gouge-in-dent, the repair should prevent its outward movement. Sleeves with fillers or pressurized sleeves are required. If the included weld is an ERW or FW seam, the defect should be removed or the stress relieved with a pressurized sleeve.

Gouges without dents in the body of the pipe or in SAW seams or girth welds may be repaired by any means except deposited weld metal. Gouges-in-dents may be repaired only by means that prevent outward movement of the dent. Sleeves without fillers are not acceptable.

When dents are involved, hot-tapping is not recommended, since it may not remove the dent entirely. For any gouge or gouge-in-dent, repair by deposited weld metal is not recommended, since concealed cracks may exist. When a dent is present, weld metal may not have sufficient ductility to withstand the severe strains that accompany outward dent movement.

Construction defects in girth weld include undercut, incomplete fusion, incomplete penetration, and cracks. Deposited weld metal may be one of the best ways to repair undercut or other externally connected defects in girth welds that can be ground for access. Sleeves, if used, should have a special shape such as a central hump to avoid interference with the girth weld reinforcement. Only sleeves with welded ends are recommended. Such sleeves tend to strengthen the defective girth joint, whereas those with nonweld ends do not.

Burn-through

Burn-through into a live pipeline would defeat the purpose of a repair, would probably cause the line to be shut down, and might create a serious safety hazard to the repair crew.

An AGA study revealed that the following parameters control burn-through: remaining wall-thickness; welding

* Present requirements of federal regulations Part 192 dictate removal of dents extending over 2% of the pipe diameter.

Table 2
Limitations on remaining wall-thickness for repairing without a burn-through

Values are the minimum recommended thicknesses in inches with natural gas as the pressurizing medium at the pressures and flows shown
Maximum welding voltage, 20 volts
Maximum welding current, 100 amps

Pressure, psia	Gas Flow Rate, Feet/Second			
	0	5	10	20
15	0.320	—	—	—
500	0.300	0.270	0.240	0.205
900	0.280	0.235	0.190	0.150

heat input; and nature and state of the pressurization media in the pipe. Consequently, it is possible to prescribe safe limits for avoiding burn-through while depositing weld metal on a defect (Table 2).

Table 2 shows a relationship between I.D. surface temperature during welding and remaining wall thickness for various pressure and flow rates of natural gas. These values were taken from AGA curves. The table is based on: heat input characteristics of the 3/32-inch or 1/8-inch low hydrogen electrode; 80 to 100 amps welding current at 20 volts DC; an average electrode travel speed of 4 to 5 in./min.

Energy input should not exceed these limits. Low hydrogen electrodes are recommended for reasons explained later. Under these conditions, repairs can safely be made in the flat, horizontal, or overhead positions provided flow or pressure is controlled. Analytic results predict that the I.D. wall temperature will not exceed 2,000°F and a burn-through is highly unlikely when repairs are made in this manner.

Values shown in Table 2 are believed to be quite conservative. Repairs were made without burn-throughs on 0.180-in. remaining wall with air at ambient pressure and no flow inside the pipe. This conservatism provides an extra safety margin and allows linear extrapolations for values between those shown in the table. The minimum wall-thickness of 0.150-in. is established on the basis of the experimental results. This thickness is also recommended by the British Gas Corp. Within these limits, repairs may be made by deposited weld metal (Table 1).

Underbead cracking

Underbead cracking can be minimized by making repairs with low hydrogen electrodes in a manner that avoids hard weld HAZs.

One way to help assure that extremely hard weld zones are not formed during repair welding is to limit repairs to well-known carbon equivalent ranges. Unfortunately, chemistries of specific samples needing repairs will seldom be known. Low hydrogen electrodes prevent hydrogen from being present in the welding atmosphere, but they do not prevent formation of hard weld zones when adverse chemistries are present. Making of crack-free weldments requires careful control of heat input to avoid rapid quenching or post-repair treatments to assure that extreme hardness does not remain.

In the case of sleeves, the critical area in which creaking most often occurs is the fillet-weld at the ends. At least two procedures for making these welds, with a minimum risk of cracking, are available.[1,4,5,8]

In the case of deposited weld metal repairs, heat input must be kept low to avoid burn-through during the first or second passes. During later passes, however, higher heat inputs can be used to soften the resulting repair metal microstructure and the HAZ of base metal. A high heat input final pass can be made with a nonfusing tungsten electrode as suggested by the British Gas Corp.[8] An alternative procedure is to make an extra pass with high heat input using a conventional electrode. This pass can be ground off, since its purpose is merely to soften the heat-affected microstructure of previous passes.

(Based on a paper, "Criteria for Repairing Pipelines in Service Using Sleeves and Deposited Weld Metal" presented by the author at the AGA Transmission Conference at Montreal, Quebec, May 8–10, 1978.)

Source

Pipe Line Industry, January 1980.

References

1. Kiefner, J. F., and Duffy, A. R., *A Study of Two Methods for Repairing Defects in Line Pipe*, Pipeline Research Committee of the American Gas Association, 1974, Catalog No. L22275.
2. Kiefner, J. F., Duffy, A. R., Bunn, J. S., and Hanna, L. E., "Feasibility and Methods of Repairing Corroded Line Pipe," Materials Protection and Performance, Vol. III, No. 10, Oct., 1972.
3. Kiefner, J. F., "Corroded Pipe: Strength and Repair Methods," *Fifth Symposium on Line Pipe Research*, Houston, Texas, 1974, American Gas Association, Catalog No. L30174.
4. Kiefner, J. F., "Repair of Line Pipe Defects by Full-Encirclement Sleeves," *Welding Journal*, Vol. 56, No. 6, June, 1977.
5. Kiefner, J. F., Whitacre, G. R., and Eiber, R. J., "Further Studies of Two Methods for Repairing Defects in Line Pipe," to Pipeline Research Committee of the American Gas Association, NG-18 Report No. 112, March 2, 1978.

TERMINOLOGY[9–11]

Defect parameters

Defect: A crack, pit, crevice, gouge, dent, metallurgical anomaly, or combination of two or more of the above that is known or suspected to reduce the effective pipe strength level to less than 100% of its specified minimum yield strength (SMYS).

O.D. defect: A defect emanating at and extending radially inward from the outside surface but not entirely through the wall of the pipe.

I.D. defect: A defect emanating at and extending radially outward from the inside surface but not entirely through the wall of the pipe.

Interior defect: A defect emanating in the interior of the pipe wall but not of sufficient radial extent to be connected with either the inside or the outside surface.

Leaking O.D. defect: A defect that was initially an O.D. defect but which has grown through the wall to become a leak.

Leaking I.D. defect: A defect that was initially an I.D. defect but which has grown through the wall to become a leak.

Superficial defect: A lap, crevice, pit, group of pits, metallurgical anomaly, or plain dent (i.e., without scratches, gouges, or cracks) that is of insufficient extent to reduce the effective strength level of the pipe below 100% of SMYS.

Kinds of defects

*(Definitions marked by asterisk are from API Bulletin 5T1 *Nondestructive Testing Terminology*, Third Edition, April, 1972.)

Not every conceivable kind of defect in pipe is covered by this list. The list is limited to those that are likely to be encountered in an in-service pipeline.

Defects originating from pipe manufacture

(a) Defects not necessarily in the seam weld (primarily in the body of the pipe)

* Lap: Fold of metal that has been rolled or otherwise worked against the surface of rolled metal but has not fused into sound metal.
* Pit: A depression resulting from the removal of foreign material rolled into the surface during manufacture.
* Rolled-in slugs: A foreign metallic body rolled into the metal surface, usually not fused.
* Seam: Crevice in rolled metal that has been more or less closed by rolling or other work but has not been fused into sound metal.
* Hard spot: An area in the pipe with a hardness level considerably higher than that of surrounding metal; usually caused by localized quenching.
* Crack: A stress-induced separation of the metal that, without any other influence, is insufficient in extent to cause complete rupture of the material.

(b) Defects in the seam weld

* Incomplete fusion: Lack of complete coalescence of some portion of the metal in a weld joint.
* Incomplete penetration: A condition where weld metal does not continue through the full thickness of the joint.
* Under-cut: Under-cutting on submerged-arc-welded pipe is the reduction in thickness of the pipe wall adjacent to the weld where it is fused to the surface of the pipe.
* Weld area crack: A crack that occurs in the weld deposit, the fusion line, or the HAZ. (Crack: A stress-induced separation of the metal that, without any other influence, is sufficient in extent to cause complete rupture of the material.)
* Upturned fiber imperfection: Metal separation, resulting from imperfections at the edge of the plate or skelp, parallel to the surface, which turn toward the I.D. or O.D. pipe surface when the edges are upset during welding.
* Penetrator: A localized spot of incomplete fusion.
* Cold weld: A metallurgically inexact term generally indicating a lack of adequate weld bonding strength of the abutting edges, due to insufficient heat and/or pressure. A cold weld may or may not have separation in the weld line. Other more definitive terms should be used whenever possible.

Defects originating from external or internal environmental degeneration of the pipe

(a) Seam weld defects

Selective corrosion: Preferential corrosion in the fusion line of an electric resistance–welded or flash-welded longitudinal seam.

Hydrogen stress cracking: Environmentally stimulated cracking of the weld metal or HAZ of the longitudinal seam.

(b) Defects not in the body of the pipe (possibly in the seam weld—but not specifically because of the seam weld).

6. Cassie, B. A., "The Welding of Hot-Tap Connections to High Pressure Gas Pipelines," J. W. Jones Memorial Lecture, Pipeline Industries Guild, British Gas Corp., October, 1974.
7. Morel, R. D., *Welded Repairs on API 5LX-52 Pipe*, 13th Annual Petroleum Mechanical Engineering Conference, Denver, Colorado, September 24, 1958.
8. Christian, J. R., and Cassie, B. A., "Analysis of a Live Welded Repair on An Artificial Defect," ERS.C.96, October, 1975.
9. Smith, R. B., and Eiber, R. J., "Field Failure Survey and Investigation," *Fourth Symposium on Line Pipe Research*, American Gas Association, 1969, Catalog No. L30075.
10. Eiber, R. J., "Field Failure Investigations," *Fifth Symposium on Line Pipe Research*, American Gas Association, 1974, Catalog No. L30174.
11. Wright, R. R., "Proper Inspection Methods Minimize Pipeline Failures," *Oil and Gas Journal*, May 23, 1977, pp. 51–56.

Cross country pipeline—vertical down electrode consumption, pounds of electrode per joint*

Wall Thickness	5/16"			3/8"			1/2"			5/8"			3/4"		
Elec. Dia.	5/32"	3/16"	Total	5/32"	3/16"	Total	5/32"	3/16"	Total	5/32"	3/16"	Total	5/32"	3/16"	Total
Nom. Pipe Dia., in.			lb			lb			lb			lb			lb
6	0.34	0.54	0.88	0.34	0.86	1.2									
8	0.45	0.71	1.2	0.45	1.1	1.6	0.45	2.2	2.7						
10	0.56	0.88	1.4	0.56	1.4	2.0	0.56	2.7	3.3						
12	0.67	1.0	1.7	0.67	1.7	2.4	0.67	3.2	3.9						
14	0.73	1.1	1.8	0.73	1.8	2.5	0.73	3.5	4.2	0.73	5.7	6.4	0.73	8.3	9.0
16	0.84	1.3	2.1	0.84	2.1	2.9	0.84	4.0	4.8	0.84	6.5	7.3	0.84	9.4	10.2
18	0.94	1.5	2.4	0.94	2.3	3.2	0.94	4.5	5.4	0.94	7.3	8.2	0.94	10.6	11.5
20	1.10	1.6	2.7	1.10	2.6	3.7	1.1	5.0	6.1	1.1	8.1	9.2	1.1	11.8	12.9
22	1.20	1.8	3.0	1.20	2.9	4.1	1.2	5.5	6.7	1.2	8.9	10.1	1.2	13.0	14.2
24	1.30	2.0	3.3	1.30	3.1	4.4	1.3	6.0	7.3	1.3	9.7	11.0	1.3	14.2	15.5
26	1.40	2.1	3.5	1.40	3.4	4.8	1.4	6.5	7.9	1.4	10.5	11.9	1.4	15.3	16.7
28	1.50	2.3	3.8	1.50	3.7	5.2	1.5	7.0	8.5	1.5	11.3	12.8	1.5	16.5	18.0
30	1.60	2.5	4.1	1.60	3.9	5.5	1.6	7.5	9.1	1.6	12.1	13.7	1.6	17.7	19.3
32	1.70	2.6	4.3	1.70	4.2	5.9	1.7	8.0	9.7	1.7	13.0	14.7	1.7	18.9	20.6
34	1.80	2.8	4.6	1.80	4.4	6.2	1.8	8.6	10.4	1.8	13.8	15.6	1.8	20.1	21.9
36	1.90	2.9	4.8	1.90	4.7	6.6	1.9	9.1	11.0	1.9	14.6	16.5	1.9	21.3	23.2
38	2.00	3.1	5.1	2.00	5.0	7.0	2.0	9.6	11.6	2.0	15.4	17.4	2.0	22.4	24.4
40	2.10	3.3	5.4	2.10	5.2	7.3	2.1	10.1	12.2	2.1	16.2	18.3	2.1	23.6	25.7
42				2.20	5.5	7.7	2.2	10.6	12.8	2.2	17.0	19.2	2.2	24.8	27.0
44				2.30	5.7	8.0	2.3	11.1	13.4	2.3	17.8	20.1	2.3	26.0	28.3
46				2.40	6.0	8.4	2.4	11.6	14.0	2.4	18.6	21.0	2.4	27.2	29.6
48				2.50	6.3	8.8	2.5	12.1	14.6	2.5	19.4	21.9	2.5	28.3	30.8

* "Electrode required." Figures in the table include 4-in. stub lengths. These figures will vary with different stub loss practices.
Quantities required for the 5/32" size will vary based on travel speeds of the stringer bead and hot pass. Slow travel speeds may increase these quantities by up to 50%.
Reprinted courtesy of the Lincoln Electric Company.

Guidelines for a successful directional crossing bid package

The Directional Crossing Contractors Association (DCCA) has been addressing the issue of what information should be made available to contractors and engineers so that future projects proceed as planned. Crossings of rivers and other obstacles using directional drilling techniques are increasingly being utilized around the world. As in any construction project, it is necessary for the contractor to have as much information as possible to prepare a competitive and comprehensive proposal and to be able to successfully install the crossing. Better preconstruction information also allows the work to be undertaken more safely and with less environmental disturbance.

Development and uses

Originally used in the 1970s, directional crossings are a marriage of conventional road boring and directional drilling of oil wells. The method is now the preferred method of construction. Crossings have been installed for pipelines carrying oil, natural gas, petrochemicals, water, sewerage and other products. Ducts have been installed to carry electric and fiberoptic cables. Besides crossing under rivers and waterways, installations have been made crossing under highways, railroads, airport runways, shore approaches, islands, areas congested with buildings, pipeline corridors, and future water channels.

Technology limits

The longest crossing to date has been about 6,000 ft. Pipe diameters of up to 48 in. have been installed. Although directional drilling was originally used primarily in the U.S. Gulf Coast through alluvial soils, more and more crossings are being undertaken through gravel, cobble, glacial till, and hard rock.

Advantages

Directional crossings have the least environmental impact of any alternate method. The technology also offers maximum depth of cover under the obstacle, thereby affording maximum protection and minimizing maintenance costs. River traffic is not interrupted, because most of the work is confined to either bank. Directional crossings have a predictable and short construction schedule. Perhaps most significant, directional crossings are in many cases less expensive than other methods.

Technique

Pilot Hole—A pilot hole is drilled beginning at a prescribed angle from horizontal and continues under and across the obstacle along a design profile made up of straight tangents and long radius arcs. A schematic of the technique is shown in Figure 1. Concurrent to drilling a pilot hole, the contractor may elect to run a larger diameter "wash pipe" that will encase the pilot drill string. The wash pipe acts as a conductor casing providing rigidity to the smaller diameter pilot drill string and will also save the drilled hole should it be necessary to retract the pilot string for bit changes. The directional control is brought about by a small bend in the drill string just behind the cutting head. The pilot drill string is not rotated except to orient the bend. If the bend is oriented to the right, the drill path then proceeds in a smooth radius bend to the right. The drill path is monitored by an electronic package housed in the pilot drill string near the cutting head. The electronic package detects the relation of the drill string to the earth's magnetic field and its inclination. These data are transmitted back to the surface where calculations are made as to the location of the cutting head. Surface location of the drill head also can be used where there is reasonable access.

Pre-ream—Once the pilot hole is complete, the hole must be enlarged to a suitable diameter for the product pipeline. For instance, if the pipeline to be installed is 36-in. diameter, the hole may be enlarged to 48-in. diameter or larger. This is accomplished by "pre-reaming" the hole to successively larger diameters. Generally, the reamer is attached to the drill string on the bank opposite the drilling rig and pulled back into the pilot hole. Joints of drill pipe are added as the reamer makes its way back to the drilling rig. Large quantities of slurry are pumped into the hole to maintain the integrity of the hole and to flush out cuttings.

Pullback—Once the drilled hole is enlarged, the product pipeline can be pulled through it. The pipeline is prefabricated on the bank opposite the drilling rig. A reamer is attached to the drill string, and then connected to the pipeline pullhead via a swivel. The swivel prevents any translation of the reamer's rotation into the pipeline string, allowing for a smooth pull into the drilled hole. The drilling rig then begins the pullback operation, rotating and pulling on the drill string and once again circulating high volumes of drilling slurry. The pullback continues until the reamer and pipeline break ground at the drilling rig.

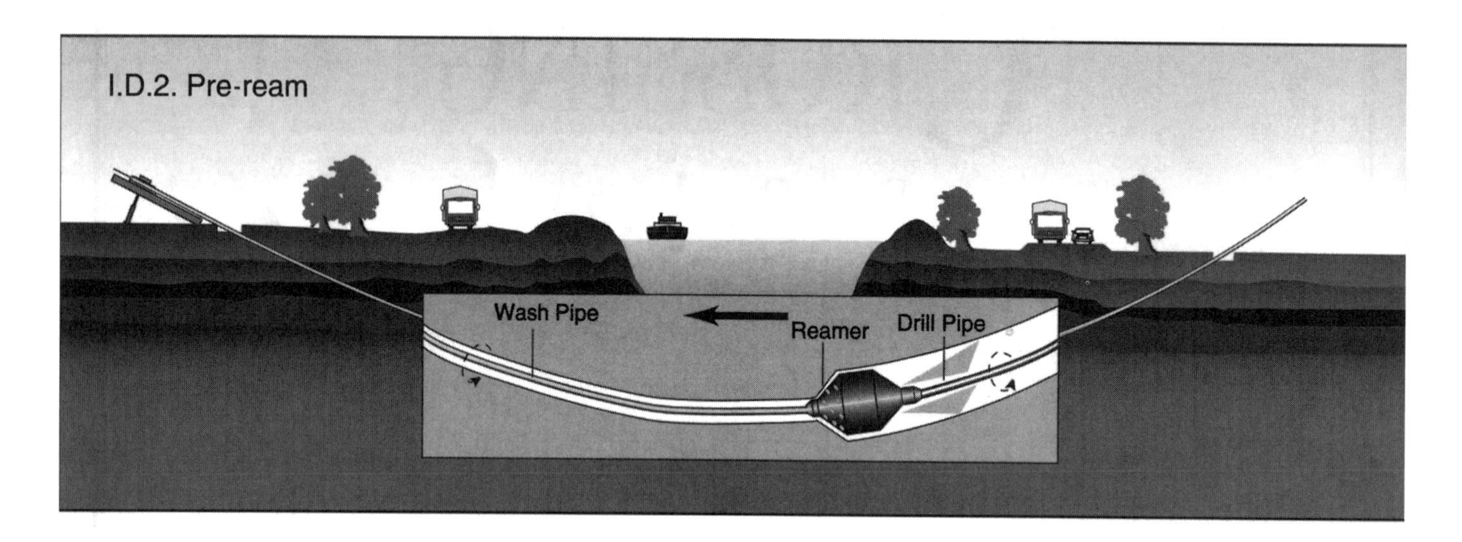

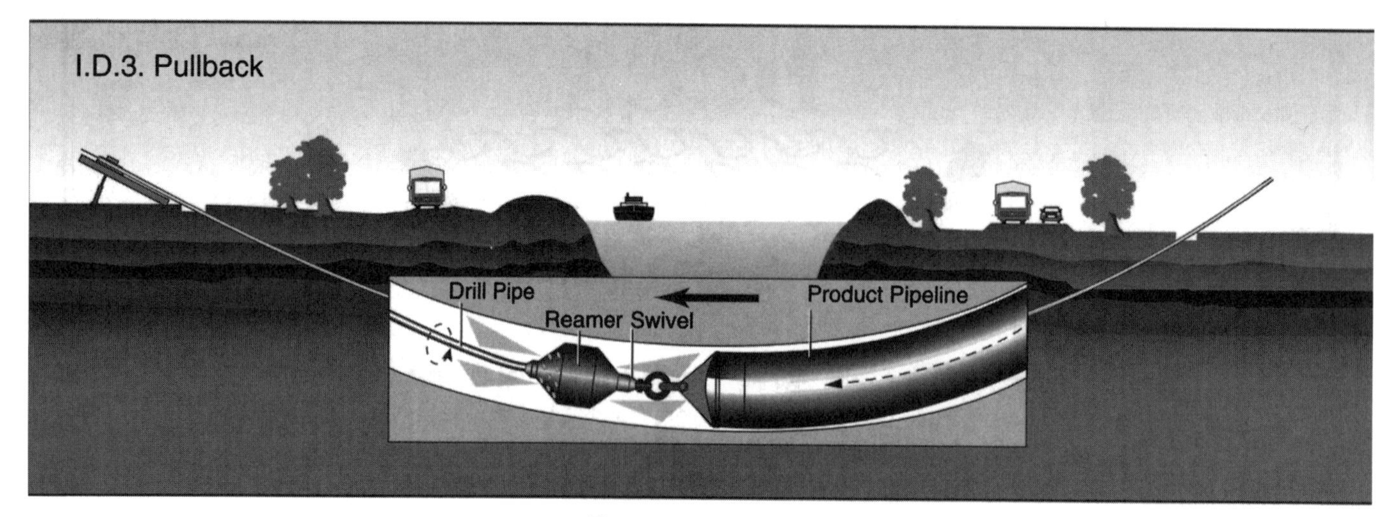

Figure 1. Technique.

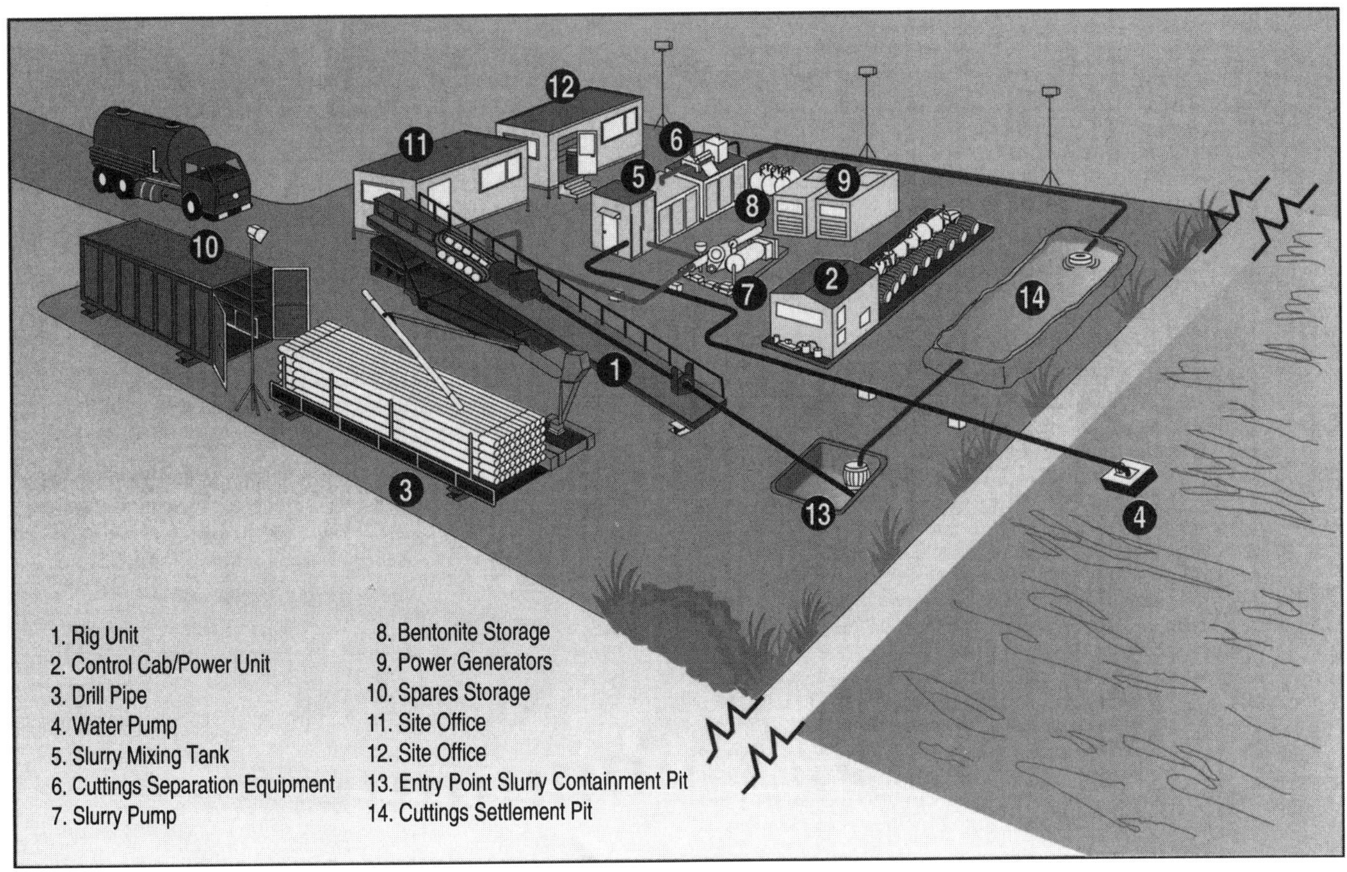

Figure 2. Rig Side Work Space.

Layout and design

Access—Heavy equipment is required on both sides of the crossing. To minimize cost, access to either side of the crossing should be provided with the least distance from an improved road. Often the pipeline right-of-way is used for access. All access agreements should be provided by the owner. It is not practical to negotiate such agreements during the bid process.

Work space

Rig Side—The rig spread requires a minimum 100-ft. wide by 150-ft. long area. This area should extend from the entry point away from the crossing, although the entry point should be at least 10 ft. inside the prescribed area. Since many components of the rig spread have no predetermined position, the rig site can be made up of smaller irregular areas. Operations are facilitated if the area is level, hardstanding, and clear of overhead obstructions. The drilling operation requires large volumes of water for the mixing of the drilling slurry. A nearby source of water is necessary (Figure 2).

Pipe Side—Strong consideration should be given to provide a sufficient length of work space to fabricate the product pipeline into one string. The width will be as necessary for normal pipeline construction, although a work space of 100-ft. wide by 150-ft. long should be provided at the exit point itself. The length will assure that during the pullback the pipe can be installed in one uninterrupted operation. Tie-ins of successive strings during the pullback operation increase the risk considerably, because the pullback should be continuous (Figure 3).

Profile survey

Once the work locations have been chosen, the area should be surveyed and detailed drawings prepared. The eventual accuracy of the drill profile and alignment is dependent on the accuracy of the survey information.

Profile design parameters

Depth of Cover—Once the crossing profile has been taken and the geotechnical investigation complete, a determination of the depth of cover under the crossing

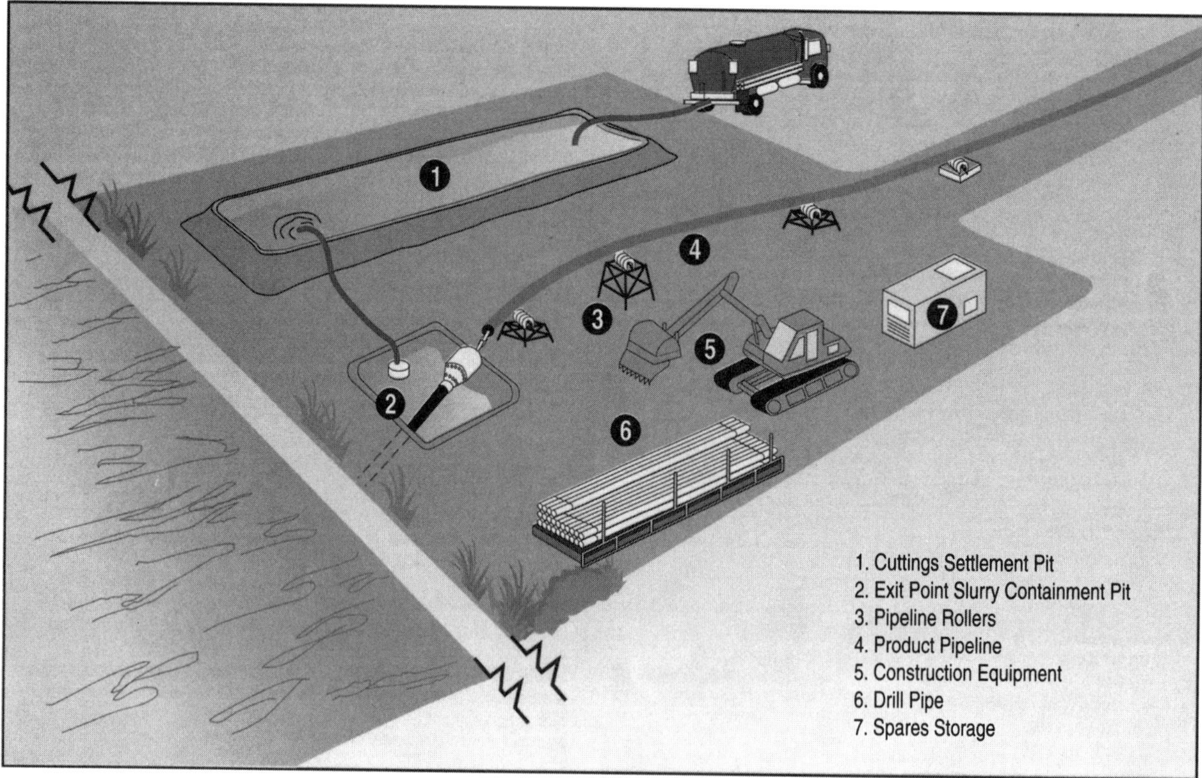

Figure 3. Pipe Side Work Space.

is made. Factors considered may include flow characteristics of the river, the depth of scour from periodic flooding, future channel widening/deepening, and whether pipeline or cable crossings already exist at the location. It is normally recommended that the minimum depth of cover be 20 ft under the lowest section of the crossing. While 20 ft is a recommended depth of cover on a river crossing, crossings of other obstacles may have differing requirements.

Penertration Angles and Radius of Curvature—An entry angle between 8 and 20 can be used for most crossings. It is preferable that straight tangent sections are drilled before the introduction of a long radius curve. The radius of the curve is determined by the bending characteristic of the product pipeline, increasing with the diameter. A general "rule-of-thumb" for the radius of curvature is 100 ft/1-in. diameter for steel line pipe. The curve usually brings the profile to the elevation providing the design cover of the pipeline under the river. Long horizontal runs can be made at this elevation before curving up toward the exit point. Exit angle should be kept between 5 and 12 to facilitate handling of the product pipeline during pullback.

Dill Survey—Most downhole survey tools are electronic devices that give a magnetic azimuth (for "right/left" control) and inclination (for "up/down" control). Surface locators can also be used in conjunction with the downhole electronic package.

Accuracy—The accuracy of the drill profile is largely dependent on variations in the earth's magnetic field. For instance, large steel structures (bridges, pilings, other pipelines, etc.) and electric power transmission lines affect magnetic field readings. However, a reasonable drill target at the pilot hole exit location is 10 ft left or right, and −10 ft to +30 ft in length.

As-Built Drawings—Normally, survey calculations are conducted every 30 ft during pilot hole operations. As-built drawings that are based on these calculations should be provided by the contractor. Alternate methods such as gyroscoping, ground penetrating radar, or "intelligent" pigs may also be used to determine the as-built position.

Geotechnical investigation

Number of Borings—The number of exploration holes is a function of the proposed crossing length and the complexity of the strata. If the crossing is about 1,000 ft, a borehole made on each side of the crossing may suffice. If an examination of these borings indicates that conditions are likely to be homogeneous on both sides, it may not be necessary to conduct further sampling. If the report indicates anomalies discontinuity in the strata or the presence of rock or large concentrations of gravel, it is advisable to make additional borings to better define the strata. Longer crossings (especially large-diameter pipelines) that indicate gravel, cobble, boulders, or rock should have samples taken about 600–800 ft apart unless significant anomalies are identified that might necessitate more borings. All borings should be located on the crossing profile along with their surface elevations being properly identified. If possible, the borings should be conducted at least 25 ft off of the proposed centerline. The bore holes should be grouted upon completion. This will help prevent the loss of drilling slurry during the crossing installation.

Depth of Borings—All borings should be made to a minimum depth of 40 ft below the lowest point in the crossing or 20 ft below the proposed depth of the crossing, whichever is greater. In some instances, it may be beneficial to the owner and the contractor to install the crossing at a greater depth than the owner requires for his permit. It is suggested that all borings be through the same elevation to better determine the consistency of the underlying material and note any patterns that may be present.

Standard Classification of Soils—A qualified technician or geologist should classify the material in accordance with the Unified Soil Classification System and ASTM Designations D-2487 and D-2488. It is beneficial to have a copy of the field-drilling log completed by the field technician or driller. These logs include visual classifications of materials, as well as the driller's interpretation of the subsurface conditions between samples.

Standard Penetration Test (SPT)—In order to better define the density of granular materials, the geotechnical engineer generally uses the Standard Penetration Test (SPT) in general accordance with ASTM Specifications D-1586. This is a field test that involves driving a 2-in. split spoon sampler into the soil by dropping a hammer of a specific weight (usually 140-lb.) a specified distance (usually 30 in.) to determine the number of blows necessary to drive the sampler 12 in. In very dense soils, the field technician may note the number of blows required to drive the sampler less than the required 12 in. (i.e., 50 blows for 3 in.). The number obtained is the standard penetration resistance value (N) and is used to estimate the in situ relative density of cohesionless soils. Some geotechnical firms will conduct these penetration tests in cohesive materials and rock, and to a lesser extent, the consistency of cohesive soils and the hardness of rock can be determined.

Thinwalled "Shelby" Tube Sampling—Most geotechnical firms prefer to use a Thinwalled Tube Sampling method for obtaining samples of cohesive materials. These tests are conducted in general accordance with ASTM Specification D-1587. This test is similar to the Standard Penetration test except the sample is collected by hydraulically pushing a thinwalled seamless steel tube with a sharp cutting edge into the ground. The hydraulic pressure required to collect the sample is noted on the field log. This produces a relatively undisturbed sample that can be further analyzed in the laboratory. These samples can be field tested with handheld penetrometers, but more accurate readings of density and consistency can be obtained by performing unconfined compressive strength tests where the results are noted in tons per square foot. Generally, for directional drilling contractors a standard penetration test using the split spoon sampler described earlier will suffice in both materials.

Sieve Analysis of Granular Materials—A sieve analysis is a mechanical test of granular materials performed on samples collected in the field during the standard penetration test with the split spoon sampler. The split spoon samples are taken to the laboratory and processed through a series of screens. The sample provides a percentage analysis of the granular material by size and weight. It is one of the most important tests undertaken.

Rock Information—If rock is encountered during the soil investigation borings, it is important to determine the type, the relative hardness, and the unconfined compressive strength. This information is typically collected by the geotechnical drilling firm by core drilling with a diamond bit core barrel. The typical core sample recovered with this process has a 2-in. diameter. The type of rock is classified by a geologist. The geologist should provide the Rock Quality Designation (RQD), which rates the quality of the rock based on the length of core retrieved in relation to the total length of the core. The hardness of the rock (Mohs' Scale of Hardness) is determined by comparing the rock to ten materials of known hardness. The compressive strength is determined by accurately measuring the core and then compressing the core to failure. This information pertaining to the underlying rock formation is imperative to determine the type of downhole equipment required and the penetration rates that can be expected.

Pipe material selection

Wall Thickness—D/t "Rule of Thumb"—The following table provides generalized recommendations for the selection of steel pipe wall thicknesses relative to pipe diameter. These recommendations are meant to be used only as a starting point in the design. It is recommended that in the final design, specific stresses be calculated and compared with allowable limits.

(For high-density polyethylene (HDPE) pipe, a standard dimension ratio of D/t, SDR, of 11 or less is recommended and the pipe manufacturer should be consulted.)

Diameter (D)	Wall Thickness (t)
6″ and smaller	0.250″
6 to 12″	0.375″
12 to 30″	0.500″
>30″	D/t <50

Stress Analysis—In finalizing the design, the stresses imposed during construction and in-service must be calculated and checked to be within allowable limits for the grade of material. The stresses at each stage must be considered acting individually and in combination. Stresses result due to spanning between rollers prior to pullback, the hydrostatic testing pressures, pulling forces during installation, radius of curvature as the pipe enters the ground, the drilling profile curvature, external pressures in the drilled hole, and the working pressure.

1. Pre-installation

   a. Hoop and longitudinal stresses resulting from hydrostatic testing are calculated.
   b. Using the known distance between rollers as the free spanning distance, the maximum hogging and sagging moments can be calculated. Considering the greater of these two moments, the maximum spanning stress is calculated. *Note:* During hydrostatic testing the pipeline will be full of water; therefore, the additional weight of water must be included in these calculations.

2. Installation

   a. The spanning stresses calculated in stage 1.b. also apply in this installation phase.
   b. The theoretical pulling force must be determined in order to provide the stresses that will result. An assumed downhole friction factor of 1.0 is recommended to provide conservative results and to include the effect of the pipeline being pulled around a curve. The maximum predicted pulling force should then be used in calculating the resulting longitudinal stress.
   c. Allowing for a 10% drilling tolerance leads to the use of a radius of curvature 90% of the design radius when calculating the longitudinal curvature stresses.
   d. External pressure from static head in the drilled hole and/or overburden pressures must be considered. It is recommended that the static head resulting from the maximum envisaged drilling fluid density should be used with a factor of safety of 1.5 to provide conservative estimations of resulting hoop and longitudinal stresses.

3. Post-installation
   a. The longitudinal curvature stresses calculated for stage 2.c. above are used again here.
   b. External pressure stresses from 2.d. apply.
   c. Hoop and longitudinal stresses resulting from the final hydrostatic test are calculated.
4. In-service
   a. Curvature—see 2.c.
   b. External pressure—see 2.d.

The maximum working pressure of the pipeline is used in calculating longitudinal and hoop stresses that will be imposed during service.

Allowable Stresses—Having determined the individual and combined stresses at each stage of construction and those for the in-service condition, they must be compared with allowable limits.

1. ASME B31.8—1992, Table A842.22 provides the following limits:

   — Maximum allowable longitudinal stress: 80% SMYS.
   — Maximum allowable hoop stress: 72% SMYS.
   — Maximum allowable combined stress: 90% SMYS.

   (Where SMYS is the Specified Minimum Yield Strength of the pipe material.)
2. Regulatory bodies may impose additional limits to those specified above—owner companies should identify any such further constraints and ensure the adequacy of the design.

Pipeline coating

Coatings are applied to provide a corrosion barrier and an abrasion barrier. Directional crossings generally encounter varying materials and often can be exposed to extra abrasion during the pullback. An outer abrasion-resistant overcoat is often warranted. To facilitate the pullback of the pipeline, the coating should bond well to the pipe to resist soil stresses and have a smooth, hard surface to reduce friction and maintain the corrosion barrier. As in any pipeline construction, the recommended external coating system should be compatible with any specifications for the field joint coating or any internal coating.

The recommended pipe coating is mill-applied fusion bonded epoxy (FBE). The recommended minimum thickness is 20 mils.

Joint Coating—The coating application of the weld area is the most critical field operation to maintain a smooth abrasion-resistant pipe string. It is recommended that the girth weld be coated with FBE powder utilizing the induction heating coil and powder application machine to a minimum dry film thickness of 25 mils. As an alternate, two component catalyzed liquid epoxy may be applied to the girth weld area to a minimum dry film thickness of 25 mils using a paintbrush or roller. Tape should never be used for joint coating on the pullback portion of a directional crossing.

Coating Repair—It is recommended that small coating damaged areas be repaired with a polymeric melt stick patching material. Holidays larger than 1 in. in diameter should be repaired utilizing the two component catalyzed liquid epoxy applied with a paintbrush or roller. Tapes should never be used for repair of coating damaged areas on the pullback portion of a directional crossing.

Abrasion-Resistant Overcoat—As an extra abrasion-resistant barrier for crossings that may encounter stones, boulders, or solid rock it is recommended the FBE-coated pipeline be overcoated with an epoxy-based polymer concrete. The material should be applied at a mill or with a portable yard coating machine to a minimum thickness of 40 mils. Girth weld and coating damaged areas should be field coated using an epoxy-based polymer concrete compatible with the overcoat material. The patch material should be applied so that the material tapers uniformly and feathers into the original coating. Stability of the pipeline in drilled crossings is not normally a concern, so a Portland cement–type concrete coating is not recommended.

Drilling slurry—containment, recycling, and disposal

The directional crossing process requires the use of large volumes of slurry that provide the following functions:

1. Hydraulic cutting with a jet.
2. Provide energy to the drill motor.
3. Lubricate the cutting head.
4. Transport drill cuttings to the surface.
5. Stabilize the hole against collapse.
6. Guard against loss of slurry into surrounding formations.

Slurry Composition—The slurries most commonly used are bentonite based. Bentonite is a naturally occurring Wyoming clay known for its hydrophilic characteristics. Often polymer extenders are also added to enhance certain characteristics. Material Safety Data Sheets (MSDS) are readily available from suppliers and can be presented to regulatory/disposal authorities.

Containment—The slurry is pumped downhole and circulates back to the surface and collected in "return pits." These pits typically have a volume of at least 500 cu. ft. Depending on the nature of the project, the slurry is pumped from the return pits to a "settling and containment pit." These pits vary in size depending on pumping rates and contain the slurry for recycling or disposal.

Recycling Slurry—Slurry that has been circulated downhole and collected in the containment pit is then passed through machinery that separates the cuttings from the slurry. This process involves a series of shaking sieves and various size hydroclones.

Slurry and Cuttings Disposal—Significant amounts of slurry are normally disposed of at the end of a project. Economics for disposal is extremely site specific. This slurry can be disposed of by:

1. Use at another drilling location.
2. Spread onto raw land for water retention improvement.
3. Evacuate to a dumpsite.

If working in an area of contaminated ground, the slurry should be tested for contamination and disposed of in a manner that meets governmental requirements.

Cost mitigation for the owner

With prebid planning and research, the owner can realize significant savings in slurry disposal. It is in the owner's interest to define and specify all disposal issues. In particular:

1. Define an approved disposal site as part of the project specifications.
2. Because it is difficult to estimate disposal quantities, disposal should be a separate bid item as either "cost plus" or on "unit rates."
3. Inadvertent returns are not uncommon and difficult to predict. The issue should be fairly represented to permitting bodies prior to construction. Contingency plans for containment and disposal of inadvertent returns should be priced as a separate bid item and agreed prior to construction.

Conditions of the contract

Always utilize a written contract to maximize communication and minimize controversy. A contract should be used to anticipate what the parties intend to do if a problem occurs on the job. The contract should be readable and understandable.

Bid package

A proposal presented by a contractor to the owner is an offer by the contractor that becomes a binding contract if accepted by the owner. The parties, price, and performance must be specified. Define the project to be undertaken by detailing the scope of work and incorporate all plans and specifications from the bid package.

Differing ground conditions and walkaway provision

Owners should accept the responsibility of performing an adequate geotechnical investigation. Despite adequate testing of ground conditions, unknown, unusual, and unexpected ground conditions may be encountered. The contract should provide solutions when the project encounters differing ground conditions. The walkaway provision in the contract should entitle the contractor to stop work and walk away from the job without the owner having the right to take over the contractor's equipment. The contractor should be entitled to receive compensation for demobilization, lost profits, and work performed prior to walkaway. If the project is completed, the contractor should be paid on a cost-plus basis. Assumption of risk of unforeseen ground conditions by the contractor affects the bid price.

Environmental concerns

Before the project begins, address environmental concerns because owners and contractors are included as potentially responsible parties when environmental damages and cleanup costs are assessed. Federal, state, and local laws must be evaluated, and licensing, permitting, and other regulations must be followed. Directional crossings that damage soil or water may cause liability.

1. Turbidity of Water and Inadvertent Returns—As these events are difficult to predict and work stoppage may occur, the contract should offer a mechanism to mutually address and mitigate the problem. Liabilities are generally shared by both the contractor and owner and many times can be insured.
2. Slurry Disposal—Comply with the regulations of the area regarding slurry disposal. Slurry disposal should be referred to in the contract and bid as a separate line item on a cost plus or unit price basis.

Allocation of risk of loss

Evaluate and allocate risks of loss that may occur during the project. Owners should share the risk of loss rather than shifting all the losses through the indemnification to the contractor, because the bid price is directly affected by contingent losses. Insurance may provide coverage by third parties for losses from differing ground conditions or environmental losses.

Dispute resolution

Provide for dispute resolution in the event of controversy by including mediation or arbitration provisions in the contract. Disputes should be resolved in the following order: 1) negotiation, 2) nonbinding mediation through a third party, 3) binding arbitration, and lastly, 4) litigation. Determine whom should be parties to the resolution, what law will be used, and where the dispute will be resolved.

A Contract Law Seminar notebook with recommended model provisions is available from the office of the DCCA.

This section is provided courtesy of:
Directional Crossing Contractors Association
One Galleria Tower, Suite 1940
13555 Noel Road, Lock Box 39
Dallas, TX 75240-6613
TEL 972-386-9545 *FAX 972-386-9547
http://www.dcca.org

3: Pipe Design

Steel pipe design

The maximum allowable design pressure stress will depend upon the intended service for the pipeline.

Pipelines to be used for transporting liquid petroleum are covered by ANSI/ASME B31.4—"Liquid Petroleum Transportation Piping Systems." Pipelines used for transporting gas are covered by ANSI/ASME B31.8—"Gas Transmission and Distribution Systems."

Pipelines that must be operated in compliance with the Federal Pipeline Safety Regulations will also need to comply with the applicable parts of these regulations: Part 192 for gas transportation systems and Part 195 for liquid transportation systems.

Gas pipelines—ANSI/ASME B31.8

The maximum allowable pressure is calculated by the following equation:

$$P = (2St/D) \times F \times E \times T$$

where: P = Design pressure, lb/in.^{2}
S = Specified minimum yield strength, lb/in.^{2} (see Table 1)
t = Nominal wall thickness, in.
D = Nominal outside diameter, in.
F = Design factor (see Table 2)
E = Longitudinal joint factor (see Table 3)
T = Temperature derating factor (see Table 4)

Class location definitions may be obtained from p. 192.111 of Part 192 of the Federal Pipeline Safety Regulations.

A typical calculation is as follows:

Pipe: 16″ OD × 0.250″ wt API 5LX X52 ERW
Location: Class 1, therefore F= 0.72 (see Table 2)
Temperature: 90°F, Temp. factor T= 1 (see Table 4)
Joint Factor: E= 1.0 (see Table 3)

$$P = (2 \times 52{,}000 \times 0.250/16.0) \times 0.72 \times 1 \times 1$$

$$P = 1{,}170 \text{lb/in.}^2 \text{ gauge}$$

Table 1
Specified minimum yield strength for steel and iron pipe commonly used in piping system

Specification	Grade	Type[1]	SMYS, psi
API 5L	A25	BW, ERW, S	25,000
API 5L	A	ERW, FW, S, DSA	30,000
API 5L	B	ERW, FW, S, DSA	35,000
API 5LS [Note (2)]	A	ERW, DSA	30,000
API 5LS	B	ERW, DSA	35,000
API 5LS	X42	ERW, DSA	42,000
API 5LS	X46	ERW, DSA	46,000
API 5LS	X52	ERW, DSA	52,000
API 5LS	X56	ERW, DSA	56,000
API 5LS	X60	ERW, DSA	60,000
API 5LS	X65	ERW, DSA	65,000
API 5LS	X70	ERW, DSA	70,000
API 5LX [Note (2)]	X42	ERW, FW, S, DSA	42,000
API 5LX	X46	ERW, FW, S, DSA	46,000
API 5LX	X52	ERW, FW, S, DSA	52,000
API 5LX	X56	ERW, FW, S, DSA	56,000
API 5LX	X60	ERW, FW, S, DSA	60,000
API 5LX	X65	ERW, FW, S, DSA	65,000
API 5LX	X70	ERW, FW, S, DSA	70,000

Table 1
Specified minimum yield strength for steel and iron pipe commonly used in piping system (Continued)

Specification	Grade	Type[1]	SMYS, psi
ASTM A53	Open Hrth. Bas. Oxy., Elec. Furn.	BW	25,000
ASTM A53	Bessemer	BW	30,000
ASTM A53	A	ERW, S	30,000
ASTM A53	B	ERW, S	35,000
ASTM A106	A	S	30,000
ASTM A106	B	S	35,000
ASTM A106	C	S	40,000
ASTM A134	—	EFW	[Note (3)]
ASTM A135	A	ERW	30,000
ASTM A135	B	ERW	35,000
ASTM A139	A	ERW	30,000
ASTM A139	B	ERW	35,000
ASTM A333	1	S, ERW	30,000
ASTM A333	3	S, ERW	35,000
ASTM A333	4	S	35,000
ASTM A333	6	S, ERW	35,000
ASTM A333	7	S, ERW	35,000
ASTM A333	8	S, ERW	75,000

Reproduced from ANSI/ASME Code B31-8-1982, Appendix D. Reprinted courtesy of The American Society of Mechanical Engineers.

Table 2
Values of design factor F

Construction Type (See 841.151)	Design Factor F
Type A	0.72
Type B	0.60
Type C	0.50
Type D	0.40

Reproduced from ANSI/ASME Code B31-8-1982, Table 841.1A. Reprinted courtesy of The American Society of Mechanical Engineers.

Table 4
Temperature derating factor T for steel pipe

Temperature, °F	Temperature Derating Factor T
250 or less	1.000
300	0.967
350	0.933
400	0.900
450	0.867

NOTE: For Intermediate temperatures, interpolate for derating factor.

Reproduced from ANSI/ASME Code B31-8-1982, Table 841.1C. Reprinted courtesy of The American Society of Mechanical Engineers.

Table 3
Longitudinal joint factor E

Spec. Number	Pipe Class	E Factor
ASTM A53	Seamless	1.00
	Electric Resistance Welded	1.00
	Furnace Welded	0.60
ASTM A106	Seamless	1.00
ASTM A134	Electric Fusion Arc Welded	0.80
ASTM A135	Electric Resistance Welded	1.00
ASTM A139	Electric Fusion Welded	0.80
ASTM A211	Spiral Welded Steel Pipe	0.80
ASTM A381	Double Submerged-Arc-Welded	1.00
ASTM A671	Electric Fusion Welded	1.00*
ASTM A672	Electric Fusion Welded	1.00*
API 5L	Seamless	1.00
	Electric Resistance Welded	1.00
	Electric Flash Welded	1.00
	Submerged Arc Welded	1.00
	Furnace Butt Welded	0.60
API 5LX	Seamless	1.00
	Electric Resistance Welded	1.00
	Electric Flash Welded	1.00
	Submerged Arc Welded	1.00
API 5LS	Electric Resistance Welded	1.00
	Submerged Welded	1.00

NOTE: Definitions for the various classes of welded pipe are given in 804.243

*Includes Classes 12, 22, 32, 42, and 52 only.

Reproduced from ANSI/ASME Code B31-8-1982, Table 841.1B. Reprinted courtesy of The American Society of Mechanical Engineers.

Liquid pipelines—ANSI/ASME B31.4

The internal design pressure is determined by using the following formula:

$P = (2St/D) \times E \times F$

where: P = Internal design pressure, lb/in.^{2} gauge
S = Specified minimum yield strength, lb/in.^{2} (see Table 5)
t = Nominal wall thickness, in.
D = Nominal outside diameter of the pipe, in.
E = Weld joint factor (see Table 6)
F = Design factor of 0.72

Note: Refer to p. 195.106 of Part 195 Federal Pipeline Safety Regulations for design factors to be used on offshore risers and platform piping and cold worked pipe.
A typical calculation of the internal design pressure is as follows:

Pipe: 26″ OD × 0.3125″ wt API 5LX X52 ERW
Weld joint factor E = 1.0 (see Table 6)
Design factor F = 0.72

$P = (2 \times 52{,}000 \times 0.3125/26) \times 1 \times 0\ 72$

$P = 900$ lb/in.^{2} in gauge

Table 5
Tabulation of examples of allowable stresses for reference use in piping systems

Allowable stress values (S) shown in this table are equal to 0.72 × E (weld joint factor) × specified minimum yield strength of the pipe.
Allowable stress values shown are for new pipe of known specification. Allowable stress values for new pipe of unknown specification, ASTM A 120 specification or used (reclaimed) pipe shall be determined in accordance with 402.3.1.
For some code computations, particularly with regard to branch connections [see 404.3.1 (d) (3)] and expansion, flexibility, structural attachments, supports, and restraints (Chapter II, Part 5), the weld joint factor *E* need not be considered.
For specified minimum yield strength of other grades in approved specifications, refer to that particular specification.
Allowable stress value for cold worked pipe subsequently heated to 600F (300C) or higher (welding excepted) shall be 75% of value listed in table.
Definitions for the various types of pipe are given in 400.2.
(Metrics Stress Levels are given in MPa [1 Megapascal = 1 million pascals])

Specification	Grade	Specified Min Yield Strength psi (MPa)	Notes	(E) Weld Joint Factor	(S) Allowable Stress Value −20F to 250F (−30C to 120C) psi (MPa)
Seamless					
API 5L	A25	25,000 (172)	(1)	1.00	18,000 (124)
API 5L, ASTM A53, ASTM A106	A	30,000 (207)	(1) (2)	1.00	21,600 (149)
API 5L, ASTM A53, ASTM A106	B	35,000 (241)	(1) (2)	1.00	25,200 (174)
ASTM A106	C	40,000 (278)	(1) (2)	1.00	28,800 (199)
ASTM A524	I	35,000 (241)	(1)	1.00	25,200 (174)
ASTM A524	II	30,000 (207)	(1)	1.00	21,600 (149)
API 5LU	U80	80,000 (551)	(1) (4)	1.00	57,600 (397)
API 5LU	U100	100,000 (689)	(1) (4)	1.00	72,000 (496)
API 5LX	X42	42,000 (289)	(1) (2) (4)	1.00	30,250 (208)
API 5LX	X46	46,000 (317)	(1) (2) (4)	1.00	33,100 (228)
API 5LX	X52	52,000 (358)	(1) (2) (4)	1.00	37,450 (258)
API 5LX	X56	56,000 (386)	(1) (4)	1.00	40,300 (278)
API 5LX	X60	60,000 (413)	(1) (4)	1.00	43,200 (298)
API 5LX	X65	65,000 (448)	(1) (4)	1.00	46,800 (323)
API 5LX	X70	70,000 (482)	(1) (4)	1.00	50,400 (347)
Furnace Welded-Butt Welded					
ASTM A53		25,000 (172)	(1) (2)	0.60	10,800 (74)
API 5L Class I & Class II	A25	25,000 (172)	(1) (2) (3)	0.60	10,800 (74)
API 5L (Bessemer), ASTM A53 (Bessemer)		30,000 (207)	(1) (2) (5)	0.60	12,950 (89)
Furnace Welded-Lap Welded					
API 5L Class I		25,000 (172)	(1) (2) (6)	0.80	14,400 (99)
API 5L Class II		28,000 (193)	(1) (2) (6)	0.80	16,150 (111)
API 5L (Bessemer)		30,000 (207)	(1) (2) (6)	0.80	17,300 (119)
API 5L Electric Furnace		25,000 (172)	(1) (2) (6)	0.80	14,400 (99)

Reproduced from ANSI/ASME Code B31-4-1979, Table 402.3.1(a). Reprinted courtesy of The American Society of Mechanical Engineers.

Table 5
Tabulation of examples of allowable stresses for reference use in piping systems (Continued)

Specification	Grade	Specified Min Yield Strength psi (MPa)	Notes	(E) Weld Joint Factor	(S) Allowable Stress Value −20F to 250F (−30C to 120C) psi (MPa)
Electric Resistance Welded and Electric Flash Welded					
API 5L	A25	25,000 (172)	(1) (7)	1.00	18,000 (124)
API 5L, ASTM A53, ASTM A135	A	30,000 (207)	(2)	0.85	18,360 (127)
API 5L, API 5LS, ASTM A53, ASTM A135	A	30,000 (207)	(1)	1.00	21.600 (149)
API 5L, ASTM A53, ASTM A135	B	35,000 (241)	(2)	0.85	21,420 (148)
API 5L, API 5LS, ASTM A53, ASTM A135	B	35,000 (241)	(1)	1.00	25,200 (174)
API 5LS, API 5LX	X42	42,000 (289)	(1) (2) (4)	1.00	30,250 (208)
API 5LS, API 5LX	X46	46,000 (317)	(1) (2) (4)	1.00	33,100 (228)
API 5LS, API 5LX	X52	52,000 (358)	(1) (2) (4)	1.00	37,450 (258)
API 5LS, API 5LX	X56	56,000 (386)	(1) (4)	1.00	40,300 (279)
API 5LS, API 5LX	X60	60,000 (413)	(1) (4)	1.00	43,200 (297)
API 5LS, API 5LX	X65	65,000 (448)	(1) (4)	1.00	46,800 (323)
API 5LS, API 5LX	X70	70,000 (482)	(1) (4)	1.00	50,400 (347)
API 5LU	U80	80,000 (551)	(1) (4)	1.00	57,600 (397)
API 5LU	U100	100,000 (689)	(1) (4)	1.00	72,000 (496)
Electric Fusion Welded					
ASTM A 134	—	—		0.80	—
ASTM A 139	A	30,000 (207)	(1) (2)	0.80	17,300 (119)
ASTM A 139	B	35,000 (241)	(1) (2)	0.80	20,150 (139)
ASTM A 155	—	—	(2) (8)	0.90	—
ASTM A 155	—	—	(1) (8)	1.00	—
Submerged Arc Welded					
API 5L, API 5LS	A	30,000 (207)	(1)	1.00	21,600 (149)
API 5L, API 5LS	B	35,000 (241)	(1)	1.00	25,200 (174)
API 5LS, API 5LX	X42	42,000 (289)	(1) (2) (4)	1.00	30,250 (208)
API 5LS, API 5LX	X46	46,000 (317)	(1) (2) (4)	1.00	33,100 (228)
API 5LS, API 5LX	X52	52,000 (358)	(1) (2) (4)	1.00	37,450 (258)
API 5LS, API 5LX	X56	56,000 (386)	(1) (4)	1.00	40,300 (278)
API 5LS, API 5LX	X60	60,000 (413)	(1) (4)	1.00	43,200 (298)
API 5LS, API 5LX	X65	65,000 (448)	(1) (4)	1.00	46,800 (323)
API 5LS, API 5LX	X70	70,000 (482)	(1) (4)	1.00	50,400 (347)
API 5LU	U80	80,000 (551)	(1) (4)	1.00	57,600 (397)
API 5LU	U100	100,000 (689)	(1) (4)	1.00	72,000 (496)
ASTM A 381	Y35	35,000 (241)	(1) (2)	1.00	25,200 (174)
ASTM A 381	Y42	42,000 (290)	(1) (2)	1.00	30,250 (209)
ASTM A 381	Y46	46,000 (317)	(1) (2)	1.00	33,100 (228)
ASTM A 381	Y48	48,000 (331)	(1) (2)	1.00	34,550 (238)
ASTM A 381	Y50	50,000 (345)	(1)	1.00	36,000 (248)
ASTM A 381	Y52	52,000 (358)	(1)	1.00	37,450 (258)
ASTM A 381	Y60	60,000 (413)	(1)	1.00	43,200 (298)
ASTM A 381	Y65	65,000 (448)	(1)	1.00	46,800 (323)

NOTES (1) Weld joint factor *E* (see Table 402.4.3) and allowable stress value are applicable to pipe manufactured after 1958.
(2) Weld joint factor *E* (see Table 402.4.3) and allowable stress value are applicable to pipe manufactured before 1959.
(3) Class II produced under API 5L 23rd Edition, 1968, or earlier has a specified minimum yield strength of 28,000 psi (193MPa).
(4) Other grades provided for in API 5LS, API 5LU, and API 5LX not precluded.
(5) Manufacture was discontinued and process deleted from API 5L in 1969.
(6) Manufacture was discontinued and process deleted from API 5L in 1962.
(7) A25 is not produced in electric flash weld.
(8) See applicable plate specification for yield point and refer to 402.3.1 for calculation of (*S*).

Reproduced from ANSI/ASME Code B31-4-1979, Table 402.3.1. Reprinted courtesy of The American Society of Mechanical Engineers.

Table 6
Weld joint factor

Specification Number	Pipe Type (1)	Weld Joint Factor E	
		Pipe Mfrd. Before 1959	Pipe Mfrd. After 1958
ASTM A53	Seamless	1.00	1.00
	Electric-Resistance-Welded	0.85 (2)	1.00
	Furnace Lap-Welded	0.80	0.80
	Furnace Butt-Welded	0.60	0.60
ASTM A106	Seamless	1.00	1.00
ASTM A134	Electric-Fusion (Arc)-Welded single or double pass	0.80	0.80
ASTM A135	Electric-Resistance-Welded	0.85 (2)	1.00
ASTM A139	Electric-Fusion-Welded single or double pass	0.80	0.80
ASTM A155	Electric-Fusion-Welded	0.90	1.00
ASTM A381	Electric-Fusion-Welded, Double Submerged Arc-Welded	—	1.00
API 5L	Seamless	1.00	1.00
	Electric-Resistance-Welded	0.85 (2)	1.00
	Electric-Flash-Welded	0.85 (2)	1.00
	Electric-Induction-Welded	—	1.00
	Submerged Arc-Welded	—	1.00
	Furnace Lap-Welded	0.80	0.80 (3)
	Furnace Butt-Welded	0.60	0.60
API 5LS	Electric-Resistance-Welded	—	1.00
	Submerged Arc-Welded	—	1.00
API 5LX	Seamless	1.00	1.00
	Electric-Resistance-Welded	1.00	1.00
	Electric-Flash-Welded	1.00	1.00
	Electric-Induction-Welded	—	1.00
	Submerged Arc-Welded	1.00	1.00
API 5LU	Seamless	—	1.00
	Electric-Resistance-Welded	—	1.00
	Electric-Flash-Welded	—	1.00
	Electric-Induction-Welded	—	1.00
	Submerged Arc-Welded	—	1.00
Known	Known	(4)	(5)
Unknown	Seamless	1.00 (6)	1.00 (6)
Unknown	Electric-Resistance or Flash-Welded	0.85 (6)	1.00 (6)
Unknown	Electric-Fusion-Welded	0.80 (6)	0.80 (6)
Unknown	Furnace Lap-Welded or over NPS 4	0.80 (7)	0.80 (7)
Unknown	Furnace Butt-Welded or NPS 4 and smaller	0.60 (8)	0.60 (8)

NOTES: (1) Definitions for the various pipe types (weld joints) are given in 400.2.

(2) A weld joint factor of 1.0 may be used for electric-resistance-welded or electric-flash-welded pipe manufactured prior to 1959 where (a) pipe furnished under this classification has been subjected to supplemental tests and/or heat treatments as agreed to by the supplier and the purchaser, and such supplemental tests and/or heat treatment demonstrate the strength characteristics of the weld to be equal to the minimum tensile strength specified for the pipe, or (b) pipe has been tested as required for a new pipeline in accordance with 437.4.1.

(3) Manufacture was discontinued and process deleted from API 5L in 1962.

(4) Factors shown above for pipe manufactured before 1959 apply for new or used (reclaimed) pipe if pipe specification and pipe type are known and it is known that pipe was manufactured before 1959 or not known whether manufactured after 1958.

(5) Factors shown above for pipe manufactured after 1958 apply for new or used (reclaimed) pipe if pipe specification and pipe type are known and it is known that pipe was manufactured after 1958.

(6) Factor applies for new or used pipe of unknown specification and ASTM A120 if type of weld joint is known.

(7) Factor applies for new or used pipe of unknown specification and ASTM A120 if type of weld joint is known to be furnace lap-welded, or for pipe over NPS 4 if type of joint is unknown.

(8) Factor applies for new or used pipe of unknown specification and ASTM A120 if type of weld joint is known to be furnace butt-welded, or for pipe NPS 4 and smaller if type of joint is unknown.

Reproduced from ANSI/ASME Code B31-4-1979, Table 402.4.3. Reprinted courtesy of The American Society of Mechanical Engineers.

Properties of pipe

D	t	d	W	A_o	V_m	A_f	I	Z
4.500	0.125	4.250	5.840	1.178097	92.644911	0.098516	4.12	1.83
	0.156	4.188	7.240		89.961575	0.095662	5.03	2.24
	0.172	4.156	7.950		88.592057	0.094206	5.49	2.44
	0.188	4.124	8.660		87.233042	0.092761	5.93	2.64
	0.219	4.062	10.010		84.629845	0.089993	6.77	3.01
STD	0.237	4.026	10.790		83.136406	0.088405	7.56	3.36
	0.250	4.000	11.350		82.066080	0.087266	7.56	3.36
6.625	0.188	6.249	12.93	1.734421	200.292532	0.212985	19.7	6.0
	0.219	6.187	14.99		196.337808	0.208779	22.6	6.8
	0.250	6.125	17.02		192.422518	0.204616	25.5	7.7
STD	0.280	6.065	18.98		188.671072	0.200627	28.1	8.5
	0.312	6.001	21.04		184.710235	0.196415	30.9	9.3
	0.375	5.875	25.03		177.035128	0.188254	36.1	10.9
8.625	0.188	8.249	16.94	2.258020	349.016785	0.371133	44.4	10.3
	0.219	8.187	19.66		343.790038	0.365575	51.1	11.9
	0.250	8.125	22.36		338.602723	0.360059	57.7	13.4
	0.277	8.071	24.70		334.116868	0.355289	63.4	14.7
	0.312	8.001	27.70		328.346391	0.349153	70.5	16.3
STD	0.322	7.981	28.56		326.706916	0.347410	72.5	16.8
	0.344	7.937	30.43		323.114514	0.343590	76.9	17.8
	0.375	7.875	33.05		318.086203	0.338243	82.9	19.2
10.750	0.219	10.312	24.63	2.814343	545.418061	0.579980	87.0	16.2
	0.250	10.250	28.04		538.879221	0.573027	113.7	21.2
	0.279	10.192	31.20		532.797939	0.566560	125.9	23.4
	0.307	10.136	34.24		526.959102	0.560352	137.5	25.6
STD	0.365	10.020	40.49		514.966704	0.547599	160.8	29.9
12.750	0.219	12.312	29.31	3.337942	777.500935	0.826770	169.3	26.6
	0.250	12.250	33.38		769.690071	0.818464	191.9	30.1
	0.281	12.188	37.43		761.918639	0.810200	214.1	33.6
	0.312	12.126	41.45		754.186639	0.801978	236.0	37.0
STD	0.375	12.000	49.57		738.594720	0.785398	279.0	43.8
	0.438	11.874	57.60		723.165661	0.768991	321.0	50.4
	0.500	11.750	65.42		708.140511	0.753014	363.0	56.7
14.000	0.219	13.562	32.24	3.665191	943.389822	1.003171	226.0	32.3
	0.250	13.500	36.72		934.783943	0.994020	255.0	36.5
	0.281	13.438	41.18		926.217495	0.984910	285.1	40.7
	0.312	13.376	45.62		917.690481	0.975843	315.0	45.0
	0.344	13.312	50.18		908.929763	0.966527	344.3	49.2
STD	0.375	13.250	54.58		900.482886	0.957545	373.0	53.3
	0.438	13.124	63.45		883.438151	0.939420	429.0	61.4
	0.500	13.000	72.10		866.822970	0.921752	484.0	69.1

D = Outside diameter, in.
t = Wall thickness, in.
d = Inside diameter, in.
W = Pipe weight #/ft
A_o = Outside surface area sq. ft/ft
V_m = Displacement, bbls/mile
A_f = Flow area, sq. ft

I = Moment of Inertia, $in.^4$
Z = Section modulus, $in.^3$
Gals/ft. = $d^2 \times 0.0408$
Gals/mile = $d^2 \times 215.4240$
Bbls/mile = $d^2 \times 5.1291$
$I = 0.0491\ (D^4 - d^4)$
$Z = 0.0982\ (D^4 - d^4)/D$

D	t	d	W	A_o	V_m	A_f	I	Z
16.000	0.219	15.562	36.92	4.188790	1242.151387	1.320864	338.0	42.3
	0.250	15.500	42.06		1232.273483	1.310360	384.0	48.0
	0.281	15.438	47.18		1222.435011	1.299899	429.0	53.6
	0.312	15.376	52.28		1212.635972	1.289479	474.0	59.3
	0.344	15.312	57.53		1202.562197	1.278766	519.0	64.8
STD	0.375	15.250	62.59		1192.843296	1.268432	562.0	70.3
	0.438	15.124	72.81		1173.213479	1.247558	649.0	81.1
	0.500	15.000	82.78		1154.054250	1.227185	732.0	91.5
18.000	0.250	17.500	47.40	4.712389	1570.796063	1.670335	549.0	61.0
	0.281	17.438	53.18		1559.685567	1.658520	614.2	68.2
	0.312	17.376	58.95		1548.614504	1.646747	679.0	75.5
	0.344	17.312	64.88		1537.227671	1.634639	744.0	82.7
STD	0.375	17.250	70.60		1526.236746	1.622952	807.0	89.6
	0.438	17.124	82.16		1504.021848	1.599329	932.0	103.6
	0.500	17.000	93.46		1482.318570	1.576250	1053.0	117.0
20.000	0.250	19.500	52.74	5.235988	1950.351683	2.073942	756.6	75.7
	0.281	19.438	59.19		1937.969163	2.060775	846.5	84.6
	0.312	19.376	65.61		1925.626075	2.047650	935.5	93.5
	0.344	19.312	72.22		1912.926185	2.034145	1026.5	102.6
STD	0.375	19.250	78.61		1900.663236	2.021105	1113.8	111.4
	0.438	19.124	91.52		1875.863256	1.994733	1288.6	128.9
	0.500	19.000	104.14		1851.615930	1.968950	1457.2	145.7
24.000	0.250	23.500	63.42	6.283185	2832.562043	3.012056	1315.7	109.6
	0.281	23.438	71.19		2817.635474	2.996184	1473.1	122.8
	0.312	23.376	78.94		2802.748338	2.980354	1629.2	135.8
	0.344	23.312	86.92		2787.422334	2.964056	1789.1	149.1
STD	0.375	23.250	94.63		2772.615336	2.948311	1942.8	161.9
	0.438	23.124	110.23		2742.645193	2.916442	2251.3	187.6
	0.500	23.000	125.51		2713.309770	2.885247	2550.0	212.5
26.000	0.250	25.500	68.76	6.806784	3335.216783	3.546564	1676.8	129.0
	0.281	25.438	77.20		3319.018190	3.529339	1878.0	144.5
	0.312	25.376	85.61		3302.859030	3.512155	2077.7	159.8
	0.344	25.312	94.27		3286.219968	3.494462	2282.3	175.6
STD	0.375	25.250	102.64		3270.140946	3.477364	2479.1	190.7
	0.438	25.124	119.59		3237.585722	3.442746	2874.5	221.1
	0.500	25.000	136.19		3205.706250	3.408846	3257.8	250.6
30.000	0.250	29.500	79.44	7.853982	4463.625383	4.746477	2585.8	172.4
	0.281	29.438	89.20		4444.882741	4.726547	2897.5	193.2
	0.312	29.376	98.94		4426.179533	4.706659	3207.1	213.8
	0.344	29.312	108.97		4406.914357	4.686173	3524.7	235.0
STD	0.375	29.250	118.66		4388.291286	4.666370	3830.4	255.4
	0.438	29.124	138.30		4350.565898	4.626254	4445.7	296.4
	0.500	29.000	157.55		4313.598330	4.586943	5043.5	336.2
32.000	0.250	31.500	84.78	8.377580	5089.379243	5.411884	3143.2	196.4
	0.281	31.438	95.20		5069.364577	5.390601	3522.7	220.2
	0.312	31.376	105.60		5049.389344	5.369360	3899.9	243.7
	0.344	31.312	116.32		5028.811111	5.347478	4287.0	267.9
STD	0.375	31.250	126.67		5008.916016	5.326322	4659.7	291.2
	0.438	31.124	147.66		4968.605547	5.283457	5410.3	338.1
	0.500	31.000	168.23		4929.093930	5.241442	6140.2	383.8

Length of pipe in bends

Radius of Pipe Bends: Inches	Feet	90° Bends	180° Bends	270° Bends	360° Bends	540° Bends
1		1½"	3"	4¾"	6¼"	9½"
2		3	6¼	9½	12½	18¾
3	¼	4¾	9½	14¼	18¾	28¼
4		6¼	12½	18¾	25¼	37¾
5		7¾	15¾	23½	31½	47¼
6	½	9½	18¾	28¼	37¾	56½
7		11	22	33	44	66
8		12½	25¼	37¾	50¼	75½
9	¾	14¼	28¼	42½	56½	84¾
10		15¾	31½	47¼	62¾	94¼
11		17¼	34½	51¾	69	103¾
12	1	18¾	37¾	56½	75½	113
12.5		19¾	39¼	59	78½	117¾
14		22	44	66	88	132
15	1¼	23½	47	70¾	94¼	141¼
16		25¼	50¼	75½	100½	150¾
17.5		27½	55	82½	110	165
18	1½	28¼	56½	84¾	113	169¾
20		31½	62¾	94¼	125¾	188½
21	1¾	33	66	99	132	198
24	2	37¾	75½	113	150¾	226¼
25		39¼	78½	117¾	157	235½
30	2½	47¼	94¼	141¼	188½	282¾
32		50¼	100½	150¾	201	301½
36	3	56½	113	169½	226¼	339¼
40		62¾	125¾	188½	251¼	377
48	4	75½	150¾	226¼	301½	452½
50		78½	157	235½	314¼	471¼
56		88	176	264	351¾	527¾
60	5	94¼	188½	282¾	377	565½
64		100½	201	301½	402	603¼
70		110	220	329¾	439¾	659¾
72	6	113	226¼	339¼	452½	678½
80		125¾	251¼	377	502¾	754
84	7	132	263¾	395¾	527¾	791½
90	7½	141¼	282¾	424	565½	848¼
96	8	150¾	301½	452½	603	904¾
100		157	314¼	471¼	628¼	942½
108	9	169½	339¼	509	678½	1017¾
120	10	188½	377	565½	754	1131
132	11	207¼	414¾	622	829½	1244
144	12	226¼	452½	678½	904¾	1357¼
156	13	245	490	735¼	980¼	1470¼
168	14	263¾	527¾	791½	1055½	1583½
180	15	282¾	565½	848¼	1131	1696½
192	16	301½	603	904¾	1206¼	1809½
204	17	320½	640¾	961¼	1281¾	1922½
216	18	339¼	678½	1017¾	1357¼	2035¾
228	19	358	716¼	1074½	1432½	2148¾
240	20	377	754	1131	1508	2262

To find the length of pipe in a bend having a radius not given above, add together the length of pipe in bends whose combined radii equal the required radius.

Example: **Find length of pipe in 90° bend of 5′ 9″ radius.**
Length of pipe in 90° bend of 5′ radius = 94¼″
Length of pipe in 90° bend of 9″ radius = 14¼″

Then, length of pipe in 90° bend of 5′ 9″ radius = 108½″

Reprinted with permission—Crane Company.

Calculation of pipe bends

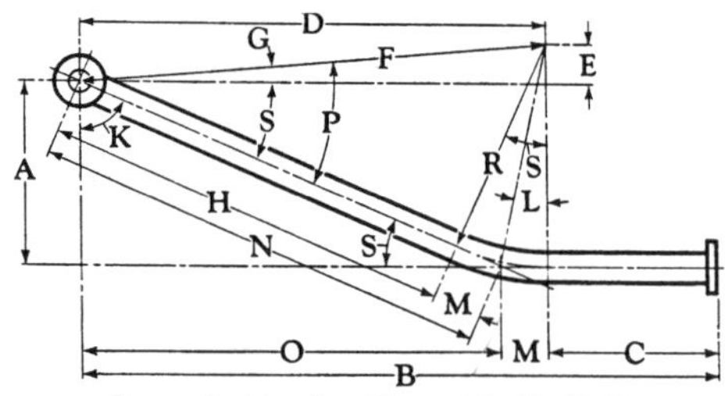

Example No. 1—**Given A, B, C, R**

$D = B - C$

$E = R - A$

$F = \sqrt{D^2 + E^2}$

$\frac{E}{F} = \sin \angle G$

$H = \sqrt{F^2 - R^2}$

$\frac{R}{F} = \sin \angle P$

$\angle S = \angle P - \angle G$

$\angle K = 90° - \angle S$

$\angle L = \frac{1}{2} \angle S$

$M = \tan \angle L \times R$

$N = H + M$

$O = B - C - M$

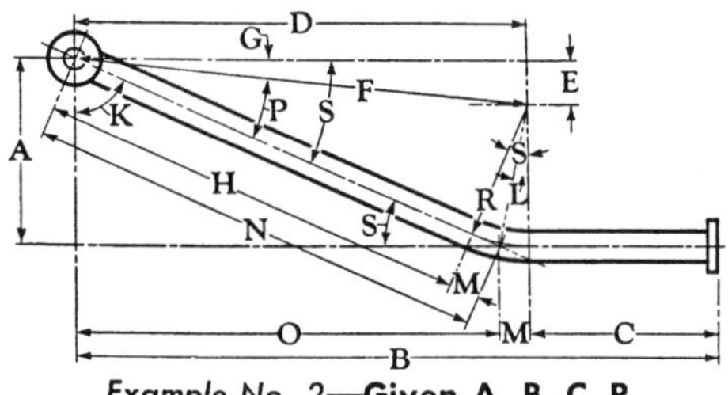

Example No. 2—**Given A, B, C, R**

$D = B - C$

$E = A - R$

$F = \sqrt{D^2 + E^2}$

$\frac{E}{F} = \sin \angle G$

$H = \sqrt{F^2 - R^2}$

$\frac{R}{F} = \sin \angle P$

$\angle S = \angle P + \angle G$

$\angle K = 90° - \angle S$

$\angle L = \frac{1}{2} \angle S$

$M = \tan \angle L \times R$

$N = H + M$

$O = B - C - M$

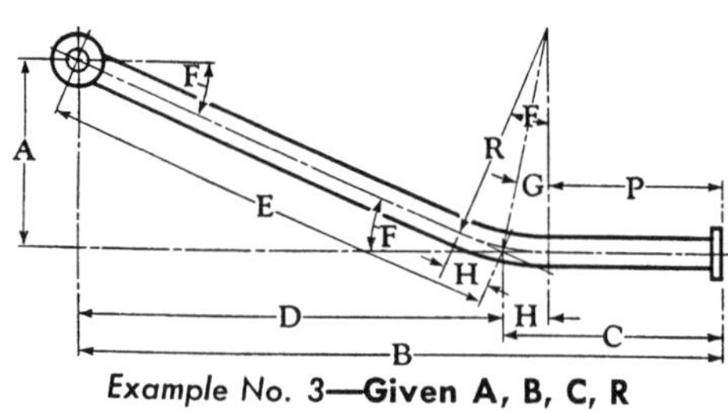

Example No. 3—**Given A, B, C, R**

$D = B - C$

$E = \sqrt{A^2 + D^2}$

$\frac{A}{E} = \sin \angle F$

$\angle G = \frac{1}{2} \angle F$

$H = \tan \angle G \times R$

$P = C - H$

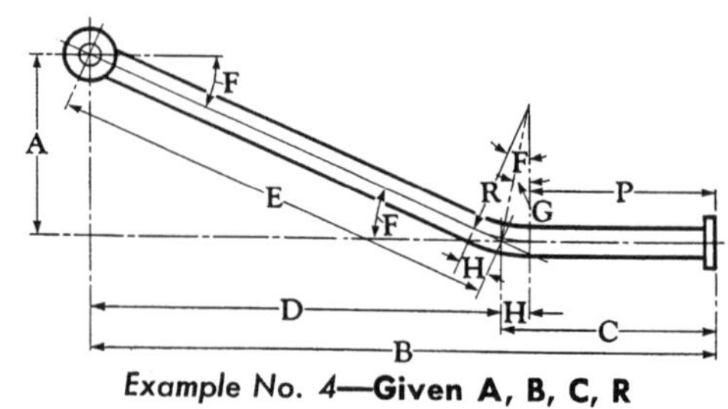

Example No. 4—**Given A, B, C, R**

$D = B - C$

$E = \sqrt{A^2 + D^2}$

$\frac{A}{E} = \sin \angle F$

$\angle G = \frac{1}{2} \angle F$

$H = \tan \angle G \times R$

$P = C - H$

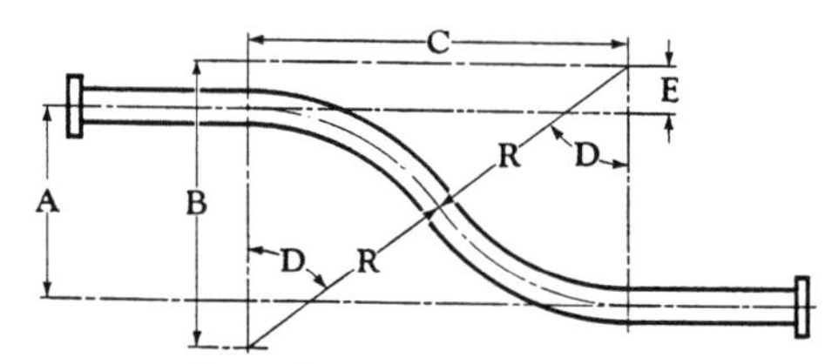

Example No. 5—**Given A, R**

$B = 2R - A$

$C = \sqrt{(2R)^2 - B^2}$

$\frac{C}{2R} = \sin \angle D$

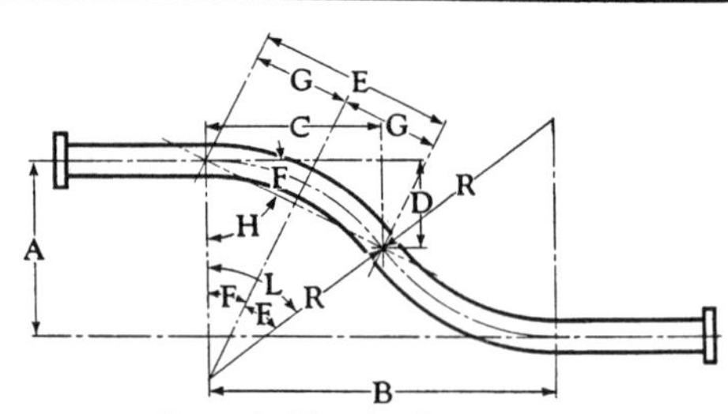

Example No. 6—**Given A, B**

$C = \frac{1}{2} B$

$D = \frac{1}{2} A$

$E = \sqrt{C^2 + D^2}$

$\frac{D}{E} = \sin \angle F$

$G = \frac{1}{2} E$

$\angle H = 90° - \angle F$

$R = \frac{A^2 + B^2}{4A}$

$L = 2F$

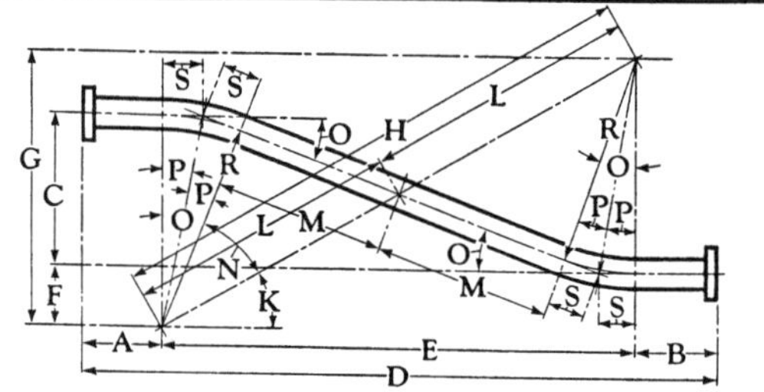

Example No. 7—**Given A, B, C, D, R**

$E = D - A - B$

$F = R - C$

$G = R + F$

$H = \sqrt{E^2 + G^2}$

$G/H = \sin \angle K$

$L = \frac{1}{2} H$

$M = \sqrt{L^2 - R^2}$

$M/L = \sin \angle N$

$\angle O = 90° - \angle K - \angle N$

$\angle P = \frac{1}{2} \angle O$

$S = \tan \angle P \times R$

Reprinted with permission—Crane Company.

Calculation of pipe bends

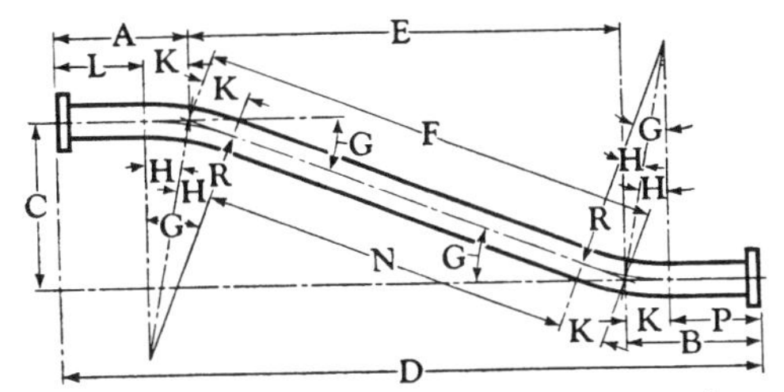

Example No. 8—**Given A, B, C, D, R**

$$E = D - A - B$$
$$F = \sqrt{E^2 + C^2}$$
$$\frac{C}{F} = \sin \angle G$$
$$\angle H = \tfrac{1}{2} \angle G$$
$$K = \tan \angle H \times R$$
$$L = A - K$$
$$P = B - K$$
$$N = F - 2K$$

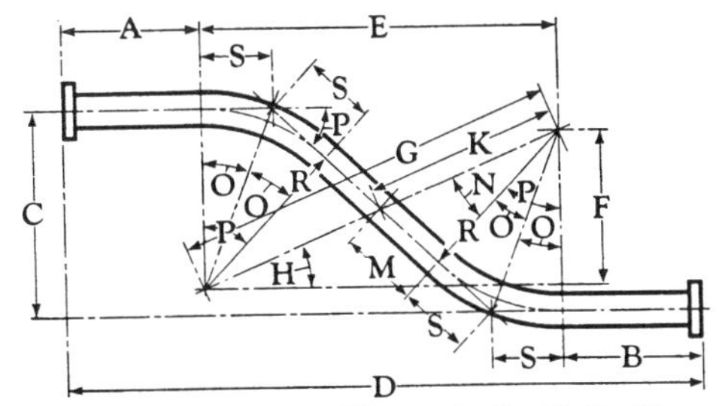

Example No. 9—**Given A, B, C, D, R**

$$E = D - A - B$$
$$F = 2R - C$$
$$G = \sqrt{E^2 + F^2}$$
$$\frac{F}{E} = \tan \angle H$$
$$K = \tfrac{1}{2} G$$
$$M = \sqrt{K^2 - R^2}$$
$$\frac{M}{K} = \sin \angle N$$
$$\angle P = 90^\circ - \angle H - \angle N$$
$$\angle O = \tfrac{1}{2} \angle P$$
$$S = \tan \angle O \times R$$

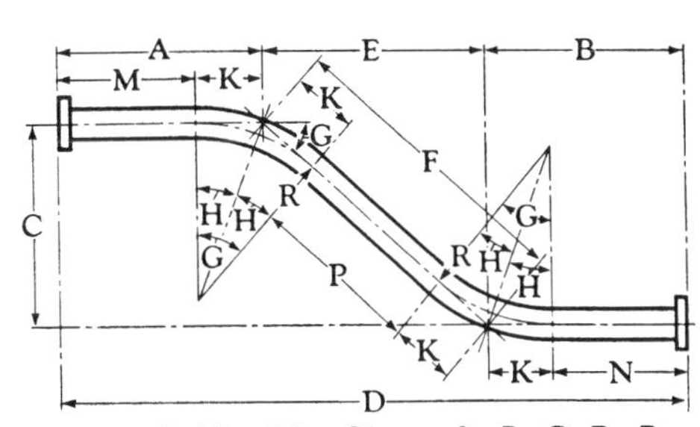

Example No. 10—**Given A, B, C, D, R**

$$E = D - A - B$$
$$F = \sqrt{C^2 + E^2}$$
$$\frac{C}{F} = \sin \angle G$$
$$\angle H = \tfrac{1}{2} \angle G$$
$$K = \tan \angle H \times R$$
$$M = A - K$$
$$N = B - K$$
$$P = F - 2K$$

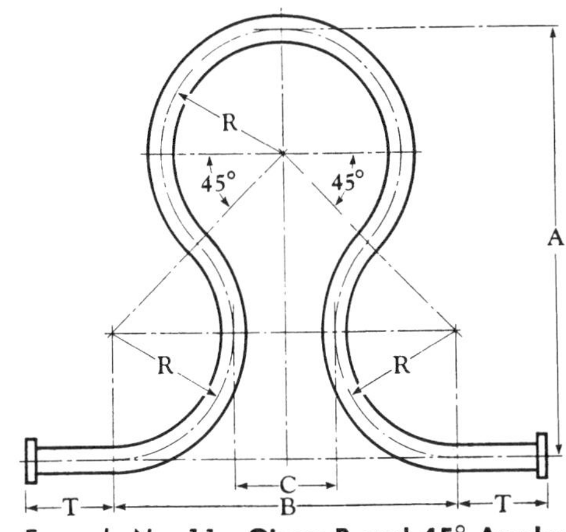

Example No. 11—**Given R and 45° Angles**

$$A = 3.414 \times R$$
$$B = 2.828 \times R$$
$$C = 0.828 \times R$$

T = Tangent

Length of pipe in bend = $9.425 \times R + 2T$

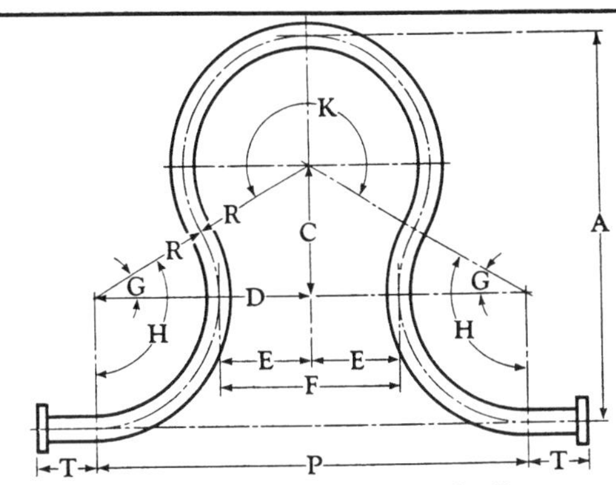

Example No. 12—**Given A, R**

$$C = A - 2R$$
$$D = \sqrt{(2R)^2 - C^2}$$
$$P = 2D$$
$$E = D - R$$
$$F = 2E$$
$$C/2R = \sin \angle G$$
$$\angle H = 90^\circ + \angle G$$
$$\angle K = 180^\circ + 2 \angle G$$

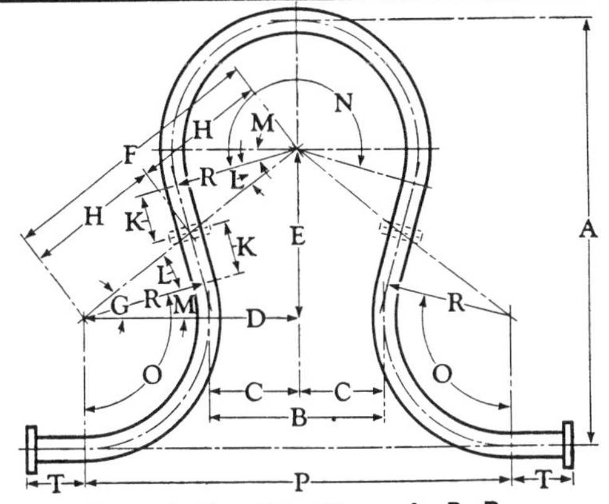

Example No. 13—**Given A, B, R**

$$C = \tfrac{1}{2} B$$
$$D = R + C$$
$$E = A - 2R$$
$$F = \sqrt{D^2 + E^2}$$
$$E/F = \sin \angle G$$
$$H = \tfrac{1}{2} F$$
$$K = \sqrt{H^2 - R^2}$$
$$K/H = \sin \angle L$$
$$\angle M = \angle G - \angle L$$
$$\angle N = 180^\circ + 2 \angle M$$
$$\angle O = 90^\circ + \angle M$$
$$P = 2D$$

Spacing of pipe supports

For chart on deflection of horizontal pipe lines, see next page

When a horizontal pipeline is supported at intermediate points, sagging of the pipe occurs between these supports, the amount of sag being dependent upon the weight of the pipe, fluid, insulation, and valves or fittings that may be included in the line. If the pipeline is installed with no downward pitch, pockets will be formed in each span, in which case condensation may collect if the line is transporting steam. In order to eliminate these pockets, the line must be pitched downward so that the outlet of each span is lower than the maximum sag.

Crane has conducted tests to determine the deflection of horizontal standard pipelines filled with water, in pipe sizes ¾ to 4″ inclusive, the results of which have indicated that for pipes larger than 2″ and with supports having center to center dimensions greater than 10 ft, the resultant deflection is less than that determined by the use of the formula for a uniformly loaded pipe fixed at both ends. For pipe sizes 2″ and smaller, the test deflection was in excess of that determined by the formula for pipe having fixed ends and approached, for the shorter spans, the deflection as determined by the use of the formula for pipelines having unrestrained ends.

Page 96 gives the deflection of horizontal standard pipelines filled with water, for varying spans, based upon the results obtained from tests for sizes 2″ and smaller and upon the formula for fixed ends for the larger sizes of pipe. The deflection values given on the chart are twice those obtained from test or calculation, to compensate for any variables including weight of insulation, etc.

The formula given below indicates the vertical distance that the span must be pitched so that the outlet is lower than the maximum sag of the pipe.

$$h = \frac{144^2 y}{36S^2 - y^2}$$

where:

h = Difference in elevation of span ends, inches
S = Length of one span, feet
y = Deflection of one span, inches

By eliminating the inconsequential term "$-y^2$" from the denominator, the formula reduces to:

$$h = 4y$$

The pitch of pipe spans, called the Average Gradient, is a ratio between the drop in elevation and the length of the span. This is expressed as so many inches in a certain number of feet.

$$\text{Average gradient} = \frac{4y}{S}$$

The dotted lines as shown on the chart on the opposite page are plotted from the above formula and indicate average gradients of 1″ in 10′, 1″ in 15′, 1″ in 20′, 1″ in 30′, and 1″ in 40′.

Example. What is the maximum distance between supports for a 4″ standard pipeline assuming a pitch or average gradient of 1″ in 30 ft?

Using the chart on the opposite page, find the point where the diagonal dotted line for an average gradient of 1″ in 30 ft intersects the diagonal solid line for 4″ pipe. From this point, proceed downward to the bottom line where the maximum span is noted to be approximately 22 ft.

Code for Pressure Piping: The Code for Pressure Piping, ASA B 31.1, makes the following statements relative to installations within the scope of the code:

"**605 (g)** Supports shall be spaced so as to prevent excessive sag, bending and shear stresses in the piping, with special consideration given to those piping sections where flanges, valves, etc., impose concentrated loads. Where calculations are not made, suggested maximum spacing of hangers or supports for carbon steel piping operating at 750°F and lower are given in Table 21a (*see the table below*).

"Where greater distance between supports, concentrated loads, higher temperatures, or vibration considerations are involved, special consideration should be given to effects of bending and shear stresses."

"**623 (b)** The design and spacing of supports shall be checked to assure that the sum of the longitudinal stresses due to weight, pressure, and other sustained external loading does not exceed the allowable stress (*S* value) in the hot condition."

Suggested maximum spacing between pipe supports for straight runs of standard wall and heavier pipe (at maximum operating temperature of 750°F)

Nominal Pipe Size Inches	Maximum Span Feet	Nominal Pipe Size Inches	Maximum Span Feet
1	7	8	19
1½	9	10	22
2	10	12	23
2½	11	14	25
3	12	16	27
3½	13	18	28
4	14	20	30
5	16	24	32
6	17	..	..

Note: The values in the table do not apply where there are concentrated loads between supports such as flanges, valves, etc.

Spacing is based on a combined bending and shear stress of 1500 psi when pipe is filled with water and the pitch of the line is such that a sag of 0.1 in. between supports is permissible.

Spacing of pipe supports

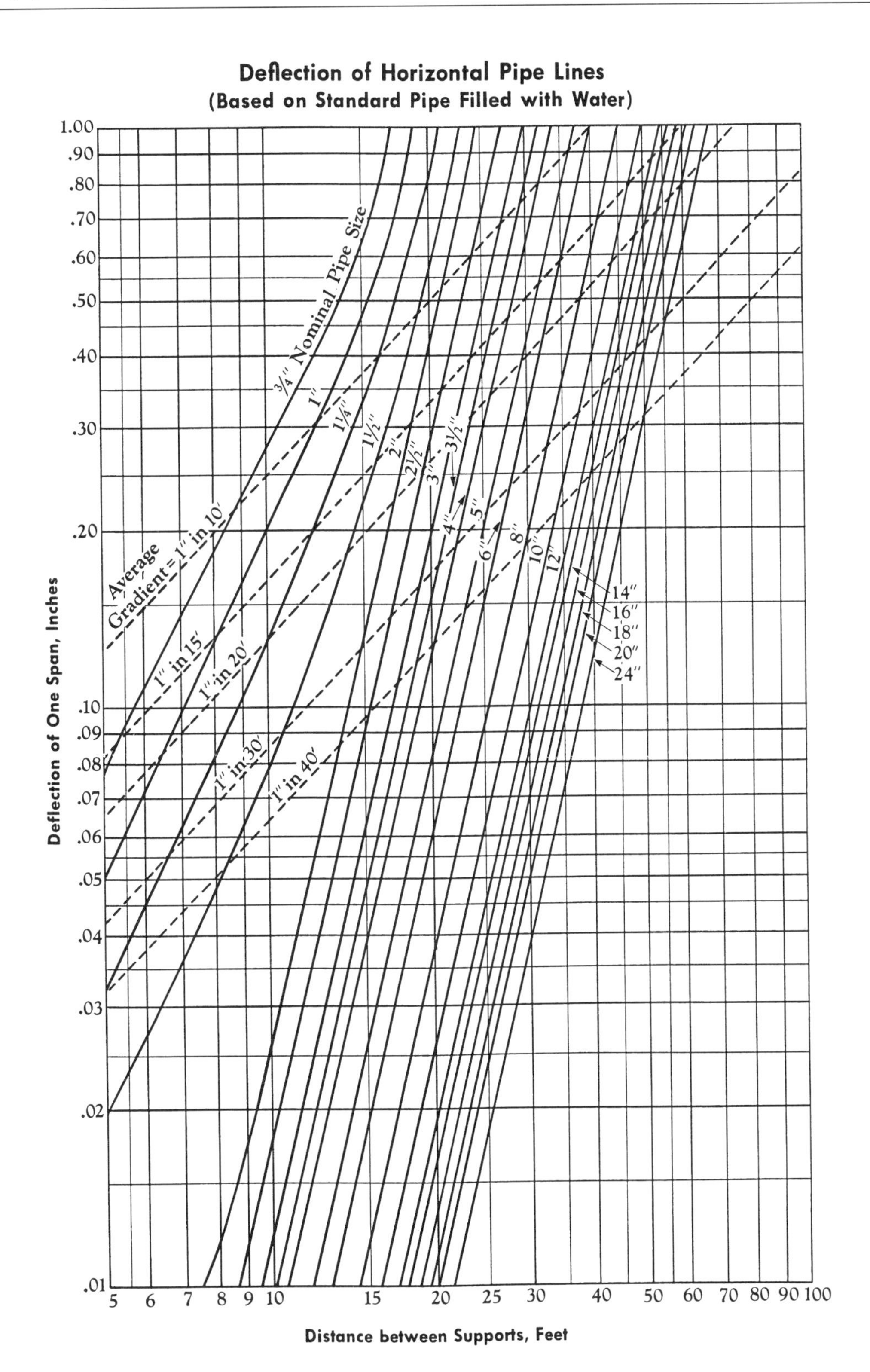

Reprinted with permission—Crane Company.

American standard taper pipe threads (NPT)

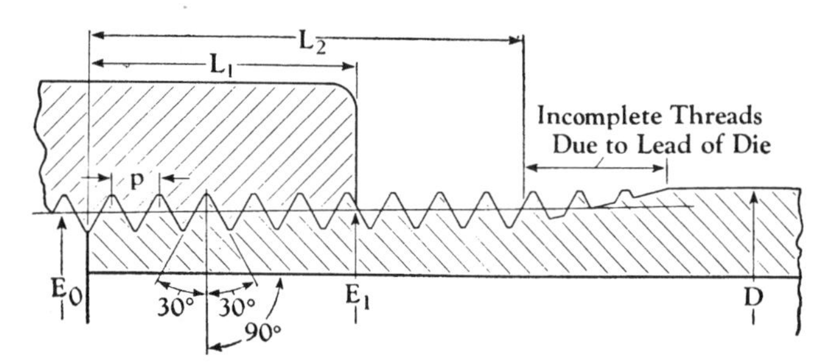

$E_0 = D - (0.050D + 1.1)p$
$E_1 = E_0 + 0.0625\ L_1$
$L_2 = (0.80D + 6.8)p$

p = Pitch
Depth of thread = $0.80p$
Total Taper ¾-in. per foot

Flush by Hand

Tolerance on Product

One turn large or small from notch on plug gauge or face of ring gauge.

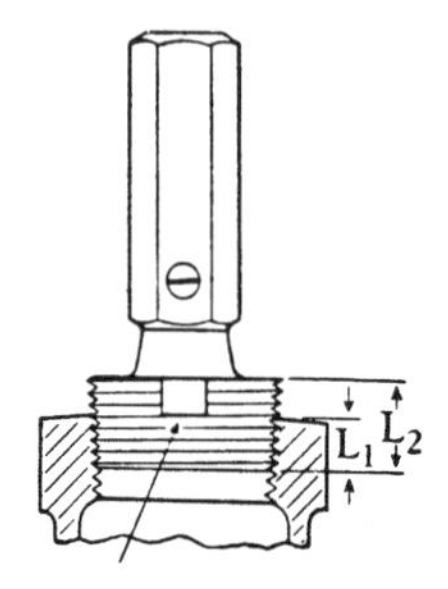

Notch flush with face of fitting. If chamfered, notch flush with bottom of chamfer.

Dimensions, in Inches

Nominal pipe size	D Outside diameter of pipe	E_0 Pitch diameter at end of external thread	E_1† Pitch diameter at end of internal thread	L_1¶ Normal engagement by hand between external and internal threads	L_2§ Length of effective thread	p Pitch of thread	Depth of thread	Number of threads Per Inch
⅛	0.405	0.36351	0.37476	0.180	0.2639	0.03704	0.02963	27
¼	0.540	0.47739	0.48989	0.200	0.4018	0.05556	0.04444	18
⅜	0.675	0.61201	0.62701	0.240	0.4078	0.05556	0.04444	18
½	0.840	0.75843	0.77843	0.320	0.5337	0.07143	0.05714	14
¾	1.050	0.96768	0.98887	0.339	0.5457	0.07143	0.05714	14
1	1.315	1.21363	1.23863	0.400	0.6828	0.08696	0.06957	11½
1¼	1.660	1.55713	1.58338	0.420	0.7068	0.08696	0.06957	11½
1½	1.900	1.79609	1.82234	0.420	0.7235	0.08696	0.06957	11½
2	2.375	2.26902	2.29627	0.436	0.7565	0.08696	0.06957	11½
2½	2.875	2.71953	2.76216	0.682	1.1375	0.12500	0.10000	8
3	3.500	3.34062	3.38850	0.766	1.2000	0.12500	0.10000	8
3½	4.000	3.83750	3.88881	0.821	1.2500	0.12500	0.10000	8
4	4.500	4.33438	4.38712	0.844	1.3000	0.12500	0.10000	8
5	5.563	5.39073	5.44929	0.937	1.4063	0.12500	0.10000	8
6	6.625	6.44609	6.50597	0.958	1.5125	0.12500	0.10000	8
8	8.625	8.43359	8.50003	1.063	1.7125	0.12500	0.10000	8
10	10.750	10.54531	10.62094	1.210	1.9250	0.12500	0.10000	8
12	12.750	12.53281	12.61781	1.360	2.1250	0.12500	0.10000	8
14 O.D.	14.000	13.77500	13.87262	1.562	2.2500	0.12500	0.10000	8
16 O.D.	16.000	15.76250	15.87575	1.812	2.4500	0.12500	0.10000	8
18 O.D.	18.000	17.75000	17.87500	2.000	2.6500	0.12500	0.10000	8
20 O.D.	20.000	19.73750	19.87031	2.125	2.8500	0.12500	0.10000	8
24 O.D.	24.000	23.71250	23.86094	2.375	3.2500	0.12500	0.10000	8

†Also pitch diameter at gauging notch.
§Also length of plug gauge.
¶Also length of ring gauge, and length from gauging notch to small end of plug gauge.

British standard taper pipe threads

Whitworth 55° Form of Thread

Total Taper ¾-inch per ft

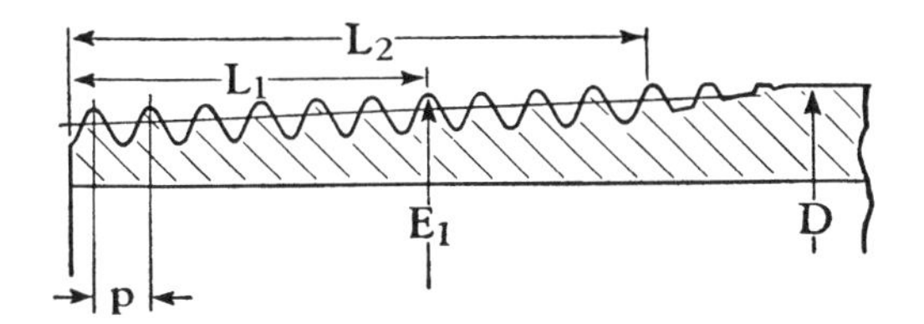

This information is taken from the British Engineering Standards Association Report No. 21-1938.

Dimensions, in Inches

Pipe size Nominal Bore	D Approximate outside diameter of pipe	E_1 Pitch diameter at end of internal thread	L_1 Normal engagement by hand between external and internal threads	L_2 Length of effective thread	p Pitch of threads	Depth of threads	Number of threads Per Inch
1/8	13/32	0.3601	0.1563	0.2545	0.03571	0.0229	28
1/4	17/32	0.4843	0.2367	0.3814	0.05263	0.0337	19
3/8	11/16	0.6223	0.2500	0.3947	0.05263	0.0337	19
1/2	27/32	0.7793	0.3214	0.5178	0.07143	0.0457	14
3/4	1 1/16	0.9953	0.3750	0.5714	0.07143	0.0457	14
1	1 11/32	1.2508	0.4091	0.6591	0.09091	0.0582	11
1 1/4	1 11/16	1.5918	0.5000	0.7500	0.09091	0.0582	11
1 1/2	1 29/32	1.8238	0.5000	0.7500	0.09091	0.0582	11
2	2 3/8	2.2888	0.6250	0.9204	0.09091	0.0582	11
2 1/2	3	2.9018	0.6875	1.0511	0.09091	0.0582	11
3	3 1/2	3.4018	0.8125	1.1761	0.09091	0.0582	11
3 1/2	4	3.8918	0.8750	1.2386	0.09091	0.0582	11
4	4 1/2	4.3918	1.0000	1.4091	0.09091	0.0582	11
5	5 1/2	5.3918	1.1250	1.5795	0.09091	0.0582	11
6	6 1/2	6.3918	1.1250	1.5795	0.09091	0.0582	11
7	7 1/2	7.3860	1.3750	1.9250	0.10000	0.0640	10
8	8 1/2	8.3860	1.5000	2.0500	0.10000	0.0640	10
9	9 1/2	9.3860	1.5000	2.0500	0.10000	0.0640	10
10	10 1/2	10.3860	1.6250	2.1750	0.10000	0.0640	10
11	11 1/2	11.3700	1.6250	2.3125	0.1250	0.0800	8
12	12 1/2	12.3700	1.6250	2.3125	0.1250	0.0800	8

Normal engagement between male and female threads to make tight joints

Normal engagement

The normal amount of engagement to make a tight joint for various types of screwed material is given in the table. These dimensions have been established from tests made under practical working conditions.

The normal engagement specified for American Standard Pipe Thread and API Line Pipe Thread joints is based on parts being threaded to the American Standard for Pipe Threads or the API Standard for Line Pipe Threads.

In order to obtain the thread engagements listed in the table, it is necessary to vary the torque or power applied according to the size, metal, and weight of material used. For example, it requires considerably less power to make up a screwed joint using a light bronze valve than a high pressure steel valve.

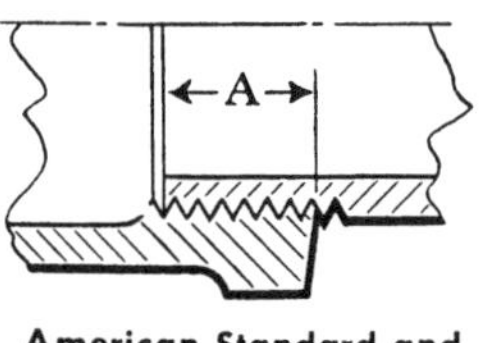

American Standard and API Line Pipe Threads

Dimensions, in Inches Dimensions Given Do Not Allow for Variations in Tapping or Threading			
Size	A	Size	A
1/8	1/4	2½	15/16
1/4	3/8	3	1
3/8	3/8	3½	1 1/16
1/2	1/2	4	1 1/8
3/4	9/16	5	1
1	11/16	6	1 5/16
1 1/4	11/16	8	1 7/16
1 1/2	11/16	10	1 5/8
2	3/4	12	1 3/4

Reprinted with permission—Crane Company.

Hand-held computer calculates pipe weight, contents, velocity

Programs, written for the HP 41 CV computer equipped with printer, work with English or metric units

Frank E. Hangs, P. E., Sovereign Engineering, Inc., Houston

Pipeline engineers, accountants, and system operators are deeply involved with data concerning line pipe such as:

How much does it weigh (lb/ft, tons/mi)? What are the corresponding metric units?

How much will it contain (bbl/mi, gal/100 ft, cu m/km)?

What are velocities at various feed rates (ft/sec, mi/hr, m/hr)?

The following program, written for the Hewlett Packard HP 41 CV hand-held computer, develops a lot of useful data for steel pipe for any outside diameter and wall thickness. Either English or metric units may be entered.

A printer is essential.

Results for the WEIGHT category are calculated by an API formula and given as: Pounds (mass) per ft; kg per m; short tons (2,000 lb) per mile; metric tons per km.

The content results are: bbl/mi, gal (U.S.)/100 ft, cu ft/mi, L/m, cu m/km.

Flowrates can be input to give: b/h or b/d; gal (U.S.)/min; or gal/hr or gal/d; cu ft/min, /hr, /d; cu m/min, /hr, /d.

Velocity results: mph; fps; km/hr; m/sec.

All results, as well as the input, are printed out, so the tape becomes a permanent record.

A good feature of the program is that, for a given size pipe (after executing PWCV and A, content routine, to store data concerning that particular size pipe), velocity routine, B, can be run as often as wanted. Press B, then key in desired flowrate after the prompt, then press the RUN key.

When a different size pipe is needed: XEQ PWCV, press A, then B, as often as needed for different flowrates.

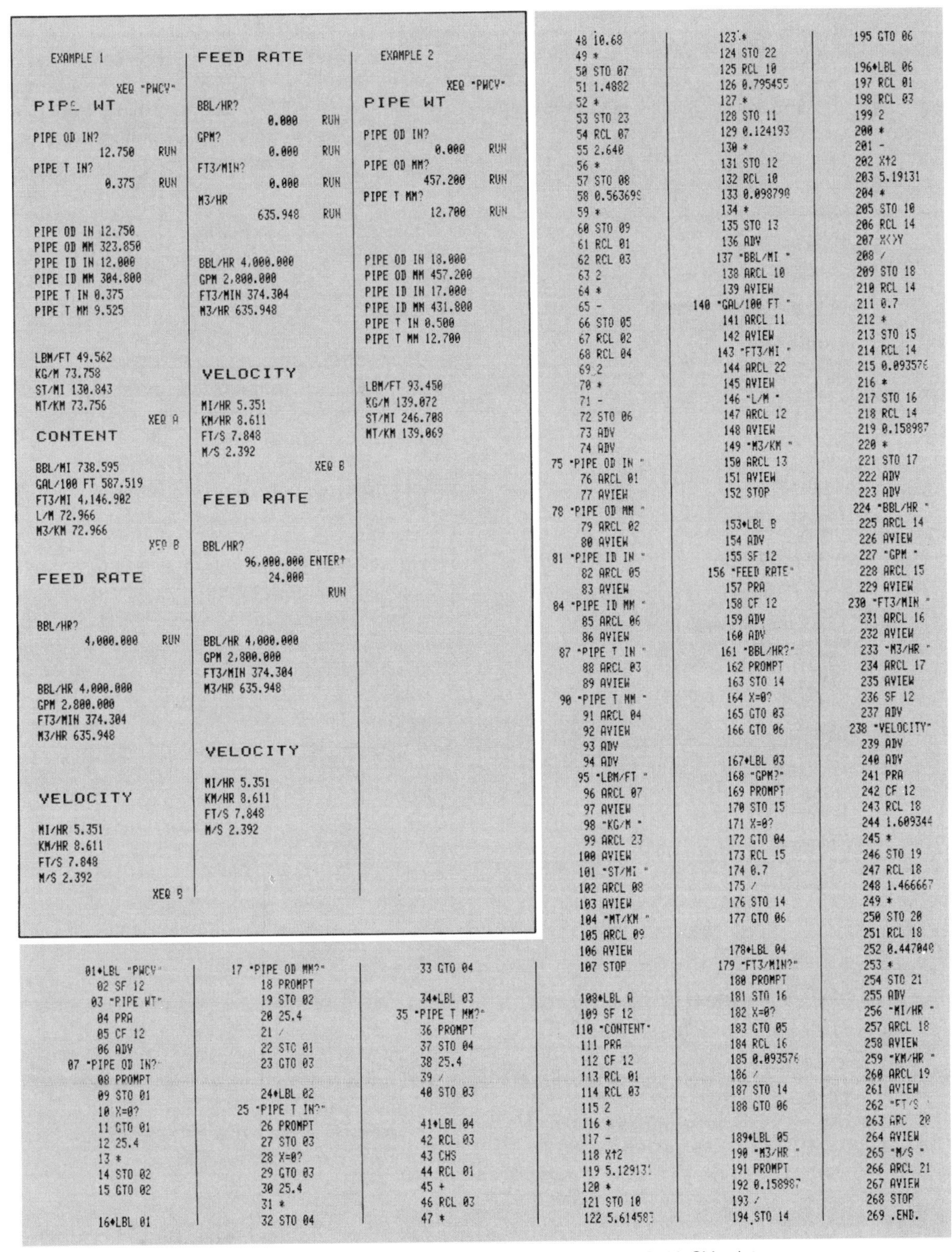

Figure 1. Examples of the program printed by the HP 41 CV printer.

Basic data calculated by this program can be used for determining the following criteria:

- Weight: Cost of construction, hauling and stringing, investments.
- Content: Amount of water for hydrostatic test, line fill, oil stocks. Amount of one grade to be displaced for another to reach a station, shut-off valve, or terminal. When a segment must be drained, barrels or gallons to be removed.
- Velocity: Time for interface or scraper to traverse distance between two points. Time to accumulate or withdraw a given quantity of fluid. Locating head of column of a grade of oil or a scraper from given station after a certain time. Determining expected time of arrival (ETA) of interface or scraper at a given point.

List of registers and legend

Register	Content
00	Not used
01	Pipe OD in.
02	Pipe OD mm
03	Pipe T in.
04	Pipe T mm
05	Pipe ID in.
06	Pipe ID mm
07	LBM/FT—lb(mass)/ft
08	ST/MI—short tons/mi
09	MT/KM—metric tons/km
10	BBL/MI—bbl(42 U.S. gal) per mi
11	GAL/100FT—gal (U.S.) per 100 ft
12	L/M—liters per m
13	M^3/KM—cum per km
14	BBL/HR—bbl (42 U.S. gal) per hr
15	GPM—gal (U.S.) per minute
16	FT^3/MIN—cuft per min
17	M^3/HR—cum per hr
18	MI/HR—mi (statute) per hr
19	KM/HR—km per hr
20	FT/S—ft per sec
21	M/S—m per sec
22	FT^3/MI—cuft per mi
23	KG/M—kg per m

LABEL	PWCV	Calls up program
LABEL	A	Calls content calculations
LABEL	B	Calls for feed rate and velocity calculations

Note. A and B are XEQ in user mode.

Numeric labels 01 to 06 are used in the program to provide necessary GTO directions. They are not stand-alone routines.

User directions. Size 25. Set user mode. Input program. If it is desired to use units as prompted, key in value and run key. If metric units are wanted, key in 0 for English units, then run key.

Example. Display: Pipe OD in.? Key in 0, press run key, see pipe OD MM? Key in value and run key. See pipe T MM?

After weights are calculated: If content is wanted, XEQ "A." Calculations can be terminated at this point.

When velocities are desired, XEQ "B" input feed rate in desired units. Label "B" can be run for several feed rates as often as desired for the given size pipe. If other pipe sizes are needed, XEQ PWCV and "A," then run "B" as many times as needed for different feed rates.

Note. If feed rates other than BBL/HR are to be input, first key in 0 and run. This causes display to move up. See GPM? No. Gallons? Key in 0 and run key. See FT3/MIN? Continue to M3/HR? If you have this figure, key it in and run. Now suppose you have M3/DAY. This *does not* agree with prompt; enter M3/DAY, divide by 24, then press run key. This routine can be used for BBL/DAY, GAL/HR (divided by 60), FT3/DAY (divide by 24), etc.

Note. When results such as OD, ID, T Values, Weight KG/M are compared with printed values in API STD 5L, small discrepancies are sometimes found, usually on the order of 0.1% or less. This is evidently due to truncation differences in making conversions.

Formulas

Pipe weight LBM/FT (plain end) is calculated by formula used by API:

$$LBM/FT = 10.68(D - T)/T$$

Weights in other units are found by multiplying by conversion factors:

BBL/MI = 5.129131 (D−2T)

Basic velocity relation:

$$\frac{BBL}{HR} \div \frac{BBL}{MI} = \frac{MI}{HR}$$

This is converted to other units.

Example 1. Size 25, Fix 3, Set user mode. Put PWCV in calculator. XEQ PWCV. Input 12.75 in. Press run key. See Pipe T in.? Input 0.375. Press run key. After calculator stops, press "A" for content. If velocities are desired, press "B." Input 4,000 BBL/HR. Press run key. Results are printed.

Now "B" can be run as often as needed.

Change feed rate units. Note zeros were put in for three feed rates, and 635.948M3/HR input was keyed in. This is to demonstrate how other feed rates are input and the accuracy of the conversions (6 decimals). Put in feed rate 96,000BPD. Enter, divide by 24, press run key.

Example 2. This shows how to input OD and T in MM. Key in 0 for pipe OD IN.? Then press run key. See Pipe OD MM? Key in metric units. Results are calculated and printed.

Questions and answers

1. How many short tons are in a 55.35-mi segment of 12.75 in. × 0.375 in. WT? (Use data from Ex. 1.)

   55.35 × 130.843 = 7.242.160ST

2. How many km are in this segment above?

   55.35 × 1.609 = 89.058 km

3. How many bbl are in (a) 150 ft? (b) gal? (c) cu ft?

   a. 150/5,280 × 738.595 = 20.983 bbl
   b. 150/100 × 587.519 = 881.279 gal
   c. 150/5,280 × 4,146.902 = 117.810 cu ft

4. Suppose, in an emergency, you have to dig a pit to hold oil from 150 m of this 12-in. line. Volume? (a) cu m (b) cu ft

   a. 150/1,000 × 72.966 = 10.945 cu m
   b. 10.945 × 34.315 = 386.519 cu ft

5. How long will it take a scraper to traverse 67.85 mi of this 12-in. line at 4,000 bbl/hr?

   67.85/5.351 = 12.680 hrs Hours min. sec? Fix 4.

   XEQ Alpha HMS Alpha = 12HR 40MIN 48SEC

Useful conversions

Meters × 3.281 = feet
Miles × 1.609 = kilometers
Cubic ft × 7.481 = gallon (U.S.)
Kilogram × 2.205 = pounds
Cubic meters × 35.315 = cubic feet
Cubic meters × 6.290 = barrels
Cubic meters × 264.172 = gallons (U.S.)
Barrels × 5.615 = cubic feet
Barrels × 42.00 = gallons (U.S.)
Barrels × 135.00 = Imperial gallons

Source

Pipe Line Industry, May 1986.

Formulas and constants of value in solving problems relating to tubular goods

Area of Metal in Tubular Section:

Area in square inches = 3.1416(D − t)t

where:
D = Outside diameter, in.
t = Wall thickness, in.

Capacity

$B_{100} = 0.1237A = 0.0972D_2$

where:
B_{100} = Barrels per 100 ft
A = Internal area of pipe, in.^{2}
D = Internal diameter, in.

$$\text{and } F_B = \frac{808.5}{A} = \frac{1029.4}{D^2}$$

where:
F_B = Number of feet filled by one barrel
A = Internal area of pipe, in.^{2}
D = Internal diameter, in.

Complete tables giving capacities of different types of tubular goods are given in chapters dealing with these types.

Steel constants* (applicable to tubular goods)

One cubic inch = 0.2833 lb
One cubic foot = 489.542 lb
Specific gravity = 7.851

Approximate coefficient of expansion for commercial casing, drill pipe, and tubing is
Coefficient = 6.9×10^{-6} per °F, over range from 0 to 400°F

Formula for expansion of pipe due to change of temperature is as follows:

$$L_t = L_o(1 + 0.0000069t)$$

where:

L_o = Length at atmospheric temperature
L_t = Length after change of temperature t

Modulus: Section Modulus $\frac{\pi}{64}(D^4 - D_1^4)$

Polar Modulus $\frac{\pi}{32}(D^4 - D_1^4)$

where:

D = Outside diameter
D_1 = Inside diameter
π = 3.1416

Relation between twisting effort, torsional strain, and stress of cylindrical shaft or tube

$$\frac{q}{r} = \frac{T}{J} = \frac{C\theta}{L}$$

*Engineering data Spang-Chalfant Division of the National Supply Co.

How to calculate the contraction or expansion of a pipeline

Steel pipe contracts or expands approximately 0.8 in. per 100°F temperature change for each 100 ft of pipe. For construction in the United States, slack is not normally needed in welded lines for contraction or expansion unless abnormal variations are encountered. However, slack is often provided near road crossings where the pipe may have to be lowered at some future time.

$$\text{Contraction} = 0.8 \times \frac{\text{length of pipe (feet)}}{100} \times \frac{\text{temp change (°F)}}{100}$$

Example. Calculate the amount a 1,000-ft section of pipeline would contract if laid at a temperature of 100°F and cooled to a low of 0°F during winter operation (assuming the line were free of soil or other resisting loads).

$$\text{Contraction} = 0.8 \times \frac{1\,000}{100} \times \frac{100}{100} = 0.8 \times 10 \times 1 = 8\,\text{in.}$$

Large temperature changes may arise on exposed line or in lines downstream from compressor stations.

When slack is desired in a line, the amount of sag or rise needed may be estimated quickly. To provide for 0.8 in. of movement longitudinally along the pipe, a sag of 21 in. would be needed in a 100-ft loop. For a 150-ft loop, 31 in. are necessary. Such slack put into the line at time of construction might allow for lowering the pipe under roadbeds without putting excessive stress upon the pipe.

Location of overbends, sags, and side bends should be considered when laying slack loops. When loops are placed, parts of the line should be backfilled to serve as a tie-down. Loops should be lowered during the coolest part of the day when the line is the shortest. After lowering in, sags should rest on the bottom of the ditch; overbends should "ride high." Side bends should rest on the bottom of the ditch and against the outside wall. The line should be lowered so that all sections of the pipe are in compression.

Estimate weight of pipe in metric tons per kilometer

To estimate the weight of pipe in metric tons per kilometer, multiply the nominal diameter by the number of sixteenths of an inch in wall thickness.

Example. Find the weight of pipe in metric tons per kilometer for 20-in. diameter pipe, wall thickness ¼-in.

$4 \times 20 = 80$ tons (metric) per kilometer

Actual answer from pipe tables is 79 metric tons per kilometer.

This rule of thumb is based on a density of 490 lb. per cubic foot for steel. For larger diameter thin wall pipe, this approximation gives an answer usually about 1% low. The accompanying table provides a comparison between actual weights in metric tons per kilometer, as compared to that calculated by this rule of thumb.

PIPE DIAMETER	WALL THICKNESS	ACTUAL TONS (METRIC)	R. O. T. TONS (METRIC)
10-in.	$^{3}/_{16}$-in.	31.5	30
12-in.	$^{5}/_{16}$-in.	62	60
16-in.	$^{7}/_{16}$-in.	109	112
20-in.	$^{1}/_{4}$-in.	79	80
20-in.	$^{5}/_{16}$-in.	99	100
20-in.	$^{7}/_{16}$-in.	137	140
24-in.	$^{5}/_{16}$-in.	118	120
24-in.	$^{8}/_{16}$-in.	189	192
28-in.	$^{5}/_{16}$-in.	139	140
30-in.	$^{5}/_{16}$-in.	148	150
32-in.	$^{7}/_{16}$-in.	221	224
36-in.	$^{4}/_{16}$-in.	143	144
36-in.	$^{8}/_{16}$-in.	285	288

How to find pipe weight from outside diameter and wall thickness

Weight (pounds per foot) $= (Dt - t^2 \times 10.68)$

where:

D = outside diameter, in.
t = wall thickness, in.

Example. Outside diameter = 4.500″ thickness = 0.250″

$W = (4.5 \times .25 - .25^2) \times 10.68$

$= (1.125 - .0625) \times 10.68$
$= 1.0625 \times 10.68$
$= 11.35$ lb. per ft

The above equation is based on a density of 490 lb. per cubic foot for the steel. High-yield-point, thin-wall pipe may run slightly heavier than indicated.

What is the maximum allowable length of unsupported line pipe?

For Schedule 40 pipe, the maximum span between supports is given by

$$S = 6.6\sqrt{P}$$

where S is the span length in feet and P is the actual outside diameter of the pipe in inches. For pipe smaller than 12-in., the nominal size may be used, and

$$S = 7\sqrt{N}$$

Example. How wide can a ditch be spanned with 4-in. line pipe without the use of intermediate supports?

$$S = 6.6\sqrt{4.5} = 6.6 \times 2.12 = 13.99 \text{ (14 ft)}$$

$$S = 7\sqrt{N} = 7 \times 2 = 14 \text{ ft}$$

Note. The above rule is approximate but conservative. It does not assume that the pipe is held fixed at either end; this condition can make longer spans safe. Also, it does not guarantee that the pipe will not be damaged by floating logs or debris! This hazard must be evaluated from the conditions encountered.

Identify the schedule number of pipe by direct measurement

Add 3 in. to the actual inside diameter; divide this by the wall thickness; the schedule number can then be identified by the following table:

Schedule	30	40 to 50
Schedule	40	29 to 39
Schedule	60	25 to 29
Schedule	80	20 to 23
Schedule	100	16 to 18
Schedule	120	13 to 15
Schedule	140	11 to 13
Schedule	160	9 to 11

Example. Pipe measures 12.5 in. inside diameter, wall thickness .75 in.; 15.5/.75 = 20.2; pipe is Schedule 80, 14 in. outside diameter.

The rule fails on small-diameter pipe (under 6 in.), and also on schedules 10 and 20 because of the use of "standard" thicknesses in this range.

Determine buoyancy of bare steel pipe

Multiply the square of the outside diameter of the pipe, in inches, by 0.341, and subtract the pipe weight in lb/ft; the answer is the buoyancy per foot in fresh water.

Example. 12.750″ OD × 0.250 w.t. (33.#/ft)

$$\text{Buoyancy} = 12.750^2 \times 0.341 - 33.4$$

$$= 162.56 \times 0.341 - 33.4$$

$$= 55.4 - 33.4 = 22 \text{ lb./ft}$$

Because all of the coating materials in common use are heavier than water (but not as much so as steel), coated pipe will have less buoyancy than indicated by the rule. For sea water, use 0.350 for the constant.

Determine buoyancy of bare and concrete-coated steel pipe in water and mud

To find the buoyancy of bare steel and concrete coated pipe in water in lb/ft:
For bare pipe:

$$\text{Buoyancy (B)} = \frac{D}{3}(D - 32t) + 11t^2$$

For coated pipe:

$$\text{Buoyancy (B)} = \frac{D}{3}(D - 32t) + t_1 D\left[\frac{63 - W_c}{48}\right]$$

where:

D = outside diameter of pipe, in.
t = wall thickness of pipe, in.
t_1 = thickness of concrete coating, in.
W_c = weight of concrete, lb/ft^3

To find the buoyancy of bare steel and coated pipe in mud in lb/ft:
For bare pipe:

$$\text{Buoyancy (B)} = 10.7D\left[\frac{DW_m}{2000} - t\right] + 11t^2$$

For coated pipe:

$$B = 10.7\ D\left[\frac{DW_m}{2000} - t\right] + t_1 D\left[\frac{W_m - W_c}{48}\right]$$

for coated pipe, where W_m is weight of mud, lb/ft^3.

Example 1. Find the buoyancy of steel pipe with 20″ OD × 0.500″ w.t. steel pipe in water. What thickness of concrete coating will be required to give this pipe a negative buoyancy of 100 lb./ft. ($W_c = 143$)?

$$\text{Buoyancy (B)} = \frac{D}{3}(D - 32t) + 11t^2$$

$$\text{Buoyancy (B)} = \frac{20}{3}(20 - 32^*0.5) + 11^*0.5^2$$

$$= 26.7 + 2.8 = 29.5 \text{ ft positive buoyancy}$$

To give this pipe a negative buoyancy of 100 lb./ft:

$$-100 = \frac{20}{3}(20 - 32^*0.5) + t_1 20\left[\frac{63 - 143}{48}\right]$$

$$-100 = 26.7 - 33.3t_1$$

$$t_1 = 3.8 \text{ in. of concrete coating}$$

The error introduced by this method is about 15% but would be less as the thickness of coating decreases.

Example 2. Bottom conditions at a certain crossing require the pipe to have a negative buoyancy of at least 50 lb./ft. Determine if 10.750″ pipe with 2″ of concrete ($W_c = 143$) has sufficient negative buoyancy

$$B = \frac{10}{3}(10 - 32^*0.5 + 2^*10\left[\frac{63 - 143}{48}\right]$$

$B = -20 - 33.3 = -53.3$ lb./ft; so coated pipe meets design requirements.

Weights of piping materials

The weight per foot of steel pipe is subject to the following tolerances:

SPECIFICATION	TOLERANCE
A.S.T.M.A – 53 A.S.T.M.A – 120	STD WT +5%, −5% XS WT +5%, −5% XXS WT+10%, −10%
A.S.T.M.A – 106	SCH 10–120 +6.5%, −3.5% SCH 140–160 +10%, −3.5%
A.S.T.M.A – 158 A.S.T.M.A – 206 A.S.T.M.A – 280	12″ and under +6.5%, −3.5% over 12″ +10%, −5%
API 5L	All sizes +65%, −3.5%

Weight of Tube $= F \times 10.6802 \times T \times (D - T)$ lb/ft

where:

T = wall thickness, in.
D = outside diameter, in.
F = relative weight factor

The weight of tube furnished in these piping data is based on low carbon steel weighing 0.2833 lb./in.^{3}

Relative weight factor F of various metals

Aluminum	= 0.35
Brass	= 1.12
Cast Iron	= 0.91
Copper	= 1.14
Lead	= 1.44
Ferritic Stainless Steel	= 0.95
Austenitic Stainless Steel	= 1.02
Steel	= 1.00
Tin	= 0.93
Wrought Iron	= 0.98

Weight of contents of a tube $= G \times 0.3405 \times (D-2T)^2$ lb/ft

where:

G = Specific gravity of contents
T = Tube wall thickness, in.
D = Tube outside diameter, in.

The weight of welding tees and laterals is for full size fittings. The weight of reducing fittings is approximately the same as for full size fittings.

The weight of welding reducers is for one size reduction and is approximately correct for other reductions.

Pipe covering temperature ranges are intended as a guide only and do not constitute a recommendation for specific thickness of material.

Pipe covering thicknesses and weights indicate average conditions and include all allowances for wire, cement, canvas, bands, and paint. The listed thicknesses of combination covering is the sum of the inner and the outer layer thickness. When specific inner and outer layer thicknesses are known, add them, and use the weight for the nearest tabulated thickness.

To find the weight of covering on fittings, valves, or flanges, multiply the weight factor (light faced subscript) by the weight per foot of covering used on straight pipe. All flange weights include the proportional weight of bolts or studs required to make up all joints.

Lap joint flange weights include the weight of the lap.

Welding neck flange weights are compensated to allow for the weight of pipe displaced by the flange. Pipe should be measured from the face of the flange.

All flanged fitting weights include the proportional weight of bolts or studs required to make up all joints.

To find the approximate weight of reducing-flanged fittings, subtract the weight of a full size slip-on flange and add the weight of reduced size slip-on flange.

Weights of valves of the same type may vary because of individual manufacturer's design. Listed valve weights are approximate only. When it is possible to obtain specific weights from the manufacturer, such weights should be used.

To obtain the approximate weight of flanged end steel valves, add the weight of two slip-on flanges of the same size and series to the weight of the corresponding welding-end valves.

Allowable working pressure for carbon steel pipe

Based on formula No. 7 of the ASA Code for Pressure Piping, Section 3B, this table was computed for $6\frac{5}{8}$-in. OD to 36-in. OD carbon steel seamless pipe, API Specifications 5L Grade A, 5L Grade B, and API 5LX Grade X-42.

The formula is:

$$t = \frac{PD}{2(S + YP)} + u + c$$

O.D. / Wall Thickness	6⅝	8⅝	10¾	12¾	14	16	18	20	22	24	26	28	30	32	34	36	O.D. / Wall Thickness
0.188	895 1044 1253	685 799 959	548 640 768	....	....	....	....	....	....	....	....	....	....	....	....	....	0.188
0.203	.....	1070	857	721	.....	.....	.....	.....	.....	.....	.....	.....	.....	.....	.....	.....	0.203
0.219	1110 1295 1554	849 991 1189	680 793 952	572 667 801	 729	 637	....	....	....	....	....	....	....	....	....	....	0.219
0.250	1327 1548 1858	1014 1184 1420	811 947 1136	683 797 956	621 725 870	543 633 760	482 562 675	434 506 607	 551	 505	 466	 433	 404	 378	 356	 336	0.250
0.281	....	....	....	794 926 1111	722 842 1010	631 736 883	560 654 784	504 588 705	 641	 587	 542	 503	 469	 439	 414	 390	0.281
0.312	1765 2059 2471	1347 1572 1886	....	905 1056 1267	823 961 1153	719 839 1007	638 745 894	574 670 804	521 608 730	478 557 669	441 514 617	 573	 534	 501	 471	 445	0.312
0.344	1994 2326 2791	1520 1774 2128	1214 1416 1700	1021 1190 1429	928 1083 1299	810 946 1135	720 839 1007	647 755 906	587 685 823	538 628 753	 695	 645	 602	 564	 530	 501	0.344
0.375	2216 2568 3103	1689 1970 2364	....	1133 1322 1586	1030 1202 1442	900 1049 1259	798 931 1118	718 837 1005	652 760 912	597 696 836	551 642 771	 715	 667	 625	 588	 555	0.375
0.406	.....	.....	.....	1744	.....	.....	1228	1104	1002	918	847	786	733	687	646	610	0.406
0.438	....	2034 2373 2848	1622 1892 2271	1362 1589 1907	1238 1444 1733	1081 1261 1513	959 1119 1342	862 1005 1206	782 913 1095	716 836 1003	661 771 925	 858	 801	 750	 706	 666	0.438
0.500	....	 3328	 2651	 2225	1444 1685 2022	1260 1470 1764	1118 1304 1565	1004 1171 1406	911 1063 1276	834 974 1168	770 898 1077	 1000	 932	 873	 822	 775	0.500
Wall Thickness / O.D.	6⅝	8⅝	10¾	12¾	14	16	18	20	22	24	26	28	30	32	34	36	Wall Thickness / O.D.

where:

- t = wall thickness, in.
- P = allowable working pressure, psi
- D = outside diameter, in.
- S = allowable stress, psi
- Y = a coefficient
- u = under-thickness tolerance, in.
- c = allowance for threading, mechanical strength, and corrosion, in.

In calculation of the tables, the values of Y, u, and c were:

$Y = 0.4$

$u = 0.125t$ ($u = 12\frac{1}{2}\%$ oft)

$c = 0.050$ in.

Substituting in formula and solving for P:

$$P = \frac{1.75S(t - 0.057)}{D + 0.04 - 0.7t}$$

The S-values are:

Spec. API 5L Grade A—25,500 psi
Spec. API 5L Grade B—29,750 psi
Spec. API 5LX Grade X-42—35,700 psi

Example. Find pipe specification for a 12-in. line operating at 1,078 psi: In the $12\frac{3}{4}$-in. column these conditions will be met using Grade X-42 pipe with 0.281-in. wall thickness, Grade B pipe with 0.330-in. wall thickness, or Grade A pipe with 0.375-in. wall thickness.

Find the stress in pipe wall due to internal pressure

Multiply ½ by diameter by pressure and divide by wall thickness to get stress in psi.

Example. Find the wall stress due to internal pressure in a 24-in. pipeline having a $\frac{1}{4}$-in. wall thickness if the pressure gauge reads 800 psi:

$1/2 \times 24 \times 800 \div 1/4 = 38\,400$ psi

What wall thickness should be specified for a 30-in. discharge line where pressure runs 1,000 psi and wall stress is to be limited to 30,000 psi?

$1/2 \times 30 \times 1{,}000 \div \text{thickness} = 30{,}000$

$\text{Wall thickness} = 1/2 \text{ in.}$

This method of calculating transverse tensile stress due to internal pressure on the pipe is the Barlow formula. Actually, there are two accepted variations, and some designers use inside diameter of the pipe in calculating this stress. This gives a slightly lower figure for the transverse tensile stress and consequently gives less safety factor. The method used here is based on the assumption that maximum stress occurs at the outside diameter of the pipe. The results give a conservative value for safety factor calculations.

How to calculate stress in aboveground/belowground transitions

Design of gas gathering/gas reinjection project in southern Iran provides solutions for deflection and anchor block forces in long underground lines

P. J. Schnackenberg, Design Engineer, Iran-Texas Engineering Co., Tehran, Iran

Stresses and deflections occur in pipelines at the transition from the belowground (fully restrained) to the aboveground (unrestrained) condition.

Analysis of the stresses and deflections in transition areas, resulting from internal pressure/temperature change, is necessary in determining anchor block requirements and design. Longitudinal deflections are used to determine whether an anchor block is required. Anchor block forces required to maintain the pipe in a fully constrained condition are then determined.

Iran-Texas Engineering Co. recently completed an analysis for a gas gathering/compression/reinjection project in southern Iran under contract to Oil Service Co. of Iran (OSCO). Each of the two fields, Bibi-Hakimeh and Rag-E-Safid have two gas/oil separation plants, and at present the gas is being flared. The broad scope of the project is as follows:

- Provision of gas gathering lines from existing and new wellhead separation to a production unit
- Design of four production units (each adjacent to the GOSP) comprising knockout facilities, compressor stations, and dehydration plant
- Provision of transfer pipelines to send gas plus condensate as a two-phase flow to a natural gas liquid separation plant (under design by Foster Wheeler)
- Provision of reinjection system to inject lean gas back into a number of wells
- Some 9.4 MMm^3/d (330 MMscfd) of gas plus 15,000 bpd condensate will be going to the NGL plant. LPG will be used as feedstock and for export.

Following is a brief review of the analysis that resulted in more accurate solutions for deflection and anchor block forces. Sample calculations are for line sizes up to 41-cm (16-in.) CD, pressure to 193 bars (2,800 psig), and temperatures to 72°C (162°F).

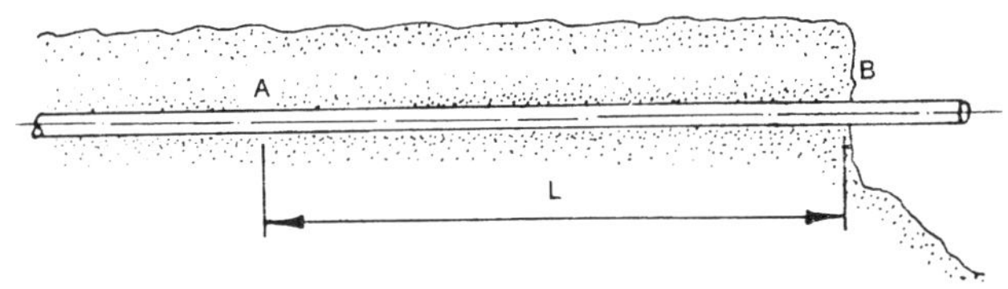

Figure 1. Area A–B depicts transition from fully restrained (left) to fully unrestrained (right).

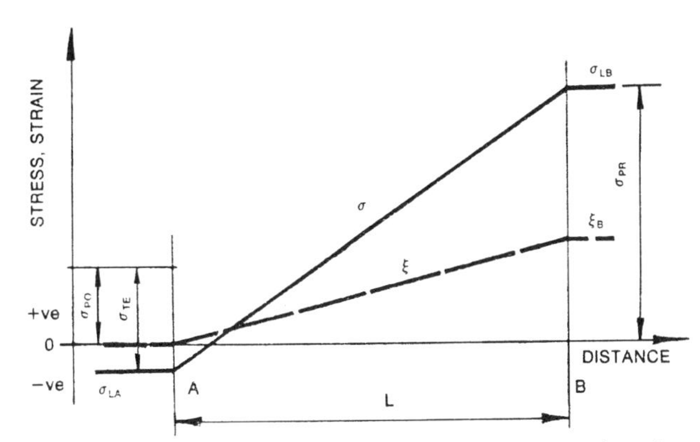

Figure 2. Transition of stress and strain between points A and B varies as linear function of length.

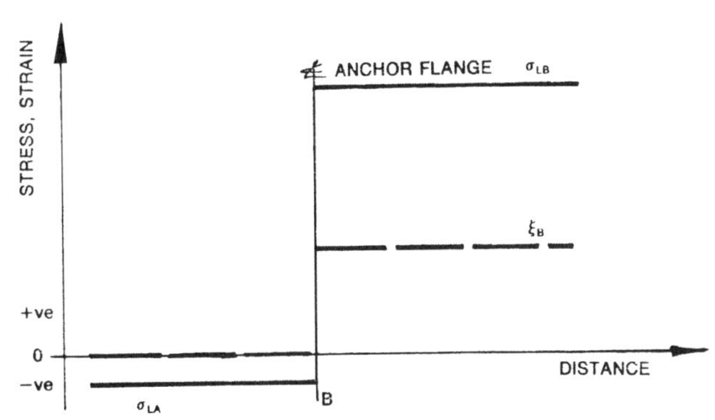

Figure 3. Distribution of stress where anchor is required to contain longitudinal deflections.

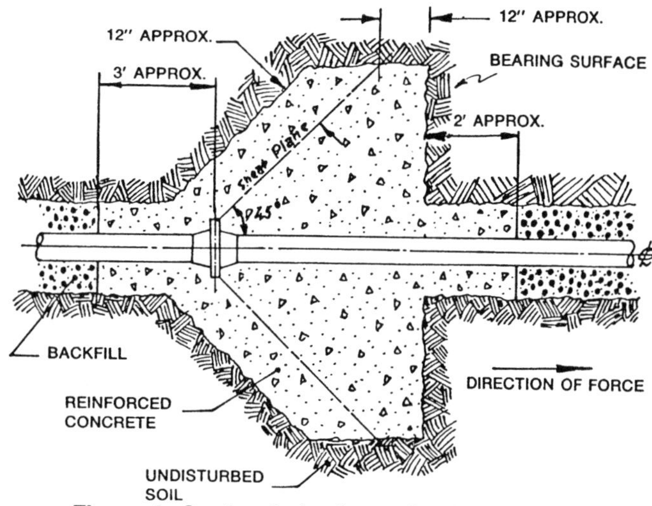

Figure 4. Sectional plan for anchor block design.

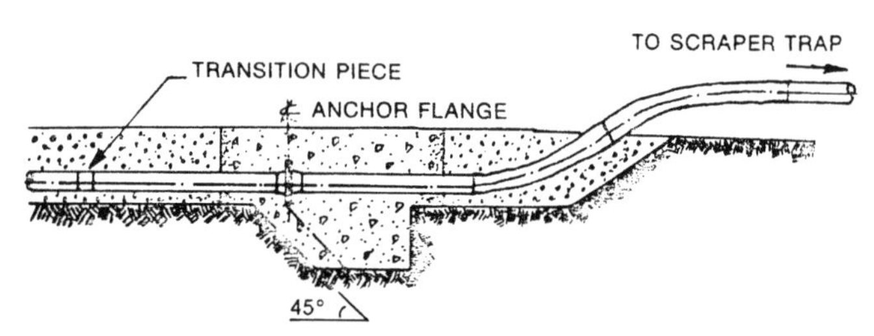

Figure 5. Sectional elevation for anchor block design.

Case I—no anchor. Considering the pipeline as shown in Figure 1, the section up to point A is fully restrained, the section A–B is a transition from fully restrained to unrestrained, and beyond point B is unrestrained. (Fully restrained is taken to be a condition of zero longitudinal strain.)

Analyzing a short section of fully restrained pipe, subject to internal pressure and a temperature change, there is a tensile stress due to a Poisson effect of

$$\sigma_{PO} = \upsilon\sigma \text{ where } \sigma_H = \frac{PDi}{2t}$$

and a compressive stress due to a temperature change of

$$\sigma_{TE} = E\alpha\Delta T$$

The net longitudinal stress, therefore, at Point A will be

$$\sigma_{LA} = \upsilon\sigma_H - E\alpha\Delta T \text{psi} \tag{1}$$

and the strain at point A, being fully restrained, will be zero

$$\varepsilon_A = 0 \tag{2}$$

In a section of unrestrained pipe at B, the longitudinal stress caused by only internal pressure will be half the hoop stress.

$$\sigma_{LB} = \frac{\sigma_H}{2} \text{psi} \tag{3}$$

The associated strain will be made up of a

(i) Temperature effect

$$\varepsilon_{TE} = \alpha\Delta T$$

(ii) Pressure effect

$$\varepsilon_{PR} = \frac{\sigma_{LB}}{E} = \frac{\sigma_H}{2E}$$

(iii) Poisson effect

$$\varepsilon_{PO} = \frac{\upsilon\sigma_H}{E}$$

The net longitudinal strain at Point B will therefore be

$$\varepsilon_B = \alpha\Delta T + \frac{\sigma_H}{E}(1/2 - \upsilon)\text{in./in.} \tag{4}$$

The transition of stress and strain between points A and B is assumed to vary as a linear function of length as shown in Figure 2.

In order to establish the length L over which the transition occurs, the longitudinal resistance of the soil needs to be known. For this part, it is assumed that any tendency to move will be counteracted by constant and opposite soil force. Wilbur[1] has recommended a design value for average soils of

$$F_s = 80\left(\frac{D_o}{12}\right)^2 \text{lbf/ft} \tag{5}$$

Between A and B equilibrium of forces exists and therefore,

$$F_s L = A_m(\sigma_{LB} - \sigma_{LA})$$

From which

$$L = \frac{A_m(\sigma_{LB} - \sigma_{LA})}{F_s} \text{ft} \tag{6}$$

Total movement at B will be average strain between A and B over length L or

$$\delta = \frac{\varepsilon_B}{2}(12L)\text{ in.} \tag{7}$$

Cell II—with anchor. Where an anchor is required to contain longitudinal deflections, the stress distribution will be as shown in Figure 3. The transition from being

Table 1
Summary of results for high pressure gas transmission lines under design in Iran

No.	Pipe (All API-5LX-60)	P psig	T_O °F	F_S lb./ft.	σ_H psi	σ_{LA} psi	σ_{LB} psi	L ft.	δ in.	F lb.
1	16″ OD × 0.312″ w.t.	1150	162	142.2	28,334	−6957	14,167	2285	9.81	324,890
2	20″ OD × 0.344″ w.t.	1000	162	222.2	28,070	−7036	14,035	2014	8.62	447,550
3	16″ OD × 0.281″ w.t.	1000	162	142.2	27,470	−7216	13,735	2045	8.71	290,800
4	16″ OD × 0.250″ w.t.	880	162	142.2	27,280	−7273	13,640	1819	7.73	258,690
5	16″ OD × 0.688″ w.t.	2800	150	142.2	29,758	−4268	14,879	4457	17.66	633,770
6	10″ NB × 0.500″ w.t.	2800	150	64.2	28,700	−4585	14,350	4728	18.60	304,850
7	6″ NB × 0.432″ w.t.	2800	150	24.4	18,670	−7594	9,335	5828	20.41	142,200

fully restrained to unrestrained will occur at the anchor. Resultant force on the anchor will simply be the difference in stress on each side times the pipe metal area (equilibrium of forces) or

$$F = (\sigma_{LB} - \sigma_{LA})A_m lb \quad (8)$$

Note that for the case of an increase in wall thickness at and beyond the anchor block, the result will be essentially the same—the decrease in stress being compensated by an increase in pipe metal area.

It can be shown that this force is equal to that to produce a deflection of 2δ in Equation 7 over length L.

Table 1 is a summary of results for a number of high pressure gas transmission lines currently under design in Iran. Lines 1 through 4 are transfer lines from a number of production units to an NGL separation plant. Lines 5 through 7 are injection lines from the NGL separation plant to a number of wellheads. The following values were used for the various parameters:

Young's modulus E 29×10^6, psi

Poisson's ratio υ, 0.3

Thermal expansion coefficient 6.5×10^{-6}, in./in.°F

Installation temperature, T_i, 80°F (minimum summer temperature when construction is in progress)

Negative stress values denote a compressive stress.

Discussion

As shown in Table 1, anchor block forces can be quite large; therefore, careful attention must be paid to the design. In order to minimize the size of the block, upper design limit values of lateral soil bearing pressure with due consideration to possible slip have been used.

For large blocks, friction between the block and the soil may also be figured in obtaining its resistance to imposed forces. The block should be reinforced concrete cast against undisturbed soil. Figures 4 and 5 indicate in broad detail the anchor block design.

Care should also be taken to ensure that connected surface piping has sufficient flexibility to absorb a degree of lateral anchor movement. Scraper traps should be installed such that they can move with the piping rather than being rigidly attached to blocks. Numerous examples of traps together with their support blocks being displaced a few inches are not uncommon.

References

1. Wilbur, W. E., *Pipe Line Industry*, February 1963.
2. Timoshenko, S., Theory of Elasticity.

Nomenclature

A_m	Pipe metal area $= \pi(D_o = t)t$, sq. in.
D_o	Pipe outside diameter, in.
D_i	Pipe inside diameter, in.
t	Pipe wall thickness, in.
E	Young's modulus of elasticity, psi
F	Anchor force, lb
F_s	Soil resistance, lb/ft
L	Transition length, ft
P	Design pressure, psig
T_i	Installation temperature, °F
T_o	Operating temperature, °F
ΔT	Temperature difference = (To − Ti), °F
α	Coefficient of thermal expansion, in./in.-°F
σ	Stress, psi
ε	Strain, in./in.
υ	Poisson's ratio
δ	Deflection, in.

Subscripts (for Greek letters only)

L Longitudinal

H Hoop

A Point A, fully restrained

B Point B, unrestrained

PR Pressure

TE Temperature

PO Poisson

How to identify the series number of flanged fittings

Divide the flange thickness, expressed in sixteenths of an inch, by the nominal pipe size + 12. If the answer is less than 1.00, the flange is Series 15; if between 1.15 and 1.33, it is Series 30; if between 1.37 and 1.54, it is Series 40; and if it is between 1.60 and 1.85, it is Series 60.

Example. A 6-in. flange has a thickness of $1^7/_{16}$ inch $= {}^{23}/_{16}$ in.

$${}^{23}/_{18} = 1.28\text{; the flange is series 30.}$$

The rule is not applicable to fittings below 4 in. in size; it also fails for a 4-in. Series 60, which comes out 1.50.

Dimensions of three-diameter ells with tangents

Here are face-to-face and center-to-face dimensions of the three-diameter radius weld ells being used in natural gas compressor station piping.

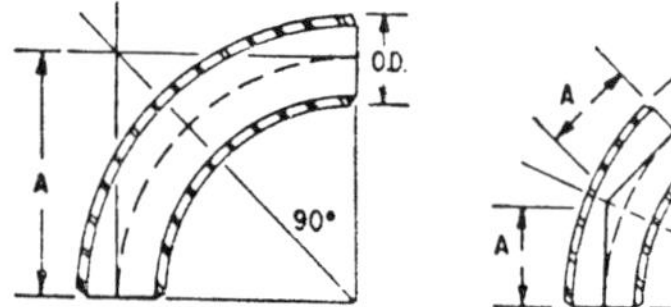

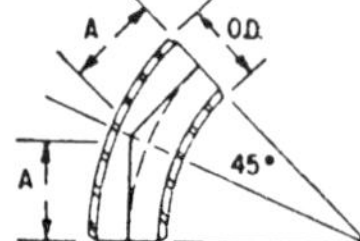

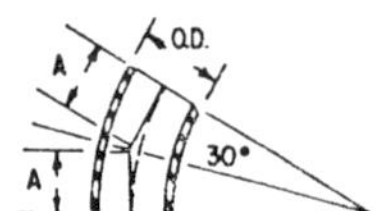

Degree Ell	Outside Diameter	A (Center To Face)	A With 4" Tan	A With 6" Tan
30°	16"	1'-0 7/8"	1'- 4 7/8"	1'- 6 7/8"
30°	20"	1'-4 1/16"	1'- 8 1/16"	1'-10 1/16"
30°	24"	1'-7 5/16"	1'-11 5/16"	2'- 1 5/16"
30°	26"	1'-8 7/8"	2'- 0 7/8"	2'- 2 7/8"
30°	30"	2'-0 1/8"	2'- 4 1/8"	2'- 6 1/8"
45°	16"	1'-7 7/8"	1'-11 7/8"	2'- 1 7/8"
45°	20"	2'-0 7/8"	2'- 4 7/8"	2'- 6 7/8"
45°	24"	2'-5 13/16"	2'- 9 13/16"	2'-11 13/16"
45°	26"	2'-8 5/16"	3'- 0 5/16"	3'- 2 5/16"
45°	30"	3'-1 1/4"	3'- 5 1/4"	3'- 7 1/4"
90°	16"	4'-0"	4'- 4"	4'- 6"
90°	20"	5'-0"	5'- 4"	5'- 6"
90°	24"	6'-0"	6'- 4"	6'- 6"
90°	26"	6'-6"	6'-10"	7'- 0"
90°	30"	7'-6"	7'-10"	8'- 0"

Spectacle blind thicknesses

The thickness of blinds is calculated using Formula 16, page 23, section 304.5.3 from Petroleum Refinery Piping ANSI B31.3-1973, sponsored by the American Society of Mechanical Engineers:

$$t = d\sqrt{\frac{3P}{16SE}} + ca$$

where:

t = Thickness, inches
D = Pipe inside diameter, inches
P = Maximum pressure blind is to contain, psig
SE = SMYS of blind material
ca = Corrision allowance

Example:

Pipe OD	8.625 in.
d	7.981 in.
P	1,440 psig
SE	42,000 psi
ca	0.125 in.

$$t = 7.981\sqrt{\frac{3 \times 1,440}{16 \times 42000}} + 0.125$$

$$t = 0.7649 \text{ in.}$$

Polypipe design data

Internal pressure

Polyethylene pipe for industrial, municipal, and mining applications is manufactured to specific dimensions as prescribed by applicable American Society for Testing and Materials (ASTM) standards. Piping outside diameters meet the IPS system. Wall thicknesses are based on the Standard Dimension Ratio (SDR) system, a specific ratio of the outside diameter to the minimum specified wall thickness. Selected SDR numbers are from the ANSI Preferred Number Series. Use of the SDR number in the ISO equation, recognized as an equation depicting the relationship of pipe dimensions'—both wall and o.d.—internal pressure carrying capabilities and tensile stress, in conjunction with a suitable safety factor (F), will give the design engineer confidence that the pipe will not fail prematurely due to internal pressurization.

A pressure differential is required to move material through a pipeline. For atmospheric systems (gravity flow), gravitational forces provide the impetus for movement of heavier-than-air mass. Internal pressure requirements must be recognized in the design stage in order to provide desired operational life. In some cases a gravity flow system must be treated comparable to the design considerations of a pressure flow system.

Polyethylene pipe under pressure is time/temperature/pressure dependent. As the temperature increases and the pressure remains constant, the creep rupture strength decreases. Conversely, as the temperature decreases, the strength increases. Varying these parameters, in conjunction with the SDR, can provide the desired service within the capabilities of polyethylene pipe.

Calculations for determining the pressure rating of polyethylene pipe is based on the ISO equation as follows:

$$P = \frac{2S}{SDR - 1} \times F$$

where:

P = Internal pressure, psi
S = Long-term hydrostatic strength, psi (1,600 psi for polyethylene pipe)
SDR = Standard dimension ratio, (D/t)
D = Outside diameter, inches
t = Minimum wall thickness, inches
F = Design safety factor (0.5 for water @ 73°F)

Use of other factors determined from Figures 1, 2, and 3 will provide for a more accurate performance characteristic for systems operating at other temperatures for shorter periods of time.

The ISO equation now becomes:

$$P = \frac{2S}{SDR - 1} \times F \times F_1 \times F_2 \times F_3$$

F_1 = Operational life factor
F_2 = Temperature correction
F_3 = Environmental service factor (Figure 3)

Other design factors may be necessary in the design of systems where abrasion, corrosion, or chemical attack may be a factor.

Figure 1. Internal pressure ratings (psi) PE pipe at 73°F for 50 years

SDR	7.3	9	11	13.5	15.5	17	21	26	32.5
Pressure Rating	250	200	160	125	110	100	80	64	50

Figure 2. Operational life

To use Figure 2 for determining operational life factor multiplier F_1, select expected operating life (less than 50 years) on the X axis and move upwards to intersect the diagonal line and then left to reading on Y axis.
Example. 5-year operational life will yield F_1 = 1.094

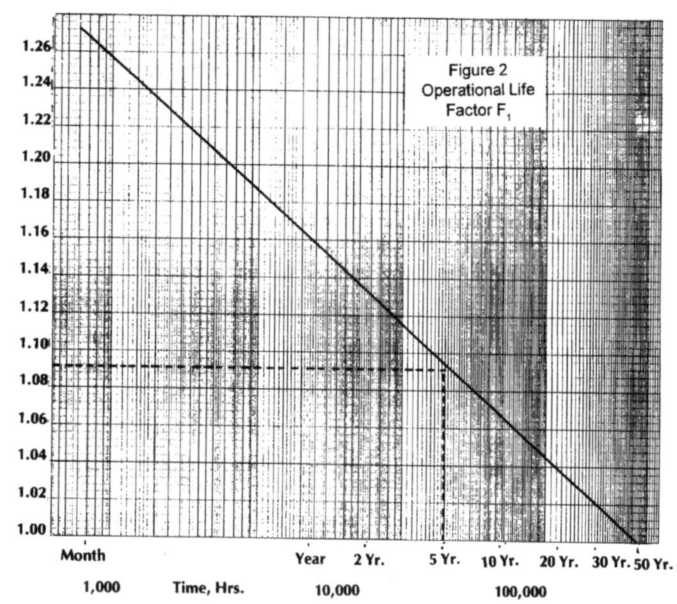

Figure 3. Environmental service factor F_3

Substance	Service Factor F_3
Crude Oil	0.5
Wet Natural Gas	0.5
Dry Natural Gas	0.64

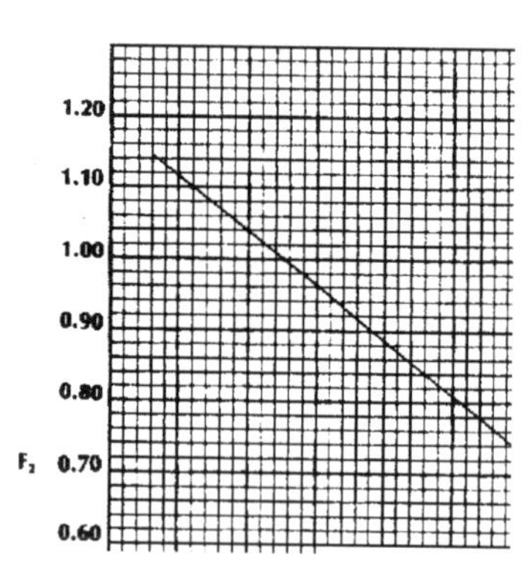

Figure 4. Temperature correction
Example: A polyethylene pipe operating at 125°F would have an F_2 factor of 0.62

Figure 5. Recommended liquid velocities

	Pipe size, inches		
	3–10	10–28	28–48
Water and similar viscosity fluids			
	Velocity, fps		
Pump suction*	1 to 4	4 to 6	5 to 8
Pump discharge	3 to 5	4 to 8	6 to 12
Gravity drain system	3 to 5	4 to 8	6 to 10
Viscous liquids			
Pump suction*	0.5 to 2	1 to 4	2 to 5
Pump discharge	3 to 5	4 to 6	5 to 7

*It is important in the selection of the pipe size that the designated pump suction velocity is lower than the discharge velocity.

Figure 6. ID to OD Ratio

SDR	F	SDR	F
7	0.706	13.5	0.852
7.3	0.719	15.5	0.873
9	0.774	17	0.884
9.3	0.781	21	0.909
11	0.817	26	0.928
		32.5	0.944

Example. A horizontal process water line 2,000 ft in length, 125°F water, 60 psig pressure, with a desired life of 5 years is proposed. Determine the standard pipe SDR for the application

$$60 = \frac{2 \times 1600}{SDR - 1} \times F \times F_1 \times F_2 \times F_3$$

where:

F = design safety factor (use 0.5 for water @ 73°F)
F_1 = operational life factor
F_2 = temperature correction
F_3 = environmental service factor (Figure 3)

Therefore, use SDR = 17.
Note: Using SDR 17 is conservative.

Pipe diameter

Volumetric flow, Q, can be determined from the continuity equation $Q = A \times V$ by selecting a suitable velocity from Figure 5 for a given pipe size. When modified for flow in gallons per minute, the equation becomes:

$$Q = 2.449 \times V \times d^2$$

where:

Q = flow rate gpm
V = velocity, in feet per second
d = inside diameter, inches

or $d = (Q \div 2.449 \times V)^{1/2}$ for pipe size determination.

Example. If the required water flow rate for a system is 2,000 gpm and the flow velocity is to be maintained below 8 fps, the pipe diameter can be found as follows:

$$d = (Q \div 2.449 \times V)^{1/2}$$

$$d = (2000 \div 2.449 \times 8)^{1/2}$$

d = 10.1 in. inside diameter
OD = ID ÷ F(F from Figure 6)

Pressure drop

Refer to Section 13: Liquids—Hydraulics and use the Hazen-Williams equation. Use a C of 150.

Fitting equivalent length

The exact pressure loss through fittings cannot readily be calculated because of the geometry of the fitting. The values shown in Figure 7 are the result of general industry consensus.

Figure 7. Equivalent Length of Fittings

Fitting type	Equivalent length of pipe
90 deg. elbow	30 D
60 deg. elbow	25 D
45 deg. elbow	16 D
45 deg. wye	60 D
Running tee	20 D
Branch tee	50 D
Gate valve, full open	8 D
Butterfly valve (3″ to 14″)	40 D
Butterfly valve ≥ 14″	30 D
Swing check valve	100 D
Ball valve full bore full open	3 D

Example. A running tee has an equivalent length of 20 D. For an 18-in. SDR 11 line, this represents 20 × ID. From Figure 6, ID = OD × F = 18 × 0.817 = 14.7″ × 20 = 294 ÷ 12 and is equal to 24.5 ft. In calculating the total pressure drop in the system, 24.5 ft of 18-in. pipe should be added to the total line length to account for the increased pressure loss due to the presence of the tee.

Reuse of polyethylene pipe in liquid hydrocarbon service

When polyethylene pipe has been permeated with liquid hydrocarbons, joining should be joined by use of mechanical connections. Joining by means of fusion bonding may produce a low-strength joint.

Gravity or nonpressure service above 180°F is not recommended.

Pressure service above 140°F is not recommended.

4: Electrical Design

Electrical design

Most pipeline facilities use electrical equipment of some sort or another, ranging from simple power circuits of a few amperes capacity to sophisticated supervisory control and data acquisition systems. It is beyond the scope of this manual to provide a complete section on electrical information for pipeline facilities. This section is intended to provide some basic data that will prove useful to field personnel responsible for electrical installations.

Electric motor selection

Based on known facts and calculations, the best selection is made after a close study of the installation, operation, and servicing of the motor. Basic steps in proper selection are numbered 1 through 8 and are briefly described.

1. Power supply

(a) Voltage-NEMA has recommended the following standards.

Nominal Power System Volts	Motor Nameplate Volts
240	230
480	460
600	575

(b) Frequency—Motors rated 200 horsepower or less can vary not to exceed 5% above or below its rated frequency.

(c) Phases—Three-phase power supplies are found in most industrial locations; for most residential and rural areas, only single-phase power is available.

2. HP and duty requirements

(a) Continuous duty—means constant load for an indefinite period (about 90% of all motor applications).

(b) Intermittent duty—means alternate periods of load and no-load, or load and rest.

(c) Varying duty—means both the load and time operation vary to a wide degree.

3. Speeds—Single speed motors are the most common, but when a range of speeds is required multispeed motors will give two, three, or four fixed speeds.

4. Service factors—Open, general purpose motors have a service factor depending upon the particular rating of the motor (usually between 1.15 and 1.25).

5. Selection of motor type

There are three main types:

(a) DC motors are designed for industrial drives requiring a controlled speed range and constant torque output on adjustable voltage control systems. Because the speed of rotation controls the flow of current in the armature, special devices must be used for starting DC motors.

(b) Single phase alternating-current motors also require some auxiliary arrangement to start rotation.

(c) Polyphase motors are alternating-current types (squirrel cage or wound rotor).

Applications—Some driven machines require a low-starting torque that gradually increases to full-load speed; others require higher-starting torque. NEMA code letters A, B, C, D, etc., on the motor nameplate designate the locked rotor kVa per horsepower of that particular motor design. The diagram shows representative speed-torque curves for polyphase and single-phase NEMA design motors Types A through D.

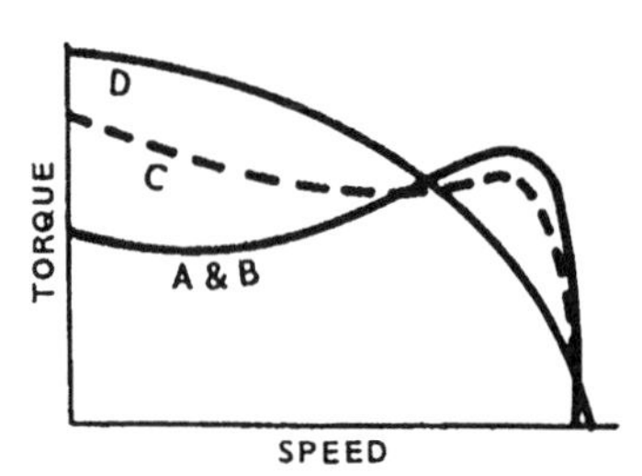

6. Torque definitions

(a) Locked rotor torque, or "starting torque," is the minimum-torque that the motor will develop at rest for all angular positions of the rotor.

Code Ltr	Locked Rotor KVa/hP
A	0–3.14
B	3.15–3.54
C	3.55–3.99
D	4.0–4.49
E	4.5–4.99
F	5–5.59

(b) Breakdown torque is the maximum-torque at rated voltage and frequency without abrupt speed drop.

(c) Full-load-torque is the torque necessary to produce rated horsepower at full-load speed.

(d) Locked-rotor current is the steady-state current of the motor with the rotor locked at rated voltage and frequency.

7. Selection of enclosure—The two general classifications are open, which permits passage of air over and around the windings, and totally enclosed, which prevents exchange of air between inside and outside of frame (but is not strictly airtight).

(a) Open Drip-Proof—means that liquid or solid particles falling on the motor at an angle not greater than 15 degrees from vertical cannot enter the motor.

(b) NEMA Type I—a weather-protected machine with its ventilating passages constructed to minimize the entrance of rain, snow, and airborne particles to the electric parts and having its ventilating openings so constructed to prevent the passage of a cylindrical rod ¾ in. in diameter.

(c) NEMA Type II—a weather-protected machine that has in addition to the enclosure defined for a Type I weather-protected machine, its ventilating passages at both intake and discharge so arranged that high velocity air and airborne particles blown into the machine by storms or high winds can be discharged without entering the internal ventilating passages leading directly to the electric parts of the machine itself. The normal path of the ventilating air that enters the electric parts of the machine shall be so arranged by baffling or separate housings as to provide at least three abrupt changes in direction, none of which shall be less than 90 degrees. In addition, an area of low velocity, not exceeding 600 ft/min., shall be provided in the intake air path to minimize the possibility of moisture or dirt being carried into the electric parts of the machine.

(b) Totally Enclosed Non-Vent—means that a motor is not equipped for cooling by external means.

(c) Totally Enclosed Fan-Cooled—means the motor has a fan integral with the motor but not external to the enclosed parts.

(d) Explosion-Proof—means the enclosure is designed to withstand an explosion of a specified gas that may occur within the motor and to prevent the ignition of gas around the motor.

8. End shield mountings—Three types of end shields with rabbets and bolt holes for mounting are standard in the industry:

(a) Type C Face provides a male rabbet and tapped holes.

(b) Type D Flange has a male rabbet with holes for through bolts in the flange.

(c) Type P base has a female rabbet and through bolts for mounting in the flange (used for mounting vertical motors).

Hazardous locations

Definition of Class Locations

Reprinted with permission from NFPA 70-1990, the *National Electrical Code®*, Copyright© 1989, National Fire Protection Association, Quincy, MA 02269. This reprinted material is not the complete and official position of the National Fire Protection Association, on the referenced subject which is represented only by the standard in its entirety.

The *National Electrical Code* and *NEC* are registered trademarks of the NFPA.

500-5. Class I Locations. Class I locations are those in which flammable gases or vapors are or may be present in the air in quantities sufficient to produce explosive or ignitible mixtures. Class I locations shall include those specified in (a) and (b) below.

(a) **Class I, Division 1**. A Class I, Division 1 location is a location: (1) in which ignitible concentrations of flammable gases or vapors can exist under normal operating conditions; or (2) in which ignitible concentrations of such gases or vapors may exist frequently because of repair or maintenance operations or because of leakage; or (3) in which breakdown or faulty operation of equipment or processes might release ignitible concentrations of flammable gases or vapors and might also cause simultaneous failure of electric equipment.

(FPN): This classification usually includes locations where volatile flammable liquids or liquefied flammable gases are transferred from one container to another; interiors of spray booths and areas in the vicinity of spraying and painting operations where volatile flammable solvents are used; locations containing open tanks or vats of volatile flammable liquids; drying rooms or compartments for the evaporation of flammable solvents; locations containing fat and oil extraction equipment using volatile flammable solvents; portions of cleaning and dyeing plants where flammable liquids are used; gas generator rooms and other portions of gas manufacturing plants where flammable gas may escape; inadequately ventilated pump rooms for flammable gas or for volatile flammable liquids; the interiors of refrigerators and freezers in which volatile flammable materials are stored in open, lightly stoppered, or easily ruptured containers; and all other locations where ignitible concentrations of flammable vapors or gases are likely to occur in the course of normal operations.

(b) **Class I, Division 2**. A Class I, Division 2 location is a location: (1) in which volatile flammable liquids or flammable gases are handled, processed, or used, but in which the liquids, vapors, or gases will normally be confined within closed containers or closed systems from which they can escape only in case of accidental rupture or breakdown of such containers or systems, or in the case of abnormal operation of equipment; or (2) in which ignitible concentrations of gases or vapors are normally prevented by positive mechanical ventilation and which might become hazardous through failure or abnormal operation of the ventilating equipment; or (3) that is adjacent to a Class I, Division 1 location, and to which ignitible concentrations of gases or vapors might occasionally be communicated unless such communication is prevented by adequate positive-pressure ventilation from a source of clean air, and effective safeguards against ventilation failure are provided.

(FPN No. 1): This classification usually includes locations where volatile flammable liquids or flammable gases or vapors are used, but which, in the judgment of the authority having jurisdiction, would become hazardous only in case of an accident or of some unusual operating condition. The quantity of flammable material that might escape in case of accident, the adequacy of ventilating equipment, the total area involved, and the record of the industry or business with respect to explosions or fires are all factors that merit consideration in determining the classification and extent of each location.

(FPN No. 2): Piping without valves, checks, meters, and similar devices would not ordinarily introduce a hazardous condition even though used for flammable liquids or gases. Locations used for the storage of flammable liquid or of liquefied or compressed gases in sealed containers would not normally be considered hazardous unless subject to other hazardous conditions also.

Group A: Atmospheres containing acetylene.

Group B: Atmospheres containing hydrogen, fuel, and combustible process gases containing more than 30% hydrogen by volume or gases or vapors of equivalent hazard such as butadiene,* ethylene oxide,** propylene oxide,** and acrolein.**

Group C: Atmospheres such as cyclopropane, ethyl ether, ethylene, or gases or vapors of equivalent hazard.

Group D: Atmospheres such as acetone, ammonia,*** benzene, butane, ethanol, gasoline, hexane, methanol, methane, natural gas, naptha, propane, or gases of vapors of equivalent hazard.

*Group D equipment may he used for this atmosphere if such equipment is isolated in accordance with Section 501-5(a) by sealing all conduits 1/2-in. or larger.

**Group C equipment may be used for this atmosphere if such equipment is isolated in accordance with Section 501-5(a) by sealing all conduit 1/2-in. or larger.

***For classification of areas involving ammonia atmosphere, see Safety Code for Mechanical Refrigeration, ANSI/ASHRAE 15-1978, and Safety Requirements for the Storage and Handling of Anhydrous Ammonia, ANSI/CGA G2.1-1981.

NEMA enclosure types[1]

Nonhazardous Locations

Type	Intended Use and Description
1	Enclosures are intended for indoor use primarily to provide a degree of protection against limited amounts of falling dirt.
2	Enclosures are intended for indoor use primarily to provide a degree of protection against limited amounts of falling water and dirt.
3	Enclosures are intended for outdoor use primarily to provide a degree of protection against rain, sleet, windblown dust, and damage from external ice formation.
3R	Enclosures are intended for outdoor use primarily to provide a degree of protection against rain, sleet, and damage from external ice formation.
3S	Enclosures are intended for outdoor use primarily to provide a degree of protection against rain, sleet, windblown dust and to provide for operation for external mechanisms when ice laden.
4	Enclosures are intended for indoor or outdoor use primarily to provide a degree of protection against windblown dust and rain, splashing water, hose-directed water, and damage from external ice formation.
4X	Enclosures are intended for indoor or outdoor use primarily to provide a degree of protection against

corrosion, windblown dust and rain, splashing water, hose-directed water, and damage from external ice formation.

5	Enclosures are intended for indoor use primarily to provide a degree of protection against settling airborne dust, falling dirt, and dripping noncorrosive liquids.
6	Enclosures are intended for indoor or outdoor use primarily to provide a degree of protection against hose-directed water, the entry of water during occasional temporary submersion at a limited depth, and damage from external ice formation.
6P	Enclosures are intended for indoor or outdoor use primarily to provide a degree of protection against hose-directed water, the entry of water during prolonged submersion at a limited depth, and damage from external ice formation.
12	Enclosures are intended for indoor use primarily to provide a degree of protection against circulating dust, falling dirt, and dripping noncorrosive liquids.
12K	Enclosures with knockouts are intended for indoor use primarily to provide a degree of protection against circulating dust, falling dirt, and dripping noncorrosive liquids.
13	Enclosures are intended for indoor use primarily to provide a degree of protection against dust, spraying of water, oil, and noncorrosive coolant.

Hazardous (classified locations)

7	Enclosures are intended for indoor use in locations classified as Class I, Groups A, B, C, or D, as defined in the National Electrical Code.
8	Enclosures are intended for indoor use in locations classified as Class I, Groups A, B, C, or D, as defined in the National Electrical Code.
9	Enclosures are intended for indoor use in locations classified as Class II, Groups E, F, or G, as defined in the National Electrical Code.
10	Enclosures are constructed to meet the applicable requirements of the Mine Safety and Health Administration.

Reference NEMA Standards Publications/No. 250-1991

Size portable electric generators

A power plant must be properly sized for the required load. Electric motors are particularly difficult for portable plants, because they require 2 to 3 times motor nameplate amps or wattage.

A portable power plant's surge capacity is limited by engine horsepower and inertia of its rotating parts. Thus, a current surge of short duration can be supplied by a power plant, but a current demand of longer duration such as a heavily loaded motor starting a high inertia load, can overload the power plant and possibly damage the power plant and the motor. A 3,450 rpm air compressor is a prime example of this type of load.

Example:

Load Requirements:

Electric Heater	1,000 watts
11–100 watt lamps	1,100 watts
Motor	600 watts
Total motor load = 600 × 3	= 1,800 watts
Total other load	= 2,100 watts
Total load	= 3,100 watts
Add 25% for future	= 975 watts
Total load for sizing power plant	= 4,875 watts

A generator capable of supplying 5,000 watts continuously will be adequate.

When more than one motor is connected to a power plant, always start the largest motor first. If the total load is connected to one receptacle on the power plant, be sure the ampere rating of the receptacle is not exceeded.

Typical wattages for tools and appliances

Equipment	Running Watts
1/2″ drill	1,000
1″ drill	1,100
6″ circular saw	800
10″ circular saw	2,000
14″ chain saw	1,100
Radio	50 to 200
Television	200 to 500
Toaster, coffeemaker	1,200
Water heater (storage type)	1,100 to 5,500
Hot plate or range (per burner)	1,000
Range oven	10,000
Skillet	1,200
Fan	50 to 200
Floodlight	1,000
Water pump	500
Vacuum cleaner	200 to 300
Refrigerator (conventional)	200 to 300
Refrigerator (freezer-combination)	250 to 600
Furnace fan blower	500 to 700

Knockout dimensions

Conduit size, Inches	Knockout Diameter, Inches		
	Min.	Nominal	Max.
1/2	0.859	0.875	0.906
3/4	1.094	1.109	1.141
1	1.359	1.375	1.406
1 1/4	1.719	1.734	1.766
1 1/2	1.958	1.984	2.016
2	2.433	2.469	2.500
2 1/2	2.938	2.969	3.000
3	3.563	3.594	3.625
3 1/2	4.063	4.125	4.156
4	4.563	4.641	4.672
5	5.625	5.719	5.750
6	6.700	6.813	6.844

Reference NEMA Standards Publications/No. 250-1991.
Used with permission—National Electrical Manufacturer's Association.

National Electrical Code tables

Table 430–150
Full-load current* three-phase alternating-current motors

HP	Induction Type Squirrel-Cage and Wound-Rotor Amperes							Synchronous Type †Unity Power Factor Amperes			
	115V	200V	208V	230V	460V	575V	2300V	230V	460V	575V	2300V
½	4	2.3	2.2	2	1	.8					
¾	5.6	3.2	3.1	2.8	1.4	1.1					
1	7.2	4.1	4.0	3.6	1.8	1.4					
1½	10.4	6.0	5.7	5.2	2.6	2.1					
2	13.6	7.8	7.5	6.8	3.4	2.7					
3		11.0	10.6	9.6	4.8	3.9					
5		17.5	16.7	15.2	7.6	6.1					
7½		25.3	24.2	22	11	9					
10		32.2	30.8	28	14	11					
15		48.3	46.2	42	21	17					
20		62.1	59.4	54	27	22					
25		78.2	74.8	68	34	27		53	26	21	
30		92	88	80	40	32		63	32	26	
40		119.6	114.4	104	52	41		83	41	33	
50		149.5	143.0	130	65	52		104	52	42	
60		177.1	169.4	154	77	62	16	123	61	49	12
75		220.8	211.2	192	96	77	20	155	78	62	15
100		285.2	272.8	248	124	99	26	202	101	81	20
125		358.8	343.2	312	156	125	31	253	126	101	25
150		414	396.0	360	180	144	37	302	151	121	30
200		552	528.0	480	240	192	49	400	201	161	40

*These values of full-load current are for motors running at speeds usual for belted motors and motors with normal torque characteristics. Motors built for especially low speeds or high torques may require more running current, and multispeed motors will have full-load current varying with speed, in which case the nameplate current rating shall be used.
†For 90% and 80% power factor the above figures shall be multipled by 1.1% and 1.25% respectively.
The voltages listed are rated motor voltages. The currents listed shall be permitted for system voltage ranges of 110 to 120, 220 to 240, 440 to 480, and 550 to 600 volts.

Table 310.16 Allowable ampacities of insulated conductors rated 0 through 2000 volts, 60°C through 90°C (140°F through 194°F), not more than three current-carrying conductors in raceway, cable, or earth (direct buried), based on ambient temperature of 30°C (86°F)

Size AWG or kcmil	Temperature Rating of Conductor (See Table 310.13.)						Size AWG or kcmil
	60°C (140°F)	75°C (167°F)	90°C (194°F)	60°C (140°F)	75°C (167°F)	90°C (194°F)	
	Types TW, UF	Types RHW, THHW, THW, THWN, XHHW, USE, ZW	Types TBS, SA, SIS, FEP, FEPB, MI, RHH, RHW-2, THHN, THHW, THW-2, THWN-2, USE-2, XHH, XHHW, XHHW-2, ZW-2	Types TW, UF	Types RHW, THHW, THW, THWN, XHHW, USE	Types TBS, SA, SIS, THHN, THHW, THW-2, THWN-2, RHH, RHW-2, USE-2, XHH, XHHW, XHHW-2, ZW-2	
	Copper			Aluminium or Copper-Clad Aluminium			
18	—	—	14	—	—	—	—
16	—	—	18	—	—	—	—
14*	20	20	25	—	—	—	—
12*	25	25	30	20	20	25	12*
10*	30	35	40	25	30	35	10*
8	40	50	55	30	40	45	8
6	55	65	75	40	50	60	6
4	70	85	95	55	65	75	4
3	85	100	110	65	75	85	3
2	95	115	130	75	90	100	2
1	110	130	150	85	100	115	1
1/0	125	150	170	100	120	135	1/0
2/0	145	175	195	115	135	150	2/0
3/0	165	200	225	130	155	175	3/0
4/0	195	230	260	150	180	205	4/0
250	215	255	290	170	205	230	250
300	240	285	320	190	230	255	300
350	260	310	350	210	250	280	350
400	280	335	380	225	270	305	400
500	320	380	430	260	310	350	500
600	355	420	475	285	340	385	600
700	385	460	520	310	375	420	700
750	400	475	535	320	385	435	750
800	410	490	555	330	395	450	800
900	435	520	585	355	425	480	900
1000	455	545	615	375	445	500	1000
1250	495	590	665	405	485	545	1250
1500	520	625	705	435	520	585	1500
1750	545	650	735	455	545	615	1750
2000	560	665	750	470	560	630	2000
Correction Factors							
Ambient Temp. (°C)	For ambient temperatures other than 30°C (86°F), multiply the allowable ampacities shown above by the appropriate factor shown below.						Ambient Temp. (°F)
21–25	1.08	1.05	1.04	1.08	1.05	1.04	70–77
26–30	1.00	1.00	1.00	1.00	1.00	1.00	78–86
31–35	0.91	0.94	0.96	0.91	0.94	0.96	87–95
36–40	0.82	0.88	0.91	0.82	0.88	0.91	96–104
41–45	0.71	0.82	0.87	0.71	0.82	0.87	105–113
46–50	0.58	0.75	0.82	0.58	0.75	0.82	114–122
51–55	0.41	0.67	0.76	0.41	0.67	0.76	123–131
56–60	—	0.58	0.71	—	0.58	0.71	132–140
61–70	—	0.33	0.58	—	0.33	0.58	141–158
71–80	—	—	0.41	—	—	0.41	159–176

*See 240.4(D)

Table C8(A) Maximum number of compact conductors in rigid metal conduct (RMC) (*based on Table 1, Chapter 9*)

Compact Conductors													
Type	Conductor Size (AWG/kcmil)	Metric Designator (Trade Size)											
		16 (½)	21 (¾)	27 (1)	35 (1¼)	41 (1½)	53 (2)	63 (2½)	78 (3)	91 (3½)	103 (4)	129 (5)	155 (6)
THW,	8	2	4	7	12	16	26	38	59	78	101	158	228
THW-2,	6	1	3	5	9	12	20	29	45	60	78	122	176
THHW	4	1	2	4	7	9	15	22	34	45	58	91	132
	2	1	1	3	5	7	11	16	25	33	43	67	97
	1	1	1	1	3	5	8	11	17	23	30	47	68
	1/0	1	1	1	3	4	7	10	15	20	26	41	59
	2/0	0	1	1	2	3	6	8	13	17	22	34	50
	3/0	0	1	1	1	3	5	7	11	14	19	29	42
	4/0	0	1	1	1	2	4	6	9	12	15	24	35
	250	0	0	1	1	1	3	4	7	9	12	19	28
	300	0	0	1	1	1	3	4	6	8	11	17	24
	350	0	0	1	1	1	2	3	5	7	9	15	22
	400	0	0	1	1	1	1	3	5	7	8	13	20
	500	0	0	0	1	1	1	3	4	5	7	11	17
	600	0	0	0	1	1	1	1	3	4	6	9	13
	700	0	0	0	1	1	1	1	3	4	5	8	12
	750	0	0	0	0	1	1	1	3	4	5	7	11
	1000	0	0	0	0	1	1	1	1	3	4	6	9
THHN,	8	—	—	—	—	—	—	—	—	—	—	—	—
THWN,	6	2	5	8	13	18	30	43	66	88	114	179	258
THWN-2	4	1	3	5	8	11	18	26	41	55	70	110	159
	2	1	1	3	6	8	13	19	29	39	50	79	114
	1	1	1	2	4	6	10	14	22	29	38	60	86
	1/0	1	1	1	4	5	8	12	19	25	32	51	73
	2/0	1	1	1	3	4	7	10	15	21	26	42	60
	3/0	0	1	1	2	3	6	8	13	17	22	35	51
	4/0	0	1	1	1	3	5	7	10	14	18	29	42
	250	0	1	1	1	2	4	5	8	11	14	23	33
	300	0	0	1	1	1	3	4	7	10	12	20	28
	350	0	0	1	1	1	3	4	6	8	11	17	25
	400	0	0	1	1	1	2	3	5	7	10	15	22
	500	0	0	0	1	1	1	3	5	6	8	13	19
	600	0	0	0	1	1	1	2	4	5	6	10	15
	700	0	0	0	1	1	1	1	3	4	6	9	13
	750	0	0	0	1	1	1	1	3	4	5	9	13
	1000	0	0	0	0	1	1	1	2	3	4	6	9
XHHW,	8	3	5	9	15	21	34	49	76	101	130	205	296
XHHW-2	6	2	4	6	11	15	25	36	56	75	97	152	220
	4	1	3	5	8	11	18	26	41	55	70	110	159
	2	1	1	3	6	8	13	19	29	39	50	79	114
	1	1	1	2	4	6	10	14	22	29	38	60	86
	1/0	1	1	1	4	5	8	12	19	25	32	51	73
	2/0	1	1	1	3	4	7	10	16	21	27	43	62
	3/0	0	1	1	2	3	6	8	13	17	22	35	51
	4/0	0	1	1	1	3	5	7	11	14	19	29	42
	250	0	1	1	1	2	4	5	8	11	15	23	34
	300	0	0	1	1	1	3	5	7	10	13	20	29
	350	0	0	1	1	1	3	4	6	9	11	18	25
	400	0	0	1	1	1	2	4	6	8	10	16	23
	500	0	0	0	1	1	1	3	5	6	8	13	19
	600	0	0	0	1	1	1	2	4	5	7	10	15
	700	0	0	0	1	1	1	1	3	4	6	9	13
	750	0	0	0	1	1	1	1	3	4	5	8	12
	1000	0	0	0	0	1	1	1	2	3	4	7	10

Definition: *Compact stranding* is the result of a manufacturing process where the standard conductor is compressed to the extent that the interstices (voids between strand wires) are virtually eliminated.

Table C9 Maximum number of conductors or fixture wires in rigid PVC conduit, schedule 80 (*based on Table 1, Chapter 9*)

Conductors													
Type	Conductor Size (AWG/kcmil)	Metric Designator (Trade Size)											
		16 (½)	21 (¾)	27 (1)	35 (1¼)	41 (1½)	53 (2)	63 (2½)	78 (3)	91 (3½)	103 (4)	129 (5)	155 (6)
RHH,	14	3	5	9	17	23	39	56	88	118	153	243	349
RHW,	12	2	4	7	14	19	32	46	73	98	127	202	290
RHW-2	10	1	3	6	11	15	26	37	59	79	103	163	234
	8	1	1	3	6	8	13	19	31	41	54	85	122
	6	1	1	2	4	6	11	16	24	33	43	68	98
	4	1	1	1	3	5	8	12	19	26	33	53	77
	3	0	1	1	3	4	7	11	17	23	29	47	67
	2	0	1	1	3	4	6	9	14	20	25	41	58
	1	0	1	1	1	2	4	6	9	13	17	27	38
	1/0	0	0	1	1	1	3	5	8	11	15	23	33
	2/0	0	0	1	1	1	3	4	7	10	13	20	29
	3/0	0	0	1	1	1	3	4	6	8	11	17	25
	4/0	0	0	0	1	1	2	3	5	7	9	15	21
	250	0	0	0	1	1	1	2	4	5	7	11	16
	300	0	0	0	1	1	1	2	3	5	6	10	14
	350	0	0	0	1	1	1	1	3	4	5	9	13
	400	0	0	0	0	1	1	1	3	4	5	8	12
	500	0	0	0	0	1	1	1	2	3	4	7	10
	600	0	0	0	0	0	1	1	1	3	3	6	8
	700	0	0	0	0	0	1	1	1	2	3	5	7
	750	0	0	0	0	0	1	1	1	2	3	5	7
	800	0	0	0	0	0	1	1	1	2	3	4	7
	1000	0	0	0	0	0	1	1	1	1	2	4	5
	1250	0	0	0	0	0	0	1	1	1	1	3	4
	1500	0	0	0	0	0	0	1	1	1	1	2	4
	1750	0	0	0	0	0	0	0	1	1	1	2	3
	2000	0	0	0	0	0	0	0	1	1	1	1	3
TW,	14	6	11	20	35	49	82	118	185	250	324	514	736
THHW,	12	5	9	15	27	38	63	91	142	192	248	394	565
THW,	10	3	6	11	20	28	47	67	106	143	185	294	421
THW-2	8	1	3	6	11	15	26	37	59	79	103	163	234
RHH*,	14	4	8	13	23	32	55	79	123	166	215	341	490
RHW*,	12	3	6	10	19	26	44	63	99	133	173	274	394
RHW-2*	10	2	5	8	15	20	34	49	77	104	135	214	307
	8	1	3	5	9	12	20	29	46	62	81	128	184
RHH*,	6	1	1	3	7	9	16	22	35	48	62	98	141
RHW*,	4	1	1	3	5	7	12	17	26	35	46	73	105
RHW-2*,	3	1	1	2	4	6	10	14	22	30	39	63	90
TW,	2	1	1	1	3	5	8	12	19	26	33	53	77
THW,	1	0	1	1	2	3	6	8	13	18	23	37	54
THHW,	1/0	0	1	1	1	3	5	7	11	15	20	32	46
THW-2	2/0	0	1	1	1	2	4	6	10	13	17	27	39
	3/0	0	0	1	1	1	3	5	8	11	14	23	33
	4/0	0	0	1	1	1	3	4	7	9	12	19	27
	250	0	0	0	1	1	2	3	5	7	9	15	22
	300	0	0	0	1	1	1	3	5	6	8	13	19
	350	0	0	0	1	1	1	2	4	6	7	12	17
	400	0	0	0	1	1	1	2	4	5	7	10	15
	500	0	0	0	1	1	1	1	3	4	5	9	13
	600	0	0	0	0	1	1	1	2	3	4	7	10
	700	0	0	0	0	1	1	1	2	3	4	6	9
	750	0	0	0	0	0	1	1	1	3	4	6	8
	800	0	0	0	0	0	1	1	1	3	3	6	8
	900	0	0	0	0	0	1	1	1	2	3	5	7
	1000	0	0	0	0	0	1	1	1	2	3	5	7
	1250	0	0	0	0	0	1	1	1	1	2	4	5
	1500	0	0	0	0	0	0	1	1	1	1	3	4
	1750	0	0	0	0	0	0	1	1	1	1	3	4
	2000	0	0	0	0	0	0	0	1	1	1	2	3

(Continued on next page)

Table C9 Maximum number of conductors or fixture wires in rigid PVC conduit, schedule 80 (*based on Table 1, Chapter 9*)

Conductors

Type	Conductor Size (AWG/kcmil)	Metric Designator (Trade Size) 16 (½)	21 (¾)	27 (1)	35 (1¼)	41 (1½)	53 (2)	63 (2½)	78 (3)	91 (3½)	103 (4)	129 (5)	155 (6)
THHN, THWN, THWN-2	14	9	17	28	51	70	118	170	265	358	464	736	1055
	12	6	12	20	37	51	86	124	193	261	338	537	770
	10	4	7	13	23	32	54	78	122	164	213	338	485
	8	2	4	7	13	18	31	45	70	95	123	195	279
	6	1	3	5	9	13	22	32	51	68	89	141	202
	4	1	1	3	6	8	14	20	31	42	54	86	124
	3	1	1	3	5	7	12	17	26	35	46	73	105
	2	1	1	2	4	6	10	14	22	30	39	61	88
	1	0	1	1	3	4	7	10	16	22	29	45	65
	1/0	0	1	1	2	3	6	9	14	18	24	38	55
	2/0	0	1	1	1	3	5	7	11	15	20	32	46
	3/0	0	1	1	1	2	4	6	9	13	17	26	38
	4/0	0	0	1	1	1	3	5	8	10	14	22	31
	250	0	0	1	1	1	3	4	6	8	11	18	25
	300	0	0	0	1	1	2	3	5	7	9	15	22
	350	0	0	0	1	1	1	3	5	6	8	13	19
	400	0	0	0	1	1	1	3	4	6	7	12	17
	500	0	0	0	1	1	1	2	3	5	6	10	14
	600	0	0	0	0	1	1	1	3	4	5	8	12
	700	0	0	0	0	1	1	1	2	3	4	7	10
	750	0	0	0	0	1	1	1	2	3	4	7	9
	800	0	0	0	0	1	1	1	2	3	4	6	9
	900	0	0	0	0	0	1	1	1	3	3	6	8
	1000	0	0	0	0	0	1	1	1	2	3	5	7
FEP, FEPB, PFA, PFAH, TFE	14	8	16	27	49	68	115	164	257	347	450	714	1024
	12	6	12	20	36	50	84	120	188	253	328	521	747
	10	4	8	14	26	36	60	86	135	182	235	374	536
	8	2	5	8	15	20	34	49	77	104	135	214	307
	6	1	3	6	10	14	24	35	55	74	96	152	218
	4	1	2	4	7	10	17	24	38	52	67	106	153
	3	1	1	3	6	8	14	20	32	43	56	89	127
	2	1	1	3	5	7	12	17	26	35	46	73	105
PFA, PFAH, TFE	1	1	1	1	3	5	8	11	18	25	32	51	73
PFA, PFAH, TFE, Z	1/0	0	1	1	3	4	7	10	15	20	27	42	61
	2/0	0	1	1	2	3	5	8	12	17	22	35	50
	3/0	0	1	1	1	2	4	6	10	14	18	29	41
	4/0	0	0	1	1	1	4	5	8	11	15	24	34
Z	14	10	19	33	59	82	138	198	310	418	542	860	1233
	12	7	14	23	42	58	98	141	220	297	385	610	875
	10	4	8	14	26	36	60	86	135	182	235	374	536
	8	3	5	9	16	22	38	54	85	115	149	236	339
	6	2	4	6	11	16	26	38	60	81	104	166	238
	4	1	2	4	8	11	18	26	41	55	72	114	164
	3	1	2	3	5	8	13	19	30	40	52	83	119
	2	1	1	2	5	6	11	16	25	33	43	69	99
	1	0	1	2	4	5	9	13	20	27	35	56	80
XHH, XHHW, XHHW-2, ZW	14	6	11	20	35	49	82	118	185	250	324	514	736
	12	5	9	15	27	38	63	91	142	192	248	394	565
	10	3	6	11	20	28	47	67	106	143	185	294	421
	8	1	3	6	11	15	26	37	59	79	103	163	234
	6	1	2	4	8	11	19	28	43	59	76	121	173
	4	1	1	3	6	8	14	20	31	42	55	87	125
	3	1	1	3	5	7	12	17	26	36	47	74	106
	2	1	1	2	4	6	10	14	22	30	39	62	89

Table C9 Maximum number of conductors or fixture wires in rigid PVC conduit, schedule 80 (*based on Table 1, Chapter 9*)

Conductors

Type	Conductor Size (AWG/kcmil)	Metric Designator (Trade Size) 16 (½)	21 (¾)	27 (1)	35 (1¼)	41 (1½)	53 (2)	63 (2½)	78 (3)	91 (3½)	103 (4)	129 (5)	155 (6)
XHH, XHHW, XHHW-2	1	0	1	1	3	4	7	10	16	22	29	46	66
	1/0	0	1	1	2	3	6	9	14	19	24	39	56
	2/0	0	1	1	1	3	5	7	11	16	20	32	46
	3/0	0	1	1	1	2	4	6	9	13	17	27	38
	4/0	0	0	1	1	1	3	5	8	11	14	22	32
	250	0	0	1	1	1	3	4	6	9	11	18	26
	300	0	0	1	1	1	2	3	5	7	10	15	22
	350	0	0	0	1	1	1	3	5	6	8	14	20
	400	0	0	0	1	1	1	3	4	6	7	12	17
	500	0	0	0	1	1	1	2	3	5	6	10	14
	600	0	0	0	0	1	1	1	3	4	5	8	11
	700	0	0	0	0	1	1	1	2	3	4	7	10
	750	0	0	0	0	1	1	1	2	3	4	6	9
	800	0	0	0	0	1	1	1	1	3	4	6	9
	900	0	0	0	0	0	1	1	—	3	3	5	8
	1000	0	0	0	0	0	1	1	1	2	3	5	7
	1250	0	0	0	0	0	1	1	1	1	2	4	6
	1500	0	0	0	0	0	0	1	1	1	1	3	5
	1750	0	0	0	0	0	0	1	1	1	1	3	4
	2000	0	0	0	0	0	0	1	1	1	1	2	4

Fixture Wires

Type	Conductor Size (AWG/kcmil)	Metric Designator (Trade Size) 16 (½)	21 (¾)	27 (1)	35 (1¼)	41 (1½)	53 (2)
FFH-2, RFH-2, RFHH-3	18	6	11	19	34	47	79
	16	5	9	16	28	39	67
SF-2, SFF-2	18	7	14	24	43	59	100
	16	6	11	20	35	49	82
	14	5	9	16	28	39	67
SF-1, SFF-1	18	13	25	42	76	105	177
RFH-1, RFHH-2, TF, TFF, XF, XFF	18	10	18	31	56	77	130
RFHH-2, TF, TFF, XF, XFF	16	8	15	25	45	62	105
XF, XFF	14	6	11	20	35	49	82
TFN, TFFN	18	16	29	50	90	124	209
	16	12	*22	38	68	95	159
PF, PFF, PGF, PGFF, PAF, PTF, PTFF, PAFF	18	15	28	47	85	118	198
	16	11	22	36	66	91	153
	14	8	16	27	49	68	115
HF, HFF, ZF, ZFF, ZHF	18	19	36	61	110	152	255
	16	14	27	45	81	112	188
	14	10	19	33	59	82	138
KF-2, KFF-2	18	28	53	88	159	220	371
	16	19	37	62	112	155	261
	14	13	25	43	77	107	179
	12	9	17	29	53	73	123
	10	6	11	20	35	49	82
KF-1, KFF-1	18	33	63	106	190	263	442
	16	23	44	74	133	185	310
	14	16	29	50	90	124	209
	12	10	19	33	59	82	138
	10	7	13	21	39	54	90
XF, XFF	12	3	6	10	19	26	44
	10	2	5	8	15	20	34

Note: This table is for concentric stranded conductors only. For compact stranded conductors, Table C9(A) should be used.

*Types RHH, RHW, and RHW-2 without outer covering.

National electrical code tables

Article 344—Rigid Metal Conduit (RMC)

Metric Designator	Trade Size	Nominal Internal Diameter		Total Area 100%		2 Wires 31%		Over 2 Wires 40%		1 Wire 53%		60%	
		mm	in.	mm^2	$in.^2$	mm^2	$in.^2$	mm^2	$in.^2$	mm^2	$in.^2$	mm^2	$in.^2$
12	3/8	—	—	—	—	—	—	—	—	—	—	—	—
16	1/2	16.1	0.632	204	0.314	63	0.097	81	0.125	108	0.166	122	0.188
21	3/4	21.2	0.836	353	0.549	109	0.170	141	0.220	187	0.291	212	0.329
27	1	27.0	1.063	573	0.887	177	0.275	229	0.355	303	0.470	344	0.532
35	1¼	35.4	1.394	984	1.526	305	0.473	394	0.610	522	0.809	591	0.916
41	1½	41.2	1.624	1333	2.071	413	0.642	533	0.829	707	1.098	800	1.243
53	2	52.9	2.083	2198	3.408	681	1.056	879	1.363	1165	1.806	1319	2.045
63	2½	63.2	2.489	3137	4.866	972	1.508	1255	1.946	1663	2.579	1882	2.919
78	3	78.5	3.090	4840	7.499	1500	2.325	1936	3.000	2565	3.974	2904	4.499
91	3½	90.7	3.570	6461	10.010	2003	3.103	2584	4.004	3424	5.305	3877	6.006
103	4	102.9	4.050	8316	12.882	2578	3.994	3326	5.153	4408	6.828	4990	7.729
129	5	128.9	5.073	13050	20.212	4045	6.266	5220	8.085	6916	10.713	7830	12.127
155	6	154.8	6.093	18821	29.158	5834	9.039	7528	11.663	9975	15.454	11292	17.495

Electrical formulas

Single Phase

KVA = VA/1,000
hp = (VA × eff × PF)/746
kw = (VA × PF)/1,000
PF = kW/kVA
Torque (ft-lb) = hp × 5,250/rpm
Motor kVA = (hp × 0.746)/eff × PF
Motor kW = (hp × 0.746)/eff
V = Line-to-line volts
A = Amperes
eff = Efficiency (decimal)
PF = Power factor (decimal)
kVA = Kilovolt amperes
kW = Kilowatts
hp = Horsepower
rpm = Revolutions per minute

Three Phase

$(3^{1/2}V \times A)/1{,}000$
$(3^{1/2}VA \times PF \times eff)/746$
$(3^{1/2}VA \times PF)/1{,}000$
$(kW \times 1{,}000)/3^{1/2} \times VA$

Full load currents—single phase transformers

kVA Rating	Voltage—Volts					
	120	240	480	2,400	4,160	7,200
1.0	8.3	4.2	2.1	0.4	0.2	0.1
3.0	25.0	12.5	6.3	1.3	0.7	0.4
5.0	41.7	20.8	10.4	2.1	1.2	0.7
7.5	62.5	31.3	15.6	3.1	1.8	1.0
10.0	83.3	41.7	20.8	4.2	2.4	1.4
15.0	125.0	62.5	31.3	6.3	3.6	2.1
25.0	208.3	104.2	52.1	10.4	6.0	3.5
37.5	312.5	156.3	78.1	15.6	9.0	5.2
50.0	416.7	208.3	104.2	20.8	12.0	6.9
75.0	625.0	312.5	156.3	31.3	18.0	10.4
100.0	833.3	416.7	208.3	41.7	24.0	13.9
125.0	1,041.7	520.8	260.4	52.1	30.0	17.4
167.5	1,395.8	697.9	349.0	69.8	40.3	23.3
200.0	1,666.7	833.3	416.7	83.3	48.1	27.8
250.0	2,083.3	1,041.7	520.8	104.2	60.1	34.7
333.0	2,775.0	1,387.5	693.8	138.8	80.0	46.3
500.0	4,166.7	2,083.3	1,041.7	208.3	120.2	69.4

Conduit size for combinations of cables with different outside diameters

The conduit size required for a group of cables with different outside diameters may be determined by calculating the equivalent outside diameter d″ and then finding the size required for this diameter in Table 1 below. For 1 to 4 cables:

$$d'' = \sqrt{\frac{n_1 d_1^2 + n_2 d_2^2 + n_m d_m^2 + \cdots}{n_1 + n_2 n_m + \cdots}}$$

where:

d″ = Equivalent diameter of same number of cables all of same outside diameter having total area equal to total area of group of cables of different sizes (a fictitious diameter appearing in the column corresponding to that number of cables ($n_1 + n_2 + n_m + \cdots$)
n_1 = number of cables of diameter d_1
n_2 = number of cables of diameter d_2
n_m = number of cables of diameter d_m, etc.

Example. Find the size conduit for two neoprene-sheathed cables having diameters of 1.20″ and one cable having a diameter of 0.63″.

$$d'' = \sqrt{\frac{2 \times (1.20)^2 + 1 \times (0.63)^2 ..}{2 + 1.}}$$

$$d'' = 1.045$$

Refer to column 3 in Table 1 and note that the diameter of 1.045″ is between 0.901 and 1.120″. Therefore 3″ conduit will be required.

To determine the size conduit required for any number (n) of equal diameter cables in excess of four, multiply the diameter of one cable by $\sqrt{n/4}$. This will give the equivalent diameter of four such cables, and the conduit size required for (n) cables may then be found by using the column for four cables.

Table 1
Maximum allowable diameter (in inches) of individual cables in given size of conduit

Nonmetallic Jacketed Cable (All Cables of Same Outside Diameter)				
Nominal Size Conduit	Number of Cables Having Same OD			
	1	2*	3	4
½	0.453	0.244	0.227	0.197
¾	0.600	0.324	0.301	0.260
1	0.763	0.412	0.383	0.332
1¼	1.010	0.542	0.504	0.436
1½	1.173	0.633	0.588	0.509
2	1.505	0.812	0.754	0.653
2½	1.797	0.970	0.901	0.780
3	2.234	1.206	1.120	0.970
3½	2.583	1.395	1.296	1.121
4	2.930	1.583	1.470	1.273
5	3.675	1.985	1.844	1.595
6	4.416	2.385	2.215	1.916

*Diameters are based on percent fill only. The Jam Ratio, Conduit OD to Cable ID should be checked to avoid possible jamming.

Minimum bending radius for insulated cables for permanent training during installation

These limits do not apply to conduit bends, sheaves, or other curved surfaces around which the cable may be pulled under tension while being installed. Larger radii bends may be required for such conditions to limit side-wall pressure. In all cases the minimum radii specified refer to the inner surface of the cable and not to the axis of the cable.

Power and control cables without metallic shielding or armor

The minimum bending radii for both single and multiple conductor cable with or without lead sheath and without metallic shielding or armor are given in Table 2.

Table 2

Thickness of Conductor Insulation, Inches	Overall Diameter of Cable, Inches		
	Minimum Bending Radius as a Multiple of Cable Diameter		
	1.000 and Less	1.001 to 2.000	2.001 and Over
0.156 and less	4	5	6
0.157 to 0.315	5	6	7
0.316 and over		7	8

Twisted pair instrumentation cable

Table 3
Twisted pair instrumentation cable

Type of Cable	Minimum Bending Radius as a Multiple of Cable Diameter
Armored, wire type, or corrugated sheath or interlocked type	7*
Nonarmored without shielded conductors	6
Nonarmored, metallized-polyester, or braid shielded	8
Nonarmored, flat, or corrugated tape shielded	12**

*With shielded conductors
**For longitudinally applied corrugated shield with PVC jacket

Power and control cables with metallic shielding or armor

Table 4
Power and control cables with metallic shielding or armor

Type of Cable	Minimum Bending Radius as a Multiple of Cable Diameter	
	Power	Control
Armored, flat tape, or wire type	12	12
Armored smooth aluminum sheath, up to		
0.75 in. cable diameter	10*	10*
0.76 in. cable diameter	12	12
>1.5 in. cable diameter	15	15
Armored corrugated sheath or interlocked type	7	7
With shielded single conductor	12	12
With shielded multiconductor	**	**
Nonarmored, flat, or corrugated		
Tape shielded single conductor	12	12
Tape shielded multiconductor	**	**
Multiconductor overall tape shield	12	12
LCS with PVC jacket	15	15
Nonarmored, flat strap shielded	8	–
Nonarmored, wire shielded	Refer to twisted pair instrumentation cable (Table 3)	

LCS = longitudinally applied corrugated shield
*With shielded conductors, 12
**12 times single conductor diameter or 7 times overall cable diameter, whichever is greater

Rubber jacketed flexible portable power and control cables used on take-up reels and sheaves

Type of Cable	Minimum Bending Radius as a Multiple of Cable Diameter
0 to 5 kV	6
Over 5 kV	8

Full load currents—three phase transformers

kVA Rating	Voltage (line-to-line)				
	240	480	2,400	4,160	12,470
3	7.2	3.6	0.7	0.4	0.1
6	14.4	7.2	1.4	0.8	0.3
9	21.7	10.8	2.2	1.2	0.4
15	36.1	18.0	3.6	2.1	0.7
30	72.2	36.1	7.2	4.2	1.4
37.5	90.2	45.1	9.0	5.2	1.7
45	108.3	54.1	10.8	6.2	2.1
50	120.3	60.1	12.0	6.9	2.3
75	180.4	90.2	18.0	10.4	3.5
100	240.6	120.3	24.1	13.9	4.6
112.5	270.6	135.3	27.1	15.6	5.2
150	360.9	180.4	36.1	20.8	6.9
200	481.1	240.6	48.1	27.8	9.3
225	541.3	270.6	54.1	31.2	10.4
300	721.7	360.9	72.2	41.6	13.9
450	1,082.6	541.3	108.3	62.5	20.8
500	1,202.8	601.4	120.3	69.4	23.2
600	1,443.4	721.7	144.3	83.3	27.8
750	1,804.3	902.1	180.4	104.1	34.7
1,000	2,405.7	1,202.8	240.6	138.8	46.3
1,500	3,608.5	1,804.3	360.9	208.2	69.5
2,000	4,811.4	2,405.7	481.1	277.6	92.6
2,500	6,014.2	3,007.1	601.4	347.0	115.8
5,000	12,028.5	6,014.2	1,202.8	694.0	231.5
10,000	24,057.0	12,028.5	2,405.7	1,387.9	463.0

Motor controller sizes

Polyphase motors

	Maximum Horsepower Full Voltage Starting	
NEMA Size	230 Volts	460–575 Volts
00	1.5	2
0	3	5
1	7.5	10
2	15	25
3	30	50
4	50	100
5	100	200
6	200	400
7	300	600

Single phase motors

NEMA Size	Maximum Horsepower Full Voltage Starting (Two Pole Contactor)	
	115 Volts	230 Volts
00	1.3	1
0	1	2
1	2	3
2	3	7.5
3	7.5	15

Voltage drop on circuits using 600 V copper conductors in steel conduit

To determine the voltage drop, multiply the length in feet of one conductor by the current in amperes and by the number listed in the table for the type of system and power factor and divide the result by 1,000,000 to obtain the voltage loss. This table takes into consideration reactance on AC circuits and resistance of the conductor. Unless otherwise noted, the table is based on 60 Hz.

Wire Size	Direct Current	Three Phase Lagging Power Factor					Single Phase Lagging Power Factor				
		100%	90%	80%	70%	60%	100%	90%	80%	70%	60%
14	6,100	5,280	4,800	4,300	3,780	3,260	6,100	5,551	4,964	4,370	3,772
12	3,828	3,320	3,030	2,720	2,400	2,080	3,828	3,502	3,138	2,773	2,404
10	2,404	2,080	1,921	1,733	1,540	1,340	2,404	2,221	2,003	1,779	1,547
8	1,520	1,316	1,234	1,120	1,000	880	1,520	1,426	1,295	1,159	1,017
6	970	840	802	735	665	590	970	926	850	769	682
4	614	531	530	487	445	400	614	613	562	514	462
3	484	420	425	398	368	334	484	491	460	425	385
2	382	331	339	322	300	274	382	392	372	346	317
1	306	265	280	270	254	236	306	323	312	294	273
0	241	208	229	224	214	202	241	265	259	247	233
00	192	166	190	188	181	173	192	219	217	209	199
000	152	132	157	158	155	150	152	181	183	179	173
0000	121	105	131	135	134	132	121	151	156	155	152
250M	102	89	118	123	125	123	103	136	142	144	142
300M	85	74	104	111	112	113	86	120	128	130	131
350M	73	63	94	101	105	106	73	108	117	121	122
400M	64	55	87	95	98	100	64	100	110	113	116
500M	51	45	76	85	90	92	52	88	98	104	106
600M	43	38	69	79	85	87	44	80	91	98	101
700M	36	33	64	74	80	84	38	74	86	92	97
750M	34	31	62	72	79	82	36	72	83	91	95
800M	32	29	61	71	76	81	33	70	82	88	93
900M	28	26	57	68	74	78	30	66	78	85	90
1000M	26	23	55	66	72	76	27	63	76	83	88

Determine the most economical size for electric power conductors

To calculate quickly the most economical size copper wire for carrying a specified current load, use the formula:

$$A = (59)(I)\sqrt{\frac{C_e \times t}{C_c \times F}}$$

where:

- A = Conductor cross-sectional area (circular mils)
- I = Current load (amperes)
- C_e = Cost of electrical power (cents per KWH)
- C_c = Cost of copper (cents per lb)
- t = Hours of service per year
- F = Factor for fixed charges (amortization, insurance, taxes, etc.)

Selection of the proper size electric conductor will depend on the load to be carried and the mechanical strength required, as well as economics. Once these are considered, the most economical size conductor is that for which the annual energy cost equals the copper cost. This method can be used for other metallic conductors.

Example. Determine the most economical size copper conductor for an installation operating 365 days a year, 8 hours per day, with a 100 ampere load. Energy costs are \$0.00875 per KWH. Fixed charges are 23%, considering a 5-year amortization with 3% for insurance and taxes. Copper costs \$0.46 per lb.

$$A = (59)(100)\sqrt{\frac{0.00875 \times 365 \times 8}{0.46 \times 0.23}}$$

$$= 91{,}686 \text{ circular mils}$$

How to find the resistance and weight of copper wires

It is easy to calculate mentally the approximate resistance and weight of standard sizes of copper wires, by remembering the following rules:

Rule 1: Number 10 wire has a resistance of 1 ohm per 1,000 ft. Number 2 wire weighs 200 lb. per 1,000 ft.
Rule 2: An increase of 10 numbers in the size is an increase of ten times in the resistance, and a decrease to one-tenth of the weight.
Rule 3: An increase of three numbers in the size doubles the resistance and halves the weight.
Rule 4: An increase of one number in the size multiplies the resistance by $\frac{5}{4}$ and the weight by $\frac{4}{5}$.

Always proceed from the known value (No. 10 for resistance, No. 2 for the weight) in the smallest number of jumps; that is, go by tens, then by threes, and finally by ones; the jumps do not all have to be in the same direction.

Example. What is the weight and resistance of 200 ft of No. 18? Apply the rules in order, as follows; resistance first:

Number 10 wire has a resistance of 1 ohm per 1,000 ft. (Rule 1)
Number 20 wire has a resistance of 10 ohms per 1,000 ft. (Rule 2)
Number 17 wire has a resistance of 5 ohms per 1,000 ft. (Rule 3)
Number 18 wire has a resistance of 6.25 ohms per 1,000 ft. (Rule 4)
Number 18 wire has a resistance of 1.25 ohms per 200 ft. (Answer)
Number 2 wire weighs 200 lb. per thousand ft. (Rule 1)
Number 12 wire weighs 200 lb. per thousand ft. (Rule 2)
Number 22 wire weighs 2 lb. per thousand ft. (Rule 2)
Number 19 wire weighs 4 lb. per thousand ft. (Rule 3)
Number 18 wire weighs 5 lb. per thousand ft. (Rule 4)
Number 18 wire weighs 1 lb. per 200 ft. (Answer)

The weight determined by the above rules is that of bare copper wire. Weight of insulated wire varies widely according to the type of insulation. Resistance values are correct for bare or insulated wire and for solid or stranded. Errors resulting from the use of the rules will rarely exceed 2%.

What you should remember about electrical formulas

Ohm's law

Easily **R**emember **I**t, or E = RI

Power

DC **P**ower **I**s **E**asy, or P = IE
It **R**eally **I**s, or P = IRI = I^2R

Other facts

The number of **watts in one hp** is equal to the year Columbus discovered America divided by two, or 1492/2 = 746 watts/hp.

The horsepower of a reciprocating engine necessary to drive a three-phase, 60-cycle electric generator can be determined by multiplying KW by 1.5.

How to calculate microwave hops on level ground

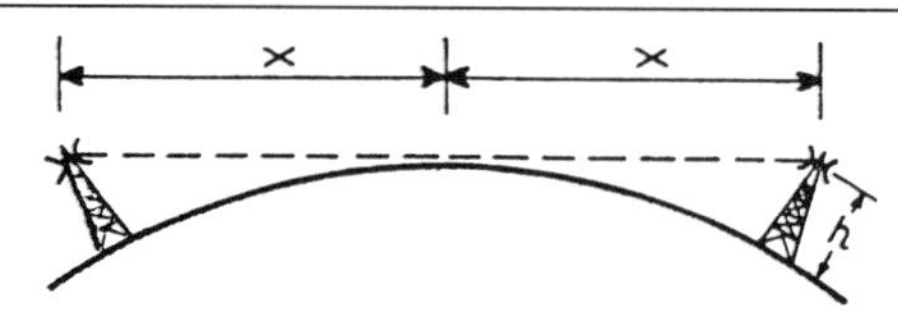

Rule. Excluding obstructions such as hills, the microwave (line-of-sight) distances between two towers will be twice the distance from the transmitter to the horizon. The line-of-sight distance in miles from the top of the tower to the horizon can be found by adding the square root of the tower height to the square root of the first square root (the latter, of course, is the fourth root).

$$x = \sqrt{h} + \sqrt{\sqrt{h}} \quad or \quad \sqrt{h} + \sqrt[4]{h}$$

Example. How far apart can 100-foot-high towers be spaced to provide line-of-sight transmission if there are no obstructions between them?

$$X = \sqrt{100} + \sqrt{10}$$

$$X = 10 + 3 + \text{or} + 13 \text{ miles}$$

Distance between towers can be 2X or 26 miles.

This same formula can be used to estimate distances in an airplane.

Example. How far is the line-of-sight distance to the horizon in an airplane flying at an altitude of 4,900 ft?

$(\sqrt{4{,}900} = 70)$

$X = \sqrt{4{,}900} + \sqrt{70}$

$X = 70 + 8$

Distance to horizon is 78 miles.

For quick determination of the horsepower per ampere for induction motors (3 phase) at different voltages

Voltage	hp per ampere
480	1
2,400	5
4,160	8

Chart of electric motor horsepower for pumping units

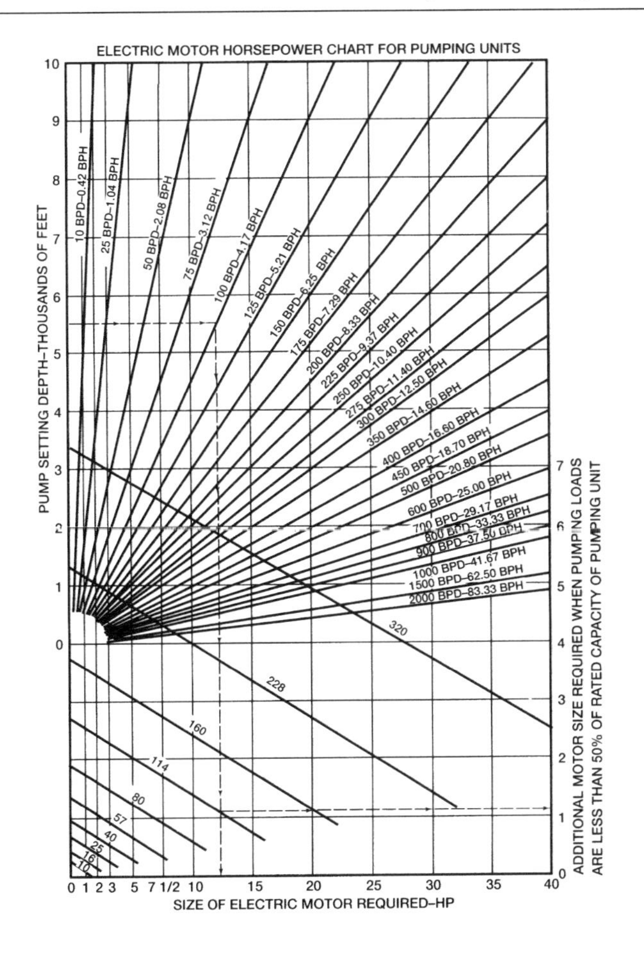

This chart provides a means of determining the power requirements for a beam pumping installation powered by an electric motor based on a fluid with a specific gravity of 1, with fluid at the pump. The power requirement determined by the chart includes a mechanical efficiency factor of 0.45 and a cyclic factor of 0.75, which are factors frequently applied to motors used in sucker rod pumping service.

An arrangement is available for correcting the power requirement in the case of an underloaded pumping unit. The example shown in the chart is self-explanatory.

After the power requirements are determined from the chart, the next higher size of commercially available motor is used.

Pumping stations

Table gives capacitor multipliers for kilowatt loads for different desired power factor improvements

Original Power Factor, Percent	DESIRED POWER FACTOR—Percent				
	100	95	90	85	80
50	1.732	1.403	1.248	1.112	0.982
51	1.687	1.358	1.203	1.067	0.937
52	1.643	1.314	1.159	1.023	0.893
53	1.600	1.271	1.116	0.980	0.850
54	1.559	1.230	1.075	0.939	0.809
55	1.518	1.190	1.035	0.898	0.769
56	1.480	1.151	0.996	0.860	0.730
57	1.442	1.113	0.958	0.822	0.692
58	1.405	1.076	0.921	0.785	0.655
59	1.369	1.040	0.885	0.749	0.619
60	1.333	1.004	0.849	0.713	0.583
61	1.299	0.970	0.815	0.679	0.549
62	1.266	0.937	0.782	0.646	0.516
63	1.233	0.904	0.749	0.613	0.483
64	1.201	0.872	0.717	0.581	0.451
65	1.169	0.840	0.685	0.549	0.419
66	1.138	0.809	0.654	0.518	0 388
67	1.108	0.779	0.624	0.488	0.358
68	1.078	0.749	0.594	0.458	0.328
69	1.049	0.720	0.565	0.429	0.299
70	1.020	0.691	0.536	0.400	0.270
71	0.992	0.663	0.508	0.372	0.242
72	0.964	0.635	0.480	0.344	0.214
73	0.936	0.607	0.452	0.316	0.186
74	0.909	0.580	0.425	0.289	0.159
75	0.882	0.553	0.398	0.262	0.132
76	0.855	0.526	0.371	0.235	0.105
77	0.829	0.500	0.345	0.209	0.079
78	0.802	0.473	0.318	0.182	0.052
79	0.776	0.447	0.292	0.156	0.026
80	0.750	0.421	0.266	0.130	....
81	0.724	0.395	0.240	0.104	....
82	0.698	0.369	0.214	0.078	....
83	0.672	0.343	0.188	0.052	....
84	0.646	0.317	0.162	0.026	....
85	0.620	0.291	0.136	....	....
86	0.593	0.264	0.109	....	....
87	0.567	0.238	0.083	....	....
88	0.540	0.211	0.056	....	....
89	0.512	0.183	0.028	....	....
90	0.484	0.155			
91	0.456	0.127			
92	0.426	0.097			
93	0.395	0.066			
94	0.363	0.034			
95	0.329	....			
96	0.292	....			
97	0.251	....			
98	0.203	....			
99	0.143	....			

Example

Assume total plant load is 100 kw at 60 percent power factor. Capacitor kvar rating necessary to improve power factor to 80 percent is found by multiplying kw (100) by multiplier in table (0.583), which gives kvar (58.3). Nearest standard rating (60 kvar) should be recommended.

Courtesy General Electric Company.

FLOODLIGHTING CONCEPTS

This material is intended as a helpful guide in the evaluation of lighting requirements and is intended to serve only as a guide.

Terms

Candela or Candlepower—Light sources do not project the same amount of light in every direction. The directional characteristic of a lamp is described by the candlepower in specific directions. This directional strength of light or luminous intensity is measured in candelas.

Lumen—Light quantity, irrespective of direction, is measured in lumens. Lamp lumens are the quantity of light produced by a lamp. Average light level calculations use total lamp lumens as a basis and then adjust for all factors that lower this quantity. The amount of useful light in a floodlight beam is measured in beam lumens.

Footcandle, fc—Specifications are usually based on density of light or level of illumination which is measured in footcandles. Footcandles are the ratio of quantity of light in lumens divided by the surface area in square feet on which the lumens are falling. A density of one lumen per square foot is one footcandle. One footcandle is equal to 10.76 lux.

Lux, lx—The SI unit of illuminance. This metric measurement is based upon the density of lumens per unit surface area similar to footcandle, except one lux is one lumen per square meter. One lux is equal to 0.09 footcandle.

Light Loss Factor, LLF—These factors are used to adjust lighting calculations from laboratory test conditions to a closer approximation of expected field results. The I.E.S. Lighting Handbook, 1984 Reference Volume, defines LLF as follows: "a factor used in calculating illuminance after a given period of time and under given conditions. It takes into account temperature and voltage variations, dirt accumulations on luminaire and room surfaces, lamp depreciation, maintenance procedures and atmosphere conditions."

Floodlighting calculations

Floodlighting encompasses many variations. Since the location of the floodlight relative to the object to be lighted can be in any plane and at any distance from the source ... floodlighting application is often considered the most complex and difficult of all lighting techniques.

The most commonly used systems for floodlight calculations are the point-by-point method and the beam-lumen method.

Point-by-point method

The point-by-point method permits the determination of footcandles at any point and orientation on a surface and the degree of lighting uniformity realized for any given set of conditions.

In such situations, the illumination is proportional to the candlepower of the source in a given direction and inversely proportional to the square of the distance from the source. See Figure 1.

$$\text{Footcandles on Plane (Normal to Light Ray)} = \frac{\text{Candlepower of Light Ray}}{\text{Distance in Feet From Source to Point-Squared}}$$

$$E = I/D^2 \qquad (1)$$

When the surface on which the illumination to be determined is tilted, the light will be spread over a greater area, reducing the illumination in the ratio of the area of plane A to the area of plane B as shown in Figure 2. This ratio is equal to the cosine of the angle of incidence; thus.

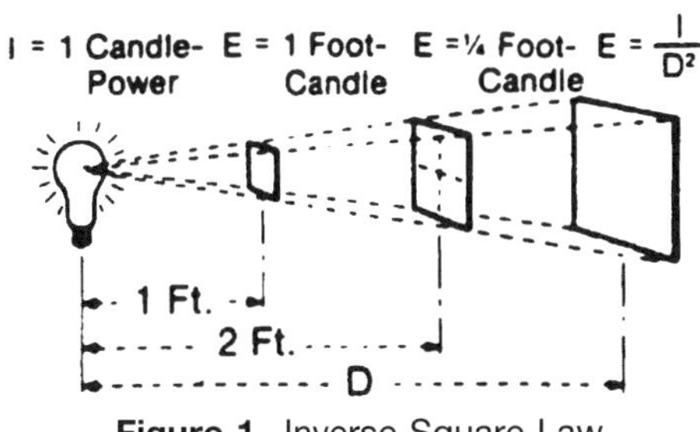

Figure 1. Inverse Square Law

$$\text{Footcandles on Plane B} = \frac{\text{Candlepower of Light Ray}}{\text{Distance in Feet From Source to Point-Squared}}$$

$\times$ Cosine of Angle β

$$E = I/D^2 \times \text{Cosine}\,\beta \qquad (2)$$

Then β equals the angle between the light ray and a perpendicular to the plane at that point.

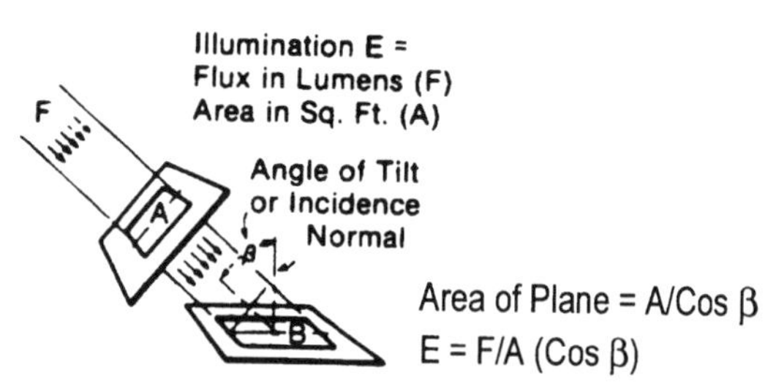

Figure 2. Cosine of Angle Law.

Beam-lumen method

The beam-lumen method is quite similar to the method used for interior lighting except that the utilization factors must take into consideration the fact that floodlights are not usually perpendicular to the surface and therefore not all of the useful light strikes the surface.

Beam lumens are defined as the quantity of light that is contained within the beam limits as described as "beam spread." Beam lumens equal the lamp lumens multiplied by the beam efficiency of the floodlight.

Coverage. It is recommended that sufficient point-by-point calculations be made for each job to check uniformity and coverage.

Light Loss Factor (LLF). The maintenance or light loss factor is an allowance for depreciation of lamp output with age and floodlight efficiency due to the collection of dirt on lamp, reflector, and cover glass. The total factor may vary from 0.65 to 0.85 depending on the type of lamp and luminaire used and may include losses due to lamp orientation, or "tilt."

Design procedure

Step 1—Determine the level of illumination. See Table 1 for some typical levels of illumination (fc). The basic formula is:

$$fc = \frac{N \times BL \times CBU \times LLF}{A} \qquad (3)$$

where:

N = quantity of luminaries
A = area in square feet
BL = beam lumens
CBU = Coefficient of beam utilization
LLF = Light loss factor

Step 2—Determine type and location of floodlights. Regardless of light source there are industry standards on beam spreads. See Table 2.

Step 3—Determine the coefficient of beam utilization. This factor, CBU, written as a decimal fraction, is expressed in the following ratio:

$$CBU = \frac{\text{Utilized Lumens}}{BL} \qquad (4)$$

The exact CBU can be determined graphically by projecting the outline of the area to be lighted upon the photometric data and totaling the utilized lumens. This procedure is detailed in the I.E.S. Lighting Handbook. See Table 3.

Table 1
Typical illumination levels

Area	fc	Area	fc
Building exteriors		Storage tanks	1
Entrances		Pump areas	2
Active	5	Parking	1
Inactive	1	Substation	2
Vital locations	5	Storage yards	
Catwalks	3	Active	20
Stairs	5	Inactive	1
Barge or truck unloading	5		

Table 2
Outdoor floodlight luminaire designations

Beam Spread, Degrees	NEMA Type
10 up to 18	1
>18 up to 29	2
>29 up to 46	3
>46 up to 70	4
>70 up to 100	5
>100 up to 130	6
>130 up	7

The following general principles apply in the choice of beam spread:

1. The greater the distance from the floodlight to the area to be lighted, the narrower the beam spread desired.
2. Since by definition the candlepower at the edge of a floodlight beam is 10% of the candlepower near the center of the beam, the illumination level at the edge of the beam is one tenth or less of that at the center. To obtain reasonable uniformity of illumination, the beams of individual floodlights must overlap each other, as well as the edge of the surface to be lighted.
3. The percentage of beam lumens falling outside the area to be lighted is usually lower with narrow-beam units than with wide-beam units. Thus, narrow-beam floodlights are preferable where they will provide the necessary degree of uniformity of illumination and the proper footcandle level.

Table 3
Typical CBUs

Area	CBU	Area	fc
Baseball			
Infield	0.65	Tennis	0.75
Outfield	0.85	Large parking lots	0.80
Football	0.60	Building facades	0.65

As an approximation, the average CBU of all the floodlights in an installation should fall within the range of 0.6 to 0.9. If less than 60% of the beam lumens are utilized, a more economical lighting plan should be possible by using different locations or narrower beam floodlights. If the CBU is greater than 0.9, it is probable that the beam spread selected is too narrow and the resultant illumination will be spotty. An estimated CBU can be determined by experience, or by making calculations for several potential aiming points and using the average figure thus obtained.

Step 4—Determine the quantity of floodlights (N) required. Rearrange the basic formula in Step 1 as follows:

$$N = \frac{A \times fc}{BL \times CBU \times LLF} \tag{5}$$

Conductor size conversion chart—Metric to AWG

Table 4

Metric Size	AWG Size
2.5 mm^2	#14
4 mm^2	#12
6 mm^2	#10
10 mm^2	#8
16 mm^2	#6
25 mm^2	#3[1]
35 mm^2	#2
35 mm^2	#1[1]
50 mm^2	#1/0
70 mm^2	#2/0
95 mm^2	#3/0
95 mm^2	#4/0[1]
120 mm^2	250 kcmil
150 mm^2	300 kcmil
185 mm^2	350 kcmil[1]
185 mm^2	400 kcmil[1]
240 mm^2	500 kcmil
300 mm^2	600 kcmil
400 mm^2	700 kcmil
400 mm^2	750 kcmil
400 mm^2	800 kcmil[1]

[1]This AWG size is the closest equivalent to the metric size.

5: Hydrostatic Testing

THE BENEFITS AND LIMITATIONS OF HYDROSTATIC TESTING

John F. Kiefner and Willard A. Maxey, Adapted from Dr. John Kiefner, Kiefner and Associates, Inc.

The purpose of this discussion is to clarify the issues regarding the use of hydrostatic testing to verify pipeline integrity. There are those who say it damages a pipeline, especially if carried out to levels of 100% or more of the specified minimum yield strength (SMYS) of the pipe material. These people assert that if it is done at all, it should be limited to levels of around 90% of SMYS. There are those who insist that pipelines should be retested periodically to reassure their serviceability. The reality is that if and when it is appropriate to test a pipeline, the test should be carried out at the highest possible level that can feasiblely be done without creating numerous test failures. The challenge is to determine if and when it should be done, the appropriate test level, and the test-section logistics that will maximize the effectiveness of the test.

The technology to meet these challenges has been known for 30 years. Nothing has arisen in the meantime to refute this technology. The problem is that people both within and outside the pipe industry either are not aware of the technology or have forgotten it, or for political reasons are choosing to ignore it.

In this discussion we show the following:

- It makes sense to test a new pipeline to a minimum of 100% of SMYS at the highest elevation in the test section.
- Pipe that meets the specified minimum yield strength is not likely to be appreciably expanded even if the maximum test pressure is 110% of SMYS.
- If hydrostatic retesting is to be conducted to revalidate the serviceability of a pipeline suspected to contain defects that are becoming larger with time in service, the highest feasible test pressure level should be used.
- If the time-dependent defects can be located reliably by means of an in-line inspection tool, using the tool is usually preferable to hydrostatic testing.

We also note the following as reminders:

- When a pipeline is tested to a level in excess of 100% of SMYS, a pressure-volume plot should be made to limit yielding.
- A test may be terminated short of the initial pressure target, if necessary, to limit the number of test breaks as long as the MOP guaranteed by the test is acceptable to the pipeline's operator.

And, we suggest that:

- Test-section length should be limited to prevent elevation differences within a test section from exceeding 300 ft.

The pressure level for verifying integrity can be higher than the level needed to validate the MOP of the pipeline, and the integrity test to a level above 1.25 times MOP, if used, needs to be no longer than ½ hour.

Background

The concept and value of high-pressure hydrostatic testing of cross-country pipelines were first demonstrated by Texas Eastern Transmission Corporation. Texas Eastern sought the advice of Battelle in the early 1950s as they began to rehabilitate the War Emergency Pipelines and to convert them to natural gas service. Prior to testing, these pipelines exhibited numerous failures in service due to original manufacturing defects in the pipe. The Battelle staff recommended hydrostatic testing to eliminate as many of these types of defects as possible. After being tested to levels of 100% to 109% of SMYS, during which time "hundreds" of test breaks occurred, not one in-service failure caused by a manufacturing defect was observed. The news of this successful use of hydrostatic testing spread quickly to other pipeline operators, and by the late 1960s the ASA B31.8 Committee (forerunner of ASME B31.8) had established an enormous database of thousands of miles of pipelines that had exhibited no in-service ruptures from original manufacturing or construction defects after having been hydrostatically tested to levels at or above 90% of SMYS.[1] These data were used to establish the standard practice and ASA B31.8 Code requirement that, prior to service, each gas pipeline should be hydrostatically tested to 1.25 times its maximum allowable operating pressure. Later, a similar requirement for liquid pipelines was inserted into the ASME B31.4 Code. When federal regulations for pipelines came along, the precedent set by the industry of testing to 1.25 times the MOP was adopted as a legal requirement.

Both field experience and full-scale laboratory tests have revealed much about the benefits and limitations of hydrostatic testing. Among the things learned were the following:

- Longitudinally oriented defects in pipe materials have unique failure pressure levels that are predictable on the basis of the axial lengths and maximum depths of the defects and the geometry of the pipe and its material properties.[2]
- The higher the test pressure, the smaller will be the defects, if any, that survive the test.
- With increasing pressure, defects in a typical line-pipe material begin to grow by ductile tearing prior to failure. If the defect is close enough to failure, the ductile tearing that occurs prior to failure will continue even if pressurization is stopped and the pressure is held constant. The damage created by this tearing when the defect is about ready to fail can be severe enough that if pressurization is stopped and the pressure is released, the defect may fail upon a second or subsequent pressurization at a pressure level below the level reached on the first pressurization. This phenomenon is referred to as a pressure reversal.[3,4]
- Testing a pipeline to its actual yield strength can cause some pipe to expand plastically, but the number of pipes affected and the amount of expansion will be small if a pressure-volume plot is made during testing and the test is terminated with an acceptably small offset volume or reduction in the pressure-volume slope.[5]

Test-pressure-to-operating-pressure ratio

The hypothesis that "the higher the test-pressure-to-operating-pressure ratio, the more effective the test," is validated by Figure 1. Figure 1 presents a set of failure-pressure-versus-defect-size relationships for a specific diameter, wall thickness, and grade of pipe. A great deal of testing of line-pipe materials over the years has validated these curves.[2] Each curve represents a flaw with a uniform depth-to-wall-thickness ratio. Nine such curves are given (d/t ranging from 0.1 to 0.9).

Consider the maximum operating pressure (MOP) for the pipeline (the pressure level corresponding to 72% of SMYS). That pressure level is represented in Figure 1 by the horizontal line labeled MOP. At the MOP, no defect longer than 10 in. and deeper than 50% of the wall thickness can exist. Any such defect would have failed in service. Similarly, no defect longer than 4 in. and deeper than 70% of the wall thickness can exist, nor can one that is longer than 16 in. and deeper than 40% of the wall thickness.

By raising the pressure level above the MOP in a hydrostatic test, the pipeline's operator can assure the absence of defects smaller than those that would fail at the MOP. For example, at a test pressure level equivalent to 90% of SMYS, the largest surviving defects are determined in Figure 1 by the horizontal line labeled 90% of SMYS. At that level, the longest surviving defect that is 50% through the wall can be only about 4.5 in. Compare that length to the length of the longest possible 50%-through flaw at the MOP; it was 10 in. Alternatively, consider the minimum survivable depth at 90% SMYS for a 10-in.-long defect (the size that fails at the MOP if it is 50% through the wall). The survivable depth is only about 32% through the wall. By a similar process of reasoning, one can show that even smaller flaws are assured by tests to 100% or 110% of SMYS (the horizontal lines drawn at those pressure levels on Figure 1).

The point is that the higher the test pressure (above MOP), the smaller will be the possible surviving flaws. This fact means a larger size margin between flaw sizes left after the test and the sizes of flaws that would cause a failure at the MOP. If surviving flaws can be extended by operating-pressure cycles, the higher test pressure will assure that it takes a longer time for these smaller flaws to grow to a size that will fail at the MOP. Thus, Figure 1 provides proof of the validity of the hypothesis (i.e., the higher the test-pressure-to-operating-pressure ratio, the more effective the test).

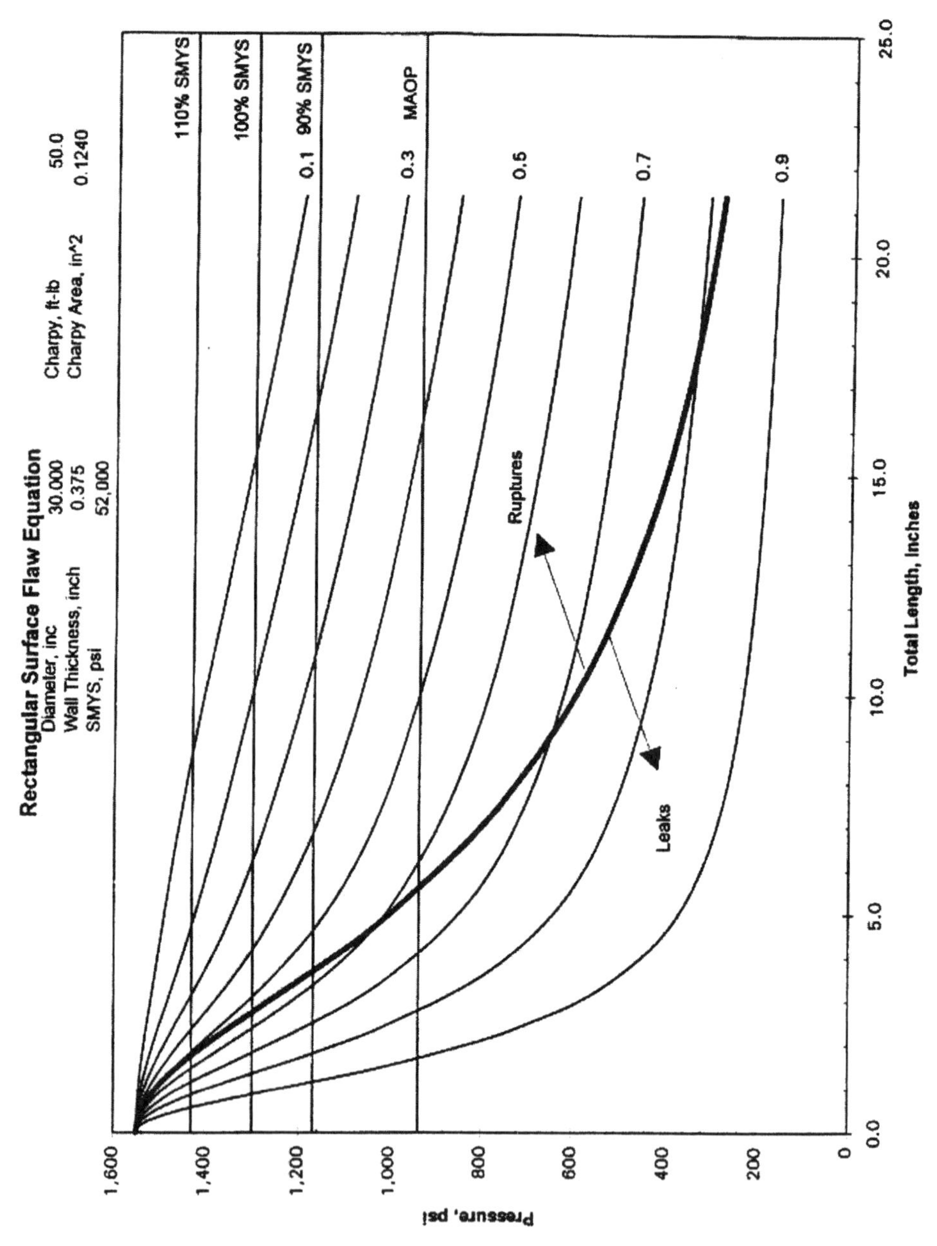

Figure 1. Impact of test pressure levels on margin of safety.

Testing to levels above 100% of SMYS

Given the previous argument for testing to the highest feasible level, one needs to consider practical upper limits. In the case of a new pipeline constructed of modern high-quality, high-toughness line pipe, the maximum test level can generally be in excess of 100% of SMYS. Reasons why this will not cause significant yielding of the pipe are as follows.

First, as shown in Figure 2, the average yield strength of an order of pipe is usually well above the mininmum specified value. Very few pieces will have yield strengths low enough to cause yielding at 100% of SMYS.

Secondly, when a buried pipeline is pressurized, it is restrained by the soil from shortening in the axial direction. This causes an axial tensile stress equal to Poisson's ratio times the hoop stress (Poisson's ratio is 0.3 for steel). What this means to testing in excess of 100% of SMYS is shown in Figure 3.

The tensile test commonly used to assess the yield strength of line pipe is a transverse, flattened uniaxial specimen designed to test the circumferential (hoop) direction tensile properties. A test of such a specimen reveals a unique value of yield strength at a certain value of applied stress. In Figure 3, that value is represented as 1.0 on the circumferential tensile stress to uniaxial yield stress ratio (vertical) axis. If one were to test the same type of specimen using a longitudinal specimen, the unique yield strength measured in the test could be plotted on Figure 3 at 1.0 on the longitudinal tensile stress uniaxial yield strength (horizontal axis). Negative numbers on the horizontal axis represent axial compressive stress. The typical line-pipe material exhibits an elliptical yield-strength relationship for various combinations of biaxial stress.[4] As shown in Figure 3, this results in yielding at a higher value of circumferential tensile stress to uniaxial yield strength ratios than 1.0. In tests of pressurized pipes, the ratio for a buried pipeline (longitudinal tensile stress to circumferential tensile stress ratio of 0.3) was found to be about 1.09.[6] So, this effect also suppresses yielding in a hydrostatic test of a pipeline to a pressure level in excess of 100% of SMYS.

To resolve how much yielding actually takes place, Texas Eastern designed a gauging pig in the mid 1960s to measure diametric expansion.[5] In 300 miles of 30-in. OD X52 pipe, tested to a maximum of 113% of SMYS, they found only 100 joints of pipe (out of 40,000) that had expanded as much as 1.0%. In 66 miles of 36-inch OD X60 pipe tested to a maximum of 113% of SMYS, they found 100 joints of pipe (out of 6,600) that had expanded as much as 1.0%, still not a lot of expanded pipe.

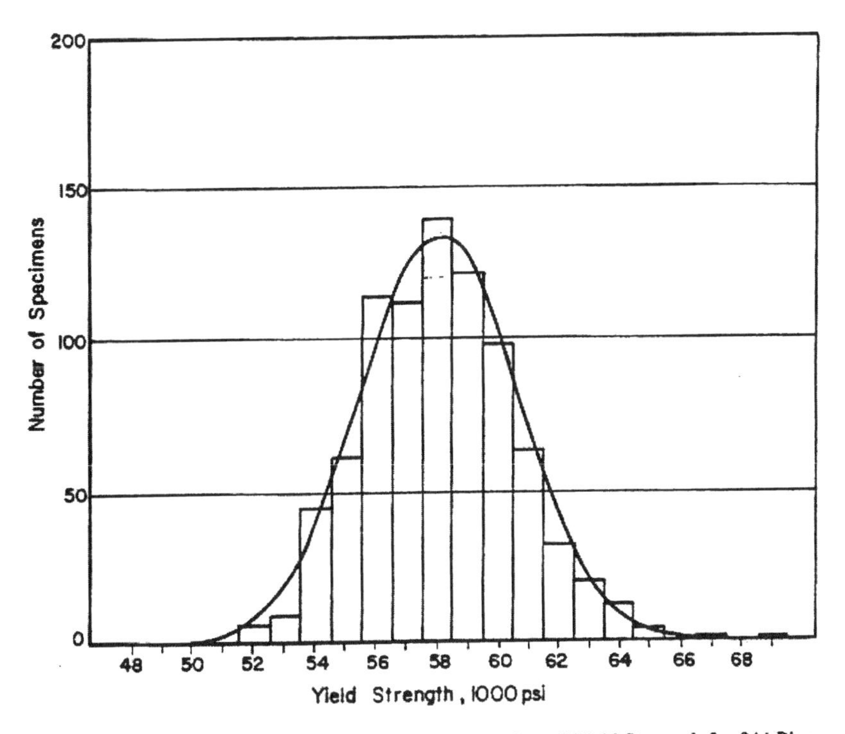

Figure 2.

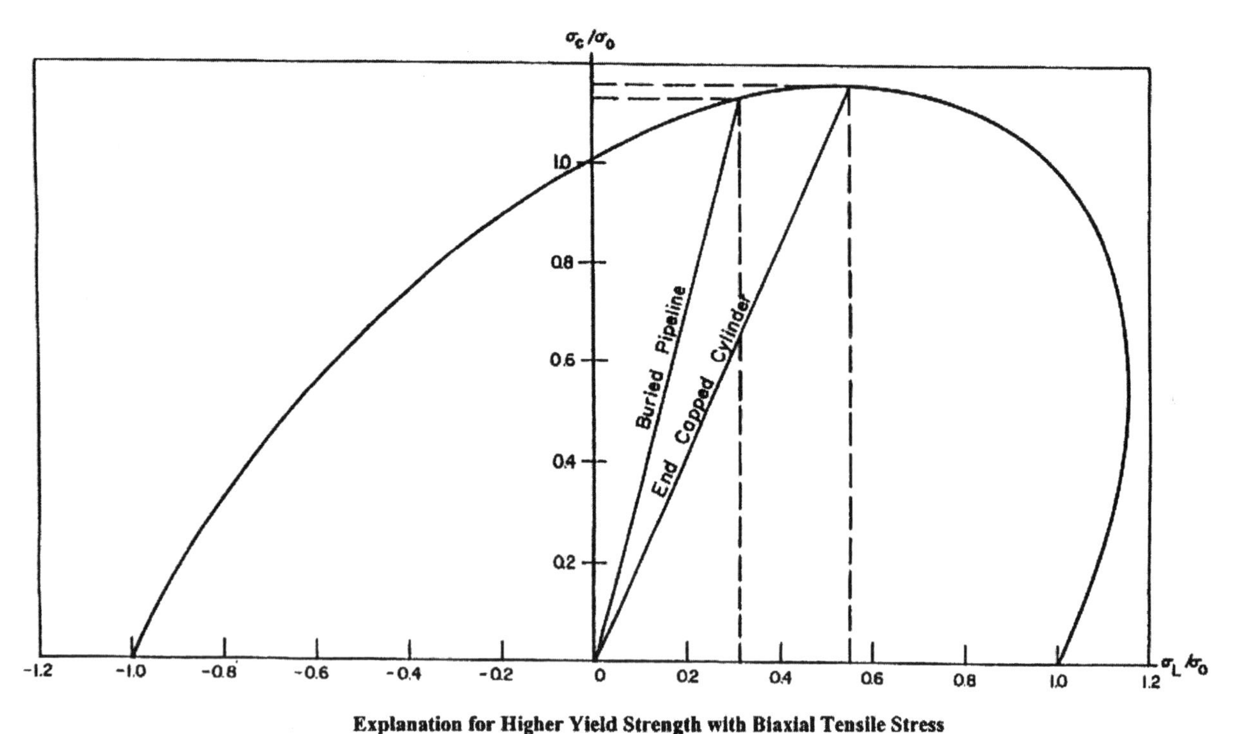

Explanation for Higher Yield Strength with Biaxial Tensile Stress

Figure 3.

Pressure-volume plots

Figure 4 shows a pressure-versus-pump-stroke plot of an actual hydrostatic test. The plot is created by recording the number of pump strokes of a positive displacement pump as each 10 psig increase of pressure is attained. Prior to beginning the plot, it is prudent to hold the test section at a constant pressure to assure that there are no leaks. After it is established that there is no leak, the plot should be started at a pressure level no higher than 90% of SMYS for the low elevation point in the test section in order to establish the "elastic" slope of the plot. By projecting the elastic slope lines across the plot as shown, one can then record pump strokes and compare the evolving plot to those slopes. If and when the actual plot begins to deviate from the elastic slope, either some pipe is beginning to yield or a leak has developed. The pressurization can be continued in any event until the "double-the-strokes" point is reached. This is the point at which it takes twice as many strokes to increase the pressure 10 psi as it did in the elastic range. Also, we suggest stopping at 110% of SMYS if that level of pressure is reached before the double-the-stroke point. Once the desired level has been reached, a hold period of 30 minutes should establish whether or not a leak has developed. Some yielding can be taking place while holding at the maximum pressure. Yielding will cease upon repeated repressurization to the maximum pressure, whereas a leak likely will not.

Testing an existing pipeline to a level at which yielding can occur may or may not be a good idea. It depends on the number and severity of defects in the pipe, the purpose of the test, and the level of maximum operating pressure that is desired. More will be said about this in the next section.

Finally, on the subject of testing to actual yield, the following statements apply:

- Yielding does not hurt or damage sound pipe. If it did, no one would be able to make cold-expanded pipe or to cold bend pipe.
- Yielding does not damage the coating. If it did, one could not field-bend coated pipe or lay coated pipe from a reel barge.
- Very little pipe actually undergoes yielding in a test to 110% of SMYS.
- Those joints that do yield do not affect pipeline integrity, and the amount of yielding is small.
- The only thing testing to a level in excess of 100% of SMYS may do is to void a manufacturer's warranty to replace test breaks if such a warranty exists.

Testing existing pipelines

Testing of an existing pipeline is a possible way to demonstrate or revalidate its serviceability. For a variety of reasons, retesting of an existing pipeline is not necessarily the

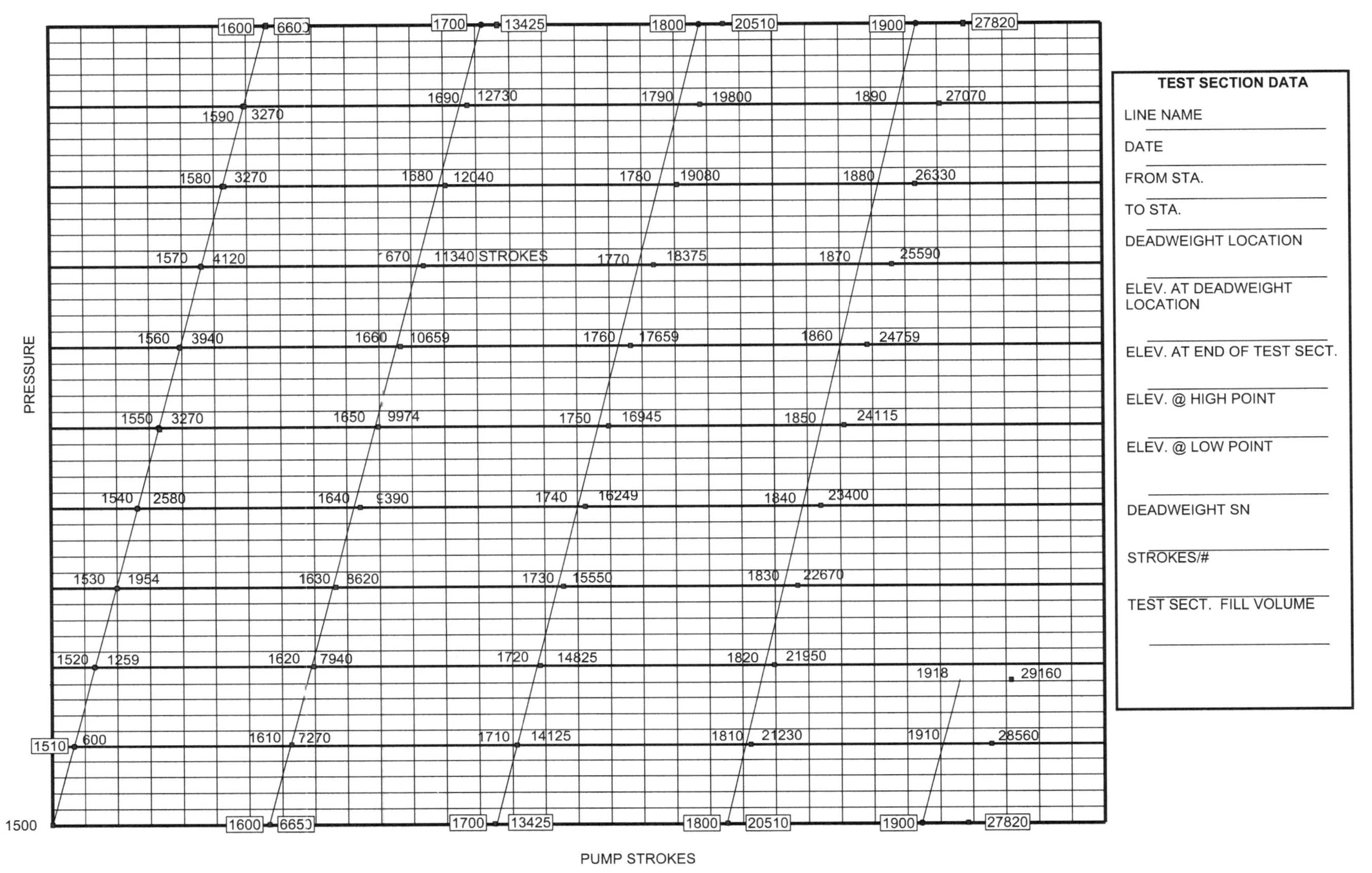

Figure 4. Pressure volume plot.

best means to achieve confidence in its serviceability, however. First, a pipeline operator who elects to retest a pipeline must take it out of service and purge it of product. The downtime represents a loss of revenue and a disruption to shippers. Second, the operator must obtain test water. To fill 30 miles of a 16-in. pipeline, an operator would need nearly 40,000 barrels of water. This is equivalent to a 100 × 100-ft pond, 22 ft in depth. For 30 miles of 36-in. pipe, the volume required would be five times as large. After the test, the water is considered a hazardous material because of being contaminated with product remaining in the pipeline. And, a test break, if one occurs, releases contaminated water into the environment. Aside from these issues, some problematic technical considerations exist.

The most important reason why a hydrostatic test may not be the best way to validate the integrity of an existing pipeline is that in-line inspection is often a better alternative. From the standpoint of corrosion-caused metal loss, this is almost certainly the case. Even with the standard resolution tools that first emerged in the late 1960s and 1970s this was true. Consider Figure 5. This figure shows the relative comparison between using a standard resolution in-line tool and testing a pipeline to a level of 90% of SMYS. The assumption is made in this case that the operator excavates and examines all "severe" and "moderate" anomalies identified by the tool, leaving only the "lights" unexcavated. In terms of the 1970s technology, the terms "light," "moderate," and "severe" meant the following:

- Light indications: metal loss having a depth less than or equal to 30% of the wall thickness.
- Moderate indication: metal loss having a depth more than 30% of the wall thickness but less than 50% of the wall thickness.
- Severe indication: metal loss having a depth more than 50% of the wall thickness.

In Figure 5, the boundary between "light" and "moderate" is nearly at the same level of failure pressure as the 90% of SMYS test for long defects, and it is well above that level for short defects. Because even the standard resolution tools have some defect–length-indicating capability, an in-line inspection on the basis represented in Figure 5 gives a better assurance of pipeline integrity than a hydrostatic test to 90% of SMYS. With the advent of high-resolution tools, the

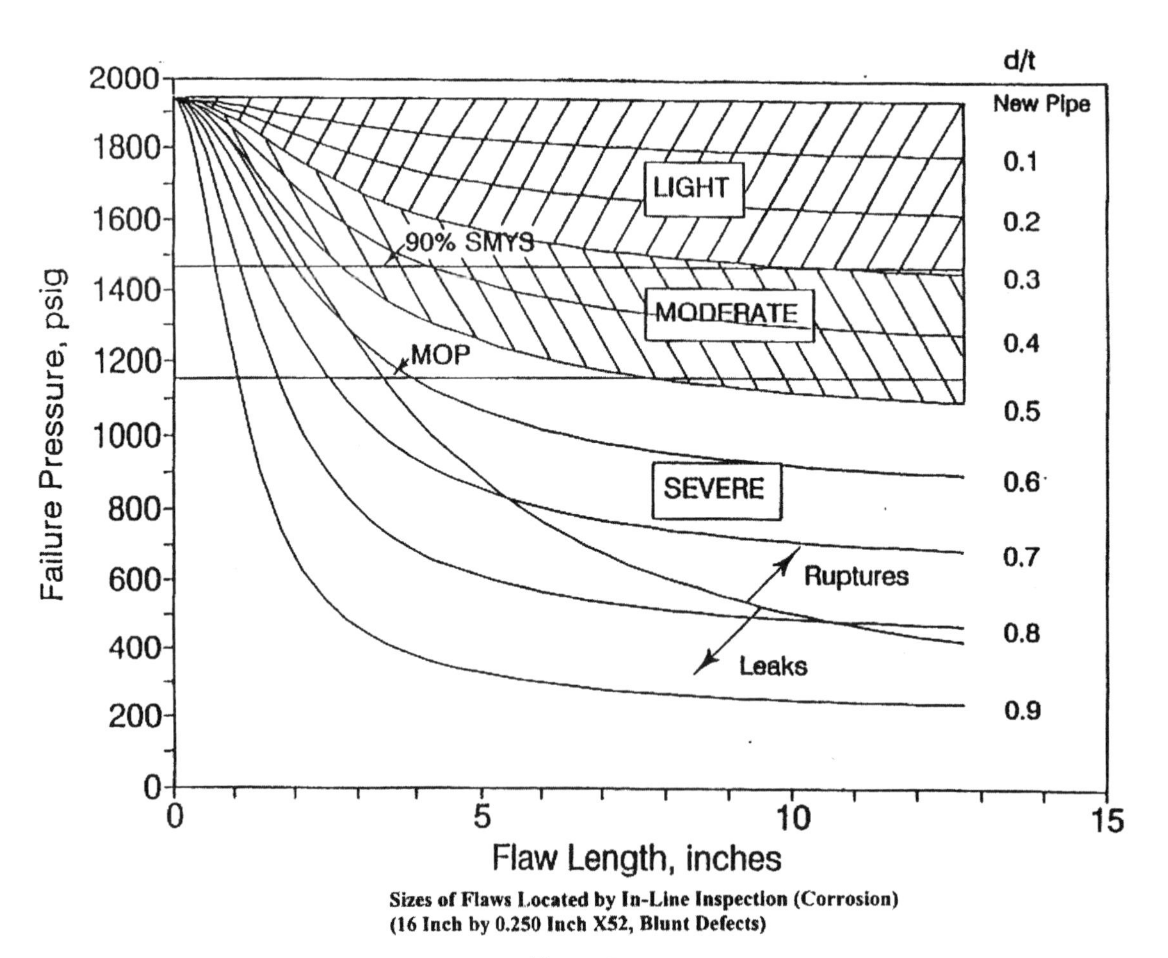

Sizes of Flaws Located by In-Line Inspection (Corrosion)
(16 Inch by 0.250 Inch X52, Blunt Defects)

Figure 5.

advantage shifts dramatically in favor of using in-line inspection instead of hydrostatic testing to validate the serviceability of a pipeline affected by corrosion-caused metal loss.

From the standpoint of other types of defects, the appropriate in-line inspection technology is evolving rapidly and, in some cases, it has proven to be more effective than hydrostatic testing. One example is the use of the elastic-wave tool for detecting seam-weld defects in submerged arc-welded pipe.[7] Another is the use of transverse-field magnetic-flux-leakage inspection to find seam anomalies alongside or in the seams of electric-resistance-welded (ERW) pipe.[8] In these cases, the particular tools revealed defects that were too small to have been found by a hydrostatic test to any reasonable level up to and including 110% of SMYS. When a tool has established this kind of track record, a pipeline operator can justify using the tool instead of hydrostatic testing.

The concept of using in-line tools to detect flaws invariably raises the question about defects possibly not being detected. The reasonable answer is that the probability of nondetection is small (acceptably small in the authors' opinion) but not zero. In the same context, one must also recognize that hydrostatic testing is not foolproof either. One issue with hydrostatic testing is the possibility of a pressure reversal (discussed later). The other issue is that because hydrostatic testing can leave behind defects that could be detected by in-line inspection, the use of hydrostatic testing often demonstrates serviceability for only a short period of time if a defect-growth mechanism exists. This possibility is discussed in our companion paper to this article.[9]

Pressure reversals

A pressure reversal is defined as the occurrence of a failure of a defect at a pressure level that is below the pressure level that the defect has previously survived due to defect growth produced by the previous higher pressurization and possible subsequent damage upon depressurization. Pressure reversals were observed long before their probable cause was identified.[3] The pipeline industry supported a considerable amount of research to determine the causes of pressure reversals. The most complete body of industry research on this subject is Kiefner, Maxey, and Eiber.[4] Figure 6, taken from that source, reveals the nature of experiments used to create and demonstrate pressure reversals. It shows photographs of highly magnified cross sections of the tips of six longitudinally oriented flaws that had been machined into a single piece of 36-in. OD by 0.390 in. w.t. X60 pipe. Each flaw had the same length, but each was of a different depth, giving a graduation in severities. When the single specimen containing all six flaws was pressurized to failure, the deepest flaw (Flaw 1) failed. By calculations based on their lengths and depths, the surviving flaws were believed to have been pressurized to the percentages of their failure pressures shown in Table 1.

Table 1

Flaw Number	Test Pressure Level at Failure of Flaw 1 as a Percentage of the Calculated Failure Pressure of the Flaw
2	97
3	94
4	91
5	89
6	87

As one can see, the tips of Flaws 2, 3, and 4 exhibit some crack extension as a result of the pressurization to failure. The nearer the defect to failure, the more crack extension it exhibited. In fact, due to its extension during the test, Flaw 2 is now deeper than Flaw 1 was at the outset. Logic suggests that if we could have pressurized the specimen again, Flaw 2 would have failed at a level below that which it experienced during the testing of Flaw 1 to failure. Indeed, in similar specimens designed in a manner to allow subsequent pressurizations, that is exactly what often occurred. This type of testing led to an understanding of pressure reversals in terms of ductile crack extension occurring at near-failure pressure levels where the amount of crack extension is so great that crack closure upon depressurization does further damage, leading to the inability of the flaw to endure a second pressurization to the previous level. The pressure reversal is expressed as a percentage:

$$\text{Pressure reversal} = \frac{(\text{original pressure} - \text{failure pressure})}{(\text{original pressure} \times 100)}$$

Once the cause was known, the next key question was: What is the implication of the potential for pressure reversals on confidence in the safety margin demonstrated by a hydrostatic test? This question has been answered in particular circumstances, and the answer comes from numerous examples of actual hydrostatic tests.*

* In actual hydrostatic tests, direct evidence that pressure reversals are the result of the type of flaw growth shown in Figure 6 has seldom been obtained. However, a few such cases have been documented and it is assumed that defect growth is responsible for all such cases.

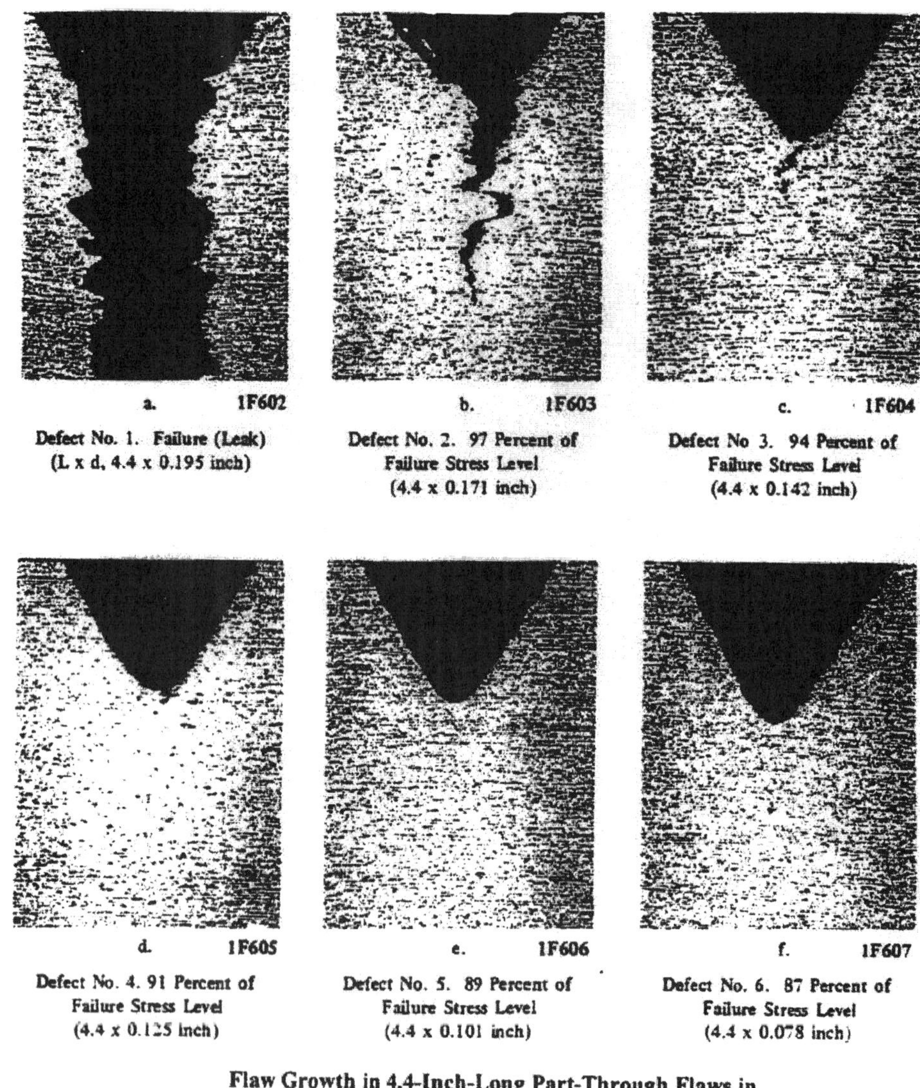

Note: Loading consisted of –
1st cycle – 0 → 1330 psig with 30 sec hold
2nd cycle – 0 → 1300 psig with 30 sec hold
3rd cycle – 0 → 1230 psig with 30 sec hold

Figure 6.

Figure 7 is an example of an analysis of pressure reversals in a specific test. This figure is a plot of sizes of pressure increases or decreases (reversals) on subsequent pressurizations versus the frequency of occurrence of that size of increase or decrease. Among other things, these data show that upon pressurization to the target test-pressure level, a 1% pressure reversal (34 psi) can be expected about once in every 15 pressurizations, a 2% pressure reversal (68 psi) can be expected about once in every 100 pressurizations, and a 3% pressure reversal (102 psi) can be expected about once in every 1,000 pressurizations. For a target test pressure level of 1.25 times the MOP, the expectation of a 20% pressure reversal (enough to cause failure at the MOP) is off the chart, that is, it is an extremely low-probability event (but not an impossible event).

There have been a handful of pipeline service failures in which a pressure reversal is the suspected but unproven

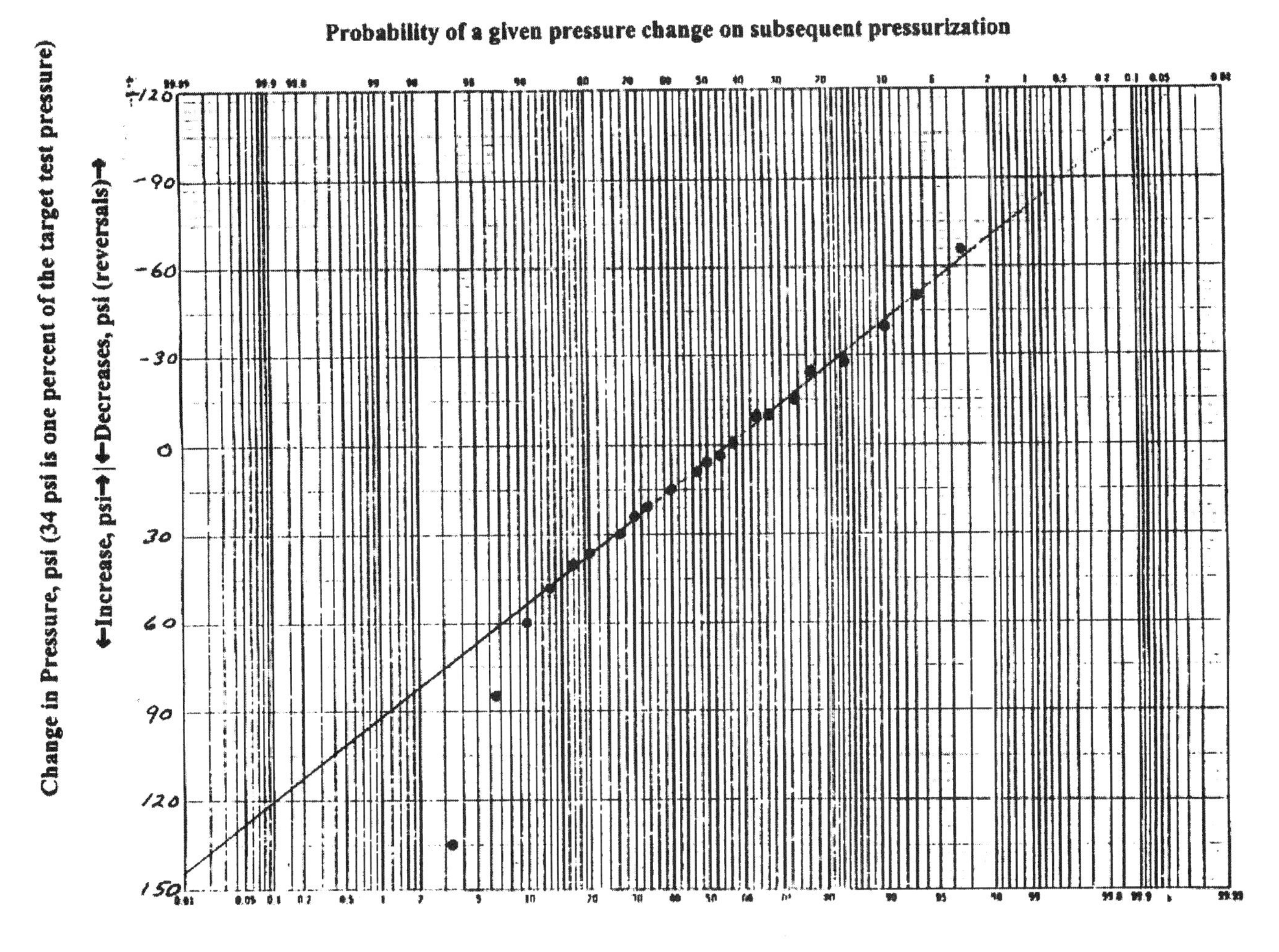

Figure 7.

cause. There is also one case of a large pressure reversal (62%) that was unequivocally demonstrated because it occurred on the fifth cycle of a five-cycle hydrostatic test.[4] It should be noted that most of the experiences of numerous and large pressure reversals in actual hydrostatic tests have involved flaws associated with manufacturing defects in or near ERW seams, particularly in materials with low-frequency welded (generally pre-1970) ERW pipe. But in most cases where numerous reversals occurred, the sizes of the actual pressure reversals observed are small (less than 5%). One thing seems clear: if a hydrostatic test can be successfully accomplished without the failure of any defect, the likelihood of a pressure reversal will be extremely small. It is the tests in which numerous failures occur that have the highest probabilities of reversals. And, when the number of reversals becomes large, the probability of a reversal of a given size can be estimated as was done on the basis of Figure 7.

Practical considerations

For new pipeline materials made to adequate specifications with adequate inspection and pipe-mill testing, one does not expect test failures even at pressure levels corresponding to 100% or more of SMYS. Therefore, there is no reason not to test a pipeline constructed of such materials to levels in excess of 100% of SMYS. As has been shown, the higher the ratio of test pressure to operating pressure, the more confidence one can have in the serviceability of a pipeline. In the case of existing pipelines, especially the older ones, such test levels may be impossible to achieve, and if numerous test failures occur, the margin of confidence may become eroded by the potential for pressure reversals. Weighing against low test-pressure-to-operating-pressure ratios, on the other hand, is the fact that such tests, by definition, generate lower levels of confidence and buy less time between retests if the issue of concern is time-dependent defect growth. A discussion of some possible responses to this dilemma follows.

First and foremost, as has already been mentioned, the use of an appropriate in-line inspection tool is always to be preferred to hydrostatic testing if there is sufficient confidence in the ability of the tool to find the defects of significance. Most of the pipe in a pipeline is usually sound. Therefore, it makes sense to use a technique that will find the critical defects and allow their repair as opposed to testing the whole pipeline when it is not necessary. The industry now has access to highly reliable tools for dealing with corrosion-caused metal loss, and tools are evolving rapidly to detect and characterize cracks. As has been noted, some uses of these tools have already proven their value and, in those cases, their use in lieu of hydrostatic testing makes good sense.

There are still certain existing pipelines for which hydrostatic testing remains the best (in some cases the only) means to revalidate their serviceability. In those cases, the following advice may be useful. Determine the mill hydrostatic test level for the pipe. The mill-test certificates will show the level applied if such certificates can be found. Also, search the records for prior hydrostatic tests at or after the time of construction. Review the pressure levels and causes of mill-test or in-place test failures if they exist. If none of these records is available, look up the API 5L specification applicable to the time the pipe was manufactured. This will reveal the standard mill-test pressure for the pipe. Do not assume, if you do not know, that the pipe was tested in the mill to 90% of SMYS. This was not always the case especially for non-X grades and smaller diameter pipe materials. If you decide to test the pipe to a level in excess of its mill-test pressure for the first time ever, anticipate test failures. If you cannot tolerate test failures, consider testing to a level just below the mill-test pressure. This may mean, of course, that the MOP you validate is less than 72% of SMYS (the minimum test pressure must be at least 1.25 times the MOP for 4 hours plus 1.10 times MOP for 4 hours for a buried pipeline; see Federal Regulation Part 195). Alternatively, if you can tolerate at least one test failure, pressurize to a level as high as you wish or until the first failure, whichever comes first. If you then conduct your 1.25-times MOP test at a level at least 5% below the level of the first failure, a second failure will be highly improbable.

It is always a good idea to conduct an integrity test as a "spike" test. This concept has been known for many years,[4] but more recently it has been advocated for dealing with stress-corrosion cracking.[10] The idea is to test to as high a pressure level as possible, but to hold it for only a short time (5 min. is good enough). Then, if you can live with the resulting MOP, conduct your 8-hour test at a level of at least 5% below the spike-test level. The spike test establishes the effective test-pressure-to-operating-pressure ratio; the rest of the test is only for the purpose of checking for leaks and for meeting the requirements of Part 195.

Summary

To summarize, it is worth reporting the following:

- Test-pressure-to-operating-pressure ratio measures the effectiveness of the test.
- In-line inspection is usually preferable to hydrostatic testing.
- Testing to actual yield is acceptable for modern materials.
- Pressure reversals, if they occur, tend to erode confidence in the effectiveness of a test but usually not to a significant degree.
- Minimizing test-pressure cycles minimizes the chance for pressure reversals.

References

1. Bergman, S. A., "Why Not Higher Operating Pressures for Lines Tested to 90% SMYS?" *Pipeline and Gas Journal*, December 1974.
2. Kiefner, J. F., Maxey, W. A., Eiber, R. J., and Duffy, A. R., "Failure Stress Levels of Flaws in Pressurized Cylinders," *Progress in Flaw Growth and Toughness Testing*, ASTM STP 536, American Society for Testing and Materials, pp. 461–481, 1973.
3. Brooks, L. E., "High-Pressure Testing—Pipeline Defect Behavior and Pressure Reversals," ASME, 68-PET-24, 1968.
4. Kiefner, J. F., Maxey, W. A., and Eiber, R. J., *A Study of the Causes of Failure of Defects That Have Survived a Prior Hydrostatic Test*, Pipeline Research Committee, American Gas Association, NG-18 Report No. 111, November 3, 1980.
5. Duffy, A. R., McClure, G. M., Maxey, W. A., and Atterbury, T. J., *Study of Feasibility of Basing Natural Gas Pipeline Operating Pressure on Hydrostatic Test Pressure*, American Gas Association, Inc., Catalogue No. L30050, February 1968.
6. Maxey, W. A., Podlasak, R. J., Kiefner, J. F., and Duffy, A. R., "Measuring the Yield Strength of Pipe in the Mill Expander," Research report to A.G.A./PRC, A.G.A. Catalog No L22273, 61 pages text plus 2 appendices, 23 pages, December 1973.
7. Maxey, W. A., Mesloh, R. E., and Kiefner, J. F., "Use of the Elastic Wave Tool to Locate Cracks Along the DSAW Seam Welds in a 32-Inch (812.8 mm) OD Products Pipeline," *International Pipeline Conference*, Volume 1, ASME, 1998.

8. Grimes, K., "A Breakthrough in the Detection of Long Seam Weld Defects in Steel Pipelines," Pipeline Integrity International.
9. Kiefner, J. F., and Maxey, W. A., "Periodic Hydrostatic Testing or In-line Inspection to Prevent Failures from Pressure-Cycle-Induced Fatigue," Paper Presented at API's 51st Annual Pipeline Conference & Cybernetics Symposium, New Orleans, Louisiana, April 18–20, 2000.
10. Leis, B. N., and Brust, F. W., *Hydrotest Strategies for Gas Transmission Pipelines based on Ductile-Flaw-Growth Considerations*, Pipeline Research Committee, American Gas Association, NG-18 Report No. 194, July 27, 1992.

Hydrostatic testing for pipelines

After construction of a pipeline is completed, or if pipe has been replaced or relocated, it is necessary to hydrostatically test the pipeline to demonstrate that the pipeline has the strength required to meet the design conditions, and to verify that the pipeline is leak free.

The U.S. Federal Safety Regulations for Pipelines require that pipelines used to transport hazardous or highly volatile liquids be tested at a pressure equal to 125% of the maximum allowable operating pressure (MAOP) for at least 4 continuous hours and for an additional 4 continuous hours at a pressure equal to 110% or more of the MAOP if the line is not visually inspected for leakage during the test. A design factor of 72% of the specified minimum yield strength (SMYS) of the pipe is used to determine the maximum allowable operating pressure. The requirement to test to 125% of the MAOP will therefore cause the pipe to be tested to a pressure equal to 90% of the SMYS of the pipe. See Section 3—Pipe Design for additional information on calculating the MAOP.

The regulations for gas lines specify design factors based on the class location of the pipeline. The class location is determined by the number of buildings in a specified area on either side of the pipeline. Refer to Part 192.111 of the Minimum Federal Safety Standards for Gas Lines for the details on how to determine the class location. See Section 3, Pipe Design for a listing of the design factors for the different class locations. The regulations specify that the test pressure must be maintained for at least 8 hours and must be equal to at least 125% of the maximum allowable operating pressure. The regulations should be consulted for the specific testing requirements, as the regulations are subject to change.

Usually, operators will specify a test pressure range from 90% to 95% of the SMYS of the pipe. Some will allow test pressures as high as 100% of the SMYS of the pipe and some will test to slightly beyond the SMYS of the pipe. Specifying a test pressure at least equal to 90% of the SMYS of the pipe will qualify it for the maximum allowable operating pressure. In some cases, the test pressure will be based on the minimum yield strength determined from the mill test reports.

Hydrostatic testing a pipeline is certainly a major operation and should be carefully planned. Most companies have hydrostatic test manuals that detail the procedures to be followed to complete the test. Usually, this work is performed by a hydrostatic testing contractor hired by the pipeline owner, or hydrostatic testing may be included as a part of the main construction contract.

One of the first steps in planning the hydrostatic test operation is to examine the elevation gradient. The gradient, along with the location of the water source, and the pipe design data, will be used to determine the length and number of test sections. Figure 1 shows a typical pipeline elevation gradient.

Where the pipeline traverses hilly terrain, the elevation gradient must be carefully considered in selecting the pipeline test segments. Different companies have differing philosophies on how to do this. Some limit the amount of elevation difference while others may specify a range of allowable percentages of the SMYS of the pipe—i.e., 90% to 95% or 90% to 100% of the SMYS of the pipe. In any case, the test gradient should be plotted to be sure the test pressure falls within the specified pressure limits.

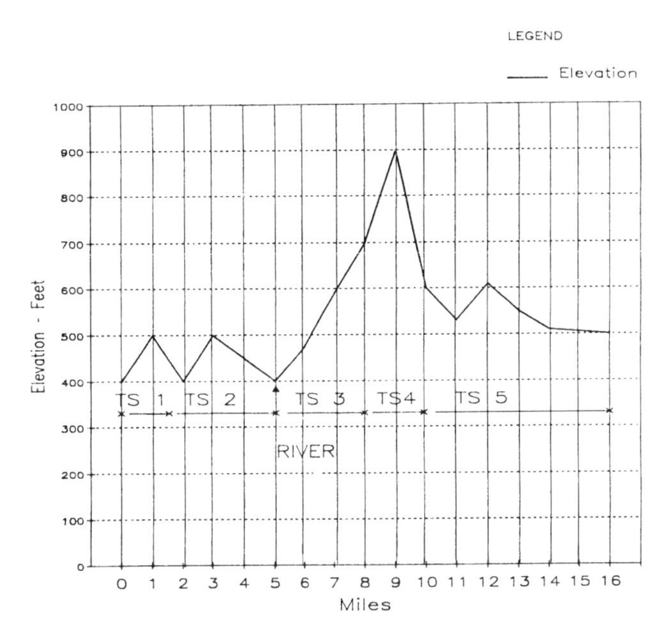

Figure 1. Pipeline elevation profile.

The test gradient must be based on water head in feet, if water is used as the test medium. Water pressure may be converted to head by dividing the pressure by 0.433. If sea water is being used as the test medium, a factor of 0.445 should be used. This assumes a specific gravity of 1.02 for sea water. In any case, the factor of 0.433 should be modified according to the specific gravity of the test medium.

Let's assume that we have a line that is to be tested and the elevation at the test site is 1,000 feet and the elevation at the end of the line is 1,200 feet. Fresh water will be used as the test medium, and it is desired to test the pipe to a minimum of 90% and a maximum of 96% of the SMYS of the pipe. The pipe is 30″ OD × 0.390″ wt API 5LXX60. There is a difference in elevation of 200 ft and this is equal to 86.6 psig (200 ft × 0.433 = 86.6 psig). A test pressure equal to 90% of SMYS is 1,404 psig. Since the test site is lower than the high end of the line, the 86.6 psig is added to 1,404 psig to obtain a test site pressure of 1,491 psig. The pressure at the end of the line will be 1,404 psig, which equates to 90% of SMYS. The pressure at the low point equates to 96% of SMYS.

Now, let's assume a line has a high point elevation of 1,100 ft, a low elevation point of 1,000 ft, and the elevation at the test site is 1,050 ft. The test pressure at the high point will need to be 1,404 psig in order to meet the 90% of SMYS requirement. The pressure at the low point will be 1,447 psig, and the pressure at the test site will be 1,426 psig.

When testing offshore lines, the pressure at or below the water surface will be the same as at the low elevation point due to the offsetting external sub-sea pressure head. If the line previously described was laid offshore, the test pressure at the water surface would be 1,404 psig. The elevations are as follows:

Top if riser	+11 feet
Test site	+7 feet
Water surface	0 feet
Pipe depth	−168 feet

The test pressure at the test site would be 1,404 − (7 × 0.445) or 1,401 psig. At the top of the riser, the test pressure would be 1,404 − (11 × 0.445) or 1,399 psig.

The typical profile shown in Figure 1 represents a pipeline that requires testing. The pipeline crosses a river at approximately MP 5, and river water will be used to test the line. Test sections 1 through 5 have been chosen as indicated. The lengths of sections 3 and 4 are limited by elevation difference. This line was designed to operate at 936 psig or 72% of the SMYS of 30″ × 0.375″ wt API 5LX X52 pipe for Class I locations, and 60% of the SMYS of 30″ × 0.390″ wt API 5LX X60 pipe for Class II locations, and 50% of the SMYS of 30″ × 0.438″ wt API 5LX X65 for Class III locations.

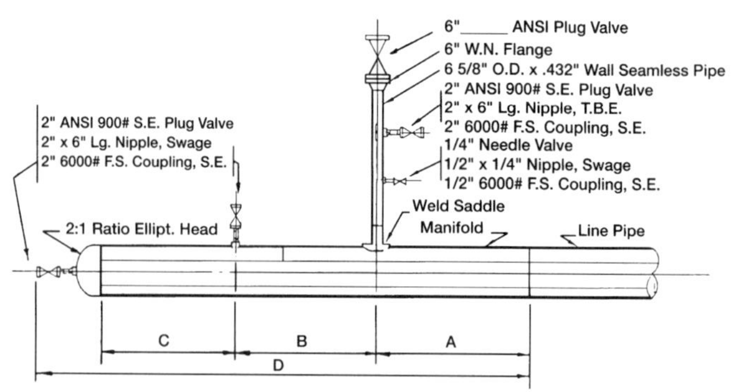

Figure 2. Typical pressure manifold. Courtesy Milbar Hydrotest.

The pipeline crosses one railroad, one highway, and the river. It also includes one main line block valve assembly. The valve assembly, river crossing, and road crossing were pre-tested before installation and are tested again after installation.

Four test manifolds were installed to facilitate filling the line and for isolating the test sections during the test operation. See Figure 2 for a typical pressure sectionalizing manifold. Two-way pigs were loaded at each of the intermediate manifolds. The pigs are moved by the fill water and are necessary to remove the air from the line.

It is generally a good idea to test the section most distant from the water source first. If it should rupture, then testing on intermediate sections can continue while repairs on the failed section are completed.

The pumps used to fill the line should have sufficient capacity to fill the line at a rate of about ½ mile per hour. Water filters should be used and are normally equipped with 100 mesh screens. See Figure 4 for a typical fill site arrangement. Some companies may specify a finer mesh screen. The filling unit should be equipped with a flow meter to measure the amount of water pumped into the pipeline. Temperature recorders will be used to record ambient temperature, the temperature of the pipe and water, and

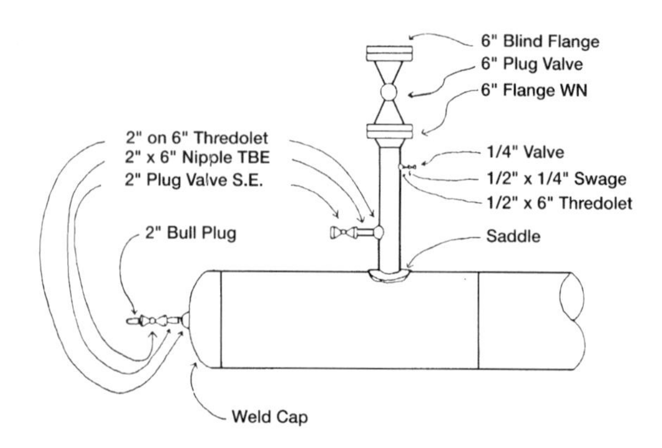

Figure 3. Typical test manifold. Courtesy Milbar Hydrotest.

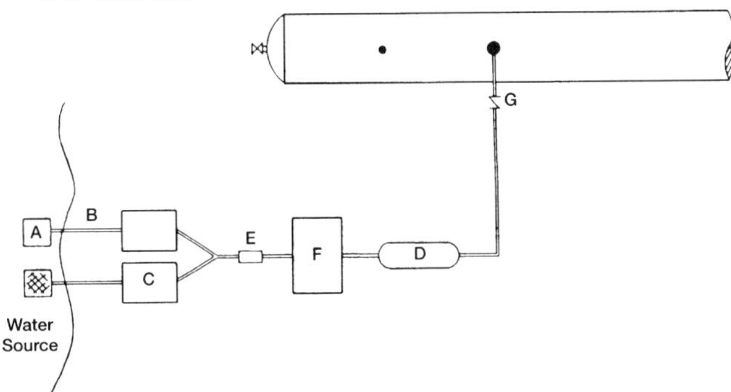

Figure 4. Typical fill site. Courtesy Milbar Hydrotest.

ground temperature. See Figure 5 for typical test thermometer placement. A pressure recorder will be used to provide a recording of the pipeline test pressure. A deadweight tester will be used to calibrate the pressure recorder. A high pressure pump capable of delivering 70 to 150 gpm at a pressure exceeding the required test pressure will be used in the final pressuring operation. Small high pressure units with a capacity of 6 to 30 gpm may be used for short sections of large diameter lines and smaller diameter lines.

Deadweight pressure and temperature readings are recorded after the prescribed test pressure has been reached and the pressures and temperatures have stabilized. A pressure vs. time plot may also be made. Readings are usually made at 15-minute intervals for the first hour and at 30-minute intervals thereafter. The procedure for accepting a leakage test will vary from company to company. Some will accept the leakage test if there is no pressure drop in a 3-hour period.

A pressure-volume plot may be required, especially if the test pressures approach or exceed the SMYS of the pipe. The plot is made manually during the pressuring operation by recording pump strokes on the X-axis and pressure on the Y-axis. A straight line will be produced until the plastic range of the pipe is reached or until a leak occurs. This type of plot is also useful should a joint of lighter wall thickness or lower yield strength pipe be inadvertently placed in the pipeline.

Small leaks during the testing operation can be difficult to locate. A change in the water/pipe temperature may give the appearance of a leak. If the temperature of the pipe/water decreases, the test pressure will decrease. An increase in water/pipe temperature will cause the test pressure to increase. The effect of a temperature change may be estimated using the equations and data contained in Appendixes A and B. To achieve any degree of accuracy in

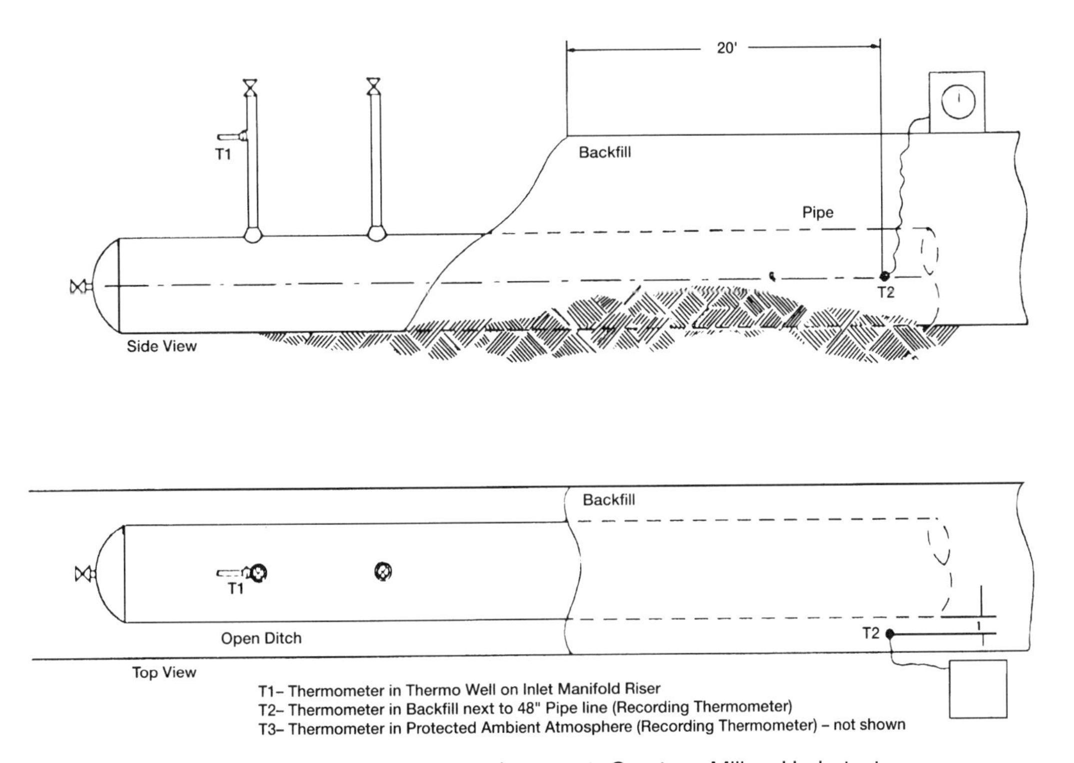

Figure 5. Test thermometer placement. Courtesy Milbar Hydrotest.

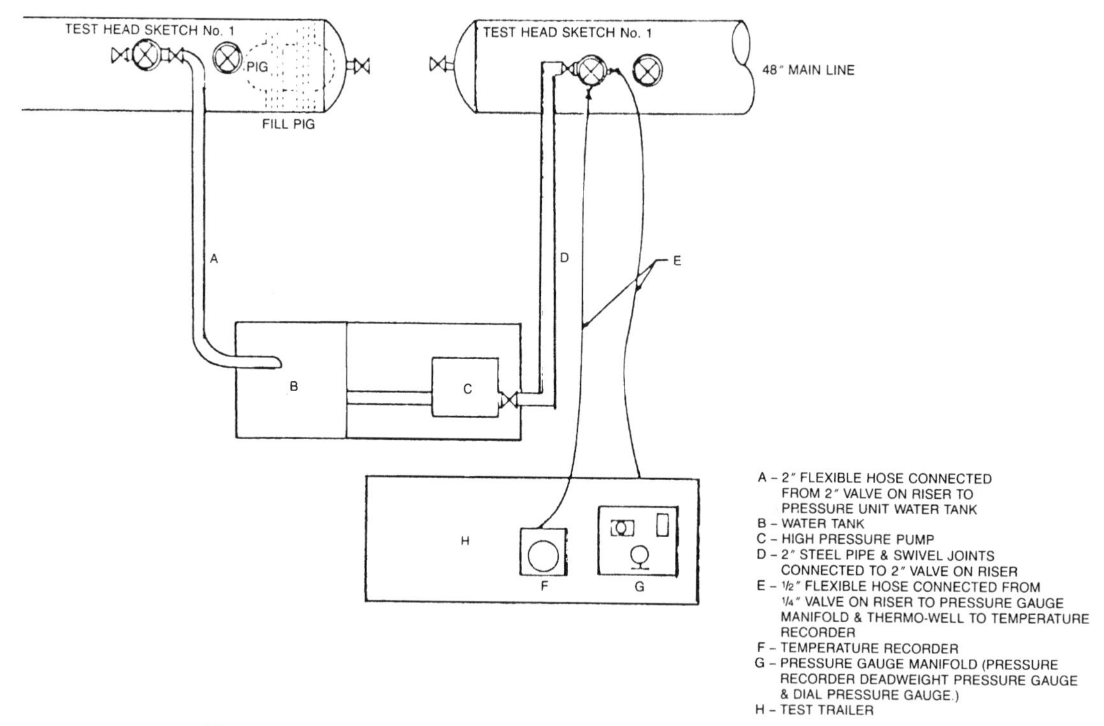

Figure 6. Typical pressure unit set-up. Courtesy Milbar Hydrotest.

these calculations, it will be necessary to have accurate temperature and pressure readings.

The pipeline test temperature may be affected by river/creek crossings along the pipeline route. Also, if cold river water is used, there will be a changing temperature gradient. Air, either trapped or entrained, will also affect the pressure-temperature calculations since the coefficient of expansion for air is not the same as for water. If the calculations are made and the decrease in pressure cannot be attributed to a temperature change, then it may be necessary to sectionalize the test section by manifolding and repressuring the shorter segments to facilitate locating the leak. Once the leak has been located and repaired, testing operations can resume.

Air compressors will be required to remove the water once the testing is complete. The maximum head for this particular project was 450 ft (195 psig). Allowance will need to be included for friction loss in the cross-over manifolds. The compressors will need to have sufficient capacity to remove the water at about the same rate as the fill rate. Most portable compressors are limited to approximately 125 psig. In this case, booster compressors will be needed for the 195+psig. See Figure 7 for a typical dewatering connection.

Upon completion of testing, the test pressure is bled off to leave approximately 10 to 20 psig at the high point in the test section. Dewatering may be accomplished using the air compressors discussed earlier. Some companies may elect to displace the test water with the material that will normally be pumped through the line. This is usually the case with crude oil and refined product pipelines. Pipelines used to transport natural gas and certain chemicals are usually dewatered using compressed air. Care must be exercised in the dewatering operation to make sure that no air is introduced into the test section, thus minimizing the possibility of air locks. An air lock is probably the most severe problem involved in dewatering. Air locks are caused by air accumulating in the downhill leg and water accumulating in the uphill leg, which creates a manometer in the pipeline. In some instances, extremely high pressures may be required to overcome the manometer. Care should be exercised to be sure the maximum allowable pipeline operating pressure is not exceeded. It may also be necessary to tap the line and vent air at the high points.

If it is necessary to pipe the water away from the right-of-way, welded pipe should be used. The pipe must be securely anchored. The use of tractors and skids will probably provide sufficient anchoring.

Always use a valve to control the amount of water being bled from the pipeline.

If there will be a delay in placing the pipeline in operation, and it is decided to leave the test water in the line until operations begin, consideration should be given to the use of a corrosion inhibitor. Long tenn storage of the line may necessitate the use of oxygen scavengers in addition to the inhibitor.

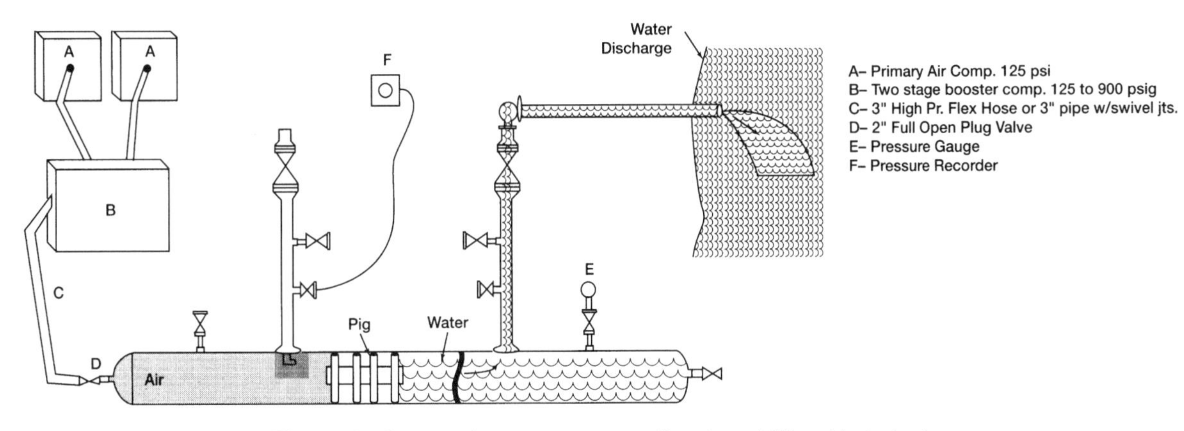

Figure 7. Dewatering arrangement. Courtesy Milbar Hydrotest.

FIELD PRESSURE & TEST REPORT

Page ______ of ______

Contractor or Company: ______

Pipe Line Description: ______

Section Tested: From: ______ To: ______

Test Section No.: ______ Length: ______

Type and Size of Pipe: ______ O. D. x ______ W.T. x Grade ______ Manufacturer: ______

Pressure Unit No.: ______ Gallons per Stroke: ______ Fill Unit No.: ______

Pressure Unit Location: ______ Water Source ______

Time and Date Test Started: ______ AM PM ______ Test Pressure (Maximum): ______ PSIG

Time and Date Test Ended: ______ AM PM ______ Pressure Volume Plot: Yes ______ No ______

Section Accepted ______ Section Leaking ______ Section Ruptured ______

DEADWEIGHT RECORDED READING (PSIG)

TIME	PRESSURE	TEMP DEG.F	TIME	PRESSURE	TEMP DEG. F	TIME	PRESSURE	TEMP DEG. F
Date: ______								
AM PM			AM PM			AM PM		

Strokes per 10 psi: ______ Remarks: ______

Milbar Supt. ______ Inspector ______

Witness: 1. ______ 2. ______

Gauges and Recorders Last Tested ______ Deadweight Serial No. ______

Figure 8. Typical test report. Courtesy Milbar Hydrotest.

If after dewatering, residual moisture in the pipeline poses a problem, it will be necessary to clean and dry the pipeline. See Section 6 for drying details.

A permit to discharge the water may be required, especially if there is a possibility the water may enter a stream of water. This should be determined and permits obtained, if required, in the planning phase of the testing operation. A "splash" plate or bales of hay may be used when dewatering to prevent erosion.

The most important aspect of any test operation is complete and accurate documentation, since it will become a permanent record that must be retained for as long as the facility tested remains in operation. Federal Pipeline Safety Regulations for pipelines transporting liquids require that these records include the following information:

- The pressure recording charts;
- Test instrument calibration data;
- The names of the operator, of the person responsible for making the test, and of the test company used, if any;
- The date and time of the test;
- The minimum test pressure;
- The test medium;
- A description of the facility tested and the test apparatus;
- An explanation of any pressure discontinuities, including test failures, that appear on the pressure recording charts; and,
- Where the elevation differences in the section under test exceed 100 ft, a profile of the pipeline that shows the elevation and test sites over the entire length of the test section.

The records required by the Federal Pipeline Safety Regulations for pipelines transporting gas are similar to those required for pipelines used to transport liquids.

A typical test report is shown in Figure 8. If failures occur during the test, a report documenting the failure and the suspected reasons for the failure should be completed. A typical form is shown in Figure 9.

A daily operating log should be used to record activities associated with the test, operating status of the test equipment in use, engine rpm's, and any unusual circumstances that occur during the test. A typical form is shown in Figure 10.

PIPE LINE FAILURE REPORT

1. Company ______
2. Section tested ______
3. Time of failure ______ AM, PM. Date ______
4. Location of failure: ______
5. Pressure, psig, at point of failure ______
6. Description of failure: Leak ______, Break ______, Length of failure ______
7. If leak, fill in blanks: ______ gallons lost per hour, ______ pounds lost per hour.
8. Describe any peculiarities or defects on failed part, such as corrosion or evidence of prior damage, etc. ______
9. Possible cause of failure ______
10. Pipe size: ______ OD x ______ Wt., Grade ______ Mfg. by ______
11. Repairs: Pipe installed ______ OD x ______ Wt., Grade ______ Mfg. by ______, Length of joint or pup ______
12. Date repaired ______ 19 ______ By ______
13. Remarks (Damage to property, persons injured, etc.) ______

Figure 9. Pipeline failure report. Courtesy Milbar Hydrotest.

DAILY OPERATING REPORT

COMPANY & CONTRACTOR ______________________ DATE __________

LOCATION (City) ______________________ CONTRACT NO. __________

PIPE LINE DESCRIPTION: ______________________

PERSONNEL: ______________________

EQUIPMENT: ______________________

ACTIVITIES: ______________________

COMPANY REP. CONTRACTOR REP. MILBAR SUPT.

Figure 10. Daily Operation Report. Courtesy Milbar Hydrotest.

APPENDIX A

Volume of water required to fill test section

$$V = L \times (3.1416 \times d^2/4) \times 12/231 \quad (1)$$

$$V = 0.0408 \times d^2 \times L$$

where: L = Length of test section, feet.
d = Inside diameter of pipe, inches.
V = Gallons required to fill @ Opsig.

Volume required at test pressure

$V_{tp} = V \times F_{wp} \times F_{pp} \times F_{pwt}$ (2)

V_{tp} = Gallons contained in test section at test pressure P, and temperature T – °F.

F_{wp} = Factor to correct for the compressibility of water due to increasing pressure from 0 psig to test pressure P, psig

F_{pp} = Factor to correct for volume change in pipeline due to pressure increase from 0 psig to test pressure P, psig

F_{pwt} = Factor to correct for the change in water volume and pipe volume due to change in pipe and water temperature from base of 60°F to pipe and water test temperature, T –°F.

$F_{wp} = 1/[1 - (4.5 \times 10^{-5}) \times (P/14.73)]$ (3)

P = Test pressure, psig.

$$F_{pp} = 1 + [(D/t) \times (0.91P/30 \times 10^6)] + [3.6 \times 10^{-6}(T - 60)] \quad (4)$$

D = Outside diameter of pipe, inches.
t = Pipe wall thickness, inches.
T = Pipe temperature, °F.

$F_{pwt} = F_{pt}/F_{wt}$ (5)

F_{pt} = Factor to correct for change in pipe volume due to thermal expansion of the pipe from base temperature of 60°F.

$F_{pt} = 1 + [(T - 60) \times 18.2 \times 10^{-6}]$ (6)

F_{wt} = Factor to correct for thermal change in specific water volume from 60°F to test water temperature. Refer to Table 1.

Sample calculation:

Pipe Size:	10.750″OD × 0.279″ w.t. X-52
Length:	5 miles
Test pressure:	2,430 psig
Temperature:	50°F

Use equation 1 to determine the initial line fill volume, V.

$$V = 0.0408 \times (10.75 - (2 \times 0.279))^2 \times 5{,}280 \times 5 = 111{,}888 \text{ gallons}$$

Volume required to achieve test pressure $= V_{tp} = V \times F_{wp} \times F_{pp} \times F_{pwt}$.

$$F_{wp} = 1/1 - [(4.5 \times 10^{-5}) \times (2430/14.73)] = 1.007479.$$

$$F_{pp} = 1 + [(10.75/0.279) \times (0.91 \times 2430/(30 \times 10^6))] + [3.6 \times 10^{-6} \times (50 - 60)] = 1.002804$$

$F_{pwt} = F_{pt}/F_{wt}$

$$F_{pt} = 1 + [(50 - 60) \times 18.2 \times 10^{-6}] = 0.999818$$

$F_{wt} = 0.9993061$ (from Table 1)

$$F_{pwt} = 0.999818/0.9993061 = 1.000512$$

$$V_{tp} = 111{,}888 \times 1.007457 \times 1.002804 \times 1.000512 = 113{,}099 \text{ gals.}$$

Incremental volume required to reach test pressure, P.

$$= 113{,}099 - 111{,}888 = 1{,}211 \text{ gals.}$$

After a period of time, the test pressure, P has decreased to 2,418 psig and the temperature of the pipe and test water has decreased to 48°F. The calculation procedure previously described may be repeated to determine if the pressure decrease is attributable to the decrease in temperature.

V = 111,888 gals.
P_1 = 2,422 psig
T_1 = 48°F.
F_{wp1} = 1.007454
F_{pp1} = 1.002796

F_{pt1} = 0.999781
F_{wt1} = 0.999217 (from Table 1)
F_{pwt1} = 0.999781/0.999217 = 1.000565
V_{tp1} = 111,888 × 1.007454 × 1.002796 × 1.00565 (@ 2,422 psig and 48°F)
V_{tpl} = 113,101 (Volume at reduced temperature—represents volume that would be required to raise test pressure to original value at 48°F)
V_{tp} = 113,099 (Volume at initial test conditions)

To restore the pressure to the original test pressure, two gallons of water could be added to the test section. If more than two gallons are required to restore the test pressure, there is the possibility that a leak may exist.

When making these calculations, care must be exercised to be sure that accurate temperature and pressure data is available. Where test sections are long, it is very likely that the temperature will not be uniform throughout the test section and thus affect the accuracy of the calculated results.

Table 1
F_{wt}—Factor to correct for the thermal change in the specific volume of water from 60°F to the test water temperature

Temp. (°F)	F_{wt}	Temp. (°F)	F_{wt}
35	0.9990777	70	1.0010364
36	0.9990590	71	1.0011696
37	0.9990458	72	1.0012832
38	0.9990375	73	1.0014229
39	0.9990340	74	1.0015420
40	0.9990357	75	1.0016883
41	0.9990421	76	1.0018130
42	0.9990536	77	1.0019657
43	0.9990694	78	1.0021222
44	0.9990903	79	1.0022552
45	0.9991150	80	1.0024178
46	0.9991451	81	1.0025561
47	0.9991791	82	1.0027251
48	0.9992168	83	1.0028684
49	0.9992599	84	1.0030435
50	0.9993061	85	1.0031919
51	0.9993615	86	1.0033730
52	0.9994112	87	1.0035573
53	0.9994715	88	1.0037133
54	0.9995322	89	1.0039034
55	0.9996046	90	1.0040642
56	0.9996683	91	1.0042601
57	0.9997488	92	1.0044357
58	0.9998191	93	1.0046271
59	0.9999074	94	1.0047972
60	1.0000000	95	1.0050043
61	1.0000803	96	1.0052142
62	1.0001805	97	1.0053915
63	1.0002671	98	1.0056067
64	1.0003746	99	1.0057884
65	1.0004674	100	1.0060090
66	1.0005823	101	1.0061949
67	1.0006811	102	1.0064207
68	1.0008031	103	1.0066108
69	1.0009290	104	1.0068417

APPENDIX B

How to use charts for estimating the amount of pressure change for a change in test water temperature

Example.

Pipe data	18 OD × 0.375″ wt
Water temp. at beginning of test	70°F
Water temp. at time T	66°F
Test pressure	1,800 psig

Calculate

D/t = 18/0.375 = 48

where:

D = Pipe OD, in.
t = Pipe wall thickness, in.

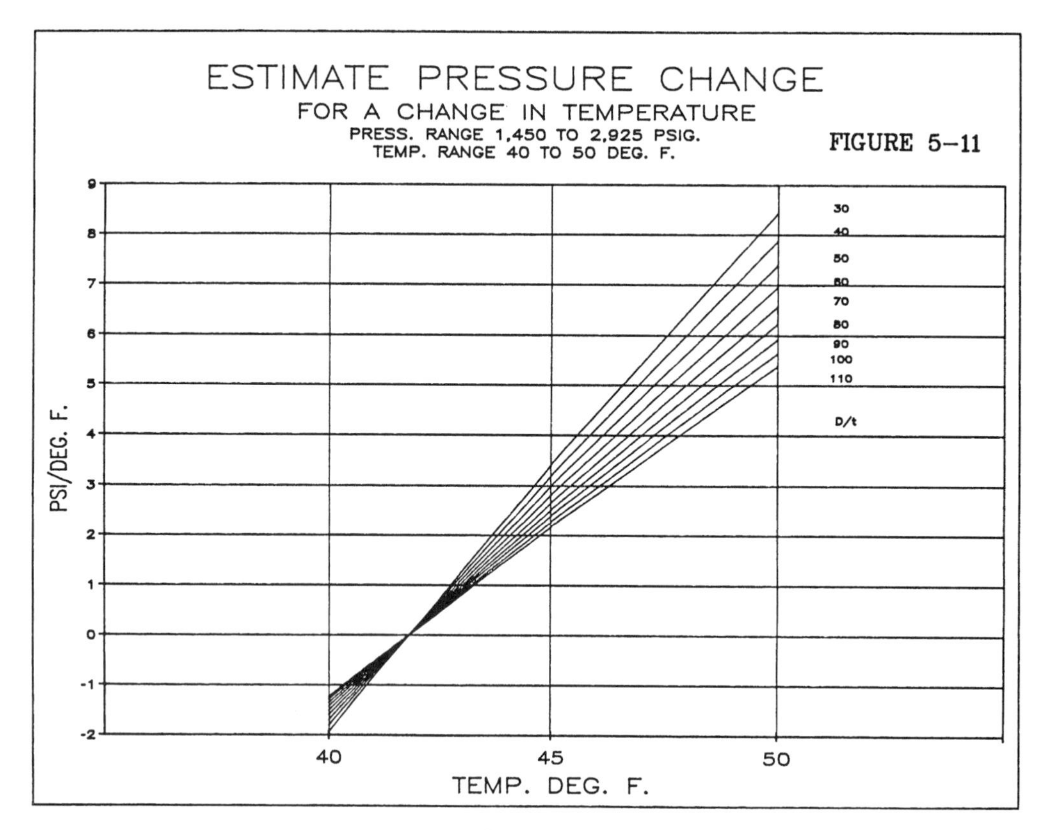

Figure 1.

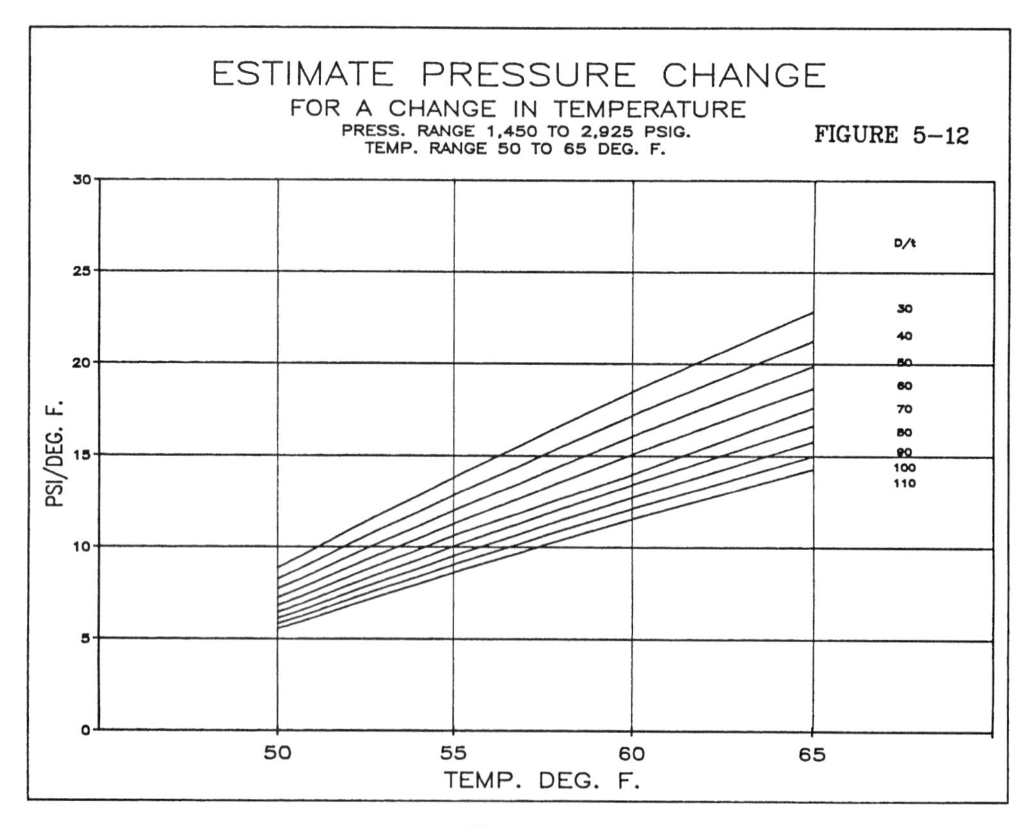

Figure 2.

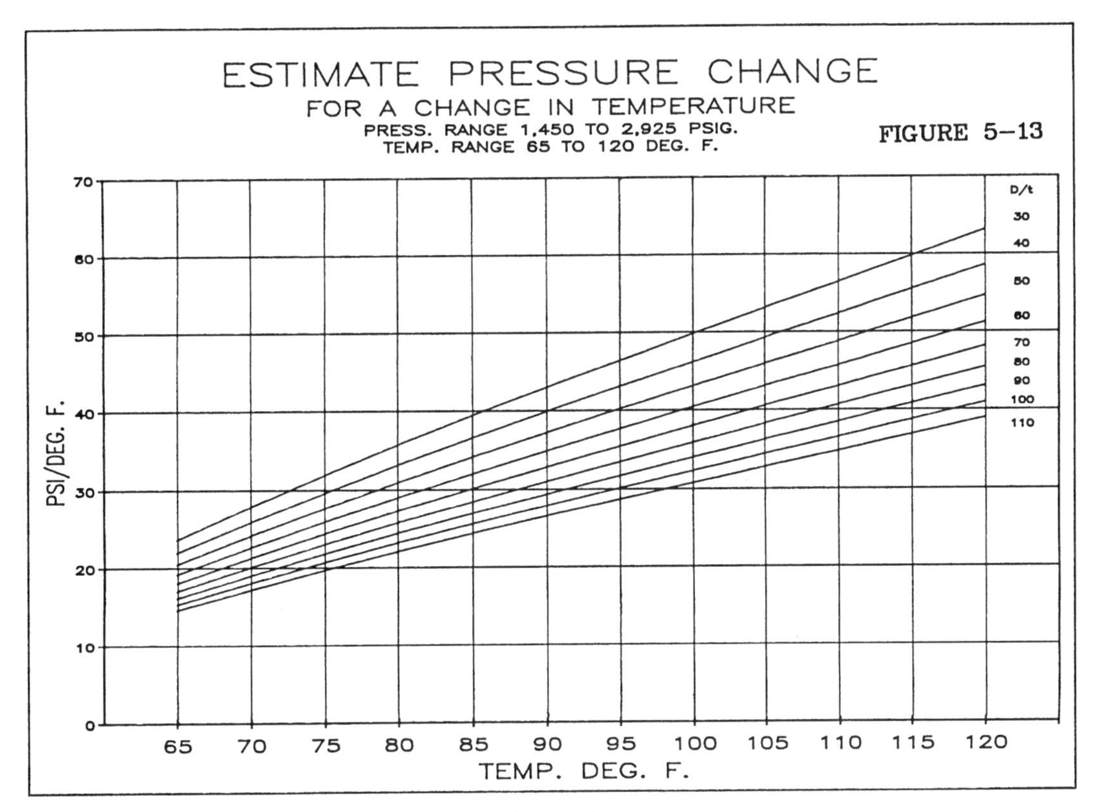

Figure 3.

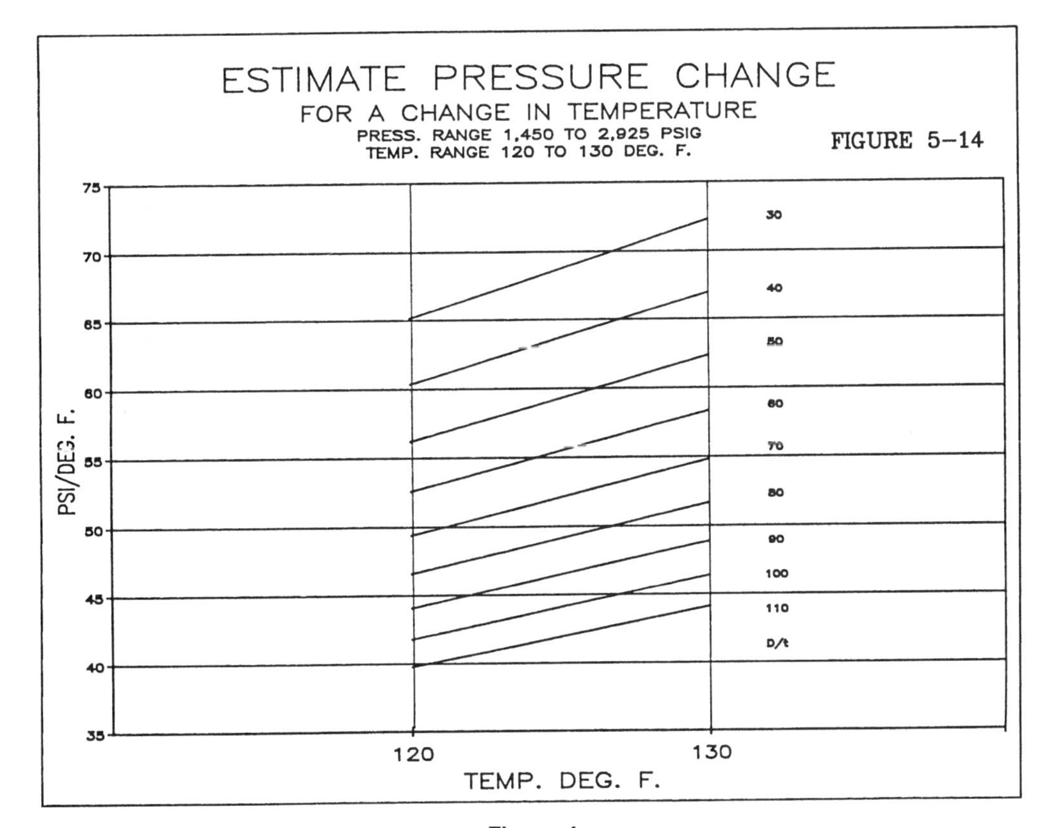

Figure 4.

Calculate

average temperature = (70 + 66)/2 = 68°F

Use the chart in Figure 3. Enter at 68°F and at the intersection with D/t line representing 48, read 23 psig/°F.

Basis for chart development

$$\Delta P = \frac{\beta - 2\alpha}{\frac{D \times (1 - v^2)}{E \times t + C}}$$

where:

DP = psig change per °C
β = Coefficient of expansion for water
α = Coefficient of linear expansion for steel 1.116×10^6 per °C
D = Pipe OD inches
v = Poisson's ratio 0.3
E = Young's modulus for steel 30×10^6
t = Pipe wall thickness, inches
C = Compressibility factor for water, cu.in./cu.in./psig

$$\beta \times 10^6 = -64.268 + (17.0105 \times T) - (0.20369 \times T^2) + (0.0016048 \times T^3)$$

T = water temperature, °C

Compressibility factor for water

Pressure range 1450 psig to 2925 psig

Temperature (°C)	Compressibility Factor cu.in./cu.in./psig
0	3.35×10^6
10	3.14×10^6
20	3.01×10^6
50	2.89×10^6

From *PIPECALC 2.0: Practical Pipeline Hydraulics*, Gulf Publishing Co., Houston, Texas.

Hydrostatic test records

At a minimum, the following records should be made for a hydrostatic test.

- Calculations supporting basis for the test pressure. This should take into account the high and low point elevations in the test section and an elevation profile of the test section.
- Pressure recording charts.
- Test instrument calibration. This will include the pressure recorder and temperature recorder. Record the serial numbers of the instruments used for the test.
- Date and time of the test.
- Temperature of the test medium.
- Test medium used.
- Explanation of any pressure or temperature discontinuities.
- Explanation of any failures that occurred during the test.
- Weather conditions during test.
- Description of pipe in the test section.
- Any permits required for the test.
- Permits for discharge of test water.
- Any other records required by agencies having jurisdiction over the testing/operation of pipelines.
- All records should be signed and dated by the engineer in charge of the test.

6: Pipeline Drying

PIPELINE DEWATERING, CLEANING, AND DRYING*

Newly constructed pipelines are typically hydrostatically tested, using water as the test medium, upon completion of construction. Older lines already in service may be re-tested either to confirm an earlier test or to qualify the line for a higher operating pressure. Once the hydrostatic testing has been completed, it is necessary to remove the water from the line and place the line in service. Dewatering can be a simple process or, if the procedure is not properly planned, a difficult one. Pipelines used to transport crude oil and/or refined products will probably only require removal of the test water before the line is placed in service. If the pipeline will be used to transport materials that must meet a specified dryness requirement, the pipeline will need to be dewatered, cleaned, and dried. Pipelines used to transport natural gas will need some drying, depending on the operating pressure and the location of the line, to prevent the formation of hydrates. Other pipelines may require drying to protect the pipe from internal corrosion caused by the formation of corrosive acids, such as carbonic acid in the case of carbon dioxide pipelines.

*Based on a paper entitled "The State of the Art of Drying Large Diameter Gas Pipelines after Hydrotest," by Marvin D. Powers, Pipeline Dehydrators, Inc., Houston, Texas.

Dewatering

Dewatering is considered to commence with the running of the first pig after hydrostatic testing is completed and begins with the insertion of a displacer, commonly referred to as a pig, in the pipeline. The dewatering pig may be pushed through the pipeline with crude oil or other petroleum product if no drying is required. If the pipeline is to be cleaned and/or dried, the pig will be pushed by either compressed air or gas. In either case, proper precautions must be taken to be sure the test water is properly disposed of and that any required water discharge permits are obtained ahead of the dewatering operation.

Several types of pigs may be used for the dewatering phase, and pig selection should depend upon the design characteristics of the pipeline and the degree of cleaning, if any, that is desired. Ideally, the pig will form a perfect seal with the inner periphery of the pipe and will not allow any of the material behind the pig to leak past the pig and co-mingle with the test water ahead of the pig.

Normally, the pig will move through the pipeline without difficulty. However, there are many opportunities for problems to develop, causing the pig to stick or even disintegrate.

Sticking may be caused by any or a combination of the following things:

- Pig is incorrect length and cannot negotiate check valves, tees, and bends;
- Pig is too large for heavy wall sections of the line;
- Pipeline may be equipped with reduced opening valves;
- Excess debris or construction material may be left in the pipeline;
- Air/gas may bypass the pig and create an air lock condition; and,
- Other unknown reasons.

Air locks are more likely to occur in hilly country than in flat land. Air locks occur when the accumulated static heads are greater than the available displacing pressure. In some cases, a pressure greater than the maximum allowable pipeline pressure would be required to overcome the air lock. Usually, the air lock occurs where air/gas has bypassed the pig and the downhill legs of the pipeline are filled with air and the uphill legs of the pipeline are filled with water. The air/gas can get in front of the dewatering pig in several ways, such as:

- Poor filling techniques;
- Poor dewatering procedure such as draining water from the line at low points;
- Air/gas bypassing the pig because the pig is too small for the pipeline;
- Air/gas bypassing the pig before it is launched;
- Air/gas bypassing the pig in a fitting such as in a tee, or a steel shaft pig in a short radius bend; and,
- Other unknown reasons.

When an air lock condition occurs, it is necessary to either increase the displacing pressure or remove air/gas through existing vents or other connections at high points in front of the pig.

For these reasons, it is important that the dewatering process be carefully planned, especially for pipelines located in hilly country.

Cleaning pipelines

It is believed there are no written specifications to define the degree of cleaning for a pipeline since there is a question of the definition of a clean pipeline, and furthermore, a method for measuring the cleanliness has not been developed. It is known, however, that cleaning a pipeline does at least four good things. Cleaning the pipeline will:

- Improve flow efficiency because of a smoother pipe wall
- Reduce product contamination and formation of hydrates
- Reduce abrasive damage to pipeline appurtenances such as valves and instruments
- Facilitate pipeline drying

Internal cleaning of the pipeline may be accomplished by any of a combination of the following methods:

- Running a brush pig with air, gas, or liquid
- Internal sand blasting
- Chemical cleaning
- Purging with air or gas followed by a liquid flush

Brush pig run with gas

One brush pig pushed by gas or air will not usually remove a significant amount of small debris from the pipeline, particularly if the debris is wet. However, the scratching action of a good brush pig will more evenly distribute the loose debris and leave a smoother pipe wall, which will improve pipeline flow efficiency.

Brush pig run with liquid

A brush pig displaced by water is a very efficient means for removing debris from the pipeline provided certain precautions are taken. Experience has shown that a pig velocity greater than 3 ft per second is desired. A pig, such as the Power Brush Pig, that allows some of the liquid to bypass the pig to keep the bristles clean and move the loose debris into suspension in the liquid in front of the pig is also desirable. The debris must be free of any sticky material such as oil, grease, or paraffin.

It is not necessary that the pipeline be completely filled with liquid. Short slugs of liquid batched between two pigs are usually preferable as long as there is enough water to keep the pigs apart and maintain a velocity greater than 3 ft per second. Short slugs will minimize the water/debris disposal problems and will allow for higher velocities with lower pressures.

Disposal of the dirty liquid may present a problem, depending upon the location. As indicated earlier, environmental aspects should be investigated thoroughly during the pre-planning phase. Usually, rust will settle out of still water in about 1 hour, leaving an almost clear liquid. If the dumping of semi-clear water is acceptable, then the problem may be solved by the use of holding tanks.

The debris-holding ability of the propelling fluid is proportional to the viscosity/density of the propelling fluid. The greater the viscosity/density, the higher the holding ability. It has been reported that gels have been developed and used that are supposedly more effective than water.

Internal sand blasting

Internal sand blasting is an extremely effective method for cleaning pipelines. This method was extensively used when natural gas was inexpensive and there were no environmental controls. Some short, small-diameter pipelines are still being internally sand blasted by using air or nitrogen to propel the sand. This procedure, if properly executed, will leave the pipeline dry upon completion.

After the pipeline has been dewatered, clay is blown into short sections of the pipeline with a dry gas to absorb the remaining water. The correct type of sand is then blown through the dust dry section at a high velocity to remove the clay, rust, and mill scale. Caution must be taken to ensure that short sections are cleaned each time so that high velocities may be maintained; that large quantities of dry air or gas are available to sustain the high velocities; that the pipeline is dust dry; that all sand has been removed from the pipeline after the sand blasting is complete; and, that the pipeline facilities will not be damaged by the high-velocity sand.

Chemical cleaning

Cleaning with chemicals is typically used for smaller diameter lines. A properly planned chemical cleaning operation can produce a very clean pipeline. Chemical cleaning will remove rust from any internal pits. Chemical cleaning is accomplished by pushing a batch of hydrochloric acid and water for rust removal, a neutralizer, and a passivator, all separated by pigs, through the pipeline. This method will leave a bright metal finish on the interior of the pipeline. If a pipeline is being converted for a different service, such as a crude oil line being converted to chemical service, a batch of detergent can be added behind the cleaning process to remove any traces of the crude oil or other product.

Typically, chemical cleaning is more involved than other cleaning processes in that some special equipment will be required to safely handle the chemicals. Disposing of the spent material will also present a problem. Care must be exercised in the selection of pigs to be sure they will not be consumed by the cleaning solution. Personnel safety must be carefully considered when choosing this method of pipeline cleaning.

Purging with gas. This method is seldom used for pipeline cleaning, because the amount of debris that will be removed is small. A high gas velocity is required.

Pipeline drying

Pipelines used to transport petrochemicals such as propylene and ethylene must be dried in order for the delivered product to meet moisture specifications. Natural gas pipelines are usually dried to a lesser extent to prevent the formation of hydrates. It is not unusual for a petrochemical line to be dried to a dew point of −80°F. A typical dew point for a propylene pipeline will be −70°F. A carbon dioxide pipeline might be typically dried to a −40°F dew point. The natural gas industry specifies dryness in pounds of water per million standard cubic feet of gas. The table at the end of this section will enable you to convert from one method to the other.

Dew point, by definition, is the temperature at which water vapor begins to condense out of a gas at atmospheric pressure. For example, at atmospheric pressure, water vapor begins to condense out of a gas that has a moisture content of 7 lb. of water per million standard cubic feet at −39°F. Therefore, a gas that has a moisture content of 7 lb. of water per million standard cubic feet has a dew point of −39°F.

Air at +70°F with a humidity of 100% has a dew point of +70°F and holds approximately 1,253 pounds of water (150 gallons) per million standard cubic feet.

The most common methods for drying pipelines are as follows:

- Drying with super dry air
- Drying with methanol
- Drying with inert gas such as nitrogen
- Internal sand blasting
- Drying with the medium to be transported
- Vacuum drying

All of these methods may be applied to pipeline drying depending on the particular line and amount of dryness required. No single method can be considered ideal for all situations. Many times, a combination of two or more methods will be used to achieve a dry pipeline at the least cost. The first three methods are probably the most economical and technically feasible for most pipeline drying applications.

Before any type of drying operation commences, it will be necessary to clean the line using one of the previously described processes. If the rust and mill scale are not removed from the pipe wall, moisture will remain trapped and will bleed out over a long period of time. A pipeline can be dried without cleaning; however, the cost and time required will be great.

In *drying with super dry air*, soft foam pigs pushed by dry air are used to absorb any free water remaining in the pipeline after dewatering. After the line is dust dry, wire brush pigs are run to remove any water bearing debris from the pipe wall. The wire brush pigs are then followed by soft foam pigs to absorb the loosened debris. Near the end of this phase, the pigs may be weighed prior to insertion and after removal to monitor the amount of debris that is being removed. The first pigs run through the line will naturally weigh much more than their clean weight. As the cleaning and drying progress, the pig weights will approach their original weight. Pig color will also give some indication of how the debris removal is proceeding.

Dew-point readings will need to be made to determine when the line has been dried to the specified dew point.

It will be necessary to give special consideration to laterals, by-passes, and valve body cavities as any free water trapped here could affect the final dew point readings.

Drying with super dry air provides internal corrosion protection if the line is to remain out of service for some period of time before it is placed in service.

Methanol drying relies on the hygroscopic effect of the methanol. Any remaining moisture in the line will be absorbed by batches of methanol pushed through the line with either gas or dry air. Pigs are used to separate the methanol batches from the displacing medium.[1] Methanol drying usually requires fewer pig runs and, consequently, less line cleaning is accomplished. Pure methanol is expensive, and sometimes a 96% methanol/water mix is used. Since the methanol mix contains water, some water will be left in the pipeline. Some of the methanol will vaporize in the pipeline and will be absorbed by the displacing medium. Toward the end of the line, the moisture content of the methanol will increase, which in turn reduces the amount of water that it can absorb. If the pressure used in the drying operation is too high, hydrate formation can occur, usually at the far end of the line. If natural gas is being used to push the methanol batches, it will probably be necessary to flare some of the line fill volume to be sure that no methanol impurities are contained in the gas.[1]

Methanol run with a dry gas will absorb most of the water and facilitate the vaporization of the remaining water. Soft swabs run through a line with a dry purge gas will accelerate the evaporation of remaining methanol/water solution.

Also of great concern with the methanol method of drying is the fact that explosive mixtures can easily be formed, whether gas or air is used to displace the methanol batches. It may be desirable to use an inert gas such as nitrogen to buffer the methanol batch from the air or gas used to displace the methanol batch. However, many pipelines have been dried without the use of nitrogen buffers with no adverse results. The air/methanol mixture is also highly poisonous and corrosive.[1]

Plans will need to be made for proper disposal of the spent methanol.

If internal corrosion protection is desired, then another drying method should be considered. If the pipeline is to be used to transport sour gas, the methanol drying method should be carefully evaluated before it is used.

Drying with nitrogen is accomplished in much the same manner as when using super dry air. Nitrogen drying will cost more than super dry air by a factor of approximately 50%. Also, super dry air is plentiful and non-polluting.

Internal sand blasting will leave the pipeline clean and dry. The procedure for internal sand blasting is described earlier in this section.

Drying with natural gas requires large volumes of gas. This method is slow and not very effective unless the line is thoroughly cleaned by one of the cleaning processes previously described. If, however, the gas being used to dry with can be blended with another dry gas stream and sold or used, then this is an economical method for pipeline drying. The cost of the gas that will be used to purge the line during the drying process should be weighed against the cost of using super dry air. The Btu value of a volume of natural gas is approximately 15 times greater than that of the fuel required to produce an equal volume of super dry air.

Vacuum drying is a slow process. All free water should be removed from the pipeline before drying begins. This method appears to be used infrequently, and perhaps only offshore.

If the pipeline has been properly cleaned by the water slug method using brush pigs run with liquid, drying can be accomplished by running soft foam pigs with dry air or gas to remove any free water left in the pipeline. This will usually produce a pipeline dry enough for natural gas operations. If additional drying is desired, it can be accomplished by using methanol or super dry air.

Reference

1. Kopp, *Pipe Line Industry*, October 1981, January 1986.

Moisture content of air

Dew Point °F @ 14.7 PSIA	Dew Point °C @ 14.7 PSIA	PPM by Volume	PPM by Weight	LBS. Water MMSCF Dry Air
-110	-78.9	0.63	0.39	.032
-109	-78.3	0.69	0.43	.034
-108	-77.8	0.75	0.47	.038
-107	-77.2	0.82	0.51	.041
-106	-76.7	0.88	0.55	.044
-105	-76.1	1.00	0.62	.050
-104	-75.5	1.08	0.67	.054
-103	-75.0	1.18	0.73	.059
-102	-74.4	1.29	0.80	.065
-101	-73.9	1.40	0.87	.070
-100	-73.3	1.53	0.95	.077
-99	-72.8	1.66	1.03	.083
-98	-72.2	1.81	1.12	.091
-97	-71.7	1.96	1.22	.098
-96	-71.1	2.15	1.34	.108
-95	-70.6	2.35	1.46	.118
-94	-70	2.54	1.58	.127
-93	-69.4	2.76	1.72	.138
-92	-68.9	3.00	1.86	.150
-91	-68.3	3.28	2.04	.164
-90	-67.8	3.53	2.19	.177
-89	-67.2	3.84	2.39	.192
-88	-66.7	4.15	2.58	.221
-87	-66.1	4.50	2.80	.225
-86	-65.6	4.78	2.97	.239
-85	-65.0	5.30	3.30	.265
-84	-64.4	5.70	3.50	.285
-83	-63.9	6.2	3.9	.31
-82	-63.3	6.6	4.1	.33
-81	-62.8	7.2	4.5	.36
-80	-62.2	7.8	4.8	.39
-79	-61.7	8.4	5.2	.42
-78	-61.1	9.1	5.7	.46
-77	-60.6	9.8	6.1	.49
-76	-60.0	10.5	6.5	.53
-75	-59.4	11.4	7.1	.57
-74	-58.9	12.3	7.6	.62
-73	-58.3	13.3	8.3	.67
-72	-57.9	14.3	8.9	.72
-71	-57.2	15.4	9.6	.77
-70	-56.7	16.6	10.3	.83
-69	-56.1	17.9	11.1	.96
-68	-55.6	19.2	11.9	1.0
-67	-55.0	20.6	12.8	1.03
-66	-54.4	22.1	13.7	1.1
-65	-53.9	23.6	14.7	1.2
-64	-53.3	25.6	15.9	1.3
-63	-52.8	27.5	17.1	1.4
-62	-52.2	29.4	18.3	1.5
-61	-51.7	31.7	19.7	1.6
-60	-51.1	34.0	21.1	1.7
-59	-50.6	36.5	22.7	1.8
-58	-50.0	39.0	24.2	2.0
-57	-49.4	41.8	26.0	2.1
-56	-48.9	44.6	27.7	2.2
-55	-48.3	48.0	30	2.4
-54	-47.8	51	32	2.6
-53	-47.2	55	34	2.8
-52	-46.7	59	37	2.9
-51	-46.1	62	39	3.1
-50	-45.6	67	42	3.4
-49	-45.0	72	45	3.6
-48	-44.4	76	47	3.8
-47	-43.9	82	51	4.1
-46	-43.3	87	54	4.4
-45	-42.8	92	57	4.6
-44	-42.2	98	61	4.9
-43	-41.7	105	65	5.3
-42	-41.1	113	70	5.7
-41	-40.6	119	74	6.0
-40	-40.0	125	80	6.4
-39	-39.4	136	85	6.8
-38	-38.9	144	89	7.2
-37	-38.3	153	95	7.7
-36	-37.9	164	102	8.2
-35	-37.2	174	108	8.7
-34	-36.7	185	115	9.3
-33	-36.1	196	122	9.8
-32	-35.6	210	131	10.5
-31	-35.0	222	140	11.1
-30	-34.4	235	146	11.8
-29	-33.9	250	155	12.5
-28	-33.3	265	165	13.3
-27	-32.8	283	176	14.2
-26	-32.2	300	186	15.0
-25	-31.7	317	197	15.9
-24	-31.1	338	210	16.9
-23	-30.6	358	223	17.9
-22	-30.0	378	235	18.9
-21	-28.4	400	249	20.0
-20	-28.9	422	212	21.1
-19	-28.3	448	278	22.4
-18	-27.8	475	295	23.8
-17	-27.2	500	311	25.0
-16	-26.7	530	329	26.5
-15	-26.1	560	348	28.0
-14	-25.6	590	367	29.5
-13	-25.0	630	391	31.5
-12	-24.4	660	410	33.0
-11	-23.9	700	435	35.0
-10	-22.3	740	460	37.0
-9	-22.8	780	485	39.0
-8	-22.2	820	509	41.0
-7	-21.7	870	541	43.5
-6	-21.1	920	572	46.0
-5	-20.6	970	603	48.5
-4	-20.0	1020	634	51.0
-3	-19.4	1087	675	54.4
-2	-18.9	1153	716	57.7
-1	-18.3	1220	758	61.0

(table continued on next page)

DEW POINT °F @ 14.7 PSIA	DEW POINT °C @ 14.7 PSIA	PPM BY VOLUME	PPM BY WEIGHT	LBS. WATER/MMSCF DRY AIR
0	-17.8	1299	807	64.9
1	-17.2	1378	856	68.9
2	-16.7	1458	906	74.3
3	-16.1	1537	955	76.9
4	-15.6	1616	1004	80.8
5	-15.0	1695	1053	84.8
6	-14.4	1775	1103	88.8
7	-13.9	1854	1152	92.7
8	-13.3	1933	1201	96.7
9	-12.8	2053	1276	103
10	-12.2	2174	1351	109
11	-11.7	2294	1425	115
12	-11.1	2415	1501	121
13	-10.6	2535	1575	127
14	-10.0	2656	1650	133
15	-9.4	2776	1725	139
16	-8.9	2897	1800	145
17	-8.3	3017	1875	151
18	-7.8	3197	1986	160
19	-7.2	3376	2098	169
20	-6.7	3556	2209	178
21	-6.1	3735	2321	187
22	-5.6	3915	2432	196
23	-5.0	4094	2544	205
24	-4.4	4274	2656	214
25	-3.9	4453	2767	223
26	-3.3	4633	2879	232
27	-2.8	4883	3034	244
28	-2.2	5133	3189	357
29	-1.7	5383	3345	269
30	-1.1	5633	3400	282
31	-0.56	5883	3655	294
32	0	6133	3811	307
33	0.56	6383	3966	319
34	1.1	6633	4121	332
35	1.7	6883	4277	344
36	2.2	7281	4524	364
37	2.8	7679	4771	384
38	3.3	8077	5019	404
39	3.9	8475	5266	424
40	4.4	8874	5514	444
41	5.0	9272	5761	464
42	5.6	9670	6008	484
43	6.1	10068	6256	503
44	6.7	10466	6503	522
45	7.2	10824	6725	541
46	7.8	11182	6948	559
47	8.3	11540	7170	577
48	8.9	11898	7393	595
49	9.4	12255	7614	613
50	10.0	12613	7837	631
51	10.6	12971	8059	649
52	11.1	13329	8282	666
53	11.7	13687	8504	684
54	12.2	14268	8865	713
55	12.8	14849	9226	742
56	13.3	15430	9587	772
57	13.9	16011	9948	801
58	14.4	16593	10310	830
59	15.0	17174	10671	859
60	15.6	17754	11031	889
61	16.1	18336	11393	917
62	16.7	18917	11754	946
63	17.2	19685	12231	984
64	17.8	20452	12707	1023
65	18.3	21220	13185	1061
66	18.9	21987	13661	1099
67	19.4	22755	14138	1138
68	20.0	23522	14615	1176
69	20.6	24290	15092	1215
70	21.1	25057	15569	1253
71	21.7	25825	16046	1291
72	22.2	26827	16668	1341
73	22.8	27829	17291	1391
74	23.3	28831	17914	1442
75	23.9	29833	18536	1492
76	24.4	30836	19159	1542
77	25.0	31838	19782	1592
78	25.6	32840	20405	1642
79	26.1	33842	21027	1692
80	26.7	34844	21650	1742
81	27.2	36138	22454	1807
82	27.8	37432	23258	1872
83	28.3	38727	24062	1936
84	28.9	40021	24866	2001
85	28.4	41315	25670	2066
86	30.0	42609	26474	2130
87	30.6	43903	27278	2195
88	31.1	45198	28083	2260
89	31.7	46492	28887	2325
90	32.2	48146	29914	2407
91	32.8	49801	30943	2490
92	33.3	51455	31971	2573
93	33.9	53109	32998	2655
94	34.4	54764	34027	2738
95	35.0	56419	35055	2821
96	35.6	58023	36083	2904
97	36.1	59728	37111	2986
98	36.7	61382	38139	3069
99	37.2	63477	39440	3174
100	37.9	65571	40741	3279
101	38.3	67666	42043	3383
102	38.9	69760	43344	3488
103	39.4	71855	44646	3593
104	40.0	73950	45948	3698
105	40.6	76045	47249	3802
106	41.1	78139	48550	3907
107	41.7	80234	49852	4012
108	42.2	83861	51484	4193
109	42.8	85489	53117	4274
110	43.3	88116	54749	4406

Commissioning petrochemical pipelines

Pipelines carrying a variety of gaseous and volatile liquid products are extensively employed in today's industry as a safe and efficient means of transportation. New pipelines must be placed into service initially (i.e., commissioned), and older lines occasionally are taken out of service in order to perform some desired maintenance and are then recommissioned. Among the reasons for taking a pipeline out of service (i.e., decommissioning) are: hydrostatic testing to recertify or upgrade the pipeline's ability to be used at higher operating pressures; performance of construction work on the pipeline; and a change in the product transported by the pipeline. Pipelines, or sections of a pipeline, may need to be relocated because of highway work, the necessity to deepen a canal, or because of increases in the population surrounding the pipeline. It may also be necessary to replace valves, fittings or a damaged section of the pipeline or add a new connection to service a customer or supplier.

Typically, decommissioning and recommissioning a petrochemical pipeline will include the steps of decommissioning by removing the product from the pipeline, and flaring any remaining residual product. Any necessary construction, upgrading, or cleaning of the pipeline can then be performed. Usually, the pipeline is then filled with water for hydrostatic testing. After pressure testing, the water is removed, the pipeline is cleaned and dried to a specific low dewpoint (to avoid the problems of water contamination of the product), and the pipeline is inerted with nitrogen for recommissioning. Refer to the section on Pipeline Drying for drying techniques.

In order to recommission the pipeline, the nitrogen must be displaced by the desired petrochemical. Before returning to service, product purity must be established and the line safely filled to operating pressure. The terms "commissioning" and "recommissioning" are used interchangeably to refer to a process whereby a first inerting gas, normally nitrogen, in a pipeline is replaced with the desired product at the desired purity and pressure.

Safety and economics are two primary concerns for any proposed pipeline operation. Thus, a commissioning process that brings the purity and pressure of potentially explosive products such as ethylene or propylene up to specification quickly, but at the risk of damage to the pipeline or reduced safety to operating personnel, is not acceptable. Similarly, a process that uses large quantities of product to push nitrogen from a line and results in wasted product/nitrogen mixtures that must be flared or otherwise disposed of would also be unacceptable. Improvement over known methods requires consideration of its safety and economic benefits (considering both the cost of wasted product and the cost attributed to the time the line must remain out of service) as well as its ability to bring the line back into service with product at desired pressures and purities.[1]

The main problems posed by recommissioning are (1) the fast and economical purging of nitrogen so as to obtain uncontaminated products in the line, and (2) possible damage to the pipeline due to cold temperatures to which it may be subjected during the process. The latter can be a problem when the pipeline is to be recommissioned with a product at a pressure significantly higher than that of the nitrogen inerted line. The pressure drop of a petrochemical product entering a pipeline can cause large drops in temperature and, consequently, potential damage to the carbon steel pipeline.

Due to the demands of the expanding petrochemical industry, and the aging of the present pipeline system, the need for decommissioning, repairing, cleaning, drying, and recommissioning pipelines is increasing. Given the value of today's petrochemical products, a recommissioning procedure, that brings the product purity to acceptable levels as quickly as possible, is desirable. Furthermore, processes and equipment which reduce the possibility for damage to the pipeline due to cold temperature conditions which can occur during the recommissioning process are also highly desirable.[2]

Due to the unusual properties of ethylene, decommissioning and recommissioning ethylene pipelines require special consideration. Ethylene is a colorless, flammable, gaseous, unsaturated hydrocarbon obtained by the pyrolysis of petroleum hydrocarbons. The relative density of ethylene is 0.9686 (air = 1), and its molecular weight is 28.054. The molecular weight of air is 28.964, and the molecular weight of nitrogen is 28.013.

Ethylene decomposition

Ethylene is subject to thermal decomposition (self propagating reaction zone, explosion, or flame) under certain circumstances. Thermal decomposition occurs when the temperature of a substance is raised above the value needed to cause it to begin to self-heat because of a change in molecular structure at a rate high enough to result in combustion. An external source of ignition is not required to precipitate this process, nor is a source of oxygen needed. Thermal decomposition is usually initiated by a source of heat from events such as sudden compression, external source of heat, or the thermal runaway of a heater.[4] Heaters are sometimes used in ethylene measurement stations to maintain the ethylene at an optimum temperature for measurement.

Ethylene is routinely handled under conditions where decomposition may be initiated. The point at which decomposition begins is dependent upon system parameters such as line or vessel size, operating pressure, and operating temperatures. Generally, ethylene decomposition is initiated in pipeline facilities by a rapid rise in temperature associated with sudden compression in the presence of a diatomic gas such as nitrogen or oxygen, exposure of the pipe to flame, possibly hot tapping operations.

The autoignition temperature (AIT) for ethylene is 914°F at atmospheric pressure.[3] Experience of others suggests that at 1,000 to 1,500 psig, the AIT could drop to as low as 400° to 700°F.[3,4] As the concentration of oxygen increases, the decomposition initiation temperature decreases.[4] From this, it may be concluded that a mixture of low oxygen concentration and ethylene could be very hazardous. For this reason, ethylene pipelines, which have been evacuated for maintenance or repairs, are usually filled with nitrogen to displace oxygen from the pipe. Compressing nitrogen with ethylene may result in high nitrogen temperatures—especially if the compression is rapid. Also, when the pressure of the ethylene source is high, refrigeration across the throttling valve used to control the ethylene may result in extremely low temperatures and could cause the valve to fail due to excessively low temperature.

Commissioning an ethylene pipeline with warm ethylene (±70°F) at a low pressure (±50 psig) is quick, easy, inexpensive, and very safe when the ethylene is heated before it is admitted to the pipeline being commissioned.

A patented process used by Pipeline Dehydrators can be used to safely commission ethylene pipelines. The process uses a shell and tube heat exchanger to heat the high pressure ethylene before the pressure is reduced to the low pressure desired for line commissioning.

The methods currently used to commission ethylene pipelines are potentially hazardous, expensive, and time consuming. The two most severe problems are potential damage to the carbon steel in the pipeline due to the cold temperature of the expanded ethylene and contamination of the ethylene from the nitrogen used to inert the pipeline during commissioning.

An ethylene pipeline that is being commissioned has usually been dried to a −70°F dew point or more and inerted with nitrogen. The nitrogen pressure left on the line for commissioning varies from 10 to 1,000 psig depending on the pressure of the ethylene source and the commissioning procedure used and the pipeline owner's preference.

The source of ethylene used for commissioning an ethylene pipeline is usually more than 900 psig and may be up to 2,200 psig.

If the pressure of the ethylene source used for commissioning is high and the nitrogen pressure in the pipeline being commissioned is low, the resulting temperature of the ethylene, due to the pressure drop, may be well below the −20°F design temperature of most carbon steel pipelines.

The carbon steel in some pipelines is subject to becoming brittle at temperatures below −20°F and may fail at a very low internal pressure because of the low temperature.

At low pressures, the very cold ethylene will be much more dense than nitrogen, thereby producing an extensive product interface during commissioning or trapping pockets of the less dense nitrogen at each high place in the pipeline.

In this case, the amount of ethylene burned to produce product purity may be expensive and time consuming and may create environmental problems in addition to subjecting the steel in the pipeline to sub-design temperatures.

If the pressure of the ethylene source used for commissioning is high, and the nitrogen pressure in the pipeline being commissioned is also high, cold temperatures will not be encountered. However, even at warm temperatures, high-pressure ethylene is still much more dense than nitrogen at the same pressure. Therefore, the extensive product interface and the tapped nitrogen at the high points will still occur, resulting in an expensive and time consuming purification process plus the cost of the large volume of nitrogen. Also, there is a large inventory of ethylene in a pipeline at high pressure. If a pipeline blow-down is required because of a leak, lack of purity, or for whatever reason, it will be a major product loss and expense.

The Pipeline Dehydrators system utilizes a tube and shell heat exchanger and a water-glycol mix that is used as the heat transfer fluid. A pump forces the heat transfer fluid

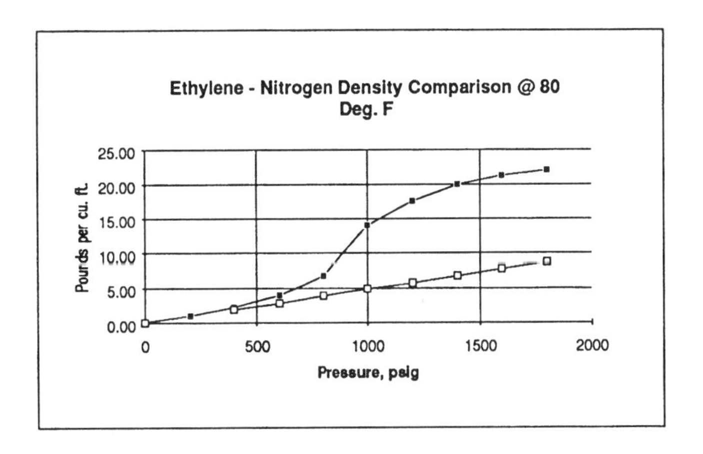

through the fired heater and then through the shell side of the exchanger. The ethylene passes through the tubes of the heat exchanger and is safely and properly warmed by the heat transfer fluid without entering the direct fired heater, thereby eliminating the possibility of an ethylene decomposition due to heater thermal runaway.

The heater will produce ethylene at ground temperature (±70°F) at ±50 psig corresponding to the conditions of the nitrogen in the pipeline. Under these conditions, ethylene

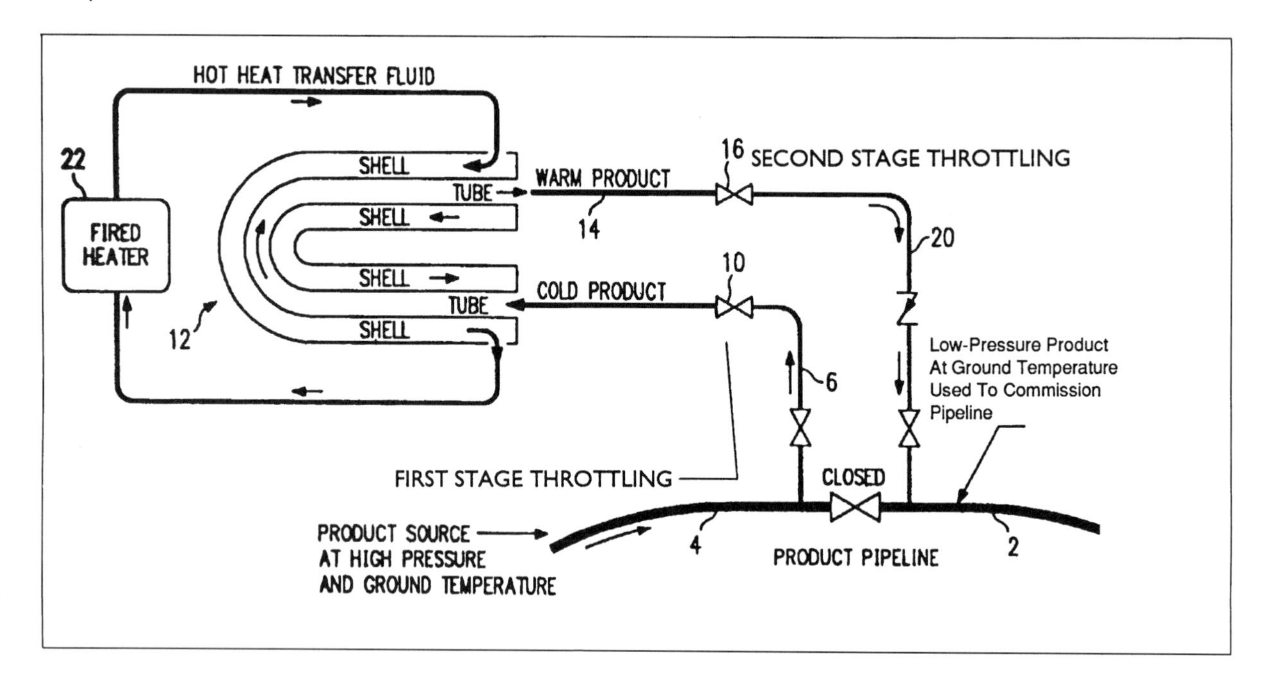

and nitrogen have basically the same density; therefore, the product interface during commissioning will be insignificant; no nitrogen will be trapped at the high points in the pipeline; the steel in the pipeline will not be subjected to cold temperatures; purification will be simple, short, and inexpensive; and the ethylene inventory in the pipeline will be minimal in case a pipeline blow-down is required. The savings on nitrogen will also be significant.

Two pressure drops will be taken on the ethylene, one on each side of the tube-shell heat exchanger. The pressure will be dropped to approximately 600 psig prior to the ethylene entering the heat exchanger. The 600 psig ethylene will be warmed to approximately 140°F. Then a second pressure drop will occur, allowing the pressure and temperature of the ethylene to equalize with the pressure and temperature of the nitrogen in the pipeline (±50 psig @ ±70°F). Under these conditions, PLD's 6 million BTU per hour heater can handle up to 45,000 lb. of ethylene per hour. This high flow rate will normally be used only during the pressuring up phase of the process and not used during the purification process. The warm ethylene at low pressure, traveling about 10 mph in the small pipelines that are typically used to transport ethylene, will produce excellent results during the purification phase of the process.

The minimum ethylene temperature encountered in this commissioning process is well above 0°F, and the maximum temperature is 160°F. This is a very safe temperature range to handle ethylene in carbon steel piping systems.

Purity can be established at 50 psig or less, with very little loss of product, and the system can be checked for leaks with ethylene in the pipeline while the pressure is low and the dollar value of ethylene in the pipeline is still small.

After purity has been established at ±50 psig and potential leak points have been checked, the major remaining effort will be to warm enough ethylene to bring the pipeline up to approximately 600 psig. Beyond this point, the ethylene can be throttled into the pipeline without heating, because the cooling due to pressure drop will no longer be great enough to cause damage to the pipeline and equipment. During this pressure-up phase of the commissioning operation, the ethylene does not have to be heated to +70°F temperature and the pipeline pressure will be increasing, both resulting in a reduction in the amount of heat required per pound of ethylene. Therefore, for this phase of the commissioning operation, PLD's new ethylene heater will handle considerably more than 45,000 lb. of ethylene per hour.

This system provides excellent conditions for commissioning an ethylene pipeline quickly, inexpensively, and safely.

This heater can also be utilized to commission other product pipelines where cold temperatures are encountered due to pressure drops such as with propylene, ethane, propane, and carbon dioxide.[2]

Acknowledgement

The editor wishes to thank Marvin D. Powers of Pipeline Dehydrators for permission to use material from a paper on *Ethylene Heater For Commissioning Pipelines*.

Literature cited

1. US Patent No. 5,027,842 July. 2, 1991 Powers.

2. Powers, Marvin D. *"Ethylene Heater For Commissioning Pipelines"* 5–89 Pipeline Dehydrators.
3. Hilado, Carlos J. and Clark, Stanley W. *"Autoignition Temperatures of Organic Chemicals"* Chemical Engineering September 4, 1972, p 76.
4. McKay, F. F., Worrell, G. R., Thornton, B. C., and Lewis, H. L. *"How To Avoid Decomposition In Ethylene Pipelines,"* Pipe Line Industry August 1978, p 80.

Vacuum drying

The vacuum drying process is shown graphically in Figure 1 and consists of three separate phases. Corrosion is generally inhibited at relative humidity (R. H.) levels below 30%, but in the presence of hygroscopic dirt (present in millscale) corrosion can occur at R.H. levels of 20%. Therefore, systems should be thoroughly drained and then vacuum dried to lower than 20% R.H.

Target Dewpoint (D.P.)

$$\text{R.H.} = \frac{\text{Water vapor pressure}}{\text{Saturated Vapor Pressure (SVP)}}$$

for a constant temperature. Assume the lowest average temperature of the system is 0°C (during the winter months), which gives

SVP = 0.6 KPaA

The vacuum level required for water vapor at 20% R.H. is:

$0.2 \times \text{SVP} = 0.2 \times 0.6 = 0.12\text{KPaA}$

This corresponds to a dewpoint of −18°C. A safety allowance should be provided for some desorption from the pipe walls; and therefore a dewpoint of −20°C would be used.

Phase 1—Evacuation

During this phase, the pressure in the pipeline is reduced to a level where the ambient temperature of the pipeline will cause the free water to boil and change to water vapor. This pressure level corresponds to the saturated vapor pressure of the free water in the pipeline, which is dependent upon the ambient temperature of the pipeline.

The approximate pressure value is calculated in advance but is easily recognized on site by a fall in the rate of pressure reduction, which is noted from the plot of pressure against time.

At some convenient point in time a "leak test" is carried out by stopping the vacuum equipment and observing the pressure, usually for a period of 4 hours. Any "air-in" leaks on flanges, fittings, or hoses are rectified at this time, although leaks are not a common occurrence.

Phase 2—Evaporation

Once the saturated vapor pressure has been reached, then evaporation of the free water into water vapor will commence. During this phase, the vacuum equipment is carefully controlled to maintain the pressure at a constant

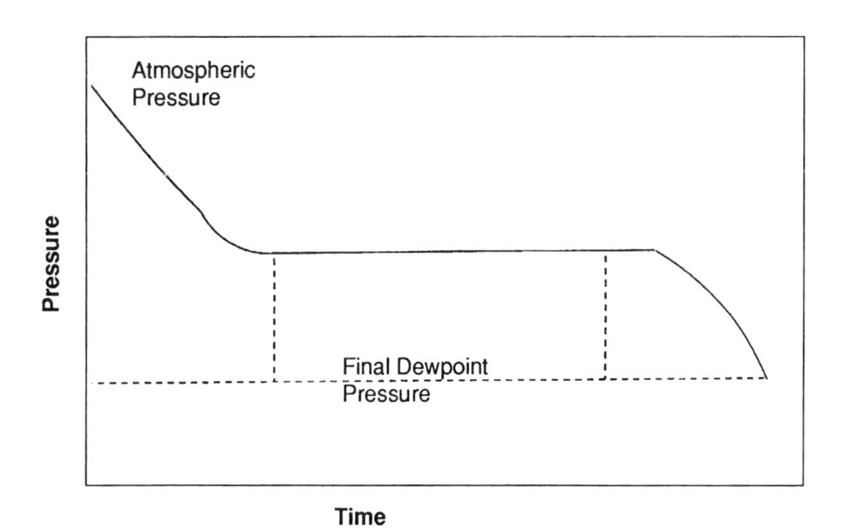

Figure 1. Vacuum drying pressure vs time curve.

level until all the free water has been converted into water vapor. This phase may take several days to complete depending on: (1) the amount of water to be evaporated; (2) the size of the vacuum equipment; and (3) the ambient temperature of the pipeline. The end of the evaporation phase will be observed on site by a noticeable decrease in pressure.

At this time it is prudent to carry out a "soak test" to ensure that all the free water has in fact evaporated. The vacuum equipment is temporarily isolated from the pipeline, usually for a period of 12 hours, and a careful note made of the pressure. If all free water has evaporated then the pressure will remain constant and the final drying phase can be commenced.

Phase 3—Final drying

Once the free water has been converted into water vapor, the majority of it must be removed from the pipeline in order to reach the required dryness level. This is achieved by reducing the pressure in the pipeline still further which has the effect of drawing the water vapor out of the pipeline through the vacuum equipment. Obviously, the more water vapor removed, the drier the pipeline will become.

During this phase a careful watch is kept on the slope of the final drying line to ensure that it follows the calculated value, since a shallower slope would indicate the continuing presence of some free water still remaining in the pipeline.

The necessary calculations to determine the time required for each phase of vacuum drying are detailed in Figure 2.

Evacuation:

$$t = \frac{V}{Se} \log e \frac{Pi - Pv}{Pf - Pv}$$

$$SE = \frac{1}{C} + \frac{1}{Sp}$$

Evaporation:

$$t = \frac{M}{SeDv}$$

Final Drying:

$$t = \frac{V}{Se} \log e \frac{Pv}{Preq}$$

Pi = Initial Pressure
Pf = Final Pressure
Pv = Vapor Pressure
Se = Effective Pump Speed
V = System Volume
Sp = Pump Speed
C = Conductance
M = Mass of Water
Dv = Vapor Density
Preq = Required Dewpoint Pressure

Figure 2. Vacuum drying calculations.

What is dryness?

The dryness of a pipeline is measured in terms of dewpoint, which is the temperature at which mist or dew will begin to form. A convenient method of measuring dewpoint is to use an instrument called a mirror hygrometer where the water vapor is passed across a polished surface that is slowly cooled until dew forms. The temperature at which the dew fonns is the dewpoint of the water vapor and is normally expressed in degrees centigrade. The drier the air, the lower the temperature at which dew will form.

In terms of a pipeline being vacuum dried, the lower the pressure in the pipeline, the lower the dewpoint will be. For example, at a pressure level of 0.26 kPaA, the equivalent dewpoint of the pipeline would be −10°C. If the pressure were further reduced to 0.104 kPaA, then the dewpoint would be −20°C.

For gas pipelines a dewpoint level of −20°C is generally considered to be adequate and the 0.10 kPaA bar pressure level required to achieve this dewpoint is readily attainable using the portable vacuum equipment described previously.

For example, consider a 100-mile-long 36-inch-diameter pipeline that, prior to drying, contained 10,000 gallons of water as a film on the inside surface of the pipe. At a dewpoint of −10°C, the quantity remaining would be reduced to only 46.5 gallons and at −20°C to 19.7 gallons. Also, this water would not be free water but rather water vapor, and it would only revert to free water if the ambient temperature of the pipeline were further reduced. This water vapor can subsequently be removed from the pipeline during the purging operation.

The relations between pressure and dewpoint are shown in Figure 3 and Table 1.

Proving the dryness

Immediately following the final drying phase, a dry gas purge using atmospheric air or nitrogen is carried out to prove the dryness of the pipeline. It is possible, under certain circumstances, for a small amount of free water to still remain in the pipeline. Usually this water will have

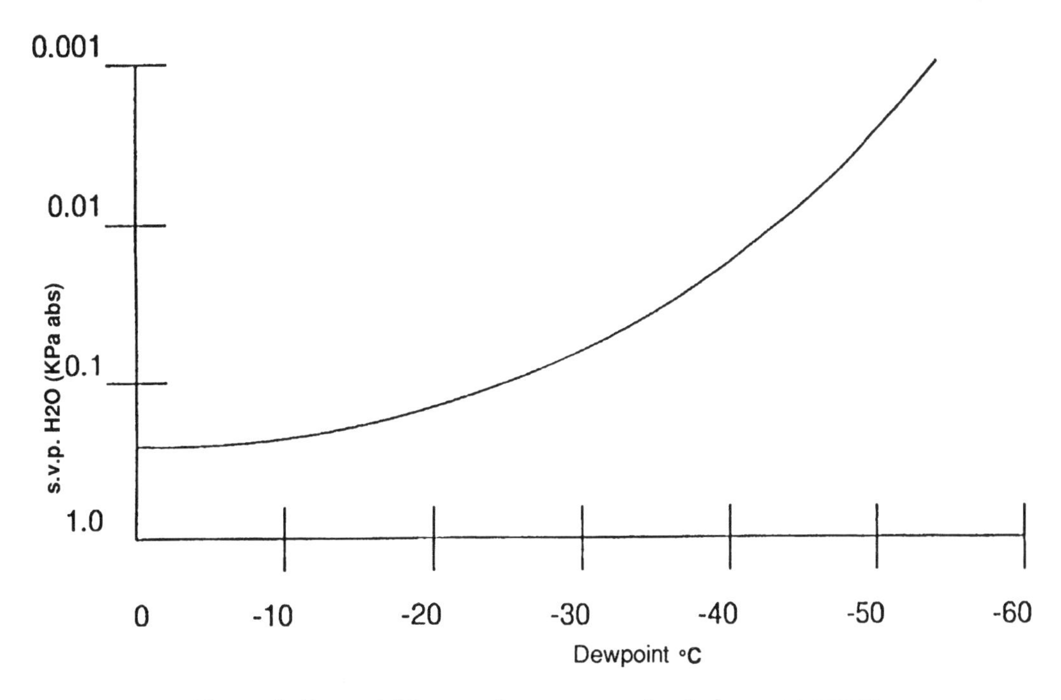

Figure 3. Dewpoint temperature versus saturated vapor pressure.

Table 1
Water Vapor Pressure Table

T °C	SVP (kPaA) (mb × 10^{-1})	Vapor density (gm m^3)
−50	0.0039 (0.039 mb)	0.038
−45	0.0128	0.119
−35	0.0223	0.203
−30	0.0308	0.339
−25	0.0632	0.552
−20	0.1043	0.884
−15	0.1652	1.387
−10	0.2597	2.139
−5	0.4015	3.246
0	0.6108	4.847
5	0.8719	6.797
10	1.2270	9.399
15	1.7040	12.830
20	2.3370	17.300
25	3.1670	23.050
30	4.2430	30.380

turned to ice due to the chilling effect of the vacuum drying process and may not be apparent during the final drying phase or soak test.

Nitrogen or atmospheric air is allowed to enter the pipeline through a valve at the end remote from the vacuum equipment until the pressure has risen to the SVP equivalent of the target dewpoint.

Once this pressure level has been reached, the vacuum equipment is started and that pressure level maintained. This has the effect of drawing gas through the pipeline under vacuum at a relatively constant dewpoint equal to the final dewpoint required.

At some point in time the purge gas, now under vacuum at a dewpoint of say, −20°C, will reach the vacuum equipment and be pulled through it. The dewpoint at both ends of the pipeline is carefully monitored and compared. If there is no free water remaining in the pipeline then the dewpoint at the vacuum equipment end will be the same as the dewpoint at the remote end. However, if there is any free water present then the dry air passing through the pipeline under vacuum will absorb the water hygroscopically. The dry gas purge operation must then continue to remove the remaining free water until the dewpoints at both ends are equal, at which time purging is discounted. The pipeline has now been vacuum dried to the required dewpoint level, and the dryness proved.

Purging and commissioning

Once the dryness has been attained and proved, the pipeline is ready for commissioning. It is possible to introduce the gas directly into the vacuum or to relieve the vacuum using dry nitrogen.

Reprinted with permission by Nowsco Well Service Ltd.

7: Control Valves

Control valve sizing formulas

Liquid formulas (noncompressible fluids).

Volume basis:

$$C_v = q\sqrt{G/\Delta P} \quad (1)$$

Weight basis:

$$C_v = W/\left(500\sqrt{\Delta P \times G}\right) \quad (2)$$

Equations 1 and 2 exclude control valve calculations where flashing, high-viscosity, two-phase flow, or high pressure drops are involved.

Gas and vapors (other than steam).

Volume basis:

$$C_v = (Q/1360)\sqrt{2GT/[\Delta P(P_1 + P_2)]} \quad (3)$$

Equation 3 does not take into consideration gas compressibility, mixed phase flow or high pressure drop. Compressibility is considered in Equations 12 and 13. Usable pressure drop for valve sizing calculations for gas, vapor, and steam flows is always limited to ½ of the absolute inlet pressure of the control valve. At that approximate value, critical (sonic) velocity is reached and any further reduction of the downstream pressure will not increase the velocity through a valve, thus there will be no increase in flow.

Steam formula.

$$C_v = \frac{W(1.0 + .0007Tsh)}{2.1\sqrt{\Delta P(P_1 + P_2)}} \quad (4)$$

Special conditions

Equations are given below to correct for flashing, cavitation, high-viscosity fluids, and mixed phase flow, where the simplified equations may lead to inaccurate sizing. These additional expressions, together with the FCI equations, provide satisfactory solutions for most control valve sizing applications.

Liquids

Flashing service. When flashing occurs, the simplified liquid sizing equation gives erroneous results. Flashing occurs when liquids enter a valve at or near their boiling points, and they begin vaporizing as their pressures decrease, due to conversions of static pressure head to velocity head. With the advent of high recovery valves, this condition has been aggravated. High recovery valves take a much higher pressure drop from inlet to valve orifice than standard control valves (see Figure 1). When the pressure at the orifice drops to the vapor pressure of the fluid at flowing temperature, the maximum valve capacity is reached. This is true, because flashing occurs across the valve orifice and a flow condition similar to the critical flow of vapors exists.

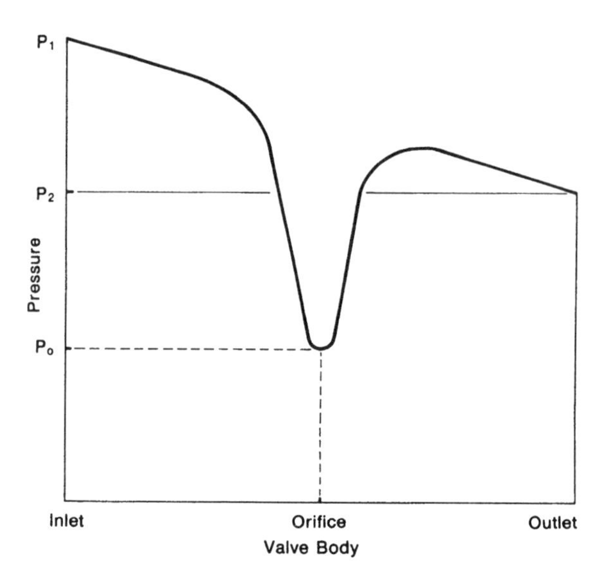

Figure 1. Typical control valve pressure gradient. The ΔP between P_1 and P_2 is much less for high recovery valves.

Two approaches take into account the flashing condition:

1. Hans D. Baumann, in a paper presented in *ISA Transaction*, Volume 2, Number 2 (April, 1963), introduced an equation using a C_f factor to help determine the valve capacity when flashing occurs. The equation, based also on the use of liquid vapor pressure (P_v), is

$$C_v = (Q/C_f)\sqrt{G/(P_1 - P_v)} \quad (5)$$

2. C. B. Schuder, Fisher Controls, in a paper, "How to Size High Recovery Valves Correctly," proposed an equation to solve for the maximum pressure drop to be used for liquid sizing:

$$\Delta P_m = K_m(P_1 - r_c P_v) \quad (6)$$

(See Figure 2 for values for r_c.)

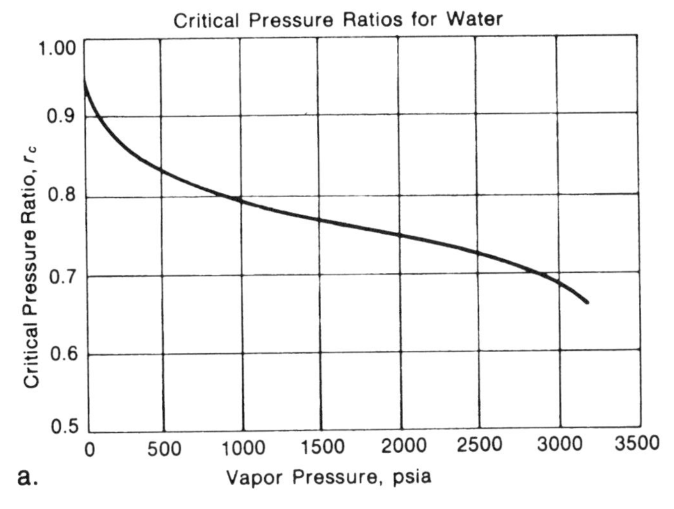

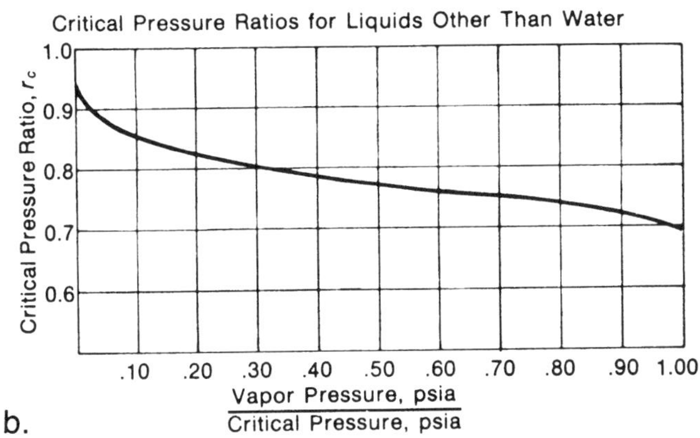

Figure 2. These critical pressure ratios (r_c) are for use with Equation 6.

When this limiting pressure drop from Equation 6 is used in the liquid flow formula, Equation 1 gives a valve size in which flashing will not occur unless the pressure drop increases above the calculated ΔP_m. When the ΔP exceeds ΔP_m, flashing and possible cavitation can occur. Cavitation is the effect caused by the sudden condensing of vapors when the static pressure increases to a point above the boiling point of the liquid.

When the possibility of cavitation exists, the process should be reviewed and the condition eliminated if at all possible. A high recovery valve should not be used in this application. This second approach is suggested as a good solution to the flashing problem.

Viscosity Correction. The liquid sizing formulas in general use have no factor applied to correct for viscosity. This correction is a complex function involving the Reynolds number, which varies with the valve size, port area, velocity, and viscosity. No correction is required for a viscosity value below 100 SSU. Above viscosity values of 100 SSU, two equations are given to obtain correction factors:

$$R = (10{,}000 \times q)/\sqrt{C_v} \times cs \tag{7}$$

This equation is used for viscosity values between 100 and 200 SSU; the viscosity must be converted to centistokes. When viscosities exceed 200 SSU, Equation 8 applies:

$$R = (46{,}500 \times q)/\sqrt{C_v} \times SSU \tag{8}$$

In either equation, the following steps should be followed:

1. Solve for C_v, assuming no viscosity effects.
2. Solve for factor R using Equation 7 or 8.
3. From the viscosity correction curve, Figure 3, read the correction factor at the intercept of factor R on the curve.
4. Multiply the C_v obtained in Step 1 by the correction factor obtained in Step 3 for the corrected C_v.

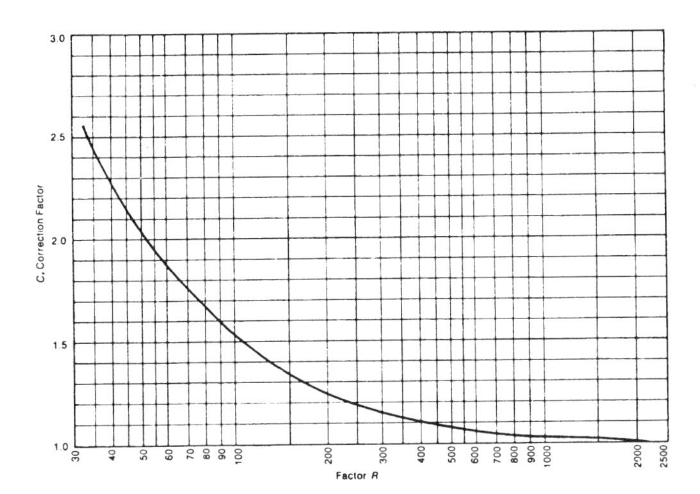

Figure 3. Viscosity correction curve for use with Equation 8.

Gases and vapors

The simplified equation for gases and vapors (Equation 3) agrees favorably with experimental data when used for conventional globe valves, but agreement with experimental data is poor when used with high recovery valves.

A universal equation proposed by C. B. Schuder, Fisher Controls, in *Instrumentation Technology*, February, 1968, reportedly showed excellent agreement with experimental results covering a wide range of body plug configurations and pressure drop ratios.

Volume basis:

$$Q = \sqrt{\frac{520}{GT}} C_1 C_2 C_v P_1 \sin\left[\frac{3417}{C_1 C_2}\sqrt{\frac{\Delta P}{P_1}}\right] \text{Deg.} \tag{9}$$

Weight basis:

$$W = 1.06\sqrt{dP_1ZC_1C_2C_v}\sin\left[\frac{3417}{C_1C_2}\sqrt{\frac{\Delta P}{P_1}}\right]\text{Deg.} \qquad (10)$$

The C_1 factor in Equations 9 and 10 is currently available from Fisher Controls for their valves. Other companies presumably do not publish this factor. (The C_2 factor is shown in Figure 4.)

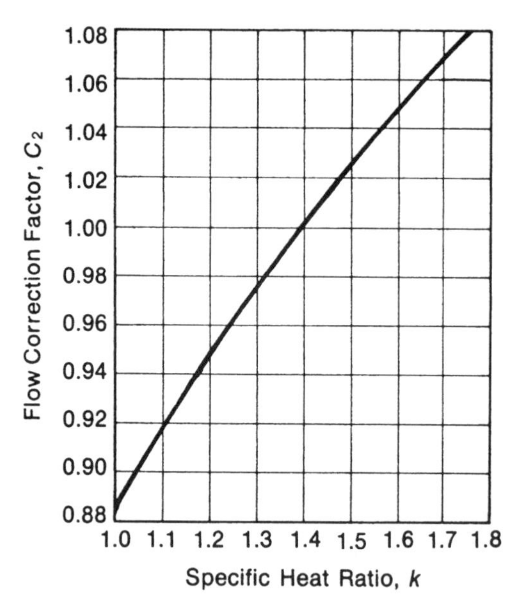

Figure 4. The C_2 factor is applicable only to Equations 9 and 10, the Fisher universal equations for gas and vapor flows.

An equation for gases and vapors proposed by the Foxboro Company (Foxboro TI Bulletin 31-4K) is identical with the FCI equation, except that P_2 is used instead of $(P_1 + P_2)/2$:

$$C_v = (Q/1360)\sqrt{GT/\Delta P(P_2)} \qquad (11)$$

A comparison of the Fisher Universal, FCI, and Foxboro equations is made in the TI Bulletin 31-4K. A plot is given showing the results of these three equations using the same data (Figure 5).

The Foxboro equation gives C_v values essentially the same as the universal equation at pressure drops up to 25% of P_1 and gives 7% larger values at a pressure drop of 50% of P_1. At the higher end of $\Delta P/P_1$ ratios, a slightly larger valve should be helpful to allow for the expanding vapors. The same agreement is claimed for steam flows, although no plot is given.

To correct for deviation from the universal gas law, the compressibility factor is inserted in Equation 11, becoming:

$$C_v = (Q/1360)\sqrt{G_bTZ/(\Delta P \times P_2)} \qquad (12)$$

On a weight basis, this equation becomes:

$$C_v = (W/104)\sqrt{TZ/(\Delta P \times P_2 \times G_b)} \qquad (13)$$

Mixed phase flow. Mixed phase flow is generally experienced when:

1. Liquid entrainment is carried by a gaseous flow.
2. Liquid at or near its boiling point vaporizes as it flows through a line and its restrictions.

In the discussion under "Flashing Service," it is explained that flashing can be reduced or eliminated by limiting the ΔP across the valve. When two-phase flow already exists or when it occurs in the control valve, two methods are given to size the valve.

1. A rule of thumb that has been used for years is to calculate the C_v based on liquid only and increase the valve body to the next size. This is satisfactory in many cases, but it obviously does not take into account varying ratios of liquid to vapor.
2. A preferred method is to determine the amount of vapor and liquid, then calculate and add the C_v of each phase to get the overall C_v requirement. The percent vaporization can be calculated by the responsible engineer if it is not furnished by the process people.

Example.

Fluid—Propane
100 gpm @ 300°psia and 130°F
s.g. @ 130°F = 0.444
s.g. @ 60°F = 0.5077
s.g. @ P_2 (200psia) = 0.466 @ 104°F
$\Delta P = 100$ psi
% vaporization = 15.05%
M.W. = 44
Z = 0.8
$G_b = 44/29 = 1.518$
Flow = 100 gpm × 0.444 = 22,200 lb/hr
Vapor = 22,200 × 0.1505 = 3,340 lb/hr
Liquid = 22,200 − 3,340 = 18,860 lb/hr

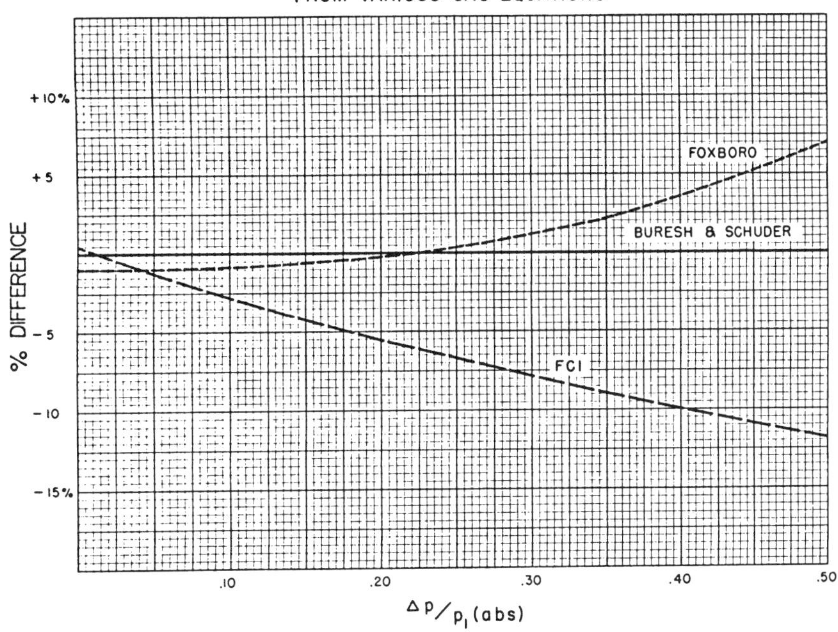

Figure 5. These curves show the close agreement (particularly for low pressure drops) of the Foxboro, FCI, and Fisher universal equations for gas flows. Courtesy of the Foxboro Co.

Using Equation 13:

$$C_v(\text{vapor}) = (W/104)\sqrt{(T \times Z)/(\Delta P \times G_b \times P_2)}$$

$$= (3340/104)\sqrt{(564)(.8)/(100(1.518)200)}$$

$$= 3.92$$

Using Equation 2:

$$C_v(\text{liquid}) = (W/500)/\left(18{,}860/\sqrt{\Delta P \times G}\right)$$

$$= 18{,}860/\left(500\sqrt{100 \times .466}\right)$$

$$= 5.53$$

$$C_v(\text{total}) = 3.92 + 5.53$$

$$= 9.45$$

The value of P must never exceed $P_2/2$ for sizing calculation.

Nomenclature

C_v = Valve flow coefficient
G = Specific gravity of fluid at flowing temp. (water = 1.0)
P = Pressure drop (P_1–P_2), psi
W = Flow of fluid, lb/hr
q = Flow rate, gpm
T = Flowing temp., °R
T_{sh} = Degrees of superheat, °F
C_f = Critical flow factor, from valve manufacturer
P_m = Maximum useable P (choked flow occurs), psi
K_m = Valve recovery coefficient
r_c = Theoretical critical pressure ratio at P_v
P_v = Fluid vapor pressure at flowing temp., psia
R = Factor for use with Figure 3
cs = Viscosity at flowing temp., centistokes
SSU = Viscosity at flowing temp., Saybolt Seconds Universal
C_1 = Valve sizing factor
C_2 = Specific heat correction factor (see Figure 4)
P_1 = Valve upstream pressure, psia
P_2 = Valve downstream pressure, psia
d = Density of vapors at valve inlet lb/ft^3
d_1 = Density of vapors at valve outlet lb/ft^3
Z = Compressibility factor
G_b = Specific gravity of gas or vapor at base conditions

Source

Applied Instrumentation in the Process Industries, Gulf Publishing Company, Houston, Texas.

Sizing control valves for throughput

These useful charts give data essential to calculation of pressure drops through regulating stations

W. G. Birkhead, Gas Measurement Engineer, United Gas Pipeline Co., Shreveport, La.

Pressure losses through piping fittings, once given little consideration in the sizing of control valves, have become important as more efficient control valves have been developed. With the development of high capacity valves and the desire for a more exacting flow formula and efficiency, designers and users look to consideration of minor losses through elbows, tees, swages, or reduced ported valves. The capacity of control valves and regulators was once calculated using the basic hydraulic head equation with an assumed efficiency. This is no longer the case.

The head loss in any closed filled conduit has been accepted as proportional to its length, diameter, and fluid velocity head, and can be calculated as

$$H_L = f\frac{L}{D}\frac{V^2}{2g} \qquad (1)$$

where:

H_L = Head loss of flowing fluid (ft of fluid)
f = Friction factor
L = Length of pipe, ft
D = Diameter of pipe, ft
V = Average fluid velocity, ft/sec
g = Acceleration of gravity, ft/sec/sec

Sometimes L/D is referred to as the "equivalent length in diameters." Valves, fittings, etc., have often been expressed in equivalent diameters. If the friction factor f is combined with L/D, a constant K is derived for any pipe, valve, or fitting. If K = f(L/D), then

$$H_L = K\left(\frac{V^2}{2g}\right) \qquad (2)$$

The term K is called the "head loss factor" or "pressure drop factor."

Most control valves are rated with a capacity term called C_v. C_v is the number of gallons of water per minute that will flow through the valve with a 1 lb/in.^{2} pressure drop across the valve.

$$C_v = Q/\sqrt{\Delta P} \qquad (3)$$

Q = Quantity of water in gpm
DP = Pressure drop in psi

Rearranging Equation 3:

$$\Delta P = Q^2/(C_v)^2 \qquad (3a)$$

This equation may be rearranged so that C_v in terms of our K factor may be expressed as:

$$C_v = \frac{29.9(d)^2}{\sqrt{K}} \qquad (4)$$

where d = inside diameter of the pipe or fitting, inches.

The head loss or pressure drop of a fluid flowing through various pipes, valves, or fittings in series can be stated:

$$(\Delta P)_{Total} = (\Delta P)_1 + (\Delta P)_2 + (\Delta P)_3 + — \qquad (5)$$

Substituting Equation 3a in Equation 5:

$$\frac{1}{(C_v)^2_{Total}} = \frac{1}{(C_v)^2_1} + \frac{1}{(C_v)^2_2} + \frac{1}{(C_v)^2_3} + — \qquad (6)$$

This equation is unwieldy if several fittings are used. By equating Equation 4, substituting in Equation 6 and dividing through by $[29.9(d)^2]^2$:

$$K_{Total} = K_1 + K_2 + K_3 + — = \sum K \qquad (7)$$

Substituting in Equation s4 for K_T an over-all C_v can be derived:

$$C_v = \frac{29.9(d^2)}{\sqrt{\sum K}}$$

K factors for various valves and fittings

For full open valves, pipes or fittings, the K factor is approximately equal to f(L/D) or Equation 1.

where:

L = Equivalent length of valve or pipe or fitting, inches (or feet)
D = Diameter of the valve, inches (or feet)
f = Friction factor

Some of the recognized K factors for various welding fittings are shown in Table 1.

Other fitting K factors can be obtained from any fluid mechanics, hydraulics, or piping handbook.

Low, average, and high values of K given in Table 1 point out caution necessary in selecting specific types of fittings. Pressure losses in valves and fittings can be more accurately calculated by consulting the references.

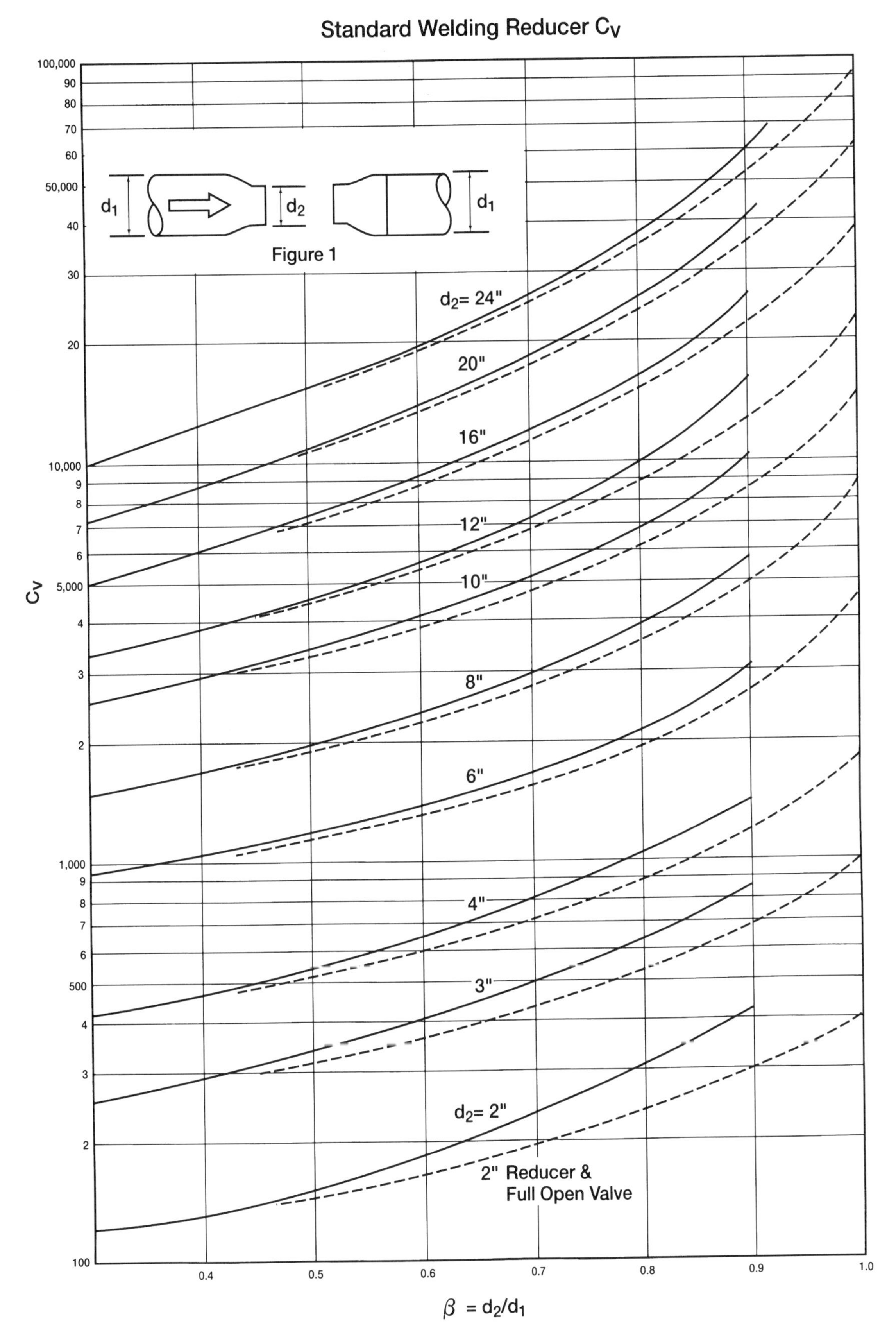

Figure 1. C_v for standard welding reducers.

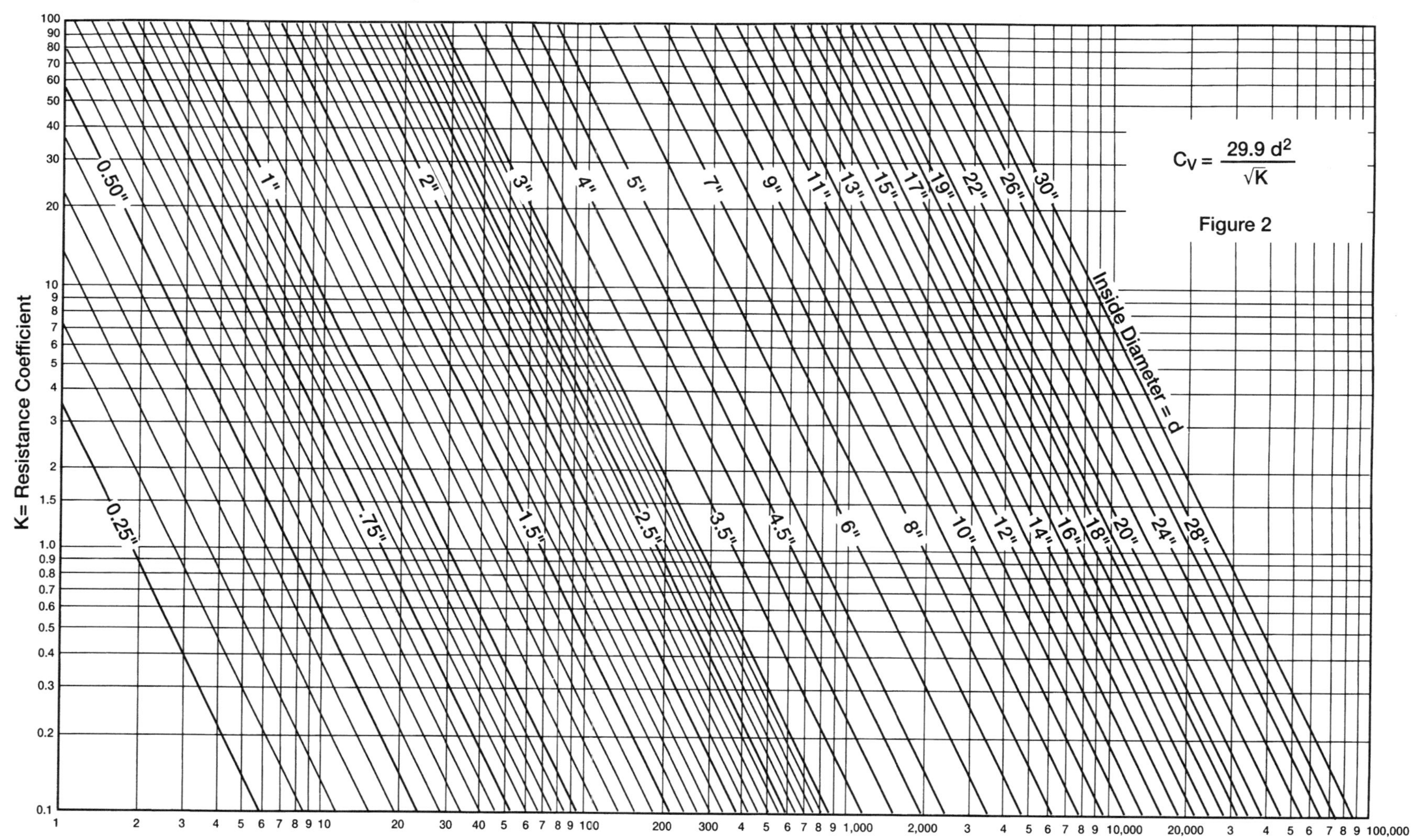

Figure 2. C_v versus K values for valves.

Table 1

K Factor			
Low	**Average**	**High**	**Type of Welding Fittings and Valves**
0.10	0.20	0.40	90° Elbow
0.05	0.12	0.25	45° Elbow
0.25	0.50	1.00	Tee (Enter run, out branch)
0.75	1.00	1.20	Tee (Enter branch, out run)
0.05	0.10	0.15	Tee (Enter run, out run)
See Figs. 1 and 2			Standard Reducer-Increaser
0.25	0.50	1.50	Plug Valves (See manufacturer's bulletin)

All K factors, L/D's or C_v's must be converted to the same basis using the same diameter d when using an overall C_v for a piping run. This can be accomplished with the following equations:

$$K_a = K_b(f_a/f_b)(d_a/d_b)^4 \quad (8)$$

$$(L/D)_a = (L/D)_b(d_a/d_b)^4 \quad (9)$$

$$(C_v)_a = (C_v)_b(d_b/d_a)^2 \quad (10)$$

Examples

Refer to Figure 3 for a schematic of a typical control valve station using an old-style control valve. Flow is from a header through a 12-in. by 6-in. reducer, a 6-in. isolating valve, 6-in. globe body control valve, 6-in. pipe spool, a second isolating valve, a 6-in. by 12-in. increaser, a 12-in. welding elbow, and into a header. Note should be made of Figure 1 in that the C_v from the solid curve is for the complete reducer-increaser combination with regard to the ratio of diameters. In this case, the Beta ratio is 6/12 or 0.5 and C_v is 1,200 for the small diameter of 6 in. Since C_v's cannot be simply added, they must be converted to K factors. To do this, refer to Figure 2, which converts C_v to K.

Also to be considered are the isolating valves and control valves. Resistance coefficients, C_v's or pressure drop factors are published by most valve manufacturers and may be converted to a K factor by using Figure 2. In this case, 6-in. K_v is 0.05.

Pipe resistance must also be considered (Figure 5). The L/D ratio must be calculated with reference to the same units of L and D. In the example, 12 ft of 6-in. pipe has a K_P, equal to 0.36.

All resistances must be converted to the same base diameter (in this case 6 in.). The entrance factor K_e is 0.05 for 12 in. Referring to Figure 6, the 6-in. K_e is about 0.04; the 12-in. elbow coefficient converted to 6-in., 6-in. K_{el} is 0.02, and the 12-in. exit coefficient, 6-in. K_x, is 0.8. A tabulation and sum of these coefficients as determined from Figure 3 would be:

12″Entrance	$6''K_e = 0.04$
6″Reducer—Increaser	$6''K_{ri} = 0.80$
2–6″ Valves	$6''K_v = 0.10$
6″Control Valve	$6''K_v = 10.00$
12-ft spool of 6″ Pipe	$6''K_P = 0.36$
12″Elbow	$6''K_{el} = 0.02$
12″Exit	$6''K_x = 0.08$
	$\sum K = 11.40$

Refer to Figure 2 for a conversion of K to C_v. With the above values noted, the overall 6-in. C_v is 320, approximately a 6% reduction of the original C_v of 340.

Now refer to Figure 4. This is an identical control valve run with Figure 3, except that a 6-in. × 4-in. × 6-in. high-capacity ball valve has been substituted for the 6-in. singleported globe body control valve. Note that the ΣK now is 1.93, which converted to an overall C_v is 775. This is a reduction of about 48% from the original 6-in. × 4-in. × 6-in. control valve C_v of 1,495.

If, in Figure 4, the 6-in. × 4-in. × 6-in. valve were substituted with a full open 6-in. ball valve, the overall C_v is 895. This represents a reduction in C_v of about 81% from the original 6-in. ball valve C_v of 4,700.

The methods and graphs presented in this article are proposed for sizing and calculating pressure drops in a control station. The assumption of assigning values to certain unusual piping configuration must be recognized. The author has some field data confirmation on similar control station designs.

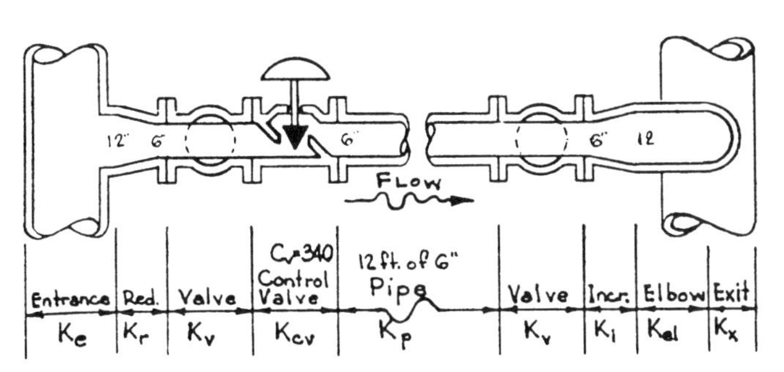

Figure 3. Piping configuration.

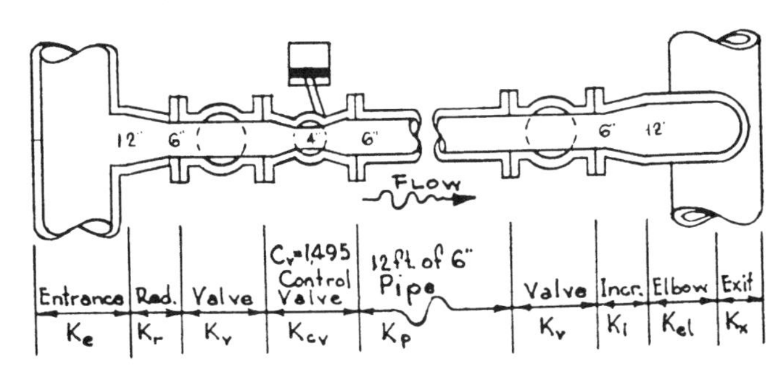

Figure 4. Piping configuration.

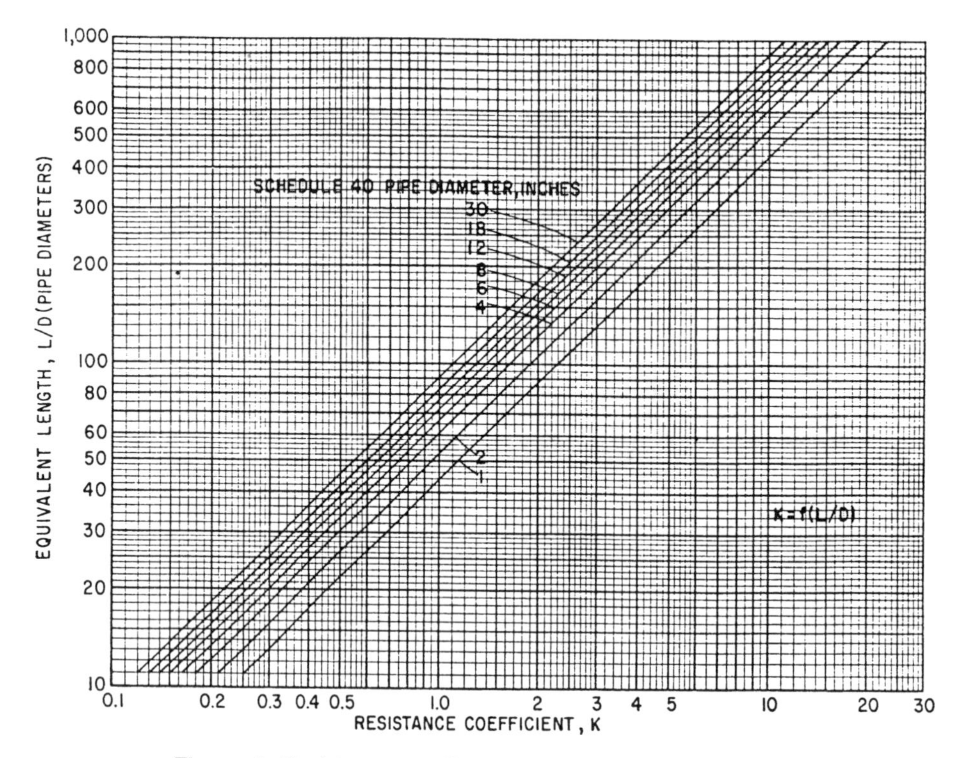

Figure 5. Resistance coefficients versus equivalent length.

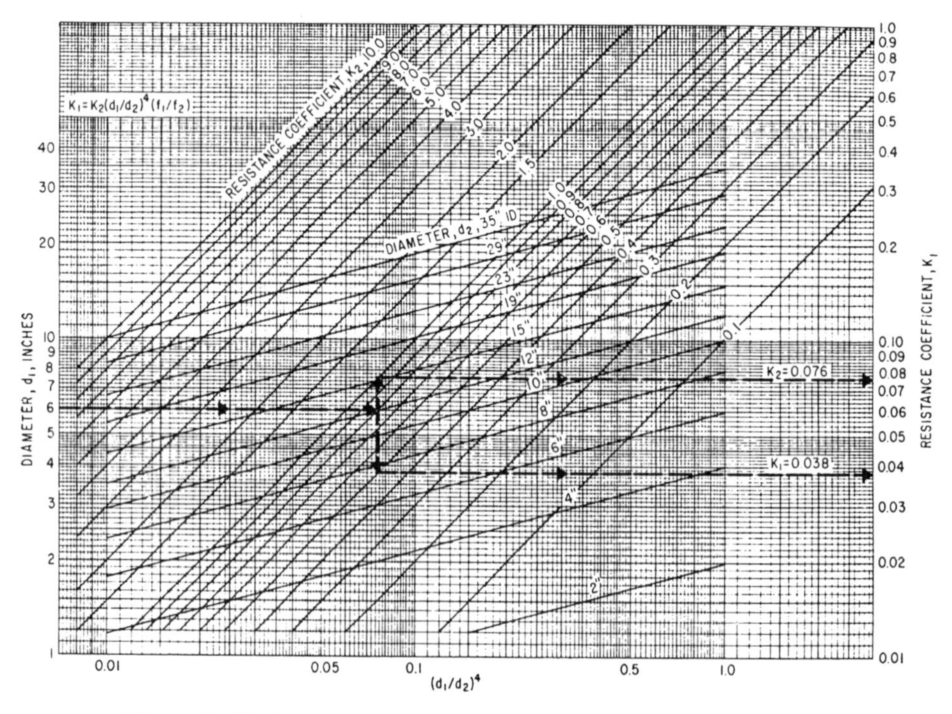

Figure 6. Equivalent resistance coefficients versus diameter ratios.

Control valve selection

A control valve may be thought of as a section of pipe with a variable resistance. Control valves are typically used to match pumps to the pipeline system characteristics. They are not normally used with reciprocating compressors but may be used with centrifugal compressors for preventing surge.

Liquid pipeline applications usually involve controlling the flow rate, suction and discharge pressure, and other variables for centrifugal pumps. Control valves may also be used with reciprocating pumps to regulate the pump discharge pressure by recycling fluid from the pump discharge to the pump suction when the discharge pressure approaches the control point. The control valve may also be used to unload the pump during the start-up by opening just prior to start-up and slowly closing after the pump driver is up to operating speed.

Other applications for control valves include:

- Pressure reduction
- Pressure relief
- Back pressure control
- Control delivery flow rates

Ideally, the pump(s) for a system will be sized such that a minimum amount of pressure throttling will be required during normal operations. System startup, system upsets, and unit startup and shutdown require that a means be provided to regulate pressures, flow rates, and in some cases motor load. This may be accomplished by using variable speed motors, variable speed couplings, or a control valve.

The control valve, when used with centrifugal pumping units, is installed in the pump discharge line. If multiple pumping units are used, only one control valve will be required.

Table 1 lists typical values of C_v for conventional and ball control valves.

When conventional globe-type control valves are used in pipeline applications for pump control, valves with equal percentage trim are the best choice. Figure 1 shows the typical control valve static characteristic curves for quick opening, linear, and equal-percentage valve trims.

Curves such as those shown in Figure 1 are based on maintaining a constant differential pressure of 1 psig across the valve for varying flow rates. In actual practice, the pressure differential across the control valve will vary according to the needs of the system. Figure 2 shows a comparison of the static and dynamic characteristics for an equal percentage control valve. The dynamic characteristic curve was developed by taking the pressure drop across a 12-in. double-ported valve with equal percentage trim in series with 10 miles of 12-in. pipeline. A maximum flow rate of 6,000 bbl/hr of 0.86 sp. gr. crude oil at a constant input pressure of 1,000 psig was used to calculate the pressure loss in the pipeline. The pressure drop across the control valve was calculated for various flow rates while maintaining a constant system input pressure. The C_v was then calculated

Table 1
Typical values of C_v for control valves*
(equal percentage trim)

Valve Size	Double Ported Globe	Single Ported Globe	Ball Valve Full Bore†
2	48	46	406
4	195	195	1,890
6	450	400	4,580
8	750	640	8,900
10	1,160	1,000	14,600
12	1,620		22,800
14	2,000		28,700
16	2,560		38,900
20			66,000
24			98,600
26			119,800

*These values are suitable for estimating only. The valve manufacturer should be consulted for the actual C_v for the valve selected.
†Courtesy: Thunderco–Houston.

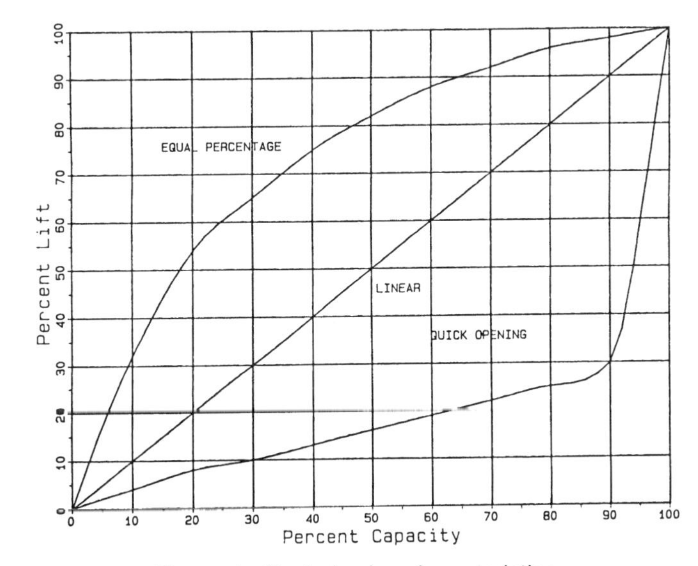

Figure 1. Control valve characteristics.

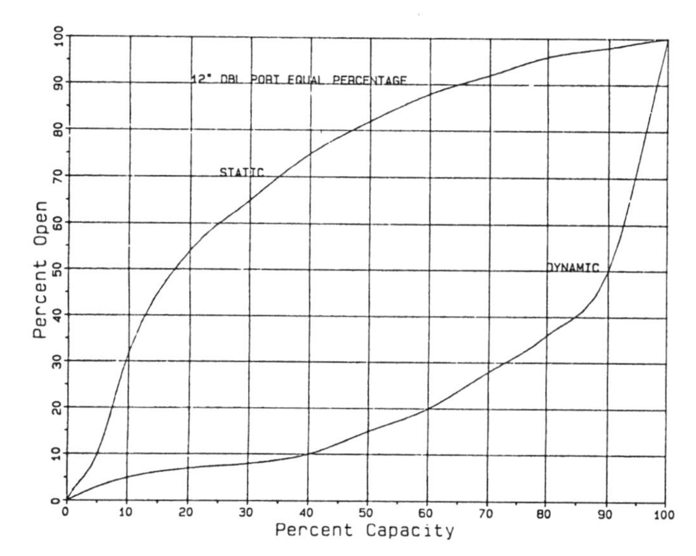

Figure 2. Control valve characteristics.

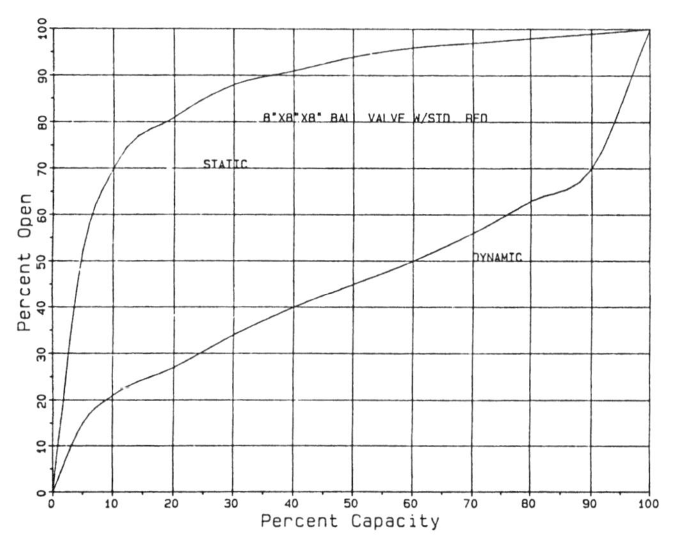

Figure 3. Control valve characteristics.

for each of the flow conditions and converted to percent of valve lift.

Figure 3 shows a comparison of the static and dynamic characteristics for an 8-in. × 8-in. × 8-in. ball valve under the same conditions of flow and pressure. A 6-in. × 6-in. × 6-in. ball valve with conical reducers has a C_v of 2,100, which would have satisfied the pressure drop requirements; however, the velocity through the 6-in. valve would have been approximately 45 ft/sec. Experience has shown that the valve velocity should be limited to about 25 ft/sec. The velocity through the 8-in. valve will be 27 ft/sec. The C_v for the 8-in. valve with standard welding reducers is 2,600, and with conical reducers, the C_v is 4,100.

The choice of using standard concentric welding reducers was made because that combination came closest to satisfying the C_v needs. When sizing and selecting valves, it is important that the valve not be oversized, as this will always result in poor system control. Reduced-size trim is available for conventional control valves, and ported ball valves are available.

In this particular application, Figures 2 and 3 show that for a 10% change in system rate, the conventional globe valve must change its position by 50% (assuming that it was operating in the full open position). The ball valve would be required to change its position by 30% under the same conditions.

In pipeline applications, it is desirable to have virtually no pressure drop across a control valve when it is wide open. Any built-in residual pressure drop across a control valve will result in added fuel costs; however, in certain situations, some residual pressure drop is unavoidable.

In process control, some pressure drop across the valve is desired to improve the control of the process. Ball valves have gained wide acceptance in pipeline operations because of their high C_v factor, which in turn results in very low pressure drop across the valve when wide open. As can be seen from Table 1, a 16-in. double ported valve (equal percentage trim) has a maximum C_v of 2,560. Based on a maximum pressure drop of 5 psig with the valve fully open, the maximum flow rate for 0.86 sp. gr. crude oil would be 212,000 bpd for a single valve. For flow rates greater than this, multiple valves may be used or one ball valve may be used. Very large ball valves of up to at least 36-in. in size have been successfully used for control valve applications.

The ideal control valve and control system will be quick to act when needed, and when controlling, will be stable at all control points. As shown by the dynamic characteristic curves, the valve must be capable of moving quickly from the wide open position to the control point without overshoot. Certain types of control systems may prove to be unstable on control when required to move the valve at high speeds.

High flow rates and high pressure drops across a control valve will require a valve operator with sufficient power to move the valve quickly and positively. Hydraulic systems have gained wide acceptance for this type of application. Air motors may also be used; however, the cost of the air compressor and air dryer will usually equal or exceed the cost of a hydraulic system. Other factors favoring a hydraulic system include the use of variable vane pumps, which reduces the power consumption during normal control operations. With a hydraulic system, a small hand pump may be used during electrical power outages. Control valves with air motors may be equipped with a handwheel for manual operation.

RELIEF VALVE SIZING, SELECTION, INSTALLATION, AND TESTING

Effectiveness of a Safety Relief Valve Depends on Proper Sizing and Type of Equipment Selected

Gary B. Emerson, Manager SVR Products, Anderson-Greenwood USA, Bellaire, Texas

A safety relief valve is an essential and important piece of equipment on virtually any pressured system. Required by the ASME-UPV Code, among others, it must be carefully sized to pass the maximum flow produced by emergency conditions.

After it is installed, the user hopes that it will never have to operate, which makes the safety relief valve rather unique as compared to other types of process equipment.

Sizing

The sizing formulas for vapors and gases fall into two general categories, based on the flowing pressure with respect to the discharge pressure. When the ratio of P_1 (set pressure plus allowable accumulation) to P_2 (outlet pressure) is greater than two, the flow through the relief valve is sonic; the flow reaches the speed of sound for the particular flowing medium. Once the flow becomes sonic, the velocity remains constant; it cannot go supersonic. No decrease of P_2 will increase the flowrate.

Sonic flow

In accordance with API RP 520, Part 1, Section 3.3, the formulas used for calculating orifice areas for sonic flow are:

$$A = \frac{W\sqrt{TZ}}{CKP_1\sqrt{M}}$$

$$A = \frac{Q\sqrt{Z}}{18.3KP_1(C/356)\sqrt{520/T}\sqrt{29/M}}$$

where:
A = Calculated orifice area (in.^{2})
W = Flow capacity (lb/h)
Q = Flow capacity (scfm)
M = Molecular weight of flowing media
T = Inlet temperature, absolute (°F + 460)
Z = Compressibility factor
C = Gas constant based on ratio of specific heats (Table 1)
K = Valve coefficient of discharge
P_1 = Inlet pressure, psia (set pressure + accumulation + atmos. pres.)

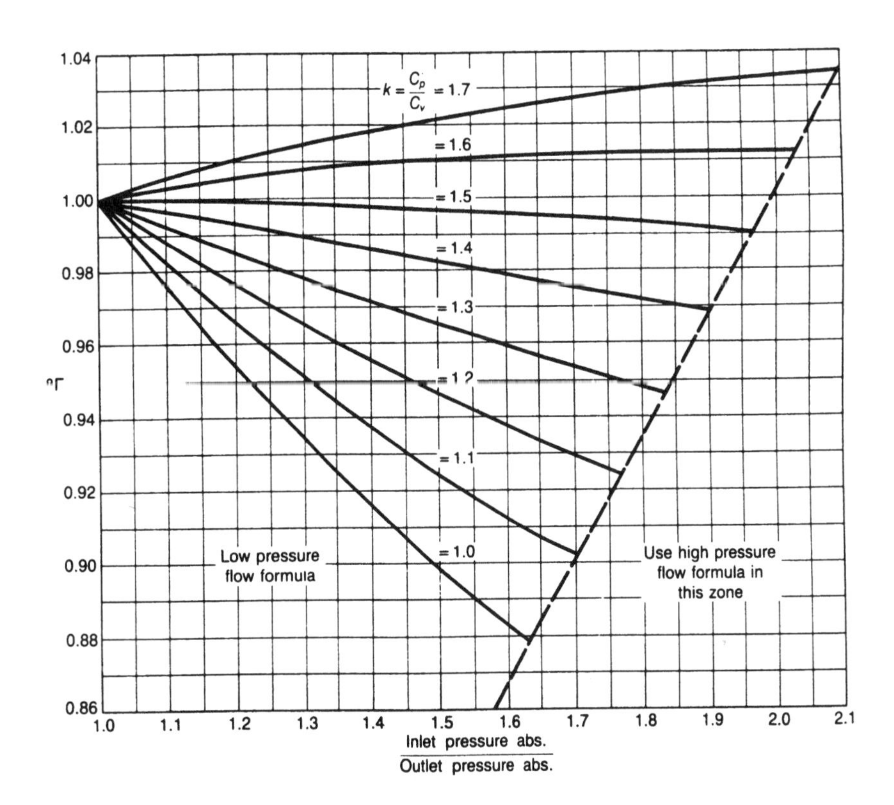

Figure 1. Low pressure and high pressure flow formulas.

Subsonic flow

The second general category for vapor or gas sizing is generally when P_2 is greater than half of P_1 (backpressure greater than half of inlet pressure).

Using k (ratio of specific heats) and P_1/P_2 (absolute), confirm from Figure 1 that the subsonic ("low pressure") flow formula is required. If so, then determine "F" factor. If not, use above sonic flow formula.

$$A = \frac{Q\sqrt{GI}}{863KF\sqrt{(P_1 - P_2) \times P_2}}$$

where: G = Specific gravity
F = Factor from Table 1
I = Temperature of relieving gas
P_2 = Outlet pressure, psia (backpressure + atmospheric pressure)

After determining the calculated orifice area, select the next largest standard orifice size from the relief valve manufacturer's catalog.

Table 1
Gas constant based on ratio of heats

Gas	Mol. Wt.	C_p/C_v	°F
Acetylene	26	1.26	343
Air	29	1.40	356
Ammonia	17	1.31	348
Argon	40	1.67	378
Benzene	78	1.12	329
Butadiene	54	1.12	324
Carbon dioxide	44	1.28	345
Carbon monoxide	28	1.40	356
Ethane	30	1.19	336
Ethylene	28	1.24	341
Freon 22	86.47	1.18	335
Helium	4	1.66	377
Hexane	86	1.06	322
Hydrogen	2	1.41	357
Hydrogen sulphide	34	1.32	349
Methane	16	1.31	348
Methyl mercapton	48.11	1.20	337
N-Butane	58	1.09	326
Natural gas (0.60)	17.4	1.27	344
Nitrogen	28	1.40	356
Oxygen	32	1.40	356
Pentane	72	1.07	323
Propane	44	1.13	330
Propylene	42	1.15	332
Sulfur dioxide	64	1.29	346
VCM	62.5	1.18	335

Selection

The fundamental selection involves the consideration of the two basic types of relief valves more commonly used: conventional spring-loaded type and pilot operated relief valves.

The advantages of the spring-loaded relief valves are as follows:

- Competitive price at lower pressures and in smaller sizes.
- Wide range of chemical compatibility.
- Wide range of temperature compatibility, particularly at higher temperatures.

The disadvantages are as follows:

- Metal-to-metal seat not tight near set pressure and usually after valve relieves.
- Sensitive to conditions that can cause chatter and/or rapid cycling.
- Protection against effects of backpressure is expensive, pressure limited, and creates possible additional maintenance problems.
- Testing of set pressure not easily accomplished.

Advantages of pilot operated type relief valves are as follows:

- Seat tight to set pressure (Figure 2).
- Ease of setting and changing set and blowdown pressures.
- Can achieve short blowdown without chatter.
- Pop or modulating action available.
- Easy maintenance.

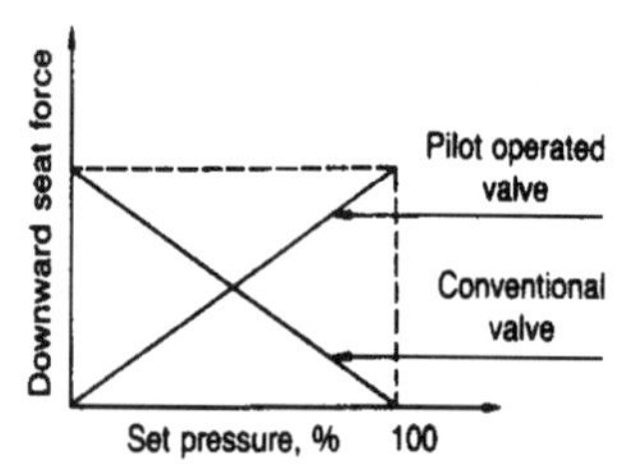

Figure 2. Setting seat pressure to set flow pressure.

- Easy verification of set pressure without removing relief valve from service.
- Flexibility: offers options for remote operation, backpressure protection, valve position indication.

The disadvantages are as follows:

- Maximum temperature limitations.
- Should not be used in extremely dirty or polymerizing type service.

Installation procedures

Safety relief valve installation requires that careful consideration be given to inlet piping, discharge piping, pressure sense lines (where used), and bracing if required. Any marginal installation practice can render the safety relief valve inoperable, at worst, or severely restrict its design capacity at best. Either condition compromises the safety of the installation.

Inlet piping

The proper design of inlet piping is extremely important. Very often, safety relief valves are added to an installation at the most physically convenient location, with little regard to flow considerations. Pressure loss during flow in a pipe always occurs. Depending upon the size, geometry, and inside surface condition of the pipe, the pressure loss may be large (20, 30 or 40%) or small (less than 5%).

API RP 520, Part 2, recommends a maximum inlet pipe pressure loss to a safety relief valve of 3%. This pressure loss shall be the sum total of the inlet loss, line loss and when a block valve is used, the loss through it. The loss should be calculated using the maximum rated flow through the pressure relief valve.

The 3% maximum inlet loss is a commendable recommendation but often very difficult to achieve. If it cannot be achieved, then the effects of excessive inlet pressure should be known. These effects are rapid or short cycling with direct spring operated valves or resonant chatter with pilot operated relief valves. In addition, on pilot operated relief valves, rapid or short cycling may occur when the pilot pressure sensing line is connected to the main valve inlet. Each of these conditions results in a loss of capacity.

Pilot operated valves can tolerate higher inlet losses when the pilot senses the system pressure at a point not affected by inlet pipe pressure drop. However, even though the valve operates satisfactorily, reduced capacity will still occur because of inlet pipe pressure losses. The sizing procedure should consider the reduced flowing inlet pressure when the required orifice area, "A," is calculated.

A conservative guideline to follow is to keep the equivalent L/D ratio (length/diameter) of the inlet piping to the relief valve inlet to 5 or less.

Rapid cycling

Rapid valve cycling with direct spring operated valves occurs when the pressure at the valve inlet decreases excessively at the start of relief valve flow. Figure 3 is a schematic of system pressures that is shown before and during flow. Before flow begins, the pressure is the same in the tank and at the valve inlet. During flow, the pressure at

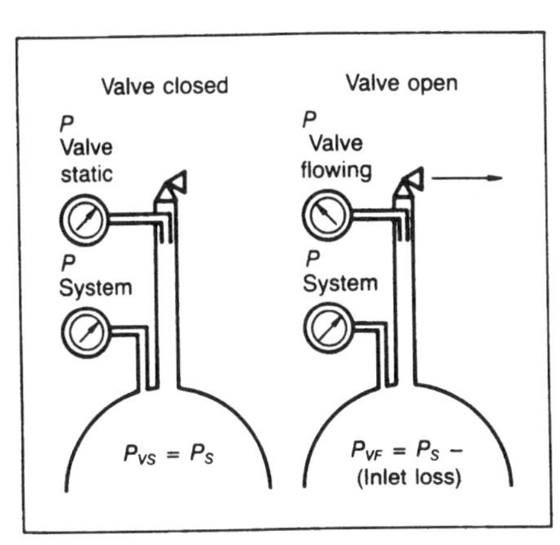

Figure 3. System pressures before and after flow.

the valve inlet is reduced due to the pressure loss in the inlet pipe to the valve. Under these conditions, the valve will cycle at a rapid rate.

The valve responds to the pressure at its inlet. When that pressure decreases during flow to below the valve's reseat point, the valve will close. However, as soon as the flow stops, the inlet pipe pressure loss becomes zero and the pressure at the valve inlet rises to tank pressure once again. If the tank pressure is still equal to or greater than the relief valve set pressure, the valve will open and close again. Rapid cycling reduces capacity and is sometimes destructive to the valve seat in addition to subjecting all the moving parts in the valve to excessive wear.

On pilot operated relief valves with the pilot valves sense line connected at the main valve inlet, rapid cycling can also occur for the same reasons described earlier; namely the pressure at the relief valve inlet is substantially reduced during flow (Figure 4). With the pilot sense line located at the main valve inlet, the pilot valve responds to the pressure at the location and therefore closes the main valve. If the pressure loss during flow is small (less than the reseat pressure setting of the pilot), the main valve may or may not partially close depending on the type of pilot design used.

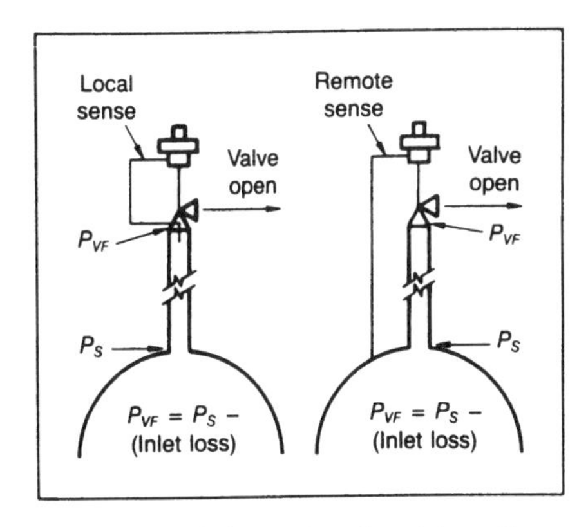

Figure 4. Schematic of rapid cycling.

In actual application, because of the long response time of a pilot operated valve compared with a direct spring operated valve, the main valve may only go into partial life since the inlet pipe pressure loss occurs before the pilot has sufficiently vented the dome area for full piston travel.

As with direct spring operated valves, capacity is reduced but under severe inlet pressure loss conditions resonant chatter may be induced in pilot operated relief valves. This chatter is extremely destructive and may result in complete valve failure.

On existing installations the corrective action is limited. For direct spring operated valves, increasing the blowdown will minimize or eliminate rapid cycling if the blowdown can be adjusted to below the initial flowing pressure.

Resonant chatter

Resonant chatter occurs with pilot operated relief valves when the inlet piping produces excessive pressure losses at the valve inlet. The higher the set pressure or the greater the inlet pipe pressure loss, the more likely resonant chatter will occur.

Unlike the rapid cycling noted earlier, resonant chatter is uncontrolled; that is, once started, it cannot be stopped unless the pressure is removed from the valve inlet. In actual application, however, the valve will self-destruct before a shutdown can take place because of the magnitude of the forces involved in a resonant mode.

Discharge piping

Discharge piping for direct spring operated valves is more critical than for pilot operated valves. As with inlet piping, pressure losses occur in discharge headers with large equivalent L/D ratios. Excessive backpressure will reduce the lift of a direct spring operated valve, and enough backpressure (15% to 20% of set + overpressure) will cause the valve to reclose. As soon as the valve closes, the backpressure in the discharge header decreases and the valve opens again with the result that rapid cycling can occur.

Pilot-operated relief valves with the pilot vented to the atmosphere or with a pilot balanced for backpressure are not affected by backpressure. However, if the discharge pressure can even exceed the inlet pressure, a back flow preventer must be used.

The valve relieving capacity for either direct or pilot operated relief valves can be affected by backpressure if the flowing pressure with respect to the discharge pressure is below critical (subsonic flow).

Balanced bellows valves (direct spring operated) have limitations on maximum permissible backpressure due to the collapse pressure rating of the bellows element. This limitation will in some cases be less than the backpressure limit of a conventional valve. Manufacturer's literature should be consulted in every case. If the bellows valve is used for systems with superimposed backpressure, the additional built-up backpressure under relieving conditions must be added to arrive at maximum backpressure.

The performance of pressure relief valves, both as to operation and flow capacity, can be achieved through the following good discharge piping practices:

- Discharge piping must be at least the same size as the valve outlet connection and may be increased when necessary to larger sizes.
- Flow direction changes should be minimized, but when necessary use long radius elbows and gradual transitions.
- If valve has a drain port on outlet side, it should be vented to a safe area. Avoid low spots in discharge piping, preferably pitch piping away from valve outlet to avoid liquid trap at valve outlet.
- Proper pipe supports to overcome thermal effects, static loads due to pipe weight, and stresses that may be imposed due to reactive thrusts forces must be considered.

Reactive force

On large orifice, high-pressure valves, the reactive forces during valve relief are substantial, and external bracing may be required (Figure 5).

API RP 520, Part 2 gives the following formula for calculating this force.

$$F = \frac{Q_h}{366}\sqrt{\frac{kT}{(k+1)M}} \tag{1}$$

where:

F = Reactive force at valve outlet centerline (lb)
Q_h = Flow capacity (lb/h)
k = Ratio of specific heats (C_p/C_v)
T = Inlet temperature, absolute (°F + 460)
M = Molecular weight of flowing media.

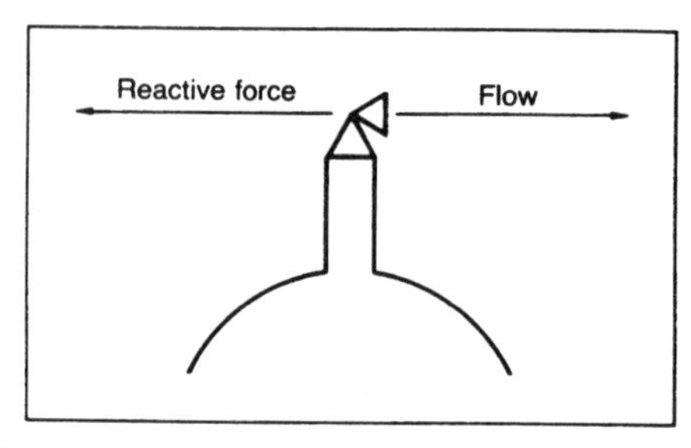

Figure 5. Reactive forces during valve relief.

If bracing is not feasible, a dual outlet valve (available in some pilot operated safety relief valves) can be used. The reactive forces are equal but opposite, resulting in zero force on the valve outlet, but if redirected can still impose loads that must be reacted in some manner.

An approximate method for calculating the reactive force is to multiply the following pressure (psig) by the orifice area (in.^{2}); the reactive force obtained will be on the high side.

Example	A-G Type 22312R68 "R" Orifice (16.00 in.^{2})
Set pressure—	1,000 psi, 10% overpressure
Flowing media—	Natural gas, 60°F
	M = 17.4(G = 0.60)
	C = 344 (k = 1.27)
Reactive force =	10,143 lb., using equation (1)
Reactive force =	17,600 lb., using approximate method.

Testing

Using DOT Title 49, Part 192.739 as a guideline, each valve should be inspected at least once a year to determine that it:

- Is in good mechanical condition
- Will operate at the correct set pressure
- Has adequate capacity and is operationally reliable
- Is properly installed and protected from dirt, liquids, or other conditions that might prevent proper operation

The effectiveness of a safety relief valve installation depends greatly on proper sizing and selection of the valve type, suitable installation conditions, and proper and timely testing of the valve.

Rupture disc sizing

A review of the ASME Section VIII requirements

Changes to Section VIII, Division 1 of the ASME Code have impacted the methods used to size pressure relief systems that include rupture discs. The objective of this article is to clarify the sizing methodologies.

Key terminology

- K_D (Coefficient of Discharge)—This is a unit-less factor used to de-rate the theoretical flowing capacity of a device or system. For rupture discs, this value is defined by the Code as 0.62.
- K_R (Resistance to Flow Factor)—This is a unit-less factor used to characterize the resistance to flow of a rupture disc device. The larger the K_R value, the more restrictive the device. The K_R value has no direct correlation to the K_D, coefficient of discharge.
- Combination Capacity Factor—This is a unit-less factor used to de-rate the capacity of a pressure relief valve, when a rupture disc is installed upstream. The combination capacity factor may be the default value of 0.9, or a higher value, if the disc/valve combination has been tested.
- MNFA (Minimum Net Flow Area)—This is a derived area based on the area of the device, and/or the piping area, and is validated during ASME certification testing. The MNFA is used only in the coefficient of the discharge method.

Sizing methodologies

The code describes three methods to provide adequate relieving capacity for a pressure vessel. These methodologies are applied based on the characteristics of the relieving system, not individual components.

When a rupture disc is installed upstream of a pressure relief valve, the code requires the capacity of the valve to be de-rated by a factor of 0.9, or by a *combination capacity factor* determined by test for that disc/valve combination. This method assumes the disc has a capacity equal to or greater than the valve, and does not interfere with the normal functioning of the valve. From the rupture disc standpoint, this means the rupture disc should be the same nominal size as the relief valve inlet (or larger) and that the disc be of a full opening design. This method is unchanged from the previous versions of Section VIII.

The *coefficient of discharge method* (K_D) is not a rupture disc sizing method but a method for sizing a relatively simple relief system. The calculations for this method remain unchanged from previous versions of the code, but guidelines for use have been added. This method calculates the theoretical capacity of system, and de-rates it by the coefficient of discharge, $K_D = 0.62$. This method is used only when each of the following conditions is true:

1) the rupture disc discharges to atmosphere
2) the rupture disc is installed within 8 pipe diameters of the vessel
3) there are no more than 5 pipe diameters of discharge piping on the outlet of the device

The reason this is considered a system analysis method is because it takes into account the entrance conditions from the vessel to the discharge piping, 8 pipe diameters of piping to the rupture disc, and 5 pipe diameters of piping, from the rupture disc to the atmosphere.

The *resistance to flow method* (K_R) is used if neither of the first two methods apply, such as systems with long runs of piping, discharging into manifolds, or flare systems. The resistance to flow method is not a rupture disc sizing method, but a system sizing method. This analysis takes into account the resistance to flow of all of the elements within the relieving system, including the resistance of the rupture disc. The code then requires that the system capacity then be derated by a factor of 0.9 to account for uncertainties in the method.

The code does not describe the detailed analysis method, other than to say "accepted engineering practices." Commonly used references for this type of analysis are: API RP521 Guide for Pressure Relieving, and Depressuring Systems of Crane Paper 410.

As a point of reference, the following table is an excerpt from API RP521 listing typical K values for various piping components. The rupture disc K_R values available from the manufacturer are used in the same way as the K values for these common piping components. In many cases the rupture disc is not a limiting or even significant factor in the capacity of a relieving system. Note the API generic rupture disc K value listed here is not necessarily conservative and actual tested values may be significantly higher or lower than 1.5.

Fitting	K
Globe valve, open	9.7
Angle valve, open	4.6
Swing check valve, open	2.3
90-degree mitered elbow	1.72
Rupture disc	1.5
Welded tee flowing through branch	1.37
15 feet of 3″ sch 40 pipe	1.04
90-degree std threaded elbow	0.93
90-degree long sweep elbow	0.59

Reprinted with permission—The Fike Corp. Blue Springs, MO.

Rupture disc sizing using the resistance to flow method (K_R)

An excerpt from Fike Technical Bulletin TB8102:

This is a sizing method for relief systems that have a rupture disc as the relief device. Rupture disc devices can be characterized as to their respective resistance to fluid flow. The K_R value represents the velocity head loss due to the rupture disc device. This head loss is included in the overall system loss calculations to determine the capacity of the relief system.

ASME PTC25 provides standardized test methods to measure the K_R of rupture disc devices. By quantifying this performance characteristic, rupture disc devices may be accounted for in piping system sizing calculations in the same way as piping and piping components (exit nozzles on vessels, elbows, tees, reducers, valves).

Sample calculation using K_R method

The example shown is based on Crane TP-410 methods. It assumes a steady-state relieving condition where the vessel volume is large relative to the relieving capacity.

Determine the Total Piping System Resistance:

Piping Component or Feature	Flow Resistance Value (K)	Reference
Entrance—Sharp Edged	$K_1 = 0.50$	Crane 410 Pg A-29
1 ft. of 3″ Sch. 40 Pipe	$K_2 = 0.07$	K = fL/D; f = .018 (Crane 410 Pg A-26) L = 1 ft, ID = 3.068″/12 ft
Fike 3″ SRX-GI Rupture Disc	$K_R = 0.99$	National Board Cert. No. FIK-M80042
20 ft. of 3″ Sch. 40 Pipe	$K_3 = 1.41$	K = fL/D; f = .018 (Crane 410 Pg A-26) L = 20 ft, ID = 3.068″/12 ft
3″ Sch. 40 Standard 90° Elbow	$K_4 = 0.54$	Crane 410 Pg A-29
40 ft. of 3″ Sch. 40 Pipe	$K_5 = 2.82$	K = fL/D; f = .018 (Crane 410 Pg A-26) L = 40 ft, ID = 3.068″/12 ft
Pipe exit—Sharp Edged	$K_6 = 1.00$	Crane 410 Pg A-29
Total System Flow Resistance	$K_T = 7.33$	$K_T = K_1 + K_2 + K_R + K_3 + K_4 + K_5 + K_6$

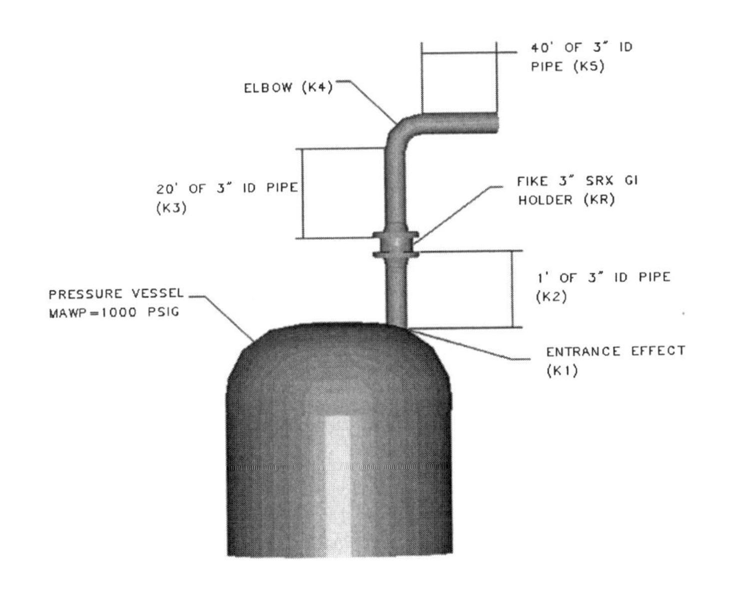

Given information:

1. Pressure vessel MAWP = 1000 psig.
2. Relieving pressure as allowed by ASME Section VII Div. 1 + 110% × MAWP = 1114.7 psia = P′1
3. Back pressure (outlet pressure) = 14.7 psia = P2
4. Working fluid – air (k = Cp/Cv = 1.4)
5. Air temperature at disc rupture = 500°F = 960°R = T1
6. Maximum flow rate into the vessel = 40,000 SCFM-Air
7. Rupture disc – Fike 3″ SRX-GI $K_R = 0.99$

The Darcy equation defines the discharge of compressible fluids through valves, fittings, and pipes. Since the flow rate into the example vessel is defined in SCFM, the following form of the Darcy equation is used:

Crane Equation 3-20

$$q'_m = 678 \times Y \times d^2 \sqrt{\frac{\Delta P \times P'_1}{KT_1 S_g}}\ \text{SCFM}$$

q'_m = rate of flow in cubic feet per minute at standard conditions (14.7 psia and 60°F)
Y = net expansion factor for compressible flow through orifices, nozzles, and pipes (Crane 410Pg A-22)
Δd = internal diameter of pipe (inches)
ΔP = Change in pressure from state 1 to 2 (entrance to exit).
P'_1 = Pressure at state 1 (entrance) psia
K = Loss coefficient
T_1 = Absolute temperature at state 1 (entrance) in degrees Rankine (°R)
Sg = Specific gravity relative to air = 1.0 for this example

To determine Y, first it must be determined if the flow will be sonic or subsonic. This is determined by comparing the *actual* $\Delta P/P'_1$ to the *limiting* $\Delta P/P'_1$ for sonic flow. Crane Table A-22 shows limiting factors for k = 1.4 for sonic flow at a known value of K_T. If $(\Delta P/P'_1)_{sonic}$ $(\Delta P/P'_1)_{actual}$, then the flow will be sonic.

Limiting Factors for Sonic Velocity ($k = 1.4$) Excerpted from Crane 410 P4 A-22

K	$\Delta P/P'_1$	Y
1.2	.552	.588
1.5	.576	.606
2.0	.612	.622
3	.622	.639
4	.697	.649
6	.737	.671
8	.762	.685
10	.784	.695
15	.818	.702
20	.839	.710
40	.883	.710
100	.926	.710

For this example : $(\Delta P/P'_1)_{actual} = \dfrac{1114.7 - 14.7}{1114.7} = 0.9868$

From table A-22 at $K_T = 7.33$

$(\Delta P/P'_1)_{sonic} = 0.754$

Since $(\Delta P/P'_1)_{sonic} (\Delta P/P'_1)_{actual}$, the flow is sonic in the piping. Thus, we will use $Y_{sonic} = 0.680$ in the flow calculations (Crane 410 Eq. 3-20).

Since $(\Delta P/P'_1)_{sonic} = .754$, then $\Delta P = .754 * P'_1 = 0.754 * 1114.7 = 840.5$ psi.

Calculating the system capacity is completed by substituting the known values into Crane 410 Equation 3-20.

$$q'_m = 678 \times Y \times d^2 \sqrt{\frac{\Delta P \times P'_1}{KT_1 S_g}}$$

$$q'_m = 678 \times 0.680 \times (3.068)^2 \sqrt{\frac{840.5 \times 1114.7}{7.33 \times 960 \times 1}}$$

$$q'_m = 50{,}074 \text{ SCFM} - \text{Air}$$

The ASME Pressure Vessel Code, Section VIII, Division 1, paragraph UG-127(a)(2)(b), also requires that the calculated system capacity using the resistance to flow method must also be multiplied by a factor of 0.90 to account for uncertainties inherent with this method.

$$q'_{m-ASME} = 50{,}074 \times 0.90 = 45{,}066 \text{ SCFM}_{Air}$$

Thus, the system capacity is greater than the required process capacity (40,000 SCFM_{Air}).

Reprinted with permission The Fike Corp. Blue Springs, MO and *Flow Control Magazine*

Variable orifice rotary control valves

Flow coefficient (C_v). The flow coefficient, which describes the flow capacity of a valve, is an empirically determined constant equal to the valve's flow capacity in U.S. gallons per minute of water at 60°F when a pressure drop of 1 psi exists across the valve in its most wide-open position. Table 1 lists C_v's for all sizes of DANTROL-270 valves, with and without trim.

The flow capacity of a valve in a partially open position may be calculated by selecting the appropriate C_v from Table 1 and multiplying it times the percent of maximum C_v at the desired position on the selected flow curve in Figure 2.

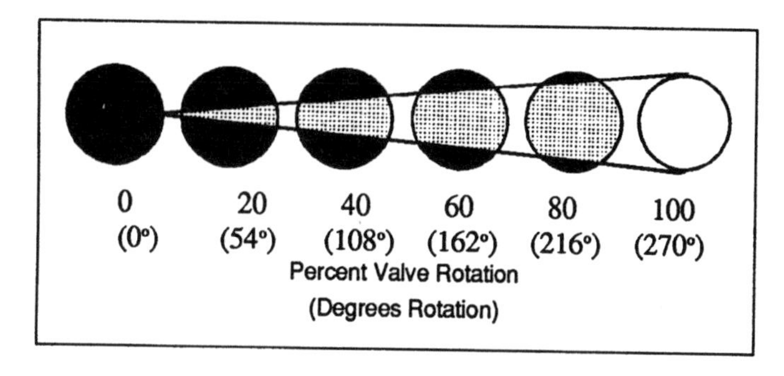

Figure 1. Orifice opening.

Table 1
Maximum Flow Coefficients (C_v)

Valve Size	Ball Dia.	Slot Length	No. Plates Each Slot	C_v w/Trim	C_v w/o Trim
1	4.25	7.40	8	45	53
1 ½	5.50	9.58	16	98	120
2	6.50	11.32	16	195	220
3	8.00	13.93	18	410	480
4	9.75	16.58	20	765	860
6	13.25	23.08	25	1900	1990

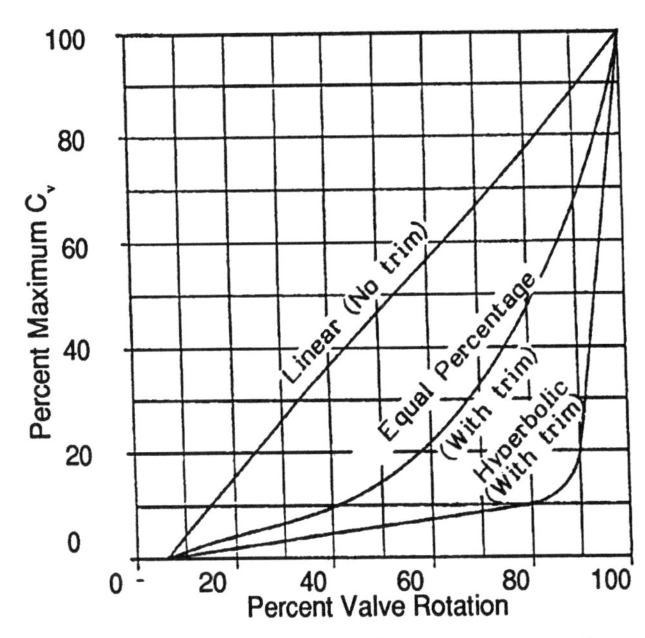

Figure 2. Percent of maximum C_v versus valve rotation.

Sizing valves for liquid flow

These equations for single-phase Newtonian liquid flow assume that the liquid is non-viscous and neither cavitating nor flashing. (Factors and constants are subsequently outlined.)

$$C_v = \frac{Q}{N_1 F_P E_R}\sqrt{\frac{G_F}{(P_1 - P_2)}}$$

or

$$C_v = \frac{W}{N_6 F_P F_R \sqrt{(P_1 - P_2)\gamma}}$$

Using the calculated RE_V, the value of F_R can be found from the graph in Figure 3. Substitute F_L for F_{LP} with line-size valves. (Values for F_L and F_{LP} are in the "Valve Reference Data"—Table 2 or 4.)

The piping geometry factor (F_P)

The piping geometry factor (F_P) is equal to 1.0 for line-size valves. With reducers, use the value of F_P from the "Valve Reference Data"—Table 2 or 4.

The Reynold's number factor (F_R)

The Reynold's number factor (F_R) has a value of 1.0 for most flows in the process industry. However, sizing errors due to low Reynold's numbers (RE_V) can occur when the fluid viscosity is high or flow velocities low. In either case, RE_V can be calculated from these equations:

$$RE_V = \frac{N_3 W}{v\sqrt{F_{LP} C_V}}\left[\frac{(F_{LP} C_V)^2}{N_2 D^4} + 1\right]^{0.25}$$

or

$$RE_V = \frac{N_4 Q}{v\sqrt{F_{LP} C_V}}\left[\frac{(F_{LP} C_V)^2}{N_2 D^4} + 1\right]^{0.25}$$

Choked flow

Choked flow is a limiting flow that occurs as a result of liquid vaporization when the internal valve pressure falls below the liquid's vapor pressure. Cavitation has then fully developed within the valve, and an increase in pressure drop across the valve will not produce an increase through the valve if

$$\Delta P \geq F_{LP}^2 (P_1 - F_F P_V)$$

(Substitute F_L for F_{LP} with line-size valves.) If the valve must be operated under choked conditions, use the following choked-flow equations in place of the basic equation for valve flow capacity:

$$C_V = \frac{Q}{N_1 F_{LP}}\sqrt{\frac{G_F}{(P_1 - F_F P_V)}}$$

or

$$C_V = \frac{W}{N_6 F_{LP}\sqrt{(P_1 - F_F P_V)\gamma}}$$

where:

$$F_F = 0.96 - 0.28\frac{P_v}{P_c}$$

(Use F_L for F_{LP} for line-size valves)

The cavitation index (K_c)

The cavitation index (K_c) relates to a condition just prior to flow choking where cavitation has only partially developed and flow rate is no longer proportional to the square root of pressure drop. This condition occurs when:

$$\Delta P_c \geq K_c(P_1 - P_V)$$

Values for the cavitation index (K_c) can be found in the Valve Reference Data—Table 2 or 4.

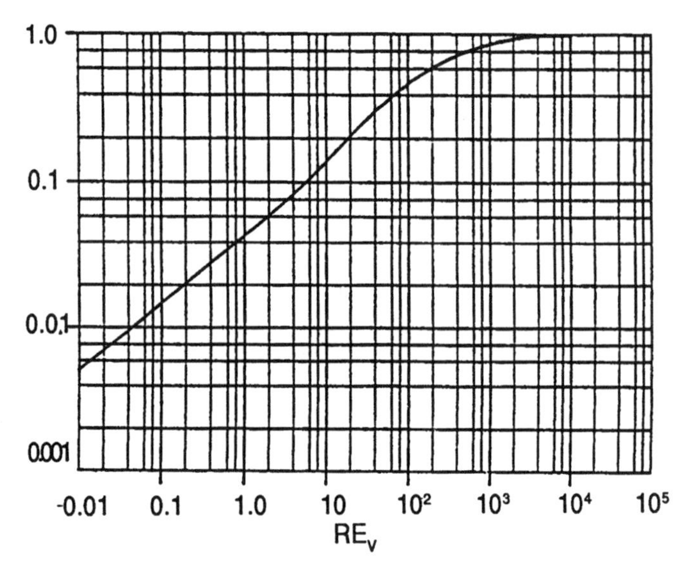

Figure 3. Reynolds Number Factor.

Anti-cavitation trim

Generally, operating a valve at conditions corresponding to K_c or choked flow will result in damaging cavitation, noise, and vibration.

If cavitation exists, a valve with anti-cavitation trim will prevent or lessen the effects of cavitation. When anti-cavitation trim is used, recalculate the flow-capacity coefficient (C_V) using factors for "With Trim" in the Valve Reference Data—Table. If noise is a concern, see the procedure under "Hydrodynamic Noise Prediction" in the manufacturer's literature.

SIZING VALVES FOR GAS AND VAPOR

Basic valve flow-capacity coefficient (C_V)

The basic valve flow coefficient (C_V) for gas or vapor flow through a variable orifice valve may be calculated using any of the following equivalent formulas for non-choked flow:

$$C_v = \frac{Q}{N_7 F_P P_1 Y}\sqrt{\frac{G_G T_1 Z}{X}}$$

or

$$C_v = \frac{W}{N_6 F_P Y \sqrt{X P_1 \gamma_1}}$$

or

$$C_v = \frac{Q}{N_9 F_P P_1 Y}\sqrt{\frac{M T_1 Z}{X}}$$

or

$$C_v = \frac{W}{N_8 F_P P_1 Y}\sqrt{\frac{T_1 Z}{X M}}$$

where: $F_k = \frac{K}{140}$ and $X = \frac{P_1 - P_2}{P_1}$

and $Y = \frac{1 - X}{3 F_K X_{TP}} \leq 0.667$*

*See "Choked Flow" for explanation.

Substitute X_T for X_{TP} with line-size valves. Values for F_P, X_T and X_{TP} are found in the "Valve Reference Data"—Table 2 or 4.

The compressibility factor (Z)

The compressibility factor (Z) can be found from Figure 4 after the reduced pressure (P_R) and reduced temperature

(T_R) have been calculated from these equations:

$$P_R = \frac{\text{Inlet Pressure (Absolute)}}{\text{Critical Pressure (Absolute)}}$$

$$T_R = \frac{\text{Inlet Temperature (Absolute)}}{\text{Critical Temperature (Absolute)}}$$

($^\circ R = ^\circ F + 460$, $^\circ K = ^\circ C + 273$)

Choked flow

Gas flow through valves can choke when the gas velocity reaches the velocity of sound. The pressure drop ratio at this point has the value:

$$X = F_K X_{TP}$$

Substitute X_T for X_{TP} with line-size valves. If this valve is to be used under choked conditions, Y is fixed at 0.667 regardless of the pressure drop. Substitute 0.667 for Y and $F_K X_T$ or $F_K X_{TP}$ for X in a basic value flow-capacity coefficient equation.

Valve outlet velocity

Sonic velocity at the valve's outlet will cause shock waves that produce vibration and noise problems. Therefore, you should make sure the outlet velocity does not reach sonic velocity for gas and steam service. The actual valve outlet diameter must be larger than the calculated diameter from the following equations:

$$d = N_{10}\sqrt{\frac{Q\sqrt{G_G T}}{P_2}} \quad \text{for gas}$$

$$d = N_{11}\sqrt{\frac{W(1 + N_{12}T_{SH})}{P_2}} \quad \text{for steam}$$

Low-noise trim

When the pressure-drop ratio (X) is near or exceeds the minimum value for choked flow, check the valve for excessive noise using the procedure under "Aerodynamic Noise Prediction." If noise is excessive, use anti-noise trim; recalculate the flow-capacity coefficient (C_V) using factors from "With Trim" in the "Valve Reference Data"—Table 4. For procedures on noise-level prediction, refer to the manufacturer's literature.

Reprinted with permission: Daniel Valve Company.

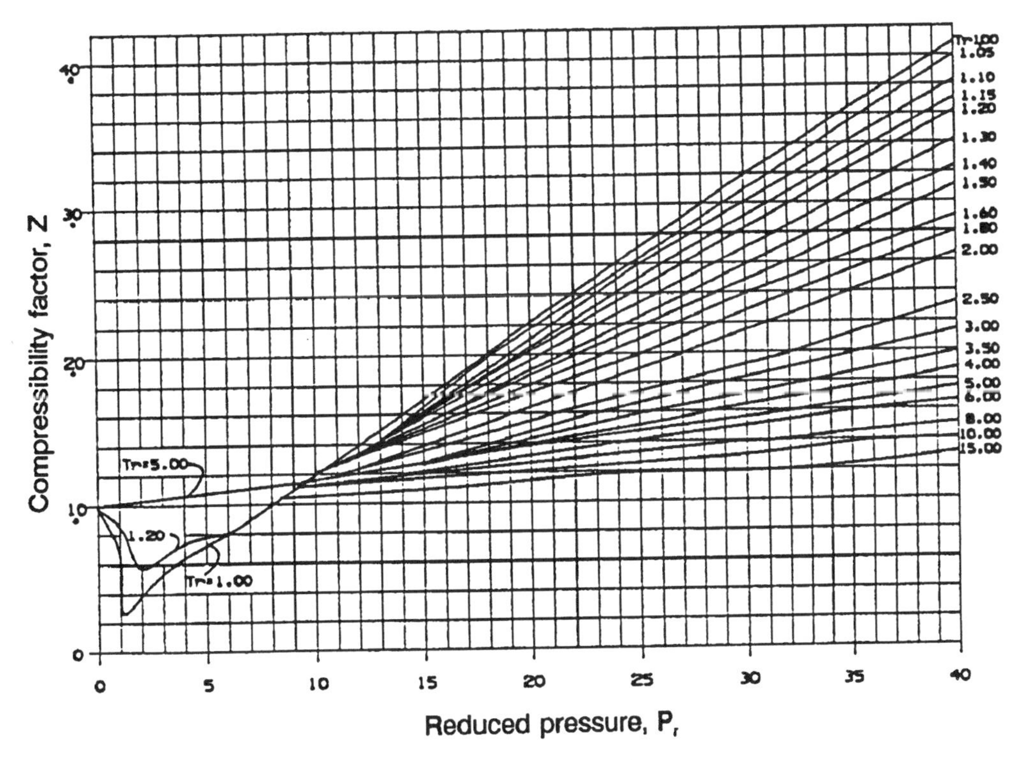

Figure 4. Compressibility versus reduced pressure.

Table 2
Valve reference data (no trim)

Value	Percent Valve Rotation												D/d
	10	20	30	40	50	60	70	80	85	90	95	100	
F_L	0.95	0.94	0.93	0.91	0.89	0.85	0.78	0.69	0.65	0.60	0.56	0.51	
X_T	0.86	0.85	0.83	0.80	0.75	0.65	0.49	0.32	0.24	0.17	0.10	0.08	1.00
K_c	0.78	0.77	0.73	0.69	0.61	0.55	0.48	0.41	0.36	0.31	0.25	0.18	
F_P	1.0	1.0	1.0	1.0	1.0	0.99	0.98	0.95	0.91	0.86	0.78	0.62	
F_{LP}	0.95	0.94	0.93	0.91	0.88	0.83	0.76	0.64	0.59	0.52	0.45	0.35	1.25
X_{TP}	0.86	0.85	0.83	0.80	0.75	0.65	0.50	0.33	0.26	0.20	0.14	0.10	
F_P	1.0	1.0	1.0	1.0	1.0	0.99	0.97	0.92	0.89	0.83	0.74	0.56	
F_{LP}	0.95	0.94	0.93	0.91	0.88	0.82	0.75	0.64	0.57	0.50	0.42	0.34	1.33
X_{TP}	0.86	0.85	0.83	0.80	0.75	0.65	0.50	0.35	0.27	0.21	0.16	0.12	
F_P	1.0	1.0	1.0	1.0	1.0	0.99	0.96	0.91	0.86	0.76	0.64	0.50	
F_{LP}	0.95	0.94	0.93	0.91	0.88	0.82	0.74	0.63	0.56	0.49	0.40	0.32	1.50
X_{TP}	0.86	0.85	0.83	0.80	0.75	0.65	0.51	0.36	0.29	0.25	0.19	0.16	
F_P	1.0	1.0	1.0	1.0	1.0	0.99	0.94	0.85	0.78	0.68	0.56	0.44	
F_{LP}	0.95	0.94	0.93	0.91	0.88	0.81	0.73	0.62	0.55	0.46	0.38	0.29	1.67
X_{TP}	0.86	0.85	0.83	0.80	0.75	0.65	0.49	0.32	0.24	0.17	0.10	0.08	
F_P	1.0	1.0	1.0	1.0	1.0	0.97	0.93	0.83	0.76	0.65	0.50	0.40	
F_{LP}	0.95	0.94	0.93	0.91	0.88	0.80	0.72	0.60	0.53	0.45	0.36	0.27	2.00
X_{TP}	0.86	0.85	0.83	0.80	0.75	0.66	0.53	0.40	0.35	0.32	0.29	0.28	
F_P	1.0	1.0	1.0	1.0	1.0	0.95	0.89	0.75	0.64	0.46	0.38	0.33	
F_{LP}	0.95	0.94	0.93	0.91	0.88	0.80	0.72	0.60	0.53	0.45	0.36	0.24	2.67
X_{TP}	0.86	0.85	0.83	0.80	0.75	0.66	0.54	0.46	0.41	0.38	0.36	0.34	

Table 3
Constants in liquid sizing equations

N		W	Q	ΔP, P	D, d	γ	v
N_1	0.865	—	M^3/HR	BAR	—	—	—
	1.000	—	GPM	PSIA	—	—	—
N_2	0.00214	—	—	—	MM	—	—
	890	—	—	—	IN	—	—
N_3	76.0	KG/HR	—	—	MM	—	cSt
	34.5	LB/HR	—	—	IN	—	cSt
N_4	76,000	—	M^3/HR	—	MM	—	cSt
	17,300	—	GPM	—	IN	—	cSt
N_6	27.3	KG/HR	—	BAR	—	KG/M^3	—
	63.3	LB/HR	—	PSIA	—	LB/FT^3	—

Table 4
Valve reference data (with trim)

Value	Percent Valve Rotation								D/d
	10–50	**60**	**70**	**80**	**85**	**90**	**95**	**100**	
F_L	1.0	1.0	1.0	0.92	0.85	0.77	0.67	0.54	
X_T	1.0	1.0	1.0	0.87	0.76	0.61	0.37	0.20	1.00
K_c	1.0	1.0	1.0	0.90	0.83	0.69	0.46	0.27	
F_P	1.0	1.0	1.0	0.98	0.97	0.95	0.92	0.87	
F_{LP}	1.0	1.0	1.0	0.90	0.84	0.75	0.63	0.48	1.25
X_{TP}	1.0	1.0	1.0	0.89	0.80	0.65	0.41	0.22	
F_P	1.0	1.0	1.0	0.98	0.97	0.95	0.92	0.87	
F_{LP}	1.0	1.0	1.0	0.90	0.84	0.75	0.62	0.47	1.33
X_{TP}	1.0	1.0	1.0	0.89	0.80	0.66	0.44	0.23	
F_P	1.0	0.99	0.98	0.95	0.93	0.89	0.85	0.78	
F_{LP}	1.0	1.0	1.0	0.89	0.83	0.74	0.61	0.46	1.50
X_{TP}	1.0	1.0	1.0	0.89	0.81	0.68	0.47	0.27	
F_P	1.0	0.99	0.98	0.94	0.91	0.86	0.80	0.71	
F_{LP}	1.0	1.0	1.0	0.89	0.83	0.74	0.61	0.45	1.67
X_{TP}	1.0	1.0	1.0	0.89	0.81	0.69	0.50	0.30	
F_P	1.0	0.99	0.98	0.94	0.90	0.84	0.77	0.66	
F_{LP}	1.0	1.0	1.0	0.89	0.83	0.73	0.60	0.44	2.00
X_{TP}	1.0	1.0	1.0	0.90	0.82	0.71	0.55	0.36	
F_P	1.0	0.99	0.98	0.94	0.89	0.83	0.75	0.64	
F_{LP}	1.0	1.0	1.0	0.89	0.83	0.73	0.59	0.43	2.33
X_{TP}	1.0	1.0	1.0	0.90	0.82	0.72	0.57	0.39	
F_P	1.0	0.99	0.98	0.93	0.88	0.81	0.73	0.61	
F_{LP}	1.0	1.0	1.0	0.88	0.81	0.71	0.57	0.42	2.50
X_{TP}	1.0	1.0	1.0	0.92	0.87	0.80	0.71	0.60	
F_P	1.0	0.99	0.98	0.92	0.86	0.79	0.70	0.59	
F_{LP}	1.0	1.0	1.0	0.88	0.81	0.70	0.55	0.41	2.67
F_{TP}	1.0	1.0	1.0	0.93	0.88	0.82	0.75	0.65	

Table 5
Constants in gas/vapor sizing equations

N	W	Q	ΔP, P	D, d	γ	T	T_{SH}
N_6 27.3 63.3	KG/HR LB/HR	— —	BAR PSIA	— —	KG/M^3 LB/FT^3	— —	— —
N_7 417 1360	— —	M^3/HR SCFH	BAR PSIA	— —	— —	°K °R	— —
N_8 94.8 19.3	KG/HR LB/HR	— —	BAR PSIA	— —	— —	°K °R	— —
N_9 2250 7320	— —	M^3/HR SCFH	BAR	— —	— —	°K °R	— —
N_{10} 0.25 63.3	— —	M^3/HR LB/HR	BAR PSIA	MM IN	— —	°K °R	— —
N_{11} 37.1 0.119	KG/HR LB/HR	— — — —	BAR PSIA	MM IN	— —	— —	°C °F
N_{12} 0.00126 0.0007	KG/HR LB/HR	— —	BAR PSIA	MM IN	— —	— —	°C °F

8: Corrosion/Coatings

Hand-held computer determines concrete coating thickness

Technical data for the engineer's file

Frank E. Hangs, Sovereign Engineering, Inc., Houston

Pipeline crossings, under water or in unstable soil, offer many challenges to engineers.

Pipe is buoyant, and an empty line may float in water. Wet silts are like viscous fluids causing inadequately weighted pipelines to pop up. There are various government entities that have jurisdiction over navigable rivers, bays, marshlands, and offshore waters. These agencies may stipulate that pipelines be buried at certain depths and be stabilized.

A good way to stabilize a pipeline is to use an adequate concrete weight coating. Determining the thickness of the concrete involves a process of balancing upward forces such as buoyancy of the mud and the downward forces—weights of pipe, protective coating, and concrete, allowing a factor of 60 (negative buoyancy). Such computations with several variables can become involved and tedious.

Computer program

The following program, written for the Hewlett Packard 41CV (Figure 1), calculates the thickness of concrete weight coating for submarine pipelines expeditiously and with satisfying results.

The prompting feature (Figure 2) is employed to aid users' data input. Important calculated values and a recap of the inputs are printed out, and each quantity is identified. Thus, the tape is a complete record. The program is flexible in that any inputs can be readily changed for a second run. Thus, many "what if" questions can be answered quickly. Suppose a heavier-weight pipe is used? What if the concrete density is changed? If the negative buoyancy is reduced? Etc.

```
01♦LBL "CONC"
02 ADV
03 SF 12
04 "  CONCRETE"
05 "⊢ COATING"
06 PRA
07 ADV
08 CF 12
09 "T TRIAL=?"
10 PROMPT
11 STO 05
12 "PIPE DIAM=?"
13 PROMPT
14 STO 15
15 "PIPE WT=?"
16 PROMPT
17 STO 16
18 "PROT CT TH=?"
19 PROMPT
20 STO 17
21 "DEN CONC=?"
22 PROMPT
23 STO 19
24 "DEN MUD=?"
25 PROMPT
26 STO 20
27 "DEN PROT CT=?"
28 PROMPT
29 STO 18
30 "T INCR=?"
31 PROMPT
32 STO 21
33 "NEG BUOY=?"
34 PROMPT
35 STO 11
36♦LBL 01
37 RCL 15
38 RCL 17
39 +
40 LASTX
41 *
42 .0218
43 *
44 RCL 18
45 *
46 STO 06
47 RCL 15
48 RCL 17
49 2
50 *
51 +
52 RCL 05
53 +
54 LASTX
55 *
56 .0218
57 *
58 RCL 19
59 *
60 STO 07
61 RCL 16
62 +
63 RCL 06
64 +
65 STO 08
66 RCL 15
67 RCL 17
68 2
69 *
70 +
71 RCL 05
72 2
73 *
74 +
75 X↑2
76 .0055
77 *
78 STO 12
79 RCL 20
80 *
81 STO 09
82 RCL 08
83 X<>Y
84 -
85 STO 10
86 FS? 01
87 GTO 03
88 RCL 11
89 X=0?
90 GTO 04
91 RCL 10
92 RCL 11
93 X<=Y?
94 GTO 03
95 GTO 02

96♦LBL 04
97 RCL 09
98 RCL 08
99 X>Y?
100 GTO 03
101 GTO 02

102♦LBL 05
103 RCL 05
104 "T=? "
105 ARCL 05
106 PROMPT
107 CLA
108 STO 05
109 RCL 15
110 "D=? "
111 ARCL 15
112 PROMPT
113 CLA
114 STO 15
115 RCL 16
116 "WP=? "
117 ARCL 16
118 PROMPT
119 CLA
120 STO 16
121 RCL 17
122 "TH PC=? "
123 ARCL 17
124 PROMPT
125 CLA
126 STO 17
127 RCL 19
128 "DC=? "
129 ARCL 19
130 PROMPT
131 CLA
132 STO 19
133 RCL 20
134 "DM=? "
135 ARCL 20
136 PROMPT
137 CLA
138 STO 20
139 RCL 18
140 "DPC=? "
141 ARCL 18
142 PROMPT
143 CLA
144 STO 18
145 RCL 21
146 "T INCR=? "
147 ARCL 21
148 PROMPT
149 CLA
150 STO 21
151 RCL 11
152 "NEG BUOY=? "
153 ARCL 11
154 PROMPT
155 CLA
156 STO 11
157 GTO 01

158♦LBL 02
159 RCL 21
160 ST+ 05
161 GTO 01

162♦LBL 03
163 RCL 08
164 RCL 12
165 /
166 STO 13
167 62.4
168 /
169 STO 14

170♦LBL "DATA"
171 ADV
172 CLA
173 "RESULTS-"
174 "⊢RECAP."
175 AVIEW
176 ADV
177 "T="
178 ARCL 05
179 AVIEW
180 CLA
181 "WPC="
182 ARCL 06
183 AVIEW
184 CLA
185 "WC="
186 ARCL 07
187 AVIEW
188 CLA
189 "WT="
190 ARCL 08
191 AVIEW
192 CLA
193 "WMD="
194 ARCL 09
195 AVIEW
196 CLA
197 "WT-WMD="
198 ARCL 10
199 AVIEW
200 CLA
201 "NEG BUOY="
202 ARCL 11
203 AVIEW
204 CLA
205 "VT="
206 ARCL 12
207 AVIEW
208 CLA
209 "BULK DEN="
210 ARCL 13
211 AVIEW
212 CLA
213 "SP GR="
214 ARCL 14
215 AVIEW
216 CLA
217 ADV
218 "D="
219 ARCL 15
220 AVIEW
221 CLA
222 "WP="
223 ARCL 16
224 AVIEW
225 CLA
226 "TH PC="
227 ARCL 17
228 AVIEW
229 CLA
230 "DC="

231 ARCL 19
232 AVIEW
233 CLA
234 "DM="
235 ARCL 20
236 AVIEW
237 CLA
238 "DPC="
239 ARCL 18
240 AVIEW
241 CLA
242 "T INCR="
243 ARCL 21
244 AVIEW
245 CLA
246 BEEP
247 END
```

Figure 1. Hewlett Packard-41CV computer program.

Nomenclature and storage registers inputs

Reg. no.

05 T = Concrete thickness, in. Assume trial value. Final answer is cumulated in Reg. 0.5.

15 D = Pipe diameter, in.

16 WP = Weight of pipe, lb/lf.

17 TH PC = Thickness of protective coating. Usually 3/32 in. or 0.0938 in.

19 DC = Density of concrete, lb/cu ft. Usually 140, 165, 190.

20 DM = Density of mud or fluid, lb/cu ft. River mud is around 90.

18 DPC = Density of protective coating, lb/cu ft. 95 is average.

21 T INCR = Increment of T. Decimal of in. 0.125 is suggested for numerical precision—not that a tolerance of this order can be constantly achieved in the pipe coating world. Increment can be made smaller, if desired.

11 NEG BUOY = Negative buoyancy. Desired minimum value, lb/lf. Can be 0 in special circumstances—e.g., to ascertain if a given pipe will float in water, or will a certain coated pipe sink in mud?

Each value keyed in is printed on the tape as a check of inputs and the program prompting feature queries user for value of next named input (Figure 2).

Calculated values

Reg. no.

06 WPC = Weight of protective coating, lb/lf.

07 WC = Weight of concrete coating, lb/lf.

08 WT = Total weight—pipe and coatings, lb/lf.

09 WMD = Weight of mud (fluid) displaced, lb/lf.

10 WT-WMD = Actual negative buoyancy, lb/lf.

12 VT = Total volume displaced by pipe and coatings, cu ft/lf.

13 BULK DEN = Bulk density, wt/vt.lb/cu ft.

14 SP GR = Specific gravity, bulk den/den water.

```
T TRIAL=?                       RESULTS-RECAP.
        3.0000    RUN           T=4.6250
PIPE DIAM=?                     WPC=4.6805
       24.0000    RUN           WC=479.3300
PIPE WT=?                       WT=578.6305
       94.6200    RUN           WMD=553.4462
PROT CT TH=?                    WT-WMD=25.1843
        0.0938    RUN           NEG BUOY=20.0000
DEN CONC=?                      VT=6.1494
      165.0000    RUN           BULK DEN=94.0954
DEN MUD=?                       SP GR=1.5079
       90.0000    RUN           D=24.0000
DEN PROT CT=?                   WP=94.6200
       95.0000    RUN           TH PC=0.0938
T INCR=?                        DC=165.0000
        0.1250    RUN           DM=90.0000
NEG BUOY=?                      DPC=95.0000
       20.0000    RUN           T INCR=0.1250
```

Figure 2. Most common questions asked when checking coating buoyancy.

Formulas

$$WPC = (D + (THPC))(THPC)0.0218(DPC)$$

$$WC = (D + 2(THPC) + T)T\ 0.0218(DC)$$

$$VT = (D + 2(THPC) + 2T)^2 0.0055$$

$$WMD = (VT)(DM)$$

Note: The user is encouraged to employ the most reliable and suitable data available, such as the density of protective coating, mud, and concrete.

Examples

Put program into calculator, XEQ Size 30. The printed RESULTS-RECAP portions of each of three examples are given. When the user XEQ "CONC" and inputs the data from the lower part of the tabulation, he or she will get the same calculated values above.

Figure 2 includes the most common questions asked when checking coating buoyancy. Always be sure flag 01 is clear XEQ "CONC" and put in new trial T and R/S. Also R/S after all other inputs. After all prompted data are keyed in, the processing starts and results are printed out. Notice T = 4.6250 in. has an actual negative buoyancy of 25.1843. If T = 4.5 in., the actual negative buoyancy = 18.4510, which is less than 20. A precise solution is between 4.5 in. and 4.625 in.

If the user wants to find T, which produces an actual negative buoyancy nearer 20: STO 4.5 in RO 5, STO $\frac{1}{16}$ in. (0.0625 in.), in R21, XEQ 01. This produces T = 4.5625 where actual negative buoyancy is 1.8113 greater than 20. This refinement can be found in choosing small T INCR. Note that the weight of the concrete coating is now 471.8269 lb/lf and we have approached closer to 20.

Calculated values. Weight of protective coating and concrete, total weight and volume of pipe, and coatings per linear foot will be of value in transporting coated pipe, etc.

Note: Results—RECAP of any run (which is in storage) can be reprinted as often as desired by XEQ "DATA."

If NEG BUOY is 0 and flag 01 is not set, then the program causes WT to approach WMD and calculates for this situation.

If a printer is not available, results can be obtained as follows:

1. XEQ 05, put in data and R/S after each item.
2. When beep sounds, RCL registers one by one in order given above (so as to identify values).

Figure 3 illustrates the changing of two input values: namely the pipe diameter and weight. The XEQ 05 routine calls up the contents of the input registers one at a time and questions them. Always put in a new T value. Then R/S.

When D = ? comes up, key in new value and R/S. Same with pipe WT = ?. Assume other data remain unchanged, R/S each time. (Any one value or any combination of values may be changed before rerunning the program.)

Figure 4 determines if a given pipe, 14 in. at 36.71 lb/lf and ½ in. of protective weight coating density of 135 lb/cu ft, will float. T, DC, T INCR, and NEG BUOY are all zero. DM becomes the density of water 62.4. Set flag 01 and XEQ 05. Key in data and R/S. When all data are in place only one computation is allowed, then printed out. WT-WMD tells us the upward forces exceed the downward forces by 19.1733 lb/lf; also the specific gravity is 0.7517, so the pipe floats. Also note that BULK DEN is *less* than the density of water.

```
                  XEQ 05              RESULTS-RECAP
T=? 4.6250
           3.0000     RUN             T=3.2500
D=? 24.0000                           WPC=3.9034
           20.0000    RUN             WC=273.9914
WP=? 94.6200                          WT=375.7248
           97.8300    RUN             WMD=352.5529
TH PC=? 0.0938                        WT-WMD=23.1720
                      RUN             NEG BUOY=20.0000
DC=? 165.0000                         VT=3.9173
                      RUN             BULK DEN=95.9154
DM=? 90.0000                          SP GR=1.5371
                      RUN
DPC=? 95.0000                         D=20.0000
                      RUN             WP=97.8300
T INCR=? 0.1250                       TH PC=0.0938
                      RUN             DC=165.0000
NEG BUOY=? 20.0000                    DM=90.0000
                      RUN             DPC=95.0000
                                      T INCR=0.1250
```

Figure 3. Changing of two input valves.

```
                XEQ "DATA"

RESULTS-RECAP.                        BULK DEN=46.9865
                                      SP GR=0.7517
T=0.0000
WPC=21.3368                           D=14.0000
WC=0.0000                             WP=36.7100
WT=58.0468                            TH PC=0.5000
WMD=77.2200                           DPC=135.0000
WT-WMD=-19.1733                       DC=0.0000
NEG BUOY=0.0000                       DM=62.4000
VT=1.2375                             T INCR=0.0000
```

Figure 4. Result of calculations to determine whether the pipe will float.

National Association of Pipe Coating Applications (NAPCA) specifications*

RECOMMENDED SPECIFICATION DESIGNATIONS FOR ENAMEL COATINGS

These are recommended enamel coating system designations for coal tar or asphalt coatings, except for specialty requirements.

SPECIFICATION SYMBOL *PART #1	**PART #2	SHOT BLAST	PRIMER	93.75 MILS MIN. ENAMEL	FIBERGLASS WRAP	#15 FELT WRAP	SEAL COAT	GLASS MAT	POLYETHYLENE WITH KRAFT	HEAVY KRAFT PAPER	ELECTRICAL INSPECTION
TF–		X	X	X		X				X	X
AF–		X	X	X		X				X	X
TG–		X	X	X	X					X	X
AG–		X	X	X	X					X	X
TGF–		X	X	X	X	X				X	X
AGF–		X	X	X	X	X				X	X
TGSF–		X	X	X	X	X	X			X	X

*NOTE: These are not all the specifications approved by NAPCA. For other specifications, consult an NAPCA member.

AGSF–	X	X	X	X	X	X			X	X
TGM–	X	X	X				X		X	X
AGM–	X	X	X				X		X	X
TGMF–	X	X	X		X		X		X	X
AGMF–	X	X	X		X		X		X	X
TGMP–	X	X	X				X	X		X
AGMP–	X	X	X				X	X		X

***PART #1 SYMBOLS– ENAMEL & WRAPPERS**

T–Coal Tar
G–Fiberglass
A–Asphalt
S–Seal Coat
F–#15 Felt
GM–Glass Mat
P–Polyethylene wrapper w/kraft paper
O–Outer wrap
To specify Outer Wrap, use letter O in place of F.

****PART #2 SYMBOLS GRADES OF ENAMEL**

3– Fully-Plasticized Coal Tar (ordered by penetration dictated by ambient temperature)
4– High Temperature Surface Coal Tar
7– Asphalt

EXAMPLE: 93.75 mils (minimum) fully-plasticized coal tar enamel wrapped with fiberglass, felt and kraft – the symbol would be TGF-3.

Reprinted courtesy of National Association of Pipe Coating Applicators.

STANDARD APPLIED PIPE COATING WEIGHTS FOR NAPCA COATING SPECIFICATIONS

Refer to NAPCA 1-65-83

Spec. Symbol		TF 3-4	AF-7	TG 3-4	AG-7	TGF -3-4	AGF-7
PIPE SIZE		Weight in Pounds–Per 100 Lineal Feet					
1/2"	I.D.	21	18	17	14	21	18
3/4"	"	26	22	21	17	26	22
1"	"	32	27	26	22	33	28
1 1/4"	"	40	34	33	27	41	35
1 1/2"	"	46	39	38	31	46	40
2 3/8"	O.D.	57	48	47	39	57	49
2 7/8"	"	68	58	57	47	69	59
3 1/2"	"	83	71	70	58	84	72
4"	"	95	81	80	66	96	82
4 1/2"	"	106	91	90	74	108	92
5 9/16"	"	130	111	111	91	132	113
6 5/8"	"	155	132	132	109	157	134
8 5/8"	"	201	171	171	141	204	174
10 3/4"	"	250	213	214	176	254	216
12 3/4"	"	296	252	253	209	300	256
14"	"	325	276	278	229	329	281
16"	"	372	317	318	262	377	322
18"	"	419	356	357	295	424	362
20"	"	463	393	397	327	469	400
22"	"	510	433	436	360	516	440
24"	"	555	472	476	392	563	480
26"	"	601	511	516	425	610	519
28"	"	648	550	555	458	656	559
30"	"	692	588	595	490	702	598
32"	"	739	627	634	523	749	638
34"	"	785	667	674	556	796	678
36"	"	831	706	714	589	842	717
42"	"	970	824	832	687	983	837

(continued on next page)

(table continued)

Spec. Symbol		TGM 3-4	AGMP 7	TGMP 3-4	AGM 7	TGMF 3-4	AGMF 7	TGSF 3-4	AGFS 7
PIPE SIZE		Weight in Pounds – Per 100 Lineal Feet							
1/2"	I.D.	17	14	17	14	21	18	26	22
3/4"	"	21	17	21	17	26	22	33	28
1"	"	26	22	26	22	33	28	41	35
1 1/4"	"	33	27	33	27	41	35	51	43
1 1/2"	"	38	31	38	31	46	40	58	50
2 3/8"	O.D.	47	39	47	39	57	49	72	61
2 7/8"	"	57	47	57	47	69	59	87	74
3 1/2"	"	70	58	70	58	84	72	106	90
4"	"	80	66	80	66	96	82	121	102
4 1/2"	"	90	74	90	74	108	92	136	115
5 9/16"	"	111	91	111	91	132	113	167	141
6 5/8"	"	132	109	132	109	157	134	198	168
8 5/8"	"	171	141	171	141	204	174	258	218
10 3/4"	"	214	176	214	176	254	216	321	271
12 3/4"	"	253	209	253	209	300	256	380	321
14"	"	278	229	278	229	329	281	417	352
16"	"	318	262	318	262	377	322	477	403
18"	"	357	295	357	295	424	362	537	453
20"	"	397	327	397	327	469	400	594	501
22"	"	436	360	436	360	516	440	654	552
24"	"	476	392	476	392	563	480	713	602
26"	"	516	425	516	425	610	519	772	651
28"	"	555	458	555	458	656	559	832	702
30"	"	595	490	595	490	702	598	889	750
32"	"	634	523	634	523	749	638	949	800
34"	"	674	556	674	556	796	678	1008	850
36"	"	714	589	714	589	842	717	1067	900
42"	"	832	687	832	687	983	837	1245	1051

All weights provide for standard cutbacks.

Weights per 100 lineal feet are figured on the following material weights per square foot of pipe surface:

	Weight Lbs.—Per Sq. Ft.
Coal Tar Primer & Enamel—1/32" Min.	.2461
Coal Tar Primer & Enamel—2/32" Min.	.4850
Coal Tar Primer & Enamel—3/32" Min.	.7238
Asphalt Primer & Enamel—1/32" Min.	.2000
Asphalt Primer & Enamel—2/32" Min.	.3950
Asphalt Primer & Enamel—3/32" Min.	.5912

	Weight Lbs.—Per Sq. Ft.
#15 Pipeline Felt (Tar or asphalt saturated)	.1271
Outerwrap	.1225
.010 Fiberglass Wrap	.0110
Kraft Paper	.0200
Glass Mat	.0114
Polyethylene	.0450

Reprinted courtesy of National Association of Pipe Coating Applicators.

NAPCA SPECIFICATIONS PIPELINE FELTS

PROPERTY	#15 TAR N.R.	#15 TAR R.F.	#15 ASPH. N.R.	#15 ASPH. R.F.
Saturated Felt – Not Perforated				
Weight (Min.) lbs. per 100 s.f. (ASTM-D-146-59 Sec. 1-9)	12	12	12	12
Saturation (Percent) (ASTM-D-146-59 Sec. 18 by extraction) MIN.	18	18	22	22
MAX.	28	28	40	40
Ash Content (MIN).) (Percent) (ASTM-D-146-59 Sec. 21)	70	70	70	70
Caliper (Min.) (Inches) (ASTM-D-645-64T Method C)	.021	.018	.021	.018
Tensile Strength (Min.) (lbs. per inch of width) (ASTM-D-146-59 Sec. 12a)				
With fiber grain	35	37	35	37
Across fiber grain	12	No. spec.	12	No. spec.
Tear Strength (Min.) (Grams) (ASTM-D-689-62)				
Across fiber grain	400	Not applicable	400	Not applicable
With fiber grain	240		240	

(continued on next page)

Loss on heating (Max.) (Percent) (AWWA C203-66 Sec. 2.6.6)	10	10	10	10
Pliability (AWWA C203-66 Sec. 2.6.4)	No cracking	No cracking	No cracking	No cracking
Saturated and Perforated Felt				
(1/16" perforations on 1" Staggered Centers)				
Tensile Strength (Min.) (lbs. per inch of width) (ASTM-D-146-59 Sec. 12a) With fiber grain Across fiber grain	30 10	35 No spec.	30 10	35 No. spec.
Tear Strength (MIN.) (Grams) (ASTM-D-689-62) Across fiber grain With fiber grain	348 190	Not applicable	348 190	Not applicable
Unsaturated Felt				
Asbestos Content (%) (MIN.) (ASTM-D-1918-67)	85	85	85	85

N.R. — NON REINFORCED
R.F. — REINFORCED LENGTHWISE WITH PARALLEL FIBER GLASS YARNS ON NOMINAL ¼" CENTERS.

Reprinted courtesy of National Association of Pipe Coating Applicators.

Minimum Test Voltages for Various Coating Thicknesses

Coating Thicknesses 32nd Inch	Mils	Test Volts
—	16	5,000
1	31	7,000
2	62	9,800
3	94	12,100
4	125	14,000
5	156	15,000
6	188	17,100
16	500	28,000
20	625	31,000
24	750	35,000

Reprinted courtesy of National Association of Pipe Coating Applicators.

How much primer for a mile of pipe?

Multiply pipe O.D. by $2\frac{3}{4}$ to get gallons of primer per mile.

Examples.

Pipe 24 in. in diameter will require $24 \times 2\frac{3}{4}$ or 66 gallons of primer per mile;

Pipe 16 in. in diameter will require $16 \times 2\frac{3}{4}$ or 44 gallons of primer per mile; and

Pipe $4\frac{1}{2}$ in. in diameter will take $4\frac{1}{2} \times 2\frac{3}{4}$ or 123/8 gallons of primer per mile (12 gallons in round figures).

The given multiplier is applicable only where the pipe is new and in good condition. For rough pipe the multiplying factor should be upped from $2\frac{3}{4}$ to 3.45. In other words, rough 20-in. pipe will require 20×3.45 or 69 gallons of primer per mile.

It is hard to estimate the amount of spillage and waste. Temperature and relative humidity affect the way the primer goes on the pipe.

How much coal-tar enamel for a mile of pipe?

For an average cover of $\frac{3}{32}$ in., multiply outside diameter of pipe by 0.6. The answer is in tons of enamel per mile. (Add 15% to take care of gate valves, drips, bends, etc.)

Examples. How many tons of coal-tar enamel will be required to apply a $\frac{3}{32}$-in. coat to a 26-in. pipeline?

$26 \times 0.6 = 15.6$ tons per mile

Add 15% to take care of gate valves, bends etc.

$15.6 + 2.3 = 17.9$ or 18 tons per mile

How many tons of coal-tar enamel will be required to apply a $\frac{3}{32}$-in. coat to a 8.625-in. pipeline?

$8.625 \times 0.6 = 5.2$ tons per mile
$5.2 \times 15\% = 0.8$
$= 6$ tons per mile

These factors strongly affect the amount of coating used per mile on pipelines: 1. Temperature of the pipe and ambient temperatures. (It is assumed that the temperature of the enamel can be controlled.) 2. The kind of dope machine. 3. Humidity. 4. The number of bends per mile. 5. The number of valves, crossovers, etc. 6. Condition of the pipe.

This latter point has a great effect on the amount of enamel required for a mile of pipeline. In fact, where reconditioned pipe is being coated, it is recommended that the amount of enamel be increased by about 20%.

How much wrapping for a mile of pipe?

For 14-in. to 48-in. pipe multiply the diameter in inches by 15 to get the number of squares (100 ft^2). For pipe sizes of 12$\frac{3}{4}$ in. and smaller, add 1 to the nominal size of pipe and multiply by 15.

Examples. How many squares of asbestos felt would be required to wrap 1 mile of 30-in. pipeline?

$30 \times 15 = 450$ squares of felt

How many squares of felt would be required to wrap 1 mile of 4-in. pipe?

$(4 + 1) \times 15 = 75$ squares of felt

These figures are conservative and do not allow for wrapping drips, for patching holidays, and for short ends that are usually discarded. They allow for machine wrapping using $\frac{3}{4}$-in. lap on diameters up to 10$\frac{3}{4}$ in. and using 1-in. lap on larger diameters.

For smaller diameters of pipe, this figure should be increased about 5% to take care of additional wastage and patching.

Estimating coating and wrapping materials required per mile of pipe

Primer and coal-tar enamel

Gallons of primer and tons of coal-tar enamel are based on covering new pipe in good condition. For rough pipe, increase primer quantities by 25% and the coal-tar quantities by 20%. No allowance is included for spillage or waste.

Wrappings

Squares are based on machine wrapping, allowing $\frac{3}{4}$-in. laps on pipe sizes through 10 in. and 1-in. laps on pipe sizes greater than 10 in. No allowance is included for wrapping drips, patching holidays, or short ends, which are usually discarded.

Nominal Pipe Size	Primer Gallons	$\frac{3}{32}''$ Coal-tar Enamel Tons	Glass Wrap Squares	Kraft Paper Squares	15-Pound Felt Squares
4	11	3	75	75	75
6	17	4	105	105	105
8	22	6	135	135	135
10	28	7	165	165	165
12	33	8	195	195	195
14	39	10	210	210	210
16	44	11	240	240	240
18	50	13	270	270	270
20	55	14	300	300	300
24	66	17	360	360	360
30	83	21	450	450	450
36	99	25	540	540	540
42	116	29	630	630	630

Coefficient of friction for pipe coating materials

Tests indicate that previous values appear to be valid for thinfilm epoxies and conservative for coal tars

J. B. Ligon, Assistant Professor, Mechanical Engineering-Engineering Mechanics Dept., Michigan Technological University, Houghton, Mich., and **G. R. Mayer,** Project Engineering Manager, Bechtel, Inc., San Francisco

A major factor in the stress analysis of buried pipelines is the movement that pipe undergoes in the presence of temperature and pressure differentials during its life. This movement is highly dependent upon friction resistance of the soil.

Although ample information is available on the static coefficient of friction for many materials, there is a lack of data on friction between soils and various coatings used in the pipeline industry. In the past, friction coefficient information was extrapolated from data in the literature that were believed to have a similarity to the external pipe coating to soil interface.

However, with the development of thinfilm epoxy resin coating systems and the increasing use of these systems in the pipeline industry, a change from a conventional coal tar felt coating to a thinfilm epoxy coating would indicate a significant change in the friction coefficient design criteria due to the extreme contrast in the surface texture of the two materials.

To evaluate the effect of the difference in surface texture on a pipeline system, test procedures were developed to determine the coefficient of friction for both coal tar felt and thinfilm epoxy to various soils and to obtain more reliable information for future pipeline designs using these coating materials.

Static friction tests were conducted to find the coefficient of friction between coal tar felt and thinfilm epoxy pipe coating and eight representative backfill soil samples from typical locations along a pipeline right of way. The results indicate that the friction coefficients are significantly larger than those previously extrapolated from literature and that coal tar has a higher friction resistance in respect to anchorage of a pipeline.

For the coal tar felt coating, the coefficient varies from 0.59 to 0.91 depending on the soil and moisture content. The thinfilm epoxy coating varies from 0.51 to 0.71 under the same conditions.

Definitions

Static coefficient of friction. The theoretical longitudinal soil force acting on the pipe surface can be calculated from the relation:

$$F = \mu \int_A p \, dA$$

where:
- F = Longitudinal soil friction force (lb)
- μ = Coefficient of friction (dimensionless)
- p = Normal soil pressure acting on the pipe surface (psi)
- dA = Soil to pipe differential contact area (in.^{2})
- $\int_A p dA$ = Total normal soil force on pipe surface (lb)

The above relation is independent of the pressure per unit area of the mating surfaces as long as very high contact pressures are not encountered, which is the case for a buried pipeline.

In terms of a buried pipeline supporting a soil burden up to three pipe diameters of depth, the soil force relation becomes:

$$F = \mu\left[2D\gamma\left(H - \frac{D}{2}\right) + Wp\right]$$

where:
- Wp = Weight of pipe and contents (lb/in.)
- D = Pipe diameter (in.)
- H = Depth of the pipe centerline (in.)
- γ = Specific weight of the soil (lb/in.^{3})

The sensitivity of the soil friction forces to the coefficient of friction is readily apparent from the above relation. Since this force is inversely proportional to the active length that a pipeline moves as a result of temperature and pressure expansion, the coefficient of friction becomes a major factor in pipeline stress design.

Test system. The theoretical soil friction force acting on the surface of a coated plate can be calculated from the relation:

$$F = \mu N$$

where:
- N = Normal force acting on the plate surfaces (lb)
- W = Total weight of plate and external load (lb)

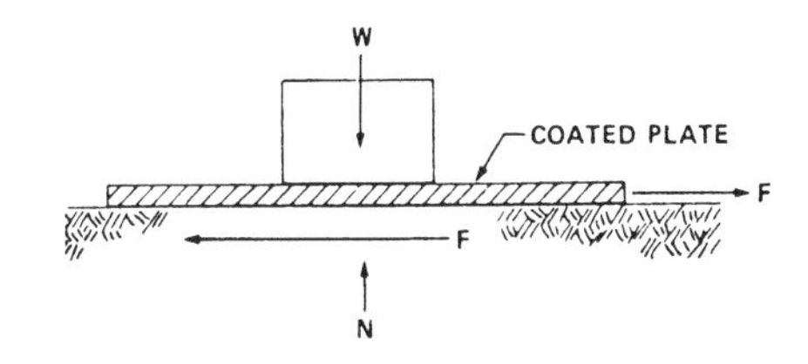

Since the coefficient of friction parameter is the same in all of the previous expressions, the plate test system can be used to simulate the soil-to-pipeline interface.

Coal tar felt. The coal tar coating, an enamel and felt combination, was prepared to simulate an actual coated pipeline.

The surface of a steel plate was coal tar coated and then wrapped in a 15-lb felt. Therefore, the soil surface contact was predominantly with the felt material.

Thinfilm epoxy. The epoxy coating simulated a thinfilm epoxy resin pipeline coating applied by the fusion method.

Plates used in the tests were coated by the fluidized bed method. However, the surface texture was the same as that expected by coating plant production equipment.

Results

Coal tar to soils. Figure 1 illustrates that the friction coefficient varies from 0.59 to 0.91 for the various soils tested at field moisture content.

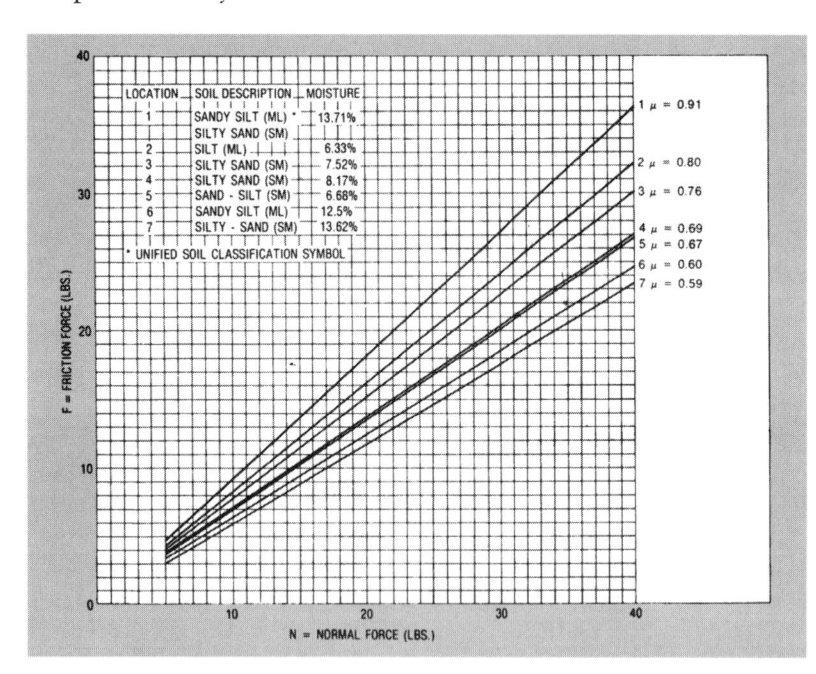

Figure 1. Coefficient of friction for coal tar coating to soil.

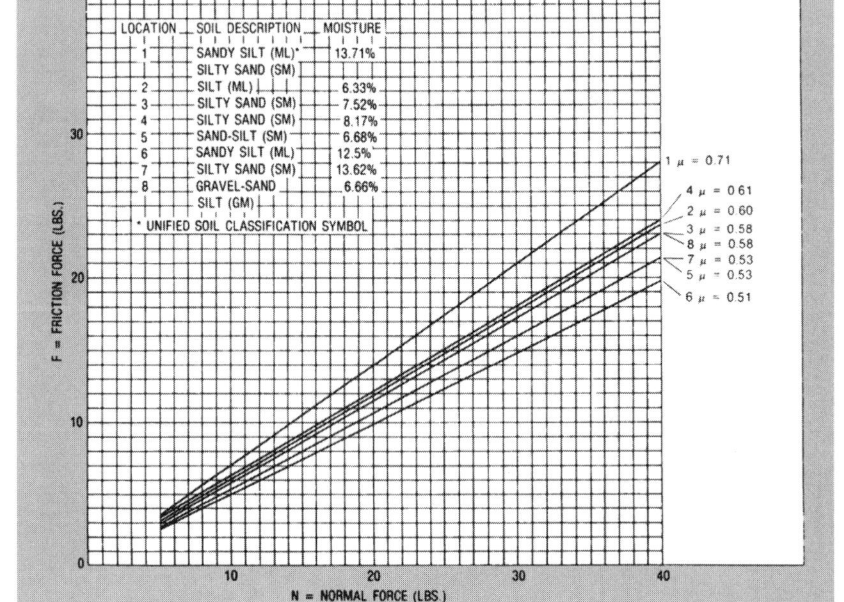

Figure 2. Coefficient of friction for thin film epoxy to soil.

The mechanism by which the friction coefficient changes among the various soils is not apparent from the results of these tests, since the soil samples used did not vary considerably in description. However, the range over which the factor changes is considerably greater than that previously extrapolated from the literature, as follows:

Soil description	Coefficient of friction commonly used
Silt	0.3
Sand	0.4
Gravel	0.5

The tests also indicate that the moisture content alters the friction factor to some extent, as would be expected.

In order to investigate the influence of temperature, tests were conducted with coal tar felt wrapping heated to 120°F. Only a slight softening beneath the coating surface was observed, and it is believed that temperatures up to this range will not significantly affect the magnitude of the friction coefficient.

Thinfilm epoxy to soils. As expected, results shown in Figure 2 indicate that the friction factor range of 0.51 to 0.71 for the epoxy is somewhat lower than that of the coal tar.

Again, the mechanism is not clear, but the results are fairly consistent with the friction coefficient increasing in most instances in the same order of magnitude as for coal tar. Since such information on this coating was not available in the literature, no comparison to similar tests could be made.

Temperatures in the 120°F range should have little or no effect on the coefficient of friction of epoxy to soil.

Conclusions

Although it is virtually impossible to precisely simulate the surface contact situation of a pipeline in a back-filled trench, the test procedure and apparatus reported here are a means of approximating it. The results indicate that coal tar coatings have a higher friction resistance than epoxy coatings as far as anchoring capabilities of the soil are concerned.

The selection of a coating, based on its soil friction resistance, could be of economic value in reducing those situations where extreme expansions call for reinforced fittings or elaborate culverts to overcome excessive stress levels in a pipeline system.

The test results show that previous values commonly used for the coefficient of friction were conservative for similar soils, and it is suggested that some conservatism still be incorporated in future analyses.

Since the tests indicated that the friction coefficient of the epoxies was similar to those previously used for coal tar coatings, their continued use for epoxy-coated pipelines should be valid. A more conservative approach to these values is recommended in soils where the presence of excessive moisture could change the friction factor substantially when interfaced with a smooth epoxy coating.

Troubleshooting cathodic protection systems: Magnesium anode system

The basic technique is the same as that outlined earlier; more measurements are required because of the multiplicity of drain points. First, the current output of stations nearest the point of low potential should be checked; if these are satisfactory, a similar check should be extended in both directions until it is clear that the trouble must be on the line. When a given anode group shows a marked drop in current output, the cause may be drying out, shrinkage of backfill, or severed or broken lead wires. If the current is zero, the pipe-to-soil potential of the lead wire will show whether it is still connected to the pipe or the anode, and thus indicate the direction to the failure. If the current is low, there may be a loss of one or more anodes by a severed wire; a pipe-to-soil potential survey over the anodes will show which are active, just as in the case of rectifier anodes.

Trouble indicated on the line, rather than at the anode stations, is tracked down in the same manner as that used for

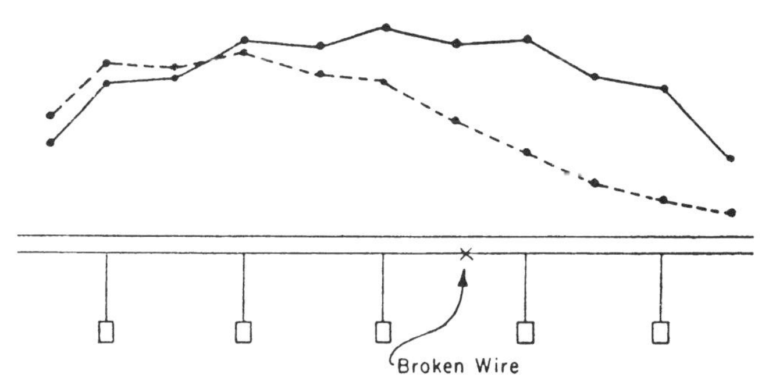

Figure 1. Locating idle anodes by surface potentials. The solid line shows the potentials found along a line of anodes when all are delivering current; the dotted line exhibits the change when there is a break in the anode lead at the point indicated. Single disconnected anodes may also be located by this method. A driven ground rod, a pipe lead, or even a rectifier terminal may be used for the reference ground; all readings should be referred to the same reference.

line protected by rectifiers—first, investigating locations where something that has been done might be responsible; next, checking potentials; and finally, making a line current survey.

Cathodic protection for pipelines

Estimate the rectifier size required for an infinite line. Refer to Figure 1. If coating conductance tests have not been performed, pick a value from Table 1.

Table 1
Typical Values of Coating Conductance

Micromhos/sq ft	Coating Condition
1–10	Excellent coating—high resistivity soil
10–50	Good coating—high resistivity soil
50–100	Excellent coating—low resistivity soil
100–250	Good coating—low resistivity soil
250–500	Average coating—low resistivity soil
500–1,000	Poor coating—low resistivity soil

Use a value of—0.3 volts for ΔE_x. This is usually enough to raise the potential of coated steel to about—0.85 volts.

Use a value of 1.5 volts for AE at the drain point. Higher values may be used in some circumstances; however, there may be a risk of some coating disbondment at higher voltages.

Calculate I at the drain point.

$$\Delta I_A = \text{amp./in.} \times \Delta E_x / 0.3 \times D$$

where D = pipe OD, inches.

Example. 30-in. OD line with coating conductivity = 100 micromhos/sq ft. What is the current change at the drain

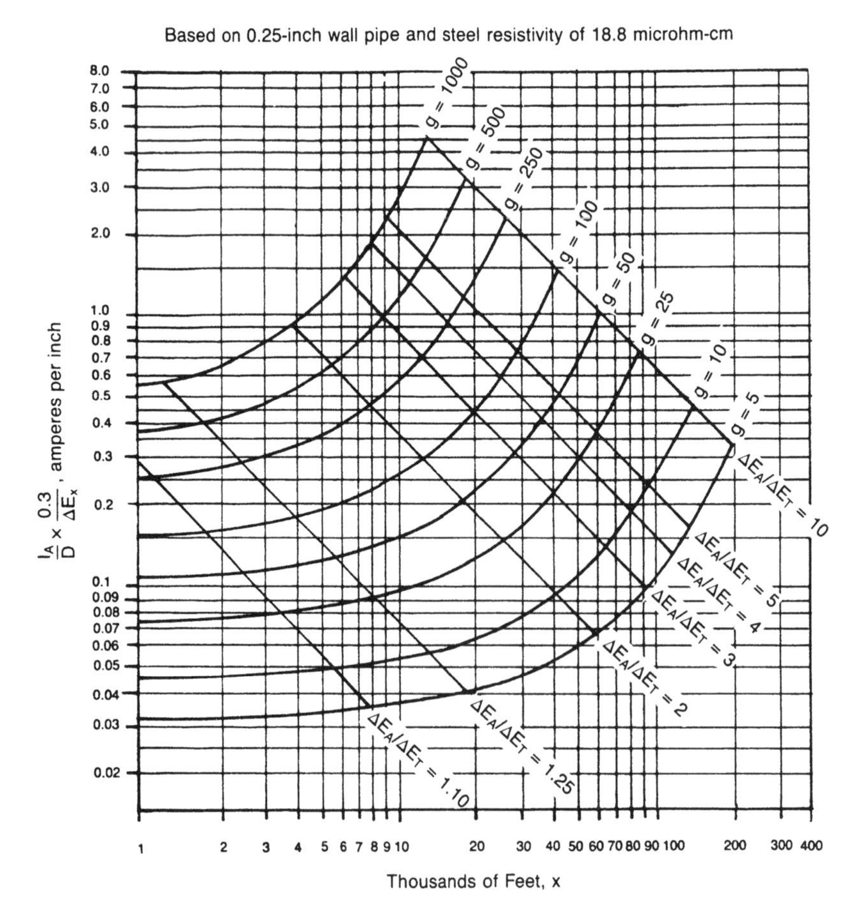

Figure 1. Drainage current versus distance for a coated infinite line.

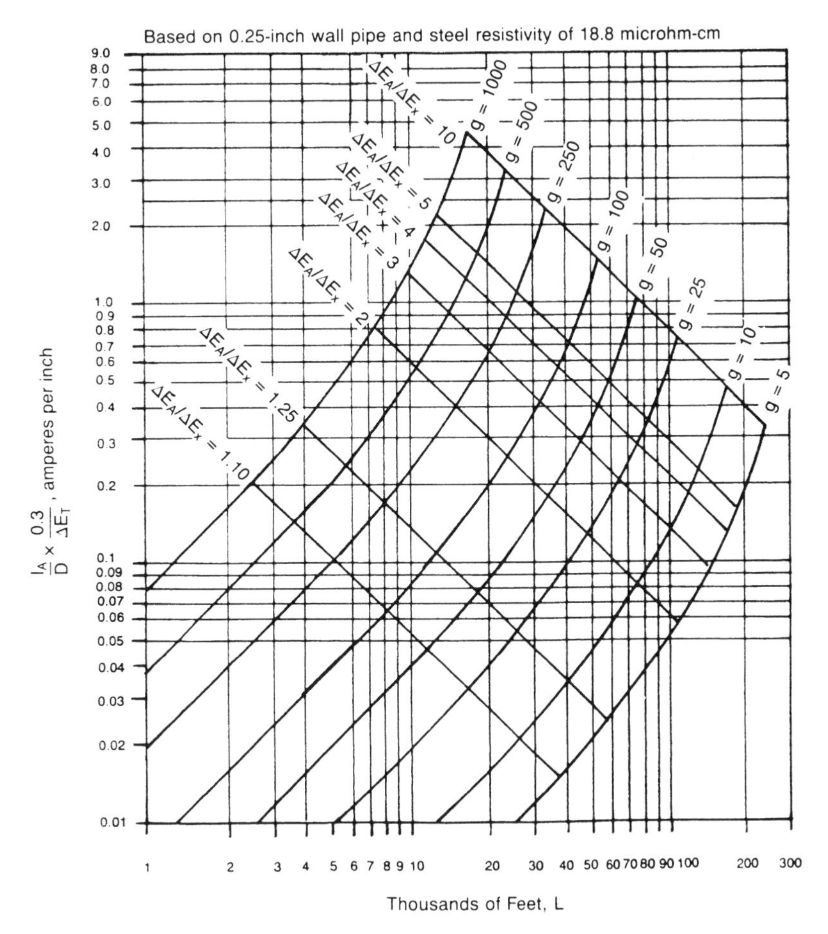

Figure 2. Drainage current versus distance for a coated finite line.

point, and how far will this current protect an infinitely long pipeline?

$\Delta E_A = 1.5$

$\Delta E_X = 0.3$

$\Delta E_A/\Delta E_x = 5.0$

Refer to Figure 1 and read

$L = 30{,}000$ ft

$\Delta I_A/D \times 0.3/\Delta E_x = 0.7$ (from Figure 1)

$\Delta I_A = 0.7 \times 30 = 21$ amp.

21 amps will protect the line for 30,000 ft in either direction from the drain point.

Estimate rectifier size for a finite line. Refer to Figure 2.

A 3½ -in. OD line 20,000 ft long is protected by insulating flanges at both ends and has a coating conductivity of 500 micromhos/sq ft. What is $\Delta E_A/E_T$ and what ΔI_A is required? Assume that

$E_T = 0.3$

From Figure 2 read

$\Delta E_A/E_T = 5.8$

$\Delta E_A = 1.74$

$\Delta I_A/D \times 0.3/E_T = 1.8$ amp./in.

$\Delta I_A = 1.8 \times 3.5 = 6.3$ amp.

Estimate ground bed resistance for a rectifier installation.

Example. A ground bed for a rectifier is to be installed in 1,000 ohm-cm soil. Seven 3-in. × 60-in. vertical graphite anodes with backfill and a spacing of loft will be used. What is the resistance of the ground bed?

Refer to Figure 3. The resistance is 0.56 ohms.

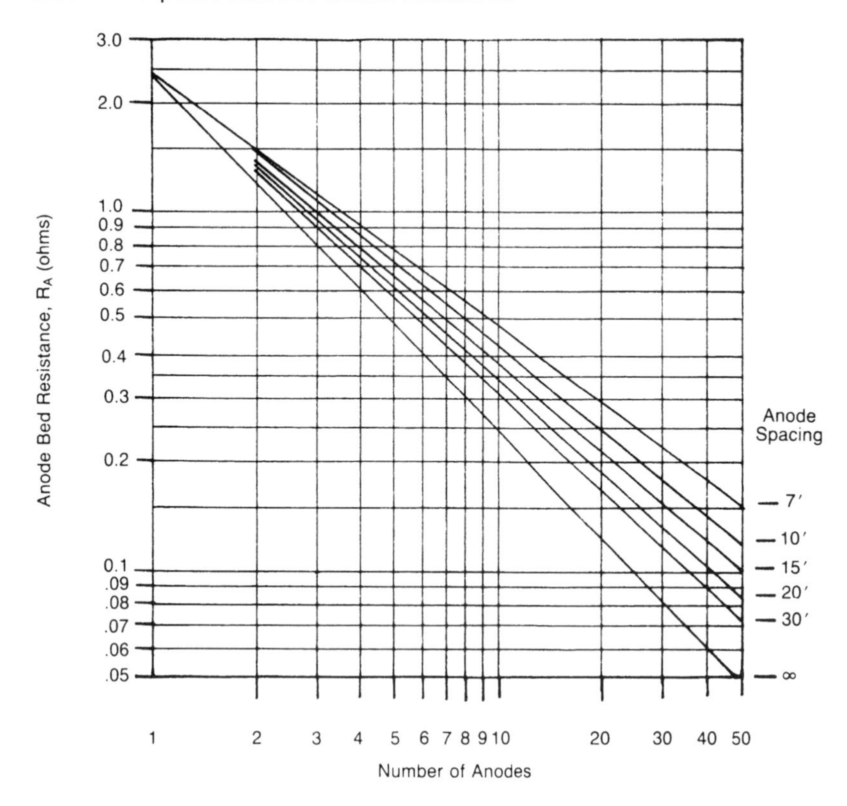

Figure 3. Anode bed resistance versus number of anodes. 3-in. × 60-in. vertical graphite anodes in backfill.

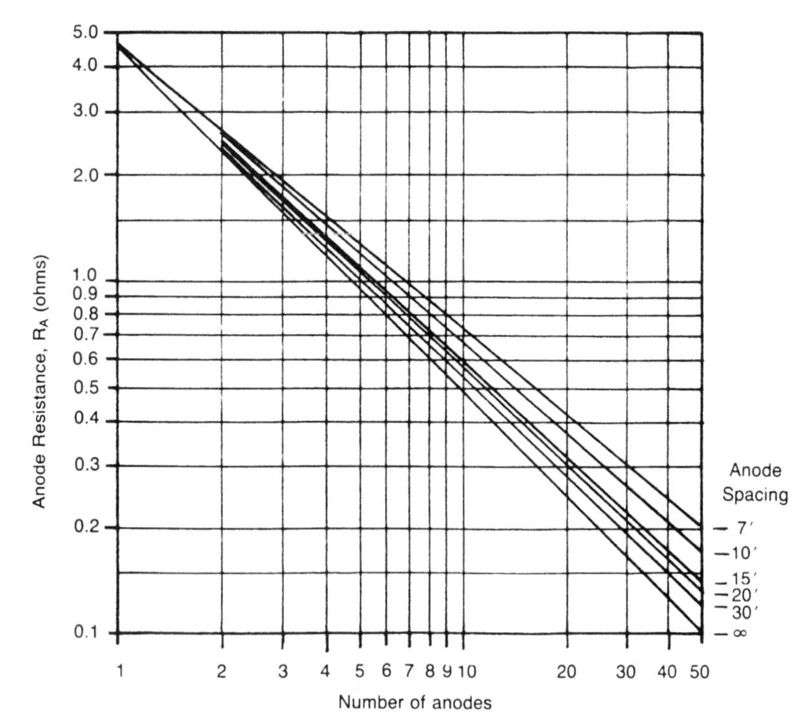

Figure 4. Anode bed resistance versus number of anodes. 2-in. × 60-in. vertical bare anodes.

Figure 3 is based on a soil resistivity of 1,000 ohms-cm. If the soil resistivity is different, use a ratio of the actual soil resistivity divided by 1,000 and multiply this by the reading obtained from Figure 3.

A rectifier ground bed is to be composed of 10 – 2 × 60-in. bare "Duriron" silicon anodes spaced 20 ft apart. The soil resistivity is 3,000 ohms-cm. What is the ground bed resistance? Refer to Figure 4. The resistivity from the chart is 0.55 ohms. Since the chart is based on soil resistivity of 1,000 ohms-cm, the ground bed resistivity is 3,000/1,000 × 0.55 or 1.65 ohms.

The resistance of multiple anodes installed vertically and connected in parallel may be calculated with the following equation:

$$R = 0.00521P/NL \times (2.3 \,\text{Log}\, 8L/d - 1 + 2L/S \,\text{Log}\, 0.656N) \quad (1)$$

where:

R = Ground bed resistance, ohms
P = Soil resistivity, ohm-cm
N = Number of anodes
d = Diameter of anode, feet
L = Length of anode, feet
S = Anode spacing, feet

If the anode is installed with backfill such as coke breeze, use the diameter and length of the hole in which the anode is installed. If the anode is installed bare, use the actual dimensions of the anode.

Figure 5 is based on Equation 1 and does not include the internal resistivity of the anode. The resistivity of a single vertical anode may be calculated with Equation 2.

$$R = 0.00521P/L \times (2.3 \,\text{Log}\, 8L/d - 1) \quad (2)$$

If the anode is installed with backfill, calculate the resistivity using the length and diameter of the hole in which the anode is installed. Calculate the resistivity using the actual anode dimensions. The difference between these two values is the internal resistance of the anode. Use the value of P, typically about 50 ohm-cm, for the backfill medium.

Figure 5 is based on 1,000 ohm-cm soil and a 7-ft × 8-in. hole with a 2-in. × 60-in. anode.

Example. Determine the resistivity of 20 anodes installed vertically in 1,500 ohm-cm soil with a spacing of 20 ft.

Read the ground bed resistivity from Figure 5.

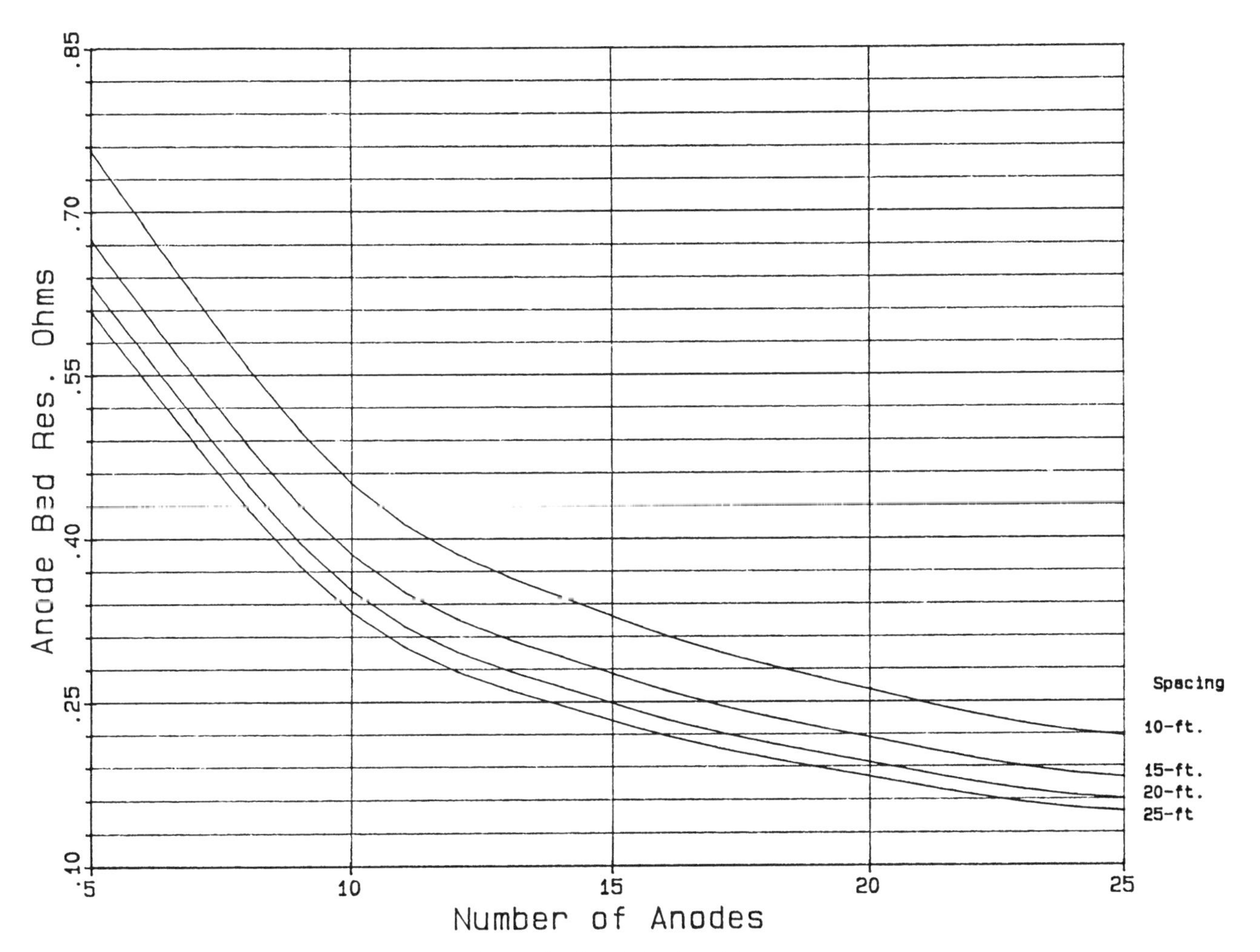

Figure 5. Anode bed resistance.

R = 0.202 ohm

Since the anodes are to be installed in 1,500 ohm-cm soil and Figure 5 is based on 1,000 ohm-cm soil, multiply R by the ratio of the actual soil resistivity to 1,000 ohm-cm.

$R = 0.202 \times 1{,}500/1{,}000$

R = 0.303 ohm

The internal resistivity for a single 2-in. × 60-in. vertical anode installed in 50 ohm-cm backfill (7 ft × 8-in. hole) is 0.106 ohm.

Since 20 anodes will be installed in parallel, divide the resistivity for one anode by the number of anodes to obtain the internal resistivity of the anode bank.

$0.106/20 = 0.005$ ohm

The total resistivity of the 20 anodes installed vertically will therefore be 0.308 ohm (0.303 + 0.005).

Galvanic Anodes

Zinc and magnesium are the most commonly used materials for galvanic anodes. Magnesium is available either in standard alloy or high purity alloy. Galvanic anodes are usually pre-packaged with backfill to facilitate their installation. They may also be ordered bare if desired. Galvanic anodes offer the advantage of more uniformly distributing the cathodic protection current along the pipeline, and it may be possible to protect the pipeline with a smaller amount of current than would be required with an impressed current system but not necessarily at a lower cost. Another advantage is that interference with other structures is minimized when galvanic anodes are used.

Galvanic anodes are not an economical source of cathodic protection current in areas of high soil resistivity. Their use is generally limited to soils of 3,000 ohm-cm except where small amounts of current are needed.

Magnesium is the most-used material for galvanic anodes for pipeline protection. Magnesium offers a higher solution potential than zinc and may therefore be used in areas of higher soil resistivity. A smaller amount of magnesium will generally be required for a comparable amount of current. Refer to Figure 6 for typical magnesium anode performance data. These curves are based on driving potentials of −0.70 volts for H-1 alloy and −0.90 volts for Galvomag working against a structure potential of −0.85 volts referenced to copper sulfate.

The driving potential with respect to steel for zinc is less than for magnesium. The efficiency of zinc at low current levels does not decrease as rapidly as the efficiency for magnesium. The solution potential for zinc referenced to a

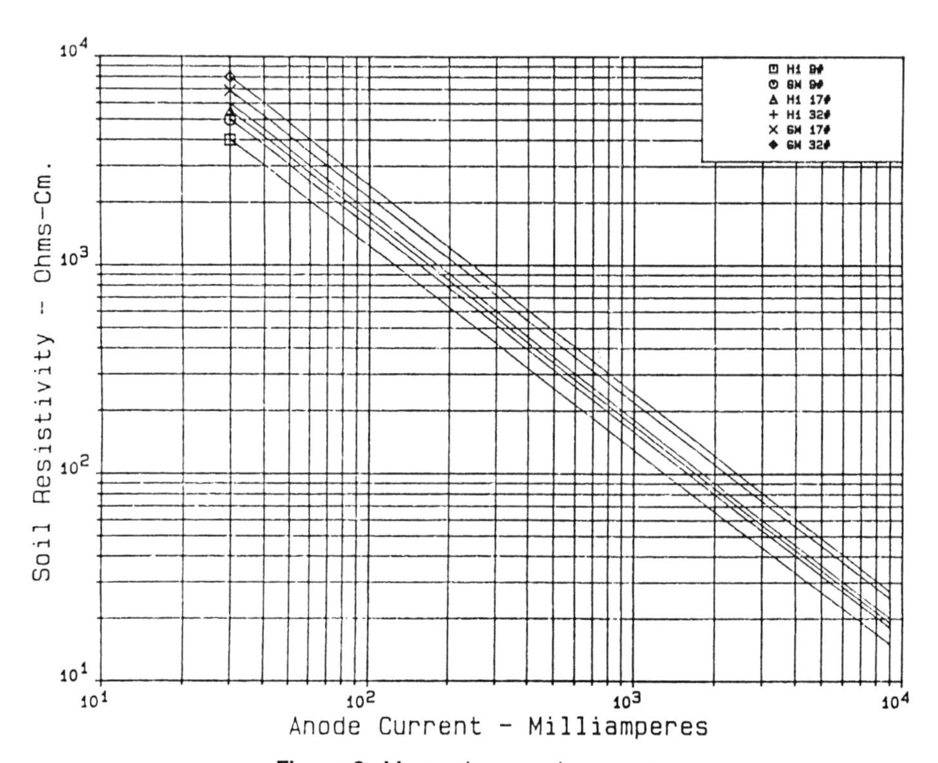

Figure 6. Magnesium anode current.

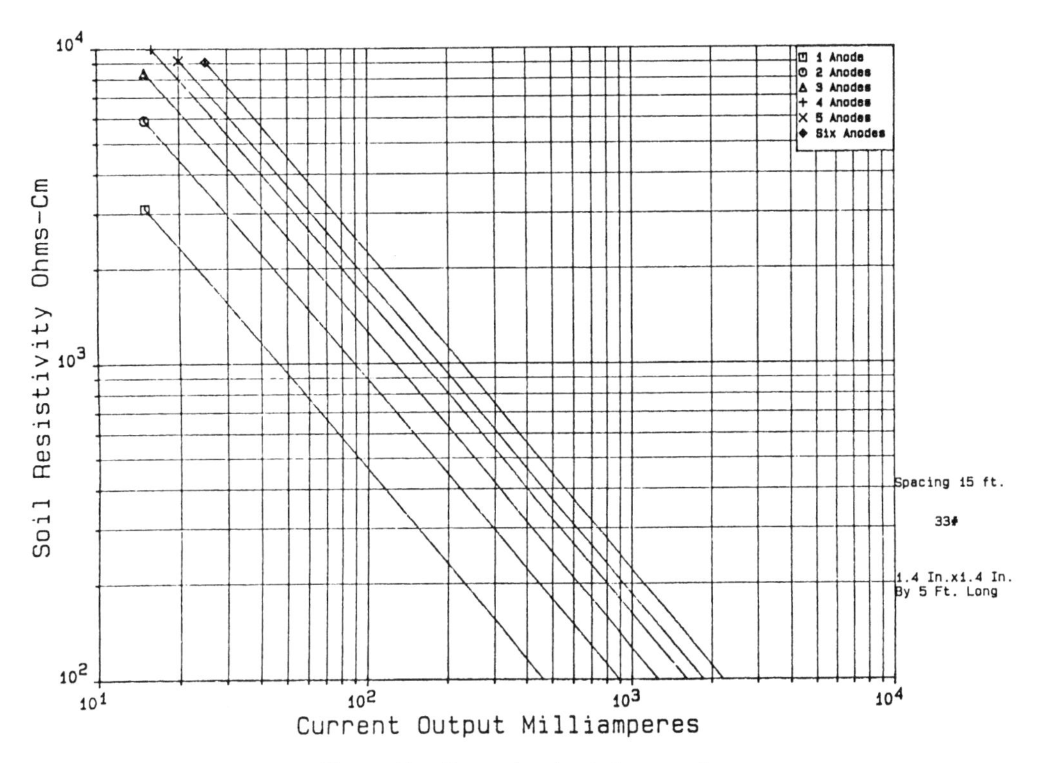

Figure 7a. Current output zinc anodes.

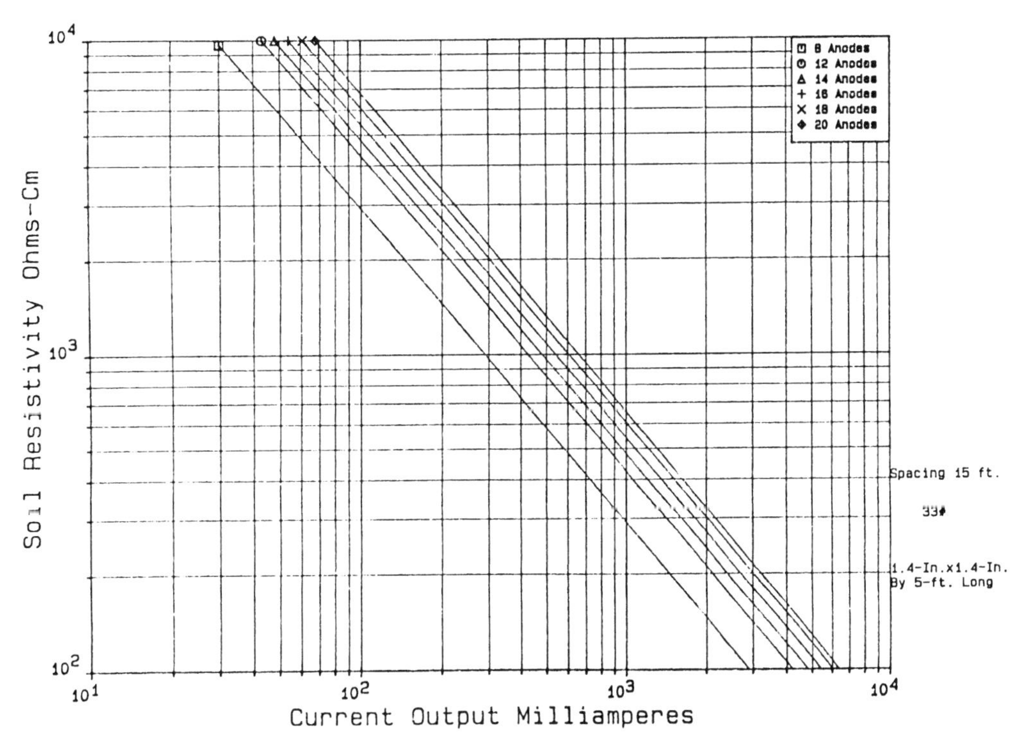

Figure 7b. Current output zinc anodes.

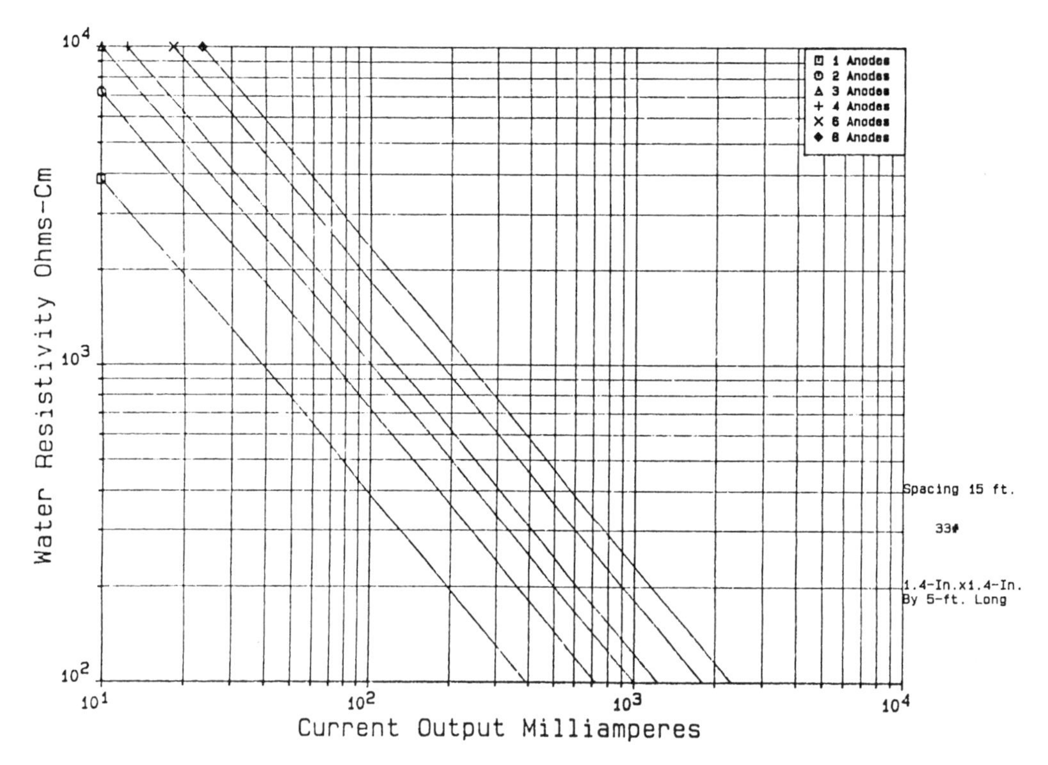

Figure 8a. Current output zinc anodes.

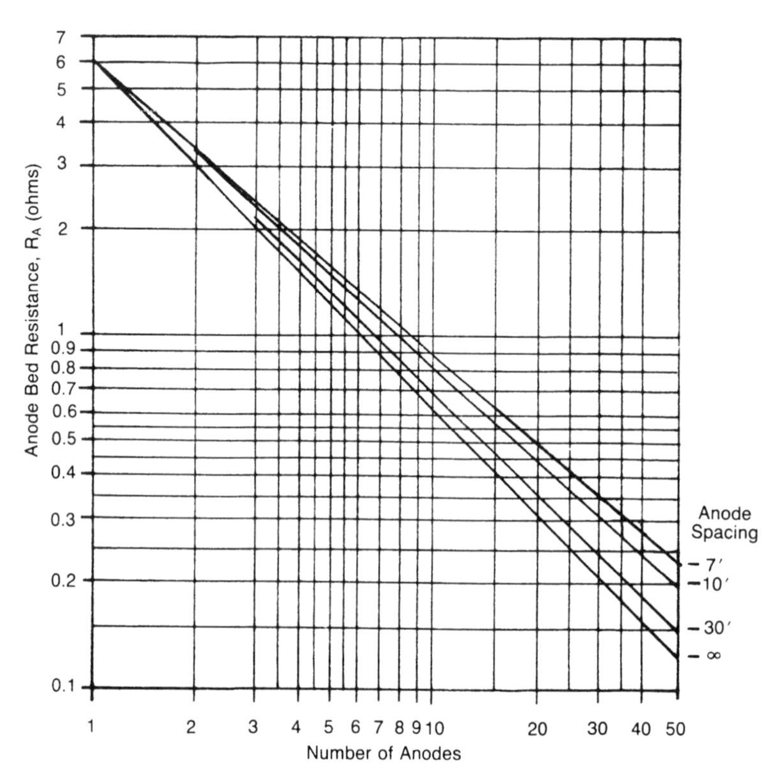

Figure 8b. Current output zinc anodes.

copper sulfate cell is −1.1 volts; standard magnesium has a solution potential of −1.55 volts; and high purity magnesium has a solution potential of −1.8 volts.

If, for example, a pipeline is protected with zinc anodes at a polarization potential of −0.9 volts, the driving potential will be −1.1 − (−0.9) or −0.2 volts. If standard magnesium is used, the driving potential will be −1.55 − (−0.9) or −0.65 volts. The circuit resistance for magnesium will be approximately three times as great as for zinc. This would be handled by using fewer magnesium anodes, smaller anodes, or using series resistors.

If the current demands for the system are increased due to coating deterioration, contact with foreign structures, or by oxygen reaching the pipe and causing depolarization, the potential drop will be less for zinc than for magnesium anodes. With zinc anodes, the current needs could increase by as much as 50% and the pipe polarization potential would still be about 0.8 volts. The polarization potential would drop to about 0.8 volts with only a 15% increase in current needs if magnesium were used.

The current efficiency for zinc is 90%, and this value holds over a wide range of current densities. Magnesium anodes have an efficiency of 50% at normal current densities. Magnesium anodes may be consumed by self-corrosion if operated at very low current densities. Refer to Figures 7a, 7b, 8a, and 8b for zinc anode performance data. The data in Figures 7a and 7b are based on the anodes being installed in a gypsum-clay backfill and having a driving potential of −0.2 volts. Figures 8a and 8b are based on the anodes being installed in water and having a driving potential of −0.2 volts.[1]

Example. Estimate the number of packaged anodes required to protect a pipeline.

What is the anode resistance of a packaged magnesium anode installation consisting of nine 32-lb. anodes spaced 7 ft apart in 2,000 ohm-cm soil?

Refer to Figure 9. This chart is based on 17# packaged anodes in 1,000 ohm-cm soil. For nine 32-lb. anodes, the resistivity will be:

$$1 \times 2{,}000/1{,}000 \times 0.9 = 1.8 \text{ ohm}$$

See Figure 10 for a table of multiplying factors for other size anodes.

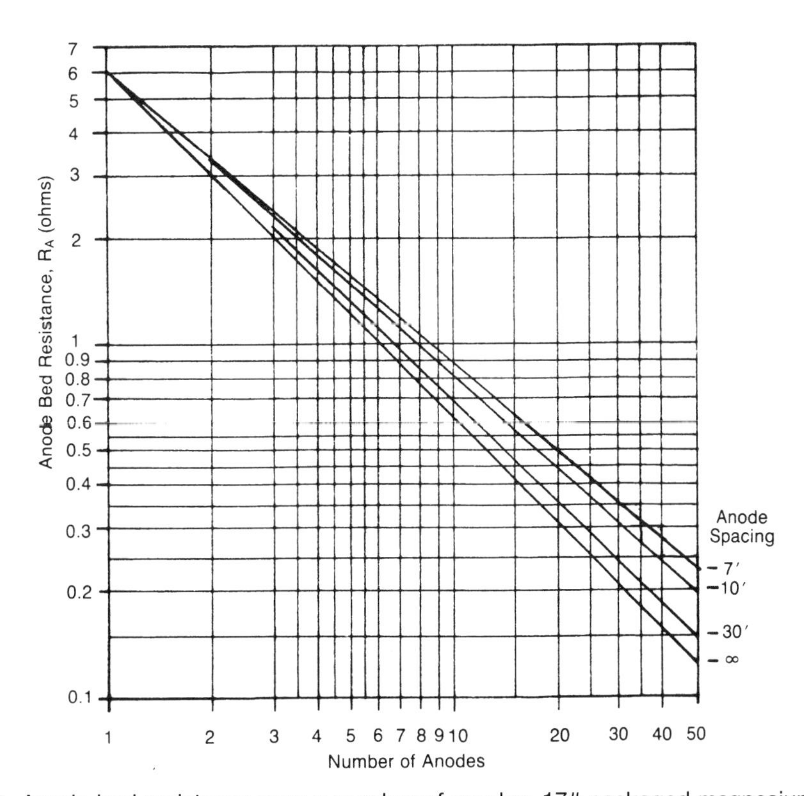

Figure 9. Anode bed resistance versus number of anodes; 17# packaged magnesium anodes.

Chart based on 17-lb. magnesium anodes installed in 1,000 ohm-cm soil in groups of 10 spaced on 10-ft centers.

For other conditions multiply number of anodes by the following multiplying factors:

For soil resistivity: $MF = \frac{p}{1{,}000}$ | For 9-lb. anodes: MF = 1.25

For conventional magnesium: MF = 1.3 | For 32-lb. anodes: MF = 0.9

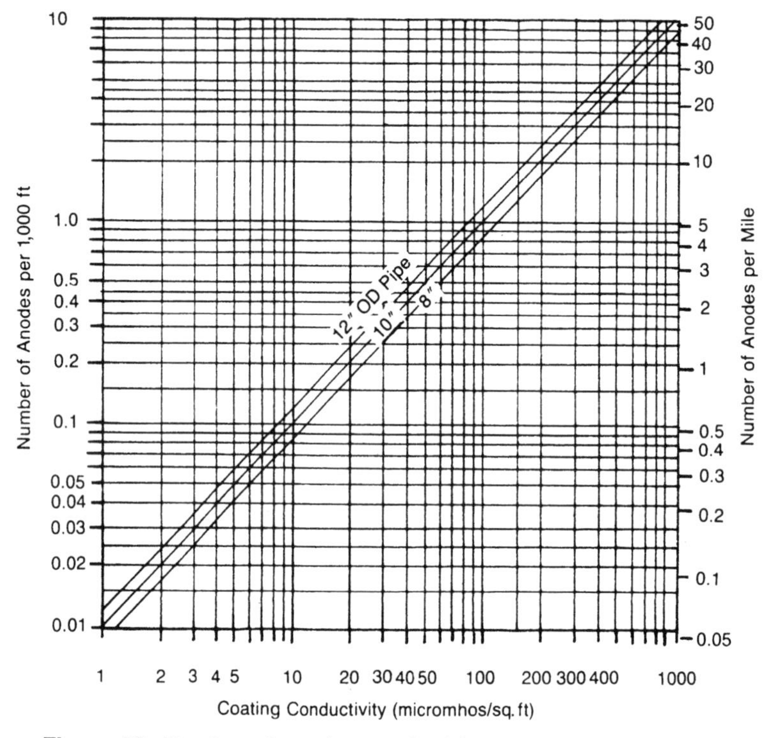

Figure 10. Number of anodes required for coated line protection.

Example. A coated pipeline has a coating conductivity of 100 micromhos/sq. ft and is 10,000 ft long, and the diameter is $10\frac{3}{4}$ in. How many 17-lb magnesium anodes will be required to protect 1,000 ft? Refer to Figure 7 and read 2 anodes per 1,000 ft. A total of twenty 17 lb. anodes will be required for the entire line.

Source

Pipeline Corrosion and Cathodic Protection, Gulf Publishing Company, Houston, Texas.

Reference

1. From data prepared for the American Zinc Institute.

Estimate the pounds of sacrificial anode material required for offshore pipelines

This rule of thumb is based on the following assumptions:

- 5% of pipeline considered bare
- 3 milliamperes/sq. ft required for protection
- 30-year design life

Basis for rule: sq. ft of surface area of pipeline/linear ft × % of bare pipe × 3.0 m.a./sq. ft × lb./amp.-year × design life in years = the lb. of galvanic anode material that will be required.

$$\text{Multiplier} = \text{OD} \times \text{lb./amp.-yr} \times (12\,\text{PI}/144) \times 30 \times 0.05 \times 0.003 \times 1{,}000$$

or:

Pipe OD × lb/amp.-yr × 1.1781 = lb of anode material/1,000 linear ft of pipeline

Example. How many pounds of zinc will be required for a 30-in. OD line installed in seawater and operating at 80°F? Pounds of zinc per ampere-year = 24.8 (seawater).

30 × 24.8 × 1.1781 = 877 lb./1,000 linear ft of pipeline

How many pounds of Galvalum® III will be required for a 30-in. OD line installed in seawater and operating at 80°F? Pounds of Galvalum® III per ampere-year in seawater = 7.62 (32–95°F).

30 × 7.62 × 1.1781 = 269 lb./1,000 linear ft of pipeline

How many pounds of Galvalum® III will be required for a 30-in. OD line installed in saline mud and operating at 80°F? Pounds of Galvalum® III per ampere-year in saline mud = 9.22 (32–95°F).

30 × 9.22 × 1.1781 = 326 lb./1,000 linear ft of pipeline

How many pounds of Galvalum® III will be required for a 30-in. OD line installed in saline mud and operated at a temperature of 160°F? Pounds of Galvalum® III per ampere-year in saline mud = 21.9 (140–170°F).

30 × 21.9 × 1.1781 = 774 lb./1,000 linear ft of pipeline

It is always a good idea to check with the bracelet vendor to determine the nearest size mold available to avoid the need for a special size mold. Also, the mold size will be a factor in determining anode spacing.

These calculations assume a utilization factor of 100%, and the designer should select the appropriate utilization factor for the specific application. For estimating, you may want to use 85%. Temperatures used in the examples are assumed to be the anode operating temperature.

Generally, zinc anodes should not be used if the pipeline will be operated at temperatures above 90°F. See ASTM B418-80 and NACE Paper No. 311 by Elwood G. Haney presented at CORROSION/85.

See Table 3 for typical anode performance data. All performance data shown here are suitable for estimating. For final design purposes, the anode manufacturer or distributor should be consulted for actual anode performance data required for a specific application.

Galvalum® III is a registered trademark of Oronzio De Nora S. A., Lugano (Switzerland). Galvomag® is a registered trademark of Dow Chemical Company.

Typical Anode Specifications

Table 1
Zinc anodes

Material	Mil. Spec. MIL-A-1800 1H	ASTM Type I	ASTM Type II
Lead	0.006 max	—	—
Iron	0.005 max	0.005 max	0.0014 max
Cadmium	0.025–0.15 Range	0.03–0.1	0.003 max
Copper	0.005 max	—	—
Aluminum	0.10–0.50 max	0.1–0.4	0.005 max
Silicon	0.125 max	—	—
Zinc	Remainder	Remainder	Remainder

Table 2*
Cast magnesium anodes

Material	H-1 Grade A Type III	H-1 Grade B Type II	H-1 Grade C Type I	Galvomag®
Al	5.3–6.7	6.3–6.7	5.0–7.0	0.010 max
Zn	2.5–3.5	2.5–3.5	2.0–4.0	0.05 max
Mn	0.15 min	0.15 min	0.10 min	0.5–1.3
Fe	0.003 max	0.003 max	0.003 max	0.03 max
Ni	0.002 max	0.003 max	0.003 max	0.001 max
Cu	0.02 max	0.05 max	0.10 max	0.02 max
Si	0.10 max	0.30 max	0.30 max	0.05 max
Others	0.30 max	0.30 max	0.30 max	0.05 max, each 0.30 max, total

*Reprinted courtesy of Dow Chemical U.S.A.

Table 3
Typical anode performance data

Anode Type	Amp-hr/lb in Seawater	Potential Negative Volts Ag/AgCl	Approx. Current Density mA/ft^2
Zinc	354	1.03	36–600
Galvalum® 111	1,150–1,204 (32–95°F)	1.10	100–1,000
	in saline mud		
	950 (32–95°F)	1.10	100
	930 (104°F)	1.08	100
	600 (140°F)	1.05	100
	400 (176°F)	1.04	100
	400 (221°F) in earth	1.02	100
Magnesium	400–500	1.53	36–600
Galvomag®	500–540	1.75	36–600

Table 4
Typical environments for sacrificial anodes

Material	Driving Potential Volts	Environment
Magnesium	0.9	Underground
		Fresh water
		200–10,000 ohm-cm
Aluminum	0.25	Sea water
		Brine
		Saline mud
		<200 ohm-cm
Zinc	0.25	Sea water
		Brine
		Underground
		<1,500 ohm-cm

Tables 1–4 are guideline-type information suitable for general use. The characteristics of a specific application must be considered, and it may be necessary to contact the anode manufacturer or distributor to obtain detailed anode performance data before finalizing a design.

Comparison of other reference electrode potentials with that of copper–copper sulfate reference electrode at 25°C

Type of comparative reference electrode	Structure-to-comparative reference electrode reading equivalent to −0.85 volt with respect to copper sulfate reference electrode	To correct readings between structure and comparative reference electrode to equivalent readings with respect to copper sulfate reference electrode
Calomel (saturated)	−0.776 volt	Add −0.074 volt
Silver–Silver Chloride (0.1 N KCl Solution)	−0.822	Add −0.028
Silver–Silver Chloride (Silver screen with deposited silver chloride)	−0.78	Add −0.07
Pure Zinc (Special high grade)	+0.25*	Add −1.10

*Based on zinc having an open circuit potential of −1.10 volt with respect to copper sulfate reference electrode.
(From A. W. Peabody, Chapter 5, p. 59, *NACE Basic Corrosion Course*, Houston, TX, 1973.)

Chart aids in calculating ground bed resistance and rectifier power cost

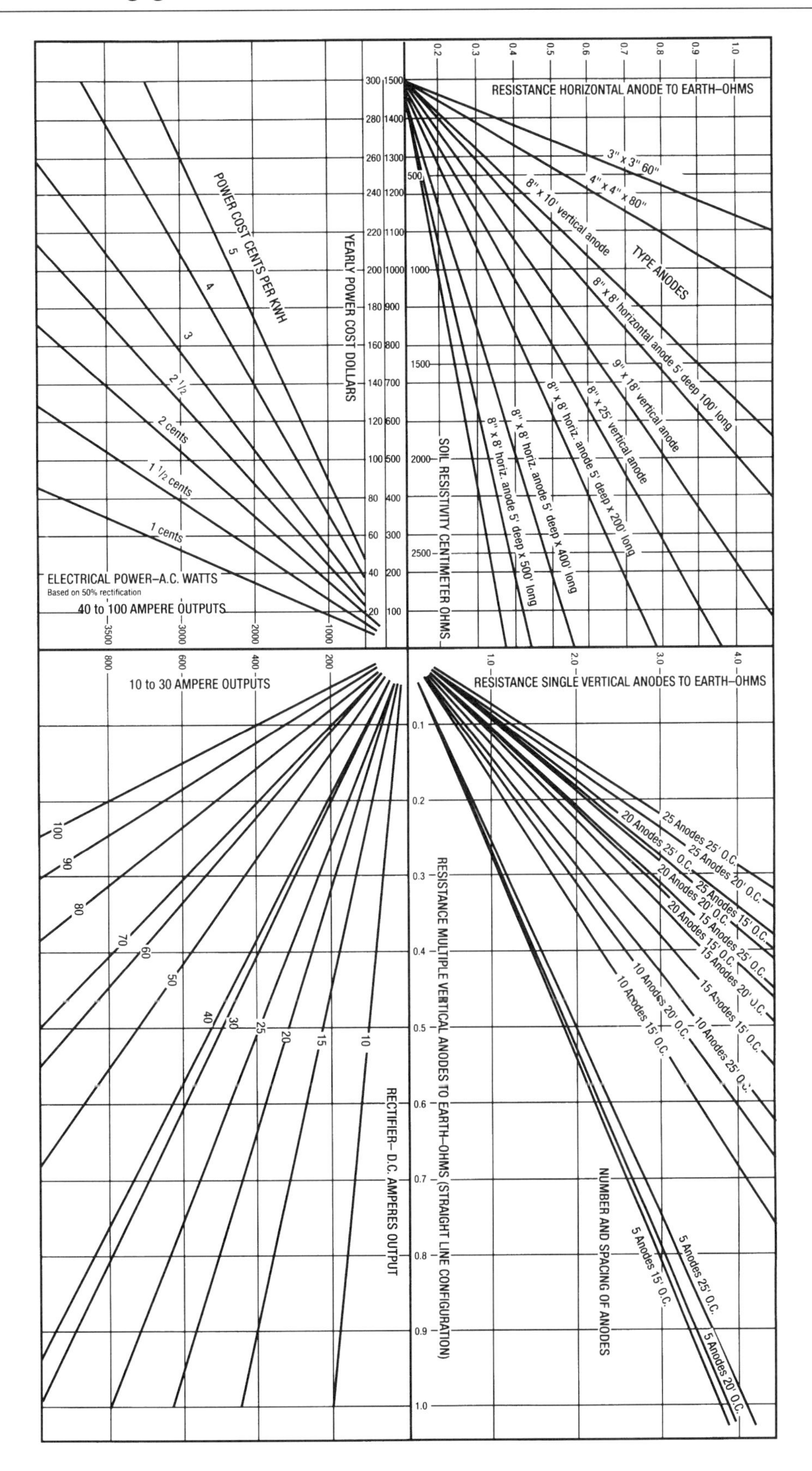

For conventional-type ground beds, this chart permits a rapid determination of the resistance to earth of various ground beds and the yearly power costs based on 50% efficient rectification. These curves are specifically designed for standard 3-in. by 60-in. and 4-in. by 80-in. graphite anodes centered in a vertical or horizontal excavation and backfilled to the dimensions indicated with well-tamped coal coke breeze.

Example.

Soil resistivity = 1,000 ohm-cm
Power rate = 3 cents per kwh
Rectifier rating = 20-ampere

To determine the optimum size ground bed, enter the chart at the "Soil Resistivity-ohm-cm" line and extend a vertical line from the 1,000-ohm-cm point to intersect all the anode types. Proceed horizontally to the left for the resistance to earth of horizontal type anodes and horizontally to the right for vertical type anodes. Various groups of vertical anodes with 15-, 20-, and 25-ft spacings are represented by curves.

Resistance values for horizontal anodes obtained from the ordinate of the upper left-hand section should be transferred to the abscissa of the upper right-hand section to determine power consumption and annual costs from the lower right-hand and lower left-hand section. In obtaining power costs for the two lower sections, the ordinate shows power consumption in AC watts based on 50% efficient rectification, and the abscissa of points on the lower left-hand section indicates annual power costs in dollars.

How can output of magnesium anodes be predicted?

For a single 17-lb. magnesium anode, working against a fully protected pipe, divide the soil resistivity in ohm-cm units into 180,000; the answer is the current output in milliamperes.

Example. Soil resistivity 3,200 ohm-cm
180,000/3,200 = 56.25

Current output will be about 56 milliamperes.

The rule works for *single* anodes against fully protected pipe; anode groups will have lower current output per anode, and if the pipe is not fully protected, the current output will be greater. For 32-lb. anodes use 195,000; for 50-lb. anodes use 204,000.

How to determine the efficiency of a cathodic protection rectifier

Multiply DC voltage output times DC amperes times 100 and divide by the AC watts input. The answer is % efficiency.

Example. A rectifier has an output of 50 amperes DC at 15 volts. The wattmeter gives a direct reading of an AC input of 1155 watts.

$$\text{Efficiency} = \frac{(\text{DC volts})(\text{DC amperes})(100)}{(\text{AC watts input})}$$

$$= \frac{(15)(50)(100)}{(1{,}155)} = 65\%$$

On those rectifier units that do not have a wattmeter, the AC input may be obtained from the watt hour meter mounted on the pole. To do this, count the number of revolutions made by the watt hour meter in so many seconds, multiply by the meter constant stamped on the meter times 3,600, and divide by the number of seconds timed. This gives AC input in watts. Then go ahead and use the formula as above.

Example. A rectifier unit does not have a watt meter. The wheel of the watt hour meter revolves 11 times in one minute. Meter constant is shown as 1.8. Output of the rectifier is the same as in the first case, 50 amperes at 15 volts DC.

$$\text{AC watts} = \frac{K \times N \times 3{,}600}{T}$$

where K is the meter constant and N is the number of meter revolutions in T seconds.

$$\text{AC watts} = \frac{(1.8)(11)(3{,}600)}{60} = 1{,}188 \text{ watts AC}$$

$$\text{Efficiency} = \frac{(15)(50)(100)}{1{,}188} = 63\%$$

How to calculate the voltage drop in ground bed cable quickly

Use the following formula for calculating the voltage drop in a copper ground bed cable:

$$V_T = \frac{R_{1000}}{1000}[I_T D_{GR} + D_A[(N-1)(I_T) - I_A(1+2+3+4....N-1)]]$$

where:

V_T = Total voltage drop in volts
R_{1000} = Cable resistance per 1,000 ft
I_T = Total current in amperes
I_A = I_T/N current per anode
D_{GB} = Distance from rectifier to first anode in ft
D_A = Distance between anodes in ft
N = Number of anodes in ground bed

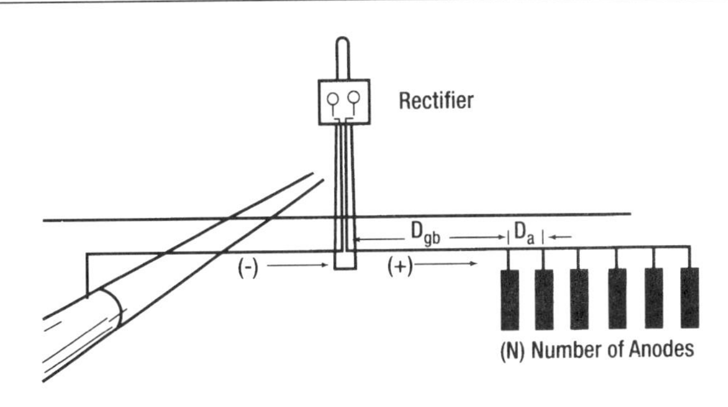

Example. What is the voltage drop in a copper wire ground bed cable in a cathodic protection system having a rectifier output of 225 amperes? The distance from rectifier to ground bed is 100 ft, and 16 anodes are spaced on 20 ft centers. The cable used is No. 4 copper having a resistance of 0.2487 ohm per 1,000 ft.

Then:

R_{1000} = 0.2487 ohm
I_T = 225 amperes
I_A = 225/16 = 14 amperes
D_{GB} = 100 ft
D_A = 20 ft
N = 16

$$V_T = \frac{.2487}{1,000}[(225)(100) + 20[(15)(225) - (14)(120)]]$$

$$V_T = \frac{.2487}{1,000}(22,500 + 20[1,695])$$

$$V_T = \left(\frac{.2487}{1,000}\right)(56,400)$$

$$V_T = (.2487)(56.4)$$

$$V_T = 14.00 \text{ volts drop}$$

What is the most economical size for a rectifier cable?

Choosing the economical size for a rectifier cable involves balancing the first cost of the cable against the power losses. A rule that does not include all of the factors, but which usually gives a very close approximation, is as follows:

$$a = 2,280\ I\sqrt{P}$$

where a is the area of the cable in circular mils (refer to a wire table for size), I is the current to be carried, in amperes, and P is the cost of power in cents per kilowatt-hour. This formula is based on the present prices of copper cables and assumes the use of a good grade of direct burial cable.

Example. Suppose

I = 60 amps
P = 4 cents

$$a = (2,280)(60)\left(\sqrt{4}\right)$$

$$a = 273,600 \text{ circular mils}$$

Cable is sized 250,000 and 300,000 circular mils. Use the latter.

How to estimate the number of magnesium anodes required and their spacing for a bare line or for a corrosion "hot spot"

Soil and pipeline conditions will vary widely, but where these given assumptions hold, an estimate can be made for test installations using this method. First find the area of pipe to be protected, multiply by the milliamperes per sq ft necessary to protect the line, and divide by design output of an anode in this type soil. This will give the number of anodes required; to find spacing, divide number of anodes into the length of line.

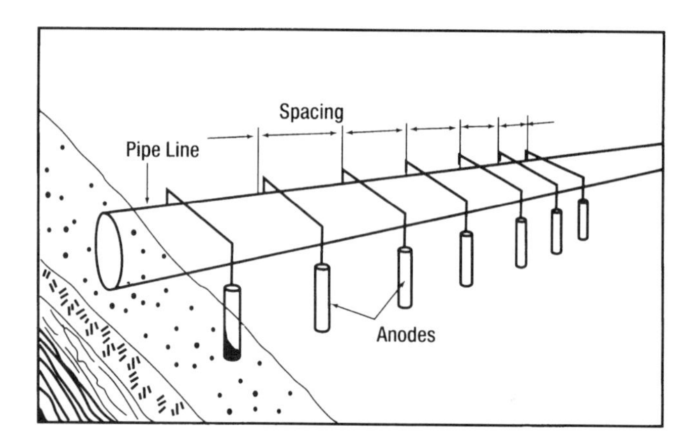

Example. Find the number of 17-lb. magnesium anodes needed and spacing used to protect 10,000 ft of bare 4-in. pipeline in a corrosive soil having a resistance of 1,000 ohm per cubic centimeter. Assume a current requirement of 1 milliampere per sq. ft, and that the anode output curve shows an output of 100 milliamperes per anode in this type soil.

$$\text{Area of pipe} = \frac{4 \times 3.14}{12} \times 10{,}000 = 10{,}500 \text{ sq ft}$$

Total current requirement $= 10{,}500 \times 1.0 = 10{,}500$ milliamperes

$$\text{Anodes required} = \frac{10{,}500 \text{ milliamperes}}{100 \text{ ma/anode}}$$

$$= 105 \text{ anodes distributed}$$

$$\text{Spacing} = \frac{10{,}000}{105} = 95 \text{ ft}$$

Similarly, find the number of anodes required to protect ten 100-ft long "hot-spots" on a 4-in. line assuming a 100-milliampere output from a 17-lb. magnesium anode. Estimate as follows:

$$\text{Area of each "hot-spot"} = \frac{4 \times 3.14}{12} = 105 \text{ sq ft}$$

$$\text{Anodes required} = \frac{105 \text{ sq ft} \times 1.0 \text{ ma per sq ft}}{100 \text{ ma per anode}} = 1.05$$

One 17-lb. magnesium anode will produce approximately enough current in 1,000-ohm soil to protect one of these "hot-spots."

Ten anodes are required for "hot-spots" on this section of line.

Note: For these rules the protective criterion is −0.85 volts pipe-to-soil potential using a copper-copper sulfate reference electrode; or a negative change of potential of as much as −0.25 volts. Soil and pipeline conditions vary widely, so consult an experienced corrosion engineer in the design of a complete cathodic protection system.

How can resistivity of fresh water be determined from chemical analysis?

Divide the total solids expressed in parts per million into 500,000; the result is the resistivity in ohm-centimeter units.

Example. Total solid content is 400 ppm.

$500{,}000 \div 400 = 1{,}250$ ohm centimeters resistivity

This rule of thumb will give very accurate results for any naturally occurring fresh waters; severely contaminated waters may differ markedly, however, depending on the kind of contamination.

What will be the resistance to earth of a single graphite anode?

For the most common size, 3-in. by 60-in., installed in 10 ft of coke breeze backfill in an 8-in. hole, multiply the soil resistivity in ohm-cm by 0.002; this will give the resistance to earth in ohm.

Example. 1,800 ohm-cm soil

$1{,}800 \times 0.002 = 3.6$ ohm

This assumes that the resistivity is uniform to considerable depth; if it increases with depth, the resistance will be slightly higher.

How to estimate the monthly power bill for a cathodic protection rectifier

Multiply the rectifier output voltage and current together, and multiply this by $1\frac{1}{2}$; the result is the power consumption for one month, in kilowatt-hours. If the unit is on a meter of its own, figure the bill by the utilities rate schedule; if it represents an additional load, use the rate per kilowatt-hour for the highest bracket reached.

Example. The terminal voltage of a rectifier is 10.2 volts, and the current output is 21 amperes.

$10.2 \times 21 \times 1.5 = 321.2$

The power consumption will be about 320 kwh per month; at 3 cents this would amount to $9.60, at 1.1 cents it would be $3.52, etc.

Note that the actual current and voltage values are used; if the rating of the unit is used, the result will be the maximum power consumption within the range of the equipment. The formula is approximate; it is based on an efficiency of about 48%.

What will be the resistance to earth of a group of graphite anodes, in terms of the resistance of a single anode?

When the spacing is 20 ft (the most common value), the resistance of the group will be equal to the resistance of a single anode, divided by the square root of the number of anodes.

Example. Resistance of 4 anodes = $\frac{1}{2}$ resistance of single; resistance of 16 anodes = $\frac{1}{4}$ resistance of single, etc.

This rule is reasonably accurate for soils that are uniform to a considerable depth; if the resistivity increases with depth, the "crowding" effect will be greater, whereas if it decreases, the individual anodes will not affect each other so much.

How can the current output of magnesium rod used for the cathodic protection of heat exchanger shells be predicted?

Divide the total dissolved solids content of the water by 13; the answer is the current output per ft of 1.315 in. (1-in. pipe size) magnesium rod to be expected.

This is an approximation to the current flow that will be found after stabilized conditions have been reached, if protection is obtained. If insufficient magnesium is installed, the current flow will probably be greater, and if too much is installed, the current will be less.

What spacing for test leads to measure current on a pipeline?

Multiply the pipe weight, in lb/ft, by 4; the result is a spacing that will give a resistance of 0.001 ohm, or 1 ampere per millivolt IR drop.

Example. Pipe weight is 40.48 lb/ft.

$40.5 \times 4 = 162$

Install test leads at 162 ft spacing.

The rule works for welded lines of grade B seamless; variations in steel may introduce errors as great as 15%. It is always better to calibrate a line section if accurate results are to be had, but if the above rule is used, the spacing will not be too small for satisfactory measurement, or too great for convenience in installation.

How many magnesium anodes are needed for supplementary protection to a short-circuited bare casing?

For 17-lb. anodes, use A $\rho/50{,}000$ to give the number needed for the casing alone; A is the area of the casing, in sq ft, and ρ is the soil resistivity in ohm-centimeters. The life of the anodes may be estimated from $\rho/180$. If the life in years, as given by this expression, is too short, refigure for 32-lb. anodes, using A $\rho/55{,}000$ for the number and $\rho/100$ for the life. If this is still too short, figure for 50-lb anodes; number is A $\rho/60{,}000$ and life is $\rho/65$.

Such an installation, grouped on both sides of the line at each end of the casing, will enable the main line protection unit to carry beyond the casing; the values obtained above are approximate only, and should be checked by potential measurements after installation.

Example. Suppose the pipeline has shorted out to 45 ft of 24-in. bare casing. Soil resistivity is 1,100 ohm-centimeters. How many 17-lb. anodes would be required for supplementary protection?

$$\text{Number of anodes} = \frac{A\rho}{50{,}000}$$

$$A = 2 \times 3.14 \times 45 = 283$$

$$\text{Number of anodes} = \frac{(284)(1{,}100)}{50{,}000}$$

$$= 6.21$$

Place three anodes at each end of casing. Their life expectancy will be:

$$\frac{1{,}100}{180} = 6 \text{ years}$$

Group installation of sacrificial anodes

When anodes are installed in groups, a spacing correction factor must be taken into account to determine the current output. The chart below shows the correction factor for multiple anode installations.

Example. Single anode 8-in. by 60-in. in 1,500 ohm soil.

Resistance = 4.849 ohm
Potential = 0.85 volts

$$\text{Current output} = \frac{E}{R} = \frac{17.5 - 0.85}{4.849} = 0.185 \text{ amps}$$

Current output of 5 anodes on 15 ft spacing under above conditions.

$$\text{Combined resistance of group} = \frac{4.849}{5} \times 1.22$$

Correction factor = 1.17 ohm

$$\text{Total current output} = \frac{1.75 - 0.85}{1.17} = 0.769 \text{ amps}$$

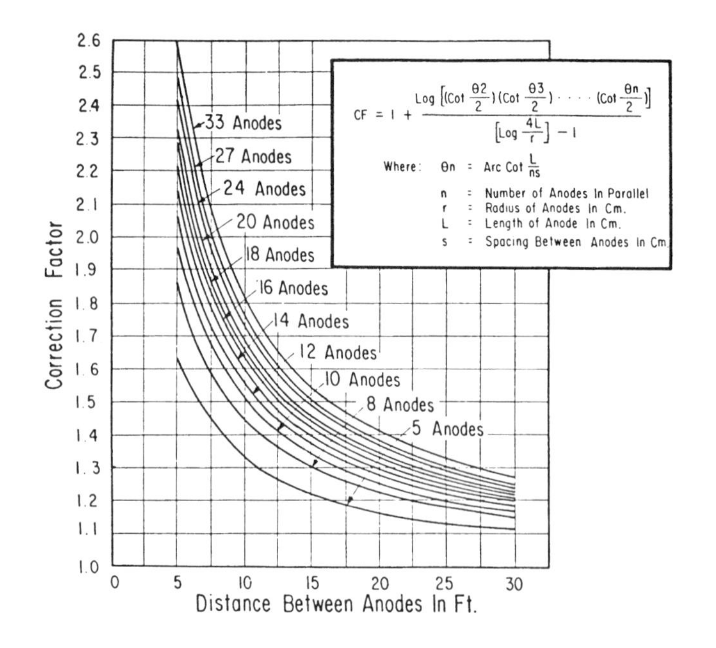

How can the life of magnesium anodes be predicted?

Multiply the anode weight in pounds by 57, and divide by the current output in milliamperes.

Example. A 17-lb anode has a current output of 83 milliamperes. What is its projected life?

$17 \times 57/83 = 11.7$

The projected life is 11.7 years.

The rule is based on 50% current efficiency, and on stabilized conditions. If more anodes are to be added, or if time has not been allowed for polarization (1–4 weeks), then the current output will probably fall off, and the actual life will be 10 to 40% longer than indicated by the rule.

How to find the voltage rating of a rectifier if it is to deliver a given amount of current through a given ground bed (graphite or carbon)

Multiply the desired current, in amperes, by the total resistance of the ground bed, in ohm (either calculated or measured); add 2 volts to overcome the galvanic difference between graphite and protected steel.

Example. A current of 20 amperes is desired from a ground bed whose resistance is 0.9 ohm.

$(20 \times 0.9) + 2 = 20$ volts

Specify the next larger size, which will probably be (depending upon the manufacturer) 24 volts.

Note: Where the resistance is measured rather than calculated, and is the total loop resistance between pipe and ground bed, this method is quite precise; for calculated values, caution should be observed.

Determining current requirements for coated lines

Survey along the line after a polarization run of 3 or 4 hours, and locate the two points whose pipe-to-soil potentials are 1.0 volt and 0.8 volt (see Figure 1). The section of line lying between these two points will then be at an average potential of about 0.9 volt, which is approximately the average of a line protected with distributed magnesium anodes. If, then, the line current is measured at each of these two points, the difference between these two line currents will be the total amount of current picked up on the line in the section. This quantity, divided by the length of the section, will then give the current requirements of the line, in amperes per mile. Naturally this figure can only be used for an entire line if the section chosen is truly representative; judgment must be used at this point.

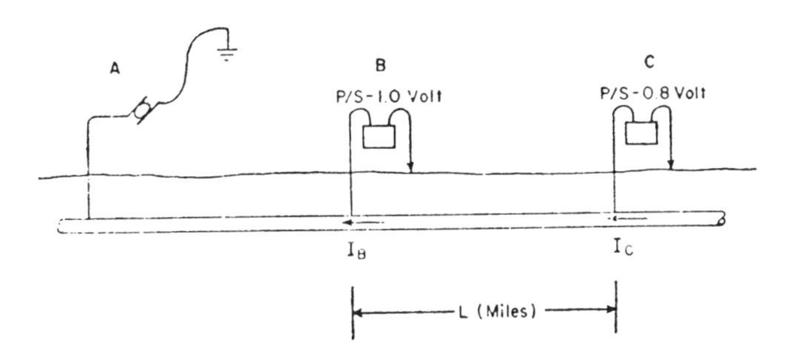

Determining current requirements for coated lines when pipe-to-soil potential values are estimated

If it is not possible or convenient to find the points that have the exact values of 1.0 and 0.8, then readings can be taken that approximate these values, and both potential and line current can then be adjusted to find the desired quantity; this is illustrated in Figure 1; readings were taken at points “A” and “B,” and, by assuming a static potential of

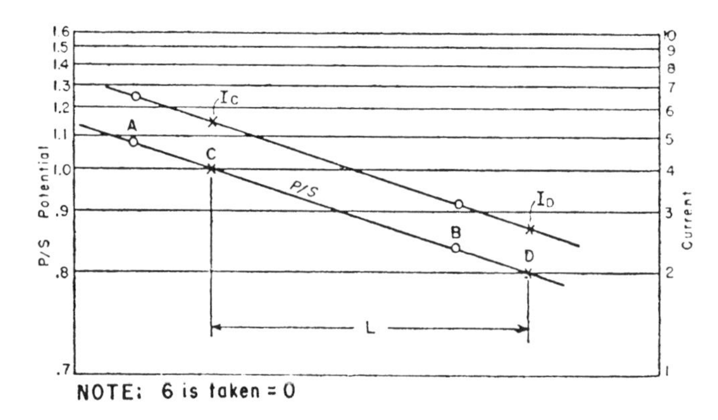

0.6 volt and linear attentuation, it is determined that points "C" and "D" have the values of 1.0 and 0.8; since line current varies according to the same laws as potential, it is possible to determine the current values at the two latter points, and so to arrive at the desired quantity.

It is not to be expected that the estimate thus made will be highly accurate; it is intended to be merely preliminary.

HVDC effects on pipelines

Latest methods to calculate quantitative characteristics of the earth's gradient and high voltage direct current effects on pipelines

Thomas B. Hays, George W. Silkworth, Robert D. Schilling, Cornell, Howell, Hayes & Merryfield, Inc.

A pipeline collects, conducts, and discharges electric current of high voltage direct current (HVDC) origin as a result of the electric field established by the DC current transmitted through the earth and by the electrical properties of the pipeline itself.

The electric field

It is assumed that the electrical properties of the earth and pipeline are uniform and that the HVDC effects are isolated from other stray or induced current effects. This assumption makes possible the analysis of this portion of the total protection problem and assists in visualizing the unique effects of HVDC earth currents.

Ultimately, the total electrical current conducted by a pipeline will be the sum of all the various components. This includes protective current, telluric current, and other stray or interference currents.

Conduction of a direct current through the earth is an electrolytic phenomena, characterized by an electric field that surrounds each of the two earth electrodes, and is directly proportional to the HVDC earth current (Figure 1). When the electrodes are spaced several hundred miles apart, the field appears to be concentric around each ground electrode.

The fields surrounding the opposite electrodes are of the opposite polarity. If the polarity of the ground current reverses, the surrounding electric fields will also reverse. The potential at any point in the electric field with respect to remote earth (V), due to an electrode current (I) at a

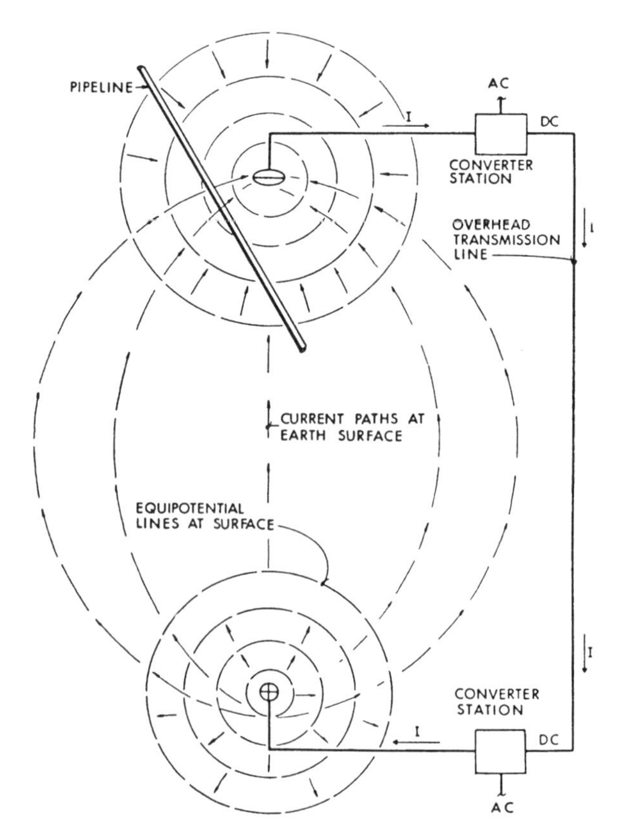

Figure 1. Conduction of a direct current through the earth is an electrolytic phenomena, characterized by an electric field that surrounds each of the two earth electrodes, and is directly proportional to the HVDC earth current.

distance (r) from an electrode in a homogeneous earth of the resistivity (ρ) is found by the relationship:

$$V = \frac{\rho I}{2\pi r}$$

This relationship shows a simplified representation of what actually occurs. The earth's resistivity is not uniform at all depths or horizontal distances, which causes irregularities to the ideal voltage distribution existing around an electrode in a homogeneous earth.

To refine the earth current analysis and to account for different resistivities, computer techniques have been developed to estimate the earth's gradients existing for conditions of multi-layers of earth of different resistivities. Pipeline effects have also been predicted by this method. Reasonable agreement between these models and field data has been reported. Techniques are available to calculate the quantitative characteristics of both the earth's gradient and HVDC effects on pipeline.

HVDC effects on a pipeline

Whether the HVDC component of the total current existing on the line causes a corrosion problem depends on the level of protection along the line and the effect of the added component to the level of protection. Figure 2a illustrates the HVDC component of the total conducted current along the pipeline for a given separation distance under the condition where the HVDC electrode is collecting current from the earth.

In determining protection requirements for the pipeline, the characteristic of the current transfer relationship is important (Figure 2b). In those regions of the pipeline where the current transfer is from the pipeline to the soil—along the center in this case—corrosion could occur if the HVDC component of the total transfer current to the soil is great enough to exceed any local level of protection against corrosion.

The pipeline HVDC arrangement shown in Figure 2a will serve as a basis to develop and illustrate the performance requirements for protecting a pipeline from the detrimental effects of HVDC earth currents. The hypothetical problem, as shown, is one that may be encountered in the field. In this example, the following assumptions are made:

- The pipeline is long, straight, electrically continuous, and uniform in wall thickness, diameter, material, and coating.
- It is approximately 30 miles away from the nearest HVDC earth electrode.
- The earth's resistivity is uniform over the general area.
- The maximum value of the HVDC earth current in either polarity causes the pipeline effects.

While these conditions may not exist in any actual pipeline HVDC system situation, they will serve to identify and evaluate the requirements for protection against HVDC effects in an actual situation.

Figure 3 shows the range or envelope of effects as the HVDC varies from its maximum value in one polarity to the maximum value of the other polarity. The coordinates on this graph are DISTANCE along the X axis and VOLTAGE on the Y axis.

The voltage corresponds to the effect that would usually be measured on an actual pipeline: pipe-to-soil voltage measured with a copper-copper sulfate half-cell. The value of the voltage effect at any point on the line is the product of the current transfer, and the electrical resistance between the pipe material and earth over the same area.

Figure 3 also shows reasonable hypothetical scale values illustrating the various combinations of protection requirements that may exist in an actual installation. The voltage values give the change in pipe-to-soil reading that would be caused by the HVDC transfer current. If this HVDC is

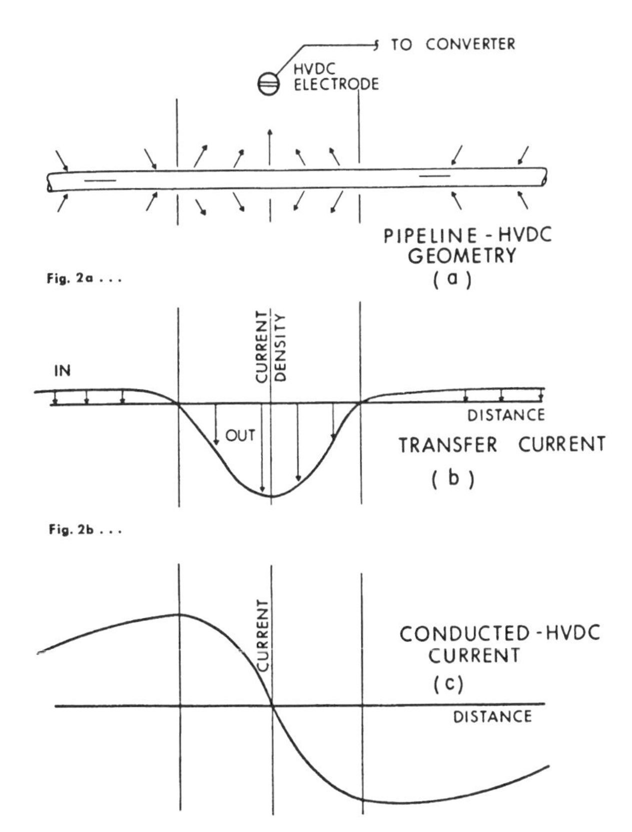

Figure 2. (a) Geometric plan between the HVDC electrode and the pipeline. Arrows indicate the direction of the HVDC component current. These would be reversed if the polarity of the electrode were reversed. (b) Curve shows characteristics of the transfer of HVDC into and out of the pipeline along its length. The transfer current is equivalent to current density. (c) Conducted current in the pipeline, which results in the summation of the transfer current along the pipe.

considered as the only source of corrosion, then a pipe-to-soil voltage reading with a copper–copper sulfate half-cell would be approximately 0.85 volts at no transfer current. This scale is shown on the left side of the figure. Corrosion occurs for any reading below this axis, since the current is leaving the pipeline and entering the soil to cause this change in the pipe-to-soil reading.

At locations on the pipeline where HVDC causes conditions that increase the volt-meter reading, current is collecting on the pipeline. The horizontal axis, therefore, represents a corrosion limit. Above this axis, a coating disbonding limit can exist where the density of the current transferred from the earth to the pipeline causes gas evolution, which may blister the coating.

In this example, a pipe-to-soil reading with a copper–copper sulfate half-cell of 1.5 volts is shown as the coating disbonding limit. The difference between these limits gives an assumed safe operating band width of 0.65 volts. Values other than these may be used for specific applications. These limits, however, will serve to illustrate the principles.

Protective system limits

The requirements for protection along the line from either corrosion or disbonding are developed from examination of Figure 3, and the results are plotted in the subsequent figures. The upper limit of protection is the maximum

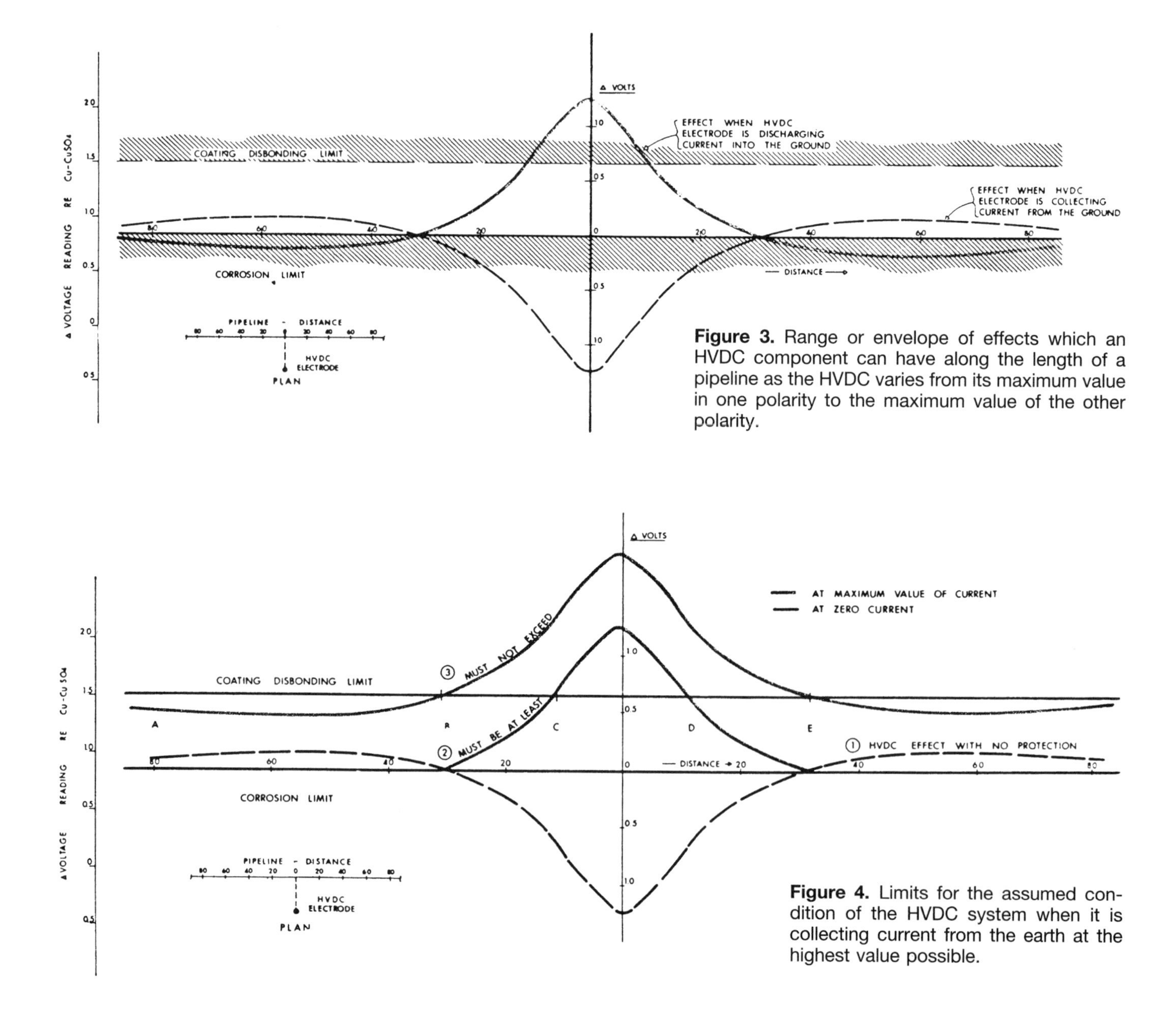

Figure 3. Range or envelope of effects which an HVDC component can have along the length of a pipeline as the HVDC varies from its maximum value in one polarity to the maximum value of the other polarity.

Figure 4. Limits for the assumed condition of the HVDC system when it is collecting current from the earth at the highest value possible.

amount of current that can be collected by the pipeline without exceeding the disbonding limit. The lower limit is the minimum amount of current that can be collected by the pipeline to keep it above the corrosion limit.

At this value of HVDC (Figure 4), the uncorrected pipeline effect is shown by Curve 1. No protection is needed along either end of the pipeline in Regions A-B or E-F. Corrosion protection is needed between B and E. The minimum protection effect along the line is shown by Curve 2. The resultant of the HVDC and this protection effect lies on the corrosion limit. The maximum protective effect that can be applied along the line, without exceeding the coating disbonding limit, is shown by Curve 3. Under this maximum value and polarity of HVDC current, the region between these two protective limits, therefore, defines the range of acceptable protective system performance at any location on the pipeline.

When the HVDC effect decreases, due to a reduction in the ground current, the required level of protection on the line also decreases. Along the pipeline length from C to D, the protection *must* be reduced to avoid exceeding the disbonding limit. At zero HVDC effect, the protective system effect must fall between the corrosion and disbonding limits all along the line.

Examination of the limits for these two values of HVDC system operation shows that the protective effect must be varied in response to the HVDC effect over the length of the pipeline from C to D: automatic regulation of the protective current is needed over this length of the pipeline. With only this polarity of HVDC current, a fixed level of protection can be applied along the pipeline from A to C and from D to F.

When the earth current can reverse polarity, as with the Pacific Northwest (PNW)-Pacific Southwest (PSW) intertie, then the same process that was used to develop the limits shown in Figure 4 can be repeated for the opposite HVDC polarity. The limits on protection along the pipeline when the HVDC electrode is discharging current, at its maximum value, into the earth are shown in Figure 5.

The combination of these two sets of limiting characteristics is shown in Figure 6. The graph shows the length along the pipeline where a fixed level of cathodic protection can be used and those regions where controlled protection is required for any value of HVDC earth current. In this particular example, variable protection is required along the pipeline for approximately 20 to 30 miles on either side of the point closest to the HVDC electrode. Cathodic protection is required when the HVDC electrode is collecting current from the earth. That is, the HVDC component of current, conducted by the pipeline, is discharging from the line near the earth electrode. The required protective current along the pipeline to counteract the HVDC effect under this condition is show as Region 1 in Figure 6.

When the HVDC polarity is reversed and the magnitude of the ground current is increased to the maximum value, a different problem is encountered. The electric field which is established around the electrode and pipeline causes current to enter the pipeline through the coating. In this example, the disbonding limit is exceeded over the length of the pipeline from C to D at the maximum value of HVDC earth current. The required correction for this detrimental effect is similar to the effect for the previous HVDC polarity: namely, the uncorrected

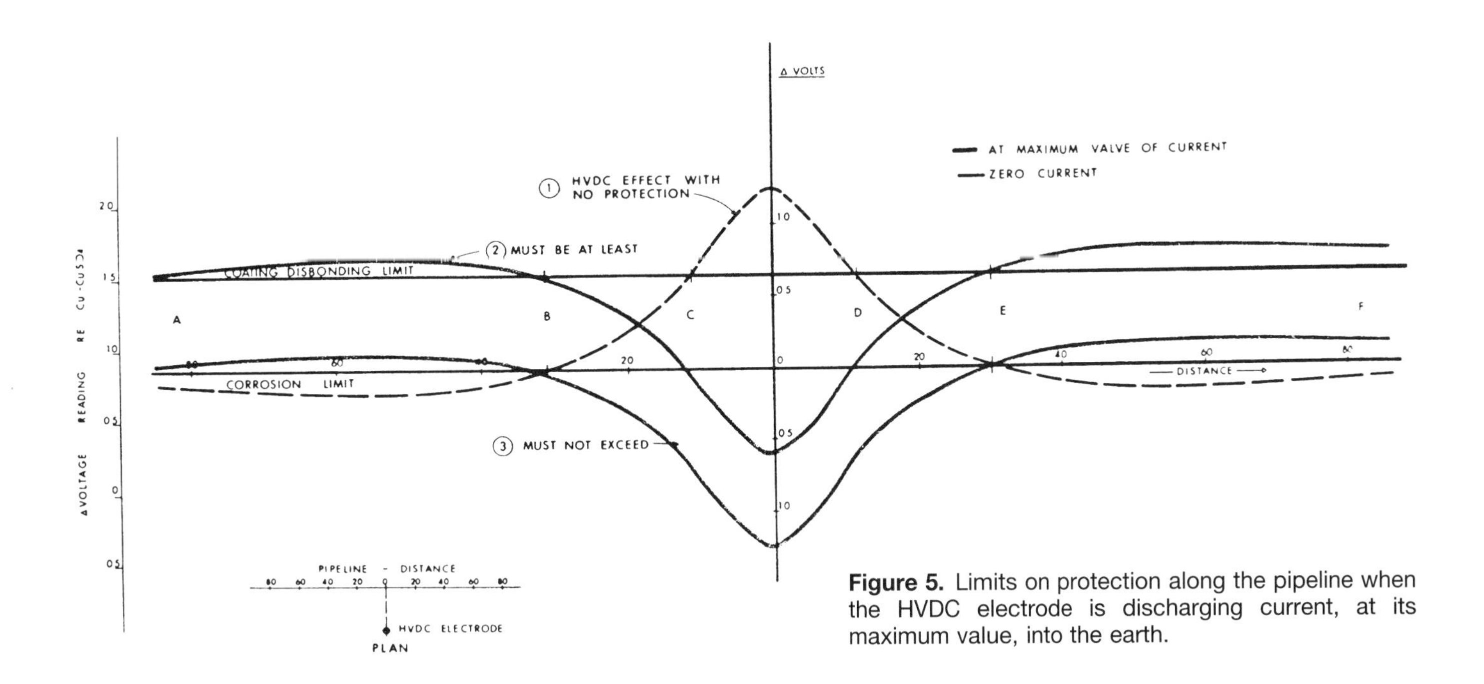

Figure 5. Limits on protection along the pipeline when the HVDC electrode is discharging current, at its maximum value, into the earth.

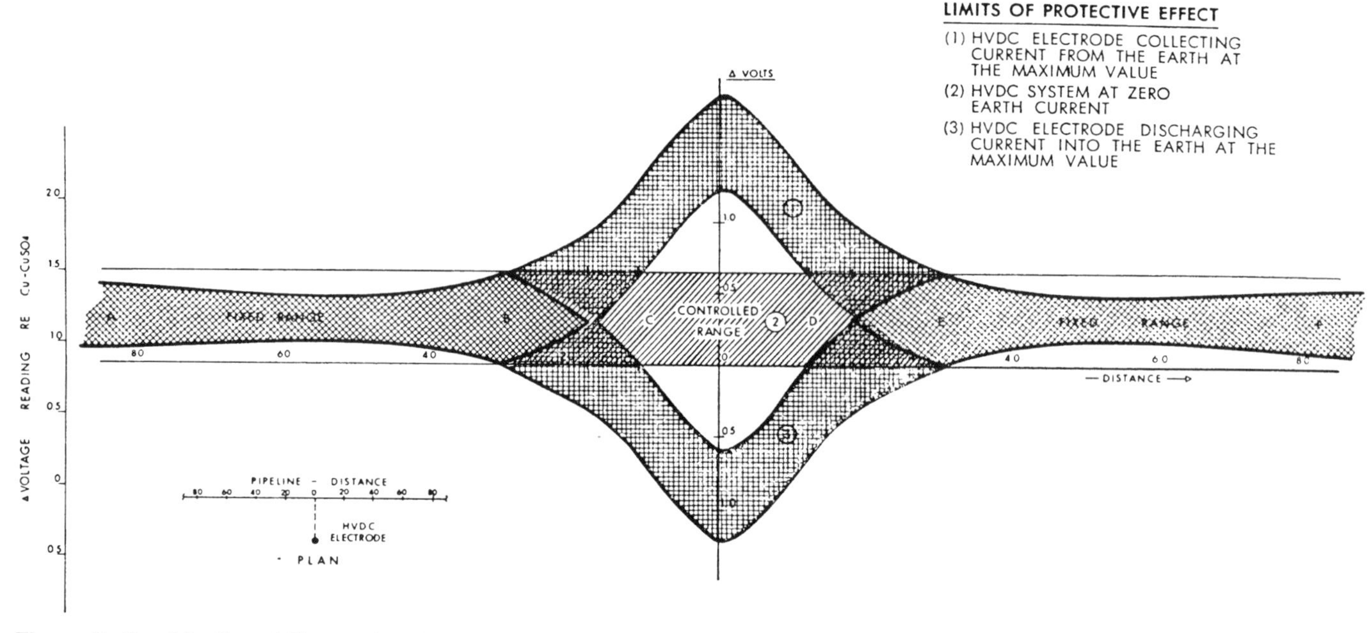

Figure 6. Combination of Figures 4 and 5. The length along the pipeline where a fixed level of cathodic protection can be used and those regions where controlled protection is required for any value of HVDC earth current. In this example, variable protection is required along the pipeline for approximately 20 to 30 miles on either side of the point closest to the HVDC electrodes.

electric field must be modified by an added field to give an acceptable total. Theoretically, in this situation, a portion of the HVDC earth current can be collected in a protective ground bed and conducted to the pipeline to satisfactorily modify the original field into an acceptable one.

Troubleshooting cathodic protection systems: Rectifier-ground bed

When the monitor, or a pipe-to-soil potential survey, indicates inadequacy of protection, the first place to look is at the protective unit. The current output of the rectifier should be checked; if it is normal, the trouble is on the line itself; if it is high, and accompanied by low voltage, the trouble is certainly on the line, and is caused either by increased current demand or by a short-circuit to parasitic metal. If the current output is low, with voltage normal or high, the trouble is in the ground bed or connecting cables. A pipe-to-soil potential over the ground bed will show a peak over every anode that is working; a disconnected anode will not show at all. This test is particularly useful if comparison can be made to a similar test made at the time of the original installation. When inactive anodes are found, only digging will uncover the cause.

If the rectifier and its anode bed appear to be performing satisfactorily, the source of the low potentials must be sought on the line itself. Any area in which work has been done recently should be investigated; for example, if a new lateral has been connected, the insulation should certainly be checked. If investigation of such suspected sources discloses nothing, then a more detailed search must be made. First, the pipe-to-soil potentials should be studied, to see if the failure seems to be localized. A more thorough, but slower, approach is that of making a detailed line current survey; find out where the drained current is coming onto the line. This will be much more easily interpreted if a similar survey, made when the line was in satisfactory condition, is on record; a comparison will often very quickly locate the offending parasite. On the other hand, if the current collected in each section is greater than on the earlier survey, with no pronounced differences, then the trouble is simply that of increased overall current requirement, probably due to coating deterioration.

How to control corrosion at compressor stations

Eight factors design engineers should consider

R. E. Hodge, Superintendent of Corrosion Control, Natural Gas Pipeline Co. of America

Corrosion failures at compressor stations result from carelessness on the part of the user, or poor choice of material/configuration by the designer.

Design engineers should consider the following:

- Coat everything underground, except ground rods, with a coating properly selected for the conditions
- Use proper coating application and inspection
- Use ground rods anodic to steel and insulated (coated) ground wires
- Avoid shielding as much as possible and allow space for corrosion control additions
- Use a minimum of insulating fittings
- Protect against water system corrosion
- Select materials carefully, with corrosion in mind, as well as the more obvious considerations
- Do not retire facilities in place—take them out

Coatings

Coatings are one of the most important considerations for control of underground corrosion. Generally, all underground metallic structures, except ground rods, should be coated. This includes gas piping, control lines, tubing, water lines, conduit, air lines, braces, etc.

There is enough buried conduit around a compressor station to soak up large amounts of cathodic protection if not coated. Non-metallic carriers can be used in some cases instead of coating.

Proper selection and use of coatings are important. The best of coatings are no good if used in the wrong place or improperly applied. The emphasis here is on the selection of coating for the temperature and environment expected.

For gas discharge lines, temperature becomes a dominant consideration. Be sure to get a coating that will withstand gas discharge temperature. Be careful, because sag temperatures listed in coating literature are not maximum operating temperatures. Very few coatings on the market can withstand operating temperatures above 120° without damage. Any bare metal underground becomes a drain on the cathodic protection system and a potential shielding problem.

There are other areas for coating consideration besides underground (e.g., tank interiors, and atmospheric coatings). These must also be matched to the situation.

Grounding

Grounds should be of materials anodic to steel. These might include magnesium or zinc anodes, or galvanized ground rods.

Copper ground rods should not be used. Copper is cathodic to steel. Some compressor stations have been installed with massive bare copper grounding systems. These are large galvanic cells with the steel gas piping as the anode (corroding element). Bare ground wires and rods also take cathodic protection current. This bad situation can be avoided by using galvanized rods, for example, and using copper ground wire with insulation. The junction of copper ground wire and galvanized ground rod must be coated. Grounding systems should be checked periodically for proper functioning.

Shielding

Shielding poses a major problem at compressor stations. It could be described as the blocking of cathodic protection due to geometric configuration.

Congested areas, where buried metal concentration is high, are difficult to cathodically protect. Piping, or other metal structures, will tend to shield cathodic protection currents from reaching all of the buried structures to be protected.

Cathodic protection ground beds of various styles have been used at compressor stations. Remote conventional beds and deep beds are popular for many applications. But NGPL has found that distributed anodes seem to be the best way to get adequate cathodic protection to all the shielded areas in a station. This type of bed can be described as a ground bed with the anodes distributed in a nongeometrical pattern, interspersed through with piping.

Designers should allow space between lines where possible; avoid congestion. This would cause less shielding and allow space for distributed anode installations where required. It would also allow space for access to work equipment for anode installations.

Insulation

Use insulating fittings (flanges, couplings, unions, etc.) only where absolutely necessary.

Hold insulating fittings to a minimum. These can often cause more problems than they solve. You might want to insulate water wells from the adjacent piping for control of cathodic protection current to the wells. This may not even be necessary if distributed ground beds are used.

Some companies insulate compressor stations from the lines entering and leaving the station. Other companies do not. This is a matter of individual company preference.

Water systems

Each water system (jacket water, piston water, boiler water, etc.) should have corrosion monitoring provisions designed into the system, such as coupons or corrosion rate probes. If there is a gas treating plant in conjunction with the compressor station, corrosion monitoring provisions should be designed into that system also.

Water treatment for corrosion control would have to be considered, depending on individual circumstances. Some treatment facilities may be designed into the systems.

Provisions should be considered for cathodic protection of the internal surfaces of storage tanks and water softeners. One problem often encountered on smaller tanks is adequate openings. Enough openings of sufficient size should be allowed on any tank for possible future work.

Water should be kept out of some systems. For instance, use air dryers on control lines.

Materials selection

Proper selection of materials is important throughout design.

Two key considerations from a corrosion standpoint are: (1) avoid the use of dissimilar metals; (2) match the material to its environment. Examples of these would include the use of a brass valve in a buried steel water line or use of mechanically strong bolts in the wrong electrolyte, resulting in subsequent failure from stress corrosion cracking.

Select materials carefully, with corrosion in mind, as well as the more obvious considerations. If design engineers do this, many subsequent failures can be eliminated.

Retirements

Many compressor station designs involve retiring something that is already there.

If something must be retired, especially underground, take it out. Many times a company pays unnecessarily to continue protection of facilities retired in place.

Even if the retired facilities can be completely disconnected from the protective systems, shielding is still a problem. Shielding is a problem by foundations also, so if they are retired, they should be removed.

Thus, attacking the corrosion problems requires a truly interdisciplinary effort. In fact, corrosion offers one of the richest fields for interdisciplinary activity.

(Based on a paper, "Construction Design to Facilitate Corrosion Control—Compressor Stations," presented by the author at AGA Distribution Conference, May 8–10, 1972, Atlanta, Ga.)

Project leak growth

The number of leaks in older pipeline systems tend to increase exponentially. Exponential growth of leaks will fit the following equation:

$$N = N_1 e^{nA}$$

where: N = Number of leaks for the study area in any year considered

N_1 = Number of leaks in the first year for which leak data are available

e = 2.718

n = nth year for which the number of leaks is to be estimated

A = Leak growth-rate coefficient

The coefficient A can be calculated from the graph drawn of the leak data for the study period. The study period should preferably be in the range of 5 to 10 years. Any data less than this will not give enough points to accurately determine the coefficient A. Before a graph is drawn, leak report quantities must be adjusted for any changes to the system during the study period. In other words, if a township was the study area, then pipe replaced, added, abandoned, or installed should be tallied. If this is not done, errors will result. For example, if leak data were collected on a township basis, and during the study period considered, 10% of the township's pipe was renewed, leak frequency in the old area will continue to rise while in the renewed area there should be none, possibly giving an overall number of leaks less than that of the year prior to the replacement. For an accurate picture and a determination of the leak growth-rate coefficient, the number of leaks in the years following the replacement should be accounted for as follows:

Let x = Percentage of replacement completed.

Let y = Actual number of leaks for the remainder of the township for the year considered

Adjusted number $N = y/(1-x)$

The leak growth rate for some pipelines may not follow the exponential curve. For example, in a particular case, a linear increase with time may be predicted. In that case, use a linear equation instead of the exponential form.

Example:

N_1 = Number of leaks in first year (i.e., 1979) = 140
N = Number of leaks in year considered (1986) = 310
n = 86−79 = 7

$310 = 140e^{7A}$

Hence, coefficient A = 0.113. The estimated number of leaks in 1990 will be:

$$N = 140e^{11\times0.113} = 485 \text{ leaks}$$

Source

Ahmad, Hayat, *Gas Pipeline Renewal Insertion Technology*, 3 and 4 (1990).

ADVANCES IN PIPELINE PROTECTION

Specialized corrosion surveys for buried pipelines: methods experience

M. D. Allen, CEng, MIM, MICorrST; **N. R. Barnes,** BTech, CEng, MIM, MiCorrST Spencer and Partners, U.K.

Synopsis

The external corrosion protection of a pipeline is commonly achieved by a system comprising an insulating coating and cathodic protection. The performance of this system is normally assessed by regular monitoring of pipe to soil potentials at selected intervals along the pipeline. Three more detailed pipeline survey procedures are available to provide additional monitoring information and the case histories described indicate the experience and value gained from their selective use.

Introduction

The external corrosion protection of a buried steel pipeline is generally achieved by a system comprising an insulating coating and cathodic protection. On modern pipelines, the coating is regarded as the primary means of protection with the cathodic protection systems providing control of corrosion where the coating has failed or has been damaged.

The performance of the cathodic protection system is usually assessed by regular monitoring of pipe-to-soil potentials at selected intervals along the pipeline.

From these results conclusions may be drawn concerning the level of cathodic protection being achieved and by inference the performance of the insulating coating. Under most circumstances, such measurements at selected locations will provide an acceptable indication of the overall level of corrosion protection being achieved on the pipeline.

These measurement locations, however, commonly at least 1 km apart, are generally selected for ease of access and thus only provide a valid assessment where pipeline coating is consistent, ensuring a constant attenuation of cathodic protection levels between each measurement point. Where the coating is variable in quality, it cannot be assumed that satisfactory levels of protection are maintained between measurement locations and a deficiency in the protection system may remain undetected allowing corrosion to occur.

Coating defect surveys have also been used to identify areas where pipeline coating quality is questionable. Historically, these surveys have almost universally taken the form of Pearson-type surveys. These surveys have entailed a two-man survey team walking the length of a pipeline identifying coating defects and subsequently marking the points with pegs for further investigation.

Surveys of this type have two inherent disadvantages, namely that the survey procedures are time consuming and no indication of cathodic protection levels at the defect location is obtained.

With these shortcomings in mind, two other survey procedures have been developed, the signal attenuation coating survey and the close interval potential survey. The former provides a rapid assessment and record of pipeline coating condition and more readily identifies areas where a defect or defects are present. Identified areas may then be surveyed in detail by close interval potential techniques that cannot only confirm coating defect locations, but also establish the status of the cathodic protection system at close intervals throughout the chosen areas.

Paper first presented at the 7th International Conference on the Internal and External Protection of Pipes. Held in London, England. Organized and sponsored by BHRA, The Fluid Engineering Centre

Methods of locating coating defects

Pearson survey

The Pearson survey is an aboveground technique used to locate coating defects on buried pipelines. An AC signal is injected onto the pipeline from a power source connected between a cathodic protection test post or valve and an earth pin. The AC frequency selected is dependent upon the type of pipeline coating in order to minimize current loss from the pipeline through sound coating. At coating defects, the AC signal leaks to earth and flows via the soil to the earth spike to complete the circuit. The localized current flow from the pipeline at the defect creates a potential gradient through the soil, and this is detected by measurement at a ground surface.

In practice, this is achieved by two surveyors walking in tandem over the entire pipeline, or section of pipeline. Earth contacts, either ski poles or foot cleats, are connected to a receiver via a cable harness. As the first surveyor approaches a coating fault, an increased voltage gradient is observed, either audibly through headphones, or visually on the receiver signal level meter. As the lead surveyor passes the defect, the signal fades, and then increases again as the rear surveyor's contacts pass over the defect.

The defect may be recorded on a preprepared record sheet complete with a measured distance from a fixed reference point or indicated by a marker peg or non-toxic paint. A recent development in Pearson equipment has been the introduction of a system that continuously records, through data logging equipment, the potential difference between each earth contact. In this way, data may be graphically presented.

Signal attenuation coating survey

This method is similar in many respects to the Pearson survey, the essential difference being the method of defect detection. The AC current injected onto the pipeline creates a magnetic field around the pipe, the strength of this field being proportional to the amplitude of the AC signal. Measurements of the field strength are made by recording the voltage generated in a sensing coil and indicated on a dB scale, when the coil is held over the pipeline. Loss of current from a coating defect creates a change in the field strength due to a change in the pipe current and this is detected as a change in output from the sensing coil.

In practice, this is achieved by one surveyor measuring field strengths at discrete, preset intervals. At each location the field strength and pipeline chainage are recorded. A graph of distance versus field strength may then be produced. Sections of pipeline having poor coating quality are indicated by the steepest gradient on the line plot. The results may also be expressed as histograms to aid interpretation. Typical results are indicated in Figure 1. Any area indicating high signal losses may then be resurveyed using the same procedures to more closely locate the coating defect(s).

Close interval potential survey

The close interval potential survey is an aboveground technique used to provide a detailed profile of the pipe-to-soil potential with distance. This profile can be used to determine detailed information concerning the performance of the cathodic protection system, the coating system, and to a lesser degree, interaction effects. The advent of portable microprocessor-based measuring equipment and computer-assisted data handling, together with accurate synchronized high-speed current interruption devices, have allowed these surveys to provide accurate, practical data and collection methods.

To eliminate errors in the pipe-to-soil potentials recorded, arising from the inclusion of volt drops caused by the flow of cathodic protection current in the soil, synchronized current interruptors are employed in the cathodic protection units. Pipe-to-soil potentials are then measured instantly after the current is interrupted, before significant chemical depolarization has occurred, thus providing an "error-free" polarized potential.

In practice, the precision current interruptors are best fitted in the DC output circuit of the cathodic protection power supplies, prior to commencing the survey to interrupt the cathodic protection current for short cyclic periods. The cycle time is chosen with regard to pipeline characteristics, although a range of "ON" and "OFF" periods and "ON/OFF" ratios are available.

For this survey a variety of purpose built-voltmeter/microprocessors are available that record and store the pipe to soil potential data with an additional facility to store coded entries of topographical or pipeline features. This information is transferred to floppy disc storage from which graphs of ON and OFF pipe-to-soil potentials against pipeline chainage are produced. A typical close interval potential graph is illustrated in Figure 2.

It will be apparent from the description of the pipeline surveys that the form of results are different for each survey:

Pearson

Results are not generally presented in graphical form. Coating defects are marked in the field; in addition, the detection of defects is dependent upon operator skill. Only if the recording Pearson technique is used can data be presented in graphical form. Experience of this technique is at present limited.

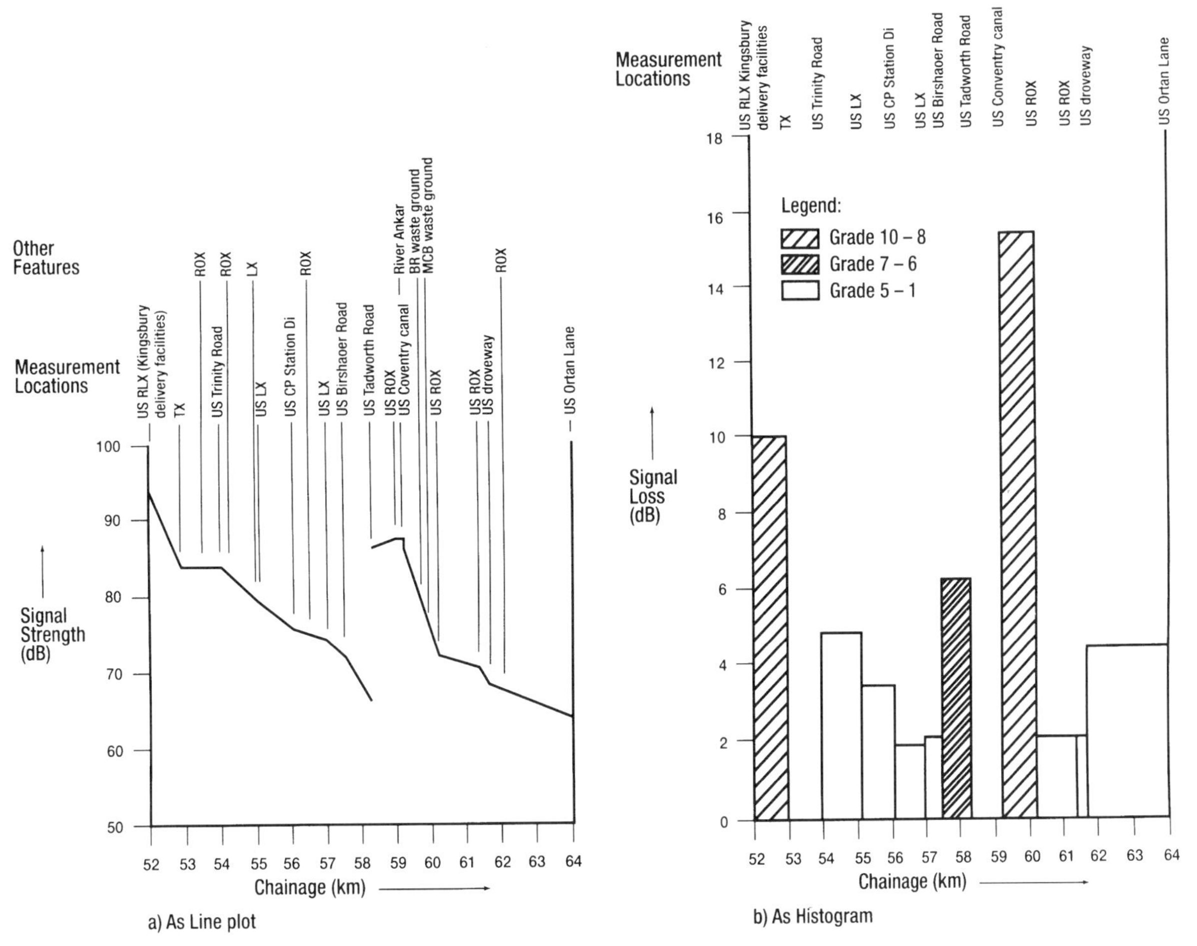

Figure 1. Typical signal attenuation coating graphs.

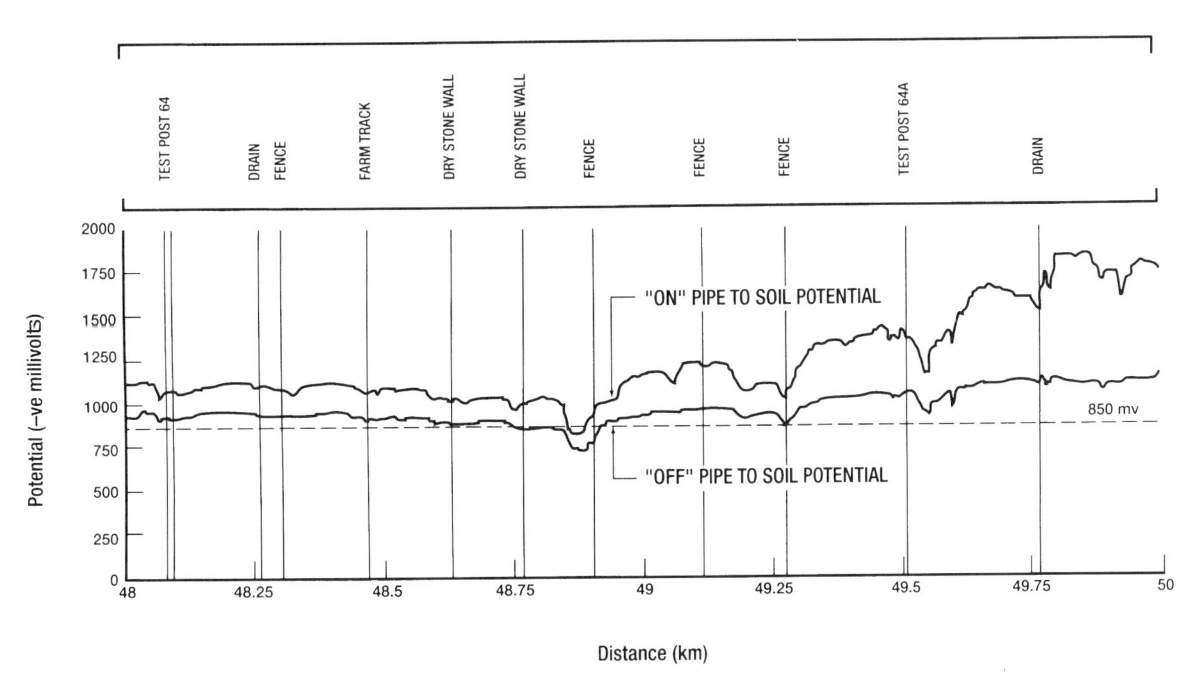

Figure 2. Typical close interval potential graph.

Signal attenuation coating survey

Results are presented in graphical form, although quick analysis may be performed in the field from raw data. Areas of poor coating quality may be readily identified from final graphs.

Following the production of graphs and histograms, an evaluation system employed in pipeline coating quality assessment may assign values to the level of signal loss and the measurement distance interval. The sum of the two values referred to as the "dB factor" and the "km factor" may be used to indicate the future action required and the level of priority for each section surveyed.

The evaluation of the coating quality is illustrated in Figure 1. It should be noted, however, that gradings assigned to one particular pipeline cannot be compared to those assigned to another.

Close interval potential survey

Results are shown in graphical form. Defect location is undertaken after the survey and requires experienced personnel to interpret the results.

The achievement of cathodic protection on the pipeline is indicated by the "ON" and "OFF" potential both being more negative than the accepted criteria.

Pipeline locations with inadequate cathodic protection will be indicated by the graph "ON" and "OFF" potentials shifting to a less negative level and become cause for concern if the "OFF" potentials are more positive than, say, −0.85 V.

It is considered that a localized coating defect would normally be indicated by a sharp positive shift of the "ON" potential plot coupled with a plot shift of smaller amplitude of the "OFF" potential, i.e., that the potential difference (IR drop) between the "ON" and "OFF" reading decreases, indicating a reduced coating resistance at that point.

The "OFF" potential plot change would not be as large as that of the "ON" potential, as the "OFF" potential would be the polarized (or partially polarized) potential of the bare steel. This differs from the case where insufficient current for the achievement of polarization is available.

In all these cases, the IR drop measured between the "ON" potentials and the "OFF" potentials includes, along with the resistance of the coating, the resistance of the soil. The resistance of the coating is assumed to be the largest component of this resistance value.

Current is also a factor of the IR value. The current density impressed on the pipeline will increase as the cathodic protection stations are approached, giving the largest "I" value of the IR drop near the location of the negative connection.

In order to interpret the data obtained fully, the anomalies observed in the plotted results may be classified into a number of categories relating to "point" anomalies or "general area" anomalies. For example, the following categories may be used.

Classification	Definition
Type I	Both the "ON" and "OFF" potentials are more positive than preset criteria. There is little or no "IR" voltage. The potentials are unsatisfactory, and the locations are not cathodically protected.
Type II	The "OFF" potential is more positive than a preset limit, the "ON" potential more negative. There is a reduced level of protection indicated at "point" locations, which may be improved by increased output from the nearest cathodic protection station, although where this occurs over long pipeline lengths, it indicates high-resistance coating or soils.
Type III	Positive change in the "ON" and "OFF" potential plot where both remain more negative than the selected criteria. This indicates increased current demand at these locations but protection is maintained at a satisfactory level.
Type IV	Unusual potential variations due to outside influences, e.g., stray currents, station earthing, AC induced voltages.
Type V	A change in level of the "ON" potential at a test post due to the change in the length of that part of the measuring circuit passing current, i.e., increasing "IR" volt drop due to increasing nine length in circuit.

Of the five categories listed above, the first three are indicative of coating defects. Types IV and V are commonly observed on close interval potential survey results and require careful interpretation.

It is important that these considerations are borne in mind when selecting the survey or combination of surveys to be employed and, in addition, the choice of which survey to employ and when to employ it is dependent on a number of factors, namely:

- The purpose of the survey
- The condition of the pipeline
- The access over the pipeline
- The time allowed

As the protection system relies on both the external coating and a cathodic protection system, these may be considered as first and second lines of defense against corrosion. It is possible to take the position that concern over damaged coating is not warranted as backup from the cathodic protection system is available, and therefore the integrity of the pipeline is not at risk. If this attitude is taken, then obviously close interval potential surveys would seem the most appropriate and perhaps the only survey system to consider.

Experience has shown, however, that the repair of significant coating defects will benefit the overall level of cathodic protection achieved, possibly permitting reduction in output levels and/or increasing the distance protected from any one protection station. In addition, the maintenance of a "two-level" protection system also ensures the highest integrity is maintained for the pipeline. With this purpose in mind, the advantages of the coating surveys become apparent.

If we consider the requirements of a newly laid pipeline, it has been considered important to ensure as near perfect a coating as is possible at the start of the pipeline's life. It has therefore been common practice to "Pearson" survey the pipeline after installation. As described earlier, this survey requires a two-man team and commonly progresses at some 4–6 km per day. This procedure may be accelerated by the use of the signal attenuation coating survey. Bearing in mind the probably high-quality coating of a new pipeline, and the comparison basis employed by the signal attenuation coating survey, even small coating defects will become readily noticeable.

At the other extreme, with poorly coated old pipeline, a Pearson survey may have an almost continuous response, making it impossible to discern individual defects or the most significant coating breakdown. The signal attenuation coating survey would, however, be able to give a rapid assessment of coating condition and indicate where limited rewrapping would be of greatest benefit.

In general, therefore, experience has shown that it is cost effective for the more detailed surveys of coating condition and cathodic protection to be undertaken in conjunction with a signal attenuation coating survey. A number of case histories are described below.

Case histories

1. Pipeline: Multiproducts UK
   Length: 115 kilometers
   Built: 1982
   Coating: Coal tar reinforced enamel with self-adhesive tape at field joints.
   Survey: Signal attenuation coating survey
   Survey period: 9 days (4 days for disconnection of all "temp" magnesium anodes, 5 days for survey).
   Survey Results:
   Survey indicated that over some 80 km of the pipeline the coating was satisfactory and further survey work was not recommended for those areas. Signal losses over these sections ranged between 1.5 dB/km and 3 dB/km. Some 9 sections, amounting to a total of some 10 km were given first priority for further investigation. Signal losses were as high as 17.6 dB over distances of 1.5 km. Two further sections, amounting to some 21 km, were classified as second-priority items.
   It was decided that this information gathered be correlated with the results obtained from a proposed intelligent pig survey before further action.
2. Pipeline: White Oil products, UK
   Length: 115 kilometers
   Built: Approx 1963
   Coating: Coal tar enamel, tape or bitumen flood field joints
   Survey: Signal attenuation coating survey
   Pearson survey
   Close interval potential survey
   Survey Period: 10 days—Signal attenuation coating survey
   1 day—Pearson survey
   4 days—Close interval potential survey
   Survey Results:
   Initial signal attenuation coating survey identified 30 locations of significantly high signal losses; of these, 12 were attributed to interpipeline bonds and sacrificial anodes, and the remaining 18 were subject to more detailed investigation.
   Close interval potential surveys showed that a satisfactory level of cathodic protection was being achieved at each of the locations where defects had been pinpointed using Pearson equipment.
   Three particular locations were excavated; in the first, a coating defect of some 800 mm associated with a welding arc burn was found within an area previously showing a signal loss of 6 dB over less than 100 m section. In the second area, showing a similar signal attenuation, a number of small coating defects were found together with a box of welding rods touching the pipe. At the third location a badly poured bitumen field joint had been indicated by a signal attenuation loss of 10 dB.
3. Pipeline: Oil, UK
   Length: 210 kilometers
   Built: Approx. 1970
   Coating: Coal tar enamel with tape-wrapped field joints.

Survey: Close interval potential survey Recording both "ON" and "OFF" potentials at 5 meter intervals.
Survey period: 40 days for 3 man team
Survey results:
Review of all the close interval potential results highlighted some 28 locations where both the "ON" and "OFF" potentials were less negative than the protection criteria. Of these 28 locations, 80% would not have been apparent from potential measurements recorded at adjacent test posts, thereby allowing the integrity of the pipeline to be put at risk where these unsatisfactory levels of cathodic protection coincided with coating defects.

4. Pipeline: Aviation Fuel 6″, UK
   Length: 5 kilometers
   Built: 1960
   Coating: Asphalt
   Survey: Signal attenuation survey
   Survey period: 2 days
   Survey results:
   The survey showed consistently high signal losses, indicating a generally deteriorated coating. Signal loss values were of the order of 40 dB per km.
   More than 50% of the total signal loss was concentrated over four areas, amounting to some 1,720 meters of pipe. Correlating close interval potential survey results over the same areas allowed 315 meters to be selected for rewrapping.
5. Pipeline: Crude Oil, Middle East
   Length: 132 kilometers
   Built: 1986
   Coating: Extruded polyethylene
   Survey: Signal attenuation coating survey
   Survey period: 4 days
   Survey results:
   Generally, signal losses were found to average less than 1 dB per kilometer. One location was found to indicate a signal loss of 14 dB over 40 m. More detailed signal attenuation survey work located two possible coating defects with signal losses of 10 dB over 20 m and 48 dB over 20 m. A subsequent Pearson survey and two excavations revealed 5 small defects in a shrink sleeve weld joint coating for the 10 dB loss and 1 defect less than 100 mm^2 in the factory-applied coating for the 4.8 dB.

Conclusion

It is considered that the discussion and case histories illustrated have shown that each of the aboveground survey procedures offer certain advantages and disadvantages, and that many of the technical and economic penalties of choosing any one specific survey may be overcome if used in conjunction with another.

References

J. Pearson, *Petroleum Engineer* 12(10) 1941.

CCEJV "Cathodic Protection Monitoring of Buried Pipelines"—State of the Art Report by Task Group E4-2.

N. Frost, "Monitoring Pipeline Corrosion." *Pipes and Pipelines International*, May 1986.

Source

Jones, Glyn, and Thron, Joanna, *Advances in Pipeline Protection*, Gulf Publishing Co., 1990.

C. J. Argent et al. "Coating defect location by above ground surveys"—proceedings of UK corrosion, 1986.

Estimate the number of squares of tape for pipe coating (machine applied)

$$\text{Squares of coating} = \frac{W \times A}{(W - O) \times 100\,\text{sq. ft/square}}$$

where: W = Width of coating, inches
A = Pipe surface area, square feet
O = Overlap, inches

Example:

Pipe OD = 6⅝″
Length = 1 mile
Tape width = 6″
Overlap = 1″

Pipe surface area = 1.73442 sq. ft/ft (Table 7 Pipe Data Section 3)

Total surface area = 1.73442 × 5280 = 9,157 sq. ft

$$\text{Squares of coating} = \frac{6 \times 9157}{(6 - 1) \times 100} = 110 \text{ squares}$$

Estimate the amount of primer required for tape

Multiply pipe surface area by 0.001047 to obtain the number of gallons of primer required per mile of pipe.

Example:

Pipe OD = $6\frac{5}{8}''$
Length = 1 mile

Pipe surface area = 1.73442 sq. ft/ft (Table 7 Pipe Data Section 3)

Total surface area = 1.73442 × 5280 = 9,157 sq. ft

Gals of primer = 9,157 × 0.001074 = 9.8 gals. (Use 10 gals.)

Tape requirements for fittings

Reducer Size	Tape Width	Linear Feet	Square Feet	Tee Size	Tape Width	Linear Feet	Square Feet	90° LR Ell Size	Tape Width	Linear Feet	Square Feet
3 × 2	2	9.2	1.5	2	2	13	2.2	2	2	9	1.5
4 × 3	3	8.4	2.1	3	2	21	3.5	3	2	11	1.8
6 × 4	4	12	4	4	2	28	4.6	4	3	14.6	3.6
8 × 4	6	14	17	6	3	38	9.5	6	4	26	8.6
10 × 6	6	18	9	8	4	45	15	8	4	40	16.5
12 × 8	6	25	12	10	6	34	17	10	6	48	24
14 × 10	6	26	13	12	6	50	25	12	6	64	32
16 × 12	6	38	19	14	6	60	30	14	6	81	45
18 × 14	6	50	25	16	6	71	35.5	16	6	101	50.5
20 × 16	6	66	33	18	6	85	42.5	18	6	123	61.5
				20	6	100	50	20	6	146	73
				22	6	126	63	22	6	173	86.5
				24	6	132	66	24	6	201	100.5

Tape requirements for fittings, continued

90° SR Ell Size	Tape Width	Linear Feet	Square Feet	45° Ell Size	Tape Width	Linear Feet	Square Feet
2	2	7	1.5	2	2	5.6	0.93
3	2	10	1.8	3	2	9	1.5
4	3	13	3.25	4	3	13	3.2
6	4	21	7	6	4	17.3	5.75
8	4	38	13	8	4	36	12
10	6	46	23	10	6	34	17
12	6	60	30	12	6	48	22
14	6	76	38	14	6	52	26
16	6	96	48	16	6	64	32
18	6	118	59	18	6	76	38
20	6	136	68	20	6	96	48
22	6	166	83	22	6	110	55
24	6	190	95	24	6	126	63

INDUCED AC VOLTAGES ON PIPELINES MAY PRESENT A SERIOUS HAZARD

by Earl L. Kirkpatrick, PE., ELK Engineering Associates, Inc., Fort Worth, TX, and Kirk Engineering Co., Oneonta, AL

The problem of induced AC voltages on pipelines has always been with us. Early pipeline construction consisted of bare steel or cast iron pipe, which was very well grounded. Bell and spigot, mechanical, or dresser-style joint couplings often were used, creating electrically discontinuous pipelines that are less susceptible to AC induction. Although induced AC affects any pipeline parallel to a high-voltage alternating current (HVAC) power line, the effects were not noticeable on bare pipelines. With the advent of welded steel pipelines, modern cathodic protection (CP) methods and materials, and the vastly improved quality of protective coatings, induced AC effects on pipelines have become a significant consideration on many pipeline rights-of-way.

In the past 2–3 decades, we have been seeing much more joint occupancy of the same right-of-way by one or more pipelines and power lines. As the cost of right-of-way and the difficulty in acquisition, particularly in urban areas, have risen, the concept of joint occupancy rights-of-way has become more attractive to many utility companies. Federal and state regulations usually insist on joint-use right-of-way when a utility proposes crossing regulated or publicly owned lands, wherever there is an existing easement. Such joint use allows the induced AC phenomena to occur and may create electrical hazards and interference to pipeline facilities. Underground pipelines are especially susceptible if they are well-coated and electrically isolated for CP.

Induced AC mechanisms

There are three coupling modes of primary concern by which voltages are induced in pipelines.

Electrostatic coupling, commonly known as "capacitive" coupling, is caused by the electrostatic field surrounding the energized conductor. This electrostatic field is proportionally produced by the voltage in the overhead conductor. The closer the pipeline is to the energized overhead conductor, the stronger the electrostatic voltage induced in the pipeline.

Electrostatic coupling is of primary concern when a pipeline is under construction near an overhead HVAC system. The length of the exposed conductor is also a factor in the electrostatic charge induced in the structure. A relatively short conductor isolated from ground, such as a single joint of line pipe supported by a nonmetallic sling or on a rubber-tired vehicle on the right-of-way, will become charged like a capacitor. (A rubber-tired vehicle on the right-of-way should have drag chains to ground the vehicle.)

Once the pipeline is backfilled, electrostatic coupling is no longer of any real concern because the pipeline coating will allow sufficient leakage to earth through pinholes or holidays, which will effectively ground out the electrostatic effects.

Electromagnetic coupling

Electromagnetic coupling is also known as "transformer action." When current flows in an energized conductor, it produces an electromagnetic field at right angles to the conductor. In AC power systems, electric current flowing in the conductor changes direction 120 times per second for 60-Hz and 100 times per second to 50-Hz systems. Thus, the electromagnetic field surrounding the energized conductor is constantly expanding and contracting.

Whenever electromagnetic lines of force cut through another conductor, a voltage is induced in that conductor. This is the principle upon which power transformers, alternators, and generators function. A well-insulated pipeline parallel to an overhead HVAC line then becomes the secondary of an air core transformer. With sufficient parallel length, very significant voltages and currents can be induced into the pipeline. These voltages can be hazardous to anyone who comes in contact with the exposed pipeline or appurtenances. The parallel power lines may induce sufficient power into the pipeline to damage the pipeline or related facilities.

Pipeline hazards

A products pipeline failed because of lightning and fault current actually penetrating the pipeline wall. This was a clear case of arc burn penetration of the pipe wall. There are other case histories where actual pipe wall penetration has occurred with or without a fire being started. People have been killed by contacting an energized pipeline under construction. We must be aware of these electromagnetic and electrostatic effects on influenced pipelines and know how to avoid injury and damage. An excellent source of information on this subject is NACE Standard RP0177-95. Additional information is contained in IEEE Standard 80.

Resistive coupling

Resistive coupling is another mode of electrical coupling between a pipeline and a parallel overhead power system. This is a concern during "fault conditions" on a power system. If lightning strikes one of the energized conductors on the overhead power line, the resulting voltage rise on the wire will exceed the breakdown insulation level (BIL) of the insulator at the nearest tower. When the BIL is exceeded, a flashover will occur from the energized conductor to the tower and then fault current will flow through the tower to the tower ground. Current will flow from the energized conductor to the tower structure via the ionized gases (plasma) generated by the lightning. Fault current will flow through the tower and the tower ground into the earth for a fraction of a second, until the circuit protection device has a chance to operate.

As a result of these fault currents, current will radiate from the tower foundation and grounds in all directions from the faulted tower. A very severe potential gradient will occur across the earth, radial to the faulted tower. If there is a nearby pipeline in the earth, the gradient field will be distorted and accentuated. This effect is greatest with bare pipelines and pipelines that are near the tower foundation. Consequently, a significant portion of the fault current may flow in the pipeline as a result of resistive coupling.

Whenever a coated pipeline and an HVAC transmission circuit are near each other, the magnetic field associated with the currents in the power transmission line will induce a voltage in the pipeline.

Physical separation between the pipeline and the power line tower is an important factor in determining the extent of the damage that will occur on the pipeline as the result of a fault.

Predicting AC voltages on a pipeline

Induced AC voltage on a pipeline can be measured by methods similar to those used to conduct a DC pipe-to-soil cathodic protection potential survey. A multimeter is placed on the appropriate AC voltage range and a steel pin is used in place of the copper/copper-sulfate reference electrode. A detailed AC pipe-to-soil potential survey should be conducted over any area where induced AC voltages are suspected. It is important to note the time of each reading because many power companies maintain a record of currents in the circuit. The voltage induced in the pipeline is directly proportional to the line currents in the overhead conductors. This can provide valuable information to assist in calculating the peak voltages that can be anticipated at each location under maximum line loading conditions.

Whenever a coated pipeline and an HVAC transmission circuit are near each other, the magnetic field associated with the currents in the power transmission line will induce a voltage in the pipeline. The actual magnitude of the induced AC voltage depends on many factors, including the overall configuration of all the structures involved, soil resistivity, pipecoating effectiveness or resistance to remote earth, pipeline propagation constant, magnitude of the line currents in the power circuit(s), and any current imbalance between the phases.

The magnitude of steady-state AC potentials induced on an underground pipeline by parallel high-voltage transmission lines can be estimated accurately using the appropriate mathematical formulas. The formulas characterize the circuit in terms of "steady-state" line currents, phase relationships, pipeline-to-conductor distances, pipeline propagation constants, characteristic impedances, soil resistivity, and other factors. The technique can predict, with reasonable accuracy, the areas where the maximum AC potentials will occur and approximate the actual induced voltage at that point. These calculations require the use of advanced mathematics, which can best be done on a computerized workstation by an experienced professional.

Mitigation techniques

Mitigation techniques that can control induced voltages on an influenced pipeline include the following.

- Supplemental grounding of the pipeline with sacrificial anodes or other grounding means.
- Bonding the pipeline to individual power line pole grounds or towers through the use of polarization cells.
- Installation of parallel mitigation wires bonded to the pipeline at regular intervals.
- Bonding the pipeline to purposefully designed "made" grounds.
- Changing phase relationships between multiple power line circuit conductors.
- Use of Faraday cages with sacrificial anodes.
- Relocation of the pipeline or power line to provide greater separation from the influencing power system.
- Installation of a nonmetallic pipeline such as high-density polyethylene pipe, if design pressures permit.
- Installation of gradient control electrodes or mats at all aboveground appurtenances.
- Security fencing around aboveground appurtenances do not mitigate induced voltages, but they do limit access to the structure.

Whenever there is a pipeline closely parallel to a power line for any significant distance (e.g., closer than 500 ft or more than 2,000 ft), further investigation is warranted. If 5 volts AC or more is measured at any point on the parallel section, this voltage should be related to line loading to estimate the pipeline-induced AC voltage at peak line-loading conditions. The induced voltage should be mitigated if the calculated values approach or exceed 15 volts AC. Fault conditions and step-and-touch voltages also must be considered.

Personnel safety

On any construction or maintenance project, safety is an attitude. This attitude is developed by proper training. A valuable source for such information is Section 4 of NACE Standard RPO177. NACE also offers an audiovisual presentation that deals with the effects of electric shock in such situations. RP0177 should be included in the construction specifications whenever a pipeline is built or exposed for maintenance on an energized HVAC right-of-way. One of the inspectors should be designated in charge of electrical safety. This individual must be familiar with and properly equipped to test for safe levels of induced voltage in the pipeline. A safety meeting should be held prior to construction with all construction employees to discuss electrical safety requirements.

The right-of-way must be vacated if thunderstorms come within 10 miles of the pipeline right-of-way or the power line(s) that are influencing the pipeline. Never work on an influenced pipeline during a thunderstorm because of the potential for a direct lightning strike. Power system fault currents will flow in the earth if lightning causes flashover of a power line insulator. A person on the right-of-way is in danger, even if they are not touching the pipeline.

Corrosion and maintenance personnel should be very cautious about stringing test lead wire on the right-of-way where the lead wire parallels the influencing HVAC line. Extremely hazardous voltages can be induced on a significant length of test lead wire laid on top of the ground.

The practice of using insulating rubber gloves should also be discouraged unless the work is being performed by a trained power company employee. "Rubber Goods" must be specially cared for and tested to ensure their reliability.

The only safe alternative in the pipeline environment is to test the level of induced AC on the pipeline before contacting anything that may be a conductor. If voltages are safe, normal measurement or repair and maintenance techniques may be employed. If unacceptably high voltages are encountered, one must work on a ground mat. A ground mat can be as simple as a piece of 6-ft road mesh laid on top of the ground and bonded to the pipeline or appurtenance with an automotive jumper cable. Since the individual, the ground mat, and the pipeline appurtenance are all at equal potential, it matters little as to how high the actual measured voltage is on the pipeline.

Reprinted with permission—*Pipeline & Gas Journal.*

MEASURING UNWANTED ALTERNATING CURRENT IN PIPE

William H. Swain, *William H. Swain Co.*

Alternating current (AC) flowing on a pipeline can be a portent of trouble, such as corrosion or harm to persons. Clamp-on ammeters represent an accurate way to measure AC. AC voltage drop measurements on a pipeline can be seriously erroneous. This article describes the use of clamp-on AC ammeters for interference and fault location, corrosion control, and personnel safety purposes.

Alternating current (AC) is best measured with a clamp-on AC ammeter. Accuracy is important in terms of pipe current and stray pickup, especially in a common corridor shared with an overhead electrical transmission line.

AC flowing in a pipe can also be detected by using the mV drop method wherein the pipe itself is used as a shunt—the voltage drop indicating current flowing in the pipe. An advantage of this method is simplicity. A disadvantage is that steel pipe is both resistive and inductive. The inductive component can cause critical errors.[1] Moreover, accurate calibration is difficult, if not impossible.

Structure of a clamp-on AC ammeter

Figure 1 is an outline of a clamp-on AC ammeter. The aperture of the clamp may range from ¾ to 60 in. (2 to 152 cm) in diameter.

Reference 1 gives the equations relating input current (i_i), winding turns (N_s), and output voltage (v_o), along with further information about clamp-on ammeters.

An undesired current will likely have an identifying characteristic (i.e., a "signature"). An oscilloscope connected across v_o in Figure 1 will show the waveform of i_i. The frequency of i_i can be measured on a portable meter, such as a Fluke 87†, or estimated with earphones. Another output is a precision rectifier. It is used for driving a data logger to record the change of i_i over time.

Some cathodic protection (CP) is pulsed at a fairly high frequency. This should have a characteristic sound when earphones are used. If the current is pulsed, this will show on an oscilloscope. Frequency and duty factor can be measured on the meter.

A foreign water pipe contact or underground residential distribution (URD) interference will likely have a strong signal—60 Hz in the U.S. or 50 Hz in some other nations. Either headphones or the meter will measure frequency.

One owner reported sharp increases and decreases in 60 Hz i_i in a large-diameter gasoline pipeline. An intermittent short to a water pipe carrying 60 Hz ground current was finally located. Every time a heavy truck went over the pipe, contact to a water line was made, then broken. A fairly deep pit was found in the gasoline pipe at the point of contact.

†Trade name.

Accuracy of a clamp-on ammeter

A typical measurement is accurate at 50 to 60 Hz to within ±1% of reading, ±3 least significant digits, and ± stray pickup. Frequency response varies with range sensitivity and clamp size. Figure 2 shows results on a 12-in. (30-cm) clamp, and Figure 3 shows results on a 49-in. (125-cm) clamp.

In the top curve of Figure 2, the sensitivity or transfer resistance (R_T) at 25 mA full-scale input is flat within ±2% (0.2 dB) from 17 to 5,500 Hz. This was measured on a 12-in. clamp with an analog indicator.

The lower curve in Figure 2 is similar, but the sensitivity is reduced so that the full-scale i_i is now 250 mA. The frequency response is flat within +1% and −3% from 17 to 5,500 Hz.

Figure 3 is similar to Figure 2 except that the clamp was much larger. When tested with an analog indictor, the

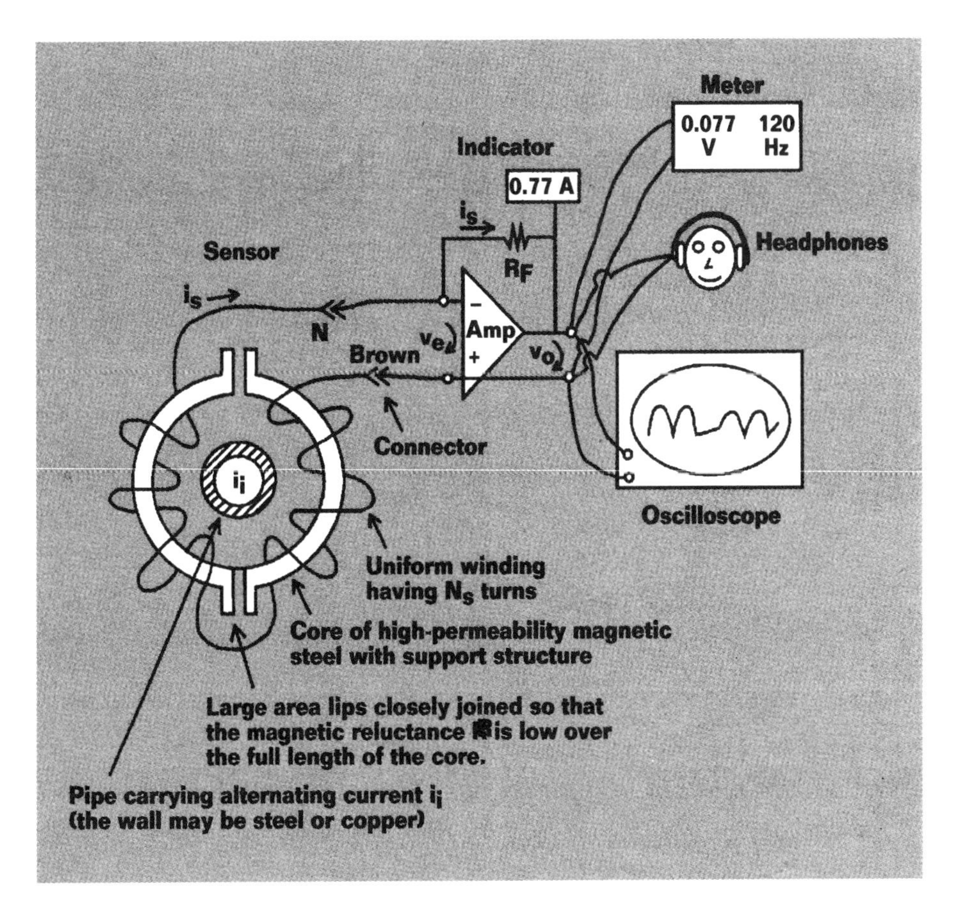

Figure 1. Clamp-on AC ammeter structure.

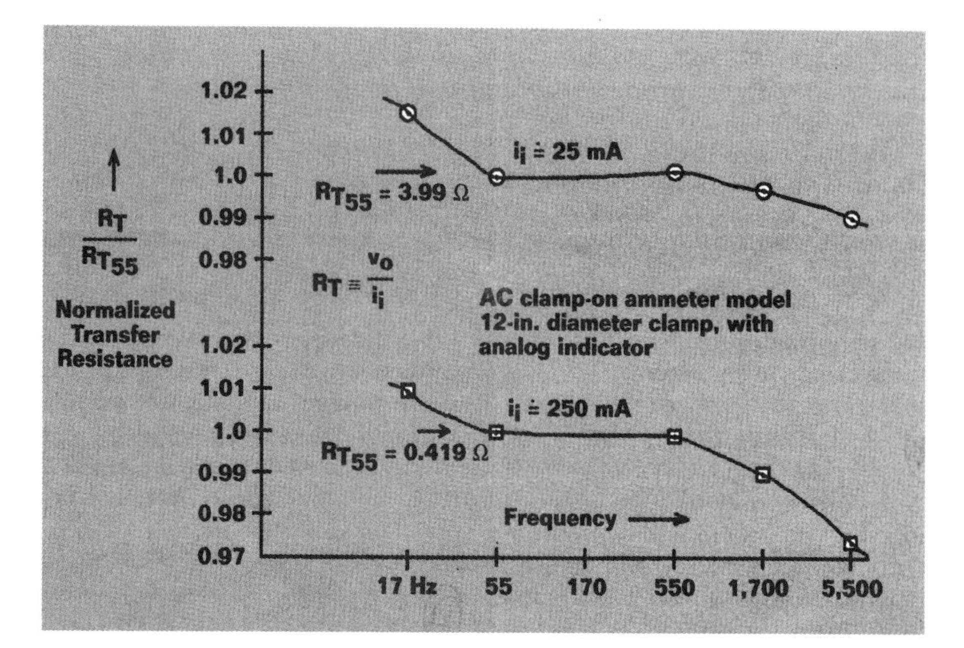

Figure 2. Transfer resistance (R_T) versus frequency, 12-in. clamp.

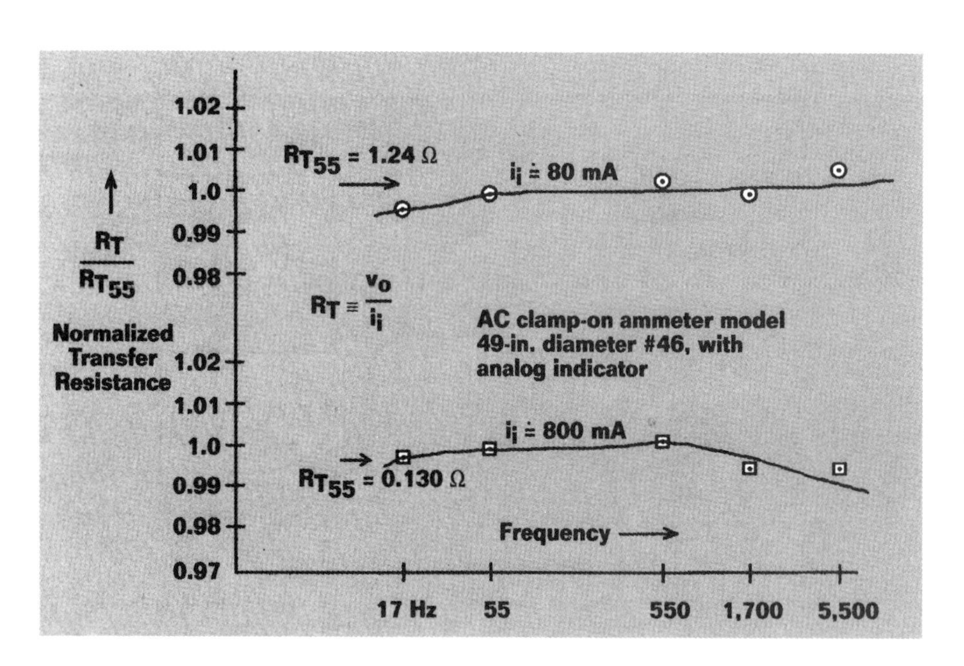

Figure 3. Transfer resistance (R_T) versus frequency, 49-in. clamp.

frequency response was flat within ±1% from 17 to 5,500 Hz.

The top curve in Figure 3 is for an i_i of 80 mA passing through the aperture of the 49-in. clamp. Flat frequency responses to within ±1% enable the user to get a more accurate picture of the current waveform with an oscilloscope, etc. The lower curve shows the response at 800 mA.

Stray pickup

The clamp-on AC ammeter shown in Figure 1 accurately measures i_i, but it is not perfect. Especially when i_i is small, the clamp may have an output current as a result of stray alternating magnetic fields. Generally, this stray pickup is small and represents an error <1% of measured current.

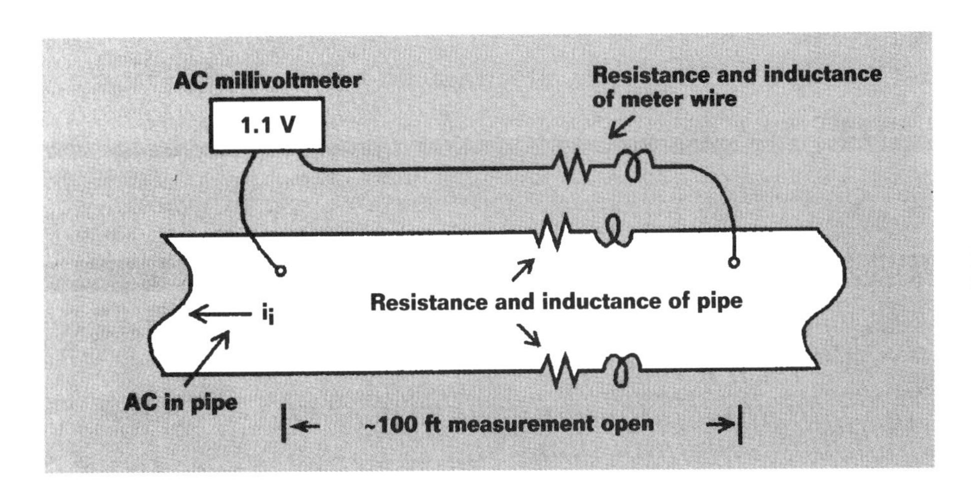

Figure 4. Structure of the mV drop method for AC.

Reference 1 provides specific data and a discussion of the likely effect of stray pickup. It is probably less than a fraction of an ampere of equivalent i_i in clamps on a pipe in a common corridor with a three-phase transmission line having up to 1,000 A in the lines.

Uses of a clamp-on AC ammeter

Interference and fault location, corrosion control, and personnel safety are important considerations when using clamp-on ammeters.

Interference and fault location. A corrosion control professional may find that there is direct current (DC) interference flowing in a line but does not know where it is coming from. The wave form or audio tone of the interfering current is a clue to the probable source.

A 24-h record of i_i will likely show a constant interference current if it is from a foreign CP system. Recurring peaks and valleys can point to interference from an electrified railroad. A 25-Hz current, for example, has been observed on a pipeline near an AC-operated railroad.

Corrosion control. Gummow has written about corrosion resulting from alternating current.[2] He notes that studies in the mid 1980s found that, above a certain minimum level of AC current density, normal levels of CP will not provide acceptable levels of AC corrosion control.

There are also concerns over corrosion problems in copper and steel freshwater piping systems associated with unwanted AC caused by grounding practices. It was reported that in a large hospital, pipe replacements were required after only 2 to 3 years because AC current was found on the piping; a pipe current >75 mA rms was deemed a hazard. The DC offset associated with nonlinear loads may have caused the problem.

Personnel safety. Gas and oil transmission pipelines are frequently laid in a corridor adjacent to high-voltage transmission lines. There can be considerable current induced in pipe in such a corridor. This can be a hazard to both property and personnel.

mV drop method

Figure 4 illustrates the mV drop method. Here, the pipe itself is used as a shunt for measuring AC flowing in the pipe. This is similar to the method that is used for measuring DC flowing in the pipe, except that the mV potential is measured using an AC millivoltmeter.

If the resistance of the pipe over the span is 0.5 mΩ and the mV meter reads 1.1 mV, one may at first think that the pipe current is 2.2 A. This would be correct for DC, but there is inductance—skin effect and stray pickup need to be considered.

Inductance

If the reactance of the ~100-ft (30.5-m) measurement span is 0.3 mΩ at 60 Hz, this adds in quadrature to the 0.5 mΩ resistance (R), so the impedance (Z) becomes 0.58 mΩ—a 16% increase. At 60 Hz, skin effect may add even more Z.

At 120 Hz, the major ripple frequency for most full-wave rectifiers, the reactance is likely doubled to 0.6 mΩ, so Z increases to 0.78 mΩ, a 56% increase. Skin effect is also greater.

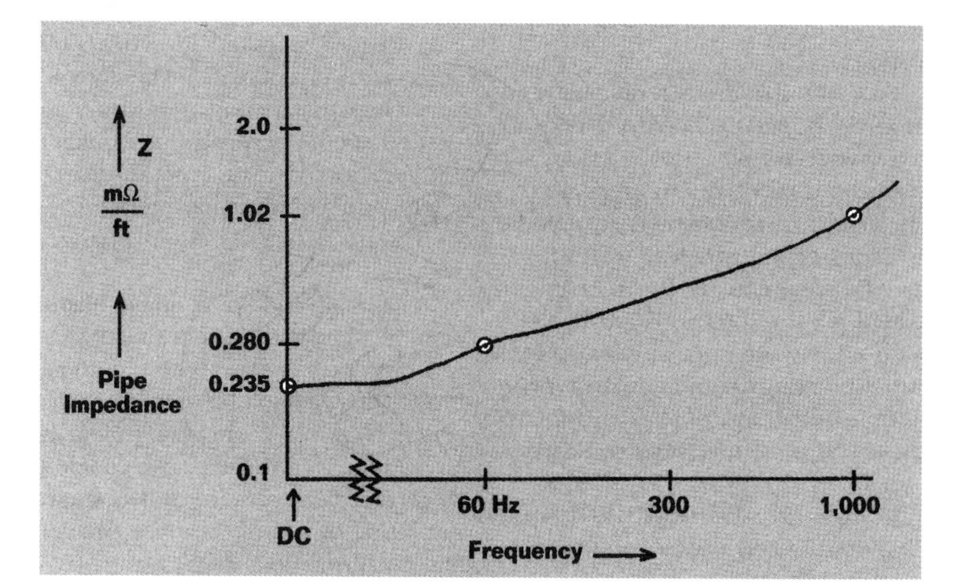

Figure 5. Impedance (Z) of cast-iron lab sample pipe versus frequency of current in pipe.

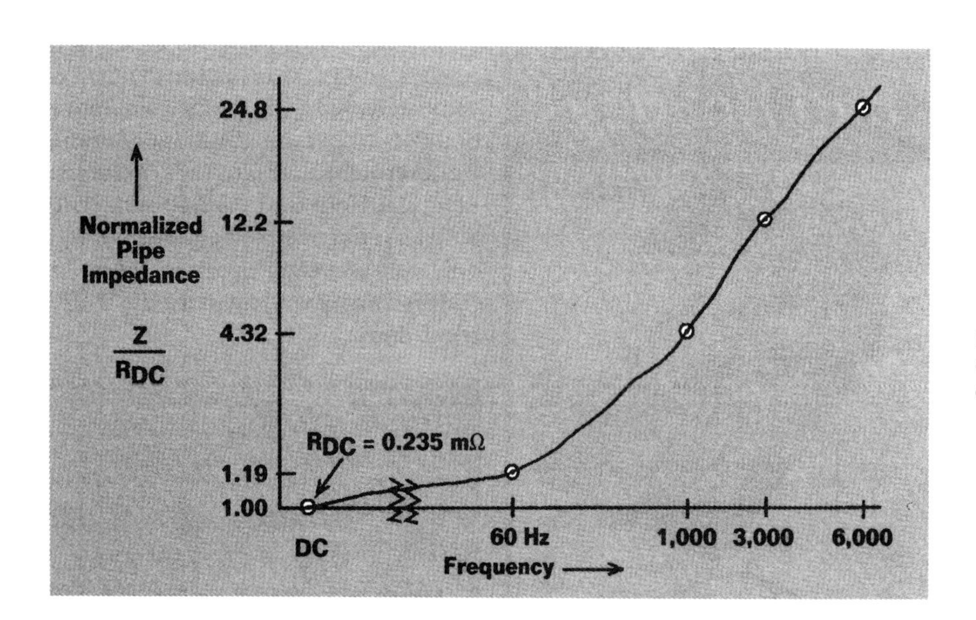

Figure 6. Normalized impedance (Z) of cast-iron lab sample versus frequency of current in pipe.

A 3-in. (76-mm) diameter cast iron pipe with $\frac{3}{16}$-in. (4.8-mm) wall thickness was tested in the laboratory. The effect of high frequency on pipe impedance was evident. Figure 5 is a log-log graph of Z against the frequency of the current flowing in the pipe.

Figure 6 is a log-log plot of normalized Z against frequency. In the cast-iron pipe sample, the 60-Hz Z was 19% greater than the DC resistance.

Sheehan[3] found that the 60-Hz normalized impedance of very large-diameter steel pipe ranged from 160 to 310 times that of the DC resistance. These numbers are within the spread calculated using several values of steel permeability.

Stray pickup

The 60-Hz magnetic field set up by the current flowing in a transmission line over a pipe will induce voltage in a properly oriented loop of wire beside the pipe. In Figure 4, the AC mV induced in the meter wire will depend on how it is laid with respect to the pipe.

General conclusion

Clamp-on AC ammeters, as described in this article, are accurate for measuring from 0.1 A to more than 20 A. The typical overall accuracy is ±3% from 17 to 5,000 Hz. This is for clips and clamps from ¾- to 60-in. diameter aperture.

Clamp-on AC ammeters are practical tools for measuring and locating current that can cause corrosion or endanger humans. The resolution is 1 mA or better, even up to 60-in. diameter pipe. Stray pickup is generally unimportant, even in a common corridor under an AC power transmission line.

The mV drop method, wherein the pipe itself is used as a current shunt, is much less sensitive and accurate, even at 60 Hz.

References

1. W. H. Swain, "Measuring Unwanted Alternating Current in Pipe," W. H. Swain Co. Web site: www.SwainMeter.com.
2. R. A. Gummow, et al., "AC Corrosion—A Challenge to Pipeline Integrity," MP 38, 2 (1999).
3. Private communication from B. Sheehan at San Diego County Water Authority, August 28, 2000.

WILLIAM H. SWAIN is President of William H. Swain Co., 239 Field End St., Sarasota, FL 34240. He founded the company in 1964 and has invented and marketed DC and AC amp clamps, meters, and other products for corrosion control on lines up to 5-ft (1.5-m) diameter. He has an M.S. degree in electrical engineering and is a professional engineer in the state of Florida. He has been a NACE member for more than 16 years. ***MP***

Minimizing shock hazards on pipelines near HVAC lines

As right-of-way for pipelines becomes more difficult to obtain, especially in congested areas, rights-of-way are being shared by pipelines and high voltage transmission lines. Sharing a common right-of-way can pose problems during operation, especially during pipeline construction unless special precautions are taken.

Considerable voltage can be induced in pipelines above ground on skids unless the section of pipeline is grounded. Without grounding, a lethal shock hazard to construction personnel can occur. To minimize the voltage on the pipe, drive ground rods at each end of the section of aboveground pipe. The ground rods should be driven at least 4 ft in the ground and connected to the aboveground section of pipe with insulated copper wire and a "C" clamp. If the aboveground section is longer than ½ mile, additional ground rods should be driven and connected to the pipe.

Work should be discontinued if thunderstorms are occurring in the vicinity (approximately 10 miles) of the pipeline construction. If lightning should strike the power line, even at a distant location, while personnel are in contact with the pipe, they could suffer a severe shock or even electrocution. To eliminate this danger, personnel should always stand on a 3-ft × 3-ft (minimum) steel mat made of 1″ square (maximum) hardware cloth that is well grounded electrically to the nearest electrical tower ground via a copper cable. The mat should be fabricated so that personnel can easily move it from tower to tower. This can be done by bolting a suitable length of flexible welding cable firmly to the metallic mat. The other end of the cable should be equipped with a suitable copper or brass clamp that can easily be connected to the tower ground. Personnel should always place the mat on the ground and stand on it while connecting the cable clamp to the tower ground. When work is complete, personnel should remain standing on the mat while disconnecting the cable clamp from the tower ground wire. This will make it safe to carry the mat to the next tower ground connection site.

When tie-ins are made, additional personnel protection is required. Use a metallic mat or driven ground rod connected to the pipe at each end of the tie-in, and connect an insulated copper cable (bond) across the tie-in point. The pipe and ground mat will be essentially at the same potential if a high-voltage surge occurs during the tie-in operation. The bond will prevent buildup of voltage between the unconnected ends at the tie-in location.

After the pipeline has been buried and placed in service, personnel should always connect a bonding cable across any location where the pipe is to be separated, i.e., a valve to be removed or a section of pipe is to be removed and replaced. The bonding cable should remain in place until all connections have been completed.

Equipment with steel tracks will not require any special grounding procedures for traveling along the power line right-of-way, but machinery with rubber tires should have a heavy-duty chain attached to the vehicle frame with a good electrical connection so that approximately 3 ft of chain will drag along the ground when moving or standing along the power line right-of-way. Any time there is a possibility of a lightning strike, personnel should not grasp any rubber-tired machinery while standing on the ground. Personnel should hop on or off when mounting or dismounting. Care should be exercised to prevent injury.

Cathodic protection test point installations

Test points provide the best means of electrically examining a buried pipeline to determine whether it is cathodically protected. They also help to ensure that other tests associated with corrosion control are working. To serve their purposes over the years, test points must be convenient to use and constructed in a manner that minimizes future damage to the test station.

Types of test points

Figure 1 illustrates test point types by function. The types shown do not necessarily represent any particular standard, but they are intended to represent the variety that may be encountered. A color code is indicated to illustrate a system whereby leads may be identified. Whatever color code is

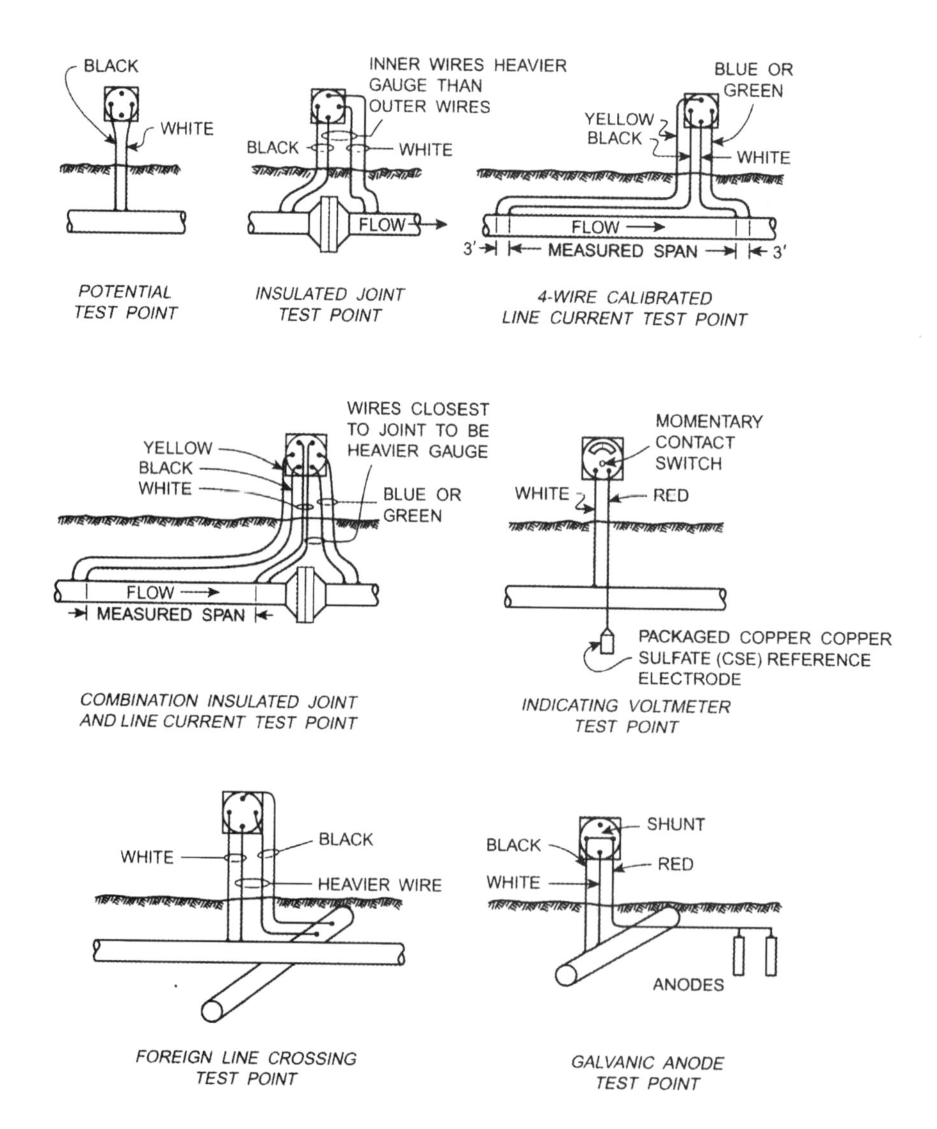

adopted should be made standard throughout a pipeline system.

The two-wire potential test point is the one used most frequently. Two wires make it possible to check pipe-to-earth potential with one while test current is being applied to the line (if desired) using the other. The four-wire insulated joint test point permits measuring pipe-to-earth potentials on each side of an insulated joint. The second pair of heavier gauge wires is available for inserting a resistance or solid bond across the insulated joint if necessary.

The four-wire calibrated line current test point permits accurate measurement of pipeline current flow. The six-wire combination insulated joint and line current test point is useful, particularly at terminal insulated flanges, because it permits positive measurement of current flow through an insulated flange should the flange become totally or partially shorted. Likewise, it will measure the current flowing through a solid or resistance bond should such measurement be necessary. One heavier-gauge wire is provided on each side of the insulated joint for bonding purposes (if required).

An indicating voltmeter test point is installed at key points on some systems. These meters may be read by operating personnel on a routine basis and the indicated values recorded and reported to the corrosion engineer. A voltmeter may be connected between the pipe and a reference electrode suitable for underground service.

The foreign line crossing test point provides two wires to each line. One wire is used for potential measurements or other tests as required and the other (the heavier-gauge wire) is available for installing a bond when needed. Test wires should never be attached to another company's pipeline unless the pipeline owner authorizes it. Further, many companies will allow such attachments only if they are made by their own personnel or if their own representatives are present while the attachments are being made.

The galvanic anode test point typically is used in connection with anodes at one location.

Adapted by Joram Lichtenstein from Peabody's Control of Pipeline Corrosion, 2 ed., *R. Bianchetti, ed. (Houston, Texas; NACE, 2001), pp. 239–241.* Published in the November 2003 issue of *Materials Performance*.

9: Gas—General

Know the gas laws

The symbols that will be used in the mathematical expression of the various gas laws are:

V_1 = Volume of the gas in cubic ft with original conditions
V_2 = Volume of gas in cubic ft under the second set of conditions
T_1 = Initial temperature of the gas in °R (°F + 460)
T_2 = Temperature of gas under second conditions °R
P_1 = Pressure of gas in psia under original conditions
P_2 = Pressure of gas in psia under second conditions

Any other symbols used in mathematically expressing the gas laws will be explained at the time of their introduction.

Boyle's law. If the temperature remains the same, the volume of a given quantity of gas will vary inversely as the absolute pressure. This may be expressed mathematically as:

$$\frac{V_1}{V_2} = \frac{P_2}{P_1}$$

Inasmuch as in the application of Boyle's Law we are generally interested in the volume at a second set of pressure conditions, a rearrangement of the formula more readily used is:

$$V_2 = V_1 \times \frac{P_1}{P_2} \qquad (1)$$

Example. A quantity of gas at a pressure of 42 psig has a volume of 1,000 cubic ft. If the gas is compressed to 100 psig, what volume would it occupy? Assume the barometric pressure to be 14.2 psia and the temperature to remain the same. Substituting in the second arrangement of the formula for Boyle's Law above would give:

$$V_2 = 1{,}000 \times \frac{42 + 14.2}{100 + 14.2}$$
$$= 492.1 \text{ cubic ft}$$

Charles' law (sometimes called Gay-Lussac's Law). If the pressure remains the same, the volume of a given quantity of gas will vary directly as the absolute temperature.

This may be expressed mathematically as:

$$\frac{V_1}{V_2} = \frac{T_1}{T_2}$$

Again, since we are usually more interested in the volume at a second set of temperature conditions than any other information, an arrangement of the formula that would be handy is:

$$V_2 = V_1 \times \frac{T_2}{T_1} \qquad (2)$$

A second part of Charles' Law is: If the volume of a quantity of gas does not change, the absolute pressure will vary directly as the absolute temperature.

Expressed mathematically this is:

$$\frac{P_1}{P_2} = \frac{T_1}{T_2}$$

In this instance, we would probably be more interested in the pressure at a second temperature condition, which could be expressed as:

$$P_2 = P_1 \times \frac{T_2}{T_1} \qquad (3)$$

Example. A given weight of gas has a volume of 450 cubic ft when the temperature is 45°F and the pressure is 10 psig. If the pressure remains the same, but the temperature is changed to 90°F, what will be the volume of the gas?

Table 1
Molecular Weights of Hydrocarbons and Other Compounds Associated with Natural Gas

Compound	Atomic Formula	Molecular Weight
Methane	CH_4	16.043
Ethane	C_2H_6	30.070
Propane	C_3H_8	44.097
Butane	C_4H_{10}	58.124
Pentane	C_5H_{12}	72.151
Hexane	C_6H_{14}	86.178
Heptane	C_7H_{16}	100.205
Carbon Dioxide	CO_2	44.011
Nitrogen	N_2	28.016
Oxygen	O_2	32.00
Water	H_2O	18.016
Air		28.967

Note: Molecular weights based on following values of atomic weights: hydrogen 1.008, carbon 12.011, nitrogen 14.008, oxygen 16.000, and argon 39.944. Air was assumed to consist of 78.09% nitrogen, 20.95% oxygen, 0.93% argon, and 0.03% carbon dioxide.

Substitution in the formula for the first part of the Charles' Law gives:

$$V_2 = 450 \times \frac{(90+460)}{(45+460)}$$
$$= 490 \text{ cubic ft}$$

It is desired to determine what the new pressure would be for the gas in the above example if the volume remains the same and the temperature changes from 45°F to 90°F as indicated. (Atmospheric pressure is 14.4 psia.)

Substitution in the formula gives:

$$P_2 = (10+14.4) \times \frac{(90+460)}{(45+460)}$$
$$= 26.6 \text{ psia or } 12.2 \text{ psig}$$

A convenient arrangement of a combination of Boyle's and Charles' laws that is easy to remember and use can be expressed mathematically as:

$$\frac{P_1V_1}{T_1} = \frac{P_2V_2}{T_2} \qquad (4)$$

One can substitute known values in the combination formula and solve for any one unknown value. In cases where one of the parameters, such as temperature, is not to be considered, it may be treated as having the same value on both sides of the formula, and consequently it can be cancelled out.

Avogadro's law. This law states that equal volumes of all gases at the same pressure and temperature conditions contain the same number of molecules.

From this it may be seen that the weight of a given volume of gas is a function of the weights of the molecules and that there is some volume at which the gas would weigh, in pounds, the numerical value of its molecular weight.

The volume at which the weight of the gas in pounds is equal to the numerical value of its molecular weight (known as the "mol-volume") is 378.9 cubic ft for gases at a temperature of 60°F and a pressure of 14.73 psia. Table 1 gives the atomic formula and molecular weights for hydrocarbons and other compounds frequently associated with natural gas. Reference to the table reveals that the molecular weight for methane is 16.043. Going back to the mol-volume explanation shows that 378.9 cubic ft of methane at 60°F and a pressure of 14.73 psia would weigh 16.043 lb.

Avogadro's law ties in closely with what is usually known as the ideal gas law.

The ideal gas law. Although expressed in many slightly different arrangements, this law is most frequently expressed as:

$$pV = nRT$$

where: p = Pressure of the gas
V = Volume of the gas
n = Number of lb-mols of gas
R = The universal gas constant, which varies depending upon the units of pressure, volume, and temperature employed

Since the number of lb-mols of a gas would be equal to the weight of the gas divided by the molecular weight of the gas, we can express the ideal gas law as:

$$pV = 10.722 \times \frac{W}{M} \times T \qquad (5)$$

where: p = Pressure of the gas, psia
V = Volume of the gas, cubic ft
W = Weight of the gas, lb
M = Molecular weight of the gas
T = Temperature of the gas, °R

The constant 10.722 is based upon the generally used value for the universal gas constant of 1,544 when the pressure is expressed in lb/sq. ft absolute.

This formula can be used in many arrangements. An arrangement that may be used to determine the weight of a quantity of gas is:

$$W = 0.0933 \times \frac{MVp}{T} \qquad (6)$$

when the symbols and units are as above.

Example. It is desired to find the weight of a gas in a 1,000-cubic-ft container if the gas is at a pressure of 150 psig and a temperature of 90°F. The molecular weight of the gas is known to be 16.816, and the barometric pressure is 14.3 psia.

Substitution in the formula gives:

$$W = \frac{.0933 \times 16.816 \times 1{,}000 \times (150+14.3)}{(90+460)}$$
$$= 468.7 \text{ lb}$$

The formula above may be used when the molecular weight of a gas is known; however, at times it is desirable to determine the weight of a given volume of gas when the molecular weight is not known or cannot be readily determined. This may be accomplished if the specific gravity of the gas is known by using the formula:

$$W = 2.698 \times \frac{GV_p}{T}$$

where G = the specific gravity of the gas (air equals 1.000) and the other symbols and units are as previously given.

Example. If the molecular weight of the gas in the preceding example was unknown, but the specific gravity was known to be 0.581, substitution in the formula would give:

$$W = 2.698 \times \frac{0.581 \times 1{,}000 \times (150 + 14.3)}{(90 + 460)}$$

$$= 468.3\ \text{lb}$$

Calculate gas properties from a gas analysis

In most practical applications involving gas calculations, the gas will consist of a mixture of components. Properties of the components are known; however, the properties of the mix must be determined for use in other calculations, such as compressor performance calculations. The gas molecular weight, K-value (isentropic exponent), and compressibility of the gas mix will be determined in the calculation to follow. Other data that may be needed for compressor performance determination will likely come from the process design and from the manufacturer's data.

Sample calculation 1

Given: A gas mixture at 30 psia and 60°F consisting of the following components:

Table 1

Gas	Mol %	Mol wt	P_c	T_c	cp lbm-mol-°R
Ethane	5	30.070	706.5	550	12.27
Propane	80	44.097	616.0	666	17.14
n-Butane	15	58.123	550.6	766	22.96
Total	100				

Table 2
Individual Component Contributions

Gas	Mol %	Mol wt	P_c	T_c	cp
Ethane	5	1.50	35	28	0.61
Propane	80	35.28	493	533	13.71
n-Butane	15	8.72	83	115	3.44
Total	100	45.5	611	676	17.76

Therefore, the properties of the mix will be:

Molecular weight = MW = 45.5
Critical pressure = P_c = 611
Critical temperature = T_c = 676
Specific heat at constant pressure = cp = 17.76

The ratio of specific heats for the mixture are calculated from Equation 1.

$$k_{mix} = cp_{mix}/(cp_{mix} - 1.986) \quad (1)$$

$$k_{mix} = 17.76/(17.76 - 1.986)$$
$$= 1.126$$

Determine the compressibility of the mix by calculating the reduced pressure and reduced temperature by using Equations 2 and 3.

$$P_r = P/P_c \quad (2)$$

$$T_r = T/T_c \quad (3)$$

$$P_r = 30/611$$
$$= 0.049$$

$$T_r = (60 + 460)/676$$
$$= 0.769$$

Refer to Figure 1 to determine the compressibility using the values of P_r and T_r. Z = 0.955.

If the mixture of components is given in mols of component or mass flow of components, these must be converted to mol % before calculating the properties of the mix.

Sample calculation 2

Mols of components are given:

Ethane	400 mols/hr
Propane	4,800 mols/hr
n-Butane	900 mols/hr
Total Flow	6,100 mols/hr

Mol % ethane = 400/6,100 = 6.56%

Table 3

Component	Mol/hr	Mol %
Ethane	400	6.56
Propane	4,800	78.69
n-Butane	900	14.75
Totals	6,100	100.00

The mol % values listed in Table 4 can be used to calculate the properties of the mixture following the procedure detailed in Sample Calculation 1.

Sample calculation 3

Mass flow of components is given.

Table 4

Component	Mass flow (kg/h)	Mol wt	Mol flow (kmol/h)	Mol %
Ethane	15,000	30.070	499	11.08
Propane	150,000	44.097	3,402	75.55
n-Butane	35,000	58.123	602	13.37
Totals			4,503	100.00

Mol flow of ethane = 15,000/30.07
= 499

Mol % of ethane = 400/4,503
= 11.08%

When a dry gas is saturated with water, it will be necessary to consider the effects of the water on the molecular weight of the mixture. The water will also affect the total mass flow of the mixture.

Sample calculation 4

Given: Compress 1,000 lbm/min dry CO_2, which is initially water saturated. MW = 44.01, $P_{inlet} = 16$ psia, $T_{inlet} = 100°F$.

Calculate the molecular weight of the mixture and the total mass flow for the required 1,000 lbm/min of dry CO_2.

$$\text{Mol fraction of water vapor} = y = P_w/P \quad (4)$$

P_w = partial pressure of water vapor. It is exactly equal to the saturation pressure if the gas is saturated.

$$MW_{mix} = yMW_{mix} + (1 - y) \times MW_{dg} \quad (5)$$

where w = water vapor
dg = dry gas

$$m_{total} = m_{dg}[1 + (MW_w \times P_w)/(MW_{dg} \times P_{dg})] \quad (6)$$

Saturated pressure of water vapor at 100°F is equal to 0.95 psia, and since the mixture is saturated, the partial pressure is also equal to 0.95 Asia.

Calculate mol % of water vapor using Equation 4.

y = 0.96/16 = 5.94 mol % water vapor

Calculate MW_{mix} using Equation 5.

$$MW_{mix} = 0.0594(18.02) + (1 - 0.059) \times 44.01 = 42.48$$

Calculate the mass flow of the mix using Equation 6.

$$M_{total} = 1,000[1 + (18.02 \times 0.95)/(44.01 \times (16 - 0.95))]$$
$$= 1,026 \text{ lbm/min}$$

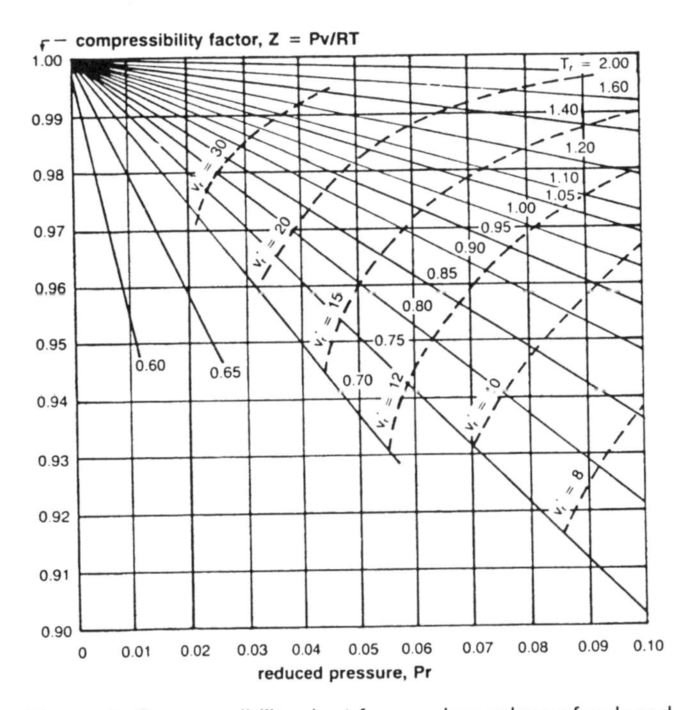

Figure 1. Compressibility chart for very low values of reduced pressure. Reproduced by permission of *Chemical Engineering*, McGraw Hill Publications Company, July 1954.

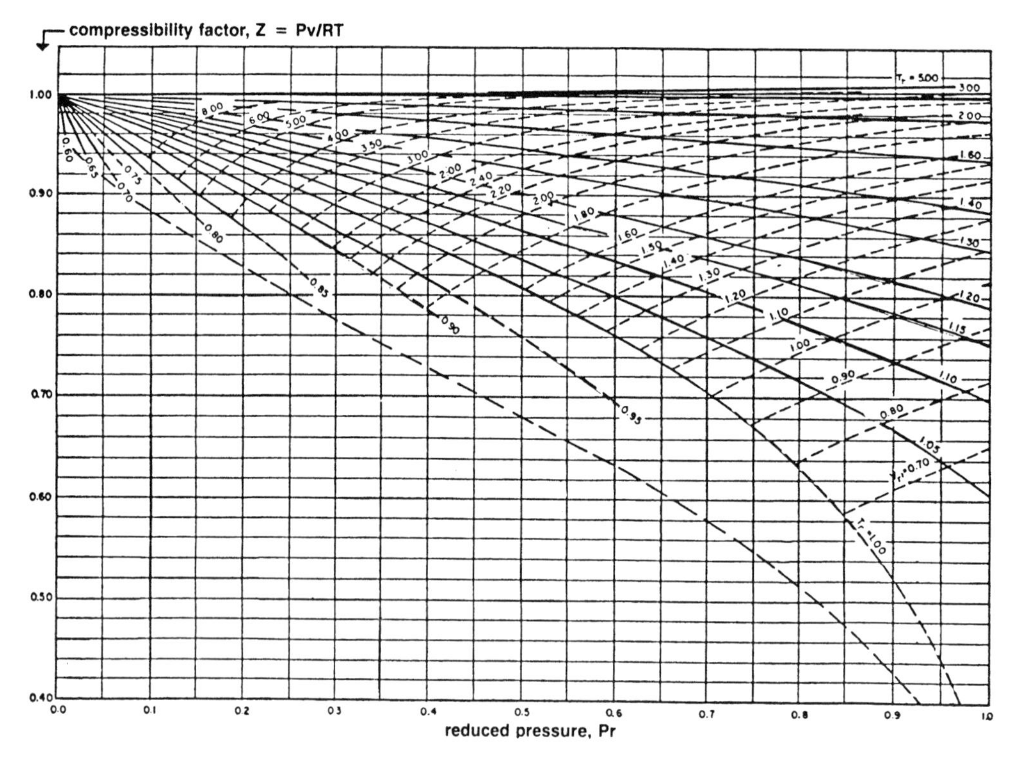

Figure 2. Compressibility chart for low values of reduced pressure. Reproduced by permission of *Chemical Engineering*, McGraw Hill Publications Company, July 1954.

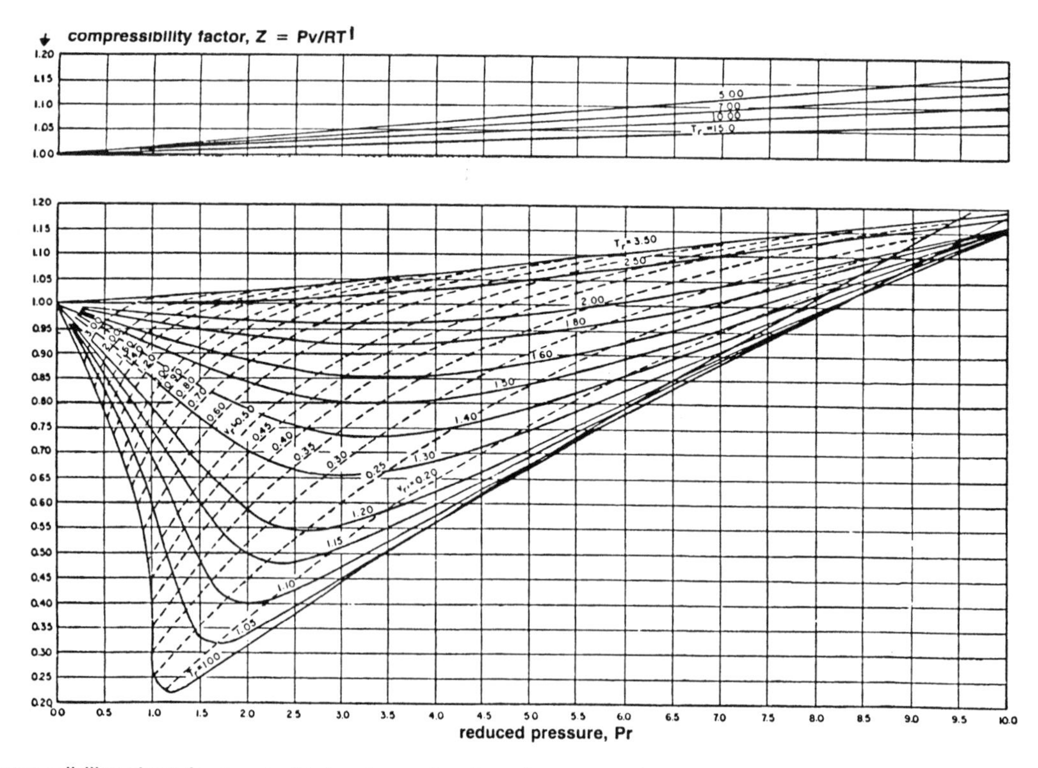

Figure 3. Compressibility chart for low to high values of reduced pressure. Reproduced by permission of *Chemical Engineering*, McGraw Hill Publications Company, July 1954.

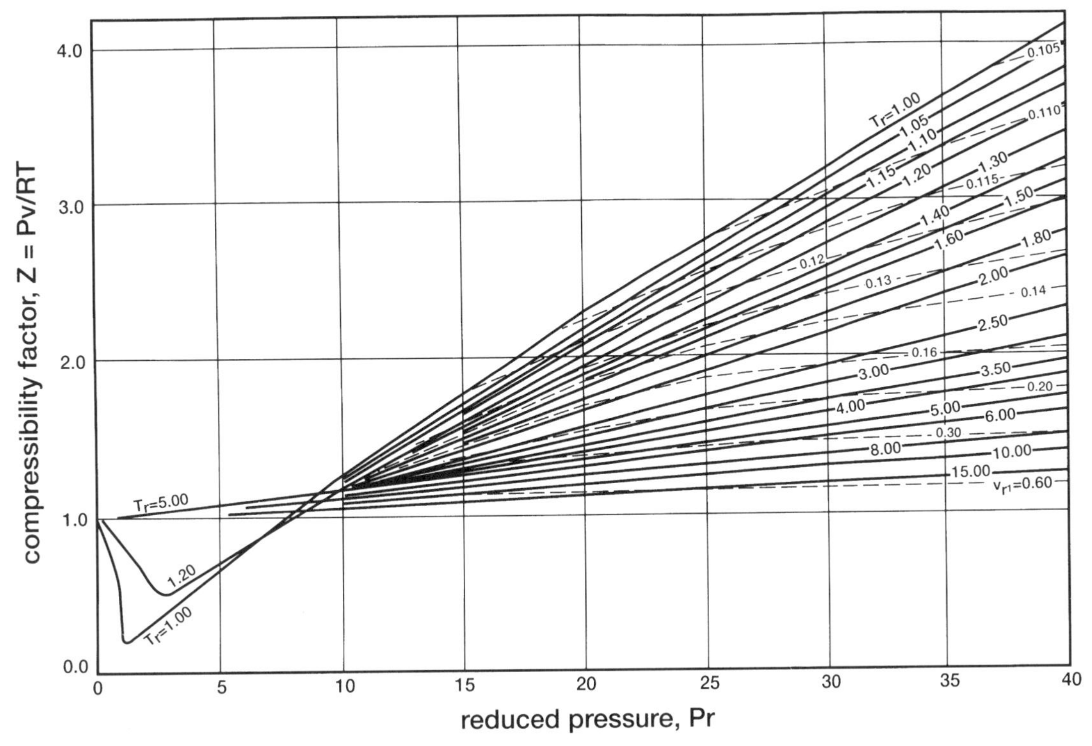

Figure 4. Compressibility chart for low to very high values of reduced pressure. Reproduced by permission of *Chemical Engineering*, McGraw Hill Publications Company, July 1954.

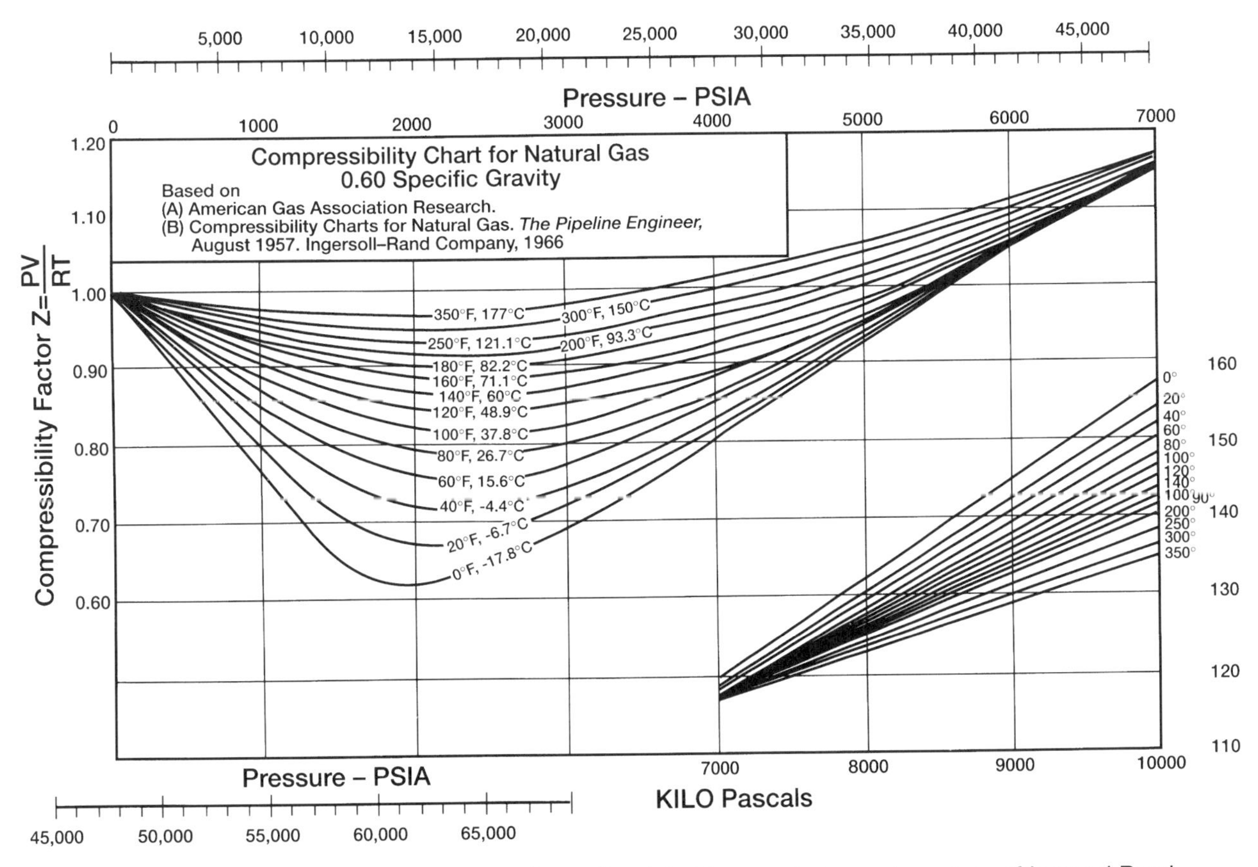

Figure 5. Compressibility chart for natural gas. Reprinted by permission and courtesy of Ingersol Rand.

Physical properties of selected hydrocarbons and other chemicals and gases

Physical Constants

*See the Table of Notes and References.

Number	See Note No. ->		A.	B.		C.	D.				
								Critical constants			
Number	Compound	Formula	Molar mass (molecular weight)	Boiling point, °F 14.696 psia	Vapor pressure, psia 100 °F	Freezing point, °F 14.696 psia	Refractive index, n_D 60 °F	Pressure, psia	Temperature, °F	Volume, ft³/lbm	Number
1	Methane	CH_4	16.043	-258.73	(5000)*	-296.44*	1.00042*	666.4	-116.67	0.0988	1
2	Ethane	C_2H_6	30.070	-127.49	(800)*	-297.04*	1.20971*	706.5	89.92	0.0783	2
3	Propane	C_3H_8	44.097	-43.75	188.64	-305.73*	1.29480*	616.0	206.06	0.0727	3
4	Isobutane	C_4H_{10}	58.123	10.78	72.581	-255.28	1.3245*	527.9	274.46	0.0714	4
5	n-Butane	C_4H_{10}	58.123	31.08	51.706	-217.05	1.33588*	550.6	305.62	0.0703	5
6	Isopentane	C_5H_{12}	72.150	82.12	20.445	-255.82	1.35631	490.4	369.10	0.0679	6
7	n-Pentane	C_5H_{12}	72.150	96.92	15.574	-201.51	1.35992	488.6	385.8	0.0675	7
8	Neopentane	C_5H_{12}	72.150	49.10	36.69	2.17	1.342*	464.0	321.13	0.0673	8
9	n-Hexane	C_6H_{14}	86.177	155.72	4.9597	-139.58	1.37708	436.9	453.6	0.0688	9
10	2-Methylpentane	C_6H_{14}	86.177	140.47	6.769	-244.62	1.37387	436.6	435.83	0.0682	10
11	3-Methylpentane	C_6H_{14}	86.177	145.89	6.103	——	1.37888	453.1	448.4	0.0682	11
12	Neohexane	C_6H_{14}	86.177	121.52	9.859	-147.72	1.37126	446.8	420.13	0.0667	12
13	2,3-Dimethylbutane	C_6H_{14}	86.177	136.36	7.406	-199.38	1.37730	453.5	440.29	0.0665	13
14	n-Heptane	C_7H_{16}	100.204	209.16	1.620	-131.05	1.38989	396.8	512.7	0.0691	14
15	2-Methylhexane	C_7H_{16}	100.204	194.09	2.272	-180.89	1.38714	396.5	495.00	0.0673	15
16	3-Methylhexane	C_7H_{16}	100.204	197.33	2.131	——	1.39091	408.1	503.80	0.0646	16
17	3-Ethylpentane	C_7H_{16}	100.204	200.25	2.013	-181.48	1.39566	419.3	513.39	0.0665	17
18	2,2-Dimethylpentane	C_7H_{16}	100.204	174.54	3.494	-190.86	1.38446	402.2	477.23	0.0665	18
19	2,4-Dimethylpentane	C_7H_{16}	100.204	176.89	3.293	-182.63	1.38379	396.9	475.95	0.0668	19
20	3,3-Dimethylpentane	C_7H_{16}	100.204	186.91	2.774	-210.01	1.38564	427.2	505.87	0.0662	20
21	Triptane	C_7H_{16}	100.204	177.58	3.375	-12.81	1.39168	428.4	496.44	0.0636	21
22	n-Octane	C_8H_{18}	114.231	258.21	0.53694	-70.18	1.39956	360.7	564.22	0.0690	22
23	Diisobutyl	C_8H_{18}	114.231	228.39	1.102	-132.11	1.39461	360.6	530.44	0.0676	23
24	Isooctane	C_8H_{18}	114.231	210.63	1.709	-161.27	1.38624	372.4	519.46	0.0656	24
25	n-Nonane	C_9H_{20}	128.258	303.47	0.17953	-64.28	1.40746	331.8	610.68	0.0684	25
26	n-Decane	$C_{10}H_{22}$	142.285	345.48	0.06088	-21.36	1.41385	305.2	652.0	0.0679	26
27	Cyclopentane	C_5H_{10}	70.134	120.65	9.915	-136.91	1.40896	653.8	461.2	0.0594	27
28	Methylcyclopentane	C_6H_{12}	84.161	161.25	4.503	-224.40	1.41210	548.9	499.35	0.0607	28
29	Cyclohexane	C_6H_{12}	84.161	177.29	3.266	43.77	1.42862	590.8	536.6	0.0586	29
30	Methylcyclohexane	C_7H_{14}	98.188	213.68	1.609	-195.87	1.42538	503.5	570.27	0.0600	30
31	Ethene(Ethylene)	C_2H_4	28.054	-154.73	(1400)*	-272.47*	(1.228)*	731.0	48.54	0.0746	31
32	Propene(Propylene)	C_3H_6	42.081	-53.84	227.7	-301.45*	1.3130*	668.6	197.17	0.0689	32
33	1-Butene(Butylene)	C_4H_8	56.108	20.79	62.10	-301.63*	1.3494*	583.5	295.48	0.0685	33
34	cis-2-Butene	C_4H_8	56.108	38.69	45.95	-218.06	1.3665*	612.1	324.37	0.0668	34
35	trans-2-Butene	C_4H_8	56.108	33.58	49.87	-157.96	1.3563*	587.4	311.86	0.0679	35
36	Isobutene	C_4H_8	56.108	19.59	63.02	-220.65	1.3512*	580.2	292.55	0.0682	36
37	1-Pentene	C_5H_{10}	70.134	85.93	19.12	-265.39	1.37426	511.8	376.93	0.0676	37
38	1,2-Butadiene	C_4H_6	54.092	51.53	36.53	-213.16	——	(653.)*	(340.)*	(0.065)*	38
39	1,3-Butadiene	C_4H_6	54.092	24.06	59.46	-164.02	1.3975*	627.5	305.	0.0654	39
40	Isoprene	C_5H_8	68.119	93.31	16.68	-230.73	1.42498	(558.)*	(412.)*	(0.065)*	40
41	Acetylene	C_2H_2	26.038	-120.49*	——	-114.5*	——	890.4	95.34	0.0695	41
42	Benzene	C_6H_6	78.114	176.18	3.225	41.95	1.50396	710.4	552.22	0.0531	42
43	Toluene	C_7H_8	92.141	231.13	1.033	-139.00	1.49942	595.5	605.57	0.0550	43
44	Ethylbenzene	C_8H_{10}	106.167	277.16	0.3716	-138.966	1.49826	523.0	651.29	0.0565	44
45	o-Xylene	C_8H_{10}	106.167	291.97	0.2643	-13.59	1.50767	541.6	674.92	0.0557	45
46	m-Xylene	C_8H_{10}	106.167	282.41	0.3265	-54.18	1.49951	512.9	651.02	0.0567	46
47	p-Xylene	C_8H_{10}	106.167	281.07	0.3424	55.83	1.49810	509.2	649.54	0.0570	47
48	Styrene	C_8H_8	104.152	293.25	0.2582	-23.10	1.54937	587.8	(703.)*	0.0534	48
49	Isopropylbenzene	C_9H_{12}	120.194	306.34	0.1884	-140.814	1.49372	465.4	676.3	0.0572	49
50	Methyl alcohol	CH_4O	32.042	148.44	4.629	-143.79	1.33034	1174.	463.08	0.0590	50
51	Ethyl alcohol	C_2H_6O	46.069	172.90	2.312	-173.4	1.36346	890.1	465.39	0.0581	51
52	Carbon monoxide	CO	28.010	-312.68	——	-337.00*	1.00036*	507.5	-220.43	0.0532	52
53	Carbon dioxide	CO_2	44.010	-109.257*	——	-69.83*	1.00048*	1071.	87.91	0.0344	53
54	Hydrogen sulfide	H_2S	34.08	-76.497	394.59	-121.88*	1.00060*	1300.	212.45	0.0461	54
55	Sulfur dioxide	SO_2	64.06	14.11	85.46	-103.86*	1.00062*	1143.	315.8	0.0305	55
56	Ammonia	NH_3	17.0305	-27.99	211.9	-107.88*	1.00036*	1646.	270.2	0.0681	56
57	Air	N_2+O_2	28.9625	-317.8	——	——	1.00028*	546.9	-221.31	0.0517	57
58	Hydrogen	H_2	2.0159	-422.955*	——	-435.26*	1.00013*	188.1	-399.9	0.5165	58
59	Oxygen	O_2	31.9988	-297.332*	——	-361.820*	1.00027*	731.4	-181.43	0.0367	59
60	Nitrogen	N_2	28.0134	-320.451	——	-346.00*	1.00028*	493.1	-232.51	0.0510	60
61	Chlorine	Cl_2	70.906	-29.13	157.3	-149.73*	1.3878*	1157.	290.75	0.0280	61
62	Water	H_2O	18.0153	212.000*	0.9501	32.00	1.33335	3198.8	705.16	0.04975	62
63	Helium	He	4.0026	-452.09	——	——	1.00003*	32.99	-450.31	0.2300	63
64	Hydrogen chloride	HCl	36.461	-121.27	906.71	-173.52*	1.00042*	1205.	124.77	0.0356	64

12/12/86

NOTE: Numbers in this table do not have accuracies greater than 1 part in 1,000; in some cases extra digits have been added to calculated values to achieve consistency or to permit recalculation of experimental values.

Physical Constants

*See the Table of Notes and References.

Number	E. Density of liquid 14.696 psia, 60°F: Relative density (specific gravity) 60°F/60°F	E. lbm/gal.	E. gal./lb mole	F. Temperature coefficient of density, 1/°F	G. Acentric factor, ω	H. Compressibility factor of real gas, Z 14.696 psia, 60°F	I. Ideal gas 14.696 psia, 60°F: Relative density (specific gravity) Air = 1	I. ft³ gas/lbm	I. ft³ gas/gal. liquid	J. Specific Heat 60°F 14.696 psia Btu/(lbm·°F): Cp, Ideal gas	J. Cp, Liquid	Number
1	(0.3)*	(2.5)*	(6.4172)*	—	0.0104	0.9980	0.5539	23.654	(59.135)*	0.52669	—	1
2	0.35619*	2.9696*	10.126*	—	0.0979	0.9919	1.0382	12.620	37.476*	0.40782	0.97225	2
3	0.50699*	4.2268*	10.433*	−0.00162*	0.1522	0.9825	1.5226	8.6059	36.375*	0.38852	0.61996	3
4	0.56287*	4.6927*	12.386*	−0.00119*	0.1852	0.9711	2.0068	6.5291	30.639*	0.38669	0.57066	4
5	0.58401*	4.8690*	11.937*	−0.00106*	0.1995	0.9667	2.0068	6.5291	31.790*	0.39499	0.57272	5
6	0.62470	5.2082	13.853	−0.00090	0.2280	—	2.4912	5.2596	27.393	0.38440	0.53331	6
7	0.63112	5.2617	13.712	−0.00086	0.2514	—	2.4912	5.2596	27.674	0.38825	0.54363	7
8	0.59666*	4.9744*	14.504*	−0.00106*	0.1963	0.9582	2.4912	5.2596	26.163*	0.39038	0.55021	8
9	0.66383	5.5344	15.571	−0.00075	0.2994	—	2.9755	4.4035	24.371	0.38628	0.53327	9
10	0.65785	5.4846	15.713	−0.00076	0.2780	—	2.9755	4.4035	24.152	0.38526	0.52732	10
11	0.66901	5.5776	15.451	−0.00076	0.2732	—	2.9755	4.4035	24.561	0.37902	0.51876	11
12	0.65385	5.4512	15.809	−0.00076	0.2326	—	2.9755	4.4035	24.005	0.38231	0.51367	12
13	0.66631	5.5551	15.513	−0.00076	0.2469	—	2.9755	4.4035	24.462	0.37762	0.51308	13
14	0.68820	5.7376	17.464	−0.00068	0.3494	—	3.4598	3.7872	21.729	0.38447	0.52802	14
15	0.68310	5.6951	17.595	−0.00070	0.3298	—	3.4598	3.7872	21.568	0.38041	0.52199	15
16	0.69165	5.7664	17.377	−0.00070	0.3232	—	3.4598	3.7872	21.838	0.37882	0.51019	16
17	0.70276	5.8590	17.103	−0.00069	0.3105	—	3.4598	3.7872	22.189	0.38646	0.51410	17
18	0.67829	5.6550	17.720	−0.00070	0.2871	—	3.4598	3.7872	21.416	0.38594	0.51678	18
19	0.67733	5.6470	17.745	−0.00073	0.3026	—	3.4598	3.7872	21.386	0.39414	0.52440	19
20	0.69772	5.8170	17.226	−0.00067	0.2674	—	3.4598	3.7872	22.030	0.38306	0.50138	20
21	0.69457	5.7907	17.304	−0.00068	0.2503	—	3.4598	3.7872	21.930	0.37724	0.49920	21
22	0.70696	5.8940	19.381	−0.00064	0.3977	—	3.9441	3.3220	19.580	0.38331	0.52406	22
23	0.69793	5.8187	19.632	−0.00067	0.3564	—	3.9441	3.3220	19.330	0.37571	0.51130	23
24	0.69624	5.8046	19.679	−0.00065	0.3035	—	3.9441	3.3220	19.283	0.38222	0.48951	24
25	0.72187	6.0183	21.311	−0.00061	0.4445	—	4.4284	2.9588	17.807	0.38246	0.52244	25
26	0.73421	6.1212	23.245	−0.00057	0.4898	—	4.9127	2.6671	16.326	0.38179	0.52103	26
27	0.75050	6.2570	11.209	−0.00073	0.1950	—	2.4215	5.4110	33.856	0.27199	0.42182	27
28	0.75349	6.2819	13.397	−0.00069	0.2302	—	2.9059	4.5090	28.325	0.30100	0.44126	28
29	0.78347	6.5319	12.885	−0.00065	0.2096	—	2.9059	4.5090	29.452	0.28817	0.43584	29
30	0.77400	6.4529	15.216	−0.00062	0.2358	—	3.3902	3.8649	24.940	0.31700	0.44012	30
31	—	—	—	—	0.0865	0.9936	0.9686	13.527	—	0.35697	—	31
32	0.52095*	4.3432*	9.6889*	−0.00173*	0.1356	0.9844	1.4529	9.0179	39.167*	0.35714	0.57116	32
33	0.60107*	5.0112*	11.197*	−0.00112*	0.1941	0.9699	1.9373	6.7636	33.894*	0.35446	0.54533	33
34	0.62717*	5.2288*	10.731*	−0.00105*	0.2029	0.9665	1.9373	6.7636	35.366*	0.33754	0.52980	34
35	0.60996*	5.0853*	11.033*	−0.00106*	0.2128	0.9667	1.9373	6.7636	34.395*	0.35574	0.54215	35
36	0.60040*	5.0056*	11.209*	−0.00117*	0.1999	0.9700	1.9373	6.7636	33.856*	0.37690	0.54839	36
37	0.64571	5.3834	13.028	−0.00089	0.2333	—	2.4215	5.4110	29.129	0.36351	0.51782	37
38	0.65799*	5.4857*	9.8605*	−0.00101*	0.2540	(0.969)	1.8677	7.0156	38.485*	0.34347	0.54029	38
39	0.62723*	5.2293*	10.344*	−0.00110*	0.2007	(0.965)	1.8677	7.0156	36.687*	0.34120	0.53447	39
40	0.68615	5.7205	11.908	−0.00082	0.1568	—	2.3520	5.5710	31.869	0.35072	0.51933	40
41	(0.41796)	(3.4842)	(7.473)	—	0.1949	0.9930	0.8990	14.574	—	0.39754	—	41
42	0.88448	7.3740	10.593	−0.00067	0.2093	—	2.6971	4.8581	35.824	0.24296	0.40989	42
43	0.87190	7.2691	12.676	−0.00059	0.2633	—	3.1814	4.1184	29.937	0.26370	0.40095	43
44	0.87168	7.2673	14.609	−0.00056	0.3027	—	3.6657	3.5744	25.976	0.27792	0.41139	44
45	0.88467	7.3756	14.394	−0.00052	0.3942	—	3.6657	3.5744	26.363	0.28964	0.41620	45
46	0.86875	7.2429	14.658	−0.00053	0.3257	—	3.6657	3.5744	25.889	0.27427	0.40545	46
47	0.86578	7.2181	14.708	−0.00056	0.3216	—	3.6657	3.5744	25.800	0.27471	0.40255	47
48	0.91108	7.5958	13.712	−0.00053	(0.2412)	—	3.5961	3.6435	27.675	0.27110	0.41220	48
49	0.86634	7.2228	16.641	−0.00055	0.3260	—	4.1500	3.1573	22.804	0.29170	0.42053	49
50	0.79626	6.6385	4.8267	−0.00066	0.5649	—	1.1063	11.843	78.622	0.32316	0.59187	50
51	0.79399	6.6196	6.9595	−0.00058	0.6438	—	1.5906	8.2372	54.527	0.33222	0.56610	51
52	0.78939*	6.5812*	4.2561*	—	0.0484	0.9959	0.9671	13.548	89.163*	0.24847	—	52
53	0.81802*	6.8199*	6.4532*	−0.00583*	0.2667	0.9943	1.5196	8.6229	58.807*	0.19911	—	53
54	0.80144*	6.6817*	5.1005*	−0.00157*	0.0948	0.9846	1.1767	11.135	74.401*	0.23827	0.50418	54
55	1.3974*	11.650*	5.4987*	—	0.2548	0.9802	2.2118	5.9238	69.012*	0.14804	0.32460	55
56	0.61832*	5.1550*	3.3037*	—	0.2557	0.9877	0.5880	22.283	114.87*	0.49677	1.1209	56
57	0.87476*	7.2930*	3.9713*	—	—	1.0000	1.0000	13.103	95.557*	0.23988	—	57
58	0.071070*	0.59252*	3.4022*	—	−0.2202	1.0006	0.06960	188.25	111.54*	3.4038	—	58
59	1.1421*	9.5221*	3.3605*	—	0.0216	0.9992	1.1048	11.859	112.93*	0.21892	—	59
60	0.80940*	6.7481*	4.1513*	—	0.0372	0.9997	0.9672	13.546	91.413*	0.24828	—	60
61	1.4244*	11.875*	5.9710*	—	0.0878	(0.9875)	2.4482	5.3519	63.554*	0.11377	—	61
62	1.00000	8.33712	2.1609	−0.00009	0.3443	—	0.62202	21.065	175.62	0.44457	0.99974	62
63	0.12510*	1.0430*	3.8376*	—	0.	1.0006	0.1382	94.814	98.891*	1.2404	—	63
64	0.85129*	7.0973*	5.1373*	−0.00300*	0.1259	0.9923	1.2589	10.408	73.869*	0.19086	—	64

12/12/86

NOTE: Numbers in this table do not have accuracies greater than 1 part in 1,000; in some cases extra digits have been added to calculated values to achieve consistency or to permit recalculation of experimental values.

Physical Constants

*See the Table of Notes and References.

Number	Compound (See Note No. ->)	K. Heating value, 60°F — Net: Btu/ft³ Ideal gas, 14.696 psia	K. Net: Btu/lbm Liquid	K. Gross: Btu/ft³ ideal gas, 14.696 psia	K. Gross: Btu/lbm Liquid	K. Gross: Btu/gal Liquid	L. Heat of vaporization 14.696 psia at boiling point, Btu/lbm	M. Air required for combustion ideal gas ft³(air)/ft³(gas)	Flammability limits, vol % in air mixture: Lower	Flammability limits: Higher	ASTM octane number: Motor method D-357	ASTM octane number: Research method D-908	Number
1	Methane	909.4	——	1010.0	——	——	219.45	9.548	5.0	15.0	——	——	1
2	Ethane	1618.7	20277.*	1769.6	22181.*	65869.*	211.14	16.710	2.9	13.0	+0.05	+1.6*	2
3	Propane	2314.9	19757.*	2516.1	21489.*	90830.*	183.01	23.871	2.0	9.5	97.1	+1.8*	3
4	Isobutane	3000.4	19437.*	3251.9	21079.*	98917.*	157.23	31.032	1.8	8.5	97.6	+0.1*	4
5	n-Butane	3010.8	19494.*	3262.3	21136.*	102911.*	165.93	31.032	1.5	9.0	89.6*	93.8*	5
6	Isopentane	3699.0	19303.	4000.9	20891.	108805.	147.12	38.193	1.3	8.0	90.3	92.3	6
7	n-Pentane	3706.9	19335.	4008.9	20923.	110091.	153.57	38.193	1.4	8.3	62.6*	61.7*	7
8	Neopentane	3682.9	19235.*	3984.7	20822.*	103577.*	135.58	38.193	1.3	7.5	80.2	85.5	8
9	n-Hexane	4403.8	19232.	4755.9	20783.	115021.	143.94	45.355	1.1	7.7	26.0	24.8	9
10	2-Methylpentane	4395.2	19202.	4747.3	20753.	113822.	138.45	45.355	1.18	7.0	73.5	73.4	10
11	3-Methylpentane	4398.2	19213.	4750.3	20764.	115813.	140.05	45.355	1.2	7.7	74.3	74.5	11
12	Neohexane	4384.0	19163.	4736.2	20714.	112916.	131.23	45.355	1.2	7.0	93.4	91.8	12
13	2,3-Dimethylbutane	4392.9	19195.	4745.0	20746.	115246.	136.07	45.355	1.2	7.0	94.3	+0.3	13
14	n-Heptane	5100.0	19155.	5502.5	20679.	118648.	136.00	52.516	1.0	7.0	0.0	0.0	14
15	2-Methylhexane	5092.2	19133.	5494.6	20657.	117644.	131.58	52.516	1.0	7.0	46.4	42.4	15
16	3-Methylhexane	5096.0	19146.	5498.6	20671.	119197.	132.10	52.516	(1.01)	(6.6)	55.8	52.0	16
17	3-Ethylpentane	5098.3	19154.	5500.7	20679.	121158.	132.82	52.516	(1.00)	(6.5)	69.3	65.0	17
18	2,2-Dimethylpentane	5079.6	19095.	5481.9	20620.	116606.	125.12	52.516	(1.09)	(6.8)	95.6	92.8	18
19	2,4-Dimethylpentane	5084.2	19111.	5486.7	20635.	116526.	126.57	52.516	(1.08)	(6.8)	83.8	83.1	19
20	3,3-Dimethylpentane	5086.4	19119.	5488.8	20643.	120080.	127.20	52.516	(1.04)	(7.0)	86.6	80.8	20
21	Triptane	5081.2	19103.	5483.5	20628.	119451.	124.21	52.516	(1.08)	(6.8)	+0.1	+1.8	21
22	n-Octane	5796.1	19096.	6248.9	20601.	121422.	129.52	59.677	0.8	6.5	——	——	22
23	Diisobutyl	5780.5	19047.	6233.5	20552.	119586.	122.83	59.677	(0.92)	(6.3)	55.7	55.2	23
24	Isooctane	5778.8	19063.	6231.7	20568.	119389.	112.94	59.677	0.95	6.0	100.0	100.0	24
25	n-Nonane	6493.2	19054.	6996.5	20543.	123634.	124.36	66.839	0.7	5.6	——	——	25
26	n-Decane	7189.6	19018.	7742.9	20494.	125448.	119.65	74.000	0.7	5.4	——	——	26
27	Cyclopentane	3512.1	18825.	3763.7	20186.	126304.	167.33	35.806	(1.48)	(8.3)	84.9*	+0.1	27
28	Methylcyclopentane	4199.4	18771.	4501.2	20132.	126467.	148.54	42.968	1.0	8.35	80.0	91.3	28
29	Cyclohexane	4179.7	18675.	4481.7	20036.	130873.	153.03	42.968	1.2	8.35	77.2	83.0	29
30	Methylcyclohexane	4863.6	18640.	5215.9	20002.	129071.	136.30	50.129	1.1	6.7	71.1	74.8	30
31	Ethene(Ethylene)	1499.1	——	1599.8	——	——	207.41	14.323	2.7	36.0	75.6	+0.03	31
32	Propene(Propylene)	2181.8	(19858.)	2332.7	(21208.)	(92113.)	188.19	21.484	2.0	11.7	84.9	+0.2	32
33	1-Butene(Butylene)	2878.7	19309.*	3079.9	20670.*	103582.*	167.96	28.645	1.6	10.	80.8*	97.4	33
34	cis-2-Butene	2871.0	19241.*	3072.2	20602.*	107724.*	178.89	28.645	1.6	10.	83.5	100.0	34
35	trans-2-Butene	2866.8	19221.*	3068.0	20582.*	104666.*	174.37	28.645	1.6	10.	——	——	35
36	Isobutene	2859.9	19182.*	3061.1	20543.*	102830.*	169.47	28.645	1.6	10.	——	——	36
37	1-Pentene	3575.0	19184.	3826.5	20545.	110602.	154.48	35.806	1.3	10.	77.1	90.9	37
38	1,2-Butadiene	2789.0	19378.*	2939.9	20437.*	112111.*	191.88	26.258	(1.62)	(10.3)	——	——	38
39	1,3-Butadiene	2729.0	18967.*	2879.9	20025.*	104717.*	185.29	26.258	2.0	12.5	——	——	39
40	Isoprene	3410.8	18832.	3612.1	19953.	114141.	163.48	33.419	(1.12)	(8.5)	81.0	99.1	40
41	Acetylene	1423.2	(20887.)	1473.5	(21613.)	(75204.)	151.90	11.935	1.5	100.	——	——	41
42	Benzene	3590.9	17256.	3741.8	17989.	132651.	169.24	35.806	1.2	8.0	+2.8	——	42
43	Toluene	4273.6	17421.	4475.0	18250.	132661.	154.83	42.968	1.2	7.1	+0.3	+5.8	43
44	Ethylbenzene	4970.5	17593.	5222.2	18492.	134387.	144.02	50.129	1.0	8.0	97.9	+0.8	44
45	o-Xylene	4958.2	17544.	5209.9	18444.	136036.	149.10	50.129	1.0	7.6	100.0	——	45
46	m-Xylene	4956.3	17541.	5207.9	18440.	133559.	147.24	50.129	1.0	7.0	+2.8	+4.0	46
47	p-Xylene	4957.1	17545.	5208.8	18444.	133131.	145.71	50.129	1.0	7.0	+1.2	+3.4	47
48	Styrene	4829.8	17414.	5031.1	18147.	137841.	152.85	47.742	1.1	8.0	+0.2	>+3.*	48
49	Isopropylbenzene	5660.9	17709.	5962.8	18662.	134792.	134.24	57.290	0.8	6.5	99.3	+2.1	49
50	Methyl alcohol	766.1	8559.	866.7	9751.	64731.	462.58	7.161	5.5	44.0	——	——	50
51	Ethyl alcohol	1448.1	11530.	1599.1	12770.	84539.	359.07	14.323	3.28	19.0	——	——	51
52	Carbon monoxide	320.5	——	320.5	——	——	92.77	2.387	12.50	74.20	——	——	52
53	Carbon dioxide	0.0	——	0.0	——	——	246.47*	——	——	——	——	——	53
54	Hydrogen sulfide	586.8	6337.*	637.1	6897.*	46086.*	235.63	7.161	4.30	45.50	——	——	54
55	Sulfur dioxide	0.0	——	0.0	——	——	167.22	——	——	——	——	——	55
56	Ammonia	359.0	——	434.4	——	——	589.48	3.581	15.50	27.00	——	——	56
57	Air	0.0	——	0.0	——	——	88.20	——	——	——	——	——	57
58	Hydrogen	273.8	——	324.2	——	——	192.74	2.387	4.00	74.20	——	——	58
59	Oxygen	0.0	——	0.0	——	——	91.59	——	——	——	——	——	59
60	Nitrogen	0.0	——	0.0	——	——	85.59	——	——	——	——	——	60
61	Chlorine	——	——	——	——	——	123.75	——	——	——	——	——	61
62	Water	0.0	*	*	0.0	0.0	970.18	——	——	——	——	——	62
63	Helium	0.0	——	0.0	——	——	——	——	——	——	——	——	63
64	Hydrogen chloride	——	——	——	——	——	190.43	——	——	——	——	——	64

12/12/86

NOTE: Numbers in this table do not have accuracies greater than 1 part in 1,000; in some cases extra digits have been added to calculated values to achieve consistency or to permit recalculation of experimental values.

(Cont'd)
NOTES AND REFERENCES FOR THE TABLE OF PHYSICAL CONSTANTS

Number	Compound	Formula	Molar mass (molecular weight)	Boiling point, °F 14.696 psia	Vapor pressure, psia 100 °F	Freezing point, °F 14.696 psia	Refractive index, n_D 60°F	Critical constants: Pressure, psia	Critical constants: Temperature, °F	Critical constants: Volume, ft³/lbm	Number
	See Note No. ->		A.	B.		C.	D.				
1	Methane	CH_4	19	4	a,b	c,57	i,57	4	4	4	1
2	Ethane	C_2H_6	19	35	a,b	c,57	h,55	35	35	35	2
3	Propane	C_3H_8	19	34	34	c,57	h,55	34	34	34	3
4	Isobutane	C_4H_{10}	19	30	32	57	h,55	32	32	32	4
5	n-Butane	C_4H_{10}	19	33	33	57	h,55	33	33	33	5
6	Isopentane	C_5H_{12}	19	57	57	57	57	43	43	43	6
7	n-Pentane	C_5H_{12}	19	57	57	57	57	43	43	43	7
8	Neopentane	C_5H_{12}	19	57	57	57	h,57	43	43	43	8
9	n-Hexane	C_6H_{14}	19	57	57	57	57	43	43	43	9
10	2-Methylpentane	C_6H_{14}	19	57	57	57	57	43	43	43	10
11	3-Methylpentane	C_6H_{14}	19	57	57	f,22,29	57	43	43	43	11
12	Neohexane	C_6H_{14}	19	57	57	57	57	43	43	43	12
13	2,3-Dimethylbutane	C_6H_{14}	19	57	57	57	57	43	43	43	13
14	n-Heptane	C_7H_{16}	19	57	57	57	57	43	43	43	14
15	2-Methylhexane	C_7H_{16}	19	57	57	57	57	43	43	43	15
16	3-Methylhexane	C_7H_{16}	19	57	57	f,38	57	43	43	43	16
17	3-Ethylpentane	C_7H_{16}	19	57	57	57	57	43	43	43	17
18	2,2-Dimethylpentane	C_7H_{16}	19	57	57	57	57	43	43	43	18
19	2,4-Dimethylpentane	C_7H_{16}	19	57	57	57	57	43	43	43	19
20	3,3-Dimethylpentane	C_7H_{16}	19	57	57	57	57	43	43	43	20
21	Triptane	C_7H_{16}	19	57	57	57	57	43	43	43	21
22	n-Octane	C_8H_{18}	19	57	57	57	57	43	43	43	22
23	Diisobutyl	C_8H_{18}	19	57	57	57	57	43	43	43	23
24	Isooctane	C_8H_{18}	19	57	57	57	57	43	43	43	24
25	n-Nonane	C_9H_{20}	19	57	57	57	57	43	43	43	25
26	n-Decane	$C_{10}H_{22}$	19	57	57	57	57	43	43	43	26
27	Cyclopentane	C_5H_{10}	19	57	57	57	57	57	57	57	27
28	Methylcyclopentane	C_6H_{12}	19	57	57	57	57	57	57	57	28
29	Cyclohexane	C_6H_{12}	19	57	57	57	57	57	57	57	29
30	Methylcyclohexane	C_7H_{14}	19	57	57	57	57	57	57	57	30
31	Ethene(Ethylene)	C_2H_4	19	63	b	c,63	h,26	63	63	63	31
32	Propene(Propylene)	C_3H_6	19	5	5	5	h,26	g,5	44	57	32
33	1-Butene(Butylene)	C_4H_8	19	57	57	57	h,26	57	57	57	33
34	cis-2-Butene	C_4H_8	19	57	25	57	h,26	57	57	57	34
35	trans-2-Butene	C_4H_8	19	57	25	57	h,26	57	57	57	35
36	Isobutene	C_4H_8	19	57	57	57	h,26	57	57	57	36
37	1-Pentene	C_5H_{10}	19	57	57	57	26	57	57	57	37
38	1,2-Butadiene	C_4H_6	19	57	57	56		a	a	a	38
39	1,3-Butadiene	C_4H_6	19	57	25	56	h,26	57	57	57	39
40	Isoprene	C_5H_8	19	57	57	56		a	a	a	40
41	Acetylene	C_2H_2	19	d,1	b	c,56		57	57	57	41
42	Benzene	C_6H_6	19	14	57	14	57	14	14	14	42
43	Toluene	C_7H_8	19	57	57	15	57	57	57	57	43
44	Ethylbenzene	C_8H_{10}	19	57	57	15	57	57	57	57	44
45	o-Xylene	C_8H_{10}	19	57	57	15	57	57	57	57	45
46	m-Xylene	C_8H_{10}	19	57	57	15	57	57	57	57	46
47	p-Xylene	C_8H_{10}	19	57	57	15	57	57	57	57	47
48	Styrene	C_8H_8	19	56	13	56	57	56	a,11	56	48
49	Isopropylbenzene	C_9H_{12}	19	57	r,57	57	57	57	57	57	49
50	Methyl alcohol	CH_4O	19	58	58	58	58	58	58	58	50
51	Ethyl alcohol	C_2H_6O	19	58	58	58	58	1	58	58	51
52	Carbon monoxide	CO	19	58	b	c,58	i,58	1	1	1	52
53	Carbon dioxide	CO_2	19	d,3	b	c,3	i,58	3	3	3	53
54	Hydrogen sulfide	H_2S	19	31	31	c,31	i,58	31	31	31	54
55	Sulfur dioxide	SO_2	19	1	23	c,58	i,58	58	58	58	55
56	Ammonia	NH_3	19	1	23	c,58	i,58	1	1	1	56
57	Air	N_2+O_2	19,40	45	b		i,60	45	45	45	57
58	Hydrogen	H_2	19	p,58	b	p,c,58	i,58	p,58	p,58	p,58	58
59	Oxygen	O_2	19	u,10	b	c,u,10	i,58	46	46	46	59
60	Nitrogen	N_2	19	6	b	c,6	i,58	6	6	6	60
61	Chlorine	Cl_2	19	1	2	c,58	h,58	2	2	2	61
62	Water	H_2O	19	u,10	37	58	58	37	37	37	62
63	Helium	He	19	7	b		i,58	7	s,7	7	63
64	Hydrogen chloride	HCl	19	1	23	c,58	i,58	58	58	58	64

12/12/86

(Cont'd)
NOTES AND REFERENCES FOR THE TABLE OF PHYSICAL CONSTANTS

	E.			F.	G.	H.	I.			J.		
	Density of liquid 14.696 psia, 60°F						Ideal gas 14.696 psia, 60°F			Specific Heat 60°F 14.696 psia Btu/(lbm·°F)		
Number	Relative density (specific gravity) 60°F/60°F	lbm/gal.	gal./lb mole	Temperature coefficient of density, 1/°F	Acentric factor, ω	Compressibility factor of real gas, Z 14.696 psia, 60°F	Relative density (specific gravity) Air = 1	ft³ gas/lbm	ft³ gas/gal. liquid	C_p, Ideal gas	C_p, Liquid	Number
1	b,a	b,a,v	b,a	b,a	4	57			b,a	8	b	1
2	h,35	h,v,35	h,35	h,35	57	57			h	8	53	2
3	h,34	h,34	h,34	h,34	34	57			h	8	34	3
4	h,n,32	h,n,32	h,n,32	h,n,32	32	57			h	30	32	4
5	h,33	h,33	h,33	h,33	33	57			h	8	33	5
6	57	57	57	57	57					8	36	6
7	57	57	57	57	57					8	47	7
8	h,57	h,57	h,57	h,57	57	57			h	8	24	8
9	57	57	57	57	57					8	47	9
10	57	57	57	57	57					8	57	10
11	57	57	57	57	57					8	57	11
12	57	57	57	57	57					8	57	12
13	57	57	57	57	57					8	57	13
14	57	57	57	57	57					57	47	14
15	57	57	57	57	57					57	57	15
16	57	57	57	57	57					57	38	16
17	57	57	57	57	57					57	57	17
18	57	57	57	57	57					57	57	18
19	57	57	57	57	57					57	57	19
20	57	57	57	57	57					57	57	20
21	57	57	57	57	57					57	57	21
22	57	57	57	57	57					57	47	22
23	57	57	57	57	57					57	57	23
24	57	57	57	57	57					57	57	24
25	57	57	57	57	57					57	47	25
26	57	57	57	57	57					57	47	26
27	57	57	57	57	57					8	22,59	27
28	57	57	57	57	r,57					57	22,59	28
29	57	57	57	57	57					8	9	29
30	57	57	57	57	57					57	22,59	30
31	b	b	b	b	57	57			b	8	b	31
32	h,5	h,5	h,5	h,5	5	57			h	8	17	32
33	h	h	h	h	57	57			h	57	24	33
34	h,r	h,r	h,r	h,r	25	57			h	57	17	34
35	h,r	h,r	h,r	h,r	25	57			h	57	r,17	35
36	h,r	h,r	h,r	h,r	57	57			h	57		36
37	57	57	57	57	57					57	57	37
38	h,r	h,r	h,r	h,r	r,57	a			h	57	57	38
39	h,r	h,r	h,r	h,r	25	a			h	57	54,59	39
40	r	r	r	r	57					57	48	40
41	b	b	b	b	r,57	57			b	8		41
42	57	57	57	57	57					57	15	42
43	57	57	57	57	57					16	15	43
44	57	57	57	57	57					16	15	44
45	57	57	57	57	57					16	15	45
46	57	57	57	57	57					16	15	46
47	57	57	57	57	57					16	15	47
48	j,50	j,50	j,50	j,52	a,51					57		48
49	57	57	57	57	57					61	42	49
50	58	58	58	58	58					62	62	50
51	62	62	62	62	58					62	62	51
52	k,39	k,39	k,39		39				t	8	b	52
53	h,3	h,3	h,3	h,3					h	8		53
54	h,31	h,v,31	h,31	h,31	31	58			h	8	31	54
55	h	h	h		23	58			h	8		55
56	h	h	h		23	58			h	58		56
57	k,45	k,45	k,45						t	45		57
58	k,p,21	k,p,21	k,p,21		p,23	58			t	8	b	58
59	k,46	k,46	k,46		q,58	58			t	8	b	59
60	k,6	k,6	k,6		6	58			t	8	b	60
61	h	h	h		2	a			h	58		61
62	37	37	37	41	r,58					27	28	62
63	k	k	k		58	58			t	8	b	63
64	h	h	h	h	58	58			h	58		64

12/12/86

(Cont'd)
NOTES AND REFERENCES FOR THE TABLE OF PHYSICAL CONSTANTS

Number	See Note No. ->	K. Heating value, 60°F — Net: Btu/ft³ Ideal gas, 14.696 psia	K. Net: Btu/lbm Liquid	K. Gross: Btu/ft³ Ideal gas, 14.696 psia	K. Gross: Btu/lbm Liquid	K. Gross: Btu/gal. Liquid	L. Heat of vaporization 14.696 psia at boiling point, Btu/lbm	M. Air required for combustion ideal gas ft³(air)/ft³(gas)	Flammability limits, vol % in air mixture: Lower	Flammability limits: Higher	ASTM octane number: Motor method D-357	ASTM octane number: Research method D-908	Number
	Compound												
1	Methane	57	b	57	b	b	4		x	x			1
2	Ethane	57	h,57	57	h,57	h,57	35		x	x	e	e,g	2
3	Propane	57	h,57	57	h,57	h,57	34		x	x		e,g	3
4	Isobutane	57	h,57	57	h,57	h,57	32		x	x		e,g	4
5	n-Butane	57	h,57	57	h,57	h,57	33		x	x	g	g	5
6	Isopentane	57	57	57	57	57	57		x	x			6
7	n-Pentane	57	57	57	57	57	57		x	x	g	g	7
8	Neopentane	57	h,57	57	h,57	h,57	57		x	x			8
9	n-Hexane	57	57	57	57	57	57		x	x			9
10	2-Methylpentane	57	57	57	57	57	57		x	x			10
11	3-Methylpentane	57	57	57	57	57	57		x	x			11
12	Neohexane	57	57	57	57	57	57		x	x			12
13	2,3-Dimethylbutane	57	57	57	57	57	57		x	x		e	13
14	n-Heptane	57	57	57	57	57	57		x	x			14
15	2-Methylhexane	57	57	57	57	57	57		x	x			15
16	3-Methylhexane	57	57	57	57	57	57		a,x	a,x			16
17	3-Ethylpentane	57	57	57	57	57	57		a,x	a,x			17
18	2,2-Dimethylpentane	57	57	57	57	57	57		a,x	a,x			18
19	2,4-Dimethylpentane	57	57	57	57	57	57		a,x	a,x			19
20	3,3-Dimethylpentane	57	57	57	57	57	57		a,x	a,x			20
21	Triptane	57	57	57	57	57	57		a,x	a,x	e	e	21
22	n-Octane	57	57	57	57	57	57		x	x			22
23	Diisobutyl	57	57	57	57	57	57		a,x	a,x			23
24	Isooctane	57	57	57	57	57	57		x	x			24
25	n-Nonane	57	57	57	57	57	57		x	x			25
26	n-Decane	57	57	57	57	57	57		a,x	a,x			26
27	Cyclopentane	57	57	57	57	57	57		a,x	a,x	g	e	27
28	Methylcyclopentane	20	20	20	20	20	57		x	x			28
29	Cyclohexane	57	57	57	57	57	57		x	x			29
30	Methylcyclohexane	20	20	20	20	20	57		x	x			30
31	Ethene(Ethylene)	57	b	57	b	b	63		x	x		e	31
32	Propene(Propylene)	57	a	57	a	a	57		x	x		e	32
33	1-Butene(Butylene)	57	h,57	57	h,57	h,57	57		x	x	g		33
34	cis-2-Butene	57	h,57	57	h,57	h,57	57		a,x	a,x			34
35	trans-2-Butene	57	h,57	57	h,57	h,57	57		a,x	a,x			35
36	Isobutene	57	h,57	57	h,57	h,57	57		a,x	a,x			36
37	1-Pentene	57	57	57	57	57	57		x	x			37
38	1,2-Butadiene	57	h,57	57	h,57	h,57	56		a,x	a,x			38
39	1,3-Butadiene	57	h,57	57	h,57	h,57	56		x	x			39
40	Isoprene	57	57	57	57	57	56		a,x	a,x			40
41	Acetylene	57	a	57	a	a	56		x	x			41
42	Benzene	57	57	57	57	57	14		x	x	e		42
43	Toluene	57	57	57	57	57	57		x	x	e	e	43
44	Ethylbenzene	57	57	57	57	57	57		x	x		e	44
45	o-Xylene	57	57	57	57	57	57		x	x			45
46	m-Xylene	57	57	57	57	57	57		x	x	e	e	46
47	p-Xylene	57	57	57	57	57	57		x	x	e	e	47
48	Styrene	20	20	20	20	20	56		x	x	e	e	48
49	Isopropylbenzene	57	57	57	57	57	57		x	x		e	49
50	Methyl alcohol	62	62	62	62	62	58		x	x			50
51	Ethyl alcohol	62	62	62	62	62	58		x	x			51
52	Carbon monoxide	8	b	8	b	b	58		x	x			52
53	Carbon dioxide						58						53
54	Hydrogen sulfide	18	h,58	18	h,58	h,58	31		x	x			54
55	Sulfur dioxide						56						55
56	Ammonia	58		58			58		x	x			56
57	Air						45						57
58	Hydrogen	58		58			p,58		x	x			58
59	Oxygen						46						59
60	Nitrogen						58						60
61	Chlorine						56						61
62	Water	8	w,8	w,8			37						62
63	Helium												63
64	Hydrogen chloride						56						64

12/12/86

(Cont'd)

NOTES FOR THE TABLE OF PHYSICAL CONSTANTS

a. Values in parentheses are estimated values.
b. The temperature is above the critical point.
c. At saturation pressure (triple point).
d. Sublimation point.
e. The + sign and number following specify the number of cm^3 of TEL added per gallon to achieve the ASTM octane number of 100, which corresponds to that of Isooctane (2,2,4-Trimethylpentane.)
f. These compounds form a glass. An earlier investigator [12] reported a value of -244°F for the freezing point of 3-Methylpentane.
g. Average value from octane numbers of more than one sample.
h. Saturation pressure and 60°F.
i. Index of refraction of the gas.
j. The temperature difference correction from [51].
k. Densities of the liquid at the normal boiling point.
m. Heat of sublimation.
n. Equation 2 of the reference was refitted to give: $a = 0.7872957; b = 0.1294083; c = 0.03439519$.
p. Normal hydrogen (25% para, 75% ortho).
q. The vapor pressure [5] at the critical temperature [43].
r. An extrapolated value.
s. The value for the critical point of helium depends upon the temperature scale chosen. See reference 7.
t. Gas at 60°F and the liquid at the normal boiling point.
u. Fixed points on the 1968 International Practical Temperature Scale (IPTS-68).
v. Densities at the normal boiling point are: Ethane, 4.540 [34]; Propane, 4.848 [33]; Propene, 5.083 [5]; Hydrogen Chloride, 9.948 [56]; Hydrogen Sulfide, 7.919 [30]; Ammonia, 5.688 [56]; Sulfur Dioxide, 12.20 [56].
w. Technically, water has a heating value in two cases: net (-1060. Btu/lbm) when water is liquid in the reactants and gross (+50.313 Btu/ft^3) when water is gas in the reactants. The value is the ideal heat of vaporization (enthalpy of the ideal gas less the enthalpy of the saturated liquid at the vapor pressure). This is a matter of definition; water does not burn.
x. Extreme values of those reported by ref. 24.

A. Molar mass (molecular weight) is based upon the following atomic weights: C = 12.011; H = 1.00794; O = 15.9994; N = 14.0067; S = 32.06, Cl = 35.453. The values were rounded off after calculating the molar mass using all significant figures in the atomic weights.
B. Boiling point: the temperature at equilibrium between the liquid and vapor phases at 14.696 psia.
C. Freezing point: the temperature at equilibrium between the crystalline phase and the air saturated liquid at 14.696 psia.
D. The refactive index reported refers to the liquid or gas and is measured for light of wavelength corresponding to the sodium D-line (589.26 nm).
E. The relative density (specific gravity): ρ(liquid, 60°F)/ρ(water, 60°F). The density of water at 60°F is 8.33718 lbm/gal.
F. The temperature coefficient of density is related to the expansion coefficient by:

$$\frac{1}{\rho}\left(\frac{\partial \rho}{\partial T}\right)_P = -\frac{1}{V}\left(\frac{\partial V}{\partial T}\right)_P, \quad \text{in units of } 1/^\circ\text{F}$$

G. Pitzer acentric factor:

$$\omega = -\log_{10}(P/P_c) - 1, \quad P \text{ at } T = 0.7\,T_c$$

H. Compressibility factor of the real gas, $Z = (PV)/(RT)$, is calculated using the second virial coefficient equation.
I. The density of an ideal gas relative to air is calculated by dividing the molar mass of the of the gas by 28.9625, the calculated average molar mass of air. See ref. 39 for the average composition of dry air. The specific volume of an ideal gas is calulated from the ideal gas equation. The volume ratio is: V(ideal gas)/ V(liquid in vacuum).
J. The liquid value is not rigorously C_P, but rather it is C_S, the heat capacity at saturated conditions, defined by:

$$C_S = C_P - T\left(\frac{\partial V}{\partial T}\right)_P \frac{dP_S}{dT}$$

When dealing with liquids far from the critical point, $C_S \approx C_P$.
K. The heating value is the negative of the enthalpy of combustion at 60°F and 14.696 psia in an ideal reaction (one where all gasses are ideal gasses). For an arbitrary organic compound, the combustion reaction is:

$$C_nH_mO_hS_jN_k(s, l, or\, g) + (n + \frac{m}{4} - \frac{h}{2} + j)O_2(g) \longrightarrow nCO_2(g) + \frac{m}{2}H_2O(g\, or\, l) + \frac{k}{2}N_2(g) + jSO_2(g)$$

where s, l,and g denote respectively solid, liquid and ideal gas. For gross heating values, the water formed is liquid; for net heating values, the water formed is ideal gas. Values reported are on a dry basis. To account for water in the heating value, see GPA 2172. The Btu/lbm or gal. liquid column assumes a reaction with the fuel in the liquid state, while the Btu/ft^3 ideal gas column assumes the gas in the ideal gas state. Therefore, the values are not consistent if used in the same calculation, e.g., a gas plant balance.
L. The heat of vaporization is the enthalpy of the saturated vapor at the boiling point at 14.696 psia minus the enthalpy of the saturated liquid at the same pressure.
M. Air required for the combustion of ideal gas for compounds of formula $C_nH_mO_hS_jN_k$ is:

$$V(air)/V(gas) = (n + \frac{m}{4} - \frac{h}{2} + j)/0.20946$$

COMMENTS

Units - all dimensional values are reported in terms of the following three units, which are defined in terms of the corresponding SI units:

mass - pound (avdp), lbm = 0.45359337 kg
length - foot, ft = 0.3048 m
temperature - degree fahrenheit, $(t/^\circ F) = 32 + [1.8(t/^\circ C)]$
Temperature of the Celsius scale is defined by the International Practical Temperature of 1968 (IPTS-68).

Other derived units are:
volume - cubic foot, $ft^3 = 0.02831685\ m^3$
gallon = 231 in^3 = 0.003785412 m^3
pressure - pound per square inch absolute = 6894.757 Pa
energy - British thermal unit (I.T.), Btu = 251.9958 cal (I.T.) = 1055.056 J

Physical constants:
gas constant, R = 1.98598 Btu(I.T.)/(°R·lb mol)
10.7316 ft^3 psia/(°R·lb mol)
8.31448 J/(K · mol)

Conversion factors:
1 ft^3 = 7.480520 gal.
1 lbm/ft^3 = 0.1336806 lbm/gal = 16.018462 kg/m^3
1 psia = 0.06804596 atm
1 atm = 14.69595 psia = 760 Torr
1 Btu (I.T.) = 252.1644 cal_{th}

REFERENCES FOR THE TABLE OF PHYSICAL CONSTANTS

1. Ambrose, D., National Physical Laboratory, Teddington, Middlesex, England: Feb. 1980, NPL Report Chem 107.
2. Ambrose, D.; Hall, D. J.; Lee, D. A.; Lewis, G. B.; Mash, C. J., J. Chem. Thermo., 11, 1089 (1979).
3. Angus, S.; Armstrong, B.; de Reuck, K. M., Eds. "Carbon Dioxide. International Thermodynamic Tables of the Fluid State-3"; Pergamon Press: Oxford, 1976.
4. Angus, S.; Armstrong, B.; de Reuck, K. M., Eds. "Methane. International Thermodynamic Tables of the Fluid State-5"; Pergamon Press: Oxford, 1978.
5. Angus, S.; Armstrong, B.; de Reuck, K. M., "Propylene (Propene). International Thermodynamic Tables of the Fluid State-7"; Pergamon Press: Oxford, 1980.
6. Angus, S.; de Reuck, K. M.; Armstrong, B., Eds. "Nitrogen. International Thermodynamic Tables of the Fluid State-6"; Pergamon Press: Oxford, 1979.
7. Angus, S.; de Reuck, K. M.; McCarthy, R. D., Eds. "Helium. International Thermodynamic Tables of the Fluid State-4"; Pergamon Press: Oxford, 1977.
8. Armstrong, G. T.; Jobe, T. L., "Heating Values of Natural Gas and its Components," NBSIR 82-2401, May 1982.
9. Aston, J. G.; Szasz, G. J.; Fink, H. L., J. Am. Chem. Soc., 65, 1135 (1943).
10. Barber, C. R., Metrologia 5, 35 (1969).
11. Boundy, R. H.; Boyer, R. F., (Eds.), "Styrene, Its Polymers, Copolymers and Derivatives," A.C.S. monograph No. 115, Reinholt, N.Y., 1952.
12. Bruun, J. H.; Hicks-Bruun, M. M., J. Res. NBS, 5, 933 (1930).
13. Chaiyavech, P.; Van Winkle, M., J. Chem. Eng. Data, 4, 53 (1959).
14. Chao, J., "Benzene," Key chemical data books, Thermodynamics Research Center, Texas A & M University, College Station, Texas (1978).
15. Chao, J., Hydrocarbon Processing, 59(11), 295 (1979).
16. Chao, J.; Hall, K. R., Thermochimica acta, 72, 323 (1984).
17. Chao, J.; Hall, K. R.; Yao, J., Thermochimica acta, 64, 285 (1983).
18. CODATA Task Group on Key Values for Thermodynamics, CODATA Special Report No. 7, 1978.
19. Commission on Atomic Weights and Isotopic Abundances, Pure and Appl. Chem. 49, 1102 (1983).
20. Cox, J. D.; Pilcher, G., "Thermochemistry of Organic and Organometallic Compounds," Academic Press, London, 1970.
21. Dean, J. W., "A Tabulation of the Properties of Normal Hydrogen from Low Temperature to 300 K and from 1 to 100 Atmospheres," NBS Tech. Note 120, November 1961.
22. Douslin, D. R.; Huffman, H. M., J. Am. Chem. Soc., 68, 1704 (1946).
23. Edwards, D. G., "The Vapor Pressure of 30 Inorganic Liquids Between One Atmosphere and the Critical Point," Univ. of Calif., Lawrence Radiation Laboratory, UCRL-7167. June 13, 1963.
24. Engineering Sciences Data Unit, "EDSU, Engineering Sciences Data," EDSU International Ltd., London.
25. Flebbe, J. L.; Barclay, D. A.; Manley, D. B., J. Chem. Eng. Data, 27, 405 (1982).
26. Francis, A. W., J. Chem. Eng. Data, 5, 534 (1960).
27. Friedman, A. S.; Haar, L., J. Chem. Phys. 22, 2051 (1954).
28. Ginnings, D. C.; Furukawa, G. T., J. Am. Chem. Soc. 75, 522 (1953).
29. Glasgow, A. R.; Murphy, E. T.; Willingham, C. B.; Rossini, F. D., J. Res. NBS, 37, 141 (1946).
30. Goodwin, R. D., "Isobutane: Provisional Thermodynamic Functions from 114 to 700K at Pressures to 700 Bar," NBSIR 79-1612, July 1979.
31. Goodwin, R. D., "Hydrogen Sulfide Provisional Thermochemical Properties from 188 to 700 K at Pressures to 75 MPa," NBSIR 83-1694, October 1983.
32. Goodwin, R. D.; Haynes, W. M., "Thermophysical Properties of Isobutane from 114 to 700 K at Pressures to 70 MPa," NBS Tech. Note 1051, January 1982.
33. Goodwin, R. D.; Haynes, W. M., "Thermophysical Properties of Normal Butane from 135 to 700 K at Pressures to 70 MPa," NBS Monograph 169, April 1982.
34. Goodwin, R. D.; Haynes, W. M., "Thermophysical Properties of Propane from 85 to 700 K at Pressures to 70 MPa," NBS Monograph 170, April 1982.
35. Goodwin, R. D.; Roder, H. M.; Straty, G. C.; "Thermophysical Properties of Ethane, 90 to 600 K at Pressures to 700 bar," NBS Tech. Note 684, August 1976.
36. Guthrie, G. B.; Huffman, H. M., J. Am. Chem. Soc., 65, 1139 (1943).
37. Haar, L.; Gallagher, J. S.; Kell, G. S., "NBS/NRC Steam Tables," Hemisphere Publishing Corporation, Washington, 1984.
38. Huffman, H. M.; Park, G. S.; Thomas, S. B., J. Am. Chem. Soc., 52, 3241 (1930).
39. Hust, J. G.; Stewart, R. B., "Thermodynamic Property Values for Gaseous and Liquid Carbon Monoxide from 70 to 300 Atmospheres," NBS Technical Note 202, Nov. 1963.
40. Jones, F. E., J. Res. NBS, 83, 419 (1978).
41. Kell, G. S., J. Chem. Eng. Data, 20, 97 (1975).
42. Kishimoto, K.; Suga, H.; Syuzo, S., Bull. Chem. Soc. Japan, 46, 3020 (1973).
43. Kudchadker, A. P.; Alani, G. H.; Zwolinski, B. J., Chem. Rev. 68, 659 (1968).
44. Marchman, H.; Prengle, H. W.; Motard, R. L., Ind. Eng. Chem., 41, 2658 (1949).
45. "The Matheson Unabridged Gas Data Book"; Matheson Gas Products; New York, 1974.
46. McCarty, R. D.; Weber, L. A., "Thermophysical Properties of Oxygen from the Freezing Liquid Line to 600 R for Pressures to 5000 Psia", NBS Technical Note 384, July 1971.
47. Messerly, J. F.; Guthrie, G. B.; Todd, S. S.; Finke, H. L., J. Chem. Eng. Data, 12, 338 (1967).
48. Messerly, J. F.; Todd, S. S.; Guthrie, G. B., J. Chem. Eng. Data, 15, 227 (1970).
49. Miller, A.; Scott, D. W., J. Chem. Phys., 68(3), 1317 (1978).
50. Miller, L.P.; Wachter, N. W.; Fried, V., J. Chem. Eng. Data, 20, 417 (1975).
51. Ohe, S., "Computer Aided Data Book of Vapor Pressure," Data Book Publishing Co., Tokyo, Japan, 1976.
52. Riddick, J. A.; Bunger, W. B., "Organic Solvents, Physical Properties and Methods of Purification," Wiley-Interscience: New York, 1970.
53. Roder, H. M., "Measurements of the Specific Heats, C_p, and C_v, of Dense Gaseous and Liquid Ethane," J. Res. Nat. Bur. Stand. (U.S.) 80A, 739 (1976).
54. Scott, R. B.; Meyers, C. H.; Rands, R. D.; Brickwedde, F. G., Bekkedahl, N., J. Res. NBS, 35, 39 (1945).
55. Sliwinski, P., Z. Physik. Chem. (Frankfurt) 63, 263 (1969).
56. Stull, D. R.; Westrum, E. F.; Sinke, G. C., "The Chemical Thermodynamics of Organic Compounds," John Wiley & Sons, Inc., New York, 1969.
57. "TRC Thermodynamic Tables - Hydrocarbons," Thermodynamics Research Center, Texas A&M University System: College Station, Texas.
58. "TRC Thermodynamic Tables–Non-Hydrocarbons," Thermodynamics Research Center, Texas A&M University System: College Station, Texas.
59. Vargaftik, N. B.; "Tables on the Thermophysical Properties of Liquids and Gases," Hemisphere Publishing Corporation, Washington D.C., 1975.
60. Weast, R. C., "Handbook of Chemistry and Physics," The Chemical Rubber Co: Cleveland, Ohio, 1969.
61. Wilhoit, R. C.; Aljoe, K. L.; Adler, S. B.; Kudchadker, S. A., "Isopropylbenzene and 1-Methyl-2-, -3-, and -4-isopropylbenzene," API Monograph Series No. 722, American Petroleum Institute, 1984.
62. Wilhoit, R. C.; Zwolinski, B. J., J. Phys. Chem. Ref. Data, 2, Suppl. no. 1 (1973).
63. Younglove, B. A., J. Phys. Chem. Ref. Data, 11, Suppl. no. 1 (1982).

Nomograph for calculating density and specific volume of gases and vapors

The density of a gas or vapor can be calculated from the equation for an ideal gas:

$$\rho = \frac{144P}{RT} = \frac{MP}{10.72\,T} = \frac{2.70\,PSg}{T}$$

where: ρ = Density of gas, lb/cu ft
P = Absolute pressure, psi (psig + 14.7)
R = Individual gas constant (10.72/M)
M = Molecular weight of gas
T = Temperature, °R
Sg = Specific gravity of individual gas relative to air (= to ratio of gas molecular weight to that of air)

The above equation and corresponding nomograph (Figure 1) can be used to calculate the density (or specific volume) of gases at low pressures and high temperatures, where the ideal gas law holds. Note that the pressure scale is calibrated in psi gauge so that the correction of 14.7 psi is not needed when using the nomograph.

Example. What is the density of dry methane if its temperature is 100°F and its pressure is 15 psig?

Connect	With	Mark or Read
M = 16	t = 100°F	Index
Index Mark	P = 15 psig	ρ = 0.08 lb/cu ft

Source

"Fluid Flow Through Pipe, Fittings and Valves," *Technical Paper No. 410*, A-11, Crane Company, Chicago, IL (1957).

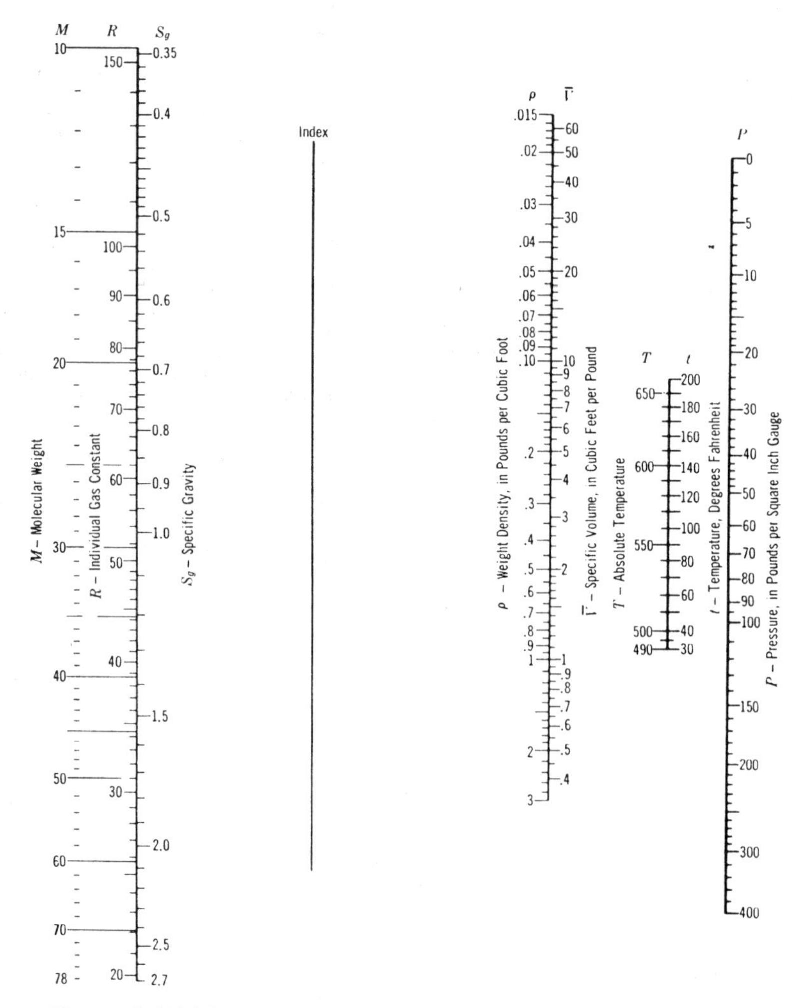

Figure 1. Weight density and specific volume of gases and vapors.

CONSIDERATIONS FOR SELECTING ENERGY MEASUREMENT EQUIPMENT

Keeping Track of Gas Heating Value is Important Phase of Today's Market

Paul E. Kizer, Applied Automation, Inc., Houston, Texas

With natural gas energy determination becoming an integral part of the transmission and distribution business in today's market, companies are placing increased emphasis on choosing the correct measuring device. Most current contracts have a Btu specification, and many others use MMBtu instead of gas volume.

Here is what happens at the burner tip. The following is the exothermic reaction of methane and oxygen (80% to 95% of natural gas):

$$CH_4 + 2O_2 \rightarrow CO_2 + 2H_2O + \text{Heat (1010 Btu/cf)}$$ [1]

for the case of ethane (up to 3% of natural gas):

$$2C_2H_6 + 7O_2 \rightarrow 4CO_2 + 6H_2O + \text{Heat (1769 Btu/cf)}$$ [1]

and for the case of propane (usually up to 0.5%, except where propane/air peak shaving is used):

$$C_3H_8 + 5O_2 \rightarrow 3CO_2 + 4H_2O + \text{Heat (2516 Btu/cf)}$$ [1]

Since natural gas is a mixture of these three hydrocarbons plus small amounts of other compounds, these three chemical reactions account for 90% to 99% of the heat generated when the gas is burned. However, one mole (a mole is a standard number of molecules [6.023×10^{23}, Avogadro's number])[17] of propane will generate more than twice the heat of one mole of methane ($CH_4 = 1010$ Btu/cf vrs $C^3H^8 = 2516$ Btu/cf at 60°F and 414.696 psia). A small error in the heavier or longer chain hydrocarbon components of natural gas leads to a larger error in the Btu calculation. Consequently, a small shift in the composition of any component can affect the heating value of a cubic foot of carefully measured gas.

The term MMBtu first came into wide use after the 1978 Natural Gas Policy Act changed the method of custody transfer from Mcf to MMBtu. This method was upheld in a court case involving FERC Order 356 (Jan. 1984).[2]

How much is it worth to keep track of the Btu? Postulate a ±5% error between doing a lab determination of the heating value on a spot sample of gas and on-line (nearly continuous) monitor of heating value, which results in a ±50 Btu difference at a 1,000 Btu/cf level. A thousand Btu/cf is fairly average for natural gas.

On a station that has 50 MMcf/d at $2.50 per Mcf or MMBtu, this is $125,000 worth of gas per day. Five percent of this is $6,250 per day. If a process chromatograph, $50,000 installed cost, is used to determine the energy content, a payout of less than 10 days is obtained on a 50 MMcf/d station. This explains why most major interconnects have Btu measurement on-line today.

General methods

First, consider the direct methods of energy determination, since this is the oldest method in use today. All direct methods involve the complete oxidation (burning) of gas to determine its energy content. There are two basic classes of direct measurement devices, calorimeter and thermtitrator.

Early on, the term representing 100,000 Btus was one therm of energy. One dekatherm (10 therms) is equal to 1,000,000 Btu or 1 MMBtu.[21] Perhaps this is where the name thermtitrator came from. However, as discussed earlier, MMBtu[2] has supplanted the therm unit of measurement for now.

The calorimeter was adapted from a laboratory physical chemistry method called "bomb calorimetry" These were laboratory instruments that yield precise results when used under controlled conditions. However, each determination required 4 to 8 hours to complete. This is fine for a research project, but in a production setting it is easy to see how gas sample cylinders would start piling up. The American calorimeter[3] provided laboratory type results in less time, but the lab conditions had to be tightly controlled with continually manned equipment. For on-line energy content measurement, the Cutler-Hammer Recording Calorimeter stood alone for decades.[3,4]

While this device still requires rigid environmental conditions, it runs unattended for long periods of time (weeks). It also provides a recorder output, which allows the correlation of the heating value with the volume of gas. The recording calorimeter's stated precision is ±5 Btu out of 1,000 Btus, although many operators can do better.

With the introduction of solid-state electronics, micro calorimetry devices have been tried, though not with much commercial success. Hart Scientific, Provo, Utah, has used a catalyst bed to perform a controlled isothermal oxidation on a very precise volume of gas. Initial results seem to show ±0.5 Btu at 1,000 Btu precision.[5]

The thermtitrator devices utilize a more elegant chemical method, stoichiomerry—that is, air to gas ratios. One instrument, the PMI thermtitrator, uses the temperature difference between two flames to maintain the gas flow ratios at stoichiometric conditions.[6,7]

Accuracy is around ±1 Btu at 1,000 Btus and compares favorably with the recording calorimeter.[8] The other stoichiometric device is the Honeywell HVT. It uses a zirconium oxide oxygen sensor to determine stoichiometric conditions and automatically adjusts air/gas ratios to minimize excess O. Both devices yield rapid reaction to step changes in Btu content and are useful in process control applications such as boiler feed forward controls.

There are two classes of indirect Btu or energy measurement, both chromatographic devices, relying on the separation, identification, and quantification of all the components that make up the natural gas mixture.[9] The two classes are: laboratory chromatographs and on-line or process chromatographs. Through the years a number of different people have compared calorimetry to chromatography with varied results.[10,11]

The laboratory chromatograph (Figure 1) is just that. It is located in a laboratory. Somehow a gas sample must be obtained and transported in a cylinder to the lab to be analyzed.

How the sample is obtained is the main difference between lab and on-line gas chromatographs (GCs). The natural gas samples analyzed in a lab today usually are called spot or grab samples. A cleaned vessel (cylinder), sometimes N_2 filled, is purged and filled at least three times for reliable sampling site. The pressured-up bottle is carried to a lab to be analyzed. More and more samples today are being collected using a continuous sampler. This device takes a small aliquot of gas at a predetermined time interval so that the bottle is filled at the end of the sampling cycle required by the gas contract.

With recent developments, these devices have been tied to electronic flow computers to give flow-weighted composite samples rather than time-weighted ones. An example of such an installation is in Figure 2.

Flow weighting generally gives even better results than time-weighted samples if there are variations in the flowrate at the sampled station. Great care should be exercised when analyzing a sample cylinder of natural gas obtained by any of the previous methods. For instance, the sample bottle should be heated to 120°F for at least 2 hours before the analysis is made to ensure that all the heavier components are analyzed. A complete discussion of the problems associated with the recommendations for bottle sample analysis have been published previously.[12,9]

The process or on-line gas chromatograph (GC) (Figure 3) is designed to continually purge a small amount of the gas flowing through the station or pipeline to provide a representative sample at the time the GC is ready to make a new analysis. Typically, this is once every 6 minutes.

The main advantage of continuous sampling and analysis is the documentation of the duration and the amount of deviation of the energy content that has occurred during the measurement period. It is possible to assign a heating value to a specific volume of gas if the on-line GC is tied directly to

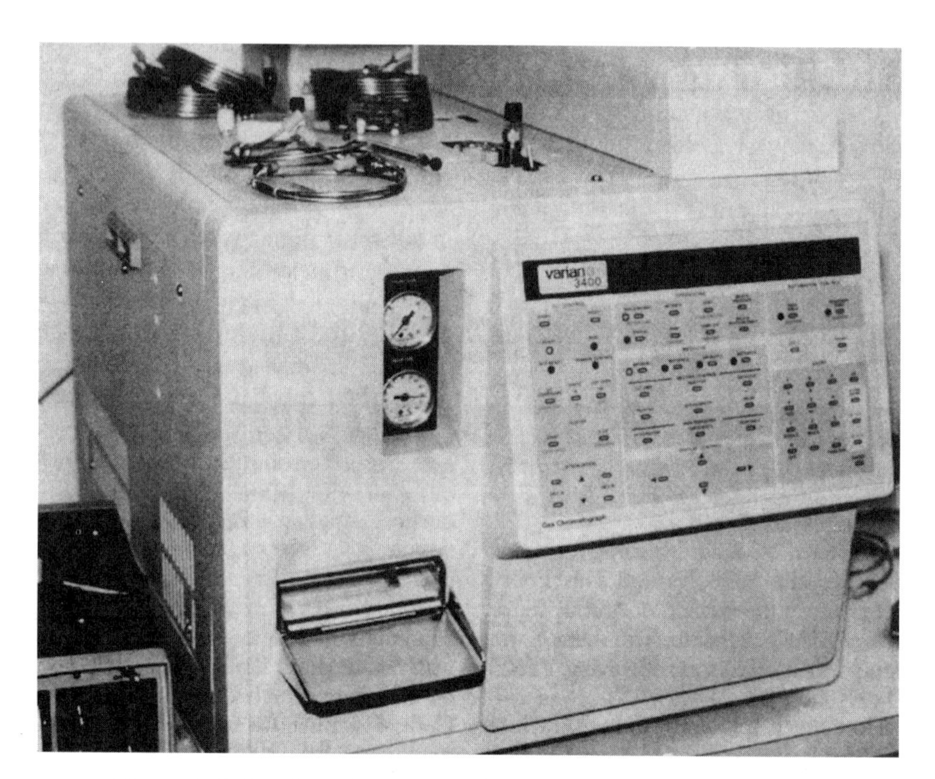

Figure 1. Laboratory chromatograph used to measure gas samples.

Figure 2. Example of on-line sampler connected to electronic flow computers.

Figure 3. Process sampler continually purges gas moving through a station or pipeline.

an electronic flow computer. This, of course, allows actual billing on an MMBtu custody transfer basis.

GCs—lab or on-line—using the computing power of today's modern microprocessor are capable of repeatability values of ±0.5 Btu at 1,000 Btu/cf. While the precision of a GC makes it well suited for custody transfer applications, the 6- to 10-minute analysis time makes it difficult to use for burner feed forward control. Generally, the on-line analyzers do not require much environmental protection at all. Most require protection from rain.

How chromatographs work

The way a modern microprocessor controlled on-line chromatograph works has been thoroughly discussed in previous papers.[13–15] However, it would be helpful to review briefly just how a GC works. Again, except for the sample conditioning system, this discussion generally applies to lab GCs as well as on-line GCs. For purposes of this discussion, the on-line GC has been divided into the following four sections:

- Sample conditioning system
- Gas chromatograph oven
- Chromatograph controller
- Input/output chassis

The sequence also is the route the sample takes from the pipeline to the printer. This makes it a logical flow for discussion as well.

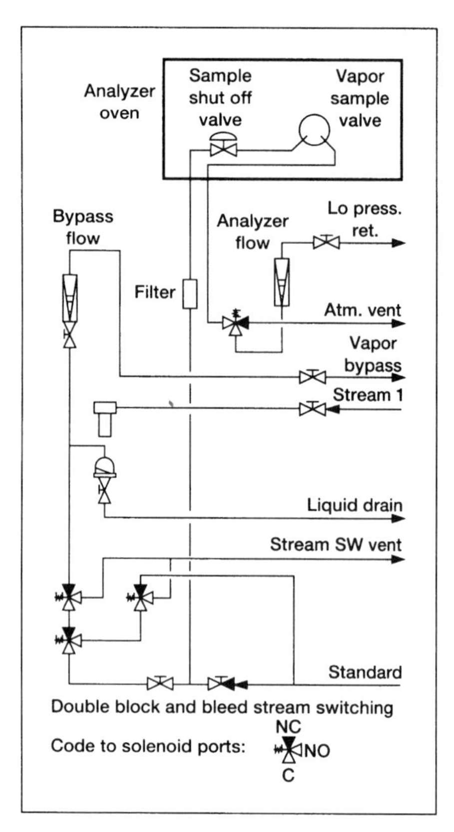

Figure 4. Bypass loop reduces sampling lag time.

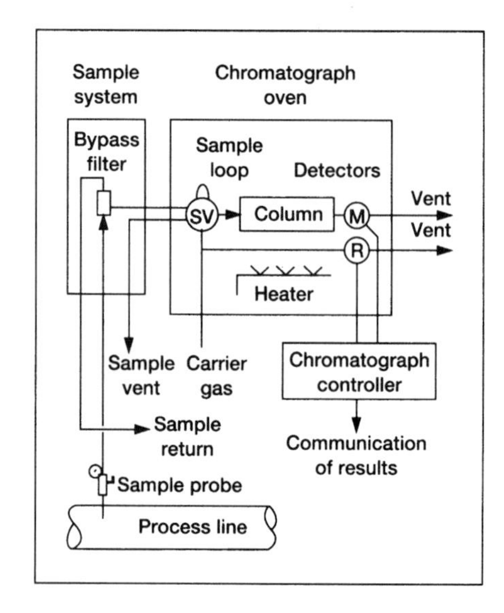

Figure 5. Block diagram illustrates gas chromatograph oven.

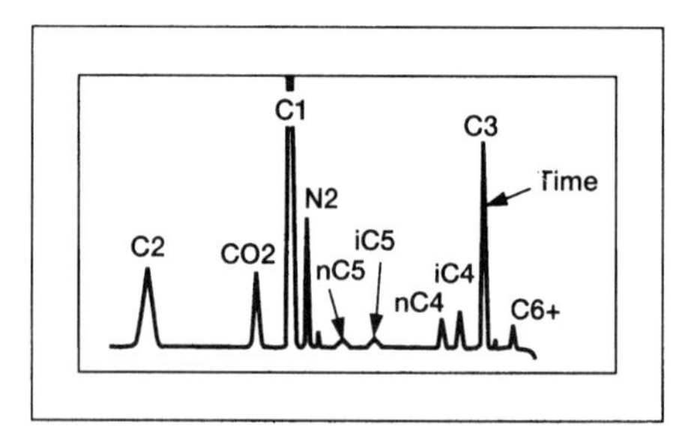

Figure 6. Separated components move through cell detectors for quantifying sample.

Sample conditioning system

To obtain a representative sample of gas, a probe should be inserted into the middle third of the pipeline. This procedure is as essential for good sampling for on-line GCs as it is for spot or composite samples. A major difference, however, is that the gas pressure is reduced to 15 psig at the probe for transport to the analyzer to ensure no heavy component (C^6+) is lost. With spot or composite samples, care is taken to maintain pipeline conditions. Sometimes it is useful to use a sample bypass loop to reduce sampling lag times, as shown in Figure 4.

Note that filtering and liquid coalescing is accomplished, if needed, to protect the integrity of the analyzer columns. Sometimes double block and bleed sample switching is used to prevent cross-contamination by the standard or other sample streams.

Gas chromatograph oven

Once the gas enters the oven, a repeatable gas sample is measured by a sample inject valve. The chromatograph controller then blocks the sample flow and allows it to equilibrate to atmospheric pressure. The sample is then swept onto a set of carefully selected chromatographic columns by a stream of helium gas for sequential separation of each component in the natural gas mixture. A block diagram is shown in Figure 5.

As each column does its specified separation job, it is switched out of the helium stream. The separated components make their way through the thermal conductivity cell detector where they can be quantified. An example of a detector, output is shown in Figure 6.

These peaks must be gated and the area under each peak integrated to be quantified, which is the job of the chromatograph controller.

Chromatograph controller

All the timing functions such as sample inject, column reverse, auto-calibration, and log results must be performed by the microprocessor controller board. In addition, this device must detect each peak, open and then close integration gates, identify the peak, and assign the correct response factor. The controller must then calculate the mole fraction of each component. Since these peaks are quite arbitrary in size due to equipment variability setup and environment, high-quality standards should be used to calibrate these instruments. This calibration assures accurate determination of mole fraction possible. All Btu determination methods require high quality standard samples.

On a Btu GC, the peaks also must be normalized to compensate for barometric pressure changes (which cause sample inject size variations). This allows the calculation of Btu to repeatability of ±0.5 Btu at 1,000 Btu/cf. This processor must also calculate the specific gravity, compressibility, condensable liquids, wobbie index, as well as the Btu. Calculation results must be made available to the outside world. This is the job of the input/output chassis.

Input/output chassis

The device houses communications option cards such as printer interface cards, analog output cards, and computer interface cards. These usually can be configured based on the customer's needs. However, caution is advised as specific computer protocols may or may not be supported by one vendor or another. One function for this input/output chassis, for example, is to hold an RS232 output card. This card would log results to an analyzer interface unit (AIU), which would then be connected to up to 30 flow computers (Fig. 7).

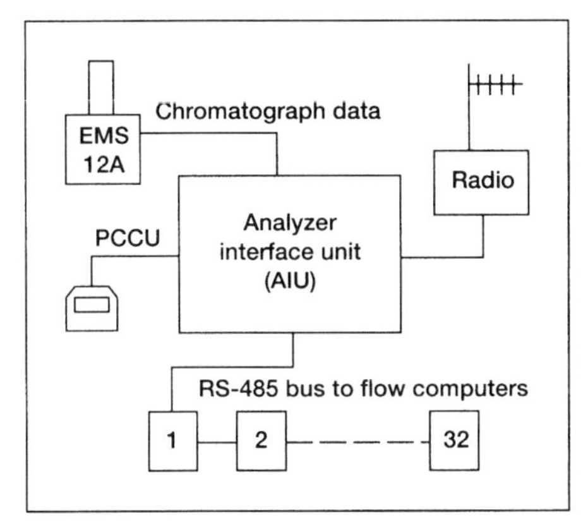

Figure 7. Output card logs results in an analyzer interface unit connected to at least 30 computers.

How Btu is calculated

The methods for calculation of the natural gas Btu from analysis data are well documented. Calculations have been published previously.[2,3,9,16,18] Nevertheless, it is good to review these methods. Although they look complicated, they really are straightforward. Below is the general formula for the Btu calculation:

$$\text{Btu/cf} = X_1 * C_1 + X_2 * C_2 + X_3 * C_3 \cdots X_{11} * C_{11}$$

where:

X_1 = Mole fraction of C_1 or methane
X_2 = Mole fraction of C_2 or ethane
X_3 = Mole fraction of C_3 or propane

" "
" "
" "

X_{11} = Mole fraction of C_6+ hexanes and heavier

and:

C_1 = Molar heating value of C_1, or methane (1010 Btu/cf)
C_2 = Molar heating value of C_2, or ethane (1769 Btu/cf)

" "
" "
" "

C_{11} = Molar heating value of C_6+ or hexanes and heavier (4943 Btu/cf)

The calculation of specific gravity and condensable liquids is handled in the same manner. For example:

$$\text{Specific gravity} = X_1 * G_1 + X_2 * G_2 + \cdots X_{11} * G_{11}$$

where: X_1, X_2, is the mole fraction of each natural gas component and: G_1, G_2, is the molar gravity contribution of each respective component.

This calculation yields gross or total heating value and is not compensated for compressibility Z. A short method for calculation of Z is:

$$Z = 1.00369 - (0.010 * \text{S.G.}) + (0.007 * X_{inert})$$

Total heating value would then be expressed by the formula:

$$\frac{\text{Btu/cf}}{Z} = \text{Btu/cf Dry}$$

in order to convert Dry Btu to saturated:

$$\text{Btu saturated} = \frac{\text{Btu Dry}}{\text{1120 saturation factor (1.0177)}}$$

Some process gas chromatographs calculate all the above values at a pressure base of 14.696 psia only. Others allow user selection of a different contract pressure base, and all the constants are then factored by this new pressure base.

Care should be exercised when comparing Btu, SG, and others that all these values are calculated at the same temperature base and pressure base.

Conclusions

There are valid reasons for choosing one energy measurement device over the other. Thermtitrators are fast if control capability is needed. They are not suited for hazardous locations and require more rigorous attention to their operational environment. The modern microprocessors used on most chromatographs make them extremely capable in terms of computer interfaces.

Chromatographs are suitable for hazardous arca installations. They are somewhat slower reacting, having 6-minute cycle times. However, they generally require less maintenance than calorimeters even though they seem more technically complicated. With the AGA-8 supercompressibility calculations, there will most likely be a requirement for a total analysis of the gas to do this part of the gas volume calculations. Thermtitrators cannot do this. GCs that only present the flow computer with four analog signals, proportional to Btu, SG, CO_2, and N_2, may not be much more help.

One company has an interface unit that will pick up and store hourly averages of all 12 components from a gas chromatograph. These hourly averages can be stored for up to three streams for 35 days each so that each hour's flow in the flow computer can be correlated with each hour's gas composition. This allows composition data for not only MMBtu computation but also AGA-8 volume calculations. From an audit standpoint, this allows downstream corrections and verification that has heretofore not been available.

References

1. GPA Standard 2145–84, p. 2.
2. GPA Reference Bulletin, Heating Value As a Basis for Custody Transfer of Natural Gas (1984 Draft).
3. AGA Gas Measurement Manual, Section 11A.2 Determination of Heating Value of Gas, p. IIA2.1 ff.
4. Installation and Operation of Recording Calorinators, A. E. Kersey ISHM #64, 1989, p. 93 ff.
5. Field Deployable Natural Gas Energy Meter, Lewis and Greefield, Sales Information 1989.
6. "Determination of Colorific Values of Natural Gas by Combustion Instruments," R. L. Howard, ISHM #64, 1989, p. 509.
7. "Fundamentals of Energy Management," R. L. Howard, GCMSC #24, 1989, p. 43.
8. "The Therm-titrator, A Comparison with You-Know-What," John Light, GPA #60, 1981.
9. "Analysis of Natural Gas by Gas Chromatography," ASTM D1945, p. 133 ff.
10. "Comparison of Calculated and Measured Heating Values of Natural Gas," Donald C. Melrose, Appalachian Gas Measurement Short Course, 1975.
11. "Measuring Gas; Chromatography Bests Calorimetry," Leisey, Potter, and McKoy, *Oil and Gas Journal*, July 1977.
12. "Sampling and Analysis of Natural Gas," Schepens, Kilmer and Bernos, GPA proceedings #61.
13. "Energy Measurement Utilizing On-Line Chromatographs," L. N. Cox, Prod. of Inter. School of Hydrocarbon Meas. #64, 1989, p. 469 ff.
14. "Btu Measurement in Natural Gas Using Process Gas Chromatography," Kizer and Sund, Proc. of Gulf Coast Meas. Soc. #24, 1989, p. 72 ff.
15. "How to Measure Btu Effectively," Staff, *Gas Industries*, August 1989.
16. "Calculations of Gross Heating Value, Relative Density and Compressibility Factor for Natural Gas Mixtures from Compositional Analysis," GPA Std. 2171–86.
17. *College Chemistry*, Bruce H. Mahan, Addison Wesley, 1966, p. 23.
18. "Gas Heating Value: What It Is and How To Measure, Calculate," D. N. McClanahan, *Oil and Gas Journal*, Feb. 20, 1967, p. 84 ff.
19. Campbell, "436 Pushes Measurement into the Electronics Age," *American Oil and Gas Report*, May 1987, p. 15 ff.
20. Hagar, "Pipelines, Producers Cope with Problems under FERC Order 436," *Oil and Gas Journal*, May 1987, p. 17 ff.
21. Paul, "A Multi Btu Approach to Fuels Procurement," Nymex Energy in the News, Spring, 1990, p. 4.

Facts about methane and its behavior

A study of the physical, transport, and thermodynamic properties of methane

Dr. Robert O. Parker and Dr. William H. Kapfer, New York University, New York, NY

Interest in liquefied natural gas (LNG) requires that the physical, transport, and thermodynamic properties of methane, the principal constituent of natural gas, be readily available. This article will present values most frequently used in process calculations and in other studies associated with the storage and processing of natural gas. In addition, sources of more detailed information are listed.

Methane, CH_4, is the hydrocarbon of lowest molecular weight (16.042). The molecule is best visualized as a regular tetrahedron with the carbon atom at the geometric center and the hydrogen atoms at each of the four apices. Methane is an end product of the anaerobic (without air) decay of plants, that is, of the breakdown of certain very complicated molecules. As such, it is the major constituent (up to 97%) of natural gas. It is the firedamp of the coal mines and can be seen as marsh gas bubbling to the surface of swamps.

The physical, transport, and thermodynamic properties required are:

Physical

1. Critical constants
2. Equation of state (PVT relationships)
3. Vapor pressure, including by and fp
4. Vapor and liquid density

Transport

1. Viscosity of liquid and gas
2. Thermal conductivity of liquid and gas
3. Diffusivity

Thermodynamic

1. Heat capacity
2. Latent heat
3. Enthalpy
4. Entropy

They are required for process calculations: material and energy balances; heat transfer, mass-transfer and pressure loss calculations.

Critical Constants

$T_c = -115.8°F\ (-82.1°C)$[1]

$P_c = 673$ psia (45.8 atm)[2]

$d_c = 10.17\ lb/ft^3$ (0.1630 g/cc)[2]

The main utility of critical constants to the engineer is related to the fact that correlations of many of the physical, transport, and thermodynamic properties of hydrocarbons are best expressed in terms of reduced temperature, T_R, and pressure, P_R:

$$\left(T_R = \frac{T}{T_c},\ P_R = \frac{P}{P_c}\right)$$

This is simply an application of the "principle of corresponding states" and is recommended only in the absence of sufficient reliable experimental data.

Equations of state (pressure–volume–temperature behavior)

For ordinary conditions, low to moderate pressures and temperatures not too close to the normal boiling point, i.e., when inspection of a compressibility factor chart shows it to be nearly unity, the ideal gas law is:

$$PV = \frac{W}{M}RT \qquad (1)$$

where: P = Pressure, psia
V = Volume, ft^3
W = Weight of gas, lb
M = Molecular weight (here 16.042)
R = Gas constant = 10.73 psia (ft^3)/lb mole (°R)
T = Temperature, °R

is sufficient. For most other conditions, Equation 1 is modified by inclusion of the compressibility factor, Z, in the right-hand member, viz:

$$PV = \frac{W}{M}ZRT \qquad (1a)$$

The writers recommend compressibility factors of Nelson and Obert.[3] If an analytical equation of state is preferred to Equation 1a the writers recommend the 8 constant equation of Benedict, Webb, and Rubin.[4]

$$P = RTd + (B_oRT - A_o - C_o/T^2)d^2 + (bRT - a)d^3 + a\alpha d^6 + \frac{cd^3(1+\gamma d^2)e^{-\gamma d^2}}{T^2} \quad (2)$$

where d = density, g-moles/l

in which the constants[4] for methane are:

$B_o = 0.042600$

$A_o = 1.85500$

$C_o = 22{,}570.0$

$b = 0.00338004$

$a = 0.0494000$

$c = 2{,}545.00$

$\gamma = 0.0060000$

$\alpha = 0.000124359$

With these constants, R = 0.0820 liter-atm/(°K) (gram mole), P is in atmospheres, T is in °K, and e is the base of natural logarithms. Constants in this equation are available for ethane, ethylene, propane, propylene, n-butane, i-butane, i-butene, n-pentane, i-pentane, n-hexane, and n-heptane.[4] Using Equation 2 to predict the pressure corresponding to 89 data points,[5,6] it was found that the average absolute and maximum deviations for methane were 0.376% and +2.25% respectively. Reference 6 discusses a generalization of the Benedict-Webb-Rubin equation.

Vapor pressure[2]

Freezing point in air at 1 atm = −296.46°F (182.48°C)

Boiling point at 29.921 in. Hg = −258.68°F (−161.49°C)

Vapor pressures are tabulated on p. 121 of Reference 2 from 10 mm to 1,500 mm Hg and are represented by the Antoine equation.

$$\log_{10} P = A - B/(C + t) \quad (3)$$

or:

$$t = B/(A - \log_{10} P) - C \quad (3a)$$

with P in mm Hg and t in °C.

$A = 6.61184$

$B = 389.93$

$C = 266.00$

Transport properties

Viscosity (liquid)[2]

t°c	μ, centipoises
−185	0.226 (subcooled)
−180	0.188
−175	0.161
−170	0.142
−165	0.127
−160	0.115 (at sat. pressure)

Viscosity (vapor)[7] at 1 atm pressure:

t°c	μ, centipoises
−181.6	0.00348
−78.5	0.00760
0	0.01026
20	0.01087
100	0.01331
200.5	0.01605
284	0.01813
380	0.02026
499	0.02264

Additional data[8] are available.

Thermal conductivity (liquid)

	t°F	K, Btu/hr ft°F
	−280	0.0841
n.b.p	−258.68	0.0790
	−200	0.0568
crit. temp	−115.8	0.029 (reference[10])

Values for liquid thermal conductivity represent estimates by the authors using the method of Viswanath,[9] which seems to be quite reliable for organic liquids (halogenated compounds excepted).

Thermal conductivity (vapor)[7]

t°F	K, Btu/hr ft°F
−300	0.00541
−200	0.00895
−100	0.01265
−40	0.0149
−20	0.0157
0	0.01645
20	0.01725
40	0.01805
60	0.0190

t°F	K, Btu/hr ft°F
80	0.01985
100	0.0208
120	0.0216
200	0.0258

Diffusivity

The transport property, mass diffusivity (D), is of importance in calculations usually entering computations in combination with viscosity and density as the Schmidt No., μ/ρ D. Reliable estimates of D_{12}, the diffusivity of species 1 through species 2, are best made by the method of Hirschfelder, Bird, and Spotz.[11]

$$D_{12} = \frac{0.001858T^{3/2}[(M_1+M_2)/M_1M_2]^{1/2}}{P\sigma_{12}^2\Omega_D} \quad (4)$$

where: D = Diffusivity, cm^2/sec
T = Temperature, °K
M = Molecular weight, M_1 of species 1, M_2 of species 2
P = Pressure, atm
$\sigma_{12} = (\sigma_1 + \sigma_2)/2$ = collision diameter, A°
Ω_D = Collision integral = function of KT/ϵ_{12}
K = Boltzmann's constant
ϵ = A force constant
$\epsilon_{12} = \sqrt{\varepsilon_1\varepsilon_2}$

Force constants and collision diameters for air and methane are:

	ϵ/K, °K	σ, A°
Air	97	3.617
Methane	136.5	3.822

These are known as the Lennard-Jones force constants, and an extensive table is given.[11] Ω_D, which depends upon the temperature through KT/ϵ, is tabulated in the reference.[12] A few values are:

KT/ϵ	Ω_D
0.30	2.662
0.50	2.066
1.00	1.439
1.50	1.198
2.00	1.075
4.00	0.8836
10	0.7424
50	0.5756
100	0.5130
400	0.4170

If the calculation requiring diffusivities is a crucial one, we recommend that the engineer consult the principal literature reference.[11]

Thermodynamic properties

The most extensive, reliable compilation of the thermodynamic properties is the one by Matthews and Hurd.[13] In addition to a pressure-enthalpy diagram, tabulations are provided for saturated methane from −280°F to −115.8°F, the critical temperature, and for superheated methane to 500°F and 1,500 psia. This tabulated information is reproduced in the Chemical Engineers' Handbook 3rd Ed.[14] In the 4th Ed. (1963) superheat tables are omitted.

Corresponding to the pressure-temperature pair, values are given for:

a. Ratio fugacity (where fugacity[13] refers to "escaping tendency") to pressure
b. Specific volume of liquid and vapor
c. Enthalpy of liquid, vaporization, and vapor
d. Entropy of liquid and vapor

A few numbers of immediate interest to the process engineer are:

a. Latent heat of vaporization at normal boiling point = 219.22 Btu/lb
b. Liquid density at normal boiling point = 26.6 lb/ft^3
c. Vapor density at normal boiling point = 0.113 lb/ft^3

The liquid density is required to compute the volume of a diked area to contain a spill. Latent heat is useful in estimating the time required for a spill to be vaporized. The vapor density points up the fact that, although methane has a specific gravity referred to air of about 0.55, vapors generated at a spill are about 1.5 times as heavy as ambient air; a matter of importance in various studies.

Heat capacities

If heat capacities are required for process calculations, they can be extracted from the thermodynamic tables mentioned above by the following procedure:

$$C_P = \frac{H_1 - H_2}{T_1 - T_2}, \text{ Btu/lb °F} \quad (5)$$

Example. Find the heat capacity for superheated methane at 15 Asia and 100°F.

Figure 14. Cold box where liquefaction of natural gas is accomplished.

From the Matthews and Hurd superheat table:[13,14]

at 15 psia

$T_1 = 110 \qquad H_1 = 426.6$

$T2 = 90 \qquad H_2 = 415.8$

Then

$$C_P = \frac{426.6 - 415.8}{110 - 90} = 0.54 \text{ Btu/lb °F}$$

The procedure given above is adequate for liquid heat capacities when enthalpy values are taken from the table for saturated methane. This works because the effect of pressure on liquid heat capacity is essentially negligible.

Alternatively, instantaneous values of heat capacity may be determined.[15]

$$C_P^o = 0.213 + 1.11 \times 10^{-3}\,T - 2.59 \times 10^{-7}\,T^2 \text{ Btu/lb °F} \qquad (6)$$

Note carefully that Equation 6 yields the heat capacity at low pressure, 1 atm; at elevated pressures a correction may be required. If Equation 5 above is used, no pressure correction is necessary.

Heat capacity ratio

The ratio $C_p/C_v = \gamma$, useful in process calculations (adapted from reference 14), is for atmospheric pressure:

t°F	$\gamma = C_p/C_v$
−150	1.39
−100	1.36
−50	1.33
0	1.31
50	1.28
100	1.27
150	1.25

Flammability characteristics

Since methane is a combustible gas, consideration of any process involving its use or storage must include studies of the associated fire hazard. Basic data are flammable limits and the autoignition temperature. For methane in air these are[16,17] at atmospheric pressure:

a. Lower Flammable, Limit @ 25°C about 5% by vol
b. Upper Flammable, Limit @ 25°C about 15% by vol
c. Autoignition Temperature 1,004°F

These figures mean that mixtures of methane and air having less than 5% methane by volume are too lean to burn, those with more than 15% are too rich. The autoignition temperature is the minimum temperature at which ignition can occur. Figures on flammable limits refer to "homogeneous" mixtures. Industrially, few mixtures are completely homogeneous. In safety computations, therefore, the engineer must establish that the "highest local concentration" is below the lower flammable limit by a suitable margin before saying there is no fire hazard. A method dealing with LNG spills has been proposed by Parker and Spata.[18]

References

1. Kobe, K. A. and R. E. Lynn, *Chem. Revs*. 52, 117 (1953).
2. American Petroleum Institute, "Selected Values of Physical and Thermodynamic Properties of Hydrocarbons and Related Compounds," Project 44, Carnegie Press, Pittsburgh, Pa. 1953 (NBS-C-461).
3. Nelson, L. C. and E. F. Obert, *Trans*. ASME 76, 1057 (1954).
4. Benedict, M., G. B. Webb, L. C. Rubin and L. Friend, *Chem. Eng. Prog*. 47, 419, 449, 571, 609 (1951).
5. Kvalnes, H. M. and V. L. Gaddy, J. *Ann. Chem. Soc*. 53, 394 (1931).
6. Edmister, W. C., J. Vairogs and A. J. Klekers, *AIChE Journal* 14, No. 3, p. 479 (1968).
7. *Handbook of Chemistry and Physics, 47th Edition*, 1966–67.
8. Huang, E. T. S., G. W. Swift and F. Kurata, *AIChE Journal* 12, p. 932 (1966).
9. Viswanath, D. S., *AIChE Journal* 13, 850 (1967).
10. Gamson, B. W., *Chem. Eng. Prog*. 45, 154 (1949).
11. Hirschfelder, J. O., R. B. Bird and E. L. Spotz, *Trans. ASME* 71, 921 (1949).
12. Hirschfelder, J. O., C. F. Curtiss and R. B. Bird, *Molecular Theory of Gases and Liquids*, John Wiley and Sons, Inc., New York, 1954.
13. Matthews, C. S. and C. O. Hurd, *Trans. AIChE* 42, p. 55 (1946); ibid., pp. 78 (1009).
14. Perry, J. H. (ed.) *Chemical Engineers Handbook, 3rd ed*., McGraw-Hill, New York, 1950.
15. Reid, R. C. and T. K. Sherwood, *The Properties of Gases and Liquids*, McGraw-Hill, New York (1958).
16. Coward, H. F. and G. W. Jones, U.S. Bureau of Mines, Bulletin 503 (1952).
17. Zabetakis, M. G., U.S. Bureau of Mines, Bulletin 627 (1965).
18. Parker, R. O. and J. K. Spata, Proc. 1st Intern. Conf. on LNG, Chicago, April 7–12, 1968, Paper No. 24.

Conversion table for pure methane

		Long tons Liquid	ft^3 Liquid	ft^3 gas	m^3 liquid	m^3 gas	U.S. bbl liquid	Therms	Thermies
long ton liquid	=	1	84.56	52,886	2.394	1,419.7	15.06	535.2	13,491
ft^3 liquid	=	0.011826	1	625.43	0.028311	16.789	0.1781	6.329	159.54
ft^3 gas $\times 10^6$	=	18.91	1,599	1×10^6	45.27	26,847	284.785	10,121	255,115
m^3 liquid	=	0.4177	35.315	22,090	1	593	6.29	223.55	5,635.2
m^3 gas $\times 10^9$	=	704,374	59.562×10^6	37.252×10^9	1.686×10^6	1×10^9	10.608×10^6	0.377×10^9	9.503×10^9
U.S. bbl liquid	=	0.0664	5.615	3,512	0.15896	94.268	1	35.54	895.8
Therm $\times 10^6$	=	1,868	157,958	98.791×10^6	4,472	2.652×10^6	28,132	1×10^6	25.201×10^6
Thermie $\times 10^6$	=	74.12	6,268	3.92×10^6	177.44	105,228	1,116.2	39,669	1×10^6

Notes

1. ft^3 measured at 60°F, 14.696 lbf/in^2, dry
2. m^3 measured at 0°C, 760 mmHg, dry
3. Gross heat content of methane is 1012 Btu/ft^3 in gaseous forms: Specific gravity 0.554 (Air = 1)
   Boiling point −258.7°F (14.696 lbf/in^2)
4. 1×10^6 ft^3 gas = 24.42 long tons (0.4417 b/d) fuel oil equivalent
5. The above factors are based on the properties of pure methane; LNG as shipped in practice will vary in composition according to the presence of heavier hydrocarbons (C_2, C_3, C_4,) in the gas stream.

Categories of natural gas and reserves terminology

Associated and non-associated natural gas

Natural gas is found in underground structures similar to those containing crude oil. There are three types of natural gas reservoirs:

1. Structures from which only gas can be produced economically—called **non-associated gas** (or unassociated gas).
2. Condensate reservoirs which yield relatively large amounts of gas per barrel of light liquid hydrocarbons. Although many condensate reservoirs, are produced primarily for gas, there are cases where gas is re-injected or "re-cycled" to improve liquid recovery, particularly if no gas market is yet available. This gas also is termed **non-associated**.
3. Reservoirs where gas is found dissolved in crude oil (solution gas) and in some cases also in contact with underlying gas-saturated crude (**gas-cap gas**). Both are called **associated gas**. (Gas-cap gas is almost never produced until most of the economically recoverable oil has been yielded.)

In such fields, gas production rates will depend on oil output, with the oil usually representing the major part in terms of energy equivalents.

Natural gas reserves

There are no generally accepted definitions of natural gas reserves, and wide variations both in terms and principles may be met. The following definitions are the most commonly used:

Expected Ultimate Recovery is the total volume of hydrocarbons, gas and/or oil, which may be expected to be recoverable commercially from a given area at current or anticipated prices and costs, under existing or anticipated regulatory practices and with known methods and equipment. This total can be expressed as the sum of the "Ultimate Recovery from Existing Fields" and the "Expectation from Future Discoveries."

Ultimate Recovery from Existing Fields is the sum of the "Cumulative Production" from such fields and the "Reserves from Existing Fields."

Reserves from Existing Fields are the sum of "Proven Reserves," "Discounted (i.e., 50% of) Probable Reserves" and "Discounted (i.e., 25% of) Possible Reserves."

Proven Reserves represent the quantities of crude oil and/or natural gas and natural gas liquids that geological and engineering data demonstrate with reasonable certainty to be recoverable in the future from known oil and/or gas reservoirs. They represent strictly technical judgments and are not knowingly influenced by attitudes of conservatism or optimism.

Discounted (i.e., 50% of) **Probable Reserves** are those quantities of crude oil and/or natural gas and natural gas liquids for which there exists a probability of 50% that they will materialize. Such reserves are usually allocated to some conjectural part of a field or reservoir as yet undrilled or incompletely evaluated where, while characteristics of the productive zone and fluid content are reasonably favorable, other uncertain features may have equal chances of being favorable or unfavorable for commercial production.

Discounted (i.e., 25% of) **Possible Reserves** are those quantities of crude oil and/or natural gas and natural gas liquids thought to be potentially producible but where the chance of their being realized is thought to be on the order of 25%. Reserves in this category are usually allocated to possible extensions of existing fields where, although geological data may be favorable, uncertainties as to the characteristics of the productive zone and/or fluid content are such as to preclude a probability greater than 25%.

Terminology of Reserves

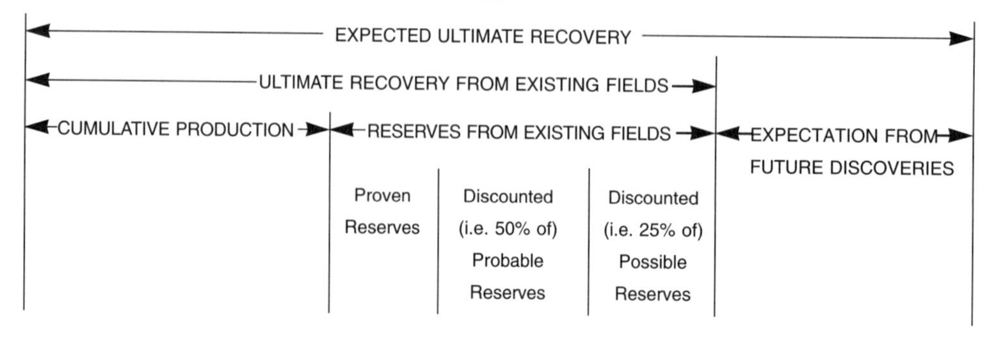

It should be noted that the terms "gas-in-place" and "oil-in-place" represent estimates of the total hydrocarbons present which, in conjunction with various engineering and economic criteria, provide the basis for the "reserve" estimates. They should never be confused with "reserves," which implies that physical recovery is possible.

Glossary of common gas industry terms

(italics denote cross reference)

absolute pressure *gauge pressure* plus *atmospheric pressure*.

associated gas see preceding page.

atmospheric pressure the pressure of the weight of air and water vapor on the surface of the earth. Approximately 14.7 lbf/in.^{2} at sea level.

blue water gas also called "water gas." Made in a cyclic process in which an incandescent bed of coke or coal is alternately subjected to blasts of air and steam. The gas consists mainly of equal proportions of carbon monoxide and hydrogen and has a gross heat content of about 300 Btu/ft^3 (see also *carbureted water gas*).

British thermal unit the heat required to raise the temperature of 1 lb of water 1°F.

burner capacity the maximum Btu/hr that can be released by a burner while burning with a stable flame.

calorie the quantity of heat required to raise the temperature of 1 gram of water 1°C (centigrade).

calorific value the quantity of heat released by the combustion, in a calora meter at constant pressure of 1 atmosphere, of a unit quantity of fuel measured under given conditions. In the case of gas, the conditions of temperature, pressure, and water vapor content under which the unit volume was measured affect the calorific value.

calorimeter an instrument for determining the heating value of a fuel.

carbureted water gas (carbureted blue gas) the gas resulting from the enrichment of *blue water gas* during its manufacture by a simultaneous process of light distillate, gas oil, or fuel oil gasification. The gas has a gross heat content of about 500 to 550 Btu/ft^3.

city gate a measuring station, which may also include pressure regulation, at which a distributing gas utility receives gas from a natural gas pipeline company or the transmission system.

coal gas or **coke oven gas** a *manufactured gas* made by destructive distillation ("carbonization") of bituminous coal in a gas retort or by-product coke oven. Its chief components are methane (20 to 30%) and hydrogen (about 50%). This gas generally has a gross heating value of 500 to 550 Btu/ft^3. When the process takes place in a closed oven (with gas as a by-product of coke production) it is generally designated as "Coke Oven Gas," and when produced in retorts it is called "Coal Gas."

commodity charge that portion of the charge or rate based upon the total volume of gas consumed or billed.

condensate a liquid/hydrocarbon mixture of 45 to 65° API gravity, which may be recovered at the surface from some *non-associated gas* reservoirs.

connection charge an amount to be paid by the customer in a lump sum, or in installments, for connecting the customer's facilities to the supplier's facilities.

contract demand the volume that the supplier of gas service agrees to deliver and, in general, the amount that the customer agrees to take or pay for.

cubic foot and **cubic meter (gas)** common units of measurement of gas volume. They are the amounts of gas required to fill a volume of 1 cubic ft or 1 cubic meter under stated conditions of temperature, pressure, and water vapor content.

customer charge a fixed amount to be paid periodically by the customer without regard to *demand or commodity* consumption.

cycling or **recycling** the process by which a *non-associated gas* reservoir is produced only for the recovery of condensate. The gas, after removal of the condensate, is reinjected into the formation.

daily average send-out the total volume of gas delivered during a period of time divided by number of days in the period.

daily peak the maximum volume of gas delivered in any one day during a given period, usually a calendar year.

degree day a measure of the extent to which the mean daily temperature falls below an assumed base, say 65°F. Thus, each degree by which the mean temperature for any day is less than 65°F would represent one degree day. (In Continental Europe, °C are used instead of °F, and the assumed base temperature is generally taken as 16°C equivalent to 60.8°F.)

demand (load) rate of flow of gas in a specified time interval, usually expressed in cubic ft (or cubic meters) per hour, per day, or per year.

demand charge that portion of the charge or rate for gas service based upon the customer's *demand* characteristics.

distribution company or **gas utility** a company that obtains the major portion of its gas operating revenues from the operation of a retail gas *distribution system* and which operates no transportation system other than incidental connections to a transportation system of another company. For purposes of American Gas Association (AGA) statistics, a distribution company obtains at least 95% of its gas operating revenues from the operation of its retail gas distribution system (see also *transmission company*).

distribution system feeders, mains, services, and equipment which carry or control the supply of gas from the point or points of local supply (usually the *city gate* station) to and including the consumer *meters*.

diversity factor the ratio of the sum of the non-coincident maximum demands of two or more loads to their coincident maximum demand for the same period.

dyne the standard centimeter-gram-second unit of force, equal to the force that produces an acceleration on one centimeter per second on a mass of one gram.

erg the centimeter-gram-second unit of work or energy, equal to the work done by a force of 1 dyne when its point of application moves through a distance of 1 centimeter in the direction of the force.

flat rate schedule a rate schedule that provides for a specified charge irrespective of the quantity of gas used or the *demand*.

gauge pressure the pressure generally shown by measuring devices. This is the pressure in excess of that exerted by the atmosphere (see *absolute pressure*).

gross calorific (heating) value the gross calorific value at constant pressure of a gaseous fuel is the number of heat units produced when unit volume of the fuel, measured under standard conditions, is burned in excess air in such a way that the materials after combustion consist of the gases carbon dioxide, sulfur dioxide, oxygen, and nitrogen, water vapor equal in quantity to that in the gaseous fuel and the air before combustion, and liquid water equal in quantity to that produced during combustion, and that the pressure and temperature of the gaseous fuel, the air, and the materials after combustion are 1 standard atmosphere and 25°C.

inch of mercury a unit of pressure (equivalent to 0.491154 lbf/in.^{2}) by which 1 lb/in.^{2} equals 2.036009 in. of mercury column at 0°C.

inch of water a unit of pressure (equivalent to 0.03606 lbf/in.^{2}) by which 1 lb/in.^{2} equals 27.68049 in. of water column at 4°C.

interruptible gas gas made available under agreements which permit curtailment or cessation of delivery by the supplier.

joule the work done when the point of application of a force of 1 newton is displaced through a distance of one meter in the direction of the force. It is equal to 107 *ergs*, and in electrical units is the energy dissipated by one watt in a second.

kilocalorie 1,000 *calories*.

kilowatt-hour a unit of energy equivalent to the energy transferred or expanded in 1 hour by 1 kilowatt of power.

line pack a method of *peak-shaving* by withdrawing gas from a section of a pipeline system in excess of the input into that section, i.e., normally the difference between the actual volume of gas in the pipeline at low flow (increased pressure) and that at normal flow.

liquefied natural gas (LNG) natural gas that has been liquefied by cooling to minus 258°F (−161°C) at atmospheric pressure.

liquefied petroleum gas (LPG) any hydrocarbon mixture, in either the liquid or gaseous state, the chief components of which consist of propane, propylene, butane, iso-butane, butylene, or mixtures thereof in any ratio.

load curve a graph in which the *send-out* of a gas system, or segment of a system, is plotted against intervals of time.

load density the concentration of gas load for a given area, expressed as gas volume per unit of time and per unit of area.

load duration curve a curve of loads, plotted in descending order of magnitude, against time intervals for a specified period. The coordinates may be absolute quantities or percentages.

load factor the ratio of the average load over a designated period to the *peak load* occurring in that period. Usually expressed as a percentage.

low pressure gas distribution system a gas *distribution system*, or the mains of a segment of a distribution system, operated at low pressures of less than 15 *inches water column*.

LPG-air mixtures mixtures of *liquefied petroleum gas* and air to obtain a desired heating value and capable of being distributed through a *distribution system* also used for stand-by and *peak-shaving* purposes by gas utilities.

manufactured gas combustible gases derived from primary energy sources by processes involving chemical reaction. For instance, gas produced from coal, coke or liquid hydrocarbons (see also *town gas*).

meter (gas) a mechanical device for automatically measuring quantities of gas. Some of the more important types are:

Diaphragm Meter a gas meter in which the passage of gas through two or more chambers moves diaphragms that are geared to a volume-indicating dial. Sizes up to

20,000 cubic ft per hour and 1,000 lbf/in.^{2} working pressure are available.

Mass Flowmeter a type of meter in which the gas flow is measured in terms of mass by utilizing the velocity and density of the gas stream to calculate the mass of a gas flowing in unit time.

Orifice Meter meter for measuring the flow of fluid through a pipe or duct by measurement of the pressure differential across a plate having a precision machined hole in its center. Ratings up to very high throughputs and pressures are available.

Rotary Displacement Meter the positive-pressure blower principle in reverse is used in this meter in which gas pressure turns twin matching impellers and the quantity of gas passing through is proportional to the number of revolutions. These meters are made in the range of 3,000 to 1,000,000 cubic ft per hour and for pressures up to 1,200 lbf/in.^{2}

Turbine Flowmeter a type of meter in which there is a turbine wheel or rotor made to rotate by the flowing gas. The rate or speed at which the turbine wheel revolves is a measure of the velocity of the gas.

Venturi Meter a meter in which the flow is determined by measuring the pressure drop caused by the flow through a Venturi throat. The flow rate is proportional to the square root of the pressure drop across the throat.

minimum charge a clause providing that the charge for a prescribed period shall not be less than a specified amount.

natural gasoline those liquid hydrocarbon mixtures containing essentially pentanes and heavier hydrocarbons that have been extracted from natural gas.

natural gas liquids a term used for those mixtures of hydrocarbon fractions that can be extracted in liquid form from natural gas. In the widest sense, they can comprise any combination of the following: ethane, propylene, propane, butylene, isobutane, n-butane, pentanes, hexanes, heptanes, and octanes.

net calorific (heating) value the net calorific value at constant pressure of a gaseous fuel is the number of heat units produced when unit volume of the fuel, measured under standard conditions, is burned in excess air in such a way that the materials after combustion consist of the gases carbon dioxide, sulfur dioxide, oxygen, nitrogen, and water vapor, and that the pressure and temperature of the gaseous fuel, the air and the materials after combustion are 1 standard atmosphere and 25°C. This value is derived from the measured gross figure.

non-associated gas see preceding section on "reserves terminology."

off-peak the period during a day, week, month, or year when the load being delivered by a gas system is not at or near the maximum volume delivered by that system.

peak or **peak load** the maximum load consumed or produced by a unit or group of units in a stated period of time.

peak shaving the practice of augmenting the normal supply of gas during peak or emergency periods from another source where gas may have either been stored during periods of low demand or manufactured specifically to meet the peak demand.

pressure maintenance (repressing) a process in which natural gas is injected into a formation capable of producing crude petroleum to aid in maintaining pressure in an underground reservoir for the purpose of assisting in the recovery of crude.

producer gas a gas manufactured by burning coal or coke with a regulated deficiency of air, normally saturated with steam. The principal combustible component is carbon monoxide (about 30%), and the gross heat content is between 120 and 160 Btu/ft^3.

recycling see *cycling*.

reforming the process of thermal or catalytic cracking of natural gas, *liquefied petroleum gas*, *refinery gas* or gas from oil, resulting in the production of a gas having a different chemical composition.

refinery gas a gas resulting from oil refinery operations consisting mainly of hydrogen, methane, ethylene, propylene, and the butylenes. Other gases such as nitrogen and carbon dioxide may also be present. The composition can be highly variable, and the heat content can range from 1,000 to 2,000 Btu/ft^3.

reserves see preceding section.

send-out the quantity of gas delivered by a plant or system during a specified period of time.

specific gravity of gas the ratio of the density of gas to the density of dry air at the same temperature and pressure.

standard metering base standard conditions, plus agreed corrections, to which all gas volumes are corrected for purposes of comparison and payment.

summer valley the decrease that occurs in the summer months in the volume of the daily load of a gas *distribution system*.

therm 100,000 Btus.

thermie the heat required to raise the temperature of 1 metric ton of water 1°C (centigrade).

town gas gas piped to consumers from a gas plant. The gas can comprise both *manufactured gas* (secondary energy) and natural gas (primary energy) used for enrichment.

transmission company a company that operates a natural gas transmission system and which either operates no retail distribution system, or, as defined by the American Gas Association (AGA), receives less than 5% of its gas operating revenues from such retail distribution system (see also *distribution company* or *gas utility*).

unaccounted for gas the difference between the total gas available from all sources and the total gas accounted for as sales, net interchange and company use. The difference includes leakage or other actual losses, discrepancies due to *meter* inaccuracies, variations of temperature, and pressure and other variants.

unassociated gas see preceding section.

utilization factor the ratio of the maximum *demand* on a system or part of a system to the rated capacity of the system or part of the system under consideration.

water gas see *blue water gas*.

watt the meter-kilogram-second unit of power, equivalent to one joule per second and equal to the power in a circuit in which a current on one ampere flows across a potential difference on one volt.

wobbe index the gross calorific value of the gas divided by the square root of the density of the gas as compared with air.

10: Gas—Compression

Compressors

The following data are for use in the approximation of horsepower needed for compression of gas.

Definitions

The "**N value**" of gas is the ratio of the specific heat at constant pressure (C_p) to the specific heat at constant volume C_v.

If the composition of gas is known, the value of N for a gas mixture may be determined from the molal heat capacities of the components. If only the specific gravity of gas is known, an approximate N value may be obtained by using the chart in Figure 1.

Piston Displacement of a compressor cylinder is the volume swept by the piston with the proper deduction for the piston rod. The displacement is usually expressed in cubic ft per minute.

Clearance is the volume remaining in one end of a cylinder with the piston positioned at the end of the delivery stroke for this end. The clearance volume is expressed as a percentage of the volume swept by the piston in making its full delivery stroke for the end of the cylinder being considered.

Ratio of compression is the ratio of the absolute discharge pressure to the absolute inlet pressure.

Actual capacity is the quantity of gas compressed and delivered, expressed in cubic ft per minute, at the intake pressure and temperature.

Volumetric efficiency is the ratio of actual capacity, in cubic ft per minute, to the piston displacement, in cubic ft per minute, expressed in percent.

Adiabatic horsepower is the theoretical horsepower required to compress gas in a cycle in which there is no transfer of sensible heat to or from the gas during compression or expansion.

Isothermal horsepower is the theoretical horsepower required to compress gas in a cycle in which there is no change in gas temperature during compression or expansion.

Indicated horsepower is the actual horsepower required to compress gas, taking into account losses within the compressor cylinder, but not taking into account any loss in frame, gear, or power transmission equipment.

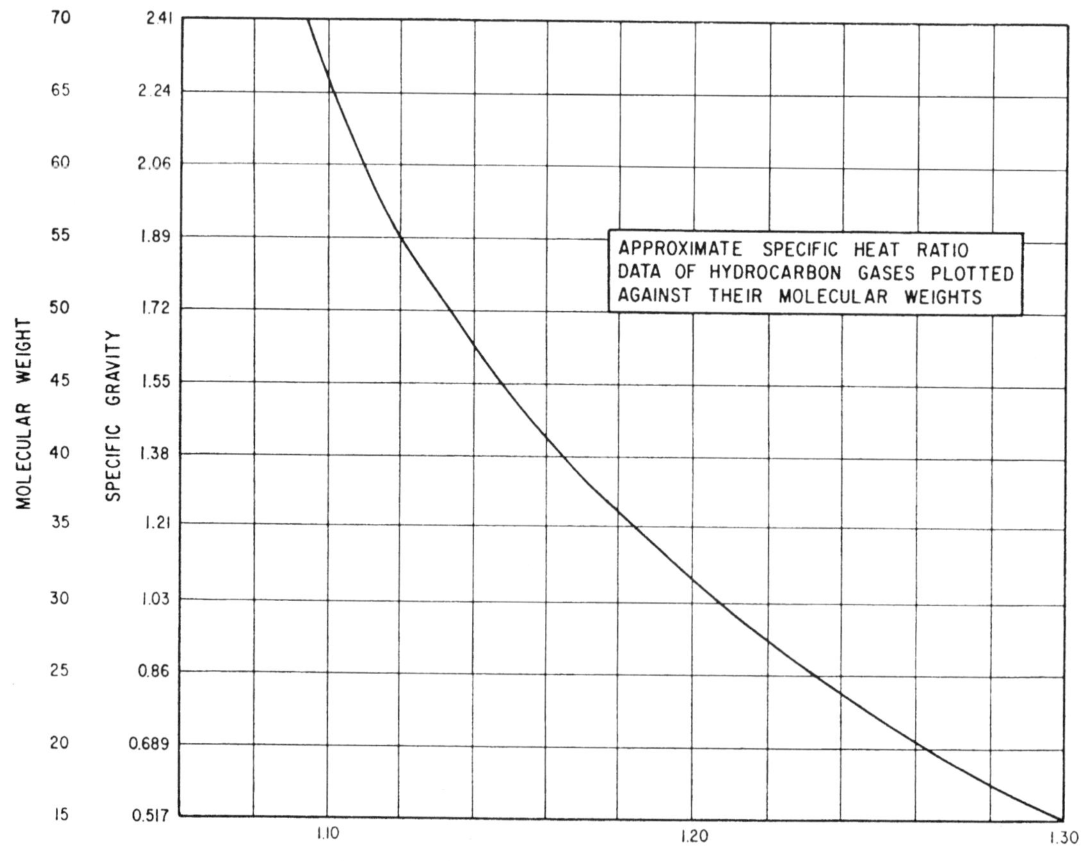

Figure 1. Ratio of specific heat (n-value).

Compression efficiency is the ratio of the theoretical horsepower to the actual indicated horsepower required to compress a definite amount of gas. The efficiency, expressed in percent, should be defined in regard to the base at which the theoretical power was calculated, whether adiabatic or isothermal.

Mechanical efficiency is the ratio of the indicated horsepower of the compressor cylinder to the brake horsepower delivered to the shaft in the case of a power-driven machine. It is expressed in percent.

Overall efficiency is the product, expressed in percent, of the compression efficiency and the mechanical efficiency. It must be defined according to the base, adiabatic isothermal, which was used in establishing the compression efficiency.

Piston rod gas load is the varying, and usually reversing, load imposed on the piston rod and crosshead during the operation, by different gas pressures existing on the faces of the compressor piston.

The maximum piston rod gas load is determined for each compressor by the manufacturer, to limit the stresses in the frame members and the bearing loads in accordance with mechanical design. The maximum allowed piston rod gas load is affected by the ratio of compression and also by the cylinder design; i.e., whether it is single or double acting.

Performance calculations for reciprocating compressors

Piston displacement

Single-acting compressor:

$$P_d = [S_t \times N \times 3.1416 \times D^2]/[4 \times 1{,}728] \quad (1)$$

Double-acting compressor without a tail rod:

$$P_d = [S_t \times N \times 3.1416 \times (2D^2 - d^2]/[4 \times 1{,}728] \quad (2)$$

Double-acting compressor with a tail rod:

$$P_d = [S_t \times N \times 3.1416 \times 2 \times (D^2 - d^2]/[4 \times 1{,}728] \quad (3)$$

Single-acting compressor compressing on frame end only:

$$P_d = [S_t \times N \times 3.1416 \times (D^2 - d^2]/[4 \times 1{,}728] \quad (4)$$

where: P_d = Cylinder displacement, cu ft/min
S_t = Stroke length, in.
N = Compressor speed, number of compression strokes/min
D = Cylinder diameter, in.
d = Piston rod diameter, in.

Volumetric efficiency

$$E_v = 0.97 - [(1/f)r_p^{1/k} - 1]C - L \quad (5)$$

where: E_v = Volumetric efficiency
f = Ratio of discharge compressibility to suction compressibility Z_2/Z_1
r_p = Pressure ratio
k = Isentropic exponent
C = Percent clearance
L = Gas slippage factor

Let L = 0.3 for lubricated compressors
Let L = 0.07 for non-lubricated compressors (6)

These values are approximations, and the exact value may vary by as much as an additional 0.02 to 0.03.

Note: A value of 0.97 is used in the volumetric efficiency equation rather than 1.0 since even with 0 clearance, the cylinder will not fill perfectly.

Cylinder inlet capacity

$$Q_1 = E_v \times P_d \quad (7)$$

Piston speed

$$PS = [2 \times S_t \times N]/12 \quad (8)$$

Discharge temperature

$$T_2 = T_1(r_p^{(k-1)/k}) \quad (9)$$

where: T_2 = Absolute discharge temperature °R
T_1 = Absolute suction temperature °R

Note: Even though this is an adiabatic relationship, cylinder cooling will generally offset the effect of efficiency.

Power

$$W_{cyl} = [144P_1Q_1/33{,}000n_{cyl}] \times [k/(k-1)] \times [r_p^{k-1/k} - 1] \quad (10)$$

where: n_{cyl} = Efficiency
W_{cyl} = Cylinder horsepower

See Figure 1 for curve of efficiency versus pressure ratio. This curve includes a 95% mechanical efficiency and a valve velocity of 3,000 ft per minute. Tables 1 and 2 permit a correction to be made to the compressor horsepower for specific gravity and low inlet pressure. While it is recognized that the efficiency is not necessarily the element affected, the desire is to modify the power required per the criteria in these figures. The efficiency correction accomplishes this. These corrections become more significant at the lower pressure ratios.

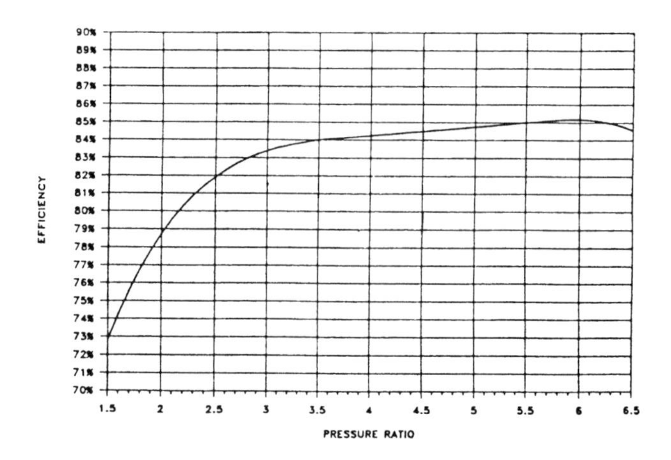

Reciprocating compressor efficiencies plotted against pressure ratio with a valve velocity of 3,000 fpm and a mechanical efficiency of 95 percent.

Figure 1. Reciprocating compressor efficiencies.

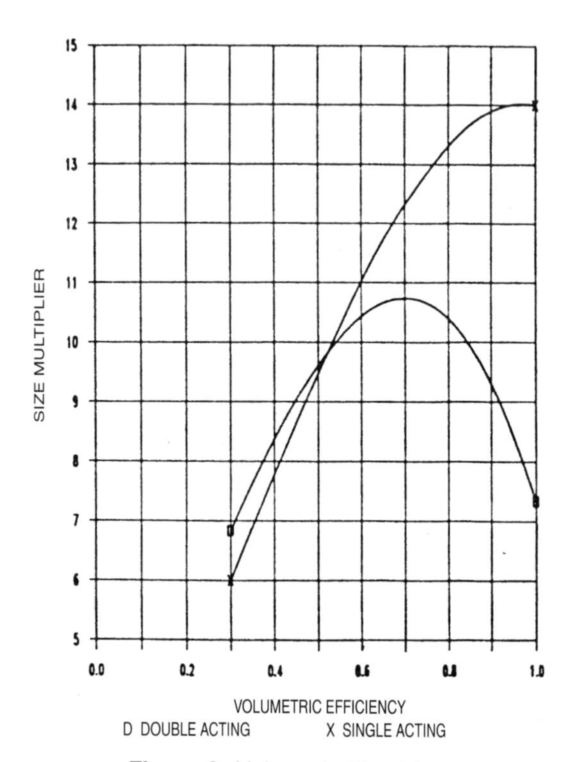

Figure 2. Volume bottle sizing.

Inlet valve velocity

$$V = 288 \times P_d/A \qquad (11)$$

where: V = Inlet valve velocity
A = Product of actual lift and the valve opening periphery and is the total for inlet valves in a cylinder expressed in square in. (This is a compressor vendor furnished number.)

Example. Calculate the following:

Suction capacity
Horsepower
Discharge temperature
Piston speed

Given:

Bore = 6 in.
Stroke = 12 in.
Speed = 300 rpm
Rod diameter = 2.5 in.
Clearance = 12%
Gas = $C0_2$
Inlet pressure = 1,720 psia
Discharge pressure = 3,440 psia
Inlet temperature = 115°

Calculate piston displacement using Equation 2.

$$P_d = 12 \times 300 \times 3.1416 \times [2(6)^2 - (2.5)^2]/1{,}728 \times 4$$
$$= 107.6\,\text{cfm}$$

Calculate volumetric efficiency using Equation 5. It will first be necessary to calculate f, which is the ratio of discharge compressibility to suction compressibility.

$$T_1 = 115 + 460 = 575°R$$

$$T_r = T/T_c$$
$$= 575/548$$
$$= 1.05$$

where: T_r = Reduced temperature
T_c = Critical temperature = 548° for CO_2
T = Inlet temperature

$$P_r = P/P_c$$
$$= 1{,}720/1{,}071$$
$$= 1.61$$

where: P_r = Reduced pressure
P_c = Critical pressure = 1,071 psia for CO_2
P = Suction pressure

From the generalized compressibility chart (page 231):

$Z_1 = 0.312$

Determine discharge compressibility. Calculate discharge temperature by using Equation 9.

$r_p = 3{,}440/1{,}720$
$= 2$

$k = cp/cv = 1.3$ for CO_2

$T_2 = 575 \times [2.0^{(13-1)/1.3}]$
$= 674.7°R$
$= 674.7 - 460 = 214°F$

$T_r = 674.7/548 = 1.23$

$P_r = 3{,}440/1{,}071 = 3.21$

From the generalized compressibility chart (page 231):

$Z_2 = 0.575$

$f = 0.575/1.61 = 1.843$

$E_v = 0.97 - [(1/1.843) \times 2.0^{1/1.3} - 1] \times 0.12 - 0.05$
$= 0.929$
$= 93\%$

Note: A value of 0.05 was used for L because of the high differential pressure.

Calculate suction capacity.

$Q_1 = E_v \times P_d$
$= 0.93 \times 107.6$
$= 100.1$ cfm

Calculate piston speed.

$PS = [2 \times S_t \times N]/12$
$= [2 \times 12 \times 300]/12$
$= 600$ ft/min

Table 1
Efficiency multiplier for low pressure

	Pressure Psia							
r_p	10	14.7	20	40	60	80	100	150
3.0	0.990	1.00	1.00	1.00	1.00	1.00	1.00	1.00
2.5	0.980	0.985	0.990	0.995	1.00	1.00	1.00	1.00
2.0	0.960	0.965	0.970	0.980	0.990	1.00	1.00	1.00
1.5	0.890	0.900	0.920	0.940	0.960	0.980	0.990	1.00

Source: Modified courtesy of the Gas Processors Suppliers Association and Ingersoll-Rand.

Table 2
Efficiency multiplier for specific gravity

	SG				
r_p	1.5	1.3	1.0	0.8	0.6
2.0	0.99	1.0	1.0	1.0	1.01
1.75	0.97	0.99	1.0	1.01	1.02
1.5	0.94	0.97	1.0	1.02	1.04

Source: Modified courtesy of the Gas Processors Suppliers Association.

Estimate suction and discharge volume bottle sizes for pulsation control for reciprocating compressors

Pressure surges are created as a result of the cessation of flow at the end of the compressor's discharge and suction stroke. As long as the compressor speed is constant, the pressure pulses will also be constant. A low pressure compressor will likely require little, if any, treatment for pulsation control; however, the same machine with increased gas density, pressure, or other operational changes may develop a problem with pressure pulses. Dealing with pulsation becomes more complex when multiple cylinders are connected to one header or when multiple stages are used.

API Standard 618 should be reviewed in detail when planning a compressor installation. The pulsation level at the outlet side of any pulsation control device, regardless of type, should be no more than 2% peak-to-peak of the line pressure of the value given by the following equation, whichever is less.

$$P_{\%} = 10/P_{line}{}^{1/3} \quad (12)$$

Where a detailed pulsation analysis is required, several approaches may be followed. An analog analysis may be performed on the Southern Gas Association dynamic compressor simulator, or the analysis may be made a part of the compressor purchase contract. Regardless of who makes the analysis, a detailed drawing of the piping in the compressor area will be needed.

The following equations are intended as an aid in estimating bottle sizes or for checking sizes proposed by a vendor for simple installations—i.e., single cylinder connected to a header without the interaction of multiple cylinders. The bottle type is the simple unbaffled type.

Example. Determine the approximate size of suction and discharge volume bottles for a single-stage, single-acting, lubricated compressor in natural gas service.

Cylinder bore = 9 in.
Cylinder stroke = 5 in.
Rod diameter = 2.25 in.
Suction temp = 80°F
Discharge temp = 141°F
Suction pressure = 514 psia
Discharge pressure = 831 psia
Isentropic exponent, k = 1.28
Specific gravity = 0.6
Percent clearance = 25.7%

Step 1. Determine suction and discharge volumetric efficiencies using Equations 5 and 13.

$$E_{vd} = [E_v \times f]/r_p^{1/k} \quad (13)$$

$$r_p = 831/514 = 1.617$$

$Z_1 = 0.93$ (from Figure 4, page 199)

$Z_2 = 0.93$ (from Figure4, page 199)

$$f = 0.93/0.93 = 1.0$$

Calculate suction volumetric efficiency using Equation 5:

$$E_v = 0.97 - [(1/1) \times (1.617)^{1/1.28} - 1] \times 0.257 - 0.03 = 0.823$$

Calculate discharge volumetric efficiency using Equation 13:

$$E_{vd} = 1 \times [0.823]/[1.617^{1/1.28}] = 0.565$$

Calculate volume displaced per revolution using Equation 1:

$$P_d/N = [S_t \times 3.1416 \times D^2]/[1,728 \times 4] = [5 \times 3.1416 \times 9^2]/[1,728 \times 4] = 0.184 \text{ cu ft or } 318 \text{ cu in.}$$

Refer to Figure 3, volume bottle sizing, using volumetric efficiencies previously calculated, and determine the multipliers.

Suction multiplier = 13.5
Discharge multiplier = 10.4

Discharge volume = 318 × 13.5
= 3,308 cu in.

Suction volume = 318 × 10.4
= 4,294 cu in.

Calculate bottle dimensions. For elliptical heads, use Equation 14.

$$\text{Bottle diameter } d_b = 0.86 \times \text{volume}^{1/3} \quad (14)$$

Volume = suction or discharge volume

$$\text{Suction bottle diameter} = 0.86 \times 4,294^{1/3} = 13.98 \text{ in.}$$

$$\text{Discharge bottle diameter} = 0.86 \times 3,308^{1/3} = 12.81 \text{ in.}$$

$$\text{Bottle length} = L_b = 2 \times d_b \quad (15)$$

Suction bottle length = 2 × 13.98
= 27.96 in.

Discharge bottle length = 2 × 12.81
= 25.62 in.

Source

Compressors—Selection & Sizing, Gulf Publishing Company, Houston, Texas.

Compression horsepower determination

The method outlined below permits determination of approximate horsepower requirements for compression of gas.

1. From Figure 1, determine the atmospheric pressure in psia for the altitude above sea level at which the compressor is to operate.
2. Determine intake pressure (P_s) and discharge pressure (P_d) by adding the atmospheric pressure to the corresponding gauge pressure for the conditions of compression.
3. Determine total compression ratio $R = P_d/P_s$. If ratio R is more than 5 to 1, two or more compressor stages will be required. Allow for a pressure loss of approximately 5 psi between stages. Use the same ratio for each stage. The ratio per stage, so that each stage has the same ratio, can be approximated by finding the nth root of the total ratio, when n = number of stages. The exact ratio can be found by trial and error, accounting for the 5 psi interstage pressure losses.
4. Determine the N value of gas (Chart, 1st ed. p. 53).
5. Figure 3 gives horsepower requirements for compression of 1 million cu. ft per day for the compression ratios and N values commonly encountered in oil-producing operations.
6. If the suction temperature is not 60°F, correct the curve horsepower figure in proportion to absolute temperature. This is done as follows:

$$\text{HP} \times \frac{460^\circ + T_s}{460^\circ + 60^\circ\text{F}} = \text{hp (corrected for suction temperature)}$$

where T_s is suction temperature in °F.

7. Add together the horsepower loads determined for each stage to secure the total compression horsepower load. For altitudes greater than 1,500 ft above sea level, apply a multiplier derived from the following table to determine the nominal sea level horsepower rating of the internal combustion engine driver.

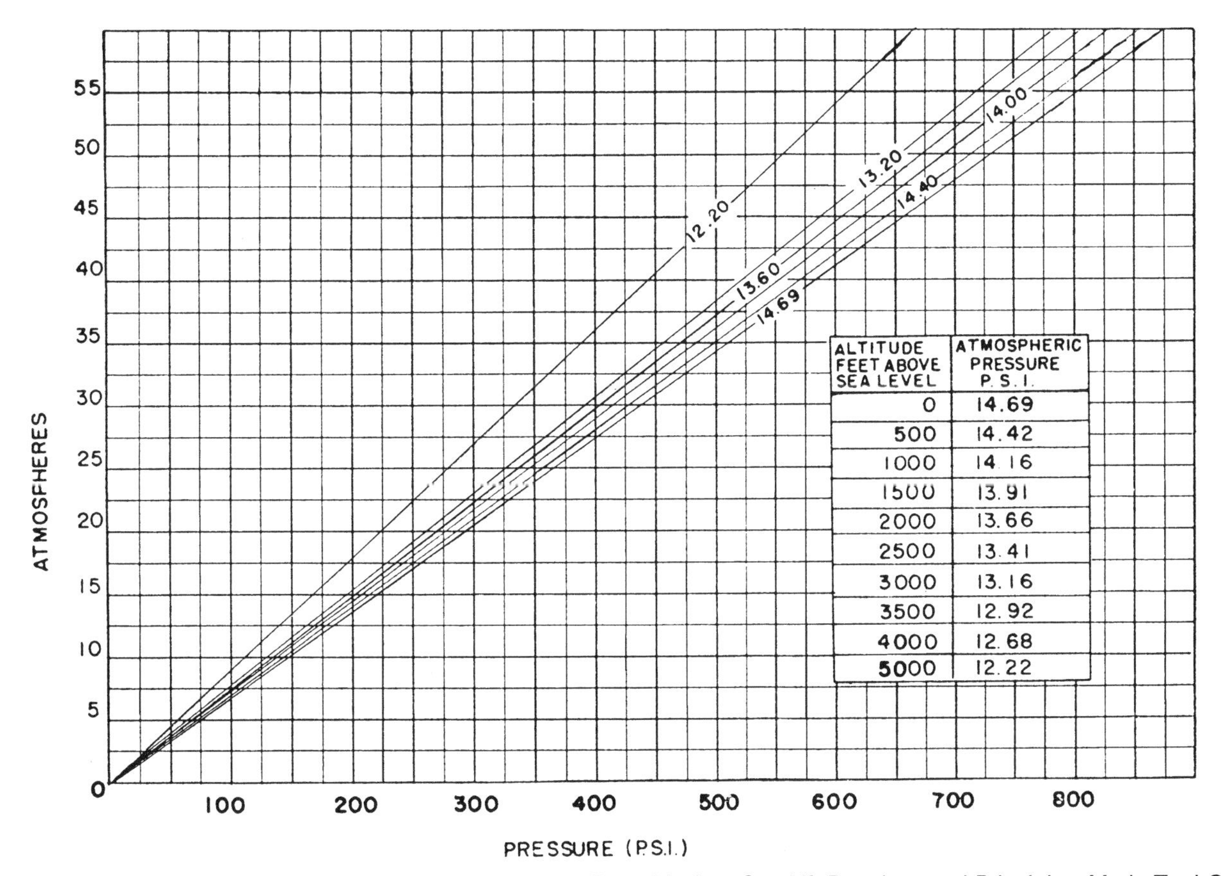

ALTITUDE FEET ABOVE SEA LEVEL	ATMOSPHERIC PRESSURE P. S. I.
0	14.69
500	14.42
1000	14.16
1500	13.91
2000	13.66
2500	13.41
3000	13.16
3500	12.92
4000	12.68
5000	12.22

Figure 1. Atmospheres at various atmospheric pressures. *From Modern Gas Lift Practices and Principles*, Merla Tool Corp.

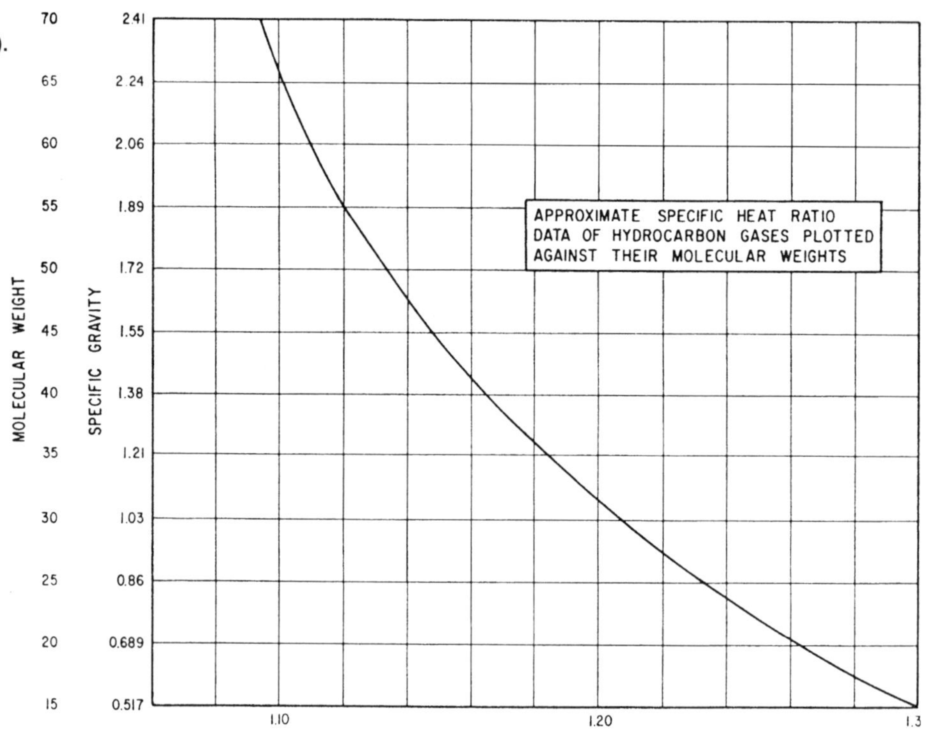

Figure 2. Ratio of specific heat (n-value).

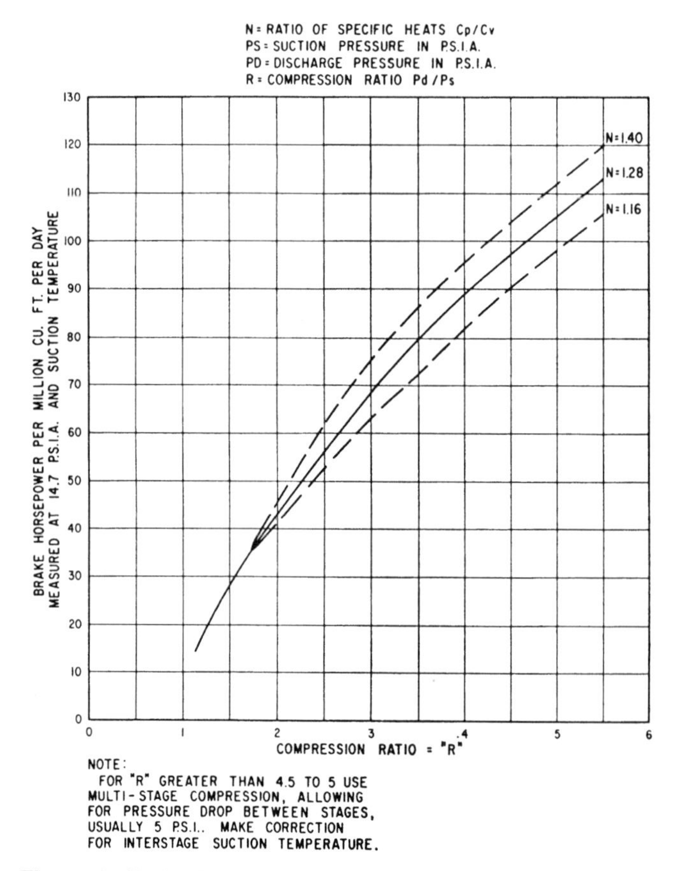

Figure 3. Brake horsepower required for compressing natural gas.

Altitude	Multiplier
1,500 ft	1.000
2,000 ft	1.03
2,500 ft	1.05
3,000 ft	1.07
3,500 ft	1.10
4,000 ft	1.12
4,500 ft	1.14
5,000 ft	1.17
5,500 ft	1.20
6,000 ft	1.22

8. For a portable unit with a fan cooler and pump driven from the compressor unit, increase the horsepower figure by 7½%.

The resulting figure is sufficiently accurate for all purposes. The nearest commercially available size of compressor is then selected.

The method does not take into consideration the supercompressibility of gas and is applicable for pressures up to 1,000 psi. In the region of high pressures, neglecting the deviation of behavior of gas from that of the perfect gas may lead to substantial errors in calculating the compression horsepower requirements. The enthalpy-entropy charts may be used conveniently in such cases. The procedures are given in references 1 and 2.

Example. What is the nominal size of a portable compressor unit required for compressing 1,600,000 standard cu. ft of gas per 24 hours at a temperature of 85°F from 40 psig pressure to 600 psig pressure? The altitude above sea level is 2,500 ft. The N value of gas is 1.28. The suction temperature of stages, other than the first stage, is 130°F.

Solution.

$P_s = 53.41$ psia

$P_d = 613.41$ psia

$R_d = P_d/P_s = 11.5 = (3.4)^2 = (2.26)^3$

Try solution using a 3.44 ratio and two stages.

1st stage : 53.41 psia × 3.44 = 183.5 psia discharge

2nd stage : 178.5 psia × 3.44 = 614 psia discharge

Horsepower from curve, Figure 3 = 77 hp for 3.44 ratio

$$\frac{77\text{ hp} \times 1{,}600{,}000}{1{,}000{,}000} = 123.1 \text{ (for } 60^\circ\text{F suction temp.)}$$

$$\text{1st stage: } 123.1\text{ hp} \times \frac{460 + 85^\circ}{460 + 60^\circ} = 129.1\text{ hp}$$

$$\text{2nd stage: } 123.1\text{ hp} \times \frac{460 + 130^\circ}{460 + 60^\circ} = 139.7\text{ hp}$$

1.05 (129.1 hp + 139.7 hp) = 282 hp

1.075 × 282 hp

Nearest nominal size compressor is 300 hp.

Centrifugal compressors

The centrifugal compressors are inherently high-volume machines. They have extensive application in gas transmission systems. Their use in producing operations is very limited.

References

1. *Engineering Data Book*, Natural Gasoline Supply Men's Association, 1957.
2. Dr. George Granger Brown: "A Series of Enthalpy-entropy Charts for Natural Gas," *Petroleum Development and Technology*, Petroleum Division AIME, 1945.

Generalized compressibility factor

The nomogram (Figure 1) is based on a generalized compressibility chart.[1] It is based on data for 26 gases, excluding helium, hydrogen, water, and ammonia. The accuracy is about 1% for gases other than those mentioned.

To use the nomogram, the values of the reduced temperature (T/T_c) and reduced pressure (P/P_c) must be calculated first.

where:

- T = Temperature in consistent units
- T_c = Critical temperature in consistent units
- P = Pressure in consistent units
- P_c = Critical pressure in consistent units

Example. $P_r = 0.078$, $T_r = 0.84$, what is the compressibility factor, z? Connect P_r with T_r and read $z = 0.948$.

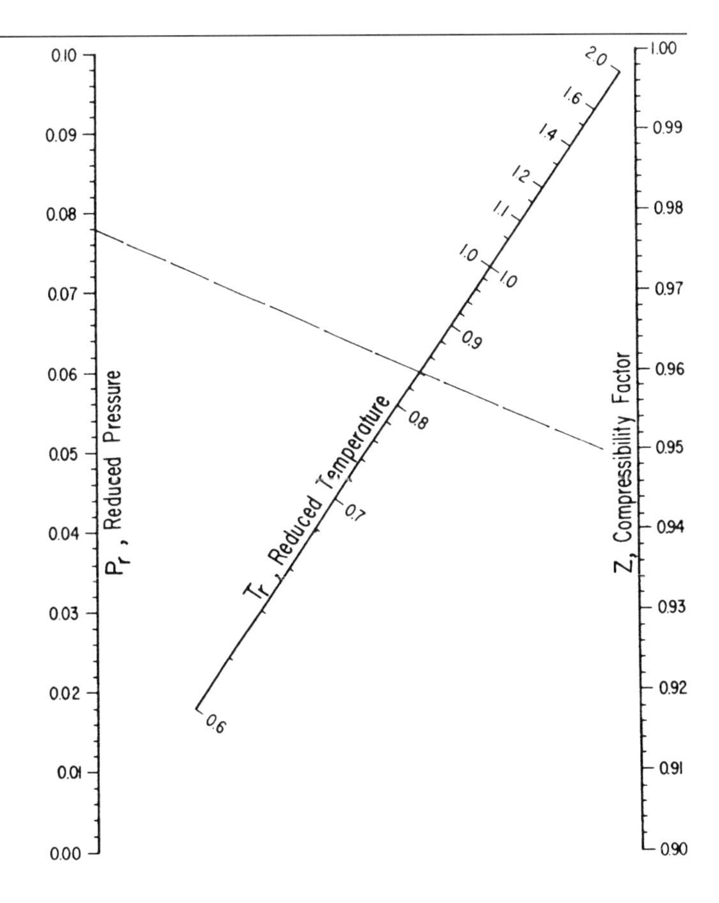

Figure 1. Generalized compressibility factor. (Reproduced by permission *Petroleum Refiner*, Vol. 37, No. 11, copyright 1961, Gulf Publishing Co., Houston.)

Source

Davis, D. S., *Petroleum Refiner*, 37, No. 11, (1961).

Reference

3. Nelson, L. C., and Obert, E. F., *Chem. Engr.*, 203 (1954).

Nomograph aids in diagnosing compressor cylinder ills

B. W. Robertson, Tennessee Gas Pipeline Co., Houston

Compressor cylinders, like people, develop a fever when they are sick. Unlike humans, however, the normal temperature (suction) is not constant. The trick of using temperatures to determine malfunctions lies in determining the correct starting position.

To determine the starting position, it is necessary to know the true k value of the product being pumped. The k value may be obtained from measured pressures and temperatures according to the formula:

$$T_2/T_1 = (P_2/P_1)\frac{k - I}{k}$$

where the subfix 2 refers to discharge conditions and the 1 to suction, and all values of P and T are absolute.[1]

The preferred time to obtain this k would be with new equipment or from data obtained when the equipment was new. The next best time would be with newly overhauled equipment. Lacking either of these data, any values of k must be considered to contain some malfunctions, and any unnecessarily high numbers should be discarded.

Nomograph 1 has been successfully used for a number of years, although it is slowly giving way to electronic equipment. Nomograph 2 may be finished by completing the suction and discharge temperature scales according to the following formulas:

Suction: $t_i = 8.557\rho \text{ Log } (t_s + 460) - 23.135\rho + 2.51$ in. above base line

Discharge: $t = 23.067\rho \text{ Log } (t_d + 460) - 62.65\rho$ in. above base line

where:

$$\rho = \frac{k}{k - 1}$$

and logs are to the base 10.

Reference

1. Virgil Moring Faires, *Applied Thermodynamics*, 11th Printing, 1946, p. 43.

Nomograph 1. To determine starting position, it is necessary to know the true "k" value of the product being pumped. The "k" value may be obtained from measured pressure and temperatures according to the formula given.

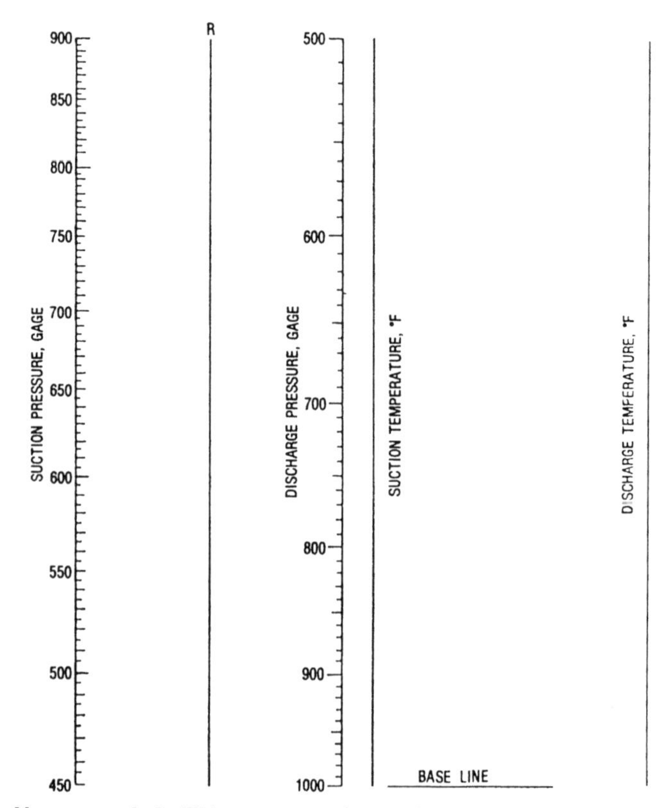

Nomograph 2. This nomograph may be finished by completing the suction and discharge temperature scales according to the formulas given.

CENTRIFUGAL COMPRESSOR DATA

Centrifugal compressor performance calculations

Centrifugal compressors are versatile, compact, and generally used in the range of 1,000 to 100,000 inlet cu. ft per minute (ICFM) for process and pipeline compression applications.

Centrifugal compressors can use either a horizontal or a vertical split case. The type of case used will depend on the pressure rating with vertical split casings generally being used for the higher pressure applications. Flow arrangements include straight through, double flow, and side flow configurations.

Centrifugal compressors may be evaluated using either the adiabatic or polytropic process method. An adiabatic process is one in which no heat transfer occurs. This doesn't imply a constant temperature, only that no heat is transferred into or out of the process system. Adiabatic is normally intended to mean adiabatic isentropic. A polytropic process is a variable-entropy process in which heat transfer can take place.

When the compressor is installed in the field, the power required from the driver will be the same whether the process is called adiabatic or polytropic during design. Therefore, the work input will be the same value for either process. It will be necessary to use corresponding values when making the calculations. When using adiabatic head, use adiabatic efficiency, and when using polytropic head, use polytropic efficiency. Polytropic calculations are easier to make even though the adiabatic approach appears to be simpler and quicker.

The polytropic approach offers two advantages over the adiabatic approach. The polytropic approach is independent of the thermodynamic state of the gas being compressed, whereas the adiabatic efficiency is a function of the pressure ratio and therefore is dependent upon the thermodynamic state of the gas.

If the design considers all processes to be polytropic, an impeller may be designed, its efficiency curve determined, and it can be applied without correction regardless of pressure, temperature, or molecular weight of the gas being compressed. Another advantage of the polytropic approach is that the sum of the polytropic heads for each stage of compression equals the total polytropic head required to get from state point 1 to state point 2. This is not true for adiabatic heads.

Sample performance calculations

Determine the compressor frame size, number of stages, rotational speed, power requirement, and discharge temperature required to compress 5,000 lb-in./min of gas from 30 psia at 60°F to 100 psia. The gas mixture molar composition is as follows:

Ethane	5%
Propane	80%
n-Butane	15%

This is the same gas mixture that was considered in Sample Calculation 1, p. 214, where the gas properties were calculated. The properties of this mixture were determined to be as follows:

$MW = 45.5$
$P_c = 611$ psia
$T_c = 676°R$
$c_p = 17.76$
$k_1 = 1.126$
$Z_1 = 0.955$

Before proceeding with the compressor calculations, let's review the merits of using average values of Z and k in calculating the polytropic head.

The inlet compressibility must be used to determine the actual volume entering the compressor to approximate the size of the compressor and to communicate with the vendor via the data sheets. The maximum value of θ is of interest and will be at its maximum at the inlet to the compressor where the inlet compressibility occurs (although using the average compressibility will result in a conservative estimate of θ).

Compressibility will decrease as the gas is compressed. This would imply that using the inlet compressibility would be conservative since as the compressibility decreases, the head requirement also decreases. If the variation in compressibility is drastic, the polytropic head requirement calculated by using the inlet compressibility would be practically useless. Compressor manufacturers calculate the performance for each stage and use the inlet compressibility for each stage. An accurate approximation may be substituted for the stage-by-stage calculation by calculating the polytropic head for the overall section using the average compressibility. This technique results in overestimating the first half of the impellers and underestimating the last half of the impellers, thereby calculating a polytropic head very near that calculated by the stage-by-stage technique.

Determine the inlet flow volume, Q_1:

$$Q_1 = m_1[(Z_1RT_1)/(144P_1)]$$

where:

m = Mass flow
Z_1 = Inlet compressibility factor
R = Gas constant = 1,545/MW
T_1 = Inlet temperature °R
P_1 = Inlet pressure

$$Q_1 = 5{,}000[(0.955)(1{,}545)(60 + 460)/(45.5)(144)(30)] = 19{,}517 \text{ ICFM}$$

Refer to Table 1 and select a compressor frame that will handle a flow rate of 19,517 ICFM. A Frame C compressor will handle a range of 13,000 to 31,000 ICFM and would have the following nominal data:

H_{pnom} = 10,000 ft – lb/lbm (nominal polytropic head)

n_p = 77% (polytropic efficiency)

N_{nom} = 5,900 rpm

Determine the pressure ratio, r_p.

$$r_p = P_2/P_1 = 100/30 = 3.33$$

Determine the approximate discharge temperature, T_2.

$$n/n - 1 = [k/k - 1]n_p = [1.126/(1.126 - 1.000)](0.77) = 6.88$$

$$T_2 = T_1\,(r_p)^{(n-1)/n} = (60 + 460)(3.33)^{1/6.88} = 619°R = 159°F$$

where T_1 = Inlet temp

Determine the average compressibility, Z_a.

$Z_1 = 0.955$ (from gas properties calculation)

where Z_1 = Inlet compressibility

$$(P_r)_2 = P_2/P_c = 100/611 = 0.164$$

$$(T_r)_2 = T_2/T_c = 619/676 = 0.916$$

Table 1
Typical centrifugal compressor frame data*

Frame	Nominal Inlet Volume Flow English (ICFM)	Nominal Inlet Volume Flow Metric (m3/h)	Nominal Polytropic Head English (ft-lbf/lbm)	Nominal Polytropic Head Metric (k-Nm/kg)	Nominal Polytropic Efficiency (%)	Nominal Rotational Speed (RPM)	Nominal Impeller Diameter English (in)	Nominal Impeller Diameter Metric (mm)
A	1,000–7,000	1,700–12,000	10,000	30	76	11,000	16	406
B	6,000–18,000	10,000–31,000	10,000	30	76	7,700	23	584
C	13,000–31,000	22,000–53,000	10,000	30	77	5,900	30	762
D	23,000–44,000	39,000–75,000	10,000	30	77	4,900	36	914
E	33,000–65,000	56,000–110,000	10,000	30	78	4,000	44	1,120
F	48,000–100,000	82,000–170,000	10,000	30	78	3,300	54	1,370

*While this table is based on a survey of currently available equipment, the instance of any machinery duplicating this table would be purely coincidental.

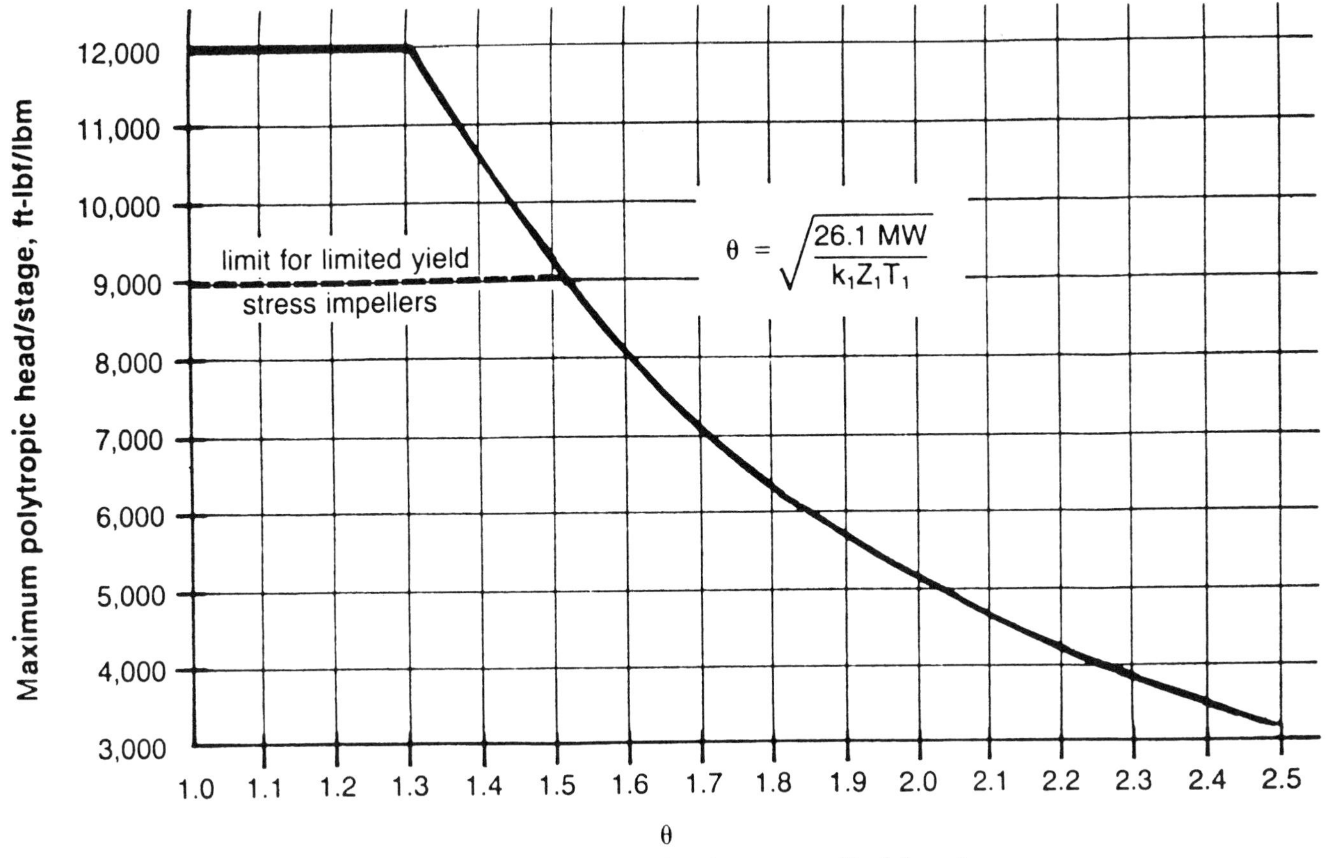

Figure 1. Maximum polytropic head per stage—English system.

Refer to Figure 2, p. 216 to find Z_2, discharge compressibility.

$$Z_2 = 0.925$$

$$\begin{aligned} Z_a &= (Z_1 + Z_2)/2 \\ &= 0.94 \end{aligned}$$

Determine average k-value. For simplicity, the inlet value of k will be used for this calculation. The polytropic head equation is insensitive to k-value (and therefore n-value) within the limits that k normally varies during compression. This is because any errors in the $n/(n-1)$ multiplier in the polytropic head equation tend to balance corresponding errors in the $(n-1)/n$ exponent. Discharge temperature is very sensitive to k-value. Since the k-value normally decreases during compression, a discharge temperature calculated by using the inlet k-value will be conservative, and the actual temperature may be several degrees higher—possibly as much as 25–50°F. Calculating the average k-value can be time-consuming, especially for mixtures containing several gases, since not only must the mol-weighted cp of the mixture be determined at the inlet temperature but also at the estimated discharge temperature.

The suggested approach is as follows:

1. If the k-value is felt to be highly variable, one pass should be made at estimating discharge temperature based on the inlet k-value; the average k-value should then be calculated using the estimated discharge temperature.
2. If the k-value is felt to be fairly constant, the inlet k-value can be used in the calculations.
3. If the k-value is felt to be highly variable, but sufficient time to calculate the average value is not available, the inlet k-value can be used (but be aware of the potential discrepancy in the calculated discharge temperature).

$$k_1 = k_a = 1{,}126$$

Determine average $n/(n-1)$ value from the average k-value. For the same reasons discussed earlier, use $n/(n-1) = 6.88$.

Table 2
Approximate mechanical losses as a percentage of gas power requirement*

Gas Power Requirement English (hp)	Gas Power Requirement Metric (kW)	Mechanical Losses, L_m (%)
0–3,000	0–2,500	3
3,000–6,000	2,500–5,000	2.5
6,000–10,000	5,000–7,500	2
10,000+	7,500+	1.5

*There is no way to estimate mechanical losses from gas power requirements. This table will, however, ensure that mechanical losses are considered and yield useful values for estimating purposes.

Determine polytropic head, Hp:

$$H_p = Z_a RT_1(n/n-1)[r_p^{(n-1)/n} - 1]$$
$$= (0.94)(1{,}545/45.5)(520)(6.88)[(3.33)^{1/6.88} - 1]$$
$$= 21{,}800 \text{ ft} - \text{lbf/lbm}$$

Determine the required number of compressor stages, θ:

$$\theta = [(26.1 \text{ MW})/(k_1 Z_1 T_1)]^{0.5}$$
$$= [(26.1)(45.5)/(1.126)(0.955)(520)]^{0.5}$$
$$= 1.46$$

max H_p/stage from Figure 1 using $\theta = 1.46$

$$\text{Number of stages} = H_p/\text{max}.\ H_p/\text{stage}$$
$$= 21{,}800/9{,}700$$
$$= 2.25$$
$$= 3 \text{ stages}$$

Determine the required rotational speed:

$$N = N_{nom}[H_p/H_{pnom} \times \text{no. stages}]^{0.5}$$
$$= 5{,}900[21{,}800/(10{,}000)(3)]^{0.5}$$
$$= 5{,}030 \text{ rpm}$$

Determine the required shaft power:

$$PWR_g = mH_p/33{,}000\ n_p$$
$$= (5{,}000)(21{,}800)/(33{,}000)(0.77)$$
$$= 4{,}290 \text{ hp}$$

Mechanical losses (L_m) = 2.5% (from Table 2)

$$L_m = (0.025)(4{,}290)$$
$$= 107 \text{ hp}$$

$$PWR_s = PWR_g + L_m$$
$$= 4{,}290 + 107$$
$$= 4{,}397 \text{ hp}$$

Determine the actual discharge temperature:

$$T_2 = T_1(r_p)^{(n-1)/n}$$
$$= 520(3.33)^{1/6.88}$$
$$= 619°R$$
$$= 159°F$$

The discharge temperature calculated in the last step is the same as that calculated earlier only because of the decision to use the inlet k-value instead of the average k-value. Had the average k-value been used, the actual discharge temperature would have been lower.

Source

Estimating Centrifugal Compressor Performance, Gulf Publishing Company, Houston.

Nomographs for estimating compressor performance*

Convert SCFM to inlet CFM

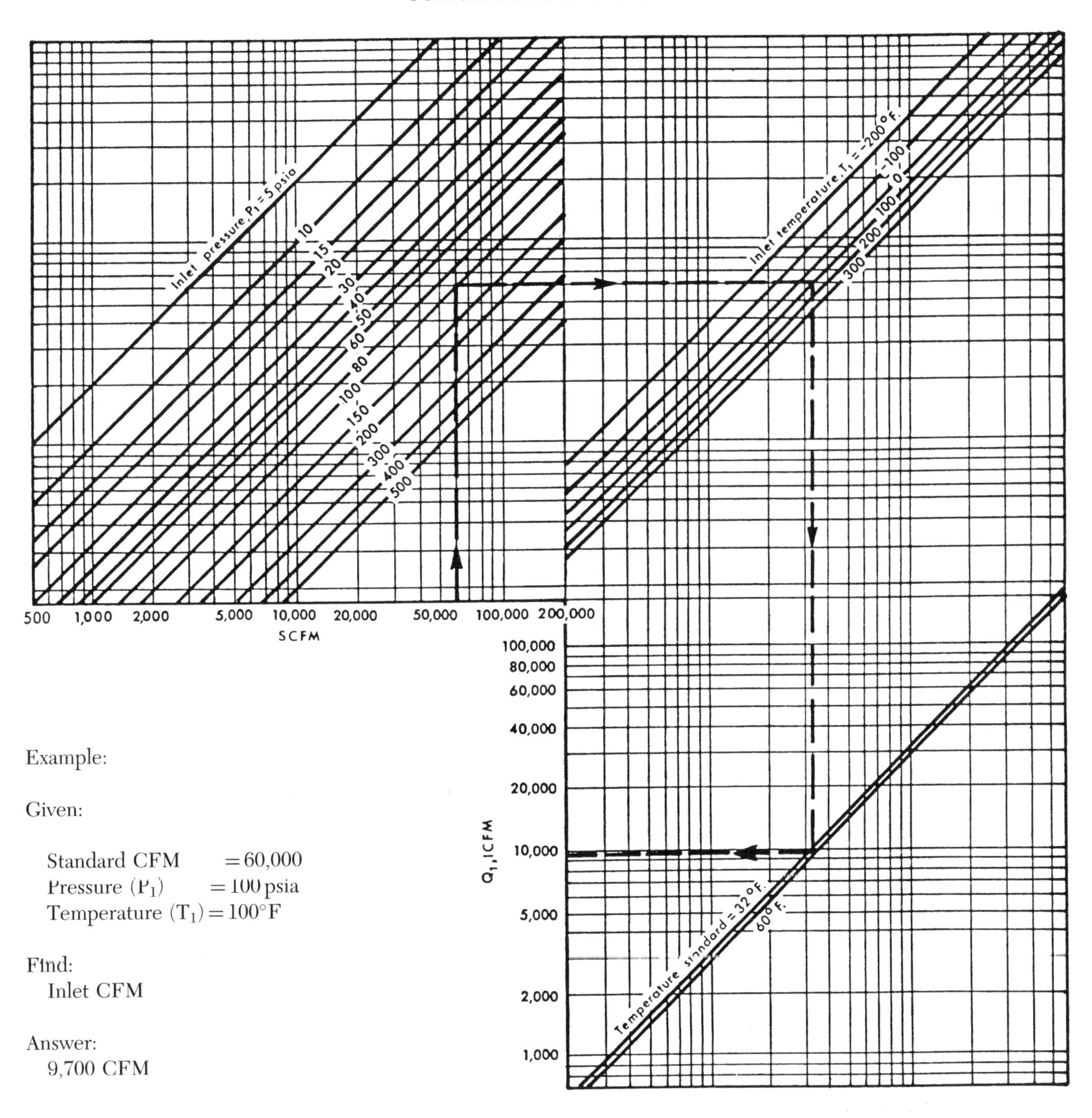

Inlet cubic feet per minute

Example:

Given:

Standard CFM $= 60{,}000$
Pressure (P_1) $= 100$ psia
Temperature (T_1) $= 100°F$

Find:
Inlet CFM

Answer:
9,700 CFM

*Reprinted by permission of Gas Processors Suppliers Association.

Convert weight flow to inlet CFN

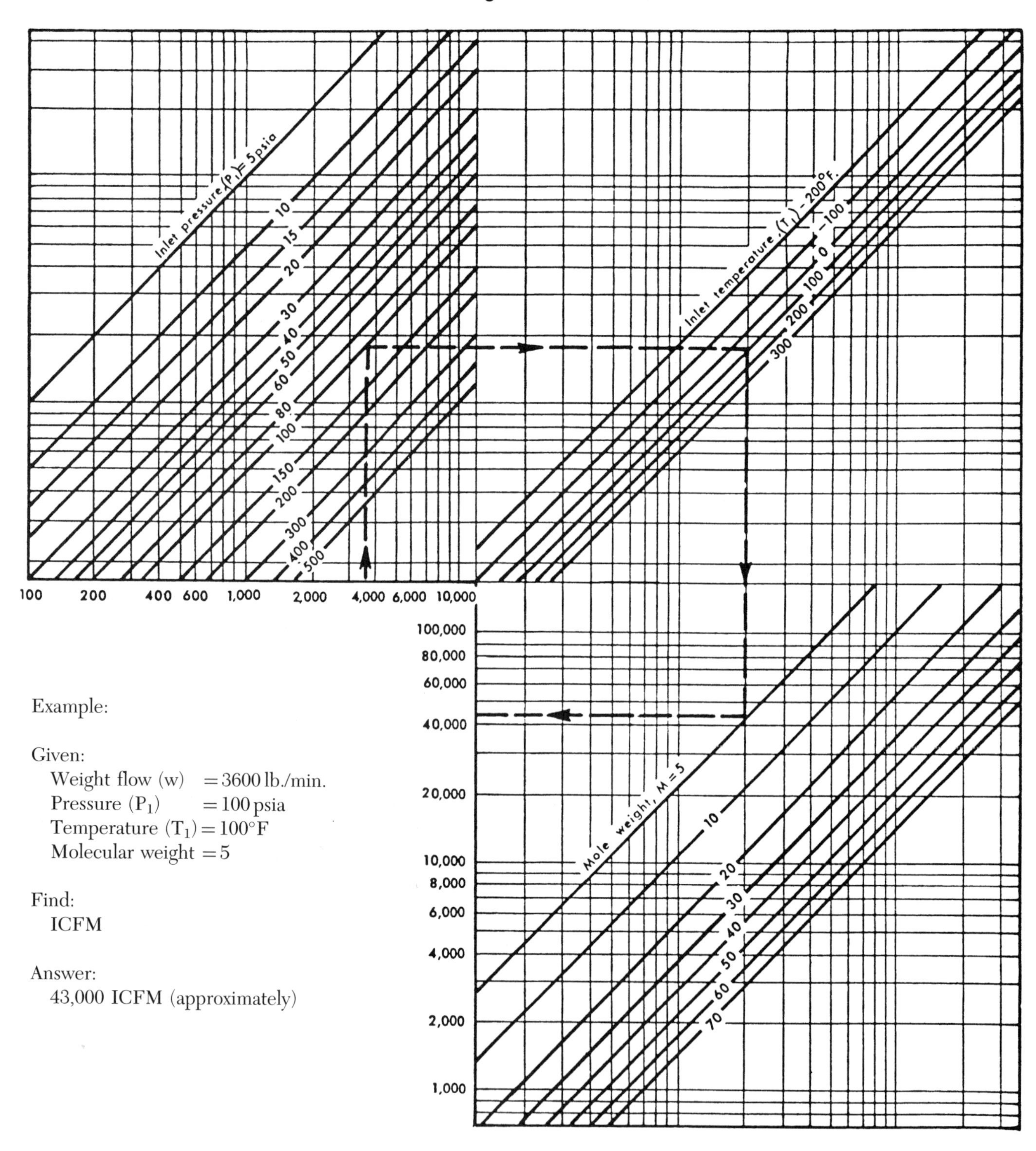

Example:

Given:
Weight flow (w) = 3600 lb./min.
Pressure (P_1) = 100 psia
Temperature (T_1) = 100°F
Molecular weight = 5

Find:
ICFM

Answer:
43,000 ICFM (approximately)

Inlet cubic feet per minute

Convert weight compression ratio to head

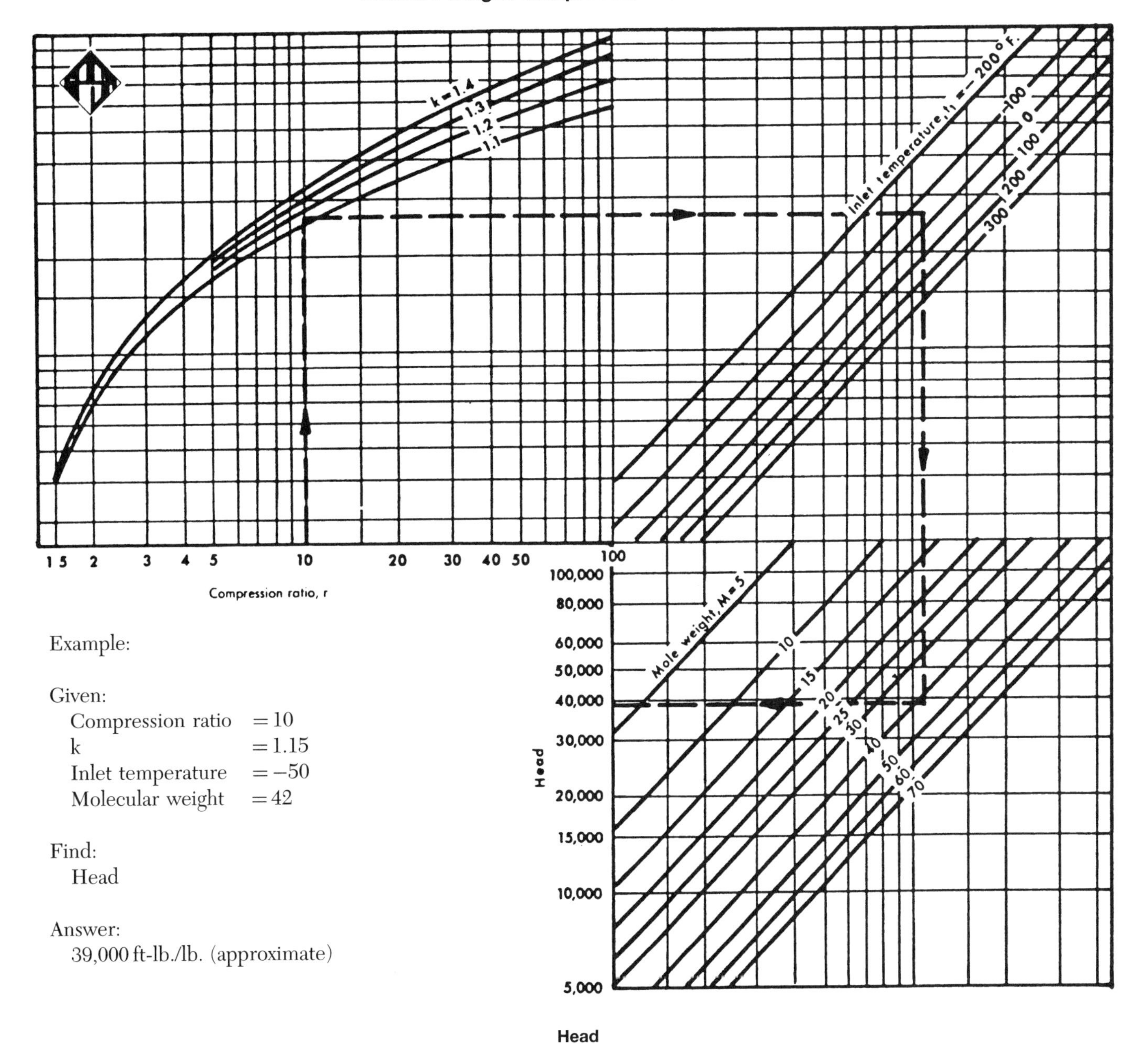

Head

Example:

Given:

Compression ratio	= 10
k	= 1.15
Inlet temperature	= −50
Molecular weight	= 42

Find:

Head

Answer:

39,000 ft-lb./lb. (approximate)

Convert compression ratio to discharge temperature

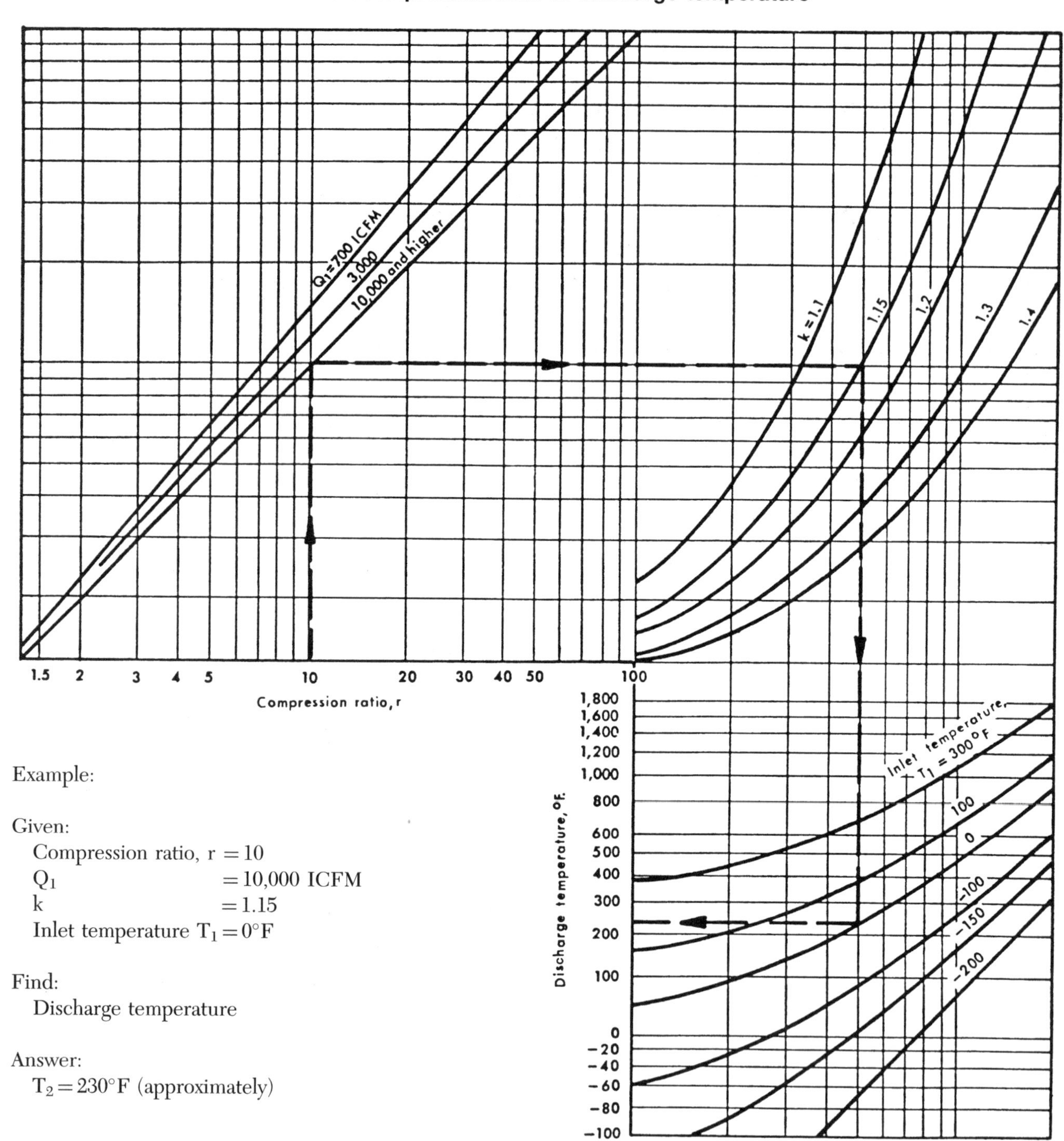

Example:

Given:

Compression ratio, r = 10
Q_1 = 10,000 ICFM
k = 1.15
Inlet temperature $T_1 = 0°F$

Find:

Discharge temperature

Answer:

$T_2 = 230°F$ (approximately)

Discharge temperature

Horsepower determination

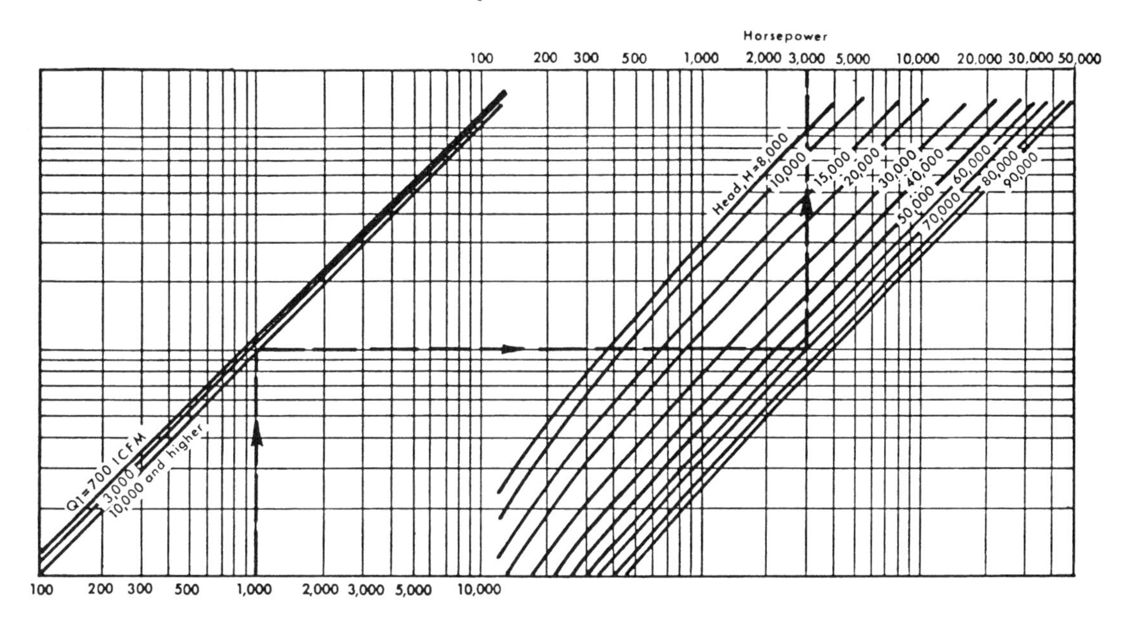

Weight flow, w, lb/min

Example:

Given:

Weight flow, w = 1,000 lb./min.

Head = 70,000 ft-lb./lb.

Find:

Horsepower

Answer:

$H_p = 3{,}000$

Efficiency conversion

Polytropic efficiency (η_p)

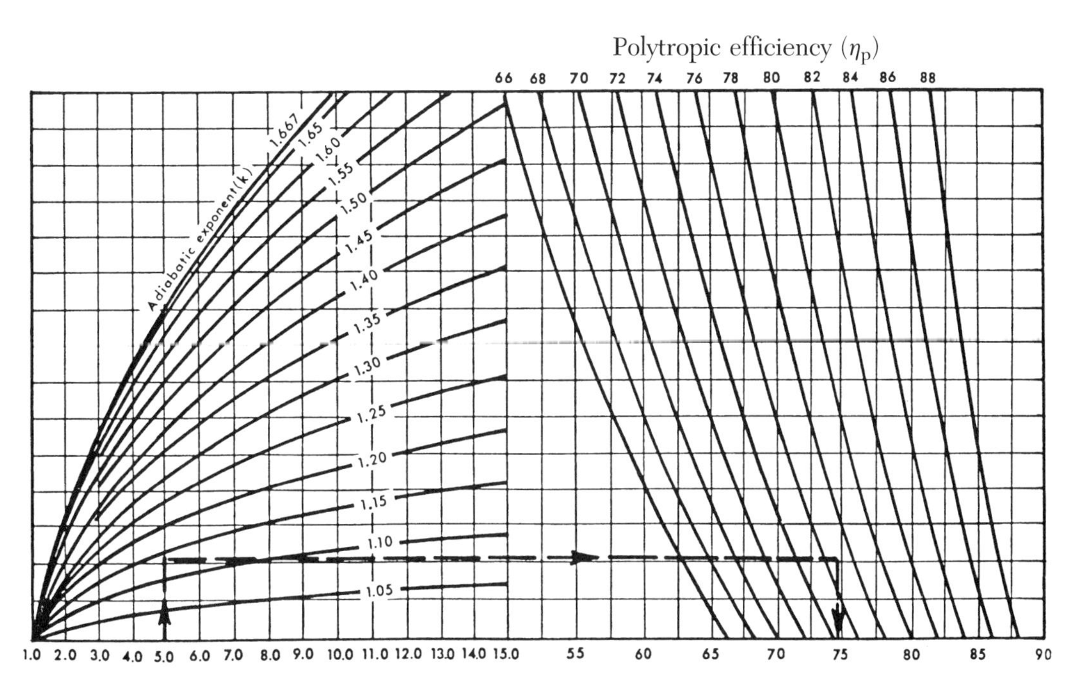

Compression ratio (r)

Adiabatic efficiency (η_{ad})

Estimate hp required to compress natural gas

To estimate the horsepower to compress a million cubic ft of gas per day (MMcfd), use the following formula:

$$\text{BHP/MMcfd} = \frac{R}{R + RJ}\left(\frac{5.16 + 124 \text{ Log}R}{0.97 - 0.03R}\right)$$

where:

R = Compression ratio. Absolute discharge pressure divided by absolute suction pressure
J = Supercompressibility factor—assumed 0.022 per 100 psia suction pressure

Example. How much horsepower should be installed to raise the pressure of 10 million cu. ft of gas per day from 185.3 psi to 985.3 psi?

This gives absolute pressures of 200 and 1,000.

$$R = \frac{1{,}000}{200} = 5.0$$

Substituting in the formula:

$$\text{BHP/MMcfd} = \frac{5.0}{5.0 + 5 \times 0.044} \times \frac{5.15 + 124 \times .699}{.97 - .03 \times 5}$$

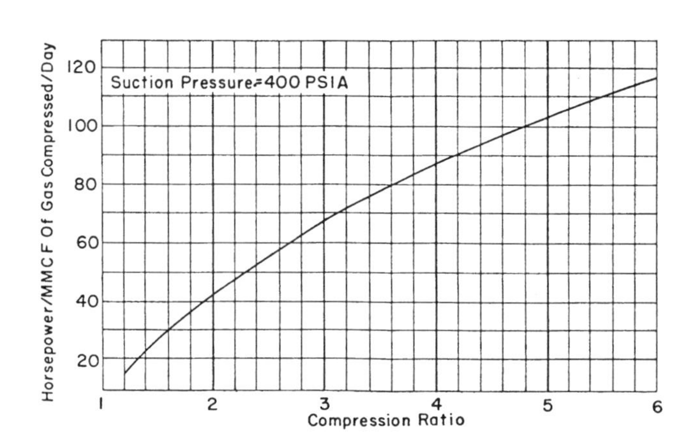

$= 106.5$ hp = BHP for 10 MMcfd
$= 1{,}065$ hp

where the suction pressure is about 400 psia, the brake horsepower per MMcfd can be read from the chart.

The previous formula may be used to calculate horsepower requirements for various suction pressures and gas physical properties to plot a family of curves.

Estimate compressor hp where discharge pressure is 1,000 psi

Rule. Use the following formula for reciprocating compressors where the compressor ratio is between 1.20 and 1.40.

Brake horsepower per MMcfd $= 11 + (42.5)(R - 1.20)$

R is compression ratio.

Example. How much horsepower would have to be installed in a field compressor station where the discharge pressure would be 1,055 psia and suction would be 800 psia? The station must handle 10,000,000 cu. ft per day.

Solution. The compression ratio is

$$\frac{1{,}055}{800} = 1.32$$

Substituting in the formula:

Brake horsepower per MMcfd $= 11 + (42.5)(1.32 - 1.20)$
$= 11 + 5.1$
$= 16.1$ hp

The station would require 161 hp.

For a centrifugal compressor the formula would be:

Brake horsepower per MMcfd $= 11 + (42.5)(R - 1.20)$

Example. How much horsepower should be installed in a centrifugal compressor station handling 100,000,000 cu. ft per day if the suction and discharge are the same as above?

Solution.

Brake horsepower per MMcfd $= 10 + (49)(0.12) = 15.9$

For 100 MMcfd the horsepower would be 1,590.

Calculate brake horsepower required to compress gas

The approximate brake horsepower per million cubic ft of gas to be compressed may be obtained from this table. The total brake horsepower required is calculated by multiplying the total million cubic ft of gas compressed. In this estimation, gas volumes are expressed at 14.7 psia and 60°F, specific gravity as 0.6, and K (the ratio of specific heats) as 1.27. The suction temperature is assumed as 80°F and interstage temperature as 130°F. The relative humidity of the air equals 100%.

Example. What is the brake horsepower required to compress 5.0 million cu. ft of gas from 200 psig suction pressure to 1,000 psig discharge pressure?

On the chart, read opposite a suction pressure of 200 and under a discharge pressure of 1,000 psig, that 100 brake horsepower is required per million cubic ft of gas compressed. Multiply 5.0 times 100 = 500 bhp.

HORSEPOWER SELECTION CHART
NATURAL GAS COMPRESSION

		DISCHARGE PRESSURE									
		250 psig	375 psig	500 psig	750 psig	1,000 psig	1,250 psig	1,500 psig	2,000 psig	3,500 psig	6,000 psig
SUCTION PRESSURE	0 psig	205	235	260	280	300	320	330	360	390	—
	10 psig	165	195	215	250	260	275	290	310	345	390
	25 psig	130	155	175	205	230	250	255	275	320	370
	50 psig	95	125	140	170	190	205	215	245	280	325
	75 psig	70	95	120	145	165	180	195	215	250	300
	100 psig	50	75	95	120	145	160	175	195	230	275
	200 psig	—	—	55	80	100	120	130	150	190	240
	300 psig	—	—	30	55	75	90	105	120	160	205
	500 psig	—	—	—	25	40	55	65	85	125	165
	750 psig	—	—	—	—	15	30	40	60	100	140
	1,000 psig	—	—	—	—	—	—	25	40	75	125
	1,500 psig	—	—	—	—	—	—	—	—	50	90
	2,000 psig	—	—	—	—	—	—	—	—	—	70
	2,500 psig	—	—	—	—	—	—	—	—	—	55

How to find the size of a fuel gas line for a compressor station

Suppose that an 8,000 hp compressor building is added to an existing compressor plant. What size fuel line should be run from the meter building 1,600 ft distant? Take-off pressure at the meter building will be 50 lb./in.^{2}, and it is desirable to maintain 25 psi pressure on the large engine fuel header in the building. Assume that the gas is heated prior to expansion so that outlet temperature is 60°F.

The first step is to estimate the ultimate compressor-building horsepower. Assume in the foreseeable future the building will be expanded to 12,000 hp.

Next, estimate the fuel needs.

$$Q = 12{,}000 \times 240 \text{ cu. ft per day}$$
$$= 2{,}880{,}000 \text{ cu. ft per day}$$

To estimate the size of line that must be run for fuel use this formula:

$$Q = \frac{(36{,}000)d^3\sqrt{P_1^2 - P_2^2}}{\sqrt{l}}$$

where: Q = Cubic feet of gas per 24 hours
d = Pipe ID in inches
l = Length of pipe in feet
P_1 = Psi (abs) at starting point
P_2 = Psi (abs) at ending point

Using this formula, try several pipe diameters to determine the approximate size. For instance, if 6-in. ID pipe were used, the approximate flow could be found as follows:

$$Q = \frac{36{,}300 \times 6^3\sqrt{(50+15)^2-(25+15)^2}}{\sqrt{1{,}600}}$$

$$Q = \frac{36{,}300 \times 216\sqrt{65^2-40^2}}{40}$$

Q = approximately 10,036,000 cu. ft per 24 hours

From this it is obvious that 6-in. pipe would be considerably oversized, especially if the fuel gas header in the compressor building is of large diameter.

Try substituting 4-in. diameter pipe in this formula to find an approximate size pipe.

$$Q = \frac{36{,}300 \times 4^3\sqrt{(50+15)^2-(25+15)^2}}{\sqrt{1{,}600}}$$

$$Q = \frac{36{,}300 \times 64\sqrt{65^2-40^2}}{40}$$

Q = 2,973,700 cu. ft per 24 hours

Apparently 4-in. pipe will be the correct size. To check it, use this more accurate formula, which is based on the Weymouth formula:

$$Q = \frac{63{,}320 \times d^{8/3}\sqrt{P_1^2-P_2^2}}{\sqrt{l}}$$

$$Q = \frac{63{,}320 \times 41\sqrt{65^2-40^2}}{\sqrt{1{,}600}}$$

Four-in. pipe is adequate.

Q = 3,322,000 cubic ft per day

Note. This latter formula is based on the Weymouth formula, for gas with a specific gravity of 0.60 and a temperature of 60°F. For every 10° variation in temperature, the answer will be approximately 1% in error. For every 0.01 variation in specific gravity, the answer will be three-fourth percent in error.

Estimate engine cooling water requirements

This equation can be used for calculating engine jacket water requirements, as well as lube oil cooling water requirements:

$$GPM = \frac{H \times BHP}{500\Delta t}$$

where: H = Heat dissipation in Btu's per BHP/hr. This will vary for different engines; where they are available, the manufacturers' values should be used. Otherwise, you will be safe in substituting the following values in the formula: For engines with water-cooled exhaust manifolds—engine jacket water = 2,200 Btu's per BHP/hr. Lube oil cooling water = 600 Btu's per BHP/hr.

For engines with dry type manifolds (so far as cooling water is concerned), use 1,500 Btu's/BHP/hr for the engine jackets and 650 Btu's/BHP/hr for lube oil cooling water requirements.

BHP = Brake Horsepower Hour

Δt = Temperature differential across engine. Usually manufacturers recommend this not exceed 15°F; 10°F is preferable.

Example. Find the jacket water requirements for a 2,000-hp gas engine that has no water jacket around the exhaust manifold.

Solution.

$$GPM = \frac{1{,}500 \times 2{,}000}{500 \times 10}$$

$$GPM = \frac{3{,}000{,}000}{5{,}000} = 600 \text{ gallons per min}$$

The lube oil cooling water requirements could be calculated in like manner.

Estimate fuel requirements for internal combustion engines

When installing an internal combustion engine at a gathering station, a quick approximation of fuel consumptions could aid in selecting the type of fuel used.

Using Natural Gas: Multiply the brake hp at drive by 11.5 to get cubic ft of gas per hour.

Using Butane: Multiply the brake hp at drive by 0.107 to get gallons of butane per hour.

Using Gasoline: Multiply the brake hp at drive by 0.112 to get gallons of gasoline per hour.

These approximations will give reasonably accurate figures under full load conditions.

Example. Internal combustion engine rated at 50 bhp—three types of fuel available.

Natural Gas: $50 \times 11.5 = 575$ cu. ft of gas per hour

Butane: $50 \times 0.107 = 5.35$ gallons of butane per hour

Gasoline: $50 \times 0.112 = 5.60$ gallons of gasoline per hour

Estimate fuel requirements for compressor installation

Rule. Multiply installation horsepower by 240 to obtain cubic ft of fuel gas required per day.

Example. Estimate the amount of fuel required to supply a 1,000-hp field compressor installation.

$1{,}000 \times 240 = 240$ Mcfd

11: Gas—Hydraulics

Gas pipeline hydraulics calculations

Equations commonly used for calculating hydraulic data for gas pipelines

Panhandle A.

$$Q_b = 435.87 \times (T_b/P_b)^{1.0788} \times D^{2.6182} \times E \times \left[\frac{P_1^2 - P_2^2 - \dfrac{(0.0375 \times G \times (h_2 - h_1) \times P_{avg}^2}{T_{avg} \times Z_{avg}}}{G^{0.8539} \times L \times T_{avg} \times Z_{avg}}\right]^{0.5394}$$

Panhandle B.

$$Q_b = 737 \times (T_b/P_b)^{1.020} \times D^{2.53} \times E \times \left[\frac{P_1^2 - P_2^2 - \dfrac{\left(0.0375 \times G \times (h_2 - h_1) \times P_{avg}^2\right)}{T_{avg} \times Z_{avg}}}{G^{0.961} \times L \times T_{avg} \times Z_{avg}}\right]^{0.51}$$

Weymouth.

$$Q_b = 433.5 \times (T_b/P_b) \times \left[\frac{P_1^2 - P_2^2}{GLTZ}\right]^{0.5} \times D^{2.667} \times E$$

$$P_{avg} = \frac{2}{3}\left[P_1 + P_2 - \frac{P_1 \times P_2}{P_1 + P_2}\right]$$

Nomenclature for Panhandle equations

Q_b = Flow rate, SCFD
P_b = – Base pressure, psia
T_b = Base temperature, °R
T_{avg} = Average gas termperature, °R
P_1 = Inlet pressure, psia
P_2 = Outlet pressure, psia
G = Gas specific gravity (air = 1.0)
L = Line length, miles
Z = Average gas compressibility
D = Pipe inside diameter, in.
h_2 = Elevation at terminus of line, ft
h_1 = Elevation at origin of line, ft
P_{avg} = Average line pressure, psia
E = Efficiency factor

E = 1 for new pipe with no bends, fittings, or pipe diameter changes.
E = 0.95 for very good operating conditions, typically through first 12–28 months
E = 0.92 for average operating conditions
E = 0.85 for unfavorable operating conditions

Nomenclature for Weymouth equation

Q = Flow rate, SCFD
T_b = Base temperature, °R
P_b = Base pressure, psia
G = Gas specific gravity (air = 1)
L = Line length, miles
T = Gas temperature, °R
Z = Gas compressibility factor
D = Pipe inside diameter, in.
E = Efficiency factor. (See Panhandle nomenclature for suggested efficiency factors

Sample calculations

Q = ?
G = 0.6
T = 100°F
L = 20 miles
P_1 = 2,000 psia
P_2 = 1,500 psia
Elev diff = 100 ft
D = 4.026-in.
T_b = 60°F
P_b = 14.7 psia
E = 1.0

$$P_{avg} = 2/3\left[2{,}000 + 1{,}500 - \frac{2{,}000 \times 1{,}500}{2{,}000 + 1{,}500}\right]$$

$$P_{avg} = 1{,}762 \text{ psia}$$

From Figure 5, p. 147, in *Pipecalc 2.0*, Z at 1,762 psia and 100°F = 0.835.

Panhandle A.

$$Q_b = 435.87 \times (520/14.7)^{1.0788} \times 4.026^{2.6182} \times 1 \times \left[\frac{(2{,}000)^2 - (1{,}500)^2 - \dfrac{(0.0375 \times 0.6 \times 100 \times (1{,}762)^2}{560 \times 0.835}}{(0.6)^{0.8539} \times 20 \times 560 \times 0.835}\right]^{0.5394}$$

Q_b = 16,577 MCFD

Panhandle B.

$$Q_b = 737 \times (520/14.7)^{1.020} \times (4.026)^{2.53} \times 1 \times \left[\frac{(2{,}000)^2 - (1{,}500)^2 - \dfrac{(0.0375 \times 0.6 \times 100 \times (1{,}762)^2}{560 \times 0.835}}{(0.6)^{0.961} \times 20 \times 560 \times 0.835}\right]^{0.51}$$

$Q_b = 17{,}498$ MCFD

Weymouth.

$$Q = 0.433 \times (520/14.7) \times \left[(2{,}000)^2 - (1{,}500)^2/(0.6 \times 20 \times 560 \times 0.835)\right]^{1/2} \times (4.026)^{2.667}$$

$Q = 11{,}101$ MCFD

Source

Pipecalc 2.0, Gulf Publishing Company, Houston, Texas. *Note:* Pipecalc 2.0 will calculate the compressibility factor, minimum pipe ID, upstream pressure, downstream pressure, and flow rate for Panhandle A, Panhandle B, Weymouth, AGA, and Colebrook-White equations. The flow rates calculated in the previous sample calculations will differ slightly from those calculated with Pipecalc 2.0, since the viscosity used in the examples was extracted from Figure 5, p. 147. Pipecalc uses the Dranchuk et al. method for calculating gas compressibility.

Equivalent lengths for multiple lines based on Panhandle A

Condition I

A single pipeline that consists of two or more different diameter lines.

Let L_E = Equivalent length
$L_1, L_2, \ldots L_n$ = Length of each diameter
$D_1, D_2, \ldots D_n$ = Internal diameter of each separate line corresponding to $L_1, L_2, \ldots L_n$
D_E = Equivalent internal diameter

$$L_e = L_1\left[\frac{D_E}{D_1}\right]^{4.8539} + L_2\left[\frac{D_E}{D_1}\right]^{4.8539} + \cdots L_n\left[\frac{D_E}{D_n}\right]^{4.8539}$$

Example. A single pipeline 100 miles in length consists of 10 miles 10¾-in. OD; 40 miles 12¾-in. OD, and 50 miles of 22-in. OD lines.

Find equivalent length (L_E) in terms of 22-in. OD pipe.

$$L_E = 50 + 40\left[\frac{21.5}{12.25}\right]^{4.8539} + 10\left[\frac{21.5}{10.25}\right]^{4.8539}$$

$$= 50 + 614 + 364$$

$= 1{,}028$ miles equivalent length of 22-in. OD

Condition II

A multiple pipeline system consisting of two or more parallel lines of different diameters and different length.

Let L_E = Equivalent length
$L_1, L_2, L_3, \ldots L_n$ = Length of various looped sections
$d_1, d_2, d_3 \ldots d_n$ = Internal diameter of the individual line corresponding to length L_1, L_2, L_3 & L_n

$$L_E = L_1\left[\frac{d_E^{2.6182}}{d_1^{2.6182} + d_2^{2.6182} + d_3^{2.6182} + \ldots + d_n^{2.6182}}\right]^{1.8539} + \ldots L_n\left[\frac{d_E^{2.6182}}{d_1^{2.6182} + d_2^{2.6182} + d_3^{2.6182} + \ldots + d_n^{2.6182}}\right]^{1.8539}$$

Let L_E = Equivalent length
L_1, L_2, L_3 & L_n = Length of various looped sections
d_1, d_2, d_3 & d_n = Internal diameter of individual line corresponding to lengths L_1, L_2, L_3 & L_n

$$L_E = L_1\left[\frac{d_E^{2.6182}}{d_1^{2.6182} + d_2^{2.6182} + d_3^{2.6182} + \ldots + d_n^{2.6182}}\right]^{1.8539} + \ldots L_n\left[\frac{d_E^{2.6182}}{d_1^{2.6182} + d_2^{2.6182} + d_3^{2.6182} + \ldots + d_n^{2.6182}}\right]^{1.8539}$$

when L_1 = Length of unlooped section
L_2 = Length of single looped section
L_3 = Length of double looped section
$d_E = d_1 = d_2$

then:

$$L_E = L_1 + 0.27664L_2 + L_3\left[\frac{d_E^{2.6182}}{2d_1^{2.6182} + d_3^{2.6182}}\right]^{1.8539}$$

when $d_E = d_1 = d_2 = d_3$

then $L_E = L_1 + 0.27664L_2 + 0.1305L_3$

Example. A multiple system consisting of a 15-mile section of 3–8⅝-in. OD lines and 1–10¾-in. OD line, and a 30-mile section of 2–8⅝-in. lines and 1–10¾-in. OD line.
Find the equivalent length in terms of single 12-in. ID line.

$$L_E = 15\left[\frac{12^{2.6182}}{3(7.981)^{2.6182} + 10.02^{2.6182}}\right]^{1.8539}$$

$$+ 30\left[\frac{12^{2.6182}}{2(7.981)^{2.6182} + 10.02^{2.6182}}\right]^{1.8539}$$

$= 5.9 + 18.1$
$= 24.0$ miles equivalent of 12-in. ID pipe

Example. A multiple system consisting of a single 12-in. ID line 5 miles in length and a 30-mile section of 3–12-in. ID lines.
Find equivalent length in terms of a single 12-in. ID line.

$$L_E = 5 + 0.1305 \times 30$$
$= 8.92$ miles equivalent of single 12-in. ID line

Determine pressure loss for a low-pressure gas system

Use the Spitzglass equation for systems operating at less than 1 psig.

$$Q_h = 3{,}550\sqrt{\frac{\Delta h_w d^5}{S_g L(1+(3.6/d)+0.03d)}}$$

where: Q_h = Rate of flow, in cubic feet per hour at standard conditions (14.7 psia and 60°F), scfh
h_w = Static pressure head, in inches of water
S_g = Specific gravity of gas relative to air = the ratio of the molecular weight of the gas to that of air
d = Internal diameter of the pipe, inches
L = Length of the pipe, feet

Example. Given the following conditions, find the flow in the system:

$h_w = 50$ in. of water
$S_g = 0.65$
$L = 500$ ft
$d = 6.187$ in.

$$Q_h = 3{,}550\sqrt{\frac{50 \times 6.187^5}{0.65 \times 500(1 + (3.6/6.187) + 0.03 \times 6.187)}}$$

$Q_h = 99{,}723$ scfh

Nomograph for determining pipe-equivalent factors

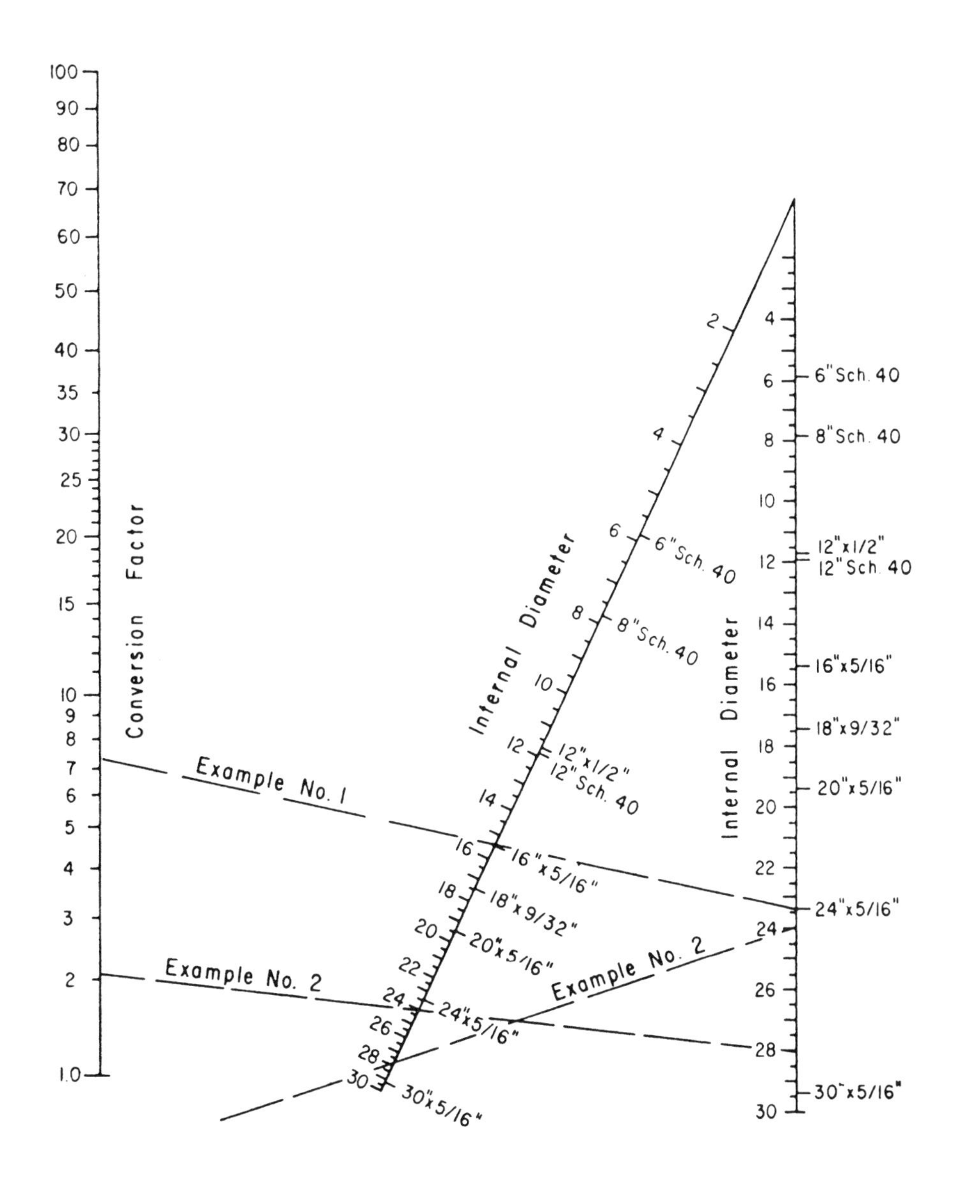

For turbulent flow, this handy nomograph will save a great deal of time in flow calculations where different sizes of pipe are connected. The advantage over a table of values is in having all sizes of pipe expressed in relation to any other size.

Example 1. 16-in. × 5⁄16-in. in terms of 24-in. × 5⁄16-in.

$$\left(\frac{23.375}{15.375}\right)^{4.735} = 1.5203^{4.735} = 7.26$$

Example 2. Reciprocal values can be found when chart limits are exceeded. 28-in. ID in terms of 24-in. ID falls beyond the lower limits of the chart; however, 24-in. ID in terms of 28-in. ID gives a conversion factor of 2.1.

How much gas is contained in a given line section?

Multiply the square of the inside diameter, in inches, by the gauge pressure, in lb/in.^{2}; multiply this by 0.372; the answer is the approximate number of cubic ft of gas (standard conditions) in 1,000 ft of line.

Example. How much gas in 1,000 ft of 16-in. schedule 30 pipe at 350-lb pressure?

$15.25 \times 15.25 \times 350 \times 0.372 = 30{,}280$ cu. ft of gas

Example. Approximately how much gas in 8 miles of .250 wall 24-in. pipe if the pressure is 400 psi?

$$\text{Cubic ft} = (23.5)(23.5)(400)(8)(5.28)(0.372) = 3{,}470{,}000 \text{ cu. ft}$$

When a section of line is blown down from one pressure to another, the total gas lost may be computed by the difference in the contents at the two pressures, using the above rule.

How to estimate equivalent length factors for gas lines

Internal Diameter	Nominal Diameter	36″ 35.250	34″ 33.250	30″ 29.375	26″ 25.375	24″ 23.375	24″ 23.000	22″ 21.500	20″ 19.500	18″ 17.500	16″ 15.500	12¾″ 12.250	10¾″ 10.250	8⅝″ 8.125
35.250″	36″	1	0.758	0.416	0.204	0.137	0.126	0.092	0.056	0.034	0.019	0.006	0.002	0.0008
33.250	34	1.32	1	0.55	0.270	0.181	0.167	0.121	0.075	0.044	0.025	0.008	0.003	0.001
29.375	30	2.41	1.82	1	0.492	0.330	0.305	0.22	0.136	0.081	0.045	0.014	0.006	0.002
25.375	26	4.90	3.71	2.03	1	0.671	0.622	0.45	0.278	0.164	0.092	0.029	0.012	0.004
23.375	24	7.25	5.54	3.04	1.49	1	0.922	0.67	0.415	0.246	0.137	0.044	0.018	0.006
23.000	24	7.90	6.00	3.29	1.62	1.08	1	0.72	0.450	0.267	0.148	0.048	0.020	0.0064
21.500	22	10.93	8.31	4.55	2.24	1.50	1.37	1	0.624	0.369	0.205	0.065	0.027	0.009
19.500	20	17.60	13.33	7.31	3.59	2.41	2.22	1.60	1	0.590	0.328	0.105	0.044	0.014
17.500	18	29.8	22.5	12.35	6.08	4.075	3.78	2.72	1.69	1	0.555	0.177	0.074	0.024
15.500	16	52.8	40.5	22.2	10.9	7.345	6.75	4.88	3.04	1.80	1	0.319	0.134	0.043
12.250	12¾	168	127	69.8	34.3	23.01	21.2	15.35	9.55	5.65	3.13	1	0.420	0.136
10.250	10¾	400	303	167	81.7	54.67	50.6	36.5	22.7	13.4	7.47	2.38	1	0.324
8.125	8⅝	1235	935	513	252	168.8	156.2	112.5	70.5	41.5	23.0	7.35	3.08	1
6.125	6⅝	4860	3680	2020	992	666	615	443	276	163	90.7	29.0	12.2	3.94

This table shows the equivalent length factors for pipe lines of different diameters. It is based on the Panhandle Formula, and the factors were obtained by varying the lengths of the pipe and keeping the other variables constant.

Example. Find the equivalent length of a 24-in. pipeline as compared with a 12¾-in. line.

Enter the chart at the top—24 in.—(23.00ID) and proceed downward to the 12.250 parameter. The equivalent length is 21.2.

This means that under the same conditions of temperatures, pressures, specific gravities, etc., 1 mile of 12¾-in. pipe will flow the same amount of gas that 21.2 miles of 24-in. pipe will flow. Another way of putting it, the pressure drop in 1 mile of 24-in. pipe will be the same as that in .048 miles of 12¾-in. pipe if the pressures, volumes, etc., are the same.

Where a line is composed of several sizes of pipe diameters or where the lines are looped, this chart should prove useful. In solving problems of this sort, the line is reduced to an equivalent length of some arbitrarily selected size pipe and treated as a single line of uniform size.

Estimating comparative capacities of gas pipelines

To estimate capacity of one diameter line in terms of another, read downward in the appropriate column to the desired diameter line.

Example. A 24-in. (23.375-in. ID) pipeline has a capacity 160% that of a 20-in. (19.500-in. ID) line.

This chart is based on diameter factors using the Panhandle equation. Equivalent inlet and outlet pressures, lengths of line section, and uniform flow are assumed for both lines.

(Figures give capacity of one diameter pipe as a percent of another.)

Internal Diameter	Nominal Diameter	36″ 35.250″	34″ 33.250″	30″ 29.3125″	26″ 25.375″	24″ 2.375″	24″ 23.000″	20″ 19.500″	18″ 17.500″	16″ 15.500″	12¾″ 12.250″
35.250″	36″	100	85.7	62.7	43.6	35.2	33.7	22.0	16.5	12.0	6.5
33.250″	34″	116.8	100	73.1	50.8	41.0	39.3	25.7	19.2	14.0	7.6
29.3125″	30″	159.7	137	100	69.6	56.1	53.8	35.7	26.3	19.2	10.4
25.375″	26″	229	197	149	100	80.7	77.5	50.2	37.8	27.6	14.8
23.375″	24″	285	244	178	124	100	96.0	62.3	46.9	34.2	18.5
23.000″	24″	298	255	186	129	104	100	65.1	49.0	35.7	19.2
19.500″	20″	458	392	287	199	160	154	100	75.3	35.0	29.6
17.500″	18″	608	519	379	264	213	204	132	100	73.2	39.3
15.500″	16″	835	715	522	363	293	281	182	127	100	53.8
12.250″	12¾″	1545	1322	968	673	543	521	238	254	185	100

Determination of leakage from gas line using pressure drop method

To determine the leakage in Mmcf per year at a base pressure of 14.4 psia and a temperature of 60°F, take as a basis a 1-mile section of line under 1 hour test. Use the formula:

$$L_y = 9 \times D^2 \left[\frac{P_1}{460 + t_1} - \frac{P_2}{460 + t_2} \right]$$

where: D = Inside diameter of the pipe, inches
P_1 = Pressure at the beginning of the test in psia
t_1 = Temperature at the beginning of the test in °F
P_2 = Pressure at end of the test
t_2 = Temperature at end of the test

The error in using this formula can be about 4%.

Example. A section of a 10-in. pipeline 1 mile in length is tested using the pressure drop method. ID is equal to 10.136 in. Pressure and temperature at the start of the test were 190.5 psig and 52°F. After 1 hour, the readings were 190.0 psig and 52°F. Calculate the leakage in Mmcf/year.

The atmospheric pressure for the area is 14.4 psia. Thus:

$$P_1 = 190.5 + 14.4 = 204.9\,\text{psia}$$

$$P_2 = 190.0 + 14.4 = 204.4\,\text{psia}$$

$$L_y = 9(10.136)^2 \left[\frac{204.9}{460 + 52} - \frac{204.4}{460 + 52} \right]$$

$$= \frac{(9)(102.7)(0.5)}{512}$$

$$= 0.903 \text{ Mmcf per mile per year}$$

A quick way to determine the size of gas gathering lines

Here is a shortcut to estimate gas flow in gathering lines. For small gathering lines, the answer will be within 10% of that obtained by more difficult and more accurate formulas.

$$Q = \frac{(500)(d^3)\sqrt{P_1^2 - P_2^2}}{\sqrt{L}}$$

where: Q = Cubic ft of gas per 24 hours
d = Pipe ID in in.
P_1 = Psi (abs) at starting point
P_2 = Psi (abs) at ending point
L = Length of line in miles

Example. How much gas would flow through four miles of 6-in. ID pipe if the pressure at the starting point is 485 psi and if the pressure at the downstream terminus is 285 psi?

Answer.

$$Q = \frac{(500)(6^3)\sqrt{(485+15)^2 - (285+15)^2}}{\sqrt{4}}$$

$$Q = \frac{(500)(216)\sqrt{500^2 - 300^2}}{2}$$

$$Q = \frac{(500)(216)(400)}{2}$$

Q = 21,600,000 cu. ft per 24 hours

Using a more accurate formula (a version of Weymouth's):

$$Q = \frac{(500)(d^{8/3})\sqrt{P_1^2 - P_2^2}}{\sqrt{L}}$$

$$Q = \frac{(851)(120)(400)}{2}$$

Q = 20,900,000 cu. ft per 24 hours

The error in this case is 700,000 cu. ft per 24 hours, or about 3%.

Energy conversion data for estimating

The accompanying table gives the high heating values of practically all commonly used fuels along with their relation one to the other.

Example. If light fuel oil will cost a potential industrial customer 18 cents per gallon and natural gas will cost him 75 cents per Mcf, which fuel will be more economical?

Solution. Assume the operator would use 10,000 gallons of light fuel oil per year at 18 cents per gallon. The cost would be $1,800. From the table, 10,000 gallons of fuel oil are equivalent to 10,000 × 134 = 1,340,000 cu. ft of natural gas. At 75 cents per Mcf this would cost 1,340 × .75 = $1,005.

The burning efficiency of both fuels is the same. The savings by using natural gas would come to $795 per year.

Fuel	Heating Value	PERCENT BURNING EFFICIENCY			Unit	NUMBER OF CUBIC FEET OF NATURAL GAS EQUIVALENT TO 1 UNIT OF OTHER FUEL		
		Domestic	Commercial	Industrial		Domestic	Commercial	Industrial
Natural Gas	1000 B.T.U. per Cubic Foot	75	75	78	Cubic Foot	1	1	1
Propane	91,600 B.T.U. per U.S. gallon	75	75	78	Gallon	91.5	91.5	91.5
Butane	102,000 B.T.U. per U.S. gallon	.....	75	78	Gallon	.....	102	102
Light Fuel Oil	134,000 B.T.U. per U.S. gallon	65–75	75	78	Gallon	117–134	134	134
Heavy Fuel Oil	146,000 B.T.U. per U.S. gallon	.....	75	78	Gallon	.....	146	146
Hydro-electricity	3,412 B.T.U. per Kilowatt Hour	100	100	100	Kilowatt Hour	4.549	4.549	4.549
*Coal	8,700–11,000 B.T.U. per pound	55–60	65	73	Pound	6.37–8.07	7.55–9.57	8.45–10.7

*Coal qualities vary considerably. The figures given are average for Western Coals; slack, run-of-the-mine; potentially available at lowest cost.

How to estimate time required to get a shut-in test on gas transmission lines and approximate a maximum acceptable pressure loss for new lines

These two rules may prove helpful for air or gas testing of gas transmission lines. They are not applicable for hydrostatic tests. Values of the constants used by different transmission companies may vary with economic considerations, line conditions, and throughputs.

To determine the minimum time required to achieve a good shut-in test after the line has been charged and becomes stabilized, use the following formula:

$$H_m = \frac{3 \times D^2 \times L}{P_1}$$

where: H_m = Minimum time necessary to achieve an accurate test in hours
D = Internal diameter of pipe in inches
L = Length of section under test in miles
P_1 = Initial test pressure in lb/in.^{2} gauge

Example. How long should a 14-mile section of 26-in. (25.375-in. ID) pipeline be shut in under a 1,050 psi test pressure to get a good test?

$$H_m = \frac{3 \times (25.375)^2 \times 14}{1{,}050}$$

$$= 25.75 \text{ hours or 25 hours and 45 minutes}$$

When a new line is being shut in for a test of this duration or longer, the following formula may be used to evaluate whether or not the line is "tight":

$$P_{ad} = \frac{H_t \times P_1}{D \times 949}$$

where: P_{ad} = Acceptable pressure drop in lb/in.^{2} gauge
H_t = Shut-in test time in hours
D = Internal diameter of pipe in in.
P_1 = Initial test pressure in lb/in.^{2} gague

Example. What would be the maximum acceptable pressure drop for a new gas transmission line under air or gas test when a 44-mile section of 30-in. (29.250-in. ID) pipeline has been shut in for 100 hours at an initial pressure of 1,100 psig?

$$P_{ad} = \frac{100 \times 1{,}100}{29.250 \times 949} = 4\,\text{lb/in.}^2$$

If the observed pressure drop after stabilization is less than 4 psig, the section of line would be considered "tight."

Note: Corrections must he made for the effect of temperature variations upon pressure during the test.

How to determine the relationship of capacity increase to investment increase

Example. Determine the relationship of capacity increase to investment increase when increasing pipe diameter, while keeping the factors P(in), P(out), and L constant.

Solution. The top number in each square represents the percent increase in capacity when increasing the pipe size from that size shown in the column at the left to the size shown by the column at the bottom of the graph. The lower numbers represent the percent increase in investment of the bottom size over the original pipe size. All numbers are based on pipe capable of 1,000 psi working pressure.

PIPE SIZE	4	6	8	10	12	14	16	18	20	22	24	26	28	30	32	34	36
34																	17/12
32																17/13	37/27
30															19/13	39/28	62/43
28														20/11	43/26	67/41	
26													23/11	47/23	75/39		
24												[illegible]/11	[illegible]/23	[illegible]/37			
22											26/14	56/26	91/40				
20										29/17	62/33	101/47					
18									34/14	72/33	117/51						
16								38/23	85/39	138/63							
14							44/14	98/39	165/58								
12						29/16	85/32	156/62									
10					61/22	108/42	198/61										
8				82/26	193/53	278/78											
6			105/22	274/54	502/87												
4	0/0	190/38	497/69	988/112													

XXX/— Top Number represents % increase in capacity when increasing pipe size from size on left to size at bottom

—/XXX Lower Number represents % increase in investment for pipe size at bottom over pipe size at left.

Estimate pipe size requirements for increasing throughput volumes of natural gas

Reflects the capacity-to–pipe size relationship when L (length), P_1 (inlet pressure), and P_2 (outlet pressure) are constant.

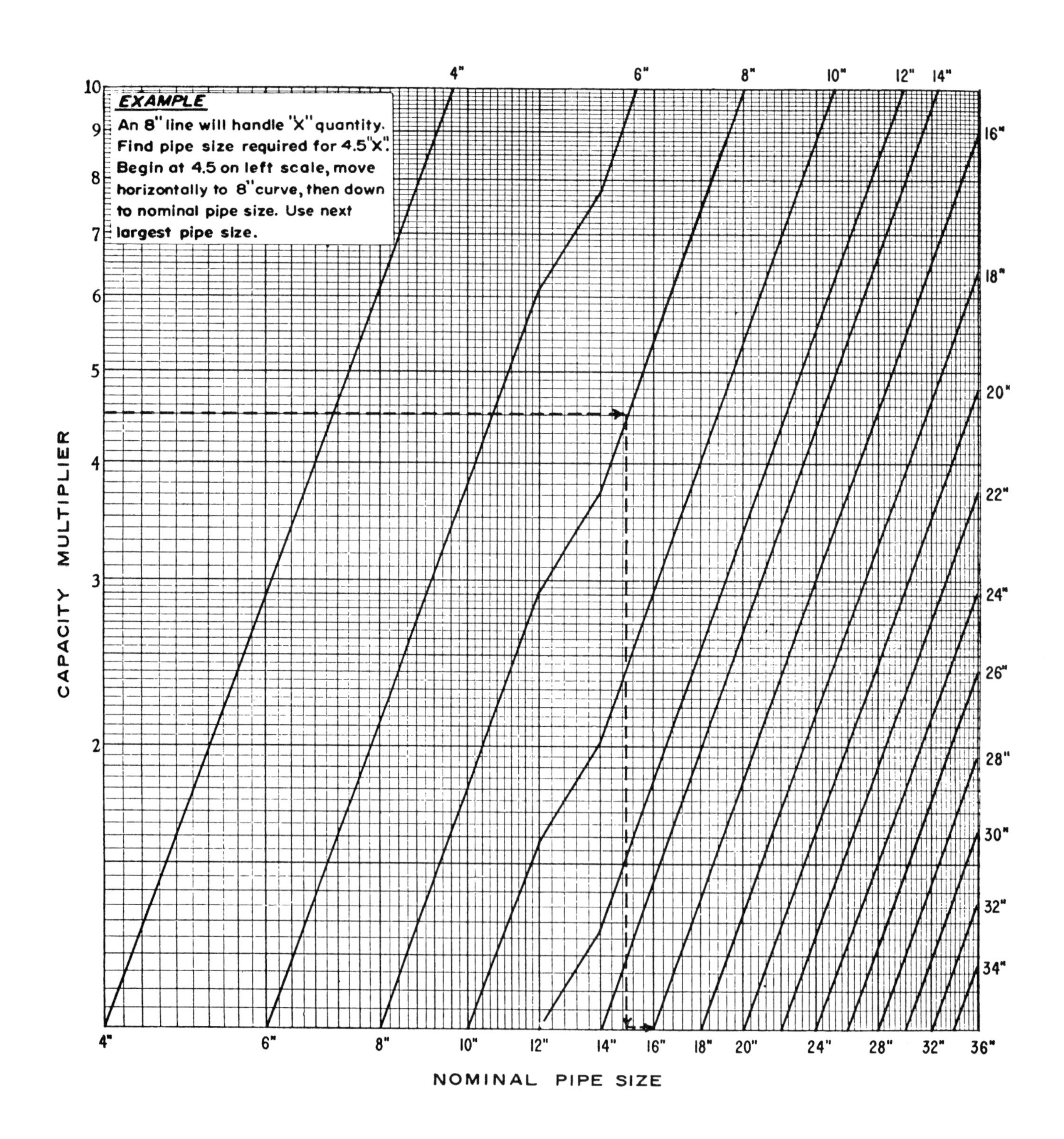

Calculate line loss using cross-sectional areas table when testing mains with air or gas

$$\text{Initial volume} = \frac{P_i}{P_B}(AL)$$

$$\text{Final volume} = \frac{P_f}{P_B}(AL)$$

where: P_i = Initial pressure, or pressure at start of test, psia
P_f = Final pressure, or pressure at end of test, psia
P_B = Base pressure, psia
A = Cross-sectional area of inside of pipe in ft^2
L = Length of line in ft

The loss then would be the difference of the volumes, found by the above formula, which could be written:

$$\text{Loss during test} = AL\frac{(P_i - P_f)}{P_B}$$

To facilitate calculations, Table 1 contains the cross-sectional areas in ft^2 for some of the more popular wall thickness pipe from 1-in. through 24-in. nominal size.

Example. Assume 3,562 ft of 8.125-in. inside diameter pipe is tested with air for a period of 26 hours. The initial pressure was 116 psig, the final pressure 112 psig, and the barometer was 14.4 psia in both instances. It is desired to calculate the volume of loss during the test using a pressure base of 14.65 psia.

If the area "A" is obtained from Table 1, and information as given above is substituted in the formula, we would have:

$$\text{Loss during test} = 0.360 \times 3{,}562 \times \frac{[(116 + 14.4) - (112 + 14.4)]}{14.65}$$

$$= 350.1 \text{ cu. ft}$$

Table 1
Characteristics of pipe

Nominal Pipe Size D_n	Outside Diameter	Wall Thickness	Inside Diameter	"A" Inside Area in Sq. Feet
1	1.315	.133	1.049	.006
		.179	.957	.005
1¼	1.660	.140	1.380	.010
		.191	1.278	.009
1½	1.900	.145	1.610	.014
		.200	1.500	.012
2	2.375	.154	2.067	.023
		.218	1.939	.021
2½	2.875	.203	2.469	.033
		.276	2.323	.029
3	3.500	.125	3.250	.058
		.156	3.188	.055
		.188	3.124	.053
		.216	3.068	.051
		.250	3.000	.049
3½	4.000	.125	3.750	.077
		.156	3.688	.074
		.188	3.624	.072
		.226	3.548	.069
		.250	3.500	.067
4	4.500	.125	4.250	.099
		.156	4.188	.096
		.188	4.124	.093
		.219	4.062	.090
		.237	4.026	.088
		.250	4.000	.087
5	5.563	.156	5.251	.150
		.188	5.187	.147
		.219	5.125	.143
		.258	5.047	.139
6	6.625	.188	6.249	.213
		.219	6.187	.209
		.250	6.125	.205
		.280	6.065	.201
8	8.625	.188	8.249	.371
		.219	8.187	.366
		.250	8.125	.360
		.277	8.071	.355
10	10.750	.188	10.374	.587
		.219	10.312	.580
		.250	10.250	.573
		.279	10.192	.567
		.307	10.136	.560
12	12.750	.219	12.312	.827
		.250	12.250	.818
		.281	12.188	.810
		.312	12.126	.802
		.330	12.090	.797
14	14.000	.250	13.500	.994
		.281	13.438	.985
		.312	13.376	.976
		.344	13.312	.966
		.375	13.250	.958
16	16.000	.250	15.500	1.310
		.281	15.438	1.300
		.312	15.376	1.289
		.344	15.312	1.279
		.375	15.250	1.268
18	18.000	.250	17.500	1.670
		.281	17.438	1.658
		.312	17.376	1.647
		.344	17.312	1.635
		.375	17.250	1.623
20	20.000	.250	19.500	2.074
		.281	19.438	2.061
		.312	19.376	2.048
		.344	19.312	2.034
		.375	19.250	2.021
22	22.000	.312	21.376	2.492
		.344	21.312	2.477
		.375	21.250	2.463
24	24.000	.312	23.376	2.980
		.344	23.312	2.964
		.375	23.250	2.948

Flow of fuel gases in pipelines

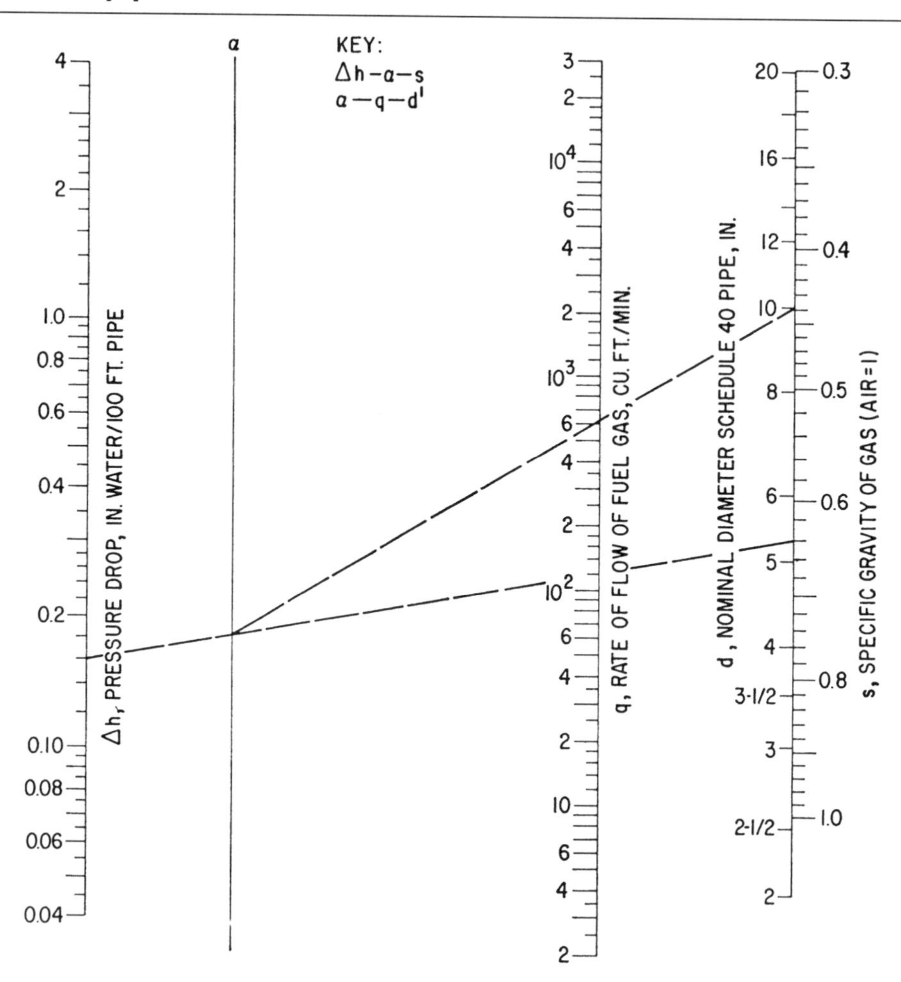

D. S. Davis, Special Projects, Inc., Bailey Island, Maine

The rate of flow of fuel gases at low pressures, ordinary temperatures, and under turbulent conditions in pipelines is reflected by an equation[1] based on sound theoretical considerations:

$$q = \frac{3.185\, \Delta h^{0.543} d^{2.63}}{s^{0.468}}$$

where: q = Rate of flow of gas, cubic ft per minute
Δh = Pressure drop, inches of water per 100 ft of pipe
d = Actual inner diameter of Schedule 40 steel pipe, inches
s = Specific gravity of gas at the prevailing temperature and pressure relative to air at 68°F and 30 inches of mercury

The equation can be solved readily and accurately by means of the accompanying monograph constructed in accordance with methods described previously.[2]

The use of the chart is illustrated as follows. At what rate will fuel gas with a specific gravity of 0.640 flow in a 10-in. Schedule 40 pipeline if the pressure is 0.160 in. of water per 100 ft of pipe? Following the key, connect 0.160 on the Δh scale and 0.640 on the s scale with a straight line and mark the intersection with the α axis. Connect this point with 10 on the d′ scale for nominal diameters and read the rate of flow at 624 cu. ft per minute on the q scale.

Some authorities give the exponent of s as 0.408 when, for the data given, the rate of flow as read from the chart would be decreased by 2.7% (17 cubic ft per minute) to attain 607 cu. ft per minute.

References

1. Perry, J. H., ed., *Chemical Engineers'* Handbook, 3rd ed., p. 1589, McGraw-Hill Book Co., Inc., New York (1950).
2. Davis, D. S., *Nomography and Empirical Equations*, 2nd ed., Chapter 6, Reinhold Publishing Corp., New York, 1962.

Calculate the velocity of gas in a pipeline

$$V = \frac{1440ZQT}{d^2P}$$

where: V = Velocity of gas, ft/sec
Z = Compressibility
Q = Volume, million cu. ft/hr at standard conditions
T = Operating temperature, °R
d = Inside diameter of pipe, inches
P = Operating pressure, psia

Example.
Q = 5 MMCFH
T = 60°F
d = 15.5 in.
P = 500 psig (514.7 psia)
Z = 0.94 (from compressibility chart)
T = 520°R

$$V = \frac{1440 \times 0.94 \times 5 \times 520}{240.25 \times 514.7}$$

V = 28 ft/sec

Consideration should be given to acceptable noise levels (60 to 80 ft/sec) and erosion, especially where the gas may contain solid particles. Refer to API RP 14E for erosion and corrosion considerations. If the gas contains liquids, the minimum velocity should be limited to about 10 ft/sec to help sweep liquids from the pipe and prevent slugging.

Determining throat pressure in a blow-down system

In designing a gas blow-down system, it is necessary to find the throat pressure in a blow-off stack of a given size with a given flow at critical velocity. The accompanying nomograph is handy to use for preliminary calculations. It was derived from the equation:

$$P_t = \frac{W}{d^2} \times \frac{\sqrt{\frac{RT_1}{K(K-1)}}}{11{,}400}$$

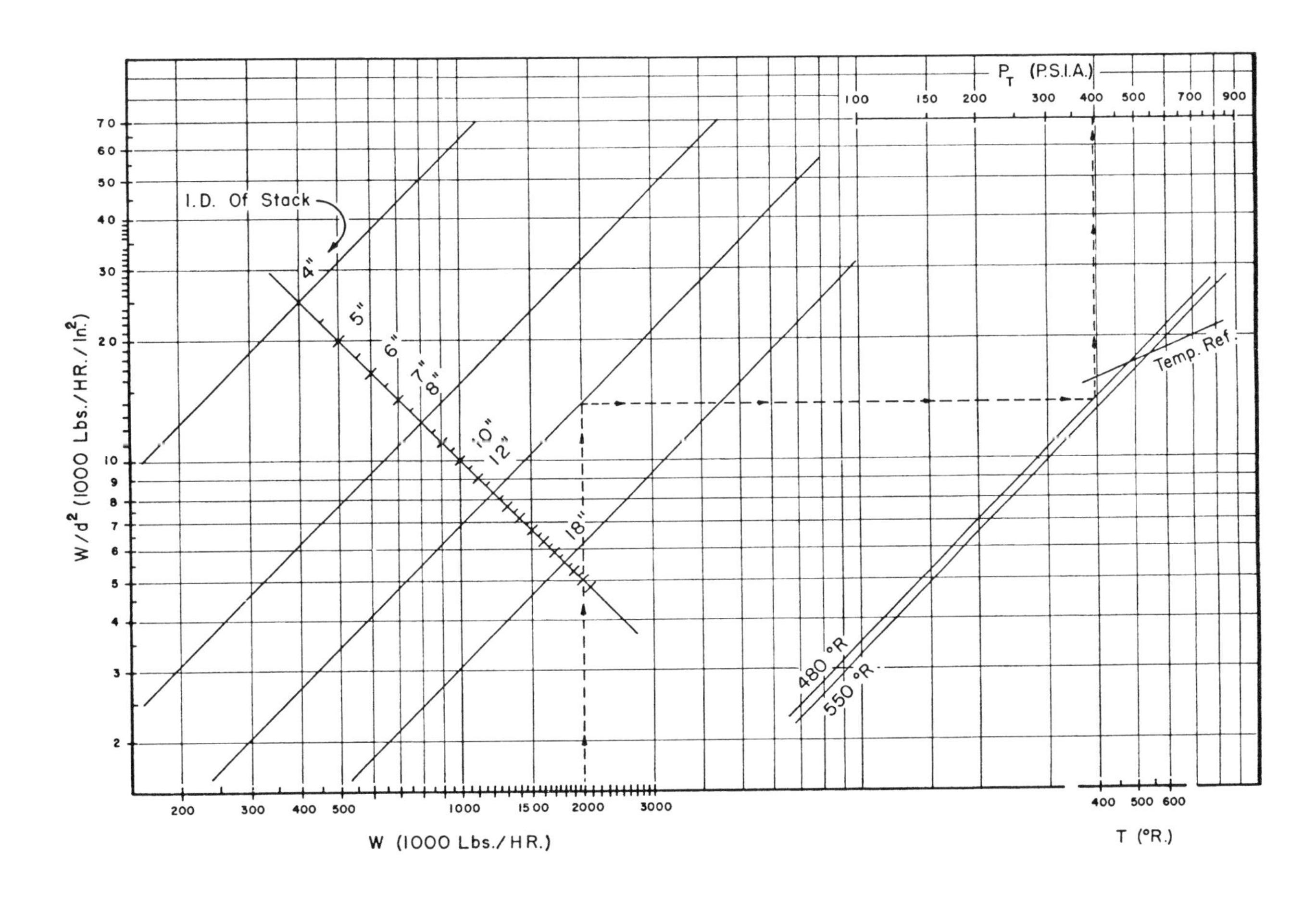

where: P_t = Pressure in the throat, psia
W = Weight of flowing gas, lb/hr
d = Actual ID of stack, in.
R = Gas constant, ft lb/°F
K = C_P/C_V
C_P = Specific heat at constant pressure, Btu/lb/°F
C_V = Specific heat at constant volume, Btu/lb/°F
T_1 = Absolute temperature of gas flowing,
°R = (°F + 460)

Using the following values for natural gas, R = 88.888 and K = 1.31, the above equation reduces to:

$$P_t = \frac{W}{d^2} \times .0012978\sqrt{T_1}$$

Example. To use the accompanying nomograph, assume that the rate of flow is known to be 2.00×10^6 lb/hr, the ID of the stack is 12 in., and the flowing temperature is 480°R. Start from the abscissa at the Rate of Flow, W, and proceed upward to the ID of the pipe. From this intersection proceed to the right to the flowing temperature of 480°R. Turn upward from this intersection to the P_t scale and read the answer 395 psia.

Estimate the amount of gas blown off through a line puncture

To calculate the volume of gas lost from a puncture or blowoff, use the equation:

$$Q = D^2 P_1$$

where: Q = Volume of gas in Mcf/hr at a pressure of 14.9psi, 60°F, and with a specific gravity of 0.60
D = Diameter of the nipple or orifice in inches
P = Absolute pressure in lb/in.² at some nearby point upstream from the opening

Example. How much gas will be lost during a five minute blowoff through a 2-in. nipple if the upstream pressure is 1,000 psi absolute?

$$Q = D^2 P_1$$

$$Q = (2)^2 \times 1{,}000$$

$$Q = 4{,}000 \text{ Mcf/hr}$$

$$Q_{5\,min} = 4{,}000 \times \frac{5}{60} = 333 \text{ Mcf}$$

A practical way to calculate gas flow for pipelines

Here is a shortcut to calculate gas flow in pipelines. It is based on the Weymouth formula. At 60°F and specific gravity of 0.60, the answer will be accurate. For every 10° variation in temperature, the answer will be 1% error. For every 0.01 variation in specific gravity, the answer will be three-fourths percent in error.

Formula:

$$Q = \frac{(871)(d^{8/3})\sqrt{P_1^2 - P_2^2}}{\sqrt{L}}$$

where: Q = Cubic ft of gas per 24 hours; 871 is a constant
d = Pipe ID in in.
P_1 = Psi (abs) at starting point
P_2 = Psi (abs) at ending point
L = Length of line in miles

Example. How much gas would flow through 1 mile of 6-in. ID pipe if the pressure at the starting point is 485 psi and if the pressure at the downstream terminus is 285 psi?

Solution.

$$Q = \frac{(871)(6^{8/3})\sqrt{(485+15)^2 - (285+15)^2}}{\sqrt{1}}$$

$$= \frac{(871)(120)\sqrt{500^2 - 300^2}}{\sqrt{1}}$$

$$= (871)(120)(400)$$

$$= 41{,}800{,}000 \text{ cu. ft per 24 hours}$$

Example. What would be the flow through 4 miles of 10¾-in. quarter-inch wall pipe using the same pressures?

$$Q = \frac{(871)(10.25^{8/3})\sqrt{500^2 - 300^2}}{\sqrt{4}}$$

$$Q = \frac{(871)(495.7)(400)}{2}$$

$$Q = 86{,}350{,}000 \text{ cu. ft per day}$$

How to calculate the weight of gas in a pipeline

Rule. Find the volume of the pipe in cubic ft and multiply by the weight of the gas per cubic ft. To find the latter, multiply the absolute pressure of the gas times 3 and divide by 1,000.

The basis for the latter is that gas with a specific gravity of .60 at 70°F weighs 3.06 lb/cu. ft at 1,000 psia. And everything else remaining equal, weight of the gas is proportional to absolute pressure. Thus, to find the weight of gas at, say, 500 psia, multiply $3 \times 500/1{,}000 = 1.5$ lb. (1.53 is more accurate.)

Example. Find the weight of gas in a 1,250-ft aerial river crossing where the average pressure reads 625.3 on the gauges; the temperature of the gas is 70°F, and specific gravity of gas is .60.

Solution. Change psig to psia:

$625.3 + 14.7 = 640$ psia

Weight of gas per cubic ft:

$$3.06 \times \frac{640}{1{,}000} = 1.959 \text{ lb/cu. ft}$$

Volume of 1,250 ft of 23¼-in. ID pipe = 3,685 cu. ft

Weight of gas = 7,217 lb

This method is fairly accurate; here is the same problem calculated with the formula:

$$W = \frac{(V)(144)(P_{abs})}{RT}$$

where: W = Weight of gas in lb
V = Volume of pipe
P_{abs} = Absolute pressure of gas
R = Universal constant 1,544 ÷ molecular weight of gas
T = Temperature of gas in °R

(To find the molecular weight of gas, multiply specific gravity × molecular weight of air, or in this case, $.6 \times 28.95 = 17.37$.)

$$W = \frac{(3{,}685)(144)(640)}{(1{,}500 \div 17.37)(70 + 460)}$$

$$W = 7{,}209 \text{ lb}$$

Estimate average pressure in gas pipeline using upstream and downstream pressures

To find the average gas pressure in a line, first divide downstream pressure by upstream pressure and look up the value of the factor X in the table shown. Then multiply X times the upstream pressure to get the average pressure.

Example. Find the average pressure in a pipeline if the upstream pressure in 775 psia and the downstream pressure is 420 psia.

$$\frac{P_d}{P_u} = \frac{420}{775} = 0.54$$

Interpolating from the table,

$X = 0.78 + 4/5(0.02) = 0.796$

$P_{average} = 0.796 \times 775 = 617$ psia

This method is accurate to within 2 or 3 psi. Here is the same problem calculated with the formula:

$$P_{average} = \frac{2}{3}\left(P_u + P_d - \frac{P_u P_d}{P_u + P_d}\right)$$
$$= \frac{2}{3}\left(1{,}195 - \frac{325{,}500}{1{,}195}\right)$$

$P_{average} = 615$ psia

P_d/P_u	X
.10	.67
.15	.68
.20	.69
.25	.70
.30	.71
.35	.725
.40	.74
.45	.76
.50	.78
.55	.80
.60	.82
.65	.84
.70	.86
.75	.88
.80	.90
.85	.925
.90	.95
.95	.975
1.00	1.00

Chart for determining viscosity of natural gas

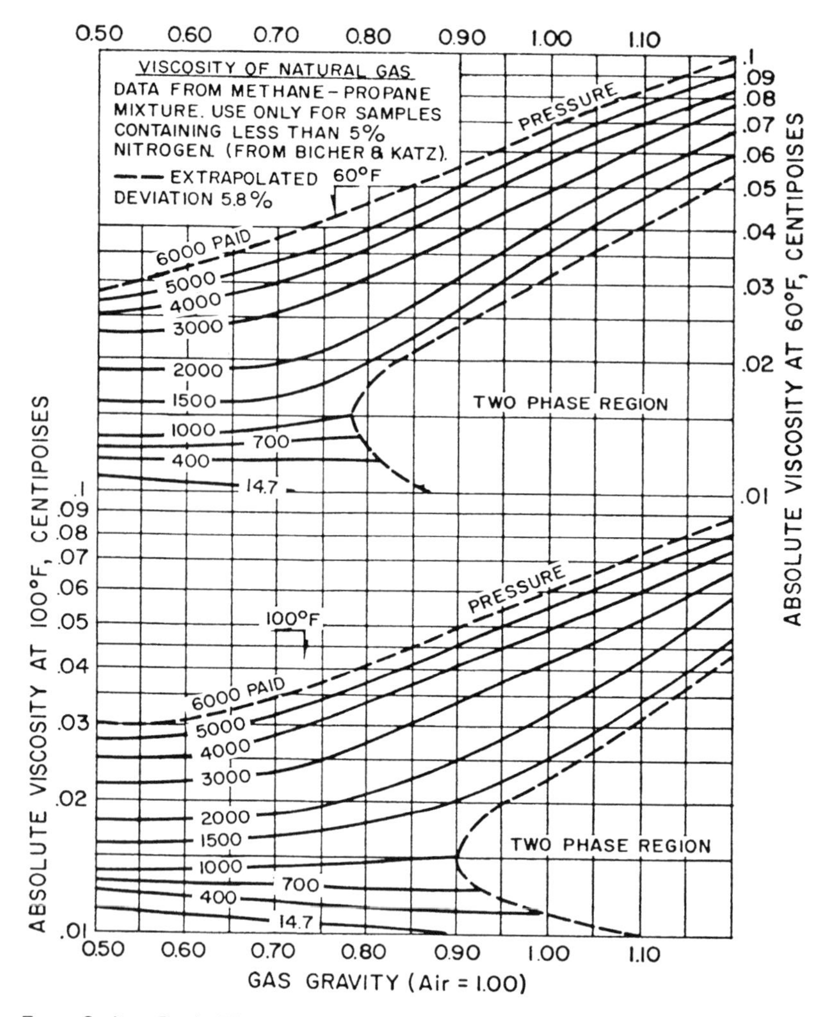

From Carlton Beal, "The Viscosity of Air, Water, Natural Gas, Crude Oil and Its Associated Gases at Oil Field Temperatures and Pressures," *Petroleum Development and Technology*, AIME, 1946.

Flow of gas

For gas flow problems as encountered in oil field production operations, the Fanning equation for pressure drop may be used. A modified form of this equation employing units commonly used in oil field practice is:[1,2,3,4,5,6]

$$W = \frac{FL(MCFD)^2STZ}{20{,}000\, d^5 P_{av}}$$

and for determining the friction factor F:

$$R_W = \frac{20.14(MCFD)S}{d\mu}$$

where:

P = Pressure drop, psia
F = Friction factor, dimensionless
L = Length of pipe, ft
MCFD = Gas flow at standard condition

S = Specific gravity of gas
T = Absolute temperature (°F + 460)
P_{av} = Average flowing pressure, psia
d = Internal pipe diameter, inches
Re = Reynolds number, dimensionless
μ = Viscosity, centipoises

In actual practice, empirical flow formulas are used by many to solve the gas flow problems of field production operations. The Weymouth formula is the one most frequently used since results obtained by its use agree quite closely with actual values. Recent modification of the formula, by including the compressibility factor, Z, made the formula applicable for calculation of high-pressure flow problems.

The modified formula is as follow:

$$Q = 433.45 \frac{T_s}{P_s} \times d^{2.667} \times \left(\frac{P_1^2 - P_2^2}{LSTZ} \right)^{1/2}$$

where:

Q_S = Rate of flow of gas in cubic ft per 24 hours measured at standard conditions
d = Internal diameter of pipe in inches
P_1 = Initial pressure, psi absolute
P_2 = Terminal pressure, psi absolute
L = Length of line in miles
S = Specific gravity of flowing gas (air = 1.0)
T = Absolute temperature of flowing gas

Table 1
Values of $d^{2.667}$ for use in Weymouth's gas-flow formula

(Increments of tenths of an inch with certain standard pipe sizes inserted)

d	$d^{2.667}$	d	$d^{2.667}$	d	$d^{2.667}$	d	$d^{2.667}$
1.0	1.0	6.2	129.9	11.5	673.8	16.9	1881
1.1	1.3	6.3	135.4	11.6	689.5	17.0	1911
1.2	1.6	6.4	141.2	11.7	705.5	17.1	1941
1.3	2.0	6.5	147.1	11.8	721.7	17.2	1971
1.4	2.4	6.6	153.3	11.9	738.1	17.250	1988
1.5	2.9	6.7	159.5	12.000	754.8	17.3	2002
1.6	3.5	6.8	166.0	12.1	771.7	17.4	2033
1.7	4.1	6.9	172.6	12.2	788.8	17.5	2064
1.8	4.8	7.0	179.3	12.3	806.1	17.6	2096
1.9	5.5	7.1	186.2	12.4	823.7	17.7	2128
2.0	6.3	7.2	193.3	12.5	841.6	17.8	2160
2.067	6.9	7.3	200.5	12.6	859.7	17.9	2193
2.1	7.2	7.4	207.9	12.7	878.0	18.0	2225
2.2	8.2	7.5	215.5	12.8	896.5	18.1	2259
2.3	9.2	7.6	223.3	12.9	915.3	18.2	2292
2.4	10.3	7.7	231.2	13.0	934.4	18.3	2326
2.5	11.5	7.8	239.3	13.1	953.7	18.4	2360
2.6	12.8	7.9	247.6	13.2	973.2	18.5	2394
2.7	14.1	7.981	254.5	13.250	983.9	18.6	2429
2.8	15.6	8.0	256.0	13.3	993.0	18.7	2464
2.9	17.1	8.1	264.6	13.4	1013	18.8	2499
3.0	18.7	8.2	273.4	13.5	1033	18.9	2535
3 068	19.9	8.3	282.4	13.6	1054	19.0	2570
3.1	20.4	8.4	291.6	13.7	1075	19.1	2607
3.2	22.2	8.5	300.9	13.8	1096	19.182	2639
3.3	24.1	8.6	310.5	13.9	1117	19.2	2643
3.4	26.1	8.7	320.2	14.0	1138	19.250	2664
3.5	28.2	8.8	330.1	14.1	1160	19.3	2680
3.6	30.4	8.9	340.2	14.2	1182	19.4	2717
3.7	32.7	8.941	344.6	14.3	1205	19.5	2755
3.8	35.2	9.0	350.5	14.4	1227	19.6	2793
3.9	37.7	9.1	361.0	14.5	1250	19.7	2831
4.0	40.3	9.2	371.6	14.6	1273	19.8	2860
4.026	41.0	9.3	382.5	14.7	1297	19.9	2908
4.1	43.1	9.4	393.6	14.8	1321	20.0	2947
4.2	45.9	9.5	404.8	14.9	1344	20.1	2987
4.3	48.9	9.6	416.3	15.0	1368	20.2	3026
4.4	52.0	9.7	428.0	15.1	1393	20.3	3067
4.5	55.2	9.8	439.8	15.2	1418	20.4	3107
4.6	58.5	9.9	451.9	15.250	1431	20.5	3148
4.7	62.0	10.0	464.2	15.3	1443	20.6	3196
4.8	65.6	10.020	467.0	15.4	1468	20.7	3230
4.9	69.3	10.1	476.6	15.5	1494	20.8	3272
5.0	73.1	10.2	489.3	15.6	1519	20.9	3314
5.1	77.1	10.3	502.2	15.7	1545	21.0	3357
5.2	81.2	10.4	515.3	15.8	1572	21.1	3400
5.3	85.4	10.5	528.7	15.9	1599	21.1	3443
5.4	89.8	10.6	542.2	16.0	1625	21.250	3468
5.5	94.3	10.7	555.9	16.1	1653	21.3	3486
5.6	98.9	10.8	569.9	16.2	1680	21.4	3530
5.7	103.7	10.9	584.1	16.3	1708	21.5	3574
5.8	108.6	11.000	598.5	16.4	1736	21.6	3619
5.9	113.7	11.1	613.1	16.5	1765	21.7	3663
6.0	118.9	11.2	627.9	16.6	1793	21.8	3709
6.065	122.4	11.3	643.0	16.7	1822	21.9	3754
6.7	124.2	11.4	658.3	16.8	1851	22 0	3800

—The Oil & Gas Journal

Standard conditions for measurements:

T_s = Standard absolute temperature
P = Standard pressure, psi absolute
Z = Compressibility factor

Values of expression $d^{2.667}$ for different inside diameter of pipe are given in Table 1.

Another formula quite commonly used is the one determined by F. H. Oliphant:

$$Q_s = 42a\sqrt{\frac{P_1^2 - P_2^2}{L}}$$

where:
Q = Discharge in cubic ft per hour
42 = A constant
P_1 = Initial pressure in lb/in.², absolute
P_2 = Final pressure in lb/in.², absolute
L = Length of line in miles
a = Diameter coefficient

This formula assumes a specific gravity of gas of 0.6. For any other specific gravity, multiply the final result by:

$$\sqrt{\frac{0.6}{\text{sp.gr.gas}}}$$

The values of diameter coefficients for different sizes of pipes are given below:

¼ in.	.0317	2½ in.	10.37	8 in.	198
½ in.	.1810	3 in.	16.50	10 in.	350
¾ in.	.5012	4 in.	34.10	12 in.	556
1 in.	1.0000	5 in.	60.00	16 in.	1160
1½ in.	2.9300	5⅝ in.	81.00	18 in.	1570
2 in.	5.9200	6 in.	95.00		

For pipes greater than 12 in. in diameter, the measure is taken from the outside and for pipes of ordinary thickness, the corresponding inside diameters and multipliers are as follows:

Outside	Inside	Multiplier
15 in.	14¼ in.	863
16 in.	15¼ in.	1025
18 in.	17¼ in	1410
20 in.	19¼ in.	1860

To determine gas flow rates for specific pressure drops for pipe sizes used in production operations, Tables 3, 4, and 5 may be used. The tables are calculated by the Weymouth formula for listed inside diameters of standard weight threaded line pipe for the sizes shown.

Tables 3, 4, and 5 give the hourly rates of flow of 0.70 specific gravity gas, flowing at 60°F and measured at a standard temperature of 60°F and at a standard pressure base of 14.4 psi + 4 oz. gauge pressure of 14.65 psi. For specific gravity of gas other than 0.70 and for flowing temperature other than 60°F, correction can be made by use of the factors shown in Table 2. For instance, to determine the flow of gas of specific gravity 0.06 flowing at the temperature of 20°F, the value obtained from the gas flow Tables 3, 4, and 5 would be multiplied by 1.12, the factor obtained from Table 2.

To find from Tables 3, 4, and 5 the volume of gas delivered through any length L, the volume found from the table should be multiplied by $1/\sqrt{L}$. L must be expressed as a multiple of the length heading the table in use. For instance, if the delivery is to be determined from a 4,500 ft line, the volume found for a 1,000 ft line in Table 3 should be multiplied by $1\sqrt{4.5}$. At the bottom of each of the two tables in question, multipliers are given to be used as correction factors.

Use of Tables 3, 4, and 5 may be illustrated by the following example.

Example. What is the delivery of a 2-in. gas line, 3,600 ft long, with an upstream pressure of 200 psi and a downstream pressure of 50 psi? The specific gravity of gas is 0.07; the flowing temperature, 70°F.

Table 2
Temperature–specific gravity multipliers

Specific Gravity	FLOWING TEMPERATURE—DEGREES F.												
	0	10	20	30	40	50	60	70	80	90	100	110	129
0.60	1.15	1.14	1.12	1.11	1.10	1.09	1.08	1.07	1.06	1.05	1.04	1.03	1.02
0.70	1.06	1.05	1.04	1.03	1.02	1.01	1.00	.991	.981	.972	.964	.955	.948
0.80	.995	.984	.974	.964	.954	.945	.935	.927	.918	.910	.901	.894	.886
0.90	.938	.928	.918	.909	.899	.891	.882	.874	.965	.858	.850	.842	.835
1.00	.890	.880	.871	.862	.853	.845	.837	.829	.821	.814	.806	.799	.792

Table 3
Volume of gas delivered—thousands of cubic feet per hour through 1000-foot and 2000-foot lines

0.7 SP. GR. 60° F.

Gage Pressure Up Stream P.S.I.	Gage Pressure Down Stream P.S.I.	1000 Foot Line, Nominal Pipe Size* 2	2½	3	4	6	2000 Foot Line, Nominal Pipe Size* 2	2½	3	4	6
		Thousands of Cu. Ft. per Hour					Thousands of Cu. Ft. per Hour				
400	10	221.5	355.8	635.0	1311.	3910.	156.6	251.6	449.0	926.8	2765.
	50	219.2	352.1	628.4	1297.	3869.	155.0	248.9	444.3	917.2	2736.
	100	213.2	342.5	611.4	1262.	3764.	150.8	242.2	432.3	892.4	2662.
	350	105.6	169.7	302.9	625.2	1865.	74.70	120.0	214.2	442.1	1319.
350	10	194.7	312.7	558.1	1152.	3436.	137.6	221.1	394.6	814.6	2430.
	50	192.0	308.5	550.6	1136.	3390.	135.8	218.1	389.3	803.6	2397.
	100	185.2	297.5	531.0	1096.	3270.	131.0	210.4	375.5	775.1	2312.
	300	98.63	158.4	282.8	583.7	1741.	69.74	112.0	200.0	412.8	1231.
300	10	167.8	269.6	481.2	993.1	2962.	118.7	190.6	340.2	702.3	2095.
	50	164.8	264.6	472.4	975.0	2908.	116.5	187.1	334.0	689.4	2056.
	100	156.8	251.9	449.5	927.9	2768.	110.9	178.1	317.9	656.1	1957.
	250	91.08	146.3	261.1	539.0	1608.	64.40	103.5	184.7	381.1	1137.
250	10	141.0	226.4	404.1	834.2	2488.	99.67	160.1	285.8	589.8	1759.
	50	137.3	220.5	393.6	812.5	2424.	97.08	155.9	278.3	574.5	1714.
	100	127.6	205.0	365.9	755.3	2253.	90.24	145.0	258.7	534.0	1583.
	200	82.84	133.1	237.5	490.3	1462.	58.58	94.09	167.9	346.7	1034.
200	10	114.0	183.2	327.0	674.9	2013.	80.64	129.5	231.2	477.2	1424.
	50	109.5	175.9	313.9	647.9	1933.	77.42	124.4	222.0	458.2	1367.
	100	97.08	155.9	278.3	574.5	1714.	68.65	110.3	196.8	406.3	1212.
	150	73.68	118.4	211.3	436.0	1301.	52.10	83.69	149.4	308.3	919.7
150	10	87.04	139.8	249.6	515.1	1537.	61.55	98.87	176.5	364.2	1086.
	25	85.45	137.3	245.0	505.7	1508.	60.42	97.06	173.2	357.6	1067.
	50	80.98	130.1	232.2	479.3	1430.	57.26	91.98	154.2	338.9	1011.
	100	63.21	101.5	181.2	374.1	1116.	44.70	71.80	128.2	264.5	789·0

Gage Pressure Up Stream P.S.I.	Gage Pressure Down Stream P.S.I.	1000 Foot Line, Nominal Pipe Size* 2	2½	3	4	6	2000 Foot Line, Nominal Pipe Size* 2	2½	3	4	6
		Thousands of Cu. Ft. Per Hour					Thousands of Cu. Ft. per Hour				
100	10	59.84	96.12	171.6	354.1	1056.	42.31	67.97	121.3	250.4	746.9
	25	57.50	92.36	164.9	340.3	1015.	40.66	65.31	116.6	240.6	717.7
	50	50.62	81.32	145.1	299.6	893.6	35.79	57.50	102.6	211.8	631.9
	75	38.22	61.39	109.6	226.2	674.6	27.02	43.41	77.5	159.9	477.0
80	5	49.46	79.45	141.8	292.7	873.1	34.97	56.18	100.3	207.0	617.4
	10	48.82	78.43	140.0	288.9	861.8	34.52	55.46	98.98	204.3	609.4
	30	44.50	71.64	127.9	264.0	787.3	31.54	50.66	90.42	186.6	556.7
	55	34.26	55.03	98.23	202.8	604.8	24.23	38.91	69.46	143.4	427.6
60	5	38.45	61.77	110.3	227.6	678.8	27.19	43.68	77.96	160.9	480.0
	10	37.63	60.45	107.9	222.7	664.3	26.61	42.74	76.29	157.5	469.7
	20	35.32	56.73	101.3	209.0	623.5	24.97	40.12	71.61	147.8	440.9
	40	27.17	43.65	77.91	160.8	479.7	19.21	30.86	55.09	113.7	339.2
40	5	27.21	43.71	78.01	161.0	480.3	19.24	30.91	55.16	113.9	339.6
	10	26.03	41.81	74.63	154.1	549.5	18.41	29.57	52.77	108.9	324.9
	15	24.50	39.36	70.26	145.0	432.6	17.33	27.83	49.68	102.5	305.9
	25	20.08	32.26	57.58	118.8	354.5	14.20	22.81	40.71	84.04	250.7
30	5	21.38	34.35	61.30	126.5	377.4	15.12	24.29	43.35	89.48	266.9
	10	19.86	31.90	56.94	117.5	350.6	14.04	22.56	40.26	83.11	247.9
	15	17.81	28.61	51.07	105.4	314.5	12.60	20.23	36.11	74.54	222.4
	20	15.03	24.14	43.09	88.94	265.3	10.63	17.07	30.47	62.89	187.6
20	5	15.21	24.43	43.61	90.00	268.5	10.75	17.28	30.83	63.65	189.8
	10	12.98	20.85	37.22	76.83	229.2	9.180	14.75	26.32	54.33	162.1
10	5	7.923	12.73	22.72	46.89	138.9	5.603	9.000	16.06	33.16	98.90

MULTIPLIERS TO CONVERT 1000 FOOT LINE VALUES TO LENGTHS LISTED:

Length of Line—Feet	200	400	600	800	1000	1200	1400	1600	1800	2000	2200	2400	2600	2800	3000	3200	3400	3600	3800	4000	4200	4400	4600	4800	5000	5200	Feet—Length of Line
Multipliers	2.24	1.58	1.29	1.12	1.00	.91	.85	.79	.75	.71	.67	.65	.62	.60	.58	.56	.54	.53	.51	.50	.49	.48	.47	.46	.45	.44	Multipliers

For remarks regarding use of the Table refer to page 78 of source.

Courtesy Armco Steel Corporation.

Table 4
1-mile and 2½-mile lines

0.7 SP. GR. 60° F.

Gage Pressure Up Stream P.S.I.	Gage Pressure Down Stream P.S.I.	1 Mile Line, Nominal Pipe Size* 2	2½	3	4	6	2½ Mile Line, Nominal Pipe Size* 2	2½	3	4	6
		Thousands of Cu. Ft. per Hour					Thousands of Cu. Ft. per Hour				
400	10	96.39	154.8	276.4	570.4	1701.	60.96	97.92	174.8	360.8	1076.
	50	95.38	153.2	273.5	564.5	1684.	60.32	96.90	173.0	357.0	1065.
	100	92.80	149.1	266.1	549.2	1638.	58.69	94.28	168.3	347.3	1036.
	350	45.98	73.85	131.8	272.1	811.6	29.08	46.71	83.37	172.1	513.3
350	10	84.71	136.1	242.9	501.3	1495.	53.58	86.06	153.6	317.1	945.8
	50	83.57	134.2	239.6	494.6	1475.	52.85	84.90	151.5	312.8	933.0
	100	80.61	129.5	231.1	477.1	1423.	50.98	81.90	146.2	301.7	900.0
	300	42.92	68.95	123.1	254.0	757.7	27.15	43.61	77.84	160.7	479.2
300	10	73.03	117.3	209.4	432.2	1289.	46.19	74.20	132.4	273.4	815.4
	50	71.70	115.2	205.6	424.3	1266.	45.35	72.84	130.0	268.4	800.5
	100	68.23	109.6	195.6	403.8	1204.	43.15	69.32	123.7	255.4	761.8
	250	38.64	63.67	113.6	234.6	699.7	25.07	40.27	71.88	148.4	442.5
250	10	61.34	98.53	175.9	363.0	1083.	38.80	62.32	111.2	229.6	684.8
	50	59.75	95.98	171.3	353.6	1055.	37.79	60.70	108.3	223.6	667.1
	100	55.54	89.21	159.2	328.7	980.4	35.13	56.42	100.7	207.9	620.1
	200	36.05	57.91	103.4	213.4	636.4	22.80	36.63	65.37	134.9	402.5
200	10	49.63	79.72	142.3	293.7	876.1	31.39	50.42	90.00	185.8	554.1
	50	47.65	76.54	136.6	282.0	841.1	30.13	49.41	86.40	178.3	532.0
	100	42.25	67.87	121.1	250.0	745.8	26.72	42.92	76.61	158.1	471.7
	150	32.06	51.51	91.94	189.8	566.0	20.28	32.58	58.14	120.0	358.0
150	10	37.88	60.85	108.6	224.2	668.7	23.96	38.48	68.69	141.8	422.9
	25	37.19	59.74	106.6	220.1	656.5	23.52	37.78	67.44	139.2	415.2
	50	35.24	56.61	101.0	208.6	622.1	22.29	35.80	63.91	131.9	393.5
	100	27.51	44.19	78.87	162.8	485.6	17.40	27.95	49.88	103.0	307.1

Gage Pressure Up Stream P.S.I.	Gage Pressure Down Stream P.S.I.	1 Mile Line, Nominal Pipe Size* 2	2½	3	4	6	2½ Mile Line, Nominal Pipe Size* 2	2½	3	4	6
		Thousands of Cu. Ft. per Hour					Thousands of Cu. Ft. per Hour				
100	10	26.04	41.83	74.67	154.1	459.7	16.47	26.46	47.22	97.47	290.7
	25	25.02	40.20	71.75	148.1	441.7	15.83	25.42	45.38	93.66	279.4
	50	22.03	35.39	63.16	130.4	388.9	13.93	22.38	39.95	82.46	246.0
	75	16.63	26.72	47.68	98.4	293.6	10.52	16.90	30.16	62.25	185.7
80	5	21.53	34.58	61.72	127.4	380.0	13.61	21.87	39.03	80.57	240.3
	10	21.25	34.13	60.92	125.7	375.1	13.44	21.59	38.53	79.53	237.2
	30	19.41	31.18	55.65	114.9	342.6	12.28	19.72	35.20	72.65	216.7
	55	14.91	23.95	42.75	88.24	263.2	9.430	15.15	27.04	55.81	166.5
60	5	16.74	26.88	47.98	99.04	285.4	10.58	17.00	30.35	62.64	186.8
	10	16.38	26.31	46.95	96.91	289.1	10.36	16.64	29.70	61.29	182.8
	20	15.37	24.69	44.07	90.96	271.3	9.721	15.62	27.87	57.53	171.6
	40	11.83	19.00	33.91	69.98	208.8	7.479	12.01	21.44	44.26	132.0
40	5	11.84	19.02	33.95	70.08	209.0	7.489	12.03	21.47	44.32	132.2
	10	11.33	18.20	32.48	67.04	200.0	7.165	11.51	20.54	42.40	126.5
	15	10.66	17.13	30.58	63.11	188.3	6.745	10.83	19.34	39.92	119.1
	25	8.740	14.04	25.06	51.72	154.3	5.527	8.879	15.85	32.71	97.57
30	5	9.315	14.95	26.68	55.07	164.3	5.885	9.453	16.87	34.83	103.9
	10	8.643	13.88	24.78	51.15	152.6	5.466	8.781	15.67	32.35	96.49
	15	7.752	12.45	22.23	45.88	136.8	4.903	7.875	14.06	29.02	86.55
	20	6.540	10.51	18.75	38.71	115.5	4.137	6.645	11.86	24.48	73.02
20	5	6.619	10.63	18.98	39.17	116.8	4.186	6.724	12.00	24.77	73.90
	10	5.650	9.076	16.20	33.44	99.74	3.573	8.740	10.25	21.15	63.08
10	5	3.448	5.539	9.886	20.41	60:87	2.181	3.503	6.253	12.91	38.50

MULTIPLIERS TO CONVERT 1 MILE LINE VALUES TO LENGTHS LISTED:

Length of Line—Miles	1.2	1.4	1.6	1.8	2.0	2.2	2.4	2.6	2.8	3.0	3.2	3.4	3.6	3.8	4.0	4.2	4.4	4.6	4.8	Miles—Length of Line
Multipliers	.91	.85	.79	.75	.71	.67	.65	.62	.60	.58	.56	.54	.53	.51	.50	.49	.48	.47	.46	Multipliers

For remarks regarding use of the Table refer to page 78 of source.

Courtesy of Armco Steel Corporation.

Table 5
5-mile and 10-mile lines

0.7 SP. GR. 60° F.

GAGE PRESSURE Up Stream P.S.I.	Down Stream P.S.I.	5 MILE LINE 2	5 MILE LINE 2½	5 MILE LINE 3	5 MILE LINE 4	5 MILE LINE 6	10 MILE LINE 2	10 MILE LINE 2½	10 MILE LINE 3	10 MILE LINE 4	10 MILE LINE 6
		Thousands of Cu. Ft. Per Hour					Thousands of Cu. Ft. Per Hour				
400	10	43.10	69.24	123.6	255.1	760.9	30.48	48.96	87.39	180.4	538.0
	50	42.66	68.52	122.3	252.4	753.0	30.16	48.45	86.48	178.5	532.4
	100	41.50	66.67	119.0	245.6	732.6	29.35	47.14	84.14	173.7	518.0
	350	20.56	33.03	58.95	121.7	363,0	14.54	23.35	41.69	86.04	256.7
350	10	37.88	60.86	108.6	224.2	668.8	26.79	43.03	76.81	158.5	472.9
	50	37.37	60.03	107.2	221.2	659.7	26.43	42.45	75.77	156.4	466.5
	100	36.05	57.91	103.4	213.3	636.4	25.49	40.95	73.09	150.9	450.0
	300	19.20	30.84	55.04	113.6	338.9	13.57	21.80	38.92	80.33	239.6
300	10	32.66	52.46	93.64	193.3	576.6	23.09	37.10	66.22	136.7	407.7
	50	32.07	51.51	91.94	189.8	566.0	22.67	36.42	65.01	134.2	400.2
	100	30.51	49.02	87.49	180.6	538.7	21.58	34.66	61.86	127.7	380.9
	250	17.73	28.47	50.82	104.9	312.9	12.53	20.13	35.94	74.18	221.3
250	10	27.43	44.07	78.65	162.3	484.3	19.40	31.16	55.62	114.8	342.4
	50	26.72	42.92	76.61	158.1	471.7	18.89	30.35	54.17	111.8	333.5
	100	24.84	39.90	71.21	147.0	438.5	17.56	28.21	50.36	103.9	310.0
	200	16.12	25.90	46.23	95.41	284.6	11.40	18.31	32.69	67.47	201.2
200	10	22.20	35.65	63.64	131.4	391.8	15.69	25.21	45.00	92.88	277.0
	50	21.31	34.23	61.10	126.1	376.2	15.07	24.20	43.20	89.17	266.0
	100	18.89	30.35	54.17	111.8	333.5	13.36	21.46	38.31	79.07	235.8
	150	14.34	23.03	41.11	84.86	253.1	10.14	16.29	29.07	60.01	179.0
150	10	16.94	27.21	48.57	100.3	299.0	11.98	19.24	34.35	70.89	211.5
	25	16.63	26.72	47.68	98.42	293.6	11.76	18.89	33.72	69.60	207.6
	50	15.76	25.32	45.19	93.27	278.2	11.14	17.90	31.95	65.96	196.7
	100	12.30	19.76	35.27	72.81	217.2	8.699	13.97	24.94	51.48	153.6

0.7 SP. GR. 60° F.

GAGE PRESSURE Up Stream P.S.I.	Down Stream P.S.I.	5 MILE LINE 2	5 MILE LINE 2½	5 MILE LINE 3	5 MILE LINE 4	5 MILE LINE 6	10 MILE LINE 2	10 MILE LINE 2½	10 MILE LINE 3	10 MILE LINE 4	10 MILE LINE 6
		Thousands of Cu. Ft. Per Hour					Thousands of Cu. Ft. Per Hour				
100	10	11.65	18.71	33.39	68.92	205.6	8.235	13.23	23.61	48.74	145.4
	25	11.19	17.98	32.09	66.23	197.5	7.913	12.71	22.69	46.83	139.7
	50	9.852	15.83	28.25	55.31	173.9	6.966	11.19	19.97	41.23	123.0
	75	7.438	11.95	21.32	44.02	131.3	5.259	8.448	15.08	31.12	92.84
80	5	9.626	15.46	27.60	56.97	169.9	6.807	10.93	19.52	40.28	120.2
	10	9.502	15.26	27.24	56.23	167.7	6.719	10.79	19.26	39.76	118.6
	30	8.680	13.94	24.89	51.37	153.2	6.138	9.860	17.60	36.32	108.4
	55	6.668	10.71	19.12	39.46	117.7	4.715	7.574	13.52	27.90	83.23
60	5	7.484	12.02	21.46	44.29	132.1	5.292	8.501	15.17	31.32	93.42
	10	7.324	11.76	21.00	43.34	129.3	5.179	8.319	14.85	30.65	91.42
	20	6.873	11.04	19.71	40.68	121.3	4.861	7.808	13.94	28.77	85.80
	40	5.288	8.495	15.16	31.30	93.36	3.740	6.007	10.72	22.13	66.01
40	5	5.296	8.507	15.18	31.34	93.48	3.745	6.015	10.74	22.16	66.10
	10	5.066	8.138	14.53	29.98	89.43	3.582	5.755	10.27	21.20	63.24
	15	4.769	7.661	13.67	28.22	84.19	3.372	5.417	9.669	19.96	59.53
	25	3.908	6.278	11.21	23.13	69.00	2.764	4.439	7.924	16.36	48.79
30	5	4.161	6.685	11.93	24.63	73.46	2.943	4.727	8.437	17.41	51.94
	10	3.865	6.209	11.08	22.87	68.23	2.733	4.390	7.836	16.17	48.25
	15	3.467	5.569	9.940	20.52	61.20	2.451	3.938	7.029	14.51	43.27
	20	2.925	4.698	8.386	17.31	51.63	2.068	3.322	5.930	12.24	36.51
20	5	2.960	4.755	8.487	17.52	52.25	2.093	3.362	6.001	12.39	36.95
	10	2.527	4.059	7.244	14.95	44.60	1.787	2.870	5.123	10.57	31.54
10	5	1.542	2.477	4.421	9.126	27.22	1.090	1.752	3.126	6.453	19.25

MULTIPLIERS TO CONVERT 10 MILE LINE VALUES TO LENGTHS LISTED.

Length of Line—Miles..........	5.2	5.4	5.6	5.8	6.0	6.2	6.4	6.6	6.8	7.0	7.2	7.4	7.6	7.8	8.0	8.2	8.4	8.6	8.8	9.0	9.2	9.4	9.6	9.8	Miles—Length of Line
Multipliers.......	1.39	1.36	1.34	1.31	1.29	1.27	1.25	1.23	1.21	1.20	1.18	1.16	1.15	1.13	1.12	1.10	1.09	1.08	1.07	1.05	1.04	1.03	1.02	1.01	Multipliers

For remarks regarding use of the Table refer to page 78 of source.

Courtesy of Armco Steel Corporation.

From Table 3 the volume for 1,000ft of line, for 0.70 specific gravity gas and 60°F flowing temperature for 200 and 50 psi up and downstream pressure, respectively, is 109.5 MCF per hour. Correction factor for the pipe length, from the same table is .53. Correction factor, from Table 2 for specific gravity and flowing temperature is .991. Therefore:

$$\text{Delivery of gas} = 109.5 \times .53 \times .991 = 57{,}512\ \text{cu. ft/hr}$$

Multiphase flow

As stated at the beginning of this chapter, the problems of the simultaneous flow of oil and gas or of oil, gas, and water through one pipe have become increasingly important in oil field production operations. The problems are complex, because the properties of two or more fluids must be considered and because of the different patterns of fluid flows, depending upon the flow conditions. These patterns, in any given line, may change as the flow conditions change, and they may be coexistent at different points of the line.[3,5,7]

Different investigators recognize different flow patterns and use different nomenclature. Those generally recognized are:

1. Bubble flow—bubbles of gas flow along the upper part of the pipe with about the same velocity as the liquid.
2. Plug flow—the bubbles of gas coalesce into large bubbles and fill out the large part of the cross-sectional area of the pipe.
3. Laminar flow—the gas-liquid interface is relatively smooth with gas flowing in the upper part of the pipe.
4. Wavy flow—same as above except that waves are formed on the surface of the liquid.
5. Slug flow—the tops of some waves reach the top of the pipe. These slugs move with high velocity.
6. Annular flow—the liquid flows along the walls of the pipe, and the gas moves through the center with high velocity.
7. Spray flow—the liquid is dispersed in the gas.

The above description of flow patterns has been given to emphasize the first reason why the pressure drop must be

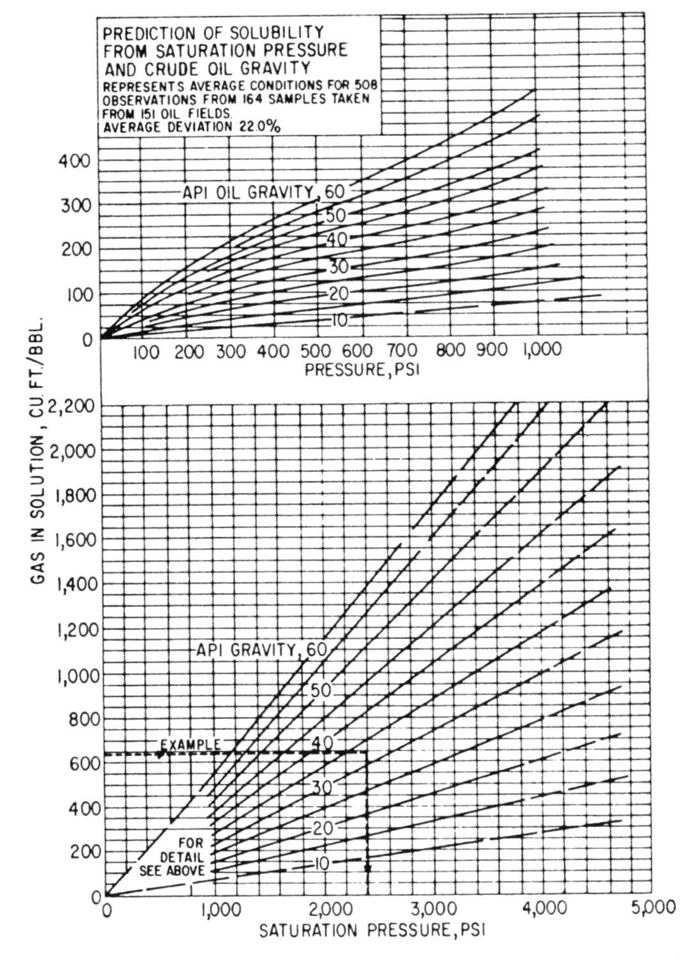

Figure 1. Prediction of solubility from saturation pressure and gravity of crude oil. (Reference 12.)

higher for the multiphase than the single-phase flow. In the latter the pressure drop is primarily the result of friction. In the multiphase flow, in addition to friction, the energy losses are due to gas accelerating the fluid, due to compression of gas and formation of waves and slugs. These losses are different for different patterns, with patterns changing as a result of changes in flow conditions. The second reason is the fact that with two phases present in the pipe, less cross-sectional area is available for each of the phases. As previously discussed, the pressure drop is inversely proportional to the fifth power of pipe diameter.

The literature on multiphase flow is quite extensive. Reference 7 gives 72 references on the subject.

Two-phase horizontal flow

More than twenty correlations have been developed by different investigators for predicting the pressure drop in the two-phase flow. Under the project, sponsored jointly by the American Petroleum Institute and American Gas Association at the University of Houston, five of these correlations were tested by comparing them with 2,620 experimental measurements culled from more than 15,000 available.

The statistical part of the report on the project concludes that of the five correlations tested, the Lockhart-Martinelli correlation shows the best agreement with the experimental data, particularly for pipe sizes commonly used in oil field production operations.[8] The method may be summarized as follows:[3,4]

1. Determine the single-phase pressure drops for the liquid and the gas phase as if each one of these were flowing alone through the pipe. This is done by use of previously given equations.
2. Determine the dimensionless parameter:

$$X = \sqrt{\frac{\sqrt{\Delta P_L}}{\Delta P_G}}$$

where ΔP_L and ΔP_G are single-phase pressure drops for liquid and gas, respectively.
3. The method recognizes four regimens of the two-phase flow as follows:

Flow Regimens

Gas	Liquid
Turbulent	Laminar
Turbulent	Turbulent
Laminar	Laminar
Laminar	Turbulent

The type of flow of each of the phases is determined by its Reynolds number (Figure 2).
4. Determine factor Φ from a chart as a function of X for the appropriate flow regimen.
5. The two-phase pressure drop is then:

$$\Delta P_{TP} = \Phi^2 P_L \text{ of } G$$

The single-phase pressure drop of either gas or liquid phase can be used. The results are the same.

For the two-phase flow problems in oil field production operations, certain simplifications can be used if only approximate estimates are desired:

1. Information that would permit determining the volume of oil and gas under flow conditions from equilibrium flash calculation is usually unavailable. The following procedure may be used:

   Gas volume is determined by use of the chart in Figure 1, which shows gas in solution for oil of different

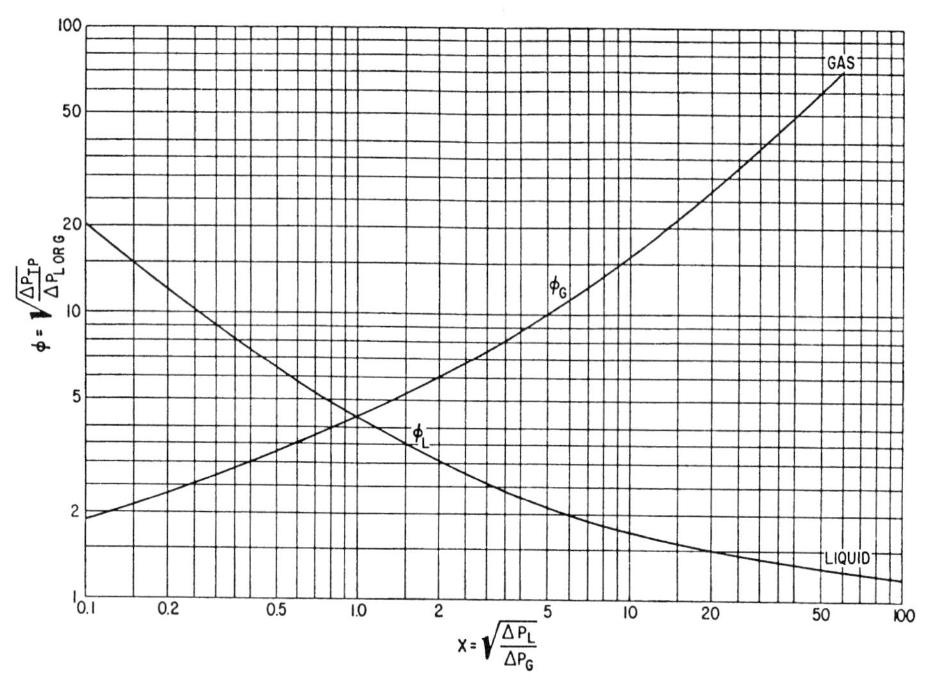

Figure 2. Parameter Φ for Lockhart-Martinelli correlation for two-phase flow. (Reference 9.)

gravities and different saturation pressures (in this case the flowline pressures). If the gas-oil ratio is known for one pressure, the ratio for another pressure can be determined from the chart as follows:

> Assume a gas–oil ratio of 600 cu. ft/bbl at 900 psi for a 30°API oil. Determine the gas oil ratio at 1,300 psi. From the chart, at 900 psi, the gas in solution is 205 cu. ft/bbl and at 1,300 psi is 300 cu. ft/bbl, an increase of 95 cu. ft/bbl. Therefore, at 1,300 psi the gas–oil ratio will be 600 – 95 = 505 cu. ft/bbl.

Increase in volume of oil because of the increase of the dissolved gas can be obtained from charts for calculation of formation volume of bubble point liquids. However, for approximate estimates in problems involving field flowlines, this step frequently can be omitted without seriously affecting the validity of results.

2. In a majority of cases here under consideration, the flow regimen is turbulent-turbulent. Therefore, the chart in Figure 2 shows the factor for only this regimen.

The procedure is illustrated by the following examples.

Example 1. Determine two-phase pressure drops for different flow ratios and different flow pressures under the following conditions:

Oil	31°API
Viscosity	5 cp at 100°F
Separator pressure	900 psig
Gas–oil ratio	670 cu. ft/bbl
Flow line	2 in.

Wellhead pressure of 1,300 psig was assumed, and calculations were made for flow rates of 100 and 50 bbl per day. The single phase pressure drops were calculated for the gas and liquid phases, each of them flowing alone in the line, according to the Fanning equations.

Calculation of pressure drop for liquid phase:

$$H_f = \frac{f\,L(\text{bbl/day})^2(\text{lb/gal})}{181{,}916D^5}$$

where: f = Friction factor
L = Length of pipe, ft
D = Inside diameter of pipe, inches

Calculation of pressure drop for gas phase: see pp. 280–281.

For the two flow rates and the two pressures, these single-phase pressure drops were calculated to be as follows:

Flow Rate bbl/day	**100**		**50**	
Pressure, psia	1,300	900	1,300	900
ΔP_G psi/100 ft	7.9×10^{-4}	2.0×10^{-3}	2.4×10^{-4}	7.7×10^{-4}
ΔP_l psi/100 ft	$2. \times 10^{-4}$	2.7×10^{-2}	1.3×10^{-2}	1.3×10^{-2}

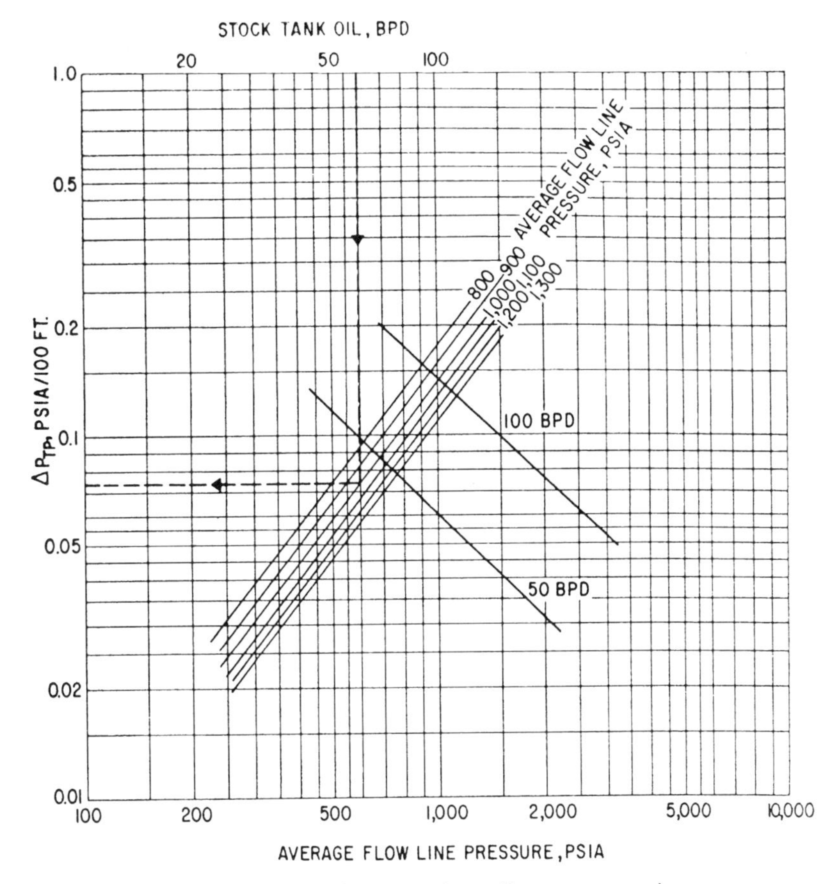

Figure 3. Determining two-phase flow pressure drops.

Then $\Delta X = \sqrt{\dfrac{\Delta P_L}{\Delta P_G}}$	5.9	3.8	7.4	4.2
From Figure 2ϕ and the two-phase pressure drop	2.0	2.4	1.9	2.3
Pressure drop $\Delta P_{TP} = \phi_L \times \Delta P_L$	0.12	0.16	0.048	0.069

With two pressure drops determined for two flow line pressures for each of the flow rates, lines can be drawn showing pressure drop for different flow line pressures for these two rates (Figure 3). From these two lines relationships can be established for pressure drops for other rates and flow line pressures. The procedure is as follows:

Start with, for instance, 800 psia line. For the rate of 50 bpd at 800 psia flow line pressure, the pressure drop from the chart is .074 psia/100 ft. From the 50 bpd rate on the scale in the upper part of the chart, go downward to the line representing .074 psia pressure drop. This gives one point of the 800 psia line.

For the rate of 100 bpd the chart indicates that at 800 psia flowline pressure, the pressure drop is .175 psia/100 ft. From the 100 bpd rate in the upper part of the chart, go downward to the .175 psia /100 ft line. This gives the second point of the 800 psia line. The line can now be drawn through these two points. Lines for other pressures can be constructed in a similar way.

The chart is used as follows: For conditions represented by this particular chart, what is the two-phase pressure drop for the rate of 60 bpd and average flow line pressure of 1,000 psia?

Start at the rate of 60 bpd on the upper scale of the chart. Go downward to the intersection with the 1,000 psia pressure, then to the left, and read the two-phase pressure drop as .075 psia/100 ft.

Example 2. If pipe diameter is to be determined for the desired pressure drop and the known length of line, flow rate, and the properties of fluids, then the problem cannot be solved directly. The procedure is as follows:

An estimate is made of the diameter of the two-phase line. The best approach is to determine, for the conditions, the diameter of the line for each of the phases flowing separately. In a majority of cases, the sum of these two diameters will be a good estimate of the diameter for the two-phase line. Calculations are then made for the two-phase pressure drop using the assumed diameter. If this drop is reasonably close to the desired one, the estimate is correct. If not, a new estimate must be made and the calculations repeated.

Inclined two-phase flow

In hilly terrain additional pressure drop can be expected in case of the two-phase flow. Little published information is available on the subject. Ovid Baker suggested the following empirical formula:[10]

$$\Delta P_{TPH} = \Delta P_{TP} + \frac{1.61 N H \rho_L}{144 V_g^{0.7}}$$

where: ΔP_{TPH} = Pressure drop, corrected for hilly terrain in psi
ΔP_{TP} = Calculated two-phase pressure drop for horizontal pipe, psi
N = Number of hills
H = Average height of hills, ft
ρ_L = Density of liquid phase, lb/ft^3
V_g = Velocity of gas, ft/sec (based on 100% pipe area)

No test data are available to determine the accuracy of this approach. Until more information is developed on the subject, the method may be considered for approximate estimates.

Horizontal three-phase flow

Very little is known about the effect on the pressure drop of addition to the gas-oil system of a third phase, an immiscible liquid, water. Formation of an emulsion results in increased viscosities. Formulas are available for approximate determination of viscosity of emulsion, if viscosity of oil and percent of water content of oil is known. The point is, however, that it is not known what portion of the water is flowing in emulsified form.

From one published reference[11] and from some unpublished data, the following conclusions appear to be well founded:

1. Within the range of water content of less than 10% or more than 90%, the flow mechanism appears to approach that of the two-phase flow.
2. Within the range of water content of from 70% to 90%, the three-phase pressure drop is considerably higher than for the two-phase flow.

For solution of three-phase flow problems of oil field flow lines, the following approach has been used. The oil and water are considered as one phase and the gas as the other phase. Previously given calculation of a two-phase drop is used. The viscosity of the oil-water mixture is determined as follows:

$$\mu_L = \frac{V \times \mu_0 + V_W \times \mu_W}{V_0 + V_W}$$

where μ_L, μ_0 and μ_W are viscosities of mixture of oil and of water, respectively, and V_0 and V_W represent the corresponding volumes of oil and water.

References

1. *Cameron Hydraulic Data*, 13th ed., Ingersoll-Rand Company, New York, 1962.
2. *Flow of Fluids Through Valves, Fittings and Pipe*, Technical Paper No. 410, Engineering Division, Crane Company, Chicago, 1957.
3. Streeter, Victor L., *Handbook of Fluid Dynamics*, McGraw-Hill Book Company, Inc., New York, 1961.
4. Prof. W. E. Durand's Discussions in *Handbook of the Petroleum Industry*, David T. Day; John Wiley & Sons, Inc., New York.
5. Baker, Ovid, "Simultaneous Flow of Oil and Gas," *The Oil and Gas Journal*, July 26, 1954.
6. Campbell, Dr. John M., "Elements of Field Processing," *The Oil and Gas Journal*, December 9, 1957.
7. Campbell, Dr. John M., "Problems of Multi-phase Pipe Line Flow," in "Flow Calculations in Pipelining," *The Oil and Gas Journal*, November 4, 1959.
8. Duckler, A. E., Wicks, III, Moye, and Cleveland, R. G., "Frictional Pressure Drop in Two-Phase Flow," *AIChE Journal*, January, 1964.
9. Lockhart, R. W., and Martinelli, R. C., "Proposed Correlation of Data for Isothermal Two-Phase, Two-Component Flow in Pipes," *Chemical Engineering Progress* 45: 39–48, (1949).
10. Ovid Baker's discussion of article "How Uphill and Downhill Flow Affect Pressure Drop in Two-Phase Pipe Lines," by Brigham, W. E., Holstein, E. D., and Huntington, R. L.; *The Oil and Gas Journal*, November 11, 1957.
11. Sobocinski, D. P., and Huntington, R. D., "Concurrent Flow of Air, Gas, Oil and Water in a Horizontal Line," *Trans. ASME* 80: 1,252, 1958.
12. Beal, Carlton, "The Viscosity of Air, Water, Natural Gas, Crude Oil and Its Associated Gases at Oil Field Temperatures and Pressures," *Petroleum Development and Technology*, AIME, 1946.
13. Switzer, F. G., *Pump Fax*, Gould Pumps, Inc.
14. *Pump Talk*, The Engineering Department, Dean Brothers Pumps, Inc., Indianapolis.
15. Switzer, F. G., *The Centrifugal Pump*, Goulds Pumps, Inc.
16. Zalis, A. A., "Don't Be Confused by Rotary Pump Performance Curves," *Hydrocarbon Processing and Petroleum Refiner*, Gulf Publishing Company, September, 1961.

Nomograph for calculating Reynolds number for compressible flow friction factor for clean steel and wrought iron pipe

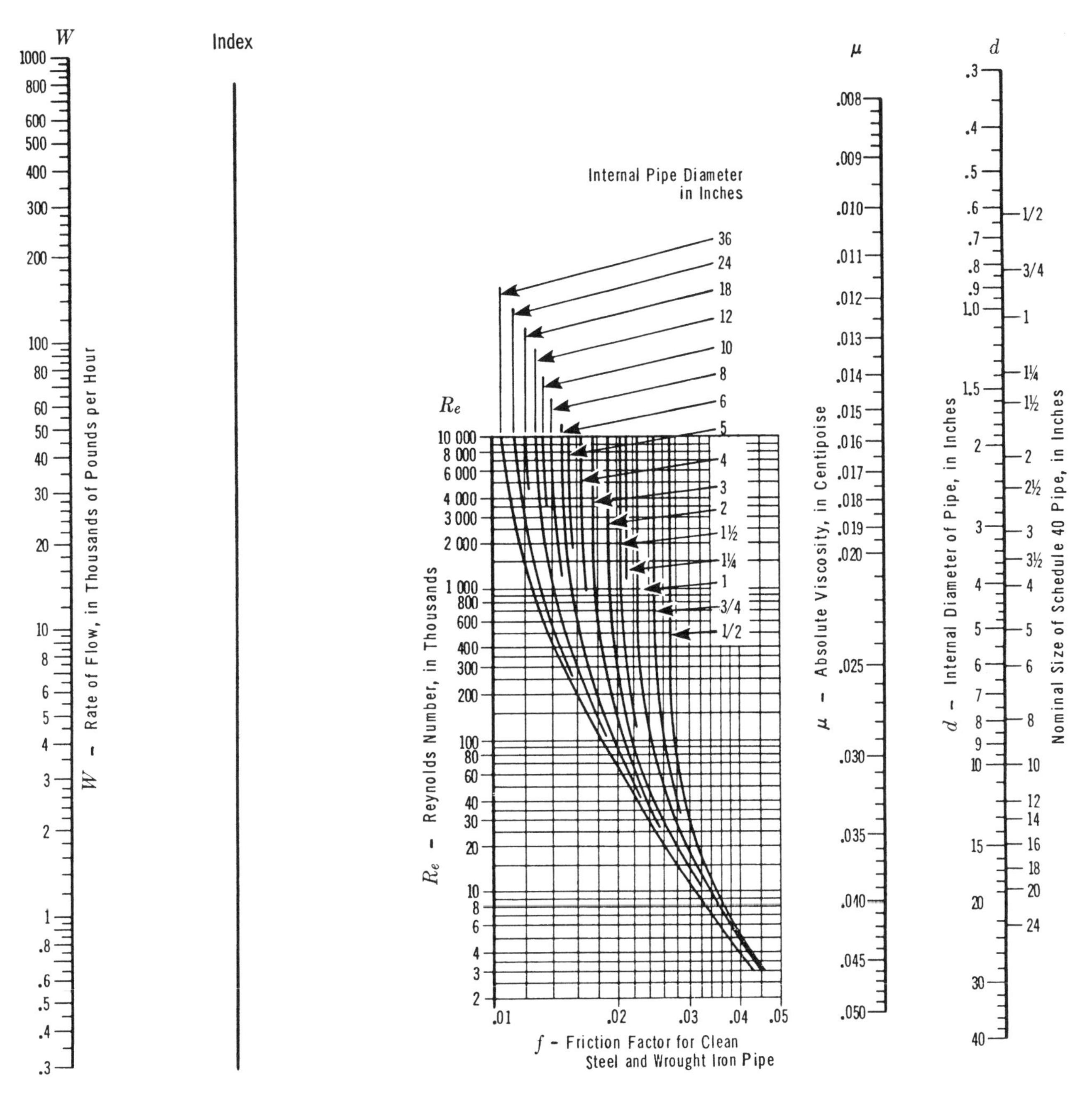

Figure 1. Reynolds number for compressible flow friction factor for clean steel and wrought iron pipe.

Background

The nomograph (Figure 1) permits calculation of the Reynolds number for compressible flow and the corresponding friction factor. It is based on the equation:

$$Re = 6.31\frac{W}{d\mu} = 0.482\frac{q_s \times Sg}{d\mu}$$

where: W = Flow rate, lb/hr
Sg = Specific gravity of gas relative to air

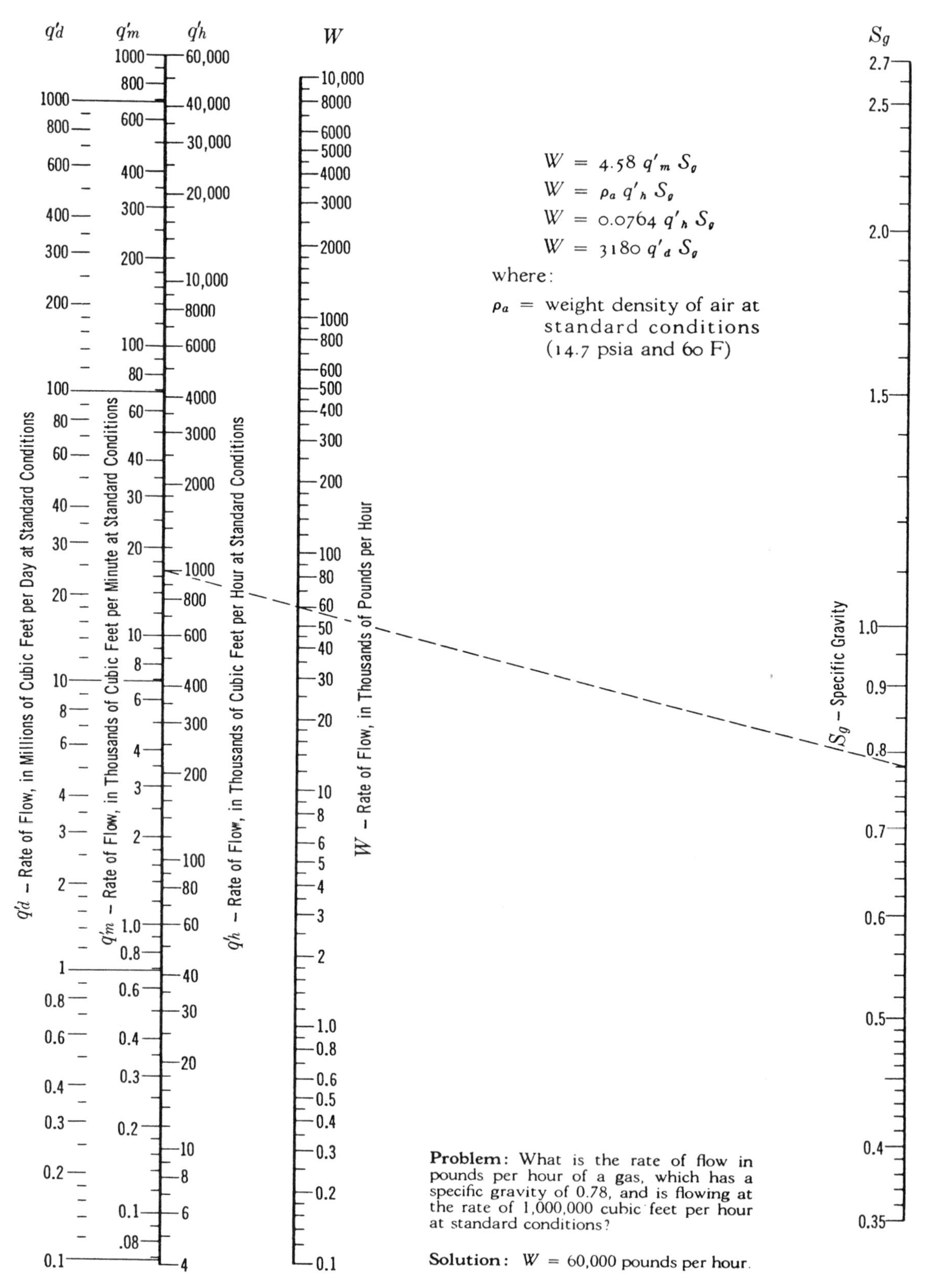

Figure 2. Equivalent volume and weight flow rates of compressible fluids.

q_s = volumetric flow rate, cubic ft/sec (at 14.7 psia and 60°F)

μ = fluid viscosity, centipoise

If the flow rate is given in lb/hr, the nomograph can be used directly without resorting to Figure 2 to obtain q_s. Figure 2 converts volume flow to weight flow rates if the specific gravity of the fluid is known.

Example. Natural gas at 250 psig and 60°F with a specific gravity of 0.75 flows through an 8-in. schedule 40 clean steel pipe at a rate of 1,200,000 cubic ft/hr at standard conditions.

Find the Reynolds number and the friction factor.

At Sg = 0.75 obtain q = 69,000 from Figure 1, $\mu = 0.011$.

Connect	With	Mark or Read
W = 69,000	$\mu = 0.011$	Index
Index	d = 8 in.	Re = 5,000,000
Re = 5,000,000	horizontally to 8 in.	f = 0.014

Source

Flow of Fluids Through Valves, Fittings, and Pipes, Technical Paper No. 410, 3–19, Crane Company, Chicago, (1957).

12: Liquids—General

Determining the viscosity of crude

If the viscosity of a gas-saturated crude oil at the saturation (bubble-point) pressure is known, using this homograph you can quickly estimate the viscosities at higher pressures.

Example. Find the viscosity at 4,200 psia for a crude oil when its viscosity is 30 cp at the saturation pressure of 1,200 psia. Notice that 4,200 psia is 3,000 psi above the saturation pressure. Connecting 3,000 on the pressure difference scale (left) with 30 on the curved scale for viscosity at the bubble-point pressure, the intersection with the scale on the right at 48 cp is the desired value.

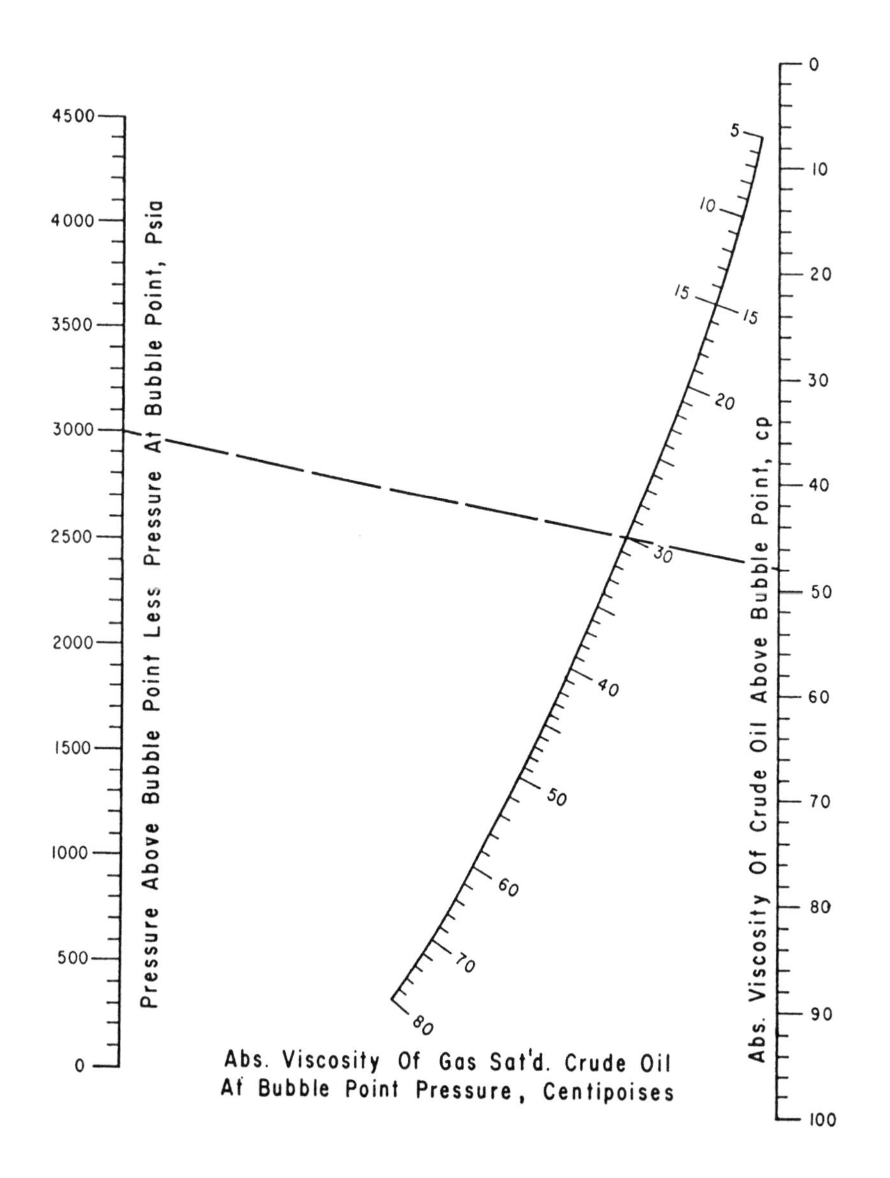

Chart gives API gravity of blends quickly

API gravity of a crude oil blend may be readily estimated from the gravity of the components and their percentage composition. It is a simple procedure to use these curves to find the resulting API gravity of the blend.

Example. If you blend a 14°API pitch (60%) with a 36°API cutting stock (40%), the resulting fuel oil has an API gravity of 22° as read from the nomograph. Calculated result from gravity tables would be 22.1°API.

Results found using this nomograph checked out within 1°API over the range of gravities and percentage of components (1) and (2) in the nomograph below. Estimates from the nomograph are based on the assumption that volumes of blends are additive and that no light components flash off in blending.

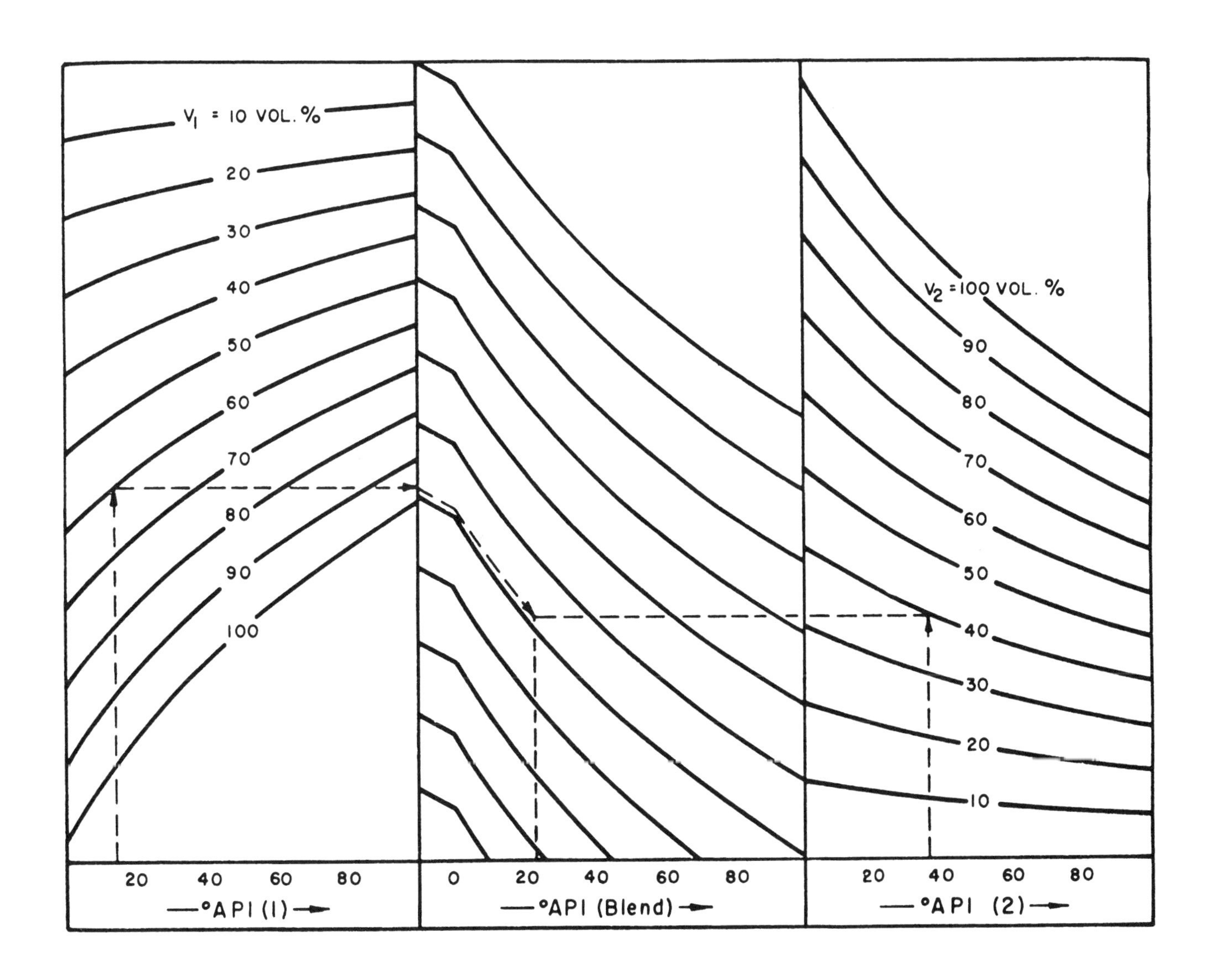

Liquid gravity and density conversion chart

This line chart provides an easy method for converting units of liquid gravity and density. Draw a horizontal line perpendicular to the scale line through a known value and read the equivalent value on any other scale.

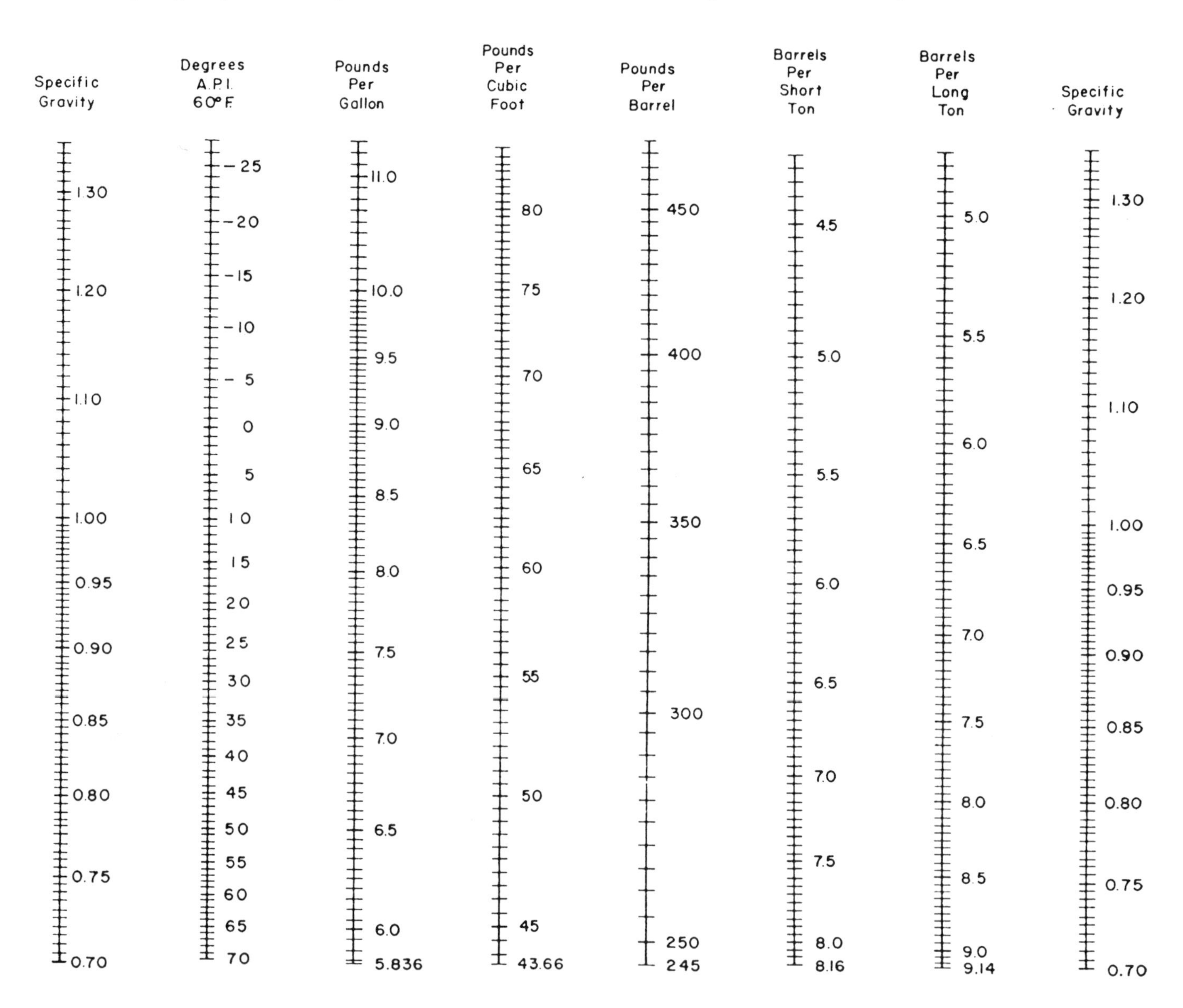

LIQUID GRAVITY AND DENSITY

Nomograph for calculating viscosities of liquid hydrocarbons at high pressure

Lockhart and Lenoir developed a graphical correlation showing the effect of pressure on viscosity of liquid hydrocarbons. This correlation is shown in Figure 1, which is based primarily on data of Hersey and Hopkins,[2] including pure hydrocarbons, lubricating oils, bright stocks, and distillates. Data from Reference 1 have also been included.

To use the nomograph, the characterization factor of Watson, K_w, and the viscosity of the liquid at atmospheric pressure are required.

The accuracy of the correlation decreases as pressure increases.

Example. What is the viscosity of an oil at 5,400 psia, if its characterization factor is 11.8 and its viscosity at atmospheric conditions is 90 centipoises?

Enter 5,400 psia in the pressure scale to the viscosity line of 90 and proceed horizontally to the middle reference scale. Follow the curve lines to intersect the vertical line drawn at $K_w = 11.8$ and read the ratio of viscosity on the extreme left scale at 2.6. The viscosity of the oil is $(2.6)(90) = 234$ centipoises.

Source

Lockhart, F. J. and Lenoir, J. M., *Petroleum Refiner*, 40, No. 3, 209 (1961).

References

1. Griest, and others, *J. Chem. Physics*, 29, 711 (1958).
2. Hersey, M. D., and Hopkins, R. F., *Viscosity of Lubricants under Pressure*, ASME, 1954.

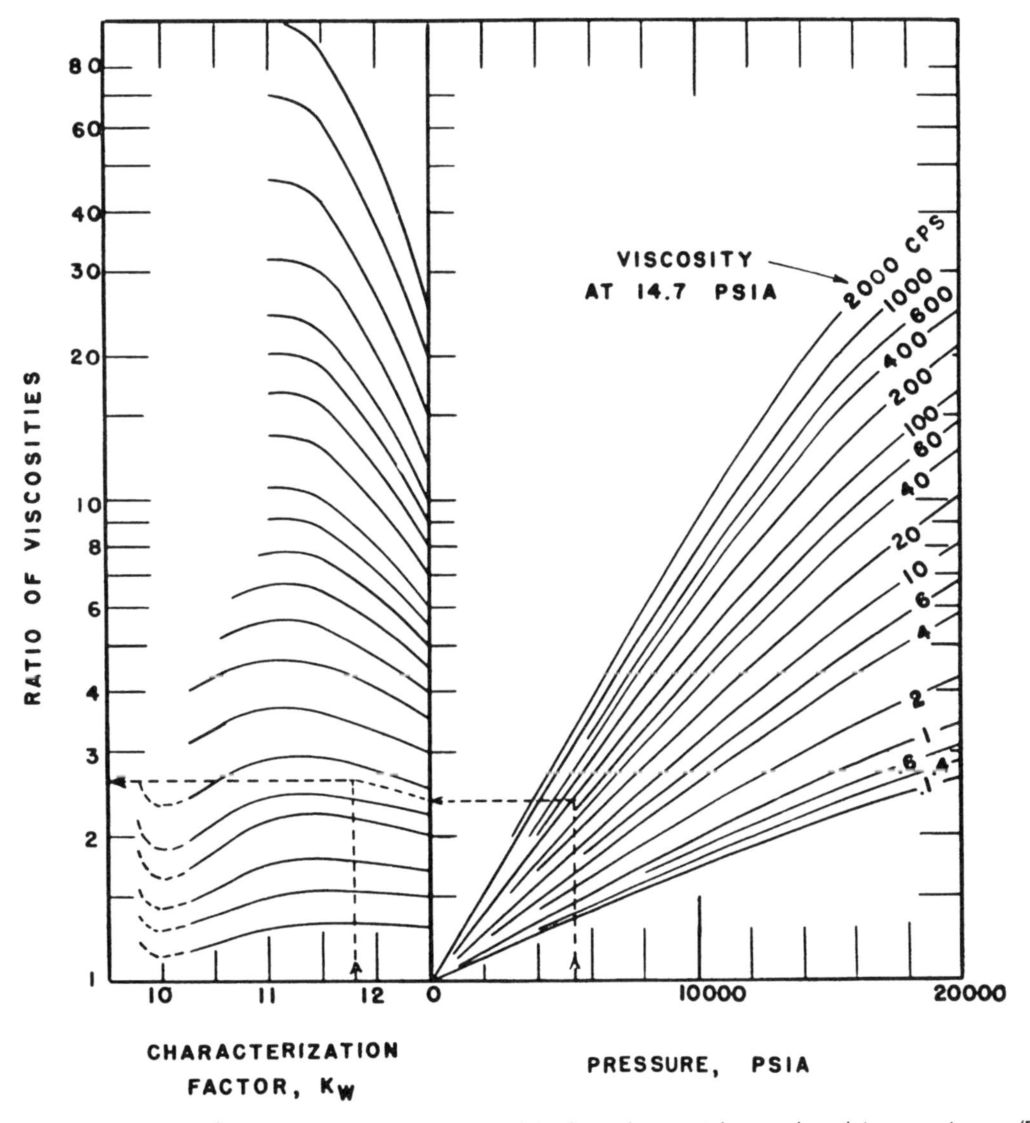

Figure 1. Isothermal effect of pressure on viscosities of liquid hydrocarbons at low reduced temperatures. (Reproduced by permission *Petroleum Refiner*, Vol. 40, No. 3 page 209, copyright 1961, Gulf Publishing Co., Houston.)

Calculate viscosity of a blend

Viscosity is not an additive function of percentage composition. The viscosity of a blend can be calculated with the following equation:

$$V_b = \left[\frac{G_1 + G_2 + G_3 \cdots\cdots\cdots}{\frac{G_1}{\sqrt{V_1}} + \frac{G_2}{\sqrt{V_2}} + \frac{G_3}{\sqrt{V_3}} \cdots\cdots\cdots\cdots} \right]^2$$

where:

V_b = Viscosity of blend
G = Volume (gals., gpm, b/d, etc.)
V = Viscosity in SUS

Example.

$G_1 = 10{,}000$ bbls $V_1 = 45$ SUS
$G_2 = 20{,}000$ bbls $V_2 = 50$ SUS
$G_3 = 5{,}000$ bbls $V_3 = 37$ SUS

$$V_b = \left[\frac{10,000 + 20,000 + 5,000}{\frac{10,000}{\sqrt{45}} + \frac{20,000}{\sqrt{50}} + \frac{5,000}{\sqrt{37}}} \right]^2$$

$V_b = 46.4$ SUS

Calculate specific gravity of a blend

The specific gravity of a blend may be calculated directly by a ratio of the gravity of each component in the blend.

Example. Assume a blend consisting of the following:

Volume, bbls	Specific Gravity	Partial Specific Gravity
10,000	0.8551	0.1900
15,000	0.8337	0.2779
20,000	0.8481	0.3769
45,000		0.8448

The specific gravity of the blend is 0.8448.

Convert viscosity units

Convert SUS to centistokes:

$$V_{CS} = 0.22V_{SUS} - \frac{135}{V_{SUS}} \text{ where } V_{SUS} > 100$$

$$V_{CS} = 0.226V_{SUS} - \frac{135}{V_{SUS}} \text{ where } V_{SUS} \leq 100$$

Convert centistokes to centipoises (cP):

$$cP = V_{CS} \times \text{density, gm/cm}^3$$

SUS = cS × 4.6673, if cS > 500, SUS = cS × 4.6347

Convert specific gravity to API gravity

$$API = \frac{141.5}{S_{60}} - 131.5$$

where:

S_{60} = sp. gr. @ 60°F

$$S_{60} = \frac{141.5}{API_{60} + 131.5}$$

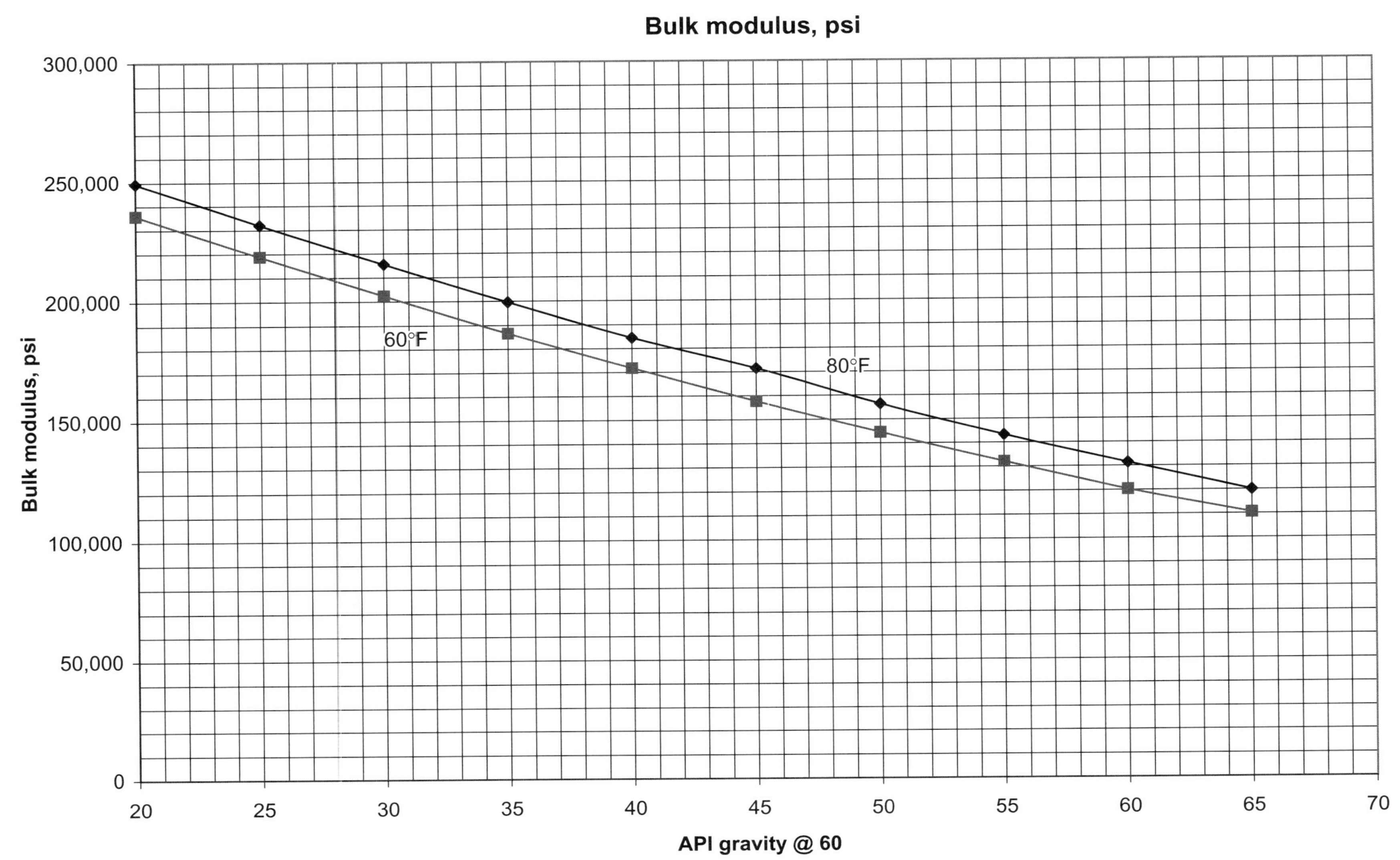

Based on Chapter 11.2.1, API MPMS, Compressibility Factors for Hydrocarbons, May 21, 1992.

Calculate bulk modulus

The bulk modulus, K, of a liquid is the reciprocal of its compressibility. The bulk modulus for water is K = 300,000 psi. The compressibility factors for most liquids can be found in Chapters 11.2.1 and 11.2.2 of the API Manual of Petroleum Measurement Standards. The following relation, known as the ARCO formula, can be used to calculate the bulk modulus for crude oil.

$$K = 1.286(10^6) + 13.55P - 4.122(10^4)T^{1/2} - 4.53(10^3)API - 10.59\,API^2 + 3.228T\,API$$

where:
P = Average pressure

Adapted from Fluid Flow Consultants, Inc., Broken Arrow, Oklahoma. Reprinted with permission.

Nomograph for calculating viscosity of slurries

This nomogram is based on the Hatschek equation for estimating the viscosity of slurries in an aqueous suspension.

$$\mu = \frac{\mu_w}{1 - x^{0.333}}$$

where:
μ = Viscosity of slurry, centipoises
μ_w = Viscosity of water at the temperature of the slurry, centipoises
x = Volume fraction of dry solids in the slurry.

The nomogram (Figure 1) shows a scale calibrated in terms of the temperature of the water instead of its viscosity. Thus, the actual water viscosity is not needed.

Example. What is the viscosity of a slurry having a volume fraction of solids of 0.06 at a temperature of 39°C?

Connect T = 39°C with x = 0.06 and read μ_w = 1.1 centipoises on the middle scale.

Source

Davis, D. S., *Brit. Chem. Eng.*, 4, 9, 478 (1959).

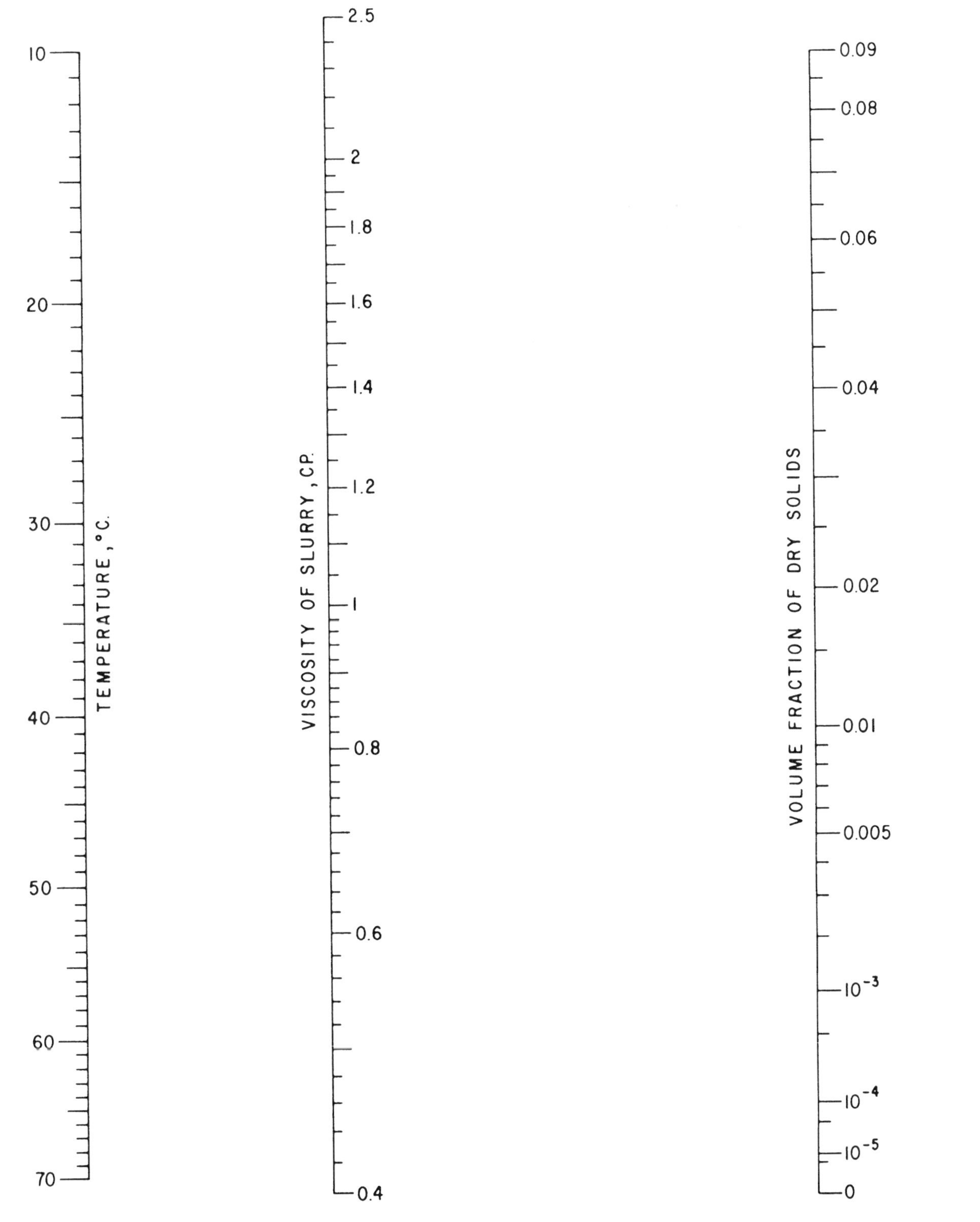

Figure 1. Viscosity of slurries. (Reproduced by permission, *British Chemical Engineering*, vol. 4, page 478, copyright 1959.)

Nomograph for calculating velocity of liquids in pipes

The mean velocity of any liquid flowing in a pipe of circular cross section can be calculated from the following formula:

$$v = 183.3\frac{q}{d^2} = 0.408\frac{Q}{d^2} = 0.0509\frac{W}{d^2\rho}$$

where: v = Average fluid velocity, feet per second
d = Inside diameter of pipe, inches
q = Rate of flow, cubic feet per second
Q = Rate of flow, gallons per minute
W = Rate of flow, pounds per hour
ρ = Fluid density, pounds per cubic foot

The Nomograph for Calculating Velocity of Liquids in Pipes (see Figure 1, p. 351) can be used to calculate the liquid velocity when the rate of flow is in cubic feet, gallons, or thousands of lb. Conversely, knowing the flow rate and velocity, the pipe diameter may be calculated.

Example. What is the velocity of fuel oil at 60°F flowing through a 2-in. schedule 40 pipe at a rate of 45,000 lb/hr. The oil density is 56.02 lb/gal.

Connect	With	Mark or Read
W = 45	ρ = 56.02	Q = 100 cu. ft/sec
Q = 100	d = 2-in. schedule 40	v = 10 ft/sec

Source

Flow of Fluids through Valves, Fittings, and Pipe, Technical Paper No. 410, 3–7, Crane Company, Chicago, (1957).

Nomograph for calculating velocity of compressible fluids in pipes

The mean velocity of a compressible fluid in a pipe can be computed by the following formula, which is obtained after dividing the rate of flow in appropriate units by the cross-sectional area of the pipe:

$$V = \frac{3.06W\overline{V}}{d^2} = \frac{3.06W}{d^2\rho}$$

where: V = Mean velocity of flow, feet per minute
W = Rate of flow, in pounds per hour
$\overline{V}$ = Specific volume of fluid, cubic feet per pound
d = Internal diameter of pipe, inches
ρ = Density of fluid, pounds per cubic foot

Example. Steam at 600 psig and 850°F is to flow through a schedule 80 pipe at a rate of 30,000 lb/hr. Find the pipe size if the velocity is to be limited to 8,000 ft/min.

Connect	With	Mark or Read
850°F	600 psig line (vertically)	600 psig
600 psig	Horizontally to	$\overline{V}$ = 1.21
$\overline{V}$ = 1.21	W = 30,000	Index
Index	V = 8,000	d = 3.7, use 4-in. schedule 80 pipe

If a 4-in. schedule 80 pipe is used, the actual velocity is found by connecting the Index with 4-in. schedule 80 to get V = 7,600 ft/min.

Note: If a different fluid is involved, the value of the density (or specific volume) needed to make the calculation can be obtained from Section 9 of this book, under "Density and specific volume of gases and vapors," p. 226.

Source

Flow of Fluids through Valves, Fittings, and Pipes, Technical Paper No. 410, 3–16, Crane Company, Chicago, (1957).

Nomograph for calculating velocity of liquids in pipes

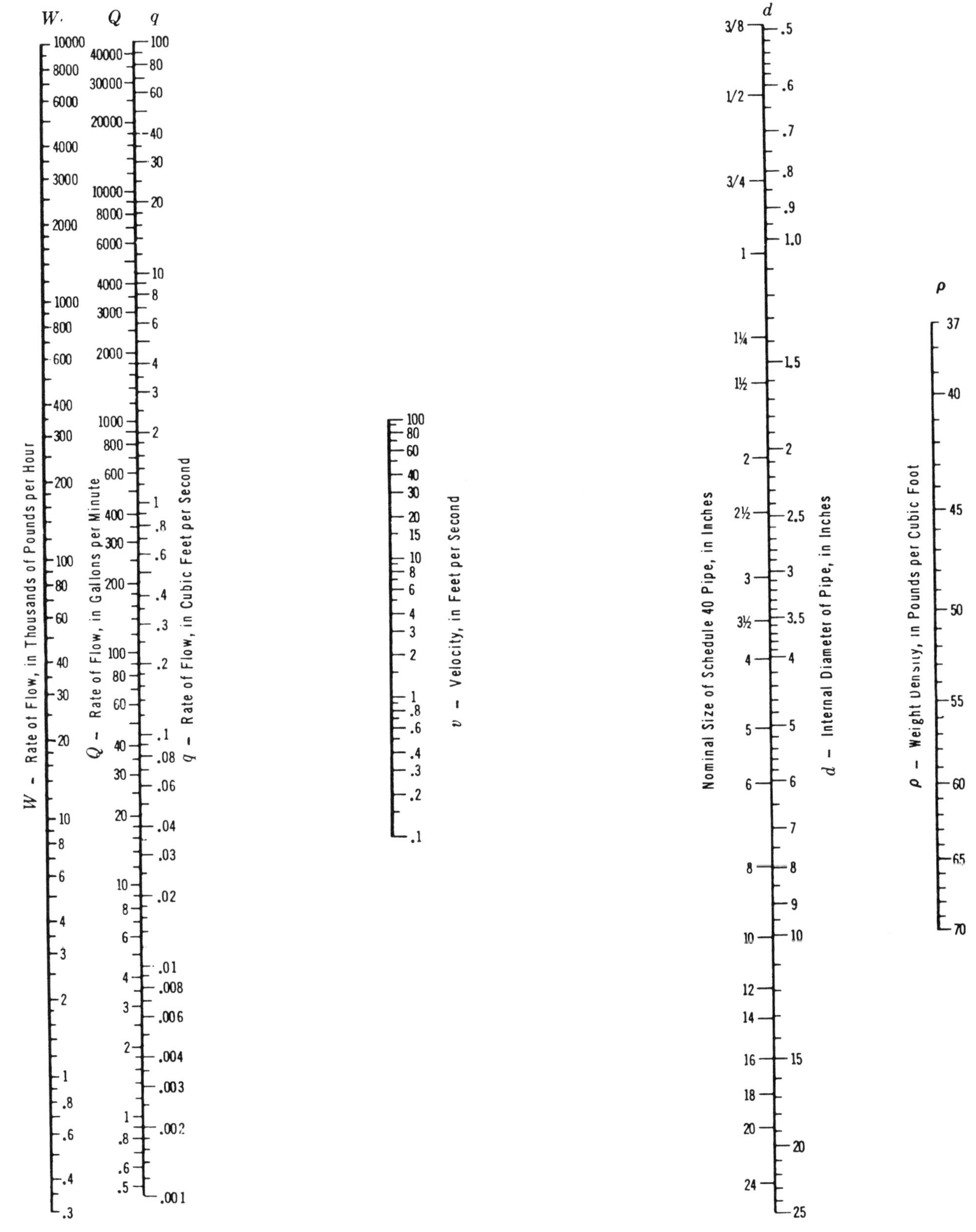

Figure 1. Velocity of liquids in pipe.

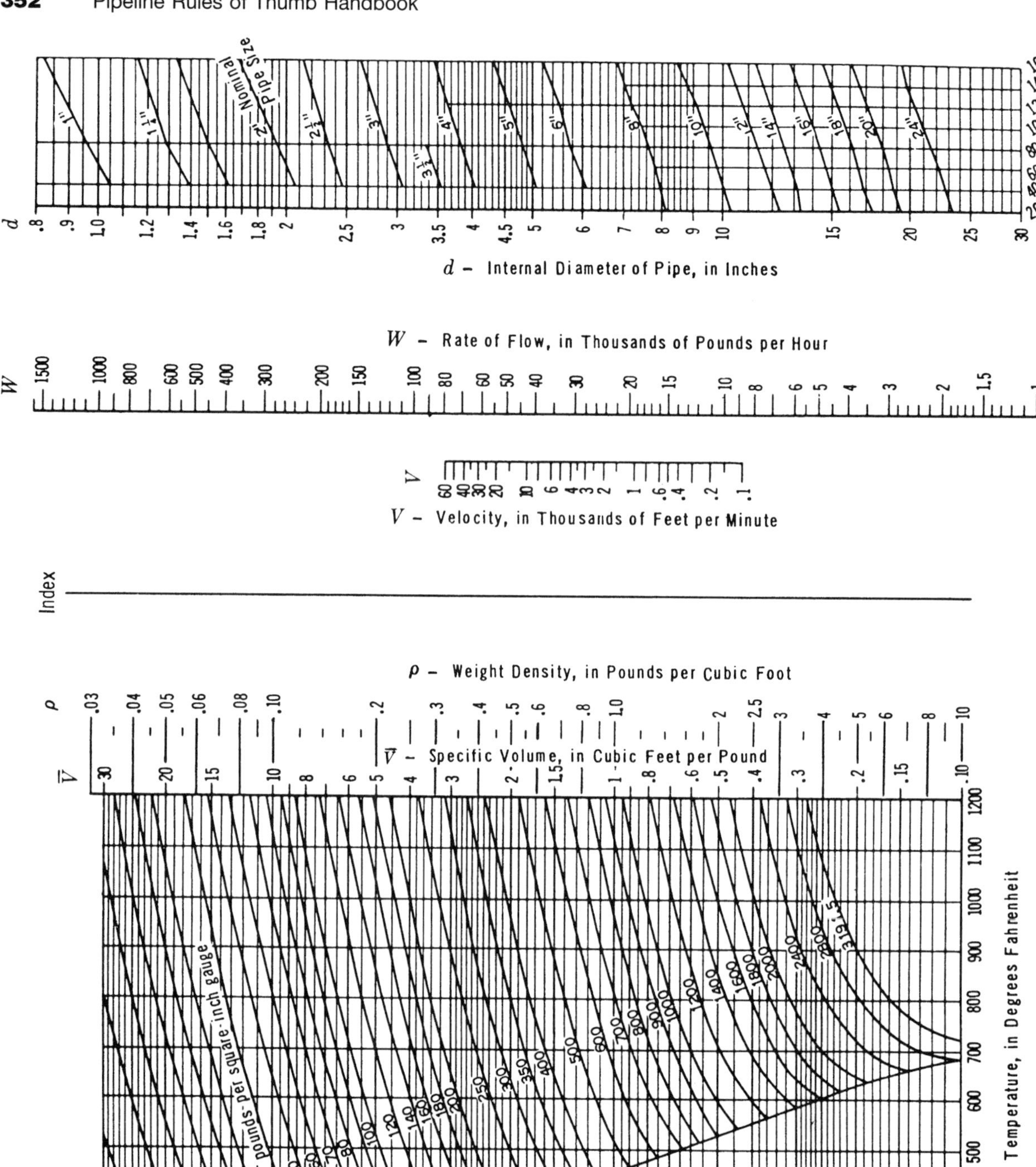

Figure 2. Velocity of compressible fluids in pipes.

Derivation of basic ultrasonic flow equations

The actual measurements performed by ultrasonic flowmeters are of flow velocity, with all other variable effects due to temperature, pressure, density, viscosity, etc., being cancelled via a differential sensing technique.

Multiplication by the cross-sectional pipe area readily enables the output to be conditioned to read in volumetric flow rate scaled to engineering units. Refinement of the technique enables density and other physical properties to be measured as well so that mass flow rates and other parameters pertaining to various process variables can be determined.

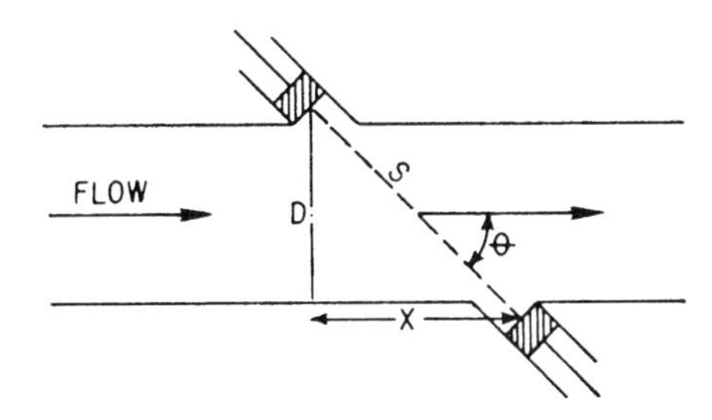

Let D = Pipe diameter
X = Axial displacement of transducers
S = Path length between transducers
θ = Angle of transducers to flow stream
c = Velocity of sound in the flowing medium
v = Velocity of flowing medium
t_1 = Flight time of upstream pulse
t_2 = Flight time of downstream pulse

$$\text{then } t_2 = \frac{S}{c + v\cos\theta} \quad (1)$$

$$t_1 = \frac{S}{c - v\cos\theta} \quad (2)$$

To cancel the effects of variations in the velocity of sound c in the medium itself, the circuitry is arranged to compute the reciprocal of each flight time and to subtract one from the other. This difference of the flight time reciprocals leads to the relationship:

$$\frac{1}{t_2} - \frac{1}{t_1} = \frac{2v\cos\theta}{S} \quad (3)$$

Solving for v

$$v = \frac{S}{2\cos\theta}\left(\frac{1}{t_2} - \frac{1}{t_1}\right) \quad (4)$$

$$\text{but } \cos\theta = \frac{X}{S} \quad (5)$$

$$\text{Therefore, } v = \frac{S^2}{2X}\left(\frac{1}{t_2} - \frac{1}{t_1}\right) \quad (6)$$

The circuitry that performs the reciprocal time computation produces an output frequency, since:

$$\frac{1}{t} = f \quad (7)$$

$$\text{Therefore, } \frac{1}{t_2} = f_2 \quad (8)$$

$$\text{and } \frac{1}{t_1} = f_1 \quad (9)$$

$$\text{and } \frac{1}{t_2} - \frac{1}{t_1} = f_2 - f_1 = \Delta f \quad (10)$$

Substituting in Equation 6 we have the basic velocity equation:

$$v = \frac{S^2}{2X}\Delta f \quad (11)$$

or

$$v = k_1\Delta f \quad (11a)$$

$$\text{The relationship } k_1 = \frac{S^2}{2X} \quad (12)$$

is the flow velocity scaling factor and is usually expressed in ft per second or ft per minute. Note that it is a function of the square of the distance between the transducers. This is an important consideration when providing for re-zeroing of the meter calibration each time the transducers are removed and re-inserted.

For volumetric flow, the velocity must be multiplied by the cross-sectional area of the pipe, and the following considerations are involved:

$$A = \pi\left(\frac{D}{2}\right)^2 \quad (13)$$

Volumetric flow

$$V - vA \quad (14)$$

Therefore, volumetric flow

$$V\pi = \frac{S^2 D^2 \Delta f}{8X} \quad (15)$$

The relationship

$$k_2 = \frac{S^2D^2}{8X}\pi \qquad (16)$$

is the volumetric flow scaling factor and is usually expressed in cubic ft per minute.

Further information on the physical properties of the flowing medium is also acquired in the process of the basic measurement. Specifically, the velocity of sound in the medium itself can be measured independent of the velocity of its motion.

Equation 3 computed the difference of the reciprocals of the upstream and downstream flight times. In the instrument this is performed in the delta computing channel. Supposing an addition of the reciprocals of the respective upstream and downstream flight times is performed in a separate computing channel, called the sigma channel. The following relationships ensue:

$$\frac{1}{t_2} + \frac{1}{t_1} = \frac{c + v\cos\theta}{S} + \frac{c - v\cos\theta}{S} = \frac{2c}{S} \qquad (17)$$

$$\text{then } c = \frac{S}{2}\sum f \qquad (18)$$

$$\text{Let } k_3 \frac{S}{2} \qquad (19)$$

be the sound velocity scaling factor.

$$\text{Then } c = k_3 \sum f \qquad (20)$$

The speed of sound c in a medium is a function of its bulk modulus and its density and is expressed by the following equation:

$$c = \sqrt{\frac{B}{\rho}} \qquad (21)$$

where: B = Bulk modulus of elasticity
ρ = Density

From this, the equation for density may be derived thus:

$$c^2 = \frac{B}{\rho} \qquad (22)$$

$$\text{and } \rho = \frac{B}{c^2} \qquad (23)$$

Substituting the value of c from Equation 20 into 23 gives

$$\rho = \frac{B}{(k_3 \sum f)^2} \qquad (24)$$

For a given gas, B is essentially constant.

$$\text{Therefore, } \frac{B}{k_3 2} = k_4 \qquad (24a)$$

is a fair representation

$$\text{Then } \rho\,(\text{density}) = k_4 \sum f^{-2} \qquad (25)$$

An interesting relationship may be derived at this point, namely, an equation for mass flow. Since

$$\text{mass flow} = \rho V \qquad (26)$$

$$\text{and since } V = k_1 \Delta f \qquad (11a)$$

$$\text{and } \rho = \frac{k_4}{\sum f^2} \qquad (25)$$

$$\text{then mass flow} = k_1 k_4 \frac{\Delta f}{\sum f^2} \qquad (27)$$

or, letting $k_1 k_4 = k_5$

$$\text{mass flow} = k_5 \frac{\Delta f}{\sum f^2} \qquad (28)$$

Further, the density of a gas is related by the equation

$$\rho = \frac{P}{T} \qquad (29)$$

where: P = Pressure
T = Temperature (°K)

Therefore, if the pressure is known, the temperature may be extracted from the sigma channel in this way:

$$\text{From Equation 25, } \frac{P}{T} = k_4 \sum f^{-2} \qquad (30)$$

$$\text{and } T = k_4 P^{-1} \sum f^{-2} \qquad (31)$$

Let us go back to Equation 24. If the chemical properties of the gas or the ratio of the mixture between two or more gases change, then B is not a constant. If separate transducers are used to measure density, say via the pressure/temperature relationship, then the variation of B can also be measured.

$$\text{Thus, } \rho = B(k_3 \sum f)^{-2} \qquad (32)$$

$$\text{and } B = \rho(k_3 \sum f)^2 \qquad (33)$$

The above relationship holds for liquids or gases.

In the case of a gas, we may further say:

$$B = \frac{P}{T}(k_3 \sum f)^2 \qquad (34)$$

How fast does oil move in a pipeline?

Divide the throughput, in barrels per day, by the square of the nominal diameter, in inches; divide this result by 100 and subtract 1 from the answer. The answer is the speed of the oil column in miles per hour.

Example. A 12-in. line has a throughput of 70,000 bpd

$70{,}000 \div 144 = 486$

$486 \div 100 = 4.86$

$4.86 - 1.00 = 3.86$

The oil moves about 4 miles per hour.

This rule is rough, but seldom is an exact answer needed for this problem. Greater accuracy can be had by using the true inside diameter of the line in the following formula:

$V = 0.0081 \times Q/d^2$

where: V = Speed, mph
Q = Throughput, bpd
d = Inside diameter of pipe, in.

Take the above example: If the wall thickness of the pipe is $\frac{3}{8}$ in.

$V = (0.0081)(70{,}000 \div 144)$

$V = 3.94$

or about 4 miles per hour.

Example. A 20-in. pipeline has a throughput of 200,000 barrels per day. How fast does the oil move inside the pipeline?

By the first method

$200{,}000 \div 400 = 500$

$500 \div 100 \div 1 = 4$ miles per hour

By the second method

$V = (.0081)(200{,}000 \div 19.25^2)$

$V = 4.37$ or about 4 miles per hour

Estimate the volume of a pipeline per linear foot using the inside diameter

To find the volume of a pipeline in gallons per ft, square the inside diameter in inches and multiply by 4%. A more accurate answer may be obtained by adding 2% of this result to the first answer.

Example. Find the volume of a pipeline having an ID of 7.981 in. expressed in gallons per ft.

$\text{Volume} = (0.04) \times (7.981)^2 = 2.548$ gallons per ft

For greater accuracy add 2% of 2.548.

$2.548 + 0.051 = 2.599$ gallon per ft

Actual volume = 2.599 gallons per ft

To find the volume of the pipeline in barrels per ft, square the inside diameter in inches and divide by 1,000. For greater accuracy subtract 3% of the first answer.

Example. Find the volume of a pipeline with an ID of 7.981 in. in barrels per ft.

$$\text{Volume} = \frac{(7.981)^2}{1{,}000} = 0.0637 \text{ barrels per ft}$$

For greater accuracy subtract 3% of 0.0637

$0.0637 - 0.0019 = 0.0618$ barrels per ft

Actual volume = 0.0619 barrels per ft

What is the linefill of a given pipe in barrels per mile?

Multiply the inside diameter of the pipe, in inches, by itself, and multiply the result by 5.13; the answer is the linefill in barrels per mile.

Example. Six-in. schedule 40 pipe; inside diameter is 6.065 in.

$6.065 \times 6.065 \times 5.13 = 188.7$ barrels per mile

If the correct inside diameter is used, the rule gives the correct linefill in standard 42-gallon barrels per mile of pipe.

Estimate leakage amount through small holes in a pipeline

The amount of fluid lost through a small hole in a pipeline can be estimated using the following equation:

$$Q = 0.61A\sqrt{2gh}$$

where:

Q = Flow in cu. ft./second
A = Cross sectional area, sq. ft.
g = Gravitational constant, ft/sec/sec
h = Head, feet

Example. Assume the following conditions:

Hole diameter = 0.125 in.
Pressure = 30 psig
Sp. gr. of fluid = 0.85

A = 0.0000852 sq. ft
h = 81.5 ft
g = 32.2

$$Q = 0.61 \times 0.0000852\sqrt{2 \times 32.2 \times 81.5}$$

$$Q = 0.002765 \text{ cu. ft/sec}$$

Table gives velocity heads for various pipe diameters and different rates of discharge

NOMINAL DIAMETER OF PIPE (Velocity Heads in Feet)

Rate of Flow in Gallons per Minute	1/8″	1/4″	3/8″	1/2″	3/4″	1″	1¼″	1½″	2″	2½″	3″	4″	5″	6″
1	.488	.148	.043	.017										
2		.590	.175	.069	.022									
3			.392	.155	.050	.020								
4			.698	.276	.090	.035	.012	.006						
5			1.089	.430	.141	.054	.018	.010						
10				1.720	.564	.215	.071	.038	.016	.007	.003			
15					1.265	.488	.160	.087	.036	.015	.007			
20					2.250	.861	.286	.154	.065	.027	.013			
25						1.345	.447	.241	.101	.041	.020			
30						1.933	.643	.346	.146	.060	.029			
35						2.635	.877	.472	.198	.082	.039			
40						3.343	1.145	.618	.259	.106	.052	.016		
45							1.548	.779	.329	.134	.065	.021		
50							1.787	.963	.406	.166	.080	.026		
70							3.503	1.888	.795	.326	.157	.050	.020	
75								2.165	.912	.373	.180	.057	.023	
100								3.851	1.621	.665	.320	.101	.041	.020
120								5.548	2.333	.956	.462	.146	.060	.029
125									2.528	1.035	.498	.158	.065	.031
150									3.649	1.493	.736	.229	.093	.045
175									4.954	2.021	.980	.309	.127	.061
200									6.483	2.656	1.281	.406	.166	.080
225										3.359	1.618	.514	.209	.101
250										4.131	1.992	.633	.259	.125
270											2.333	.740	.302	.146
275											2.429	.769	.313	.151
300											2.883	.912	.373	.180
350												1.242	.509	.246
400												1.621	.665	.320
450												2.053	.839	.406
470												2.239	.917	.443
475												2.276	.934	.453
500												2.535	1.038	.502
550													1.256	.606
600													1.493	.721
650													1.752	.846
700													2.035	.980
750													2.337	1.123
800														1.281
850														1.548
900														1.618
950														1.813
1000														1.999
1050														2.202
1100														2.421
1150														2.648
1200														2.880
1250														3.135

Rate of Flow in Gallons per Minute	8″	10″	12″	14″	15″	16″	20″	24″	30″	36″	42″
250	.040										
270	.047										
275	.048										
300	.057										
350	.078										
400	.102										
450	.129	.053									
470	.141	.057									
475	.144	.059									
500	.159	.065	.031								
550	.193	.079	.038								
600	.229	.093	.045								
650	.269	.109	.053	.026							
700	.312	.127	.062	.031							
750	.358	.146	.071	.035							
800	.407	.166	.080	.040							
850	.462	.187	.090	.045							
900	.516	.209	.102	.050							
950	.574	.234	.113	.056	.043						
1000	.637	.259	.125	.062	.048						
1050	.703	.296	.138	.069	.053						
1100	.771	.313	.152	.075	.058						
1150	.841	.343	.166	.082	.064						
1200	.917	.373	.181	.090	.069						
1250	.995	.406	.196	.097	.075	.062					
1500	1.433	.582	.282	.145	.108	.088					
2000	2.547	1.035	.502	.249	.193	.158					
2500		1.618	.784	.391	.301	.246					
3000		2.329	1.128	.562	.434	.355	.147				
3500		3.224	1.540	.700	.590	.502	.199	.096			
4000			2.007	1.000	.771	.633	.261	.125	.051		
4200			2.209	1.102	.851	.698	.288	.137	.056		
4500			2.539	1.265	.975	.799	.330	.158	.065		
5000			3.135	1.561	1.204	.985	.407	.195	.079		
5500				1.884	1.460	1.196	.493	.227	.096		
6000				2.239	1.733	1.421	.586	.281	.115	.056	
6500				2.627	2.035	1.669	.690	.329	.134	.065	
7000				3.056	2.360	1.936	.797	.384	.156	.076	
7200					2.709	2.056	.846	.404	.165	.080	
7500					3.047	2.224	.917	.440	.180	.087	
8000							1.045	.498	.204	.099	
8500							1.177	.564	.230	.112	
9000							1.319	.633	.259	.125	
9500							1.469	.702	.289	.140	
10000							1.630	.779	.319	.154	.083
10500								.858	.352	.169	.092
11000								.943	.387	.186	.101
11500								1.033	.423	.204	.110
12000								1.123	.460	.222	.120
12500								1.218	.500	.240	.130
13000								1.316	.541	.260	.141
13500								1.421	.582	.281	.152

F. G. Switzer "Pump-Fax," Goulds Pumps, Inc.

Viscosities of hydrocarbon liquids

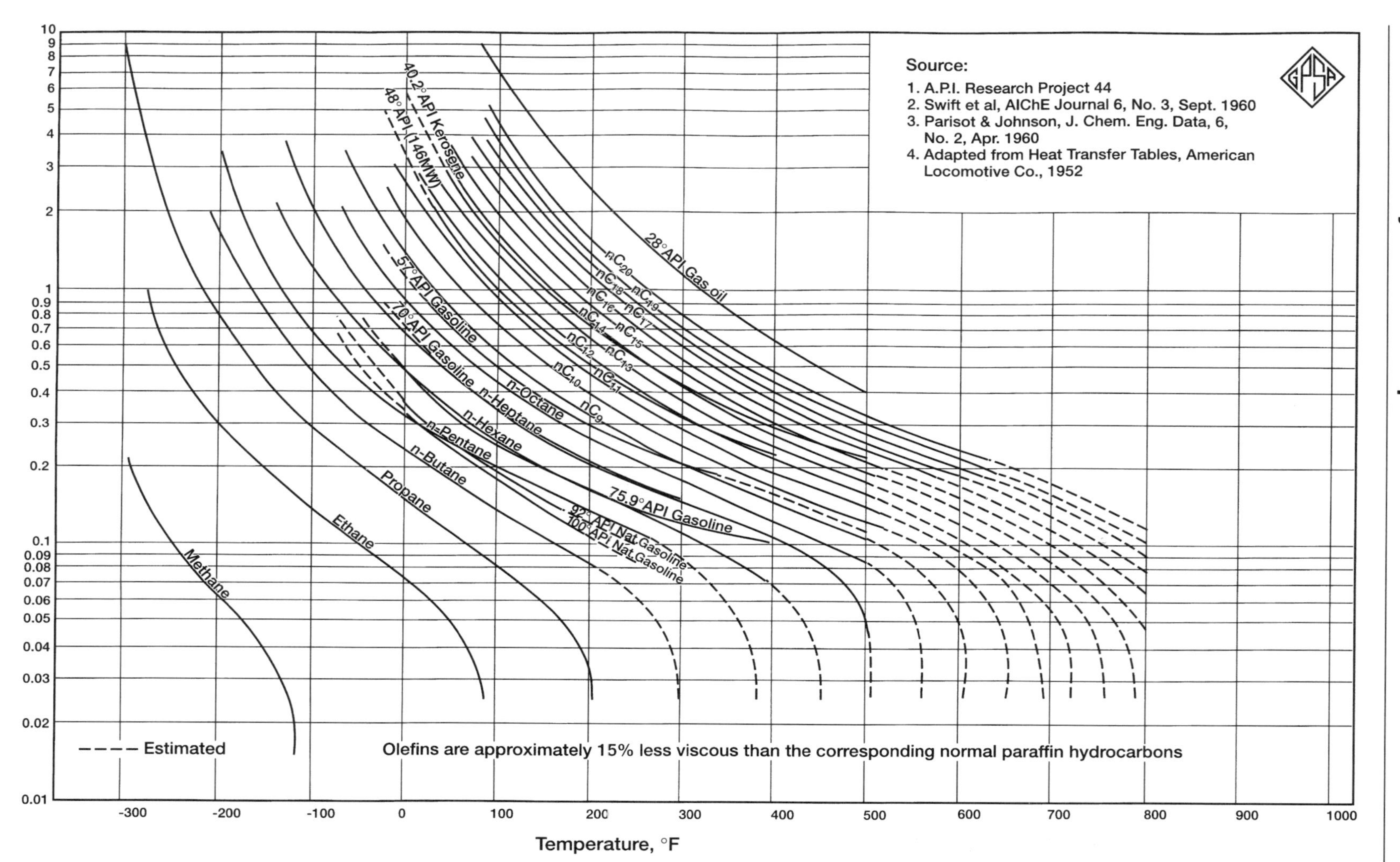

Reprinted with Permission—Gas Prrocessors Suppliers Association, *Tenth Edition Engineering Data Book.*

13: Liquids—Hydraulics

MARINE HOSE DATA

This section presents data on how to calculate the length of hose, and pressure loss, through marine hose.

CALM system CATENARY ANCHOR LEG MOORING

The overall length of the floating hose string is designed as follows:

$$TL = C + RX(1 + E) + B/2 + L + H + M + G \qquad (1)$$

where:

TL = Required overall length
C = Diameter of buoy
R = Length of mooring hawser
E = Max. elongation of mooring hawser
B = Beam of tanker
L = Distance between bow and manifold
H = Maximum (light ballast) freeboard (approx. 0.8 × tanker depth)
M = Distance between rail and manifold
G = Height of manifold from deck

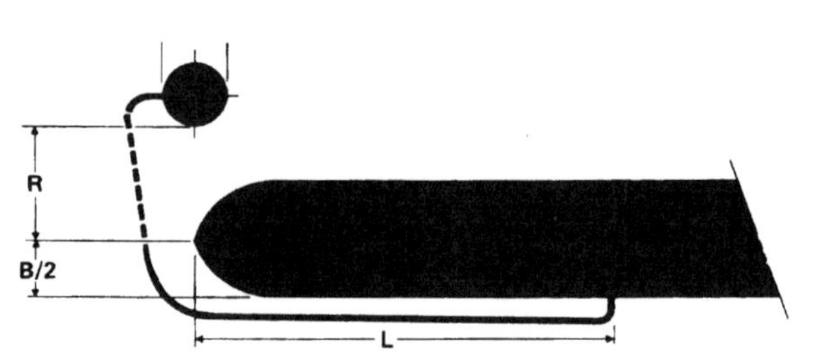

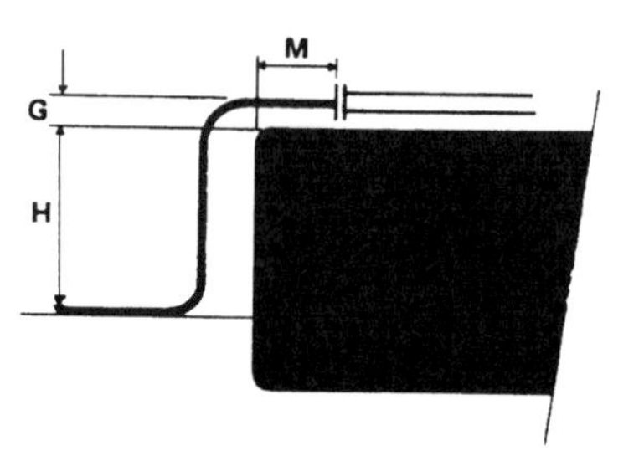

SALM system SINGLE ANCHOR LEG MOORING

The required overall length of hose string is designed as follows:

$\ell_1 < R$

where:

ℓ_1 = Distance between hose arm* and floating hose
R = Length of mooring hawser
$TW = \sqrt{\ell_1 + W^2}$
W = Maximum water depth at hose arm*

$$TL = HM + RX(1 + E) + TW + (B/2) + L + H + M + G + \ell_1 \qquad (2)$$

where:

TL = Required overall length of hose string
HM = Max. horizontal movement of buoy
R = Length of mooring hawser
E = Max. elongation of mooring hawser
B = Beam of tanker
L = Distance between bow and manifold
H = Maximum light (ballast) freeboard (approx. 0.8 × tanker's depth)
M = Distance between rail and manifold
G = Height of manifold from deck

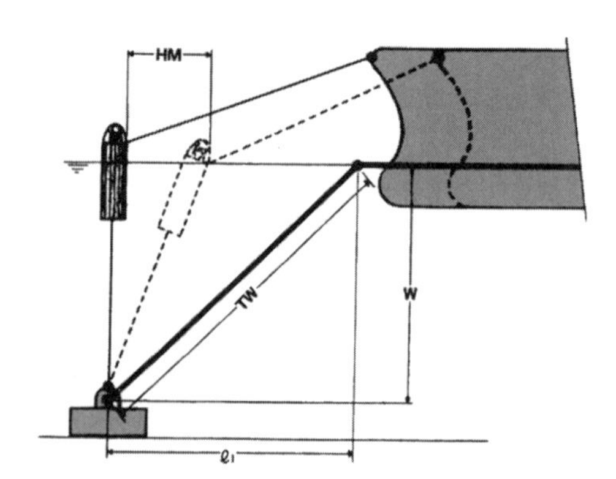

* Hose arm—the horizontal (I_1) or the vertical (W) distance between the underwater hose to swivel connection and the commencement of the main floating line.

Reprinted with permission Bridgestone Engineered Products Co.

Tandem system

The required overall length of hose string is designed as follows:

$$TL = LF + HF + R + (1 + E) + LT + HT + MT + GT \quad (3)$$

where:

TL = Maximum water depth at hose arm
LF = Distance between manifold and stern of floating storage vessel (FSV)
HF = Max. height of manifold of FSV from sea-surface
R = Length of mooring hawser
E = Max. elongation of mooring hawser
LT = Distance between bow and manifold of off-take-tanker (OTT)
HT = Maximum light (ballast) freeboard of OTT (approx. 0.8 × tanker's depth)
MT = Distance between rail and manifold of OTT
GT = Height of manifold from deck of OTT

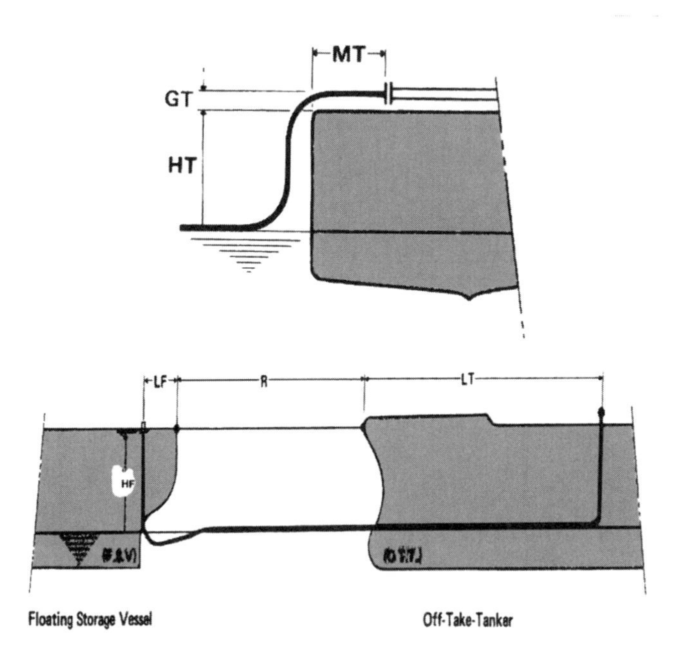

Multi-point mooring system

The required overall length of hose string is designed as follows:

$$TL = 1.5 \times L + D + H + G + M + \sqrt{(FA/2)^2 + (SW/2)^2} \quad (4)$$

where:

TL = Required overall length
L = Distance between PLEM and tanker hull
D = Maximum water depth
H = Maximum light (ballast) freeboard (approx. 0.8 × tanker's depth)
M = Distance between rail and manifold
G = Height of manifold from deck
FA = Max. fore and aft excursion of tanker
SW = Max. sideways excursion of tanker

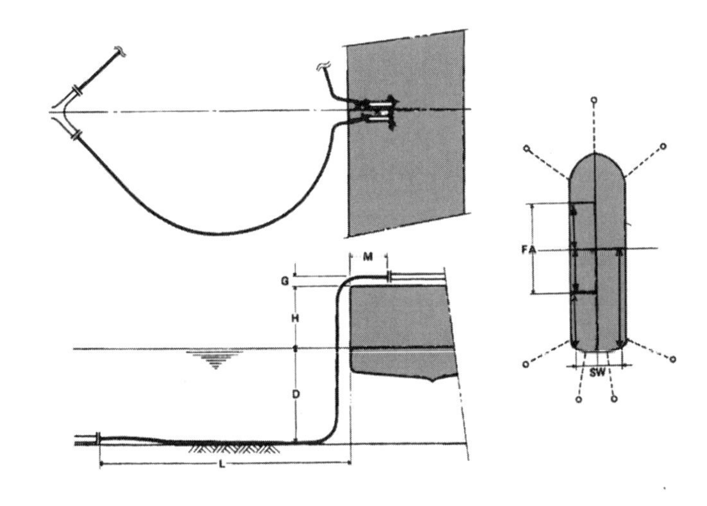

Reprinted with permission Bridgestone Engineered Products Co.

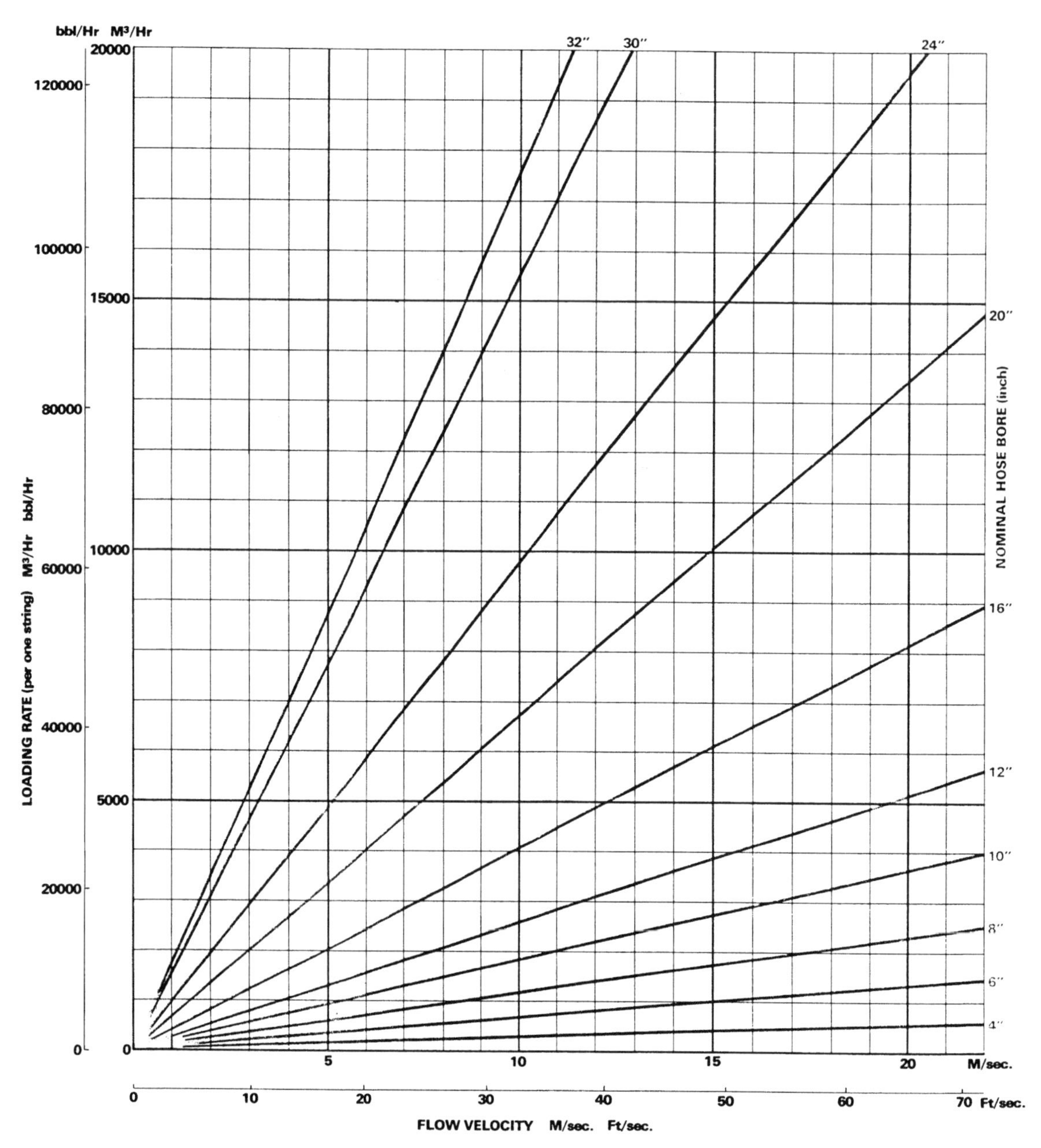

Figure 1. Loading rate and flow velocity.

Pressure loss in hose string

The pressure loss in a hose string is calculated as follows:

$$\Delta P = \frac{\lambda \times L \times V^2 \times \gamma \times 10^{-4}}{2 \times g \times D} \quad (1)$$

where:

ΔP = Pressure loss (Kg/cm^2)
λ = Friction factor
L = Length of hose string (m)
V = Flow velocity (in/sec)
γ = Specific weight of oil (Kg/m^3)
= (1000 × specific gravity)
g = Acceleration of gravity (in/sec^2)
D = Inside diameter of hose (m)

λ is calculated by using the following Mises' experimental formula:

$$\lambda = 0.0096 + 5.7 \times \sqrt{K/D} + 1.7 \times \sqrt{1/Re} \quad (2)$$

where:

K = Coefficient of wall roughness by Mises ' experiment (m)
K = 0.3×10^{-6} for rubber hose
Re = Reynolds number

Re is calculated as follows:

$$Re = \frac{D \times V}{\nu} \quad (3)$$

where:

ν = Kinematic viscosity (m^2/sec)

Input Data:

Viscosity and specific weight of oil depend on temperature, and the values at the operating temperature sometimes need to be converted from values at other given temperatures.

Specific weight can be converted from other units as follows:
Specific weight (Kg/m^3) = 1000 × specific gravity

$$\text{Specific gravity} = \frac{1415}{1315 + \text{Deg.API}} \quad (4)$$

Specific gravity at operating temperature can be calculated as follows:

S.G. operating = S.G. measured − A × (Temp. operating − Temp. measured) *Note:* Temp: °C. Coefficient of S.G. change "A" depends on the specific gravity of the oil and is given in Table 1.

Kinematic viscosity is converted from other units as follows:

1 cSt = 1 CP/S.G. (cP is Absolute Viscosity)
1 cSt = 0.245 × R (Redwood Sec.)
1 cSt = 7.60 × E (Engler Deg.)
1 cSt = 0.220 × S (Saybolt Universal Sec.)

Conversion of kinematic viscosity at a different temperature depends on the characteristics of the cargo and can be calculated from data at two different temperatures.

$$\log_{10} \log_{10}(n + 0.6) = \nu - m \times \log T \quad (5)$$

where:

T = oK
ν = cSt

Flow velocity is determined by the internal diameter of the hose and the loading rate as follows:

$$V = \frac{4 \times Q}{3600 \times \Pi \times D^2} \quad (6)$$

where:

V = Flow velocity (m/sec)
Q = Loading rate (m^3/hr)
D = Inside diameter of hose (m)

Table 1
Change of specific gravity of oil

S.G. at meas. temp.	Coeff. "A"	S.G. at meas. temp.	Coeff. "A"
0.640–0.643	0.00095	0.754–0.759	0.00078
0.644–0.648	0.00094	0.760–0.765	0.00077
0.649–0.654	0.00093	0.766–0.771	0.00076
0.655–0.661	0.00092	0.772–0.777	0.00075
0.662–0.667	0.00091	0.778–0.783	0.00074
0.668–0.674	0.00090	0.784–0.790	0.00073
0.675–0.681	0.00089	0.791–0.799	0.00072
0.682–0.688	0.00088	0.800–0.808	0.00071
0.689–0.696	0.00087	0.809–0.818	0.00070
0.697–0.703	0.00086	0.819–0.828	0.00069
0.704–0.711	0.00085	0.829–0.838	0.00068
0.712–0.719	0.00084	0.839–0.852	0.00067
0.720–0.726	0.00083	0.853–0.870	0.00066
0.727–0.734	0.00082	0.871–0.890	0.00065
0.735–0.741	0.00081	0.891–0.970	0.00064
0.742–0.747	0.00080	0.971–1.000	0.00063
0.748–0.753	0.00079		

Example:

Conditions:

Loading rate:	40,000 bbl/hr.
Hose size:	20″ & 16″
Hose length:	100 m for 20″ hose 50 m for 16″ hose
Operating temp.:	30°C
Specific gravity of oil	34.4° API at 60°F
Kinematic viscosity:	6.90 cSt at 30°C

Calculations:

Loading rate: $Q = 40{,}000 \times 0.15899 = 6{,}360\ m^3/hr$
Flow velocity for 20″ hose:

$$V_{20''} = \frac{4 \times 6{,}360}{3600 \times \Pi \times 0.486^2} = 4.83\ (m/sec)$$

$$V_{16''} = \frac{4 \times 6{,}360}{3600 \times \Pi \times 0.380^2} = 5.92\ (m/sec)$$

Specific weight:

$$S.G._{60°F} = \frac{141.5}{131.5 + 34.4} = 0.853\ (\text{at } 15.56°C)$$

$$S.G._{30°C} = 0.853 - 0.00066X(30 - 15.56) = 0.843$$

Specific weight = 843 (Kg/m^3)

Kinematic viscosity:

$$\nu = 6.9\ cSt = 6.9 \times 10^{-6} \times (m^2/sec)$$

Reynolds number:

$$Re = \frac{D \times V}{\nu}$$

$$Re_{20''} = \frac{0.486 \times 9.52}{6.9 \times 10^{-6}} = 670{,}539$$

$$Re_{16''} = \frac{0.380 \times 15.6}{6.9 \times 10^{-6}} = 859{,}130$$

Friction factor:

$$\lambda = 0.0096 + 5.7 \times \sqrt{\frac{K}{D}} + 1.7 \times \sqrt{\frac{1}{Re}}$$

$$\lambda_{20''} = 0.0096 + 5.7 \times \sqrt{\frac{0.3 \times 10^{-6}}{0.486}} + 1.7 \times \sqrt{\frac{1}{670539}} = 0.01615$$

$$\lambda_{16''} = 0.0096 + 5.7 \times \sqrt{\frac{0.3 \times 10^{-6}}{0.380}} + 1.7 \times \sqrt{\frac{1}{859130}} = 0.1650$$

Pressure loss:

$$\Delta P = \frac{\lambda \times L \times V^2 \times \gamma \times 10^{-4}}{2 \times g \times D}$$

$$\Delta P_{20''} = \frac{0.01615 \times 200 \times 9.52^2 \times 843 \times 10^{-4}}{2 \times 9.81 \times 0.486} = 2.59\ (Kg/cm^2)$$

$$\Delta P_{16''} = \frac{0.01650 \times 50 \times 15.6^2 \times 843 \times 10^{-4}}{2 \times 9.81 \times 0.380} = 2.27\ (Kg/cm^2)$$

Total pressure loss = $2.59 + 2.27 = 4.86\ (Kg/cm^2)$

Pressure drop calculations for rubber hose

Fluid flow

The general flow formula is as follows:

$$\Delta P = \frac{0.0134 \times f \times S \times L \times Q^2}{d^5}$$

where: ΔP = Pressure drop, psi
f = Friction factor
S = Specific gravity
L = Length of line, ft
Q = Flow, gpm
d = Inside diameter of line, in.

$$N_R = \frac{3{,}160 \times Q}{V_C \times d}$$

where: N_R = Reynolds number
Q = Flow, gpm
V_C = Fluid viscosity, centistokes
d = Inside diameter of line, in.

Friction factor, f.

For laminar flow, $f = \dfrac{64}{N_R}$.

Source

Gates Rubber Company.

Examples of pressure drop calculations for rubber hose

Flow rate = Q = 1,000 gpm
Hose ID = d = 6.0 in.
Length = L = 100 ft
Fluid = Gasoline
Specific gravity = S = 0.71
Viscosity = V_C = 0.476 centipoise/0.71
= 0.6704 centistokes

$$\begin{aligned}\text{Reynolds number} = N_R &= [3{,}160 \times Q]/[V_C \times d] \\ &= [3{,}160 \times 1{,}000][0.6704 \times 6] \\ &= 785{,}600\end{aligned}$$

From Figure 1, f = 0.012 for N_R = 785,600

$P = [0.0134 \times f \times S \times L \times Q^2]/d^5$
$P = [0.0134 \times 0.012 \times 0.71 \times 100 \times 1{,}000{,}000]/7{,}776$
P = 1.468 psi per 100 ft

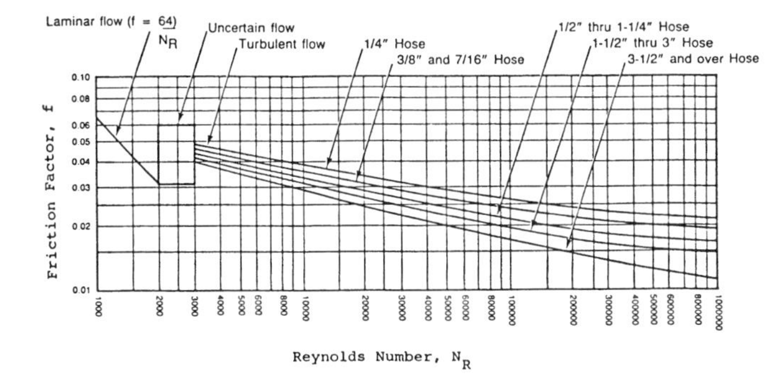

Figure 1. Friction factors for rubber hose. Reproduced with permission of Gates Rubber Company.

As a rule of thumb, loss for each set of couplings equals 5% of the loss per 100 ft of hose. For two sets of couplings,

Pressure loss = $2 \times 0.05 \times 1.468 = 0.15$ psig

Typical formulas used for calculating pressure drop and flow rates for pipelines

Shell/MIT

$P = (0.241\, fsQ^2)/d^5$

For viscous flow,

$f = 0.00207 \times (1/r)$

For turbulent flow,

$f = 0.0018 + 0.00662 \times (1/r)^{0.355}$
$r = 0.0119Q/dv$

Viscous flow,

r = 0.1 to 0.135

Turbulent flow,

$r > 0.4$

Indeterminate region,

$r = 0.135$ to 0.4

Benjamin Miller

$$B = (0.1692)(d^5P/s)^{0.5} (\log(d^3sP/z^2) + 4.35)$$

T. R. Aude

$$B = 0.871K(1/z^{0.104})((P^{0.552}d^{2.656})/s^{0.448})$$

$$P = ((Bz^{0.104}s^{0.448})/(0.871\, Kd^{2.656}))^{1.812}$$

Hazen-Williams

$$Q = (0.148\, d^{2.63}P^{0.54})/s^{0.54}$$

Notation.

Q = Barrels per day
B = Barrels per hour
P = Pressure loss, psi per mile
d = Inside diameter of pipe, in.
z = Absolute viscosity, centipoises
s = Specific gravity
K = Efficiency factor (Aude's formula)
f = Friction factor
r = Reynolds number modified Re/7,742
C = Hazen-Williams coefficient (use 120 for new steel pipe; 100 for ordinary 15-year-old steel pipe; and, 150 for fiberglass pipe)
SUS = Saybolt Universal Seconds
v = Kinetic viscosity, centistokes

Examples

Shell/MIT. Find the pressure loss, psi/mile, for a crude oil line where the flow rate is 30,000 barrels per day with the following conditions:

Pipe: 10¾ -in. × 0.307-in. wt
Specific gravity = 0.85
Viscosity = 80 SUS

Convert viscosity (80 SUS) to centistokes:

$$v = 0.22(\text{SUS}) - 180/\text{SUS}$$
$$= 0.22(80) - 180/80$$
$$= 15.35 \text{ centistokes}$$

$$r = 0.0119Q/dv$$
$$= (0.0119 \times 30{,}000)/(10.136 \times 15.35)$$
$$= 2.295$$

$$f = 0.0018 + 0.00662(1/r)^{0.355}$$
$$= 0.0018 + 0.00662(1/2.295)^{0.355}$$
$$= 0.00673$$

$$P = (0.241fsQ^2)/d^5$$
$$= (0.241 \times 0.00673 \times 0.85 \times 30{,}000^2)/10.136^5$$
$$= 11.6 \text{ psi/mile}$$

T. R. Aude. Find the pressure loss, psi/mile, when pumping gasoline with the following conditions:

B = 2,000 barrels per hour
Pipe: 8.625-in. × 0.322-in.wt
z = 0.59 centipoise
s = 0.733
K = 0.94 (pipe roughness/efficiency factor)

$$P = ((Bz^{0.104}s^{0.448})/(0.871\, Kd^{2.656}))^{1.812}$$
$$= ((2{,}000 \times 0.59^{0.104} \times 0.733^{0.448})/(0.871 \times 0.94 \times 7.981^{2.656}))^{1.812}$$
$$= 44.1 \text{ psi/mile}$$

Note: Aude's formula is based on data taken from 6-in. and 8-in. lines, and no extension data are available.

Benjamin Miller. Find the pressure loss, psi/mile, when pumping gasoline with the following conditions:

B = 4,167 barrels per hour
Pipe: 16.0-in. × 0.3125-in.wt
z = 0.59 centipoise
s = 0.733
d = 15.375 – in.

$$B = (0.169176)(d^5P/s)^{0.5}(\log(d^3sP/z^2) + 4.35)$$
$$4{,}167 = (0.169176)(15.375^5P/0.733)^{0.5} (\log (15.375^3 \times 0.733 \times P)/(0.59^2) + 4.35)$$

Let P = 6.4 psi/mile for first try:

$$4{,}167 = (0.169167)(15.375^5 \times 6.4)^{0.5}(\log(15.375^3 \times 0.733 \times 6.4)/(0.59^2) + 4.35)$$
$$4{,}167 = 0.169167 \times 2{,}738 \times (4.69 + 4.35)$$
$$4{,}167 \neq 4{,}187$$

Let P = 6.34 psi/mile for second try

$$4{,}167 = 0.169167 \times 2{,}726 \times (4.68 + 4.35)$$

$$4{,}167 = 4{,}167$$

Therefore, P = 6.34 psi/mile.

Hazen-Williams. (Typically used for water system hydraulic calculations.) Find the pressure drop, psi/mile, in a water system where the flow rate is 20,000 barrels per day.

The pipe is 4.5-in. × 0.226-in. wt new steel. Since the flowing medium is water, s = 1.0.

$$20{,}000 = (0.148 \times 4{,}048^{2.63} \times 120 \times P^{0.54})$$

$$P^{0.54} = 20{,}000/702.23$$

$$= 28.48$$

$$P = 493 \text{ psi/mile}$$

All of the preceding examples have assumed no difference in elevation. Where a difference in elevation exists, this must be considered in determining the total pressure required for a given flow rate.

Hydraulic gradients

Some of the pressure drop formulas used to calculate pressure drop and flow rates for pipelines are shown earlier in this section. In order to use these formulas, it is first necessary to determine the characteristics of the fluid to be transported—primarily the specific gravity, viscosity, and either the flow rate or allowable pressure drop.

Before the system design is completed, it is necessary to know the elevation of the terrain to be traversed by the pipeline. Once the elevation of the system is known and a pipe size has been selected, the pressure loss data may be used to calculate or plot a hydraulic gradient. A hydraulic gradient is the height that a column of fluid will rise to if small tubes were installed along the pipeline as shown in Figure 1. Feet of head may be converted to pressure by dividing by 2.31 and multiplying by the specific gravity. Conversely, pressure is converted to feet of head by multiplying pressure by 2.31 and dividing by the specific gravity of the fluid.

The slope of the hydraulic gradient will represent the amount of pressure drop per unit of linear distance. The pressure drop per unit of length is uniform as long as there is no change in pipe size or specific gravity, and therefore

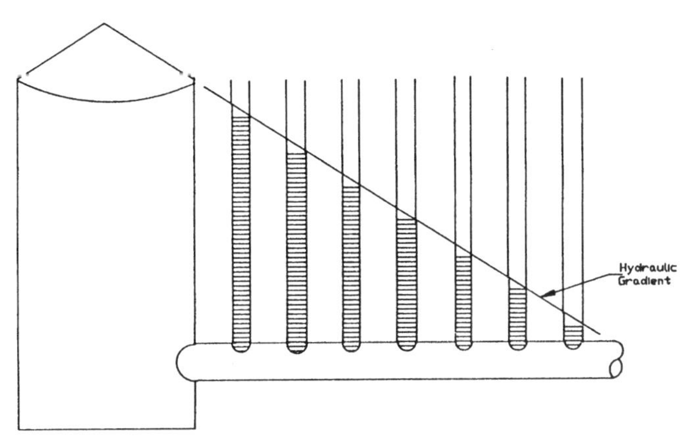

Figure 1. Schematic for hydraulic gradient.

the hydraulic gradient will be a straight line with a slope proportional to the system flow rate.

Since the pipeline elevation profile is usually plotted in feet, it is helpful to plot the hydraulic gradient in feet rather than pressure. The hydraulic gradient may be plotted by converting pressure at the origin and end points to feet using the specific gravity of the fluid in the pipeline. The end point in this case will be the end of the pipeline if there are no intermediate booster stations. If the system is equipped with intermediate booster stations, then the suction of the booster station in the first segment will be the end point for that segment of the line. Don't forget to consider the suction pressure requirement for the booster station. These two points are then connected with a straight line, which will represent the pressure (in feet) at any intermediate point along the pipeline. The pressure at any point may be determined by taking the difference between the gradient and pipeline elevation and converting feet to psi.

Figure 2 shows a typical elevation profile plot with two hydraulic gradients—one for no booster station and one for an intermediate booster station and no change in elevation. In Figure 3 there is an elevation change, and for a hydraulically balanced system, it was necessary to locate the booster station at a point other than the geographic mid-point of the line. The location of the booster station in this example corresponds to the hydraulic mid-point of the system. A hydraulically balanced system is one in which all stations develop the same differential head. In Figure 2 where there was no elevation change, the booster station location is at the geographic mid-point of the system, which also corresponds to the hydraulic mid-point of the system. This is not meant to imply that a system must absolutely be hydraulically balanced. A booster station may be located off the hydraulic center; however, its differential head requirement will differ from that of the origin station and system throughput capacity will be affected. The amount of difference in head required, for a booster station which is

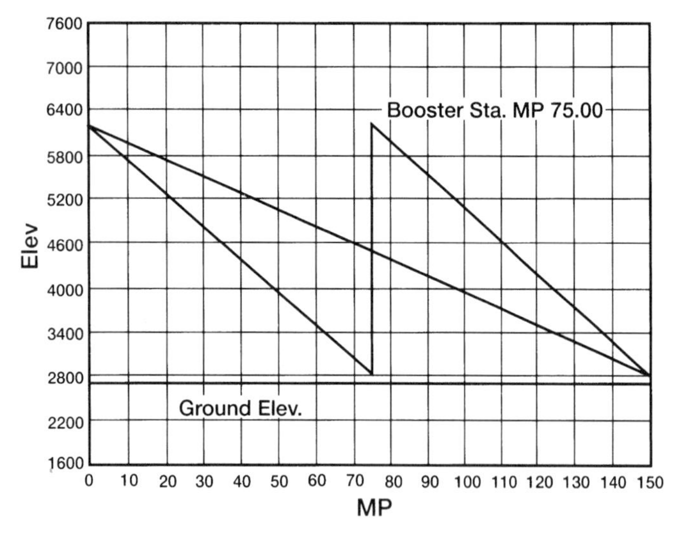

Figure 2. Profile with no elevation change.

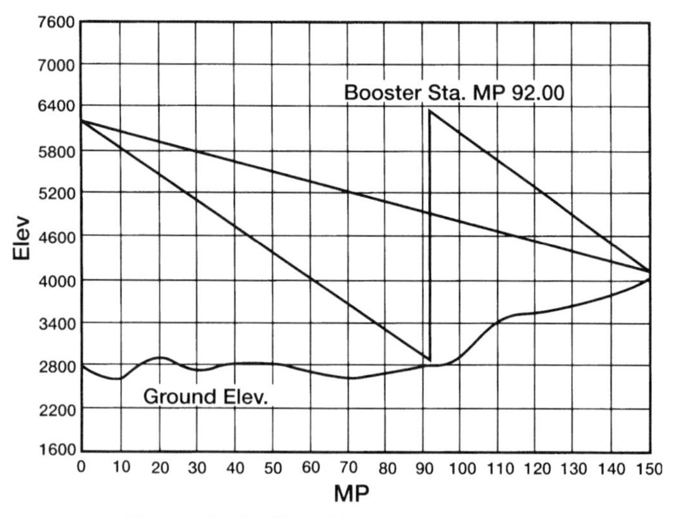

Figure 3. Profile with elevation change.

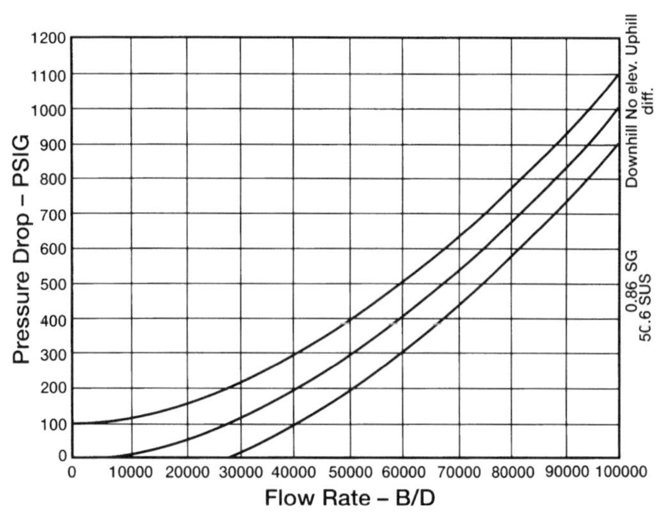

Figure 4. System curve. 30 mi., 12.750″ × 0.250″ pipe.

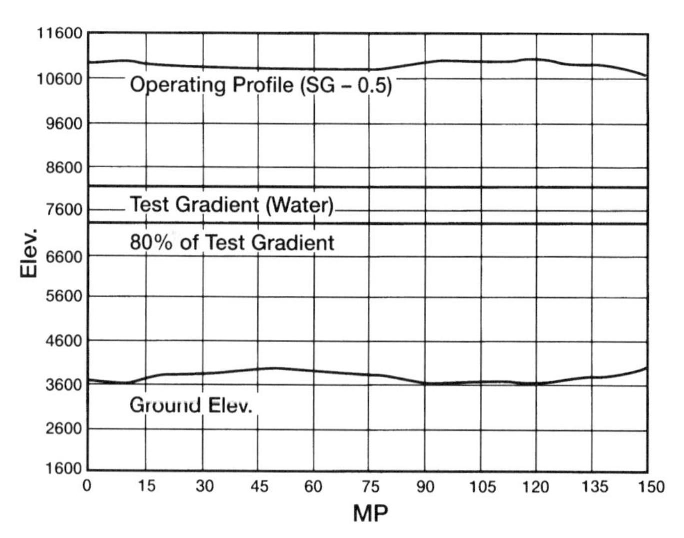

Figure 5. Pipeline elevation and test gradients and operating profile.

located off hydraulic center, will depend upon the amount of system elevation change and back pressure maintained at the end of the line.

The total system friction loss may be divided equally among an optimum number of booster stations. The discharge pressure at each station may be calculated with the following equation:

$$Pdp = (Pt + S + (N - 1) \times Ps + Pr)/N$$

where:

Pdp = Station discharge pressure
Pt = Total system friction loss
S = System static pressure
N = Number of booster stations
Ps = Station suction pressure
Pr = Residual or back pressure at end of line

In Figure 4, three system curves are shown to illustrate how elevation will affect the capacity of a system. With a constant input pressure of 750 psig, the capacity will be 77,500 barrels per day pumping uphill (272 ft); 84,000 barrels per day with no elevation change; and 90,000 barrels per day pumping downhill (272 ft). Note that where there is a difference in elevation, the amount of displacement of the curves, 100 psi in this case, is uniform.

The hydraulic gradient may also be used to analyze a system to assure that it is operated within the pressure limits determined by the hydrostatic test and the pipe design parameters. Figure 5 shows the test and maximum allowable operating gradients for a system. The hydrostatic test point was located at MP 150, and the test pressure was 1,800 psig, which in this case corresponds to 90% of the specified minimum yield strength (SMYS) of the pipe. Water was used as the test medium, and therefore 1,800 psig equates to 4,157 ft of head. The test gradient is drawn as a straight line across the profile from 8,157 ft (4,157 ft of head + 4,000 ft of pipeline elevation). This system may be operated at 80% of the test pressure, since this represents 72% of the SMYS of the pipe. See Section 3 for details on pipe design and limits on maximum allowable operating pressure.

The 80% gradient (still based on water) is drawn as a straight line across the profile from 7,325 ft. The maximum allowable operating profile is now developed by taking the difference between the 80% of test pressure gradient, in feet, and the pipeline elevation and converting this difference to feet of head for the fluid to be transported. For this example, a specific gravity of 0.5 was used. At MP 150, the difference between pipeline elevation and the 80% of test gradient is 3,325 ft of head (water), and for a specific gravity of 0.5, this will convert to 6,650 ft of head. Therefore, the maximum operating pressure in ft of head for a fluid with a specific gravity of 0.5 will be 10,650 ft (6,650 ft of head + 4,000 ft pipeline elevation) at MP 150.

The maximum allowable operating profile will not be a straight line but will be a displaced mirror image of the pipeline elevation profile. The remainder of the maximum allowable operating profile may be developed by following this procedure at other points along the pipeline. When fluids with different specific gravities are transported, an operating profile will need to be developed for each specific gravity. If the line is hydrostatically tested in several segments rather than in one continuous length, an operating profile will need to be developed for each of the test segments.

Once the maximum allowable operating profile has been established, the hydraulic gradient may be superimposed. If the hydraulic gradient crosses the maximum allowable operating profile, the pipe will be overpressured in the segments where the gradient crosses the profile.

Figure 6 illustrates the effect hills have on the operation of a system. In this case, a hill is located near the end of the line and, due to its elevation, it will control the system capacity, since adequate back pressure must be maintained (if the vapor pressure of the fluid is above atmospheric pressure) at the end of the line to assure the gradient will "clear" the hill. If no back pressure is maintained at the end of the line, the gradient will intersect the side of the hill and, theoretically, a siphon action will occur if the elevation difference is not too great. In reality, the system flow rate will change of its own

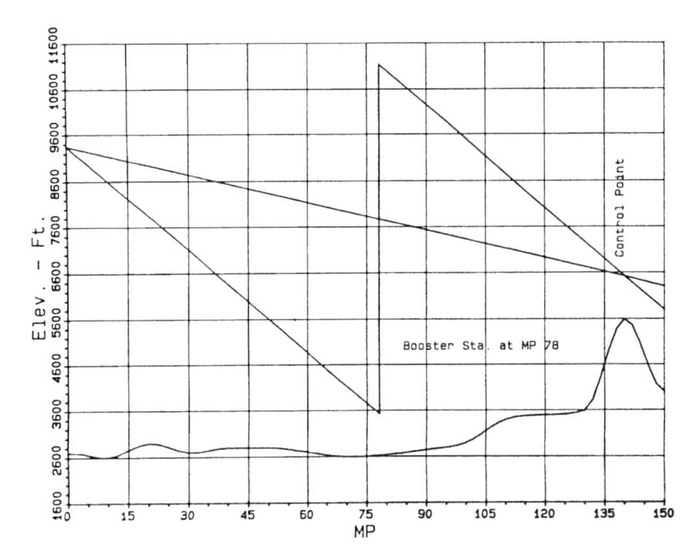

Figure 6. Pipeline elevation profile and hydraulic gradients.

accord to cause the gradient to intersect the top of the hill. The flow from that point to the end of the line will cascade or run away from the flow on the other side of the hill. If the vapor pressure of the fluid is very low, this condition can probably be tolerated. If segregated batches of material are handled or if the fluid vapor pressure is above atmospheric pressure, adequate back pressure must be maintained to keep the batches from mixing or forming vapor. Inadequate back pressure will cause difficulties in getting the flow in versus flow out to balance on a short-term basis as would be required for a leak detection system.

Two conditions are illustrated in Figure 6 to show the effect the flow rate will have on the amount of back pressure required to clear the hill. Some safety factor should be included when determining the amount of pressure to maintain in the system—especially when handling high vapor pressure materials. The flowing temperature will also need to be considered in making this determination.

For this example, the fluid has a specific gravity of 0.5 and its vapor pressure requires a minimum of 200 psig in the system. Therefore, it will be necessary to maintain 200 psig (924 ft) at the top of the hill located at MP 140. All gradients must be drawn through this point. If the specific gravity changes, the control point will move up or down depending upon the direction of the specific gravity change. The gradient connecting MP 0 to the end of the line passes through the control point and intersects the end of the line at 6,300 ft or 498 psig. The second condition, which assumes a booster station at MP 78, illustrates what happens to the back pressure requirement when the flow rate changes. The gradient from the discharge of the booster station must also pass through the control point, and since the flow rate is greater, the slope of the gradient is steeper and intersects the end of the line at 5,800 ft or 390 psig. It follows that as the system flow rate changes, it will be necessary to adjust the amount of back pressure. If the back pressure is permanently set to the higher value, system capacity will be penalized. See Section 16: Instrumentation, for a discussion of control systems.

Equivalent lengths

Some pipeline systems are not made up of one single uniform pipe size. As the pressure in a system decreases, it may be desirable to use pipe with a thinner wall thickness to reduce the investment cost, or where deliveries are made along the pipeline, it is sometimes practical to use a smaller diameter pipe for the lower flow rate. Conversely, where injections are made into a system, it may be necessary to use a larger diameter pipe for the increased flow rate.

The capacity of a system may be increased by installing loop lines. Loop lines are two or more pipelines that are operated in parallel. Loop lines are not practical where segregated batches are pumped, since there will be a problem in getting the batches to arrive simultaneously. Also, if there are injections or deliveries along such a system, it will be almost impossible to balance the amount injected or delivered between the two lines. The most practical method for operating loops where segregated batches are handled is to operate each line separately. A product system might, for example, use one line for distillates and the other line for gasoline. If the gasoline volume is too large for one line, the excess could be batched through the other system. One other caution on loops is that the lengths of the legs may not be exactly the same.

Hydraulic systems are analogous to electrical circuits as shown in Figure 1, and this sometimes helps in analyzing a pipeline system. In a series electrical circuit, the same current flows through all of the circuit elements, and the voltage drops across the elements are additive. This same logic may be applied to a pipeline system with segments of different pipe sizes connected in a series. The pressure drops through the various segments are additive, and the sum of the pressure drops will equal the total system pressure drop.

A loop system is analogous to a parallel electrical circuit as shown in Figure 1. Since a common pressure is applied to all the legs of the loop system, the pressure drop across each leg of a loop system must be the same and the flow will divide in proportion to the friction loss in each leg.

An examination of the Shell/MIT flow formula reveals that the friction factor is dependent upon flow rate, pipe ID, and viscosity. Since the flow through each leg of a loop system will change as the total system flow changes, this will have an impact on the calculated equivalent length factor. For best accuracy, the equivalent length factor should be determined as near as possible to the expected operating parameters.

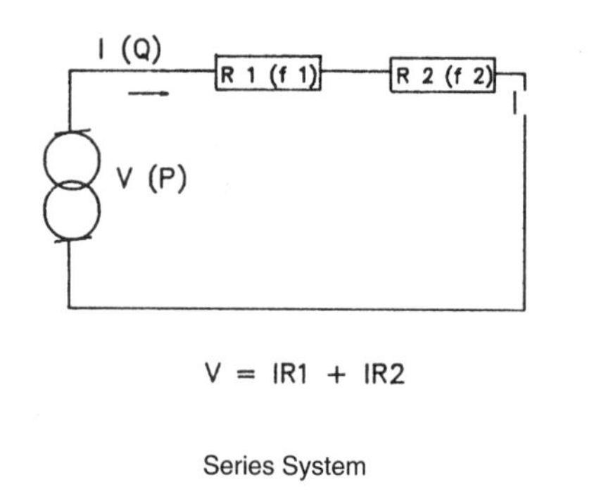

Series System

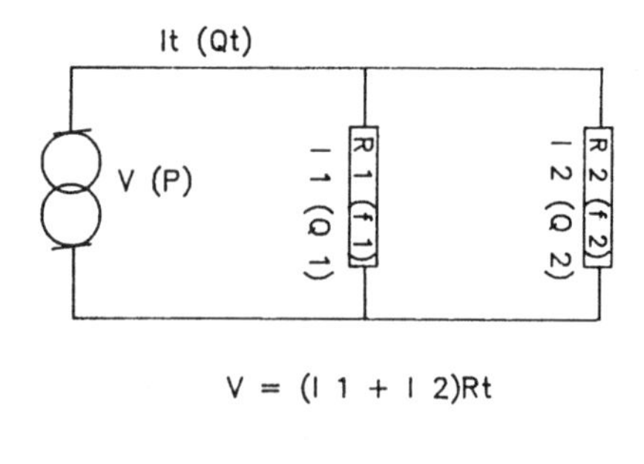

Loop System

Note: Hydraulic equivalents shown in ( )
P = pressure, Q = flow, f = friction
V = voltage, I = current, R = resistance

Figure 1. Series and loop system analogy.

Series systems

Figure 1 shows a system curve for a pipeline system 30 miles in length and consisting of 15 miles of 12.750-in. OD × 0.250-in. pipe in series with 15 miles of 10.750-in. OD × 0.250-in. pipe. System friction curves were developed for each pipe size and for the combination of the two sizes. Inspection shows that the curve for the combination is the sum of the 12.750-in. OD curve and the 10.750-in. OD curve. Pressure drop is additive on lines of constant flow for different pipe sizes connected in series.

The step of developing the individual curves for 12.750-in. OD and 10.750-in. OD pipe may be eliminated by calculating the equivalent length of 12.750-in. OD or 10.750-in. OD pipe. A system friction loss curve may then be calculated using the equivalent length of one pipe size or the other. For this example, the fluid is crude oil with a specific gravity of 0.86 and a viscosity of 50.6 SUS.

At a flow rate of 40,000 barrels per day, the pressure drop through 12.750-in. OD × 0.250-in. pipe is 6.618 psi/mile, and the pressure drop through 10.750-in. OD × 0.250-in. pipe is 15.469 psi/mile. The ratio of these two pressure drops is 6.618/15.469 or 0.428. This ratio times the length of the 12.750-in. OD pipe equals the equivalent amount of 10.750-in. OD pipe that will be required to yield the same pressure drop as 15 miles of 12.750-in. OD pipe. Therefore, the equivalent length of 10.750-in. OD pipe is 15 + (0.428 × 15) or 21.41 equivalent miles of 10.750-in. OD pipe. If it is desired to use 12.750-in. OD pipe as the base size, the same procedure may be followed except that the pressure ratio will be reversed and the reciprocal of 0.428 or 2.3364 multiplied by 15 miles of 10.750-in. OD pipe to obtain 35.046 equivalent miles of 12.750-in. OD pipe. The total equivalent length will be 15 miles + 35.046 miles, or 50.046 equivalent miles of 12.750-in. OD pipe.

This may be checked as follows:

At a flow rate of 40,000 b/d, the pressure drop through 15 miles of 10.750-in. OD is

$$15.469 \times 15 = 232 \text{ psig}$$

The pressure drop through 15 miles of 12.750-in. OD is

$$6.618 \times 15 = 99 \text{ psig}$$

The total pressure drop will be

$$232 + 99 = 331 \text{ psig}$$

The pressure drop through 21.42 equivalent miles of 10.750-in. OD is

$$21.42 \times 15.469 = 331 \text{ psig}$$

The pressure drop through 50.046 equivalent miles of 12.750-in. OD is

$$50.046 \times 6.618 = 331 \text{ psig}$$

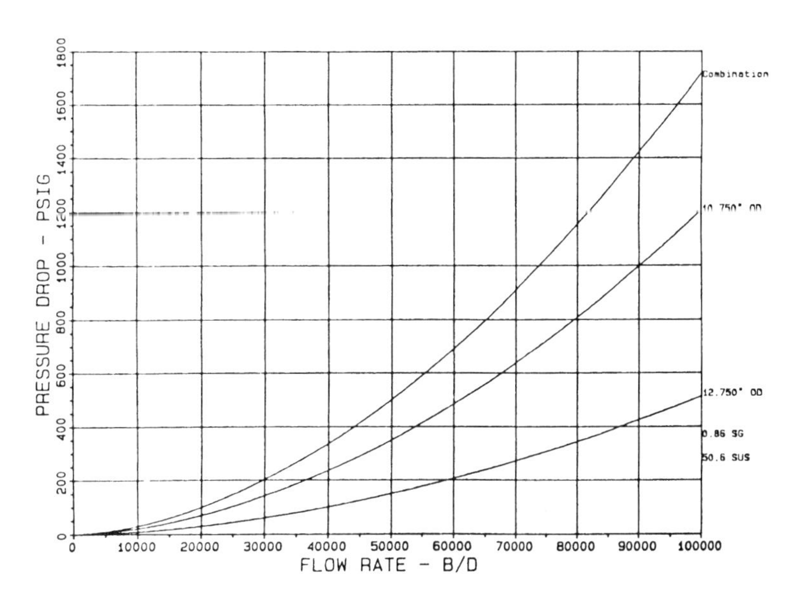

Figure 1. System curve. 15 mi., 12.750″ × 0.250″ pipe in series with 15 mi., 10.750″ × 0.250″ pipe.

Looped systems

The ratio of pressure drops may also be used for calculating the equivalent lengths of pipeline loops; however, the procedure is slightly different from that used for series systems.

Figure 1 shows the friction curves for a looped system, which is shown schematically in Figure 2. For reference, individual curves are shown for each pipe size and the combination of the two lines in parallel. The total friction loss for the system may be determined graphically using a system of curves such as shown in Figure 1, or the ratio of pressure drops may be used to determine an equivalent length for the looped system.

To use the ratio of pressure drops to convert the loop to a single line of equivalent size, determine the pressure drop through the larger size pipe (if this is to be the base size) using a chosen flow rate. In this example, a flow rate of 60,000 b/d of crude oil with a specific gravity of 0.86 and a viscosity of 50.6 SUS is used. The pressure drop through the 12.750-in. OD pipe at 60,000 b/d is 13.54 psi/mile. Next, find a flow rate through the 10.750-in. OD pipe that will yield the same pressure drop of 13.54 psi/mile. The required flow rate through the 10.750-in. OD pipe is 37,100 b/d for a drop of 13.54 psi/mile. The sum of these two rates is 97,100 b/d. Use this rate, 97,100 b/d, and determine the pressure drop/mile through the 12.750-in. OD pipe. The pressure drop/mile is 31.91 psi/mile. The factor for converting to an equivalent length of 12.750-in. OD × 0.250-in. pipe is 13.54/31.91 or 0.4243. The equivalent length of the loop combination will be 15 miles × 0.4243 or 6.3645 equivalent miles of 12.750-in. OD × 0.250-in. pipe.

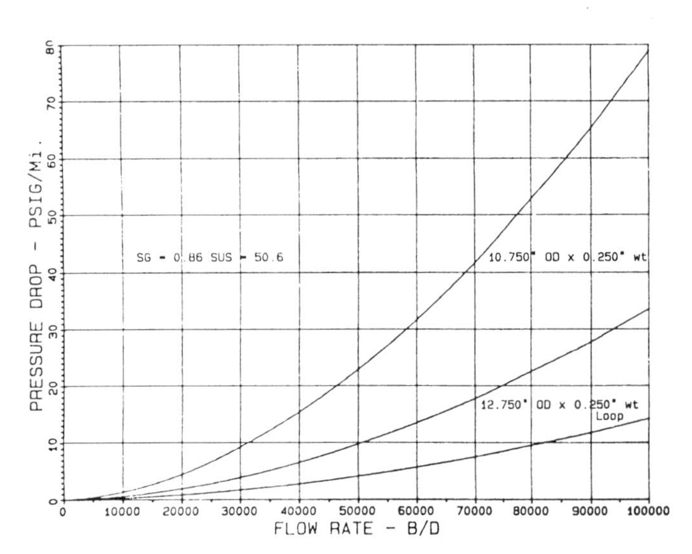

Figure 1. Friction curves. 10.750″ × 0.250″ and 12.750″ × 0.250″ and looped combination.

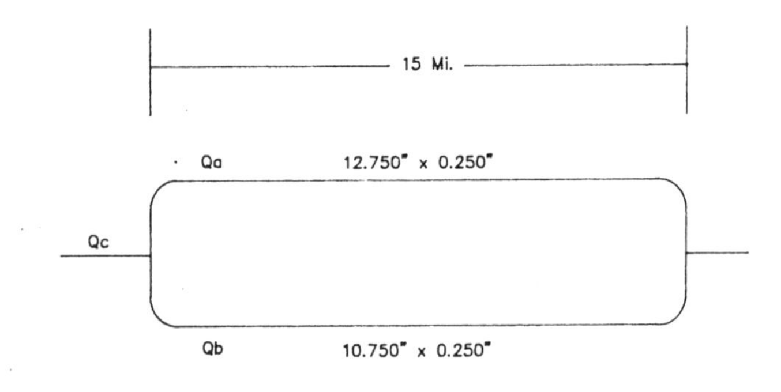

Figure 2. Looped pipelines.

If it is desired to convert the loop combination to a base diameter of 10.750-in. OD, select a flow rate and determine the pressure drop in the 10.750-in. OD leg. At 30,000 b/d, the pressure drop in the 10.750-in. OD leg will be 9.32 psi/mile. Determine the flow rate through the 12.750-in. OD leg to produce a pressure drop of 9.32 psi/mile. The rate will be 48,560 b/d. The sum of these two rates is 78,560 b/d. This rate through the 10.750-in. OD leg will produce a pressure drop of 51.29 psi/mile. The equivalent length factor for 10.750-in. OD pipe is 9.32/51.29 = 0.1817. The equivalent length will be 0.1817 × 15 or 2.7257 equivalent miles of 10.750-in. OD pipe.

This may be checked as follows:

The pressure drop through the loop combination at a flow rate of 80,000 b/d is

$$9.62 \text{ psi/mile} \times 15 \text{ miles} = 144 \text{ pisg}$$

The flow through the 10.750-in. OD leg will be 30,550 b/d and 49,450 b/d through the 12.750-in. OD leg.

The pressure drop for 6.3645 equivalent miles of 12.750-in. OD pipe at a flow rate of 80,000 b/d is

$$6.3645 \text{ equivalent miles} \times 22.584 \text{ psi/mile} = 144 \text{ pisg}$$

The pressure drop for 2.7257 equivalent miles of 10.750-in. OD pipe at a flow rate of 80,000 b/d is

$$2.7257 \text{ equivalent miles} \times 52.982 \text{ psi/mile} = 144 \text{ pisg}$$

The ratio-of-pressures method will work in conjunction with any of the flow formulas listed in this section.

Calculate pressure loss in annular sections

$P = 0.000027\ flpv^2/M$

where:

f = Friction factor (from Figure 2, p. 384)
l = Length of section, ft
p = Density of fluid, #/ft^3
v = Average velocity, fps
M = Mean hydraulic radius, ft
P = Pressure loss of length, psi
u = Absolute viscosity, ft/lb-sec. (0.000672* centipose)
R_n = Reynolds number
Q = Volume, cu. ft per second
e/d = Relative roughness

$v = Q/0.7854d_A^2$

$M = (\text{area/wetted perimeter}) = (d_1 - d_2)/4$

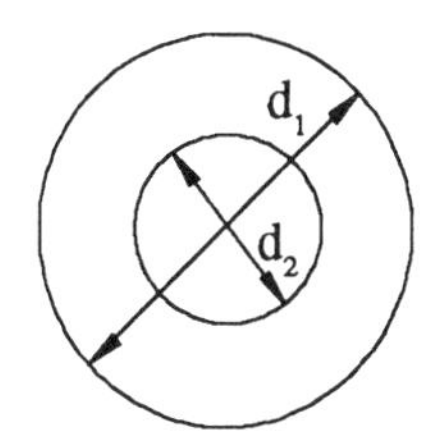

Example:

Throughput 60,000 barrels per day = 6.44 fps

Specific gravity 0.68

Viscosity 0.50 centipoise

$d_1 = 12.615''$
$d_2 = 7.000''$
$1 = 2{,}000$ ft
$d_A = (d_1^2 - d_2^2)^{0.5} = (12.615^2 - 7.0^2)^{0.5}$
$M = (d_1 - d_2)/4 = (1.051 - 0.583)/4 = 0.117$ ft.
$v = 6.44$ fps
$p = 62.4 \times 0.68 = 42.432\#/ft^3$
$u = 0.000672 \times 0.50 = 0.000336$ ft.lb.-sec.
$Rn = (4 \times 0.1178 \times 6.44 \times 42.432) \div 0.0000336 = 383{,}200$
$e/d = 0.00017$
$f = 0.015$ (from Figure 2, page 333)
$P = (0.000027 \times 0.015 \times 2000 \times 42.432 \times 6.44^2) \div 0.1178$
$P = 12.1$ psi

Calculate pressure and temperature loss for viscous crudes $\geq$ 1,000 cP

When calculating pressure and temperature losses for highly viscous Newtonian crude oils, accurate flowing temperature data are required. Usually, crudes of this nature are flowing in the laminar region, and a change of a few degrees in temperature can change the viscosity by as much as 50% or more and thus cause a significant change in the friction factor.

Figure 1 illustates a typical pressure and temperature profile for an uninsulated pipeline.

The data typically required for determining a pressure/temperature profile for a viscous crude oil pipeline is shown in Table 1.

Once the data have been tabulated, a computer program such as BATCH[1] can be used to calculate the temperature/pressure profile of the pipeline. The program performs a rigorous energy balance, i.e., static head, friction, and kinetic mechanical energy, plus friction, compression, and stored and transferred heat are included. Thermal resistances of the inside film, pipe, any insulation, and the outside environment (soil, air, or water) are calculated.

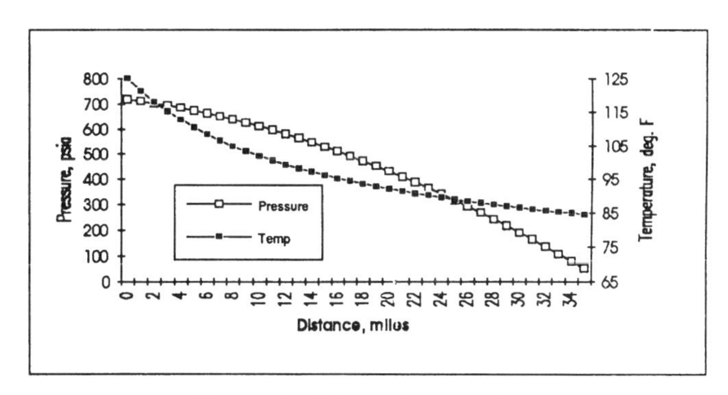

Figure 1. Uninsulated pipeline.

$$P_2 = P_1 - s\Delta x\left[\frac{C_Q Q^2 f}{d^5} + \frac{C_H(e_2 - e_1)}{\Delta x}\right]$$

where:

P_1 = Pressure at the upstream end of the calculation segment

Table 1
Data required for pressure/temperature loss calculation

Length	Heat transfer coefficient (fluid)	Inlet temperature
Size, OD	Insulation thickness	Inlet pressure
Wall thickness	Thermal conductivity of insul.	Throughput
Pipe wall roughness	Thermal conductivity of soil	Sp. gr. at two temp.
Thermal conductivity of pipe	Pipe burial depth	Viscosity at two temp.
Upstream elevation	Soil temperature	Specific heat of liquid at two temp.
Downstream elevation	Vapor pressure at two temp.	Thermal conductivity of liquid at two temp.
Number of calculation segments	Soil K factor	

P_2 = Pressure at the downstream end of the calculation segment
Δx = Length of the calculation segment
f = Darcy-Weisbach friction factor
d = Inside diameter of the pipe
s = Liquid-specific gravity
e_1 = Elevation at the upstream end of the calculation segment
e_2 = Elevation at the downstream end of the calculation segment
C_Q = Flow conversion constant
C_H = Head conversion constant

An energy balance is performed on each calculation segment and its surroundings, resulting in the next equation:

$$t_2 = [wc_{p1}t_1 - UA(LMTD) + w(h_1 - h_2)/J] \div (wC_{p2})$$

where:

t_1 = Flowing temp. at the upstream end of the calculation segment
t_2 = Flowing temp. at the downstream end of the calculation segment
h_1 = Head at the upstream end of the calculation segment
h_2 = Head at the downstream end of the calculation segment
C_{p1} = Heat capacity at constant pressure evaluated at t_1
C_{p2} = Heat capacity at constant pressure evaluated at t_2
w = Total mass flow rate
U = Overall heat transfer coefficient
A = Surface area of the pipe in calculation segment
LMTD = Log mean temp. difference between the average flowing temp. and the average ambient temp.
J = Mechanical equivalent of heat

Example:

Pipeline length	35 miles
Size	14.000″
Wall thickness	0.250″
Pipe wall roughness	0.0018″
Thermal conductivity (pipe)	29 Btu/hr-ft-F
Upstream elevation	1,500 feet
Downstream elevation	1,650 feet
Length	35 miles
Number of calculation segments	35
Heat transfer coefficient (calculated by program	Btu/hr-ft2-F)
Insulation thickness	0″ (Case 1)
Insulation thermal conductivity	0 Btu/hr-ft-F
Burial depth	36″
Soil temperature	50°F
Soil K factor	1.2
Inlet temperature	125°F
Inlet pressure	calculate
Outlet pressure	50 psia
Throughput	8,000 bpd

Liquid properties:

	Value	Temp.
Specific gravity	0.95	77
	0.90	210
Viscosity, cP	3260	80
	90	180
Specific heat	0.5	60
Btu/Lbm-F	0.5	80
Thermal conductivity	0.2	60
Btu/hr-ft-F	0.22	100
Vapor pressure	0.4	100
psia	0.3	70

The pressure/temperature profile for an uninsulated line, based on the above conditions, is shown in Figure 1 and in Table 3.

If the pipeline crosses areas where the parameters such as soil thermal conductivity, pipe size, etc. change, BATCH will allow the system to be segmented into additional pipes (up to 99 pipes) with each pipe being configured with its own input data.

Assume that 2″ of urethane foam insulation is added to the pipeline previously described. The new pressure/temperature profile is shown in Figure 2 and in Table 4.

Typical soil K factors are shown in Table 2. BATCH may also be used to calculate pump requirements and heater duty.

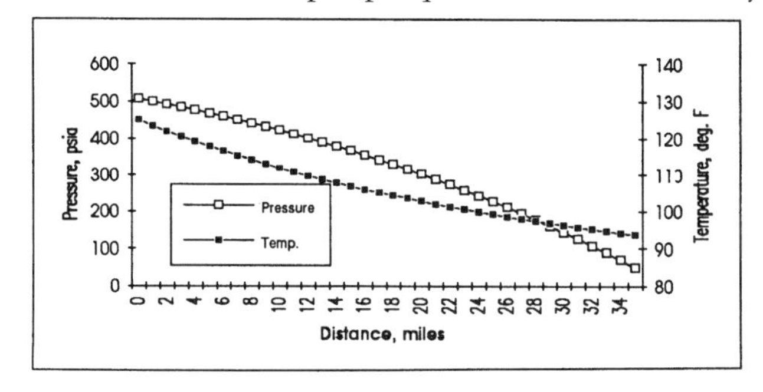

Figure 2. Insulated pipeline—2″ polyurethane.

Table 2
Soil K factors Btu/hr/ft^2/°F

Soil	Condition	Cover, In.	K
Sandy	Dry	24	0.25 to 0.40
Sandy	Moist	24	0.50 to 0.60
Sandy	Soaked	24	1.10 to 1.30
Clay	Dry	24	0.20 to 0.30
Clay	Moist	24	0.40 to 0.50
Clay	Moist to wet	24	0.60 to 0.90
Sandy	Dry	8	0.60 to 0.70
Sandy	Moist to wet	8	1.20 to 2.40
River	Water	60	2.00 to 2.50
Air	Exposed	—	2.0 (approx.)

Table 3
Uninsulated pipe

Distance miles	Pressure psia	Temp. deg. F
0	718	125
1	710.59	121.3
2	702.57	118.1
3	693.79	115.3
4	684.25	112.8
5	673.93	110.5
6	662.83	108.4
7	650.95	106.5
8	638.3	104.8
9	624.89	103.3
10	610.73	101.9
11	595.82	100.6
12	580.18	99.3
13	563.81	98.2
14	546.74	97.2
15	528.97	96.2
16	510.51	95.2
17	491.38	94.4
18	471.6	93.6
19	451.18	92.8
20	430.13	92.1
21	408.46	91.4
22	386.21	90.7
23	363.37	90.1
24	339.97	89.5
25	316.02	88.9
26	291.55	88.4
27	266.55	87.9
28	241.06	87.4
29	215.09	86.9
30	188.65	86.5
31	161.76	86
32	134.43	85.6
33	106.68	85.2
34	78.54	84.9
35	50	84.5

Table 4
Insulated pipe

Distance miles	Pressure psia	Temp. deg. F
0	506.76	125
1	499.69	123.4
2	492.32	121.9
3	484.64	120.5
4	476.65	119.1
5	468.33	117.8
6	459.68	116.5
7	450.7	115.3
8	441.38	114.1
9	431.71	112.9
10	421.68	111.8
11	411.3	110.7
12	400.56	109.7
13	389.45	108.7
14	377.98	107.8
15	366.13	106.8
16	353.9	105.9
17	341.3	105.1
18	328.32	104.3
19	314.96	103.5
20	301.22	102.7
21	287.1	101.9
22	272.59	101.2
23	257.71	100.5
24	242.44	99.8
25	226.8	99.2
26	210.77	98.5
27	194.38	97.9
28	177.6	97.3
29	160.46	96.7
30	142.95	96.2
31	125.08	95.6
32	106.84	95.1
33	88.25	94.6
34	69.3	94.1
35	50	93.6

Determine batch injection rate as per enclosure

In the operation of a pipeline system, it may be necessary to inject a specific volume into a passing block of liquid. When it is desirable to maintain the same flow rate downstream of the injection and inject at the minimum rate or into most of the passing block, the following equations can be used to determine the upstream injection rates.

$$Q_U = \frac{B + H_O Q_D + Q - [(B + H_O Q_D + Q)^2 - 4 \times H_O \times B \times Q]^{0.5}}{2 \times H_O}$$

$$Q_I = Q_D - Q_U$$

where:

Q_D = Downstream rate that is to be held constant, bph
Q_I = Injection rate to be determined, bph
Q_U = Upstream pumping rate to be determined, bph
Q = Volume to be injected, barrels
B = Volume of passing block to be injected into, barrels
H_O = Block outage in hours. This is the time that the head and tail end of the block interface is passing the injection location, during which time injection is not allowed.

Q_I = INJ. RATE
Q_U
BLOCK TO BE
INJECT INTO
73,500 BBLS
Q_D = 15,600 BPH

Example:

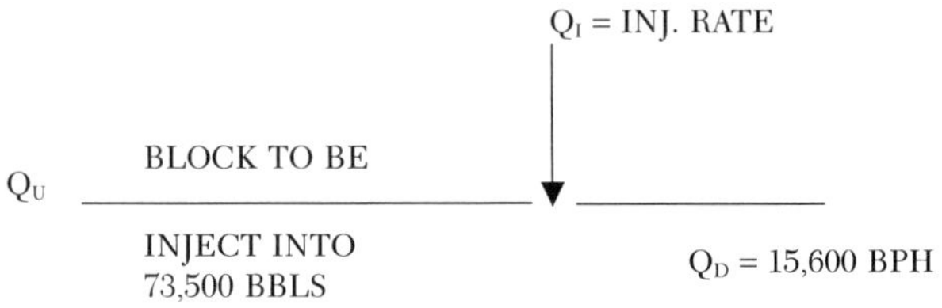

Q = 15,000 bbls for injection
Q_I = Injection rate, bph
B = 73,500 bbls
Q_D = 15,600 bph (to be held constant)
H_O = 0.5 hours

$$Q_U = \frac{73{,}500 + 0.5 \times 15{,}600 + 15{,}000 - [(73{,}5000 + 0.5 \times 15{,}600 - 15{,}000)^2 - 4 \times 0.5 \times 73{,}500 \times 15{,}600]^{0.5}}{2 \times 0.5}$$

Q_U = 12,751 bph
Q_I = 15,600 − 12,751 = 2,849 bph

PRESSURE LOSS THROUGH VALVES AND FITTINGS

"K" FACTOR TABLE—SHEET 1 of 4

Representative Resistance Coefficients (K) for Valves and Fittings

("K" is based on use of schedule pipe as listed on page 2-10)

PIPE FRICTION DATA FOR CLEAN COMMERCIAL STEEL PIPE WITH FLOW IN ZONE OF COMPLETE TURBULENCE

Nominal Size	½″	¾″	1″	1¼″	1½″	2″	2½, 3″	4″	5″	6″	8-10″	12-16″	18-24″
Friction Factor (f_T)	.027	.025	.023	.022	.021	.019	.018	.017	.016	.015	.014	.013	.012

FORMULAS FOR CALCULATING "K" FACTORS* FOR VALVES AND FITTINGS WITH REDUCED PORT

(Ref: Pages 2-11 and 3-4)

- **Formula 1**

$$K_2 = \frac{0.8\left(\sin\frac{\theta}{2}\right)(1-\beta^2)}{\beta^4} = \frac{K_1}{\beta^4}$$

- **Formula 2**

$$K_2 = \frac{0.5\,(1-\beta^2)\sqrt{\sin\frac{\theta}{2}}}{\beta^4} = \frac{K_1}{\beta^4}$$

- **Formula 3**

$$K_2 = \frac{2.6\left(\sin\frac{\theta}{2}\right)(1-\beta^2)^2}{\beta^4} = \frac{K_1}{\beta^4}$$

- **Formula 4**

$$K_2 = \frac{(1-\beta^2)^2}{\beta^4} = \frac{K_1}{\beta^4}$$

- **Formula 5**

$$K_2 = \frac{K_1}{\beta^4} + \text{Formula 1} + \text{Formula 3}$$

$$K_2 = \frac{K_1 + \sin\frac{\theta}{2}\left[0.8\,(1-\beta^2) + 2.6\,(1-\beta^2)^2\right]}{\beta^4}$$

- **Formula 6**

$$K_2 = \frac{K_1}{\beta^4} + \text{Formula 2} + \text{Formula 4}$$

$$K_2 = \frac{K_1 + 0.5\sqrt{\sin\frac{\theta}{2}}\,(1-\beta^2) + (1-\beta^2)^2}{\beta^4}$$

- **Formula 7**

$$K_2 = \frac{K_1}{\beta^4} + \beta\,(\text{Formula 2} + \text{Formula 4}) \text{ when } \theta = 180^\circ$$

$$K_2 = \frac{K_1 + \beta\left[0.5\,(1-\beta^2) + (1-\beta^2)^2\right]}{\beta^4}$$

$$\beta = \frac{d_1}{d_2}$$

$$\beta^2 = \left(\frac{d_1}{d_2}\right)^2 = \frac{a_1}{a_2}$$

Subscript 1 defines dimensions and coefficients with reference to the smaller diameter.

Subscript 2 refers to the larger diameter.

*Use "K" furnished by valve or fitting supplier when available.

SUDDEN AND GRADUAL CONTRACTION

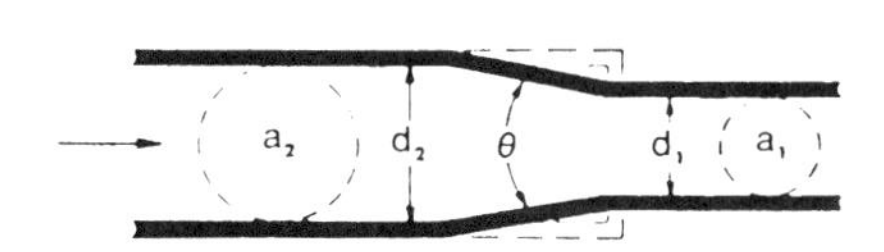

If: $\theta \leq 45^\circ$ .... K_2 = Formula 1

$45^\circ < \theta \leq 180^\circ$ ... K_2 = Formula 2

SUDDEN AND GRADUAL ENLARGEMENT

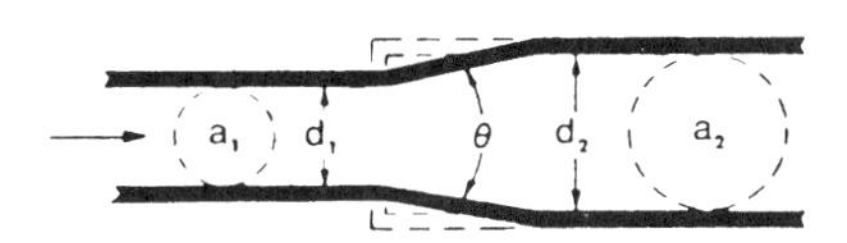

If: $\theta \leq 45^\circ$ .... K_2 = Formula 3

$45^\circ < \theta \leq 180^\circ$ ... K_2 = Formula 4

Reprinted by permission and courtesy of Crane Co. from Technical Report 410

"K" FACTOR TABLE—SHEET 2 of 4

Representative Resistance Coefficients (K) for Valves and Fittings

(for formulas and friction data, see page A-26)

("K" is based on use of schedule pipe as listed on page 2-10)

GATE VALVES
Wedge Disc, Double Disc, or Plug Type

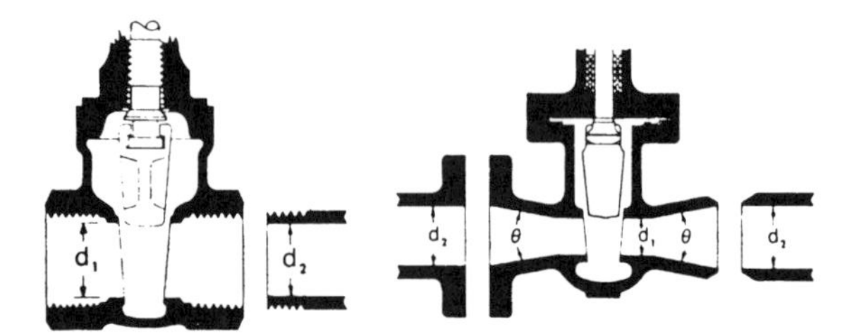

If: $\beta = 1$, $\theta = 0$ . . . . $K_1 = 8 f_T$
$\beta < 1$ and $\theta \leq 45°$ . . . . K_2 = Formula 5
$\beta < 1$ and $45° < \theta \leq 180°$ . . . K_2 = Formula 6

SWING CHECK VALVES

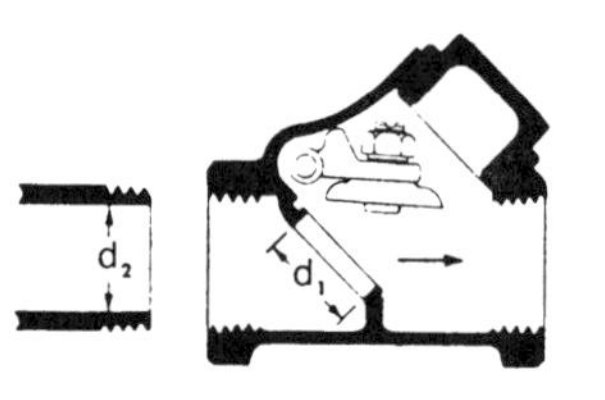

$K = 100 f_T$

Minimum pipe velocity (fps) for full disc lift $= 35 \sqrt{\bar{V}}$

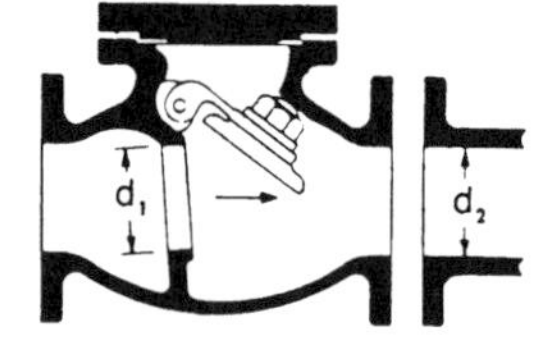

$K = 50 f_T$

Minimum pipe velocity (fps) for full disc lift $= 60 \sqrt{\bar{V}}$ except
U/L listed $= 100 \sqrt{\bar{V}}$

GLOBE AND ANGLE VALVES

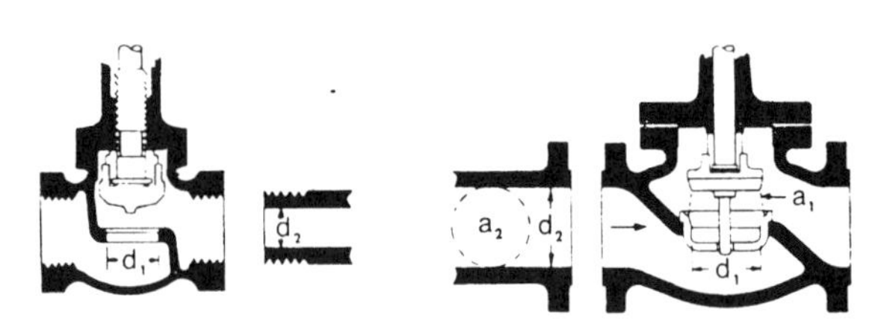

If: $\beta = 1 \ldots K_1 = 340 f_T$

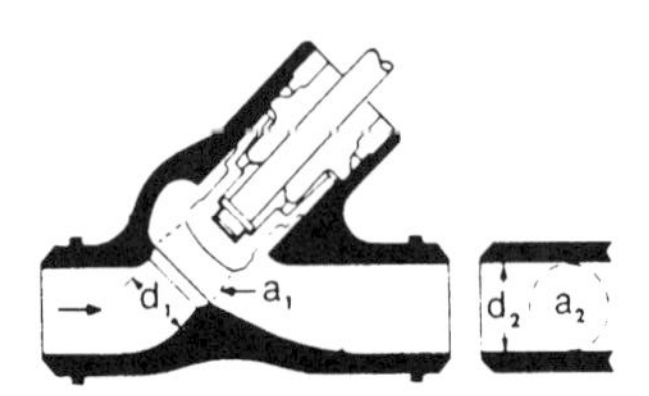

If: $\beta = 1 \ldots K_1 = 55 f_T$

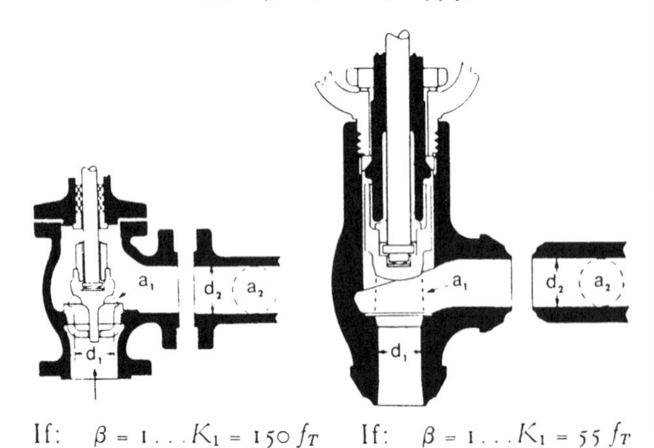

If: $\beta = 1 \ldots K_1 = 150 f_T$ If: $\beta = 1 \ldots K_1 = 55 f_T$

All globe and angle valves, whether reduced seat or throttled,
If: $\beta < 1 \ldots K_2$ = Formula 7

LIFT CHECK VALVES

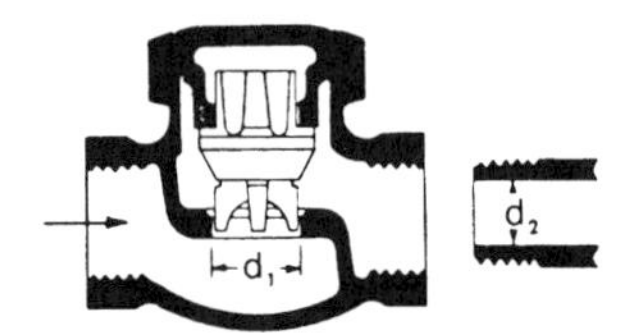

If: $\beta = 1 \ldots K_1 = 600 f_T$
$\beta < 1 \ldots K_2$ = Formula 7

Minimum pipe velocity (fps) for full disc lift $= 40 \beta^2 \sqrt{\bar{V}}$

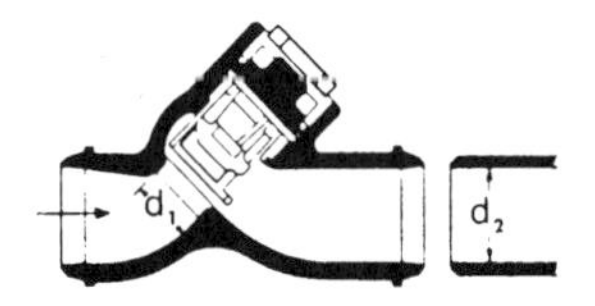

If: $\beta = 1 \ldots K_1 = 55 f_T$
$\beta < 1 \ldots K_2$ = Formula 7

Minimum pipe velocity (fps) for full disc lift $= 140 \beta^2 \sqrt{\bar{V}}$

TILTING DISC CHECK VALVES

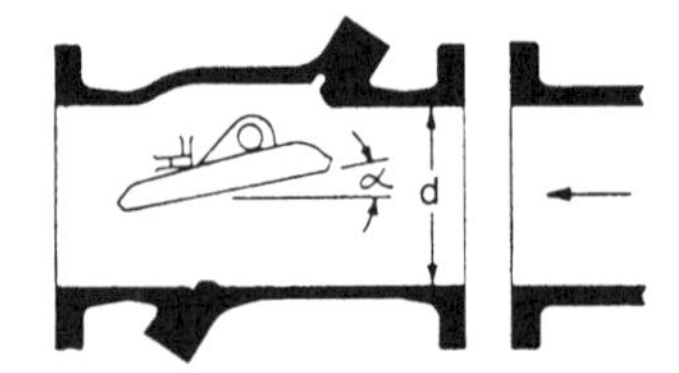

	$\alpha = 5°$	$\alpha = 15°$
Sizes 2 to 8" . . . K =	$40 f_T$	$120 f_T$
Sizes 10 to 14" . . . K =	$30 f_T$	$90 f_T$
Sizes 16 to 48" . . . K =	$20 f_T$	$60 f_T$
Minimum pipe velocity (fps) for full disc lift =	$80 \sqrt{\bar{V}}$	$30 \sqrt{\bar{V}}$

"K" FACTOR TABLE—SHEET 3 of 4

Representative Resistance Coefficients (K) for Valves and Fittings

(for formulas and friction data, see page A-26)

("K" is based on use of scheduled pipe as listed on page 2-10)

STOP-CHECK VALVES (Globe and Angle Types)

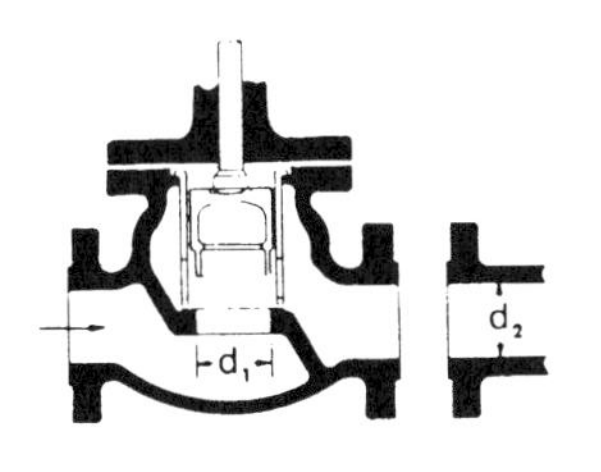

If:
$\beta = 1 \ldots K_1 = 400\ f_T$
$\beta < 1 \ldots K_2 =$ Formula 7

Minimum pipe velocity for full disc lift
$= 55\ \beta^2 \sqrt{\bar{V}}$

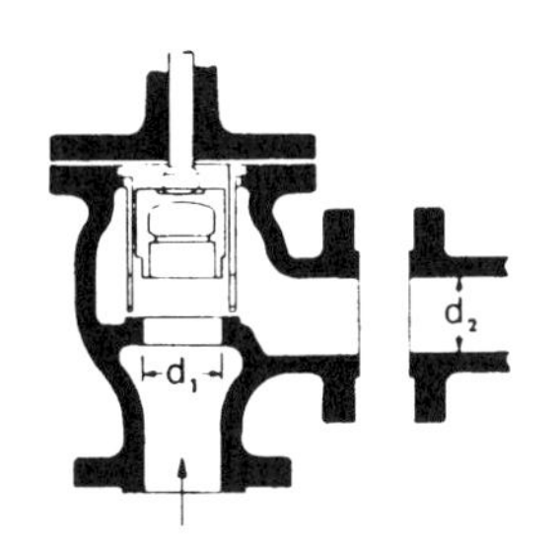

If:
$\beta = 1 \ldots K_1 = 200\ f_T$
$\beta < 1 \ldots K_2 =$ Formula 7

Minimum pipe velocity for full disc lift
$= 75\ \beta^2 \sqrt{\bar{V}}$

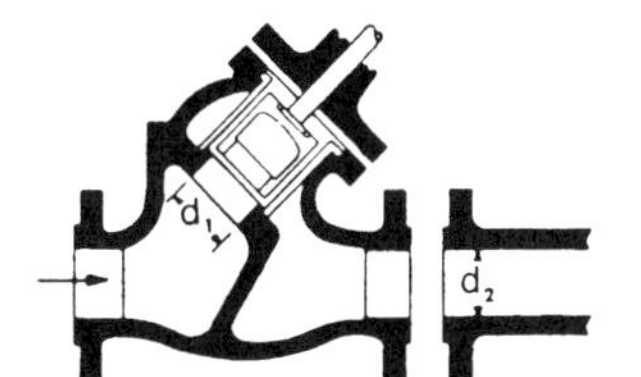

If:
$\beta = 1 \ldots K_1 = 300\ f_T$
$\beta < 1 \ldots K_2 =$ Formula 7

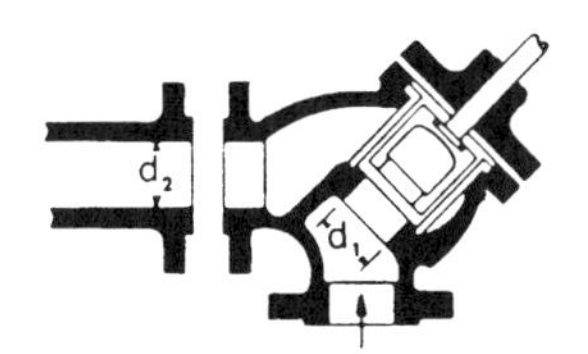

If:
$\beta = 1 \ldots K_1 = 350\ f_T$
$\beta < 1 \ldots K_2 =$ Formula 7

Minimum pipe velocity (fps) for full disc lift
$= 60\ \beta^2 \sqrt{\bar{V}}$

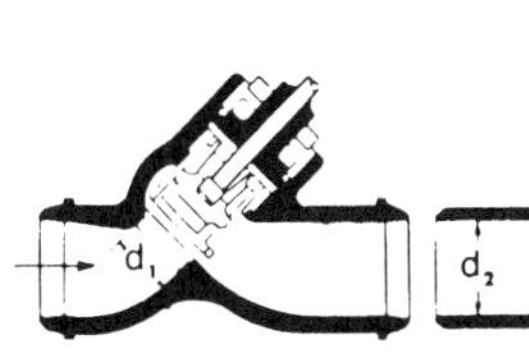

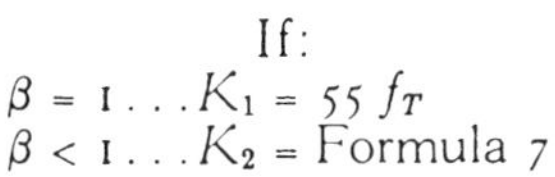

If:
$\beta = 1 \ldots K_1 = 55\ f_T$
$\beta < 1 \ldots K_2 =$ Formula 7

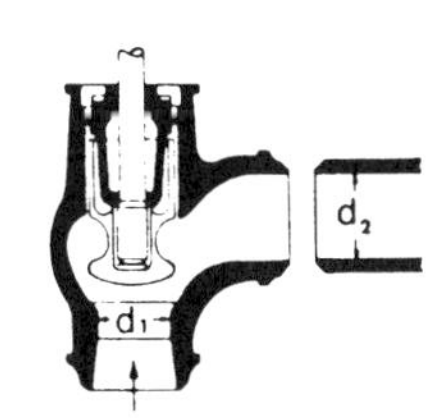

If:
$\beta = 1 \ldots K_1 = 55\ f_T$
$\beta < 1 \ldots K_2 =$ Formula 7

Minimum pipe velocity (fps) for full disc lift
$= 140\ \beta^2 \sqrt{\bar{V}}$

FOOT VALVES WITH STRAINER

Poppet Disc

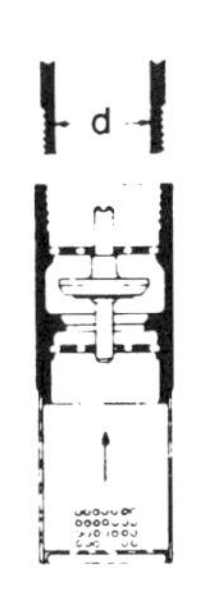

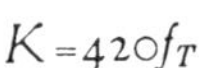

$K = 420 f_T$

Minimum pipe velocity (fps) for full disc lift
$= 15 \sqrt{\bar{V}}$

Hinged Disc

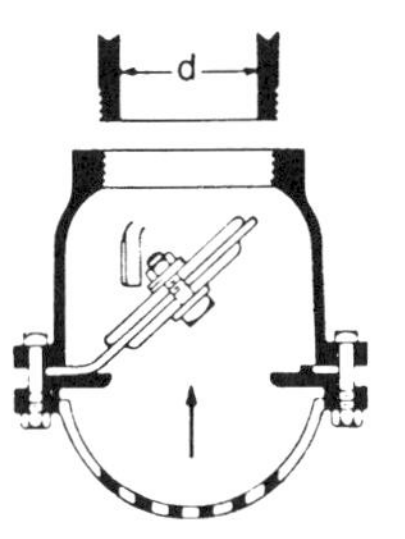

$K = 75 f_T$

Minimum pipe velocity (fps) for full disc lift
$= 35 \sqrt{\bar{V}}$

BALL VALVES

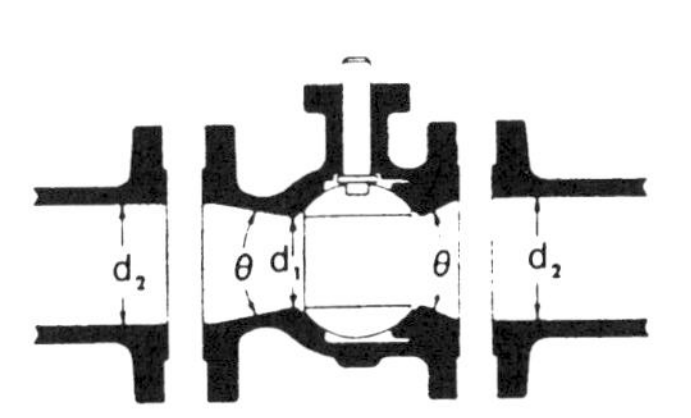

If: $\beta = 1,\ \theta = 0 \ldots\ldots\ldots\ldots\ldots K_1 = 3\ f_T$
$\beta < 1$ and $\theta \leqq 45° \ldots\ldots\ldots K_2 =$ Formula 5
$\beta < 1$ and $45° < \theta \leqq 180° \ldots K_2 =$ Formula 6

BUTTERFLY VALVES

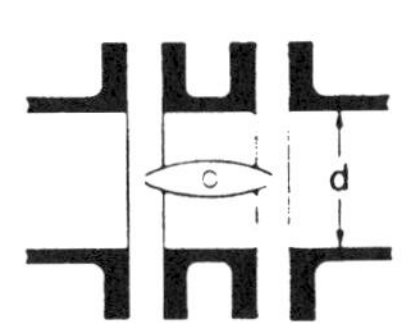

Sizes 2 to 8"... $K = 45\ f_T$
Sizes 10 to 14"... $K = 35\ f_T$
Sizes 16 to 24"... $K = 25\ f_T$

"K" FACTOR TABLE—SHEET 4 of 4

Representative Resistance Coefficients (K) for Valves and Fittings

(for formulas and friction data, see page A-26)

("K" is based on use of schedule pipe as listed on page 2-10)

PLUG VALVES AND COCKS

Straight-Way

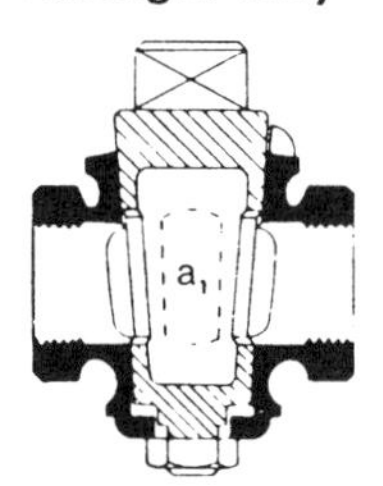

If: $\beta = 1$, $K_1 = 18\,f_T$

3-Way

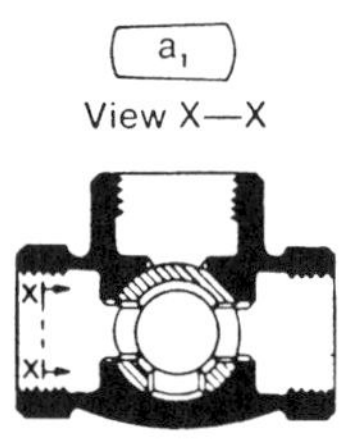

If: $\beta = 1$, $K_1 = 30\,f_T$

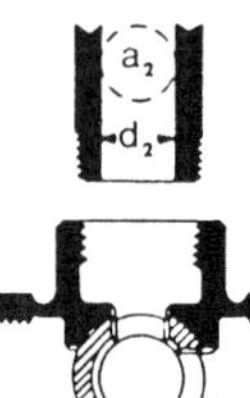

If: $\beta = 1$, $K_1 = 90\,f_T$

If: $\beta < 1 \ldots K_2$ = Formula 6

MITRE BENDS

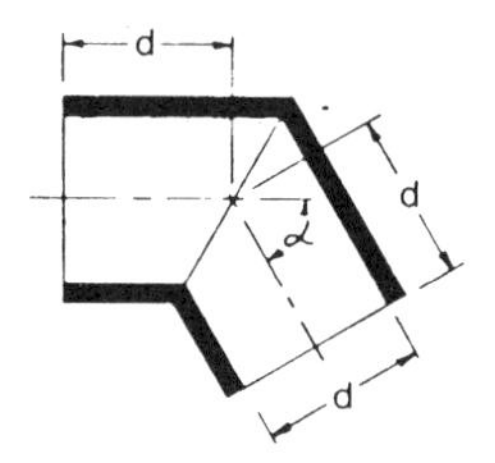

α	K
0°	2 f_T
15°	4 f_T
30°	8 f_T
45°	15 f_T
60°	25 f_T
75°	40 f_T
90°	60 f_T

90° PIPE BENDS AND FLANGED OR BUTT-WELDING 90° ELBOWS

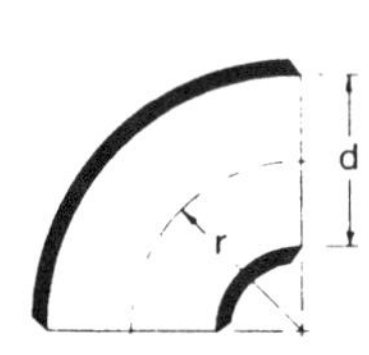

r/d	K	r/d	K
1	20 f_T	8	24 f_T
1.5	14 f_T	10	30 f_T
2	12 f_T	12	34 f_T
3	12 f_T	14	38 f_T
4	14 f_T	16	42 f_T
6	17 f_T	20	50 f_T

The resistance coefficient, K_B, for pipe bends other than 90° may be determined as follows:

$$K_B = (n - 1)\left(0.25\,\pi\,f_T\frac{r}{d} + 0.5\,K\right) + K$$

n = number of 90° bends
K = resistance coefficient for one 90° bend (per table)

CLOSE PATTERN RETURN BENDS

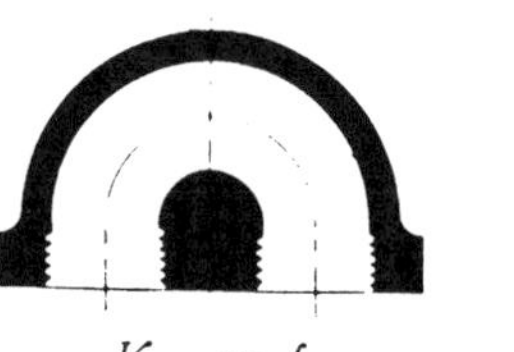

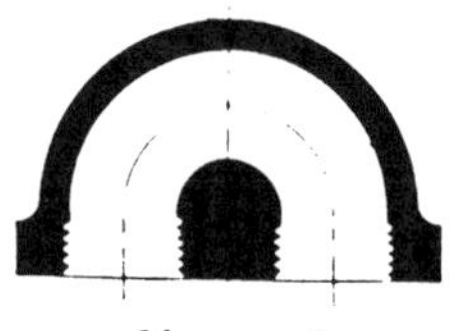

$K = 50\,f_T$

STANDARD ELBOWS

90°

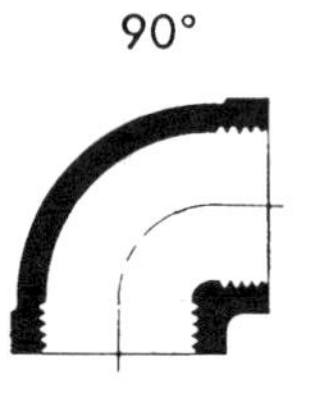

$K = 30\,f_T$

45°

$K = 16\,f_T$

STANDARD TEES

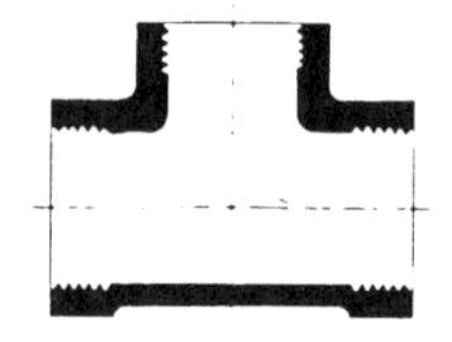

Flow thru run....... $K = 20\,f_T$
Flow thru branch.... $K = 60\,f_T$

PIPE ENTRANCE

Inward Projecting

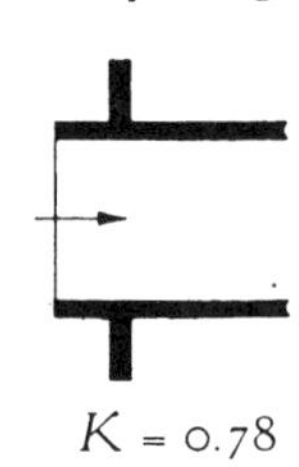

$K = 0.78$

r/d	K
0.00*	0.5
0.02	0.28
0.04	0.24
0.06	0.15
0.10	0.09
0.15 & up	0.04

*Sharp-edged

Flush

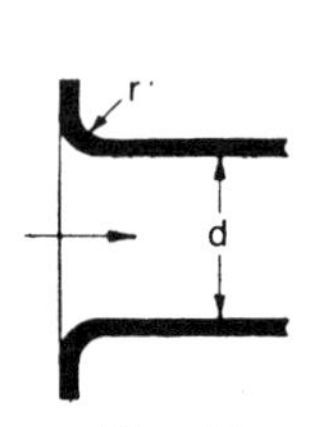

For K, see table

PIPE EXIT

Projecting **Sharp-Edged** **Rounded**

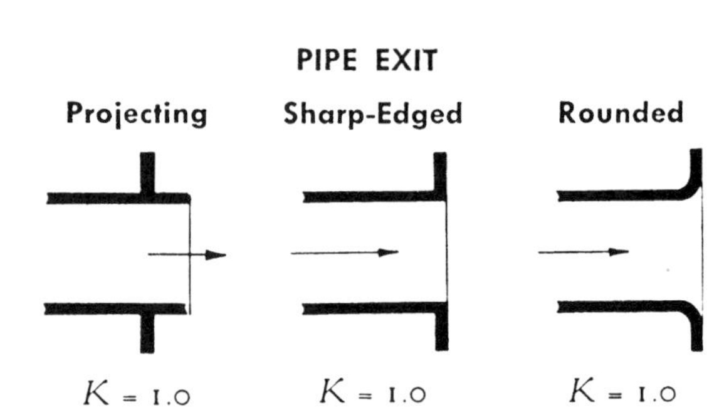

$K = 1.0$ $K = 1.0$ $K = 1.0$

Reprinted by permission and courtesy of Crane Co. from Technical Report 410

Equivalent Lengths L and L/D and Resistance Coefficient K

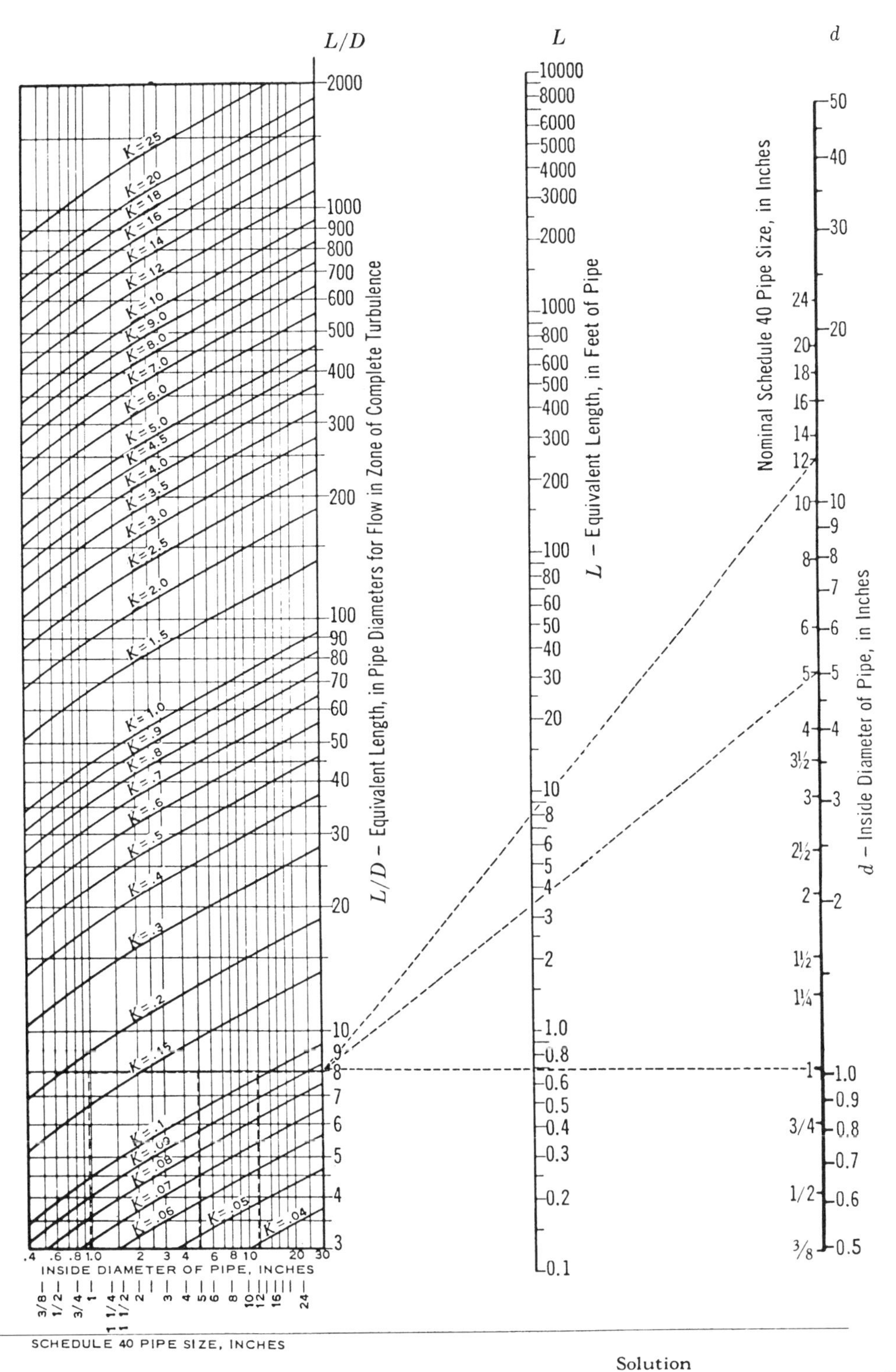

Problem: Find the equivalent length in pipe diameters and feet of Schedule 40 clean commercial steel pipe, and the resistance factor K for 1, 5, and 12-inch fully-opened gate valves with flow in zone of complete turbulence.

Solution

Valve Size	1″	5″	12″	*Refer to*
Equivalent length, pipe diameters	8	8	8	Page A-27
Equivalent length, feet of Sched. 40 pipe	0.7	3.4	7.9	Dotted lines on chart.
Resist. factor K, based on Sched. 40 pipe	0.18	0.13	0.10	

Reprinted by permission and courtesy of Crane Co. from Technical Report 410.

Equivalents of Resistance Coefficient *K* And Flow Coefficient C_v

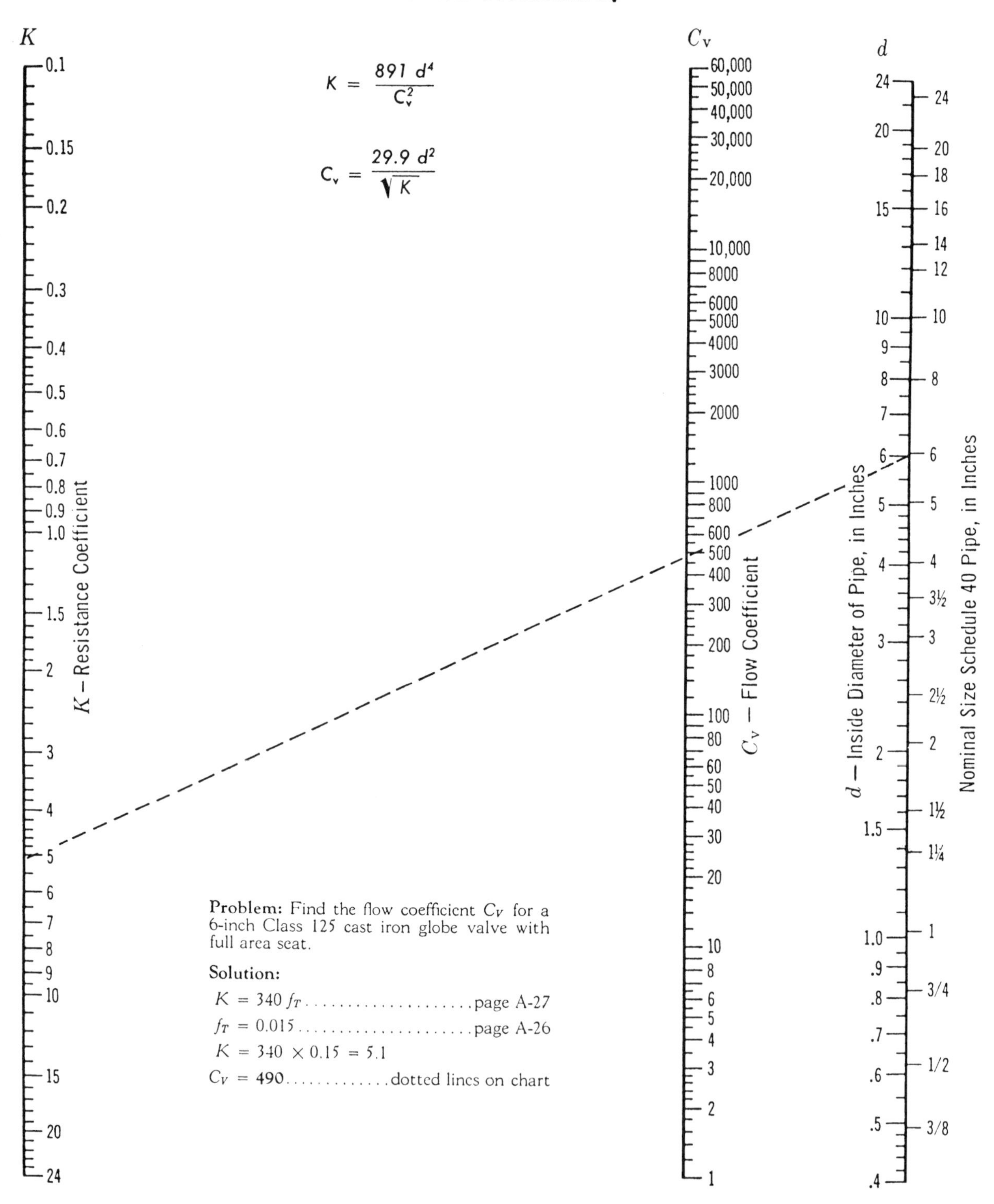

Reprinted by permission and courtesy of Crane Co. from Technical Report 410.

Nomograph for calculating Reynolds number for flow of liquids and friction factor for clean steel and wrought iron pipe

The nomograph (Figure 1) permits calculation of the Reynolds number for liquids and the corresponding friction factor for clean steel and wrought iron pipe.

The Reynolds number is defined as:

$$Re = 22{,}700\frac{q\rho}{d\mu} = 50.6\frac{Q\rho}{d\mu} = 6.31\frac{W}{d\mu} = \frac{dv\rho}{\mu a}$$

where: Re = Reynolds number
d = Inside pipe diameter, in.
q = Volumetric flow rate, cu. ft/sec

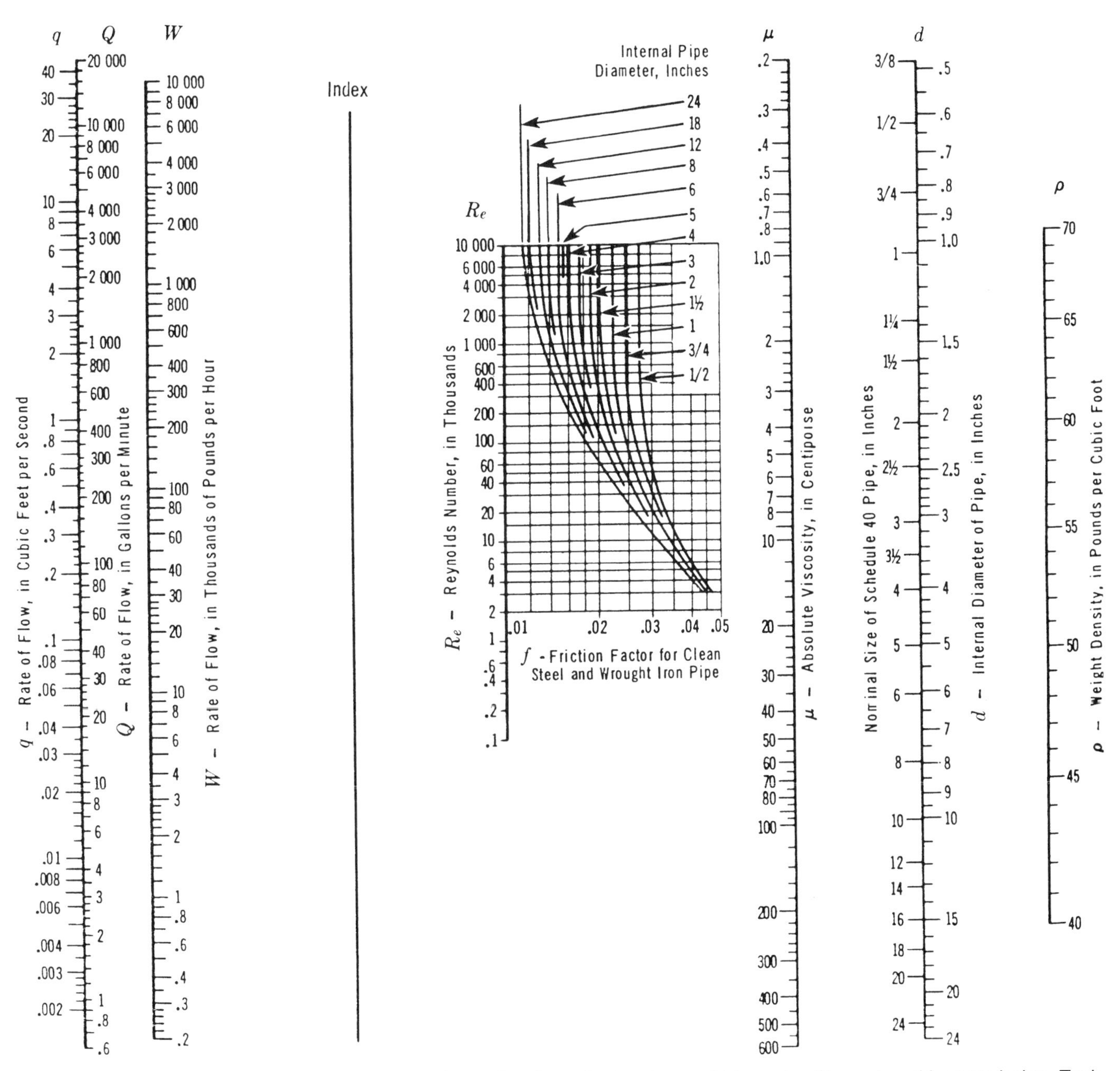

Figure 1. Reynolds number for liquid flow friction factor for clean steel and wrought iron pipe (Reproduced by permission, *Tech. Paper, 410*, Crane Co., copyright 1957).

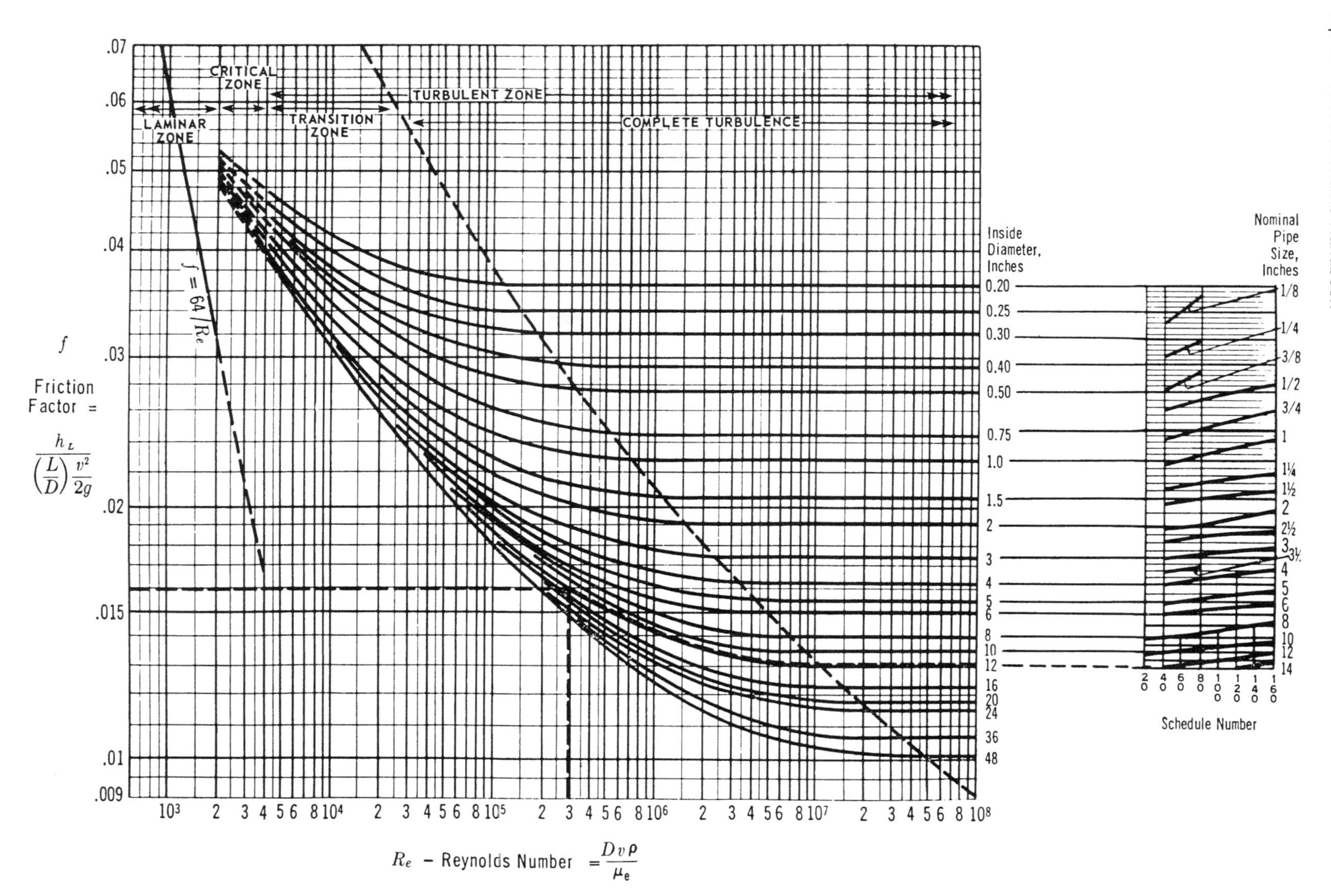

Figure 2. Friction factors for clean commercial steel and wrought iron pipe. (Reproduced by permission, *Tech. Paper, 410*, Crane Co., 1957.)

Q = Volumetric flow rate, gal/min
ρ = Fluid density, lb/cubic ft
μ = Fluid viscosity, centipoise
μa = Fluid viscosity, lb/(ft)(sec)
v = Average fluid velocity, ft/sec
w = Mass rate of flow, lb/hr

Example.

Given:
water at 200°F
$d = 4$-in. schedule 40 steel pipe
$Q = 415$ gal/min
$\rho = 60.13$ lb/cubic ft
$\mu = 0.3$ centipoises

Obtain the Reynolds number and the friction factor.

Connect	With	Mark or Read
$Q = 415$	$\rho = 60.13$	$W = 200{,}000$
$W = 200{,}000$	$d = 4$-in. schedule 40	Index
Index mark	$\mu = 0.3$	$Re = 1{,}000{,}000$

At $Re = 1{,}000{,}000$ read $f = 0.017$ from the graph on the center of the nomograph.

Figure 2 is an enlarged version of the friction factor chart, based on Moody's data[1] for commercial steel and wrought iron. This chart can be used instead of the smaller version shown in Figure 1.

Reference

1. Moody, L. F., *Transactions, American Society of Mechanical Engineers*, 66, 671 (1944).

Nomograph for calculating pressure drop of liquids in lines for turbulent flow

The nomograph (Figure 1) is based on a modified version of the Fanning equation. It gives the pressure drop per 100 ft of pipe for liquids flowing through a circular pipe. The basic equation is:

$$P_{100} = \frac{0.0216\, f\rho Q^2}{d^2} = 4{,}350\frac{f\rho q^2}{d^2} = 0.000336\frac{fW^2}{d^5\rho}$$
$$= 0.1294\frac{fv^2\rho}{d}$$

where:

P_{100} = Pressure loss per 100 ft of pipe, psi
f = Friction factor ($4 \times f'$ in Fanning equation)
ρ = Fluid density, lb/cu. ft
d = Tube internal diameter, in.
W = Flow rate, lb/hr
Q = Flow rate, gal/min
q = Flow rate, cu. ft/sec

The Moody[1] friction factor and relative roughness generalized charts (Figures 2 and 3) are also included for use with the nomogram.

Example. The estimated Reynolds number is 100,000 when water is flowing at a rate of 100 gal/min and a temperature of 70°F. Find the pressure drop per 100 ft of 2-in. diameter pipe, the relative roughness from Figure 3 is 0.0009.

At $e/d = 0.0009$ and $Re = 100{,}000$, the friction factor, f, is 0.022 from Figure 2.

Using Figure 1:

Connect	With	Mark or Read
$f = 0.022$	$\rho = 62.3$ lb/cu. ft	Intersection at Index 1
Index 1	$Q = 100$ gal/min	Intersection at Index 2
Index 2	$d = 2$-in.	$P = 8$ psi per 100 ft of pipe

Source

Flow of Fluids Through Valves, Fittings and Pipe, Technical Paper No. 410, 3-11, Crane Company, Chicago (1957).

Anon., *British Chemical Engineering*, 2, 3, 152 (1957).

Reference

1. Moody, L. F., *Transactions, American Society of Mechanical Engineers*, 66, 671 (1944).

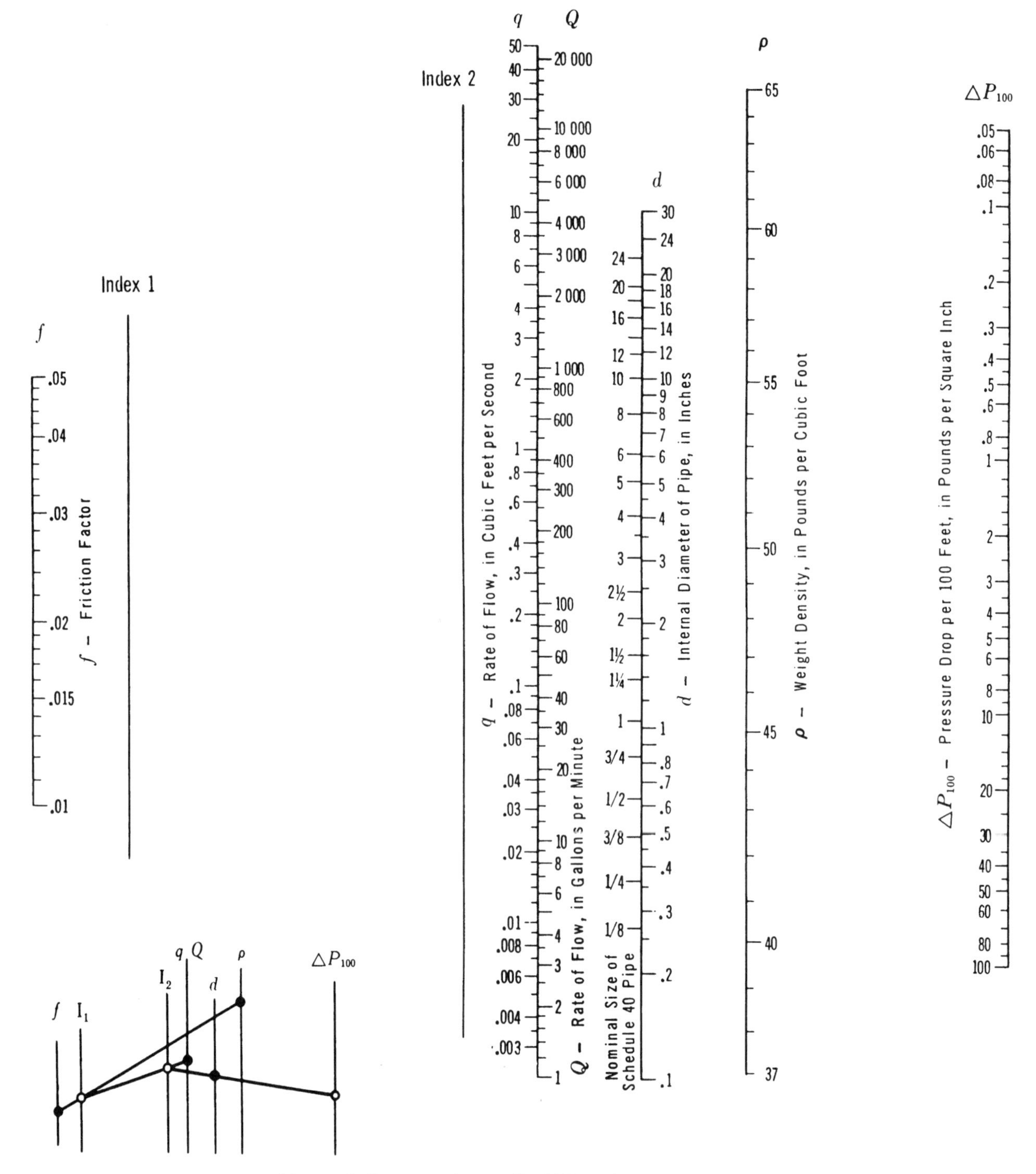

Figure 1. Pressure drop in liquid lines for turbulent flow.

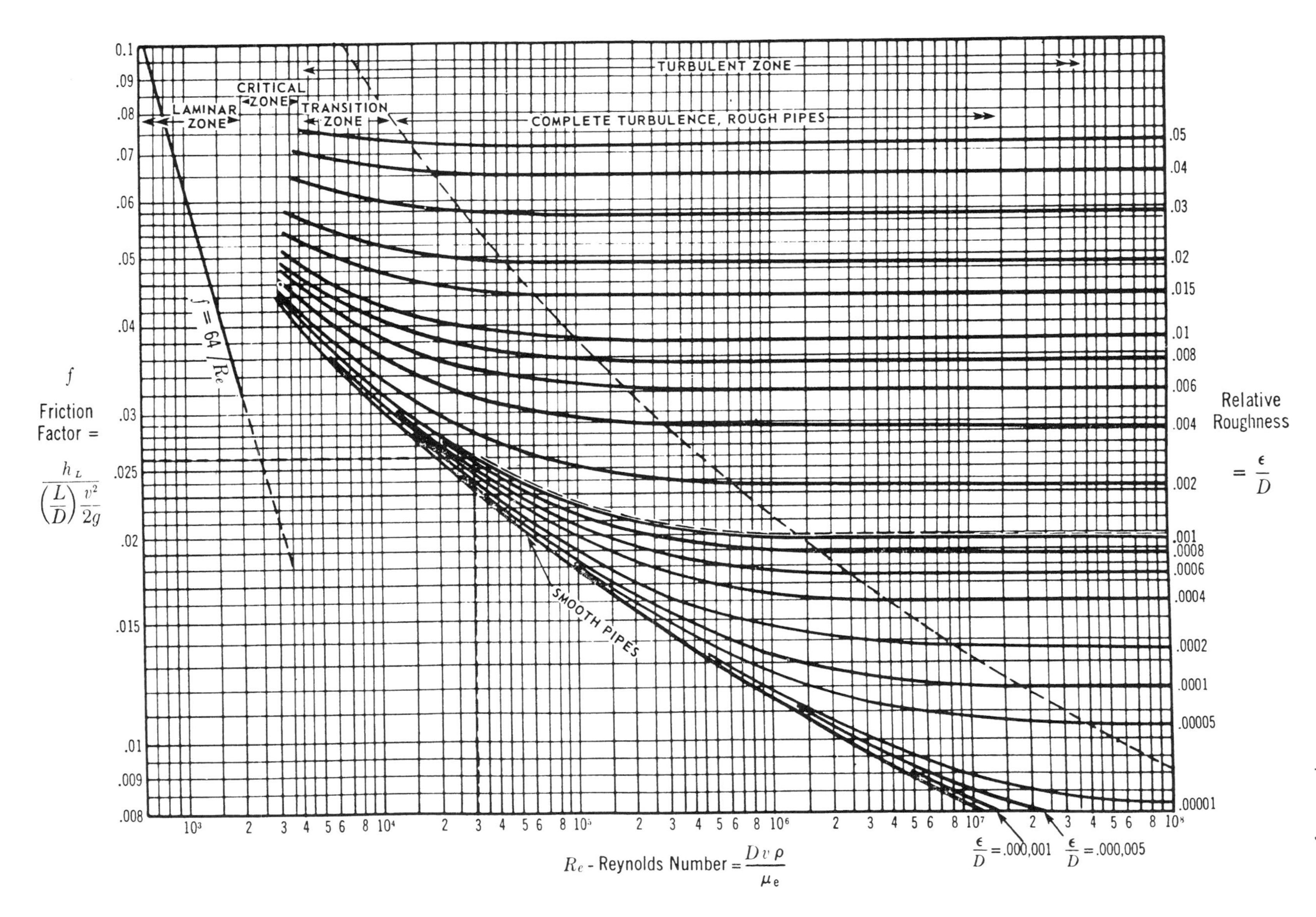

Figure 2. Friction factors for any type of commercial pipe.

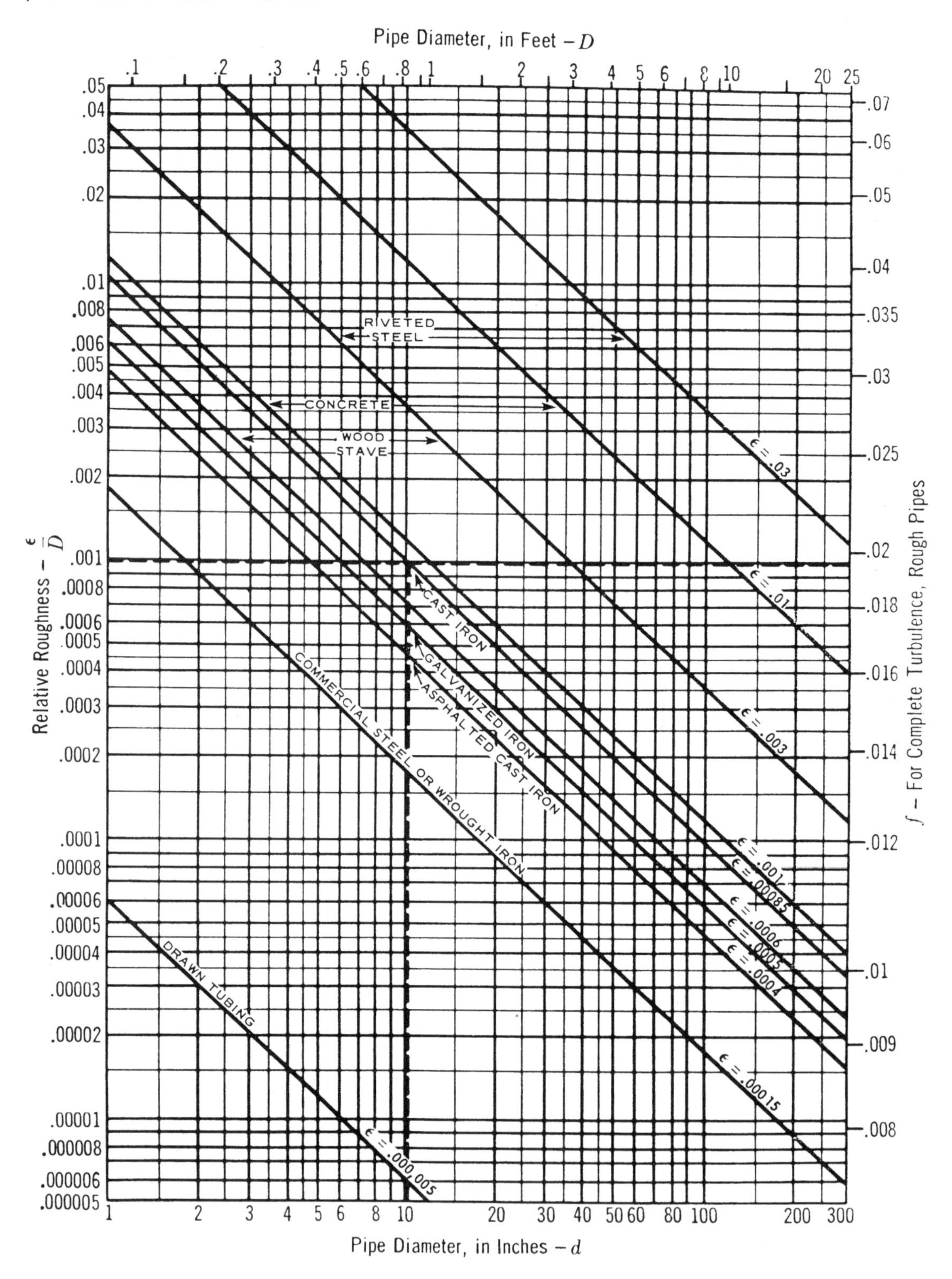

Figure 3. Relative roughness of pipe materials and friction factors for complete turbulence.

Drag-reducing agents

Drag-reducing agents (DRAs) can substantially reduce friction loss in most pipelines when flowing most hydrocarbon liquids.

When a DRA is dissolved in a solvent, it produces a solution that in laminar flow has the same pressure drop, and therefore the same Newtonian viscosity, as the solvent. But when the solution is in turbulent flow, a DRA produces a pressure drop smaller than that which would occur with untreated solvent moving at the same flow rate. The turbulent flow friction of the solution is less than that of the solvent.

DRAs are highly viscoelastic and thixotropic. They may have a viscosity at low shear rates in excess of 3,000,000 cP, with a viscosity 100 to 1,000 times less at higher rates. The viscosity is somewhat affected by temperature and increases with decreasing temperature. DRA-solvent solutions are viscoelastic, time-independent, shear degradable, non-Newtonian fluids. A DRA-solvent solution handles as an ordinary hydrocarbon except in turbulent flow, when the reduced friction becomes evident.

DRAs in present use in oil and products pipelines are themselves hydrocarbons and thus should have no effect on refining processes or refined products. Millions of barrels of crude treated with DRAs on the Alyeska system have been refined without incident. Drag-reduction installations need not involve large-diameter pipelines. The 200,000 b/d increase in the Alyeska system resulting from the injection of some thousands of gallons per day of DRA in this 48-in. line is dramatic.

However, the offshore operator who obtains a 20% increase in capacity in his 8-in. flowline by injecting a few dozen gallons of DRA per day is happy: Without the DRA, he would have to shut in production or loop the flowline. Either alternative would be orders of magnitude more expensive than the injection of a DRA.

As production decays over time, the amount of DRA injected will be reduced, with the eventual result that production will match the untreated flowline capability and operations can proceed as before but with no shut in wells in past history and no sterilized investment in a loop to be carried on the books in future years.

Early applications

The drag-reduction effect as applied to solutions was accidentally discovered by B. A. Toms in the United Kingdom in 1947.[1] It had been known for years that concentrated water–coal slurries and wood pulp–water slurries could produce a friction factor lower than that of water alone. But Dr. Toms's 1948 publication was the first indication of drag reduction in otherwise Newtonian solutions.

The first use of DRAs in the oil industry was to reduce pressure loss during pumping of fluids downhole to fracture tight formations. The quantities of DRA used are proportionally large, but the savings are substantial. Data published on one DRA-kerosene solution with 600 weight ppm of DRA in kerosene showed an 80% drag reduction over untreated kerosene.

The ongoing use of DRAs to debottleneck the Alyeska 48-in. crude line has been described in several published articles.[2] There are many other uses on commercial pipelines—one considerably larger than the Alyeska example—and the list of operators using DRA is growing. Use of DRAs in multistation pipelines is a special application, because the drag-reduction effect does not survive shear stresses raised by centrifugal pumps and appears to be badly degraded going through most positive displacement pumps.

The general rule is that DRAs are always injected on the discharge side of a pump station. If drag reduction is required along a pipeline with N pump stations, the DRA must be injected at N points.

The effectiveness of a DRA depends to a certain extent on the viscosity of the untreated oil and decreases as viscosity increases. It varies slightly with diameter and decreases as diameter increases. The effect varies greatly with flow velocity, which, for a given oil and line diameter, means it varies greatly with Reynolds number. Drag reduction increases as Reynolds number increases. However, dilute solutions are believed to disintegrate at shear rates found in commercial pipelines.

There are some as yet little understood practical aspects of DRA application. For instance, there seems to be some relationship between the physical characteristics of the oil and of the DRA. A certain DRA will produce good results in a given crude yet produce poor results in a seemingly similar oil.

Untreated non-Newtonian oils may produce non-Newtonian DRA–oil solutions, and a particular DRA may increase, rather than decrease, flowing friction. Sometimes an additional, specially compounded additive is required to obtain the desired drag reduction. In addition, there appears to be something in certain DRA–solvent combinations that has to do with the solubility of the DRA in the oil. Test loop studies verify that for these combinations drag reduction appears to increase asymptotically as the distance from the point of injection increases. Specially designed DRA injection nozzles help in these cases.

DRA activity

The precise mechanism describing how a DRA works to reduce friction is not established. It is believed these agents work by directly reducing turbulence or by absorbing and later returning to the flowing stream energy that otherwise would have been wasted in producing the crossflows that comprise turbulence. The absorb-and-return theory has a certain intuitive attractiveness, because one of the characteristics of a viscoelastic fluid is its ability to do this on a physically large scale. Whether the effect exists at the molecular level is unknown.

Pipe liners generally work with the rational flow formula. The friction factor therein, which is a scaling factor applied to velocity head to yield friction loss, is a macro factor in that it lumps all of the friction-producing effects acting in a flowing system into a single numerical value. The friction factor concept does not provide an insight into how a DRA might work. It is necessary to look at the velocity distribution pattern in a pipeline in turbulent flow to understand the bases that support the concept of the f-factor and understand better how drag reduction might take place.

The concept of turbulent flow in pipelines provides that the point flow may be directed in any direction while maintaining net pipe flow in the direction of decreasing pressure. All flow not in the direction of net pipe flow absorbs energy; the more turbulent the flow becomes, the more energy absorbed in these crossflows. In flow in hydraulically rough pipes, the f-factor does not vary with increasing flow, but the unit friction loss increases as the square of the velocity of net pipe flow.

The model of turbulent pipe flow we will use is based on the Universal Law of the Wall, which applies to both smooth and rough pipe flow. It presumes two general modes of flow in a pipe: the wall layer and the turbulent core.

The wall layer is presumed to contain all of the flow where the point velocity varies with the distance of the point from the pipe wall. The turbulent core carries all of the flow where variations in point velocity are random and independent of this distance. The wall layer itself is presumed to be composed of three sublayers: the laminar sublayer, the buffer zone, and the turbulent sublayer.

The laminar sublayer has as its outer boundary the wall of the pipe, and flow in the entire sublayer is laminar. This definition requires that immediately at the wall lies a film of liquid that is not moving.

The increase in point velocity as a point moves away from the wall is a linear function of the distance from the wall and directed parallel to the wall in the direction of pipe flow. There are no crossflows in this sublayer. The contribution of the laminar sublayer to total friction arises in viscous shear generated in the sublayer.

Pipe roughness acts to inhibit laminar flow; therefore, the definition is only statistically true for rough or partially rough pipe.

In the turbulent sublayer, which has as its inner boundary the turbulent core, it is presumed that the velocity varies logarithmically with distance from the wall, but that crossflows in the sublayer may be statistically large and nearly as great as in the core. The contribution of the turbulent sublayer to total friction lies in the turbulent shear generated in this sublayer.

In the buffer zone, which lies between the laminar and turbulent sublayers, variation of point velocity with point position is not established. Ideally, it should be the same as the laminar sublayer at the outer boundary and the same as the turbulent sublayer at the inner boundary. Any number of mathematical functions can be written to approximate these two requirements. Both viscosity and density contribute to shear losses in the buffer zone.

The contribution of the turbulent core to total friction loss lies entirely in turbulent shear; DRAs are presumed not to act on this component of total friction. Therefore, the action of a DRA must take place in the laminar sublayer, the buffer zone, or the turbulent sublayer.

That the action of a DRA increases with increasing flowing velocity would indicate that DRA action could not be primarily in the laminar sublayer. It sometimes would appear that this is so: Liquids treated with a DRA flowing under conditions in which the pipe is hydraulically rough and viscosity presumed to make no contribution to total friction can yield excellent drag reduction. This kind of result would lead to the presumption that a DRA works in the buffer zone.

On the other hand, large doses of a DRA have been observed to keep a liquid in laminar flow at Reynolds numbers much higher than those expected of the same liquid without DRA, which leads to the conclusion that a DRA acts in the laminar sublayer.

This conclusion suggests that a DRA may inhibit the action of crossflows, whether generated in the laminar sublayer or the buffer zone, which move inward into the turbulent sublayer and core and absorb energy that otherwise would go toward producing pipe flow. It appears this is the case.

Reduction in the energy absorbed by crossflows originating at the pipe wall can be accomplished by reducing their number, size, or velocity. The energy in the crossflows may also be absorbed by a DRA and later returned to the flowing stream for reuse, which, in effect, means reducing their effect if not their size, number, or velocity. It is here that theory breaks down; no one really knows which effect predominates, or if either does.

The following is known:

- That DRAs act to reduce turbulence.
- That to do so, they seem to act in the laminar sublayer or the buffer zone by effectively enlarging them or their effect.
- That the net effect is to reduce total friction.
- That the f-factor is less than would be calculated from the pipe liner's usual sources.

Experimental data taken on liquids flowing without a DRA do not translate directly to liquids that have been dosed with a DRA. The rational flow formula still applies, but the usual f-factor relationship does not.[3]

At a given flow rate, the percentage of drag reduction is defined as follows:

$$\%DR = \frac{\Delta Po - \Delta Pi}{Po} \times 100$$

where:
%DR = Percentage drag reduction
Po = Pressure drop of untreated fluid
Pi = Pressure drop of fluid containing drag reducer.

Percentage throughput increase can be estimated as follows:

$$= \left[\left(\frac{1}{1 - \frac{\%DR}{100}}\right)^{0.55} - 1\right] \times 100$$

where %DR is the same as previously defined.

Given a constant Q (flow rate), elevation 1 = elevation 2, and baseline pressures P_1 and P_2, $DPf = P_2 - P_1$ (without DRA).

The volumetric capacity and/or horsepower of the existing pumps may limit the amount of throughput increase that can be achieved. The installation of additional pump capacity may be required in order to obtain the full benefits of DRA use.

Example.

Pipe OD = 10.750 in.
Pipe w.t. = 0.250 in.
Length = 50 miles
Upstream elev. = 500 ft
Downstream elev. = 500 ft
Fluid crude oil
Oil temperature = 85°F
Discharge (source) pressure = 1,440 psig
Residual pressure (sink) = 50 psig
Base throughput (untreated) = 54,000 b/d

Oil properties

Temp.	60°F	85°F	90°F
Visc.	60 SUS	49 SUS	48 SUS
SG	0.8528	0.8423	0.8398
Visc.	8.70 cP	5.99 cP	5.72 cP
Visc.	10.20 cSt	7.11 cSt	6.81 cSt

Desired throughput increase = 20% or 64,800 b/d.

$$20\% = \left[\left(\frac{1}{1 - \frac{\%DR}{100}}\right)^{0.55} - 1\right] \times 100$$

Solve for %DR
%DR = 28.25%

Using a program such as GENNET M,[4] the calibration factor on the pipe data input screen can be adjusted as follows:

1 − 0.2825 = 0.7175. Once this is done, run the program to find the new flow rate of 64,900 b/d.

Contact the DRA vendor to determine the amount of DRA that needs to be added to achieve the desired %DR.

References

1. Toms, B. A., "Some observations on the flow of linear polymer solutions through straight pipe at large Reynolds numbers," *Proceedings of the International Congress on Rheology*, Scheveningen, The Netherlands, vol. 2, pp. 135–141 (1948).
2. Burger, Edward D., Monk, Werner R., and Wahl, Harry A., "Flow increase in the Trans Alaska pipeline using a polymeric drag reducing additive," *Society of Petroleum Engineers*, 9419 (1980).
3. *Oil and Gas Journal*, February 18, 1985.
4. Refer to the multiphase flow calculation example for information on GENNET M (Fluid Flow Consultants, Inc., Broken Arrow, Oklahoma).

How to estimate the rate of liquid discharge from a pipe

A quick and surprisingly accurate method of determining liquid discharge from an open-ended horizontal pipe makes use of the equation:

$$gpm = \frac{2.56 \times d^2}{\sqrt{Y}}$$

where:

gpm = Gallons per minute
d = Diameter of pipe, in.
X = Inches
Y = Inches

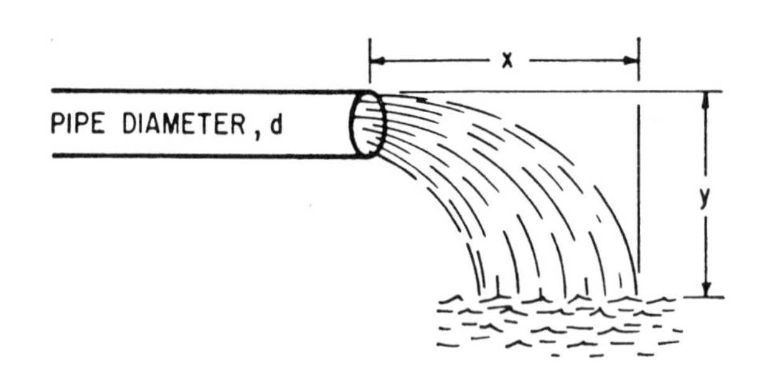

Example. From the drawing:

X = 24 in.
Y = 25 in.
d = 2 in.

$$gpm = \frac{(2.56)(24)(2^2)}{\sqrt{25}}$$

$$= \frac{245.76}{5}$$

$$= 49.15$$

Predict subsurface temperature ranges

Ground temperatures are vital to pipeline design for two reasons. First, the heat exchange between the contents of the line and the ground are greatly affected by the differential temperatures of the pipe and the soil. Second, the temperature of the pipe at times of installation will have a considerable effect on stresses, particularly where the subsequent ground temperatures will vary considerably from the installation temperatures.

For example, in the recent construction of a 34-in. line that is to operate at temperatures well over 100°F, care was exercised to limit bends when it was installed at temperatures ranging from 50°F to 70°F and the underground temperature will range from about 50°F to 80°F over a year's time.

Generally speaking, most areas will have somewhat similar curves from 3 ft on down. Of course, with depth, they approach the temperature of underground water, and the summer-winter variation ceases.

To adapt for U.S. work. As a comparison, the following temperature ranges are encountered in winter and summer at depths of 3 ft in various areas in the United States.

California	51°F to 81°F
Colorado plains	35°F to 65°F
Texas (Panhandle)	42°F to 75°F
Texas (Coast)	60°F to 80°F
Mississippi	57°F to 80°F

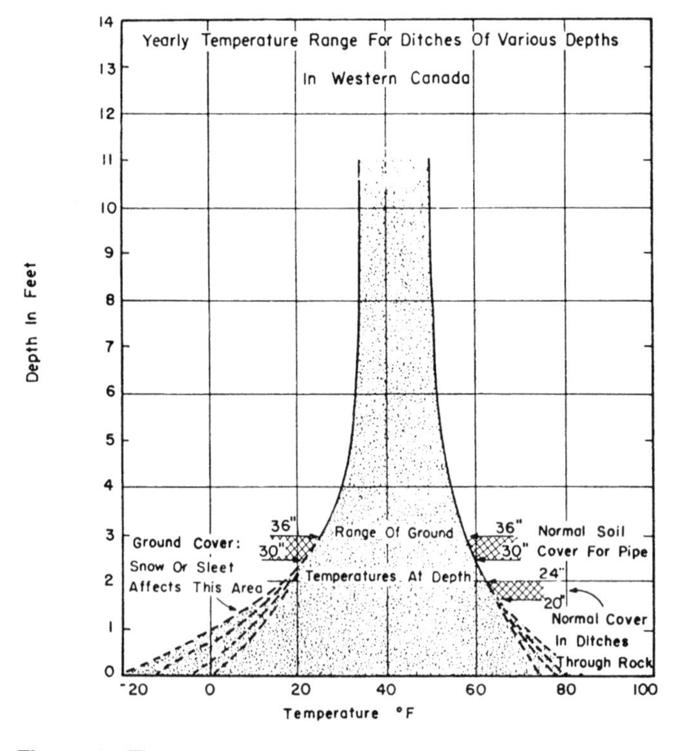

Figure 1. The accompanying curve and table shows the range of underground temperatures you can expect in the prairie regions of Canada. At shallow depths, there is some variation and uncertainty as to temperature because of snow or sleet covering. But from 3 ft down, the temperature stabilizes.

Idaho .... 37°F to 56°F
Illinois .... 41°F to 56°F
Montana .... 33°F to 54°F
Michigan .... 38°F to 60°F
Wyoming .... 26°F to 65°F

Generally, a curve plotted for any one of the above will closely parallel the curve for Canada.

Sizing pipelines for water flow

This form gives a reasonable degree of accuracy.

$$d^5 = \frac{(gpm)^2(D)}{(10)(H)(100)}$$

where:

d = Inside diameter of pipe in inches
gpm = Gallons per minute
D = Equivalent distance (in ft) from standpipe (or pump) to outlet
H = Total head lost due to friction in feet

Example. What size line should be run from a water tower to a cooling tower for a minimum of 160 gpm if the equivalent distance (actual distance in ft plus allowances for valves and fittings) is 1,000 ft and the minimum head supplied by the tower will always be at least 40 ft?

Solution.

$$d^5 = \frac{(160)^2(1{,}000)}{(10)(40)(100)} = 640 \text{ in.}$$

Rather than extract the fifth root of 640, simply raise inside diameters of logical pipe sizes to the fifth power and compare them with 640.

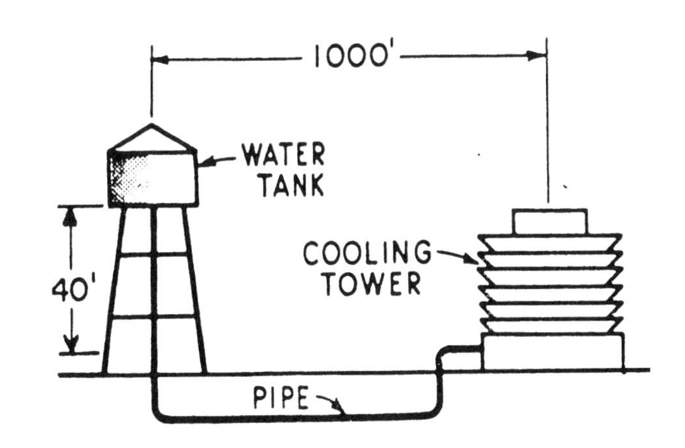

If 3-in. ID pipe were used

$$d^5 = (3)(3)(3)(3)(3) = 243$$

This is obviously too small. Trying 4-in. pipe

$$d^5 = (4)(4)(4)(4)(4) = 1{,}024$$

How approximate throughput of a line can be estimated from pipe size

Multiply the square of the nominal pipe size by 500.

Example. The throughput of a 12-in. line is approximately 144 by 500 = 72,000 barrels per day.

Naturally, the results of this rule are at best only a very rough approximation, since no account is taken of gravity, viscosity, pressures, or even of spacing between pump stations. Nevertheless, almost all of the oil and products lines now in operation have throughput capacities that will be given by the above rule, with the constant (given as 500) ranging between 300 and 600. The rule does *not* apply to short lines, such as loading lines, or runs within a refinery yard, but only to cross-country pipelines.

Gauge liquid flow where no weir or meter is available

Rule. Measure number of seconds required to fill 5-gallon pail and divide this number into the factor 10,000 to get the answer in barrels per day.

Actually, the factor is 10,286, but it is so close to 10,000 that 10,000 can be used for a rule of thumb number since it is easy to remember.

Example. It requires 100 seconds to fill a 5-gallon bucket. 10,286 divided by 100 equals 102.86 bbl per day, or for practical purposes:

$10{,}000 \div 100 = 100$ bpd

Proof: 100 seconds to fill 5-gallon pail equals 20 seconds per gallon equals .05 gallons per second.

$60 \times 60 \times 24 = 86{,}400$ seconds in a day

$86{,}400 \times .05 = 4{,}320$ gallons per day

$4{,}320 \div 42$ gal per bbl $= 102.86$

Estimate crude gathering line throughput for a given pipe diameter

To select the proper size line or estimate throughput for crude oil gathering lines, an engineer with a leading oil company has devised this handy rule: **Throughput in barrels per hour is roughly equal to a constant times the square root of pressure drop per mile (psi/mile).** Constants were derived from field data for various diameter lines between 2 and 10 in.

Pipe diameter (in.)	Constant (K)
2	7.8
3	19.2
4	37.8
6	105
8	209
10	385

Example. What size gathering line would be required to move 240 barrels per hour of crude through a 10-mile line using a pump discharge pressure of 400 psi? Assuming no significant changes in head due to elevation and a pressure drop of 40 psi per mile, for the first trial calculation a 4-in. line appears to be the best choice, so try the constant 37.8.

Throughput (4-in.)

$$= 37.8\sqrt{40} = (37.8)(6.4) = 241 \text{ barrels per hour}$$

So a 4-in. gathering line will be adequate.

How to determine head loss due to friction in ordinary iron pipeline carrying clear water

The form below gives a reasonable degree of accuracy:

$$S = \frac{G^2}{10\,d^5}$$

where:

S = Head loss due to friction in ft per 100 ft length of pipeline
G = Rate of flow in gallons per minute
d = Nominal diameter of pipe in in.

The exact head loss due to friction in pipelines carrying clear water is given by the Williams and Hazen formula:

$$S = \left[\frac{V}{1.318\,C\,R^{.63}}\right]^{1.85}$$

where:

S = Loss of head due to friction in ft per 100 ft
V = Average velocity in ft per second
C = Coefficient depending on the roughness inside the pipe
R = The mean hydraulic radius $= \frac{D}{4}$

where:

D is the diameter in ft

The average velocity V is given by Q/a where Q is the rate of flow in ft/sec and a is the cross-sectional area of the pipe.

C is a function of the material and age of the pipe and the type of water. For smooth new iron pipe C = 120. For

15-year-old ordinary pipe C = 100. For 25-year-old ordinary pipe C = 90.

Example. What will be the head loss per 100 ft if a 30-in. line flows 10,000 gpm?

Solution.

$$S = \frac{(10{,}000)^2}{10(30)^5} = \frac{100}{243} = 0.41\ \text{ft}$$

Exact answer using the Williams and Hazen formula:

$$S = 0.34\ \text{ft}$$

Remarks. This rule of thumb could be very satisfactorily used for pipe diameters 3 in. and larger and a discharge greater than 400 gpm.

This rule applies only to wrought and cast iron pipes (20 years service). If new smooth brass or steel pipe is used, a factor should be used = 0.6.

For 30-year-old ordinary pipe factor = 1.2. These are empirical factors.

For pipe diameters 2 in. and less and discharges less than 200 gpm, this rule of thumb does not give satisfactory results.

How to size lines, estimate pressure drop, and estimate optimum station spacing for crude systems

This table will give "horseback" estimates as to optimum line size, pressure drop, and pump station spacing for crude oil of approximately 40°API gravity and 60 SSU viscosity.

The depreciation was assumed to be 6% straight line, return on the investment 8%, and capital and operating costs were based on Mid-Continent averages:

For	Use		
Throughputs (bbl per day)	Pipe Diameter (in. OD)	Pressure Drop (psi per mile)	Pump Station Spacing (miles) (for 800 psi disch. press.)
0 to 2,000	3½	–	–
2,000 to 3,000	4½	32	25
3,000 to 7,500	6⅝	16	50
7,500 to 16,500	8⅝	10.5	76
16,500 to 23,500	10¾	8.5	94
23,500 to 40,000	12¾	7	112

Estimate the optimum working pressures in crude oil transmission lines

The approximate economic optimum working pressure in psi = 1250 − 15 × OD of line.

Example. Estimate the maximum economic optimum working pressure—which is usually station discharge pressure—for a 20-in. crude oil pipeline.

Solution.

Approximate maximum working pressure = 1,250 − 15 × 20 = 950 psi.

If the line profile is approximately horizontal, the maximum working pressure occurs at the discharge of a pump station. This method is useful only for approximations.

How to size crude oil and products lines for capacity increases

A new pipeline for crude oil or products service should be sized to carry ⅔ of its ultimate capacity.

Example. Design a pipeline for an initial capacity of 100,000 barrels per day. This could be done by laying a 16-in. line and spacing 14 pump stations along its length. To increase the capacity by 50% or to 150,000 barrels per day would require adding 14 intermediate booster stations, resulting in excessive operating costs. If the line had been designed so that the 100,000 barrels per day was ⅔ of its ultimate capacity, a 20-in. line with five stations would be installed. To increase the capacity to 150,000 barrels per day would require adding five stations. This is based on the theory that when expanding the capacity of a line, install a loop line if the capacity is to be greater than 50% of the original capacity. If the expansion is to be less, add horsepower. Both of these approaches are based on statistical results rather than economic studies.

How to determine the maximum surge pressure in liquid-filled pipeline when a valve is suddenly closed

The maximum surge pressure that occurs in a liquid pipeline when a valve is suddenly closed may be quickly determined by using the simple rule of thumb: Line pressure equals 0.8 times the weight per cubic foot of liquid times its velocity in feet per second. This compares favorably to the results when using the following equation.

$$P = \frac{12wV\sqrt{\dfrac{32.2}{w}\left(\dfrac{1}{\frac{1}{K}+\frac{D}{tE}}\right)}}{144 \times 32.2}$$

where:

P = Surge pressure, psi
w = Weight of liquid, lb./ft^3
V = Velocity change, fps
K = Bulk modulus of the liquid, psi
D = Outside diameter of the pipe, inches
t = Pipe wall thickness, inches
E = Young's modulus of the pipe material, psi

Example. An 8.625-in. × 0.250-in. w.t. products pipeline transports kerosene at 4.5 fps. The weight of the kerosene is 51.12 lb./ft^3, and the API gravity is 41.1 (rel. density = 0.8198). K = 171,000 psi.

Using the simplified equation, $P = 0.8\,wV = 0.8 \times 51.12 \times 4.5 = 184$ psi

Using the full equation, P = 178 psi; the difference is approximately 3%.

What is the hydrostatic pressure due to a column of liquid H feet in height?

$$P = \frac{H \times 61.41}{API + 131.5}$$

Example. A column of 37° oil is 20 ft high; what is the pressure at the base of the column? P – 20 × 61.41

$$P = \frac{20 \times 61.41}{37 + 131.5}$$

$$P = 7.28\,\text{psi}$$

Transient pressure analysis

Transient pressure analysis is the study of a pipeline system's response to a time-dependent upset condition such as a valve closing or a pump shutting down. When a valve is closed downstream of the source pressure, the pressure will increase on the upstream side of the valve while the downstream pressure decreases. The valve closure time will determine how fast the upstream pressure increases and the downstream pressure decreases. The high upstream pressure "surge" will propagate upstream while the low pressure surge moves downstream. If the downstream side is open to a tank, the downstream surge will quickly dissipate due to the surge being absorbed in the tank. Similarly, when a pump shuts down, the pressure downstream of the pump decreases and the pressure upstream of the pump will increase. Other causes of pressure surges include tank changes, manifold changes, or leaks.

Steady state versus transient conditions

Steady-state means that the pressure gradients and flow rates in the pipeline are not changing. Transient means the gradients and flows are a function of time. The event that causes conditions to be time dependent is the upset condition such as valve closure or pump failure. Transient conditions must be considered when determining if the pipeline will be subjected to abnormally high operating pressures.

Fluid properties

All pressures and flows, whether steady state or transient, depend on the fluid properties. This example deals with a pipeline filled with batches of motor gasoline and diesel fuel. The program used for the analysis is designed for almost incompressible fluids such as crude oil, products, and water.

The properties required for surge analysis include: specific gravity, viscosity, bulk modulus, and vapor pressure. All properties should be at the same base temperature. Multiple batches can be handled and PVT data will be needed for each batch.

Pipe

The pipe data, such as outside diameter, wall thickness, specified minimum yield strength (SMYS), Poisson's ratio, and Young's modulus of elasticity are required. The length and elevation profile of the pipeline are also required.

Pipe environment

Since an isothermal surge analysis is being performed, the only environmental information required is how the pipeline is anchored. Is the pipe aboveground or underwater? Does it have expansion loops throughout, or is it just constrained at the source? Multiple pipes can be handled, and each pipe may be constrained differently.

Equipment

The data required for a centrifugal pump includes rated flow, rated head differential, shutoff head differential (PSOH), maximum allowed discharge pressure, minimum allowed suction pressure, discharge pressure trip setting, initial speed, final speed, time it takes to go from initial to final speed (ramp-down time), time when the pump starts to ramp down, and whether the pump is protected by a check valve in the pump discharge line.

The data required for valves includes valve closure time, time or pressure when the valve starts to close, closure curve shape exponent (the exponent is different for gate valves, ball valves, etc.), and the amount of valve closure.

Data required for pressure relief valves include the set pressure and the back-pressure on the relief line.

Initial conditions

Since surge is a time-dependent phenomenon, we need the conditions that exist before any upset occurs. The required initial conditions are: the pressure at the delivery end of the pipeline. The program will establish the initial steady-state flow rate automatically.

Perturbations that cause surges

Any time-dependent change, such as a leak, valve closure, or pump shutdown will cause surges in the system. Cascading upset events, such as internal pumps tripping, may cause other pumps to trip and can be modeled.

The fluid-pipe-equipment interaction

The liquid and the pipe, according to its environment, expand and contract as the surges travel. The pumps, valves, and other equipment change the pressures and/or flows as the surges reach them. The surges travel at the speed of sound in the fluid-pipe system. The fluid itself is what relays the surge energy to the various pipe sections and the equipment. The fluid, pipe, and equipment are inextricably linked, and their behavior must be determined by a model that employs simultaneous fluid pipe and equipment solution procedures. The Method of Characteristics, which NATASHA PLUS© uses, is a simultaneous technique and therefore models "surge" phenomenon rigorously.

To illustrate the use of the program, the pipeline system shown in Figure 1 is equipped with a valve at the end of the pipeline that is caused to close at time $T = 300$ seconds.

The valve travels from full open to full closed in 5 seconds. A pressure relief valve is connected to the system and is set to relieve at 275 psig. The system consists of six pipes configured as shown in Table 1.

Simulation

The operation of the system was simulated for 1,200 seconds, and the results are shown graphically in Figure 2. The program outputs a detailed tabular listing showing pressures and flow rates at various locations along the pipeline at selected time steps. A graphics file containing selected node pressures and flow rates is also produced by the program. Figure 2 was prepared from the graphics file produced by the program. A typical time page report is shown in Table 2. The program also prints a maximum pressure summary which is shown in Table 3. A summary of the relief valve operation is shown in Table 4, indicating that the total volume released was 855 bbls during the simulation.

The pump is not shut down in this simulation because of the action of the relief valve. A review of the maximum pressure summary indicates that the pipeline MOP is exceeded at two locations. In order to prevent overpressure, it will be necessary to make changes to the system such as changing the relief valve setting or slowing the closing time for the valve. NATASHA-PLUS© can then be used to simulate the changes to confirm that pipeline MOP is not exceeded.

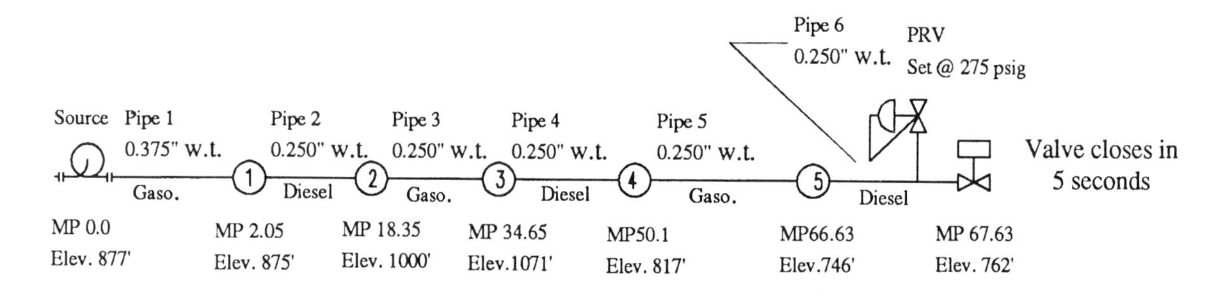

Product parameters:

Product	sp. gr.	visc. cSt	bulk modulus psi
Diesel	0.79	2.15	156,495
Gasoline	0.741	0.622	121,505

Initial flow rate is 92,500 BPD
Downstream pressure is 40 psig.

Figure 1. Sample pipeline for transient analysis study.

Table 1

Data Required By The Program

Pipe Design Data

Pipe No.	Length Mi.	OD in.	W. T. in.	Roughness in.	SMYS psig	Design Fact.
1	2.05	12.75	0.375	0.0018	35000	72%
2	16.3	12.75	0.25	0.0018	52000	72%
3	16.3	12.75	0.25	0.0018	52000	72%
4	15.45	12.75	0.25	0.0018	52000	72%
5	16.53	12.75	0.25	0.0018	52000	72%
6	1.0	12.75	0.25	0.0018	52000	72%

Table 1 continued

Liquid Properties

Pipe No.	Sp. Gr.	Vis. cSt	Bulk Mod. psi	Vapor Press psia
1	0.741	0.622	0.122	5.7
2	0.79	2.15	0.156	10
3	0.741	0.622	0.122	5.7
4	0.79	2.15	0.156	10
5	0.741	0.622	0.122	5.7
6	0.79	2.15	0.156	10

Initial Conditions

Pipe No.	Flow BPD	Upstream Press psig
1	92500	1398.22
2	92500	1444.19
3	92500	958.58
4	92500	659.75
5	92500	363.1
6	92500	68.76

Pipe No.	Friction Factor	Seg. Length Length, mi.	Press. Drop psi/mi.	Head Drop ft/mi
1	0.014	1.03	21.5907	67.2098
2	0.0157	1.02	23.2816	67.978
3	0.014	0.91	19.4468	60.5359
4	0.0157	1.03	23.2816	67.978
5	0.014	0.92	19.4468	60.5359
6	0.0157	1.00	23.2816	67.978

PUMP PARAMETERS

PUMP AT NODE 7 THAT DISCHARGES INTO PIPE 1

RATED FLOW	=	92,500 bbl/day
ZERO FLOW HEAD DIFFERENTIAL	=	4520.00 ft-head
RATED HEAD DIFFERENTIAL	=	4228.02 ft-head
SUCTION PRESSURE	=	40.00 psig
MAXIMUM DISCHARGE CONTROLLED PRESSURE	=	1440.00 psig
DISCHARGE TRIP PRESSURE	=	1490.00 psig
INITIAL SPEED	=	3565.00 rpm
FINAL SPEED	=	3565.00 rpm
CURVE EXPONENT	=	0.5000

Table 1 continued

VALVE PARAMETERS

PRESSURE RELIEF VALVE AT NODE 5

MAX PRESS. BEFORE RELIEF	=	275.00 psig
BACK PRESSURE ON RELIEF LINE	=	0.00 psig

VALVE AT NODE 6

VALVE STARTS TO CLOSE AT T	=	300.00 sec
VALVE CLOSES IN	=	5.00 sec
VALVE STARTS TO CLOSE ABOVE P	=	99999.00 psig
FINAL FRACTIONAL AREA OPEN TO FLOW	=	0.00
CLOSURE SHAPE EXPONENT	=	1.00

Elevation Data

Pipe No.	Point	MP mi.	Elev. ft.	MOP psig
1	1	0	877	1440
	2	1.59	885	1440
	3	2.05	875	1440
2	1	0	875	1440
	2	1.13	890	1440
	3	1.27	860	1440
	4	4.37	940	1440
	5	4.96	890	1440
	6	10.96	990	1440
	7	11.98	1060	1440
	8	13.03	980	1440
	9	16.3	1000	1440
3	1	0	1000	1440
	2	1.71	1065	1440
	3	4.66	1010	1440
	4	6.42	1085	1440
3	5	9.9	930	1440
	6	10.21	1040	1440
	7	11.46	970	1440
	8	14.21	1040	1440
	9	16.3	1071	1440
4	1	0	1071	1440
	2	3.76	1050	1440
	3	5.53	1005	1440
	4	5.92	1020	1440
	5	6.32	1020	1440
	6	15.45	817	1440

Pipe No.	Point	MP mi.	Elev. ft.	MOP psig
5	1	0	817	1440
	2	4.05	770	1440
	3	5.45	875	1440
	4	5.68	810	1440
	5	6.26	915	1440
	6	7.13	805	1440
	7	8.31	930	1440
	8	12.36	755	1440
	9	16.53	746	1440
6	1	0	746	275
	2	0.28	760	275
	3	0.62	762	275
	4	1	762	275

Table 2

Typical time page report for T = 0 seconds

Pipe No.	Loc. mi.	Elev Ft.	Up Flow M bbl/day	Down Flow M bbl/day	Head Ft.	Press. psig
1	0	877	93	93	5230	1398
1	1	882	93	93	5161	1374
1	2	875	93	93	5092	1355
2	0	875	93	93	5092	1444
2	1	889	93	93	5023	1416
2	2	880	93	93	4953	1395
2	3	906	93	93	4884	1362
2	4	932	93	93	4815	1330
2	5	892	93	93	4746	1320
2	6	909	93	93	4676	1290
2	7	926	93	93	4607	1261
2	8	943	93	93	4538	1231
2	9	960	93	93	4468	1202
2	10	977	93	93	4399	1172
2	11	1007	93	93	4330	1138
2	12	1041	93	93	4261	1103
2	13	981	93	93	4191	1099
2	14	988	93	93	4122	1074
2	15	994	93	93	4053	1048
2	16	1000	93	93	3984	1022
3	0	1000	93	93	3984	959
3	1	1034	93	93	3929	930
3	2	1063	93	93	3874	903
3	3	1046	93	93	3820	891
3	4	1029	93	93	3765	879
3	5	1012	93	93	3710	867
3	5	1043	93	93	3655	839
3	6	1082	93	93	3600	809
3	7	1048	93	93	3545	802
3	8	1008	93	93	3491	798
3	9	968	93	93	3436	793
3	10	952	93	93	3381	780
3	11	1003	93	93	3326	746
3	12	978	93	93	3271	737
3	13	1001	93	93	3217	712
3	14	1024	93	93	3162	687
3	14	1044	93	93	3107	663
3	15	1058	93	93	3052	641
3	16	1071	93	93	2997	619

Table 2 continued

Typical time page report for T = 0 seconds

Pipe No.	Loc. mi.	Elev Ft.	Up Flow M bbl/day	Down Flow M bbl/day	Head Ft.	Press. psig
4	0	1071	93	93	2997	660
4	1	1065	93	93	2927	638
4	2	1059	93	93	2857	616
4	3	1054	93	93	2787	594
4	4	1041	93	93	2717	574
4	5	1015	93	93	2647	559
4	6	1020	93	93	2577	533
4	7	1000	93	93	2507	516
4	8	977	93	93	2437	500
4	9	954	93	93	2367	484
4	10	932	93	93	2297	468
4	11	909	93	93	2227	452
4	12	886	93	93	2157	435
4	13	863	93	93	2087	419
4	14	840	93	93	2017	403
4	15	817	93	93	1947	387
5	0	817	93	93	1947	363
5	1	806	93	93	1892	349
5	2	796	93	93	1836	334
5	3	785	93	93	1781	320
5	4	774	93	93	1725	305
5	5	811	93	93	1669	276
5	6	858	93	93	1614	243
5	6	894	93	93	1558	213
5	7	828	93	93	1503	217
5	8	925	93	93	1447	168
5	9	892	93	93	1391	160
5	10	853	93	93	1336	155
5	11	813	93	93	1280	150
5	12	773	93	93	1225	145
5	13	754	93	93	1169	133
5	14	752	93	93	1113	116
5	15	750	93	93	1058	99
5	16	748	93	93	1002	82
5	17	746	93	93	947	64
6	0	746	93	93	947	69
6	1	762	93	93	879	40

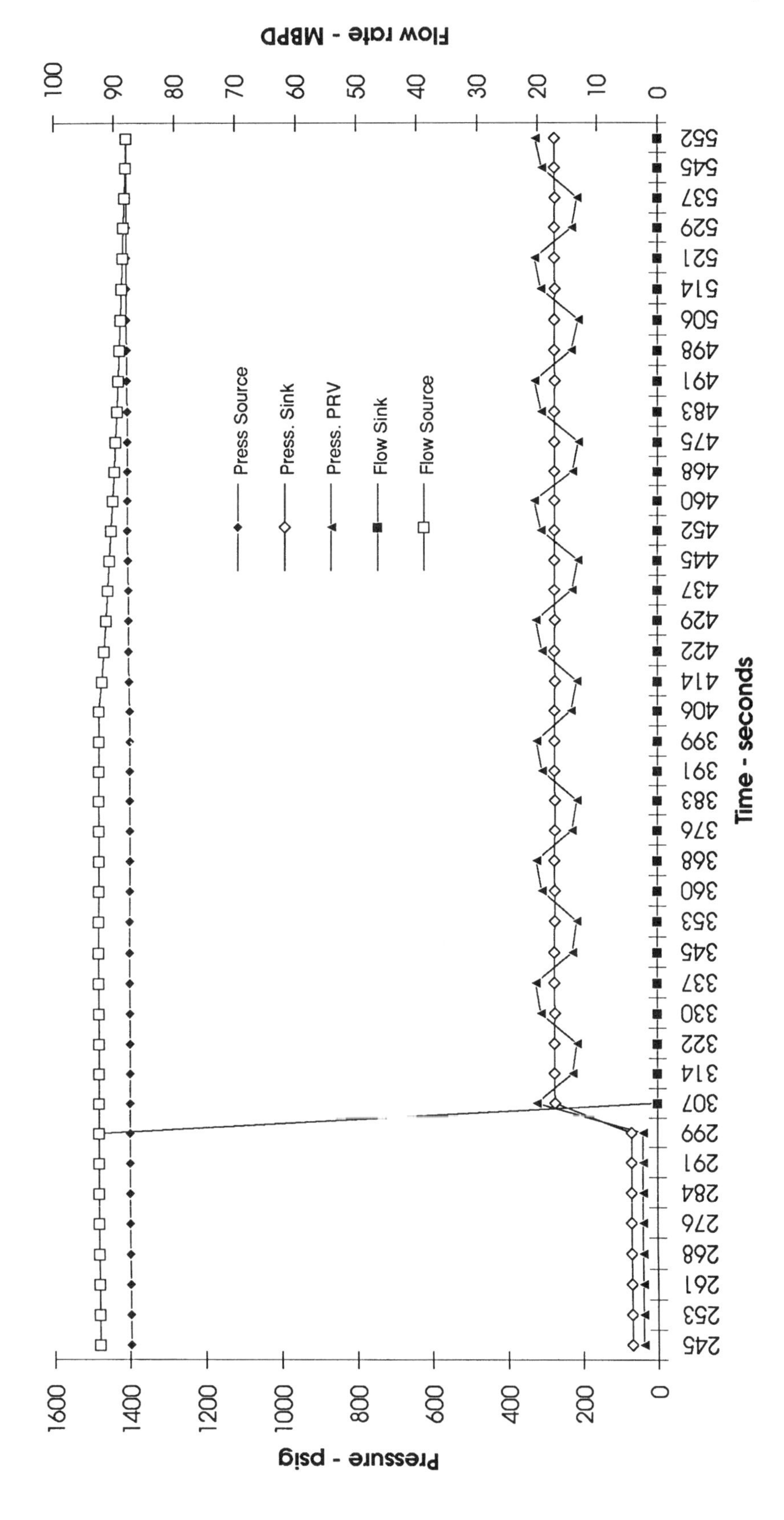

Figure 2. Transient analysis.

Table 3
Maximum pressure summary

Pipe No.	MP mi.	Elev. ft.	Hmax ft.	Pmax psig	MOP psig	Time sec.	
1	0	877	5269.51	1411.07	1440	1199.44	
1	1.03	882.16	5209.78	1390.22	1440	1200.98	
1	2.05	875	5150.04	1373.33	1440	1199.44	
2	0	875	5150.04	1464.14	1440	1199.44	× Exceeds MOP)
2	1.02	888.52	5089.72	1438.85	1440	1200.98	
2	2.04	879.81	5029.4	1421.18	1440	1199.44	
2	3.06	906.1	4969.08	1391.52	1440	1200.98	
2	4.07	932.39	4908.75	1361.85	1440	1199.44	
2	5.09	892.23	4848.44	1354.95	1440	1200.98	
2	6.11	909.21	4788.11	1328.47	1440	1199.44	
2	7.13	926.19	4727.8	1302	1440	1200.98	
2	8.15	943.17	4667.47	1275.53	1440	1199.44	
2	9.17	960.15	4607.16	1249.05	1440	1200.98	
2	10.19	977.13	4546.84	1222.58	1440	1199.44	
2	11.21	1006.9	4486.53	1191.73	1440	1200.98	
2	12.22	1041.33	4426.21	1159.27	1440	1199.44	
2	13.24	981.31	4365.9	1159.18	1440	1200.98	
2	14.26	987.54	4305.58	1136.38	1440	1199.44	
2	15.28	993.77	4245.27	1113.6	1440	1200.98	
2	16.3	1000	4184.95	1090.8	1440	1199.44	
3	0	1000	4184.95	1023.15	1440	1199.44	
3	0.91	1034.42	4137.43	996.82	1440	1200.98	
3	1.81	1063.11	4089.9	972.34	1440	1199.44	
3	2.72	1046.23	4042.38	962.49	1440	1200.98	
3	3.62	1029.35	3994.85	952.65	1440	1199.44	
3	4.53	1012.47	3947.33	942.81	1440	1200.98	
3	5.43	1042.95	3899.8	917.74	1440	1199.44	
3	6.34	1081.54	3852.29	890.08	1440	1200.98	
3	7.24	1048.28	3804.76	885.5	1440	1199.44	
3	8.15	1007.95	3757.25	883.2	1440	1200.98	
3	9.06	967.61	3709.72	880.89	1440	1199.44	
3	9.96	951.68	3662.21	870.74	1440	1200.98	
3	10.87	1003.23	3614.69	838.92	1440	1199.44	
3	11.77	977.95	3567.18	831.77	1440	1200.98	
3	12.68	1001	3519.66	809.1	1440	1199.44	
3	13.58	1024.05	3472.15	786.44	1440	1200.98	
3	14.49	1044.14	3424.63	764.72	1440	1199.44	
3	15.39	1057.57	3377.13	745.15	1440	1200.98	
3	16.3	1071	3329.61	725.57	1440	1199.44	
4	0	1071	3329.61	773.55	1440	1199.44	
4	1.03	1065.25	3268.67	754.64	1440	1200.98	
4	2.06	1059.49	3207.72	735.74	1440	1199.44	

Table 3
Maximum pressure summary

Pipe No.	MP mi.	Elev. ft.	Hmax ft.	Pmax psig	MOP psig	Time sec.	
4	3.09	1053.74	3146.78	716.84	1440	1200.98	
4	4.12	1040.85	3085.82	700.38	1440	1199.44	
4	5.15	1014.66	3024.89	688.48	1440	1200.98	
4	6.18	1020	2963.94	665.77	1440	1199.44	
4	7.21	1000.21	2903	651.68	1440	1200.98	
4	8.24	977.31	2842.06	638.65	1440	1199.44	
4	9.27	954.41	2781.13	625.63	1440	1200.98	
4	10.3	931.51	2720.18	612.6	1440	1199.44	
4	11.33	908.61	2659.25	599.57	1440	1200.98	
4	12.36	885.7	2598.31	586.55	1440	1199.44	
4	13.39	862.8	2537.39	573.52	1440	1200.98	
4	14.42	839.9	2476.45	560.5	1440	1199.44	
4	15.45	817	2415.53	547.47	1440	1200.98	
5	0	817	2415.53	513.52	1440	1200.98	
5	0.92	806.34	2367.36	501.47	1440	1199.44	
5	1.84	795.69	2319.21	489.42	1440	1200.98	
5	2.76	785.03	2271.05	477.37	1440	1199.44	
5	3.67	774.37	2222.9	465.33	1440	1200.98	
5	4.59	810.63	2174.74	438.21	1440	1199.44	
5	5.51	858.04	2126.59	407.51	1440	1200.98	
5	6.43	893.72	2078.43	380.58	1440	1199.44	
5	7.35	827.95	2030.29	386.24	1440	1200.98	
5	8.27	925.23	1982.13	339.52	1440	1199.44	
5	9.18	892.26	1933.99	334.65	1440	1200.98	
5	10.1	852.58	1885.84	331.93	1440	1199.44	
5	11.02	812.9	1837.7	329.21	1440	1200.98	
5	11.94	773.22	1789.55	326.49	1440	1199.44	
5	12.86	753.93	1741.41	317.22	1440	1200.98	
5	13.78	751.95	1693.26	302.39	1440	1199.44	
5	14.69	749.96	1645.13	287.57	1440	1200.98	
5	15.61	747.98	1596.98	272.73	1440	1199.44	
5	16.53	746	1548.9	257.93	1440	742.37	
6	0	746	1548.9	274.98	275	742.37	
6	1	762	1721.75	328.7	275	460.14	× (Exceeds MOP)

Table 4
Pressure relief valve summary

TOTAL VOLUME RELEASED AT NODE 5 = 855.4117 bbl

MAX VOLUME RATE RELEASED AT NODE 5 = 102.958 1000 bbl/day

Tank farm line sizing

By: J. C. Stivers

The designer must locate the tanks, size the tank lines, and select pumps. The designer must know the receiving flow rates, discharge flow rates, liquids to be handled, along with the properties of the liquid such as viscosity, specific gravity, and vapor pressure. The minimum and maximum tank levels must be determined, which will set the available volume that can be pumped. Other factors include determining the number of tanks that will depend upon the type of material being stored, tank turnover frequency, system load factor (which takes into consideration system outages), and environmental considerations.

If possible, the tanks should have combination suction and fill lines. To size the lines, it is necessary to consider the fill and discharge flow rates. If a tank farm can receive from a number of small systems and has one outgoing pipeline, the size of the tank lines will probably be determined by the outgoing flow rate. If a tank farm is fed by one stream and delivers to a number of smaller systems, the line sizes will probably be determined by the incoming stream rate. It is necessary to check both to confirm this generalization.

The criteria for sizing a tank line based on filling rates is somewhat arbitrary. When there are a number of sources, the operator may specify a maximum pressure of 25 to 50 psig required at the connection. The piping manifold, meters, and other equipment must then be sized so that the pressure will not exceed 25 to 50 psig at the maximum flow rate. If a metering system is to be used on the incoming stream and a back pressure valve is required for the meters, the tank line size may be sized to produce the minimum back pressure at the maximum flow rate. When the tank farm owner develops the pressure required to fill tanks, an economic analysis should be made to determine line sizes. Large-diameter pipelines require less pressure and therefore less power cost but require greater investments than smaller-diameter lines.

Determine pipeline size

The equation for determining the line size from the tanks and to fill a pump based on discharge flow rates is as follows:

NPSH available $\geq$ NPSH required by the pump

$$\frac{\text{Atoms. Press (In. Hg)} \times 1.135}{\text{s.g.}} - \frac{\text{Vapor Press.} \times 2.31}{\text{s.g.}} \pm \text{Static}$$

$$\text{Press (Ft. of Head)} - \text{Pipe Friction Loss (Psi/mi.)} \times$$

$$\frac{\text{equiv. feet of suction line} \times 2.31}{5{,}280 \times \text{s.g.}}$$

$$\geq \text{NPSH required by the pump.}$$

Atmospheric Pressure—The lowest atmospheric pressure for the area should be considered. This is particularly important for tank farms where product is transferred from one tank to another using the pump. This usually occurs when the atmospheric pressure is low.

Vapor Pressure—Vapor pressure increases with temperature and therefore reduces the NPSH available.

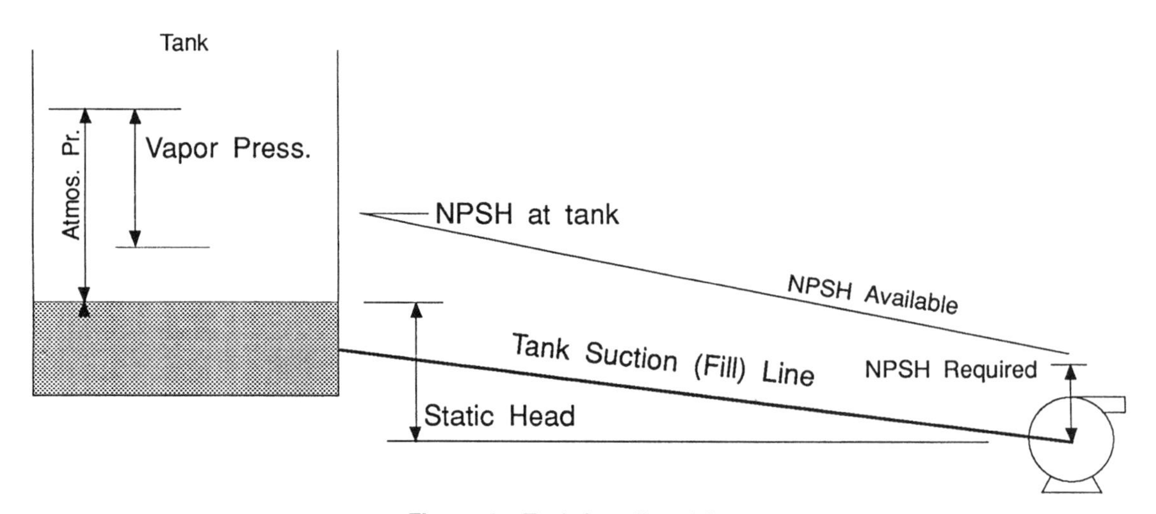

Figure 1. Tank farm line sizing.

Conversely liquid viscosity decreases with temperature increases and therefore reduces pipe friction losses. Therefore, it becomes necessary to determine the liquid temperature that produces the lowest NPSH available.

Static Pressure—This is the elevation difference between the minimum liquid level in the tank and the centerline of the pump suction. If the ground elevation in the tank farm changes, the designer can take advantage of the terrain and locate the tanks accordingly.

Pipe Friction Loss—There are many equations used to determine pipe friction loss. See Section 13.

Most tank lines are not designed to be pigged. Some crudes, particularly sour crudes, may cause a buildup in the tank line, in turn reducing the pipe i.d. Any buildup will cause the friction factor to be higher. If buildup is expected, the designer may wish to increase the pipe size—particularly if NPSH available to the pump is marginal.

NPSH required by pump

The pump manufacturer supplies the NPSH requirement for each pump. Some positive displacement pump manufacturers may not publish these requirements due to a wide range of operating speeds. They will, however, provide requirements for specific conditions. If the NPSH requirements of the pump are not met, the pump will cavitate, and a loss of pressure and flow will occur. Mechanical damage may also be possible.

Types of booster pumps

The pumps most typically used include:

Horizontal split case
Vertical can
Vertical inline

A horizontal split case pump may be used in some instances and will be preferred by most maintenance and operations personnel because it is easy to maintain. The NPSH requirement for a horizontal pump may be 7 to 10 ft, and the suction nozzle may be 2 to 4 ft above grade. The vertical pump can have an NPSH requirement with the suction nozzle 3 or more ft below grade. The NPSH advantage for the can pump can range from 12 to 17 ft, which may permit a size reduction of the tank lines. NPSH requirements for the vertical inline pump are similar to the horizontal pump. The advantage of the inline pump is pump and installation cost.

Design considerations

The designer should not automatically assume that a booster pump is required. Most positive displacement pumps have low NPSH requirements and therefore a booster pump is required unless metering is installed between the tank and mainline pump(s). If the terrain permits the tanks to be located considerably higher than the main pumps, a booster pump may not be required for centrifugal pumps.

If a series of main line pumps are used and the pumping rate is determined by the number of pumps operating, it may be possible to obtain a pump with a lower NPSH requirement for the first pump and thus eliminate the need for a booster pump. This is shown by the following example.

Assume three main line pumps are connected in a series.

Pump combinations produce the following flow rates:

No. of Pumps Operating	Flow Rate
3	100%
2	75%
1	50%

NPSH available

Flow Rate	NPSH Available
100%	15 feet
75%	20 feet
50%	27 feet

is seldom required unless metering is installed between the tank and mainline pump(s). If the terrain permits the tanks to be located considerably higher than the main pumps, a booster pump may not be required for centrifugal pumps.

Low NPSH requirement booster pumps are normally less efficient than mainline pumps. When this is the case, a booster pump should be selected to develop just enough head to feed the main line or delivery pumps.

If a metering system is required, it would normally be installed between the tanks and mainline pumps. In this case, a booster pump would almost certainly be required to overcome the pressure losses in the metering system. The booster pump then must develop enough head to produce the minimum back pressure required by the meters or the minimum NPSH required by the mainline pumps.

If the booster pump is used to transfer liquid from one tank to another, it will probably develop more head than required. This allows the designer to reduce the size of the transfer header and valves.

Once the minimum atmospheric pressure and vapor pressure of the liquid to be pumped is determined, there are three variables on the left side (NPSH available) and one on the right (NPSH required by the pump). The solution is trial and error and the following procedure is suggested:

Select a tentative location for the tanks and determine the minimum liquid level elevation.

Determine the equivalent pipe length of the suction line.

Locate the booster pump and assume it can operate at a specific NPSH.

This allows the solution of the equation for maximum pipe friction loss. Application of pipe pressure loss formula will give the minimum size pipe that will result in a specific NPSH. Select the next larger standard pipe size and rework the equation to determine the actual minimum NPSH available at the pump. Generally, the cost of the booster pump will decrease as the NPSH requirement increases. Higher NPSH will also allow the designer to select a different style pump. For these reasons, the equation should be solved using the next two larger pipe sizes. Pumps can now be selected for three different pipe sizes. The costs of the three systems should be estimated, and an economic analysis made to select the system.

The designer must be aware that a low NPSH pump can pull a vacuum, which must be considered in determining pump differential. (Pipe collapse?) Frequently, the required pump discharge pressure is used to specify a pump differential head, which assumes zero pressure at the pump suction. The following example assumes the booster pump is designed to operate at 7 ft NPSH and 7 ft NPSH is available. The liquid is fuel oil with a vapor pressure of 0.2 psig and 0.85 s.g. and atmospheric pressure is 14.7 psia.

$$(2.31/\text{s.g.}) \times (\text{press. avail.} + \text{atmos. press} - \text{vapor press}) = \text{NPSH}$$

$$\text{Press. avail.} = [(\text{s.g.} \times \text{NPSH})/2.31] - \text{atmos. press.} + \text{vapor press.}$$

$$\text{Press. avail.} = [(0.85 \times 7)/2.31] - 14.7 + 0.2 = -11.92 \text{ psig}$$

If a pump discharge pressure of 25 psig is required, the pump differential is $25 - (-11.92) = 36.92$ psig or 102 ft.

Hydraulics calculations for multiphase systems, including networks

The calculation of pressure, temperature, and volumes in multiphase systems (pipelines, wells, and pipe networks), is a complicated function of fluid composition, pressure, temperature, flow rate, fluid properties, pipe properties, angle of inclination (elevation differences), properties of the medium surrounding the pipes, and pipe connectivity. Therefore, for all but the simplest of systems, a computer program is required.

GENNET-M is a powerful Multiphase Flow program from Fluid Flow Consultants, Inc., of Broken Arrow, Okla. GENNET-M handles pipeline, well, and pipe NETWORK models (one or more sources, one or more sinks, and zero or more loops) of up to 500 pipes, has a Liquids Removal ("pigging") option, a pressure-drop calibration option, a viscosity curve input option, and "DP+DT" equipment (pressure and/or temperature change device, i.e., a pump, compressor, heater, cooler, minor loss across a valve or fitting, pressure drop from an IPR curve, etc.). GENNET seeks to solve for the flow rates in each pipe of an arbitrarily connected network subject to constant boundary (source/sink) pressures and/or flows and the resistance (friction and elevation differences) in the pipes.

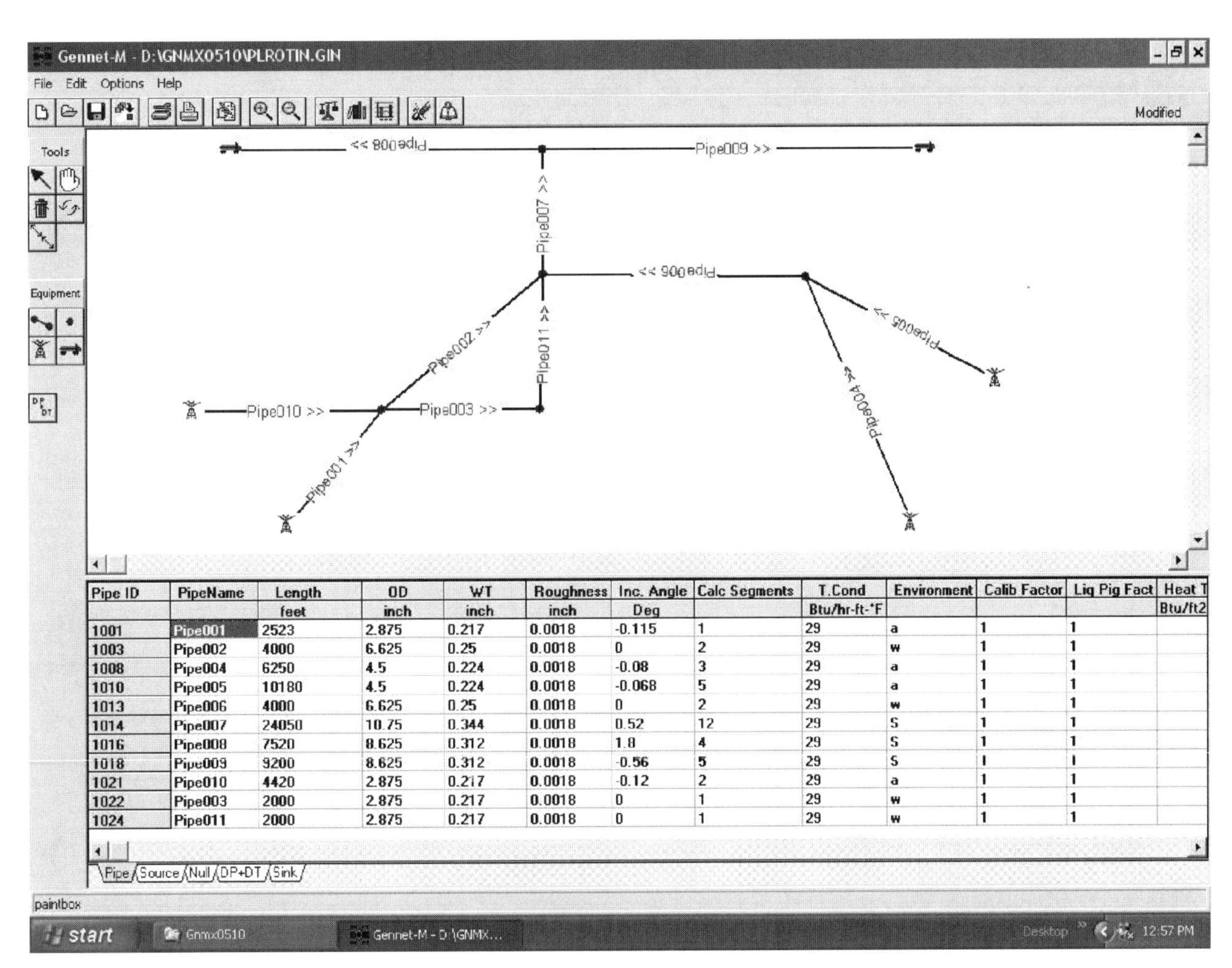

Pipe ID	PipeName	Length	OD	WT	Roughness	Inc. Angle	Calc Segments	T.Cond	Environment	Calib Factor	Liq Pig Fact	Heat T
		feet	inch	inch	inch	Deg		Btu/hr-ft-°F				Btu/ft2
1001	Pipe001	2523	2.875	0.217	0.0018	-0.115	1	29	a	1	1	
1003	Pipe002	4000	6.625	0.25	0.0018	0	2	29	w	1	1	
1008	Pipe004	6250	4.5	0.224	0.0018	-0.08	3	29	a	1	1	
1010	Pipe005	10180	4.5	0.224	0.0018	-0.068	5	29	a	1	1	
1013	Pipe006	4000	6.625	0.25	0.0018	0	2	29	w	1	1	
1014	Pipe007	24050	10.75	0.344	0.0018	0.52	12	29	S	1	1	
1016	Pipe008	7520	8.625	0.312	0.0018	1.8	4	29	S	1	1	
1018	Pipe009	9200	8.625	0.312	0.0018	-0.56	5	29	S	1	1	
1021	Pipe010	4420	2.875	0.217	0.0018	-0.12	2	29	a	1	1	
1022	Pipe003	2000	2.875	0.217	0.0018	0	1	29	w	1	1	
1024	Pipe011	2000	2.875	0.217	0.0018	0	1	29	w	1	1	

Figure 1. Diagram of network and pipe data input screen for the example.

Example.

The GENNET-M input file name is PLROT.GIN. The text input file is PLROTIN.TXT. The text output file name is PLROTOUT.TXT.

This example (PLROT) is a multiphase (oil, gas, and water) problem with four sources, two sinks, and one loop. The source flows are known (GENNET-M calculates the source pressures), and the sink pressures are known (GENNET-M calculates the sink flows). Each source has a different flow rate and composition (oil API; oil viscosity curve; gas gravity; and CO_2, N_2, and H_2S content). GENNET-M mixes the fluids according to the flow rate in each pipe and then calculates the various fluid properties it needs as it calculates the flowing pressures and temperatures. All source pipes (Pipe001, Pipe010, Pipe004, and Pipe005) are in air exposed to a 10 mph wind. The loop pipes (Pipe002, Pipe003, and Pipe011) and Pipe006 are submerged in water exposed to a 10 mph cross-current. The sink pipes (Pipe008 and Pipe009) and Pipe007 are buried in soil at a depth of 36 in. The complete input data can be found in PLROTIN.TXT, which is shown below.

Input

GENNET-M 5.4.23

GENeral NETwork Multiphase flow analyzer
DATE: 01/30/2004 TIME: 12:13

General Data

Title:
Company: Fluid Flow Consultants
Number of Pipes: 11
Fluid Type: Oil
Global Units: English
Pipe Length Units: Feet

PIPE PROPERTIES 1

Pipe Number	Pipe Name	Outside Diameter (in)	Wall Thickness (in)	Incl Angle (Deg)	Rough (in)
1	Pipe010	2.875	0.217	-0.1	0.0018
2	Pipe002	6.625	0.250	0.0	0.0018
3	Pipe004	4.500	0.224	-0.1	0.0018
4	Pipe005	4.500	0.224	-0.1	0.0018
5	Pipe006	6.625	0.250	0.0	0.0018
6	Pipe007	10.750	0.344	0.5	0.0018
7	Pipe008	8.625	0.312	1.8	0.0018
8	Pipe009	8.625	0.312	-0.6	0.0018
9	Pipe001	2.875	0.217	-0.1	0.0018
10	Pipe003	2.875	0.217	0.0	0.0018
11	Pipe011	2.875	0.217	0.0	0.0018

PIPE PROPERTIES 2

Pipe Number	Pipe Name	Thermal Conductivity (Btu/'-h-F)	Length (ft)	Calc Segs	Env	Calibration Factor	Liq Holdup Factor
1	Pipe010	29.00000	4420.0	2	Air	1.00000	1.00000
2	Pipe002	29.00000	4000.0	2	Water	1.00000	1.00000
3	Pipe004	29.00000	6250.0	3	Air	1.00000	1.00000
4	Pipe005	29.00000	10180.0	5	Air	1.00000	1.00000
5	Pipe006	29.00000	4000.0	2	Water	1.00000	1.00000
6	Pipe007	29.00000	24050.0	12	Soil	1.00000	1.00000
7	Pipe008	29.00000	7520.0	4	Soil	1.00000	1.00000
8	Pipe009	29.00000	9200.0	5	Soil	1.00000	1.00000
9	Pipe001	29.00000	2523.0	1	Air	1.00000	1.00000
10	Pipe003	29.00000	2000.0	1	Water	1.00000	1.00000
11	Pipe011	29.00000	2000.0	1	Water	1.00000	1.00000

PIPE CONNECTIVITY

Pipe Number	Upstream Node	Pipe Name	Downstream Node
1	8	Pipe010	1
2	1	Pipe002	2
3	6	Pipe004	3
4	7	Pipe005	3
5	3	Pipe006	2
6	2	Pipe007	4
7	4	Pipe008	9
8	4	Pipe009	10
9	5	Pipe001	1
10	1	Pipe003	11
11	11	Pipe011	2

PIPE ENVIRONMENT

Pipe Number	Pipe Name	Heat Transfer (Btu/ft2-hr-F)	Insul Thick (in)	Insul T.Cond (Btu/'-h-F)	Amb Temp (F)	A/W Vel (mph)	Burial Depth (in)	Soil T.COND (Btu/'-h-F)
1	Pipe010	0.00000	0.00	0.30	60.0	10.0	36.00	0.80
2	Pipe002	0.00000	0.00	0.30	60.0	10.0	36.00	0.80
3	Pipe004	0.00000	0.00	0.30	60.0	10.0	36.00	0.80
4	Pipe005	0.00000	0.00	0.30	60.0	10.0	36.00	0.80
5	Pipe006	0.00000	0.00	0.30	60.0	10.0	36.00	0.80
6	Pipe007	0.00000	0.09	0.30	60.0	10.0	36.00	0.80
7	Pipe008	0.00000	0.09	0.30	60.0	10.0	36.00	0.80
8	Pipe009	0.00000	0.09	0.30	60.0	10.0	36.00	0.80
9	Pipe001	0.00000	0.00	0.30	60.0	10.0	36.00	0.80
10	Pipe003	0.00000	0.00	0.30	60.0	10.0	36.00	0.80
11	Pipe011	0.00000	0.00	0.30	60.0	10.0	36.00	0.80

SOURCE CONDITIONS

```
Source Name: Source004,    Pipe Number:  1,    Pipe Name: Pipe010
  Flowing Temperature     (F):                     124.0
  Flowing Pressure        (psig):                  1200.49
  Oil Rate                (    bbl/day    ):       4350.000
  Gas/Oil Ratio           (  std ft3/bbl    ):      448.000
  Water Cut               (    Volume %     ):        8.200
  Gas Gravity:                                        0.695
  Separator Pressure      (psig):                     0.000
  Separator Temperature   (F):                       60.000
  Nitrogen, N2            (mole%):                    2.120
  Carbon Dioxide, CO2     (mole%):                    3.210
  Hydrogen Sulfide, H2S   (mole%):                    0.082
  Oil Gravity             (API):                     24.550
  Bub Pt Press            (psig):                     0.000
  Water Spec Grav         (FRAC):                     1.060

Source Name: Source002,    Pipe Number:  3,    Pipe Name: Pipe004
  Flowing Temperature     (F):                     136.0
  Flowing Pressure        (psig):                  721.52
  Oil Rate                (    bbl/day    ):       4400.000
  Gas/Oil Ratio           (  std ft3/bbl    ):     1000.000
  Water Cut               (    Volume %     ):       22.300
  Gas Gravity:                                        0.685
  Separator Pressure      (psig):                     0.000
  Separator Temperature   (F):                       60.000
  Nitrogen, N2            (mole%):                    0.125
  Carbon Dioxide, CO2     (mole%):                    1.080
  Hydrogen Sulfide, H2S   (mole%):                    0.015
  Oil Gravity             (API):                     32.800
  Bub Pt Press            (psig):                     0.000
  Water Spec Grav         (FRAC):                     1.100

Source Name: Source003,    Pipe Number:  4,    Pipe Name: Pipe005
  Flowing Temperature     (F):                     142.0
  Flowing Pressure        (psig):                  851.61
  Oil Rate                (    bbl/day    ):       4500.000
  Gas/Oil Ratio           (  std ft3/bbl    ):     1026.000
  Water Cut               (    Volume %     ):       23.800
  Gas Gravity:                                        0.669
  Separator Pressure      (psig):                     0.000
  Separator Temperature   (F):                       60.000
  Nitrogen, N2            (mole%):                    0.095
  Carbon Dioxide, CO2     (mole%):                    0.098
  Hydrogen Sulfide, H2S   (mole%):                    0.006
  Oil Gravity             (API):                     34.600
  Bub Pt Press            (psig):                     0.000
  Water Spec Grav         (FRAC):                     1.090

Source Name: Source001,    Pipe Number:  9,    Pipe Name: Pipe001
  Flowing Temperature     (F):                     128.0
  Flowing Pressure        (psig):                  1238.40
  Oil Rate                (    bbl/day    ):       5100.000
  Gas/Oil Ratio           (  std ft3/bbl    ):      750.000
  Water Cut               (    Volume %     ):        6.300
  Gas Gravity:                                        0.723
  Separator Pressure      (psig):                     0.000
  Separator Temperature   (F):                       60.000
  Nitrogen, N2            (mole%):                    1.430
  Carbon Dioxide, CO2     (mole%):                    4.230
  Hydrogen Sulfide, H2S   (mole%):                    0.065
  Oil Gravity             (API):                     28.600
  Bub Pt Press            (psig):                     0.000
  Water Spec Grav         (FRAC):                     1.060
```

```
                         OIL VISCOSITY

 Source Name: Source004,     Pipe Name: Pipe010:
   Pipe      Temperature       Viscosity
                 (F)           (  cp  )
-----------  -----------      ---------
       1       100.000           1.62000
       1       210.000           0.88000
 Power: Visc(cp) = 0.7154E+02 * Temp(F) ^ -.8225E+00

 Source Name: Source002,     Pipe Name: Pipe004:
       3       100.000           1.20000
       3       210.000           0.44500
 Power: Visc(cp) = 0.5666E+03 * Temp(F) ^ -.1337E+01

 Source Name: Source003,     Pipe Name: Pipe005:
       4       100.000           1.05000
       4       210.000           0.22600
 Power: Visc(cp) = 0.1451E+05 * Temp(F) ^ -.2070E+01

 Source Name: Source001,     Pipe Name: Pipe001:
       9       100.000           1.45000
       9       210.000           0.95000
 Power: Visc(cp) = 0.2001E+02 * Temp(F) ^ -.5699E+00
```

```
                                     SINK CONDITIONS

Sink       Pipe     Pipe      Flowing           Flow
Name      Number    Name      Pressure          Rate
                                (psig)     (    bbl/day     )
          ------  --------   --------     -----------
Sink001      7    Pipe008     100.00        11274.102
Sink002      8    Pipe009     200.00         7075.899
```

Output

The calculated output can be found in PLROTOUT.TXT.

```
                                  GENNET - M 5.4.23

                       GENeral NETwork Multiphase flow analyzer
                          DATE: 01/30/2004          TIME: 12:12


                     PROGRAM and its DOCUMENTATION Copyrighted by:
                            Fluid Flow Consultants 1995-2004
                              --- All Rights Reserved ---
                   (Photocopies of Input, Output, User's Manual may be
                   made as required by authorized users of the PROGRAM)

Pipe   Pipe      Dist       Flow          Pressure  Temp  Liq Holdup Vel mix  Flow
 No    Name      (ft)    (bbl/day)         (psig)   (øF)    (Dbl )   (ft/sec) Patt
---- -------- -------- -----------      ---------- ----- ---------- -------  ----

  9  Pipe001       0.0    0.510E+04      1238.40 128.00
  9  Pipe001    2523.0    0.510E+04       465.89 109.24       6.9       30.3   Intr
                                                          ---------
                                                              6.9

  3  Pipe004       0.0    0.440E+04       721.52 136.00
  3  Pipe004    2083.3    0.440E+04       654.40 115.02      13.6       15.3   Intr
  3  Pipe004    4166.7    0.440E+04       583.04  99.37      13.1       16.2   Intr
  3  Pipe004    6250.0    0.440E+04       504.91  87.90      12.5       17.6   Intr
                                                          ---------
                                                             39.2

  4  Pipe005       0.0    0.450E+04       851.61 142.00
  4  Pipe005    2036.0    0.450E+04       790.08 120.64      14.4       14.0   Intr
  4  Pipe005    4072.0    0.450E+04       726.25 104.37      14.1       14.4   Intr
  4  Pipe005    6108.0    0.450E+04       658.77  92.18      13.7       15.1   Intr
  4  Pipe005    8144.0    0.450E+04       585.88  83.18      13.1       16.2   Intr
  4  Pipe005   10180.0    0.450E+04       504.91  76.64      12.3       17.8   Intr
                                                          ---------
                                                             67.7
```

```
  1  Pipe010      0.0    0.435E+04    1200.49 124.00
  1  Pipe010   2210.0    0.435E+04     880.72 106.14      7.9        16.6   Liq
  1  Pipe010   4420.0    0.435E+04     465.89  92.76      6.7        21.5   Intr
                                                       ---------
                                                         14.6

  2  Pipe002      0.0    0.869E+04     465.89 101.70
  2  Pipe002   2000.0    0.869E+04     444.63  60.00     29.9        11.2   Intr
  2  Pipe002   4000.0    0.869E+04     423.41  60.00     30.1        11.0   Intr
                                                       ---------
                                                         60.0

  5  Pipe006      0.0    0.890E+04     504.91  82.17
  5  Pipe006   2000.0    0.890E+04     465.45  60.00     26.5        16.8   Intr
  5  Pipe006   4000.0    0.890E+04     423.41  60.00     25.7        17.5   Intr
                                                       ---------
                                                         52.2

 10  Pipe003      0.0    0.760E+03     465.89 101.70
 10  Pipe003   2000.0    0.760E+03     444.74  60.00      4.8         6.2   Intr
                                                       ---------
                                                          4.8

 11  Pipe011      0.0    0.760E+03     444.74  60.00
 11  Pipe011   2000.0    0.760E+03     423.41  60.00      4.8         6.0   Intr
                                                       ---------
                                                          4.8


  6  Pipe007      0.0    0.184E+05     423.41  60.00
  6  Pipe007   2004.2    0.184E+05     409.86  60.00     73.2        11.6   Intr
  6  Pipe007   4008.3    0.184E+05     396.04  60.00     72.0        11.9   Intr
  6  Pipe007   6012.5    0.184E+05     381.94  60.00     70.7        12.3   Intr

Pipe   Pipe      Dist      Flow        Pressure   Temp  Liq Holdup Vel mix  Flow
 No    Name      (ft)   (bbl/day)       (psig)    (øF)   (Bbl )   (ft/sec) Patt
---- -------- -------- -----------  ---------- ----- ---------- -------  ----

  6  Pipe007   8016.7    0.184E+05     367.52  60.00     69.3        12.7   Intr
  6  Pipe007  10020.8    0.184E+05     352.75  60.00     67.9        13.1   Intr
  6  Pipe007  12025.0    0.184E+05     337.60  60.00     66.4        13.7   Intr
  6  Pipe007  14029.2    0.184E+05     321.99  60.00     64.9        14.2   Intr
  6  Pipe007  16033.3    0.184E+05     305.89  60.00     63.2        14.9   Intr
  6  Pipe007  18037.5    0.184E+05     289.21  60.00     61.5        15.6   Intr
  6  Pipe007  20041.7    0.184E+05     271.86  60.00     59.6        16.4   Intr
  6  Pipe007  22045.8    0.184E+05     253.71  60.00     57.6        17.4   Intr
  6  Pipe007  24050.0    0.184E+05     234.60  60.00     55.5        18.6   Intr
                                                       ---------
                                                        781.7

  7  Pipe008      0.0    0.113E+05     234.60  60.00
  7  Pipe008   1880.0    0.113E+05     206.43  60.00     31.1        19.8   Intr
  7  Pipe008   3760.0    0.113E+05     175.85  60.00     28.9        22.5   Intr
  7  Pipe008   5640.0    0.113E+05     141.48  60.00     26.3        26.6   Intr
  7  Pipe008   7520.0    0.113E+05      99.99  60.00     22.9        33.7   Intr
                                                       ---------
                                                        109.3

  8  Pipe009      0.0    0.708E+04     234.60  60.00
  8  Pipe009   1840.0    0.708E+04     228.19  60.00     31.0        11.9   Intr
  8  Pipe009   3680.0    0.708E+04     221.55  60.00     30.6        12.2   Intr
  8  Pipe009   5520.0    0.708E+04     214.65  60.00     30.1        12.6   Intr
  8  Pipe009   7360.0    0.708E+04     207.47  60.00     29.6        12.9   Intr
  8  Pipe009   9200.0    0.708E+04     199.99  60.00     29.1        13.3   Intr
                                                       ---------
                                                        150.3

END
```

Flow pattern

Some correlations define flow patterns, and others do not. The modified Beggs and Brill-Moody correlations, which GENNET-M uses, define the following patterns:

1. Segregated (usually Stratified or Stratified-Wavy for near horizontal segments; usually Annular for near vertical segments)
2. Intermittent (Slug or Plug)
3. Distributed (Bubble for high liquid loadings; Mist for low liquid loadings)
4. Transition (somewhere between Segregated and Intermittent)

Because any optimization method requires continuity across flow pattern boundaries, three more "transition" zones were added to the Beggs and Brill flow pattern map:

5. D-S (transition zone between Distributed and Segregated)
6. D-T (transition zone between Distributed and Transition)
7. D-I (transition zone between Distributed and Intermittent)

It is important to note that the Beggs and Brill correlations do not necessarily predict the correct flow pattern if the angle of inclination is more than a few degrees from horizontal. If the segment is not close to being horizontal, the flow pattern is used only as a correlating parameter for the holdup.

Adapted from Fluid Flow Consultants, Inc. Broken Arrow, Oklahoma; ©1997 Fluid Flow Consultants. Reprinted with permission.

14: Pumps

Centrifugal pumps

Centrifugal pumps are best suited for large volume applications or for smaller volumes when the ratio of volume to pressure is high. The selection of the proper pump will depend on the system throughput, viscosity, specific gravity, and head requirements for a particular application. Where a variety of products with different characteristics are handled, it may be necessary to utilize multiple pumps either in series or parallel, or a series-parallel combination, or pumps with variable speed capabilities to achieve the desired throughput at least cost. Economics will ultimately determine the type and number of pumps to be used.

The first step in selecting centrifugal pumps is to analyze the pipeline system and determine its characteristics such as the pressure required to move the desired flow rate. Initial and future conditions should be evaluated so that pumps with sufficient flexibility to handle both conditions with minor changes may be selected. Refer to Section 13—Liquid Hydraulics for data on calculating pipeline pressure drop.

Pump manufacturers publish pump performance maps similar to the one shown in Figure 1 to aid in pump selection. The performance map shows the range of differential head and capacity that will be developed by a family of pumps. Once the system data have been

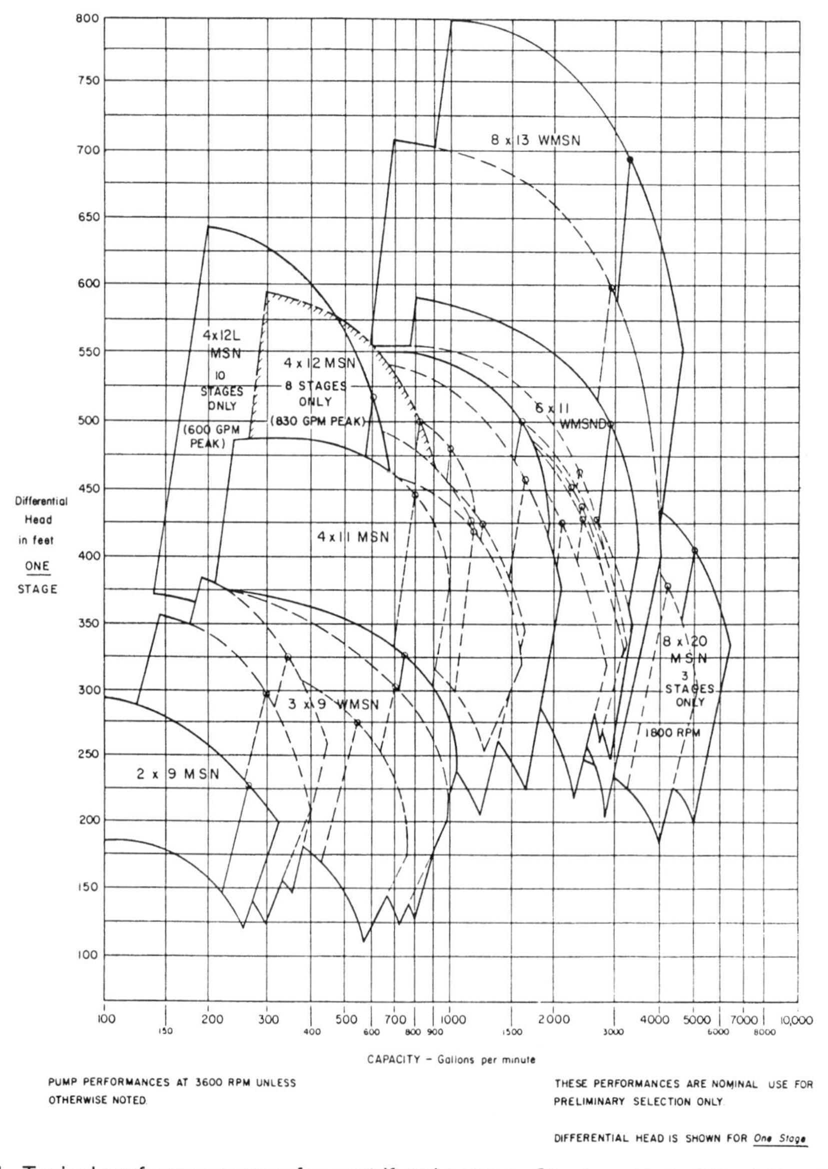

Figure 1. Typical performance map for centrifugal pumps. Courtesy United Centrifugal Pumps.

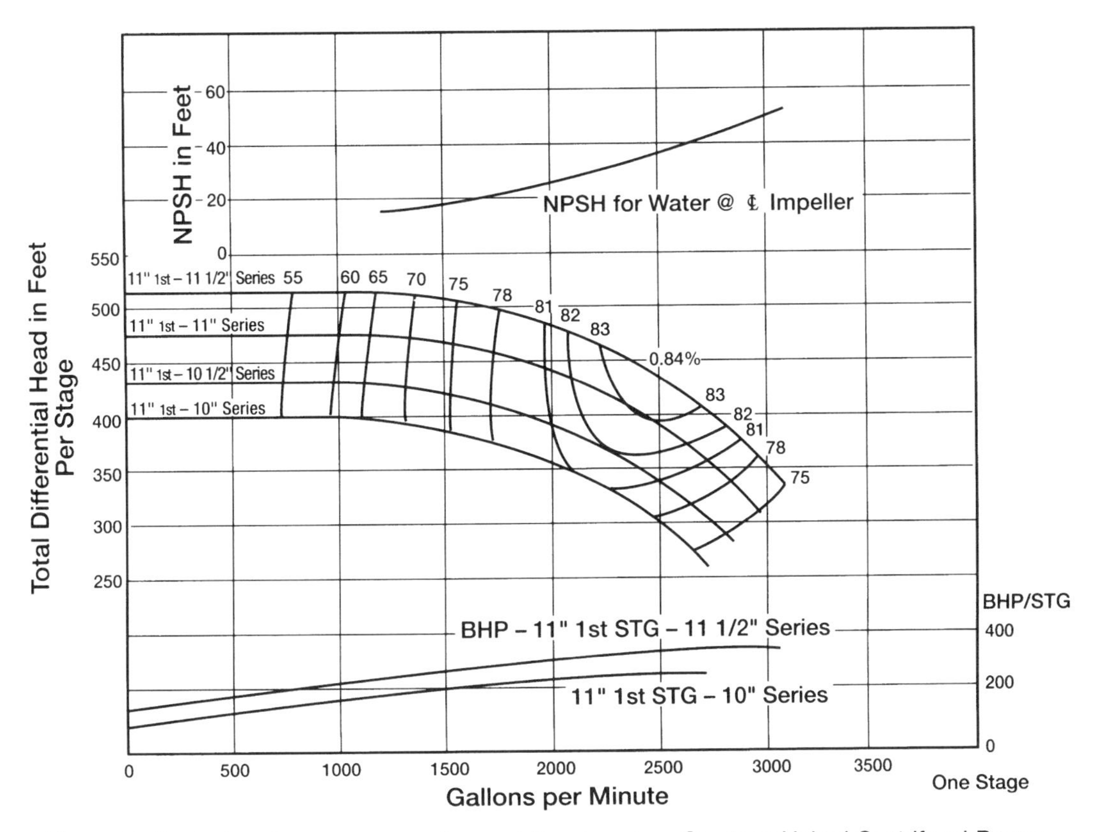

Figure 2. Typical performance curve for centrifugal pumps. Courtesy United Centrifugal Pumps.

established, a pump may be selected from the map. The selection is further refined by referring to a curve such as that shown in Figure 2, which shows a typical pump manufacturer's catalog curve for a specific pump showing the performance for various impeller size, as well as the brake horsepower, NPSH requirement, and pump efficiency. Pump performance curves are plotted in feet and for a specific gravity of 1.0. When pumping a nonviscous liquid, the pump will develop the same head shown on the performance curve; however, the *pressure* will change with specific gravity, as will the horsepower required to drive the pump. Pressure may be calculated by multiplying the pump differential head by the specific gravity and by 0.4329. The horsepower required to drive the pump will also vary directly with the specific gravity of the fluid being pumped.

The viscosity of the fluid will influence the performance of the centrifugal pump. This is discussed in more detail later.

Centrifugal pumps are available in a number of different configurations such as horizontal, vertical, radial split, and axial split casings. The final choice will depend on the system hydraulic conditions (pressure and flow rate), the ease of maintenance desired, and amount of space available for installing the pump. The pressure rating of the case and material selection will be determined by the system requirements and the fluid to be pumped.

It is extremely important that sufficient net positive suction head (NPSH) above the vapor pressure of the liquid being pumped is available for safe and reliable pump operation. NPSH is always specified in feet of head above vapor pressure required at the center line of the impeller. The NPSH required will increase as the capacity of the pump increases, which makes it necessary for the user to make a best effort to furnish the pump manufacturer with an accurate range of flow rate conditions. The pump suction nozzle geometry, pump speed, and the impeller inlet will determine the NPSH that will be required. If too much margin of safety is included in the pump specifications, the pump will probably be oversized to obtain a low NPSH and will not operate at its best efficiency point. Figures 8, 9, and 10 show three examples of how to calculate available NPSH. As a rule of thumb, the NPSH available should be at least 10% greater than the NPSH required by the pump. It is difficult and costly to correct NPSH problems after the pump is placed in service. NPSH is discussed in detail in *Centrifugal Pumps: Design and Application* by Ross and Lobanoff, Gulf Publishing Co., and it's recommended

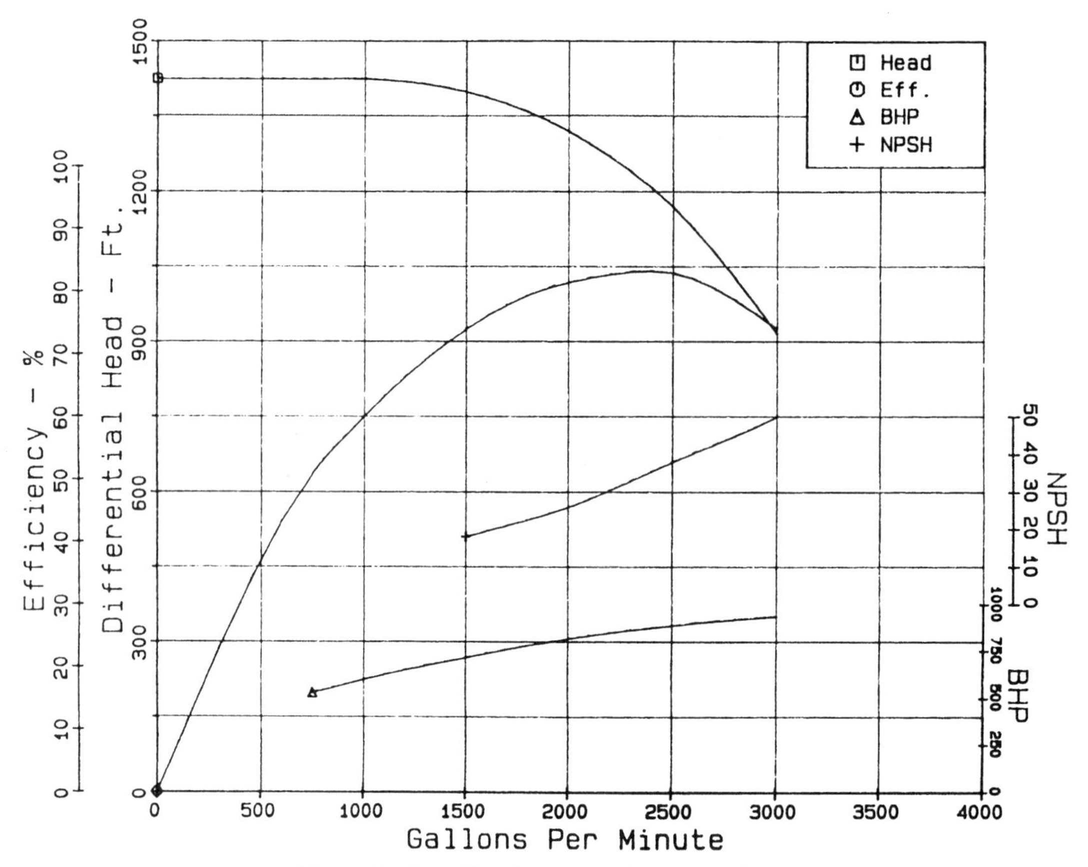

Figure 3. Centrifugal pump performance data.

reading for those who are involved in the selection and application of centrifugal pumps.

Table 1 gives some conversion factors and formulas that will be useful in problems dealing with pumps.

All centrifugal pumps, except those pumps that are specifically designed for low suction pressure applications such as tank farm booster pumps, should be protected by a low suction pressure shut down switch. The value of pressure to be used to set the switch will be determined by the amount of NPSH required at the maximum expected operating flow rate, the vapor pressure of the fluid being pumped, and the atmospheric pressure at the pump location. Once the maximum expected operating flow rate is known, the amount of NPSH required may be obtained from the pump performance curve. The NPSH curve is plotted in feet of head, and this may be converted to psi for switch setting purposes as follows:

$$\text{PSIG} = \text{NPSH} \times 0.433 \times \text{specific gravity} + \text{vapor pressure} - \text{atmospheric pressure}$$

Figure 3 shows the performance data for a centrifugal pump that was chosen from the performance map. The impeller diameter was chosen from the curve shown in Figure 2. In this case, an impeller diameter of 11-in. was chosen.

Centrifugal pumps may be operated singly, in series, in parallel, or in a combination of series and parallel. When pumps are operated in series, the heads are additive at a given rate of flow. When pumps are operated in parallel, flow is additive at a given value of head. This is illustrated in Figure 3. When pumps are operated in a series-parallel combination, these same rules will apply to determine the total head developed by the combination of pumps. Figure 4 shows performance data for one pump and for series and parallel operation of two pumps.

Pumps connected in series have the discharge of one pump connected to the suction of the downstream pump. Pumps connected in parallel use a common suction line and common discharge line. (See Figure 4a.) Pumps that are to be operated in parallel must be matched so that both develop the same head at the same flow rate. They should also have constantly rising head curves. Pumps with a head curve that has a hump (one that develops less head at shutoff than at some other point on the head curve) should be avoided since

Table 1*
Conversion factors and formulas

gpm = .07 × boiler hp

gpm = 449 × cfs

gpm = 0.0292 × bbl/day

gpm = 0.7 × bbl/hr

gpm = 4.4 × cubic meters/hr

$$\text{gpm} = \frac{\text{lb per hr}}{500 \times \text{specific gravity}}$$

$$H = \frac{P \times 2.31}{\text{specific gravity}}$$

$$V = \frac{Q \times .321}{A}$$

$$U = \frac{\text{Diameter (in.)} \times N}{229}$$

$$h_v = \frac{V^2}{2\,g}$$

$$\text{whp} = \frac{Q \times H \times \text{specific gravity}}{3{,}960}$$

$$\text{bhp} = \frac{Q \times H \times \text{specific gravity}}{3{,}960 \times e}$$

$$\text{bhp} = \frac{Q \times P}{1{,}715 \times e}$$

$$T = \frac{\text{bhp} \times 5{,}250}{N}$$

$$N_s = \frac{N\sqrt{Q}}{H_1^{3/4}} = \frac{N\sqrt{\sqrt{H_1}} \times \sqrt{Q}}{H_1}$$

$$S = \frac{N\sqrt{Q}}{h_{sv}^{3/4}} = \frac{N\sqrt{\sqrt{h_{sv}}} \times \sqrt{Q}}{h_{sv}}$$

$$t_r = \frac{H\left(\frac{1}{e} - 1\right)}{780 \times C}$$

where
N = speed in rpm
N_s = specific speed in rpm
S = suction specific speed in rpm
Q = capacity in gpm
P = pressure in psi
H = total head in ft
H_1 = head per stage in ft
h_{sv} = net positive suction head in ft
h_v = velocity head in ft
whp = water horsepower
bhp = brake horsepower
U = peripheral velocity in ft per second
g = 32.16 ft per second (acceleration of gravity)
mgd = million gallons per day
cfs = cubic ft per second
bbl = barrel (42 gallons)
C = specific heat
psi = lb/in.^{2}
gpm = gallons per minute
e = pump efficiency in decimal
D = impeller diameter in in.
V = velocity in ft per second
T = torque in ft-lb
t = temperature in °F
t_r = temperature rise in °F
A = area in in.^{2}

*Courtesy United Centrifugal Pumps.

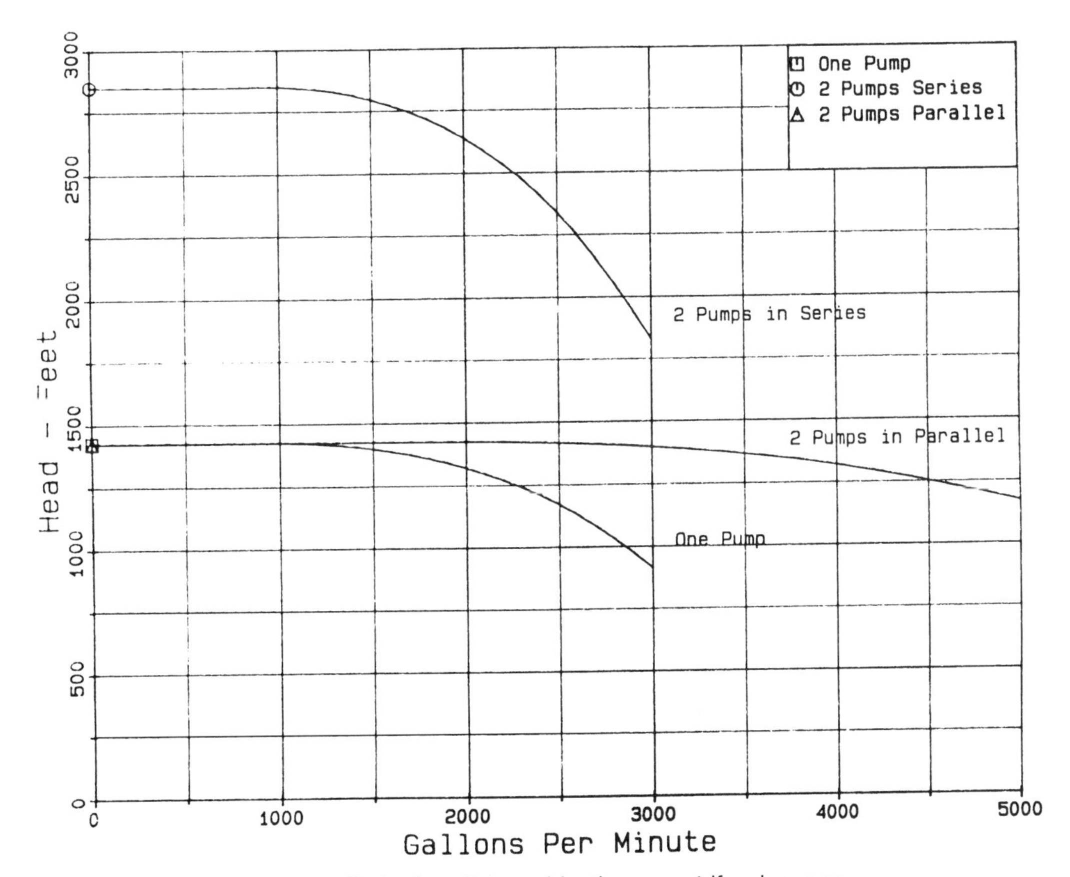

Figure 4. Series/parallel combinations—centrifugal pumps.

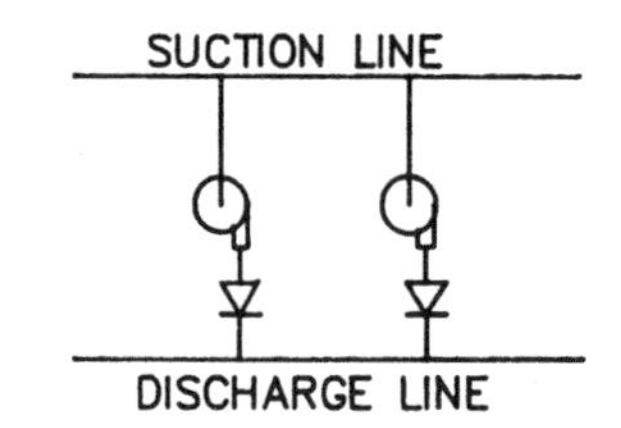

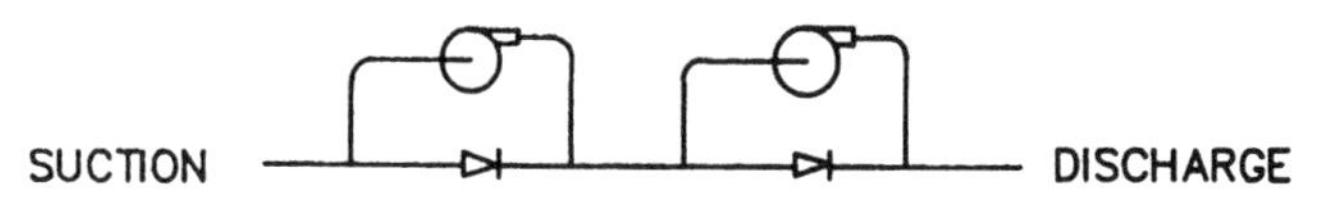

Figure 4a. Pumps connected in parallel/in series.

this could possibly preclude the second pump from pumping.

The high cost of energy will in almost all cases justify the additional cost of multiple pumping units or variable speed drivers. The use of multiple units will allow most pumping conditions to be met without throttling and therefore wasting power. If as many as three series units can be justified, the units should be sized as follows:

One unit—full head
One unit—$\frac{3}{4}$ of the head of the full head unit
One unit—$\frac{1}{2}$ of the head of the full head unit

This will allow the following pump combinations to be used:

Units	Total head	Units	Total head
$\frac{1}{2}$	$\frac{1}{2}$	$\frac{1}{2}$, 1	$\frac{1}{2}$
$\frac{3}{4}$	$\frac{3}{4}$	$\frac{3}{4}$, 1	$1\frac{3}{4}$
1	1	$\frac{1}{2}$, $\frac{3}{4}$, 1	$2\frac{1}{4}$
$\frac{1}{2}$, $\frac{3}{4}$	$1\frac{1}{4}$	$\frac{3}{4}$	

Increments of $\frac{1}{4}$ head can be achieved up to the last $\frac{1}{2}$ head increment. This results in maximum flexibility when constant rpm drivers are used.

Figure 5 shows the performance of series versus parallel operation of pumps. In this particular example, series

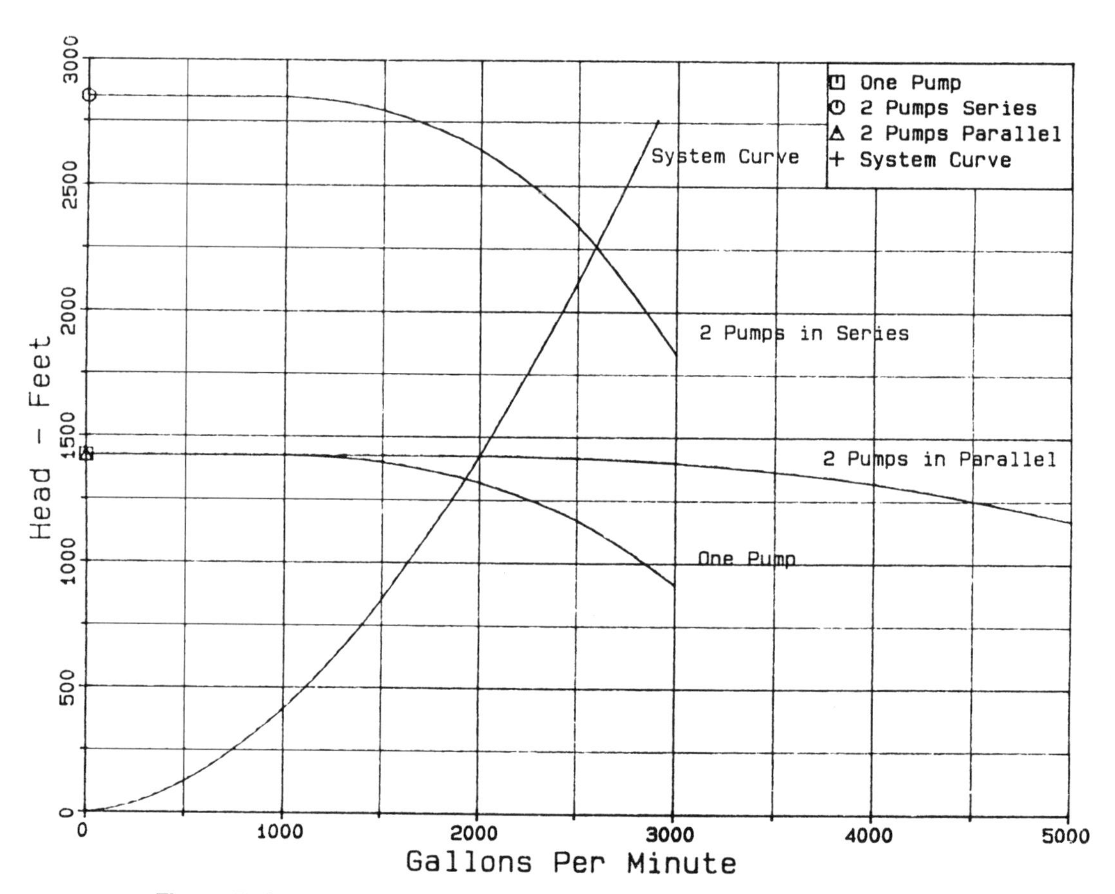

Figure 5. Series/parallel combinations with system curve—centrifugal pumps.

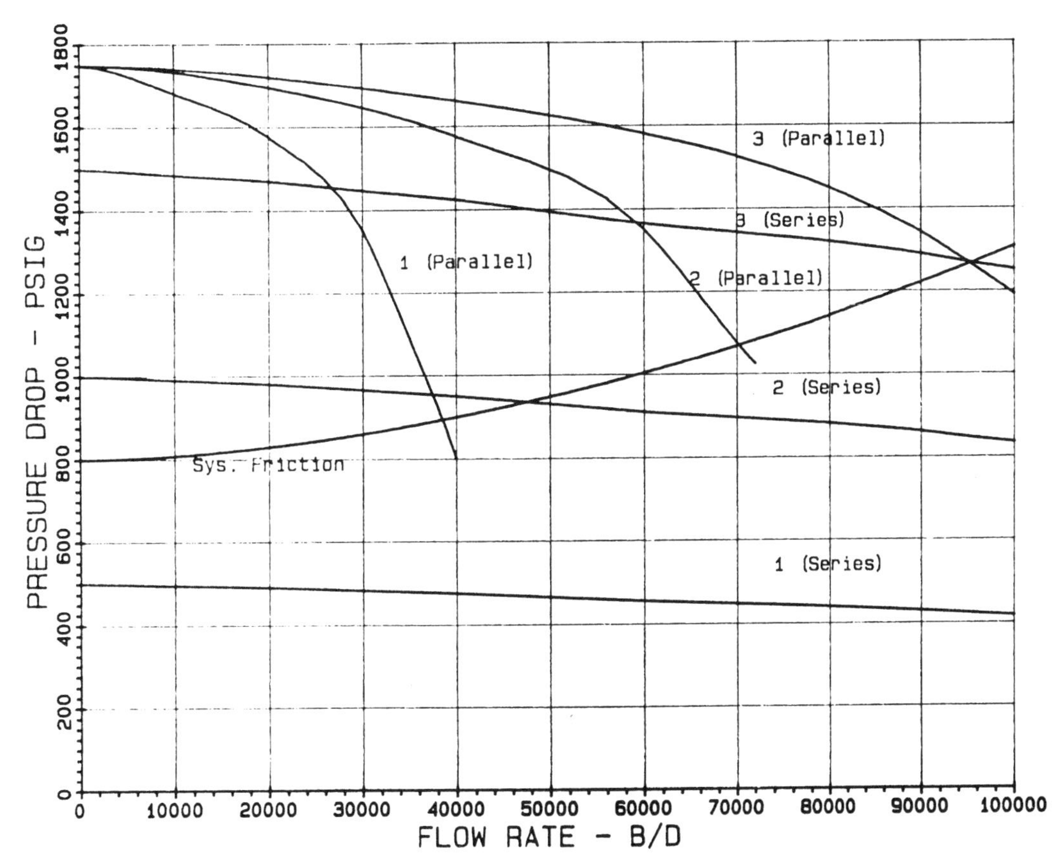

Figure 5a. Series and parallel operation—centrifugal pumps.

operation will provide a higher pumping rate than provided with a parallel operation, and two pumps in parallel will pump only slightly more than one single pump. The maximum flow rate with two pumps in series will be 2,594 gpm or 89,439 b/d. The maximum flow rate for one pump will be 1,917 gpm or 66,092 b/d.

For system rates between 66,092 b/d and 89,439 b/d, pressure throttling will be necessary to match the pump performance to the system curves, unless the pumps are equipped with variable speed drives. If throttling is necessary, it may be minimized by operating at a high flow rate for part of the day and at a lower rate for the balance of the day. If a system rate of 70,000 b/d is desired with the system shown in Figure 5, one pump operating 20 hours per day and two pumps operating 4 hours per day would achieve this rate; however, this may not be the most economical method of operation. It will be necessary to calculate and compare the cost of throttling pressure versus the cost to operate the second unit for 4 hours. System rates less than 66,092 b/d will require pressure throttling unless the pump speed may be varied. It is not uncommon for pipeline pump stations to contain several pumps, each of which may be sized differently to increase the amount of flow rate flexibility.

The use of multiple pumps affords some protection against throughput reductions due to one or more pumps being out of service.

Parallel pump operation is usually best suited for systems where the residual static pressure is high compared to the line friction loss. Such systems usually have a flat system curve. An example of this is shown in Figure 5a.

For this system, three parallel pumps would offer the most flexibility, since one unit will always be able to overcome the system friction and static and would pump almost as much as two series pumps. One could make the argument that one large pump could be selected to meet the 2 (series) condition shown in Figure 5a and a ½-head pump could be used for the maximum volume condition. This argument is valid as long as the large pump is available. When it is not available, the system will be down, since the ½-head unit will not develop enough head to overcome the static and friction losses. Three pumps operating in parallel will offer the most flexibility for the system shown in Figure 5a.

Changes in system throughputs may require resizing the pump impellers to meet the new conditions. Some types of multi-stage pumps may be destaged or upstaged by adding or taking out impellers, depending upon the number of

Table 2*
Formulas for recalculating pump performance with impeller diameter or speed change

Q_1, H_1, bhp_1, D_1 and N_1 = Initial Capacity, Head, Brake Horsepower, Diameter, and Speed.

Q_2, H_2, bhp_2, D_2 and N_2 = New Capacity, Head, Brake Horsepower, Diameter, and Speed.

Diameter Change Only	Speed Change Only	Diameter & Speed Change
$Q_2 = Q_1\left(\frac{D_2}{D_1}\right)$	$Q_2 = Q_1\left(\frac{N_2}{N_1}\right)$	$Q_2 = Q_1\left(\frac{D_2}{D_1} \times \frac{N_2}{N_1}\right)$
$H_2 = H_1\left(\frac{D_2}{D_1}\right)^2$	$H_2 = H_1\left(\frac{N_2}{N_1}\right)^2$	$H_2 = H_1\left(\frac{D_2}{D_1} \times \frac{N_2}{N_1}\right)^2$
$bhp_2 = bhp_1\left(\frac{D_2}{D_1}\right)^3$	$bhp_2 = bhp_1\left(\frac{N_2}{N_1}\right)^3$	$bhp_2 = bhp_1\left(\frac{D_2}{D_1} \times \frac{N_2}{N_1}\right)^3$

* Courtesy United Centrifugal Pumps.

stages in operation and the amount of head change required. If it is contemplated that it may be necessary in the future to destage or upstage, this should be specified when the pump is purchased. It is also possible to optimize the efficiency point by chipping the pump volutes and under-filing the pump impeller(s) or adding volute inserts. The pump manufacturer should be contacted for details on how to accomplish this.

The pump affinity laws shown in Table 1 may be used to predict the performance of the pump at different operating conditions. These laws are commonly used to determine new impeller diameters required for nominal changes in flow rates or performance at different speeds. The affinity laws will allow the translation of a point on the performance curves but will not define the operation of the pump in the system. A new pump performance curve may be constructed using these laws and then applied to the system curve to determine the actual operation of the pump.

The system shown in Figure 5, with both pumps configured with 11-in. diameter impellers, will pump a maximum of 89,439 b/d. The system throughput requirements have now changed such that the maximum desired throughput is 87,931 b/d, and it is desired to trim the impellers in one of the pumps to meet this new condition. The new H–Q condition for one pump will be $H_2 = 955$ ft and $Q_2 = 2{,}550$ gpm. In order to establish the new impeller diameter, it will be necessary to translate this point to the existing pump curve to determine H_1 and Q_1. Select flow rates on both sides of the new rate and calculate the

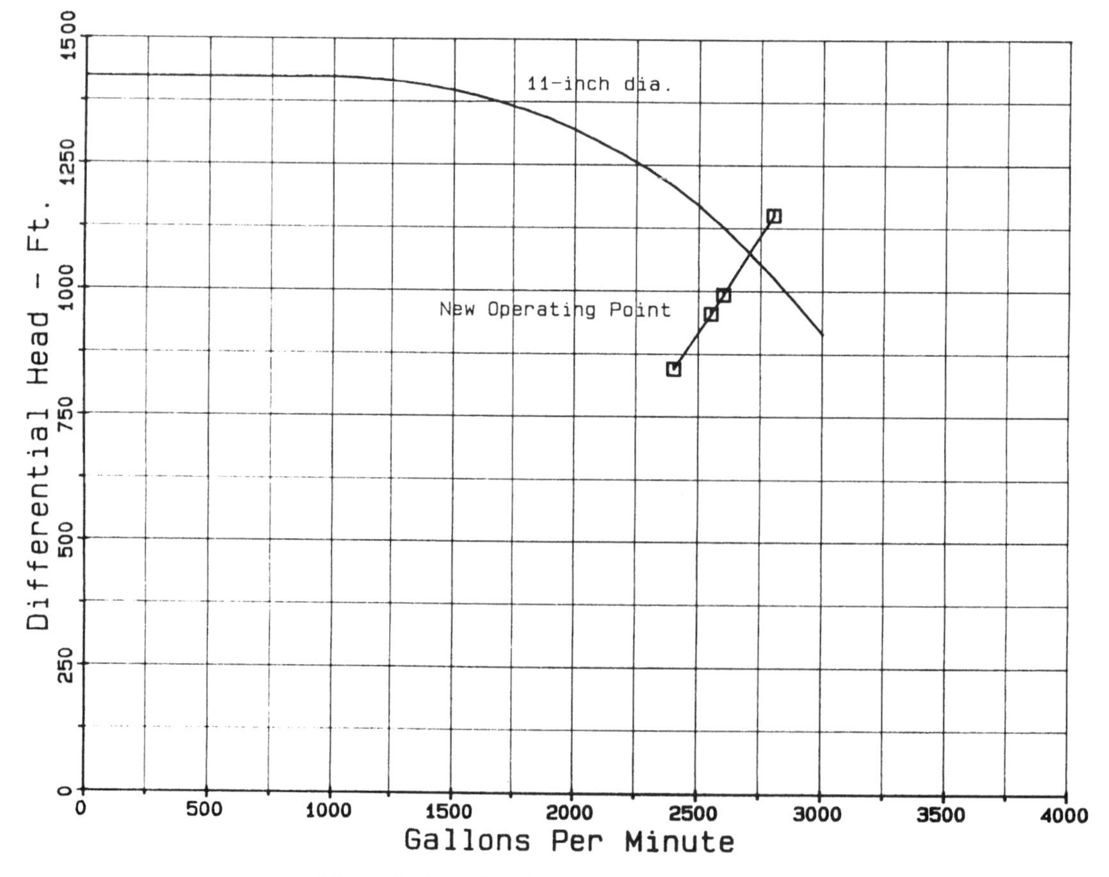

Figure 6. Impeller sizing–centrifugal pumps.

corresponding head. These data are plotted and the resulting line will cross the existing pump head curve as shown in Figure 6. Determine the head and flow rate at the point where this line crosses the existing head curve, and these values will be H_1 and Q_1. For this calculation, flow rates were selected as follows:

Selected Q	Calculated H
2,800	1,151
2,600	993
2,400	846

The head was calculated as follows:

$$H = H_2 \times (Q/Q_2)^2$$

This line crosses the existing head curve at 2,700 gpm and 1,080 ft, and these values will be the values to use for Q_1 and H_1, respectively.

Use the pump affinity law to calculate the new diameter as follows:

$$H_2/H_1 = (D_2/D_1)^2$$
$$D_2 = D_1 \times (H_2/H_1)^{1/2}$$
$$D_2 = 11.0 \times (955/1{,}080)^{1/2}$$
$$D_2 = 10.34 \text{ in.}$$

Since the pump affinity laws are not quite exact for a diameter change, it will be necessary to apply a correction factor to the calculated value of D_2 to determine the final diameter. The correction factor is determined by calculating the ratio of the calculated diameter to the original diameter. This value, expressed as a percentage, is then used to enter Figure 7 to determine the actual required diameter as a percentage of the original diameter.

$$D\ calc/D\ orig = 10.34/11 = 94\%$$

Enter the chart with 94% and read 94.5% as the actual required diameter as a percentage of the original diameter.

The new diameter will therefore be $0.945 \times 11.0 = 10.395$ in. The impellers would probably be trimmed to $10\frac{7}{16}$-in.

Once the new diameter is established, the new pump head curve may be calculated by using the affinity laws to calculate new H–Q values.

The pump affinity laws are exact for a speed change. Figure 7a shows the effect of a speed change on the pump performance. The pump manufacturer should be consulted for data on the range of speeds over which the pump may be efficiently operated.

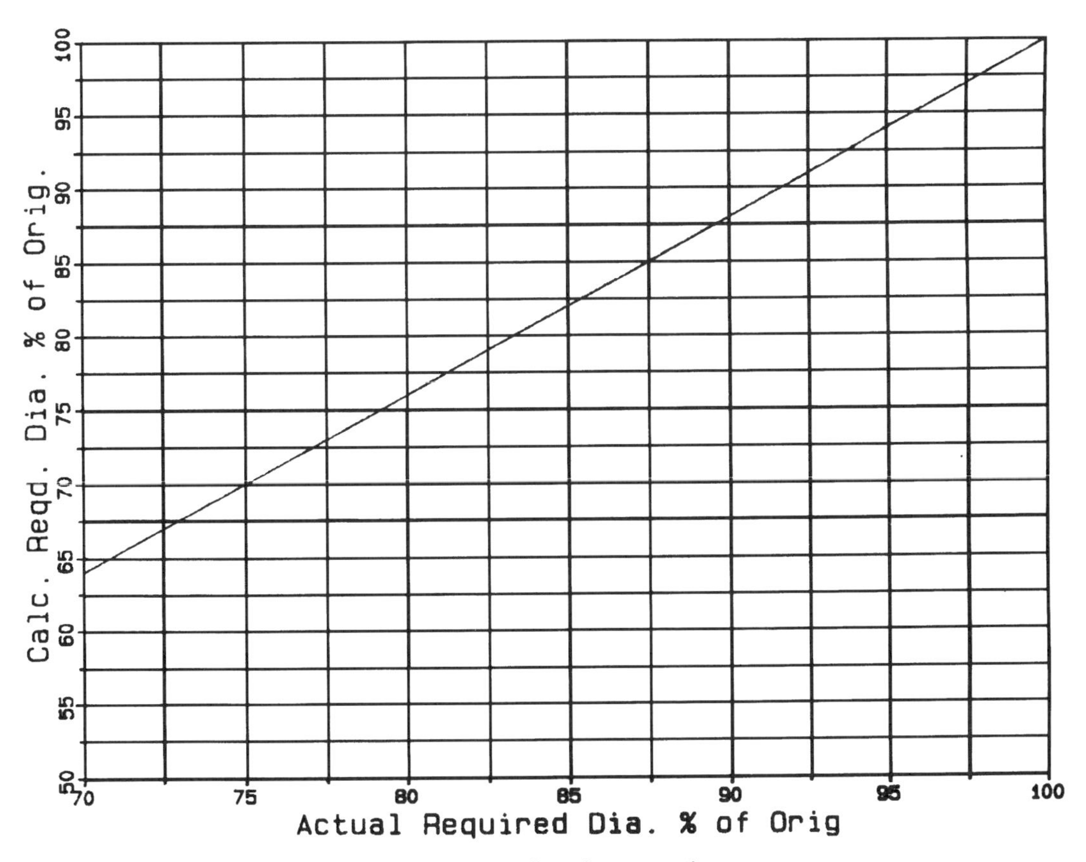

Figure 7. Impeller trim correction.

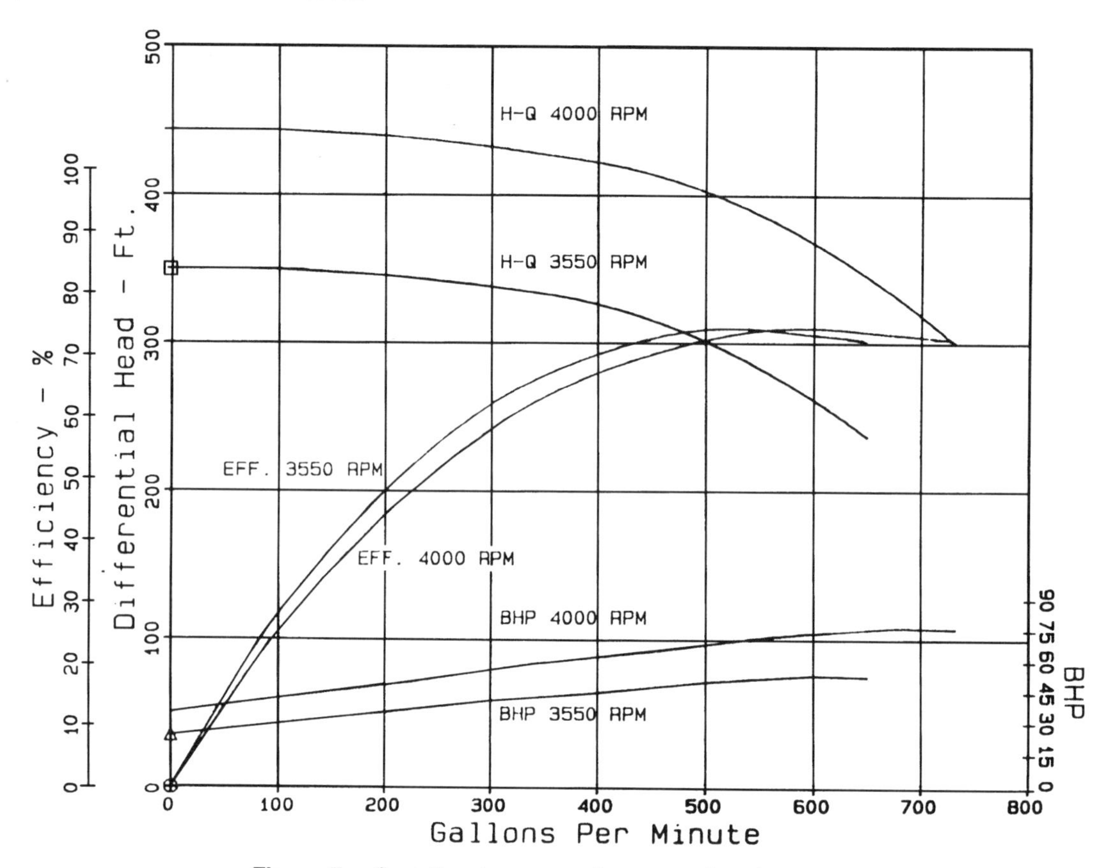

Figure 7a. Centrifugal pump performance data for speed.

Very viscous liquids will have an impact on the performance of centrifugal pumps. Figure 11 compares pump performance when pumping water to pumping a liquid with a viscosity of 6,000 SUS. The effects of viscosity on pump performance may be predicted by the use of Figures 12 and 13. These charts are used as follows. Refer to Figure 11 and select the water capacity at bep (best efficiency point). In this example, the water capacity at bep is 23,000 gpm. Refer to Figure 12 and enter the chart with 23,000 gpm and move vertically to intersect the desired impeller diameter, 27 in., and then horizontally to intersect the 6,000 SUS line. Move vertically to intersect the line representing the water efficiency (85%), then horizontally to read the viscous efficiency 62% (not a correction factor). The previous vertical line is extended to intersect the head correction and capacity correction curves. The capacity correction factor is 93%, and the head correction factor is 95%. The viscous head at shutoff will be the same as the shutoff head for water. The viscous head will be 617 ft at 21,400 gpm. Plot this point and then draw the head-capacity

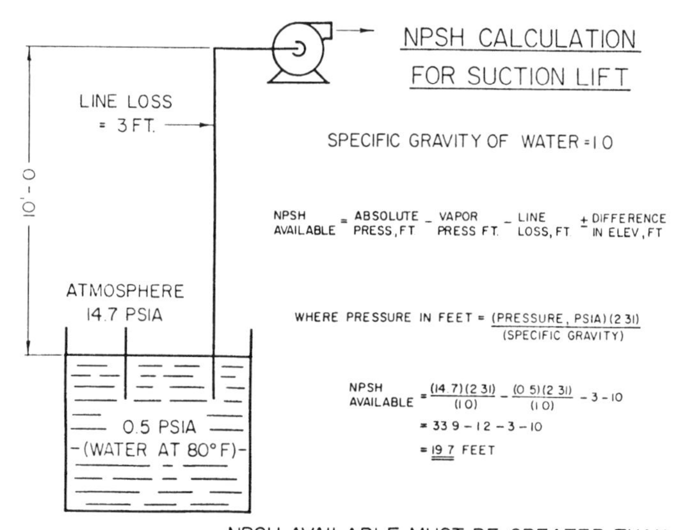

Figure 8. NPSH calculation for suction lift.

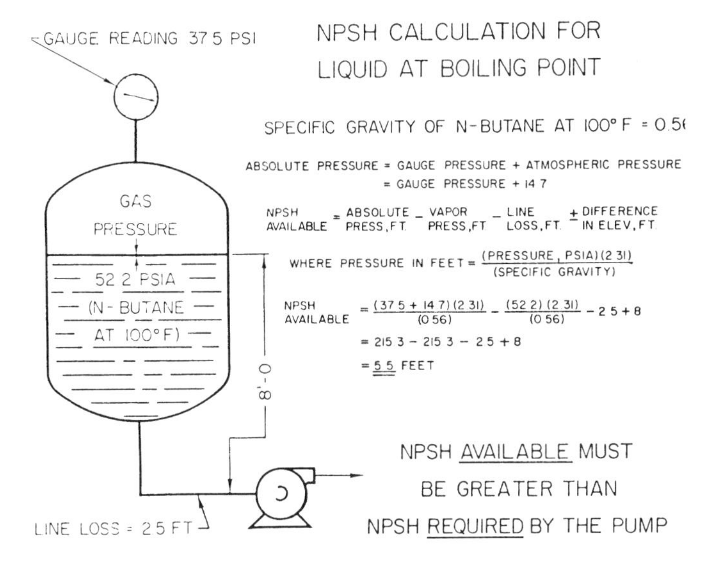

Figure 9. NPSH calculation for liquid at boiling point.

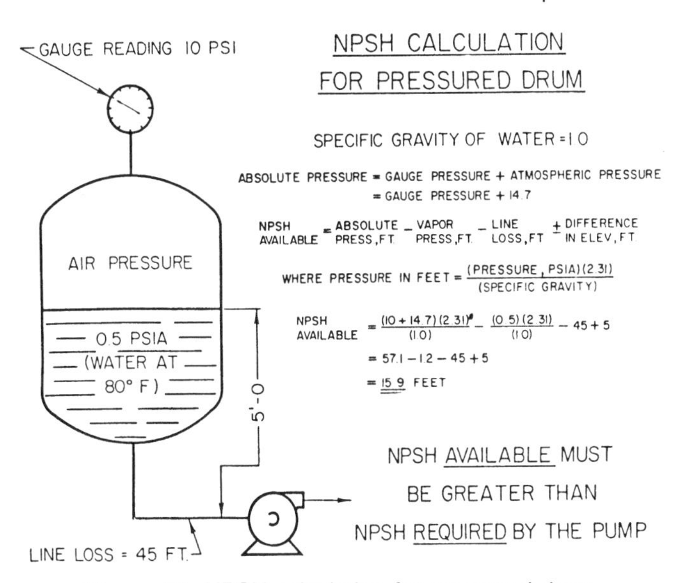

Figure 10. NPSH calculation for pressured drum.

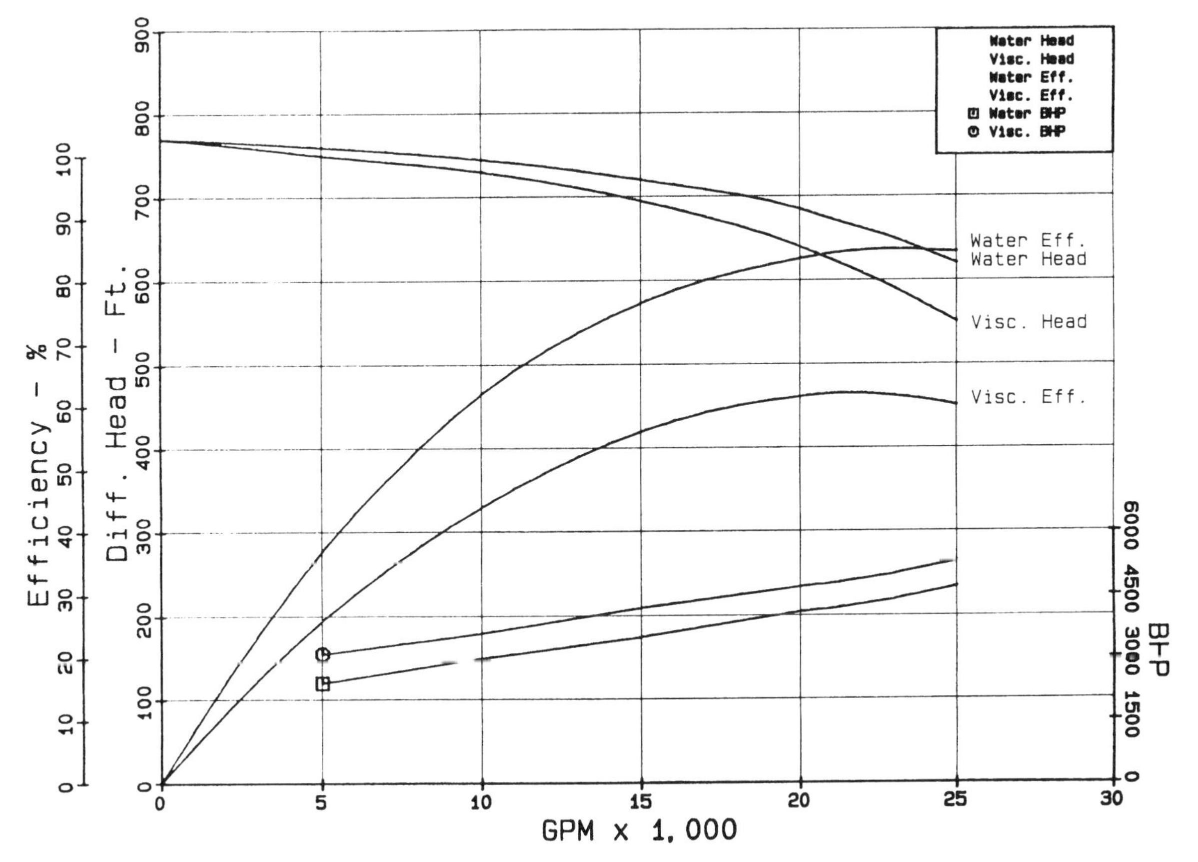

Figure 11. Effects of viscosity on centrifugal pump performance.

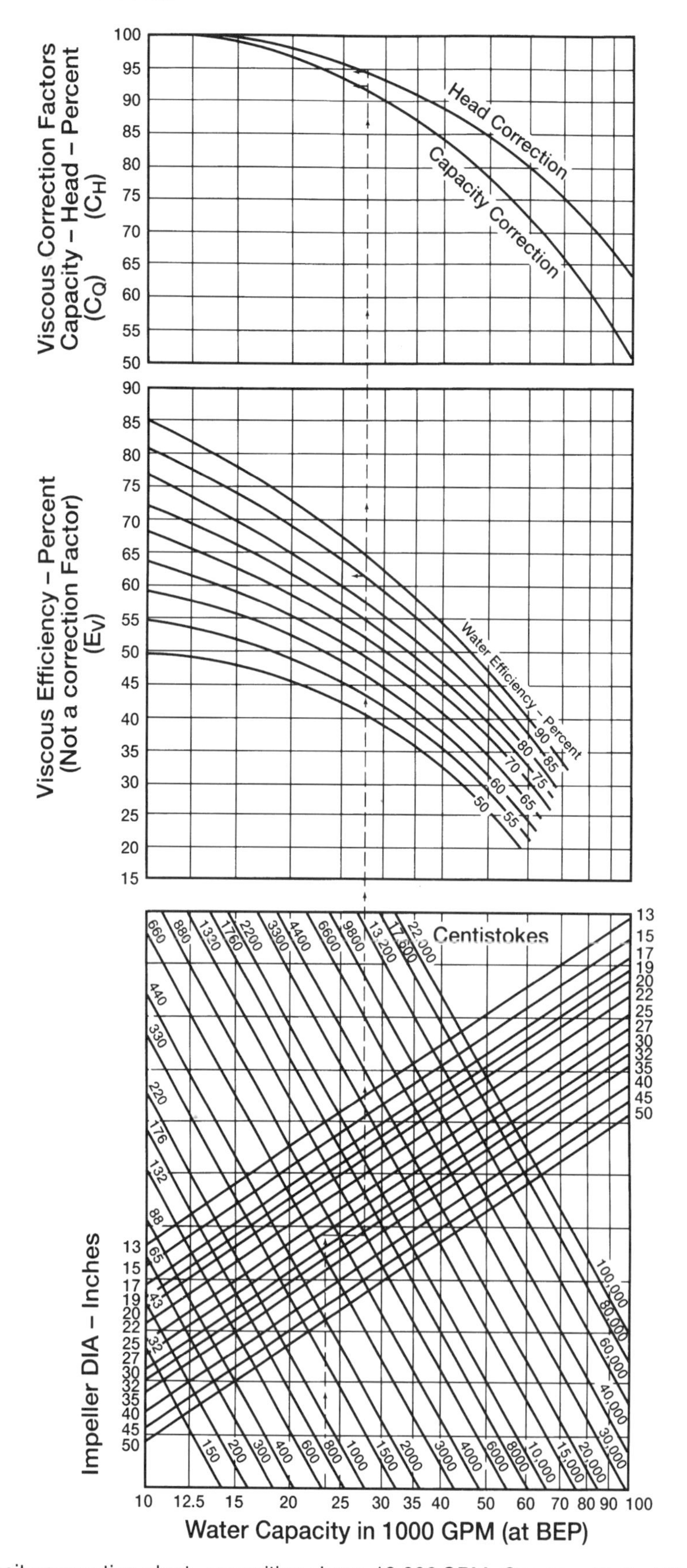

Figure 12. Viscosity correction chart—capacities above 10,000 GPM. Courtesy United Centrifugal Pumps.

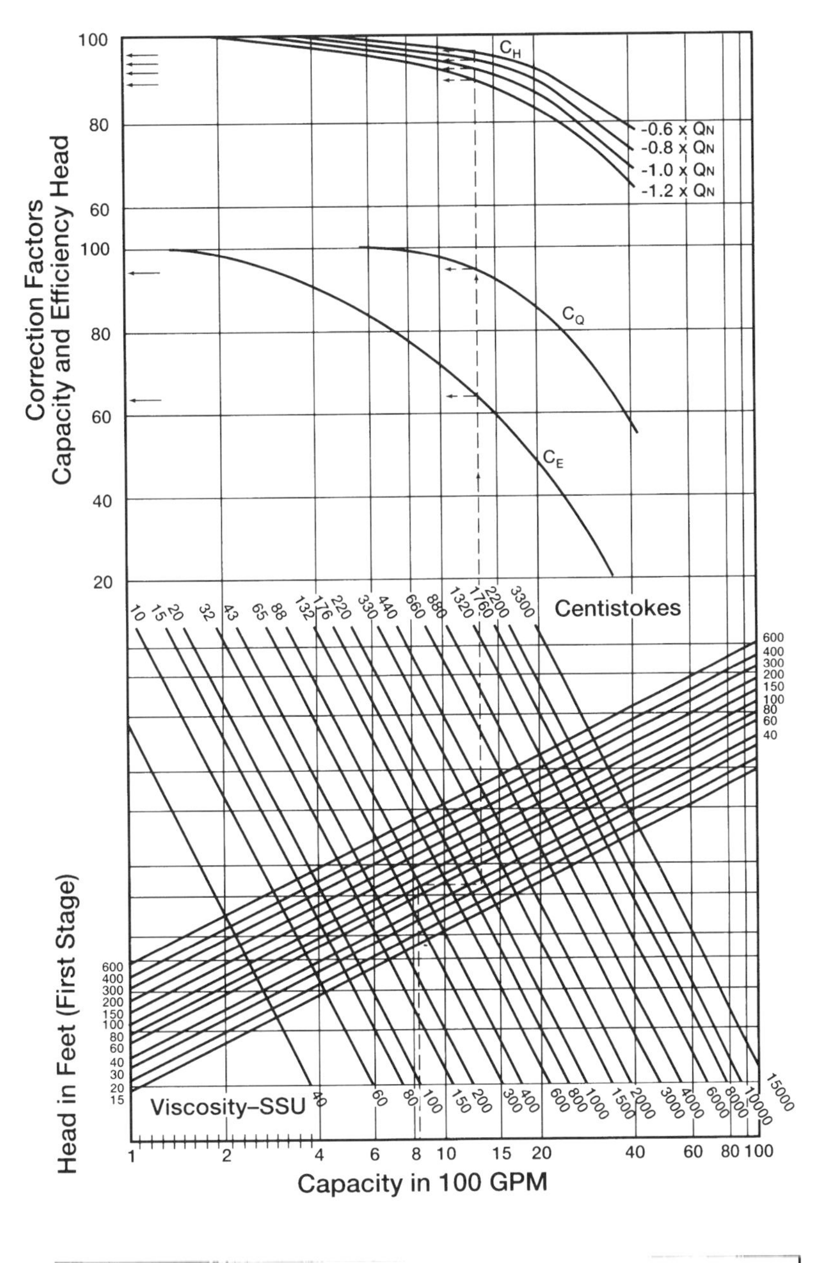

CORRECTION APPLIES TO PEAK EFFICIENCY ONLY. FOR OTHER CAP DRAW CORRECTED H.P. CURVE PARALLEL TO WATER H.P. CURVE AND CALCULATE EFFICIENCIES BASED ON ASSUMED HEAD - CAP CURVE. ADJUST AS REQUIRED TO OBTAIN A SMOOTH CURVE. USE CORRECTIONS OBTAINED FOR MAXIMUM DIAMETER FOR CUT DIAMETERS.

Figure 13. Viscosity correction chart—capacities up to 10,000 GPM. Courtesy United Centrifugal Pumps.

curve through this point. The viscous brake horsepower is calculated at the bep (62% at 21,400 gpm). Draw the viscous brake horsepower curve through this point. The slope of the line will follow the slope of the water brake horsepower curve. The viscous efficiency will be 62% at 21,400 gpm, and the viscous efficiency curve may now be calculated using the viscous brake horsepower, viscous head and capacity, and the viscous specific gravity, which in this example is 0.9.

Figure 13 is used for pumps with capacities up to 10,000 gpm, and Figure 12 is used for pumps with capacities greater than 10,000 gpm.

There may be instances where it is necessary to calculate the pump speed torque requirements to determine if the pump driver will be able to accelerate the pump to operating speed. Example 1 illustrates this calculation procedure, and the results are shown in Figure 14.

Speed torque calculation

The formula for calculating the speed-torque relation for a pump with a given specific gravity and viscosity is:

$$\text{Torque (lb.-ft.)} = \frac{5250 \times HP}{RPM}$$

The following data is required to plot the speed-torque curve:

1. HP @ Shutoff.
2. HP @ Open Valve (use one of the following):
   2a. HP @ 120% of peak capacity.
   2b. HP rating of driver if greater than HP @ 120% of peak capacity.
   2c. Depending on pump application maximum horsepower on performance curve which is greater than driver rated horsepower.
   NOTE: Using "2a," or "2b" will cover most of the applications. However on some applications "2c" must be considered.
3. Pump RPM.
4. Torque @ Shutoff.
5. Torque @ "2a," "2b" or "2c".

From the above data the torque with the valve closed (using HP @ Shutoff), and the torque with the valve open (using HP @ "2a," "2b" or "2c") at full speed can be computed.

The torque varies as the square of the speed; therefore to obtain the torque at:

3/4 Speed—multiply full speed torque by—0.563
1/2 Speed—multiply full speed torque by—0.250
1/4 Speed—multiply full speed torque by—0.063
1/8 Speed—multiply full speed torque by—0.016

From the figures a speed-torque curve for both closed and open valve, covering the complete speed range, can be plotted.

Example

HP @ Shutoff = 80
HP @ 120% of peak capacity = 150
Pump RPM = 3550

$$\text{Torque @ Shutoff} = \frac{5250 \times 80}{3550} = 118 \text{ lb.-ft. @ 3550 RPM}$$

$$\text{Torque @ 120\% of peak cap.} = \frac{5250 \times 150}{3550} = 222 \text{ lb.-ft. @ 3550 RPM}$$

Speed-RPM	Torque-Valve Open (150 HP)	Torque-Valve Closed (80 HP)
Full–3550	Full–222	Full–118
3/4–2660	0.563–125	0.563–67
1/2–1775	0.250–56	0.250–30
1/4–885	0.063–14	0.063–8
1/8–445	0.016–4	0.016–2
0–0	0.050–11	

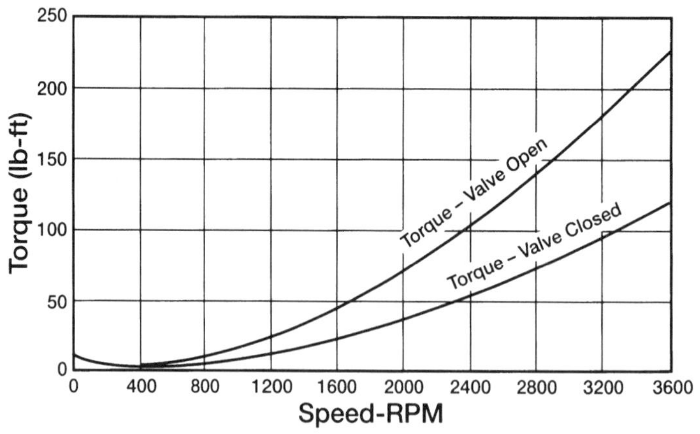

Figure 14. Courtesy United Centrifugal Pumps.

PULSATION CONTROL FOR RECIPROCATING PUMPS

Suppress the surge in the suction and discharge systems of your positive displacement pumps

By V. Larry Beynart, Fluid Energy Controls, Inc.

This article reviews the specific requirements for the selection and sizing of pulsation dampeners to reduce the fluid pressure pulsations generated by positive displacement reciprocating pumps. It includes analyses of the nature of pressure pulsation and explains the advantages of using pulsation dampeners and suction stabilizers. Sizing procedures for dampeners illustrate how their application improves suction and discharge control in reciprocating pumps, reduces pressure pulsations, increases Net Positive Suction Head (NPSH), and makes the service life of pumps longer and more efficient.

Combating pulsation

The primary purpose of pulsation control for reciprocating pumps is to attenuate or filter out pump-generated pressures that create destructive forces, vibration, and noise in the piping system.

Every reciprocating pump design has inherent, built-in pressure surges. These surges are directly related to the crank-piston arrangement (Figure 1). Fluid passing through a reciprocating pump is subjected to the continuous change in piston velocity as the piston accelerates, decelerates, and stops with each crankshaft revolution. During the suction stroke, the piston moves away from the pump's head, thereby reducing the pressure in the cylinder. Atmospheric pressure, which exists on the surface of a liquid in a tank, forces liquid into the suction pipe and pump chamber. During the discharge stroke, the piston moves toward the pump's head, which increases the pressure in the cylinder and forces the liquid into the discharge pipe. This cycle is repeated at a frequency related to the rotational speed of the crank.

At every stroke of the pump, the inertia of the column of liquid in the discharge and suction lines must be overcome to accelerate the liquid column to maximum velocity. At the end of each stroke, inertia must be overcome and the column decelerated and brought to rest. Figure 2 compares pump pulsation frequencies to pipe span natural frequencies of vibration.

A key benefit of pulsation control is the reduction of fatigue in the pump's liquid end and expendable parts. Reducing the pressure peaks experienced by the piston will in turn reduce the power-end peak loading.

The source of fluid pressure pulsations

It is almost mandatory that multiple pump installations have well-designed individual suction and discharge pulsation control equipment. While multiple pumps may be arranged to run at slightly different speeds, it is impossible to prevent them from frequently reaching an "in-phase" condition where all the pump flow or acceleration disturbances occur simultaneously. Having more than one pump can multiply the extent of the pipe vibration caused by such disturbances because the energy involved is similarly increased.

The reduction of the hydraulic pressure pulses achieved by using pulsation control equipment is usually reported in total pulsation pressure "swing" as a percentage of average pressure. This method is used widely within the industry and is often recommended as a standard. Referring to Figure 3,

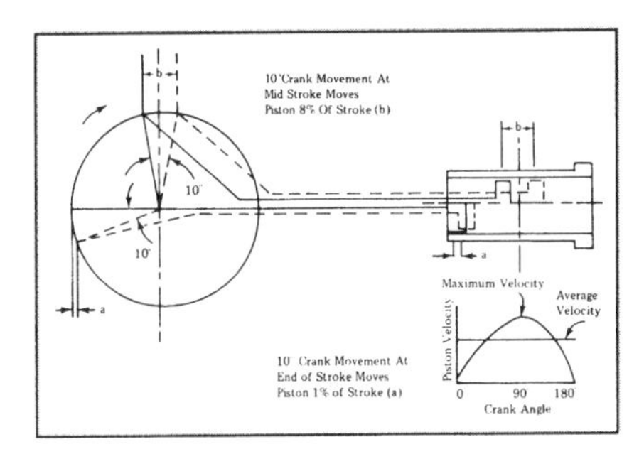

Figure 1. The rotary motion of the crank movement is transformed into the reciprocating motion of the piston.

RPM	Frequency, Hz Pump	Duplex Pulse[a]	Triplex Pulse[a]	Quint Pulse[a]
50	0.8	3.2	2.4	4.0
100	1.7	6.8	5.1	8.5
150	2.5	10.0	7.5	12.5
200	3.3	13.2	9.9	16.5

[a] Pulse = pump RPM x number of cylinders/60

(a)

30'

	2", 5 lb Pipe	6", 19 lb Pipe
	2.9 Hz	8.5 Hz
	4.3 Hz	12.5 Hz
	1.9 Hz	5.5 Hz

(b)

Figure 2. (a) Typical pump and pulse frequencies. (b) Typical piping natural frequencies.

the percent of residual pulsation pressure can be calculated as follows:

Case I: $\Delta P/Ave \times 100 = 460/1000 \times 100 = 46\%$

Case II: $\Delta P/Ave \times 100 = 70/100 \times 100 = 7\%$

It is very important that any reference to degree of pulsation should apply to the total excursion in terms of pressure and percentage. In Case I, the total pulsation of 460 psi (46%) implies that the excursion is from 680 psi (32% below the average) to 1,140 psi (14% above the average).

Reciprocating pumps introduce into the suction and discharge systems three apparently unrelated pressure disturbances, which are illustrated in Figure 4. These include:

- A low-frequency disturbance at A, based on the rate at maximum flow velocity pressures
- A higher frequency due to maximum acceleration pressure at the beginning of each piston stroke at B
- A pressure disturbance at the point of flow velocity change (valley) at C

Major problems caused by pressure pulsations

Unsteady flow. The most obvious problem caused by pulsation is that flow is not constant, which can lead to problems in processes where a steady flow rate is required, such as applications involving spraying, mixing, or metering. Flow rates can be difficult to measure with some conventional types of flowmeters.

Noise and vibration. Many reciprocating pump installations suffer from problems that can lead to excessive maintenance costs and unreliable operation. Typical examples are noise and vibration in the piping and the pump. Vibration can lead to loss of performance and failure of valves, crossheads, crankshafts, piping, and even pump barrels. High levels of pulsation can occur when the pulsation energy from the pump interacts with the natural acoustic frequencies of the piping. A reciprocating pump produces pulsation at multiples of the pump speed, and the magnified pulsation in the system is generally worse at multiples of the plunger frequency. Most systems have more than one natural frequency, so problems can occur at differing pump speeds.

Shaking. Shaking forces result from the pulsation and cause mechanical vibration of the piping system. These forces are a function of the amplitude of the pulsation and the cross-sectional area of the piping. When the exciting frequency of the pulsation coincides with a natural frequency of the system, amplification and excessive vibration occur. Amplification factors can be as high as 40 for pulsation resonance and 20 for mechanical resonance. If the mechanical resonance coincides with the acoustic resonance, a combined amplification factor of 800 is possible.

Wear and fatigue. Pulsation can lead to wearing of pump valves and bearings and, if coupled with mechanical vibration, will often lead to loosening of fittings and bolts and thus to leakage. In severe cases, mechanical fatigue can lead to total failure of components and welded joints, particularly when the liquid being pumped is corrosive.

Cavitation. Under certain pumping conditions, areas of low system pressure can occur. If this pressure falls below the liquid's vapor pressure, local boiling of the liquid will occur and bubbles of vapor will form. If the liquid has dissolved gases, they will come out of solution before the liquid boils.

Pressure waves summary

The earlier analysis of the source and nature of fluid pressure pulsations enables us to build the idealized pressure wave pattern (Figure 5). Although identical forces are acting on the liquid in the suction and discharge sides of a pump, the effects of these forces vary greatly. The major difference is that acceleration effects in the suction at B tend to separate the liquid into vapor or "cavitation" pockets that collapse with a high-magnitude pressure pulse. Acceleration on the discharge side of the piston at B, however, tends to compress the liquid and create an impulse pressure pulse. This requires separate consideration of the suction and discharge systems when choosing and installing pulsation control and suction stabilizing equipment.

Pulsations in suction systems

The friction losses in a suction system are usually low because of the relatively short length and large diameter of the piping involved. Accordingly, the flow-induced A-type pressures are of low magnitude compared to the accepted

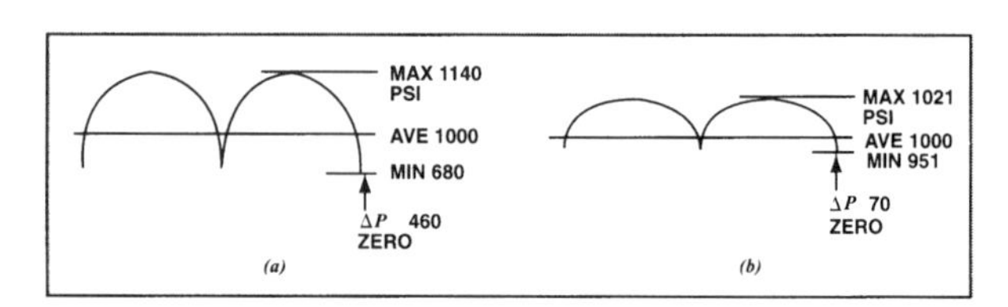

Figure 3. Reporting the degree of pulsation and control. (a) Non-dampened waveform from flow variation (Case I). (b) Dampened waveform from flow variation (Case II).

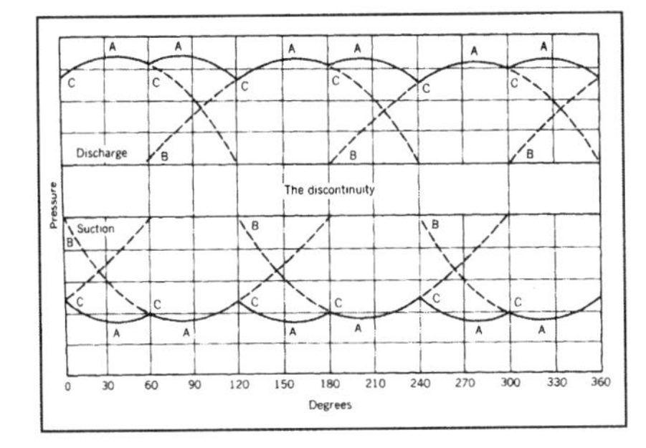

Figure 4. Triplex single-acting pump—points of induced pressure disturbances.

percentage of change (Figure 4). For example, if the suction pressure is a static 20 psi, the 23% variation of a triplex single-acting pump would generate a theoretical A-type pressure variation of only 9.2 psi. This is hardly enough energy to set pipes in motion. The same flow variation at 1,400 psi working pressure, however, creates more than enough energy—a 644 psi A-type pressure variation in the discharge line.

Even so, the forces of acceleration become overwhelming disturbances in the suction. Pressure pulses of more than 25 psi are often encountered in pumps—even in systems with short suction pipes. A small amount of dampening of the flow-induced pressure can reduce pulsation to a negligible amount, leaving the 25-psi C-type acceleration pulsations. These remaining pulsations can create damage by enabling cavitation (Figure 4).

Cavitation. In practice, cavitation occurs at a pressure slightly higher than the vapor pressure of the liquid. Because vapor pressure is a function of temperature, a system may run without cavitation during the winter months but be subject to cavitation problems during the summer when temperatures are higher. Cavitation is connected with the presence of microscopic gas nuclei, which are in the solid materials bonded in the liquid. It is because of these nuclei that liquids cannot withstand tension. Without them, it is estimated that water could transmit a tension of approximately 150,000 psi.

During the beginning stages of cavitation, the nuclei give rise to the formation of gas bubbles, which are swept along to an area where the pressure is higher. The bubbles then collapse, producing high-intensity pressure waves. The formation of successive bubbles quickly follows. This repeating cycle lasts for only a few milliseconds, but during this time the localized pressures can be as high as 60,000 psi and temperatures can rise as much as 1500°F.

Cavitation can occur at a pump's inlets, inside the barrels, at the draft tubes of turbines and on propellers, mixing blades, Venturi meters, and other places where changes in fluid velocity occur. The cavitation effects in rotary positive displacement pumps are similar to those in rotodynamic pumps. The usual effects are wear of surfaces in contact due to erosion, slow increase in clearances, reduction in performance, noise, and vibration. Cavitation damage in reciprocating pumps can happen very quickly and be catastrophic. A cavitating 60 hp pump can break a major component every 350 hours.

Suction control. Suction control is an essential step in the design of any pump system and is equally important for reciprocating and centrifugal pumps. This process requires the following major steps:

1. Compare Net Positive Suction Head available (NPSHa) to Net Positive Suction Head required (NPSHr). (NPSH is a feature of the system layout and liquid properties. It is often referred to simply as NPSHa or NPSH available. Pump manufacturers usually quote NPSHr. This is a function of the pump's performance and is related to the pump's susceptibility to cavitation. To calculate the NPSHa to the pump, it is necessary to take into account the friction losses of the suction piping. NPSHa is then the difference between the total pressure on the inlet side of the pump and the sum of the vapor pressure of the liquid and the friction losses of the suction pipes.)
2. NPSHa is greater than NPSHr, the general suction conditions are under control, and their improvement (with respect to NPSHa) by installation of a suction pulsation dampener is not required. In this situation, the installation of a pulsation dampener would be required only for pressure fluctuation and noise reduction purposes.
3. If NPSHa is less than NPSHr and the piping rearrangement of a suction system is impossible or insufficient, installation of a pulsation dampener before the pump's inlet is essential.
4. Choose the appropriate dampener design based on steps 1–3. Using this process will help you achieve maximum reduction of pressure fluctuation.

Pulsations in discharge systems

The same forces at work on the suction side also affect discharge systems, but the pressure at A due to flow-induced pulsations becomes overwhelming at 460 psi (Figure 4). The 25 psi contribution from acceleration at C is a small percentage (2.5%) of the total discharge pressure. One exception is when the pump is delivering into a low-friction, high-pressure system such as a short vertical discharge system (Figure 6(a)). An example is a mine dewatering application, where the pump is delivering to an already pressurized system—a pressurized pipeline—through a short connecting pipe. Another exception is when the pump is delivering to already pressurized systems such as hydraulic press accumulators (Figure 6(c)). In those two cases, the acceleration pressures can become the overwhelming disturbance, particularly if the piping system is relatively long compared to a suction system (but considerably shorter than a "pipeline").

Control of pulsation with dampeners

Successful control of damaging pulsation requires careful selection of pulsation dampeners and proper location of those dampeners in a piping system. The earlier pressure pulsation analysis showed that dampeners will reduce the level of pulsation and consequently pressure fluctuations. With lower pipe vibration and noise, wear at the pump, risk of cavitation, and metal fatigue will be less as well. (There are several OSHA standards dedicated to the control of industrial noise in the workplace. One of the most practical and inexpensive ways to meet OSHA noise level requirements and achieve the overall performance improvement of reciprocating pumps is installation of pulsation dampeners on suction and discharge systems).

Dampener types. After evaluating the causes and types of pressure pulsation effects in reciprocating pumps, we can review which available fluid pressure pulsation control equipment is best for specific conditions.

Hydro-pneumatic pulsation dampeners are gas-type energy-absorbing design that uses a gas-filled bladder to absorb the "A" flow on peaks and to give back that flow on the "lows," thereby reducing flow variations beyond the pump and consequently reducing the flow-induced pressure pulsations. This is the most efficient, low-cost, and widely used type of pulsation control equipment.

This discussion is limited to the two major designs of bladder-type hydro-pneumatic pulsation dampeners: appendage and flow-through.

Appendage dampeners. Figure 7 shows a typical appendage design suction pulsation dampener. It consists of a homogenous shell and a nitrogen-filled flexible rubber bladder. The fluid port can be threaded (NPT or SAE) or flanged. This design is recommended for installation on pumping systems where suction is flooded. The dampener should be installed as close as possible to the pump inlet, and the nitrogen in the bladder should be precharged to 60% of the pump's inlet pressure. Figure 8 shows a large capacity version of this design.

The same type of suction pulsation dampener (150 psi) was installed on a 120 gpm and 30 psi Wheatley triplex pump with 3″ piston bore, 3-1/2″ stroke. Another two discharge 1,500 psi pulsation dampeners were used with a 263 gpm and 642 psi Wheatley quintuplex pump with 3-3/8″ bore, 4-1/2″ stroke.

Flow-through dampeners. This design should be installed on a pump's suction system when inlet pressure (NPSH) is very low and the fluid can contain entrained gases such as air, carbon dioxide, sulphur dioxide, etc. Flow-through units have significant advantages over the appendage design for a few reasons. First, the inlet and outlet diameters of the dampener are equal to the pipe diameter. Second, the in-line installation enables reduction of the inlet fluid velocity into the pump. In addition, the bladder response time is faster than an appendage dampener, due to the minimum distance between the shell inlet and pipe centerline. The flow-through design can also be used to improve the NPSHa of systems with low NPSH because it acts as a suction stabilizer.

Figure 9 shows an actual installation of a discharge flow through 2.5 gallon capacity dampener. This carbon steel, 3,000 psi, 3″ flanged unit was installed on the discharge

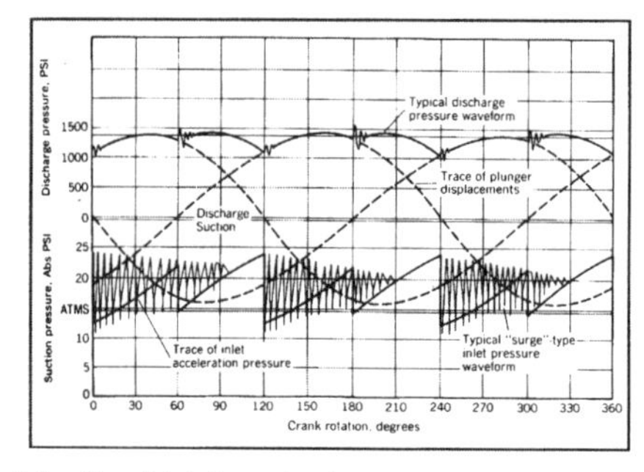

Figure 5. Idealized triplex single-acting pump pressure waveform showing the effect of surge-like acceleration pressures.

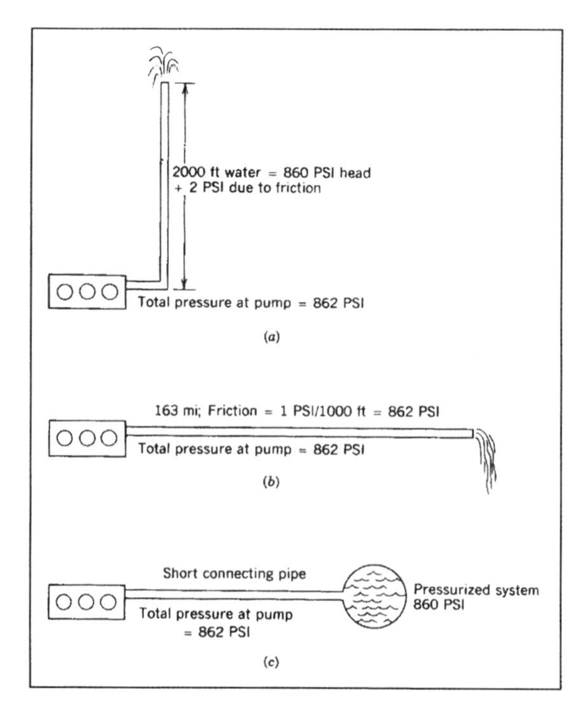

Figure 6. Dynamic differences in discharge systems. (a) vertical discharge. (b) horizontal discharge. (c) pressurized systems.

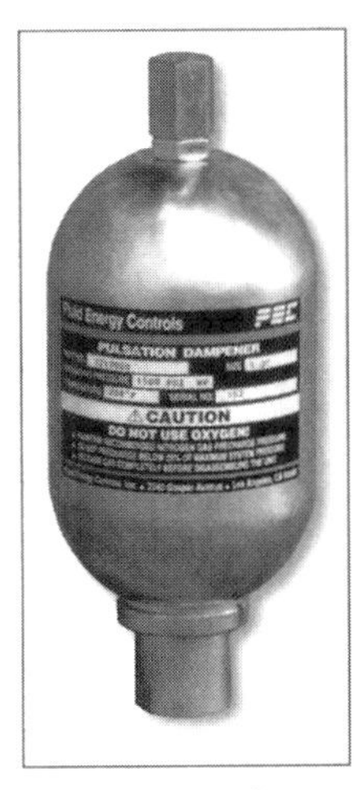

Figure 7. Appendage type 60 in^3 1,500 psi stainless steel pulsation dampener with threaded fluid port.

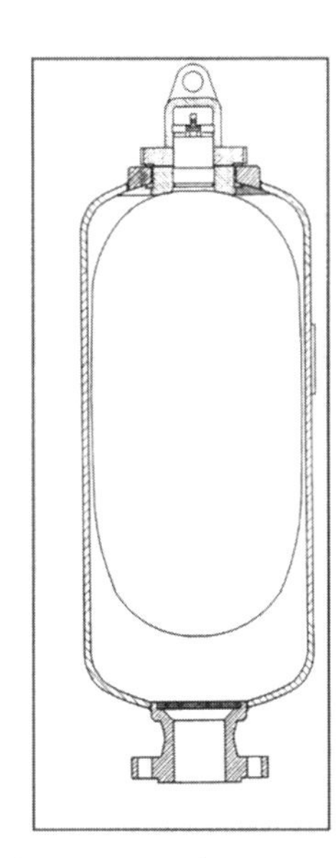

Figure 8. Typical large capacity appendage type flanged pulsation dampener.

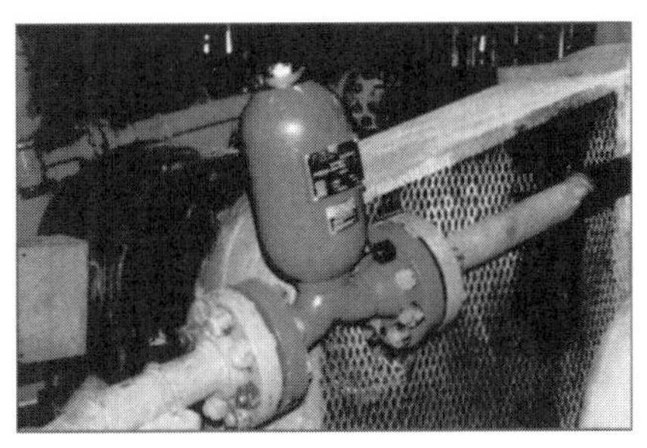

Figure 9. Flow-through pulsation dampener (Courtesy of Ormet Aluminum Mills Products Corporation, a Hannibal Rolling Mill Division).

system of an existing triplex single-acting reciprocating pump with 2.88″ piston bore diameter and 5″ stroke—one of seven such units. The owner, Ormet Aluminum Mill Products Corporation, is a Hannibal Rolling Mill Division, Hannibal, Ohio. Currently, the company is scheduled to install another two pulsation dampeners of the same type. The installation of pulsation dampeners for the remaining pumps is scheduled for sometime in 2000.

The abilities of pulsation dampeners

Hydro-pneumatic bladder pulsation dampeners are able to prevent the generation of the most destructive low-frequency pulses such as those generated by the pump's rotary motion and the combination of flow from each of the pump's cylinders. (This is basically rpm multiplied by the number of cylinders.) Accordingly, based on a maximum of 500 rpm for most small pumps, the maximum frequency involved should not be more than about 50 Hz.

Performance comparisons. Figure 10 compares the performance of the two types of pulsation dampeners on a typical pump. In this particular case, installation of hydro-pneumatic dampeners offers overall improved performance.

An analysis of reciprocating pump pulsations is based on the premise that pressure or pressure variation (pulsation) due to flow velocity (flow-velocity variation) is a function of the square of such velocity. And the system must present enough resistance to flow—due to pipe friction, bends, fittings, chokes—to enable the generation of a pressure or pulsation in the first place. The magnitude of that pressure is directly related to the resistance. The analysis is also based on the premise that the acceleration of the liquid at the beginning of an increase in velocity (start of stroke) generates a pressure pulse, and the magnitude is directly related to the length of pipe connected to the pump suction or discharge. It is also inversely related to the diameter of the pipe and directly related to the rate of change of velocity.

Because the inherent flow variations of a reciprocating pump are of the same intensity (but out of phase) in both the suction and discharge, for many years it was believed that a pulsation dampener of the same size as the discharge should be used on the suction, and many such installations can be found. However, from the standpoint of flow-induced pulsations alone, it is logical that a smaller dampener can be used on the suction side. For example, a typical triplex single-acting pump has a flow variation of about 23% in both the suction and the discharge. If all of this flow variation were converted to pressure pulsations, the pulsations would be a maximum of 46% of the average pump pressure. In our example, the maximum change in pressure at the discharge

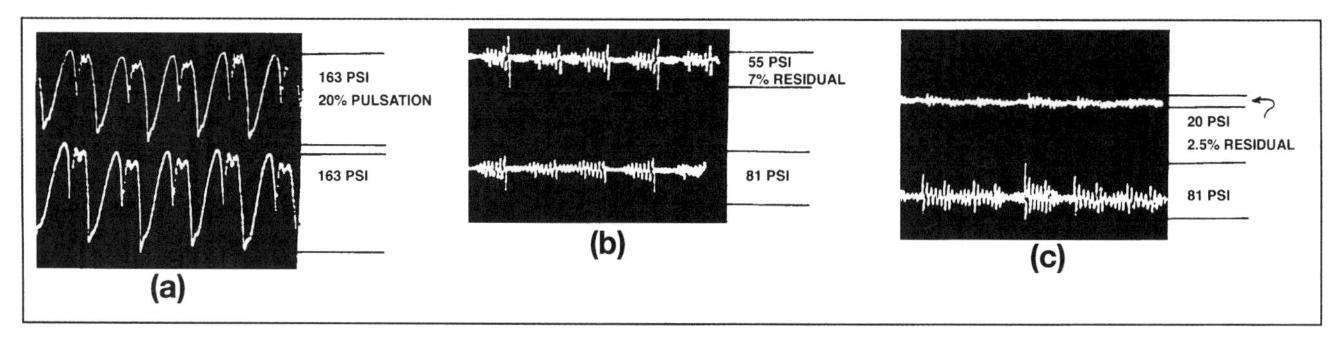

Figure 10. (a) No dampener installed. (b) appendage type hydro-pneumatic dampener. (c) flow-through type dampener. Top chart: downstream of dampener. Bottom charts: upstream of dampener.

would be about 460 psi at 1,000 psi average pressure. If there is a 50 psi suction pressure, the change in pressure there would be only 23 psi; that is, there would be about 20 times less energy available to set the piping in motion.

In addition, the acceleration due to velocity variations is the same on the suction and discharge sides. In the suction system, however, such disturbances usually overwhelm the disturbances due to flow variation, and conversely, in the discharge system the flow-induced pressure pulses usually overwhelm the acceleration pressure pulses.

There is evidence that only a flow disturbance or an acceleration disturbance (not both) can exist during any suction or discharge stroke of a pump. One will overwhelm and reduce the effect of the other. A plausible explanation is the theory that in pumps of greater than about 300 fpm piston speed, a disturbance of either type will momentarily reduce the volumetric efficiency to such an extent that it cannot recover fast enough to enable the flow rate to reach an adequate value to affect another disturbance in the short period of time (about 3 ms) between the disturbances. There is no such worry with slow speed pumps because few if any disturbances are generated.

It is possible for a pump with high inherent flow variations to actually deliver an almost pressure-pulseless liquid into an open-ended (in effect), short, relatively large-diameter horizontal discharge pipe. Some typical cases are pumps with short connecting pipes feeding a large-diameter already-pressurized system such as an existing pipeline or hydraulic press supply. If the discharge pipe is vertical, as in a mine dewatering system, the acceleration of the liquid column due to gravity becomes the predominant generator of pressure pulses. This great dynamic and hydraulic difference is shown in Figure 6, where the pump "sees" the same discharge pressure under three different physical arrangements. Energy-absorbing or hydro-pneumatic pulsation dampeners may not solve pulsation problems in cases (a) and (c) of Figure 6.

Application of a large-capacity, flange type of surge suppressors should be carefully considered (Figure 8).

Advantages of bladder type pulsation dampeners

- Positive sealing between the gas and fluid chambers, such as the bladder, serves as a permanent wall between them
- Zero leakage between both chambers due to bladder barrier
- Quick pressure response times for pressure control pump pulsation and shock-dampening applications are facilitated by the light weight of the bladder
- Simplicity of design reduces overall maintenance and operating cost of pumping system

Sizing considerations

Suction pulsation dampeners and suction stabilizers. The following sizing procedure will serve two purposes:

- Evaluation of the pump suction system and subsequent sizing of the pulsation dampener and suction stabilizer with respect to the reduction of pressure fluctuation
- Improvement of pump suction conditions represented by NPSHa

Sample data

Piping
- The maximum flow of 200 gpm originates from the storage tank.
- The tank is open to atmosphere.
- The length of the 8″ cast iron suction pipe is 106 ft.
- The difference of elevation between the storage tank and centerline of pump suction inlet is 23 ft.
- The fittings are: (1) 8″-90° elbows, (1) 8″-45° elbows, (4) 8″ Tees.

Pump (4 each)

- Single action triplex
- Max speed = 45 rpm
- Flow rate = 200 gpm (800 gpm for all 4 pumps)
- Min NPSH = 23 ft. of column of water
- Bore = 9.0 in.
- Stroke = 6.0 in.

Fluid

- Anaerobically digested municipal sludge conditioned with 10% lime and 5% ferric chloride.
- Fluid temperature = 100°F

Analysis. In order to determinate the Net Positive Suction Head Available (NPSHa), the acceleration head and the friction losses along the suction piping must be calculated. Acceleration head can be calculated as follows:

$$h_a = \frac{LVnC}{Kg}$$

where:

h_a = Acceleration head, ft. of fluid
L = Length of suction line, ft
V = Velocity in the suction line, fps
n = Pump speed, rpm
C = Pump type constant, for triplex single-acting pump C = 0.066
K = 1.4 for hot water
$g = 32.2\ \text{ft./sec}^2$

Pipe friction loss factor

$$K_1 = f\frac{L}{D} = \frac{0.020 \times 106}{0.336} = 6.31$$

Valve and fitting friction loss factor K_2 = (4 valves) × 1.8 + (1–90° elbow) × 1.5 + (1–45° elbow) × 0.8 = 9.50

Total valve and fitting frictional loss factor

$K_t = 15.81$

Total pipe frictional losses

$$h_t = \frac{K_t V^2}{2g} = \frac{(15.81) \times (5.12)^2}{2 \times 32.2} = 6.44\ \text{ft. of w.c.}$$

Acceleration head

$$h_a = \frac{(106) \times (5.12)^2 \times 45 \times 0.066}{1.4 \times 32.2} = 34.96\ \text{ft. of w.c.}$$

The maximum static elevation head hs = 23.0 ft. of water column.

Net Positive Suction Head Required NPSHr = 9.87 psig @ 45 rpm = 22.8 ft. of w.c., say 23 ft. of w.c.

P_{tank} = 0 psig = 0 ft. of w.c.

Net Positive Suction Head Available

$$\begin{aligned} NPSHa &= h_s + P_{tank} - h_a + h_f \\ &= (23 + 0) - (34.96 + 6.44) \\ &= -18.40\ \text{ft. of w.c.} \end{aligned}$$

Since NPSHr = 23.0 ft. of w.c., the NPSHa = 18.40 ft. of w.c. and is less than NPSHr. The actual suction head is not adequate to operate the pump at the 23 ft. of w.c. NPSH requirement.

This analysis shows the necessity of installing the suction stabilizer to improve suction conditions. Such stabilizers can reduce the acceleration head, thereby increasing the NPSHa. The unit will also eliminate the acceleration head by minimizing the mass of fluid in the suction pipeline that must be accelerated. The suction stabilizer installed at the pump inlet will reduce the fluid column that must be accelerated to the fluid between the bladder and the pump suction manifold. This fluid mass is normally very small and requires a low-pressure fluctuation. The losses correspond to the acceleration effects and are about 67.5% of the operating pressure.

Acceleration losses

h_a = (0.075) × (23 ft. of w.c.) = 1.73 ft. of w.c.

NPSHa for the system with the suction stabilizer is

NPSHa = (23.0 + 0) − (1.73 + 6.44) = 14.84 ft. of w.c.

As can be seen from the earlier analysis, the installation of a suction stabilizer will effectively reduce the acceleration head and provide a much greater NPSHa to the pump. This will prevent local cavitation and increase pump efficiency. To provide the required Net Positive Suction Head in our example, the sludge conditioning tank should be raised only 8.0 ft. (versus to 18.4 feet) to compensate for the friction losses between the tank and pump suction manifold.

The actual volume of pulsation dampener/suction stabilizer can be calculated using the following sizing procedure:

Volume = 10 × displaced flow/revolution
Displaced flow = (0.7854 × B^2 × stroke × no. pistons × action)/231 gallons

where:
B = Pump cylinder bore, in.
stroke = Pump piston stroke, in.
no. piston = 1 for simplex, 2 for duplex, 3 for triplex.
suction = 1 for single acting pump, 2 for double acting pump.

By installing the pump data in the formula, we get the required actual volume of pulsation dampener/suction stabilizer.

$$\text{Volume} = 10 \times (0.7854 \times 9^2 \times 6 \times 3 \times 1)/1 \times 231 = 49.6 \text{ gallons}$$

The 50-gallon suction stabilizer installed at the pump inlet will reduce the pressure fluctuation on the inlet pipeline to 1.73 ft. of w.c., thereby providing an NPSHa of 33.24 ft., greater than without a suction stabilizer.

Sizing of discharge pulsation dampeners

In order to achieve optimum results from the dampener, its size (capacity) must be properly calculated. The following simple formula can be applied for sizing a pulsation dampener for either a simplex, duplex, or triplex single-acting pump.

Size of dampener = plunger area (sq. in.) × plunger stroke (in.) × 50 × pump type factor

Notes:
1. 50 = pressure factor for about 65% pulsation control
2. pump type factor = 0.25 for duplex single-acting pump

Example. Determine the size of a pulsation dampener to control pulsations caused by a duplex single-acting plunger chemical metering pump handling 50% sodium chloride.

Data

- Plunger diameter (D) = $1\frac{3}{4}$ in.
- Plunger stroke (L) = 3 in.
- Max. pressure = 150 psig
- Pump type factor = 0.25

Size of pulsation dampener = plunger area (sq. in.) × plunger stroke (in.) × 50 × pump type factor = $\pi D^2/4 \times 3 \times 50 \times 0.25 = 90.2$ cu. in. Therefore, a 0.5-gallon size dampener should be selected.

Material and design specification guidelines for pulsation dampeners and suction stabilizers

The next step is proper specification of the dampener's material of construction with respect to working conditions. The diaphragm or bladder and all other seals in the dampener shall be EPR (Ethylene Propylene) compound compatible with 50% sodium hydroxide (NaOH). The metal parts of the dampener shall be of 316 stainless steel. The selected dampener in this example shall dampen pulsations within 5% to 10% of the mean line pressure.

Conclusion

The following factors must be thoroughly considered during the selection process of pulsation dampeners.

1. Maximum design pressure rating of the dampener must be within the operating pressure limit of the system.
2. Liquid port size of the dampener should be equal to the discharge port of the pump or equal to the discharge line size.
3. Dampener should be charged with a neutral gas (usually nitrogen) to approximately 60% of the mean system pressure.
4. Appendage type dampeners: pipe or nipple connecting the dampener to the line should be as short as possible. The diameter of connecting pipe should not be smaller than the size of the dampener liquid port.
5. The dampener should be installed as close as possible to the pump discharge port.
6. A periodic check of nitrogen gas pressure in the dampener should be made to make sure the dampener is in operable condition.
7. The dampener's rubber and metal parts must be compatible with the pumped liquid. ■

V. Larry Beynart is an Application Engineer with Fluid Energy Controls, Inc., a Los Angeles manufacturer of hydropneumatic accumulators and pulsation dampeners. He holds a Bachelor of Science Degree in Mechanical Engineering and Hydraulic Certification from Fluid Power Society.

Rotary pumps on pipeline services

James R. Brennan

Imo Pump, Monroe, NC

Abstract

Pipelines handling liquids in excess of about 500 SSU viscosity will normally use rotary, positive displacement pumps both for optimum efficiency and, frequently, lower initial cost. This section provides some insight into the application of rotary pumps to some of the unique requirements facing users of pipelines in crude oil transport, electric utility power generation, and general industrial needs. Coverage includes pumping system designs, flow control on variable flow systems, serial pump operation on multistation pipelines, and key components that help ensure reliable pump operation and maintainability.

Pipelines are used to transport a broad range of liquids, including crude oils, fuel oils, and refined products such as gasoline, as well as water, specialty chemicals, and others. Lines can be as short as several dozen feet to hundreds or even thousands of miles long. They can cross state or country boundaries. The modern world relies upon safe, efficient operation of all manner of pipelines, both liquid and gas. Oil producers, pipeline transport companies, refineries, and power plants are probably the largest operators of pipelines.

Types of pipeline services

Pipeline services can be broadly categorized as loading/unloading, local transport, and long-distance transport. Examples of load/unload include pumping a relatively short distance from or to rail cars, tank trucks, barges, or tanker ships. These services are typically intermittent, perhaps 8 or fewer hours per day, perhaps not every day of the week. Differential pump pressures are usually low, less than 125 psid. The combination of low power and intermittent use makes pump operating efficiencies less important than in other services.

Local transport would be defined as a single pumping station installation with line lengths less than 50 miles. Design operating pressures can reach 1450 psig or more. Local or national regulation may dictate maximum pressures. These pipelines can be intermittent or continuous duty services and can range from modest power levels to many hundreds of horsepower or more.

Long-distance transport is usually more than 50 miles and includes multiple pumping stations along the length of the pipeline. These pipelines normally operate around the clock at partial or full capacity. Controls and instrumentation tend to be more sophisticated and automated. Long-distance pipelines will frequently operate at variable station flow rates, depending on product demand at the end terminal. Pumping stations will normally have three or more pumps with one in standby mode (three half-capacity pumps). If pipeline capacity is to be increased over time, more pumps may be added to each station.

Pumping stations

Pumping stations for liquid pipelines can be as simple as a single pump located adjacent to a supply tank pumping liquid down the pipeline to a receiving tank (Figure 1). The supply and receiving tank level switches can control the pumping cycle by starting or stopping the pump at appropriate minimum and maximum tank liquid levels. Longer pipelines needing intermediate pumping stations can be handled in one of two basic arrangements. The first is to use storage tank(s) at each station (Figure 2). The upstream station pumps to the downstream station's tanks. The downstream station pump(s) take suction from these tanks and send the liquid farther downstream. This is a simple,

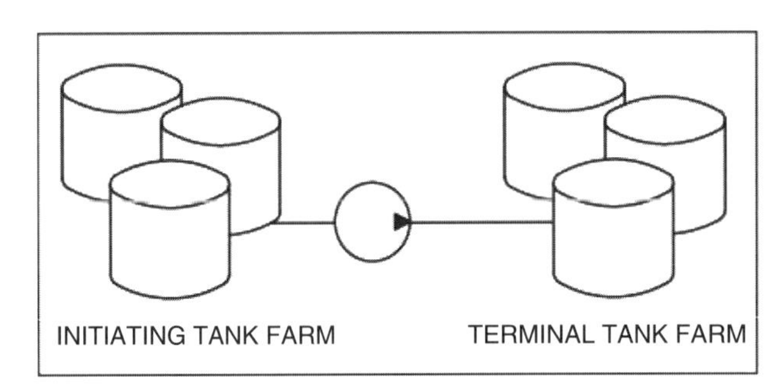

Figure 1. Simple pipeline system.

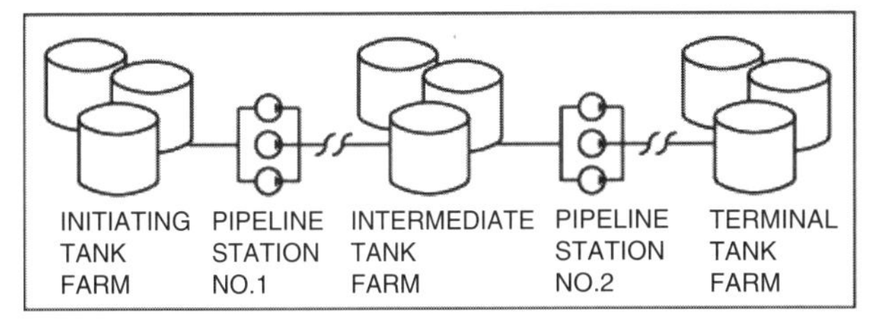

Figure 2. Multistation, tank system.

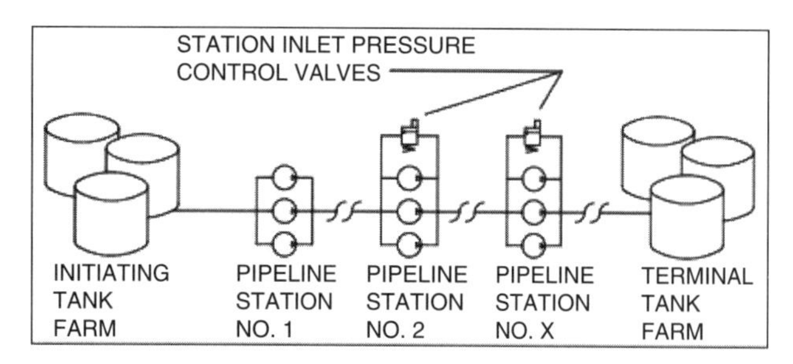

Figure 3. Multistation, tight system.

reliable method of long-distance transport. Pumps are started or stopped depending on liquid level in tanks. This has several disadvantages. The cost of storage tanks, dykes, and tank condition monitoring are not small. There is also product in "inventory" in the tanks that does not help a business's cash flow. The alternate arrangement is a "tight" system, leaving the pumping stations in series with each station's pump(s) taking suction from the pipeline (Figure 3). This has the advantage of eliminating the need for storage tanks but requires control over each pumping station in the pipeline such that exactly the same flow rate is going through each station—more on that later.

Types of pumps applied

Most pipeline pumps are centrifugal units. When the product being transported is always a low viscosity liquid like water, gasoline, diesel oil, or very light crude oil, centrifugal pumps are cost-effective, reliable, and efficient. However, as the liquid viscosity increases, the frictional losses within a centrifugal pump quickly reduce pumping efficiency dramatically (Figure 4). For this reason, rotary, positive displacement pumps are often used when products such as heavy crude oil, bunker fuels (no. 6 fuel oil), low sulfur fuels, asphalt, Orimulsion®* (a manufactured boiler fuel emulsion of 30% water and 70% bitumen), and similar need to be transported. The energy costs of operating a pipeline can represent close to 50% of total operating costs, so pump operating efficiencies play a critical role in profitability. Table 1 illustrates differential annual operating costs for two real examples of centrifugal pumps versus rotary pumps.

Although there are a great many types of rotary pumps, the most common on pipeline services are two- and three-screw pumps, gear pumps (internal and external), and vane pumps for low pressure services. Two- and three-screw pumps are used for higher pressure services, especially at medium to high flow rates. Table 2 shows the approximate maximum flow and pressure capability for each type, not necessarily achievable concurrently.

Orimulsion is a registered trademark of Petroleos de Venezuela S. A.

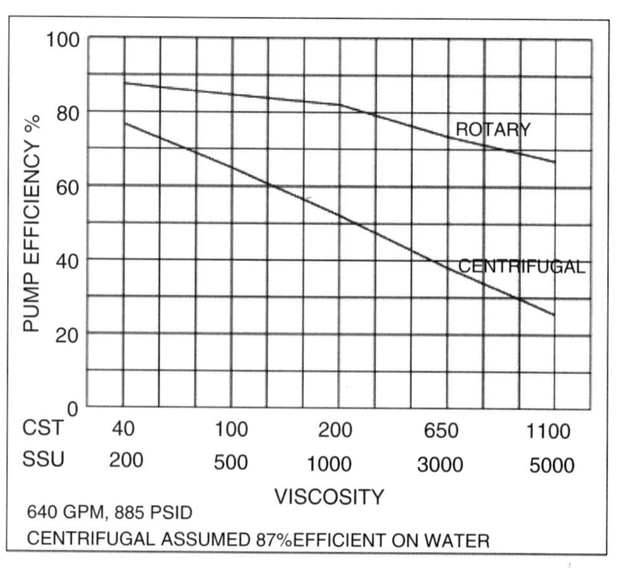

Figure 4. Rotary versus centrifugal efficiency.

Table 1
Operating costs comparison

Parameter	Centrifugal	Rotary
Type	6 stage	3 screw
rpm	3500	1750
Cst	200	200
psid	765	765
gpm	410	410
bhp	395	229
Efficiency %	45	81
Annual energy cost at $0.07/kw-h	$180,449	$104,615

Table 2
Rotary pump flow and pressure ranges

Type	Maximum gpm	Maximum psid
Gear	1500	250
Vane	1000	250
2 Screw	5500	1500
3 Screw	3400	2000

Gear and vane pumps (Figure 5) tend to have the lower first cost and may or may not have the best lifecycle cost. They will normally operate at relatively low speeds, requiring speed reducers on all but the very smallest sizes.

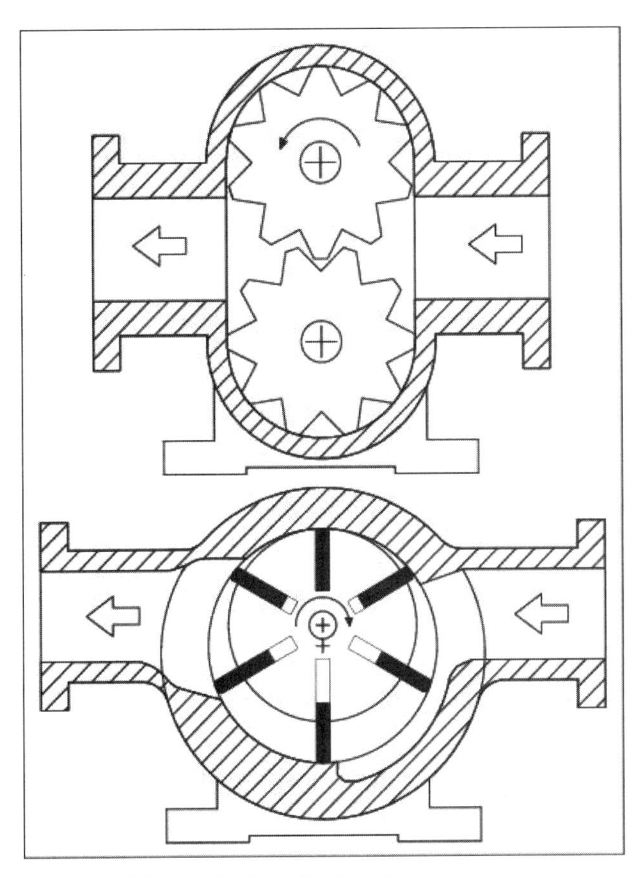

Figure 5. Gear (top) and vane pump.

Multiple-screw pumps (Figure 6) move liquid axially rather than in a radial direction, so fluid velocities are relatively low even when directly driven at four- and six-pole motor speeds (1800 and 1200 rpm at 60 Hz, 1500 and 1000 rpm at 50 Hz).

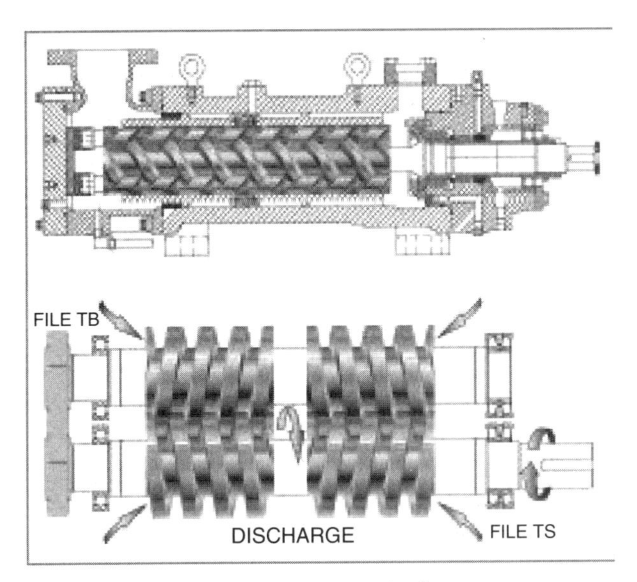

Figure 6. Multiple-screw pipeline pumps.

Figure 7. Fifteen twin-screw pumps for SUBSEA pipeline.

Efficiencies remain high over a much broader range of differential pressure than centrifugal pumps. Figure 7 shows 12,000 hp of installed two-screw pumps (15 pumps in all) that are used to load *Orimulsion* tankers 8 miles offshore. Figure 8 illustrates a diesel engine–driven three-screw crude oil pipeline pump operating in Guatemala at 1400 psig.

Specified operating requirements

Obviously, pump and driver specifications will be needed to acquire equipment that will perform to requirements. Unless your organization has these talents in house, an engineering firm will be needed to explore the pipeline

Figure 8. Diesel-driven three-screw pump at 1400 psi.

requirements. As it relates to rotary pumps, at a minimum the following will need to be defined:

1. Number of pumping stations
2. Number of pumps per station
3. Possible range of
   - Pump inlet pressures
   - Pump discharge pressures
   - Liquid temperature range
   - Liquid viscosity range
   - Per pump flow rate range
   - Particulate content
   - Upset conditions
   - Static conditions
4. Driver requirements
5. Pump instrumentation
6. Pump leak detection

Particulate content can range from unknown (take your chances) to specific weight percentage of sediment, sand, etc. For liquids containing abrasives, pump life is inversely proportional to something between the square and the cube of pump speed. Operating larger pumps at lower speeds will increase initial cost but can save on maintenance costs over many years. Another proven technique to extend pump life in abrasive environments is to select pumps with effectively more stages than needed just for pressure capability. For example, every wrap in a screw pump acts as a pressure stage, similar to a multistage centrifugal pump. More stages reduce the pressure rise per stage, reduce the slip flow (volumetric inefficiency), and spread out the wear. The result is extended pump service life and, frequently, higher average efficiency over time.

Product leak detection could be a critical element in both pumping station safety and environmental contamination. Unless there are observant attendants on site at all times, automatic alarm or shutdown should be considered in the event of product leakage from any source. Rotary pumps most commonly use mechanical seals to contain pumped product within the pump casing. A mechanical seal will wear and can fail catastrophically, and product under pressure can leak large volumes in short time periods. Many pumps used for pipeline services can be equipped with various seal-leak detection schemes to alert/alarm or shut down in the event of excessive seal leakage.

Driver selection depends on available power sources. In remote areas, diesel engines are usually chosen. They provide some range of speed and, therefore, flow control of the pumps. By far, the most common drive is a fixed speed electric motor, although variable frequency drives are becoming more popular. They can improve station efficiency by avoiding the necessity of drawing power to pump flow that is to be bypassed around the station (wasted energy). Spacer couplings are highly recommended between the pump and its drive. They allow access to the pump shaft-end seal and bearing, as well as the motor shaft-end bearing.

Upset conditions and static conditions need to be assessed during pipeline design. Most rotary, high-pressure pumps cannot tolerate anything close to discharge pressure at their inlet (suction) port. In a tight system, if an upstream pumping station is started and the downstream station is not started in very close synchronization, the downstream station may see excessive pressure on its inlet side. A low pressure set inlet side relief valve may be necessary, together with an overspill tank and appropriate instrumentation.

Instrumentation tends to be a very owner-specific issue. Some systems have only a few pressure gauges. Others include bearing temperature detectors, vibration detection, inlet and outlet pressure transducers, and temperature transducers, all sufficient to get a good picture of what is happening at a pumping station, sometimes from many hundreds of miles away.

Pump heating and/or cooling

If the product being pumped is so viscous that heating it before pumping is necessary, then it will probably be prudent or essential to provide a means of preheating the pumps so they are not faced with sudden introduction of hot liquid into cold pumps. Depending on available utilities, pumps can be jacketed for use of heating mediums such as steam or hot oil. Heat tracing with electric thermal wire is also an option. The outside of the pump will need to be insulated to contain the heat. Heating and insulation systems must take into account pump maintainability if the heating or insulating system must be removed to service the pump. Be sure to avoid applying heat to timing gear cases or pump bearings. In some instances, these pump components may actually need cooling to operate for extended periods.

Pump protection

Like most other pumps, rotary pumps are vulnerable to premature wear or failure from too little or too much inlet pressure, too high a discharge pressure, excessive inlet temperature, dry running, and ingestion of foreign material (weld rod, wire pieces from cleaning brushes, etc). In addition, if the liquid pumped is inherently "dirty," like many crude oils and bunker fuels, then pump wear will be a normal part of operating pipeline pumps and one should prepare for that eventuality.

Systems need to be instrumented to at least alarm if any of these parameters approach a limiting value. If pumps are unattended most of the time, then automatic shutdown is the only protection from catastrophic pump failure. Use of inlet strainers is highly desirable, provided they are instrumented to detect excessive pressure drop and cleaned when needed. Without regular monitoring and cleaning, strainers eventually will accumulate debris to the point that the pressure loss of liquid going through the strainer will allow pump inlet pressure to fall below the minimum allowable. Worse still, that pressure drop can cause the strainer element to collapse. On collapse, all the accumulated debris, as well as the strainer element, will arrive at the pump inlet. Nothing good will come of this.

As with any positive displacement pump, a pressure relief device is essential to safe operation. The "shutoff head" of a positive displacement pump can be enormous, far above the pressure rating of any system component. A relief valve is normally installed at each pump discharge with the valve outlet connected back to the supply tank (preferred) or to the pump inlet piping as far from the pump as practical.

System considerations

When starting a rotary pipeline pump with the pipeline already full, the system must provide for the time required to accelerate the liquid from zero velocity to final velocity. There may be 50 miles of product that needs to get to 10 ft/sec. This is normally accomplished with an electric-operated valve around each pump that bypasses discharge flow either back to the supply tank (preferred) or back to pump suction. The valve is open on pump startup and closed gradually as the flow velocity in the pipeline builds. If bypassing back to the pump inlet piping, go as far away from the pumps as practical. This will maximize the mass of liquid available to absorb the driver power draw that is not yet going down the pipeline while the bypass is not fully closed.

Flow control can be achieved in several ways. Variable-frequency drives are becoming more affordable in larger sizes. Because rotary pump flow is directly proportional to speed, control is fairly straightforward. In tight systems with fixed-speed drivers, pumping station variable flow is achieved first by how many of the station's pumps are in operation and second by bypassing flow around the station to control how much total flow leaves the station. Figure 9 is a simplified schematic of such a system. The bypass valve senses station inlet pressure. It bypasses whatever flow is necessary to control station inlet pressure at set point. If the station inlet pressure rises, less flow will be bypassed and more flow will therefore leave the station.

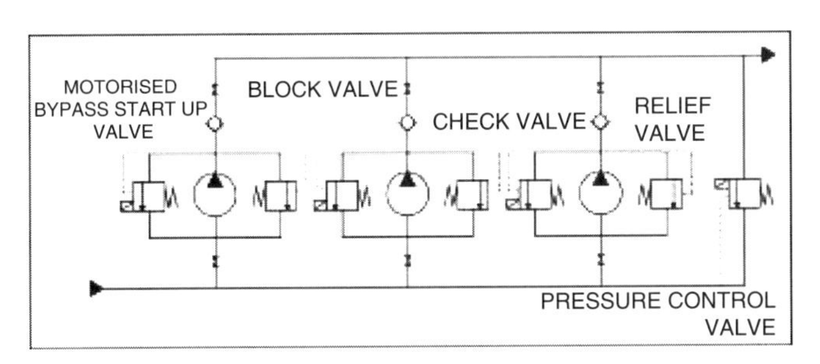

Figure 9. Tight series pump operation.

Maintenance

Maintenance for rotary pumps is not very different from any other pump. Hopefully, the installation provided both space to access the pump for service as well as some shelter from the elements. Because rotary pumps have relatively close internal running clearances, be especially careful to achieve and maintain good pump-to-motor shaft alignment. Avoid pipe strain at all costs for the same reason. If pumping station downtime is a serious matter (usually), then have an installed standby pump ready to take over, as well as critical spare parts. If a pump "goes down," try to identify the cause before starting another one. Sadly, too many times a series of failures is allowed to occur before enough time and effort is expended on root cause analysis.

Many pipeline pumps have service parts that need to be removed from both ends of the pump. Spacer couplings go a long way in easing the drive end service. Keeping walls or objects away from the pump end opposite the driver is usually necessary to avoid the need to disconnect piping and remove pumps to a bench area.

Some pumps are, in fact, better maintained on a bench in a maintenance area, where space to perform the work is more available, components can be more easily assessed for their condition, and new parts can be installed more correctly.

Conclusions

Rotary, positive displacement pumps can play key roles in reliable, efficient, effective transport of liquids over short or very long distances. They are an optimum choice where efficiency is a strong driver of pump selection and operation. There is an extensive history of successful applications of rotary pumps to pipeline transport.

Source

This tutorial was presented at the Pump Users Expo 200, Louisville, KY, September 2000.

James R. Brennan is market services manager with Imo Pump, Monroe, NC. He is a 1973 graduate of Drexel University in Philadelphia, a member of the Society of Petroleum Engineers, SPE, and has more than 30 years of service with Imo Pump.

KEY CENTRIFUGAL PUMP PARAMETERS AND HOW THEY IMPACT YOUR APPLICATIONS—PART 1

Pumps and Systems: They Go Together

By Doug Kriebel, PE, Kriebel Engineered Equipment

The purpose of this article is to familiarize the reader with centrifugal pump manufacturers' performance curves, show how a pump system head curve is constructed, and demonstrate how to analyze the effect of changes in the system.

A pumping system operates where the pump curve and the system resistance curve intersect. This important fact is one of the most overlooked concepts when designing or troubleshooting a pumping system.

The pump curve has a head/capacity portion, which indicates the change in flow with respect to head at the pump. It is generated by tests performed by the pump manufacturer. We will review the key facts that this curve offers.

The system resistance curve is the change in flow with respect to head of the *system*. It must be plotted by the user based upon the conditions of service. These include physical layout, process conditions, and fluid characteristics.

Anyone operating a pumping system must know where the pump is operating. Otherwise, it is like driving a car and not knowing how fast it's going.

The pump curve

The pump curve consists of a head/capacity curve (curve A of Figure 1) that shows the Total Differential Head (H is expressed in feet of pumped liquid). This is the difference between the Total Discharge Head (Hd) at the pump discharge and the Total Suction Head (Hs) at the pump inlet at any given flow rate (Q), expressed in gpm (gallons per minute).

$$H = Hd - Hs \tag{1}$$

All pressure is in feet of liquid

At this point, it's important to re-visit some pressure units:

psi $=$ pounds/in^2
psia $=$ pound/in^2 absolute
psig $=$ pound/in^2 gage
psia $=$ psig $+$ barometric pressure

The conversion from psi to feet of liquid is to divide the pressure, in psi, by its density:

$$(\text{lbs/in}^2)/(\text{lbs/ft}^3) \times (144\ \text{in}^2/\text{ft}^2) = \text{feet of pumped fluid}$$

For water with a specific gravity (S.G.) of 1, you substitute its density (62.4 lbs/ft^3) into the above equation and the result is:

psi $\times$ 2.31/S.G $=$ feet of liquid. (S.G. is the ratio of the density of a liquid to water at 60°F). See Figure 2 for illustration of feet, psia/psig/S.G.

The pump curve also includes the pump efficiency (in percent-curve B), BHP (brake horsepower–curve C), and NPSHr (Net Positive Suction Head required–curve D) at

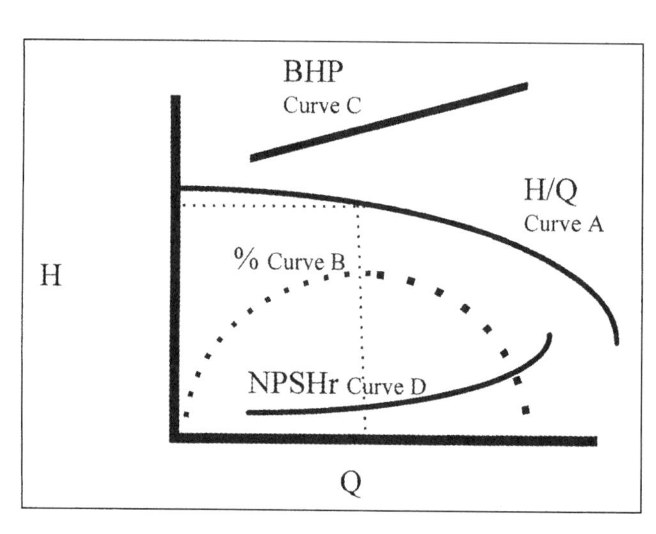

Figure 1. Performance curve data.

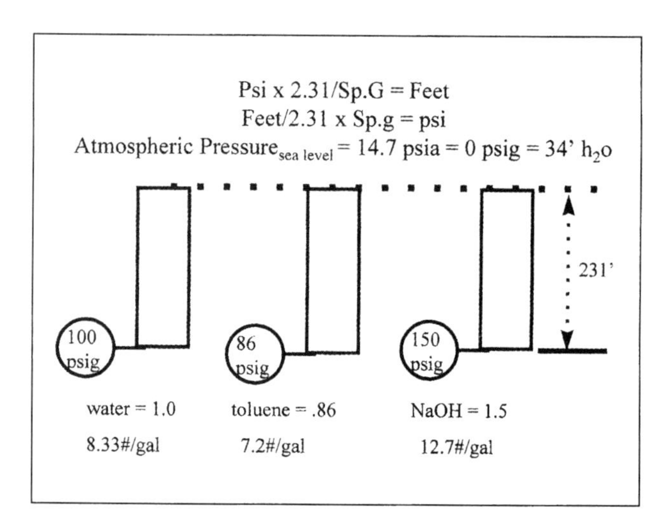

Figure 2. Pressure equivalent.

any flow rate. This curve is plotted for a constant speed (rpm) and a given impeller diameter (or series of diameters).

Figure 1 is a single line curve; Figure 3 is an actual manufacturer's performance curve showing the same information but expressed as a series of curves for different impeller diameters (from 12-5/8″ to 15-15/16″). The pump impeller can be trimmed to any dimension between the maximum and minimum diameters.

The H, NPSHr, efficiency, and BHP all vary with flow rate, Q. The best efficiency point (BEP) is the point on the curve where the efficiency is at a maximum. For Figure 3, it's 82% at 1300 gpm and 100′ TDH.

BHP can be read from the curve at any flow rate; however, it is usually calculated from the efficiency, head, and flow using Eq. 2. Pump curves are based on a specific gravity of 1.0. Other liquids' specific gravity must be considered.

$$BHP = Q \times H \times S.G./(3960 \times \%\ eff) \quad (2)$$

The NPSHr is read from the bottom curve from flows, and the NPSHr in feet is read on the lower right.

The following examples apply to Figure 3.

Example 1. For an impeller diameter of 15-15/16″ (maximum), what are the head, efficiency, NPSHr, and BHP at 900 gpm? (Answer: 110′; 76.5%, 5′ NPSHr, and 32 BHP)

Example 2. What are the head, efficiency, NPSHr, and BHP at 1600 gpm? (Answer: 85′, 78%, 7.5′ NPSHr, and 44 BHP)

In a pumping system, the capacity of the pump moves along the pump curve based upon the differential head across the pump. For instance, for Figure 3, with the maximum impeller diameter, if the flow is 900 gpm and you want to increase it, you reduce the head to 85′, and the flow will increase to 1600 gpm. It's by controlling the pump differential head that you control the flow of the pump. If you wanted 1300 gpm, you would have to throttle the head back to 100′.

The slopes of pump curves vary depending on the pump-specific speed (Ns). How they vary depends on Ns and is shown in Figure 4. The Ns is a dimensionless number calculated as:

$$Ns = rpm \times (Q^{1/2})/H^{3/4} \text{ all at BEP} \quad (3)$$

The Ns for Figure 3 is: 1311

NPSHr is Net Positive Suction Head required by the pump—based on test.

NPSHa is Net Positive Suction Head available by the system—must be calculated.

$$NPSHa = P1 + hs - hvp - hfs \quad (4)$$

P1 = Pressure of suction vessel (atmospheric in open tanks)
hvp = Vapor pressure of liquid at its pumping temperature, t

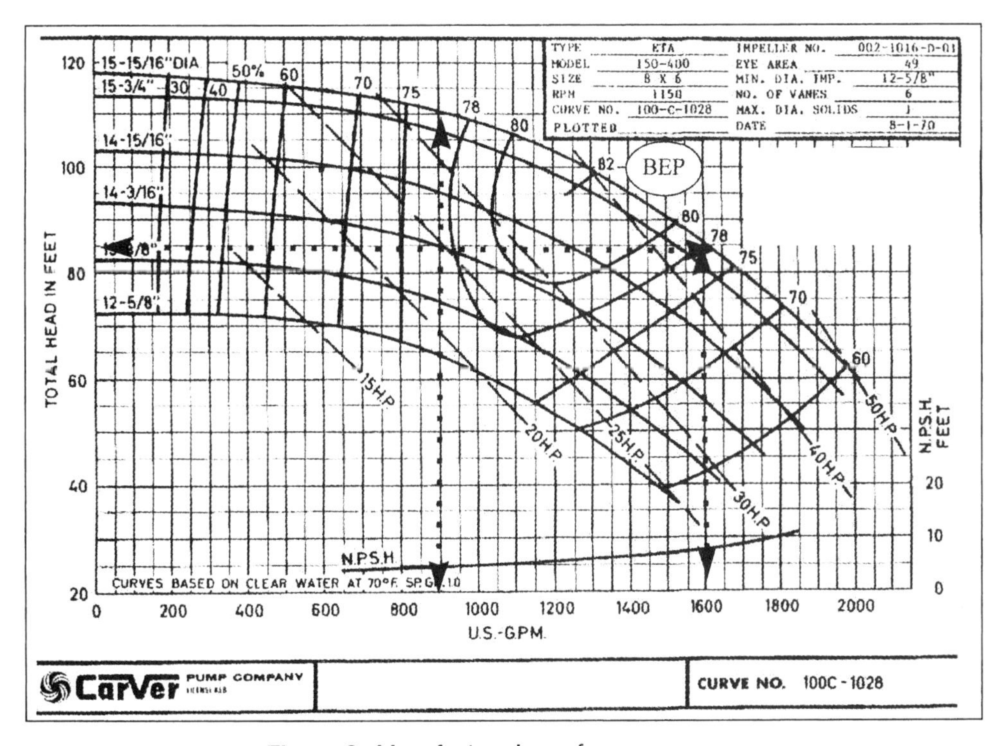

Figure 3. Manufacturer's performance curve.

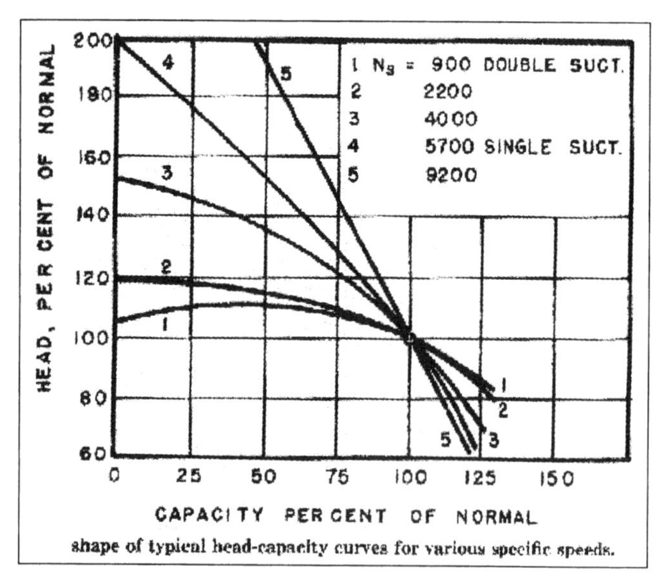

Figure 4. Head versus capacity versus Ns.

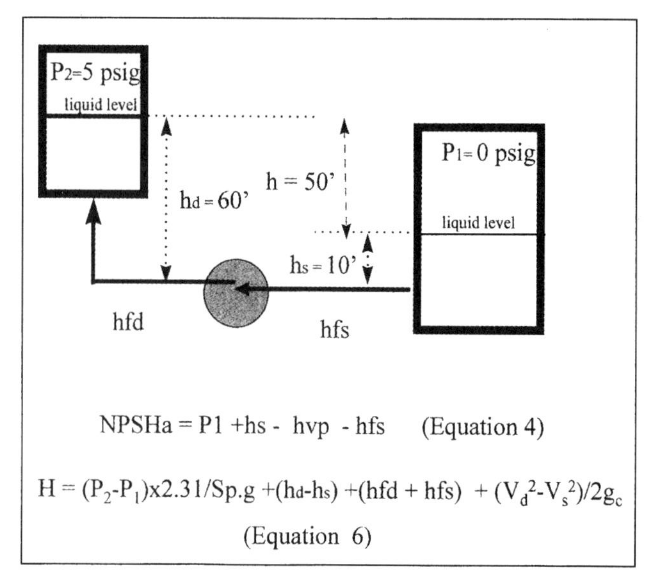

Figure 5. Atypical pump system.

hs = Static suction head, the vertical distance between the eye of the first stage impeller centerline and the suction liquid level. (This may be a positive or negative number, depending on whether the suction liquid is above the pump suction eye or below it, as on a "lift.")

hsf = Friction and entrance losses on the suction side

Pump curves are generated by manufacturers and show a range of impeller diameters from minimum to maximum size for the casing size selected. These are based on a single speed (rpm). Head/capacity relationship changes with speed or impeller diameter per the affinity laws. Affinity laws are discussed in all the references. Other information often found is suction specific speed (Nss), which is a dimensionless number calculated at the BEP as

$$Nss = rmp \times (gpm_{eye})^{1/2}/NPSHr^{3/4} \quad (5)$$

(*Note*: The gpm is per impeller eye, so for a double suction impeller, the flow is half the flow of single suction impeller.)

The Nss in Figure 3 is 10,800.

Nss is a function of NPSHr. To reduce NPSHr, the impeller eye is modified. If the eye is modified too much, it causes a reduction in the low flow capabilities of the pump and limits the low end point of operation.

System head curves (Figure 5)

Step one in constructing a system/head curve is to consider the fluid energy requirement.

At any point in a pumping system, there are four components making up its total energy (head in feet):

- Pressure
- Velocity
- Elevation
- Heat (ignored, since pumping is usually at a constant temperature across the pump)

The difference in energy levels (differential head, H) between the suction (Hs) head and discharge (Hd) head of a pump can be expressed using Bernoulli's equation and adding the friction loss.

The friction loss is the loss in energy due to resistance to flow as it moves through a system. This resistance is due to viscous shear stresses within the liquid and turbulence at the walls of the pipe and fittings. It varies with liquid properties (viscosity, solids entrained); size of the pipe (velocity); roughness of the pipe's internals; length of travel; and losses at fittings, control valves, and the entrance and exits.

$$H\ system = Hd - Hs = (P_2 - P_1) + (hd - hs) + (V_d^2 - V_s^2)/2g_c + (hfs + hfd) \quad (6)$$

where:

P_1 and P_2 = The entrance and exit pressures

hd and hs = The elevations from the pump impeller eye to the entrance and exit liquid levels

V_s and V_d = The entrance and exit velocities (velocities are given in ft/sec in this article)

hfs and hfd = The frictional losses in the suction and discharge pipe system, expressed in feet

To simplify:

$$Hf = hfs + hfd \text{ is the total friction loss (at each flow rate) through the system} \quad (7)$$

$$h = hd - hs \text{ is the difference in elevation form dischsrge level to suction level} \quad (8)$$

$$hv = V_d^2 - V_s^2/2g_c \text{ is the velocity head component} \quad (9)$$

Velocity head is often ignored, since in most cases the velocity is low (in tanks maintained at constant level, it's 0). However, if the discharge is coming out through a nozzle or other orifice, its velocity could be high and, therefore, significant if the other components are low. This should not be confused with entrance and exit head losses, which are accounted for in the Hf calculations.

Velocity head ($hv = V_2/2gc$) is a confusing concept to many—it's the head (distance) required for a liquid to fall in order to reach a velocity, v. It costs energy to move (accelerate) from some low velocity to a higher velocity.

All components of Eq. 6 must be consistent and usually in feet of liquid. Refer to Figure 5 for the following example.

For the pressure component

$$P_2 - P_1 = \frac{(5\,\text{psig} - 0\,\text{psig}) \times \frac{2.31\,\text{ft}}{\text{psi}}}{1.0} = 11.5'$$

It could have been:

$$29.7\,\text{psia} - 14.7\,\text{psia} = 5\,\text{psia} \times \frac{2.31}{1.0} = 11.5.$$

Use either; just be consistent. For open tanks there is 0 psig, 14.7 psia (at sea level).

For the static head component

$$h = hd - hs = 60 - 10 = 50'$$

Note that this is from liquid level to liquid level. This layout was for illustration; most industrial applications do not have the discharge pipe located beneath the liquid level because it would back siphon when the pump stops. If the pipe goes overhead and empties above the liquid line (or more often uses a vacuum breaker), use the centerline elevation of the discharge pipe at the exit for the hd. The pressure and static head components are constant with flow and plot as a straight line.

For the velocity head components. Assume the tanks are either being filled/emptied at the pumping rate or they are large enough that the downward and upward velocities are very low, or they are the same diameter and the velocity heads cancel. Therefore, the velocity head components is 0.

The friction component. Hf includes pipe losses and minor losses, as follows.

Pipe losses. There are several ways to calculate these losses:

1. Published data such as the Crane Handbook, the Cameron Hydraulic Data Book, or KSB's Centrifugal Pump Design.
2. Using the Hazen Williams formula:

$$Hf = \left[3.022(v)^{1.85}L\right]/\left[C^{1.85} \times D^{1.165}\right] \quad (10)$$

where:

L = Length of pipe in feet
v = Velocity ft/sec
D = Pipe diameter in feet
C = Constant for various pipe materials

This is a simple way to go. The disadvantage is that it assumes constant NRe (Reynolds Number) in the turbulent range and a viscosity of 1.13 cs (31.5 SSU), which is water at 60°F. This can yield errors of 20% or more for the range of 32 to 212°F.

3. Using the Darcy formula:

$$Hf = fLv^2/2Dg_c \quad (11)$$

where f = the friction factor based on Reynolds Number (NRe)

$$NRe = D \times v/vis \quad (12)$$

where:

vis = Viscosity in ft^2/sec

$$\text{In laminar flow (less than Nre} = 2000)\ f = 64/Nre \quad (13)$$

In turbulent flow (NRe greater than 4000), f varies with Nre and the roughness of the pipe. This is expressed as "relative roughness" = e/D; where e is absolute roughness (in feet) found in tables and f is found in a Moody Friction Factor Chart using NRe and e/D.

If the range of NRe is between 2000 and 4000 (the transition zone), use turbulent flow design.

4. Using computerized flow simulators. (You need to know limits and assumptions upon which the program is based.)

Minor losses. These include obstructions in the line, changes in direction, and changes in size (flow area). The two most common methods of calculating these losses are:

1. *Equivalent lengths:* uses tables to convert each valve or fitting into an "equivalent length," which is then added to the pipe length in L, and Hf is calculated in Darcy (Eq. 11 above).
2. *Loss coefficients:* uses published resistance coefficients (K) in the formula: $hf = Kv^2/2g_c$. K values for many types of fittings are published. Loss coefficients are the way to determine entrance and exit losses and losses due to sudden reductions or enlargements.

Control valve losses. These losses are calculated using the Cv for the valve:

Cv = gpm pressure drop (psi)

The Hf increases as flow increases and is a parabolic curve. The Hf for the system is calculated using one of the above methods. Assume the suction pipe is 10″ diameter, 20′ long, with two elbows and one gate valve. The discharge pipe is 8″, 1000′ long with eight elbows, one gate valve, and one check valve. The calculations yield Hf values of:

1. Hf = 4′ @ 400 gpm
2. Hf = 12′ @ 800 gpm
3. Hf = 27′ @ 1200 gpm
4. Hf = 37′ @ 1400 gpm

With the system head calculated at several flows (the pressure and height components are not affected by flow), we can plot the results (Figure 6).

The Total Differential Head at 1200 gpm is H = 11.5′ + 50′ + 27′ = 88.5′.

The pump's head capacity curve is plotted on the same graph. Where they intersect is the operating point ("A").

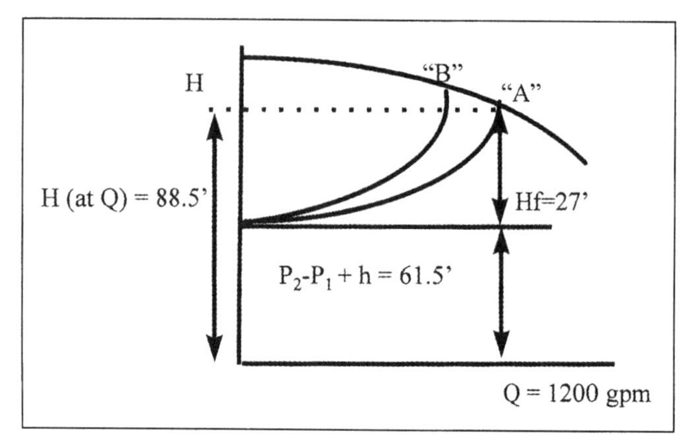

Figure 6. System head curve.

Note the pressure and static head components (66.5′) are constant with flow, while the Hf component increases as flow increases.

If a control valve in the discharge piping were used, the system head curve could be throttled back to a higher head and therefore lower flow ("B").

What can go wrong?

A centrifugal pump will run where the system head curve tells it to run. The point where the centrifugal pump performance curve and system head curve intersect is the operating point. It's important to understand where the operating point is at all times and to avoid the dangerous areas of pump operation. What are some of the dangerous areas to consider?

Horsepower. A pump must always have a driver with more hp than it consumes. With most centrifugal pumps, as the pump flow increases, the BHP increases to a maximum, and there must always be enough hp available from the driver to supply the needs of the pump. This means as TDH falls, the flow goes up and so does the BHP. Under-sizing overloads the driver.

NPSH. As the pump flow increases, the NPSHr increases. As flow increases, NPSHa decreases (hfs increases with flow). There must always be more NPSHa than NPSHr or the pump cavitates. Cavitation is where the pressure at a given point in the pump, prior to entering the impeller eye, falls below the vapor pressure of the liquid pumped, turning it to vapor. As the pressure increases through the impeller, the bubble of vapor collapses. This causes capacity reduction, noise, vibration, pitting of the impeller and, eventually, failure of a component.

No flow. It is possible for the TDH to rise to a point where there is no *flow*. This is called *shutoff* and means no flow is going through the pump. At this point, the temperature will rise in the pump because the inefficiencies of the pump are causing heat to go into the pumpage. This raises the temperature and will eventually cause the pumpage to vaporize, and the close clearances such as ring fits and mechanical seal faces will not receive lubrication. Eventually the seals fail or the pump seizes. High temperatures also cause the pump to distort and change clearances. Failure of some component will occur.

Low flow. It is possible for the TDH to rise to a point of low flow where damage other than no flow occurs. Such circumstances include:

- Low flow temperature rise—As the flow decreases, the efficiency falls and there is more BHP needed to drive

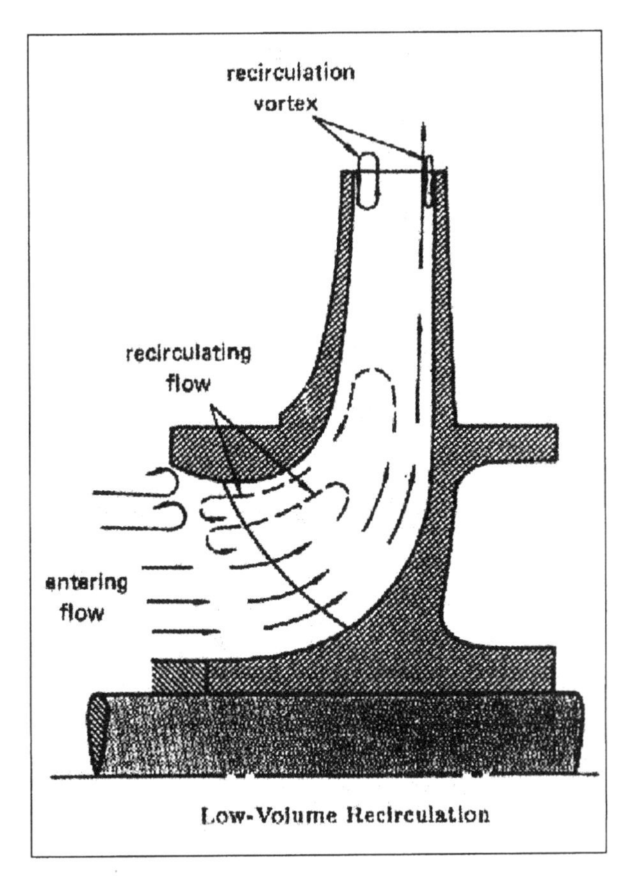

Figure 7. Low flow recirculation.

the pump than is needed to move the liquid through the TDH. The excess BHP goes into the pump as heat. The flow is lower (less to carry away the heat), and therefore the temperature rise is greater as flow drops. It is possible to reach a point where the temperature rises to the point that it reduces the NPSHa below the NPSHr and flashing occurs as above. This point is the "thermal minimum flow" and is a function of not only the pump, but also the system characteristics.

- Low flow recirculation/separation damage—As flow decreases, the liquid begins to recirculate in both the impeller eye and impeller discharge (at the volute cut-water or diffuser). This recirculation will cause separation, eddy currents, and turbulence that will cause the same noise, vibration, and damage as cavitation (Figure 7). The minimum flow required to prevent low flow re-circulation separation damage is called the "continuous minimum flow." It is specified by the pump manufacturer based on the impeller eye geometry and Nss.
- Radial or axial thrust damage—All pumps use bearings and balancing devices to prevent the rotating parts from contacting the stationary parts. These are pressure and flow related, and they may limit the flow of the pump to a minimum or maximum. The pump manufacturer must make known the limits based on the pump design. ■

References

1. Carver Pump Company. *Performance Curves*, Muscatine, IA.
2. Lindeburg, PE, Michael. *Mechanical Engineering Reference Manual, 8th Ed.*, Professional Publications, Belmont, CA (1990).
3. Karassik, Igor. *Pump Handbook*, McGraw-Hill, New York, (1976).
4. Westaway CR. *Cameron Hydraulic Data Book, 16th Ed.*, Ingersoll-Rand, Woodcliff Lake, NJ (1984).

Doug Kriebel has presented pump seminars and workshops throughout his 30-year industrial career. He has extensive experience with pumping applications in a wide range of industries including positions in the electric utility industry, as well as with a chemical equipment manufacturer and major pump OEM. Mr. Kriebel is currently president of a pump and equipment manufacturer's representative/distributor organization. He holds a B.S. degree in chemical engineering. He is an active member of AIChE and is a registered Professional Engineer in Pennsylvania.

KEY CENTRIFUGAL PUMP PARAMETERS AND HOW THEY IMPACT YOUR APPLICATIONS—PART 2

Pumps and Systems: They Go Together

By Doug Kriebel, PE, Kriebel Engineered Equipment

Editor's note: This article is the second installment of a feature that first appeared in the August 2000 issue of Pumps & Systems Magazine. Part 1 explained the commonly applied parameters of head/capacity, NPSH and brake horsepower. This month's conclusion examines what can go wrong in a system and how to fix common problems. If you've lost track of your August issue, you can read the full text of Part 1 online at www.pump-zone.com.

How to examine a system

Each system must be analyzed to determine what happens to the pump at not only design, but also every other possible point of operation—start-up, off peak, unusual modes, "best case," and "worst case." The following examples illustrate how a system change can create either low flow, high flow, or NPSH problems. Refer to Figure 1, which shows a pump system. We will first look at the suction side to determine the NPSHa. The suction side is 20′ of 5″ pipe with two elbows and a gate valve. At 300 gpm the hfs is 1′. The pump selected is shown in Figure 2.

Example 1. The tank is open to atmosphere, 0 psig, 34′ water = P_l; pumping water at 70°F, hvp = .36 psi = .8′; from hs1 = 10′ then:
The NPSHa = Pl + hs − hvp − hfs = 34 + 10 − .8 − 1 = 42.2′

Referring to the pump curve Figure 2, at 300 gpm, the NPSHr = 23′ so we have adequate NPSHa.

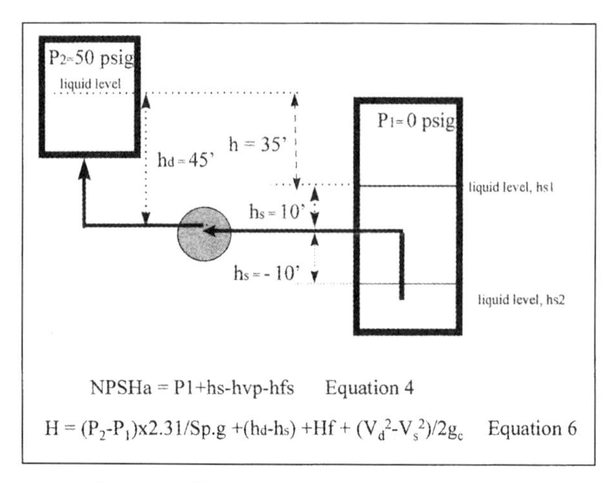

Figure 1. Pump system for Examples 1–6.

Example 2. All conditions the same except the tank is drawn down to level hs2 = −10′

Then the NPSHa = 34 − 10 − .8 − 1 = 22.2′ and the NPSHa is less than the NPSHr and the pump cavitates.

Example 3. All conditions the same as Example 1 except the water is now 190°F; hvp = 9.34 psi = 9.34 × 2.31/.966 = 22.3′.

Then NPSHa = 34 + 10 − 22.3 − 1 = 22.7′ and the pump cavitates.

Example 4. Same as Example 3 except from $hs_2 = -10'$

Then NPSHa = 34 − 10 − 22.3 − 1 = 0.7′ and the pump cavitates.

Example 5. Conditions same as any above except the suction pipe is installed as 3″ instead of 5″: hfs = 10′ (instead of 1′). All the above NPSHa would be reduced by 9′!

The lesson to be learned on the *suction side*: Have as much static suction head as possible (hs), locate the pump as close as possible to the suction vessel (with a minimum suction side frictional pressure drop (hfs)), and be careful of liquids near their vapor pressure (hvp).

Let's go further and use the Example 1 conditions and look at the discharge side: Assume 150′ of 4″ pipe with six elbows, one gate valve, and one check valve: hfd = 14′ @ 300 gpm.

Example 6. Assume the levels in both suction tank and discharge tank remain constant with suction at hs1 then:

$$
\begin{aligned}
H &= (P1 - P2) + (hd - hs) + (V22 - V21)/2gc + (hfd + hfs) \\
&= (50\,\text{psig} - 0\,\text{psig})2.31/1.0 + (45' - 10') + 0 + (14' + 1') \\
&= 115.5' + 35' + 15' = 165.5'
\end{aligned}
$$

The pump selected is shown in Figure 3. The operating point is "A".

What happens if the levels don't remain constant and the suction tank pulls down from level hs1 = +10′ to level $hs_2 = -10'$ and the discharge tank fills up an equal 20′?

The "h" component (hd − hs) would go up from 35′ to 75′ and the "P" + "h" components would rise from 150.5 to (115.5 + 75) 190.5′. This would cause the pump to "walk" back on the curve to some lower flow, point B.

All this means is that the pump will pump a lower flow rate. This is not necessarily a bad thing as long as it meets the process needs and is above the minimum flow for the pump. If it does go below minimum flow, damage will occur.

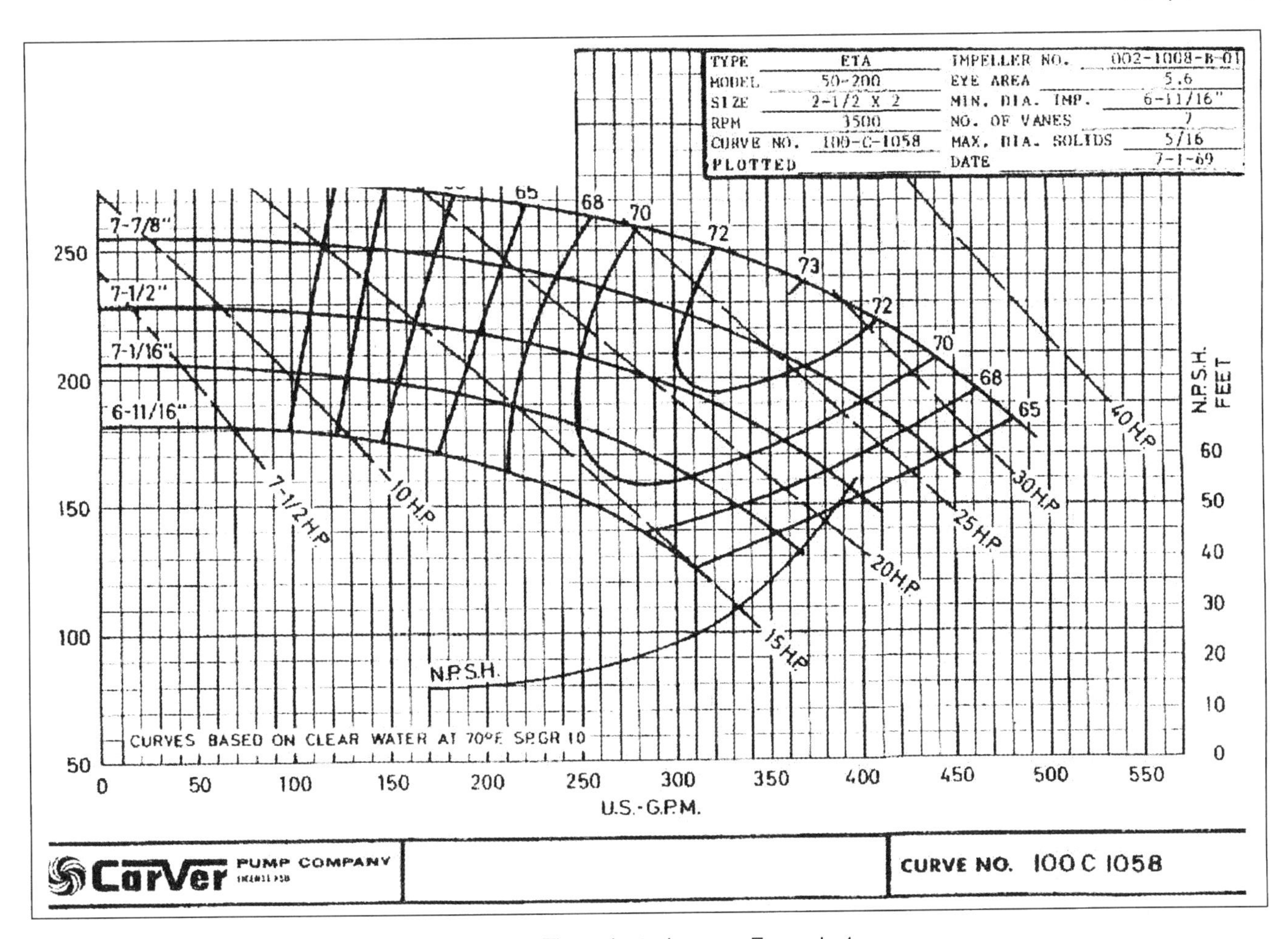

Figure 2. The selected pump, Example 1.

If the pressure (P2) is increased to 75 psig, the head goes up $(75-0)2.31/1.0=173$ plus $h=35'$ to $H=208$, which is above the shutoff of the pump (205′). The pump would be running with no flow. This could be an upset condition that causes over-pressure to the discharge tank. It is also a situation where an inline filter is installed into an existing installation and the pressure drop is underestimated or the filter is not maintained (cleaned). This example cost an East Coast utility company more than $45,000 in damage to a pump when maintenance installed a filter to protect some heat exchangers without taking into account the pressure drop when fouled!

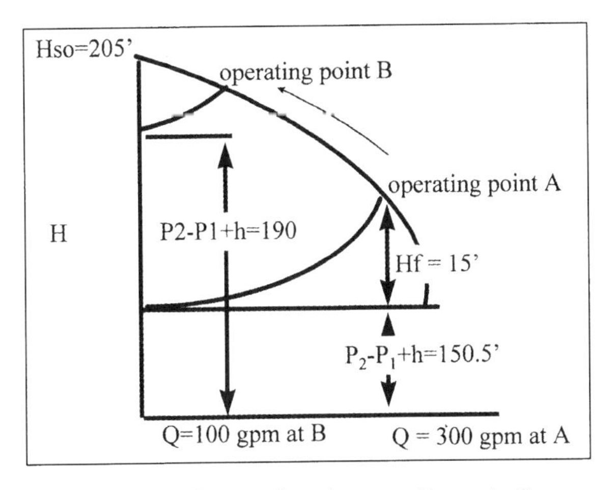

Figure 3. System head curve—Example 6.

Example 7. More often the pump is selected for a "worst case" condition, (sometimes called the "design condition"). This example uses the same system as above except use 75 psig for the design P2, assuming "worst case" situation (Figure 4, Point A). This accounts for an upset, over-pressure situation in the discharge tank or other circumstance.

However, when the system actually operates at 50 psig, the pump runs out to point B. This could overload the motor, since more BHP is required. In addition, the increased flow may cause the NPSHr to exceed the NPSHa (which decreases due to increased hfs). This may cause the pump to cavitate.

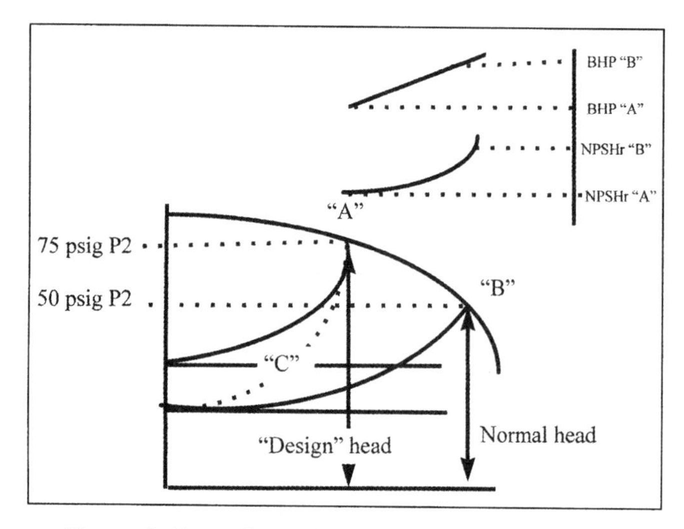

Figure 4. Example 7—changes in head and throttling.

Most systems that are designed for higher P2 than normal have a control valve that will increase the hfd and "walk" the pump back on the curve to the correct operating point (back to "A" via line C).

A major chemical company using sealless pumps and no control valves (they were handling toxic material and did not want to risk leaks) designed a system for a specific H based on a P2 set by the process. After startup, which went without a hitch, the reactor pressure (P2) was decreased to improve yield. Without control valves, the pump ran out on the curve and destroyed itself by cavitation. It took $70,000 in repairs until someone ran the calculations to determine how the system change caused the pump failures!

A similar situation occurs if excess safety factor is utilized (Figure 5). If the actual design point is calculated as 300 gpm at 165′ (point A) and the pump is selected for 345 gpm and 200′ head (point B) and the calculations are correct, the pump will run out to point C. This would pump more than required. So the pump would be throttled back to point D to pump the 300 gpm. This means 215 − 165 = 50′ f head is throttled out, which costs 5.2 BHP. Since you pay for BHP, this is wasted energy. In this case it would amount to $3,100 per year at a rate of $.085/kW-hr for service 24 hrs × 360 days. Higher heads or flows would magnify the costs.

There is nothing wrong with safety factors, and it is always better safe than sorry, but after installation, check the hydraulics. If gross over-sizing was the case, trimming the impeller will save considerable money in operations.

Example 8 (variable speed operation). An economical alternative to throttling with a valve is to use variable speed to change the pump curve to meet variable demands of a system head curve. This is illustrated in Figure 6. The system head curve is plotted, a pump is picked to meet the worst case condition, and reduced speed is used to ride down the system curve to meet varying flows at reduced pressure. In some cases this can be very economical when compared to the wasted BHP of throttling control.

If the normal point is the same as above (300 gpm at 165′ Point A on Figure 6) and with safety factors (345 gpm at 200′ Point B) the pump operates on the uppermost curve for 3500 rpm, where it intersects the system head curve at point C. By slowing down to 3200 rpm, it goes through point A without having to waste energy by throttling the head. Any point can be reached on the system head curve by reducing speed. This is a valuable method of operating when flows or heads have large changes in magnitude.

The danger comes if you reduce the pump speed to a point where the system head curve causes the pump to

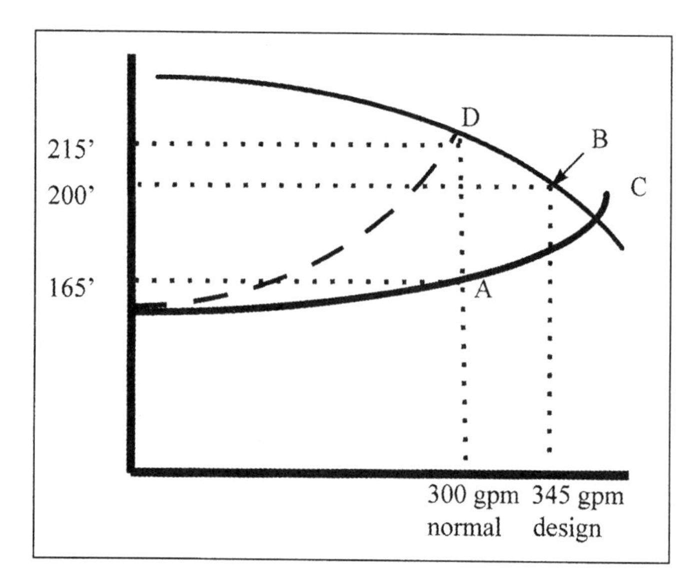

Figure 5. System head curve: design/normal/runout/throttle.

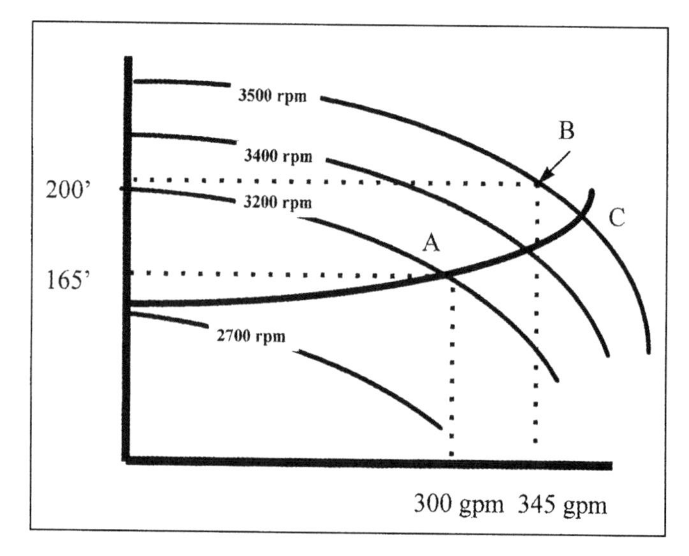

Figure 6. Example 8—variable speed operation.

operate at a point of too low flow or no flow (i.e., beneath the system head curve, such as 2700 rpm). It is possible to operate too high flow where BHP or NPSH becomes a factor, or at a speed where other damage could occur. Examples of other speed-related damages are if the pump reaches a critical speed, a speed where auxiliary equipment no longer functions (a shaft driven lube oil pump may not provide proper lubrication at too low a speed, or a motor fan motor may turn too slowly to provide adequate cooling).

Example 9 (parallel pumping applications). It's not uncommon to use two or more pumps in parallel. The flows are additive at a constant head (Figure 7). (The flow, Q, at head H for one pump at A is added to a second pump at the same head to get twice the flow, 2Q, at the intersection of the two pump curves, the system head curve, B. It's important to draw the system head curve to make sure that if one pump operates alone it does not run out to point which will overload the motor or exceed the NPSH requirements. Point B is the operating point of two pumps in parallel, and Point C is for one pump at runout (estimated here as 1.2Q). The one pump runout must be checked to provide sufficient BHP, NPSHa, submergence against vortexing, and mechanical design for the higher flow.

This is more common in vertical cooling water service where several pumps are used in parallel and a control valve is not used on each pump. If one pump trips, the system allows the remaining pump(s) to run out.

Another potential for parallel pump problems is where two dissimilar pumps (dissimilar curves) are used (Figure 8). The pump curve for pump A is added to pump B at constant head. The system head is drawn. Q1 is the flow with only pump A running, Q2 is the flow with only B running, Q3 is the flow with both A + B running. *Note:* This is not Q1 + Q2. If the system is throttled back (by any means), the system head could cause the pump with the lower head to run below a minimum flow. Pump A is shut off at point C.

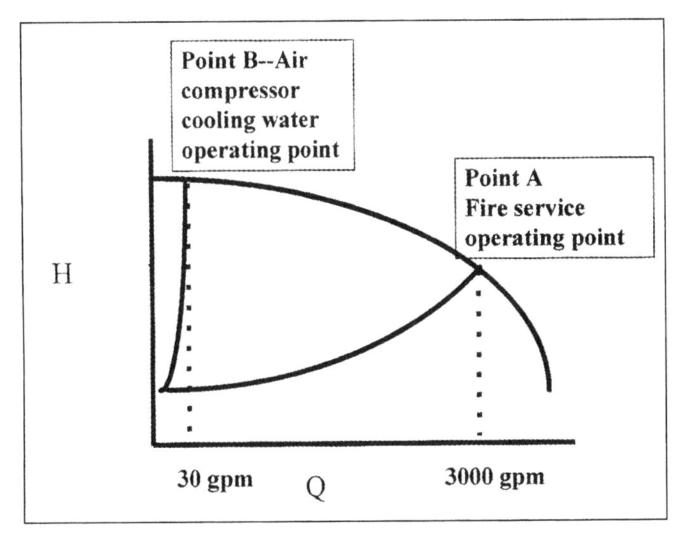

Figure 8. Parallel pumps.

Example 10 (a pump for a given service is used for an alternate service). Figure 9 is the performance curve for a horizontal split case pump put in as a fire protection pump that is to run at 3000 gpm. The plant engineer lost his air compressor cooling water pump and decided to use the fire pump to supply cooling water for the compressor inner and after coolers. (If you need 300 gpm, a 3000 gpm pump will certainly do!)

The pump had six failures of the radial bearings. The maintenance group replaced them, then redesigned the pump and added angular contact back-to-back bearings in place of the double row bearings. They then added oil lubrication to the existing grease bearings and were about to

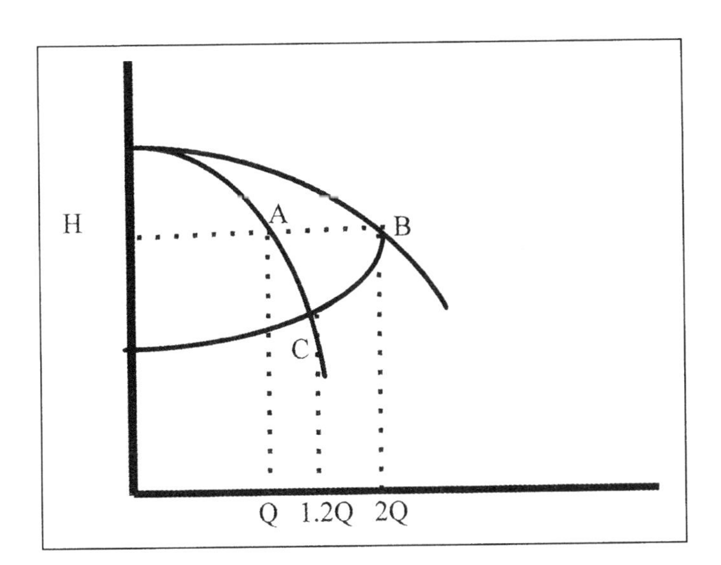

Figure 7. Example 9—parallel pumps.

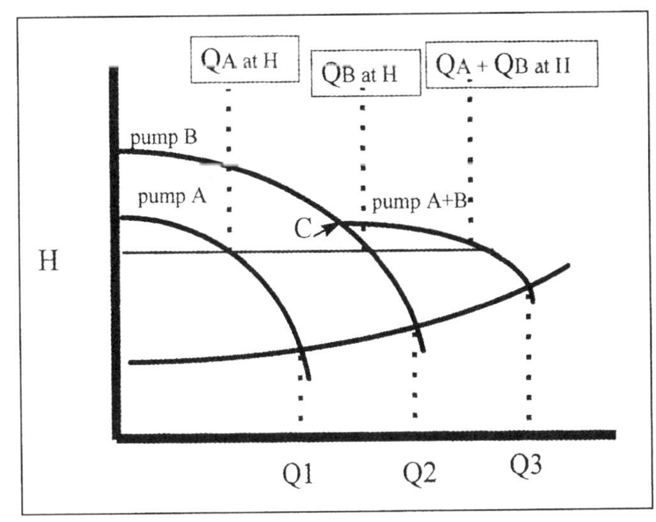

Figure 9. Example 10.

change to oil mist when someone finally asked why the fire pump was running when there was no fire!

Point A is the design point, but point B is where it operated to supply the air compressor's cooling water. This was too far back on the curve and caused high radial loads that were the cause of failure. There was no way any bearing or lubrication upgrade would enable this pump to operate at point B.

In conclusion, the purpose of this article is to provide operations and maintenance people who do not normally do process calculations with the tools to analyze the pump systems they have to operate. This will enable them to operate the systems more economically and to determine the "real" root cause of failures. Failures are mechanical in nature, but the cause of the mechanical failure is a "system" incompatibility failure. In today's business world, systems are often designed by process engineers who are not available when the system becomes operational, and the system may not have been installed, operated, or maintained the way it was intended to be, or the correct method of operation is no longer the way it was originally designed. ■

References

1. Carver Pump Company. *Performance Curves*, Muscatine, IA.
2. Lindeburg, PE, Michael. *Mechanical Engineering Reference Manual, 8th Ed.*, Professional Publications, Belmont, CA (1990).
3. Karassik, Igor. *Pump Handbook*, McGraw-Hill, New York, (1976).
4. Westaway CR. *Cameron Hydraulic Data Book, 16th Ed.*, Ingersoll-Rand, Woodcliff Lake, NJ (1984).

Nomenclature/symbols:

H = Total differential head
Hd = Total discharge head
Hs = Total suction head
S.G. = Specific Gravity
Q = Flowrate
NPSH = Net Positive Suction Head (r required, a available)
BHP = Brake horsepower
Ns = Specific speed
Nss = Suction specific speed
P = Pressure (1 suction, 2 discharge)
hvp = Vapor pressure of liquid
hs = Static suction head
hd = Static discharge head
h = (hd − hs)
hfs = Friction losses, suction side
hfd = Friction losses, discharge side
Hf = (hfs + hfd)
vd = Discharge velocity
vs = Suction velocity
v = Velocity (ft/sec)
hv = Celocity head
g_c = Gravitational constant (32.2 ft/sec^2)
L = Pipe length in feet
D = Pipe diameter in feet
C = Hazen Williams constant for pipe materials
NRe = Reynolds number
f = Friction factor
vis = Viscosity
e = Absolute roughness
K = Resistance coefficient for pipe fittings

Doug Kriebel has presented pump seminars and workshops throughout his 30-year industrial career. He has extensive experience with pumping applications in a wide range of industries, including positions in the electric utility industry, as well as with a chemical equipment manufacturer and major pump OEM. Mr. Kriebel is currently president of a pump and equipment manufacturers' representative/distributor organization. He holds a B.S. degree in chemical engineering, is an active member of AIChE, and is a registered Professional Engineer in Pennsylvania.

Estimate the discharge of a centrifugal pump at various speeds

Within the design limits of a pump, the discharge will be approximately proportionate to the speed.

$$Q_2 = Q_1 \frac{N_2}{N_1}$$

where Q_1 is the gallons per minute at a particular speed N_1, and Q_2 is the gallons per minute at N_2.

Example. If a pump delivers 350 gpm at 1,200 rpm, what will it deliver at 1,800 rpm?

$$Q_2 = 350 \times \frac{1{,}800}{1{,}200} = 525\,\text{gpm}$$

Similarly, for a given pump, design head varies as the square of the impeller speed and power as the cube of the speed.

How to estimate the head for an average centrifugal pump

By slightly modifying the basic pump formula:

$$H = \left(\frac{DN}{1{,}840\,K_u}\right)^2$$

it is possible to find a means of estimating the head a pump will develop for each impeller at various speeds. The terms are defined as follows:

H = Head, ft
D = Diameter of the impeller, in.
N = Rpm of the impeller
K_u = Pump design coefficient

The values of this expression may vary from 0.95 to 1.15. For estimating purposes, substitute 1,900 for 1,840 K_u. The formula becomes:

$$H = \left(\frac{DN}{1{,}900}\right)^2 \times \text{(the number of impellers)}$$

Example. What head will be developed by a well-designed centrifugal pump with four 11.75-in. impellers turning at 1,200 rpm?

$$H = 4 \times \left(\frac{11.75 \times 1{,}200}{1{,}900}\right)^2 = 4 \times 7.4^2 = 220 \text{ ft of head}$$

Find the reciprocating pump capacity

Multiply the diameter of the plunger by itself, multiply this by the stroke, then by the number of strokes per minute, and finally by 0.11; the result is the capacity in barrels per day.

$$C = B^2 \times S \times N \times 0.11$$

Example. Four-in. bore, 6-in. stroke, 80 strokes per minute.

$4 \times 4 \times 6 \times 80 \times 0.11 = 845$ barrels per day

Note. The rule is approximate only; it allows about 5% for "slip," which is actually somewhat variable. Also note that it is applicable only to single-acting pumps; for a double-acting piston type pump, N must be twice the rpm, and a reduction must be made for the area of the piston rod on the crank end of the pump.

How to estimate the hp required to pump at a given rate at a desired discharge pressure

Ignoring effects of temperature, viscosity, etc.

$$hp = psi \times \frac{\text{barrels per day}}{1{,}000} \times .022$$

The factor .022 is arrived at by using a pump efficiency of 80% and a gear efficiency of 97%.

Example. What horsepower would be required to pump 15,000 barrels per day with an 800 psi discharge?

$$hp = 800 \times 15 \times .022$$
$$= 264 \text{ horsepower}$$

The factor .022 was derived from the formula:

$$hp = \frac{(psi)(bpd)}{58{,}800 \times \text{pump efficiency} \times \text{gear effciency}}$$

Nomograph for determining reciprocating pump capacity

The theoretical capacity of reciprocating pumps can be calculated by the following equation:

$$C = 6/55\ D^2\ L\ N$$

where:

C = Theoretical pump capacity (delivery), gal/min
D = Diameter of cylinder, ft
L = Length of stroke, ft
N = Number of strokes/min

The nomograph (Figure 1) is based on this equation. The diameter and length-of-stroke scales are calibrated in in. units instead of ft. The nomograph also gives the delivery in British imperial gallons.

Example. What is the theoretical capacity of a reciprocating pump having the following characteristics?

D = 3 in.
L = 3 in.
N = 60 strokes/min

Connect	With	Mark or Read
D = 3 in.	L = 3 in.	Reference Scale
Ref. Point	N = 60	C = 5 gal/min

Source

Macdonald, J. O. S., *British Chemical Engineering*, 1, 6, 301 (1956).

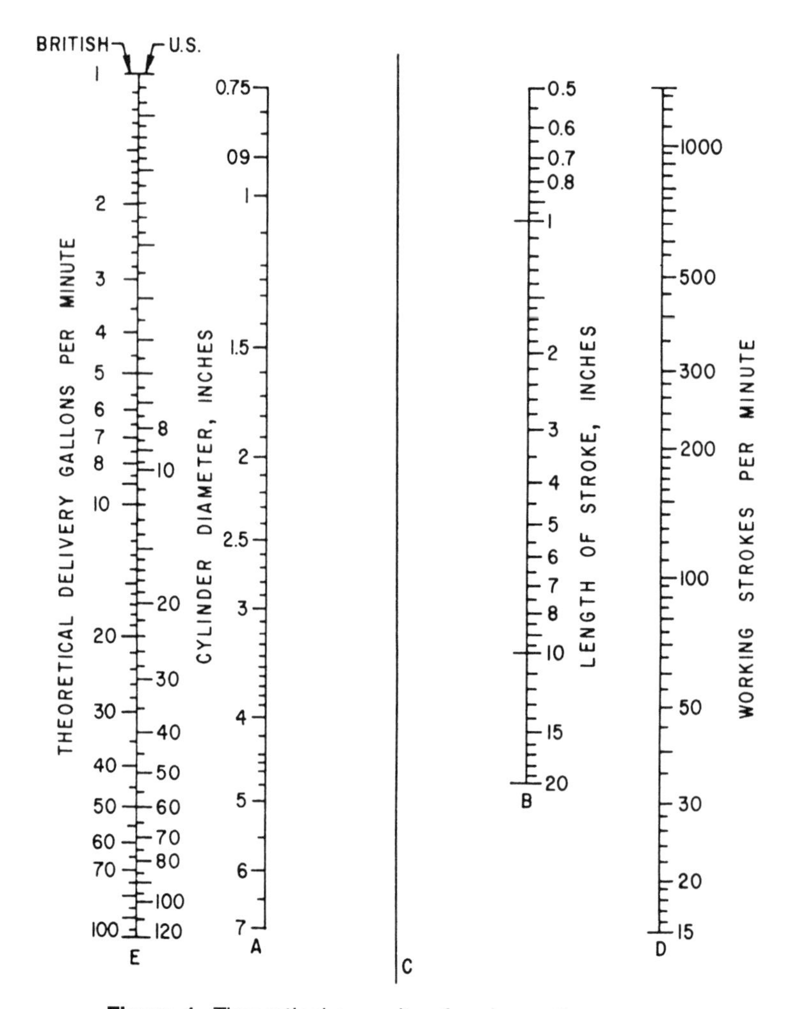

Figure 1. Theoretical capacity of reciprocating pumps.

Nomograph for determining specific speed of pumps

The specific speed is a criterion used in the selection of pumps for a given service. It can be used to (a) avoid cavitation troubles, (b) select the most economic pump for a given process layout, and (c) assist in redesigning process conditions.

For single-stage, side-entering pumps, the NPSH is defined by the following equation[1,2]:

$$S = N\frac{Q^{1/2}}{H^{3/4}} \quad (1)$$

where: S = Specific speed, no units
Q = Pump capacity
N = Pump speed, rev/min
H = Net positive suction head above the liquid vapor pressure (NPSH) measured at the pump suction flange, ft of fluid

H (in ft of liquid) is given by:

$$H = h_a \pm h_h - h_p - h_v \quad (2)$$

where: h_a = Ambient pressure at the liquid surface
h_h = Static head between liquid surfaces
h_p = Combined entrance, piping and fitting friction loss
h_v = Proper pressure at pumping temperature

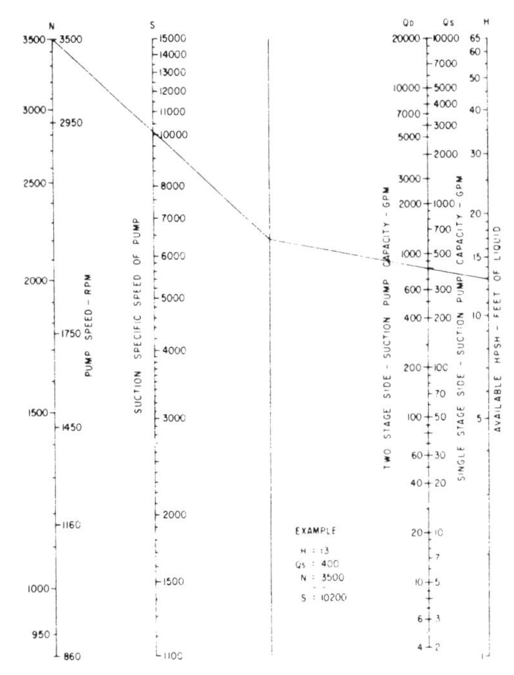

Figure 1. Pump-specific speed from net positive suction head (Reproduced by permission, *Petroleum Refiner*, Vol 36, No. 10, copyright 1957).

The specific speed of centrifugal pumps may vary between 1,100 and 15,000. Values above 12,000 are not considered practical and, where possible, the use of a specific speed below 8,500 is recommended. Specific speed limits for each type of pump are usually available from manufacturers of pumps.

The nomograph (Figure 1) solves Equation 1.

The specific speed is also defined in terms of a dimensionless number, using consistent units, by the following equation:

$$S\frac{Q^{0.5}N}{(\Delta H \times g_c)^{0.75}} \quad (3)$$

where g_c = gravitational conversion constant, 32.2 (lb mass/lb force) (ft/sec^2)

The range for this equation is 0.03 to 0.87. Results of Equation 1 can be converted to these units by dividing by 17,200.

Example. The following data are given for a single-suction reboiler pump:

$Q = 400$ gal/min
$N = 3{,}500$ rev/min

The reboiler liquid level is 18 ft above the pump, and friction from entrance, pipe, and fittings loss add up to 5 ft of liquid.

For a reboiler, the liquid is at its boiling point, so $h_a = h_v$, using Equation 2, $H = 13$ ft.

Connect	With	Mark or Read
NPSH = 13	Q = 400 gal/min	Reference line
Ref. line pivot	N = 3,500	S = 10,200

Source

Jacobs, J. K., *Petroleum Refiner*, 36, 10, 145 (1957).

References

1. *Hydraulic Institute Standards*, 10th Edition, New York.
2. Brown, et al., *Unit Operations*, 190, John Wiley and Sons, New York (1950).

Nomograph for determining horsepower requirement of pumps

In the selection of pumps it is necessary to calculate the horsepower requirement to handle a specified delivery rate of fluid against the specific head. The formula for the horsepower required by a pump is as follows:

$$hp = \frac{G \times H \times \text{specific gravity}}{3{,}960 \times e}$$

where:

hp = Pump horsepower
G = Flow rate, gal/min
H = Total dynamic head (change in head), ft
e = Pump efficiency

Figure 1 solves this equation.

Example. Given:

G = 200 U.S. gal/min
H = 60 ft
Specific gravity = 1.1
e = .80

What is the required horse power?

Connect	With	Mark or Read
e = .80	H = 60 ft	C
C	G = 200 gal/min	hp = 0.78 horsepower

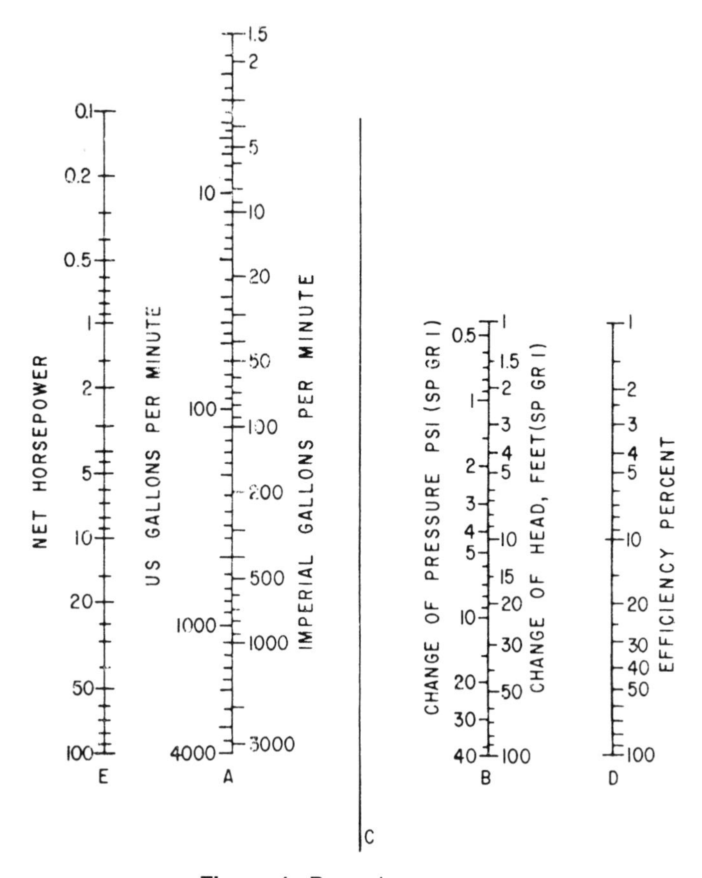

Figure 1. Pump horsepower.

Source

Macdonald, J. O. S., *British Chemical Engineering*, 1, 6, 290 (1956).

How to select motors for field-gathering pumps

The following simple formulas will provide a reasonably close approximation of the motor hp necessary to drive duplex double-acting pumps normally used in the field.

Rule.

1. $hp = \frac{bpd \times psi}{48{,}000}$

2. $hp = \frac{bpd \times psi}{2{,}000}$

3. $hp = \frac{gpm \times psi}{1{,}400}$

Motors selected with these formulas will fit delivery rates used in the formula if they are within capacity of the pump. If anticipated rate is below pump capacity, a savings may be made in first cost and power bills by selecting a motor corresponding to the actual rate rather than to full pump capacity.

It should be noted that net bpd values are to be used in Formula 1. That is, if 600 barrels are to be pumped each day, but it is desired to accomplish this in only 8 hours, the rate to be used in the formula is 1,800 bpd.

Reciprocating pumps

A. How a reciprocating pump works

A reciprocating pump is a positive displacement mechanism with liquid discharge pressure being limited only by the strength of the structural parts. Liquid volume or capacity delivered is constant regardless of pressure and is varied only by speed changes.

Characteristics of a GASO reciprocating pump are 1) positive displacement of liquid, 2) high pulsations caused by the sinusoidal motion of the piston, 3) high volumetric efficiency, and 4) low pump maintenance cost.

B. Plunger or piston rod load

Plunger or piston "rod load" is an important power and design consideration for reciprocating pumps. Rod load is the force caused by the liquid pressure acting on the face of the piston or plunger. This load is transmitted directly to the power frame assembly and is normally the limiting factor in determining maximum discharge pressure ratings. This load is directly proportional to the pump gauge discharge pressure and proportional to the square of the plunger or piston diameter.

Occasionally, allowable liquid end pressures limit the allowable rod load to a value below the design rod load. IT IS IMPORTANT THAT LIQUID END PRESSURES DO NOT EXCEED PUBLISHED LIMITS.

C. Calculations of volumetric efficiency

Volumetric efficiency (E_v) is defined as the ratio of plunger or piston displacement to liquid displacement. The volumetric efficiency calculation depends upon the internal configuration of each individual liquid body, the piston size, and the compressibility of the liquid being pumped.

D. Tools for liquid pulsation control, inlet and discharge

Pulsation Control Tools ("PCT", often referred to as "dampeners" or "stabilizers") are used on the inlet and discharge piping to protect the pumping mechanism and associated piping by reducing the high pulsations within the liquid caused by the motions of the slider-crank mechanism. A properly located and charged pulsation control tool may reduce the length of pipe used in the acceleration head equation to a value of 5 to 15 nominal pipe diameters. Figure 1 is a suggested piping system for power pumps. The pulsation control tools are specially required to compensate for inadequately designed or old/adapted supply and discharge systems.

E. Acceleration head

Whenever a column of liquid is accelerated or decelerated, pressure surges exist. This condition is found on the

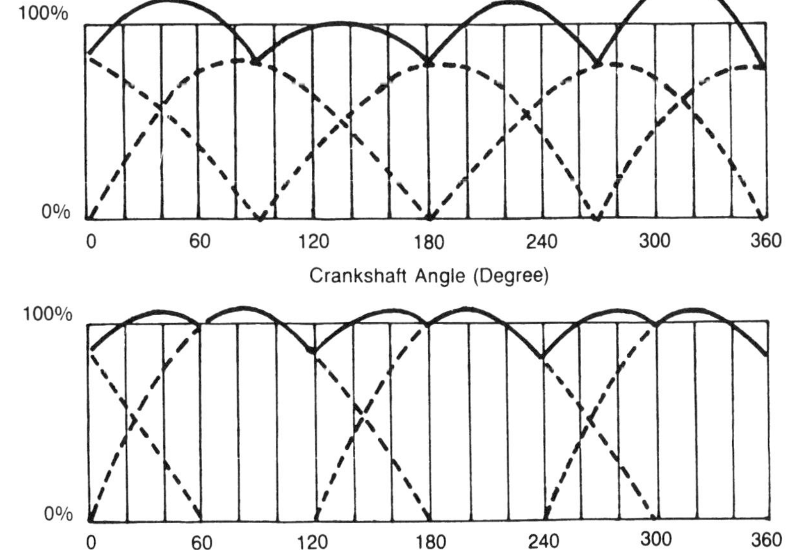

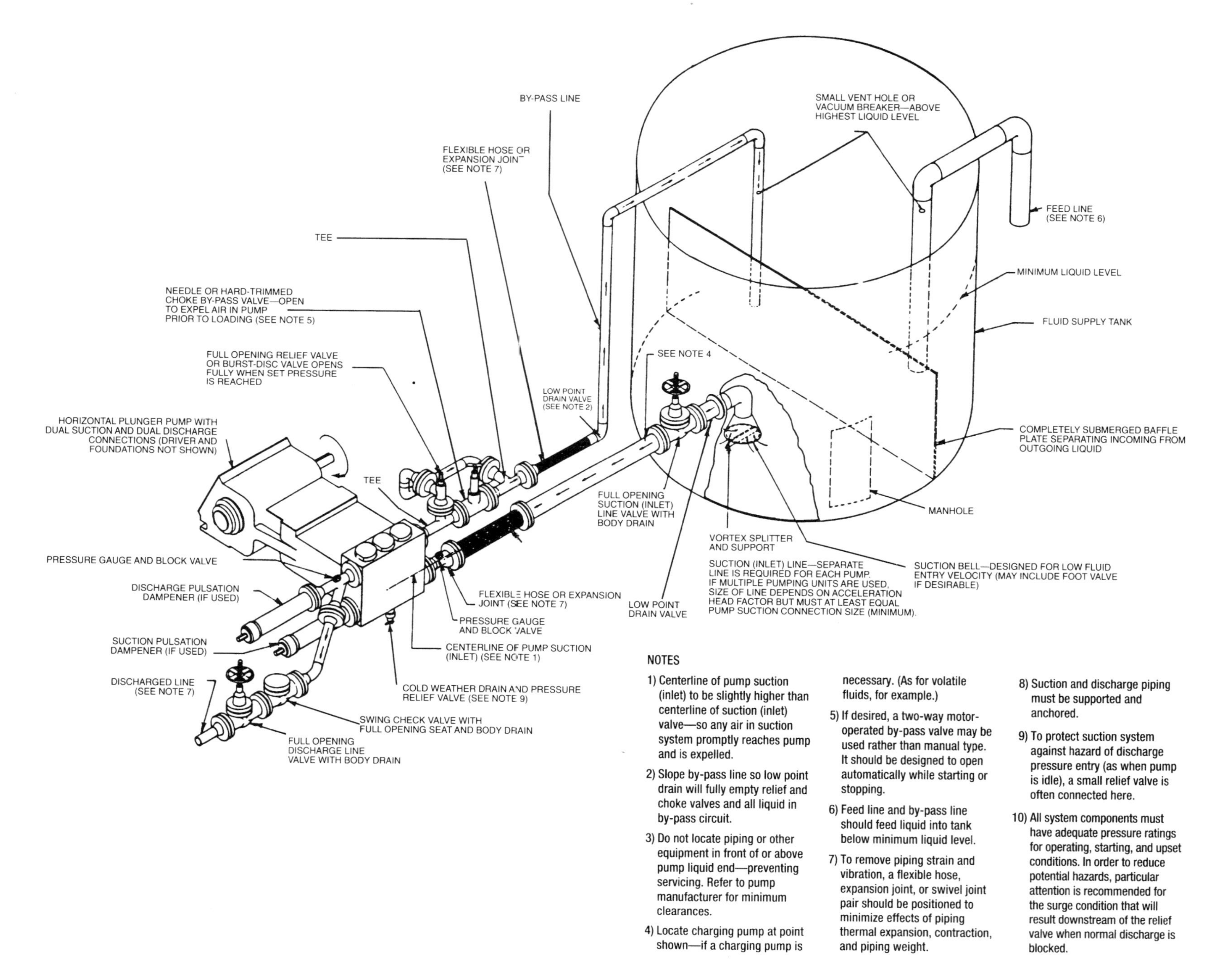

NOTES

1) Centerline of pump suction (inlet) to be slightly higher than centerline of suction (inlet) valve—so any air in suction system promptly reaches pump and is expelled.

2) Slope by-pass line so low point drain will fully empty relief and choke valves and all liquid in by-pass circuit.

3) Do not locate piping or other equipment in front of or above pump liquid end—preventing servicing. Refer to pump manufacturer for minimum clearances.

4) Locate charging pump at point shown—if a charging pump is necessary. (As for volatile fluids, for example.)

5) If desired, a two-way motor-operated by-pass valve may be used rather than manual type. It should be designed to open automatically while starting or stopping.

6) Feed line and by-pass line should feed liquid into tank below minimum liquid level.

7) To remove piping strain and vibration, a flexible hose, expansion joint, or swivel joint pair should be positioned to minimize effects of piping thermal expansion, contraction, and piping weight.

8) Suction and discharge piping must be supported and anchored.

9) To protect suction system against hazard of discharge pressure entry (as when pump is idle), a small relief valve is often connected here.

10) All system components must have adequate pressure ratings for operating, starting, and upset conditions. In order to reduce potential hazards, particular attention is recommended for the surge condition that will result downstream of the relief valve when normal discharge is blocked.

Figure 1. Suggested piping system for power pumps.

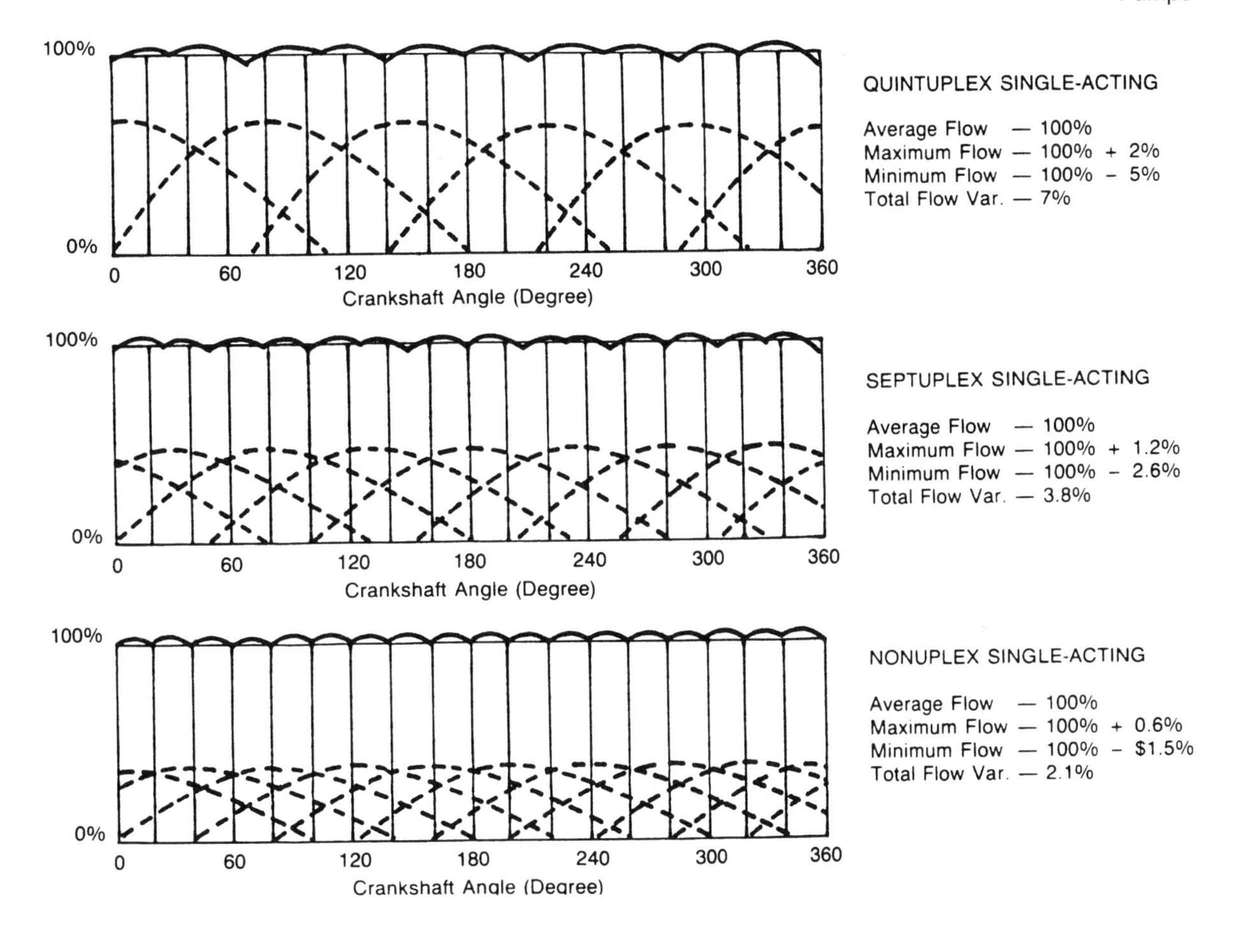

inlet side of the pump as well as the discharge side. Not only can the surges cause vibration in the inlet line, but they can restrict and impede the flow of liquid and cause incomplete filling of the inlet valve chamber. The magnitude of the surges and how they will react in the system is impossible to predict without an extremely complex and costly analysis of the system. Since the behavior of the natural frequencies in the system is not easily predictable, as much of the surge as possible must be eliminated at the source. Proper installation of an inlet pulsation control PCT will absorb a large percentage of the surge before it can travel into the system. The function of the PCT is to absorb the "peak" of the surge and feed it back at the low part of the cycle. The best position for the PCT is in the liquid supply line as close to the pump as possible or attached to the blind flange side of the pump inlet. In either location, the surges will be dampened and harmful vibrations reduced.

REQUIRED FORMULAE AND DEFINITIONS

Acceleration head

$$h_a = \frac{LVNC}{Kg} \qquad V = \frac{GPM}{(2.45)(ID)^2}$$

where:

h_a = Acceleration head (in feet)
L = Length of liquid supply line (in line)
V = Average velocity in liquid supply line (in fps)
N = Pump speed (revolutions per minute)
C = Constant depending on the type of pump
C = 0.200 for simplex double-acting
= 0.200 for duplex single-acting
= 0.115 for duplex double-acting
= 0.066 for triplex single- or double-acting
= 0.040 for quintuplex single- or double-acting
= 0.028 for septuplex, single- or double-acting
= 0.022 for nonuplex, single- or double-acting
K = Liquid compressibility factor
K = 2.5 For relatively compressible liquids (ethane, hot oil)
K = 2.0 For most other hydrocarbons
K = 1.5 For amine, glycol, and water
K = 1.4 For liquids with almost no compressibility (hot water)
g = Gravitational constant = 32.2 ft/sec^2
d = Inside diameter of pipe (inches)

Stroke

One complete uni-directional motion of piston or plunger. Stroke length is expressed in inches.

Pump capacity (Q)

The capacity of a reciprocating pump is the total volume throughput per unit of time at suction conditions. It includes both liquid and any dissolved or entrained gases at the stated operating conditions. The standard unit of pump capacity is the U.S. gallon per minute.

Pump displacement (D)

The displacement of a reciprocating pump is the volume swept by all pistons or plungers per unit time. Deduction for piston rod volume is made on double-acting piston type pumps when calculating displacement. The standard unit of pump displacement is the U.S. gallon per minute.

For single-acting pumps:

$$D = \frac{ASnm}{231}$$

For double-acting piston pumps:

$$D - \frac{(2A - a)snm}{231}$$

where:

A = Plunger or piston area, square inch
a = Piston rod cross-sectional area, square inch (double-acting pumps)
s = Stroke length, inch
n = RPM of crankshaft
m = Number of pistons or plungers

Plunger or piston speed (v)

The plunger or piston speed is the average speed of the plunger or piston. It is expressed in feet per minute.

$$v = \frac{ns}{6}$$

Pressures

The standard unit of pressure is the pound force per square inch.

Discharge pressure (Pd)—The liquid pressure at the centerline of the pump discharge port.

Suction pressure (Ps)—The liquid pressure at the centerline of the suction port.

Differential pressure (Ptd)—The difference between the liquid discharge pressure and suction pressure.

Net positive suction head required (NPSHR)—The amount of suction pressure, over vapor pressure, required by the pump to obtain satisfactory volumetric efficiency and prevent excessive cavitation.

The pump manufacturer determines (by test) the net positive suction head required by the pump at the specified operating conditions.

NPSHR is related to losses in the suction valves of the pump and frictional losses in the pump suction manifold and pumping chambers. Required NPSHR does not include system acceleration head, which is a system-related factor.

Slip (S)

Slip of a reciprocating pump is the loss of capacity, expressed as a fraction or percent of displacement, due to leaks past the valves (including the back-flow through the valves caused by delayed closing) and past double-acting pistons. Slip does not include fluid compressibility or leaks from the liquid end.

Power (P)

Pump power input (Pi)—The mechanical power delivered to a pump input shaft, at the specified operating conditions. Input horsepower may be calculated as follows:

$$Pi = \frac{Q \times Ptd}{1714 \times \eta p}$$

Pump power output (Po)—The hydraulic power imparted to the liquid by the pump, at the specified operating conditions. Output horsepower may be calculated as follows:

$$Po = \frac{Q \times Ptd}{1714}$$

The standard unit for power is the horsepower.

Efficiencies (n)

Pump efficiency (ηp) (also called pump mechanical efficiency)—The ratio of the pump power output to the pump power input.

$$\eta p = \frac{Po}{Pi}$$

Volumetric efficiency (ηv)—The ratio of the pump capacity to displacement.

$$\eta v = \frac{Q}{D}$$

Plunger load (single-acting pump)

The computed axial hydraulic load, acting upon one plunger during the discharge portion of the stroke, is the plunger load. It is the product of plunger area and the gauge discharge pressure. It is expressed in pounds force.

Piston rod load (double-acting pump)

The computed axial hydraulic load, acting upon one piston rod during the forward stroke (toward head end), is the piston rod load.

It is the product of piston area and discharge pressure, less the product of net piston area (rod area deducted) and suction pressure. It is expressed in pounds force.

Liquid pressure $\left(\frac{\text{pounds}}{\text{square inch}}\right)$ or (psi)

= **Cylinder area** (square inches)

$$= (3.1416) \times (\text{Radius(inches)})^2$$

$$= \frac{(3.1416) \times (\text{diameter(inches)})^2}{4}$$

Cylinder force (pounds)—pressure (psi) × area(square inches)

Cylinder speed or average liquid velocity through piping (feet/second) $= \frac{\text{flow rate(gpm)}}{2.448 \times (\text{inside diameter(inch)})^2}$

Reciprocating pump displacement (gpm)

$$= \frac{\text{rpm} \times \text{displacement (cubic in/revolution)}}{231}$$

Pump input horsepower—see horsepower calculations

Shaft torque (foot-pounds) $= \frac{\text{horsepower} \times 5252}{\text{shaft speed (rpm)}}$

Electric motor speed (rpm) $= \frac{120 \times \text{frequency(HZ)}}{\text{number of poles}}$

Three phase motor horsepower (output)

$$= \frac{1.73 \times \text{amperes} \times \text{volts} \times \text{efficiency} \times \text{power factor}}{746}$$

Static head of liquid (feet)

$$= \frac{2.31 \times \text{static pressure (psig)}}{\text{specific gravity}}$$

Velocity head of liquid (feet) $= \frac{\text{liquid velocity}^2}{g = 32.2\ \text{ft/sec}^2}$

Absolute viscosity (centipoise) = specific gravity × Kinematic viscosity (centistrokes)

Kinematic viscosity (centistrokes)
= 0.22 × Saybolt viscosity (ssu)

$$= \frac{180}{\text{Saybolt viscosity (ssu)}}$$

Absolute pressure (psia)
= Local atmospheric pressure + gauge pressure (psig)

Gallon per revolution

$$= \frac{\text{Area of plunger (sq in)} \times \text{length of stroke(in)} \times \text{number of plungers}}{231}$$

Barrels per day = gal/rev × pump speed (rpm) × 34.3

Example: Find the volumetric efficiency of a reciprocating pump with the following conditions:

Type of pump	3 in diam plunger × 5 in stroke triplex
Liquid pumped	Water
Suction pressure	Zero psig
Discharge pressure	1785 psig
Pumping temperature	140 F
c	127.42 cu in
d	35.343 cu in
S	.02

Find βt from Table of Water Compressibility (Table 1).

Table 1 Water Compressibility
Compressibility Factor $\beta t \times 10^{-6}$ = Contraction in Unit Volume Per Psi Pressure
Compressibility from 14.7 Psia, 32°F to 212°F and from Saturation Pressure Above 212°F

Pressure	Temperature																		
Psia	0°C 32°F	20°C 63°F	40°C 104°F	60°C 140°F	80°C 176°F	100°C 212°F	120°C 248°F	140°C 284°F	160°C 320°F	180°C 356°F	200°C 392°F	220°C 428°F	240°C 464°F	260°C 500°F	280°C 536°F	300°C 572°F	320°C 608°F	340°C 644°F	360°C 680°F
200	3.12	3.06	3.06	3.12	3.23	3.40	3.66	4.00	4.47	5.11	6.00	7.27							
400	3.11	3.05	3.05	3.11	3.22	3.39	3.64	3.99	4.45	5.09	5.97	7.21							
600	3.10	3.05	3.05	3.10	3.21	3.39	3.63	3.97	4.44	5.07	5.93	7.15	8.95						
800	3.10	3.04	3.04	3.09	3.21	3.38	3.62	3.96	4.42	5.04	5.90	7.10	8.85	11.6					
1000	3.09	3.03	3.03	3.09	3.20	3.37	3.61	3.95	4.40	5.02	5.87	7.05	8.76	11.4	16.0				
1200	3.08	3.02	3.02	3.08	3.19	3.36	3.60	3.94	4.39	5.00	5.84	7.00	8.68	11.2	15.4				
1400	3.07	3.01	3.01	3.07	3.18	3.35	3.59	3.92	4.37	4.98	5.81	6.95	8.61	11.1	15.1	23.0			
1600	3.06	3.00	3.00	3.06	3.17	3.34	3.58	3.91	4.35	4.96	5.78	6.91	8.53	10.9	14.8	21.9			
1800	3.05	2.99	3.00	3.05	3.16	3.33	3.57	3.90	4.34	4.94	5.75	6.87	8.47	10.8	14.6	21.2	36.9		
2000	3.04	2.99	2.99	3.04	3.15	3.32	3.56	3.88	4.32	4.91	5.72	6.83	8.40	10.7	14.3	20.7	34.7		
2200	3.03	2.98	2.98	3.04	3.14	3.31	3.55	3.87	4.31	4.89	5.69	6.78	8.33	10.6	14.1	20.2	32.9	86.4	
2400	3.02	2.97	2.97	3.03	3.14	3.30	3.54	3.85	4.29	4.87	5.66	6.74	8.26	10.5	13.9	19.8	31.6	69.1	
2600	3.01	2.96	2.96	3.02	3.13	3.29	3.53	3.85	4.28	4.85	5.63	6.70	8.20	10.4	13.7	19.4	30.5	61.7	
2800	3.00	2.95	2.96	3.01	3.12	3.28	3.52	3.83	4.26	4.83	5.61	6.66	8.14	10.3	13.5	19.0	29.6	57.2	238.2
3000	3.00	2.94	2.95	3.00	3.11	3.28	3.51	3.82	4.25	4.81	5.58	6.62	8.08	10.2	13.4	18.6	28.7	53.8	193.4
3200	2.99	2.94	2.94	3.00	3.10	3.27	3.50	3.81	4.23	4.79	5.55	6.58	8.02	10.1	13.2	18.3	27.9	51.0	161.0
3400	2.98	2.93	2.93	2.99	3.09	3.26	3.49	3.80	4.22	4.78	5.53	6.54	7.96	9.98	13.0	17.9	27.1	48.6	138.1
3600	2.97	2.92	2.93	2.98	3.09	3.25	3.48	3.79	4.20	4.76	5.50	6.51	7.90	9.89	12.9	17.6	26.4	45.4	122.4
3800	2.96	2.91	2.92	2.97	3.08	3.24	3.47	178	4.19	4.74	5.47	6.47	7.84	9.79	12.7	17.3	25.8	44.5	110.8
4000	2.95	2.90	2.91	2.97	3.07	3.23	3.46	3.76	4.17	4.72	5.45	6.43	7.78	9.70	12.5	17.1	25.2	42.8	101.5
4200	2.95	2.90	2.90	2.96	3.06	3.22	3.45	3.75	4.16	4.70	5.42	6.40	7.73	9.62	12.4	16.8	24.6	41.3	93.9
4400	2.94	2.89	2.90	2.95	3.05	3.21	3.44	3.74	4.14	4.68	5.40	6.36	7.68	9.53	12.2	16.5	24.1	40.0	87.6
4600	2.93	2.83	2.89	2.94	3.05	3.20	3.43	3.73	4.13	4.66	5.37	6.32	7.62	9.44	12.1	16.3	23.6	38.8	82.3
4800	2.92	2.87	2.88	2.94	3.04	3.20	3.42	3.72	4.12	4.64	5.35	6.29	7.57	9.36	12.0	16.0	23.2	37.6	77.7
5000	2.91	2.87	2.87	2.93	3.03	3.10	3.41	3.71	4.10	4.63	5.32	6.25	7.52	9.28	11.8	15.8	22.7	36.6	73.9
5200	2.90	2.85	2.87	2.92	3.02	3.18	3.40	3.69	4.09	4.61	5.30	6.22	7.47	9.19	11.7	15.6	22.3	35.6	70.3
5400	2.90	2.85	2.86	2.91	3.01	3.17	3.39	3.68	4.07	4.59	5.27	6.19	7.41	9.12	11.6	15.3	21.9	34.6	66.9

$\beta t = .00000305$ at 140°F and 1800 psia. Calculate volumetric efficiency:

$$\text{Vol. Eff.} = \frac{1 - \left[P_{td\beta t}\left(1 + \frac{c}{d}\right)\right]}{1 - P_{td\beta t}} - S$$

$$= \frac{1 - \left[(1785 - 0)(.00000305)\left[1 + \frac{127.42}{35.343}\right]\right]}{1 - (1785 - 0)(.00000305)} - .02$$

$$= .96026 = 96 \text{ percent}$$

Specific gravity (at 60°F)

$$= \frac{141.5}{131.5 + \text{API gravity (degree)}}$$

Bolt clamp load (lb)
$= 0.75 \times$ proof strength (psi) $\times$ tensile stress area (in^2)

Bolt torque (ft-lb) $= \dfrac{0.2(\text{or } 0.15)}{12}$

$\times$ nominal diameter in inches $\times$ bolt clamp load (lb)

0.2 for dry
0.15 for lubricated, plating, and hardened washers

Calculating volumetric efficiency for water

The volumetric efficiency of a reciprocating pump, based on capacity at suction conditions, using table of water compressibility, shall be calculated as follows:

$$\text{Vol. Eff.} = \frac{1 - \text{Ptd}\,\beta t\left(1 + \frac{c}{d}\right)}{1 - \text{Ptd}\,\beta t} - S$$

where:

βt = Compressibility factor at temperature t (degrees Fahrenheit or centigrade). (See Tables 1 and 2).
c = Liquid chamber volume in the passages of chamber between valves when plunger is at the end of discharge stroke in cubic inches
d = Volume displacement per plunger in cubic inches
Ptd = Discharge pressure minus suction pressure in psi
S = Slip, expressed in decimal value

Calculating volumetric efficiency for hydrocarbons

The volumetric efficiency of a reciprocating pump based on capacity at suction conditions, using compressibility factors for hydrocarbons, shall be calculated as follows:

$$\text{Vol. Eff.} = 1 - \left[S - \frac{c}{d}\left(1 - \frac{\rho d}{\rho s}\right)\right]$$

where:

c = Fluid chamber volume in the passages of chamber between valves, when plunger is at the end of discharge strike, in cubic inches
d = Volume displacement per plunger, in cubic inches
P = Pressure in psia (P_s = suction pressure in psia; P_d = discharge pressure in psia)
P_c = Critical pressure of liquid in psia (See p. 218).
P_r = Reduced pressure

$$\frac{\text{Actual pressure in psia}}{\text{Critical pressure in psia}} = \frac{P}{P_C}$$

P_{rs} = Reduced suction pressure $= \dfrac{P_S}{P_c}$

P_{rd} = Reduced discharge pressure $= \dfrac{P_d}{P_c}$

S = Slip expressed in decimal value

t = Temperature, in degrees Rankine
= Degrees F + 460 (t_s = suction temperature in degrees Rankine; t_d = discharge temperature in degrees Rankine)
T_c = Critical temperature of liquid, in degrees Rankine (See Table 3)
T_r = Reduced temperature

$$= \frac{\text{actual temp. in degrees Rankine}}{\text{critical temp. in degrees Rankine}}$$

$= \dfrac{t}{T_C}$ (See Fig.2)

T_{rs} = Reduced suction temperature

$$= \frac{t_s}{T_C}$$

T_{rd} = Reduced discharge temperature

$$= \frac{t_d}{T_C}$$

Vol. Eff. = Volumetric efficiency expressed in decimal value.

$$= \frac{1}{1} \times \bullet\bullet \times 62.4 = \text{density of liquid in lb per cu ft}$$

s = Density in lb per cu ft at suction pressure

Table 2 Water Compressibility
Compressibility Factor $\beta t \times 10^{-6}$ = Contraction in Unit Volume Per Psi Pressure
Compressibility from 14.7 Psia at 68°F and 212°F and from Saturation Pressure at 392°F

Pressure Psia	Temperature 20°C 68°F	100°C 212°F	200°C 392°F
6000	2.84	3.14	5.20
7000	2.82	3.10	5.09
8000	2.80	3.05	4.97
9000	2.78	3.01	4.87
10000	2.76	2.96	4.76
11000	2.75	2.92	4.66
12000	2.73	2.87	4.57
13000	2.71	2.83	4.47
14000	2.70	2.78	4.38
15000	2.69	2.74	4.29
16000	2.67	2.69	4.21
17000	2.66	2.65	4.13
18000	2.65	2.60	4.05
19000	2.64	2.56	3.97
20000	2.63	2.51	3.89
21000	2.62	2.47	3.82
22000	2.61	2.42	3.75
23000	2.59	2.38	3.68
24000	2.58	2.33	3.61
25000	2.57	2.29	3.55
26000	2.56	2.24	3.49
27000	2.55	2.20	3.43
28000	2.55	2.15	3.37
29000	2.54	2.11	3.31
30000	2.53	2.06	3.26
31000	2.52	2.02	3.21
32000	2.51	1.97	3.16
33000	2.50	1.93	3.11
34000	2.49	1.88	3.07
35000	2.49	1.84	3.03
36000	2.48	1.79	2.99

d = Density in lb per cu ft at discharge pressure = Expansion factor of liquid

$\frac{1}{1}$ = Characteristic constant in grams per cubic centimeter for any one liquid that is established by density measurements and the corresponding values of (See Table 3).

Table 3

Carbon Atoms	Name	T_c Degrees Rankine	P_c Lb Per Sq In	pl/wl Grams Per cc
1	Methane	343	673	3.679
2	Ethane	550	717	4.429
3	Propane	666	642	4.803
4	Butane	766	544	5.002
5	Pentane	847	482	5.128
6	Hexane	915	433	5.216
7	Heptane	972	394	5.285
8	Octane	1025	362	5.340
9	Nonane	1073	332	5.382
10	Decane	1114	308	5.414
12	Dodecane	1185	272	5.459
14	Tetradecane	1248	244	5.483

Example. Find volumetric efficiency of the previous reciprocating pump example with the following new conditions:

Type of pump	3 inch dia plunger × 5 inch stroke triplex
Liquid pumped	Propane
Suction temperature	70 F
Discharge temperature	80 F
Suction pressure	242 psig
Discharge pressure	1911 psig

Find density at suction pressure:

$$T_{rs} = \frac{t_s}{T_c} = \frac{460 + 70}{666} = .795$$

$$P_{rs} = \frac{P_s}{P_c} = \frac{257}{642} = .4$$

$$\frac{\rho 1}{\omega 1} = 4.803 \text{ (From Table 3, prapane)}$$

$$\omega = .1048 \text{ (From Figure 2)}$$

$$\rho_s = \frac{\rho 1}{\omega 1} \times \omega \times 62.4$$

$$= 4.803 \times .1048 \times 62.4$$

$$= 31.4 \text{ lb per cu ft}$$

Find density at discharge pressure:

$$T_{rd} = \frac{t_d}{T_c} = \frac{460 + 80}{666} = .81$$

$$P_{rd} = \frac{P_d}{P_c} = \frac{1926}{642} = 3.0$$

$$\omega = .1089 \text{ (From Fig. 2)}$$

$$\rho d = \frac{\rho 1}{\omega 1} \times \omega \times 62.4$$

$$= 4.803 \times .1089 \times 62.4$$

$$= 32.4 \text{ lb per cu ft}$$

Therefore

$$\text{Vol. Eff.} = 1 - \left[S - \frac{c}{d}\left(1 - \frac{\rho d}{\rho s}\right)\right]$$

$$= 1 - \left[.02 - \frac{127.42}{35.343}\left(1 - \frac{32.64}{31.4}\right)\right]$$

$$= .8376$$

$$= 83.76\%$$

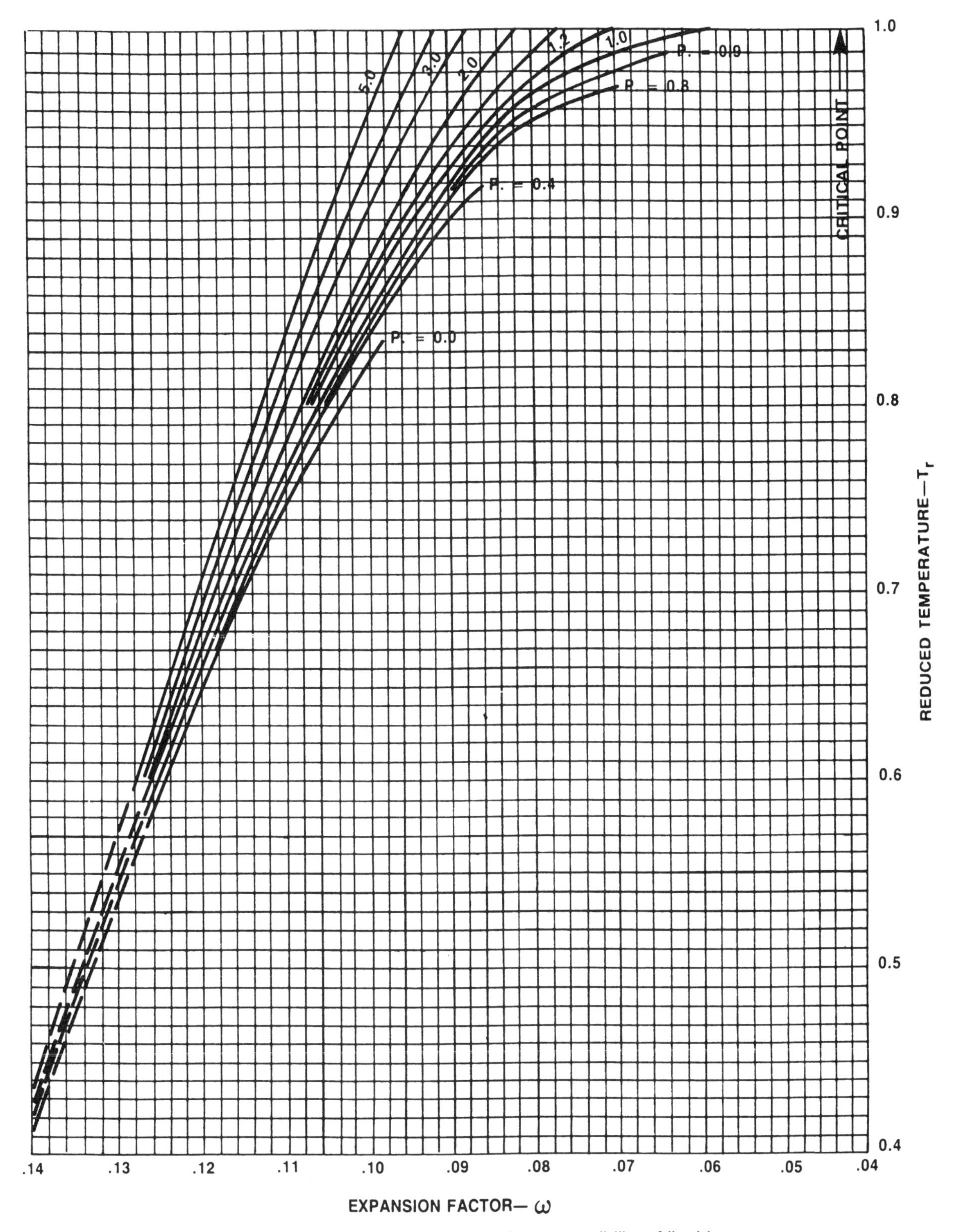

Figure 2. Thermal expansion and compressibility of liquids.

Reprinted with permission—Goulds Pumps.

Understanding the basics of rotary screw pumps

. . . operating principles, application factors, and performance characteristics

ALBERT A. ZALIS, Vice President Engineering, Warren Pumps, Inc., IMO Industries, Inc.

Basically only two categories of pumps need be considered when a pumping application is evaluated: centrifugal pumps and positive-displacement pumps. Centrifugal pumps depend on rotational speed to establish head and capacity characteristics and are capable of moderately high discharge heads and very high pumping capacities. Centrifugal pump design configurations include single and multistage, enclosed and open impeller, single and double suction, and horizontal and vertical.

When the application of a centrifugal pump is considered, in addition to the normal requirements of head and capacity, the pumped medium fluid characteristics must be known in detail:

- Abrasiveness
- Consistency and viscosity
- System head/capacity variations
- Net positive suction head available (NPSHA)
- Corrosiveness
- Required construction materials

Although centrifugal pumps successfully handle all types of slurries and viscous fluids, a suitable relationship between head, capacity, and fluid characteristics must exist. A simple end suction centrifugal pump with open impeller can be used for handling both clear water and slurries, Fig. 1. Typical centrifugal pump performance curves are shown in Fig. 2. Centrifugal pump performance can be evaluated using the affinity laws: With the impeller diameter held constant, capacity Q varies as the ratio of the change in speed N:

$$\frac{Q_2}{Q_1} = \frac{N_2}{N_1}$$

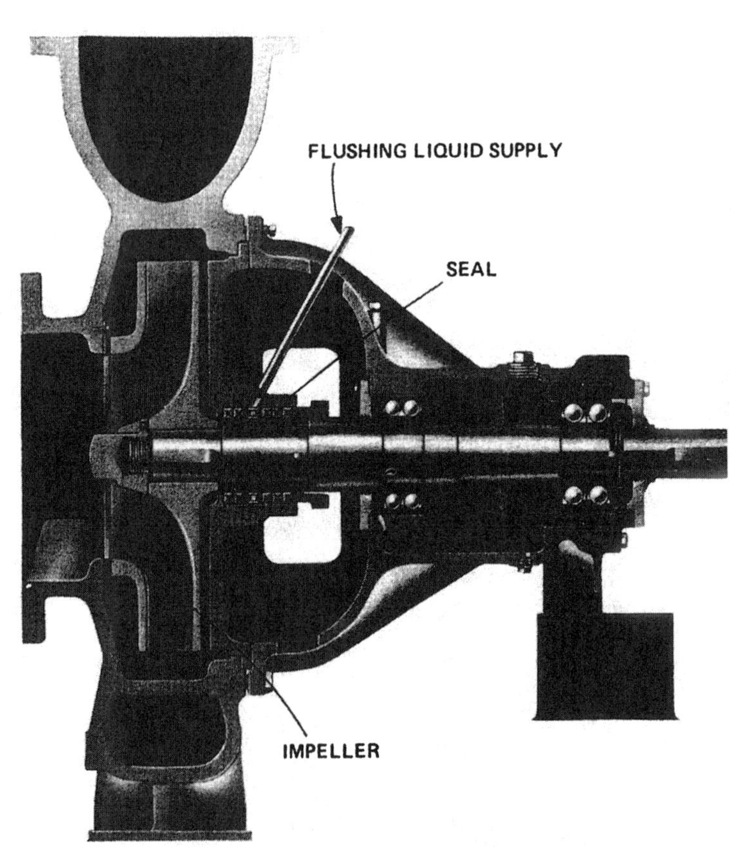

Figure 1. The end suction centrifugal pump is a general-purpose pump that is widely used for pumping a broad range of liquids. Seal may be flushed with clean liquid when corrosive or abrasive slurries are pumped.

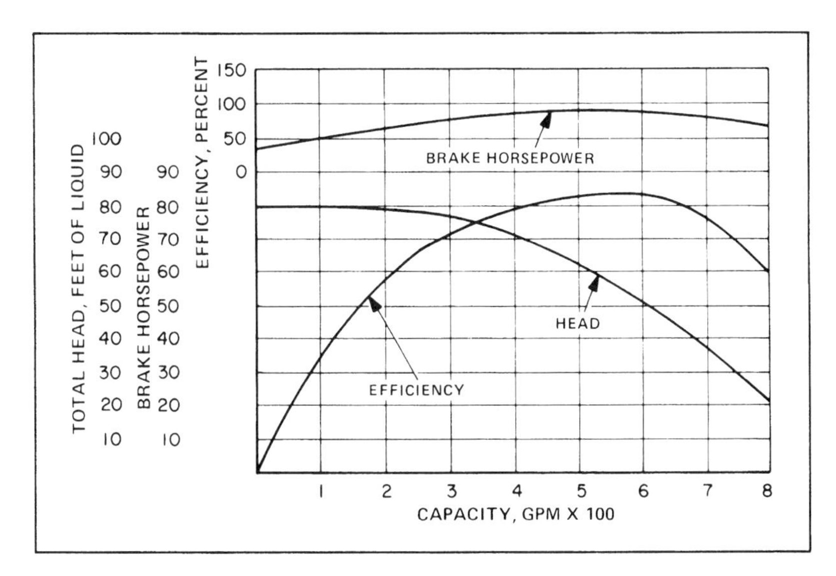

Figure 2. Typical performance curves for a centrifugal pump plot the brake horsepower, efficiency, and head (discharge pressure) as functions of capacity. The plots of horsepower and head are not straight lines, a fact that increases the difficulty of accurately predicting performance.

Head H varies as the square of the speed change ratio:

$$\frac{H_2}{H_1} = \left(\frac{N_2}{N_1}\right)^2$$

Brake horsepower bhp varies as the cube of the speed change ratio:

$$\frac{bhp_2}{bhp_1} = \left(\frac{N_2}{N_1}\right)^3$$

These laws make it simple to correct for varying speeds. However, these relationships assume that efficiency remains constant. Tests have shown that the correction for head, with regard to speed, is not precisely correct when head is reduced more than the amount calculated to result from a speed increase. Also, a speed increase yields a reduction in efficiency exceeding that calculated using the brake horsepower and head formula. However, the deviation of values is not great, and the affinity laws are a valuable tool when used with good engineering judgment.

Positive-displacement pumps include several pump designs. The major classifications are reciprocating and rotary pumps. Reciprocating pumps include power and steam driven, single and double acting, and plunger and piston.

Reciprocating pumps have high head capability, easily handle all types of slurries, and have high volumetric efficiency, permitting lowcapacity/high-head relationships.

When the application of a reciprocating pump is considered in addition to the usual head and capacity data, the following conditions related to the fluid must be known:

- NPSHA
- Allowable pulsation levels
- Corrosiveness or required construction materials

Rotary pumps cover a broad range including screw, gear, vane, and radial piston types.

Screw pumps can be further classified as single (progressive cavity), two screw (timed), and three screw (untimed). Gear pumps include external and internal gear designs.

Rotary pumps operate in generally the same pressure ranges as centrifugal pumps, except at lower capacities. Rotary pumps are not particularly suitable for handling severely abrasive fluids, except for the single-screw progressive cavity pump, which performs well in pumping abrasive slurries. Rotary pumps are generally capable of handling a wide range of fluid viscosities, at medium pressures, over a broad range of flow rates.

The remainder of this article is confined to a discussion of screw pumps, primarily the two-screw timed pump that utilizes timing gears to drive the screws and prevent driving contact between the intermeshing screws. This pump uses bearings to support the shafts and absorb some or all of the radial load, giving it a broad field of application suitable for pumping corrosive and noncorrosive fluids, lubricating and nonlubricating fluids, clean fluids and slurries, and Newtonian and nonNewtonian fluids.

Relatively high capacities, for positive-displacement pumps, can be handled by screw pumps. The primary limitation is the production of screws having the displacement required to produce a high discharge rate.

Performance characteristics of a screw pump are functions of screw design, provided it is used with properly designed inlet and outlet ports. In a particular pump size, performance characteristics can be radically changed by altering

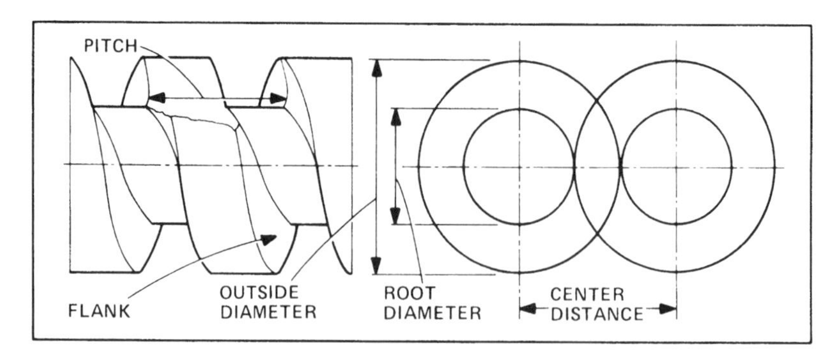

Figure 3. Four basic dimensions establish the pump displacement characteristics. One of the flanks must have an epicycloidal profile to establish a seal with the mating screw. The mating flank may have an epicycloidal curve or any form of convex curve that will achieve a sealed running mesh.

screw design. When a screw pump performs poorly in a specific application, the application parameters should be rechecked and the screw design examined.

The capacity of a screw pump is a function of four critical screw dimensions, which control pump displacement (Figure 3). For the pump to deliver maximum displacement efficiently, the profiles of the intermeshing threads must be given careful attention. The outside diameter of the screw and the root diameter are cylindrical. Clearances of these dimensions average 0.0015 in. per in. of diameter. The flanks (thread sides) are profiled and must be carefully controlled to obtain good pumping efficiency. One of the two flanks must have a profile defined as an epicycloid. The other flank may have an epicycloidal curve or any form of convex curve that will form a continuous seal when it is mated with the meshing screw. The total axial or flank clearance normally equals the total diametrical clearance between the screw outside diameter and the casing bore.

There must be small clearances between the screws and between the screws and casing. These clearances allow a small amount of leakage from the pump outlet cavity to the pump inlet cavity. This leakage, identified as slip, is a function of pump outlet pressure, fluid viscosity, and fluid characteristics. If the screws are designed with epicycloidal curves on both flanks, the slip will be 70% greater than if the design provides for one epicycloidal curve and one convex curve. However, for certain applications the double epicycloid has distinct advantages.

In a pump design utilizing four pumping screws, the arrangement usually allows the fluid to enter the screws on each end and discharge in the center (Figure 4). When the screws are all the same diameter, axial thrust forces are equal and opposite and therefore cancel each other. Although

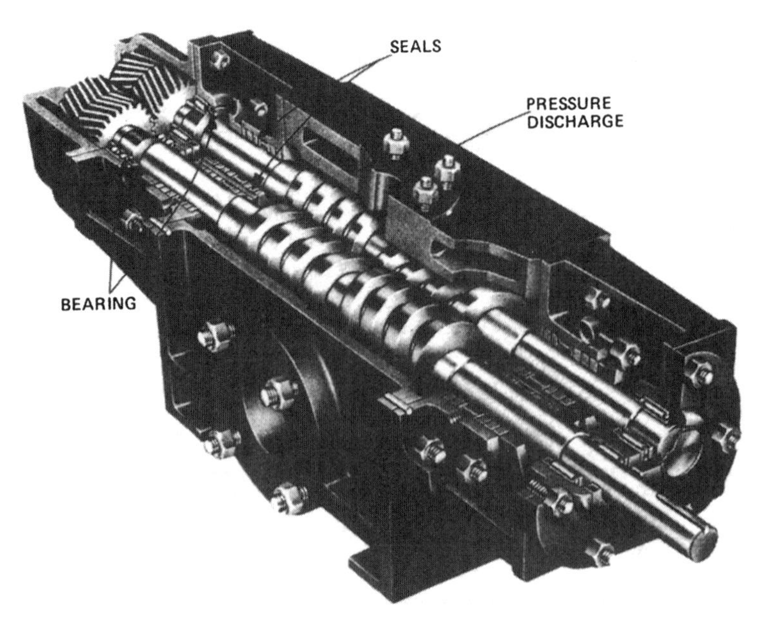

Figure 4. In a two-shaft, four-screw pump, the material enters at each end and discharges in the center. End thrust forces are balanced. Radial forces are functions of outlet pressure and screw dimensions and must be carried by the bearings at each end.

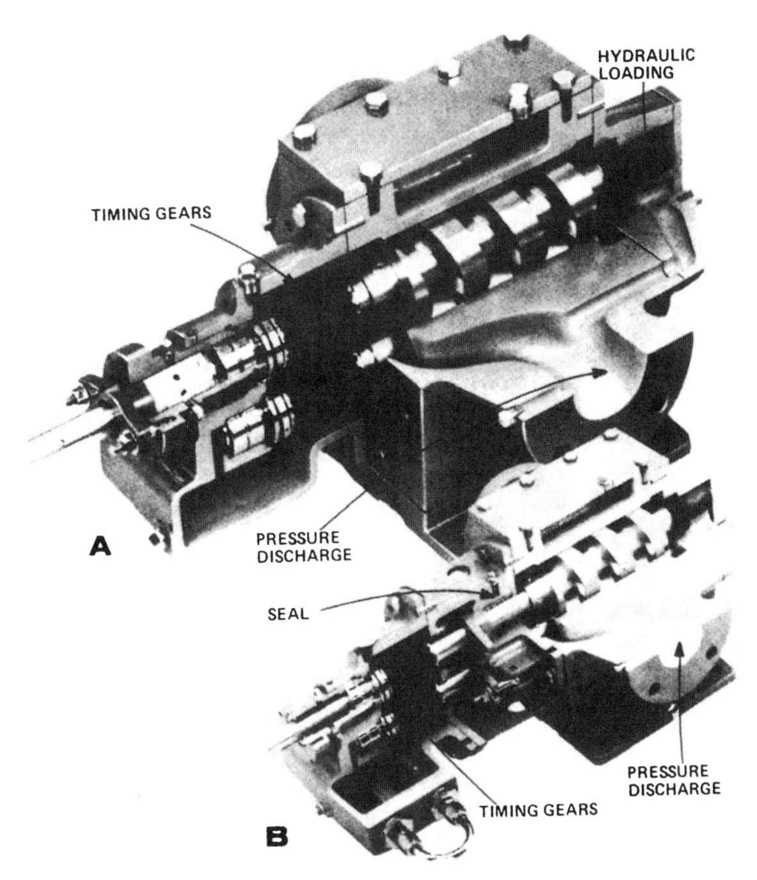

Figure 5. Pumps with two screws require timing gears to maintain constant relationship between screws. Axial forces are normally counterbalanced by hydraulic loading. The gear set may be product lubricated (a) or provided with an external lubrication system (b).

other arrangements for balancing radial forces are available, they are, at best, only partly effective and sometimes introduce other problems. Radial forces in screw pumps are functions of outlet pressure and screw dimensions. If the epicycloid/convex screw flank design is used, proper location of the profiles for the two screws on each shaft will partially balance radial loads.

Timed screw pumps are available in which only two pumping screws are used (Figure 5). In these pumps, strategic location of the flank profiles has no effect on radial forces. Axial forces are of appreciable magnitude and are normally counterbalanced hydraulically.

Screw pumps are capable of discharge pressures from atmospheric to more than 300 psig. However, the same screw design cannot be used throughout this range of pressures. In a centrifugal pump, high discharge heads are achieved either by high speeds or, more often, by staging a number of impellers to operate in series; each impeller carries a proportionate share of the discharge head.

Discharge head in a screw pump is regulated by controlling the number of screw turns. The minimum number of screw turns that can be used is two. Fewer than two turns results in erratic performance and excessive slip. Unlike the centrifugal pump impeller, each screw turn does not generate a proportionate share of the outlet pressure. The additional screw turns assist in reducing slip, thereby increasing volumetric efficiency.

"... lower volumetric efficiency could result in a higher mechanical efficiency."

A higher volumetric efficiency can mean higher mechanical efficiency, and with today's power costs it would appear desirable to use screws having many turns to achieve a high volumetric efficiency. This is not necessarily true. The greater the number of screw turns is, the longer the screw (screw length equals the number of turns multiplied by the screw pitch) is. Screw length has a proportional bearing on the horsepower required to drive the pump at zero outlet pressure.

If a low-viscosity fluid is being pumped, this horsepower will be of small magnitude. However, the horsepower

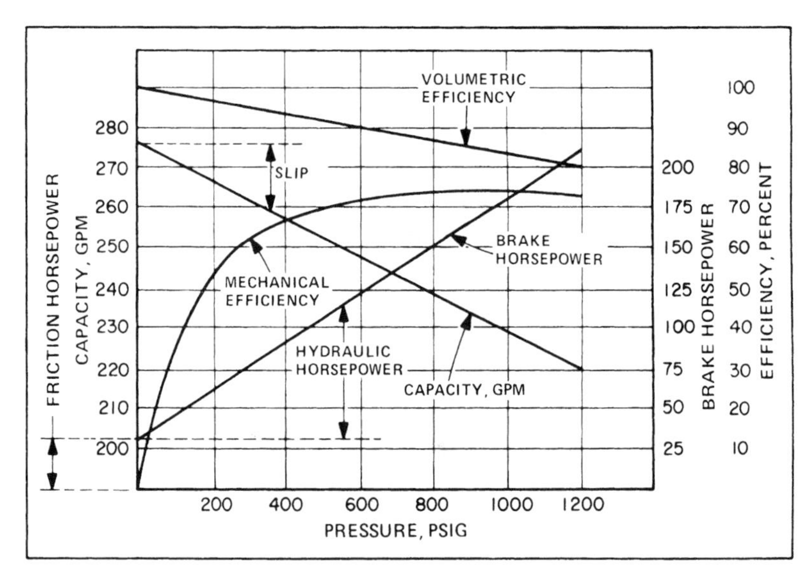

Figure 6. In contrast to performance curves for a centrifugal pump, the screw pump performance curves for capacity and brake horsepower are normally straight lines. Capacity, brake horsepower, mechanical efficiency, and volumetric efficiency are plotted as functions of discharge pressure.

requirement increases rapidly as viscosity increases. Therefore, a slightly lower volumetric efficiency could result in a higher mechanical efficiency. Mechanical efficiency controls power costs; consequently, the application engineer, when selecting a screw pump, must consider both slip and friction horsepower to determine the best arrangement.

The remaining parts of a screw pump are subject to conventional mechanical design: Shafts must be designed for bending and torsional loads, and bearings must match the center distance between the two parallel shafts. Timing gears should be designed in accordance with American Gear Manufacturers Association standards. If spur gears are used, both shafts should be rigidly located axially. If herringbone gears are used to reduce noise level, the driven shaft should be allowed to float axially and be positioned by the apex of the gear. Velocities through the fluid inlet passages should be kept low, normally not more than 5 fps for average viscosities of 150 to 1000 SSU and decreased as the viscosity increases.

The performance characteristics of a screw pump are similar to those of other positive-displacement pumps. Capacity, horsepower, and efficiency are plotted as functions of discharge pressure. For centrifugal pumps, horsepower, efficiency, and discharge pressure are plotted as functions of capacity. With centrifugal pumps, plots of horsepower and head against capacity are not straight lines. Therefore, accurately predicting performance is difficult, and more testing and modeling are required. For positive-displacement pumps, plots of capacity and brake horsepower against pressure normally result in straight lines when coordinate paper is used. However, there are instances in which the capacity versus pressure line is slightly curved. This characteristic usually results from excessive clearance in the pump permitting deflection and eccentric operation of the rotating screw.

Performance conditions can be manipulated to reflect operation at speeds or pressures other than those shown in Figure 6, using these fundamental relationships:

Friction horsepower is the power required to overcome losses including packing seal friction, timing gear inefficiencies, bearing friction, and pumped fluid viscous shear. All of the losses are so small when compared to the viscous drag of the pumped fluid that it can be safely assumed that friction horsepower is entirely dependent on fluid viscosity and pump speed. Even pump rotor clearances, within the limits of efficient pump operation, have such a minor influence on the power consumed that extremely sensitive instrumentation is required for taking measurements.

Figure 7 shows the relationship between friction horsepower and viscosity at 1200 rpm. If, for example, it is necessary to determine friction horsepower fhp at 1800 rpm, the following relationship exists:

$$\text{fhp}_2 = \text{fhp}_1\left(\frac{1800}{1200}\right)^{1.5}$$

With today's programmable calculators and computers, the curves shown in Figure 7 may be easier to work with if they are reduced to mathematical formulas. Because the equation will take the form $x = ay^n$, by means of simultaneous equations the friction horsepower equation becomes:

$$\text{fhp}_2 = 2.9541 \times (\text{viscosity in SSU})^{0.42065}$$

Hydraulic horsepower is the measure of the total hydraulic work done by the pump. This term should not be confused with water horsepower or work horsepower, the measure of the useful work done by the pump.

These terms can be defined by the formulas:

Hydraulic horsepower

$$= \frac{(\text{displaced gpm})(\text{psi})}{1714}$$

Water horsepower

$$= \frac{(\text{displaced gpm})(\text{psi})}{1714}$$

Displaced gpm is the capacity at zero psi discharge pressure. Delivered gpm is the capacity at the rated or required discharge pressure, or the displaced gpm minus the slip gpm.

Referring to Figure 6, the brake horsepower is the sum of the friction horsepower and the hydraulic horsepower. This approach yields a fairly accurate brake horsepower through a wide range of operating conditions.

Slip is the loss of capacity from the higher pressure area through the internal clearances. Slip is a function of viscosity and differential pressure. Figure 7 is a slip/viscosity curve for 500 psi operating pressure. Slip varies in direct proportion to the change in pressure. Similarly, with the friction horsepower curve, slip can be expressed by the formula:

$$\text{Slip, gpm} = 294.95 \times (\text{viscosity in SSU})^{-0.4660}$$

Mechanical efficiency is a calculated quantity that until recently received very little attention. The energy crisis has made it more important. Mechanical efficiency is calculated by:

Mechanical efficiency

$$= \frac{(\text{displaced gpm})(\text{psi})}{1714(\text{bhp})} \times 100$$

Volumetric efficiency is an infrequently used measure of internal slip and can be calculated by:

Volumetric efficiency

$$= \frac{\text{delivered gpm}}{\text{displaced gpm}} \times 100$$

It may be well to review a suggested approach to the pumping of non-Newtonian fluids. One fairly common application for screw pumps is pumping pseudoplastic polymers of various types. A pseudoplastic fluid is one whose viscosity decreases as the shear rate in the pump increases. Data on pseudoplastic fluids are usually presented as a curve plotting apparent viscosity versus shear rate in reciprocal seconds. The determination of shear rate in a positive-displacement pump is a widely discussed topic. It is

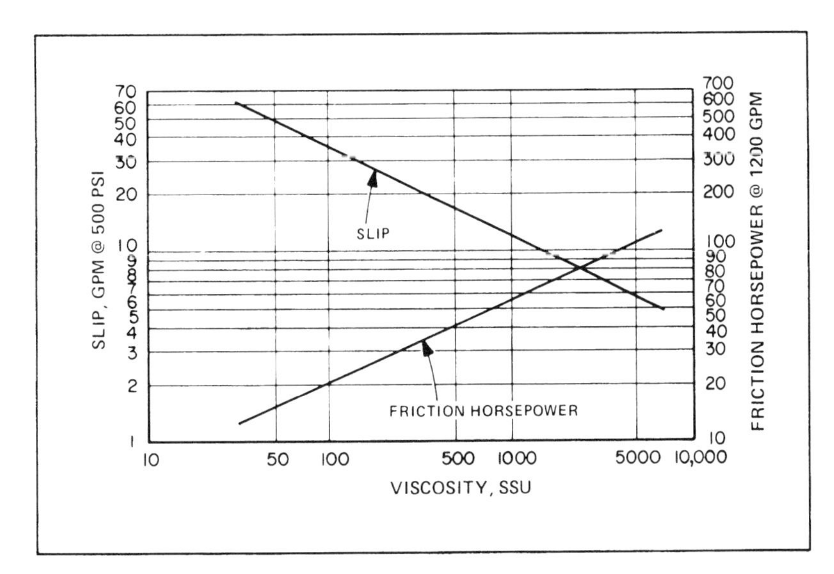

Figure 7. Slip is a loss of capacity from the higher pressure area to the lower pressure area and is a function of viscosity and pressure differential. Slip varies in direct proportion to the change in pressure. At a specific pressure as the viscosity increases, slip decreases, while the friction horsepower increases.

difficult, if not impossible, to arrive at one formula that is applicable to all pumps. For the timed screw pump, successful application has been made using the formula:

Shear rate, sec^{-1}

$$= \frac{\pi(\text{rotor diameter})(\text{rpm})}{(\text{radial clearance})(60)}$$

where:

rotor diameter = Pumping screw diameter, inches
radial clearance = Clearance between screw diameter and casing bore, inches

This formula gives consideration to only the area of maximum shear in the pump. Experience has shown that refining this formula to include all shear surfaces yields values that are grossly incorrect. The shear rate calculated from the formula will permit a satisfactory pump selection; however, care should be taken to consider the shear rate in the inlet piping. This relatively small quantity results in a much higher viscosity and thereby influences the inlet pressure requirements.

Sometimes it becomes necessary to take a surplus pump and apply it to a new application. This often happens in chemical plants where processes are frequently subject to change. Earlier discussions dealt in some detail with screw profile variations, importance of screw turns, slip, and friction horsepower. All these factors must be considered for the new application. Performance characteristics can be changed radically by installation of a new rotating element in an existing casing.

This article has not covered all there is to learn about screw pumps and their applications but has discussed the more important areas on which there is little, if any, literature published.

15: Measurement

Multiphase flow meter

Principle of operation

The major parts of the TopFlow meter are the Venturi insert and the electrodes inside the Venturi throat. The flow rates of oil, water, and gas are calculated based on the measurements performed by the electrodes and the measurement of the differential pressure across the Venturi inlet.

The TopFlow meter utilizes a capacitance sensor for oil-continuous liquid mixtures, where the capacitance caused by the dielectric constant of the fluid mixture is measured in the throat of the Venturi. Similarly it utilizes a four-electrode conductance sensor for water-continuous liquid mixtures, where the conductance is caused by the conductivity of the fluid mixture. All the necessary electrodes are incorporated within a Venturi that is somewhat modified. Advanced data processing is utilized to have continuous readings of the flow rates of oil, water, and gas.

The Venturi has a minimum of pressure drop. Nearly all multiphase meter vendors use a Venturi in one way or the other, and the reason is that this is probably the most reliable device for multiphase measurement because it measures the mass flow rate independent of what happens. One objection is the relatively limited turndown ratio that is specified, especially for single-phase measurements within fiscal uncertainties. However, because most wells produce in the high GVF range, where the total mass flow rate does not vary much, there is no general problem connected to the use of this device. Especially when installing one meter at each single well and not having to deal with varying flow rates, the Venturi is very suitable for multiphase measurement. Erosion will not give any major difficulties, because the TopFlow meter uses a Venturi in terms of an insert into the pipe. This means that the insert itself can be made of a wear-resistant material like high-quality steel or even ceramics in cases when erosion is considered to be a problem. In extreme cases, due to the simplicity of the TopFlow meter, the insert itself can easily be replaced. However, in the latter case the other parts of the production pipeline, such as bends and valves, would also be eroded to destruction and would have to be replaced. A Venturi is able to give a good picture of the dynamics of the flow (i.e., it is not sensitive to the flow regime effects for measuring the instant mass flow rate accurately). Modern differential pressure transmitters are also able to follow the fast variations through high-frequency response.

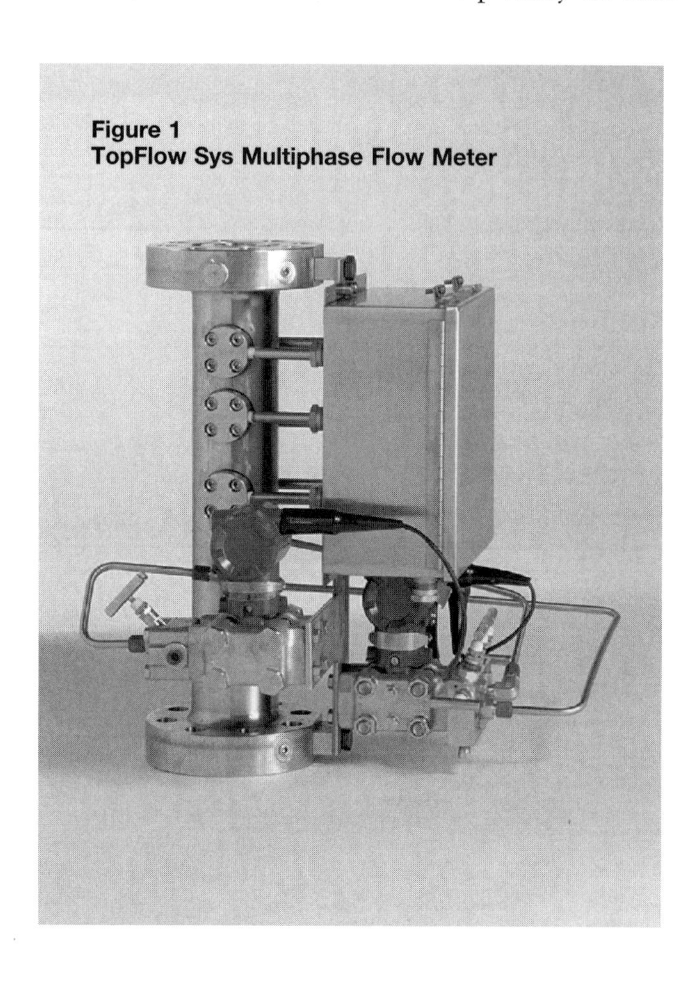
Figure 1
TopFlow Sys Multiphase Flow Meter

The most important flow regime effects are mainly due to the ratio between gas and liquid. One is the longitudinal effect along the axis of the flow conduit, represented by the slug regime, where longer or shorter liquid plugs are separated by longer or shorter gas pockets. One challenge with this is that the liquid plug is not entirely liquid but contains varying amounts of small gas bubbles. In addition, the gas pockets are not entirely gas but contain larger or smaller amounts of liquid. In general the liquid and gas are also flowing with different velocities. It is of very high importance to employ sensors with fast time response to follow the rapid changes in slug flow. It is likewise important to measure the velocity often, because there is neither a constant, distinct gas velocity nor a corresponding constant liquid velocity. Thereby the velocity "components" are measured in real time and the slip between liquid and gas is handled automatically, without the need for theoretical or empirical slip models.

Another flow regime aspect is the radial effect, concerning the distribution of the liquid and gas over the cross-section of the flow conduit. In general this plays a part in all multiphase flow conditions, but especially at high gas volume fractions, where the slip can be significant, the gas tends to gather in the center of the cross section, along the axis of the flow conduit. This is because of the lower flow resistance in the turbulent zone in this center. In plain words, the gas races through this channel, whereas the liquid is pushed aside and flows in contact with the pipe wall. The velocity difference is quite substantial and increases at increasing gas volume fraction. However, a part of the liquid flows as droplets following the gas phase. Taking these matters into consideration, it is obvious that the main source of uncertainty in multiphase flow rate measurement can be found in the complexity of the flow itself and not necessarily in the type of sensor used for the purpose. The crucial matter is therefore to choose appropriate technology and develop good physical models and reliable interpretation tools to solve the difficult tasks.

Adapted from FlowSys AS, Norway. Reprinted with permission.

Pipeline flow measurement—the new influences

E. L. Upp, Flow Measurement Consultant

Introduction

There are many considerations that influence our choice of methods and equipment in the measurement of liquids and gases in pipelines. A review of these problems is timely, since we have changing influences that are affecting our choices to accomplish the job of flow measurement. These are accuracy, electronic equipment, economics, fluid characteristics, standards and governmental influences, and testing or proving practices. All of these factors have had importance to some degree in the past; however, their relative importance to flow measurement and more importantly to a pipeline's overall operation has changed considerably as we have seen the prices of the fluids handled make dramatic changes along with supply and demand fluctuations of considerable magnitude versus a steady growth pattern of just a few years back.

Accuracy

Today's demand—better flow measurement accuracy. In analyzing the economics of the flow measurement industry, we find the customers are no longer in an expansion mode. With the turndown in their economies, they are more concerned with improving efficiencies in their present installations. Better measurement accuracy is one way to achieve this. The users are asking the manufacturer to improve his present devices as well as develop new meters and new technology to do the job "better." No one can fault these desires. However, measurement is a very conservative, "tradition-following" market. Industry metering standards are slow to change, and it has been difficult in the past to get the users to adopt new meters and products.

In the gas industry, contracts are written by lawyers who are not measurement experts and who tend to follow a contract form that was successful in the past. Manufacturers tried to switch users to electronic readout devices some 20 years ago. Electronic computers and flow recorders have been available to industry all this time, though admittedly not of today's quality. Yet today in the United States—for pipeline large-volume natural gas measurement—less than 2% of the stations installed utilize electronic computer controls and readouts. The industry has been more successful in applying electronic digital readout devices to liquid metering in conjunction with the turbine meter and is beginning to see production fields being automated for both liquid and gas measurement.

If the industry had a new meter today that offered absolute accuracy; infinite rangeability; met all the environmental requirements, codes, and governmental regulations; could be installed and operated by present personnel with minimal maintenance; was priced low compared to present devices on the market—it would still be difficult to sell it in sufficient quantities in the marketplace to make it profitable to manufacturers. The companies would at best be willing to try the meter for a year and compare it with present meters. Then and only then would they add it to their accepted list. The manufacturers are asking the user to change his cash register. The user has to be convinced that he is on solid legal, engineering, and procedural grounds.

Costs of accuracy. Many companies are taking a hard look at all capital expenditures in this time of high interest rates. Management people feel that the high price of the fluids they are handling allows a larger investment in order to get accurate measurement. Simple expenditure of capital does not eliminate the fact that normally the more expensive the device, the "better answer" we will get. However, careful consideration must be given to what the "better answer" will mean to our operations.

Quite often the flow measurement industry is called upon to solve flow measurement problems. Sometimes the end result is not exactly what the users had in mind. A new seller of natural gas or fluids that has a meter reading low by 1% would be delighted to use the new "more accurate meter." However, a purchaser who was just as interested in his balance sheet and who had his flow figure raised by 1%

would not be too happy with a decision to use this "more accurate meter."

Many times, users have evaluated electronic flow computers, and when compared with the existing meter will read differently. There are good solid technical arguments why the computer is more accurate, but there are also political and economic reasons why customers cannot use more accuracy "if it lowers my billing." When two parties and a contract are involved, you must keep in mind that you can't have a more accurate metering device that will raise measurement for both sides of the equation. As manufacturers, the only path that can be followed is to sell the most accurate metering device they know how to make and to let the customer work with the political and economic consequences. These influences certainly affect the industry's ability to bring out new products and have them accepted based on accuracy.

Their place in flow measurement. There is no denying that the high technology age is upon us. This basically refers to electronic calculating and data handling systems as far as the pipeline flow measurement problems are concerned. These tools have been available for several decades; however, we are finally seeing the beginning of the significant use of electronics to replace our previous mechanical methods of converting meter readings to custody transfer volumes. This is basically due to the availability of low-cost, reliable electronic devices that will meet the field conditions where flow measurement takes place and a growing awareness of the user on the requirements to successfully apply these devices. We have moved through stages of applications from "flow computers can do anything," to "flow computers don't operate in the field," to the stage we are in now where properly applied flow computers are a viable consideration for accomplishing the job we have done in the past, plus additional timely uses of the flow data to improve operations. However, this decision is not made without a serious look at the economic consequences.

Economics of flow measurement

Accuracy versus cost. As the cost of the fluids handled goes up, there is a demand by all parties involved in its purchase and sale to "reduce the tolerances," "get more accurate measurement," or "get more sophisticated equipment to get a better answer." Capital investment required to upgrade measurement is getting a hard look by pipeline management, as are all of their expenditures. So our desire for better measurement is tempered by available capital and the resulting savings to be expected. The definition of "more accurate answers" of various measuring devices, both in terms of theoretical and practical use, must be reviewed to improve our ability to get the best measurement for the dollars invested.

If you are producing a well at .25 MMcfd, one level of equipment and procedure investment is justified, whereas a station delivering 25 MMcfd or 250 MMcfd may require another. Similar thinking applies to liquid measurement. Physical limitations such as available power, cost of space to install equipment, and effect on the environment must also be considered. On offshore platforms, where space costs a premium, meters that can be installed with a minimum of piping have been used because of the savings realized. A large portion of the new metering devices' electronic readouts require good quality and dependable power. In some remote locations where commercial power is not available, the cost of getting the proper power may exceed the cost of the metering. In this case, the cost of power may dictate the use of mechanical or pneumatic readout devices or some local generated source such as solar or thermoelectric. These power sources are normally low-wattage devices, which may restrict the capabilities of the electronics or limit the choices of devices.

Personnel requirements. A very definite consideration affecting the economics of flow measurement is the personnel required to accomplish a given job. This may mean at least a retraining, if not a replacing of present measurement people. Generally they will be in an upgraded position that requires both electronic and mechanical skills along with a solid understanding of flow measurement. Otherwise, you may end up with an electronics-oriented person chasing the details of proper programs and round-off procedures of the measurement readout system while not recognizing that the basic meter may be improperly applied or functioning and the total system is in error. The reverse can be true if a mechanically oriented person is lacking in electronic knowledge. Fortunately, in the past few years most new workers have had good exposure to electronic devices and are more comfortable with electronic flow approaches than some of our older personnel. This problem exists at all levels of our company's operations and to a large degree will be the deciding factor of the final success of our system. It has a definite economic impact on our choice of changing our flow measurement systems.

Improving present flow measurement. Since most pipelines are in a nonexpansion mode, if not a reduction in throughput, operating people are spending a large portion of their efforts in improving efficiencies of their systems. This includes much closer attention to the efficiencies of compressor or pump stations. We are not talking about large percentage reductions in efficiencies being significant cost savings.

As an example, some gas pipelines have as much as 80% of their total cost of operation coming from fuel costs, so even a 5% reduction relates to a reduction in operation cost of 4%. To study these efficiencies for real results requires very accurate flow measurement including individual measurement on each compressor or pump and engine. This is a case where the use of flow measurement to reduce costs is the driving factor rather than the custody transfer service that we generally worry about.

Fluid characteristics

A changing problem. For years we have had the luxury of handling fairly simple fluids with well-defined and/or accepted correction factors to reduce fluids at flowing conditions to fluids at base conditions. We now are being faced with pipelines handling "dense fluids" such as carbon dioxide and ethylene and liquids such as non-stabilized crudes, light hydrocarbons operating at their vapor pressures, and refrigerated LPG. In these cases, the choice of the basic meter is less of a concern than the proper preparation of the fluid prior to passing through the meter, and the proper factors for making corrections to base conditions.

Sometimes we are trapped by our previous practices established with well-behaving fluids and can find very large errors when these practices are applied to ill-defined fluids. In these cases the problems are not in the basic flow meter but in the flow meter system application of the correcting factors based on inadequate knowledge. The proper answer to these problems has not been worked out to give the accuracy of flow measurement we desire, but with the recognition of the problem we have been able to minimize some of the errors while continuing to seek new flow meters or methods of correcting for the fluid characteristics to solve the problem.

Standards and governmental effects on flow measurement

Outside influences on flow measurement. Industry practices, standards, and contracts, along with the ever increasing governmental requirements, must be considered to complete the picture of present day pipeline flow measurement. It is the belief of some of the bodies that measurement can be accomplished by legislation.

All orifice plates in some countries, for example, must be stamped by a government inspector, which makes them acceptable and correct. The manufacturer or user then can adopt the attitude that if the standards body or government finds it correct, it is not his worry and he really does not need to know much about flow measurement. This fundamentally has a serious flaw, since it removes the one item effective in obtaining accuracy—personal commitment and responsibility to accomplish the job. The manufacturers and users in countries where this is acceptable practice have passed the responsibility back to the government. They have reduced the caring as to whether or not their metering is accurate or properly installed and whether or not it maintains this accuracy in long-term service.

As we get further into the governmental agencies controlling our business, we have seen that the cost of doing the job is of secondary importance to meeting DOE, EPA, or OSHA standards, as an example. Unfortunately, the governmental bodies are looking at getting further in the measurement of fluids business with every passing day. Experience tells us that present economics of fluid measurement will be less of a controlling factor as this continues to occur.

Testing or proving flow meters

Present practices. The term "proving" in flow measurement means a throughput test and comes to us from the liquid measurement people who developed the mechanical displacement prover (now termed the "pipe prover") for proving turbine and positive displacement meters on crude oil and petroleum products. Prior to this development, we had critical flow and low flow provers used on PD meters on gas, but these had relatively limited use. On the other hand, "proving head meters" was never done. We substituted mechanical inspection of the primary devices and calibration of the secondary devices for the throughput test.

Proving is the checking of the throughput of a meter versus a standard determination of volume from another device. The purpose of all inspection and calibration or proving is to establish a meter's performance accuracy to some tolerance. This tolerance is dependent on the basic accuracy of the standard used plus the conditions of the test. The allowable tolerance is dependent on the value and quantity of the product handled. Because of this, some metering installations cannot justify an extra few dollars in cost. In those cases we continue to use inspection and calibration to determine the meter's accuracy.

Proving meters can be quite expensive in terms of permanently installed provers. The decision is based on "cost considerations" and is balanced by the degree of "accuracy needed" and operations and maintenance money available to be spent in obtaining and maintaining the "accuracy." Here, then, are several of the influences that affect our measurement testing and proving: present practice, standards, accuracy desired, and economics of the measurement job.

Improvement of present practices. Present practices will be continued in concept; however, equipment development has improved our tools such as air deadweight devices for differential pressure tests and precision test gauges for static pressure for gas meters. Liquid meters have improved equipment available for sampling and fluid characteristic testing.

The standards for flow measurement have had a great deal of study done recently to improve the accuracy and application range with such programs as the American Petroleum Institute work on the fluid characteristics, the development of the small provers for liquids, and the basic flow coefficient research on the orifice by the American Petroleum Institute and the Gas Processors Association. This work of upgrading standards to improve accuracy is continuing at a rapid pace. The accuracy requirements for all flow measurement is improved because of the high cost of the measured fluids, and we anticipate that this demand will increase.

Summary

The business of pipeline flow measurement has assumed a more important role in the overall pipeline operation picture because the economics of the job to be done requires: more accuracy; better knowledge of the fluids being handled; wider application of electronics to accomplish the basic flow measurement plus other uses of these data to improve operations; the adaptation to changing standards and governmental rules; and the use of new equipment and techniques to determine the accuracy of a flow meter. The people who are responsible for flow measurement have a challenge to respond to these changing influences to improve their company's flow measurement.

Liquid measurement orifice plate flange taps

This program, designed for the HP-41CV calculator, will calculate the bore required for an orifice plate for a given rate of flow, or the flow rate for a given bore size, when the following variables are entered into the program.

Initiate program by pressing A:

Enter:

Flow rate, lb/hr	Press R/S
Flowing temperature, °F	Press R/S
Flowing specific gravity, dimensionless	Press R/S
Differential pressure range, in. of water	Press R/S
Pipe inside diameter, in.	Press R/S
Absolute viscosity, centipoises	Press R/S

Program calculates beta ratio (dimensionless) and orifice bore (in.).

If an even-sized bore is required, press C and enter:

Required orifice bore, in.	Press R/S

Program calculates beta ratio (dimensionless) and new flow rate (lb/hr).

Notes

1. The E subroutine is available to reenter data and recalculate.
2. Viscosity correction is automatic.
3. If the orifice plate material is not 316SS or 304SS, see Table 1.

Sequence of calculations

1. Calculate the S value.

$$S = \frac{W_m}{D^2(2{,}831 + .0567T_f)\sqrt{G_f h_m}}$$

2. Calculate Reynolds number.

$$R_D = \frac{4.42\,W_m}{D_\mu}$$

Table 1
Table of equal thermal expansion temperatures, °F, for various orifice plate materials

Type 316 or Type 304	Alu-minum	Copper	Type 430	2% CRMO	5% CRMO	Bronze	Steel	Monel
	−298					−360		
	−230	−378				−273		
−325	−178	−266				−216		
−226	−131	−194				−163		−281
−152	−85	−132				−111		−191
−86	−41	−72				−60		−108
−24	+3	−15				−8		−33
+38	+47	+42	+24	+28	+27	+43	+31	+38
+101	+91	+99	+112	+108	+110	+95	+106	+104
+158			+202	+184	+191	+149	+181	+168
213			290	260	272	200	254	230
266			374	331	352	249	324	290
318			456	400	426	297	392	348
367			531	468	493	345	457	405
414			604	534	560	393	520	461
463			677	597	628	440	542	516
512			749	660	696		644	571
560			817	720	763		705	626
607			884	780	827		765	680
655			952	842	888		823	733
703			1020	899	948		879	785
750			1087	956	1009		934	835
796			1155	1011	1071		988	886
840			1223	1065	1133		1041	935

The program is written for 316SS and 304SS. If another material is used, move down the scale for the selected material, then across (left) to the equivalent 316SS/04SS temperature. Enter this value. For example, if the material selected for orifice plate was Monel, with a flow temperature of 405°F, enter 367°F. (The temperature is only entered in the program for orifice plate thermal expansion. It does not affect any other calculation.)

HP 41CV Data Sheet

```
************

ORIFICE LIQD
FLANGE   TAPS

************

ENTER FLOW  PPH
    100,000.0000    ***
ENTER TEMP DEG F
        150.0000    ***
ENTER SG
          0.9800    ***
ENTER DIFF RANGE INS H20
        100.0000    ***
ENTER PIPE ID INS
          4.0260    ***
ENTER VISCOSITY cP
          0.8000    ***
CALC   BETA
RATIO
          0.5810    ***
CALC ORIFICE
 BORE  INS
          2.3391    ***
************
                  2.3391
ENTER REQD  ORIFICE INS
          2.3750    ***
CALC   BETA
RATIO
          0.5899    ***
CALC   FLOW
PPH
    103,700.4948    ***
************
```

HP 41CV Register Loading

SIZE 080

```
R00=  0.000000
R01=  0.000000
R02=  0.000000
R03=  0.000000
R04=  0.000000
R05=  0.000000
R06=  0.000000
R07=  0.000000
R08=  0.000000
R09=  0.000000
R10=  0.000000
R11=  0.000000
R12=  0.000000
R13=  0.000000
R14=  0.000000
R15=  0.000000
R16=  0.000000
R17=  0.000000
R18=  0.000000
R19=  1,200.000000
R20=  3.328200
R21=  5.425000
R22=  1.196000
R23=  0.000010
R24=  0.518040
R25=  6.425000
R26=  0.056700
R27=  2,831.000000
R28=  0.000000
R29=  0.000000
R30=  72.450000
R31=  46.470000
R32=  57.484900
R33=  57.505100
R34=  57.520000
R35=  57.535460
R36=  57.616200
R37=  57.636400
R38=  0.000000
R39=  56.656600
R40=  71.676869
R41=  57.707273
R42=  0.000000
R43=  56.656600
R44=  56.484900
R45= "E LIQD"
R46= "FLANGE"
R47= "  TAPS"
R48= "FLOW  "
R49= "PPH   "
R50= "TEMP D"
R51= "EG F  "
R52= "SG    "
R53= "DIFF R"
R54= "ANGE I"
R55= "******"
R56= "CALC  "
R57= "ENTER "
R58=  44.000000
R59=  0.000000
R60= "NS H20"
R61= "PIPE I"
R62= "D INS "
R63= "VISCOS"
R64= "ITY cP"
R65= "BETA  "
R66= "RATIO "
R67= "RIFICE"
R68= " BORE "
R69= "INS   "
R70= "REQD  "
R71= "CALC O"
R72= "ORIFIC"
R73= "E INS "
R74=  0.000000
R75=  0.000000
R76=  0.000000
R77=  0.000000
R78=  0.000000
R79=  0.000000
```

HP 41CV Listing

SUBROUTINES

```
01♦LBL "FT LIQ"
02 GTO A
```

PRINT LINE OF STARS

```
03♦LBL a
04 FC? 55
05 RTN
06 55.55
07 ADV
08 SF 12
09 XEQ b
10 ADV
11 RTN
```

PRINT LINES

```
12♦LBL d
13 SF 12
14 1
15 ST+ 58
16 RCL IND 58
17♦LBL b
18 STO 59
19 CLA
20♦LBL c
21 ARCL IND 59
22 RCL 59
23 FRC
24 100
25 *
26 STO 59
27 X≠0?
28 GTO c
29 FC? 55
30 GTO 01
31 PRA
32 CF 12
33 RTN
34♦LBL 01
35 CF 12
36 AVIEW
37 RTN
```

PRINT & STORE VARIABLES

```
38♦LBL e
39 1
40 ST+ 58
41 RCL IND 58
42 X=0?
43 GTO d
44 XEQ b
45 1
46 ST+ 01
47 RCL IND 01
48 TONE 9
49 FS? 00
50 GTO 01
51 STOP
```

```
52 STO IND 01
53 .5
54 STO 12
55 RCL IND 01
56♦LBL 01
57 XEQ J
58 GTO e
```

RE-ENTER VARIABLES

```
59♦LBL "REENTER"
60 ADV
61 1
62 +
63 STO 01
64 RCL IND 01
65 XEQ J
66 STOP
67 XEQ J
68 STO IND 01
69 ADV
70 SF 00
71 SF 02
72 STOP
73 GTO 02
```

BYPASS PRINTER

```
74♦LBL J
75 FC? 55
76 GTO 01
77 PRX
78 RTN
79♦LBL 01
80 VIEW X
81 RTN
```

CALC S

```
82♦LBL G
83 RCL 12
84 PSE
85 X↑2
86 .598
87 *
88 RCL 12
89 3
90 Y↑X
91 .01
92 *
93 +
94 RCL 12
95 RCL 25
96 Y↑X
97 RCL 24
98 *
99 +
100 RTN
```

ITERATE β (ENDS AT 218)

```
101♦LBL D
102 RCL 12
103 RCL 22
104 *
105 RCL 12
106 X↑2
107 .03
108 *
109 +
110 RCL 12
111 RCL 21
112 Y↑X
113 RCL 20
114 *
115 +
116 1/X
117 XEQ G
118 RCL 11
119 RCL 10
120 /
121 -
122 *
123 ST- 12
124 ABS
125 RCL 23
126 X<>Y
127 X>Y?
128 GTO D
```

CALC F_c WHEN $R_D \leq 700\beta^{1.3}$

```
129 RCL 12
130 1.3
131 Y↑X
132 700
133 *
134 STO 18
135 STO 15
136 2
137 /
138 STO 14
139 RCL 09
140 /
141 1/X
142 SQRT
143 90
144 *
145 SIN
146 RCL 12
147 .15
148 Y↑X
149 /
150 STO 17
151 RCL 09
152 RCL 14
153 X>Y?
154 GTO I
155 RCL 17
156 ABS
157 SQRT
158 STO 17
```

CALC F_c WHEN $R_D > 700\beta^{1.3}$

```
159 RCL 09
160 RCL 18
161 X>Y?
162 GTO I
163 1
164 STO 17
165 RCL 09
166 STO 18
167 STO 15
168 RCL 09
169 RCL 19
170 X>Y?
171 GTO I
172 STO 18
```

CALC F_c WHEN $R_D = 1200$

```
173♦LBL I
174 0
175 RCL 12
176 .2
177 -
178 X>Y?
179 0
180 X↑2
181 RCL 12
182 3.9
183 *
184 E↑X
185 8.7
186 *
187 RCL 15
188 -.89
189 Y↑X
190 *
191 +
192 STO 28
193 RCL 14
194 RCL 18
195 /
196 RCL 12
197 SQRT
198 SQRT
199 .12
200 -
201 Y↑X
202 *
203 1
204 RCL 18
205 RCL 19
206 /
207 -
208 SQRT
209 *
210 CHS
211 RCL 28
212 +
213 RCL 17
214 *
215 .997
216 +
217 STO 10
218 RTN
```

MAIN PROGRAM ENTER DATA

```
219♦LBL A
220 CF 02
221 1
222 STO 01
223 STO 10
224 29
225 STO 58
226 XEQ a
227 XEQ d
228 XEQ d
229 XEQ a
230 XEQ e
231 CF 00
```

CALC S

```
232♦LBL 02
233 39
234 STO 58
235 RCL 02
236 RCL 06
237 X↑2
238 /
239 RCL 27
240 RCL 26
241 RCL 03
242 *
243 +
244 /
245 RCL 04
246 RCL 05
247 *
248 SQRT
249 /
250 STO 11
```

CALC R_D

```
251 4.42
252 RCL 02
253 *
254 RCL 06
255 /
256 RCL 07
257 /
258 STO 09
```

CALC β

```
259 XEQ D
260 XEQ D
261 XEQ D
262 RCL 12
263 XEQ J
```

CALC d

```
264 XEQ d
265 RCL 06
266 RCL 12
267 *
268 STO 29
269 GTO 01
```

ENTER REQD BORI

```
270♦LBL "EVEN SZ"
271 7
272 STO 01
273 40
274 STO 58
275 XEQ e
```

CALC β

```
276 RCL 08
277 RCL 06
278 /
279 STO 12
280 XEQ J
281 XEQ G
282 STO 16
```

CALC FLOW

```
283 XEQ d
284 RCL 02
285 RCL 11
286 /
287 RCL 16
288 *
289 STO 29
290♦LBL 01
291 XEQ J
292 FS? 00
293 STOP
294 XEQ a
295 RCL 29
296 VIEW X
297 BEEP
298 ADV
299 STOP
300 .END.
```

3. Iterate to find uncorrected beta ratio.

$$S = 0.598\beta^2 + .01\beta^3 + .518\beta^{6.425}$$
$$fx = 0.598\beta^2 + .01\beta^3 + .518\beta^{6.425} - S$$
$$f^1x = 1.196\beta + .03\beta^2 + 3.3282\beta^{5.425}$$

Start with $\beta = 0.5$ and calculate fx/f^1x. If greater than 0.0001, subtract fx/f^1x from current value of β and repeat until the error is less than 0.00001.

4. Calculate Reynolds number correction factor. If $R_D < 350\,\beta^{1.3}$:

$$F_c = \left[A - \left(A\left(.5^{(\beta^{.25} - .12)}\right)\sqrt{1 - \frac{700\beta^{1.3}}{1{,}200}}\right)\right] \left[\frac{\mathrm{SIN}\left(90\sqrt{\frac{R_D}{350\beta^{1.3}}}\right)}{\beta^{.15}}\right] + .997$$

where $A = (\beta - .2)^2 + (8.7e^{3.9\beta})(700\beta^{1.3})^{-.89}$

when $\beta - 2 > 0$

$(\beta - .2)^2 = 0$

If $350^{1.3} < R_D < 700\beta^{1.3}$:

$$F_c = \left[A - \left(A\left(.5^{(\beta^{.25} - .12)}\right)\sqrt{\frac{1 - 700\beta^{1.3}}{1{,}200}}\right)\right] \left[\sqrt{\frac{\mathrm{SIN}\left(90\sqrt{\frac{R_D}{350\beta^{1.3}}}\right)}{\beta^{.15}}}\right] + .997$$

If $700\beta^{1.3} < R_D < 1{,}200$:

$$F_c = \left[A - \left(A\frac{(350\beta^{1.3})^{(\beta^{.25} - .12)}}{R_D}\sqrt{1 - \frac{R_D}{1{,}200}}\right)\right] + .997$$

When $R_D > 1{,}200$:

$F_c = (A + .997)$

5. Divide original S by F_c for viscosity correction.
6. Repeat step 3 for new β for corrected S.
7. Repeat step 4 for final F_c for corrected β.
8. Repeat step 5 for final S.
9. Repeat step 3 for final β.
10. Calculate orifice bore.

$d = \beta D$ in.

11. Calculate flow rate for an even-sized orifice plate.

$$\beta = \frac{d}{D}$$

$$S = .598\beta^2 + .01\beta^3 + .518\beta^{6.425}$$

$$\text{new flow} = \frac{\text{new S}}{\text{original S}} \times \text{original flow, lb/hr}$$

Nomenclature

A	Partial calculation of F_c dimensionless
D	Inside diameter of pipe, in.
d	Diameter of orifice bore, in.
F_c	Reynolds number correction factor, dimensionless
G_f	Specific gravity of liquid at flowing temperature, dimensionless
h_m	Maximum differential range, in. of water
R_D	Reynolds number, dimensionless
S	Beta ratio index, dimensionless
T_f	Flowing temperature, °F
W_m	Maximum rate of flow, lb/hr
μ	Absolute viscosity, centipoises
β	Beta ratio d/D, dimensionless

Source

Instrument Engineering Programs, Gulf Publishing Company, Houston.

Reference

12. Spink, L. K., *The Principles and Practice of Flowmeter Engineering*, The Foxboro Company.

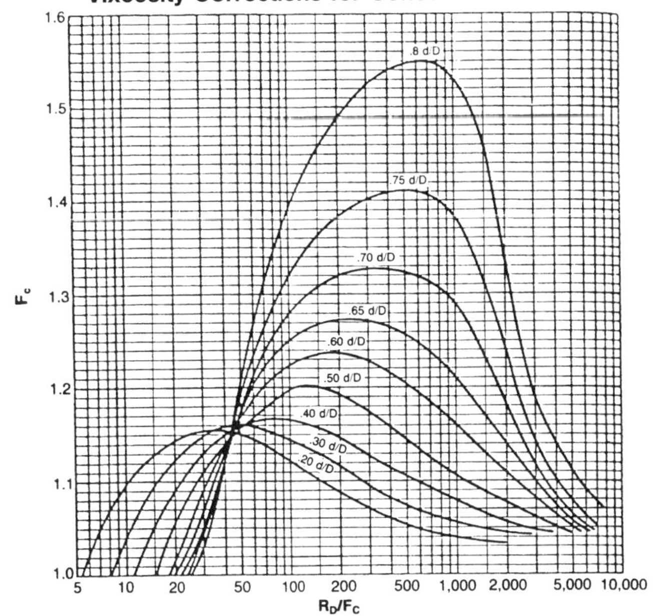

Mass measurement light hydrocarbons

Streams that transport mixed natural gas liquids require the use of mass measurement to accurately account for the volume of the components which make up the mix. When natural gas liquid components are mixed, non-ideal mixing occurs. Compressibility factors are difficult to determine for mixes. Mass measurement eliminates the effects of non-ideal mixing and the need for compressibility factors.

Certain precautions must be noted in using the mass measurement method. The density and volume must be measured at the same flowing conditions. An accurate composite sample of the flowing stream must be collected and analyzed with care. On systems where there are a number of input points that will be compared to an output point, it may be desirable to have all composite samples analyzed at the same laboratory.

The following data are needed for the calculation of quantities of natural gas liquid components:

- Metered volume at operating conditions
- Meter factor, if applicable
- Density at flowing conditions
- Composite sample and sample analysis for metered volume
- Component lb mass for each component in the stream
- Density—lb mass per U.S. barrel for each component in the stream

Sample calculation

Metered volume = 25,000 barrels
Meter factor = 1.0015
Density = 0.480

Composite sample analysis

Component	mol %
Carbon dioxide	0.08
Methane	1.90
Ethane	41.30
Propane	38.00
Isobutane	5.00
Normal butane	9.30
Isopentane	1.50
Normal pentane	1.10
Hexanes+	1.82
Total	100.00

See Table 1 for the details of the calculation.

Table 1
Mass measurement—light hydrocarbons

Component	Mol % 1	Mol Wt 2	Mol % × Mol Wt 3	Wt Frac. of Comp. 4	# Mass Comp. 5	Density #/bbl 6	U.S. bbl 7
CO_2	0.08	44.010	3.5208	0.000852	3,588	286.4358	13
C_1	1.90	16.043	30.4817	0.007381	31,063	105.0000	296
C_2	41.30	30.070	1,241.891	0.300739	1,265,556	124.7232	10,147
C_3	38.00	44.097	1,675.686	0.405787	1,707,617	177.5256	9,619
IC_4	5.00	58.123	290.615	0.070375	296,153	197.0934	1,503
NC_4	9.30	58.123	540.5439	0.130899	550,844	204.4980	2,694
IC_5	1.50	72.150	108.225	0.026208	110,287	218.7444	504
NC_5	1.10	72.150	79.365	0.019219	80,877	220.9914	366
C_6^+	1.82	87.436	159.1335	0.038536	162,166	244.2300	664
	100.00		4,129.461		4,208,151		25,805

Total lb mass = 25,000 × 1.0015 × 0.480 × 8.337 × 42 = 4,208,151
Column 1—From Sample Analysis
Column 2—From Table of Hydrocarbon Properties (Figure 1)
Column 3—Column 1 × Column 2
Column 4—Mol % × Mol Wt divided by sum of Column 3
Column 5—Column 4 × total lb mass
Column 6—From Table of Hydrocarbon Properties (Figure 1)
Column 7—Column 5 divided by Column 6

Reference: GPA Standard 8173-83, Method for Converting Natural Gas Liquids and Vapors to Equivalent Liquid Volumes. GPA Standard 8182-82, Tentative Standard for the Mass Measurement of Natural Gas Liquids.

Pipeline measurement of supercritical carbon dioxide

G. W. Marsden, Shell Pipeline Corporation and **D. G. Wolter**, Shell Oil Company

Introduction

The Shell Companies have been involved in carbon dioxide (CO_2) systems used for enhanced oil recovery for a number of years. Pilot programs existed in the late 1970s using recovered combustion generated carbon dioxide. In 1981 a plan was in place for a large-scale commercial pipeline transportation system connecting carbon–dioxide producing fields in southwestern Colorado and the oil fields of the Permian Basin in West Texas (see Figure 1). The Cortez Pipeline Company was set up as the ownership company, and Shell Pipeline Corporation was established as the operator.

The 810-km (502 mi) long, 760-mm (30-in.) diameter Cortez pipeline was designed to operate under supercritical conditions of 9,000 to 19,000 kPa (1,300 to 2,700 psig), 286 to 322°K (55 to 120°F), and with the nominal stream composition shown in Table 1. Design flow rates were 4.6 to 13.83 kmol/s (330 to 1,000 MMscfd). Line fill was completed in May 1984, and the system has been in continuous operation since then. There are 10 receipt meters and 9 delivery meters currently in operation, each with a custody transfer measurement facility.

Table 1
Stream composition

Major Components	m^3/m^3	% Vol
CO_2 minimum	.95	95
Nitrogen, maximum	.04	4
Methane	.01–.05	1–5
Contaminants		
Hydrogen sulfide	.00002	0.002
Water	0.48 g per m^3 at standard conditions	30 lb/MMscf

Definitions

"Carbon dioxide," as referred to here and as used in the pipeline industry, is a CO_2-rich mixture normally having nitrogen and methane as principal volumetric contaminants. A minimum quality of 95% carbon dioxide purity is used by many pipeline carriers, while Cortez experience thus far is circa 98% CO_2 purity. The measured quantity is the total of all mixture constituents.

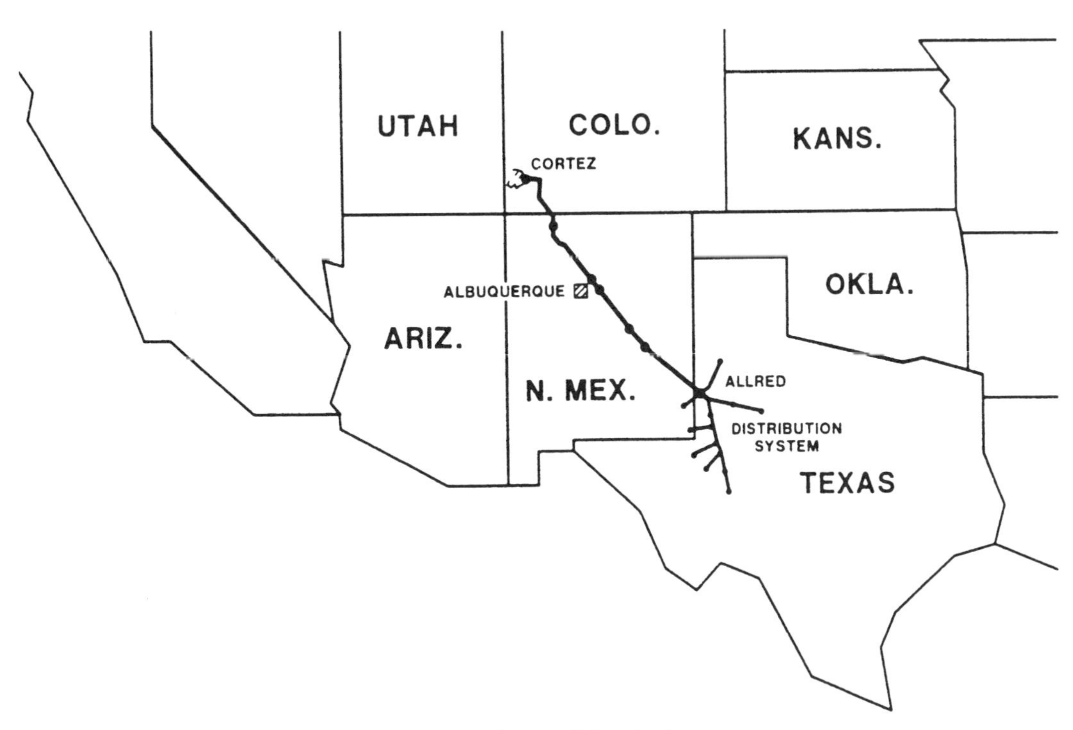

Figure 1. Cortez CO_2 pipeline.

Another term used by liquid pipeline operators is "custody transfer." Liquid pipeline operators typically do not own the fluid being transported but instead accept custody at pipeline origin (receipt) and transfer that custody back to the owner or other party at the pipeline destination (delivery).

Pipelines thus have custody transfer measurement facilities, which are composed of generally accepted, commercially available, and accurate measurement equipment. Additionally, the operation of this equipment is accomplished in a uniform manner acceptable to all parties to the transaction: buyers, transporters, and sellers. The API Manual of Petroleum Measurement Standards is widely used in the design and operation of these facilities.

These brief definitions give some insight into the selection of custody transfer measurement methods and equipment. Let us return now to the CO_2-rich fluid we want to address.

Physical properties

CO_2, in its pure form, has a critical point of 304 K (88°F) and 7,415 kPa (1,071 psia). Within practical limits, the characteristics of pipeline-quality CO_2-rich mixtures will approximate those of pure CO_2. Refer to the pressure enthalpy diagram for CO_2 (Figure 2), which illustrates the area of normal measurement for Cortez Pipeline.

The most economical pipeline operation is in the supercritical single-phase area. Fortunately, this operating area also coincides with our preferred measurement envelope. There is no pressure letdown for the delivery; in practice, the receiving party prefers the fluid with a pressure level similar to the pipeline operating pressure. This leads to measurement systems where pressure conditions are quite similar at both the inlet to and the outlet from the pipeline.

Product temperature does vary from the inlet measurement facility, which is downstream of compression equipment, to the delivery stations where the CO_2 stream approaches the ambient ground temperature. Future increases in throughput will require additional pumping/compression equipment, which will tend to create measurement operating temperatures that are similar at both pipeline inlet and outlet.

The upshot of these temperature and pressure conditions is the measurement envelope shown in Figure 2. This envelope is considerably more confined during normal measurement operations than is the pipeline operating envelope. The operating envelope includes the changes in hydraulic conditions that occur along the route of the pipeline.

Looking again at the P-E diagram (Figure 2), observe that the density of the CO_2-rich stream is in the range of 640 to 960 kg/m^3 (40 to 60 lb/ft^3). Additionally, note that the prime determinant of a change in density will be a change in temperature. This sensitivity of change in density with temperature is of concern in the design of the measurement facilities, and it has led to special attention being given to temperature loss/gain between elements of the measuring system. There is an approximate 0.9% volume change for each °C (0.5%/°F).

The change of density with pressure change in the measurement envelope is considerably less prominent, being on the order of one tenth as severe as temperature.

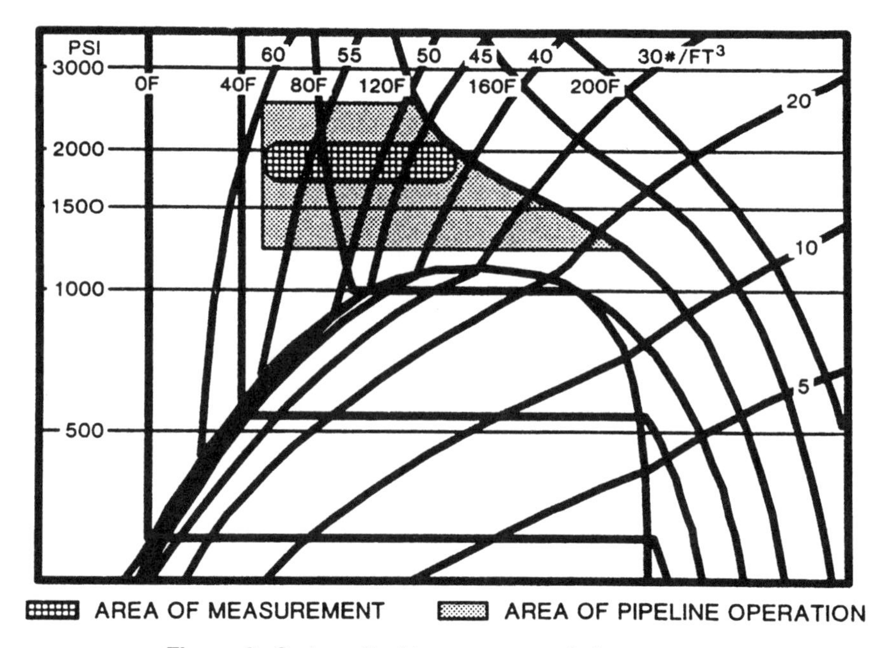

Figure 2. Carbon dioxide pressure enthalpy diagram.

Measurement methods in use today

The commercial planning and development of the Cortez carbon dioxide project preceded the engineering activity, and the supply and sales contract negotiations were well advanced before custody transfer flow measurement methods were investigated. The resulting contracts, based on existing natural gas pipeline systems, dictated the use of MMscfd volumetric units for gas phase equivalent volume of carbon dioxide.

Custody transfer measurement of carbon dioxide in large volumes at high pressures presented significant technological hurdles in the methods, performance, and materials areas. First, the probable stream compositions and the variations of the source field carbon dioxide were uncertain. This uncertainty of composition, and thus the physical properties, led to the selection of mass measurement techniques that would provide flexibility and operability when product conditions changed. Shell's decade-long experience with mass measurement of variable mix hydrocarbon streams and with supercritical streams suggested mass measurement as one of the strongest contenders. Mass measurement, however, would not directly provide the volumetric quantities that had been established in the preceding business negotiations.

The selection of meter systems to meet the above requirements resulted in two systems that are used today in the carbon dioxide transportation industry. The orifice meter, common in the natural gas industry, and the turbine meter, similar to those used in liquid measurement, were the dominant choices. As with most other choices, there are advantages and disadvantages (see Table 2) to each method. Both are operated as mass meter systems with subsequent conversion to volumetric terms for contract compliance and accounting requirements.

Fortunately, Shell Pipeline had several years of operating experience with mass measurement of petrochemical products, i.e., ethane and ethylene, as background in the orifice measurement system. The orifice meter was selected based on analysis of the factors shown in Table 2. The units conversion process, however, required additional facilities and methods for routinely and rapidly determining the composition of the streams, the calculation of molecular weight for the equivalent gas, and the conversion to volume at the selected reference pressure and temperature conditions. The resulting methods and facilities in service today are discussed below.

Table 2
Carbon dioxide pipeline metering options

	Orifice Meter	Turbine Meter
Advantages	Industry acceptance	Accuracy ±0.5%
	Proven equipment and operations	Rangeability ±10:1
	Least sensitive to density determination accuracy	
Disadvantages	Limited rangeability	Performance unknown with no experience in CO_2
	Accuracy ±1%	No experience with proving in CO_2
	Costly installation	

Orifice measurement practices

The measurement system that provided basic simplicity, reliability, wide acceptance, and the capability of handling variable mix streams without breaking new frontiers in measurement methods was resolved to be the orifice flow measuring element with online density meter and microprocessor flow computer. This basic assembly of three proven devices provides real time-mass flow output. The orifice is the key to the success of the measurement system.

The orifice plate has a 50-year record of widespread use in the measurement of all manner of fluids. It is a static device generally inert to the measured fluid conditions, and calibration consists of simple dimensional measurement and conformance to specified physical tolerances. This, along with the periodic calibration of the associated differential pressure transmitter, completes the verification of this part of the measurement system. The orifice meter is one of the few measurement devices accepted for custody transfer that does not require in-service proving.

The second element, the density meter, has been selected to be an external unit that is easily isolated from the flowing stream for calibration, inspection, and maintenance. The relative insensitivity of CO_2 density to small changes in pressure in the primary flow to the orifice meters permits locating the density meter upstream of the meter manifold, thereby serving several meters. All installations are insulated to assure that pipeline density is maintained in the density meter.

The microprocessor flow computer, or third element of the system, is essential to achieving the advantages of integrated mass flow. This advantage comes from the ability of the computer to make computations in essentially "real time." How this advantage comes about is seen upon examination of the flow equation.

Orifice flow equation

The API Manual of Petroleum Measurement Standards, Chapter 14, Section 3 (also API 2530, AGA Report No. 3,

ANSI/API 2530-1985, and GPA 8185-85) provides the widely accepted formulation for mass flow. Mass flow is considered by this source to be the fundamental orifice meter flow equation.

Equation (38) of this publication, "Equation for Mass Flow Rate when Density of Fluid Flowing is Known," is shown in Figure 3.

The terms used in Figure 3 are:

q_m is the mass flow rate.

K and Y, the flow coefficient and the expansion factor, are the empirical elements of the orifice equation, based upon the work reported by Buckingham and Bean in 1934.

The flow coefficient, K, is in part dependent upon the viscosity of the fluid. Pipeline CO_2 viscosity ranges from approximately 0.05 to 0.08 mPa's (0.05 to 0.08 centipoise) in the measurement envelope.

The expansion factor, Y, is a function of the ratio of the orifice differential pressure to static pressure and to the specific heat ratio of the CO_2. The specific heat ratio ranges from 2.9 to 4.0 for CO_2 in the measurement envelope.

ρ_{fl} is the density output of the density meter.

ΔP is the differential pressure across the orifice plate.

Two correction factors, F_a and F_{pwl}, from Appendix E of Chapter 14, Section 3 are also utilized. F_a provides for the thermal expansion of the orifice plate, and F_{pwl} provides for local gravitational correction for deadweight testers.

Note that errors in density are taken at half their real value by integrating the density value with the orifice differential before taking the square root.

Figure 4 illustrates the accumulation of individual tolerances to arrive at the mass measurement system uncertainty tolerance. Individual element tolerances are modified by an

14.3.5.4.3 "Equation for Mass Flow Rate when Density of Fluid Flowing is Known"

$$q_m = \frac{\pi}{4}[2g_c]^{0.5}KY_1d^2[\rho_{fl}\Delta P]^{0.5}$$

Correction Factors in Accordance with Appendix E

E.3: Orifice Thermal Expansion Factor

$$F_a = 1 + [0.0000185(t_f - t_{meas})]$$

E.7: Local Gravitational Correction for Deadweight Tester

$$F_{pwl} = \left[\frac{g}{g_o}\right]^{0.5}$$

Figure 3. CO_2-rich mixture orifice measurement equation. From the *API Manual of Petroleum Measurement Standards*, 2nd Edition, September 1985.

VARIABLE	TOLERANCE %	EFFECT	SQ OF TOL & EFF	SUM OF SQUARES	SQ. RT. OF SUM %
Basic Orifice, F(B)	0.50	1.00	.2500	.2500	.5000
Diffntl Press, H	0.50	0.50	.0625	.3125	.5590
Density	1.00	0.50	.2500	.5625	.7500
Processor Error	0.01	1.00	.0001	.5626	.7500
Reynolds, F(R)	0.03	1.00	.0009	.5635	.7506
Expansion, Y	0.02	1.00	.0004	.5639	.7509
Orifice, F(A)	0.05	1.00	.0025	.5664	.7523

Individual factors with tolerance as above will provide system measurement with tolerance of (plus or minus) 0.75%

Figure 4. Orifice meter system tolerance (root-sum-square method).

effect factor, then the square root of the sum of the squares of these products is computed. The resulting overall measurement uncertainty for the Cortez mass measurement system is ±0.75%.

Pipeline operation

Full Cortez pipeline operation is accomplished using the mass flow output of the measurement system. This system also forms the basis for all custody transfer measurements in volumetric units.

It is necessary to introduce a mass-to-volume conversion step to satisfy the volumetric contracts. This conversion is accomplished using an analysis of the composite stream with a gas-liquid chromatograph maintained at each measurement area. The CO_2, N_2, and CH_4 composition of the CO_2-rich stream is determined for each measurement ticket period. An off-line equation of state computer program then calculates the appropriate standard volumetric units. A custody transfer measurement ticket (see Figure 5) is produced to document all pertinent data, including mass units, three-component composition, and the standard volumetric units.

The equation of state computer program utilized by Cortez Pipeline for this conversion was developed specifically for CO_2-rich mixtures. It is based upon the IUPAC (International Union of Pure and Applied Chemists) publication, "CO_2 International Thermodynamic Tables of the Fluid State," Volume III, compiled by Angus, Armstrong, and DeReuch. The method of extended corresponding states is utilized for the stated properties of the CO_2-rich mixture.

Turbine metering

The authors do not have firsthand experience or knowledge with turbine metering of carbon dioxide and therefore will not delve into the details of the application. Turbine metering was considered in 1981 as one of the two major alternatives and rejected for a variety of reasons. Interestingly, other companies did select turbine metering, and experience is now being accumulated on large-scale carbon dioxide pipelines.

Turbine meters were not selected for Cortez or subsequent Shell systems, because the experience with mechanical displacement provers at such high pressure with compressible fluids was non-existent. Likewise, the application of liquid or gas turbine meters in supercritical carbon dioxide service was a complete unknown.

The volumetric accuracy of turbine meters was attractive and no doubt led others to select them. However, the direct influence of density in the mass flow calculation is a disadvantage (as compared to orifice metering where density enters the equation as the square root of density and effectively halves the sensitivity of mass flow accuracy to density determination errors). (See Figure 6.)

The proving of turbine meters was a concern and has been a significant commissioning problem on systems utilizing turbine meters. High-pressure provers in the 900 to 1,500 ANSI flange ratings were scarce and expensive. The seals, design, and materials were considered potential problems. Significant progress has been made by those utilizing the equipment. In the long term, turbine meters and provers will likely be more widely applied as seal systems are

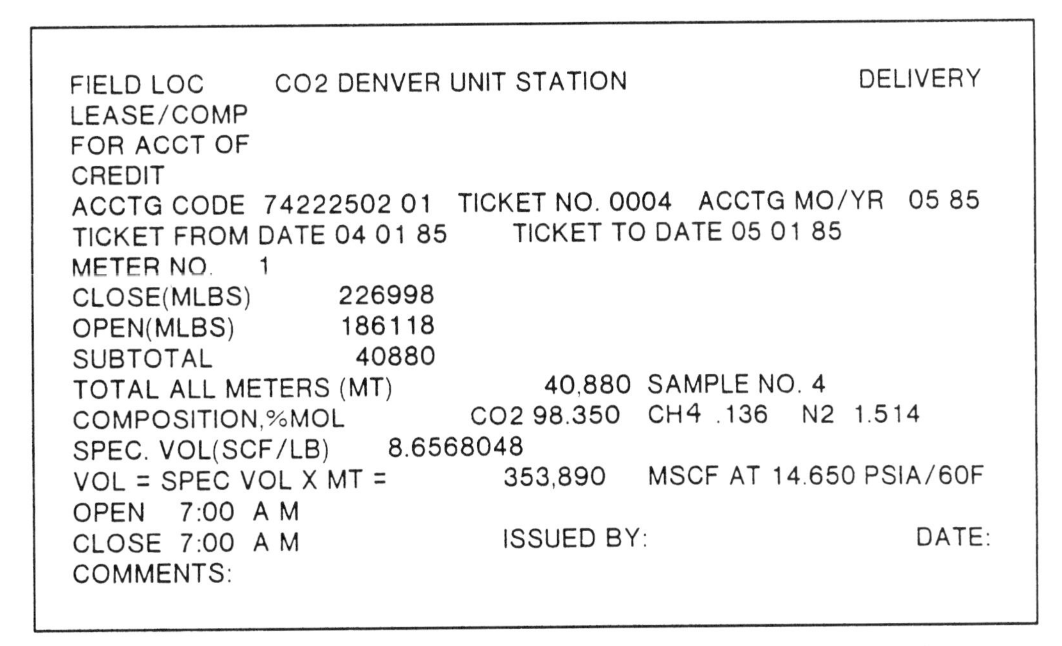
FIELD LOC CO2 DENVER UNIT STATION DELIVERY
LEASE/COMP
FOR ACCT OF
CREDIT
ACCTG CODE 74222502 01 TICKET NO. 0004 ACCTG MO/YR 05 85
TICKET FROM DATE 04 01 85 TICKET TO DATE 05 01 85
METER NO. 1
CLOSE(MLBS) 226998
OPEN(MLBS) 186118
SUBTOTAL 40880
TOTAL ALL METERS (MT) 40,880 SAMPLE NO. 4
COMPOSITION,%MOL CO2 98.350 CH4 .136 N2 1.514
SPEC. VOL(SCF/LB) 8.6568048
VOL = SPEC VOL X MT = 353,890 MSCF AT 14.650 PSIA/60F
OPEN 7:00 A M
CLOSE 7:00 A M ISSUED BY: DATE:
COMMENTS:

Figure 5. Cortez Pipeline Company. CO_2 meter ticket.

```
FOR ORIFICE METERING:

MASS = (CONSTANT) (DENSITY * DIFFERENTIAL) 0.5

FOR TURBINE OR VORTEX METERING:

MASS = (VOLUME) (DENSITY)
```

Figure 6. Mass flow calculation for orifice metering.

demonstrated to be reliable and as the other operational aspects are demonstrated to be manageable.

Potential measurement methods

Two additional meter types appear attractive and are coming into use in carbon dioxide service for various applications. Cortez Pipeline now utilizes the vortex shedding meter for line integrity/leak detection purposes, and the coriolis meter is attractive due to its simplicity and apparent accuracy.

The vortex shedding meter offers advantages in higher rangeability, requiring fewer meter runs than orifice meters, and thus lower installation costs. We believe the rangeability to be between 10 and 20 to 1 and the off-the-shelf accuracy to be 1% or better. Operating personnel have made online comparisons with adjacent, though several miles distant, orifice metering systems, and observe agreement within less than 1% for extended periods. No detailed studies have been made, however.

Currently, no custody transfer applications of the vortex shedding meter are known, although consideration is becoming more common. Application will first occur on low-volume service between agreeable parties or at the opportunity that assures a proving system will be available. (Proving via master meter and master prover techniques is one possibility that has been tested on other similar products.) Once the proving methods are applied, it is expected that accuracies better than 0.5% mass basis would be achievable.

Application of the coriolis meter lags the vortex shedding meter since it is a newer development. The expected accuracy of better than 1% and rangeability of 10 to 1 are attractive. The coriolis meter also provides direct mass measurement, which removes the need for a separate density determination and should reduce both the installed and operating costs significantly. The meter currently lacks widespread industry acceptance for custody transfer measurement applications, and the high-pressure operation of carbon dioxide systems currently limits selection to the smaller meter sizes. No custody transfer metering applications are known at this time.

Future application of the coriolis meter in carbon dioxide service will be much the same as described earlier for the vortex shedding meter. Industry experience, acceptance, application of proving methods, and availability in larger sizes with higher pressure ratings will open the routes for increased use.

The future

Carbon dioxide measurement technology is advancing rapidly, and the major difficulties in accuracy and operation are now identified. One accuracy enhancement that is progressing is the proprietary development of a flow computer with an accurate equation of state for carbon dioxide density determination from on-line/real time pressure and temperature inputs and keypad, or possibly analog, input of stream composition.

The Shell equation of state is a proprietary unit at this time and is being used in its mainframe form for conversion of mass to volumetric units on Cortez Pipeline. Industry acceptance and verification in field practice has been excellent. The equation of state is a modification of the IUPAC equation for pure carbon dioxide and incorporates the effects of nitrogen and methane impurities over a range of 80% to 100% carbon dioxide. The equation of state is considered to be accurate to ±0.2% for density determination in the supercritical regions. The program is being applied to a flow computer with density update calculations every 1 to 2 minutes. Pressure and temperature are real-time analog inputs, while composition will be manually input via the keypad on a periodic basis dependent upon compositional stability of the fluid at each location. The planned operation region is shown in Table 3.

It is recognized that the National Bureau of Standards and others are pursuing similar equation-of-state expressions. Early investigation, however, determined that an industry-accepted equation of state was in the distant future and the earliest implementation could be achieved on a proprietary

Table 3
Region of operation for CO_2 equation of state flow computer

Fluid	Single Phase CO_2		
		m^3/m^3	Vol %
Composition	CO_2	0.8–1.0	80–100
	Methane	0–0.02	0–2
	Nitrogen	0–0.20	0–20
Pressure	3,400–20,500 kPa absolute (500–3,000 psia)		
Temperature	273.16–333.16 K (32–140°F)		

development with industry acceptance to follow. This decision was also supported from the view that numerous inhouse uses could be made of the computers where outside acceptance was not required. These applications included process operations at the carbon dioxide production, separation, and reinjection plants where density meter operation is a considerable burden and optimized accuracy is not justified. Likewise, the computer is attractive for comparison against critical density meters and for line integrity/leak detection flow measurement facilities. We are hoping to implement some of these applications in 1986.

In addition to the equation of state, it is our intent to remain flexible and consider the use of the potential meter types discussed earlier. The meter manifolds on the Cortez system have been designed with flexibility for other meter types and proving operations if these become sufficiently attractive.

The technology of carbon dioxide pipeline measurement is gaining uniformity via conferences such as this and joint industry development projects such as the equation-of-state work at NBS. In the commercial area, it is desirable that similar efforts be directed toward the use of mass measurement for operation and custody transfer purposes and to do away with the processes necessary for conversion to volumetric units. Mass units are obtained with simpler systems, at a lower cost, and more accurately. We should all work toward converting the contracts and agreements to mass units.

Source

Reprinted from *Oil & Gas Journal*, November 3, 1986 by permission and courtesy of Perm Well Publishing Co.; G. W. Marsden, Shell Pipeline Corporation; and D. G. Wolter, Shell Oil Company.

GAS MEASUREMENT

Master meter proving orifice meters in dense phase ethylene

James E. Gallagher, Shell Oil Company, Houston

Abstract

Shell Pipeline Corporation, cognizant of the importance of highly accurate orifice measurement for polymer-grade ethylene, embarked on a development program to improve the overall uncertainty of orifice metering. The program's goal was to successfully apply master meter proving techniques to dense phase ethylene orifice meter facilities operated on a mass basis.

This paper presents the techniques and results of mass proving concentric, square-edged, flange-tapped orifice meters. All data were obtained on commercially operated metering facilities that are in compliance with ANSI 2530 requirements.

Results indicate an overall orifice meter uncertainty tolerance of approximately ±0.31% is attainable when orifice meters are mass proved in situ using master meter proving techniques.

Goal of development program

During the past decade, increasing commercial activity for ethylene demanded improvement of measurement technology. Significant measurement uncertainties were being experienced on ethylene pipeline systems. In response, a development program to improve the overall uncertainty of orifice metering was initiated by Shell Pipeline Corporation.

The program's goal was to develop an economical method for proving ethylene orifice meters under actual operating conditions. Master meter proving on a mass basis was selected as the only viable approach.

Pipeline transportation and measurement systems

An ethylene distribution system consists of a pipeline network and salt dome storage facility linking producers and consumers. Since producers and consumers are not equipped with on-site storage facilities, the pipeline is designed with maximum flexibility to satisfy the continually changing demands on the operations.

Shell's ethylene systems are operated in the dense phase fluid region due to lower transportation costs. The ethylene meter stations operate in two regions depicted in Figure 1—the dense-phase fluid and single-phase gas regions.

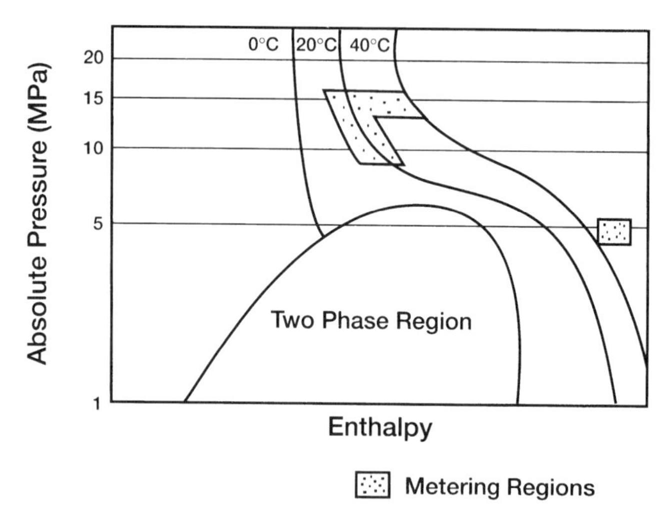

Figure 1. Ethylene pressure-enthalpy diagram.

The quantity of ethylene measured by concentric, square-edged, flange-tapped orifice meters, shown in Figure 3, is calculated through the combined use of the mass orifice equation and API's ethylene density equation. In the United States, the applicable standards are API, Chapter 14, Section 3—Orifice Metering of Natural Gas and Other Related Hydrocarbon Fluids, Second Edition, September 1985 (ANSI 2530), and API Chapter 11.3.2.1—Ethylene Density.

In Shell, the concentric square-edged flange-tapped orifice predominates as the primary element. Most installations are equipped with a dual chamber orifice fitting to allow retrieval of the orifice plate under pressure. For the most part, the orifice plates are sized to maintain a 0.3 to 0.6 beta ratio. Solid state electronic flow computers and transmitters are installed to maximize metering accuracy. Parallel meter runs and stacked "dp" transmitters maximize rangeability while maintaining a high degree of accuracy.

Historically, the industry has presumed that compliance with the reference standard piping conditions infers accuracy. No consideration was given to deviation from the empirical equation's criteria upon initial installation or further deviation over the life of the facility. The major

Velocity Profile
- Obstructions, Deposits, Etc.
- Piping Configuration
- Pre & Post Run Lengths
- Pipe Condition
  - Wall Roughness
  - Roundness
- Plate Condition
  - Concentric
  - Edge Sharpness
  - Nicks or Burrs
  - Flatness
  - Roughness
- D.P. Sensing Taps
  - Burrs

Operating Conditions
- Pulsating Flow
- Widely Varying Flow

Fluid Properties
- Density
  - Equation of State Uncertainty
  - Transmitters' Sensitivity
- Expansion Factor
- Viscosity

Instrumentation
- Ambient Effects
- Static Pressure on "DP" Sensing
- Calibration Accuracy

Figure 2. Influence quantities on orifice meters.

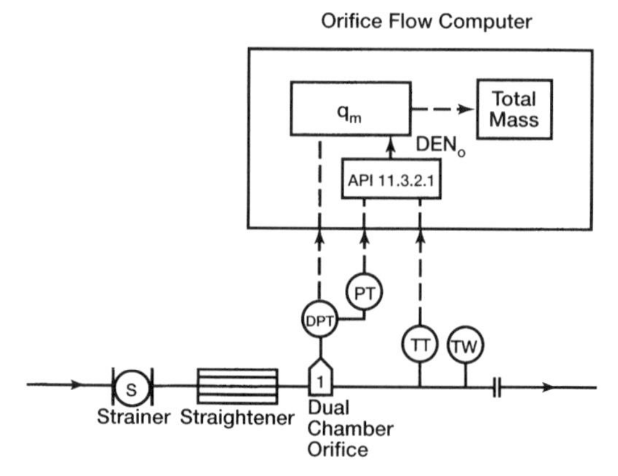

Figure 3. Ethylene orifice meter.

weakness of orifice metering has been the inability to prove the meter under actual operating conditions.

In 1982, the Industry Ethylene Measurement Project conducted at Mt. Belvieu, Texas, demonstrated the feasibility of gas turbine meters, conventional piston pipe provers, and API's equation of state for custody transfer. The project also demonstrated the turbine meter and pipe prover approach produced a lower uncertainty than orifice meters. The ability to accurately prove the turbine meter, shown in Figure 4, on a mass basis under actual operating conditions eliminated the reliance on reference conditions.

However, installation costs could not justify this technique for most custody transfer installations. As a result, the majority of Shell's ethylene custody transfers continues to be based on the traditional orifice meter.

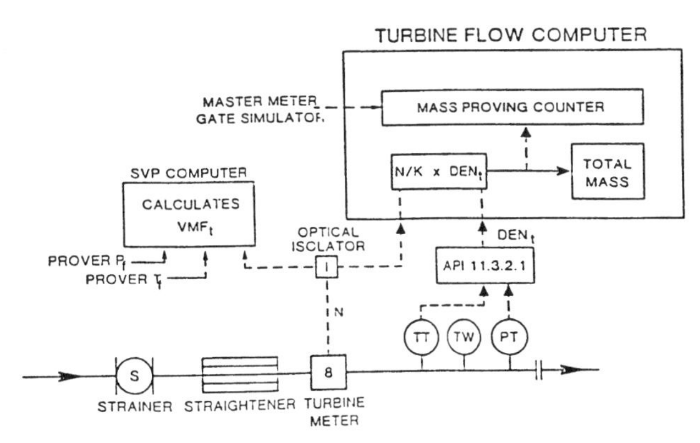

Figure 4. Ethylene turbine meter.

Development program

Development of the master meter proving technique was conducted on commercially operating ethylene pipelines by field personnel. Orifice meters associated with these systems were proven under actual operating conditions.

Fluid properties. The thermodynamic and physical properties of polymer-grade ethylene have been the subject of intensive investigations. Comprehensive treatises on the density of pure ethylene have been published by two highly reputable sources—API and the NBS. The pipeline industry in the United States presently utilizes API.

Chapter 11.3.2—Ethylene Density (API 2565). Ethylene densities are calculated using measured temperatures and pressures. The accuracy of the calculated density depends on the accuracy of the equation of state, the accuracy of the measuring devices, and the sensitivity of the calculated density to small temperature and pressure variations.

Portable proving facility. A truck-mounted 900 ANSI proving system was developed to allow maximum utilization of capital investment. The truck is equipped with a small volume prover, master turbine meter, and associated instrumentation, as shown in Figures 5 and 6.

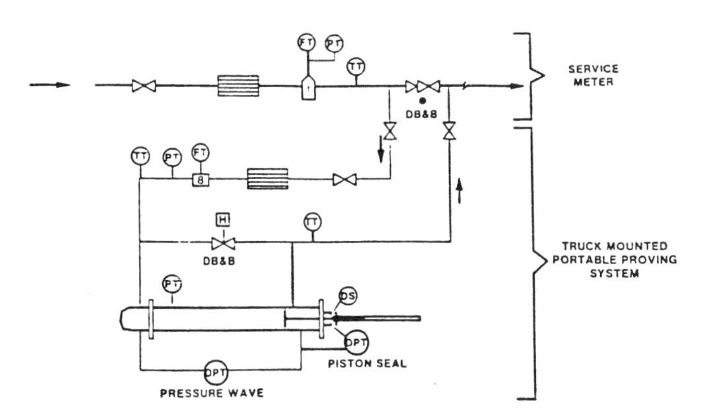

Figure 5. Master meter method—mechanical.

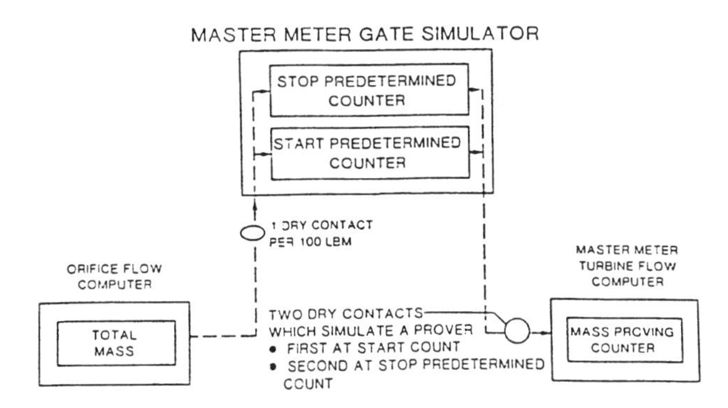

Figure 6. Master meter method—electronic.

The small volume prover, SVP, operates on the principle of repeatable displacement of a known volume of fluid in the calibrated section. The mass of the fluid in the prover is determined through the combination of the prover's base volume, the fluid's pressure and temperature, and the API ethylene density equation.

The SVP was selected based on the following criteria:

- Minimize capital investment by using portable system
- Unidirectional flow design
- Honed and calibrated pipe section
- Dynamic piston seal leak detection
- Single fiber optic piston position detector
- Control of launching pressure wave
- Double chronometry pulse interpolation

Principal parts include an outer housing, a honed inner measuring cylinder, a piston displacer with an Invar extension rod, and an automatic hydraulic system. The hydraulic system returns the piston displacer to the upstream launch position after each proving run and actuates the prover bypass valve to allow the fluid to bypass the prover while the piston is being returned to the launch position.

The pressure wave that occurs when launching the prover's piston is monitored by a high-speed recorder, which was added during initial testing. The differential pressure, or pressure wave, across the prover is recorded to assure the piston does not act as a compressor or brake. The hydraulic system controls the piston to maintain a constant velocity through the calibrated section.

A gas turbine meter specially modified by the manufacturer for ethylene service was selected for the master turbine meter. The meter is equipped with a Lexan rotor, ball bearings, external lubricator, and conventional preamplified electronic pulse output. The master meter has a stated linearity of ±0.25% over a 5:1 turndown. The truck is equipped with two master turbine meters to cover normal operating flowrates—10 mm (4 in.) and 50 mm (2 in.) turbine meters.

The SVP is controlled by a microprocessor-based computer. The computer monitors repeatability, piston position, and prover bypass valve position; performs dual chronometric pulse interpolation; and calculates/prints a hard copy report for each successful proving.

The computer receives input from the prover's temperature and pressure transmitters, the turbine meter's pulse train, the fiberoptic gate detector, and various hydraulic controls.

Two microprocessor-based flow computers are used to calculate the turbine meter's mass flow. The first computer calculates the API ethylene density based on the turbine meter's pressure and temperature transmitters. The second computer calculates the turbine's mass flow based on the turbine meter's pulse input and contains an internal mass proving counter.

Solid state temperature transmitters, equipped with high-accuracy 100 ohm platinum RTD sensors, are installed 4–5 pipe diameters downstream of the turbine meter and at the exit of the prover. The calibrated accuracy is ±0.08°C (0.15°F).

Solid state pressure transmitters, equipped with a sealed capacitance sensing element, are installed 4–5 pipe diameters downstream of the turbine meter and at the inlet to the prover. The calibrated accuracy is ±6.89 kPa (1.0 psi).

Solid state differential pressure (dp) transmitters, equipped with a sealed capacitance sensing element, are installed across the prover and the piston's seal. The calibrated accuracy is ±12 Pa (0.05 in. H_2O).

A high-speed chart recorder monitors the prover's "dp" transmitters:

- To assure that the prover does not induce a pressure wave greater than 6.89 kPa (1 psi) while the piston is in the calibrated section
- To assure that the piston's seal does not leak in the dynamic state

A solid state counter, equipped to receive a contact closure from the orifice's computer and send a dry contact closure output to the turbine meter's flow computer, is installed to start and stop the master meter proving run. This device triggers the running start/stop master meter proving.

Master meter proving principle. The proving is based on the concept that a "master meter" can act as a transfer standard between the primary standard and another meter.

Meter proving by the master meter method requires two distinct steps:

- Mass prove the master turbine meter on site under operating conditions
- Master meter prove the orifice meter on a mass basis using the master turbine meter

To assure the devices are experiencing identical conditions, the orifice meter, turbine meter, and SVP are placed in series, as shown in Figure 5, during the entire calibration. Before proving the "master meter," the flow and/or density conditions through the orifice meter, turbine meter, and SVP must approach steady state conditions.

In the "master meter" proving method, the turbine meter is calibrated on a mass basis under actual operating conditions by the SVP. Once the turbine meter has been proved, a comparison of the two meters' mass outputs forms the basis for determining the orifice meter factor as shown in Figure 6.

Proving procedures

In order to assure accuracy, the operator must be certain the system approaches steady state conditions (flow rate, density, pressure, temperature) and that no leakage is occurring between the orifice, turbine, and the prover. If steady state conditions do not exist, accuracy and/or repeatability will be extremely difficult to obtain. The orifice meter's instrumentation is calibrated prior to proving.

Preparation.

1.1 Initiate flow through the SVP. Check for leakage on all critical valves, relief valves, and piston displacer seal.

1.2 Calibrate all instrumentation associated with the prover truck (pressures, temperatures, flow computers).

1.3 Check operation of assembled instrumentation by performing preliminary turbine meter proving runs.

1.4 If necessary, adjust hydraulic system to assure the prover does not act as a compressor or a brake. Adjust the hydraulic system if the piston displacer's differential pressure exceeds ±6.89 kPa (1 psi).
1.5 Allow ethylene to circulate through the prover truck until the outlet temperature is stabilized.

Proving run.

2.1 Initiate the SVP proving sequence by commanding the prover computer/controller to obtain a VMFt. (Acceptable repeatability for the turbine meter is five consecutive runs in which the volumetric meter factor, VMFt, does not exceed a tolerance of 0.05%.) The computer will print a proving report when this criterion has been met.
2.2 Calculate the turbine meter's mass meter factor, MMFt.
2.3 Clear the turbine meter flow computer's internal mass prover counter.
2.4 Reset and program the master meter gate simulator to allow the master meter proving to occur for a minimum of 1 hour. Record the turbine meter's mass flow rate. (The simulator initiates a master meter proving after receiving five dry contacts from the orifice computer. The simulator stops the proving after approximately one hour. Example—for flow at 50,000 lb/hr, the gates would be set at 5 and 505.)
2.5 When the master meter run is complete, record the turbine meter's mass counts.
2.6 Calculate the orifice's mass meter factor (MMFo).
2.7 Repeat steps 2.1 thru 2.6 until two consecutive master meter runs satisfy repeatability criteria. (Acceptable repeatability for orifice meter is two consecutive runs in which the MMFo does not exceed a tolerance of 0.10%.)
2.8 Average the results of the master meter runs. Enter the average MMFo into the orifice flow computer for use in continuing operations.

Figure 7 is a typical orifice proving report.

Analysis of data

Analysis of the data is contained in this section, along with the calculation methods for reduction of the data.

Proving calculations. The prover base volume, PBV, is stated at 101.325 kPa (14.696 psia) and 15.6°C (60.0°F). Therefore, correction for the effect of operating pressure and temperature shall be applied to determine the prover's volume at operating conditions. These corrections are as follows:

Correction for Effect of Pressure on Steel Prover (Cpsp)

$Cpsp = 1.0000$ (since the prover is equipped with a double wall design)

Correction for Effect of Temperature on Steel Prover (Ctsp)

$Ctsp = [1 + 0.000024(Tf - 15.6)]$ SI Units

$Ctsp = [1 + 0.0000134(Tf - 60)]$ IP Units

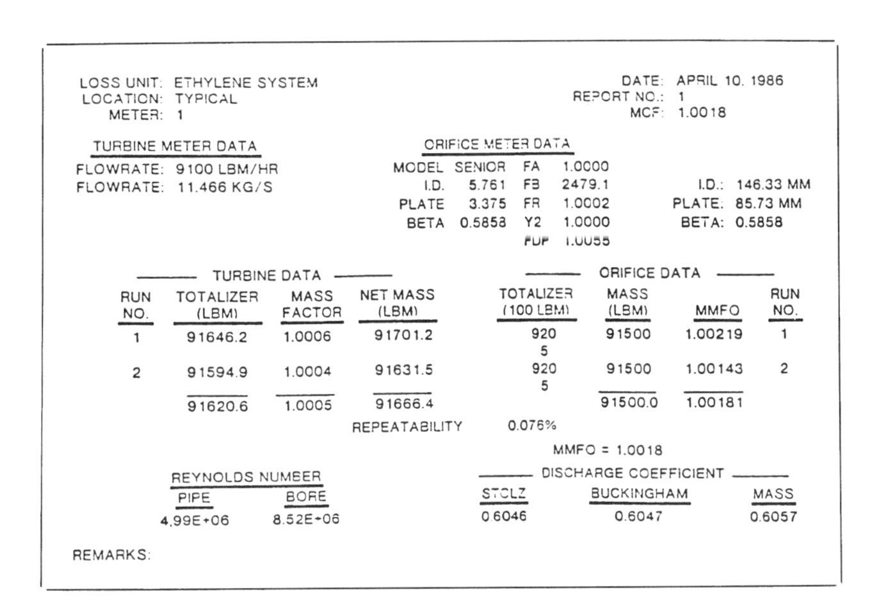

LOSS UNIT: ETHYLENE SYSTEM
LOCATION: TYPICAL
METER: 1

DATE: APRIL 10. 1986
REPORT NO.: 1
MCF: 1.0018

TURBINE METER DATA
FLOWRATE: 9100 LBM/HR
FLOWRATE: 11.466 KG/S

ORIFICE METER DATA

MODEL	SENIOR	FA	1.0000		
I.D.	5.761	FB	2479.1	I.D.:	146.33 MM
PLATE	3.375	FR	1.0002	PLATE:	85.73 MM
BETA	0.5858	Y2	1.0000	BETA:	0.5858
		FDP	1.0055		

TURBINE DATA				ORIFICE DATA			
RUN NO.	TOTALIZER (LBM)	MASS FACTOR	NET MASS (LBM)	TOTALIZER (100 LBM)	MASS (LBM)	MMFO	RUN NO.
1	91646.2	1.0006	91701.2	920 5	91500	1.00219	1
2	91594.9	1.0004	91631.5	920 5	91500	1.00143	2
	91620.6	1.0005	91666.4		91500.0	1.00181	

REPEATABILITY 0.076%

MMFO = 1.0018

REYNOLDS NUMBER	
PIPE	BORE
4.99E+06	8.52E+06

DISCHARGE COEFFICIENT		
STOLZ	BUCKINGHAM	MASS
0.6046	0.6047	0.6057

REMARKS:

Figure 7. Shell Pipeline Corporation orifice master meter proving report version 86.2A.

The calibrated volume of the prover at operating conditions is:

PVtp = [PBV * Cpsp * Ctsp]

The mass of fluid in the prover's calibrated section at operating conditions is:

PMtp = [PVtp * DENp]

In the "master meter" proving method, the turbine meter is calibrated on a mass basis under actual operating conditions. The master meter's Mass Meter Factor, MMFt, can be calculated using the following equations:

TVtp = [(Proving Run Pulses)/(K Factor)]
TMtp = [TVtp * DENt]
VMFt = [PVtp/TVtp]
MMFt = [PMtp/TMtp]

The average mass meter factor, MMFt, for the turbine meter is calculated along with the repeatability for five consecutive runs. Acceptable repeatability is based on a maximum tolerance of 0.05% for the turbine's volumetric meter factor, VMFt. The master turbine meter is reproved at the start of each master meter proving run. This mass factor is applied forward.

In the "master meter" proving method, the orifice meter is proved on a mass basis under actual operating conditions. The average mass meter factor for the orifice meter, MMFo, is calculated along with the repeatability for two consecutive runs. Acceptable repeatability is based on a maximum deviation of 0.15%. An MMFo is calculated for each master meter proving run using the following equation:

MMFo = [(Turbine Mass) * (MMFt)]/[(Orifice Mass)]

Discharge coefficients. When an orifice proving report was calculated, the empirically derived mass discharge coefficient was compared with two empirical predictions—the Stolz and Buckingham equations. The Mass Discharge Coefficient was calculated from the master meter proving results using the following formula:

Mass Cd = [(MMFo) * (Buckingham's Cd)]

Influence quantities. Orifice meter applications are based on steady flow of a homogeneous, single-phase, Newtonian fluid with a velocity profile that follows the reference standard piping conditions. Departures from these reference piping conditions, the effect of the dynamic properties of the fluid, instrumentation effects, and operating conditions are termed influence quantities. Influence quantities are summarized in Figure 2.

Repeatability and reproducibility. Repeatability is the ability to obtain the same result when measuring the same quantity several times in succession. The turbine meter's VMFt has a required repeatability range of 0.05%. The orifice meter's MMFo has a required repeatability range of 0.10%. Repeatability has been expressed as a percentage on each orifice proving report.

Reproducibility is the ability to obtain the same result when measuring the same quantity under different conditions (i.e., time, observer, operating conditions, etc.). With respect to the turbine meter, reproducibility is not significant, since the meter is proved immediately prior to proving an orifice meter. The orifice meter's MMFo reproducibility was demonstrated over a time period of several days at different flow rates and was generally within 0.10%.

Uncertainty of orifice calibrations. The generally accepted statistical method of the quadrature sum (square root of the sum of the squares) has been applied in deriving the overall measurement accuracy.

In calculating the uncertainties, the following assumptions were made:

- The universe is homogeneous (statistically controlled) with respect to MMFo.
- The distribution is approximately Normal or Gaussian.
- The sample is random.
- A 95% confidence limit is required.

The prover base volume, PBV, was established to within ±0.031% by water draw calibration using NBS certified standards.

When proved separately as a volume meter, the turbine meter's VMFt demonstrated a standard deviation of ±0.025% for five consecutive proving runs. An assigned uncertainty of ±0.05%, or two standard deviations, was used in calculating the overall uncertainty.

The flow computer has a stated uncertainty of 0.10%. The density of ethylene was determined using API Chapter 11.3.2.1—Ethylene Density, using measured pressures and temperatures. The uncertainty of the equation of state is stated as ±0.20%.

Additional uncertainties when calculating density using an equation of state, no matter how accurate the equation of state might be, are errors resulting from the uncertainties of the measured temperature and pressure. Transmitter uncertainties of ±0.08°C (0.15°F) and ±6.89 kPa (1 psi) at metering conditions of 8 MPa (1,160 psia) and 20°C (68°F) would result in a calculated density uncertainty of ±0.15%.

Following is a summary of the master meter uncertainty equation:

	Tolerance (%)	Effect	Square of tol & eff
Prover calibration	0.03	1.0000	0.0009
Turbine repeatability	0.05	1.0000	0.0025
Turbine computer	0.10	1.0000	0.0100
Density computer	0.10	1.0000	0.0100
Equation of state	0.20	1.0000	0.0400
Density sensing	0.15	1.0000	0.0225
		Sum	0.0861
	Square root of sum ±0.2934%		

When the orifice meters were proved on a mass basis, they demonstrated a standard deviation of ±0.05% for two consecutive proving runs. An assigned uncertainty of ±0.10%, or two standard deviations, was used in calculating the overall uncertainty.

Following is a summary of the overall orifice meter uncertainty equation as a result of proving on a mass basis using the master meter method:

	Tolerance %	Effect	Square of tol & eff
Master turbine	0.2934	1.0000	0.0861
Master meter proving	0.10	1.0000	0.0100
		Sum	0.0961
	Square root of sum ±0.3100%		

The uncertainties calculated are indicative at the time of proving and do not reflect additional uncertainties associated with measurements occurring beyond the proving period.

Many factors contribute an element of uncertainty to the flow rate indicated by the flow meter in relation to the actual or true flow. Orifice metering systems are made of an assembly of parts originating from different manufacturers. To determine the uncertainty with which the total assembly can measure, the tolerances on the different components and the way they affect the metering accuracy have to be considered. Since the orifice meter has been proved in situ, the practical working formula for the uncertainty can be expressed as follows:

$$\% = \left(\left[(\text{orifice calibration})^2\right] + 0.25\left[(\text{dp error})^2\right] + 0.25\left[(\text{density sensing error})^2\right]\right)^{0.5}$$

Conclusions

The results of the orifice calibrations were rather interesting. Installations that were not in compliance with ANSI 2530 requirements were discovered, as well as human errors associated with flow computer programming.

The master meter method identified and measured errors associated with the following problems:

- Deformed orifice plates
- Eccentrically mounted orifice plates
- Orifice plates installed backwards
- Improperly programmed computers
- Effect of compressor pulsations

Additional information was no doubt masked by the influence quantities. From an application standpoint, identifying the source of error is not as important as measuring and correcting for the error.

Based upon the results, the majority of orifice meters have indicated a general understatement of measured quantities prior to proving.

In summary, our original goal was achieved—to develop an economically viable method for proving ethylene orifice meters under actual operating conditions. The master meter proving method has demonstrated to be a reliable measurement tool in identifying the quantifying measurement problems.

References

1. *Orifice Metering of Natural Gas and Other Related Hydrocarbon Fluids*, API MPMS Chapter 14.3, Second Edition, September 1985 (AGA Report No. 3), (GPA 8185-85), (ANSUAPI 2530).
2. *Ethylene Density*, API MPMS Chapter 11.3.2.1, 1974.
3. *Shell Flow Meter Engineering Handbook*, Second Edition, McGraw-Hill Book Company (UK) Limited, 1985.
4. Miller, R. W., *Flow Measurement Engineering Handbook*, McGraw-Hill Book Company, 1983.
5. Stolz, J., *A Universal Equation for the Calculation of Discharge Coefficients of Orifice Plates*, Flow Measurements of Fluids, IMEKO Conference on Flow Measurement of Fluids, Groningen, The Netherlands, 1978.
6. Kolb, R. H., *Ethylene Mass Measurement System with Prover*, The Second Petroleum Measurement Seminar, 1980.
7. Pfrehm, R. H., "Improved Turbine Meter System Measures Ethylene Accurately," *Oil & Gas Journal*, April, 1981.
8. Bellinga, H., Kooi, A., Hoeks, C. P., van Laak, F. A. L., and Orbons, P. J., "Piston Prover Used to Calibrate Gas Meters," *Oil & Gas Journal*, August, 1985.
9. Boncher, F. S., Shipman, L. M., and Yen, L. C., *Hydrocarbon Processing*, 54 (2), 89 (1975).
10. *ASTM Manual on Presentation of Data and Control Chart Analysis*, ASTM Special Technical Publication 15D, 1976.

11. *Measurement Uncertainty for Fluid Flow in Closed Conduits*, ANSI/ASME MFC-2M-1983.

Notation

β	ratio of orifice plate bore to internal meter diameter
Cd	orifice discharge coefficient
Cpsp	correction for effect of pressure on prover
Ctsp	correction for effect of temperature on prover
d	orifice plate bore, in. (mm)
dp	differential pressure across orifice plate, in H_2O @ 60 (Pa)
D	meter internal diameter, in. (mm)
DENo	density of fluid at orifice, lbm/ft^3 (kg/m^3)
DENp	density of fluid at prover, lbm/ft^3 (kg/m^3)
DENt	density of fluid at turbine, lbm/ft^3 (kg/m^3)
Fa	orifice thermal expansion factor
Fb	basic orifice factor
Fdp	correction for static pressure on dp transmitter
Fr	Reynolds number factor
IP	Inch-Pound System of Units
MMFt	mass meter factor for turbine
MMFo	mass meter factor for orifice
PBV	prover base volume, ft^3 (m^3)
PMtp	prover mass, lbm (kg)
PVtp	prover volume, ft^3 (m^3)
qm	mass flow rate, lbm/s (kg/s)
Qm	mass flow rate, lbm/hr (kg/hr)
Rd	orifice bore Reynolds number
RD	pipe Reynolds number
SI	International System of Units
SVP	small volume prover
Tf	flowing temperature, °F (°C)
TMtp	turbine meter's mass, lbm (kg)
TVtp	turbine meter's volume, ft^3 (m^3)
VMFt	volumetric meter factor for turbine meter
Y2	expansion factor based on downstream static pressure

Source

Reprinted from *Oil & Gas Journal*, December 15, 1986 by permission and courtesy of Perm Well Publishing Co. and James E. Gallagher, Shell Oil Company.

Gas or vapor flow measurement—orifice plate flange taps

HP 41CV Data Sheet

```
************

ORIFICE  GAS
FLANGE  TAPS

************

ENTER FLOW  PPH
     2,500.0000   ***
ENTER TEMP DEG F
      366.0000    ***
ENTER PRSS PSIG
      150.0000    ***
ENTER DENSITY PPCF
        0.3640    ***

ENTER DIFF RANGE INS H20
      100.0000    ***
ENTER PIPE ID INS
        2.0670    ***
ENTER CP/CV
        1.3320    ***
ENTER VISCOSITY cP
        0.0150    ***
CALC  BETA
RATIO
        0.6366    ***
CALC ORIFICE
 BORE    INS
        1.3160    ***

************

        1.3160

ENTER REQD ORIFICE  INS
        1.3750    ***
CALC  BETA
RATIO
        0.6652    ***
CALC  FLOW
PPH
     2,731.2064   ***

************
```

HP 41CV Register Loading

```
SIZE 080
R00= 0.000000    R08= 0.000000    R17= 0.000000
R01= 0.000000    R09= 0.000000    R18= 0.000000
R02= 0.000000    R10= 0.000000    R19= 0.000000
R03= 0.000000    R11= 0.000000    R20= 0.000000
R04= 0.000000    R12= 0.000000    R21= 0.000000
R05= 0.000000    R13= 0.000000    R22= 0.000000
R06= 0.000000    R14= 0.000000    R23= 0.000000
R07= 0.000000    R15= 0.000000    R24= 0.000000
                 R16= 0.000000    R25= "ORIFIC"
```

```
R26= "E  GAS"
R27= "FLANGE"
R28= "  TAPS"
R29= "FLOW  "
R30=  0.000000
R31=  25.262728
R32=  57.295000
R33=  57.515200
R34=  57.535400
R35=  57.606100
R36=  57.626364
R37=  57.656600
R38=  57.670000
R39=  57.686900
R40=  0.000000
R41=  56.707100
R42=  75.767273
R43=  57.747673
R44=  0.000000
R45=  56.707100
R46=  56.295000
R47=  0.000000
R48=  0.000000
R49=  0.000000
R50= "PPH   "
R51= "TEMP D"
R52= "EG F  "
R53= "PRSS P"
R54= "SIG   "
R55= "******"
R56= "CALC  "
R57= "ENTER "
R58=  0.000000
R59=  0.000000
R60= "DENSIT"
R61= "Y PPCF"
R62= "DIFF P"
R63= "ANGE I"
R64= "NS H2O"
R65= "PIPE I"
R66= "D INS "
R67= "CP/CV "
R68= "VISCOS"
R69= "ITY cP"
R70= "BETA  "
R71= "RATIO "
R72= " BORE "
R73= "  INS "
R74= "REQD O"
R75= "CALC O"
R76= "RIFICE"
R77=  0.000000
R78=  0.000000
R79=  0.000000
```

HP 41CV Listing

SUBROUTINES

```
01♦LBL "FT GAS"
02 GTO A
```

PRINT LINE OF STARS

```
03♦LBL a
04 FC? 55
05 RTN
06 55.55
07 ADV
08 SF 12
09 XEQ b
10 ADV
11 RTN
```

PRINT LINES

```
12♦LBL d
13 SF 12
14 1
15 ST+ 58
16 RCL IND 58
17♦LBL b
18 STO 59
19 CLA
20♦LBL c
21 ARCL IND 59
22 RCL 59
23 FRC
24 100
25 *
26 STO 59
27 X≠0?
28 GTO c
29 FC? 55
30 GTO 01
31 PRA
32 CF 12
33 RTN
34♦LBL 01
35 CF 12
36 AVIEW
37 RTN
```

PRINT & STORE VARIABLES

```
38♦LBL e
39 1
40 ST+ 58
41 RCL IND 58
42 X=0?
43 GTO d
44 XEQ b
45 1
46 ST+ 01
47 RCL IND 01
48 TONE 9
49 FS? 00
50 GTO 01
51 STOP
52 STO IND 01
53 .7
54 STO 14
55 RCL IND 01
56♦LBL 01
57 XEQ J
58 GTO e
```

RE-ENTER VARIABLES

```
59♦LBL "REENTER"
60 ADV
61 1
62 +
63 STO 01
64 RCL IND 01
65 XEQ J
66 STOP
67 XEQ J
68 STO IND 01
69 ADV
70 SF 00
71 SF 02
72 STOP
73 GTO 02
```

BYPASS PRINTER

```
74♦LBL J
75 FC? 55
76 GTO 01
77 PRX
78 RTN
79♦LBL 01
80 VIEW X
81 RTN
```

ITERATE FOR β

```
82♦LBL B
83 .076
84 RCL 07
85 SQRT
86 /
87 .364
88 +
89 STO 13
90♦LBL F
91 .007
92 RCL 07
93 /
94 .5993
95 +
96 STO 12
97 .5
98 RCL 07
99 /
100 .07
101 +
102 RCL 14
103 PSE
104 -
105 0
106 X<>Y
107 X<=Y?
108 GTO 01
109 2.5
110 Y↑X
111 1.6
112 RCL 07
113 1/X
114 -
115 5
116 Y↑X
117 .4
118 *
119 *
120 ST+ 12
121♦LBL 01
122 .5
123 RCL 14
124 -
125 0
126 X<>Y
127 X<=Y?
128 GTO 01
129 1.5
130 Y↑X
131 .09034
132 RCL 07
133 /
```

```
134 +
135 *
136 ST- 12
137♦LBL 01
138 RCL 14
139 .7
140 -
141 0
142 X<>Y
143 X<=Y?
144 GTO 01
145 2.5
146 Y↑X
147 65
148 RCL 07
149 X↑2
150 /
151 3
152 +
153 *
154 ST+ 12
155♦LBL 01
156 RCL 12
157 2
158 *
159 RCL 14
160 *
161 RCL 14
162 5
163 Y↑X
164 6
165 *
166 RCL 13
167 *
168 +
169 STO 23
170 RCL 12
171 RCL 14
172 X↑2
173 *
174 RCL 13
175 RCL 14
176 6
177 Y↑X
178 *
179 +
180 830
181 RCL 14
182 5000
183 *
184 -
185 9000
186 RCL 14
187 X↑2
188 *
189 +
190 4200
191 RCL 14
192 X↑2
193 *
194 RCL 14
195 *
196 -
197 530
198 RCL 07
199 SQRT
200 /
201 +
202 STO 24
203 .000015
204 *
205 1
206 +
207 /
208 STO 11
209 FS? 01
210 GTO G
211 RCL 16
212 -
213 RCL 23
214 /
215 ST- 14
216 ABS
217 .000015
218 X<>Y
219 X>Y?
220 GTO F
221 RCL 14
222 RTN
```

CALC F_R

```
223♦LBL D
224 RCL 14
225 4
226 Y↑X
227 .35
228 *
229 .41
230 +
231 RCL 06
232 *
233 39.6
234 /
235 RCL 08
236 /
237 RCL 04
238 15
239 +
240 /
241 CHS
242 1
243 +
244 ST/ 16
245 RCL 14
246 RCL 07
247 *
248 X↑2
249 RCL 24
250 *
251 RCL 09
252 *
253 4.42
254 /
255 RCL 02
256 /
257 1
258 +
259 ST/ 16
260 RTN
```

MAIN PROGRAM ENTER DATA

```
261♦LBL A
262 1
263 STO 01
264 30
265 STO 58
266 XEQ a
267 XEQ d
268 XEQ a
269 XEQ e
270 CF 00
271♦LBL 02
272 41
273 STO 58
```

CALC $K_o\beta^2$

```
274 RCL 02
275 RCL 07
276 X↑2
277 /
278 358.5
279 RCL 03
280 .007
281 *
282 +
283 /
284 RCL 05
285 RCL 06
286 *
287 SQRT
288 /
289 STO 16
290 STO 10
```

CALC β

```
291 XEQ B
292 XEQ D
293 XEQ B
294 RCL 10
295 STO 16
296 XEQ D
297 XEQ B
298 XEQ J
```

CALC d

```
299 XEQ d
300 RCL 14
301 RCL 07
302 *
303 STO 19
304 GTO 01
```

ENTER REQD BORE

```
305♦LBL "EVEN SZ"
306 SF 01
307 14
308 STO 01
309 42
310 STO 58
311 XEQ e
```

CALC β

```
312 RCL 15
313 RCL 07
314 /
315 STO 14
316 XEQ J
```

CALC FLOW

```
317 GTO B
318♦LBL G
319 CF 01
320 XEQ d
321 RCL 11
322 RCL 16
323 /
324 RCL 02
325 *
326 STO 19
327♦LBL 01
328 XEQ J
329 FS? 00
330 STOP
331 XEQ a
332 CF 02
333 RCL 19
334 VIEW X
335 BEEP
336 ADV
337 STOP
338 .END.
```

Sequence of Calculations

1. Calculate $K_o\beta^2$ value.

$$K_o\beta^2 = \frac{W_m}{D^2(358.5 + .007T_f)\sqrt{\gamma_f h_m}}$$

2. Iterate to find uncorrected beta ratio.

$$K_o\beta^2 = \beta^2\left[\left(0.5993 + \frac{0.007}{D}\right) + \left(0.364 + \frac{0.0076}{\sqrt{D}}\right)\beta^4 + 0.4\left(1.6 - \frac{1}{D}\right)^5 \left(0.07 + \frac{0.5}{D} - \beta\right)^{5/2}\left(0.009 + \frac{0.034}{D}\right) (0.5 - \beta)^{3/2} + \left(\frac{65}{D^2} + 3\right) (\beta - 0.7)^{5/2}\right] \div \left[1 + 0.000015(830 - 5{,}000\beta + 9{,}000\beta^2 - 4{,}200\beta^3 + \frac{530}{\sqrt{D}})\right]$$

Any negative factor, such as $(\beta - 0.7)^{5/2}$ when β is less than 0.7, shall be taken as equal to zero, and thus the whole term $(65/D^2 + 3)(\beta - 0.7)^{5/2}$ will drop out of the equation.

Let fx equal the equation in step 2—$K_o\beta^2$, and let:

$$f^1x = 2\left(\left(.5993 + \frac{.007}{D}\right) + 4\left(1.6 - \frac{1}{D}\right)^5 \left(.07 + \frac{5}{D} - \beta\right)^{5/2} - \left(.009 + \frac{.034}{D}\right) (0.5 - \beta)^{3/2} + \left(\frac{65}{D^2} + 3\right)(\beta - .7)^{5/2}\beta + 6\left(.364 + \frac{.076}{\sqrt{D}}\right)\right)\beta^6$$

Start with $\beta = 0.7$ and calculate fx/f^1x. If greater than 0.000015, subtract fx/f^1x from current value of β and repeat until the error is less than 0.000015.

3. Calculate the expansion factor.

$$Y = 1 - \frac{(.41 + .35\beta^4)h_m}{39.6kP_f}$$

4. Divide $K_o\beta^2$ by Y.
5. Calculate the Reynolds number correction factor.

$$F_r = 1 + \left(830 - 5{,}000\beta + 9{,}000\beta^3 - 4{,}200\beta^3 + \frac{530}{\sqrt{D}}\right) \times \frac{(D\beta)^2\mu_{cp}}{4.42W_m}$$

6. Divide $K_o\beta^2$ by F_r.
7. Repeat step 2 to find corrected beta ratio.
8. Repeat steps 3, 4, 5, and 6 using corrected beta ratio.
9. Repeat step 2 for final beta ratio.
10. Calculate orifice bore.

$$d = BD \text{ in.}$$

11. Calculate flow rate for an even-sized orifice bore.

$$\beta = \frac{d}{D}$$

Calculate $K_o\beta^2$ using the equation in step 2.

$$\text{new flow} = \frac{\text{new } K_o\beta^2}{\text{original } K_o\beta^2} \times \text{original flow, lb/hr}$$

Notation

D	inside diameter of pipe, in.
d	bore diameter of primary device, in.
F_r	Reynolds number correction factor for steam, vapor, or gas, dimensionless
h_m	maximum differential or range of differential gauge, in. of water, dry calibrated
K_o	coefficient of discharge, including velocity of approach at a hypothetical Reynolds number of infinity, dimensionless
k	ratio of specific heats C_p/C_r, dimensionless
P_f	flowing pressure, psia
T_f	flowing temperature, °F
W_m	maximum rate of flow, lb/hr
Y	expansion factor, dimensionless
β	diameter ratio d/D, dimensionless
γ_f	specific weight of flowing conditions, lb/ft^3
μ_{cp}	absolute viscosity, centipoises

Source

Instrument Engineering Programs, Gulf Publishing Company, Houston, Texas.

Reference

1. Spink, L. K., *The Principles and Practice of Flowmeter Engineering*, The Foxboro Company.

Properties of gas and vapors

This program, designed for the HP-41CV calculator, will provide:

1. Compressibility factor Z, dimensionless
2. Specific weight (or density), lb/ft^3
3. Mixture molecular weight M, dimensionless
4. Mixture critical temperature T_c, °R
5. Mixture critical pressure P_c, psia
6. Mixture ratio of specific heats C_p/C_v, dimensionless
7. Mixture normal boiling point latent heat of vaporization HV, Btu/lb

Caution must be exercised when using the information in number 7. For mixtures having components with widely different values, the answer may be of limited value.

The program defaults to a compressibility factor and specific weight calculation if the first component volume % entry is 100. Any number of components can be handled by the program. It loops until 100% value is reached.

User instructions. Two programs are available. Program A calculates items 1–5 and program B calculates items 6–7.

Initiate desired program by pressing **A** or **B**.

Enter:

Flowing pressure, psig — Press **R/S**
Flowing temperature, °F — Press **R/S**
Component-number volume % (*Note*: When the volume % is 100, the program goes immediately to the calculation. When the volume % is not 100, it continues.) — Press **R/S**
Critical pressure, psia — Press **R/S**
Critical temperature, °R — Press **R/S**
Molecular weight, dimensionless — Press **R/S**

And, if program B is selected:

Ratio of specific heats C_P/C_v, dimensionless — Press **R/S**
Normal boiling point latent heat of vaporization, Btu/lb — Press **R/S**

Program calculates remaining volume %. If less than 100, it loops back to 3. If equal to 100, it continues and calculates.

- Mixture critical pressure, psia
- Mixture critical temperature, °R
- Mixture molecular weight, dimensionless

And, if program B was selected, it calculates:

- Mixture ratio of specific heats C_p/C_v, dimensionless
- Mixture normal boiling point latent heat of vaporization, Btu/lb

Both programs calculate:

- Compressibility factor Z, dimensionless
- Specific weight (or density), lb/ft^3

See Table 1 for table of physical properties.

Table 1
Physical properties

No.	Compound	Formula	Critical Constants Pressure, PSIA (P_c)	Critical Constants Temperature, °R (T_c)	Molecular Weight (M)	Ratio of Specific Heats C_p/C_v at 14.7 PSIA and 60°F (k)	Heat of Vaporization, 14.696 PSIA at Boiling Point, Btu/lb ($NBP\ \Delta HV$)
1	Methane	CH_4	667.8	343.4	16.043	1.311	219.22
2	Ethane	C_2H_6	707.8	550.0	30.070	1.193	210.41
3	Propane	C_3H_8	616.3	666.0	44.097	1.136	183.05
4	*n*-Butane	C_4H_{10}	550.7	765.0	58.124	1.094	165.65
5	Isobutane	C_4H_{10}	529.1	735.0	58.124	1.097	157.53
6	*n*-Pentane	C_5H_{12}	488.6	845.7	72.151	1.074	153.59
7	Isopentane	C_5H_{12}	490.4	829.1	72.151	1.075	147.13
8	Neopentane	C_5H_{12}	464.0	781.0	72.151	1.074	135.58
9	*n*-Hexane	C_6H_{14}	436.9	913.7	86.178	1.062	143.95
10	2-Methylpentane	C_6H_{14}	436.6	895.8	86.178	1.063	138.67
11	3-Methylpentane	C_6H_{14}	453.1	908.3	86.178	1.064	140.09
12	Neohexane	C_6H_{14}	446.8	860.1	86.178	1.064	131.24
13	2,3-Dimethylbutane	C_6H_{14}	453.5	900.3	86.178	1.065	136.08

Table 1
Physical properties (continued)

No.	Compound	Formula	Critical Constants Pressure, PSIA (P_c)	Critical Constants Temperature, °R (T_c)	Molecular Weight (M)	Ratio of Specific Heats C_p/C_v at 14.7 PSIA and 60°F (k)	Heat of Vaporization, 14.696 PSIA at Boiling Point, Btu/lb ($NBP\ \Delta HV$)
14	*n*-Heptane	C_7H_{16}	396.8	972.8	100.205	1.054	136.01
15	2-Methylhexane	C_7H_{16}	396.5	955.0	100.205	1.054	131.59
16	3-Methylhexane	C_7H_{16}	408.1	963.8	100.205	1.054	132.11
17	3-Ethylpentane	C_7H_{16}	419.3	973.5	100.205	1.054	132.83
18	2,2-Dimethylpentane	C_7H_{16}	402.2	937.2	100.205	1.053	125.13
19	2,4-Dimethylpentane	C_7H_{16}	396.9	936.0	100.205	1.053	126.58
20	3,3-Dimethylpentane	C_7H_{16}	427.2	965.9	100.205	1.053	127.21
21	Triptane	C_7H_{16}	428.4	956.4	100.205	1.055	124.21
22	*n*-Octane	C_8H_{18}	360.6	1024.2	114.232	1.046	129.53
23	Diisobutyl	C_8H_{18}	360.6	990.4	114.232	1.049	122.80
24	Isooctane	C_8H_{18}	372.4	979.5	114.232	1.049	116.71
25	*n*-Nonane	C_9H_{20}	332.0	1070.7	128.259	1.042	123.76
26	*n*-Decane	$C_{10}H_{22}$	304.0	1112.1	142.286	1.038	118.68
27	Cyclopentane	C_5H_{10}	653.8	921.5	70.135	1.117	167.34
28	Methylcyclopentane	C_6H_{12}	548.9	959.4	84.162	1.085	147.83
29	Cyclohexane	C_6H_{12}	591.0	996.7	84.162	1.089	153.00
30	Methylcyclohexane	C_7H_{14}	503.5	1030.3	98.189	1.068	136.30
31	Ethylene	C_2H_4	729.8	506.6	28.054	1.243	207.57
32	Propene	C_3H_6	669.0	656.9	42.081	1.154	188.18
33	1-Butene	C_4H_8	583.0	755.6	56.108	1.111	167.94
34	*cis*-2-Butene	C_4H_8	610.0	784.4	56.108	1.121	178.91
35	*trans*-2-Butene	C_4H_8	595.0	771.9	56.108	1.107	174.39
36	Isobutene	C_4H_8	580.0	752.6	56.108	1.106	169.48
37	1-Pentene	C_5H_{10}	590.0	836.9	70.135	1.084	154.46
38	1,2-Butadiene	C_4H_6	653.0	799.0	54.092	1.118	181.00
39	1,3-Butadiene	C_4H_6	628.0	766.0	54.092	1.120	174.00
40	Isoprene	C_5H_8	558.4	872.0	68.119	1.089	153.00
41	Acetylene	C_2H_2	890.4	555.3	26.038	1.238	—
42	Benzene	C_6H_6	710.4	1012.2	78.114	1.117	169.31
43	Toluene	C_7H_8	595.9	1065.6	92.141	1.091	154.84
44	Ethylbenzene	C_8H_{10}	523.5	1111.2	106.168	1.072	144.00
45	*o*-Xylene	C_8H_{10}	541.4	1135.0	106.168	1.069	149.10
46	*m*-Xylene	C_8H_{10}	513.6	1111.0	106.168	1.072	147.20
47	*p*-Xylene	C_8H_{10}	509.2	1109.6	106.168	1.072	144.52
48	Styrene	C_8H_8	580.0	1166.0	104.152	1.076	151.00
49	Isopropylbenzene	C_9H_{12}	465.4	1136.4	120.195	1.060	134.30
50	Methyl Alcohol	CH_4O	1174.2	923.0	32.042	1.237	473.0
51	Ethyl Alcohol	C_2H_6O	925.3	929.6	46.069	1.149	367.0
52	Carbon Monoxide	CO	507.0	240.0	28.010	1.403	92.7
53	Carbon Dioxide	CO_2	1071.0	547.9	44.010	1.293	238.2
54	Hydrogen Sulfide	H_2S	1306.0	672.7	34.076	1.323	235.6
55	Sulfur Dioxide	SO_2	1145.1	775.5	64.059	1.272	166.7
56	Ammonia	NH_3	1636.1	730.3	17.031	1.304	587.2
57	Air	N_2O_2	547.0	238.7	28.964	1.395	92.0
58	Hydrogen	H_2	188.1	60.2	2.016	1.408	193.9
59	Oxygen	O_2	736.9	278.9	31.999	1.401	91.6
60	Nitrogen	N_2	493.0	227.4	28.013	1.402	87.8
61	Chloride	Cl_2	1118.4	751.0	70.906	1.330	123.8
62	Water	H_2O	3208.0	1165.6	18.015	1.335	970.3
63	Helium	He	—	—	4.003	1.660	—
64	Hydrogen Chloride	HCl	1198.0	584.5	36.461	1.410	185.5

Note: C_p/C_v calculated by the formula $k = C_p/(C_p - 1.1987)/M$.

HP 41CV Data Sheet

```
************

   *GAS*
*PROPERTIES*

************



ENTER PRSS PSIG
     100.0000000    ***
ENTER TEMP DEG F
     60.00000000    ***
ENTER % COMPNT NO 1.
     100.0000000    ***
ENTER CRIT  PRSS PSIA
     200.0000000    ***
ENTER CRIT  TEMP DEG R
     500.0000000    ***
ENTER MW
     20.00000000    ***
CALC     Z
     0.801732805    ***
CALC DENSITY
 PPCF
     0.512814969    ***



************
```

```
************

   *GAS*
*PROPERTIES*

************

ENTER PRSS PSIG
     100.0000000    ***
ENTER TEMP DEG F
     60.00000000    ***
ENTER % COMPNT NO 1.
     25.00000000    ***
ENTER CRIT  PRSS PSIA
     200.0000000    ***
ENTER CRIT  TEMP DEG R
     500.0000000    ***
ENTER MW
     20.00000000    ***
ENTER CP/CV
     1.400000000    ***
ENTER NBP dHV BTU/P
     200.0000000    ***
CALC % TO GO
     75.00000000    ***

ENTER % COMPNT NO 2.
     75.00000000    ***
ENTER CRIT  PRSS PSIA
     200.0000000    ***
```

```
ENTER CRIT  TEMP DEG R
     500.0000000    ***
ENTER MW
     20.00000000    ***
ENTER CP/CV
     1.400000000    ***
ENTER NBP dHV BTU/P
     200.0000000    ***
CALC MIXTURE
CRIT  PRSS
     200.0000000    ***
CALC MIXTURE
CRIT  TEMP
     500.0000000    ***
CALC MIXTURE
MW
     20.00000000    ***
CALC MIXTURE
CP/CV
     1.400000000    ***
CALC MIXTURE
NBP dHV
     200.0000000    ***

CALC     Z
     0.801732805    ***
CALC DENSITY
 PPCF
     0.512814969    ***

************
```

HP 41CV Register Loading

SIZE 080

```
R00=  0.000000
R01=  0.000000
R02=  0.000000
R03=  0.000000
R04=  0.000000
R05=  0.000000
R06=  0.000000
R07=  0.000000
R08=  0.000000
R09=  0.000000
R10=  0.000000
R11=  0.000000
R12=  0.000000
R13=  0.000000
R14=  0.000000
R15=  0.000000
R16=  0.000000
R17=  0.000000
R18=  0.000000
R19=  0.000000
R20=  0.000000
R21=  0.000000
R22=  0.000000
R23=  0.000000
R24= "   *GA"
R25= "S*    "
```

```
R26= "*PROPE"
R27= "RTIES*"
R28= "CALC M"
R29=  24.252627
R30=  57.505100
R31=  57.525300
R32=  0.000000
R33=  57.546260
R34=  0.000000
R35=  57.635064
R36=  57.635265
R37=  57.660000
R38=  0.000000
R39=  57.670000
R40=  57.686970
R41=  0.000000
R42=  79.710000
R43=  28.726373
R44=  28.726374
R45=  28.726600
R46=  28.726700
R47=  28.726861
R48=  56.750000
R49=  76.777800
R50= "PRSS P"
R51= "SIG   "
R52= "TEMP D"
R53= "EG F  "
```

```
R54= "% COMP"
R55= "******"
R56= "CALC  "
R57= "ENTER "
R58=  49.000000
R59=  0.000000
R60=  1.000000
R61= "V     "
R62= "NT NO "
R63= "CRIT  "
R64= "SIA   "
R65= "EG R  "
R66= "MW    "
R67= "CP/CV"
R68= "NBP dH"
R69= "V BTU/"
R70= "P     "
R71= " TO GO"
R72= "IXTURE"
R73= "PRSS  "
R74= "TEMP  "
R75= " Z    "
R76= "CALC D"
R77= "ENSITY"
R78= " PPCF "
R79= "CALC %"
```

HP 41CV Listing

SUBROUTINES

```
01♦LBL "GAS PRP"
02 GTO A
```

PRINT LINE OF STARS

```
03♦LBL a
04 FC? 55
05 RTN
06 55.55
07 ADV
08 SF 12
09 XEQ b
10 ADV
11 RTN
```

PRINT LINES

```
12♦LBL d
13 SF 12
14 1
15 ST+ 58
16 RCL IND 58
17♦LBL b
18 STO 59
19 CLA
20♦LBL c
21 FIX 0
22 ARCL IND 59
23 FIX 9
24 RCL 59
25 FRC
26 100
27 *
28 STO 59
29 X≠0?
30 GTO c
31 FC? 55
32 GTO 01
33 PRA
34 CF 12
35 RTN
36♦LBL 01
37 CF 12
38 AVIEW
39 RTN
```

PRINT & STORE VARIABLES

```
40♦LBL e
41 1
42 ST+ 58
43 RCL IND 58
44 X=0?
45 RTN
46 XEQ b
47 1
48 ST+ 01
49 RCL IND 01
50 TONE 9
51 FS? 00
52 GTO 01
53 STOP
54 FC? 01
55 GTO 02
56 XEQ J
57 RCL 04
58 *
59 100
60 /
61 ST+ IND 01
62 GTO e
63♦LBL 02
64 STO IND 01
65♦LBL 01
66 XEQ J
67 GTO e
```

RE-ENTER VARIABLES

```
68♦LBL "REENTER"
69 ADV
70 1
71 +
72 STO 01
73 RCL IND 01
74 XEQ J
75 STOP
76 XEQ J
77 STO IND 01
78 ADV
79 SF 00
80 SF 02
81 100
82 STO 04
83 STOP
84 GTO C
```

BYPASS PRINTER

```
85♦LBL J
86 FC? 55
87 GTO 01
88 PRX
89 RTN
90♦LBL 01
91 VIEW X
92 RTN
```

MAIN PROGRAM INCLUDES CP/CV & dHV

```
93♦LBL "CP/CV"
94 SF 03
```

ENTER DATA

```
95♦LBL A
96 1
97 STO 01
98 STO 21
99 STO 60
100 28
101 STO 58
102 FS? 00
103 GTO 01
104 0
105 STO 05
106 STO 06
107 STO 07
108 STO 08
109 STO 09
110 100
111 STO 20
112 STO 04
113♦LBL 01
114 XEQ a
115 XEQ d
116 XEQ a
117 XEQ e
118 CF 02
119♦LBL B
120 3
121 STO 01
122 32
123 STO 58
124 FS? 00
125 GTO G
126 XEQ e
127 RCL 04
128 ST- 20
129 100
130 -
131 X=0?
132 GTO G
133 SF 01
134 XEQ e
135 FS? 03
136 XEQ e
137 RCL 20
138 0
139 X≠Y?
140 GTO 01
141 CF 00
142 CF 01
```

PRINT MIXTURE VALUES

```
143 42
144 STO 58
145 XEQ d
146 RCL 05
147 XEQ J
148 XEQ d
149 RCL 06
150 XEQ J
151 XEQ d
152 RCL 07
153 XEQ J
154 FC? 03
155 GTO C
156 XEQ d
157 RCL 08
158 XEQ J
159 XEQ d
160 RCL 09
161 XEQ J
162 GTO C
163♦LBL 01
164 RCL 20
165 STO 04
166 41
167 STO 58
168 XEQ d
169 CF 01
170 RCL 20
171 XEQ J
172 1
173 ST+ 60
174 ADV
175 GTO B
176♦LBL G
177 34
178 STO 58
179 4
180 STO 01
181 XEQ e
```

CALC BP_f

```
182♦LBL C
183 47
184 STO 58
185 XEQ d
```

```
186 .0866403
187 RCL 06
188 *
189 RCL 05
190 /
191 RCL 02
192 14.7
193 +
194 *
195 RCL 03
196 460
197 +
198 /
199 STO 11
```

CALC A²/B

```
200 RCL 06
201 RCL 03
202 460
203 +
204 /
205 1.5
206 Y↑X
207 4.93396
208 *
209 STO 12
```

ITERATE FOR Z

```
210•LBL F
211 RCL 21
212 X↑2
213 3
214 *
215 RCL 21
216 2
217 *
218 -
219 RCL 12
220 RCL 11
221 -
222 1
223 -
224 RCL 11
225 *
226 STO 22
227 +
228 STO 23
229 RCL 21
230 3
231 Y↑X
232 RCL 21
233 X↑2
234 -
235 RCL 22
236 RCL 21
237 *
238 +
239 RCL 12
240 RCL 11
241 X↑2
242 *
243 -
244 RCL 23
245 /
246 ST- 21
247 ABS
248 PSE
249 .00001
250 X<>Y
251 X>Y?
252 GTO F
253 RCL 21
254 XEQ J
```

CALC DENSITY

```
255 XEQ d
256 RCL 02
257 14.7
258 +
259 RCL 03
260 460
261 +
262 /
263 RCL 07
264 *
265 RCL 21
266 /
267 10.73
268 /
269 STO 13
270 XEQ J
271 FS? 02
272 STOP
273 XEQ a
274 CF 00
275 CF 03
276 CF 04
277 RCL 21
278 RCL 13
279 VIEW X
280 BEEP
281 ADV
282 STOP
283 END
```

Sequence of calculations

1. Mixture values are calculated by the mixture law, which assumes that the property of the mixture is the sum of the partial fractions.
2. The compressibility factor is iterated from the Redlich-Kwong equation:

$$fx = Z^3 - Z^2 + BP_f\left[\left(\left(\frac{A^2}{B}\right) - BP_f - 1\right)Z - \left(\frac{A^2}{B}\right)BP_f\right]^2$$

$$f^1x = 3Z^2 - 2Z + BP_f\left(\left(\frac{A^2}{B}\right) - BP_f - 1\right)$$

$$\text{where } A = \frac{.42748T_c^{2.5}}{P_cT_f^{2.5}}$$

$$BP_f = .0866403\frac{T_cP_f}{P_cT_f}$$

$$\frac{A^2}{B} = 4.93396\left(\frac{T_c}{T_f}\right)^{1.5}$$

Start with Z = 1 and calculate fx/f^1x. If greater than 0.00001, subtract fx/f^1x from the current value of Z and repeat until the error is less than 0.00001.

3. Calculate the specific weight:

$$\gamma_f = \frac{P_f}{T_f}\frac{M}{1.314Z}\text{lb/ft}^3$$

Notation

A, B	partial calculations for Z
M	molecular weight, dimensionless
P_f	flowing pressure, atmospheres
P_c	critical pressure, atmospheres
T_f	flowing temperature, °K
T_c	critical temperature, °K
Z	compressibility, dimensionless
γ_f	specific weight, lb/ft^3

Note: These units do not correspond with the calculator input units. The program converts as necessary.

Source

Instrument Engineering Programs, Gulf Publishing Company, Houston.

Determine required orifice diameter for any required differential when the present orifice and differential are known in gas measurement

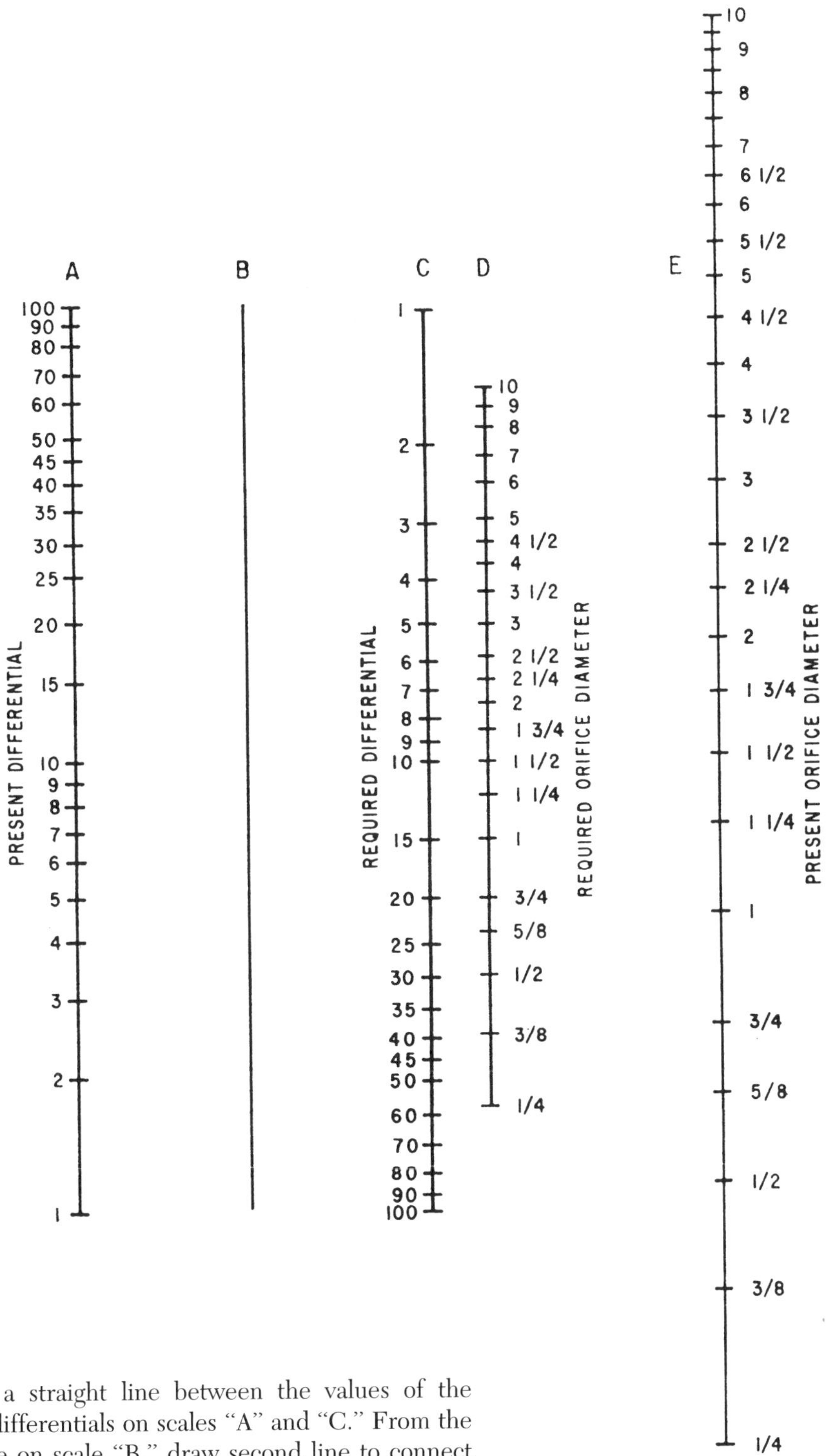

Procedure. Draw a straight line between the values of the present and required differentials on scales "A" and "C." From the intersection of this line on scale "B," draw second line to connect with the size of the present orifice diameter on scale "E." The size of the required orifice will be found at the intersection of his second line with scale "D."

Estimate the temperature drop across a regulator

The temperature drop that will be experienced when gas is passed through a regulator may be estimated as 1° drop for each atmosphere (15 psi) of pressure drop.

Example. Determine the temperature drop that will be experienced when reducing a gas from 750 psig to 600 psig.

Solution

$750 - 600\,\text{psig} = 150\,\text{psig}$

$\frac{150}{15} = 10\,\text{atmospheres}$

$10 \times 1°/\text{atm} = 10°$ of drop

Estimate natural gas flow rates

This chart provides a good estimate of natural gas flow rates through 4-in. meter runs equipped with flange taps. No calculations are required to obtain a flow rate. Orifice plate size and differential and static pressures from the orifice meter chart are used to make the estimate.

To use the Gas Flow Rate Estimator, enter the chart at the proper differential value at the indicated lower horizontal scale. Project this value vertically upward to the appropriate static pressure line, and then horizontally to the proper plate size line. Drop vertically from the plate size line to the lowermost horizontal rate scale at the base of the chart to obtain the estimated flow rate.

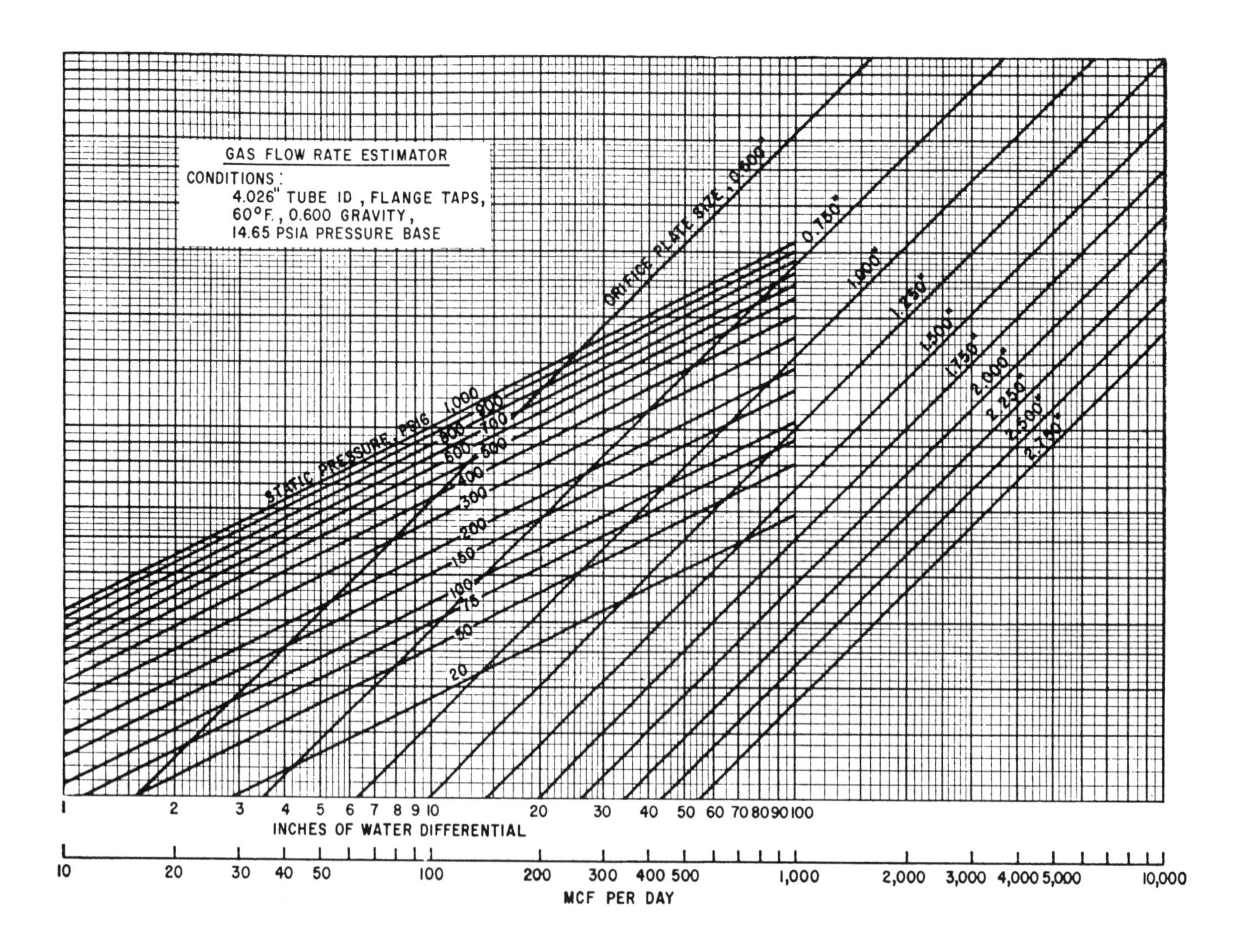

How to estimate the average pressure differential on the remaining meter runs of a parallel system when one or more runs are shut off

On gas systems where inferential meters are installed in parallel, it may at times be necessary to shut down one or more of the runs temporarily for repair or maintenance work. When this occurs, it is important to know beforehand what the pressure differential on the remaining runs will be so that the meters will not be over-ranged. The equation is:

Average differential for runs remaining in service (A) =

$$\left[\frac{(\text{No. of runs in service})\sqrt{(\text{Average differential})}}{(\text{No. of runs remaining in service})}\right]^2$$

Example. A measurement engineer wishes to shut off one meter run on a four-meter parallel system to inspect the orifice plate. The average differential on the four runs is 49 in. and his meters have a 100-in. range. He wishes to determine whether shutting down one meter will over-range the three remaining meters.

$$(\text{A}) = \left(\frac{4 \times \sqrt{49}}{3}\right)^3 = (9.3)^2 = 86.5\,\text{in.}$$

Since the average differential for the remaining runs is 86.5 in., the operator may safely shut the meter down as planned to inspect its orifice plate.

Sizing a gas metering run

Multiply the absolute pressure by the desired differential pressure (the differential in inches of water should not be greater than the absolute pressure in psia). Take the square root of the product. Divide this figure into the maximum flow rate in cubic ft/hr to obtain the approximate coefficient. Then divide the maximum coefficient by 250 and take the square root to determine the approximate inside diameter of the orifice in inches. The metering run diameter should not be less than one-third larger than the orifice diameter.

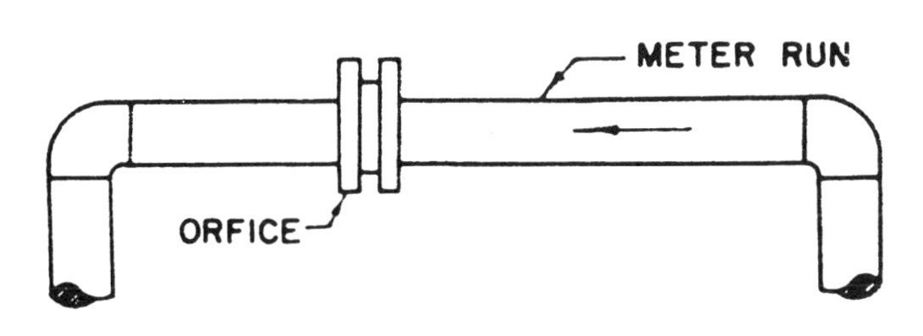

Example. Find the orifice diameter and meter run for a line designed to carry a maximum flow of 2 MMcf/hr at a pressure of 600 psia. Use a differential pressure of 50 in. of water.

$$\sqrt{600 \times 50} = \sqrt{30{,}000} = 173.2$$

$$\frac{2{,}000{,}000}{173.2} = 11{,}540$$

$$\frac{11{,}540}{250} = 46.2$$

$\sqrt{46.2} = 6.8$ in., approximateorificediameter

Meter run should not be less than 10 in. in diameter.

List of typical specifications for domestic and commercial natural gas

1. **Heating value:** Not less than 950 Btu's per std cubic ft.
2. **Purity:** The gas to be delivered shall be free from dust, gums, crude oils, and hydrocarbons liquefiable at temperatures in excess of 15°F at 800 psig.
3. **Sulfur content:** The gas shall contain not more than one grain of hydrogen sulfide per 100 cu. ft of gas, and not more than 20 grains of total sulfur per 100 cu. ft of gas.
4. **CO_2 content:** The gas shall contain not more than 2% carbon dioxide.
5. **Moisture content:** The gas shall contain not more than 4 lb. of water vapor per MMcf of gas measured at a pressure of 14.4 psi and a temperature of 60°F.
6. **Temperature:** Maximum temperature at point of delivery shall be 120°F.

Determine the number of purges for sample cylinders

To determine the proper number of times a sample cylinder should be pressured to ensure removal of the original gas, multiply the reciprocal of each filling pressure (in atmospheres) by each succeeding reciprocal. The final filling will determine what fraction of the original gas is still in the cylinder.

Example. The problem is to fill a gas sample for analysis to obtain an accuracy of 0.05%. Source pressure is 450 psia.

Solution. First filling to line pressure and emptying:

$$\frac{1}{450/15} = \frac{1}{30} \times 100 = 3.33\%$$

Second filling to line pressure and emptying:

$$\frac{1}{30} \times \frac{1}{30} \times 100 = 0.11\%$$

Third filling to line pressure:

$$\frac{1}{30} \times \frac{1}{30} \times \frac{1}{30} \times 100 = 0.0037\%$$

Find the British thermal units (Btu) when the specific gravity of a pipeline gas is known

Use the graph below prepared from the Headlee formula:

Btu @ 30 in. Hg and 60° saturated = 1,545 (specific gravity) + 140 − 16.33 (% N_2)

Example. Assume a gas having a gravity of 0.670, which contains 1.0% nitrogen. From the chart, read up from 0.670 gravity on the base line to the intercept with the curve marked 1% nitrogen. Reading horizontally across, the total Btu of the gas is 1,160 Btu at 30 in., 60° saturated.

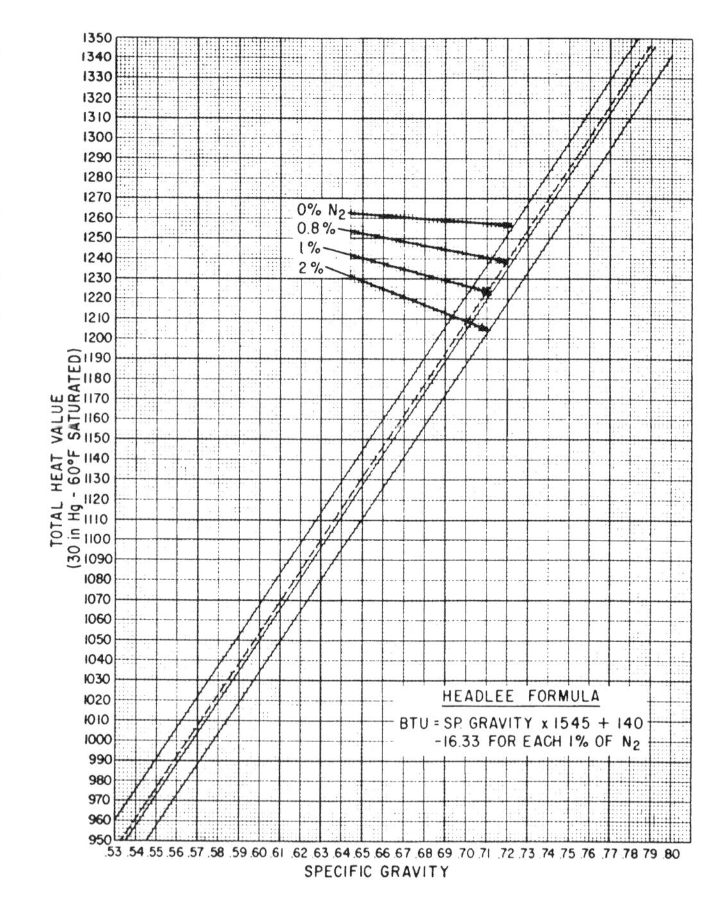

Estimate for variations in measurement factors

Orifice measurement

1. Each 10° change in flowing temperature will cause a 1% change in flow calculated.
2. Each 0.01-change in specific gravity of the gas measured will result in a 1% change in passing flow calculated.

3. Each 10-psi error in static pressure will make the following error in flow calculated:

   at 100 psia—5%
   at 500 psia—1%
   at 1,000 psia—0.5%

4. Each 0.1 in. of differential pressure error will make the following error in flow calculated.

   at 5 in. differential—1%
   at 25 in. differential—0.2%
   at 50 in. differential—0.1%
   at 100 in. differential—0.05%

5. Supercompressibility affects volume measured approximately 0.5% per 100 psi pressure in flowing temperature range of 60–90°F.

Rules of measurement of gas by orifice meter

Where the static pressure remains constant, to reduce the flow by one half, lower the differential reading to one fourth. For example, if the differential is reading 64 in., to reduce the flow by one half, reduce the differential to 16 in.

For better measurement, keep the beta ratio between 0.2 and 0.6.

Headers on meter stations should be twice the internal diameter of the meter runs they manifold together.

Measurement stations should be installed on the suction side of a compressor station rather than the discharge side where possible, to reduce effect of pulsation.

Use the wine size orifice plate in meter runs that are manifolded together where possible; recordings can then be used to check against the other, since their differentials should be approximately the same.

Make quick calculation of the gas flow through orifice meters

On meter runs up to and including 10-in., this can be accomplished by multiplying the coefficient for a 1-in. orifice times the actual orifice installed squared times the square root of the extension (the product of the differential in in. of water times the static pressure in lb/in.^{2} absolute). AGA-3 coefficients for a 1-in. orifice in 2, 3, 4, 6, 8, and 10 in. standard orifice meter runs using flange taps should be committed to memory. Base conditions would be for a 0.60 specific gravity, 60°F, and 14.73 psia pressure base. The answer would be in standard cubic ft/hr or day depending upon the coefficient used.

AGA Committee Report Number 3 coefficients at the conditions shown are:

2 in. meter run—1 in. orifice—271.4 cfh, 0.6 G, 60°F, and 14.73 psia
3 in. meter run—1 in. orifice—262.8
4 in. meter run—1 in. orifice—260.8
6 in. meter run—1 in. orifice—259.3
8 in. meter run—1 in. orifice—258.7
10 in. meter run—1 in. orifice—258.5 (heavy)

Example. Find the hourly gas flow through an 8-in. meter run having a 2.00-in. orifice when the differential pen is recording 16 in. of water and the static pressure is 900 lb/in.^{2} absolute.

Solution. From memory, the coefficient for an 8-in. meter run, 1-in. orifice is 258.7. Multiplying 258.7 times 4 (or 2 squared) times 120 (the square root of the product of 900 times 16) equals 124,176 cfh, the approximate flow in cubic ft of gas per hour at 0.6 gravity, 60°F temperature, and 14.73 psia base pressure.

How to measure high pressure gas

Accurate gas measurement is obtained only through a good maintenance and operating schedule

T J. Lewis, Measurement Engineer, Columbia Gulf Transmission Co., Houston

The rapidly increasing value of natural gas, along with rising operating costs, has made accurate gas measurement vital for the natural gas industry.

Accurate gas measurement is obtained only through an efficient maintenance and operating schedule, adhered to rigidly, after a properly designed measuring station has been installed. Any error in the measurement of gas at high pressure and large volumes magnifies itself, so that there can be either a considerable gain or loss in revenue.

Field measurement at high pressure is defined as the measurement of gas at a specific point in a gas producing area at a pressure higher than that of contract base pressure.

Contracts

The gas purchase contract is of prime importance in determining the requirements for gas measurement at a given location.

Information such as gas analysis, gas/oil ratios, specific gravity determinations, flowing temperatures, pressures, etc., should be gathered before a gas purchase contract is written. This collection of data should be used to eliminate any assumptions regarding the requirements for performing accurate measurement. It also provides information for the engineering and accounting functions, so that they are made aware of their responsibilities and limitations in helping provide good measurement.

The contract should also state as clearly as possible all parameters of the gas measurement operation to prevent possible disputes by having a clear understanding between the buyer and the seller.

Design and planning

In the planning and designing of a natural gas measuring station, many items of equal importance must be considered if accurate measurement is to be obtained.

The design and fabrication of the orifice meter runs must meet the specifications of the AGA Gas Measurement Committee Report No. 3, latest revision. In addition to these requirements, consideration must be given to safety, volume and quality of gas, term of contract, flexibility of station design, and physical location of the station. The most important point in this whole procedure is remembering that the measuring station is like a large cash register with money flowing both ways for the company.

A meter tube should be designed to allow easy internal inspection. Shutoff valves should be placed on each end of the meter, so that it may be depressured independently of any other piping. The instrumentation and/or control piping should be so located and supported to furnish maximum protection against accidental breakage.

Meter runs, piping, etc., should be designed as a Class III area facility. If possible, a properly vented walk-in type meter building should be used, with the total installation laid out with the safety of the operating personnel firmly in mind (Figure 1). As periodic orifice plate inspections are necessary, provisions should be made so that these inspections can be made without interruption of the gas flow. This can be done by installing a bypass line or by using a senior type orifice fitting.

The flow and temperature recording instruments should be located in a walk-in type meter building. This arrangement provides a year-round working area for chart changing, meter tests, maintenance work, etc.

Installation

The measuring or metering components in a measuring station are divided into two elements: primary and auxiliary.

The *primary element* of an orifice-type measuring station consists of the meter tube, the orifice plate fitting, the orifice plate, pressure taps and straightening vanes. While the

Figure 1. Meter runs, piping, etc., should be designed as a Class III area facility. If possible, a properly vented walk-in type meter building should be used, with the total installation laid out with the safety of the operating personnel firmly in mind.

primary element is being fabricated, its components should be checked against the specifications listed in AGA Gas Measurement Committee Report No. 3. Any discrepancies found then can be corrected before the meter tube is finally assembled and shipped to the field.

When the meter tube is being installed in the field, the orifice plate is checked to make sure it is centered in the meter tube, and that it meets AGA Report No. 3 specifications. Meter tubes are normally installed in a horizontal or vertical plane. Pressure taps in a vertical plane are preferred, since this allows the gauge lines to be installed from the top of the orifice fitting. The pressure taps located on the bottom of the orifice fittings then can be used for installing drips or drains.

The *auxiliary element* consists of the meters (recording devices) used for measuring the static pressure, differential pressure, and temperature of the gas in the meter tube. Meters are placed as close as possible to the orifice or flange fitting. Gauge lines are connected from the meter to the orifice fitting in a way that will best provide a drain to the meter tube. Liquids trapped in a gauge line can cause significant errors in the recorded pressures.

There are two principal types of orifice meter gauges: the mercury Manometer-type and the bellows-type meters. Each of these meters is suitable for field measurement, with the bellows-type meter becoming the favorite for wet gas measurement.

Installation of these elements in a measuring station should be under the supervision of experienced measurement personnel.

Operation and maintenance

After a measuring station has been properly installed, accurate measurement is obtainable only by conscientious care and maintenance by the field measurement technicians.

Any discrepancy noticed on the chart should be explained on the chart, and a correction should be made immediately. Since immediate correction is not always possible, a report should then accompany the chart describing the condition of the meter and when the correction was completed. These reports, along with test and inspection reports, are necessary for the calculation of accurate volumes.

The accuracy of recording instruments depends on the men who test and operate the measuring station equipment. Good test equipment, properly maintained, is absolutely necessary in checking the calibration of temperature and pressure recording instruments and the quality of the gas being measured. The frequency of meter tests, gas quality tests, and meter tube inspections depends on the individual company and its contractual agreements with other companies.

Monthly calibration tests and inspections on the flow and temperature recorders are normally adequate for accurate measurement. In special cases, however, a weekly inspection may be required. Monthly calibration tests and inspections are vital to accurate measurement practices. Figures 2 and 3 show what happens to volume calculations when there are

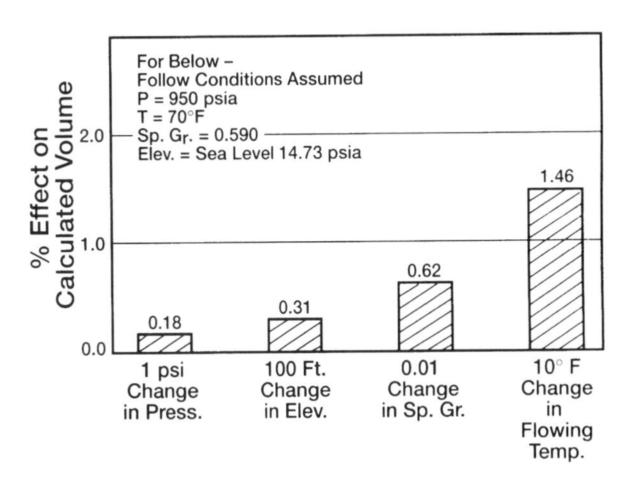

Figure 2. Relative effect of various factors on calculated flow volume (also Figure 3).

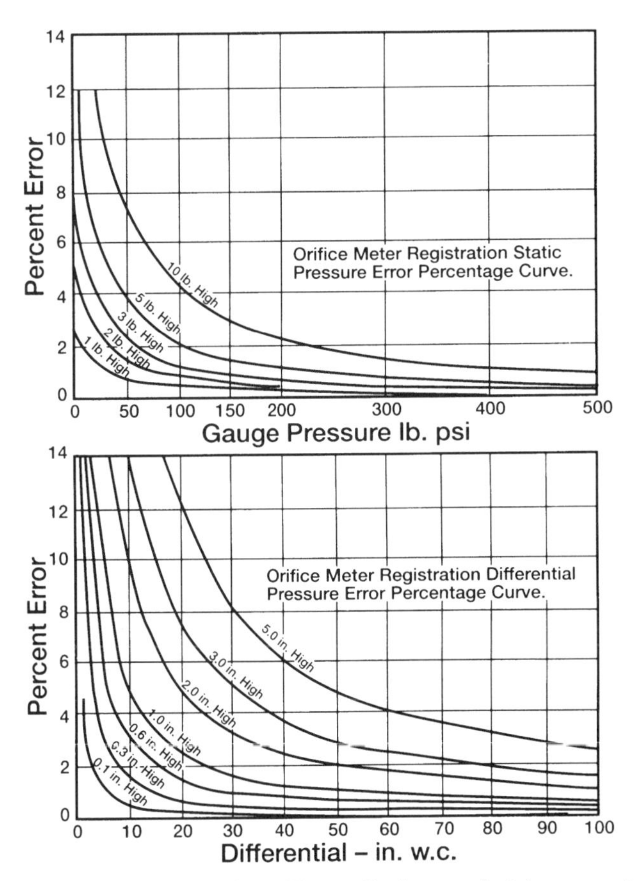

Figure 3. The above (also Figure 2) shows what happens to volume calculations when there are small errors in static pressure, differential pressure, temperature and specific gravity measurements.

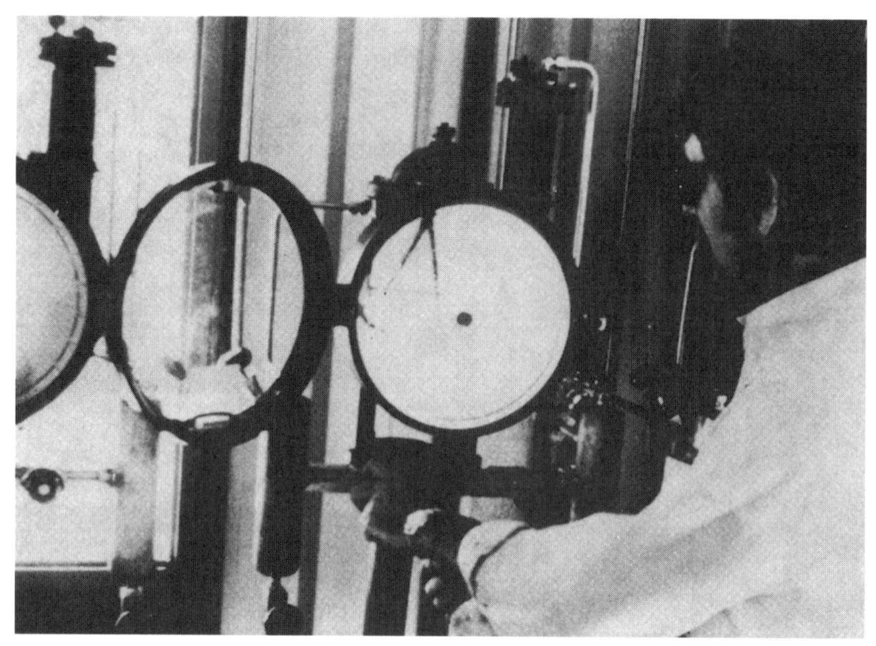

Figure 4. Charts are shown being changed, and clocks wound and adjusted. Each measuring station has its own flow pattern, which is recognizable by experienced chart changers.

small errors in static pressure, differential pressure, temperature, and specific gravity measurements.

Since external tests and inspections do not reveal the internal condition of a meter tube, the meter tube and its components must be inspected internally on an annual basis for possible buildups and internal corrosion. This inspection ensures continuing accuracy.

Chart changing

Charts must be changed on time, and clocks must be kept wound and adjusted (Figure 4).

Each measuring station has its own flow pattern, which is recognizable by experienced chart changers. Any change in this flow pattern indicates a possible error, which should be corrected if possible. If the correction cannot be made by the chart changer, he should notify the field measurement personnel who are qualified to perform required adjustment and/or repairs.

Each chart must be properly dated and identified along with field information necessary for calculating the required volumes. Automatic chart changers have become increasingly popular, especially in remote areas such as offshore Louisiana.

Periodic pulsation test should be made at each measuring station, particularly where a measuring station is located on the suction or discharge side of a compressor station. In this instance, pulsation dampeners must be installed between the compressor and the measuring station. Errors caused by pulsation can be very large, but they will not necessarily show in calibration tests.

A comprehensive maintenance program, faithfully followed, can detect and prevent costly errors before they occur. Leaking valves, gauge lines, etc., will cause errors in an increase or decrease in company revenue. Such leaks should be corrected as soon as possible.

Wet gas measurement

The problems encountered in measuring wet gas are considerably different from dry gas measurement.

Wet gas contains entrained liquids consisting of water and hydrocarbon condensates that can cause hydrate formations in the orifice plate fitting, flow recorders, and instrument lines. These hydrates are formed with the aid of free water under certain pressure and temperature conditions. Formation of hydrates can be reduced or eliminated in several different ways.

One successful solution to prevent hydrate formation is to use drip or condensate bottles in instrument and gauge lines. Infrared gas heaters can be used in meter buildings for heating gauge lines and flow recorders. This type of heater has become very popular lately, since it does not produce an open flame.

Errors caused by hydrate formations in the meter tube and around the orifice plate are common in wet gas measurement. The formation of these hydrates can be prevented by continuously injecting methanol into the gas stream ahead of the meter run. Experience and experimentation will determine the amount of methanol to be injected.

The use of methanol-containing inhibitors is not recommended as the inhibitor often solidifies in the meter tubes.

Four ways to calculate orifice flow in field

Rogers G. Thompson, Measurement Technician, Texas Eastern Transmission Corp., Shreveport, Louisiana.

Several methods are available for orifice flow field calculations.

The particular method used is determined by such factors as company policy, advance warning, time available, individual preference, accuracy required, and the information and equipment available at the time of need. A good rule to follow is to calculate as precise a volume as possible at all times.

Any volume calculated manually will be considered a field calculation.

Most companies now use some type of computer system, and rarely will a manually calculated volume match a computer volume. Field calculations are usually made under conditions that eliminate the use of adding machines and calculators. Accuracies can vary from an extremely accurate

Table 1
Combined specific gravity and temperature correction factors—orifice meters combined factor $= \left(\sqrt{\frac{1}{G}}\sqrt{\frac{520}{460°F}}\right)$

Flow Temp. °F	Specific Gravity .590	.591	.592	.593	.594	.595	.596	.597	.598
35	1.3343	1.3332	1.3321	1.3309	1.3298	1.3287	1.3276	1.3264	1.3254
36	1.3330	1.3319	1.3208	1.3296	1.3285	1.3274	1.3263	1.3251	1.3241
37	1.3317	1.3306	1.3295	1.3283	1.3272	1.3261	1.3250	1.3238	1.3228
38	1.3303	1.3292	1.3280	1.3269	1.3258	1.3247	1.3235	1.3224	1.3214
39	1.3290	1.3279	1.3267	1.3256	1.3245	1.3234	1.3222	1.3211	1.3201
40	1.3277	1.3266	1.3254	1.3243	1.3232	1.3221	1.3209	1.3198	1.3188
41	1.3264	1.3253	1.3241	1.3230	1.3219	1.3208	1.3197	1.3185	1.3175
42	1.3251	1.3240	1.3228	1.3217	1.3206	1.3195	1.3184	1.3172	1.3162
43	1.3238	1.3227	1.3215	1.3204	1.3193	1.3182	1.3171	1.3159	1.3149
44	1.3223	1.3212	1.3201	1.3190	1.3179	1.3168	1.3156	1.3145	1.3135
45	1.3210	1.3199	1.3188	1.3177	1.3166	1.3155	1.3143	1.3132	1.3122
46	1.3197	1.3186	1.3175	1.3164	1.3153	1.3142	1.3130	1.3119	1.3109
47	1.3184	1.3173	1.3162	1.3151	1.3140	1.3129	1.3118	1.3106	1.3096
48	1.3171	1.3160	1.3149	1.3138	1.3127	1.3116	1.3105	1.3093	1.3083
49	1.3158	1.3147	1.3136	1.3125	1.3114	1.3103	1.3092	1.3080	1.3070
50	1.3147	1.3135	1.3124	1.3113	1.3102	1.3091	1.3080	1.3069	1.3059
51	1.3134	1.3122	1.3111	1.3100	1.3089	1.3078	1.3067	1.3056	1.3046
52	1.3121	1.3109	1.3098	1.3087	1.3076	1.3065	1.3054	1.3043	1.3033
53	1.3108	1.3096	1.3085	1.3074	1.3063	1.3052	1.3041	1.3030	1.3020
54	1.3095	1.3083	1.3072	1.3061	1.3050	1.3039	1.3028	1.3017	1.3007
55	1.3081	1.3070	1.3059	1.3048	1.3037	1.3026	1.3015	1.3004	1.2994
56	1.3070	1.3059	1.3048	1.3037	1.3026	1.3015	1.3004	1.2992	1.2982
57	1.3057	1.3046	1.3035	1.3024	1.3013	1.3002	1.2991	1.2980	1.2970
58	1.3044	1.3033	1.3022	1.3011	1.3000	1.2989	1.2978	1.2967	1.2957
59	1.3032	1.3021	1.3010	1.2999	1.2988	1.2977	1.2966	1.2955	1.2945
60	1.3019	1.3008	1.2997	1.2986	1.2975	1.2964	1.2953	1.2942	1.2932
61	1.3006	1.2995	1.2984	1.2973	1.2962	1.2951	1.2940	1.2929	1.2919
62	1.2994	1.2983	1.2972	1.2961	1.2950	1.2939	1.2928	1.2917	1.2907
63	1.2981	1.2970	1.2959	1.2948	1.2937	1.2926	1.2915	1.2904	1.2894
64	1.2970	1.2959	1.2948	1.2937	1.2926	1.2915	1.2904	1.2893	1.2883
65	1.2957	1.2946	1.2935	1.2924	1.2913	1.2902	1.2891	1.2880	1.2870
66	1.2945	1.2934	1.2923	1.2912	1.2901	1.2890	1.2879	1.2868	1.2858
67	1.2932	1.2921	1.2910	1.2899	1.2888	1.2877	1.2866	1.2855	1.2845
68	1.2920	1.2909	1.2898	1.2887	1.2876	1.2865	1.2855	1.2844	1.2834
69	1.2908	1.2897	1.2887	1.2876	1.2865	1.2854	1.2843	1.2832	1.2822
70	1.2895	1.2884	1.2874	1.2863	1.2852	1.2841	1.2830	1.2819	1.2809
71	1.2884	1.2873	1.2862	1.2851	1.2840	1.2829	1.2818	1.2807	1.2798
72	1.2872	1.2861	1.2850	1.2839	1.2828	1.2818	1.2807	1.2796	1.2786
73	1.2859	1.2848	1.2837	1.2826	1.2815	1.2805	1.2794	1.2783	1.2773
74	1.2847	1.2836	1.2825	1.2815	1.2804	1.2793	1.2782	1.2771	1.2761
75	1.2835	1.2825	1.2814	1.2803	1.2792	1.2781	1.2770	1.2760	1.2750
76	1.2824	1.2813	1.2802	1.2791	1.2780	1.2770	1.2759	1.2748	1.2738
77	1.2811	1.2800	1.2789	1.2778	1.2767	1.2757	1.2746	1.2735	1.2725
78	1.2799	1.2788	1.2777	1.2767	1.2756	1.2745	1.2734	1.2723	1.2713
79	1.2787	1.2776	1.2766	1.2755	1.2744	1.2733	1.2722	1.2712	1.2702

spot check to a very basic calculation. External factors usually determine what procedure to use.

Uses for the calculations determine the degree of accuracy required for each particular instance. Since time is usually at a premium, it would be absurd to calculate to the cubic ft when calculating to the Mcf would suffice. A good example is automatic control of a compressor station where rates are given in terms of MMcf.

Field calculations are used for such things as determining purge gas volumes, relief valve losses, set point determination and sizing of plates. In general, volume, rate, differential, and plate size are the form variables that are determined by field calculations.

This discussion will be based on the AGA Committee Report No. 3 as it is the most commonly used. It is also the most complicated due to the large number of corrections applied.

Four categories have been devised to aid in discussing field calculations: the AGA formula, graphs, computers, and the quick method.

AGA formula

In making field calculations a good rule to follow is—the simpler, the better. Detailed calculations are difficult to perform under usual field conditions.

Another good rule to follow is cool and check. Always allow some time to elapse between calculating and checking. This can be from 5 minutes to 5 hours. Even better is to have someone else check your work. Following these rules will eliminate most errors.

The gas flow equation as stated in the AGA Committee Report No. 3 is:

$$Qh = Fb\ Fr\ Y\ Fm\ Fpb\ Ftb\ Fm\ Fa\ Fl\ Ftf\ Fg\ Fpv\ \sqrt{h_w pf}$$

Obviously the most accurate method of volume calculation would use all of the above factors. Most field calculations will use some modified form of the AGA equation.

In order to present a logical discussion, these correction factors are broken down into three groups—major, minor, and intermediate. This grouping is based upon the effect each factor has on the volume.

The major factors are the basic orifice factor Fb and the specific gravity factor Fg. These two factors have to be included in all computations.

The intermediate factors are the pressure base correction Fpb, flowing temperature correction Ftf, and supercompressibility correction Fpv. The value of these factors is variable. Usually they are determined by operating conditions at each specific location or are set by contract as in the case of pressure base.

Supercompressibility can become a large correction with transmission companies and producers where gas is measured at elevated pressures. For field calculation purposes, intermediate factors will be considered as either major or minor.

The Fr, Y, Fm, Ftb, Fa, and Fl corrections are minor factors for field calculation because of their minimal effect upon the final volume. Most of these factors are usually ignored.

Condensed formula

As previously stated, the most accurate method of field calculation is the AGA formula using all corrections. However, this is usually not possible because of insufficient time, advance warning, or operating information.

If a little work is done prior to the time of use, the formula can be condensed to three or four calculations. However, some operating information must be available. The static pressure, differential, run and plate size, specific gravity, and gas composition must either be known or assumed.

Assumptions should be based on knowledge of normal operating procedures and characteristics of similar M&R stations. As an example, the inert composition of gas is assumed to be 0% nitrogen and 0% carbon dioxide until an analysis is performed. This procedure permits volume calculations until more definite information is available.

By using known and assumed data, an hourly coefficient composed of Fb, Fr, Y, Fm, Ftb, Fpb, Fa and Fl can be computed. The only corrections left to be made are Fg, Ftf and Fpv. By giving some thought, time, and prior work, the formula can be reduced to:

$$Qh = C\ Fg\ Ftf\ Fg\sqrt{h_w pf}$$

Tables are available that further reduce the amount of calculation that has to be done. Combined specific gravity and temperature tables are available from different sources. Also available are extension values for $\sqrt{h_w pf}$ or the individual values of $\sqrt{h_w}$ and $\sqrt{pf}$. These tables eliminate what is probably the hardest part of a field calculation.

When following these guidelines, calculations become fairly simple. One note of caution: always make computations in a legible and logical order. This makes everything fairly simple to check. As many errors are caused by illogical layout of calculations as any other factor.

All methods from this point on are an adaptation of the AGA Report No. 3 equation.

Flow graphs

Usually the next most accurate method is the company-developed flow graphs.

Although the graphs necessarily have to be based on average data, they are tailored to each particular company's

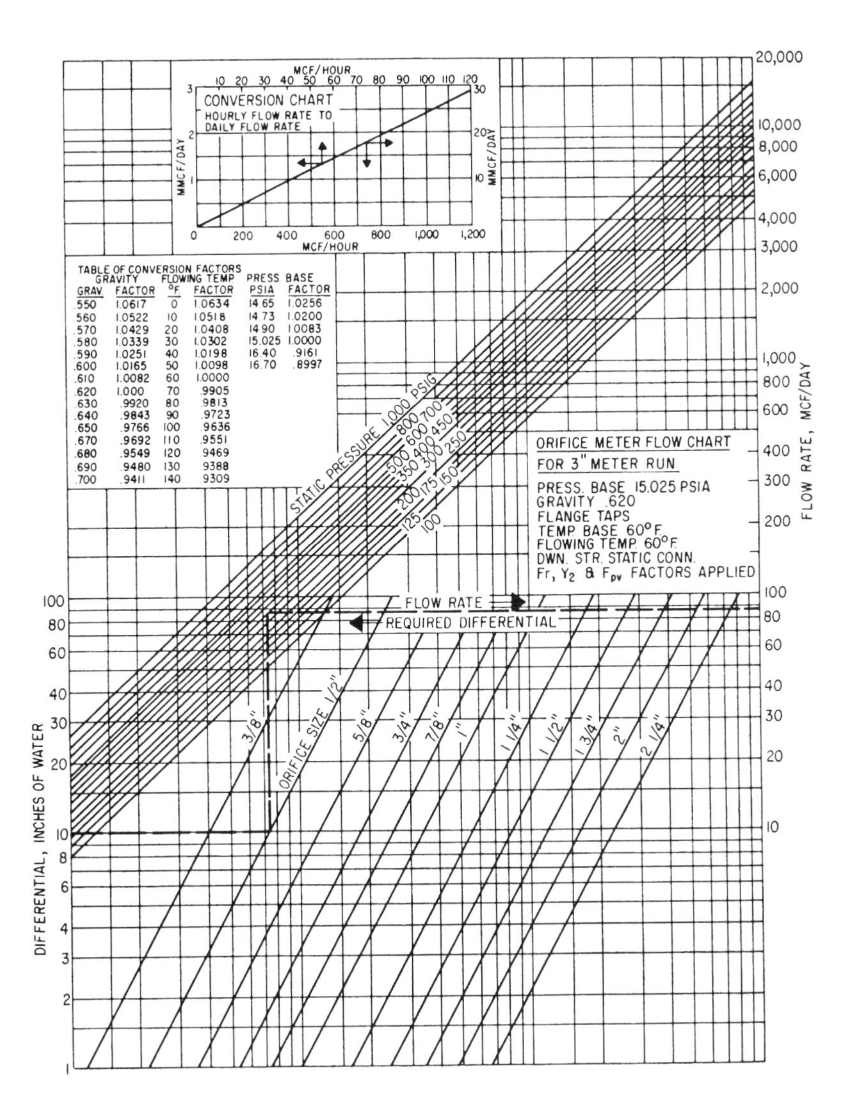

TABLE OF CONVERSION FACTORS

GRAVITY		FLOWING TEMP		PRESS BASE	
GRAV.	FACTOR	°F.	FACTOR	PSIA	FACTOR
.550	1.0617	0	1.0634	14.65	1.0256
.560	1.0522	10	1.0518	14.73	1.0200
.570	1.0429	20	1.0408	14.90	1.0083
.580	1.0339	30	1.0302	15.025	1.0000
.590	1.0251	40	1.0198	16.40	.9161
.600	1.0165	50	1.0098	16.70	.8997
.610	1.0082	60	1.0000		
.620	1.000	70	.9905		
.630	.9920	80	.9813		
.640	.9843	90	.9723		
.650	.9766	100	.9636		
.670	.9692	110	.9551		
.680	.9549	120	.9469		
.690	.9480	130	.9388		
.700	.9411	140	.9309		

Figure 1. Example of orifice meter flow graph used by Texas Eastern.

operating conditions. Thus, you get more accurate results for your particular requirements. Corrections such as gravity can be made to the final figure if desired, but usually are not.

The graph has the advantage of giving considerable information quickly. All plate size and differential combinations are shown at a glance. Daily and hourly flow rates are also given. This is the fastest and easiest of all methods to use, and accuracy is usually good enough to use under most conditions. No calculations are necessary under most conditions.

Most paper graphs are bulky, requiring a straight edge and space to use. Humidity will destroy them. However, replacements usually are available upon request. Oddly enough, this is probably the least used of all methods.

Most companies use the AGA equation to develop their graphs. The newer factors Fa and Fl probably are not included in current data. Most factors are based on a particular company's average data.

Some factors are given a value of 1.000 and ignored. Supercompressibility corrections are usually by average specific gravity and inert composition.

Computers

Flow computers and slide rules are probably the most widely used method of field calculation. This is due to their compactness and simplicity of operation. Accuracy is generally in the acceptable range for field calculations.

Most manufacturers of measurement equipment furnish some type of flow computer either free or at a nominal cost. Construction usually is of paper and plastic.

All of these flow computers are based on the AGA Report No. 3 flow equation. Most of the minor corrections are not used, with the omissions listed in the instructions. Since computers are designed for use throughout the industry, they must be kept standard. Consequently, they will not fit a particular set of conditions without making additional corrections.

Most flow computers incorporate Fb, Fg, Ftf, Ftb, and sometimes a generalized form of Fpv. The minor factors are usually ignored. The pressure base is usually set at 14.7 or 14.73, but some computers do incorporate a pressure base correction feature.

The biggest single drawback to computers is the fact that they must be general in scope and not slanted to one particular method of calculating volumes. Corrections are kept too general to fully correct for a specific set of conditions. There must be enough use of a computer for a person to maintain speed and accuracy of operation. If not, the instruction booklet will have to be studied each time the computer is used.

Flow computers are compact, easy to carry and store, and are very quick to use. As long as their limits are recognized, they can do an excellent job.

Quick method

Primarily, the quick method uses the major factors Fb and Fg only.

Inclusion of the intermediate factors is desirable if sufficient information is available. These intermediate corrections are made in a general way only. For example, if the pressure base is 15.025, subtract 2% of the volume. Algebraically, add 1% of the volume for each 10° variation from the temperature base. For each 100 psi of static pressure, add 8% of the volume.

This procedure will suffice for a lot of field calculation requirements. Keep in mind that this method is a rough calculation and is not intended to be used as a precise quantity.

Some mention must be made of the extension that probably causes more trouble than any other part of a volume calculation. In determining average differential and pressure, spot readings are the easiest to get and the most commonly used method. Sometimes predetermined values are used. Quite often differential and pressure are averaged over several hours. When averaging, care should be exercised for differentials under 25 in. and especially for differentials under 10 in. Due to the square root effect, errors are magnified in these ranges.

Extension books are available either in the form $\sqrt{h_w pf}$ or the individual $\sqrt{h_w}$ and $\sqrt{pf}$ tables. Although the combined tables are bulky, the individual tables can be limited to a few small sheets. Use of these tables will improve accuracy and speed.

Practical maintenance tips for positive displacement meters

Here are the most common malfunction causes and points to watch on the widely used meters

J. A. Wilson, Meter Shop Manager, Northern District, Skelly Oil Co., Pampa, Texas

Thorough understanding of a meter's function is necessary to properly repair any mechanism within it.

There is a definite reason for each meter part, and each part has a definite relation to the other. The alignment of any one part with another has been precisely established for each type and size of meter.

The functions of a positive displacement meter are similar to a four-cylinder engine. For one revolution of the tangent, each chamber must fill and exhaust once.

One revolution of the tangent is 360°. Therefore, to make a smooth running meter with uniform displacement, one of the chambers must start to exhaust every 90°. Since the valves control the time at which the chamber starts to fill and exhaust, they must be coordinated, one with the other, so that this 90° relationship is maintained.

Similarly, the diaphragms must also be connected to assure this relationship, at least in their extreme positions. Both valves and diaphragms must be connected to coordinate both 90° relationships.

Following are practical tips for PD meter repair and testing.

PDM repairs

The repairman cannot always carry parts for every type meter. Also, testers who only do an occasional repair job are not as efficient as men performing repairs daily. Therefore, most major repairs and all major overhauls are more efficiently handled in the meter shop.

Operating conditions in the field vary and are not always the best. Some of the operating problems in the field are gas saturated with water and hydrocarbons, iron oxide, sour gas, and lubrication.

Shop repair. Meter shops, no matter how well equipped, cannot perform satisfactory meter repair without qualified and dedicated personnel.

When an experienced man disassembles a meter for repair, he inspects each part as he tears it down, considering each as a possible cause of the malfunction. When the trouble is found, it is corrected before proceeding, if possible.

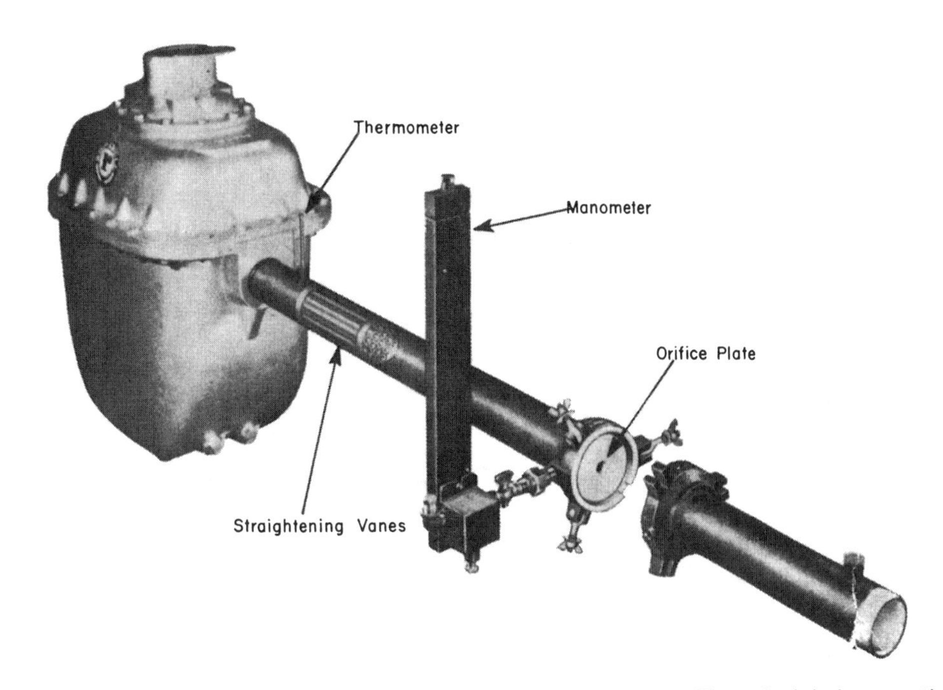

Figure 1. Low pressure flow provers for PD meters can be used in the field or shop. The principle is essentially that of the orifice meter, except the orifice discs are calibrated for different flow rates instead of using various coefficients. Shown above is nomenclature for the Rockwell prover.

Some of the most common causes of malfunction and points to watch on the more widely used meters are:

- Exhaust port gaskets should be inspected to make sure the seal is good. Some meters have metal spacers under the seal and must be in place when reassembled.
- The clearance of the flag rods at the top of the meter should be checked. After a meter has been time checked and balanced, it will sometimes run rough. Look for a bright spot right over the top of the flag rods in the top cover. This is not a common problem but is difficult to locate, and a few strokes with a file will take care of it.
- The reassembly of the tangent gear between the valve gears of meters, as well as other gear-driven valves, should be centered carefully. There is usually a small amount of side play as well as end play on the tangent shaft bracket. A little extra care here will contribute to a quiet operating meter.
- The body seal O-rings on the later models should be placed carefully in the slot when the top cover is put into place.

Flag arms. Because new materials are being used in meters and the new aluminum-case meters are designed differently, there have been changes in some repair procedures. Some meters use an aluminum, long flag arm with a brass bushing cast in the center, so the flag arm can be soldered to the flag rod.

Most repairmen use a wet sponge to rapidly cool a solder joint on flag rods when flag arms are attached, but because of the different expansion coefficients of brass and aluminum, the repairman must use extra precaution in cooling these solder connections.

Tests have proven that cooling the aluminum part of the flag arms too rapidly will cause fractures in the aluminum surrounding the brass bushing. Most of these fractured flag arms will go undetected and later break down in operation. However, if the repairman cools the solder joints by easing a wet sponge slowly up the flag rod, the joint is cooled from the inside out and no damage is done.

Valves and seats. Careful grinding of the valves and valve seats is very important. Number 80 grit may be used to clean up bad valves and seats, finishing with number 120 grit. Most repairmen use the figure eight motion in grinding, letting the weight of the glass block control the pressure on the valve seat. Use a very light touch on the valves. Where it is practical, both valve seats should be ground together. Grinding paper that is self adhesive is certainly recommended over grinding paper that has to be cemented to a glass block.

There are thousands of ironcase meters that have been in service for a great number of years. The valve seats on these meters should be carefully checked to make sure they are still bonded to the meter body. Remove and

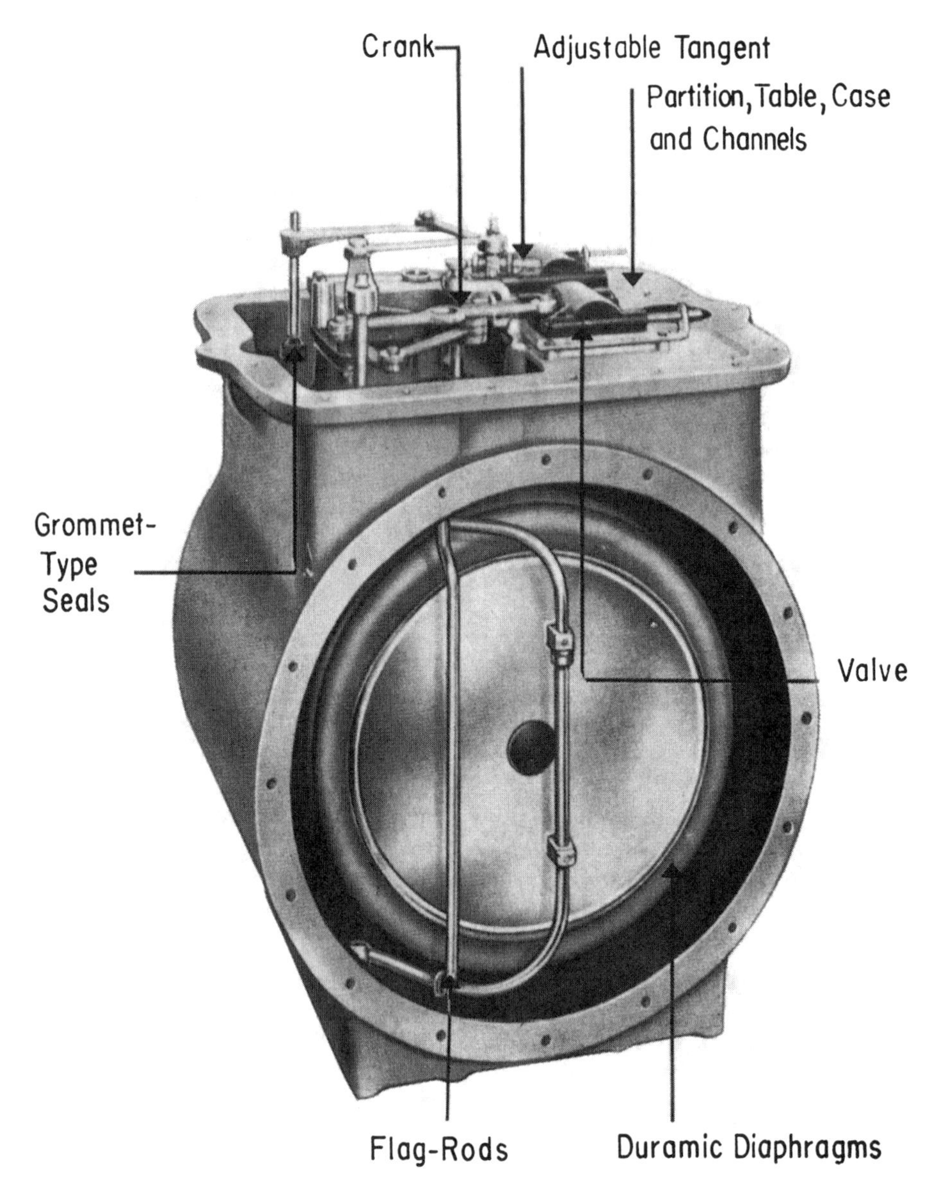

Figure 2. Thousands of ironcase PD meters, such as the American 25B and 35B shown above, have been in service for many years. Valve seats should be carefully checked to make sure they still are bonded to the meter body.

reseal the seats if there is any doubt about the strength of the bond. Use an old grinding block to true up the bottom of the valve seats and buff the meter case clean to ensure a good bond.

Clean all surplus cement from the intersurfaces of the valve seat opening. Be careful not to use screws that are longer than standard in valve guides that screw into the valve seat. An overlength screw can jack the valve seat off the body and break the seal.

Watch the clearance on the new plastic parts when they are installed in old style meters. The raised reinforcing rib on some of these parts may pose clearance problems in many meters. Soldering acid will dissolve some plastic, so if you must use acid core solder, be careful to keep it off the plastic.

Balance and time check. Take plenty of time to balance and time the valves and diaphragms to ensure a smooth operating meter. Use care in setting the tangent in proper relation to the valves. Time spent here will be regained manyfold when the meter is calibrated.

Most shops have a utility-type vacuum sweeper. This is an excellent means of rotating a meter before the top cover is installed. This gives the repairman a good chance to watch the valve action and make a final check on timing and balance, gauge, and depth of the undergear driving arm in relation to the tangent. Make sure the tangent wrist is engaged, but do not touch the short flag arm.

Installing a long pointer on the index driver dog will help proof test the meter, since the circumference of the circle will be much longer than the test-hand on the index. It is a good practice to rotate the meter at least two rotations of the test-hand after the index is installed. This gives a final check to make certain the meter is operating smoothly. A vacuum cleaner can also be used to good advantage here.

Pan-type diaphragms can be rebuilt with a minimum amount of investment in jigs and other equipment. Many parts, such as crank brackets and flag arms that have pressed-in bushings, are easily repaired. Rebuilding parts can save up to 75% of the cost of a new item.

When it comes to repairing meters, "about right" isn't good enough. The meter repairman who wants to make his company money will strive to make all checks and adjustments as accurately as possible.

PDM testing

Positive displacement meters may be tested by any of the following methods: (1) bell type meter prover, (2) low pressure flow prover, (3) critical flow prover, (4) test meter. All of these methods, when correctly used under proper conditions, will give similar results.

Bell prover. Meter shop personnel are more familiar with the bell prover than with the other three types.

When a bell of sufficient size is available, large-capacity meters can be calibrated very rapidly in the shop. The use of a slope gauge will certainly reveal if the repairman has assembled a well-balanced, timed meter with a minimum of internal friction.

Bell provers should be in a temperature-controlled area, and all meters should be at room temperature before testing. Normally both a 20% rate and capacity rate are run when proof testing flow rate on the meter. 2 in. of water pressure on the bell is used instead of the 1½ in. used. To test small domestic meters, 2 in. of water pressure on the bell is used instead of the 1½ in. used.

Low pressure flow prover. The low pressure flow prover is used both in the field and in the shop. It is very compact and portable, consisting of two pieces of aluminum pipe with straightening vanes in the upstream sections.

The principle is essentially that of the orifice meter, except the orifice discs are calibrated for different flow rates instead of using various coefficients. Many shops use the low pressure flow prover to test large capacity meters when a bell prover of sufficient size is not available. One or two blowers may be used to supply the necessary volume of air, or low pressure gas may be used where safe.

Critical flow prover. Critical flow is a condition occurring in the gas flow when the downstream pressure is approximately one half the upstream pressure.

The critical flow prover consists of a piece of 2-in., extra heavy pipe approximately 18 in. long. One end is threaded to screw into a test connection placed as close to the meter outlet as possible. The other end has a special orifice plate holder. The critical flow prover can be used to test positive displacement meters operating on pressure in excess of 15 psi by passing gas or air through the meter and the prover and discharging it into the air.

The orifice plates of a critical flow prover are stamped on the downstream side with the time (in seconds) required for one cubic ft of air to pass through the orifice at critical flow conditions. Plates used in critical flow provers have rounded edges on the upstream side, whereas orifice plates in a low pressure flow prover are sharp on the upstream edge.

It is vitally important that both the critical flow and low pressure orifice plate be handled with care to prevent damage, which would change the time stamped on the plate.

The stopwatch, which is important in timing prover tests, should be serviced and checked at regular intervals. Like many pieces of mechanical equipment, the stopwatch will operate for a long time before it completely gives out, and there is a very good possibility that during this time it will be operating inaccurately. Remember that tests taken with a defective timer are inaccurate.

Test meter method. Test meters have become popular in recent years with larger companies and contract testers. Continued development of rotary displacement has made the test meter a practical method for field use.

Although the shop tester normally runs only two rates—check rate and capacity rate—on a bell prover, most testers run three rates on the critical flow, low pressure, and test meter methods. These rates are usually about 20%, 50%, and 100%.

Testing frequency is a much-discussed subject. Sales through a large-capacity meter represent a great deal of money; thus, it is important to regularly check the meter's accuracy. Some of the larger meters should be tested at 90-day intervals. Others of intermediate size could perhaps operate for 2, 5, or 7 years without being tested and at the end of the scheduled time, simply be changed out.

Many factors affect a testing and change-out schedule. The meter's mileage, its type of service, the type and amount of gas the meter must measure, and the length of time the meter is in service are important factors in setting up a testing and change-out schedule. Only after all of these factors have been considered can an adequate testing and change-out program be established.

Sizing headers for meter stations

Calculate the velocity of gas in meter headers for proper sizing to achieve good measurement

Glerm E. Peck, Senior Engineer, Pipe Line Division, Bechtel Corp., San Francisco

Meter and regulator station design for natural gas transmission systems continues to grow more complex and demand more of the design engineer. Basic elements of today's station remain about the same, but higher pressures and ever-increasing capacities have replaced the positive displacement meter with multi-tube orifice meters and automatic chart changers. Buildings are often pre-fabricated structures designed specifically to house telemetering transmitters and supervisory and communications equipment in addition to basic recording instruments. The multi-run setting may have a computer giving the dispatcher instantaneous rate and totalized flow for the station. Electrical equipment may cost several times that of the entire station of a few years ago.

There is a sparsity of published criteria for sizing of orifice meter headers in the multi-run station. A discussion of factors that affect the sizing of the header and arrangement of component parts for satisfactory operating conditions should be in order.

We shall assume that the orifice tube and its component parts are designed in accordance with AGA Gas Measurement Report No. 3, that piping meets all the requirements of USASI B31.8 for pressure, materials, and class of construction, that all valves and fittings meet appropriate codes or standards, and that the differential pressure across the orifice plate will not exceed 100 in. of water.

The design goals may be listed as follows:

1. Commercially available equipment will be used that is specified to the closest reproducible economic tolerances.
2. The arrangement of the equipment shall conform to accepted practice for good measurement and allow fullest use of the available meter tubes.
3. Fittings and connections to the header shall minimize the flow turbulence in so far as is practical.

Meeting the above goals allows considerable latitude in the choice of components according to the judgment and experience of the designer without contradicting code, contract, or economic requirements.

In sizing the header a number of factors must be considered, namely:

- The number of orifice tubes that will eventually be connected to the header, or the ultimate flow capacity
- The setting layout, or arrangement, which governs the resistance flow paths the gas must take from entering the inlet header and dividing through the meter tubes before recombining in the outlet header and exiting
- The type of connections which the tubes make with the header

Station layout

There are as many opinions of what constitutes good meter station layout as there are engineers working with gas measurement. Four of the more common arrangements are illustrated in Figures 1–4 and will be discussed relative to the flow-resistance paths offered.

Header connections

Not many years ago, the header was a common piece of pipe with "orange-peel" ends and notched-miter nipple connections. The next step was a shaped nipple reinforced by a welding saddle or pad. Higher pressures brought forged steel tees into use. On larger-sized tees the spacing between runs or tubes depended on the measurement for the tee and required a number of girth welds. With the introduction of the extruded nozzle header, the limitation of tees was eliminated for a generally more economical header.

The radius of curvature or flaring of the nozzle as extruded has superior flow characteristics over the reduced branch of a forged tee, and both are vast improvements over the square cut opening for a shaped nipple. Some designers have gone a step further and reduced the 90° header nozzle by the use of a 45° or 60° lateral nipple to the axis of the header. This reduces the turbulence entering or leaving the header and could reduce the diameter of the header. This method can be much more costly for fabrication, is suspect for structural integrity at higher pressures and extreme ambient conditions, and goes beyond the configurations used in establishing the proportions, beta ratios, and coefficients for the orifice meter.

Runs per header

The number of runs, or meter tubes, for a given header is limited by the velocity criterion selected, the operating pressure, the size of fittings obtainable, and the relationship of the peak load to the normal load or capacity for which the meter setting is designed. The operating pressure has an important bearing on the number of orifice tubes used with a given header, because the capacity of the tube is

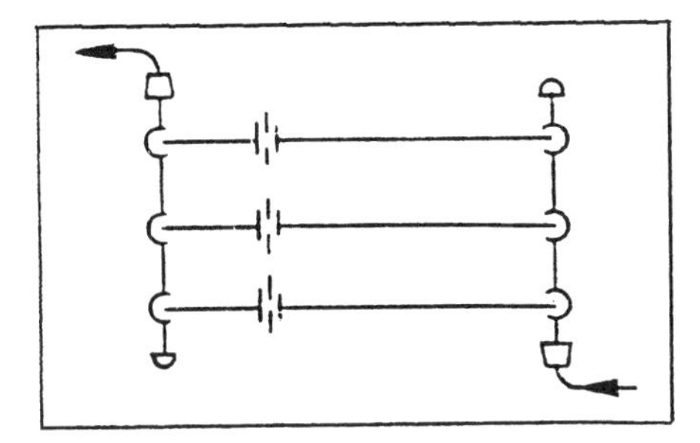

Figure 1. The flow-resistant paths are identical for this layout and each of the two runs will furnish its full capacity with near identical differential readings on the meters. However, when additional capacity is required, an additional run will have a higher or longer resistance-flow path and will furnish a lower differential reading and capacity as compared with the first two runs.

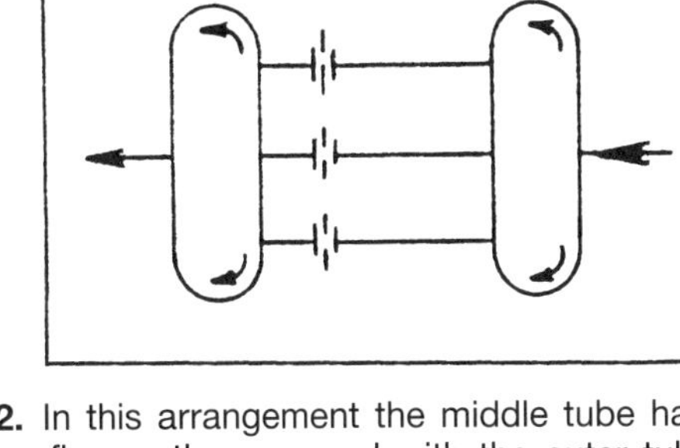

Figure 2. In this arrangement the middle tube has the lower resistance-flow path compared with the outer tubes and will affect the meter reading accordingly. The difference between meter readings will be more noticeable if the headers are undersized.

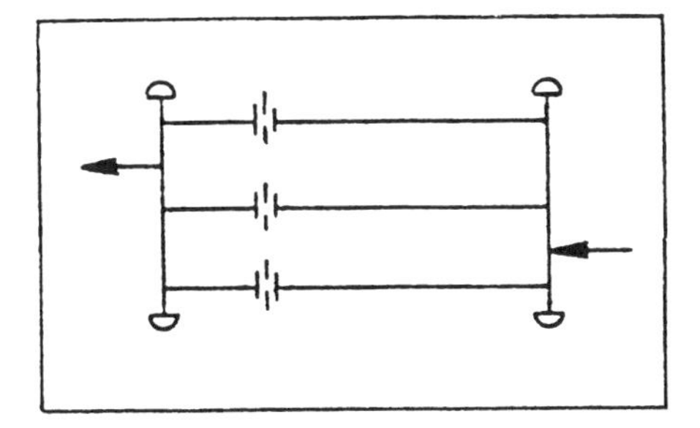

Figure 3. The "donut" header has been used frequently when large headers could be utilized. At higher pressures and capacities this layout places a high economic burden on the station and requires more space, which can be a problem when a limited site is available.

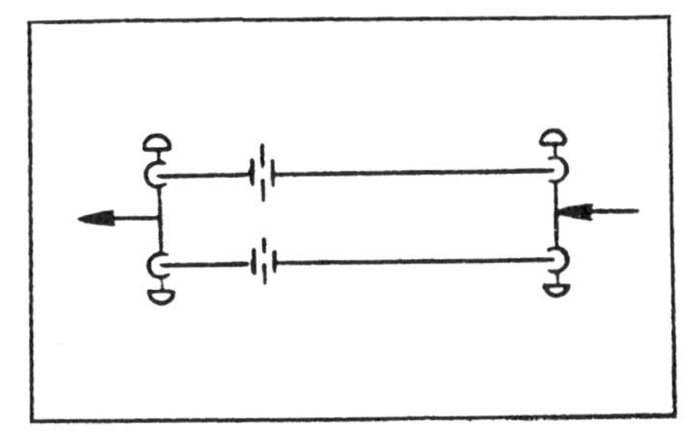

Figure 4. This layout has been a favorite for many since the advent of the extruded or drawn nozzle header and because the resistance-flow paths are as nearly identical as is possible using commercially available equipment. It is limited to three or four meter tubes before the header begins to be "out-sized."

proportional to the square root of the pressure while the capacity (velocity) of the header is proportional to the pressure in atmospheres. At lower pressures the number of runs might be limited to three while at higher pressures the limit would be four runs. Size of fittings is generally less limiting than other factors unless the meter tube is greater than 12 in. in diameter. For the very large tube, the extruded nozzle header is required, since reducing branch tees are not generally available. The relationship of the peak load to the sizing of the header is discussed under the velocity sizing criterion.

Velocity sizing

The following velocity criterion is suggested for sizing of headers for the sales orifice meter setting (expressed at ft per minute):

Target	Normal Limit	Maximum Limit
1,500	2,000	2,500

The relationship of the peak hour load to the average hour for the peak day designed for the station determines the range of velocity used in sizing the header. For example, the peak hour which is 10% of the peak day and one that is only 1% higher than the average hour for the peak day obviously demands that different limits be used for economic justification. The 10% peak hour suggests that for higher fluctuation in demand, a higher velocity be allowed in the header for the short time involved.

Keeping the velocity as low as possible within this criterion will provide the basis for good measurement and minimize the difference between adjacent meter readings. The way that gas enters and leaves the header connections to the meter tube is important in considering the velocity criterion that is chosen.

Sizing procedure. The normal procedure for sizing the station follows:

1. Take the peak day forecast or contract load and calculate the appropriate peak hour figure—provided the

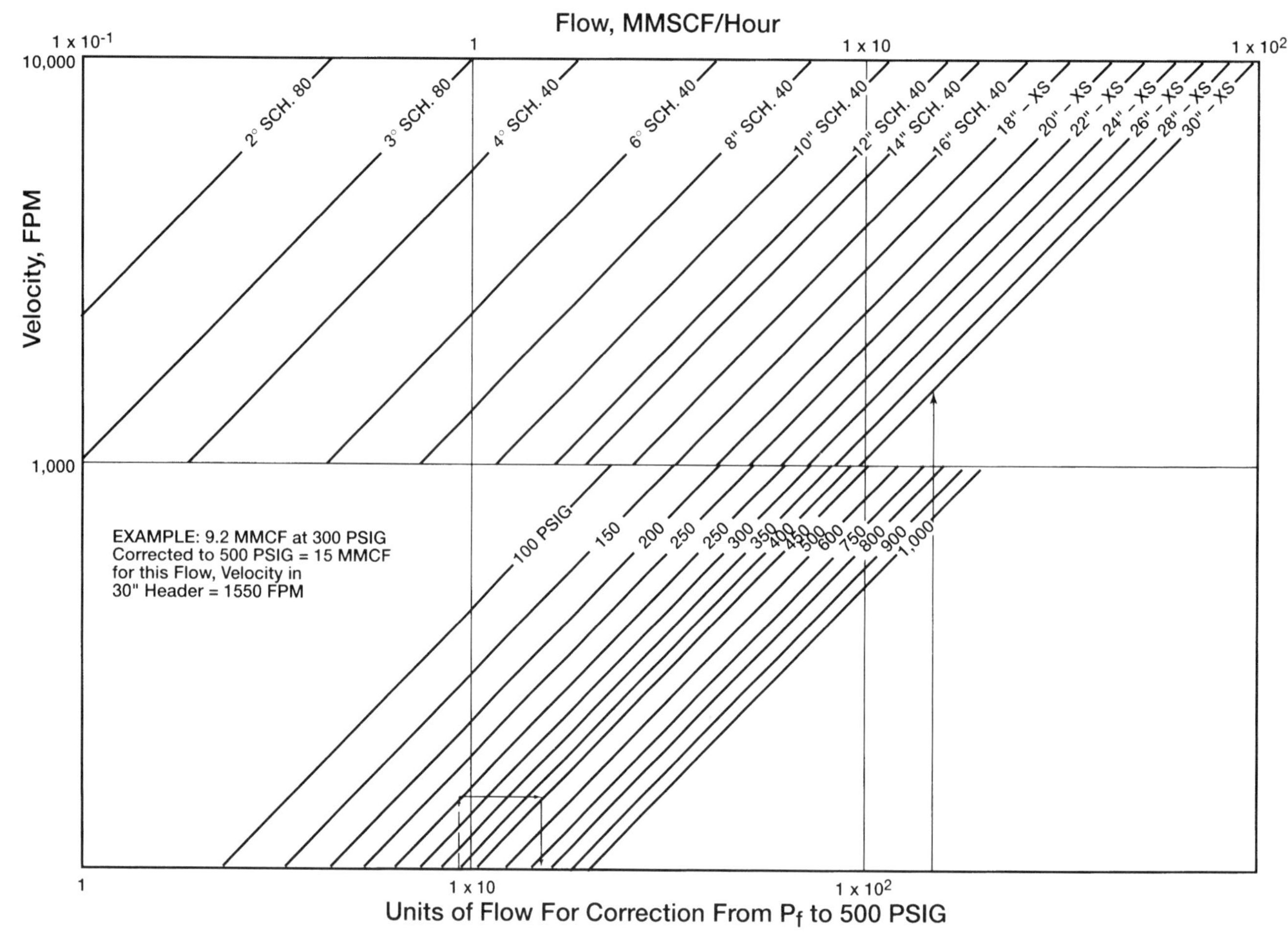

Figure 5. Pipe and header velocity sizing chart.

peak hour figure has not been fixed as a part of the forecast or contract. (Suggested period: 10 years, minimum.)

2. Divide the peak hour by the number of parallel tubes desired.
3. Use a flow rule or chart to size the orifice tube. Check the initial minimum hour against the minimum beta ratio and differential desirable, since this may affect the size and number of tubes.
4. Use a chart, such as Figure 5, to calculate the velocity in the header for the proper criterion. Figure 5 has been found useful since the capacity can be traced up to the desirable velocity and the nearest appropriate size of header chosen.

Figure 5 was prepared for 500 psig or 514.65 psia at a temperature of 60°F in order that sizing of pipe or headers be made against velocity and capacity in Mscf/h on the upper part of the chart. The lower part of the chart is used to correct the flow at other pressures to the proportionate flow at 500 psig. Since the sizing procedure is an approximate one, it is felt that corrections for temperatures other than the standard condition and for compressibility are not necessary. Use of Figure 5 or a chart prepared to your own physical conditions will allow sizing of headers for the orifice meter setting quickly and accurately within the present state of the art.

16: Instrumentation

Types of control systems

Early pipeline systems depended upon manual control and the use of tanks to maintain a pipeline system in balance. On long pipelines that required multiple pump stations, a surge tank was used to catch excess flow or make up the deficiency. Engine speed was varied to maintain the level in the surge tank at an acceptable level. This type of operation was fine as long as one grade of product was being pumped but was not satisfactory for batching different products unless there was a tank available for each grade of product handled.

It was obvious that to build a tight line system, one that would not require surge tankage, some kind of automatic control system was needed to keep the booster stations in balance. Before the widespread use of centrifugal pumps, engine speed (for reciprocating pumps) was controlled to maintain the suction and discharge pressure within limits by sensing discharge and suction pressure and directing a control signal to either the engine governor or a control valve. When control valves are used with reciprocating pumps, they are connected to recycle fluid from the discharge of the pump to the pump section.

When centrifugal pumps came into use, the same type of control system was used except that the final control element was a control valve installed in the pump discharge line as shown in Figure 1. The control system at origin pump stations normally controls discharge pressure or system flow rate. Intermediate booster stations require a control system that will control suction or discharge pressure and in some cases, motor load. An examination of the system line friction and pump head curve will show that simultaneous control of suction and discharge pressure and high flow rate is not required since the pipeline system characteristics will not permit the coexistence of these events except in case of extreme system upset. As will be seen later, the control system is configured so that only one controller will be in control at a given time.

There are several modes of control available to the pipeline engineer. Mode of control is the manner in which control systems make corrections relative to the deviation of the controlled variable. The modes of control commonly used by pipeliners include:

- Proportional position
- Proportional plus reset
- Proportional-speed floating

Other modes such as proportional plus reset plus derivative are more commonly used in the process industries. This mode of control may, however, be used in pipeline measurement stations where it is necessary to control the temperature of a heater used to heat the product being pumped or measured.

In the proportional-position mode, there is a constant linear relation between the controlled value and the position of the final control element. With this type of control system,

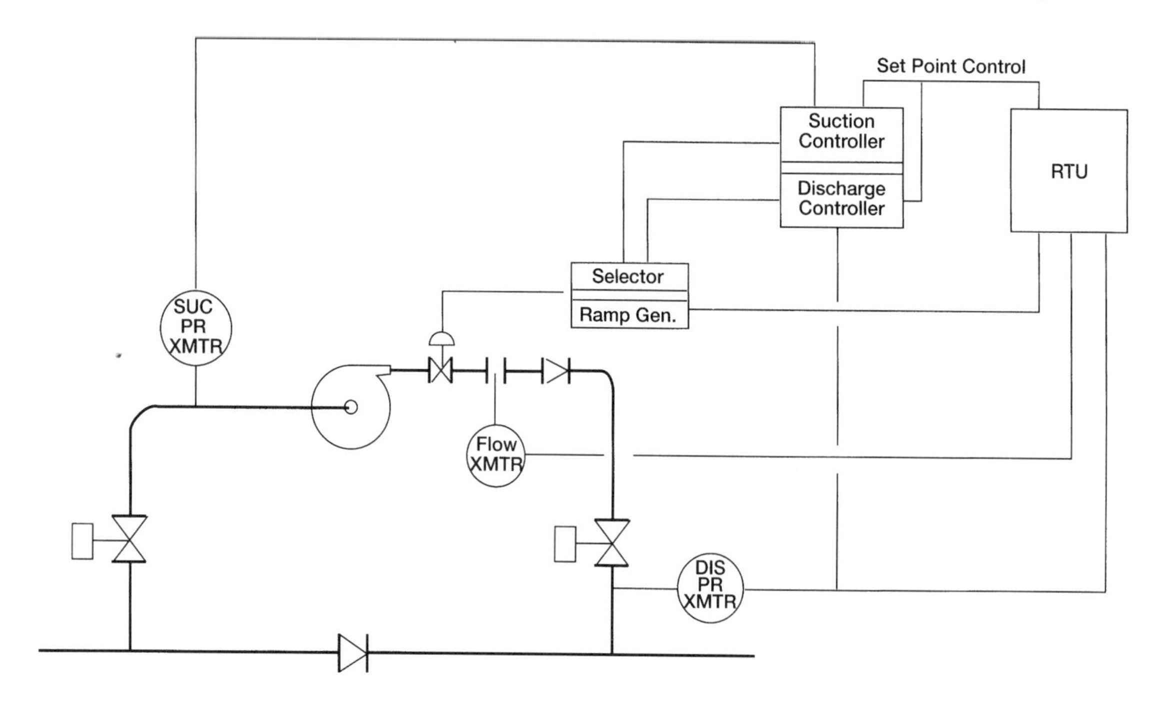

Figure 1. Typical control scheme for a pipeline booster station.

the amount of correction will be the same regardless of the amount of deviation of the controlled variable. A proportional-position control system will not necessarily return the controlled variable to the exact value of the set point when an upset occurs. This may be satisfactory for some types of control systems, but not for pipeline pump station control.

The offset deviation noted in the proportional-position system can be overcome by adding reset to the controller. The reset action will continue to correct valve position until no deviation occurs between the controlled variable and controller setpoint.

Proportional-speed floating causes the final control element to change at a rate proportional to the amount of deviation between the setpoint and the measured variable. Correction to the final control element will continue until there is no deviation between the setpoint and the measured variable. The amount of the correction signal is proportional to the amount of deviation between the setpoint and the measured variable. This mode of control will produce an exact correction for any system load change.

Proportional-position and proportional-speed floating are the control modes most commonly used for pipeline control.

Figure 1 shows a typical control scheme for a single unit booster station. A minimum control system will include a suction pressure and discharge pressure controller. Other controllers such as motor current controllers and a ramp generator may be connected in the control loop.

The suction pressure controller is set to maintain the suction pressure above the low suction pressure shut down setting while the discharge pressure controller limits the station discharge pressure to a value that will not exceed the maximum allowable operating pressure. The output from each of the controllers is fed to a signal selector that will select either the high or low signal, depending upon the control action selected, and direct this signal to the control valve. Some control systems are designed so that a selector module may not be required, but they will include a means for selecting the proper signal. In a properly designed system, the transfer of control from one controller to another will be smooth and bumpless.

A ramp generator is used for ramping the control valve to a new position. The control valve may be closed when the station is at rest, and upon station startup the ramp generator will cause the valve to open slowly until one of the controllers is in control of the system or until the pump is "floating"—that is, operating at a condition where neither suction nor discharge pressure control is required. It may also be used to ramp the valve partially closed prior to unit shut down to minimize system upset. This feature is desirable in a multi-unit booster station since when one unit is shut down, the remaining operating units will be operating momentarily at a no-flow condition. This condition will continue to exist until the system pressure decays to a value equal to the discharge pressure of the remaining units. Ramp speed should be adjustable so that it can be fine-tuned to the system requirements.

The control valve action, i.e., valve opens on increasing signal or valve closes on decreasing signal, will determine how the controller action is set. The suction pressure controller will be set opposite of the discharge controller.

Although not shown, the control system is usually equipped with a manual/automatic switch so that the control valve may be controlled in case one of the controllers fails or during system maintenance.

The controllers may also be equipped with remote setpoint control so that controller setpoint may be varied from a remote location to accommodate a changed condition in the system.

Figure 2 shows a control system for a multi-unit booster station. The control systems for single- and multi-unit stations are identical except that in a multi-unit station the station pressure transmitters are located so that one control system will suffice regardless of the number of units installed. It should also be noted that one control valve serves both units.

The control valve may be operated electrically, pneumatically, or hydraulically. Hydraulic systems offer the advantage of being able to operate the valve at a high rate of speed at virtually any system load. Most hydraulic operators are equipped with dual cylinders and may be equipped with quad cylinders where the system loads are high. The selection of control valves is covered in Section 7.

Variable speed drives or controls are becoming an economically attractive choice for the operation of pumps. The use of these devices will virtually eliminate throttling excess head and the attendant energy costs. Few systems are operated at a constant rate, and therefore some throttling is inevitable. Also, systems transporting multiple grades will require throttling to keep the system in balance as the batches move through the system.

Most control systems used for pipeline control are made up of standard control system elements. Some applications may, however, have unique characteristics that will require a system that is just a little bit different from a standard control system. Most of these types of systems can be put together using standard modules, or a programmable logic controller may be used.

In Section 13—Liquid Hydraulics, the discussion on hydraulic gradients illustrated a case where a hill required maintaining back pressure at the end of the line. It was shown that the amount of back pressure required varied with the system flow rate. A conventional back pressure controller and control valve installed at the end of the line and set for the worst case (maximum back pressure) would work; however, it would not be economical since more pressure than required would be maintained at the higher flow rates. The pressure at the hill could be telemetered to the end of

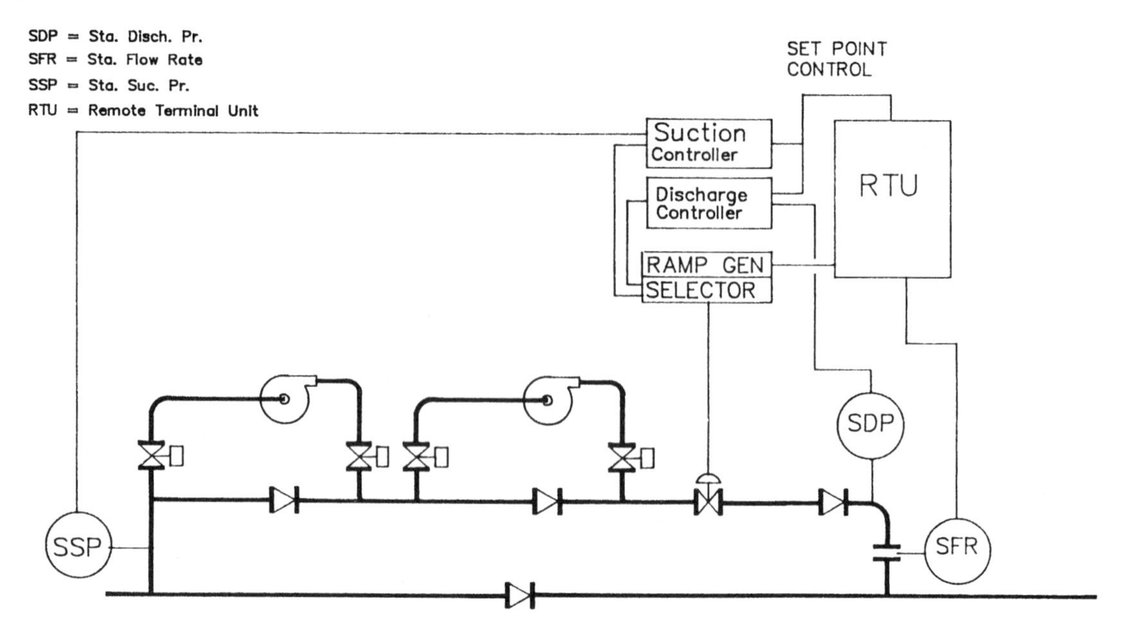

Figure 2. Multi-unit booster station.

the line, where it could be used as the measured variable for the controller. The setpoint would be set to the amount of pressure required at the hill which in this case is 200 psig. This type of control system would not penalize system economics since it would always maintain the required 200 psig at the hill. However, it would require the installation and maintenance of a telemetry link.

Another option is to use a relationship controller that would not require telemetry of the hill pressure. The relationship controller would use the system flow rate to drive the setpoint of a pressure controller. The controller would need to be equipped with limit stops so that the setpoint could only be changed between the upper and lower limits. High system flow rates would drive the controller setpoint to the minimum allowable value of back pressure and low system flow rates would drive the controller to the allowable maximum limit. Intermediate system flow rates would drive the setpoint to an intermediate pressure setting.

With today's technology and hardware, the control engineer is limited only by his imagination in solving simple as well as complex control problems. Programmable logic controllers (PLC) and self-tuning controllers are being used more and more in the control of pipelines and pipeline stations. Control systems are being incorporated as a part of the PLC. With the addition of a modem and a suitable communication protocol, the PLC may also be used as the remote terminal unit (RTU) for the supervisory control and data acquisition system (SCADA).

The following article, "Developments in Pipeline Instrumentation," by Dr. R. A. Furness, is an excellent discussion on the state-of-the-art in pipeline instrumentation.

DEVELOPMENTS IN PIPELINE INSTRUMENTATION

R. A. Furness, Center for Flow Measurement, Department of Fluid Engineering and Instrumentation, Cranfield Institute of Technology

Abstract

Pipelines are an integral part of the world's economy and literally billions of lb worth of fluids are moved each year in pipelines of varying lengths and diameters. As the cost of some of these fluids and the price of moving them has increased, so the need to measure the flows more accurately and control and operate the line more effectively has arisen. Instrumentation and control equipment have developed steadily in the past decade but not as fast as the computers and microprocessors that are now a part of most large-scale pipeline systems. It is the interfacing of the new generation of digital and sometimes "intelligent" instrumentation with smaller and more powerful computers that has led to a quiet but rapid revolution in pipeline monitoring and control. This paper looks at the more significant developments from the many that have appeared in the past few years and attempts to project future trends in the industry for the next decade.

Introduction

Our need to transport fluids from the point of production to the area of end use has led in the past 25 years to a large increase in the number of pipelines being designed and constructed. Many of these carry hazardous or toxic products, often through areas of high environmental sensitivity or close to centers of population. The cost of oil production rose sharply to more than $35 per barrel before the recent sudden fall, and the cost of manufactured chemicals and industrial gases has also risen in the same period. With the need to safeguard the cost of the transported fluid and also the population and the environment close to the pipeline, on-line monitoring and in some cases 24-hour surveillance is now becoming routine in many systems throughout the world.

When one sees maps of the pipeline network in some parts of the world, the number and length of pipelines is staggering. At the recent Aberdeen conference, one speaker gave just two of the many statistics that can be produced to show how vast the network is. One line runs from central Canada eastwards to the Toronto/Montreal area. This line alone has 54 million square meters of pipe surface to survey. Also in 1984 some 6 billion barrels of oil and 4.5 billion barrels of manufactured liquids were transported in pipelines in the USA. The actual world volume of pipeline fluids is many times this amount and the value immense.

Pipelines are not new, indeed almost 50% of the world's total network is 30 or more years old. These have been subject to corrosion and wear, and some have carried fluids for which they were not designed. Others have been operated outside design limits. Monitoring and control systems therefore are not just required for measurement purposes but also for integrity surveillance and on line fault diagnostics. Some of the many developments that can be applied in modern day line operation are briefly reviewed, and some design pointers for their integration into complete systems are given.

Flow measurements

Flow measurement is the most important process variable in the operation and control of pipelines. Simple line flow balances are used to check for discrepancies on hourly, daily, or weekly bases. This may identify that a problem exists, but flowmeters at the end of a line will not identify the location of the problem. For this, pressure transducers along the pipeline will be required together with the flowmeters to establish where the pressure gradient has deviated from that predicted from the basic equations of fluid flow. These equations use pressure/flow values at inlet and outlet to confirm the line is in balance. Such a simple system is described in more detail later in this paper.

Most flowmeters sold are of the differential pressure type, and of these the predominant meter is the sharp-edged orifice plate. Many hundreds if not thousands of these are installed on pipelines around the world to measure gas, liquid, and mixed fluid streams. With the arrival of the modern flow computer these types of flowmeter have almost received a new lease on life. One fact not appreciated, though, is the total measurement uncertainty of such devices. Frequently we read performance claims of ±0.5% and better for these pipeline metering systems. This is very difficult to believe when it must be remembered there is an uncertainty of ±0.6% on the data used to derive the flow coefficient in the basic calculation. When all the other variables that can affect the uncertainty of measurement are included, such as fluid composition changes, ambient and process temperature and pressure variations, errors in temperature and pressure measurement, A/D conversion and computational errors, to name but a few, then it is unreasonable to expect uncertainties much better than ±1% FSD. If wide flow rates are also experienced, then the uncertainty of the low flow determination could be ±5% FSD or worse. This makes their use as accurate pipeline flow devices somewhat questionable, although they are very common in the metering of natural gas in this country.

Turbine flowmeters also find widespread use in pipeline applications, particularly hydrocarbons. They all suffer from bearing limitations, however, and this means they require regular calibration to ensure good performance. Recent developments have included twin rotor designs, where measurements of the speed of both rotors simultaneously has led to the meter becoming self diagnostic. These meters can now detect their own bearing wear and also deviations from initial calibration due to blade damage. The same on-board microprocessors can also check the integrity of the data generated by the turbine meter, as well as providing alarm output for loss of signal and other problems.

Two different designs have been applied in pipeline systems with good success. The first design, from Quantum Dynamics and shown in Figure 1, has two rotors with two independent sets of bearings. The shaft and slave rotor ride on one set of bearings with the measurement rotor upstream of the slave motor riding on a second set of bearings on the rotating shaft.

Relative motion between the indicating turbine and the shaft therefore remains close to zero. This plus a drag-free RF pick-off results in an almost frictionless flowmeter.

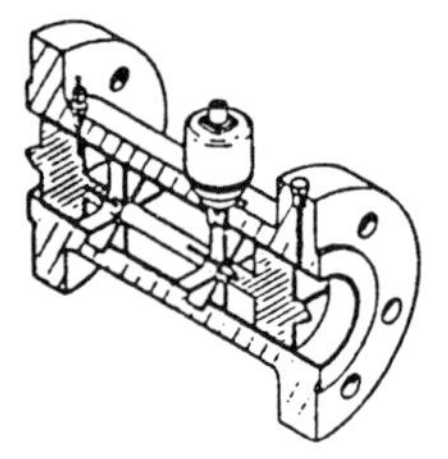

Figure 1. Twin rotor turbine meter (courtesy Quantum Dynamics Inc).

Should the shaft or measuring motor bearings begin to wear, this will result in a change in the ratio of the rotor speeds for a fixed flow rate. Thus, bearing wear can be detected on-line and action taken. This design of meter is also different from all other turbines because of the slender center body and axially short blades. Because friction is very low, the fluid does not have to be accelerated past the rotor to obtain sufficient torque to overcome retarding forces. Thus, pressure drop is also low and pressure and temperature measurements can be carried out at the point of volume flow rate determination. They have been shown to be outstanding meters for leak detection purposes on gas lines.

The Auto Adjust turbo-meter from Rockwell International is also self-checking but is in addition self-adjusting. The design shown in Figure 2 is quite different from the Quantum meter. The rotors are next to each other with the measuring rotor downstream of the main rotor. Constant

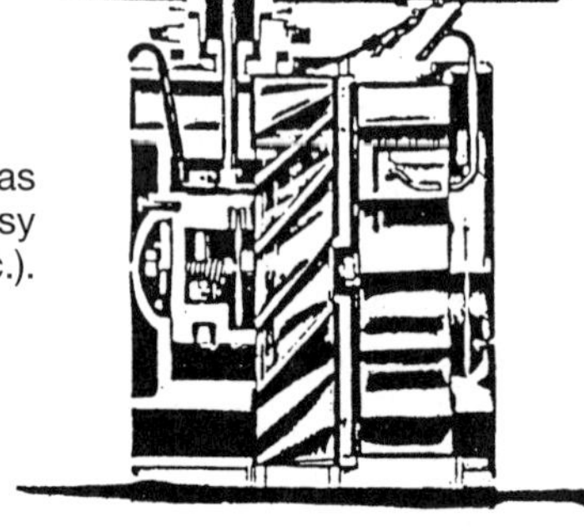

Figure 2. Auto adjust gas pipeline meter (courtesy Rockwell International, Inc.).

performance is obtained because the sensor rotor detects deviations in blade angle of fluid leaving the main rotor. With a very small angle, the sense rotor runs at a much lower speed than the upstream rotor. Performance is then given by the difference in the speed of the two rotors rather than from a single rotor. When bearings begin to wear, the main rotor will spin slightly slower and so the exit angle will also change. This is sensed by the second rotor, which adjusts speed by an equal amount. Bearing wear is therefore automatically adjusted out. These meters have been installed in many gas transmission lines in North America with generally very good performance. The Quantum meter, on the other hand, is a newcomer to pipeline applications, but personal firsthand experiences in a gas pipeline were favorable.

Pipelines are generally operated on material balances (mass), but most flowmeters are volumetric devices. This means that for pure fluids measurements of temperature and pressure plus an appropriate equation of state are required to enable density to be determined. For mixtures of gas or liquids or indeed as an alternative to the above method, on-line densitometers are used. The product of volume flow and density gives mass flow. Until recently pipeline mass flowmeters were not available, but recently a number of Coriolis type direct mass meters have appeared. These have found slow but steady acceptance in industrial applications and are beginning to find applications in pipeline monitoring. One such design is shown in Figure 3, and recently four other designs were launched at the INTERKAMA trade show in West Germany. This type of meter is applicable to both liquids and gases, and some have been used on liquid/liquid and gas/liquid mixtures but with mixed success. They are currently undergoing rigorous evaluation and analysis in many chemical, process, and small pipeline applications, and

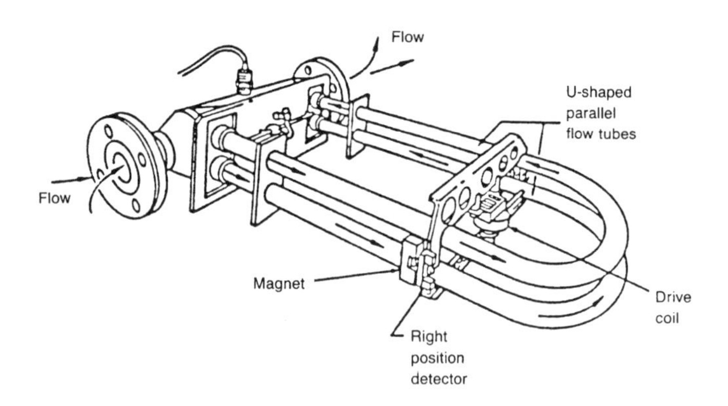

Figure 3. Micro motion direct mass flowmeter (courtesy Brooks Instruments).

indications are that they are standing up well in some applications but not others. Accuracy is around ±0.5% of reading or better, and they do not need additional measurements to determine flowing density. Indeed, by monitoring the output signal, they can be used to indicate density as well as mass flow. The output is in digital and/or analog form making them compatible with many types of flow computer.

Direct mass measurement has one main advantage over volume measurement, since an integration of the signal over a period of time will indicate the pipeline fluid inventory directly. The disadvantage for pipeline operators is the current size range that is available, the largest being 150 mm bore. As most major pipelines are in the range of 500–2,000 mm bore, these Coriolis meters must be installed in parallel to meet the larger applications. Mass metering is relatively new, and the next decade should see a steady increase in sales and acceptance of these meters.

The ideal flowmeter is one that is installed on the outside of the line but gives the performance of the best flowmeters installed inside the line. Ultrasonic time-of-flight meters are slowly developing toward these apparently conflicting but demanding criteria. Multiple beam systems have been installed on many pipelines, the most notable of which is the Alaskan oil pipeline. Some 23 meters each 4 ft in diameter look for gross imbalances along sections of this important installation. They permit the early detection of product spillage and also indicate the section of line giving the problem. This system has been operating since the late 1970s with millions of barrels of fluid being shipped through the line in that time. Experience of these meters has been very favorable in this application.

Ultrasonic meters, like most modern types of meter, were initially oversold, but the main problem has been to convince users that the two types available, the Doppler and transit time meters, are completely different animals that both use ultrasonics to infer flow. When ultrasonic meters are mentioned, people tend to think of Doppler types. Most of these are relatively crude flow indicating devices. The modern clamp-on transit time meter can indicate flow to around ±2% of reading or better depending on design, compared to the average Doppler meter, which is a ±5% FSD device. The advantage of ultrasonic methods is the negligible headloss and the ability to install other portable or dedicated systems without line shutdown. It is surprising that more have not been installed in pipeline systems, but as discussed later in this paper, this is an example of the conservatism often shown in the application of new technology.

Flow measurement has really been transformed by the application of digital techniques to on-line signal analysis. Much more data than just flow rate is now available, and with the latest systems, operators are taking advantage of the new generation of micro and flow computers to control and run their pipelines. It is unusual these days not to install a local flow computer with each flowmeter, with temperature, pressure, and differential pressure signals being fed into the machine for local "number crunching."

Proving devices

The problem with all flowmeters is that they are affected to some degree by the fluid they are metering. For the least uncertainty of measurement, the flowmeter, whatever type is chosen, should be calibrated on-line under actual operating conditions. In the past this was performed with large and expensive pieces of hardware called ball provers. These had to be large to provide adequate volume within the calibrated section of pipe to ensure sufficient resolution and accuracy for the flowmeter they were intended to calibrate.

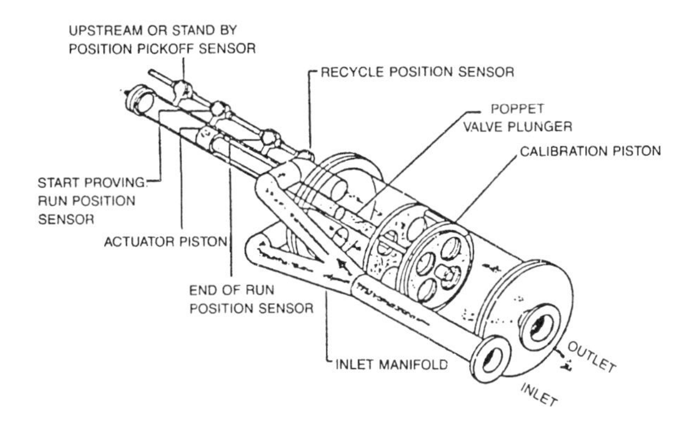

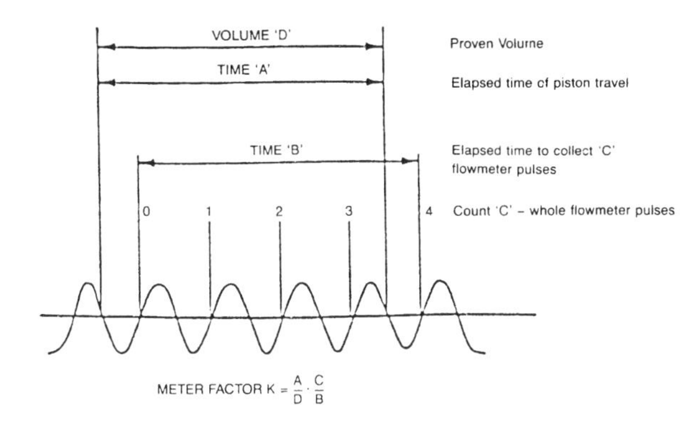

Figure 4. Modern compact meter prover (courtesy Brooks Instruments).

The latest generation of proving devices, called compact provers, are essentially time/volume calibrators that eliminate the ±1 count error present on their "big brothers." It is physically much smaller but with the same or better performance. There are currently 10 different versions on the market, of which five are available in the UK. The prover, one design of which is shown in Figure 4, uses pneumatics to launch a piston, which is then driven by the fluid through a calibrated chamber, displaying a known volume. The key to their excellent performance is the data acquisition system illustrated in the lower part of Figure 4. As the piston passes through the calibrated section, pulses from a high frequency internal clock are counted to determine the time, A, to displace the calibrated volume D. Once the counting has started, a second internal timer is gated to respond to the first input pulse from the flowmeter. This timer then records the time interval B to collect C whole flowmeter pulses.

During the proving run there were then C × A/B meter pulses. The meter factor is found by dividing this count by the volume displaced:

$$K = \frac{A}{B} \times \frac{C}{D}$$

Some designs are uni-directional and some bi-directional. In the uni-directional types pneumatics or hydraulics launch and retrieve the piston; in the bi-directional designs the process fluid is used to drive the piston. Other versions incorporate double-barrelled liners to remove the need for pressure compensation and also reduce the effect of temperature variations.

In one proving run there may be as few as 100 pulses collected, so timing error must be minimized. If the pulse output is frequency modulated, from a positive displacement bulk meter, for example, then these few pulses may not represent a true statistical sample of the pulse train and multiple passes may be required. Alternatively, pulse interpolation techniques may be required, several of which are applicable to give the accuracy required. These are currently the subject of intense activity in standards committees who are producing a new general code of practice on meter provers.

The compact prover is better suited to offshore platform applications and to field use on operating pipelines. Being portable, it can be easily coupled in series with the meter to be proved. A suitable choice of materials allows proving on a vast range of fluids to be carried out. Operational experience has generally been favorable, and several direct comparison tests with ball provers have shown the new generation of compact provers to be superior in most respects. However, their acceptance in custody transfer applications has been slower than had been expected, even though several countries have certified some designs for Weights and Measures approval. Some international calibration centers are now using this technology for routine traceable calibrations, and confidence is growing as operational experience is accumulated.

By far the most economic method of measuring flows in pipelines is to install pairs of flowmeters in custom-designed metering stations. By monitoring the ratio of the outputs, a measure of wear or drift is obtained. Provided the outputs of the meters remain within preset tight limits, uncertainty of flow rate measurement is low. The disadvantage of this on pipeline operation is that there are two meters to be maintained and checked, and should faults develop, the pipeline may have to be shut down more frequently, unless by-pass loops are provided when the metering station is designed. Turbine meters and PD meters with pulse outputs give the best performance, particularly if they are regularly proved with either a dedicated or portable prover of the type described.

For large bore lines, particularly gas lines, the proving of flow meters is very expensive if it can be performed at all. The majority of installations rely on orifice plates built to the standards. Recently a new high-pressure gas proving station was completed by British Gas at Bishop Auckland. Here turbine and other meters can be calibrated in actual operating conditions on natural gas. Such an installation has previously only been available in Holland, where gas calibration standards have been developed to a high level of confidence over many years of patient data collection. With the new British Gas facility we should see the ability to cross check our own standards against international standards.

Valves

Valves are an important part of any pipeline system. They come in many shapes and sizes and perform numerous functions. Some act as primary flow regulators, some as check valves for flow direction control, and some as pressure reducers for safety purposes. Developments have centered in two main areas, materials and actuators (both simple and intelligent). One of the development sources is again the North Sea, where the demands for better reliability, better in-field performance, and reduced cost have forced many manufacturers to look critically at designs. As one of the most expensive components in the valve is the body, effort has been concentrated in this area, both in terms of new materials to meet more demanding specifications and of quality and price. These last two items are critical to the survival of many companies, and in a market sector that is growing slowly or even declining, as suggested by one leading instrument journal, valve body cost savings, however small, may mean the difference between life and death.

Slimline butterfly valves have recently been introduced by a number of manufacturers to meet the demand for lower cost reliable devices. These can be installed between existing pipeline flanges. They can be manufactured in a wide range of materials with special seal materials chosen for the most demanding of duties. With suitable actuators, even valves as large as 60 in. in diameter do not take up much space.

Other designs of control valve provide straight-through flow patterns to minimize flow turbulence, vibration, and noise. Should a pipeline be fitted with acoustic line break detectors (see section following), line noise will be important in setting the sensitivity of the detectors, and hence the type of valve fitted will be important to system operation. Other operational problems such as "hunting" in globe valves have prompted the designers to work more closely with the operators to define full specifications. Pipeline operators often run their systems with rapid transients present. Demand changes of both a positive and negative value can subject the control valves to very severe step changes in flow and pressure, and initially many conventional valves could not withstand such treatment.

This is where intelligent sensors on pressure relief valves have come into their own. They are able to give better speed of response and meet some stringent operating criteria. Pressure may also rise gradually due to a steady rise in line temperature, as well as the rapid transient caused by block valve closure or pump failure/surge. These transients travel at sonic velocity and even tenths of a second's advantage in operating the valves can affect safety on the line.

Microprocessor-based detectors fitted to the valves can be used to sense line conditions and provide suitable inputs to the valve. In some advanced designs manufactured in the United States, it is possible to program intelligence into the unit to begin closure/opening based on measurements of an anticipated condition such as pump surge further down the line. If the signals from one part of the line can be transmitted to other sectors, corrective action can be taken before safety blowoff valves are activated, releasing product into the environment.

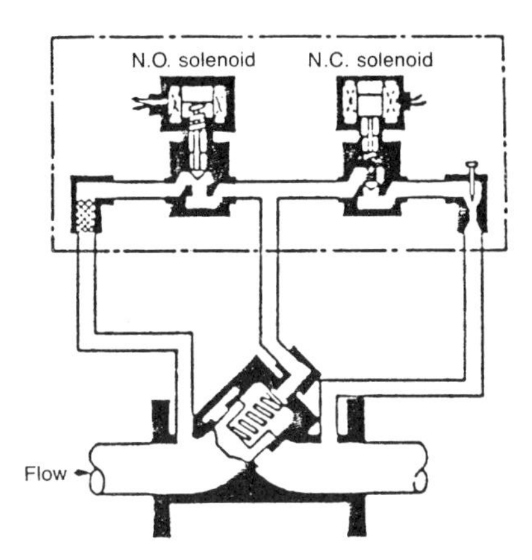

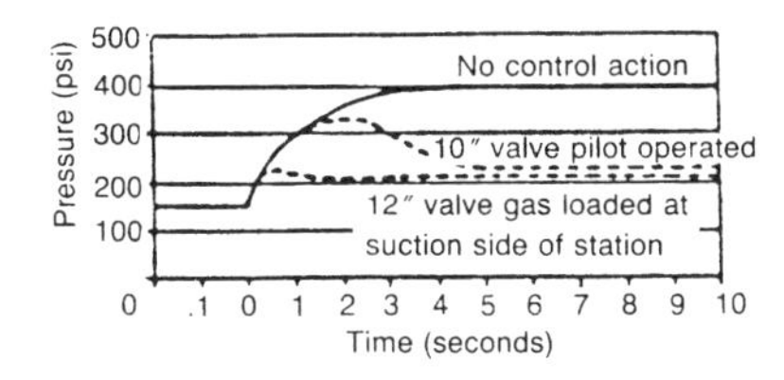

Figure 5. Intelligent pressure relief valve.

A pipeline pressure relief system based on this principle is shown in Figure 5. A strain gauge sensor fitted on the suction side of the compressor is monitored by a microprocessor that linearizes and conditions the signal. The microprocessor operates a pair of solenoid valves acting as the pilot mode with the line pressure acting as the actuating medium. By switching solenoids, the relief valve can be opened, closed, or programmed to act as a regulator at the pressure trapped in the valve body. Such a system can provide adequate response to step changes on the order of 70 bar in 3 milliseconds. In this system the status of the pipeline pumps is provided as an additional input to the microprocessor to trigger action before the pressure surge occurs. Tests on large lines have shown such intelligence to be vital in the safe operation of systems when 5,000 hp pumps were suddenly lost.

Line block valves can also be actuated by output signals from line break detectors. This is important in isolating problem sections of line carrying toxic products. The wave detector will sense problems in seconds, and the line can be closed to minimize the amount of toxic product released. That section of the line may well depressure, but the release effects will be reduced. Such systems are fitted as anti-terrorist devices on lines in problem areas of the world such as the Middle East and Central Africa, where all too often pipelines are political targets.

The choice of materials for valve bodies and the styles of valve available is now very wide. A survey in *Processing* (January 1986) listed 15 different types supplied by 110 companies. These range from the cryogenic applications up to line temperatures in excess of 700°C. Selecting the correct valve for the duty is now the biggest problem facing the user, and where it is necessary to take outputs into computers for safety and control purposes, the choice becomes more difficult. In such a competitive market, developments are bound to come thick and fast, and it will be interesting to watch the development of the next generation of intelligent pipeline valves.

Acoustic line break detectors

Problems with the pipeline are frequently accompanied by the generation of noise, either from problem valves (see above) or line breaks. If the flow becomes too high from a surge or a leak occurs through corrosion or third-party intervention, then there are instances where vibrations in excess of 20 kHz are produced. These frequencies are in the ultrasonic range and can be detected with suitable transducers. Such devices have been made portable so that pipeline crews can clamp them at any point along the line to check for noise. By noting the signal strength, the source of the problem can be pinpointed. British Gas has used this principle to develop one of two types of pipeline monitoring tools that are described later.

A similar technique, though based on a different principle, is the acoustic "wavealert" monitor, though more correctly called a negative pressure wave detector. The central element is a piezo-electric sensor that gives an output when stressed. When a line rupture or leak occurs, there is a sudden drop in line pressure at the source of the problem, followed by line repressurization a few milliseconds later. The rarefaction wave generated by these events moves away from the leak in both directions at the speed of sound. The pipewalls act as wave guides enabling this wave to travel for great distances, attentuating in amplitude as it does so. Sensors placed at known points along the line can detect the passage of this pressure transient. From the triggering of internal timing circuits and a knowledge of the prevailing acoustic velocity for the line, the leak location can be calculated.

Such devices are particularly useful for detecting large breaks in lines very rapidly. The transient wave moves at 1 mile per second in liquids and 1 mile in 5 seconds in gases. If the

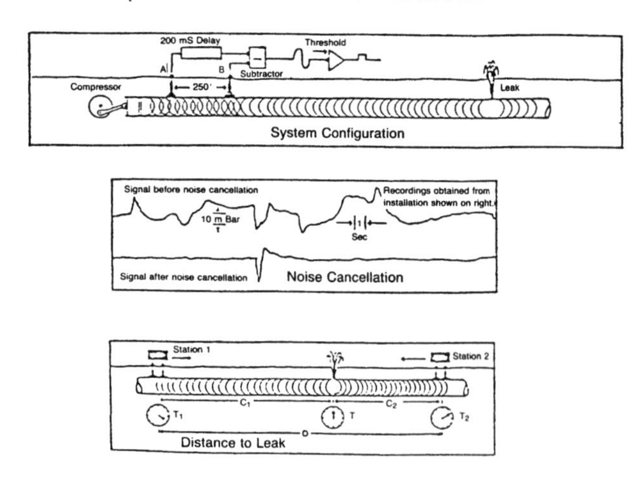

Figure 6. Negative pressure wave "wavealert" monitor (courtesy Spectra-Tek).

detectors are positioned at intervals of 2 to 5 miles in critical areas, then the break is detected in a few seconds. If the output signal is fed into a computer, then this can be used to trigger block valves to isolate the problem section of the line very easily. Figure 6 shows one further development where the instrument can be made to cancel line noise and use the full measuring capacity of the sensor for signal detection. There is, however, the practical problem in setting the background threshold correctly. Pipeline noise is generated by bends, valves, pumps, regulators, etc., and it is crucial to be able to remove this background noise to avoid unnecessary line shutdowns when the line noise exceeds the detector setting. The Reynolds number of the flow has been shown to be the important parameter in installations both in the United States and the United Kingdom. Experience has also shown that the installation of these devices is critical to reliable performance. It is important to have the sensor as close as possible to the line to avoid attenuation effects and improve sensitivity. Both this and the background line noise will influence the number of "false alarms" from the instrument.

ICI has been experimenting with strain gauge–type sensors on their cross country ethylene line, and these are showing more promise. These experiments will be closely monitored in the next few years to see whether we can improve the reliability and sensitivity of line break detectors.

"Smart" pressure sensors

One area where the impact of microprocessors has been felt is in the development of smart sensors, particularly pressure. Many physical effects respond directly or indirectly to pressure, and as a result, the first generation of "smart" process transducers appeared on the market some years ago. Surprisingly, however, market penetration has been slow, mainly, I suspect, because hand-picked conventional pressure and differential pressure cells perform almost as well over the limited range often required. There is virtually no limit to the amount of intelligence and diagnostics that can be incorporated into pressure transducers, but so far only those that are reasonably linear or that have well defined characteristics have been commercially exploited. The need for linearity is now no longer there because all signal processing the digital and algorithms can be written to cope with any signal/pressure curve, provided it is repeatable. Even hysteresis can be taken account of.

It is in the communications area that the full potential of intelligent transducers is likely to be felt. Hand held field communicators enable the user to rerange, configure, and display process output values without the need to be close to the transmitters. On-line reranging is particularly useful when the pressure transducer is located in a very inaccessible place. It is possible to use two-way communication links to examine the outputs from a variety of intelligent transducers located at remote outstations, but the cost of such a feature should be carefully considered as this is likely to be expensive. Intelligent transducers are more expensive than the commercial instrumentation, but in certain applications cost is not a consideration.

The general features of the new generation of smart sensors, in addition to those mentioned above, include:

- On-line temperature compensation
- Built in-diagnostics
- Claimed accuracies of ±0.1% of span

Leading companies in the field include Honeywell, Bailey, and SattControl, but other established companies are working on similar devices. Most of these sensors use the properties of silicon plus the addition of micro electronics to give the performance. Silicon is an elastic drift-free material, and pressure sensing elements are diffused onto it. A cavity is usually etched on the reverse side to form the diaphragm. Where the line fluid is very aggressive, a second insulating diaphragm is required to protect the silicon from corrosion. Generally speaking, the cost of these sensors is coming down, and it appears we can look forward to many more instruments appearing in the more difficult applications. Provided we can overcome the general conservatism about applying new technology, this is one section of the instrumentation industry where microelectronics will have a great future influence, particularly when used with the computer and real-time models. This will give the pipeline operator more confidence and more information on both the condition of the instrument and the accuracy of the measurement. Interfacing problems should also reduce.

Another type of pressure transducer that looks promising for future applications in pipelines is the vibrating wire sensor. One such instrument from Foxboro has shown considerable reduction in size and manufacturing costs from other sensors. A tungsten wire encapsulated inside a cell is vibrated at its natural frequency. As pressure is applied to the silicon barrier diaphragms, this changes the tension in the wire and so alters the resonant frequency. This is sensed and amplified and gives the transducer excellent resolution characteristics. Pressure and temperature compensation is accomplished within the cell, and although the instrument is not strictly a "smart" transducer, it does possess basic characteristics that should be attractive to pipeline applications.

Densitometers

Continuous high accuracy on-line density measurement has become increasingly important in pipeline operations. The composition of natural gas and hydrocarbon streams are different from each production well, and assumptions about constant composition cannot be made. In fiscal measurement and where the energy content of the stream is required, a knowledge of density is essential, particularly for gases.

Recently great strides in the development of rugged on-line densitometers have been made with the introduction of the vibrating element transducer. This type of instrument is fast becoming the industry standard for pipeline applications. One such instrument is shown in Figure 7. The operation is very simple, being based on the spring mass principle. A flat plate, cylinder, or fork is maintained at resonance by an electronic circuit. A fluid flowing over or through the vibrating element adds to the vibrating mass and causes a decrease in natural frequency. The higher the fluid density, the lower the frequency and vice versa. By calibration using fluids of precisely known density, the relationship between frequency and density for each element can be determined very accurately. The characteristic equation for most vibrating element densitometers is of the form:

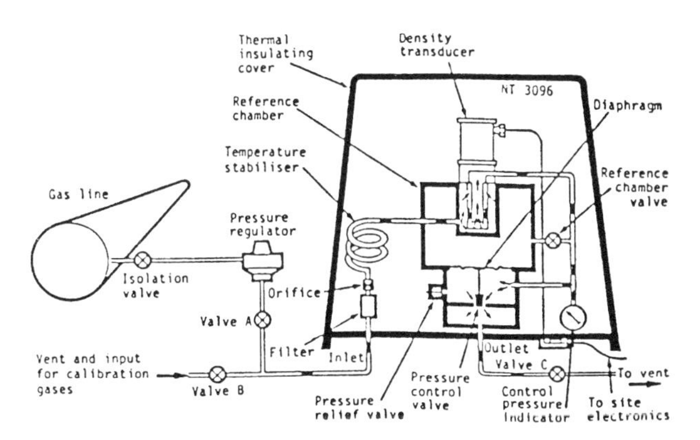

Figure 7. Pipeline densitometer (courtesy Solartron Inst and Control and Instrumentation).

$$\rho = At^2 + Bt + C$$

where A, B, and C are calibration constants, t is the period of oscillation, and ρ the fluid density. Once the constants have been determined by the laboratory tests, the characteristic equation can be programmed into the system computer for on-line calculation of density. In some cases it may be possible to check the performance *in situ* by passing a fluid of known density through the transducer. This will establish if there has been any drift from the original manufacturer's calibration and will help to minimize the uncertainty of the operating system. The latest generation of instruments can approach accuracies of ±0.1% of reading under ideal conditions and are typically ±0.2% of reading under actual operating conditions.

When used in conjunction with orifice meters, for example, the flow through the densitometer is maintained by taking fluid from a tapping downstream of the orifice and feeding this back at the downstream tapping of the plate. This method, known as pressure recovery, is very popular in pipeline applications, and a typical installation is shown in Figure 8. Other methods include pumping a small percentage of the stream flow through a flowmeter and densitometer before feeding back into the main stream.

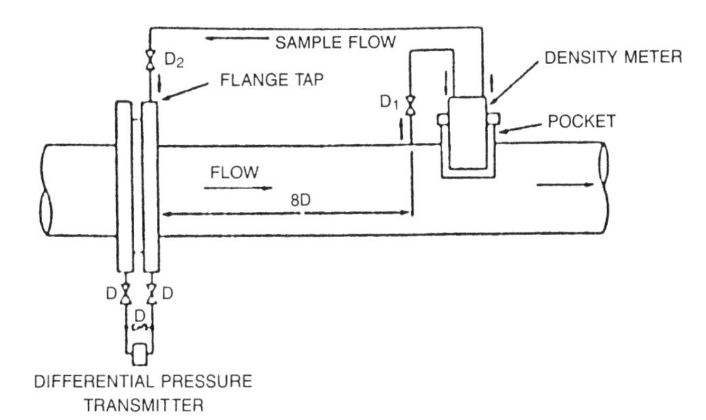

Figure 8. Standard installation for pipeline density measurement (courtesy IP manual section 6.2 Nov 1983).

There are other types of fluid densitometer such as nuclear or force balance, but these do not find the same widespread use on pipelines as the vibrating element devices. The choice of transducer will depend on the application, since fluid properties, flow rate, sensor element design, and materials of construction all influence the type of densitometer used. For example, if the fluid contains solid particles or is of a "dirty" nature, then straight pipe vibrating cylinder elements are more likely to give better long-term performance. The actual elements are manufactured in a wide range of materials. Ni-Span-C is used for high accuracy metering, and this material also has a low temperature coefficient. Stainless steel, on the other hand, has better corrosion characteristics and is cheaper to use. Materials such as hastelloy C-276 offer excellent resistance to the most corrosive of fluids.

It is probable that the use of these instruments will increase as general experience on pipelines has been favorable. They must, however, be operated in single-phase fluids, as the presence of either droplets in gas lines or vapor bubbles in liquid lines will produce erroneous outputs of unknown magnitudes. The straight tube cylinder-type designs could be used in low second-phase concentration applications if the densitometer is installed vertically with upward flow. This will ensure closer approximation to homogeneous flow. Experience has shown, however, that two-phase flows are significantly more troublesome to handle, and the output of the transducer should be used with some caution.

Pipeline samplers

Oil lines, particularly crude oil lines, often contain significant amounts of water. This has to be monitored and allowed for when the oil is being metered for custody transfer or taxation purposes. It is a growing and important area of pipeline technology to install on-line samplers for the measurement of water content, and work is also under way to use statistical methods to estimate how many samples are representative of the composition of the flowing stream. There are currently two different basic methods being used on pipelines to monitor the water content. These are the capacitance method and grab samplers.

The first method is based on the fact that the difference in the relative permittivities of water and oil makes a capacitive transducer sensitive to the water content flowing between two electrodes. In the past 3 years, commercial instruments using this method have appeared and have attracted a great deal of interest, particularly in North Sea applications where water content could be a problem. They are basically also non-intrusive sensors with the potential for in-line installation. There are two basic designs, by-pass and main line types. By placing electrodes in a by-pass line from the main flow, a measure of water content in the main stream can be made. The accuracy of these meters, however, is highly dependent on the extent to which the flow in the by-pass loop is representative of the main stream. With the main line type, this source of error is reduced, but the problem of flow regime effects still remains. Both types are also affected by the excitation frequency, the salinity of the water content, the water droplet size, the thickness of the isolation material between electrode and fluid, and the homogeneity of the electric field.

Field experience has shown that meaningful measurements can only be made if the flow regime is constant and the best results are obtained when the water is evenly dispersed and in droplet form in the liquid. Such criteria often do not exist, with slugs of water passing down the pipeline at irregular intervals. One such commercial instrument is shown in Figure 9. The important feature of the design is the short distance between the electrodes, which results in high sensitivity. The electrodes are intrusive and therefore liable to either coating or blockage effects if the oil is particularly dirty.

Grab samplers are devices that take small volumes of fluid from the line at regular intervals to build up a representative sample of the mixture over a period of time. An international standard is currently being reassessed in the light of operating experiences by many oil producers. The current practice states that 10,000 "grabs" of around 1 cc each should be taken, and there are guidelines laid down for the sampling frequency and other criteria to be considered.

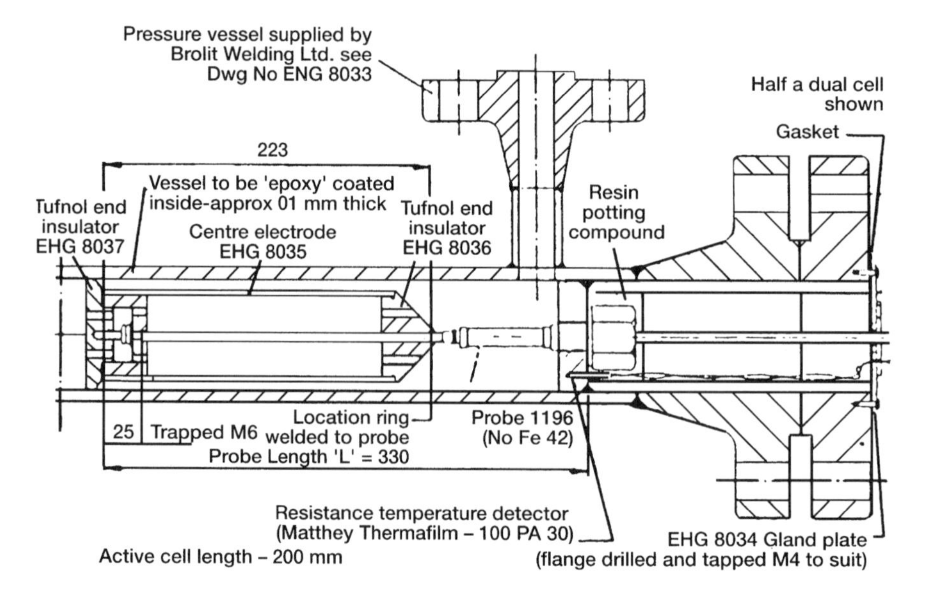

Figure 9. Capacitance-type water content transducer (coutesy Endress and Hauser).

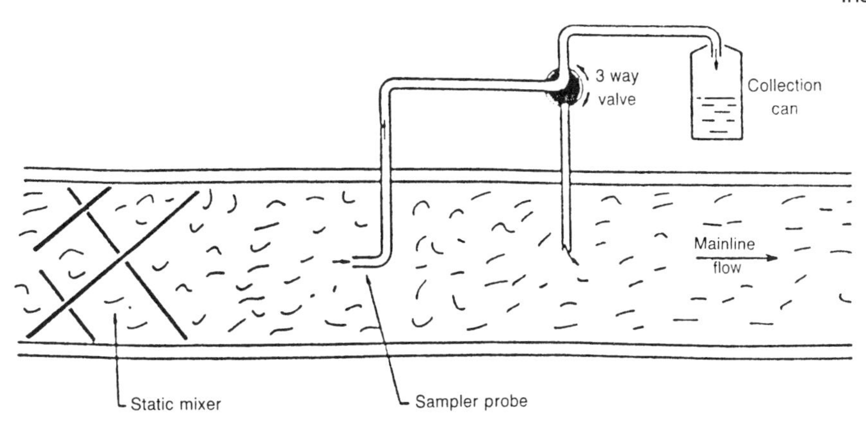

Figure 10. Fast loop pipeline sampling system.

Work at Cranfield and elsewhere has been evaluating this standard, and some of the limitations have been identified. As with the capacitance devices discussed previously, these may be installed either in the main stream or in a by-pass loop. Often signals from flowmeters in the line are used to set the sampling frequency, and the use of jet mixers, bends, pumps, and other devices that produce a more homogeneous stream is desirable to improve performance and accuracy.

Another variation, the fast loop sampler, is used for continuous monitoring of water content, and such a system is shown in Figure 10. The main advantage of the fast loop sampler is that flow is continuously extracted, and the sampler itself seems to have a smaller perturbing effect on the stream characteristics. A great deal of work is currently under way to examine the effect that the presence of the sampling probe has on the fluid dynamics of the stream. Early work has indicated that the flow streamlines around the probe head do have an influence on the droplet distribution and hence the oil/water ratio. This is developing technology, and we should see increased activity in this area as some of the older North Sea fields begin to produce greater concentrations of water.

Pipeline monitoring systems

On-line inspection techniques have recently been developed to a very high level of sophistication. British Gas has spent some 12 years and more than £100 M producing intelligent monitoring tools and the data analysis software to inspect line sizes from 8-in. to 42-in. bore. There is a demand worldwide for their expertise, and with a potential 1 million miles plus of line to survey, they should be busy for years to come! These inspection tools are commonly called "pigs" but should be distinguished from the pipeline pigs used, for cleaning, dewaxing, batching, filling, and dewatering of pipelines. I feel the term *monitoring tool* should be used, because these devices are very advanced indeed.

Two types of monitoring tools have been developed, one based on magnetic sensors to detect metal loss and the second based on ultrasonics to sense defects and cracks in pipe walls. The two used together give a complete picture of the mechanical state of the pipeline. They are not continuous monitors like pressure, temperature, and flow transducers but are useful in supplementing the on-line measurements. If the lines are inspected at regular intervals, then changes can be identified and possible trouble spots pinpointed. The data can provide a very useful historical record to be used when maintenance, cleaning, or line repairs are needed.

The magnetic pipeline inspection vehicle operates by magnetizing a short section of line as it travels down the pipe, driven by the process fluid. If a section of pipe has suffered metal loss, it will cause a local distortion of the magnetic field. The data are recorded on 14 channel tape systems, and onboard software reduces the amount of data saved to only that where significant signals are noted. The information from the clean or undamaged pipe is not saved. Once the inspection run is complete, the data are analyzed to identify the problem areas. The first offshore run was completed in early 1986 when the Rough and Morecambe gas field lines were surveyed. This run took around 5 hours. The method can be used to survey sections of line around 300 km long, although a more typical single pass would be around 80 km (50 miles) (*Monitor*, No. 2, 1986).

The ultrasonic system is virtually identical except for the transducers and detectors. The sensors are ultrasonic, operating in the megahertz region. They are housed in wheels that press against the pipe walls. They are pulsed from an on-board electronic control system, and the reflected waves are stored in the same way as in the magnetic system. Both inspection vehicles travel at around 3 meters per second. If a line problem is detected by an imbalance between flowmeters, for example, the location could be confirmed by the acoustic detectors on board the inspection vehicle. This could be arranged to alarm when the detector output reaches a maximum, and the precise location can be given by radio transmitters mounted on board.

Other problems that can be identified include corrosion, fatigue, ground movement (causing pipe buckling), and third-party damage. The inspection center at Cramlington has already undertaken a number of overseas surveys and has experience in gas, crude oil, kerosene, and gasoline lines. The impressive part of these pieces of equipment is the engineering. They have to negotiate bends, valves, crossovers, etc. and still protect the relatively fragile electronic recording equipment from damage. They have been used both in on-shore and off-shore lines with equal success. This is one area of pipeline instrumentation development that I will be watching over the next 2 years as new on-line vehicles are developed for pipe coat-less monitoring.

Computer systems

It is in computer systems that the greatest amount of data can be gathered, analyzed, and acted upon in the shortest amount of time. Most operators now have some form of computer-based monitoring system employing either commercially available or custom-developed software to run the system. Programs can calculate the inventory of the line at any point in time and compare this with line measurements to provide independent cross checking of pipeline measurement data. These data can be transmitted by telemetry, radio, or telephone links to a central computer that continuously monitors the "health" of the line. The central operations room is now more likely to contain VDUs rather than banks of process indicators and enables the engineers to have rapid and accurate updates of conditions along the entire line or network. By changing programs and subroutines, a large number of tasks and functions can be performed easily, simultaneously, and cost effectively.

The many functions that can be performed by computer-based systems include:

- Leak detection and location
- Batch tracking of fluids
- Pig tracking
- On-line flow compensation
- Real-time pipeline modeling
- Instrument data and malfunction checking
- Inventory and material balance accounting

Such systems have very rapid responses but are dependent on the accuracy of the input data to be effective. Most computer-based systems consist of two major elements, namely:

1. A supervisory computer plus associated software
2. A number of independent pipeline monitoring stations plus data transmission equipment for communication with the computer

The complexity of the software, the number of input variables, and the pipeline instrumentation selected all influence the size and the choice of computer. Some systems are capable of running on the compact powerful micros such as the IBM AT or XT, while the more complex pipeline networks will require a mainframe machine, possibly communicating with satellite micros installed close to the measurement stations. The overall design time varies with each system but should not be underestimated, especially if SCADA systems are being developed in house (see section following).

At the most basic level the computer system can take flowmeter inputs to perform gross balances along the pipeline. Such a method is very useful in identifying small but consistent loss of product from corrosion pits in the pipe wall. This will arise if the integrated flowmeter outputs diverge with time when the flow in the line is maintained constant. If the line flow rate varies with time, then imbalances are more difficult to detect, since the meters at each end may have different characteristics or their outputs may vary nonlinearly with flow rate. A loss of product will be identified simply as the difference between the steady state inventory of the system and the instantaneous inlet and outlet flows. Mathematically, this is:

$$\Delta V = V_{in} - V_{out} - V_l$$

where ΔV is the leakage volume, V_{in} and V_{out} are the metered inlet and outlet flows, and V_l is the inventory of the pipeline. This last term can be calculated as the average of the integrated flowmeter signals but is dependent on variations in temperature, pressure, and many other variables. It is the use of computers that enables this last term to be calculated on line and as a function of pipeline elevation, pipe material characteristics, process variations, pipeline fitting loss data among the many variables. Temperature and pressure variations change both the density of the fluid

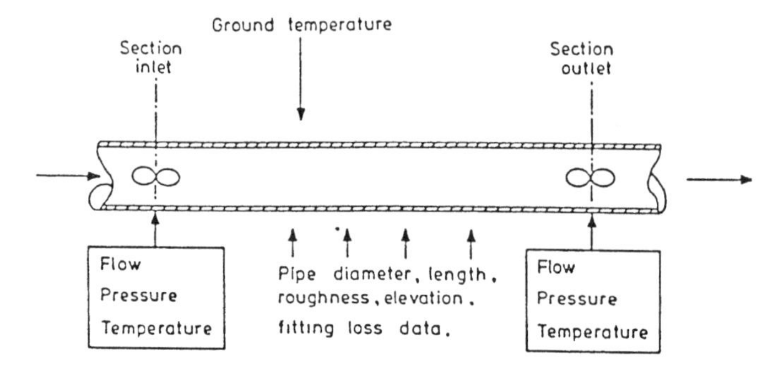

Figure 11. Input data for computer inventory calculation.

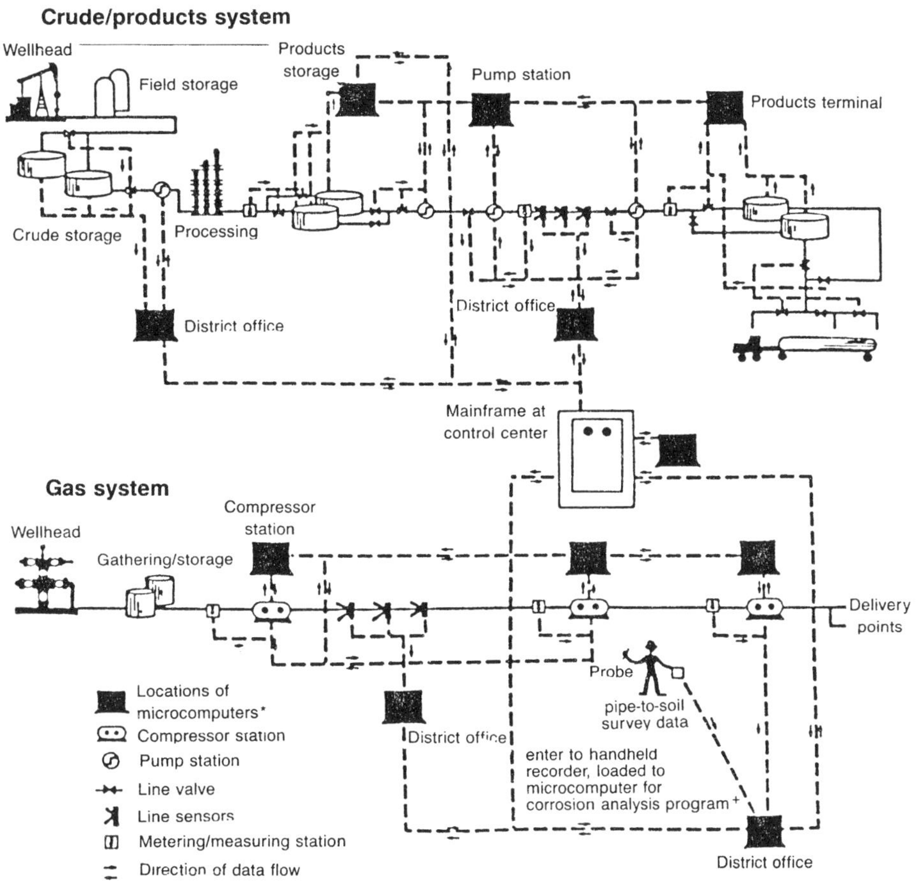

Crude/products system

Gas system

* A 16 or 32-bit microprocessor supporting a CRT (cathode-ray tube) screen and full-function keyboard.
† Data physically returned to station or communicated via modem.

Figure 12. Examples of modern computer-based-pipeline systems (courtesy Oil and Gas Journal, 1986).

and the actual dimensions of the line. Both will give rise to sources of errors in pipeline inventory if compensation is not included in the calculations. Figure 11 shows typical input data requirements for simple on-line inventory calculations.

At the other extreme, computer-based systems can control entire networks with almost no need for the operators to intervene unless serious problems arise. Figure 12, taken from *Oil and Gas Journal*, March 3, 1986, shows two examples from the many hundreds that exist of computer-based monitoring. The benefits from the addition of real-time pipeline models cannot be overemphasized. Even when operating transients occur, some commercially available programs enable problems to be detected. An example is shown in Figure 13. Here a pipeline leak was induced following pump failure and restart, and although the line was operating under highly transient conditions, a material imbalance was detected. Real-time modeling also allows density profiles to be calculated along the line. It is fairly common practice to have amounts of different fluids in the same pipeline separated by a pig or sphere. From the pressure drop/flow characteristics, the real time can track the progress of these batches of fluids along the pipe.

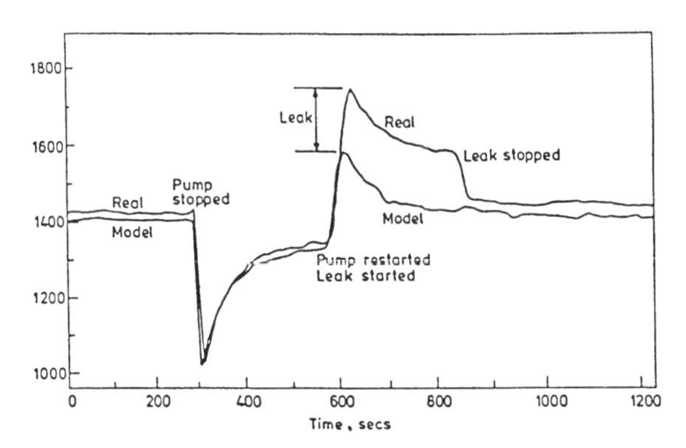

Figure 13. Example of transient real time modelling of pipelines (data from Ref 9).

SCADA systems

Real time computer modeling of pipelines is a technique that uses the full data gathering capabilities of modern digital systems and the computational power of computers to give accurate "snapshots" of the line. These enable the process operator to act quickly when problems develop. The whole system is under the control of the SCADA (Supervisory Control And Data Acquisition) system. This suite of programs polls the data stations on the line in turn, processes the data, controls the running of any pipeline model, and activates alarm and other monitoring routines. Most modern SCADA systems now include checking routines to determine the validity of the data before any calculations are made. It has been found that the use of continuous quality and cross checking of data improves the sensitivity of the system. In addition to these basic functions, some of the more advanced SCADA systems might include predictive models to analyze "what if" operating scenarios, include automatic startup and shutdown procedures, or provide optimization routines for least-cost operating strategies.

There are a number of companies supplying standard SCADA packages, but the majority of systems use custom written master routines to enable the commercially available programs to be used. Each pipeline is different, in line characteristics, instrumentation installed, fluid properties, and need of the user to name but four requirements. Experience has shown that the adaptation of standard software modules to specific needs takes much longer than is first estimated and also gives rise to the greatest number of problems when pipeline monitoring and control systems are being commissioned.

It is very important for the software specialist to communicate directly with the instrument engineer who specified the hardware, to ensure proper integration of the various parts of the system. All too frequently these two key members of the design team do not meet until the project is well under way.

Cathodic protection

Protecting a pipeline from corrosion is an important and expensive part of building and running a pipeline system safely. Every year, operators spend millions of pounds on pipe coatings, gas detectors, and sacrificial anodes to lengthen the effective life of pipelines by reducing the effects of corrosion. Whenever a steel structure of any type comes into contact with the soil and/or water there is almost certainly going to be corrosion. The structure, in our case the pipeline, divides itself into anodes and cathodes with corrosion taking place at the anode. This chemical process is accompanied by the flow of electric current from one point on the line to another via the ground or water. There are two basic preventative methods, sacrificial anodes and impressed currents (more commonly called cathodic protection). This second method is the more common and preferred method for long pipelines to identify the problem spots.

If a voltage is applied to a section of line, a profile of the current flow can be drawn up by measuring the current flow at intermediate points. This gradient will show areas where current is escaping, indicating a pipe defect. Recently, techniques using AC current have been perfected where the magnitude of the current can be sensed inductively from above ground.

This method can be used by pipeline crews walking the line, and compact lightweight handheld units store data on the mechanical condition of the line in lengths as short as 5 meters. On long cross-country lines the profile can be plotted at intervals of a few kilometers, and from historical cross-comparisons, line changes can be identified.

The object of cathodic protection is to prevent current flow from the pipe. The operating principle is to apply a voltage in the opposite direction to the natural potential of the pipe. The pipeline voltage is made negative with respect to earth so that in theory the voltage and current flowing from the cathodic protection system balance the corrosion current that would otherwise flow. Recent developments include the use of solar panels and converters so that energy from the sun is used to run the system. One of the largest examples of this new technique is the Sarir to Tobruk in Libya, where a 32-in. diameter, 319-mile pipeline is protected by such an installation.

Remote station monitoring is easily accomplished with the data being telemetered back to central control rooms many miles from the location of the actual installation. Continuous automatic control is often performed by means of thyristors. Such continuous monitoring is now routine on new lines that are being constructed, and many older lines are being fitted with some form of cathodic protection system. In most cases it is probably more important for the older lines to be surveyed than the newer ones, especially where different fluids have been transported over the lifetime of the line.

System design guidelines

The aims and requirements of a project should be specified as fully as possible as early as possible when designing a pipeline monitoring system. Based on personal experience, the following points should be borne in mind. They are not in any special order but do form a useful checklist when starting system development.

1. Make the system as simple as is practical but do not underspecify the needs for the sake of simplicity.
2. Select instrumentation and pipeline equipment based on performance and not economic grounds. It is better to

install fewer high-quality pieces of equipment than numerous poor ones.

3. Equipment compatibility is important. Use transducers, interface modules, and other pieces or hardware that have standard communications protocol. Custom-designed systems invariably put up development costs and do lead to endless sources of trouble and frustration.
4. When calibration and maintenance are due on the system, do not put them off. Wherever possible use computer software to cross check data to aid in the identification of problems. This will help to reduce operating and maintenance costs.
5. Where system availability is important, install devices that are self checking, self diagnosing, or dual systems where appropriate. The initial cost may be higher but the benefits may be more economical in the longer term.
6. Seek independent references, user experience, or validation of the equipment chosen. Most pieces of hardware perform differently in real applications to the specification published under ideal conditions.

Future trends

Future trends are in one respect easy to assess, but it is very difficult to estimate where we will be in 10 years' time. There is no doubt that the application of computer-based technology is going to proceed at an ever-increasing speed to both the instrumentation and pipeline monitoring systems. There will be reliable and rugged intelligent flow, pressure, valve, and other hardware components appearing, and once the initial testing over the application has occurred, the second-generation systems will give the desired performance, if previous history is anything to go on.

By far the biggest problem facing the whole spectrum of control and instrumentation is to develop standard protocols to facilitate communication between instruments. The situation at present means that most, if not all, monitoring computers must have a multitude of interface boxes scattered through the system to enable some elements of the system to "talk" to others. It is sincerely hoped that manufacturers will develop microcomputers that are able to communicate directly with each other, with full-scale industrial process monitoring and control systems and with the individual process sensors and associated controllers. The more elements there are in the system, the greater the potential of line downtime. Some moves are being made in this area, but it could be some time before we see commonality of process protocol.

The most rapid area of advancement has been in the microchip. Oil pipeline operators are among the many people who have been watching the development of 32-bit machines. The Intel 80386 chip for example, now contains more than 275,000 transistors in one piece of silicon. This device has an operating speed of 4 million instructions per second and has the capability to address over 4 gigabytes of main memory. One other manufacturer is working on a computer that will perform 10 billion calculations per second, more than 10 times faster than any other computer on the current market. Such machines will permit easier real-time modeling of pipelines and will present the line operator with many extra pieces of data on the condition of the line. Of course, complete new suites of software will be required for these fifth generation machines.

Conclusion

In such a short space, the paper cannot do justice or even mention the vast number of recent instrumentation developments that have had an impact on pipeline operation and control. The technology of measurement and control is proceeding at an ever-increasing pace, and it is now quite difficult to keep up with developments in one area. Pipeline instrumentation covers a very wide spectrum from pressure and surge control to intelligent instrumentation and advanced software. Certainly our "friend" the microchip has totally transformed our thinking in the past decade, and the poor pipeline operator now needs specialized help in specifying a system to meet his needs. There are many options open: control using PLCs at sources, control using software in the process computer, and manual override if required. With the development of expert systems, systems can now be programmed to almost preempt problems before they arise and take appropriate corrective action. The flexibility in the industry is vast, but one thing is sure, you cannot control unless you measure accurately, and all too frequently the primary elements of a system are under-specified compared to the rest of the system. It is of very little use, for example, to pressure and temperature compensate transducer outputs when the basic measurement uncertainty is poor.

Source

Measurement + Control, Vol. 20, February 1987.

References

1. Adams, R. B., "Buckeye modernizes computer operations and SCADA system," *Oil and Gas Journal*, 11, 55–59, August 1986.
2. Anderson, C. C., "The Ekofisk pipeline control system," *Pipes and Pipelines International*, 30, No. 3, 21–25, 1985.

3. Baker, R. C. and Hayes, E. R., "Multiphase measurement problems and techniques for crude oil production systems," *Petroleum Review*, 18–22, November 1985.
4. Chester, J. R. and Phillips, R. D., "The best pressure relief valves: simple or intelligent," *In Tech*, 51–55, January 1985.
5. Covington, M. T, "Transient models permit quick leak identification," *Int Pipeline Industry*, August 1979.
6. Davenport, S. T, "Forecast of major international pipeline projects for the 1986–2000 period," *Pipes and Pipelines International*, Vol. 31, No. 1, 9–14, 1986.
7. Furness, R. A., "Modern pipeline monitoring techniques—Parts 1 and 2," *Pipes and Pipelines International*, Vol. 30, No. 3, 7–11, and Vol. 30, No. 5, 14–18, May and September 1985.
8. Kershaw, C., "Pig tracking and location—Parts 1 and 2," *Pipes and Pipelines International*, Vol. 31, No. 2, 9–15 and Vol. 31, No. 3, 13–16, 1986.
9. Stewart, T L., "Operating experience using a computer model for pipeline leak detector," *Journal of Pipelines*, Vol. 3, 233–237, 1983.
10. Taffe, P., "Intelligent pigs crack pipe corrosion problems," *Processing*, 19–20, September 1986.
11. Tinham, B., "Micros and sensors make smart transmitters," *Control and Instrumentation*, 55–57, September 1986.
12. Williams, R. I., "Fundamentals of pipeline instrumentation, automation and supervisory control," ISA *Transactions*, Vol. 21, No. 1, 45–54, 1982.

CHOOSING THE RIGHT TECHNOLOGY FOR INTEGRATED SCADA COMMUNICATIONS

Service was consolidated for both current and projected computer environments

Vernon J. Sterba, Director, System and Telecommunications, Enron Gas Pipeline Group, Group Technical Services, Houston

Consolidation of overlapping telecommunication systems for Enron subsidiaries Florida Gas Transmission, Houston Pipeline, Northern Natural Gas, and Transwestern Pipeline eliminated redundant facilities and integrated information and communication required for upcoming open transportation services.

A wide-area communications study group (WAC) was formed to define and document operations/SCADA communications required for both current and projected computer environments. A functional and technical design would be developed and documented, as well as supporting costs, an implementation timetable, and equipment specifications.

The WAC project was to meet the following requirements:

- Pipeline monitoring and control (SCADA)
- Electronic flow measurement/custody transfer (EFM)
- Real-time pipeline modeling and optimization
- Compressor automation

Since installing an advanced, integrated communication system required a considerable capital investment, the WAC team chose not to take a "SCADA-only" view, since this could miss opportunities to effectively develop the existing communications network.

Mainframe data and voice communications were considered for consolidation. This later became a key factor in technical decision making. The project emphasis was to identify where a cost-effective solution could simultaneously meet SCADA and other voice and data requirements.

The WAC team consisted of Enron members representing both organizational and technical experience. In addition, a telecommunications specialist served as project leader and expediter for formulating project methodology and imparting a "third party" perspective to team discussions.

WAC methodology

The main sequence of tasks was as follows:

- Define existing communication systems
- Define communication requirements
- Analyze adequacy of existing systems
- Identify and evaluate technology alternatives
- Establish communication "direction"
- Develop functional and technical design
- Develop project implementation schedule

This approach allowed the team to fulfill project objectives in a scheduled time frame. Next, an inventory of communication circuits and facilities was entered into a PC database. The database included the latitude and longitude of all facilities, type and name of facility, communication media, circuit number (where applicable), and monthly cost of leased communication.

The database was used to develop communication system maps by interfacing with a computer-automated design package. The maps identified redundant communication paths, duplication of facilities, high-cost communication routes, and other details. Without the maps' visual assistance, it is unlikely the team could have quickly developed an understanding of the extensive and complex communication system.

Existing communication facilities included the following technology combinations:

- Multi-drop and point-to-point leased lines. These lines represented a major cost component; many redundant circuits were identified, and some surprising cost anomalies were discovered. For example, the cost of a 40-mi hop from an RTU to the nearest microwave tower could cost more than a 300-mi hop from the RTU to Houston.
- Company-owned 2/6-GHz microwave system. The system provides voice and data communications in Texas inland and Gulf Coast areas and incorporates voice circuits for toll-free calling at 30 field offices along the microwave system.
- Radio telemetry. This system was used by Northern Natural Gas and Transwestern for gathering SCADA data at master control units (MTUs) or central data concentration points in major supply basins. The MTUs, in turn, relayed data to the SCADA master via leased phone lines. Meteor burst technology. The system "bounces" radio waves off the ionized trails of micro-meteorites entering the earth's atmosphere. Northern Natural Gas used this technology, as a prototype, for nine RTU sites.
- VSAT technology. The VSATs (very small aperture terminals) use satellites in orbit for relaying radio signals. Dish antennas of 1.8 to 2.4-meter diameter (6 to 8 ft) were required. In 1986, Transwestern Pipeline installed 11 satellite terminals to eliminate high construction costs involved with leased line service at remote RTU sites.

Analysis of technology

Sending data to the SCADA master required both short and long-haul communication segments. The segments could be established by separate and independent technologies.

For example, to aggregate data from several RTUs at a central hub, use radio telemetry or local leased phone lines. The long-haul technology (central hub to SCADA master) could use either leased phone line, microwave, or VSAT. This requires a serial byte-oriented, SCADA protocol standard, which had been in place since 1987.

To determine the best communication direction, various technology "mixes" were applied system-wide. This provided comparative technology costs for the entire communication system.

Table 1
Communication guidelines

Short-haul choices
1. Use radio, if available, in a radio hub arrangement.
2. Use leased lines if economically available, connected to a hub or the home office site.
3. Use VSAT only at remote sites when the above options are infeasible.
Long-haul choices
1. Use existing microwave, when available.
2. Use VSAT for sites with multiple applications.
3. Use single-application leased lines when the monthly cost is less than $450; otherwise, use VSAT.

Amortized capital cost and O&M cost, including leased lines, were estimated for each design case. The cost analysis was conducted for both current and projected data density. The technology mixes considered were:

Case	Short-haul	Long-haul
1	radio	leased line
2	radio	microwave
3	radio	VSAT
4	leased line	leased line
5	leased line	microwave
6	leased line	VSAT

In both current and future data density environments, the combination of radio telemetry and long-haul leased lines (Case no. 1) yielded the minimum monthly cost. This was followed closely by radio telemetry and VSAT (Case no. 3). When VSAT was used also as a mainframe connection, Case no. 3 became the minimum cost option.

In all cases, radio telemetry was the most economical technology for the short-haul communication segment. Radio systems bypass local telephone-company leased lines to eliminate escalating costs and service problems found in remote areas. Radio is not always available, due to frequency availability.

For existing facilities, the microwave network was a clear choice for long-haul systems, with no capital investment required and a small incremental cost for adding circuits. However, when considering microwave expansion, the situation changed significantly. In the current data density environment, increased microwave coverage (Cases no. 2 and 5) did not justify capital costs involved, due to insufficient data traffic.

In the projected environment, higher data densities began to support the capital and O&M costs as microwave system capacity was fully utilized. This moved microwave technology from seventh to third most favorable system, compared to other technologies.

Generalizations shown in Table 1 apply to one system's economics and topography and are not intended to suggest the same options should be chosen by other gas pipeline companies.

For example, the short-haul radio option was not viable in some cases. In one case studied, the cost of a high radio tower exceeded the cost of a VSAT system. Frequency availability was a problem as well.

In addition, the long-haul segment was affected by site-specific factors such as data density, leased-line installation and operating costs, and VSAT line-of-sight availability. VSATs require a clear view to the south.

The technical evaluation recommended combining short haul radio telemetry with existing microwave or VSAT for the long-haul segment. Economics were especially favorable when the VSAT could provide mainframe connections, a situation already found at a majority of sites.

Another decision was whether to recommend C-band or Ku-band VSAT technology. The company was using C-band VSATs in the Transwestern Pipeline but at the time had no experience with Ku-band VSATs.

C-band VSAT advantages

- More cost effective for low data rates; C-band hardware is generally 25%-50% less expensive.
- Capabilities meet existing SCADA transmission speed requirements.
- C-band SCADA communication services were available and had been used since 1986, a "known-quantity" advantage.

C-band VSAT disadvantages

- Not amenable to moderate or high-speed data transmission rates, voice, or video.
- Shares the radio wave spectrum with terrestrial microwave services, increasing the possibility of interference, a rare occurrence in rural areas, but a major concern in metropolitan areas.
- A single vendor was the main provider of VSAT transponder capacity.
- Existing C-band satellites have low utilization.
- C-band is older technology. The expected life span of a satellite, both C and Ku-band, is about 7 years, usually set by fuel limitations for keeping the satellite onstation in geostationary orbit.
- Projected shortages of C-band satellite capacity were expected beyond 1991, as existing satellites reach lifespan limits. However, the projected shortage did not develop.
- When conducting the WAC project, there were no launch plans for replacement C-band satellites.

Ku-band advantages

- Growth trend in satellite communications is toward Ku-band, with future satellite launches planned.
- More suppliers are expected in the Ku-band marketplace.
- Ku-band VSATs can be used for high-speed applications such as voice, video, computer-to-computer data communications, and SCADA and mainframe workstations. Local networks can be extended into wide-area networks (WANs).
- Although the value of satellite voice communications is limited by the inherent 0.5-second delay, receive-only video and data communications will become more valuable in the future, if costs decrease.
- Television receive-only (TVRO) can be used to communicate with a dispersed workforce and provides cost-effective training when compared to increasing airfare costs.
- Computer-to-computer data communications will become necessary components as microcomputer installations increase at field locations.

Ku-band disadvantages

- Cost usually exceeds C-band cost for one-application installations.
- Can be affected by heavy rainfall, such as rainfall rates sufficient to restrict driving. This occurs because water droplets absorb the satellite frequency; however, this can be mitigated by proper antenna design and other equipment. Most users report actual outage rates less than statistically predicted.

VSAT decisions

Because a distributed mainframe application was planned, the Ku-band scenario was cost effective at a majority of sites. A secondary decision was made not to add more C-band VSAT sites. Also, because of the economics mentioned earlier, radio became the first choice for the short-haul option. The combined radio/VSAT or radio/microwave became the "radio hub."

Operating costs, for a Ku-band VSAT at each SCADA site, were $375 to $450 per month. The exact cost varied depending on vendor, data traffic costs, and capital amortization period. This estimated cost agreed with other published analyses.[1-3]

One disadvantage of VSAT technology was the delay introduced by equipment and related store-and-forward message handling. With one vendor, this delay ran as long as 12 seconds, reducing the effective circuit speed.

Do's and don'ts of VSAT technology installation:

- Do install bumper guards around VSATs in compressor station yards.
- Do install a security fence around VSATs in or near public areas and be prepared to install camouflage, such as shrubbery, to hide the VSAT from public view.
- Do allow extra time for installing VSATs mounted on building roofs, especially if the building is not company-owned.
- Do use dish heaters where ice storms are common.
- Do install antennas high enough to avoid snow drifts in northern latitudes.
- Don't install dish heaters in snowy regions where ice storms are frequent.
- Don't install VSAT antennas just outside office buildings; this may obstruct view or reflect sunlight into windows.

Because of inherent VSAT delays, special features were mitigated into the SCADA system. A report-by-exception feature was implemented at data concentration (MTU) sites. Delay was reduced also by using single-hop rather than double-hop VSAT technology.

In a single-hop system (Figure 1), the message path is:

- Outbound poll—SCADA master to vendor master earth station (MES) via leased lines.
- Inbound response-VSAT at RTU to MES via space segment; MES to SCADA master via leased lines.

The double-hop system (Figure 2) replaces the SCADA master to an MES leased-line link with another pair of space-segment hops, increasing time delay, but still cost-effective for smaller systems.

A dial-up backup to key radio hubs and other critical VSAT sites was chosen. The SCADA system supports this failure mode by automatically dialing up failed VSAT sites. This is cost effective, since the business line's monthly cost is low relative to other options. Dial-up on-line costs are high, but rarely needed.

Implementation

Because of very close bids and a WAC team recommendation to use multiple VSAT vendors, two vendors were awarded contracts to install Ku-band VSATs across the nationwide 38,000-mi pipeline grid. Initial construction was completed in 1989.

Enron currently has 125 VSAT sites throughout the United States and has changed over to primarily Ku-band technology. Not all sites use the same VSAT options, as shown below:

Ku-band application area	Number of sites
SCADA	104
Mainframe	60
TV RO	63
Other (ASCII/X.25)	14
Total number of sites	125

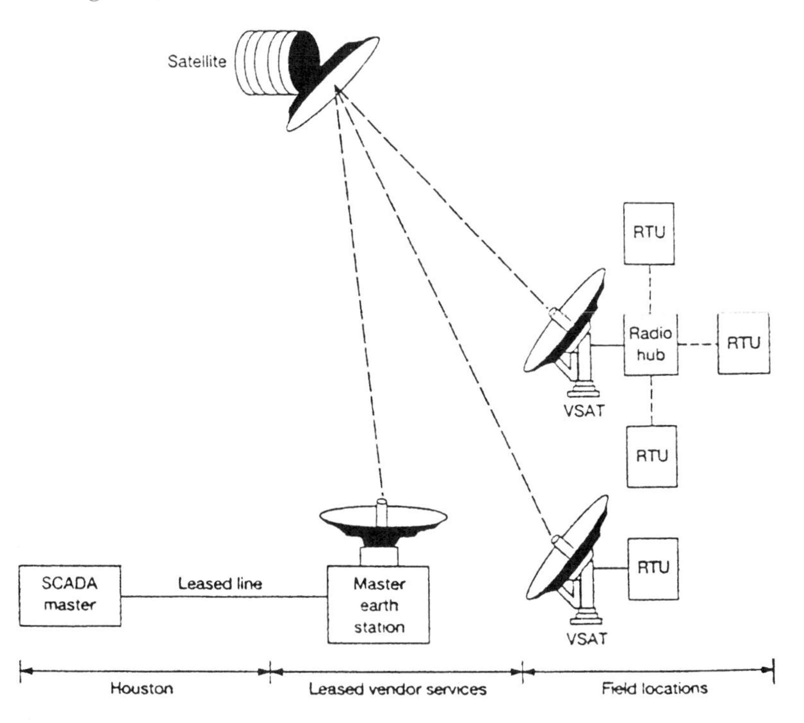

Figure 1. Single-hop VSAT.

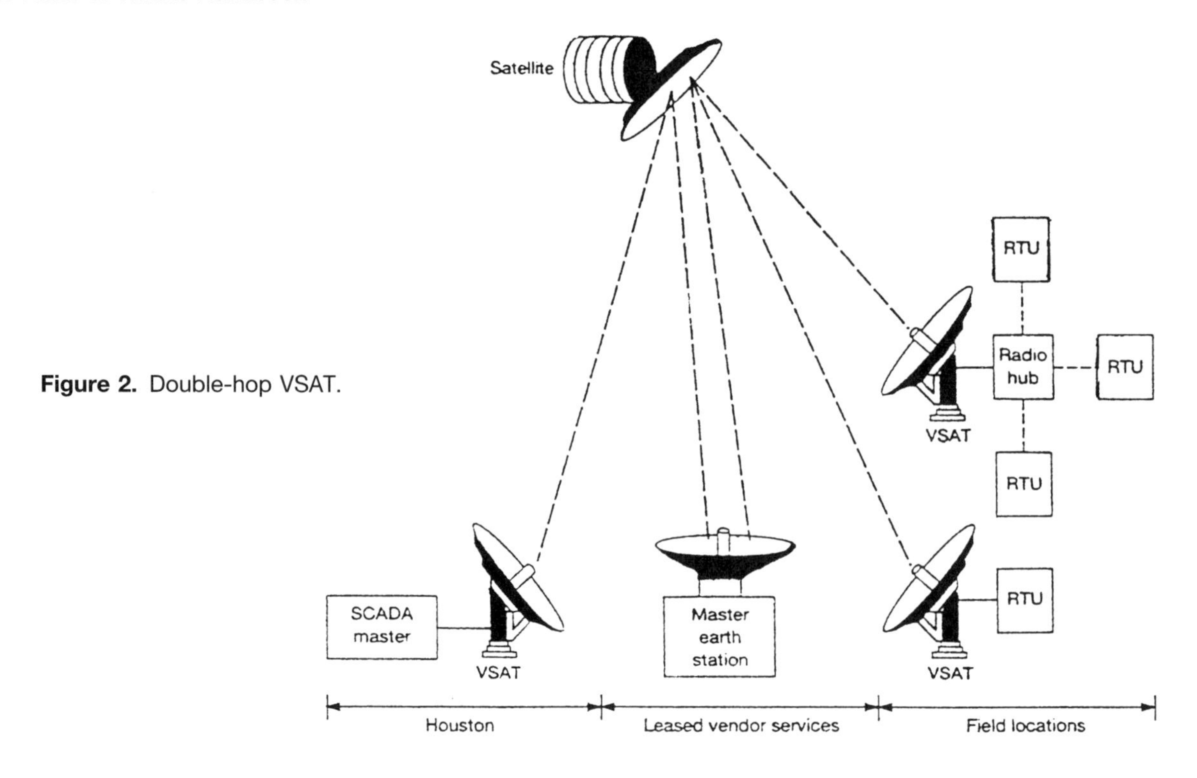

Figure 2. Double-hop VSAT.

The television receive-only (TVRO) application currently is being installed. The X.25 application permits linkup between field and home office systems, including local area network, electronic mail, and remote access to centralized DEC VAX applications without any dial-up cost. Because many sites are in remote areas, the dial-up costs are often dwarfed by the costs and frustrations of poor quality telephone connections.

Acknowledgment

Based on a paper, "VSAT Communications Technology for SCADA System Communications," presented by the author at the AGA Distribution Conference, Nashville, Tenn., April 1991. The author received the Acker Medal Award 1991–92 for this technical paper, sponsored by Central Hudson Gas and Electric Corp., in recognition of excellent contribution to gas industry.

Literature cited

1. Sharma, R., "VSAT Network Economics: A Comparative Analysis," IEEE Communications, Feb. 1989.
2. Simo, E., "VSATs and the Pipeline Industry," Satellite Communications, Aug. 1988.
3. Coulter, J., "Design and Installation of a Gas Control SCADA Network using Satellite, Microwave and Radio Systems," Entelec News, Vol. 61, no. 2, Aug. 1988.

17: Leak Detection

Pipeline leak detection techniques

R. A. Furness, Cranfield Institute of Technology, Cranfield, Bedford, UK, and **J. D. van Reet**, Scientific Software Intercomp, Houston, Texas

Summary

Pipelines are now an integral part of the world's economic structure, and literally billions of dollars worth of products are now moved annually in pipelines. Both economic and environmental factors are influential in pipeline operation, and therefore integrity monitoring is vitally important in the control and operation of complex systems.

Leak detection systems range from simple, visual line walking and checking to complex arrangements of hardware and software. No one method is universally applicable, and operating requirements dictate which method is the most cost effective. The aim of this paper is to review the basic techniques of leak detection that are currently in use. The advantages and disadvantages of each method are discussed, and some indications of applicability are outlined.

Modern pipeline computer modeling and control is then reviewed in detail. These systems are the most flexible and versatile and are steadily being adopted. The influence of instrument accuracy on system design and performance is discussed, and the basic modeling equations are reviewed.

Introduction

Our need to transport fluids from the point of production to the area of end use has led to a rapid increase in the number of pipelines being designed and constructed. Many of these carry toxic or hazardous products, often close to centers of high population or through areas of high environmental sensitivity. With the need to safeguard these lines, online monitoring is becoming routine and in some cases 24-hour surveillance is mandatory. With the increase in world terrorism, the need for rapid and reliable pipeline measurement and control systems will increase.

The review begins with a summary of causes of leaks and the implications of the failure to detect them. Basic techniques are covered, and the features of each are briefly discussed. The bulk of the paper deals with modern computer-based techniques. Basic flow equations are covered, and the on-line dynamic calculations required are listed together with the input data required to enable the monitoring with minimal downtime. The latest systems are capable of resolving down to 1% of the maximum rated flow with a response time of a few minutes. Practical experience verifies this figure, but the total costs of such a system could be high. The system therefore becomes a compromise among response, performance, and alarm availability, and choice of instrumentation is crucial. The integration of good quality instruments with advanced real-time models seems to be the current trend, and the paper closes with some personal thoughts on future trends.

Causes and economic aspects of leaks

There are four main categories of pipeline failures. These are:

- Pipeline corrosion and wear
- Operation outside design limits
- Unintentional third-party damage
- Intentional damage

Many pipelines are operated for a number of years with no regard to any possible mechanical changes occurring in the line. Some of the products may be corrosive, the pipeline may be left partially full for periods of time, or atmospheric effects may cause external damage. These three reasons are responsible for pipeline corrosion, and this may give rise to corrosion "pits" developing along the line. These are small in nature and could be responsible for material imbalances over a period of time. Very accurate flowmetering can be used to detect this, as discussed in the next section. Abrasive fluids or dust-laden gas streams can give rise to pipeline wear. Again, this is a slow process, but should a weak spot develop (more often than not close to a change in direction or section) then a pipe break may occur very rapidly and totally unexpectedly.

Operation outside design guidelines is more common than is realized, as operators seek to use the line for as many fluids as possible. If the line is designed for a certain maximum temperature and pressure, then operation at higher pressure and/or higher temperature could lead to spontaneous failure. The problem could be compounded if the line has a large but unknown amount of corrosion.

Unintentional third-party damage may occur if excavation or building occurs near buried lines. More often than not the right-of-ways are not clearly marked, and lines are sometimes broken by bulldozers or similar plant machinery, possibly with fatal results. An example occurred in West Virginia in 1984 when a 20-in. natural gas line was punctured by an excavator, and product leaked slowly into a nearby supermarket during the night. The building was totally destroyed early next morning by a massive explosion caused by a staff member lighting a cigarette. Such an occurrence could have been avoided by some form of leak detection on the line.

Intentional damage is unfortunately on the increase, and pipelines carrying flammable or high-value products make ideal targets. Alarm systems linked to block valves can help to minimize the amount of product released as a result of sabotage, so again certain lines are instrumented with the intention of reducing the effects of planned terrorism.

The costs of failure to detect leaks also fall into four main areas:

- Loss of life and property
- Direct cost of lost product and line downtime
- Environmental cleanup costs
- Possible fines and legal suits

These are all self explanatory, with the most costly of these being the last, although any of the four areas could be very expensive. The size of claims can run into many millions of dollars, so the cost of fitting and operating leak detection systems is often insignificant compared with the costs of failure of the line.

It is this background that is causing operators and designers to turn to on-line integrity monitoring systems, and the next section looks at the more basic methods.

Simple leak detection systems

The most basic method of leak detection involves either walking, driving, or flying the pipeline right-of-way to look for evidence of discoloration of vegetation near the line or actually hear or see the leak. Often "unofficial" pipeline monitoring is performed by people living nearby who can inform the operator of a problem with the line.

The most cost-effective way to detect leaks in nonflammable products is to simply add an odorant to the fluid. This requires some care in selection, as frequently the odorant has to be removed before the transported fluid can be used. Organic compounds make the most useful odorizers, especially when the fluid being carried has no natural smell of its own. A good example is carbon monoxide, a highly toxic but odorless gas that is often pumped in large quantities between chemical plants. Chemicals such as mercaptans (rotten egg smell) or Trimethylamine (rotten fish smell) can be added in small quantities to enable any leak to be located by smell. The disadvantage of such a method is that if the leak occurs in an area of no population, the leak will go undetected unless the line is walked regularly by pipeline surveillance crews carrying suitable "sniffer" detectors. Thus, to the apparent low costs of this method have to be added the costs of removing the odorant and maintaining staff to check the line at frequent intervals. The location of a leak is also dependent on prevailing weather conditions. Strong winds may disperse the smell, and atmospheric inversions may give an incorrect location of the leak. The uncertainty of relying on this method alone is high. Nevertheless, it is a useful method if used in conjunction with other techniques.

Simple line flow balances are frequently used to check for gross imbalances over hourly or daily bases. This method may identify that a leak is present, but flowmeters at each end of the line will not identify the leak location. A line pressure measurement system will be required in conjunction with the flowmeters to establish that the pressure gradient has changed from the no-leak situation. The method is useful, however, in identifying the existence of corrosion pits, as the outputs of the flowmeters at each end of the line will consistently diverge if flow in the line is maintained constant. If line flow rate varies with time, then imbalances are more difficult to detect, since the flowmeter outputs may vary nonlinearly with flow rate or may have different flow characteristics from each other.

A loss of product will be identified simply as the difference between the steady state inventory of the system and the instantaneous inlet and outlet flows. Mathematically this is:

$$\Delta V = V_{in} - V_{out} - V_1$$

where: ΔV = Leakage volume
V_{in} = Metered inlet flow
V_{out} = Metered outlet flow
V_1 = Pipeline fluid inventory

This last term can be calculated as the average of the integrated inlet and outlet flows in simple systems, but as will be seen later, the value of this term can be calculated more accurately and easily in real time as a function of several variables.

Another method is based on detecting the noise associated with or generated by a leak. There are many instances where fluid flow can generate vibrations at frequencies in excess of 20 kHz. These frequencies are in the ultrasonic range but can be detected with suitable transducers. The device can be made portable so that pipeline crews can clamp a transducer at any point along the line to check for noise. By noting the signal strength, the source of the leak can be pinpointed.

A similar technique, though based on a different principle, is the acoustic "wavealert" monitor, more correctly called a negative pressure wave detector. This is a piezoelectric sensor that gives an output when dynamically stressed. When a leak occurs, there is a sudden drop in pressure at the leak followed by rapid line repressurization a few milliseconds later. The low pressure wave moves away from the leak in both directions at the speed of sound. The pipe walls act as a waveguide so that this rarefaction wave can travel for great distances, attenuating in amplitude as it does so. Sensors placed at distances along the line can be triggered as the wave passes, and the location of the leak can be calculated from the line conditions and the internal timing devices in the instrument.

Such devices are particularly useful in identifying large breaks in lines very rapidly, since the transient wave typically moves 1 mile in 5 seconds in gases and almost 1 mile per second in liquids. Response is therefore on the order of a few seconds depending on the positioning of the transducers. Figure 1 shows one adaptation where the instrument can be made to cancel line noise and use the full measuring capability of the sensor for signal detection. There is the problem of setting the background threshold correctly, as this may be affected by the location of the instrument in relation to bends, valves, pumps, regulators, etc. Experience has shown that the installation is also critical to reliable performance, and there is also a dependence on the Reynolds number of the flow. Both of these affect the number of "false alarms" from the instrument.

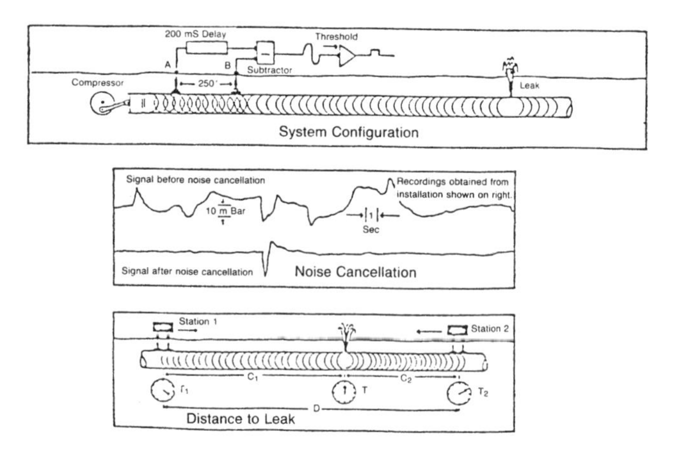

Figure 1. Negative pressure wave detector (acoustic monitor).

Pig-based monitoring systems

Pipeline pigs are frequently used for pipeline commissioning, cleaning, filling, dewaxing, batching, and more recently pipeline monitoring. This last type of pig can be designed to carry a wide range of surveillance and monitoring equipment and can be used at regular intervals to check internal conditions rather than continuously monitoring the line. Data, however, can be built up over a period of time to provide a history of the line. This information can be used to predict or estimate when maintenance, line cleaning, or repairs are required. If a leak is detected, for example, by flow meter imbalance, the location can be found by using a pig with acoustic equipment on board. This will alarm when the detection equipment output reaches a maximum, and the precise location of the pig can be confirmed by radio transmitters also mounted on board.

Pigs require tracking because they may become stuck, at a point of debris build-up, for example. Pigging should be carried out at a steady speed, but occasionally the pig may stop and start, particularly in smaller lines. Information on when and where the pig stops is therefore important in interpreting the inspection records. Pig tracking is not new, and many such proprietary systems exist. In the best systems, however, a picture of the line is often programmed in so that outputs from junctions, valves, cross-overs, and other geometries act as an aid to location. Pig tracking can make use of the acoustic methods discussed earlier. When the sealing cups at the front of the pig encounter a weld, vibrational or acoustic signals are generated. Each pipeline therefore has its characteristic sound pattern. When a crack occurs, this pattern changes from the no-leak case, and the location can be found from direct comparison. The technology has become so advanced that information on dents, buckles, ovality, weld penetration, expansion, and pipeline footage can be generated.

The equipment is often simple, consisting of sensor, conditioning, and amplifier circuits and suitable output and recording devices. Such a device developed by British Gas is shown in Figure 2. The range of detection is dependent on the pipeline diameter and the type of pig. Operational data have shown that light pigs in a 200 mm line can be detected at a range of 8 km, increasing to 80 km for a heavy pig in a 900 mm line. As the signals travel at acoustic velocity, this means a signal from a pig at 80 km range will take 190 seconds to be picked up. Such technology is now becoming routine in both offshore gas and onshore liquid lines.

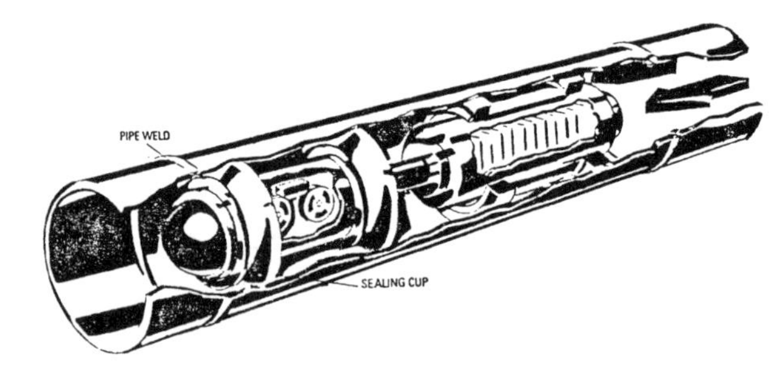

Figure 2. "Intelligent" pipeline monitoring pig. Courtesy British Gas Corp.

Computer-based monitoring systems

It is in computer-based systems that the greatest amount of data can be gathered, processed, analyzed, and acted upon in the shortest period of time. Programs can calculate the inventory of the line at any time and compare this with accurate measurements at any section in the system. The effects of pressure and temperature on line dimensions, for example, can be calculated to provide an accurate estimate of the mass of fluid in the line. Data from a wide range of instruments can be transmitted by telemetry, radio, or phone links to a central computer that monitors the "health" of the line continuously. By changing programs and subroutines, a vast amount of functions and tasks can be accomplished very easily and cost effectively.

The many functions that can be performed by computer-based systems include not only leak detection but also:

- Pig tracking
- Batch tracking of fluids
- Inventory accounting
- On-line flow compensation
- Instrument data and malfunction checking, etc.

Such systems have very rapid response and have the advantage of multiple inputs being required before leaks are declared. Thus, some systems can run for short outage periods with no loss of integrity. They are often complex and costly to install, but once the initial capital investment has been made, running costs are low. The first section in the detailed review of such systems looks at the phenomena that need to be modeled when a leak occurs.

Pipeline leak phenomena

When a leak occurs in a pipeline, the measured pressure downstream of the leak falls, but the pressure at the same location is predicted to rise. The first is not difficult to understand as the line is depressurizing as mass leaves through the leak. The second effect can be explained as follows. The equations presented later in the paper predict pressure based on measured flow or flow based on measured pressure. As mass leaves the system through the leak hole, a reduced flow at the downstream end is compared to the inlet flow. This may not have changed, and so to balance the system, the equations predict a downstream pressure rise. In physical terms the model thinks the line is "packing" and total system inventory is increasing. There is therefore a divergence between measured and modeled pressure.

The same is true of flow changes, but here the inlet flow could increase due to lower pipe flow resistance between meter and leak, while the section outlet flow will fall as mass leaves through the leak instead of passing through the meter. Thus, a real imbalance will result. The model, however, will show an inconsistency since the pressure comparison will indicate line packing and the flow comparison a line unpacking. If selected pressure and flow imbalance limits are exceeded, then a leak is declared. The magnitude of the leak is predicted from the flow imbalance and the location from the pressure profile imbalance and the flow leak indicators. The impact of instrument accuracy on the predicted location is discussed later in the paper. It is vitally important to good leak sizing and location to have the best pipeline instrumentation possible to minimize uncertainty.

Background philosophy of pipeline modeling

Real-time modeling is a technique that uses the full data-gathering capabilities of modern digital systems and the computational power of small computers to give accurate "snapshots" of the pipeline. The whole system is under the control of a SCADA package of programs, which poll the data stations on the line, process the data, control the running of the transient pipeline model, and activate the alarm and leak location routines. In addition to these basic software modules, more complex systems might include a predictive model to analyze "what if" operating scenarios, provide an optimization routine for least-cost operating strategies, or include a separate man/machine interface for the model system.

The SCADA interface is responsible for acquiring the data from the SCADA system and relating them to the model representation of the line. As a point in the system where two large and independently developed systems join, this can be the source of many problems in the implementation of the real-time modeling. Once the measurement data have been obtained, noise filtering and plausibility checking can be performed prior to running of the model. The model is the mathematical representation of the pipeline and will include such features as elevation data, diameters, valve and pump locations, changes of direction, and the location of cross-overs and junctions. The model provides data on the flow conditions within the line at intervals between seconds and minutes, depending on operational needs.

With the data available from both the measurement system and the pipeline model, the real-time applications modules are run. These are the leak detection and location routines in the context of integrity monitoring. The leak detection module functions by computing the difference between the modeled flows and pressures and the measured values at all points where measurements not already used as boundary conditions are located. Because the model accounts for normal transient operations, these differences will be small under normal conditions. When a leak is present, the differences become larger since the model system does not account for leakage. When these differences exceed preselected values, a leak alarm is declared. Sophisticated voting schemes that require multiple leak indicators to be in alarm for several consecutive time intervals are used to reduce false alarms while maintaining low thresholds. Often a simple pipeline balance of the type discussed earlier is used as a back-up to verify the transient model. Response characteristics are, however, much slower than the real time model.

Once the leak detection module declares a leak, the location routine is activated. The location is calculated from the magnitude and distribution of the leak indicators. As an example, in a straight pipeline with an upstream flow discrepancy and a downstream pressure discrepancy as leak indicators, it is an easy calculation to determine where the leak must be such that the leak flow when added to the modeled flow will produce the additional pressure drop observed at the downstream end. Solutions for pipe networks are more complicated, and unique locations do not always exist. This might be the case with parallel looped lines, for example. In this case all the calculated locations should be checked.

The components of the real-time modeling system work together to reduce the large volumes of raw data from the data acquisition system to a much smaller number of parameters and alarms that are more meaningful to pipeline operations. In the case of integrity monitoring, this means leak events that could not be detected by inspection of the measured data can be found and isolated quickly and reliably.

Basic pipeline modeling equations

The transient pipeline flow model is the heart of a pipeline modeling system. The model computes the state of the pipeline at each time interval for which data are available. The state of the pipeline is defined as a set of pressures, temperatures, flows, and densities that describe the fluids being transported at all points within the system. These quantities are found as the solution to a set of equations that describe the behavior of the pipeline system. These basic equations are the Continuity equation, the Momentum equation, the Energy equation, and an equation of state.

The continuity equation enforces the conservation of mass principle. Simply stated, it requires that the difference in mass flow into and out of a section of pipeline is equal to the rate of change of mass within the section. This can be expressed mathematically by the relation:

$$\frac{d(\rho A)}{dt} + \frac{d(\rho AV)}{dx} = 0$$

The momentum equation describes the force balance on the fluid within a section of pipeline. It requires that any unbalanced forces result in an acceleration of the fluid element. In mathematical form, this is:

$$\frac{dV}{dt} + V \times \frac{dV}{dx} + \frac{1}{\rho} \times \frac{dP}{dx} + g \times \frac{dH}{dx} + \frac{fV|V|}{2 \times D} = 0$$

The energy equation states that the difference in the energy flow into and out of a section equals the rate of change of energy within the section. The equation is:

$$\frac{dV}{dt} + V \times \frac{dT}{dx} + \frac{T}{\rho \times c} \times \frac{dP}{dT} \times \frac{dV}{dT} - \frac{f|V^3|}{2cD} + \frac{4U}{\rho c D}(T - T_g) = 0$$

These three are the basic one-dimensional pipeflow equations and are present in one form or another in all transient pipe models. What is needed to solve them, however, is a relation between the pressure, density, and temperature for the fluid—an equation of state.

The state equation depends on the type of fluid being modeled, as no one equation fully describes the variety of products that are shipped in pipelines. Some of the forms in use include a bulk modulus type of relation of the form:

$$\rho = \rho_o \left[1 + \frac{P - P_o}{B} + \alpha(T - T_o)\right]$$

This is normally used for liquids that can be regarded as incompressible. The bulk modulus B and the thermal expansion coefficient α can be constant or functions of temperature and/or pressure depending on the application. For light hydrocarbon gases a basic equation such as $P = \rho \times R \times Z \times T$ is appropriate, where Z (the compressibility) is a known function of temperature and pressure. For reasonable ranges of temperature and pressure, a function of the form:

$$\frac{1}{Z} - 1 \cong \frac{P}{T^y}$$

may be adequate. For conditions where fluids are transported at or near the critical point, a more sophisticated correlation is required to obtain the required accuracy, but there is still a large uncertainty in the true density under these operating conditions, and they should be avoided wherever possible. Many real-time systems have been installed on lines carrying ethylene, butane, propane, and LNG/LPG products and have used the Benedict-Webb-Rubin correlation as modified by Starling (commonly called the BWRS equation) with reasonable results. Alternatively, tables such as NX-19 or special correlations such as the NBS ethylene equation can be used, but this increases the complexity of the programming.

A further complication arises if the product is not uniform throughout the system. This can occur due to batching of fluids or from varying inlet conditions. The first is more common where different products are shipped in a common line. The properties are essentially discontinuous across the interface of the two fluids but can be considered as uniform within batches. The basic problem here is to keep track of the location of the interface. Systems with continuous variations in inlet conditions occur in both liquid and gas systems. The variations can result from mixing of fluids of slightly different composition or from large variations in supply conditions.

The governing equations presented are non-linear partial differential equations, which are not suitable for machine computation. They have to be solved by implicit or explicit finite difference techniques or the method or characteristics. Of these, the implicit method seems the most appropriate, as the other two methods could give rise to mathematical instabilities if the wrong timestep or distance interval is used.

In order for the transient pipeline model to compute the state along the line at the end of each time interval, a set of initial conditions and a set of boundary conditions are required. The initial conditions specify the state at the beginning of the time interval and are normally the last set of data from the model. A steady state model must be used to generate an initial state when the model is started from rest, a so-called "cold start." In this case a period of time must elapse before the pipeline model truly represents the actual state of the line. This time period allows any transient conditions present and not represented by the steady state model to die out. Generally, less compressible systems will cold start faster than the more compressible ones, but the actual time for the transient model to be activated depends on the application. This may typically be on the order of 30 minutes for a gas pipeline.

The boundary conditions required by the model are taken from measured data along the line. For each point where fluids enter the system, its temperature, fluid type or composition, and either a flow or a pressure is required. For any equipment in the system that affects or controls the line, a suitable boundary condition must also be given. For a gas compressor, for example, either its suction pressure, flow rate, or discharge pressure must be specified. Additional measurements are used by the applications modules, generally by comparing their values to the corresponding model calculations.

Impact of instrument accuracy

The performance of real-time pipeline monitoring systems is limited primarily by the accuracy of the instrumentation installed on the line. To estimate the performance of the leak detection and location routines, it is important to understand the effect of measurement uncertainty on the model and the real-time applications module. Measurement uncertainty is composed of bias and random components. The first is usually a fixed error between the indicated and true values, but this could change with time as components wear. The second is temporal and possibly spatial fluctuation of the output about its mean value. Fortunately there are techniques that can be employed within the software to largely mitigate the effects of bias, but care should be exercised as this is not always the case.

Leak detection and location both use differences between the measured values and the modeled values to discern leak characteristics. The measured values could contain both bias and/or random errors as discussed. The model values, because they are driven by the measured values as boundary conditions, also include error terms that are less obvious. Because of these errors, the differences in the model and measured values, or leak indicators, will not normally be zero but will fluctuate about some non-zero mean. This mean value is determined by observation during periods when no leak is present and is attributed to bias errors in the measuring system. By subtracting this from the leak indicators, the bias component can be eliminated. The leak detection then will be a function of the measurement precision errors.

Problems with this technique arise when dealing with pipelines whose operations change substantially from time to time. An example would be a liquid line operating intermittently. By monitoring differences in this way, the instruments contributing to the error are not identified. Because fluid flow in a pipeline is governed by highly non-linear relationships, fixed errors in the boundary conditions can cause variable differences in the leak indicators when the pipeline operation changes. As an example, consider a steady state pipeline that is driven by pressure difference between upstream and downstream boundary points. The leak indicator is the difference in the measured and modeled flow. (Note that a transient model would have a flow difference at each end of the pipeline.) The true value of the data for this pipeline is an upstream pressure of 1,000 units with a pressure drop of 100 units for a flow of 100 units. The model of this system is then described by the equation:

$$P1 - P2 = 0.01Q^2$$

If a 1% pressure error is introduced into the upstream pressure, the modeled flow becomes 105 and a difference of 5 units between measured and modeled flow would show up in the leak indicators. As long as the flow stays near this value, the error in the leak indicator will remain nearly constant. For instance, an actual flow of 80 units would result in a modeled flow of 86 units. Thus, the effects of the bias error in the pressure measurement can be substantially mitigated by subtracting 5 units from the leak indicator. This is termed the leak indicator "offset."

Now assume the line is shut down. The 10-unit pressure measurement error causes the flow to compute a flow of 31 units. After applying the offset, the value of the leak indicator still remains at 25. In general, a line that undergoes large and rapid changes in operation will be affected by instrument bias errors. The more common case of lines that operate within relatively narrow bounds, or that change operations slowly so that the offset can be automatically adjusted, will only be affected by the precision error of the measurements.

Measurement errors impact leak detection by limiting the size of leak that can be detected by the monitoring system. The problem comes in finding a threshold value for each alarm in the system. For a simple system that operates within narrow bounds, this can be as simple as the offset previously discussed. The values of the leak indicators (now after the offset has been removed) can be observed during normal operation and the appropriate alarm values set. For more complicated systems, the thresholds can be set in a more rigorous way. The pipeline hydraulic equations can be used to determine the sensitivity of each leak indicator to the measurement at each boundary point. The error, whether bias or precision, of each boundary point can be estimated from knowledge of the transducers and the data-gathering equipment installed on the pipeline. The error in the boundary instruments times the sensitivity of the leak indicator to the boundary point will give the threshold required to prevent normal noise in the measurement value from being interpreted as a leak. Using the root sum square as a result of combining the threshold for each boundary point with the error for the leak indicator's comparison measurement, a threshold for the leak indicator can be calculated online. This "auto tuning" of the leak indicator is found in the more advanced systems commercially available.

As an example, consider the steady state pipeline used in the earlier paragraph. The sensitivity of the modeled flow to the upstream and downstream pressure are:

$$\frac{dQ}{dP1} = -\frac{dQ}{dP2} = \frac{50}{Q}$$

Thus, at a flow of 100 units, the sensitivity of the flow to either pressure would be 0.5, with a decrease in the downstream pressure being equal to an increase in the upstream pressure. For a 1%, or 10-unit error, in the pressure, a 5-unit error in the flow would be expected, which is consistent with the previous results. The threshold required for this system would then be:

$$TL = \left[\left(\frac{50}{Q} \times EP1\right)^2 + \left(\frac{50}{Q} \times EP2\right)^2\right]^{1/2}$$

This shows that the required threshold for the system increases with decreasing flow, with leak detection being impossible at zero flow. The overly stringent requirement at zero flow is due to the simplified model used. Other than that, the results are representative of the manner in which the leak thresholds must be adjusted when large flow variations occur in a pipeline.

Measurement uncertainty affects leak location by increasing the uncertainty in the calculated leak position. Leaks are located in a line by discovering where a leak of a given size would need to be located to best match the observed discrepancies in the leak indicators. Consider our steady state model again. If there is a leak of 20 units halfway down the line so that the flow is 120 units before and 100 after the leak, then the pressure drop would be 122 units. This pressure drop would correspond to a modeled flow of 110. This would result in discrepancies in the leak indicator at both ends of the pipe at 10. The leak location for a pressure–pressure boundary condition is given by:

$$X = L \times \frac{Q2}{Q1 + Q2}$$

For the conditions given above, the correct location of half the pipe length is obtained. If, however, the flowmeters have a 2-unit error such that the upstream leak indicator is 12 and the downstream indicator is 8, the leak is located 40% down the pipe, not halfway. In a 10-mi line this is an error of 100 miles, which is very significant.

When evaluating the effect of measurement uncertainty on leak detection or location, it is useful to compare the uncertainty to the magnitude of the hydraulic events of interest. Thus, a pressure transducer that is 1% accurate over a 1,500 psi span is only 30% accurate when the pressure drop between two closely spaced valve sites is 50 psi. This is because the flow is governed by pressure differences and not absolute pressures. Likewise, a flowmeter that is 2% accurate in comparison to its span is 70% accurate for sizing a 3% leak.

System design aspects and guidelines

The availability of leak alarm uptime depends heavily on the system design and the choice of hardware. Generally the more complex the system, the greater the risk of leak indicator loss, but the more accurate the location of the leaks. Design is therefore a compromise between cost, performance, and reliability. Consider a section of line shown in Figure 3. For the three simple alarms of flow imbalance, pressure imbalance, and acoustic alarms between

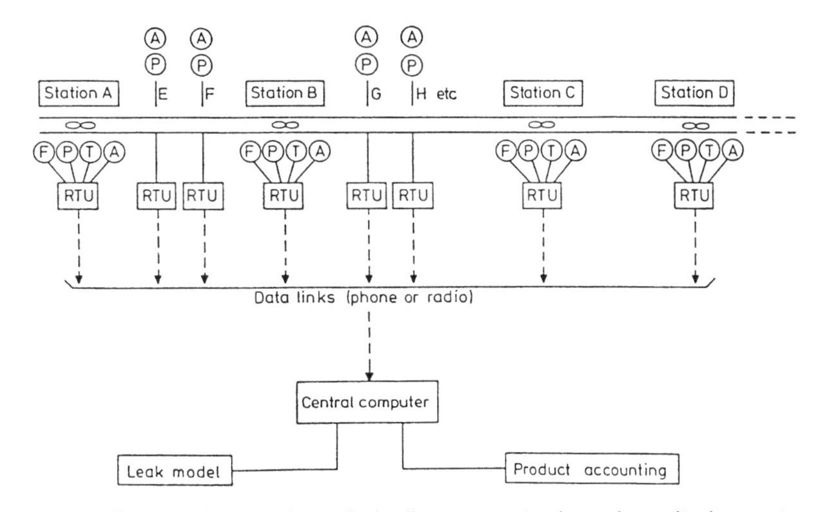

Figure 3. Schematic overview of pipeline computer-based monitoring system.

stations A and B or B and C, the following components are needed:

Mass flow: 2 flowmeters, 2 pressure sensors, 2 temperature sensors, 2 RTUs, 2 communications links, 1 computer (11 elements)

Pressure: 4 pressure sensors, 4 RTUs, 4 communications links, and 1 computer (13 elements)

Acoustic: 2 acoustic monitors, 2 RTUs, 2 communications links, and 1 computer (7 elements)

If combined hybrid alarms of flow/acoustic, flow/pressure, or pressure/acoustic are used, then the number of components in the chain is increased. By summing the component availabilities for each element, an uptime for each alarm can be estimated. For example, a flowmeter with a failure of once in 3 years with a repair time of 4 hours has an availability factor of 0.99985 (or 99.985%).

From such an analysis, the conclusion can be drawn that the system should be made as simple as possible or instruments should be installed in duplicate to maximize alarm uptime. Section 9 also showed that instruments should be selected on performance and not on economic grounds. It is better to install fewer high-performance instruments than numerous poor ones. Digital conversion also requires attention. Often 12- or 14-bit conversion is required to give the necessary accuracy of data processing, and usually double-precision computation is also required. The use of standard outputs should be made wherever possible. Custom-designed electronics invariably lead to problems. The best guideline, however, is to seek users of pipeline monitoring systems to ask their advice and experience with instruments and system components. Independent validation of all information should be made wherever possible. Companies that supply such complete systems usually have a client list, and it is worth spending time talking to these clients before the final design specification is fixed.

With regard to instrumentation, flowmeters with the highest accuracy are required for mass balance functions. Suitable types include turbine and displacement meters with pulse outputs. Orifice meters are not really suitable, since the best accuracy that can be obtained from a well-maintained system is around ±1% of full scale. The correct choice of turbine by comparison is ±0.25% of reading or better.

Development of pipeline monitoring systems

The speed of instrumentation development generally is rather frightening. The impact of micro-electronics is still being felt some 10 years after they first appeared, and new and improved transducers with on-board "intelligence" are being sold in increasing numbers. At the same time the size of computers is decreasing and the computating capability is increasing. Software is also advancing rapidly, and the performance of modern flow monitoring systems is becoming dependent on the accuracy of the modeling equations. Equations of state and the behavior of hydrocarbon mixtures are not particularly advanced or well understood, and fundamental research is required before the next advance in this type of technology can proceed.

All of these points indicate that computer-based monitoring systems will become the standard technique of operating and controlling pipelines in the future. Control algorithms can be integrated with the applications modules to produce a semi-intelligent complete integrity monitoring scheme. As experience in the design and operation of such systems grows, they will be applied with increasing confidence.

Future systems will use a combination of the new technology discussed. This could include internal monitoring pigs and advanced pipeline models, both run from a central control room. Thus, the internal and external state of the line could be checked simultaneously. Such technology can enable safer and more economic operation of pipelines to be carried out.

Conclusion

This section cannot do justice in such a short space, to the complex and diverse subject of leak detection. Such systems have been in operation in many forms all over the world, but it is only recently that environmental as well as economic factors have influenced their development. Modern digital systems are transforming operation and design, with many parallel functions being possible with a single system. There is now a clear need to ensure complete integration of all components in the system to guarantee safer and more accurate pipeline management. Instrument selection is critical, as is the need to develop better thermodynamic models for the next generation of systems to become more reliable and accurate.

References

1. Furness, R. A., "Modern Pipe Line Monitoring Techniques," *Pipes and Pipelines International*, May–June, 7–11 and September–October, 14–18, 1985.
2. Mailloux, R. L. and van Reet, J. D., "Real time Transient Flow Modeling Applications," PSIG Annual Meeting, 1983.
3. Covington, M. T., "Pipe Line Rupture Detection and Control," ASME Paper 78-PET-54,1978.
4. Bernard, G. G., "Energy Balance Derivation," CRC Bethany Internal Report, 1982.
5. Anon, "Pinpointing pigs in pipe lines," *R & D Digest*, 8, 14–15, British Gas Corporation, 1986.
6. Fujimori, N. and Sugaya, S., "A study of a leak detection based on in-out flow difference method," Proceedings of IMEKO Symposium on Flow Measurement and Control, pp. 205–209, Tokyo, Japan, 1979.
7. Stewart, T. L., "Operating experience using a computer model for pipeline leak detection," *Journal of Pipelines*, 3, 233–237,1983.

Notation

A cross-sectional area
B fluid bulk modulus
c specific heat at constant volume
D pipeline diameter
EP1 uncertainty in upstream pressure measurement
EP2 uncertainty in downstream pressure measurement
EQ uncertainty in flow measurement
g gravitational acceleration
h elevation
L pipe length
P pressure
P_o reference or base pressure
P_1 section upstream pressure
P_2 section downstream pressure
Q pipeline flow rate
Q_1 upstream measured vs. modeled flow discrepancy
Q_2 downstream measured vs. modeled flow discrepancy
R gas constant
T temperature
T_g ground temperature
TL leak detection threshold level
T_o reference or base temperature
t time
U heat transfer coefficient
V velocity
X leak location
x incremental distance along the pipeline
y Z factor correlation coefficient
Z gas compressibility factor
α coefficient of thermal expansion
ρ density
ρ_o reference or base density

18: Tanks

Charts give vapor loss from internal floating-roof tanks

S. Sivaraman, Exxon Research & Engineering Co., Florham Park, N.J.

Nomographs, based on the guidelines presented in American Petroleum Institute (API) Publication No. 2519, have been constructed to estimate the average evaporation loss from internal floating-roof tanks.[1] Loss determined from the charts can be used to evaluate the economics of seal conversion and to reconcile refinery, petrochemical plant, and storage terminal losses.

The losses represent average standing losses only. They do not cover losses associated with the movement of product into or out of the tank.

The average standing evaporation loss from an internal floating-roof tank depends on:

- Vapor pressure of the product
- Type and condition of roof seal
- Tank diameter
- Type of fixed roof support

The nomographs (Figures 1–4) can estimate evaporation loss for product true vapor pressures (TVP) ranging from 1.5 to 14 psia, the most commonly used seals for average and tight fit conditions, tank diameters ranging from 50-250 ft, welded and bolted designs, and both self-supporting and column-supported fixed roof designs. The charts are purposely limited to tank diameters 250 ft and less, because internal floating-roof tanks are generally below this diameter.

Typical values of the deck fitting loss factors presented as a function of tank diameters in the API Publication 2519 have been used in the preparation of these nomographs. In addition, for the calculations of the evaporation loss for the bolted deck design, a typical deck seam loss factor value of 0.2 has been assumed.

Table 1 gives the proper axis to use for various seal designs and fits.

Table 1
Selection of seal axis

Seal type	Seal axis: Average fit	Seal axis: Tight fit
Vapor-mounted primary seal only	H	G
Liquid-mounted primary seal only	F	E
Vapor-mounted primary seal plus secondary seal	D	C
Liquid-mounted primary seal plus secondary seal	B	A

Use of these nomographs is illustrated by the following example.

Example. Determine the evaporation loss for an internal floating roof tank given the following:

- Tank diameter 200 ft

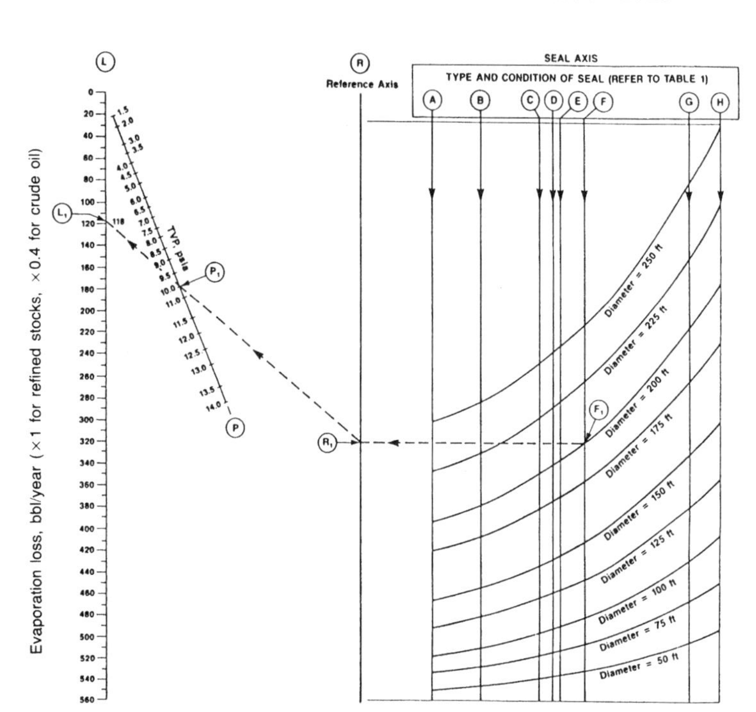

Figure 1. Loss from welded deck, self-supporting fixed roofs.

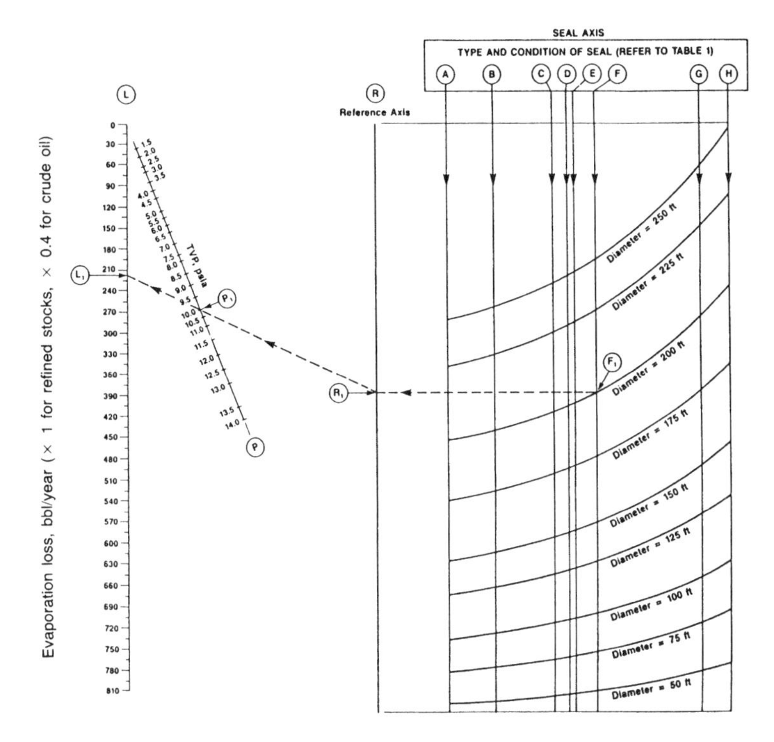

Figure 2. Welded deck, column-supported fixed roofs.

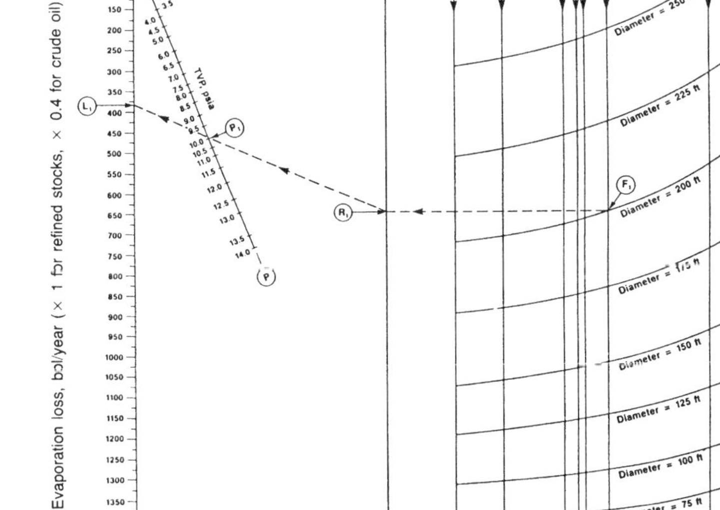

Figure 3. Bolted deck, self-supporting fixed roofs.

- Liquid-mounted primary seal only and an average seal fit
- Product true vapor pressure of 10 psia
- Welded deck with self-supporting fixed roof

Solution

1. Use Figure 1 for the welded deck and self-supporting fixed roof.

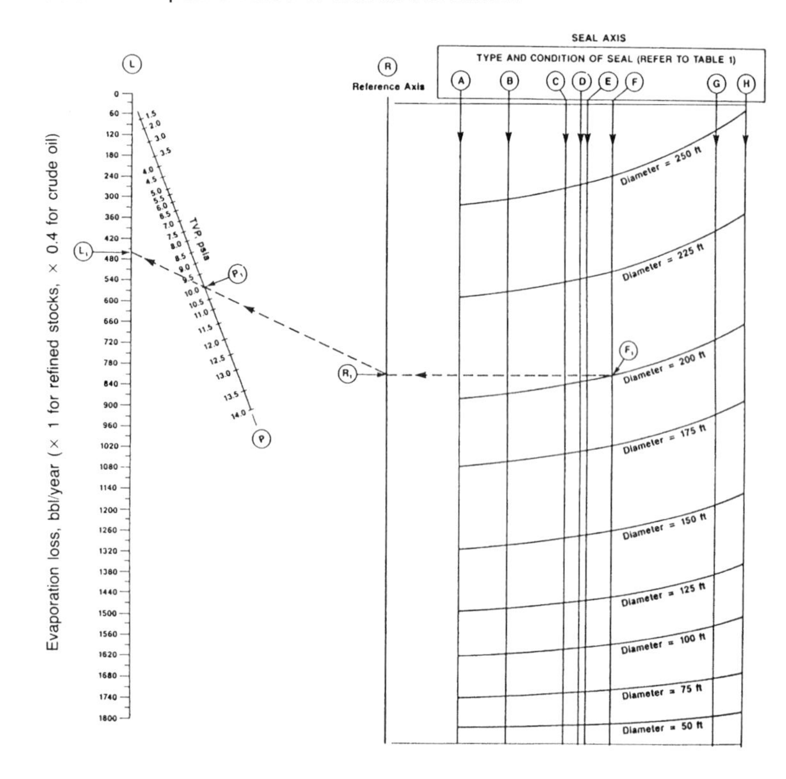

Figure 4. Bolted deck, column-supported fixed roofs.

2. From Table 1 select the seal axis. The seal axis for the example problem is F.
3. Locate the point of intersection F1 between the seal axis F and the tank diameter contour for the 200-ft diameter tank.
4. From the point F1 traverse horizontally to intersect the reference axis R at R1.
5. Locate the true vapor pressure point P1 corresponding to 10 psia on the pressure axis P.
6. Connect the point R1 on the reference axis R and the point P1 on the pressure axis P and extend in to intersect the evaporation loss axis L at L1.

Read the evaporation loss in bbl/year at L1. The average evaporation loss is 188 bbl/year for this example. The same example is shown in Figures 2, 3, and 4 for other deck designs and roof supports.

Source

Oil & Gas Journal, March 9, 1987.

Reference

1. "Evaporation Loss from Internal Floating-Roof Tanks," American Petroleum Institute Publication No. 2519.

Estimating the contents of horizontal cylindrical tanks

Horizontal cylindrical tanks are frequently used for water and fuel storage, and in many cases it is important to be able to gauge these vessels to determine the volume of liquid contained in them. However, it is normally much more difficult to establish a volume-per-inch scale for a horizontal tank than for one in a vertical position. The accompanying nomograph simplifies this problem.

To use the nomograph, it is necessary to gauge the tank and determine the ratio of the depth of liquid in the tank to the tank diameter. After this is found, draw a straight line from the point on the "ratio" scale through the known point on the "diameter of tank" scale and read the intercept on the "gallons per ft of length" scale. From this point, draw a second line through the known point on the "length of tank" scale and read the intercept on the "gallons (or barrels) in total length" scale.

Example. Find the volume of liquid contained in a horizontal cylindrical tank 7 ft in diameter and 20 ft long when the depth of the liquid is 4 ft, 10.8 in.

The ratio of depth of liquid to tank diameter is:

$$58.8/84 = 0.70$$

Connect 0.70 on the ratio scale with 7 ft on the diameter scale and continue the straight line to obtain the intercept 215 on the gallons per ft of length scale. Draw a second line

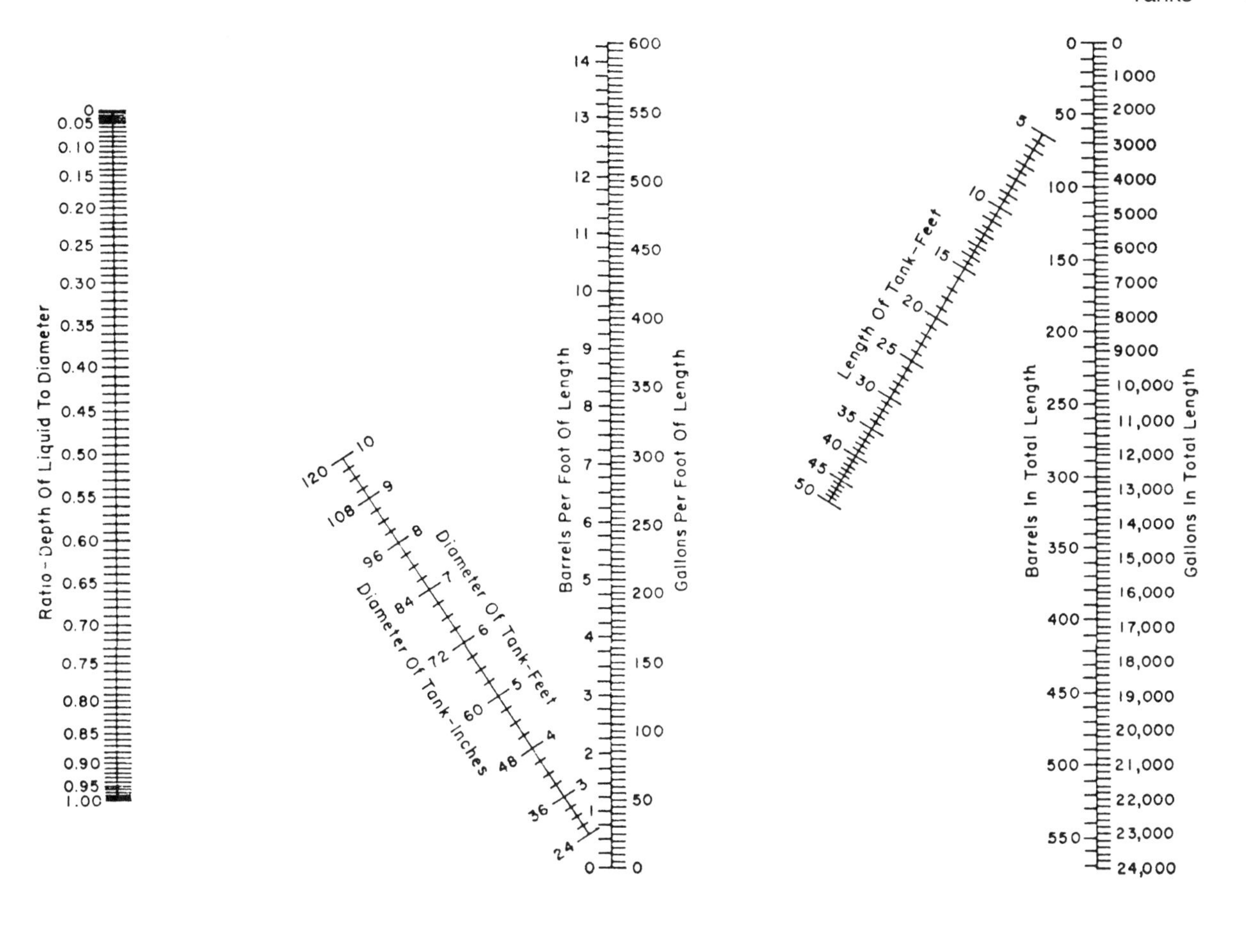

from the point 215 through the 20 on the length of tank scale and continue the line to obtain the intercept 4,300 on the gallons in total length scale. Total liquid content of the tank is 4,300 on the gallons.

How to gauge a horizontal cylindrical tank

Express the depth in % of the diameter; then the result will be given in % of total capacity of the tank.

Rule 1. For depth up to 30; multiply the square root of the depth by the depth, and then by 0.155.

Example. Liquid depth is 16% of tank diameter

$16 \times 16 \times 0.155 = 4 \times 16 \times 0.155 = 9.9\%$

The correct answer is 10.3%; error is about .4%.

Rule 2. For depth between 30 and 50; subtract 10 from the depth, multiply by 1.25.

Example. Liquid depth is 44% of tank diameter

$(44 - 10) \times 1.25 = 34 \times 1.25 = 42.5\%$

The correct answer is 42.4%.

The maximum error for depths less than 5% may be as great as 10%; gauging in this range is always difficult, and a very small slope can introduce a much larger error than this. When the depth is greater than 50%, apply the same rule to get the volume of the empty space above the fluid, and subtract.

Use nomograph to find tank capacity

This simple nomograph can be used to find the capacity of your vertical cylindrical tanks. Here's how it works:

Draw a straight line from the "height" scale through the "diameter" scale and to the first "capacity, barrels" scale.

Read directly the capacity of the tank in barrels. (*Note:* The "height" scale may be used to indicate the overall height of the tank or the depth of liquid in the tank.)

Draw a second straight line connecting the two "capacity, barrels" scales at the same reading on each scale. Read the capacity of the tank in gallons and cubic ft on the proper scales.

The nomograph was constructed as follows:

1. The "height" scale is based on two log cycles per 10 in. with a range of 1-60 ft.
2. The "capacity, barrels" scale is based on four log cycles per 10 in. with a range of 20-150,000 barrels.
3. The "diameter" scale is based on three log cycles per 10 in. with a range of 4-150 ft.
4. The distance between the height and diameter scales is exactly two-thirds the distance between the height and "capacity, barrels" scale.
5. Determine points to locate the diameter scale from the following equation:

$$\text{Capacity, barrels} = 0.1399\,(\text{diameter})^2\,(\text{height}),\ \text{units in ft}$$

6. The "capacity, gallons" scale is based on four log cycles per 10 in. The initial point on the scale is determined as follows:

$$20 \text{ barrels} \times 42 \text{ gallons per barrel} - 840 \text{ gallons}$$

The range of the scale is 900 to 6 million gal.

7. The "capacity, cubic feet" scale is based on four log cycles per 10 in. The initial point on the scale is determined as follows:

$$20 \text{ barrels} \times 5.6146 \text{ cu. ft per barrel} = 112.292 \text{ cu. ft}$$

The range of the scale is 120 to 800,000 cu. ft.

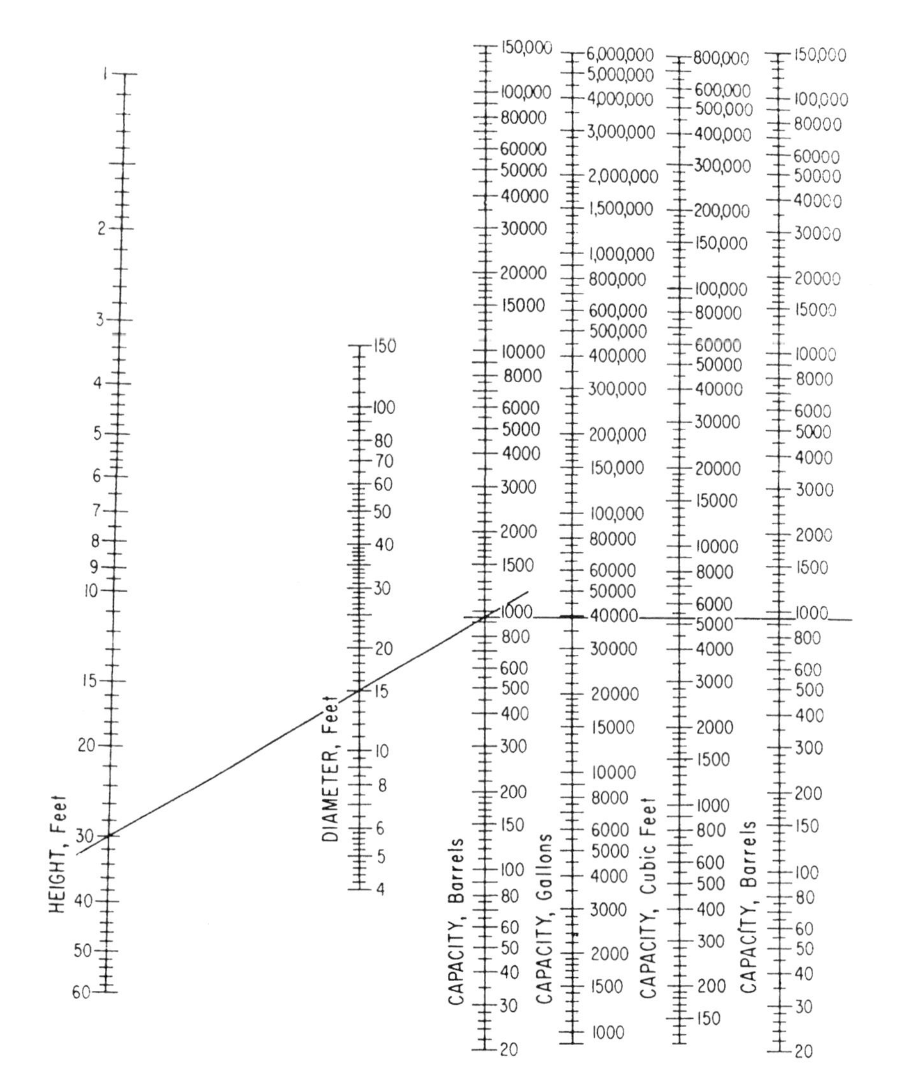

Correct the volume of light fuels from actual temperature to a base of 60°F

To approximate quickly the volume of gasoline or other light liquid fuel at 60°F from a known volume at any temperature in the atmospheric range, use the formula:

$$V_a - V_{60} = 0.0006(T - 60)V_{60}$$

where: V_a = Volume at actual temperature
V_{60} = Volume corrected to 60°F
T_a = Actual temperature of fuel

Example. A tank contains 5,500 gallons of gasoline at 46°F. Correct the volume to a base of 60°F.

$(5{,}500 - V_{60}) = 0.0006(46 - 60)V_{60}$
$(5{,}500 - V_{60}) = 0.0006(-14)V_{60}$
$5{,}500 = V_{60} - 0.0084V_{60}$
$5{,}500 = 0.9916V_{60}$
Volume at 60°F = 5,546.6 gallons

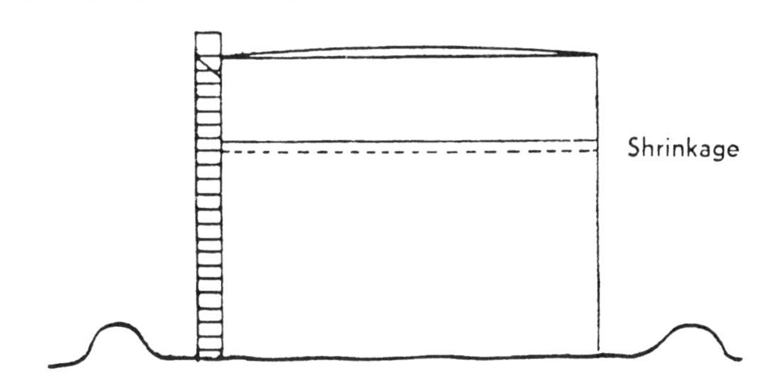

To approximate the shrinkage or expansion, obtain the difference between the actual volume measured and the corrected volume. In this case:

Shrinkage = 5,546.6 − 5,500 = 46.6 gallons

Volume of liquid in vertical cylindrical tanks

Measure the depth of the liquid and either the diameter or circumference of the tank, then the volume in:

Gallons $= 0.0034\,d^2h$ or $0.00034\,c^2h$
Barrels $= 0.000081\,d^2h$ or $0.00082\,c^2h$
Gallons $= 5.88\,D^2H$ or $0.595\,C^2H$
Barrels $= 0.140\,D^2H$ or $0.0142\,C^2H$

where: d = Diameter, inches
c = Circumference, inches
h = Depth, inches
D = Diameter, feet
C = Circumference, feet
H = Depth, feet

If the circumference is measured on the outside, then three times the thickness of the tank wall should be subtracted before using the formula. Naturally, these rules cannot supplant the results of accurate tank strapping, which take many other factors into account.

Example. How many gallons will a tank 12 ft in diameter and 16 ft high hold when full?

$$\text{Gallons} = 5.88\,D^2H = (5.88)(144)(16) = 13{,}548 \text{ gallons}$$

Example. How many barrels will a tank 8 ft in diameter and 16 ft high hold when full?

$$\text{Barrels} = 0.140\,D^2H = (0.140)(64)(16) = 143 \text{ barrels}$$

Chart gives tank's vapor formation rate

When sizing the vapor piping for a manifolded expansion-roof tank system, the rate of vapor formation must be known. While the rate of vapor formation can be computed by longhand methods, the calculation is tedious and takes much valuable time.

Example. Determine the rate of formation of vapor in a 140,000 barrel capacity tank when it is filled at the rate of 8,000 barrels per hour.

Solution. Enter the chart on the left at a capacity of 140,000 barrels and draw a straight line through the filling rate of 8,000 barrels per hour on the right. At the intersection

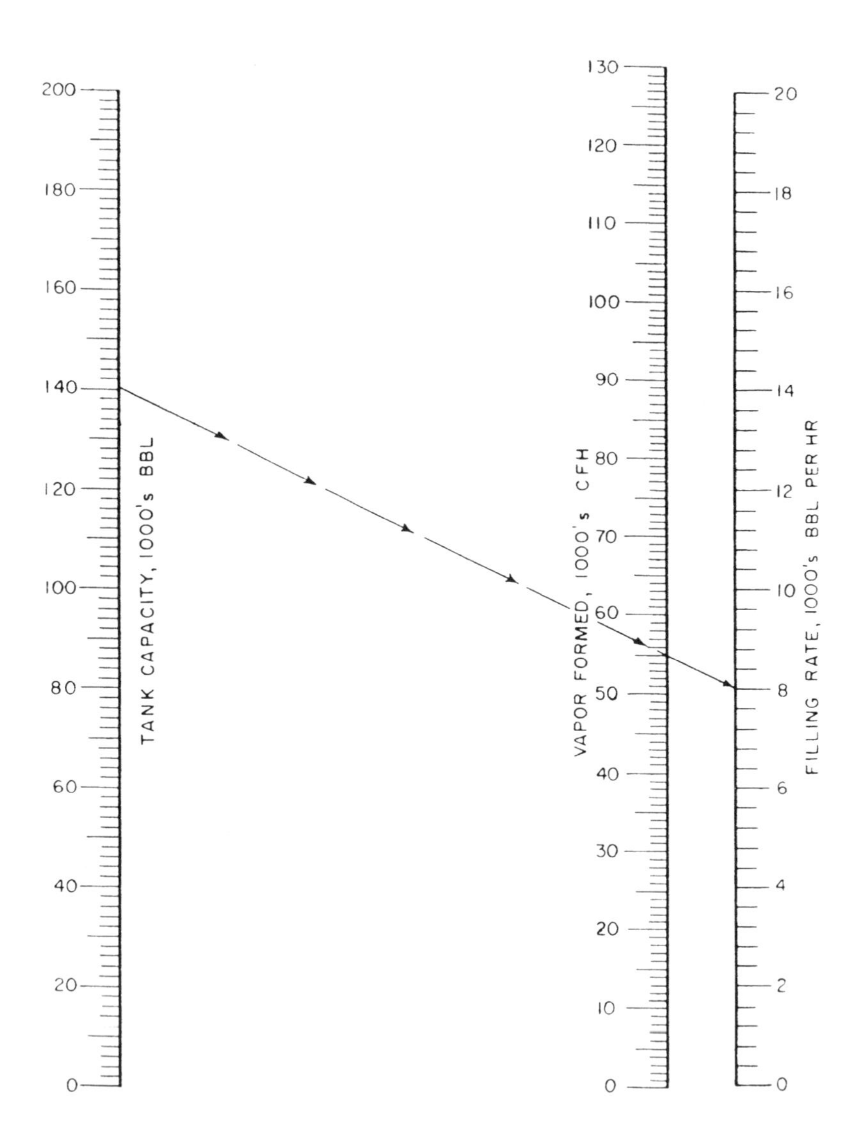

with the central scale read the vapor formation rate as 55,000 cu. ft per hour. The vapor piping for this tank would have to be designed for this formation rate if the maximum filling rate anticipated were 8,000 barrels per hour. But if a great filling rate were expected, the vapor formed would have to be computed for the higher rate. The chart could, of course, also be used for this computation.

This chart is based on the following equation:

$$V = \frac{\text{tank capacity, bb1}}{14.3} + \frac{\text{filling rate, bb1 per hr}}{0.178}$$

where V = vapor formed, cubic feet per hour

Hand-held calculator program simplifies dike computations

Calculating height of earthen dikes around above-ground storage can be done easily with a program for a portable calculator

Frank E. Hangs, Sovereign Engineering Co., Houston

Earthen dikes are widely used all over the world to contain flammable volumes of above-ground storage. They perform two vital functions: to prevent loss of fluid into the environment and to reduce the likelihood of fire spreading from one tank to another.

Sizing dikes by conventional methods is a time-consuming, trial-and-error process. A complete assessment of the problem involves: applicable codes and regulations; land

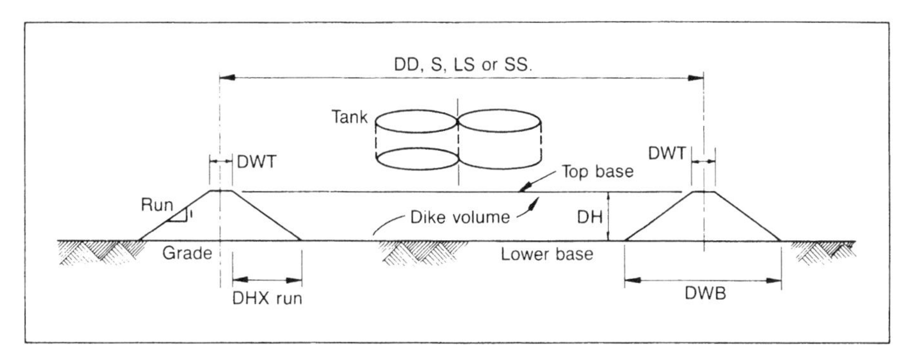

Figure 1. Cross section of a typical dike.

area available; topography of the area; soil characteristics; and the stipulated volume contained by dike and other dimensions of the dike section.

The following program for the HP-41CV hand-held calculator enables one to enter required data at a prompt and to calculate the height of the dike to retain the required volume of fluid, cross section of dike, width of the base, and the cubic yards of earth required, quickly. When a printer is available, a record of the input and output (results) is made. Without a printer, the input and output items (all identified) can be displayed one at a time and advanced at will.

Many "what if" questions can be answered readily, and different configurations compared as desirable. This is explained in detail in text and examples.

The Flammable and Combustible Liquids Code, as promulgated by the National Fire Protection Association, NFPA No. 30, is used as a basis for this program. Important stipulations are:

- Volume contained in dike area shall not be less than the full tank. (We have taken one tank per dike.)
- For crude petroleum with boilover characteristics, stored in fixed roof tanks, the contained volume above shall be calculated by deducting the volume of the tank below the height of the dike.
- Earthen dikes 3 ft or more in height shall have a flat section at the top not less than 2 ft wide.
- The slope of the earth wall shall be consistent with the natural angle of repose of the material of construction.
- The walls of the diked area shall be restricted to an average height of 6 ft above interior grade.

Dikes are constructed in circular, square, or rectangular configurations. For the purposes of this program, the volumes contained in the dikes are calculated as invented frustums of a cone or pyramid. The dike volume (converted to barrels) is compared to the total volume

```
EXAMPLE 1

            XEQ "DIKE"
TV=?
     54,200.00    RUN
TD=?
        110.00    RUN
TH=?
         32.00    RUN
RUN=1.5?
          1.50    RUN
DWT=2?
          2.00    RUN
DH INCR=?
          0.05    RUN
BOIL OVER?Y? N?
N                 RUN
TRIAL DH=?
          3.00    RUN
A? B? C?
                XEQ A
DD=3TD?
        330.00    RUN


TV=54,200.00
TD=110.00
TH=32.00
RUN=1.50
DWT=2.00
DH INCR=0.05
DD=330.00
SQ=0.00
LS=0.00
SS=0.00


DV=54,521.90
DH=3.75
X-SECT=28.59
DWB=13.25
EARTH YD3=1,097.92
B O BBL=0.00
TOT BBL=54,200.00
```

```
EXAMPLE 2

                SF 00
          4.00 STO 07
                XEQ A
DD=3TD?
       330.00     RUN

TV=54,200.00
TD=110.00
TH=32.00
RUN=1.50
DWT=2.00
DH INCR=0.05
DD=330.00
SQ=0.00
LS=0.00
SS=0.00

DV=61,506.12
DH=4.25
X-SECT=35.59
DWB=14.75
EARTH YD3=1,366.70
B O BBL=7,193.62
TOT BBL=61,393.62

EXAMPLE 3

                CF 00
                SF 01
               RCL 00
               STO 01
          0.00 STO 24
          4.50 STO 07
                XEQ A
DD=3TD?
       300.00     RUN


TV=54,200.00
TD-110.00
TH=32.00
RUN=1.50
DWT=2.00
DH INCR=0.05
DD=300.00
SQ=0.00
LS=0.00
SS=0.00

DV=53,406.62
DH=4.50
X-SECT=39.38
DWB=15.50
EARTH YD3=1,374.45
B O BBL=0.00
TOT BBL=54,200.00
```

Figure 2. Examples of the dike computation program for the HP-41CV hand-held calculator.

DIKE PROGRAM

Legend and storage registers

		REG.
TV	= Tank vol. (bbl)	00
TOT BBL	= Total bbl TV + boilover (if needed)	01
TD	= Tank dia (ft)	02
TH	= Tank height (ft)	03
Run	= For angle of repose of dike earth expressed as bevel Rise/Run = 1/Run, 1/1.5 is widely used	04
DWT	= Dike width top—ft Min. 2 ft for dikes 3 ft and higher—NFPA No. 30.	05
DH INCR	= Dike height increment Suggest 0.10 ft for prelim. run; Use 0.05 ft (0.60-in.) to finalize	06
TRIAL DH	= Trial dike height (ft) Try 3 ft for tanks 10,000 bbl and larger	07
DD	= Dike dia (ft)	10
EARTH YD3	= Dike vol. (cu yd)	20
SQ	= Square side (ft)	21
LS	= Long side (rectangle) ft	22
SS	= Short side (rectangle) ft	23
BO BBL	= Boilover barrels	24

Registers 8, 9, 13 and 17 not used.
Registers 11, 12, 14 and 15 scratch.

FLAGS

00—Boilover calculation
01—Single calculation
02—Circular dike
03—Square dike
04—Rectangular dike
21—Printer enables following: Allows data and results to be displayed one by one without printer. (Must be set each time calculator is turned on.)

Note: Flags 01 and 21 must be manually set. Flag 01—Clear manually—other flags set and cleared in program. See Example 3 for "short cut" exceptions.

PRINCIPAL LABELS

01—Calculates X-SECT and DWB; directs program to proper EARTH VOLUME routine, i.e., SQ ? etc.

03, 11 and 13 Bypass incrementation in a loop for a single calculation

06, 07 and 08 Calculates EARTH VOLUME for A, B, and C, respectively.

09 Summarizes data and results for display or printout.

A, B and C Subroutines for circular, square and rectangular dikes, respectively, calculating dike volumes for DH increments to converge dike volume and total volume.

a. Subroutine for calculating boilover volume.
b. Sets flag 00 for boilover calculations.

DV	= Dike vol. (bbl)	16
X-SECT	= Cross sect. dike (sq ft)	18
DWB	= Dike width base (ft)	19

FORMULA: For right truncated cone or pyramid:

$$V = \frac{H}{3}(A + \sqrt{AB} + B)$$

H = Height; A = Area of larger base, B = Area of smaller base.

(See Fig. 1.)

Dia or side of top base = (DD, S, LS or SS) − DWT

Dia or side of lower base = (DD, S, LS or SS) − DWT − 2 (DH × Run.)

USER INSTRUCTIONS

1) Put dike program in calculator.
2) XEQ size 25 and set User Mode.
3) XEQ "Dike."
4) Key in data according to prompts and R/S. Notice Run = 1.5? and DWT = 2? If these values are acceptable, key in and R/S. DH INCR = ? Can be 0.10 ft for preliminary runs, otherwise, use 0.05 ft (0.60-in.).

   Trial DH = ? Try 3 ft. If this is too much, machine will stop, STO smaller value in R07 and try again ("A," "B" or "C"). Likewise, if it is known that DH is much greater than 3—try larger value: This saves iterations!

A? Key in "A" for circular dikes (Be sure User Mode!)
B? Key in "B" for square dikes. (Be sure User Mode!)
C? Key in "C" for rectangular dikes. (Be sure User Mode!)

Boilover? Y? N?
This routine is available when needed. Usually answer is no. Press "N" and R/S.

Whichever key A, B, or C is pressed, a dia or side(s) will be called for; key in appropriate data and calculator will converge DV with total barrels for solution of DH (See Fig. 2).

Example 1: Illustrates a 54,200-bbl tank in a circular dike. Notice how data are entered according to prompts. A printer is a great convenience but is not indispensible. Without printer, SF 21 each time calculator is turned on, now data and results will be displayed one at time and advanced by pressing R/S key. Notice formating of data and results—zeros are shown for inactive dimensions or boilover volume.

Example 2: Demonstrates an approach, where most input data do not change: Calculate DH for above tank for boilover crude. (This avoids going through entering routine XEQ Dike.) SF 00 and a new trial DH try 4, STO 07. (Result of Example 1 is DH = 3.75 ft. It is evident dike will be higher to contain larger volume.) Press "A," re-enter DD (330). Results: DH = 4.25 ft. larger X-SECT, MORE EARTH, a boilover volume is shown and included in total volume.

Example 3: Same tank, no boilover, what is DD for 4.5 ft. dike? CF 00 (Notice it appears in annunciator when set). SF 01 (This is not a looping routine!) RCL 00 STO 01, 0.00 STD 24. STO 4.5 in 07. Press "A"; put some value for DD less than 330 (as above). Try 300 key in and R/S. One calculation will be made. Compare total volume and dike volume. Press "A" and key new DD. Repeat until satisfactory convergence is achieved.

The following table shows convergence for Example 3:

Total vol 54,200 bbl DH = 4.5 ft.

Trial DD ft	*DV bbl*
300	53,406.62
305	55,255.73
302	54,142.48
302.5	54,727.24
302.3	54,253.30

It is evident that with a few iterations, one can home-in on a precise solution.

SUMMARY

For a new tank size, one should XEQ "Dike" and enter data according to prompts.

Subroutines "B" and "C" are similar to "A." One can build up a dike height for given center line distances (square or rectangle) with and without boilover.

Consider same tank as Example 1. (Let 3 = Trial DH STO 07.)

Square 300 ft (No B.O)
Square 300 ft B.O
DH = 3.60
DV = 54,894.66
DH = 4.05
DV = 61,473.78
Total volume = 61,055.09

Rectangle LS 450: SS 200 (3.00 STO 07)
LS 450: SS 200 B.O.
H—3.60 DV = 54,655.16 DH = 4.10
DV = 61,900.04
Total volume = 61,139.72

One can "free-wheel" with "B" and "C" as for "A" for a fixed dike height. SF 01 and try different dike centerline distances. Compare DV and TOT. Val. and continue to desired convergence.

Warning: Be sure Flag 00 is clear when boilover routine is not required. When clearing Flag 00 STD 0.00 in 24 to prevent extraneous volume from being involved in program. BO VOL should be 0.00 when there is no boilover. Flat 01 must be clear for incrementing routines. Always STO new trial DH in 07 each run.

Note: XEQ 09 for printout or display of input and results in storage registers.

```
PRP "DIKE"

01♦LBL "DIKE"
02 "TV=?"
03 PROMPT
04 STO 01
05 STO 00
06 "TD=?"
07 PROMPT
08 STO 02
09 "TH=?"
10 PROMPT
11 STO 03
12 "RUN=1.5?"
13 PROMPT
14 STO 04
15 "DWT=2?"
16 PROMPT
17 STO 05
18 "DH INCR=?"
19 PROMPT
20 STO 06
21 "BOIL OVER?"
22 "Y? N?"
23 AON
24 PROMPT
25 ASTO Y
26 AOFF
27 "Y"
28 ASTO X
29 X=Y?
30 GTO b
31 FS? 00
32 CF 00
33 CLX
34 STO 24

35♦LBL 14
36 "TRIAL DH=?"
37 PROMPT
38 STO 07
39 "A? B? C?"
40 PROMPT

41♦LBL A
42 "DD=3TD?"
43 PROMPT
44 STO 10
45 STO 11
46 RCL 05
47 ST- 11
48 RCL 11
49 X↑2
50 STO 14
51 SF 02
52 FS? 03
53 CF 03
54 FS? 04
55 CF 04
56 FS? 01
57 GTO 03

58♦LBL 02
59 RCL 06
60 ST+ 07

61♦LBL 03
62 RCL 10
63 STO 11
64 RCL 05
65 ST- 11
66 RCL 07
67 RCL 04
68 *
69 2
70 *
71 ST- 11
72 RCL 11
73 X↑2
74 STO 15
75 RCL 14
76 *
77 SQRT
78 RCL 14
79 +
80 RCL 15
81 +
82 RCL 07
83 *
84 21.4461
85 /
86 STO 16
87 FS? 00
88 XEQ a
89 RCL 01
90 X<=Y?
91 GTO 01
92 FS? 01
93 GTO 01
94 GTO 02

95♦LBL B
96 "SQ=?"
97 PROMPT
98 STO 21
99 STO 11
100 RCL 05
101 ST- 11
102 RCL 11
103 X↑2
104 STO 14
105 SF 03
106 FS? 02
107 CF 02
108 FS? 04
109 CF 04
110 FS? 01
111 GTO 11

112♦LBL 10
113 RCL 06
114 ST+ 07

115♦LBL 11
116 RCL 21
117 STO 11
118 RCL 05
119 ST- 11
120 RCL 07
121 RCL 04
122 *
123 2
124 *
125 ST- 11
126 RCL 11
127 X↑2
128 STO 15
129 RCL 14
130 *
131 SQRT
132 RCL 14
133 +
134 RCL 15
135 +
136 RCL 07
137 *
138 16.8458
139 /
140 STO 16
141 FS? 00
142 XEQ a
143 RCL 01
144 X<=Y?
145 GTO 01
146 FS? 01
147 GTO 01
148 GTO 10

149♦LBL C
150 "LS=?"
151 PROMPT
152 STO 22
153 STO 11
154 "SS=?"
155 PROMPT
156 STO 23
157 STO 12
158 RCL 05
159 ST- 11
160 ST- 12
161 RCL 11
162 RCL 12
163 *
164 STO 14
165 SF 04
166 FS? 02
167 CF 02
168 FS? 03
169 CF 03
170 FS? 01
171 GTO 13

172♦LBL 12
173 RCL 06
174 ST+ 07

175♦LBL 13
176 RCL 22
177 STO 11
178 RCL 23
179 STO 12
180 RCL 05
181 ST- 11
182 ST- 12
183 RCL 07
184 RCL 04
185 *
186 2
187 *
188 ST- 11
189 ST- 12
190 RCL 11
191 RCL 12
192 *
193 STO 15
194 RCL 14
195 *
196 SQRT
197 RCL 14
198 +
199 RCL 15
200 +
201 RCL 07
202 *
203 16.8458
204 /
205 STO 16
206 FS? 00
207 XEQ a
208 RCL 01
209 X<=Y?
210 GTO 01
211 FS? 01
212 GTO 01
213 GTO 12

214♦LBL a
215 RCL 02
216 X↑2
217 7.1487
218 /
219 RCL 07
220 *
221 STO 24
222 RCL 00
223 STO 01
224 RCL 24
225 ST+ 01
226 RCL 16
227 RTN

228♦LBL b
229 SF 00
230 GTO 14

231♦LBL 01
232 RCL 07
233 X↑2
234 RCL 04
235 *
236 RCL 05
237 RCL 07
238 *
239 +
240 STO 18
241 RCL 07
242 RCL 04
243 *
244 2
245 *
246 RCL 05
247 +
248 STO 19
249 FS? 02
250 GTO 06
251 FS? 03
252 GTO 07
253 FS? 04
254 GTO 08
255♦LBL 06
256 RCL 18
257 RCL 10
258 *
259 PI
260 *
261 27
262 /
263 STO 20
264 CLX
265 STO 21
266 STO 22
267 STO 23
268 GTO 09

269♦LBL 07
270 RCL 21
271 4
272 *
273 RCL 18
274 *
275 27
276 /
277 STO 20
278 CLX
279 STO 10
280 STO 22
281 STO 23
282 GTO 09

283♦LBL 08
284 RCL 22
285 2
286 *
287 RCL 23
288 2
289 *
290 +
291 RCL 18
292 *
293 27
294 /
295 STO 20
296 CLX
297 STO 10
298 STO 21
299 GTO 09

300♦LBL 09
301 ADV
302 ADV
303 "TV="
304 ARCL 00
305 AVIEW
306 "TD="
307 ARCL 02
308 AVIEW
309 "TH="
310 ARCL 03
311 AVIEW
312 "RUN="
313 ARCL 04
314 AVIEW
315 "DWT="
316 ARCL 05
317 AVIEW
318 "DH INCR="
319 ARCL 06
320 AVIEW
321 "DD="
322 ARCL 10
323 AVIEW
324 "SQ="
325 ARCL 21
326 AVIEW
327 "LS="
328 ARCL 22
329 AVIEW
330 "SS="
331 ARCL 23
332 AVIEW
333 ADV
334 ADV
335 "DV="
336 ARCL 16
337 AVIEW
338 "DH="
339 ARCL 07
340 AVIEW
341 "X-SECT="
342 ARCL 18
343 AVIEW
344 "DWB="
345 ARCL 19
346 AVIEW
347 "EARTH YD3="
348 ARCL 20
349 AVIEW
350 "B O BBL="
351 ARCL 24
352 AVIEW
353 "TOT BBL="
354 ARCL 01
355 AVIEW
356 .END.
```

Example 3.

(volume of tank or volume of tank plus boilover, if applicable). The calculations begin with given dike centerline, dike width at top, repose angle of soil, and trial DH. As long as the dike volume is less than the total volume the program loops, increment DH for the next calculation. When the two volumes converge, calculations stop and input data and results are displayed or printed. The DH value, when the volumes converge, is the solution.

In some cases, it will be required to ascertain dike diameter or sides for a fixed dike height (DH). This is accomplished by storing DH Value in 07, setting Flag O1 (for single calculation). Press "A," "B," or "C" key in trial centerline distances. The results of any calculation give one an opportunity to compare total barrels with dike volume. Then alter centerline distances to fit trend and continue. See Example 3.

This program is based on the site being essentially level.

Source

Pipe Line Industry, August 1986.

19: Maintenance

How to plan for oil pipeline spills (part 1)

Here are ideas to consider in development of a comprehensive contingency plan

J. D. Sartor, vice president, and **R. W. Castle,** senior project engineer, Woodward-Clyde Consultants, Environmental Systems Division, San Francisco, California

Federal regulations and permit stipulations require the preparation of written plans and procedures for dealing with accidental spills of materials carried by liquids pipelines. These requirements can be met through the preparation of comprehensive spill contingency plans.

Proper contingency planning involves analysis of the material(s) carried and the environments crossed by the pipeline. Based on company personnel and operating procedures, a response organization and necessary equipment are developed. The planning concept assumes that preplanned emergency actions are appropriate for protection of human life, property, and the environment.

In addition, however, the planning must provide the flexibility to respond to unanticipated situations. The final planning element is the design of a comprehensive training program to ensure the safety and efficiency of the response teams.

Regulatory requirements

Historically, federal requirements pertaining to liquids pipeline construction and operation have contained no specific requirements or guidelines for the preparation of spill contingency plans.

In spite of this, such plans have been required in conjunction with various permits and right-of-way grants. Cases where contingency plans have been stipulated include the Trans Alaska, SOHIO West Coast to Midcontinent, Northern Tier, and the Northwest Alaska Gas pipelines.

The National Environment Policy Act (NEPA) of January 1, 1970 and ensuing amendments have led to federal and state requirements for analysis of the potential for spills, their probable impact, and the degree to which spill impacts may be mitigated through the Environmental Impact Statement—Environmental Impact Report process. Spill contingency is often the major method of spill impact mitigation, and its association with the environmental review process is becoming increasingly common.

Effective July 15, 1980, the Department of Transportation (DOT) amended regulations for the transportation of liquids by pipeline. Title 49 of the Code of Federal Regulations, Part 195 (revised October 1, 1979) describes these regulations and amendments. Subpart—Operation and Maintenance—requires the preparation of written emergency plans and procedures for dealing with accidental release of commodity, operational failures, and natural disasters affecting facilities. The same regulations also require assignment of personnel to emergency functions and the design and implementation of emergency procedures programs. These requirements summarize the essential elements of a spill contingency plan.

Contingency plan objectives

A rapid response capability is essential to adequately control and remove spilled material.

To provide this capability, adequate advance planning is mandatory. Such pre-spill planning should recognize the various types of spill situations that could occur and incorporate a response strategy for each.

These strategies should outline methods to be used and describe the required types and amounts of equipment, materials, and manpower.

In addition to pre-spill planning for anticipated spill types and situations, a comprehensive plan must allow the flexibility to respond effectively to unanticipated spill situations.

Related studies

The development of a spill contingency plan is based on consideration and analysis of a wide variety of factors. Included are geographical elements (location of the spill, drainage characteristics, surface conditions, soil type, accessibility, trafficability), environmental elements (weather conditions, hydrology, etc.), and ecological elements (sensitive and vulnerable areas, rare and endangered species, etc.).

For special environments, additional studies may be required. For example, a series of special studies on reclamation of spill affected areas was required for tundra during the implementation of the Alaska pipeline. These studies included evaluation of plant types that could be used for rehabilitation, effects of spilled oil on native plants, effects of various types of cleanup measures, and residual effects of oil over different weathering periods.

Engineering elements necessary for contingency planning include pipeline pumping and drainage characteristics, valve placements, monitoring equipment, operating procedures, and communications and control systems. Typically much of this information is available through project-related design engineering and environmental studies at an early phase in project implementation.

In some cases, however, all of the required information may not be available and may require generation prior to the completion of the final spill plan. Early versions of spill contingency plans are considered "draft" plans and are submitted to satisfy the requirements of regulatory agencies. The "draft" plans are submitted with the stipulation that final spill plans are prepared and implemented prior to the pipeline becoming operational.

Planning concepts

Pipelines characteristically extend considerable distances and encounter a variety of environmental conditions. The resulting spill planning must respond to these factors and is commonly either overly simplified or so complex that it is unusable in practice. To counter either tendency and yet provide sufficient information in a usable format, it is advantageous to design the contingency plan as a series of volumes, each specific to a particular need. Using this approach, the master volume or General Provisions contains organizational and operational information common to the entire pipeline system. It also forms a basis for personnel training.

The second volume or volumes delineate specific response actions and personnel for specific sections of the pipeline. Commonly the areas covered by these volumes are defined by major drainage patterns. If necessary as a result of drainage basin size or other unusual factors, each of these sections may be divided into further subdivisions until workable size field response plans are achieved.

Typical contents of successive volumes of a pipeline contingency plan are shown in Figure 1. Information pertinent to the entire system is contained only in the general provisions.

An effective contingency plan indicates the actions to be taken, their sequence, and timing in relation to other events. During or immediately following a spill, the following representative actions should be taken:

	GENERAL PROVISIONS		DISTRICT PLAN		CONTENGENCY AREA
100	Introduction	100	Introduction	100	Introduction
101	Background	101		101	
102	Policy	102		102	
103	Scope	103	Scope	103	Scope
104	Plan Concept	104	Plan Concept	104	
105	Plan Format	105	Plan Format	105	
200	Contingency Response Organization	200	Contingency Response Organization - District	200	Contingency Response Organization - Area
300	Immediate Response Actions	300	Immediate Response Actions - District	300	Immediate Response Actions - Area
301	Detection	301	Detection	301	Detection
302	Alert Procedures	302	Alert Procedures	302	Alert Procedures
303	Containment	303	Containment	303	Containment
304	Equipment	304	Equipment	304	Equipment
305	Pipeline Operational Actions	305	Pipeline Operational Actions	305	Pipeline Operational Actions
306	Location Cross References	306		306	
400	Contingency Plan	400	Contingency Plan	400	Contingency Plan
401	System Description	401	District Description	401	Area Discription
402	Oil Spill Detection	402	Oil Spill Detection	402	
403	Assessment	403	Assessment	403	
404	Containment	404	Containment	404	Containment
405	Exclusion	405	Exclusion	405	Exclusion
406	Cleanup	406	Cleanup	406	Cleanup
407	Disposal	407	Disposal	407	
408	Restoration	408	Restoration	408	
409	Documentation	409	Documentation	409	
500	Environmental Parameters	500	Environmental Parameters	500	
501	Biological Environment	501	Biological Environment	501	
502	Meterology	502	Meteorology	502	
503	Oceanography	503	Oceanography	503	
504	Hydrology	504	Hydrology	504	
600	Training Program	600	Training Program	600	
700	Response Plan Annexes	700	Response Plan Annexes	700	Response Plan Annexes
701	Governmental Relations	701	Governmental Relations	701	
702	Third Party Contractors	702	Third Party Contractors	702	
703	Logistics	703	Logistics	703	
704	Communications	704	Booms	704	
705	Oil Spill Task Force Directory	705	Skimmers	705	
706	Pipeline Drainage Calculations	706	Sorbents	706	
707	Booms	707	Revegetation	707	
708	Skimmers	708	Wildlife Care & Rehabilitation	708	
709	Sorbents	709	Communications	709	
710	Manpower	710	Oil-On-Water Sensors	710	
711	Revegetation	711	Portable Lightering Systems	711	
712	Oil-On-Water Sensors	712	Public Relations Guidelines	712	
713	Public Relations Guidelines	713	Telephone Addresses - Company Personnel - District	713	
714	Population Centers	714		714	
715	Airports & Landing Strips	715		715	
716	Commercial Food & Lodging Facilities	716		716	
717	Academic Institutions & Laboratories	717		717	
718	Oil Spill Cooperatives	718	Airports & Landing Strips	718	
719	Wildlife Care & Rehabilitation	719		719	
720	Lightering Systems	720	Commercial Food & Lodging Facilities	720	
800	Bibliography				

Figure 1. Contingency plan content.

a. Verify and locate the spill and notify the spill response team to initiate emergency procedures.
b. Control or limit the amount of material spilled. Exclude or prevent the spread of material to sensitive areas. Evaluate threats to the general public and act accordingly.
c. Determine the extent of the spill and predict its subsequent dispersion as a function of time.
d. Plan and direct overall operations; provide administrative support, liaison with government/local officials, and public information.
e. Remove and clean material from contaminated land areas, subsoils, shorelines, and water surfaces.
f. Determine procedures and site locations for the disposal of contaminated materials.
g. Document all phases of spill incident and subsequent cleanup activities.

Selection of the action to take at different times during a spill depends on: (a) the location and nature of the spill; (b) the quantity and type of material; (c) hydrology, topography, and soil types; (d) ice, weather, and currents. Supervision will start immediately, and documentation must be conducted simultaneously with all other operations.

Spills require prompt action whether they are on land or on water. Detailed response plans must be prepared for immediate action to be taken wherever and whenever spills occur.

In general, the contingency plan describes the logical and sequential order of what-to-do actions. Some actions are taken in all situations. Other actions are contingent on and initiated only by special circumstances. The basic plan is supported by appendices that provide how-to-do-it instructions for basic actions. These appendices also provide information on regulations, sources of equipment, logistics, and general background.

Contingency response

The nature of each individual response organization is dependent on the nature of the pipeline system and the area through which it crosses. In some remote areas such as Alaska, any effective response organization must rely heavily on internal resources. In more populated areas, a spill response organization can be designed more around the use of local resources.

Any response organization must, however, incorporate certain functions. Typically these functions may be grouped as management, advisory, support, and local field response. The management, advisory, and support roles generally perform staff functions and are usually staffed by company personnel. The local field response functions are normally staffed by local operating personnel stationed along the pipeline route. As necessary, these personnel may be supplemented by local contractors.

The assemblage of the above roles is known as the Spill Task Force (Figure 2).

Source

Pipe Line Industry, February 1982.

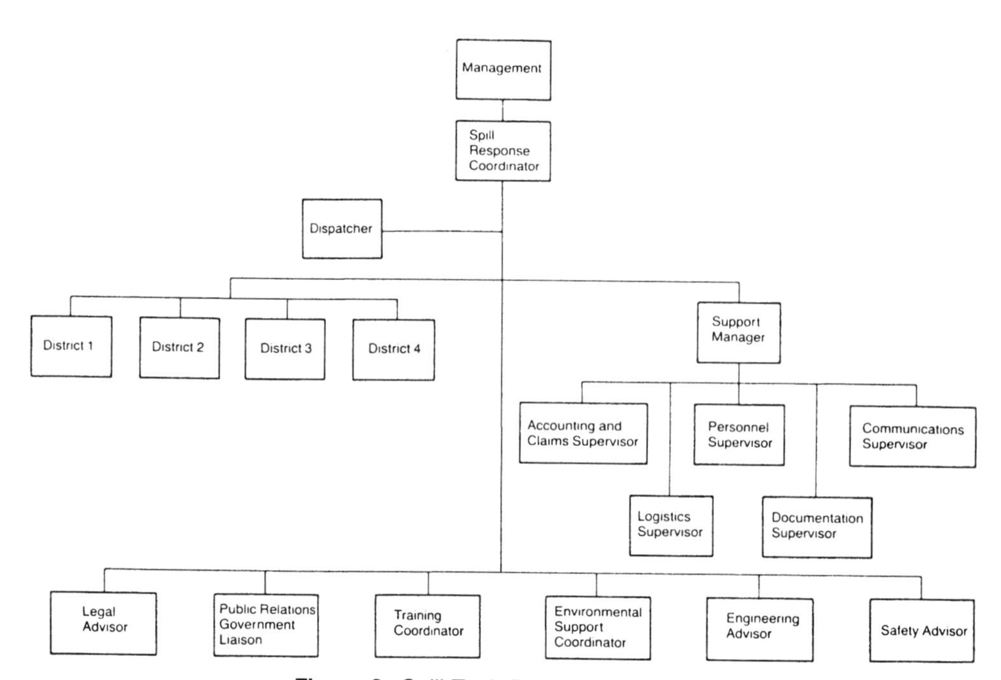

Figure 2. Spill Task Force organization.

How to plan for oil pipeline spills (part 2)

What to consider for initial and followup actions for spills and task force training

J. D. Sartor, vice president, and **R. W. Castle,** senior project engineer, Woodward-Clyde Consultants, Environmental Systems Division, San Francisco, California

Within the Spill Task Force, each district uses a two-level organization to provide fast and effective initial and followup response. A typical district organization is shown in Figure 1. The first level of response, the Immediate Response Team (IRT), reacts immediately to any spill within its area of concern. This team takes pre-planned response actions, limiting the spills' impacts until further actions can be activated.

The IRT is trained and equipped to handle minor spills without additional assistance. Should this team be unable to completely control a spill, it will take pre-planned control actions and notify the spill control coordinator. The spill coordinator will then activate company and contractor resources as necessary. The second level, the remainder of the District Response Team, is then activated as necessary to complete the control and recovery of the spill. For sufficiently large spills, the overall Spill Task Force may require activation.

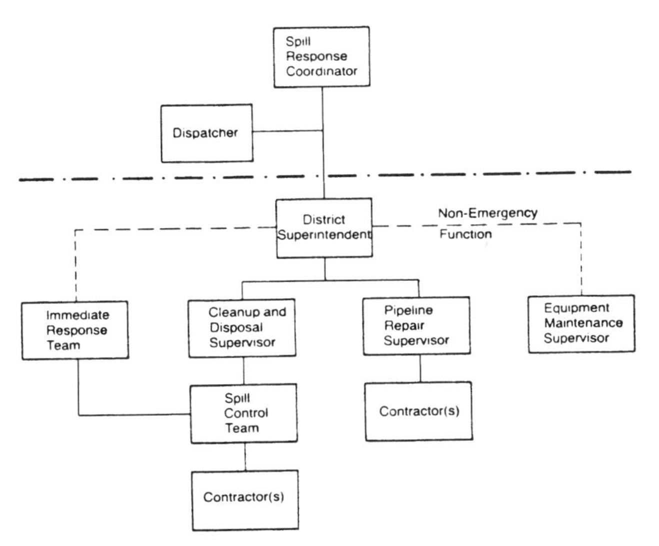

Figure 1. District Response Team organization.

Spills will normally be detected by or reported directly to the pipeline dispatcher, who will ensure proper communications within the Spill Task Force and who will shut down pumping operations if necessary. The dispatcher will ensure that the district superintendent responsible for the area in question is notified and that the appropriate Immediate Response Team has been activated. The dispatcher will then notify the spill coordinator, who will activate additional resources as necessary after consulting with the local area superintendent. Figure 2 illustrates how the contingency response organization functions.

Immediate response

The composition of any individual IRT will vary with the size and nature of the pipeline system. In many cases several company personnel will be available along a specific pipeline segment. With proper training and equipment, however, even two- or three-man teams can be constructive in implementing immediate actions.

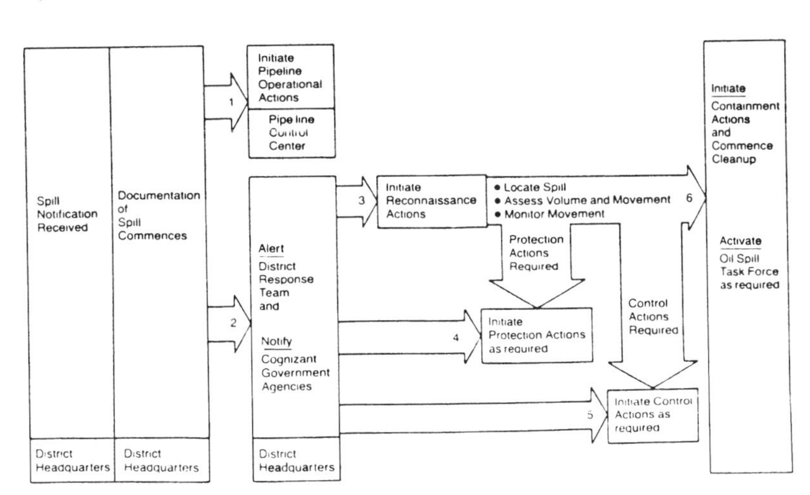

Figure 2. Immediate response action flow chart.

The duties of the IRT may need to be taken simultaneously. These duties include:

- Verification and location of the leak
- Leak control
- Conducting prescribed initial actions to ensure the safety of the general public, contain the spill, and protect sensitive areas
- Assess the situation and provide initial inputs to the district supervisor and the spill coordinator
- Assume assigned positions in the spill task force when activated

IRT positions can be manned 24 hours per day or on a 24-hour standby basis depending on the nature of the facility and available along-route personnel. Typically, the dispatcher position will be manned on a 24-hour basis. Individuals filling response positions are listed in the contingency plan by name and method of notification.

Immediate response actions

When a spill is detected by monitoring equipment or reported through some other mechanism, the dispatcher immediately notifies the IRT. A typical alert and notification procedure is shown in Figure 3. The primary functions of the IRT are to provide for public safety, contain the spill, and exclude the spill from sensitive or vulnerable areas. For small spills, the IRT may be capable of complete control and cleanup. For larger spills, they provide stop-gap measures until greater resources can be mobilized.

The type of response is dependent on the specific spill area and the nature of the material. The intention of all actions is to locate the leak and begin emergency actions immediately. To facilitate such actions, the contingency plan must provide predesignated control sites and actions, equipment recommendations, and sufficient information on the physical and ecologic characteristics of the threatened area to allow flexible and defendable response actions. Figure 4 depicts a presentation of pre-planned response actions and support information for a typical pipeline segment.

Directions of probable spill movement, access points, and potential control sites are indicated on the pipeline segment map. Supporting information is tabularized and includes predicted maximum spill sizes at selected points, location of equipment storage, recommendations for primary control actions under varying conditions (letter codes refer to a contingency plan appendix that details respective technique implementation), special features of concern during spill control, and general hydrologic characteristics. Ecologic characteristics of the area are provided in an additional supporting appendix.

Flexible response actions

While it is a basic premise in spill planning that pre-planned response actions are appropriate, it is also accepted that no two spills are ever identical.

As a result, spill contingency planning must provide the flexibility to respond to unpredicted situations. This can be accomplished through the design of decision guides that equate the types and ranges of environmental parameters to appropriate control and recovery techniques. A representative decision guide for the protection of inland waters is shown in Figure 5. A pre-planned decision making process can be particularly useful in designing cleanup and disposal programs.

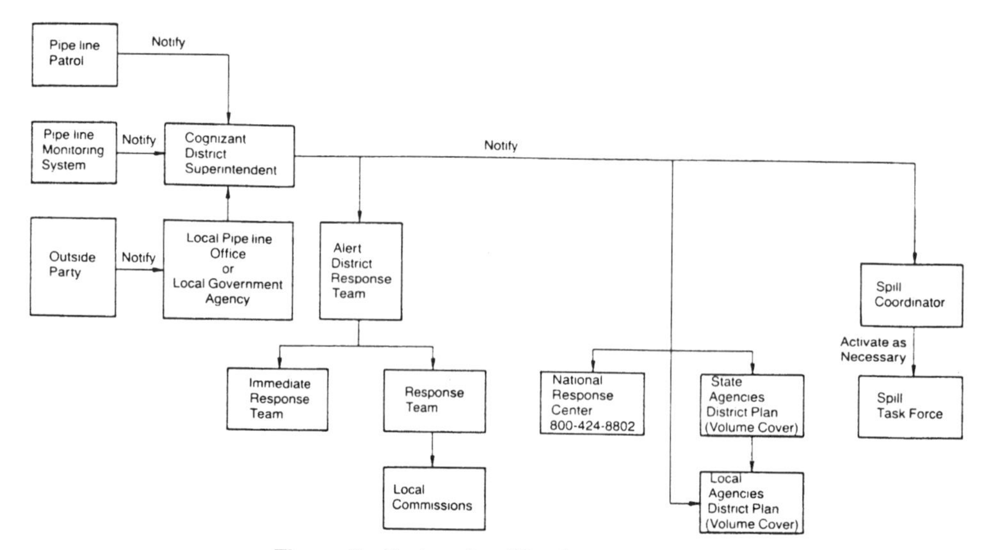

Figure 3. Alert and notification procedure.

PIPELINE MILEPOSTS 5.8 to 9.4 COUNTY County C MAP REF. 404-5

MAXIMUM SPILL SIZE River 1: 1,682 bbl (16,148 bbl west of river)
Creek A:

NEAREST EQUIPMENT SITE Site X:

SPILL CONTROL RECOMMENDATIONS

Location	Low Flow	High Flow	Terrestrial	Remarks
CS-1	F,G,H	G,H	-	Private property
CS-2	F,G,H	G,H	-	
CS-3	F,G,H	G,H	-	
CS-4	F,G	-	ABC	
CS-5	F,G,H	I,J,K,L	-	
CS-6	F,G,H	I,J,L	-	Limited access, if oil passes this point, contain in river at River 1
CS-7	F,G,H	I,J,L	-	
CS-8	F,G,H	G,H	-	
CS-9	F,G,H	G,H	-	Limited access, private property
CS-10	F,G,H	I,J,K	-	
CS-11	F,G,H	I,J,K	-	
CS-12	F,G,H	I,J	-	
CS-13	F,G,H	I,J	-	
CS-14	F,G,H	I,J	-	Private property

SPECIAL FEATURES/SPILL CONTROL PROCEDURES

1 Stream velocity during moderate to peak flow may preclude effective booming. During lower flow periods bury boom skirt on banks and bars

- River 1 is a major salmon migration and spawning watercourse
- See Table C-29, Appendix C for an inventory of the Natural Resources of River 1.

WATER CROSSINGS/HYDROLOGIC CHARACTERISTICS

- Depth to water table generally exceeds 75 feet
- Irrigation ditch (mp 6.0)
- River 1 (mp 7.0)

Condition	Discharge	Velocity	Max. Width	Max. Depth	Occurrence
Approx. 100 yr. flow	8,800 cfs	8-9 fps	270-350 ft.	9-10 ft.	Dec. - Feb.
Low flow	70 cfs	3 fps	20 ft.±	2 ft.	Jun. - Aug.

- Unnamed Creek (mp 7.75): perennial; water quality classification A
- Irrigation ditch (mp 7.9): underground
- Creek A (mp 8.6): perennial; water quality classification A
- Unnamed Creek (mp 8.75): perennial; water quality classification A

Figure 4. Typical pipeline segment information and response actions.

Training

Training of all spill response personnel, including simulated spill and equipment drills, is critical. Accordingly, a comprehensive training program is an integral part of the contingency plan. The objectives of the training program are to:

- Maintain the plan as a fully operable and working document
- Inform Task Force members of their respective duties and standard communication procedures
- Allow Task Force members to become familiar with the use of all equipment and to update the plan to reflect the current state of the art, including new equipment and procedures
- Modify the plan in the light of information gained from the field exercises
- Incorporate experience gained from response to spills into the plan

The training program is structured according to the level of responsibility of the participants and may include:

- Classroom instruction of field demonstrations
- Oil spill response drills to test notification, alert, and mobilization procedures
- Drills to practice specific containment measures and general containment and cleanup equipment and techniques

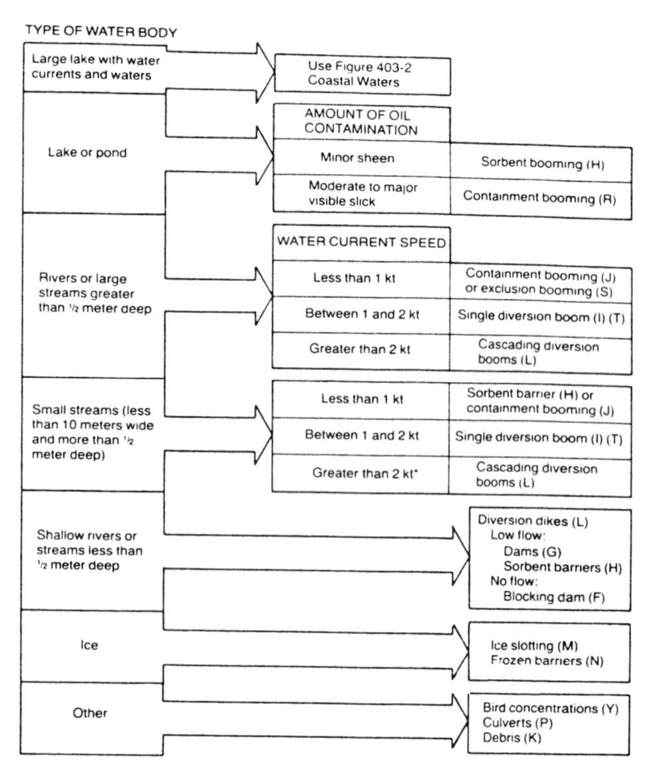

Figure 5. Decision guide for the protection of inland waters.

Conclusion

Accidental pipeline spills have and will continue to occur. Recognizing this fact, regulatory agencies have required emergency spill planning to minimize resulting impacts.

Comprehensive spill contingency planning satisfies these requirements and provides quick and effective response to environmental emergencies. Included in the preparation of such plans are the development of a response organization and response actions for each individual system. Also important in the development of a plan is the inclusion of a training and proficiency program.

(Based on a paper, Contingency Planning for Liquid Pipe Line Spills, presented by the authors as a contribution of the American Society of Mechanical Engineers at the Energy-Resources Conference and Exhibition, January 18–22, 1981, Houston, Texas.)

Source

Pipe Line Industry, March 1982.

20: Economics

Rule of thumb speeds payroll estimates

Hewlett Packard HP41-CV computation provides an estimate of the weekly expense of a "manpower spread" for pipeline construction in a given year

R. D. Green, Energy Center Fellow, University of Pennsylvania

Pipeline construction payroll expenses can be estimated in a rule-of-thumb computation that indexes 1982 survey data to any year of interest when an inflation figure is provided (software is provided in the appendix, p. 591).

An estimate of the weekly payroll expense of a "manpower spread" for pipeline construction in a given year is obtained as follows (Table 1):

$$P(y) = \left(\sum_{j=1}^{15} P_j E_j \right) e^{i(y-1982)}$$

where: P_j = Number of employees in category j
E_j = Compensation expense per employee in category j
i = Rate of inflation
y = Year for the construction estimate

The present value (PV) of a stream of payroll expenses is found as follows:

Table 1
Compensation categories of manpower

Category* (j)	Position	Weekly compensation** (E_j)	Category* (j)	Position	Weekly compensation** (E_j)
I	Welding crew foreman	1,750	VII	Field office & overhead spreadman	1,100
II	Welding crew welders	1,670		Stringing foreman	
	Pipe crew foreman			Over-the-ditch, cleaning & priming crew foreman	
	Pipe crew stringer bead welder		VIII	Testing crew operator	1,050
	Pipe crew hot pass welders		IX	Field office & overhead asst. spreadman	1,000
	Tie-in welders		X	R.O.W. clearing & fencing power saw operators	995
III	Field office & overhead heavy-duty mechanics	1,250		Pipe crew skid truck drivers	
IV	Bending foreman	1,200	XI	Field office & overhead lowboy truck driver	980
	Bending engineer			Field office & overhead float truck driver	
V	Field office & overhead fuel truck driver	1,175		Pipe crew welder helpers	
	Field office & overhead grease truck driver			Tie-in welder helpers	
VI	R.O.W. & fencing dozer operator	1,125	XII	R.O.W. clearing & fencing laborers	960
	R.O.W. grading foreman			R.O.W. grading laborers	
	R.O.W. grading dozer operators			Ditching (earth) laborers	
	Ditching (earth) foreman			Stringing truck drivers	
	Ditching (earth) machine operators			Stringing laborers	
	Ditching (earth) backhoe operators			Bending laborers	
	Road crossing road boring machine operator			Pipe crew laborers	
	Road crossing backhoe operator			Over-the-ditch, cleaning & priming crew laborers	
	Stringing crane operator			Tie-in laborers	
	Stringing sideboom operator			Backfilling & clean-up laborers	
	Bending sideboom operator			Lower-in laborers	
	Pipe crew sideboom operator		XIII	R.O.W. clearing & fencing foreman	900
	Pipe crew tack rig operators		XIV	Field office & overhead office manager	800
	Tie-in foreman			Field office & overhead payroll clerk	
	Tie-in sideboom operators		XV	Field office overhead night watchman	350
	Lower in foreman				
	Lower in sideboom operators				
	Testing crew foreman				
	Backfilling & clean-up foreman				
	Backfilling & clean-up dozer operators				

$$PV = \sum_{n=1}^{N} \frac{P(y)}{\left(1+\frac{i}{52}\right)^n}$$

where: N = number of weeks
i = decimal "interest" rate

The above estimates are computed by the program CSTM (see appendix); the compensation data presented in Table 1 are incorporated into CSTM.

Estimate

If an estimate for P_j is needed, example manpower spreads for 8- and 30-in. cross-country pipelines are provided in Table 2. In this case, an estimate of the number of weeks the spreads will be in operation is obtained as follows:

$$PD = \frac{L}{S \times d}$$

where: PD = Project duration in weeks
L = Length of the pipeline in miles
S = Average daily spread movement rate in miles per day
d = Number of days in the work week (e.g., d = 6 if the crews work 6 days per week)

For the manpower spreads of Table 2, S is probably in the range expressed by these inequalities:
1 mile/day $\leq$ S $\leq$ .35 miles/day for 8-in. size

Table 2
Typical manpower spreads

Category (j)	Numbers of employees in category j for 8 & 30-in. pipeline*	
	(P_j for 8-in. pipeline)	(P_j for 30-in. pipeline)
I	1	1
II	6	25
III	2	7
IV	1	3
V	1	5
VI	29	74
VII	3	4
VIII	1	3
IX	1	1
X	14	26
XI	11	36
XII	40	144
XIII	1	2
XIV	2	4
XV	1	1

*Cross-country pipeline construction.

Source

Pipe Line Industry, March 1985.

Rule of thumb estimates optimum time to keep construction equipment

Hewlett Packard HP41-CV computation estimates the optimum retention period for construction equipment

R. D. Green, Energy Center Fellow, University of Pennsylvania

Construction firm executives prefer cost-wise decisions concerning whether and when equipment will be replaced. The problem is extensive since, as depicted in Table 1, equipment spreads are both extensive and varied.

Equipment needed to construct a pipeline deteriorates with use, and maintenance costs rise over time. These maintenance costs might include, among other things, (1) funds for a preventive maintenance program; (2) funds to be set aside for costs incurred due to equipment failures; (3) replacement parts; and (4) labor expenses for repairs.

The problem of how to decide when to replace equipment is simplified with a "life cycle analysis," which compares (over an infinite time horizon) any number of alternatives of maintaining or replacing machines and then identifies the optimum choice. A "rule of thumb" approach to life cycle analysis for construction equipment can be performed by making some assumptions to simplify the usually complex calculations. Taking this approach, the assumptions and method are as follows.

Procedure

Let the original cost of the equipment item (e.g., a machine) be K. The problem is to find the number of years, N, that the machine should be kept such that the Net Present Worth (NPW) of operating costs is a minimum. This optimum number of years is called N* (read "N Star"). Suppose that the machine maintenance costs increase geometrically by a fraction, G, over the previous year and that the first year maintenance cost is C dollars. To simplify the equations, assume that all costs are paid as they are incurred; hence all costs are at present value. The relevant discount rate is R.

Table 1
Equipment spread

Typical spread—equipment type and number for 8-in. and 30-in. cross-country lines

Equipment	8″	30″
Field Office and Overhead		
Cars	—	2
Pickups	2	7
Welding rig	—	1
Fuel truck	1	1
Grease truck	1	1
Utility truck	1	1
Tractor w/lowboy	1	1
Tractor w/float	1	1
Right of way, clear, burn, and fence		
Pickups	3	2
Dozers	2	2
Winch truck	—	1
Flatbed truck	1	1
Power saws	8	6
Right of way, grade		
Pickup	2	1
Dozers	4	3
Ditch (Earth)		
Pickups	2	7
Ditching machines	1	2
Dozer w/ripper	—	1
Backhoes	3	4
Clamshell	—	1
Ditch (rock)		
Sidebooms	—	2
Wagon drills	—	2
Compressors	—	2
Trucks	—	2
Jackhammers	—	2
Road crossings		
Pickup	1	1
Boring machine	1	1
Sideboom	—	1
Winch truck	—	1
Tar kettles		1
Back hoe	1	1
Stringing		
Pickup	1	1
Crane hoist	1	1
Tractors w/pipe trailers	3	7
Sideboom	1	1
Tow tractors	1	3
Bending		
Pickups	1	2
Sideboom	1	1
Bending machine	1	1
Transit & level rod	—	1
Pipe crew		
Pickup	1	1
Tractor welders	1	2
Sidebooms	—	2
Trucks	—	2
Set inside clamps	—	1
Buffing rig	1	1
Welding crew		
Pickup	—	1
Welders	3	11
Tow tractor	—	1
Crew bus	—	1
Coating crew		
Pickups	1	4
Sidebooms	—	3
Tow tractor	—	1
Clean & prime machine	—	1
Coat & wrap machine	—	1
Tar kettles	—	4
Winch truck	—	1
Flatbed truck	—	1
Sets, cradles & strongbacks	2	2
Holiday detectors	—	2
Lower in		
Pickups	—	2
Sidebooms	—	6
Clamshell	—	1
Dozer	—	1
Trucks	—	2
Tar kettles	—	1
Water pump	—	5
Tie in		
Pickups	2	4
Sidebooms	3	4
Welders	1	4
Truck	—	1
Clamshell	—	1
Tar kettle	—	1
Water pump	—	1
Testing		
Special truck	1	1
Pumps, high pressure	—	2 sets
Utilities		
Pickup	—	1
Clamshell	—	1
Clamps and cutters	—	1 set
Welders	—	2
Backfill and clean up		
Pickups	2	4
Backfiller	—	1
Dozers	3	3
High lift	—	1
Farm tractor	1	1
Dragline	—	1
Winch truck	—	1
Flatbed truck	—	1
Other* **		
Base station	1	3
Two-way radios	11	33
Comfort station/toilet	3	9
Skids	5000	15000
Helicopters	—	—
Internal line-up clamp	—	—
External line-up clamp	5	15
Tamper tractor	—	—
External plant coating machine	—	—
Vacuum-lift unit	—	—
Laybarge	—	—
Motor patrol/grader	—	—
Crane	—	—
Kitchen	—	—
Trailer/sleeping	—	—
Medical station	—	—
First aid kits	12	36

*Survey data, 1982.
***Source: A Manual of Pipeline Construction, *3rd Ed*., Dallas, Energy Communications, 1979, except for "other" category.
***In "Other" category only: 30″ data is estimated by multiplying 8″ data by three.
Note: For more information on construction spreads, see *Cost Estimating Manual For Pipelines And Marine Structures*, John S. Page, Houston, Gulf Publishing Co., 1977.

The solution to this problem is to derive an equation that allows a comparison for machines with different lifetimes. This is done by finding the NPW of an infinite stream of constant scale replications. The analysis is performed by using the (algorithm) equations below to find an optimum replacement period for each alternative considered:

Find N* such that:

$$f(N_i) - f(N_{i+1}) \leq 0 \text{ and } f(N_{i-1}) - f(N_i) > 0$$

where:

$$f(N) = K + C\left(\frac{1-\left(\frac{1+G}{1+R}\right)^{N}}{R-G} \times \frac{(1+R)^{N}}{(1+R)^{N}-1}\right)$$

and

$$f(N+1) = K + C\left(\frac{1-\left(\frac{1+G}{1+R}\right)^{N+1}}{R-G} - \frac{(1+R)^{N+1}}{(1+R)^{N+1}-1}\right)$$

and where: K = The initial cost of the machine
C = The first-year maintenance cost
G = The fraction of increased maintenance costs (e.g., G = .20 when maintenance costs, C, are rising at 20% per year compounded annually)
R = The rate (e.g., R = .15 for a discount rate of 15%)

A stopping rule for operating the algorithm is to halt the computation when the value of N is no longer decreasing (e.g., the computation is over when, as N decreases, an N for which $f(N_i) - f(N_{i+1}) \leq 0$ is finally obtained). An example program and program run is given in the Appendix (written in the language of the Hewlett Packard HP41-CV). Figure 1 is a flowchart that is useful for interpreting the algorithm and translating the program into another computer language.

The procedures for choosing among the technical and managerial options (e.g., alternatives of repairing, replacing, rehabilitating, abandoning, or decreasing the work load of machines) for managing the construction spread are: (1) Given i options, use the algorithm to identify the optimum replacement period N* for each equipment replacement alternative (that is, given i alternatives, find all N_i^*); (2) For each N_i^*, there is an associated NPW_i. Compare these net present worths and identify the least expensive one (e.g., of several N* values, each with an associated cost and equipment replacement scheme, the least expensive one is the smallest NPW_i); (3) Usually, the equipment replacement plan associated with this minimum NPW_i will be selected and then to implement the decision, that machine will be replaced every N* years (where the N_i^* of interest is the one

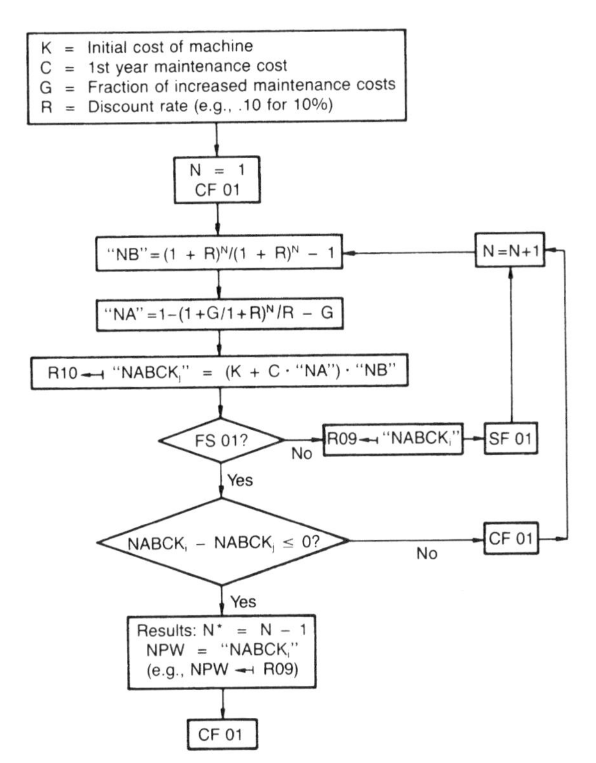

Figure 1. Flow chart for interpreting the algorithm and translating the program into another computer language.

for which NPW_i is a minimum); (4) A simple sensitivity analysis may be performed by reworking items 1 through 3 for various values (e.g., for checking the sensitivity to interest rates, compute N* for various R values).

"Cutoff rates" may be used to decide when equipment failures are too high despite favorable cost curves for repairs since the pipeline construction firm must be reliable in serving customers, but these cutoff rates should be selected in recognition of the N* values for the various alternatives to avoid short-circuiting the whole decision process.

The result of the above procedure is a cost-wise strategy for the management of the equipment used in pipeline construction spreads.

If the assumptions of the life cycle analysis are modified to more accurately reflect the rising maintenance costs over time, then a new equation must be derived. If the analytical problem becomes unwieldy, an option is to solve the problem via a simulation; the saddle of the cost curve can then be found using a numerical searching method.

Another perspective is to view N* as the optimum retention period, e.g., the length of time that a machine should be kept if costs are to be minimized. Costs would be greater if the machine were kept shorter or longer than the ideal duration expressed by N*.

Source

Pipe Line Industry, August 1985.

Acknowledgments

Prof Zandi, Civil and Urban Engineering Department, University of Pennsylvania; Dr. Pollack-Johnson, Department of Mathematics, Oberlin College, Ohio; and Mr. V James Joyce, Joyce Western Corp., Andover, N.Y.

How to estimate construction costs

Two hand-held calculator programs divide pipeline costs into four component areas: right of way, material, labor, and miscellaneous

R. D. Green, Energy Center Fellow, University of Pennsylvania

Pipeline cost models are numerous, and their uncertainty is associated with the degree of generality.

Costs vary with "distances from manufacturers, terrain, locality, labor costs, season of the year, and other variables."[1] Hence, it is vital to the interests of construction firms to maintain historical records of costs and to update these records routinely. A well-developed model may be used as a management tool during construction to identify *either* discrepancies between the model and "reality" or to encourage project superintendents to avoid cost over-runs and outperform previous time and motion constants.

A detailed account of cost estimating has been written by J. S. Page (1977)[2]; the work is similar to the approach taken by contractors seeking to estimate costs of manpower, materials, etc., based on historical time frames and the like. Pipeline cost models that identify cost parameters and the inter-relationship of these parameters have been developed by I. Zandi (1982)[3] and K. Gimm and I. Zandi (1976).[4]

A rule of thumb calculation has also been computed (1964),[5] and it is updated here. Table 1, "Pipeline rule of thumb cost estimate," presents a means of estimating the cost based on costs of pipelines in the United States. The "1964" figures are

Table 1. Pipeline rule of thumb cost estimate, Cost = D × A × Miles of pipeline

Year	Diameter of pipe, inches	$\left(\frac{\text{Average cost US\\$}}{\text{inch diameter} \times \text{mile}}\right)$	$\left(\frac{\text{Range US\\$}}{\text{in. dia.} \times \text{mile}}\right)$ Low	high	Average % inflation since 1964*
1964**	10	2,059	–	–	–
	16	2,849	–	–	–
	20	2,632	–	–	–
	24	3,325	–	–	–
	30	4,085	–	–	–
	36	4,260	–	–	–
1980***	8	16,920	6,953	24,125	–
	12	18,446	13,882	16,644	–
	16	10,823	9,200	12,368	8.3
	20	14,226	12,039	16,698	10.5
	24	14,738	8,184	28,263	9.3
	30	16,730	13,100	28,756	8.8
	36	25,468	19,087	33,929	11.2

*Compounded continuously; from $100\% \times \left(\ln\left(\frac{A\ 1980}{A\ 1964}\right)\right) \div 16$.

**Source is *Pipeline Rules of Thumb Handbook*,[1] and these figures are based on 1964 data from the Federal Power Commission.

***Based on averages of costs for onshore projects from FERC Construction Permit applications as reported in the *International Petroleum Encyclopaedias*.[5]

Table 2. A real division of pipeline costs

Size (inches)	ROW	Material	Labor	Misc.	Total
Part I: By area, % of total pipeline costs over 10-year period:*					
8	2.8	37.2	49.2	10.8	100.0
12	5.7	37.3	44.3	11.2	98.5
16	3.9	46.6	38.8	10.5	99.5
20	3.1	46.0	41.8	9.4	100.3
24	3.1	46.3	38.0	11.8	99.2
30	3.7	49.6	37.9	8.5	99.7
36	4.1	48.1	37.2	10.6	100.0
Average	3.8	44.4	41.0	10.4	99.6
Part II: Simplified version of Part I.					
8	3.5	37.	49.	10.5	100.0
12	5.7	37.	44.3	10.5	97.5
16	3.5	47.	38.75	10.5	99.75
20	3.5	47.	38.75	10.5	99.75
24	3.5	47.	38.75	10.5	99.75
30	3.5	47.	38.75	10.5	99.75
36	3.5	47.	38.75	10.5	99.75
Average	3.8	44.1	41.0	10.5	99.46
Part III: Version used in hand-calculator program "CSTA." (*D* inches)					
$D \leq 8$	3.5	37	49.	10.5	100.0
$8 < D \leq 12$	5.7	37	44.3	10.5	97.5
$12 < D \leq 36$	3.5	47	38.75	10.5	99.75
Average	–	–	–	–	–

*Tabulated from FERC based data presented in the *International Petroleum Encyclopaedias.*[5]

listed for comparison only and are from the (1978) *Pipeline Rules of Thumb Handbook.*[1]

Adjusting for inflation

The pipeline rule of thumb cost estimate is made more useful by including a means for adjusting the "1980" based formula of Table 1 to other years:

$$C(y) = DAme^{i}\ (y - 1980)$$

where: $C(y)$ = Cost of the pipeline in year y
D = Diameter of the pipeline
A = Average cost of pipeline construction per inch-diameter-mile
m = Length of pipeline in miles

i = Continuously compounded rate of inflation (e.g., % inflation from Table 1 divided by 100)
y = Year for which the cost estimate is desired

Hand calculator program CST performs this cost calculation when the user inputs 1980-based data from Table 1; an example run is included in the appendix.

Estimate for cost areas

The cost of a pipeline may be divided into four component cost areas: Right of way (ROW), material, labor, miscellaneous.

An estimate of any one of these component costs (C_i) is obtained from the following:

$$C_i = a_i C(y)$$

where: a_i = 10-year average proportion of total cost allocated for one of the four areas
$C(y)$ = Total cost of the pipeline (or an estimate)

The a_i's are taken from Table 2, Part III, in the program CSTA (see appendix), according to the size (diameter) of the pipeline given the following divisions:

$D \geq 8$
$8 < D \leq 12$
$12 < D \leq 36$

Appendix—Software for Hewlett Packward HP41-CV

CSTA

Given both the cost and size of a pipe line, this program estimates the division of expenses between four categories: Right of way (ROW), material, labor and miscellaneous. Table 2 forms the basis of the program, and in Part III of the table it is apparent that the program user may elect to use pipe line sizes (diameters) for which no data is supplied. For example, a 10-inch line is viewed as a 12-inch line by the program.

```
PROGRAM "CSTA"
01 ▲ LBL "CSTA"
02 CF 01                Clear flags.
03 CF 02
04 "COST?"              User inputs pipe line construction cost
                        and diameter of line (D).
05 PROMPT
06 STO 03
07 "D?"
08 PROMPT
09 STO 04
10 8                    Is D≤8?, 8<D≤12?, or 12<D≤36?
11 X( )Y
12 X( = Y?
13 SF 01                Set flags accordingly (1, 2, or neither).
14 FS? 01
15 GTO A
16 RCL 04
17 12
18 X( )Y?
19 X( = Y?
20 SF 02
21 FS? 02
22 GTO "AA"
23 ▲ LBL A              ROW cost calculation for D ≠ x
                        such that 8 < x ≤ 12.
24 RCL 03
25 .035
26 *
27 STO 08
28 GTO "AAA"
29 ▲ LBL "AA"           "ROW" cost calculation for D such
                        that 8<D≤12.
30 RCL 03
31 .057
32 *
33 STO 08
34 ▲ LBL "AAA"
35 "R.O.W. = "          output
36 PRA
37 RCL 08
38 PRX
39 FS? 01               "Material" calculation.
40 GTO B
41 FS? 02
42 GTO B
43 RCL 03
44 .47
45 *
46 STO 05
47 GTO C
48 ▲ LBL B
49 RCL 03
50 .37
51 *
52 STO 05
53 ▲LBL C
54 "MATERIAL = "        output
55 PRA
56 RCL 05
57 PRX
58 FS? 01               "Labor" calculation.
59 GTO D
60 FS? 02
61 GTO E
62 RCL 03
63 .3875
64 *
65 STO 06
66 GTO F
67 ▲ LBL D
68 RCL 03
69 .49
70 *
71 STO 06
72 GTO F
73 ▲ LBL E
74 RCL 03
75 .43
76 *
77 STO 06
78 ▲ LBL F
79 "LABOR = "           output
80 PRA
81 RCL 06
82 PRX
83 "MISC. = "           "Misc." calculation and output.
84 PRA
85 RCL 03
86 .105
87 *
88 STO 07
89 PRX
90 "TOTAL = "           "Total" = ROW + Material + Labor + Misc.
                        Due to assumptions of model (see Table 2), "Total"
                        might be different from input cost.
91 PRA
92 RCL 08
93 RCL 05
94 +
95 RCL 06
96 +
97 RCL 07
98 +
99 PRX                  output
100 END
```

EXAMPLE RUN

```
                        XEQ "CSTA"
COST?
   5,605,238.00         RUN
D?
            8.0         RUN
R.O.W. =
     196,183.33         ...
MATERIAL =
   2,073,938.06         ...
LABOR =
   2,746,566.62         ...
MISC. =
     588,549.99         ...
TOTAL =
   5,605,238.00         ...
```

CST

This program is to be used with the "1980" data of Table 1. The program estimates the cost of a pipe line from the equation,

$$\text{Cost} = D \times A \times \text{Miles} \times \exp\left(\frac{\%\ \text{inflation}}{100} \times \#\ \text{years}\right).$$

The exponential in this cost equation is merely a means for adjusting the "1980" figures to other years. User input is as follows: "Year" (e.g., 1983; the program also accepts years prior to the base year 1980), "INFLATION" (e.g., 8.3%; this will be compounded continuously), *"D"* (e.g., 16; diameter of pipe line from Table 1, "1980" category); *"A"* (e.g.,10,823; average cost from Table 1, "1980" category); "MILES" (e.g., 20.5; length of pipe line in miles).

Program output is as follows: "COST" estimate of pipe line in U.S. dollars, based on U.S. data (e.g., 4,553,660); "IN YEAR", is the year for which the cost estimate has been computed (e.g., 1983).

```
PROGRAM "CST"
01▲ LBL "CST"
02  "YEAR?"             1980 is the base year; the program ac-
                        cepts any year prior to, after or includ-
                        ing 1980.
03  PROMPT
04  1980
05  -
06  STO 00
07  "INFLATION?"        Inflation rate (in percent) is any figure
                        selected by the user; it is continuously
                        compounded.
08  PROMPT
09  100
10  /
11  STO 01
12  "D?"                Diameter of pipe line.
13  PROMPT
14  "A?"                Average cost in units of US$ per (inch
                        diameter × mile).
15  PROMPT
16 *
17  "MILES?"            Length of pipe line in miles.
18  PROMPT
19 *
20  STO 02
21  RCL 01
22  RCL 00
23 *
24  E↑X                 Compound factor is exp (rate × time).
25  RCL 02
26 *
27  STO 03
28  "COST = "           Cost estimate of pipe line in U.S. dol-
                        lars for a specified year. The cost esti-
                        mate is for construction and encom-
                        passes an average of material, labor,
                        right of way, and miscellaneous ex-
                        penses, as presented in Table 1.
29  PRA
30  RCL 03
31  PRX
32  "IN YEAR-"
33  PRA
34  RCL 00
35  1980
36  +
37   PRX
38   END
```

EXAMPLE RUN

```
                  XEQ "CST"
YEAR?
          1983    RUN
INFLATION?
           8.3    RUN
D?
            16    RUN
A?
        10,823    RUN
MILES?
          20.5    RUN
COST =
    4,553,660.    ...
IN YEAR =
          1983    ...
```

After program has been entered, user executes the program and responds to the prompts; after each data entry, the user presses the "RUN" button. User assignable keys may be defined to facilitate rapid program execution.

The percentages listed in Table 2 are simple averages of pipeline construction cost data for a specific pipeline size:

$$a_i \times 100 = \sum_{y=1971}^{1980} \frac{A_y}{C_y}$$

where: A_y = Average pipeline cost for a given component cost area (e.g., ROW) in a given year
C_y = Average total pipeline cost in a given year (e.g., sum of average costs for a given year in the four categories)

Source

Pipe Line Industry, September 1984.

References

1. *Pipeline Rules of Thumb Handbook*, Gulf Publishing Company, 1978, p. 163.
2. *Cost Estimating Manual For Pipelines and Marine Structures*, Gulf Publishing Company, 1977.
3. "Coal Slurry Pipeline Cost Model," *Journal of Pipelines*, Vol. 2, No. 3, 1982, pp. 53–66.
4. "Freight Pipeline Cost Model," *Transportation of Solid Commodities Via Freight Pipeline*, Department of Transportation (DOT/RSPA/DPB-50/78/33–39), 1978.
5. "Pipelining Continues To Set New Records," *International Petroleum Encyclopaedia, 14th ed.*, Petroleum Publishing Co., 1981, pp. 368–373.

Cost estimating strategies for pipelines, stations, and terminals (part 1)

Microcomputers and up-to-date estimating software are more efficient means for cost studies

R. D. Green, Energy Center Fellow, University of Pennsylvania, Folcroft, Pa.

Cost estimating strategies vary in the accuracy of representing cost factor data. Cost factors for pipelines, stations, and terminals are presented in Tables 1 and 2.

In the pipeline industry, the practice has been to assume that the most accurate cost estimates for constructing pipelines, stations, and terminals will come from a very detailed estimating procedure that sums the expected value or best guess for each cost factor involved. Such an assumption ignores more accurate cost-estimating strategies that require many more computations and have therefore traditionally been avoided.

Historically, cost factor information has been organized in files, tables, and ledgers. Estimating has been a lengthy procedure taking days, if not weeks, of work effort by a full staff of estimators. Now, with microcomputers and up-to-date estimating software, we have a more efficient means of storing, retrieving, and handling data. It is time to reconsider the use of more accurate estimating strategies.

By accurate, I mean representing data in a form that identifies what is and is not known about the cost factor.

Current, future estimating practices

Currently, spreadsheet based-cost models for pipelines, stations, and terminals sum the simply represented costs for the cost factors given in Tables 1 and 2. Future construction estimating practices will likely involve two changes from the current practices.

These changes will be a response to the two problems estimators face. One problem is how to speed the computational procedure that yields the final or grand total project cost estimate, and the other is how best to represent cost factor data. This article considers these two problems.

Speeding computational procedures

Estimating computations are often lengthy and redundant. Estimating can be performed quickly either by simplifying the computations (often with an associated loss in accuracy) or by computerization, in which case accuracy need not be sacrificed. Current and future construction estimating procedures involve the use of spreadsheet-based cost models. There are three basic types of spreadsheets.

- *Spreadsheets* are typically large ledger-like sheets of paper upon which are spread the cost factors and costs that go in to the cost-estimating procedure in estimating the costs of a construction project.
- *Electronic spreadsheets* are a computerized version of the spreadsheet in which cost factors are listed and the estimator fills in the costs for the appropriate items and the computer automatically tallies the results.
- *Automated electronic spreadsheets* are an automated version of the electronic spreadsheet in the sense that

Table 1

Cost factors in pipeline estimate

Design engineering
Electrical & mechanical; engineering-supervisor; engineer, drafting; corrosion-engineer, drafting; planning engineering-supervisor; engineer; survey & drawing; aerial photography; stake out; as-built information & drawing; data processing-supervisor; programmer; travel; supplies.

Right of way, legal, environmental & safety
Acquisition labor; damages labor; permit/legal labor; environmental & safety labor; travel; field office rental; supplies; considerations-existing line; rights, new R/W rights; damages-damage payments, land; purchases, permit payments-permit fees; cost of deposits/bonds; first year rentals/leases.

Construction
Labor–electrical, mechanical; corrosion, inspection labor, expenses–regular, temporary; field office rental, equipment; rentals, supplies; inspection–corrosion coating mill; pipeline, survey, mechanical-pipe mill; pipeline, X-ray, tuboscope, caliper pig; contract–pipeline installation–open area, moderate area, congested; install casing; supplementary work–install valves, purge line, caliper pig (contractor's labor), river crossing, hydrostatic test, gravitometer, station work, buildings, ground bed, rectifier, anodes, environmental requirements, performance bond; extra work–rock removal–blasting, air hammer; sand padding; contingencies.

Materials
Instrumentation–transmitters, gravitometers, miscellaneous; mainline pipe (size, grade, & wall), vent & marker, freight charges; casing pipe (size, grade & wall), freight charges; valves (size, series, type); fittings (description, size, series); valve operators; buildings, furnishings & fencing (size, type); coating (pipe size, type); cathodic protection–rectifiers; test stations; insulation joints; anodes (type); seals and insulators; data processing.

Overheads
Indirect labor (managers and above); travel; supplies; interest; fringe benefits.

much of the data, e.g., costs and cost factors, are provided by the computer rather than by having the estimator fill in the blanks with costs. The computer has a database to turn to for cost and material information, thus the estimator need not look up such information in compiling the estimate.

The time and effort required to produce routine estimates may be reduced by using an automated electronic spreadsheet. Given the tremendous data-handling abilities of even low-cost computers vis-a-vis manual computation, cost estimating may be performed using various strategies ranging from the traditional to the more complex.

A fully integrated cost estimating approach enables the estimator to rely on data that have as their basis historical, current, and forecast aspects of the project. Since such an approach is conveniently packaged as a (high-speed) computerized estimating program, the estimator may base estimates on cost factor data in any of the strategies for representing data.

Three estimating strategies

The three basic strategies for obtaining an estimate differ in the way uncertainty is viewed and represented. Each strategy is characterized by the type of estimates that are utilized:

- A point estimate views factors in the estimate as quantifiable without regard to error, range, and uncertainty (such estimates are rarely viewed as conclusive since managers often ask estimators, "How good are these numbers?").
- A range estimate is generally more sophisticated than the point estimate and may yield, for example, an "expected cost of pipeline construction, ranging as low as . . . and as high as. . ."
- A probability estimate is the most sophisticated estimate since it does not ignore uncertainty and is based on an evaluation of probability distributions for (a) historical data, (b) current conditions, and (c) projections of future conditions.

Probability estimates may be depicted in simplified forms without regressing to a range or point estimating strategy. In particular, a probability distribution can be represented with some loss of accuracy as any of several simplified distributions (such as a histogram or a triangular distribution or some smooth distribution that is beta or normal).

One problem with point estimates is that these do not convey an awareness of any uncertainty behind the estimate. Point and range estimates can be misleading. As Hertz

Table 2
Cost factors in station and terminal estimate

Design engineering
Electrical & mechanical; engineering supervisor; engineer, drafting; corrosion-engineer, drafting; planning engineering-supervisor; engineer; supplies; civil-engineer, drafting; surveying-property, topograph, aerial, soil borings; stake out/as-built-crew; draftsman; data processing-supervisor; programmer; travel; expenses; supplies.

Right of way, legal, environmental & safety
Acquisition/damages labor; permit/legal labor; environmental & safety; travel; field office rental; supplies; payments-considerations; damages; land purchase; title searches; permits & fees–first year rental/leases.

Construction
Labor-electrical, mechanical; corrosion, expenses regular & temporary; office & equipment rental; supplies; inspection mechanical; electrical; X-ray; soil testing; concrete testing; contracts; corrosion-anodes; test stations; deep well; groundbed; rectifier; set pole; insulating joint; painting (including sandblasting, primer, all coats); tanks; piping; pipe coating; excavation, machine, backfill; cable, substation-install; transformers; install switchgear; install fuse/circuit breakers; structure work; motors-set motor, install motor; control; starters; instrumentation-install transmitters, recorders, etc.; tank gauges; ticket printers; meter totalizers; electrical cabinets; valve operators; gravitometers; miscellaneous; conduit & wire; lighting-set poles, mount lights; install computer; other-concrete, excavation; modular building; mechanical-welds, valves, cuts & bevels; pipe, civil-excavation, backfill; concrete; stoning; fencing; gates; water well; septic system; building; oil/water separator; seeding; extra work–corrosion, electrical, mechanical, civil; electrical & mechanical–manpower; equipment.

Material
Motors; motor control center; starters; substation and transformers–transformer, substation structure, lightning arrestors, miscellaneous fittings; switchgear; instrumentation-general–gravitometers, transmitters, receivers, power supply, alarm annunciator, recorders, totalizers, ticket printers; switches-pressure, temperature, level; vibration monitor; cabinets; micro pumps; other & miscellaneous; instrumentation–tanks, tank gauges, transmitters, other; lighting–pole, lights; miscellaneous items; circuit breakers; wire & conduit; miscellaneous special equipment; mechanical–pipe (size, grade, & wall), valves (size, series, type); fittings (description, size, series); valve operators; pumps & pump/motor combinations–spare rotating element; meters; prover loop; meter/prover skid; strainers; tanks–large storage tanks, other; communication–radios, other; fire protection–pipe, valves, fittings, fire extinguishers, foam proportioners, foam tank, foam; environmental control equipment–boat, spill boom, vapor recovery system; miscellaneous and special equipment; civil-buildings, building furnishings; oil water separator; structural steel supports; stone; special equipment; corrosion-coating, anodes; insulating joint; test stations; cable; rectifier; other; paint; data processing-computers, CRT displays, printers, disks, RTU; special equipment.

Overheads
Indirect labor-directors & managers of engineering, planning & analysis, construction, right of way, other; travel; supplies; interest; fringe benefits.

Administrative overheads
Administrative labor; travel expenses; supplies.

Indirect costs
Labor; materials, pipe, other.

describes, the "estimated probability distribution paints the clearest picture of all possible outcomes,"[1] and he gives examples in the area of capital investment projects. Risk analysis is designed to provide more information than conventional techniques that fail, as Hespos and Strassman note, for not taking into account the "skewed factors," and for being "influenced by the subjective aspects of best guesses."[2]

To illustrate the subjective aspects, note that an estimator typically places the "best point estimate" in the low end if a high outcome is more serious than a low outcome. And the estimator's "best point estimate" is in the high end if a low outcome is more serious than a high outcome. The point estimate is a result of the estimator's judgments about uncertainties, outcomes, and risks or values of the consequences of outcomes. Probability estimates are more useful than point estimates since they more accurately represent what is and is not known about cost factors. There is less need or motivation to distort probability estimates (e.g., these are less vulnerable to subjective aspects).

Although the range estimate conveys the notion that a precise estimate is not available, it does not convey any probability of where the actual cost will fall. Range estimates are often misleading since the actual cost may fall not only within but outside of the range specified. Range estimates are usually defined by two point estimates (e.g., low and high).

The probability estimate conveys the most information and offers a means of expressing the uncertainty of the estimate.

Probability estimates of uncertain parameters are made by converting point forecasts to probabilistic forecasts. This is done generally according to either the historical records of the firm or the "guess-timates" provided by experienced personnel.

For example, where the future is statistically like the past (e.g., no identifiable characteristics such as cycles or trends), probability estimates for future cost factors can be based on historical relative frequency distributions that may have been "refined" by interpolation or smoothing. Concisely stated, the probability estimate conveys information concerning all that is and is not known about the estimated cost.

Therefore, it can visually represent the likeliness that the real value will be, for example, below a given cost level. This is especially important for construction contractors who engage in bidding to perform work on a project.

For the contract-engineering and construction firms, the variability in construction costs must be closely monitored in view of the competitive nature of the construction industry where profits are related to the accuracy of forecasting engineering and construction costs. As for pipeline operating firms that often have work performed by outside contractors, there is a need to pay attention to the variability in construction costs for both large and small projects.

The database ought to reflect the variability in cost factors so that uncertainties for large projects, which can impact the firm's capital position, are more clearly comprehended. Also, for an operating firm that has a large proportion of small construction projects relative to large projects, most of the data gathering opportunities will be missed if the management posture is one of ignoring variability for small projects.

Cost estimating strategies for pipelines, stations, and terminals (part 2)

Developing a cost model for automated electronic spreadsheets may involve three estimating strategies

R. D. Green, Energy Center Fellow, University of Pennsylvania, Folcroft, Pennsylvania

Developing a cost model (e.g., an automated electronic spreadsheet) for pipelines, stations, and terminals, which provides a probability estimate of cost, may involve the use of three estimating strategies. For example, the variability in cost for many cost factors may be so limited that it is feasible to use point estimates. This is often the case for material estimates.

Similarly, range estimates may be used where probability estimates will not add any significant information (e.g., where a probability distribution is approximately "mesa" shaped). But for the cost factors that exhibit great variability, it is essential to use probability estimates (estimated probability distributions).

Manpower (cost) estimates are often in this category. Whether or not historical data are available, probability estimates can be created. For example, based on the experience of an estimator, a three-point estimate can be obtained. Or if adequate data are available, histograms can be used to estimate the probability density functions (probability estimates). Some commonly used approaches for estimating the probability density functions are:

- The three-point estimate approach (which may involve any of several distributions, e.g., beta or the triangular distribution)
- The histogram approach (which may closely approximate the "smooth curve" of a probability distribution).

Three-point estimate approach

The three-point estimate, e.g., low (L), expected (E), and high (H), can be used to estimate the probability distribution for the cost factor. Given the simplifying assumption that the probability distribution for the cost factor is approximately beta and that the standard deviation equals 1/6 of the range of reasonably possible costs for factors, then the estimated expected value of a given cost factor is:

$$C_e = (1/3) \times [2E + (1/2) \times (L + H)]$$

The variance is the square of the standard deviation:

$$s^2 = ((1/6) \times (H - L))^2$$

Given many cost factors that have been estimated in this way, the expected value of the final or grand total cost is:

$$G = \sum_{i=1}^{n} C_{ei}$$

The variance for G is:

$$V_G = \sum_{i=1}^{n} s_i^2$$

The final or grand total cost may be viewed as a sum of many independent random variables. By invoking the general version of the central limit theorem, the probability distribution of such a sum is approximately normal. Given the expected value G and the variance V_G, it is then straightforward to find the probability that this normal random variable (grand total project cost, G) will be less than a given (e.g., "bid," B) price. This is done by finding the area under the standard normal curve from negative infinity to Z where:

$$Z = (B - G)/\sqrt{V_G}$$

That is, the estimated probability that the actual grand total cost is less than the given bid price is:

$$P = .5 + \int_0^{x=Z} 1/\sqrt{2\pi} - x^2/e^2 dx$$

This equation is solved by invoking Simpson's rule for several intervals and yields the following approximation (if more accuracy is needed, more intervals can be used):

$$P = .5 + \frac{\frac{Z}{6}}{3\sqrt{2\pi}} \times \left(1 + \left[\frac{4}{\frac{Z^2}{e^{72}}}\right] + \left[\frac{2}{\frac{Z^2}{e^{18}}}\right] + \left[\frac{4}{\frac{Z^2}{e^{8}}}\right] + \left[\frac{2}{\frac{2Z^2}{e^{9}}}\right] + \left[\frac{4}{\frac{25Z^2}{e^{72}}}\right] + \left[\frac{1}{\frac{Z^2}{e^{2}}}\right]\right)$$

Histogram approach

The histogram approach can be a more accurate means of representing the estimated probability distributions for cost factors (rather than using the three-estimate approach) since bimodal and skewed characteristics, etc., can be portrayed. Depending on the mix of estimating strategies and the types of probability estimates involved, the final cost estimate will be obtained in part via a simulation by Monte Carlo sampling. Further details of this approach are beyond the scope of this article.

Observations and conclusion

The advantage of a well-developed electronic spreadsheet is that it can automatically maintain historically based data in the form of, for example, histograms that may be sampled later on to extract information. This supports future efforts to obtain probability estimates for cost factors. The system may accommodate inflation rates that vary among the cost factors. These probability estimates enable the estimator to communicate information about the project risks and the uncertainty of the estimate so that the right risks can be taken by the executive or managerial component of pipeline construction and operating firms.

An implication for the future is related to the current "data policy" of the pipeline organization. In view of the estimating tools that are needed by the pipeline industry, data currently being gathered and preserved ought to reflect the variability of costs for the cost factors. Such data are important to the future of the pipeline organization.

	A	B	C	D	E	F	G
1	CST						
2	DATA:	PIPE DIA.	AVG. COST		RANGE	AVG. % INFLATION	
3		D INCHES	A	LOW TO	HIGH	SINCE 1964	
4		8	$16920	$6953	$24125	8.3	
5		12	$18446	$13882	$16644	8.3	
6		16	$10823	$9200	$12368	8.3	
7		20	$14226	$12039	$16698	10.5	
8		24	$14738	$8184	$28263	9.3	
9		30	$16730	$13100	$28756	8.8	
10		36	$25468	$19087	$33929	11.2	
11	YEAR?	1983	expected	lo	hi		
12	D?	16.00					
13	MILES?	20.50					
14	A	$10823	GIVEN D=	16.00	24.00		
15	INFLATION	8.3					
16	--------	AVERAGE	LOW	HIGH			
17	COST=	$4,553,662.26	$3,870,802.25	$5,203,704.60			
18	--------	--------------	--------	--------			
19	CSTA						
20	DATA:						
21	8	2.8	37.2	49.2	10.8	100.0	
22	12	5.7	37.3	44.3	11.2	98.5	
23	16	3.9	46.6	38.8	10.5	99.5	
24	20	3.1	46.0	41.8	9.4	100.3	
25	24	3.1	46.3	38.0	11.8	99.2	
26	30	3.7	49.6	37.9	8.5	99.7	
27	36	4.1	48.1	37.2	10.6	100.0	
28	FOR COST=	$4,553,662.26	$3,870,802.25	$5,203,704.60			
29	FOR D =	16					
30	--------	POINT ESTIMATE	-------------				
31	ROW	$178,485.25	GIVEN D=	16	24		
32	MATERIAL	$2,132,669.95					
33	LABOR	$1,775,699.44					
34	MISC.	$480,537.22					
35	TOTAL	$4,567,391.86					
36	---------------	RANGE	ESTIMATE -----				
37		LO	HI				
38	ROW	$151,719.89	$203,964.30				
39	MATERIAL	$1,812,858.13	$2,437,111.89				
40	LABOR	$1,509,418.36	$2,029,183.29				
41	MISC.	$408,476.62	$549,134.66				
42	TOTAL	$3,882,472.98	$5,219,394.13				
43	________________	PROBABILITY	ESTIMATE______	________________	________	________________	________
44	B =	$6,000,000		Ce		s square	
45			ROW	$178,270.87		75,818,855	
46			MATERIAL	$2,130,108.30		10,824,798,700	
47			LABOR	$1,773,566.56		7,504,321,729	
48			MISC.	$479,960.03		549,574,535	
49			G =	$4,561,905.74	VG =	18,954,513,700	
50							Z =10.445543
51							
52							0.2315
53						THE PROBABILITY	1.0000
54						OF THE ACTUAL	0.8789
55						GRAND TOTAL COST	0.0047
56						BEING LESS THAN	0.0000
57						A GIVEN COST "B"	0.0000
58						IS APPROXIMATELY	0.0000
59						EQUAL TO "P"	0.0000
60						WHERE	0.5000
61						P =	0.9361

Figure 1. Worksheet example of an electronic spreadsheet.

```
A1 CST
A2 DATA:
A11 YEAR?
A12 D?
A13 MILES?
A14 A
A15 INFLATION
A16 ________
A17 COST=
A18 ________
A19 CSTA
A20 DATA:
A21 8
A22 12
A23 16
A24 20
A25 24
A26 30
A27 36
A28 FOR COST=
A29 FOR D =
A30 ________
A31 ROW
A32 MATERIAL
A33 LABOR
A34 MISC.
A35 TOTAL
A36 _________
A38 ROW
A39 MATERIAL
A40 LABOR
A41 MISC.
A42 TOTAL
A43 ________________PROBABILITY ESTIMATE_____________
A44 B =
B2 PIPE DIA.
B3 D INCHES
B4 8
B5 12
B6 16
B7 20
B8 24
B9 30
B10 36
B11 1983
B12 16
B13 20.5
B14 @LOOKUP(D14,B4:F10,1)
B15 @LOOKUP(D14,B4:F10,4)
B16 AVERAGE
B17 B12*B13*B14*@EXP((B15/100)*(B11-1980))
B18 _____________
B21 2.8
B22 5.7
B23 3.9
B24 3.1
B25 3.1
B26 3.7
B27 4.1
B28 B17
B29 B12
B30 POINT ESTIMATE
B31 @LOOKUP(D31,A21:A27,1)*B28/(@LOOKUP(D31,A21:A27,5))
B33 @LOOKUP(D31,A21:A27,3)*B28/(@LOOKUP(D31,A21:A27,5))
B34 @LOOKUP(D31,A21:A27,4)*B28/(@LOOKUP(D31,A21:A27,5))
B35 @SUM(B31,B34)
B36 _______ RANGE    ESTIMATE _____
B37 LO
B38 @LOOKUP(D31,A21:A27,1)*C17/(@LOOKUP(D31,A21:A27,5))
B39 @LOOKUP(D31,A21:A27,2)*C17/(@LOOKUP(D31,A21:A27,5))
B40 @LOOKUP(D31,A21:A27,3)*C17/(@LOOKUP(D31,A21:A27,5))
B41 @LOOKUP(D31,A21:A27,4)*C17/(@LOOKUP(D31,A21:A27,5))
B42 @SUM(B38,B41)
B43 _______PROBABILI
B44 1E+10
C2 AVG. COST
C3 A
C4 16920
C5 18446
C6 10823
C7 14226
C8 14738
C9 16730
C10 25468
C11 expected
C14 GIVEN D=
C16 LOW
C17 B12*B13*(@LOOKUP(D14,B4:B10,2))*@EXP((B15/100)*(B11-1980))
C18 ________
C21 37.2
C22 37.3
C23 46.6
C24 46
C25 46.3
C26 49.6
C27 48.1
C28 C17
C30 _____________
C31 GIVEN D=
C36  ESTIMATE ____
C37 HI
C38 @LOOKUP(D31,A21:A27,1)*D17/(@LOOKUP(D31,A21:A27,5))
C39 @LOOKUP(D31,A21:A27,2)*D17/(@LOOKUP(D31,A21:A27,5))
C40 @LOOKUP(D31,A21:A27,3)*D17/(@LOOKUP(D31,A21:A27,5))
C41 @LOOKUP(D31,A21:A27,4)*D17/(@LOOKUP(D31,A21:A27,5))
C42 @SUM(C38,C41)
C43 TY ESTIMATE___
C45 ROW
C46 MATERIAL
C47 LABOR
C48 MISC.
C49 G =
D2              RANGE
D3    LOW TO
D4 6953
D5 13882
D6 9200
D7 12039
D8 8184
D9 13100
D10 19887
D11 lo
D14 @IF B12<=8 THEN 8 ELSE ( @IF B12<=12 THEN 12 ELSE
( @IF B12<=16 THEN 16 ELSE ( @IF B12<=20 THEN 20 ELSE E14)))
D16 HIGH
D17 B12*B13*(@LOOKUP(D14,B4:B10,3))*@EXP((B15/100)
*(B11-1980))
D18 ________
D21 49.2
D22 44.3
D24 41.8
D25 38
D26 37.9
D27 37.2
D28 D17
D31 @IF B29<=8 THEN 8 ELSE ( @IF B29<=12 THEN 12 ELSE
( @IF B29<=16 THEN 16 ELSE ( @IF B29<=20 THEN 20 ELSE E31)))
D36 _
D43 ______________
D44 Ce
D45 (1/3)*(2*B31+(1/2)*(B38+C38))
D46 (1/3)*(2*B32+(1/2)*(B39+C39))
D47 (1/3)*(2*B33+(1/2)*(B40+C40))
D48 (1/3)*(2*B34+(1/2)*(B41+C41))
D49 @SUM(D45,D48)
E2 NGE
E3  HIGH
E4 24125
E5 16644
E6 12368
E7 16698
E8 28263
E9 28756
E10 33929
E11 hi
E14 @IF B12<=24 THEN 24 ELSE ( @IF B12<=30 THEN 30 ELSE 36)
E21 10.8
E22 11.2
E23 10.5
E24 9.4
E25 11.8
E26 8.5
E27 10.6
E31 @IF B29<=24 THEN 24 ELSE ( @IF B29<=30 THEN 30 ELSE 36)
E43 _________
E49 VG =
F2 AVG. % INFLATION
F3 SINCE 1964
F4 8.3
F5 8.3
F6 8.3
F7 10.5
F8 9.3
F9 8.8
F10 11.2
F21 100
F22 98.5
F23 99.5
F24 100.3
F25 99.2
F26 99.7
F27 100
F43 __________________
F44 s square
F45 ((C38-B38)/6)*((C38-B38)/6)
F46 ((C39-B39)/6)*((C39-B39)/6)
F47 ((C40-B40)/6)*((C40-B40)/6)
F48 ((C41-B41)/6)*((C41-B41)/6)
F49 @SUM(F45,F48)
F50 Z =
F53 THE PROBABILITY
F54 OF THE ACTUAL
F55 GRAND TOTAL COST
F56 BEING LESS THAN
F57 A GIVEN COST "B"
F58 IS APPROXIMATELY
F60 WHERE
F61 P =
G43 _________
G50 (B44-D49)/@SQRT(F49)
G52 (G50/6)/(3*@SQRT(2*@PI))
G53 1/@EXP(0)
G54 4/@EXP(G50*G50/72)
G55 2/@EXP(G50*G50/18)
G56 4/@EXP(G50*G50/8)
G57 2/@EXP(2*G50*G50/9)
G58 4/@EXP(25*G50*G50/72)
G59 1/@EXP(G50*G50/2)
G60 0.5
G61 (G52*(@SUM(G53,G59)))+G60
```

Figure 2. Program listing of an electronic spreadsheet.

Example spreadsheet

A simple automated electronic spreadsheet example is provided (with software). It illustrates the use of "lookup tables" as a means of accessing a "built-in" database. In this example, a rule of thumb estimate[3] is used for estimating the probability distribution of pipeline construction costs in several cost areas. An overall distribution is obtained by using the procedure described earlier (three-point estimate approach). The estimated probability distribution for the grand total cost of the project is used in a step to assess the probability that the project cost is less than a given "bid" price. Given relatively few inputs (year of construction, pipeline diameter, and length) and a limited database, this simple automated electronic spreadsheet finds point, range, and probability estimates for several cost areas. The expected grand total cost G = $4,561,905.74 is surrounded by considerable uncertainty. In a full-scale automated electronic spreadsheet, the variance surrounding the cost factors and expected grand total cost will be greatly reduced relative to this simple example.

A possible improvement upon this example is as follows. If more representative distributions for the cost factors are obtained and presented as histograms, then the estimated probability distribution for grand total cost will be obtained through a simulation that performs a summation involving random sampling of histograms (histogram approach).

Overall, the example provides an introductory automated electronic spreadsheet. Given some project-specific information and the historical record held within the program, cost projections are effortlessly produced. A large-scale model would work in a similar way but with more detailed data to represent the many cost factors given in Figures 1 and 2.

Source

Pipe Line Industry, September 1986.

Acknowledgments

Mr. B. H. Basavaraj, Facilities Planning Supervisor, Buckeye Pipe Line Co., Emmaus, Pa.; Prof I. Zandi, Civil and Urban Engineering Dept., University of Pennsylvania; Mr. Andy Anderson and Mr. K. N. Wheelwright, Weyher/Livsey Constructors, Inc., Salt Lake City, Utah.

References

1. Hertz, D. B., "Risk Analysis in Capital Investment," *Harvard Business Review*, Sept.–Oct. 1979, pp. 169–180.
2. Hespos, R. F. And Strassmann, P. A., "Stochastic Decision Trees for The Analysis of Investment Decisions," *Management Science*, August 1965, p. 11.
3. Green, R. D., "Rules of thumb, How to estimate construction costs," *Pipe Line Industry*, Vol. 61, No. 3, Sept. 1984, p. 47–48.

APPENDIX

A worksheet and program listing for the example spreadsheet discussed above are provided in Figures 1 and 2. The software and computer used is Syncalc on the Atari 800XL. (Note the similarities to VisiCalc and Lotus-123 electronic spreadsheets.) For industrial scale work, a more powerful system such as the Atari 520ST or IBM XT/AT computer offers an ideal system for developing a full-scale automated electronic spreadsheet for pipeline construction estimating. For more information on estimating, see *Cost Estimating Manual for Pipelines and Marine Structures* by John S. Page (Gulf Publishing Co., Houston, 1977).

Economics

The goal of all economic evaluations should be to eliminate all unprofitable projects. Some borderline, and perhaps a few good projects, will be killed along with the bad ones. However, it is much better to pass up several good projects than to build one loser.

Every company should develop a standard form for their economic calculations, with "a place for everything and everything in its place." As management becomes familiar with the form, it is easier to evaluate projects and choose among several based on their merits, rather than the presentation.

Components of an economic study

Most economic studies for processing plants will have revenue, expense, tax, and profit items to be developed. These are discussed in the following sections.

Revenues. The selling price for the major product and the quantity to be sold each year must be estimated. If the product is new, this can be a major study. In any case, input is needed from sales, transportation, research, and any other department or individual who can increase the accuracy of the revenue data. The operating costs may be off by 20% without fatal results, but a 20% error in the sales volume or price will have a much larger impact. Further, the operating costs are mostly within the control of the operator, while the sales price and volume can be impacted by circumstances totally beyond the control of the company.

Depending on the product and sales arrangement, the revenues calculation can take from a few man-hours to many man-months of effort. For instance, determining the price for a synthetic fuel from coal can be done in a very short time, based on the cost of service. However, if the price is tied to the price of natural gas or oil, the task becomes very difficult, if not impossible. On the other hand, determining the sales growth and selling price for a new product requires a great deal of analysis, speculation, market research, and luck, but projections can be made.

Most revenue projections should be on a plant net back basis with all transportation, sales expense, etc., deducted. This makes the later evaluation of the plant performance easier.

Operating Costs. The operating costs, as developed in the previous section, can be presented as a single line item in the economics form or broken down into several items.

Amortization. Items of expense incurred before start-up can be accumulated and amortized during the first 5 or 10 years of plant operations. The tax laws and corporate policy on what profit to show can influence the number of years for amortization.

Such items as organization expense, unusual prestart-up expenses, and interest during construction, not capitalized, can be amortized. During plant operations this will be treated as an expense, but it will be a non-cash expense similar to depreciation, which is discussed later.

Land lease costs. If there are annual payments for land use, this item of expense is often kept separate from other operating costs, since it does not inflate at the same rate as other costs. If the land is purchased, it cannot be depreciated for tax purposes.

Interest. The interest for each year must be calculated for the type of loan expected. The interest rate can be obtained from the company's financial section, recent loan history, or the literature. The *Wall Street Journal* publishes the weekly interest rates for several borrowings. The company's interest rate will probably be the banks' prime rate, or slightly higher.

The calculation of interest is usually based on the outstanding loan amount. For example, if semi-annual payments are expected, the amount of the first year's interest (I) would be:

$$I = \left(\frac{i}{2}\right)P + \left(\frac{i}{2}\right)(P - PP) \quad (1)$$

where: i = Annual interest rate
P = Original principal amount
PP = First principal payment

The equation for any annual payment would be:

$$I_y = \sum_{n=1}^{n=N} \left(\frac{i}{N}\right)(P_{y,n-1} - PP_{y,n-1}) \quad (2)$$

where: Iy = Interest for any year
$P_{y,n-1}$ = Principal or loan balance during the previous period
n = Period for loan payment
N = Total number of payments expected each year
y = Year during the loan life
$PP_{y,n-1}$ = Principal payment made at the end of the previous period

Debt payments. The method of making principal and interest payments is determined by the terms of the loan agreement.

Loans can be scheduled to be repaid at a fixed amount each period, including principal and interest. This is similar to most home mortgage payments. The size of level payments can be calculated using the formula:

$$I + PP = P\frac{\left(\frac{i}{n}\right)\left(1+\frac{i}{n}\right)^{n\times y}}{\left[\left(1+\frac{i}{n}\right)^{n\times y}\right]-1} \quad (3)$$

In this type of loan, the principal payment is different each period, so the calculation of interest using Equation 2 must take into account the variable principal payment. Even though both interest and principal are paid together, they must be kept separate. Interest is an expense item for federal income taxes, while principal payments are not.

Alternatively, the debt repayment can be scheduled with uniform principal payments and varying total payments each period. The principal payment can be calculated by the formula:

$$PP_{y,n-1=} = PP = \frac{P}{n \times y} \quad (4)$$

In some loans a grace period is allowed for the first 6 to 12 months of operation, where no principal payments are due. The economics must reflect this.

Depreciation. This is a noncash cost that accumulates money to rebuild the facilities after the project life is over. There are usually two depreciation rates. One is the corporation's book depreciation, which reflects the expected operating life of the project. The other is the tax depreciation, which is usually the maximum rate allowed by tax laws. To be correct, the tax depreciation should now be called "Accelerated Capital Cost Recovery System" (ACRS).

Economics can be shown using either depreciation rate. If book depreciation is used, a line for deferred income tax must be used. This is a cash-available item that becomes negative when the tax depreciation becomes less than the book depreciation. At the end of the book life, the cumulative deferred income tax should be zero.

The 1984 allowable ACRS tax depreciation is shown in Table 1. Most chemical plants have a 5-year tax life.

Federal income tax. The calculation of the federal income tax depends on the current tax laws. State income taxes are often included in this line item, since many are directly proportional to the federal taxes.

The federal tax laws are complicated and ever-changing. It is recommended that expert tax advice be obtained for most feasibility studies. If no current tax advice is available, many tax manuals are published each year that can be purchased or borrowed from a library.

Table 1
ACRS schedule in % per year of depreciable plant

Year	Years for Recovery			
	3	5	10	15
1	25	15	8	5
2	38	22	14	10
3	37	21	12	9
4		21	10	8
5		21	10	7
6			10	7
7			9	6
8			9	6
9			9	6
10			9	6
11–15				6

Current tax laws allow deductions from revenues for sales expenses, operating costs, amortizations, lease costs, interest payments, depreciation, and depletion.

The net income before federal income tax (FIT) is calculated by subtracting all of the deductions from the gross revenues. If there is a net loss for any year, it can be carried forward and deducted the next year from the net income before FIT. If not all of the loss carried forward can be used in the following year, it can be carried forward for a total of 7 years.

At times, tax laws allow a 10% investment tax credit for certain qualifying facilities. Thus, 10% of the qualifying investment can be subtracted from the first 85% of the FIT due. If the taxes paid in the first year are not enough to cover the investment tax credit, it can sometimes be carried forward for several years.

Exhaustible natural deposits and timber might qualify for a cost depletion allowance. Each year, depletion is equal to the property's basis, times the number of units sold that year, divided by the number of recoverable units in the deposit at the first of the year. The basis is the cost of the property, exclusive of the investment that can be depreciated, less depletion taken in previous years.

At times, mining and some oil and gas operations are entitled to an alternate depletion allowance calculated from a percentage of gross sales. A current tax manual should be consulted to get the percentage depletion allowance for the mineral being studied.

Percentage depletion for most minerals is limited to a percentage of the taxable income from the property receiving the depletion allowance. For certain oil and gas producers the percentage depletion can be limited to a percentage of the taxpayer's net income from all sources.

The procedure for a feasibility study is to calculate both the cost and percentage depletion allowances, then choose the larger. Remember, either type reduces the basis.

However, percentage depletion can still be deducted after the basis reaches zero.

The cash taxes due are calculated by multiplying the book net income before tax, times the tax rate, then adjusting for deferred tax and tax credits, if any.

Economics form

As stated at the start of this chapter, it is desirable to establish a form for reporting economics. Table 2 shows such a form. By following this form, the procedure can be seen for calculating the entire economics.

As can be seen in Table 2, all revenues less expenses associated with selling are summed in Row 17. All expenses including non-cash expenses such as depreciation, amortization, and depletion are summed in Row 30. The net profit before tax, Row 32, is obtained by subtracting Row 30 from Row 17 and making any inventory adjustment required. Row 34 is the cash taxes that are to be paid unless offset by investment or energy tax credits in Row 36. The deferred income tax is shown in Row 35. The deferred tax decreases the net profit after tax in the early years and increases the net profit after tax in later years. The impact on cash flow is just the other way around, as discussed later. Row 37, profit after tax, is obtained as follows:

Row 32 (profit before tax)
− Row 34 (income tax)
− Row 35 (deferred tax)
+ Row 36 (tax credits)
= Row 37 (profit after tax)

Row 33, loss carried forward, is subtracted from Row 32 before calculating the income tax.

The cash flow, Row 39, is calculated as follows:

Row 37 (profit after tax)
+ Row 35 (deferred tax)
+ Row 29 (depletion)
+ Row 28 (amortization)
+ Row 27 (depreciation)
− Row 38 (principal payments)
= Row 39 (cash flow)

If one of the rows is a negative number, such as deferred tax, the proper sign is used to obtain the correct answer. Any required inventory adjustment is also made.

Table 2
Economic calculation example

	Periods/Year = 1		
	Fiscal years	1992	1993
12	Sales revenue	239,093	248,657
13	Byproduct revenue	17,031	17,713
14	Advert'g & market'g	(1,638)	(1,703)
15	Shipping & freight	(1,310)	(1,363)
16	Other sales expense	(983)	(1,022)
17	Income from sales	252,194	262,282
18	Raw material cost	76,075	78,738
19	Operating labor	26,508	27,701
20	Maintenance labor	23,984	25,063
21	Maint'nce material	37,869	39,573
22	Utilities	25,246	26,382
23	Administr. & general	9,467	9,893
24	Unchanging expense	5,000	5,000
25	Temporary expense	0	0
26	Interest	7,614	7,188
27	Depreciation	5,500	5,500
28	Amortization	3,639	2,139
29	Depletion	12,621	19,541
30	Total Cost	233,524	246,719
31	Inventory adjust't	0	0
32	Profit before tax	18,671	15,563
33	Loss carried forw'd	12,621	4,031
34	Income tax	0	7,835
35	Deferred tax	2,783	(2,530)
36	Tax credits	0	7,835
37	Profit after tax	15,888	18,093
38	Principal payments	4,105	4,531
39	Cash flow	36,325	38,212

Economic analysis

Once the cash flow has been calculated, various yardsticks can be applied to determine the desirability of the project. The years for payout (payback) is often used. This is obtained by adding the cash flow each year, from the first year of construction, until the sum of the negative and positive cash flows reaches zero. The payout is usually reported in years to the nearest tenth, i.e., 5.6 years.

Each company has its own cutoff period for payout, depending on the economy, the company position, alternate projects, etc. One company used 5 years as the maximum payout acceptable for several years. As conditions changed, this was reduced to 3 years, then 1 year. It was hard to find projects with 1 year payout after taxes.

Some companies feel they should earn interest on their equity as well as get a payout. To calculate this, the cash flows are discounted back to year 0 using some arbitrary interest (discount) rate, then the positive discounted cash flows are summed each year of operation until they equal the sum of the discounted negative cash flows. This yardstick also works, but consistency must be maintained from one project to another. The discount (interest) rate and payout both become variables that must be set and perhaps changed from time to time depending on economic conditions inside and outside the company.

A simple return on equity can be used as a yardstick. Utilities are required by regulators to set their rates to obtain a given rate of return. This return is usually constant over the years and from project to project, so management must choose among projects on some basis other than profit.

In free enterprise projects, the return on equity will probably vary from year to year, so an average would have to be used. However, use of an average would consider return in later years to be as valuable as income in the first year of operation, which is not true.

The best way to evaluate free enterprise projects is to use discounted cash flow (DCF) rate of return, sometimes called internal rate of return.

This calculation can be expressed as follows:

$$\text{SDCF} = \sum_{y=1}^{y=N} \frac{\text{CF}_y}{(1+i)^y} = 0 \qquad (5)$$

where: SDCF = Sum of the equity discounted cash flows
CF_Y = Cash flow, both negative and positive for each year y
i = Discount rate in %/100 that results in the sum being zero
N = Total number of years

Since this calculation uses trial and error to find a discount rate at which all discounted negative and positive cash flows are equal, it can be tedious without a computer. Table 3 demonstrates this calculation for a simple set of cash flows. The DCF rate of return results in a single number that reflects the time value of money. It is certainly one of the yardsticks that should be used to measure projects.

Table 3
Discounted Cash Flow Calculations

Discount Rate (%)	Discounted Cash Flows			
Year	0	20	30	25.9
1	−20	−17	−15	−16
2	−20	−14	−12	−12
3	−10	−6	−5	−5
4	10	5	3	4
5	15	6	4	4
6	20	7	4	5
7	25	7	4	5
8	30	7	4	5
9	40	8	4	5
10	50	8	4	5
Totals	+140	+11	−5	00

Sensitivity analysis

Once the economic analysis has been completed, the project should be analyzed for unexpected as well as expected impacts on the economics. This is usually done through a set of "what if" calculations that test the project's sensitivity to missed estimates and changing economic environment. As a minimum, the DCF rate of return should be calculated for ±10% variations in capital, operating expenses, and sales volume and price.

The impact of loss of tax credits, loss carry forward, and other tax benefits should be studied. Such things as start-up problems that increase the first year's operating costs and reduce the production should also be studied. With a good economic computer program, such as described earlier, all of these sensitivity analyses can be made in an afternoon.

Additional reading on economics can be found in the References.

References

1. Perry, R. H. and Chilton, C. H., *Chemical Engineers Handbook*, Fifth Edition, McGraw-Hill, Inc., New York, section 25, p. 3.
2. Constantinescu, S., "Payback Formula Accounts for Inflation," *Chemical Engineering*, Dec. 13, 1982, p. 109.
3. Horwitz, B. A., "How Does Construction Time Affect Return?" *Chemical Engineering*, Sept. 21, 1981, p. 158.
4. Kasner, E., "Breakeven Analysis Evaluates Investment Alternatives," *Chemical Engineering*, Feb. 26, 1979, pp. 117–118.
5. Cason, R. L., "Go Broke While Showing A Profit," *Hydrocarbon Processing*, July, 1975, pp. 179–190.
6. Reul, R. I., "Which Investment Appraisal Technique Should You Use," *Chemical Engineering*, April 22, 1968, pp. 212–218.
7. Branan, C. and Mills, J., *The Process Engineer's Pocket Handbook*, *Vol. 3*, Gulf Publishing Co., 1984.

TIME VALUE OF MONEY: CONCEPTS AND FORMULAS

The value of money is not constant but changes with time. A dollar received today is worth more than a dollar received a month from now, for two reasons. First, the dollar received today can be invested immediately and earn interest. Second, the purchasing power of each dollar will decrease during times of inflation, and consequently a dollar received today will likely purchase more than a dollar next month. Therefore, in the capital budgeting and decision-making process for any engineering or financial project, it is important to understand the concepts of *present value, future value,* and *interest*. Present value is the "current worth" of a dollar amount to be received or paid out in the future. Future value is the "future worth" of an amount invested today at some future time. Interest is the *cost* of borrowing money or the *return* from lending money. Interest rates vary with time and are dependent upon both risk and inflation. The rate of interest charged, as well as the rate of return expected, will be higher for any project with considerable risk than for a "safe" investment or project. Similarly, banks and investors will also demand a higher rate of return during periods of monetary inflation, and interest rates will be adjusted to reflect the effects of inflation.

Simple interest versus compound interest

Simple interest is interest that is only earned on the original principal borrowed or loaned. Simple interest is calculated as follows:

$$I = (P)(i)(n)$$

where: I = Simple interest
P = Principal (money) borrowed or loaned
i = Interest rate per time period (usually years)
n = Number of time periods (usually years)

Compound interest is interest that is earned on both the principal and interest. When interest is compounded, interest is earned each time period on the original principal *and* on interest accumulated from preceding time periods. Exhibits 1 and 2 illustrate the difference between simple and compound interest and show what a dramatic effect interest compounding can have on an investment after a few years.

Exhibit 1
Comparison between $5,000 Invested for 3 Years at 12% Simple and 12% Compound Interest

Simple Interest			
End of Year	Interest Earned	Cumulative Interest Earned	Investment Balance
0	$0.00	$0.00	$5,000.00
1	$5,000.00 × 0.12 = $600.00	$600.00	$5,600.00
2	$5,000.00 × 0.12 = $600.00	$1,200.00	$6,200.00
3	$5,000.00 × 0.12 = $600.00	$1,800.00	$6,800.00

Compound Interest (Annual Compounding)			
End of Year	Interest Earned	Cumulative Interest Earned	Investment Balance
0	$0.00	$0.00	$5,000.00
1	$5,000.00 × 0.12 = $600.00	$600.00	$5,600.00
2	$5,600.00 × 0.12 = $672.00	$1,272.00	$6,272.00
3	$6,272.00 × 0.12 = $752.74	$2,024.64	$7,024.64

Exhibit 2
Comparison of Simple vs. Compound Interest

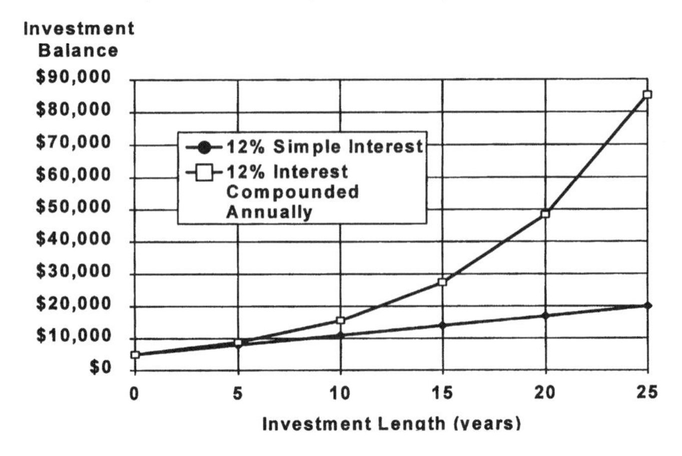

Nominal interest rate versus effective annual interest rate

Almost all interest rates in the financial world now involve compound rates of interest, rather than simple interest. Compound interest rates can be quoted either as a *nominal rate* of interest or as an *effective annual rate* of interest. The nominal rate is the stated annual interest rate compounded periodically (at the stated compounding time interval). The effective annual rate is the rate that produces the same final value as the nominal rate, when compounded only once per year. When the compounding period for a stated nominal rate is one year, then the nominal rate *is the same* as the effective annual rate of interest. The following formula can be used to convert nominal interest rates to effective annual rates:

$$i_{eff} = \left(1 - \frac{i_{nom}}{c}\right)^{c} - 1$$

where:

i_{eff} = Effective annual rate
i_{nom} = Nominal rate
c = Number of compounding periods per year

Note: Interest rates are expressed as fractions (.12) rather than percentage (12%).

For *continuous* compounding, natural logarithms may be used to convert to an effective annual rate:

$$i_{eff} = e^{i_{cont}} - 1$$

where:

i_{cont} = Continuous compounding rate

Note: Interest rates are expressed as fractions (.12) rather than percentage (12%).

Present value of a single cash flow to be received in the future

How much is an amount of money to be received in the future worth today? Stated differently, how much money would you invest today to receive this cash flow in the future? To answer these questions, you need to know the *present value* of the amount to be received in the future. The present value (PV) can easily be calculated by *discounting* the future amount by an appropriate compound interest rate. The interest rate used to determine PV is often referred to as the *discount rate, hurdle rate*, or *opportunity cost of capital*. It represents the opportunity cost (rate of return) foregone by making this particular investment rather than other alternatives of comparable risk.

The formula for calculating present value is:

$$PV = \frac{FV}{(1 + r^{n})}$$

where: PV = Present value
FV = Future value
r = Discount rate (per compounding period)
n = Number of compounding periods

Example. What is the present value of an investment that guarantees to pay you $100,000 three years from now? The first important step is to understand what risks are involved in this investment, and then choose a discount rate equal to the rate of return on investments of comparable risk. Let's say you decide that 9% effective annual rate of interest (9% compounded annually) is an appropriate discount rate.

$$PV = \frac{\$100,000}{(1+.09)^{3}} = \$77.218$$

Thus, you would be willing to invest $77,218 today in this investment to receive $100,000 in 3 years.

Future value of a single investment

Rearranging the relationship between PV and FV, we have a simple formula to allow us to calculate the future value of an amount of money invested today:

$$FV = PV(1 + r)^n$$

where: FV = Future value
PV = Present value
R = Rate of return (per compounding period)
N = Number of compounding periods

Example. What is the future value of a single $10,000 lump sum invested at a rate of return of 10% compounded monthly (10.47 effective annual rate) for 5 years? Note that monthly compounding is used in this example.

$$FV = \$10{,}000(1 + .10/12)^{60} = \$16{,}453$$

The $10,000 investment grows to $16,453 in 5 years.

The importance of cash flow diagrams

Many financial problems are not nearly as simple as the two previous examples, which involved only one *cash outflow* and one *cash inflow*. To help analyze financial problems, it is extremely useful to diagram cash flows, since most problems are considerably more complex than these previous examples and may have multiple, repetitive, or irregular cash flows. The *cash flow diagram* is an important tool that will help you understand an investment project and the timing of its cash flows and help with calculating its PV or FV. The cash flow diagram shown represents the earlier example (in which PV was calculated for an investment that grew to $100,000 after annual compounding at 9% for 3 years).

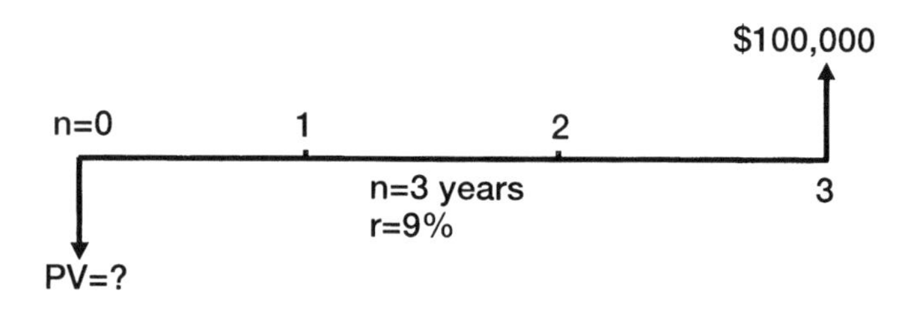

The cash flow diagram is divided into equal time periods, which correspond to the compounding (or payment) periods of the cash flows. Note that this diagram is also presented from the investor's (lender's) point of view, since the initial investment is a cash outflow (negative) and the final payoff is a cash inflow (positive) to the investor. The diagram could also have been drawn from the borrower's point of view, in which case the PV would be a cash inflow and therefore positive. Before drawing a cash flow diagram, you must first choose which viewpoint (lender or borrower) the diagram will represent. It is then extremely important to give each cash flow the correct sign (positive or negative), based upon whichever point of view is chosen. Other examples of cash flow diagrams will be presented in additional sample problems in this chapter.

Analyzing and valuing investments/projects with multiple or irregular cash flows

Many investments and financial projects have multiple, periodic, and irregular cash flows. A cash flow diagram will help in understanding the cash flows and with their analysis. The value of the investment or project (in current dollars) may then be found by discounting the cash flows to determine their present value. Each cash flow must be discounted individually back to time zero and then summed to obtain the overall PV of the investment or project. This *discounted cash flow formula* may be written as:

$$PV = \sum_{i=1}^{n} \frac{C_i}{(1+r)^i}$$

where:

PV = Present value
C_i = Cash flow at time i
n = Number of periods
r = Discount rate of return (opportunity cost of capital)

Example. How much would you be willing to invest today in a project that will have generated positive annual cash flows of $1,000, $1,500, $2,400, and $1,600, respectively, at the end of each of the first 4 years? (Assume a 10% annual rate of return can be earned on other projects of similar risk.)

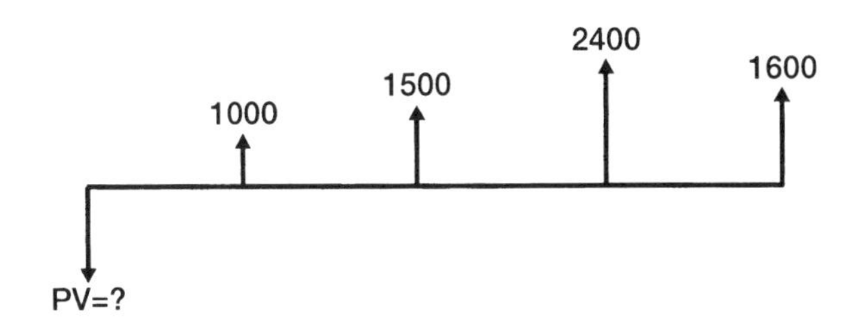

Step 1: Cash flow diagram:

Step 2: Discount cash flows into current dollars (present value):

$$PV = \frac{1,000}{(1+.1)^1} + \frac{1,500}{(1+.1)^2} + \frac{2,400}{(1+.1)^3} + \frac{1,600}{(1+.1)^4}$$
$$= \$5,044.74$$

Thus, you should be willing to invest a maximum of $5,044.74 in this project.

Perpetuities

A *perpetuity* is an asset that provides a periodic fixed amount of cash flow in perpetuity (forever). Endowments and trust funds are often set up as perpetuities, so that they pay out a fixed monetary amount each year to the beneficiary, without depleting the original principal. For valuation purposes, some physical assets such as an oil well, natural gas field, or other long-lived asset can also be treated as a perpetuity, if the asset can provide a steady cash flow stream for many years to come. The cash flow diagram for a perpetuity is represented here:

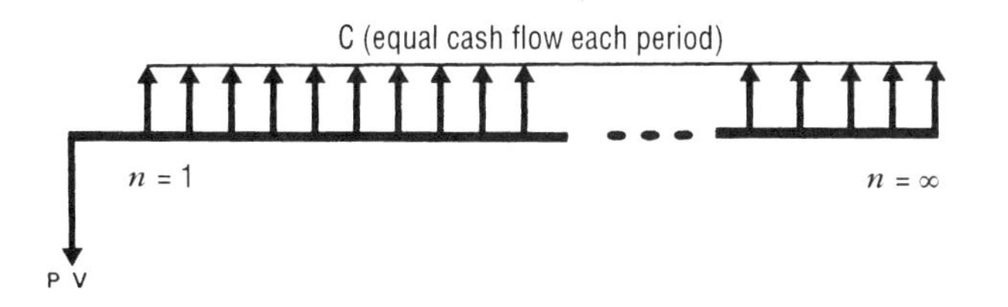

The value of a perpetuity can be found by this simple formula:

$$PV_{propetuity} = \frac{C}{r}$$

where: C = Cash flow per time period
r = Discount rate of return (per time period)

A *growing perpetuity* is a cash flow stream that grows at a constant rate. The formula to value a growing perpetuity expands the basic perpetuity formula to include a growth rate for the cash flow stream:

$$PV = \frac{C_1}{r-g}$$

where: C_1 = Initial cash flow
r = Discount rate of return (per time period)
g = Growth rate of cash flows (per time period)

Example. An existing oil well has been producing an average of 10,000 barrels per year for several years. Recent estimates by geologists show that this well's oil field is not large enough to justify additional drilling but can maintain the existing well's production rate for hundreds of years to come. Assume that the recent price of oil has averaged $20 per barrel oil, prices are historically estimated to grow 3% a year forever, and your discount rate is 10%. Also assume that maintenance costs are insignificant, and the land has no value other than for the oil. What is the current monetary value of this oil field and well, if the owner were to sell it?

Solution:

$$PV = \left(\frac{10{,}000 \times 20}{.10 - .03}\right) = \$2{,}857{,}143$$

The value of the oil field and well should be $2,857,143 based upon these assumptions.

Future value of a periodic series of investments

Both individual investors and companies often set aside a fixed quantity of money every month, quarter, or year, for some specific purpose in the future. This quantity of money then grows with compound interest until it is withdrawn and used at some future date. A cash flow diagram for a periodic investment is shown.

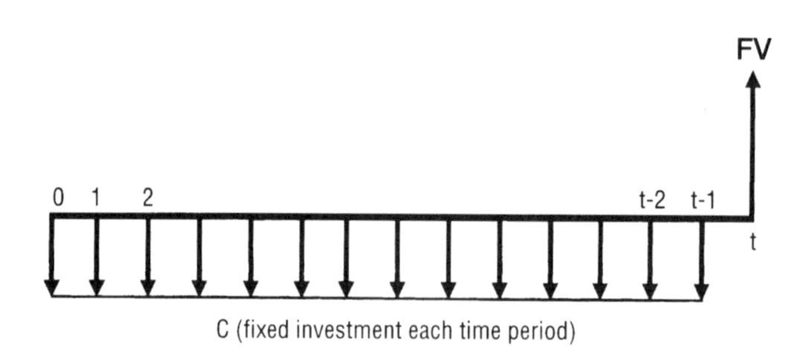

It can easily be calculated how much these periodic, fixed investments each time period will accumulate with compounding interest to be in the future. In other words, we want to determine the future value (FV), which can be calculated using the following formula:

$$FV = \frac{C}{r}\left[(1 + r)^t - 1\right]$$

where:

C = Cash flow invested each time period
r = Fractional compound interest rate (each time period)
t = Number of time periods (length of investment)

Example. A self-employed contract engineer has no company-paid retirement plan, so he decides to invest $2,000 per year in his own individual retirement account (IRA), which will allow the money to accumulate tax-free until it is withdrawn in the future. If he invests this amount every year for 25 years, and the investment earns 14% a year, how much will the investment be worth in 25 years?

Solution:

$$FV = \frac{\$2{,}000}{.14}\left[(1+.14)^{25} - 1\right] = \$363{,}742$$

The periodic investment of $2,000 each year over 25 years will grow to $363,742 if a 14% rate of return is achieved. This example illustrates the "power of compounding." A total of only $50,000 was invested over the 25-year period, yet the future value grew to more than seven times this amount.

Annuities, loans, and leases

An *annuity* is an asset that pays a constant periodic sum of money for a fixed length of time. Bank loans, mortgages, and many leasing contracts are similar to an annuity in that they usually require the borrower to pay back the lender a constant, periodic sum of money (principal and interest) for a fixed length of time. Loans, mortgages, and leases can therefore be considered "reverse" annuities, as shown by the cash flow diagrams.

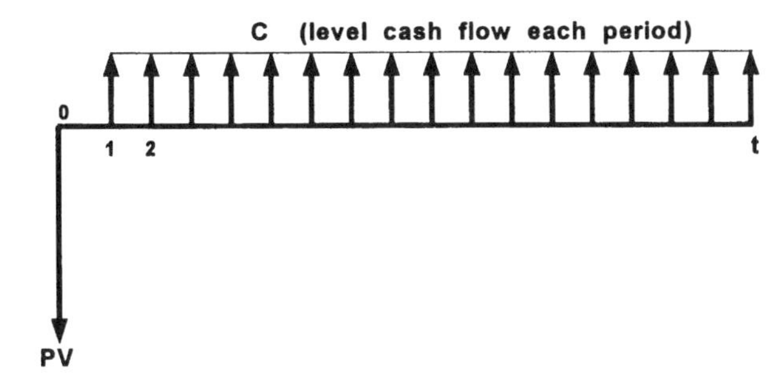

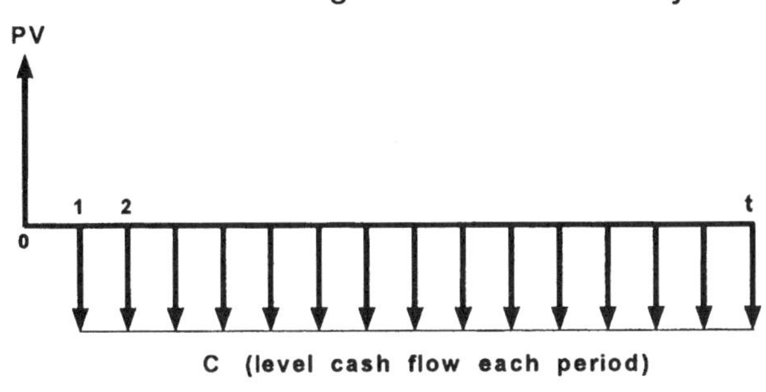

The relationship between the value (PV) of the annuity/loan/lease, its periodic cash flows, and the interest or discount rate is:

$$PV = \frac{C}{r}\left[1 - \frac{1}{(1+r)^t}\right]$$

where: C = Cash flow received/paid each time period
r = Interest or discount rate (per time period)
t = Number of time periods (time length of the annuity)

This formula may be used to determine the PV of the annuity, or conversely, to determine the periodic payouts or payments when the PV is known. It should be noted that the future value (FV) at time "t" will be zero. In other words, all the principal of the annuity will have been expended, or all the principal of the loan or lease will have been paid off, at the final time period. Financial calculators are also pre-programmed with this formula and can simplify and expedite annuity, loan, and other cash flow problem calculations.

Example. Oil pipeline equipment is to be purchased and installed on a working pipeline in a remote location. This new equipment will have an operational life of 5 years and will require routine maintenance once per month at a fixed cost of $100. How much would the oil company have to deposit now with a bank at the remote location to set up an annuity to pay the $100 a month maintenance expense for 5 years to a local contractor? Assume annual interest rates are 9%.

Solution:

$$PV = \frac{100}{(.09/12)}\left[1 - \frac{1}{(1+.09/12)^{60}}\right] = \$4,817.34$$

Example. Apex Corp. borrows $200,000 to finance the purchase of tools for a new line of gears it's producing. If the terms of the loan are annual payments over 4 years at 14% annual interest, *how much are the payments*?

$$\$200,000 = \frac{C}{.14}\left[1 - \frac{1}{(1+.14)^4}\right]$$

Solution:

C = $68,640.96 per year

Gradients (payouts/payments with constant growth rates)

A *gradient* is similar to an annuity, but its periodic payouts increase at a constant rate of growth, as illustrated by its cash flow diagram. A gradient is also similar to a growing perpetuity, except that it has a fixed time length. The cash flow diagram for a gradient loan or lease is simply the inverse of this cash flow diagram, since the loan/lease payments will be negative from the borrower's point of view. The PV of a gradient may be found by:

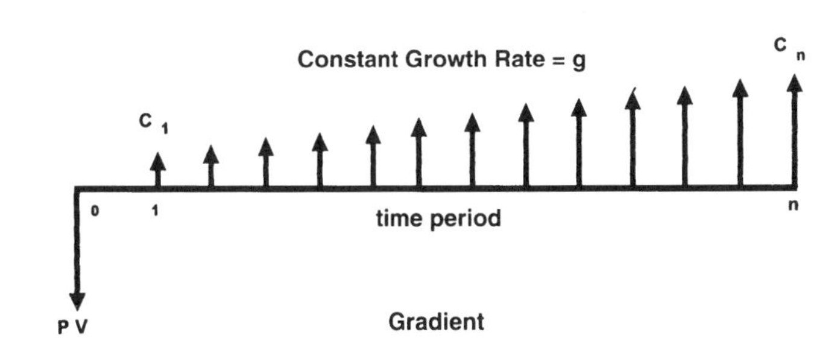

$$PV = \frac{C_1(1 - x^n)}{(1+r)(1-x)} \quad \text{when} \quad g \neq r$$

or

$$PV = \frac{C_1 n}{(1 + r)} \qquad \text{when} \quad g = r$$

where: C_1 = Initial cash flow at the end of first time period
R = Rate of return (discount rate)
n = Number of time periods
x = (1 + g)/(1 + r)
g = Fractional growth rate of cash flow period

Analyzing complex investments and cash flow problems

No matter how complex a financial project or investment is, the cash flows can always be diagrammed and split apart into separate cash flows if necessary. Using the discounted cash flow (DCF) formula, each cash flow can be discounted individually back to time zero to obtain its present value (PV), and then summed to determine the overall valuation of the project or investment. It may also be possible to simplify things if you can identify a series of cash flows in the complex investment. Cash flow series can be discounted using either annuity or perpetuity formulas, and then added together with the PV of any singular cash flow.

When analyzing any complex cash flow problem, it is important to remember the basic time value of money (TVM) concept (i.e., *the value of money changes with time*). Therefore, if you wish to add or compare any two cash flows, they both must be from the same time period. Cash flows in the future must be discounted to their present value (PV) before they can be added or compared with current cash flows. Alternatively, current cash flows can be compounded into the future (FV) and then added or compared with future cash flows.

Example. An automotive components company has been asked by a large automobile manufacturing company to increase its production rate of components for one particular car model. Sales of this car model have been much higher than expected. In order to increase its production rate, the components company will have to purchase additional specialized machinery, but this machinery will easily fit into its existing facility. The automobile manufacturer has offered to sign a contract to purchase all the additional output that can be produced for 4 years, which will increase annual net revenues by $300,000 without any additional administrative overhead or personnel expense. The added machinery will cost $450,000 and will require maintenance every 6 months at an estimated cost of $20,000. The machinery will be sold at the end of the 4-year period, with an estimated salvage value of $80,000. *If the company requires a minimum rate of return of 20% on all investments, should it sign the contract, purchase the machinery, and increase its production rate?* (Depreciation and tax effects are not considered in this example.)

Step 1: From the point of view of the automotive components company, the cash flow diagram will be:

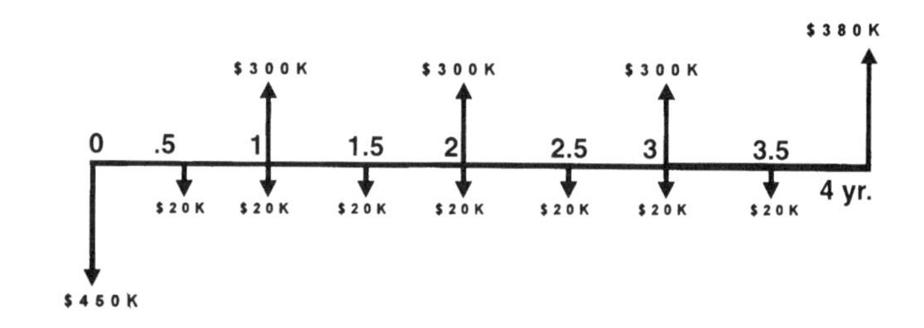

Step 2: Separate the overall cash flow diagram into individual components and series, and then determine the PV of each future cash flow or series of flows by discounting at the required annual rate of return of 20%:

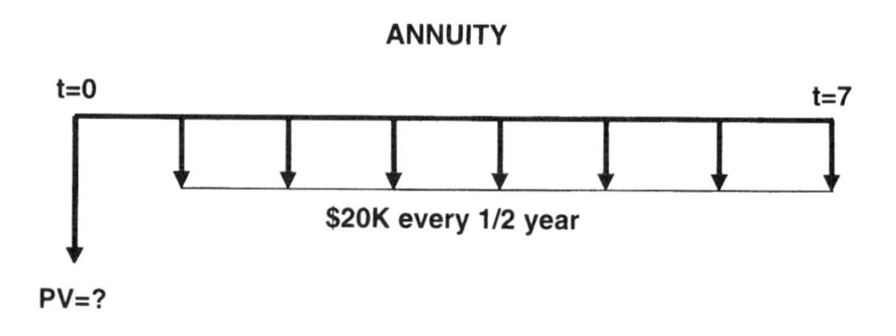

$$PV = \frac{\$20{,}000}{.20/2}\left[1 - \frac{1}{(1+.20/2)^7}\right] = \$97{,}368 \quad \text{(negative)}$$

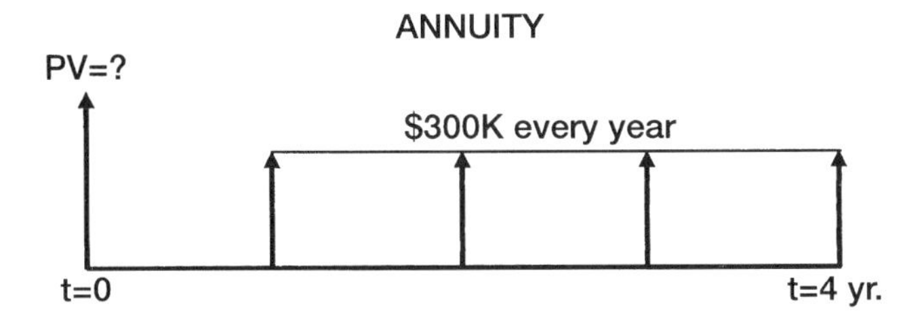

$$PV = \frac{\$30,000}{.2}\left[1 - \frac{1}{(1+.2)^4}\right] = \$776,620 \quad \text{(positive)}$$

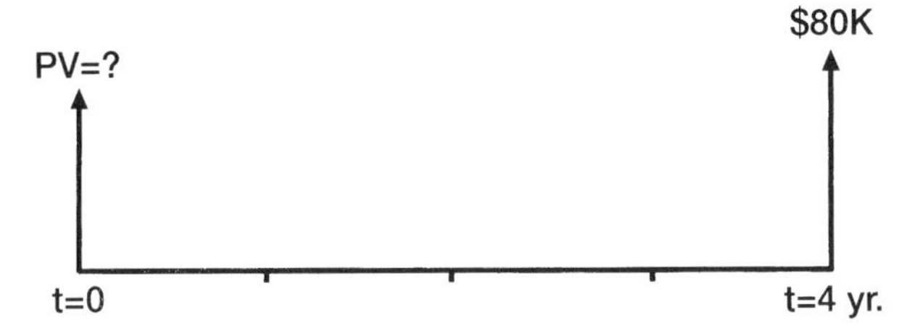

$$PV = \frac{\$80,000}{(1+.2)^4} = \$38,580 \quad \text{(positive)}$$

Step 3: Sum the PV of all cash flows to obtain the project's overall PV from all *future* cash flows:

$$PV = \$776,620 + \$38,580 - \$97,368$$
$$= \$717,832 \text{ (positive)}$$

Step 4: Compare the PV of the future cash flows with the project's *initial* cash flow (initial investment):

The $717,832 PV of the project's future cash inflows (positive) is clearly greater than the $450,000 (negative) initial investment. Thus, this project has a positive *net present value* (NPV), because the PV of the future cash flows are greater than the initial investment. A positive NPV means that the rate of return on the investment is greater than the rate (20%) used to discount the cash flows.

Conclusion: The company should buy the additional machinery and increase its production rate.

DECISION AND EVALUATION CRITERIA FOR INVESTMENTS AND FINANCIAL PROJECTS

How should you decide whether or not to make an investment or go forward with a financial project? If you also have a number of alternative investment options, how do you decide which is the best alternative? How do you evaluate an investment's performance some time period after the investment has been made and the project is up-and-running? There are a number of analysis techniques that can help with these questions. Four separate methods are presented here, although each method varies in its effectiveness and complexity. There are also additional methods and techniques that can be found in other engineering economics, finance, and accounting textbooks. However, most modern books and courses in these fields of study recommend the *net present value* (NPV) method as the most effective and accurate technique of evaluating potential investments and financial projects. Accounting measures such as *rate of return* (ROR) on an investment or *return on equity* (ROE) are useful to evaluate a project's financial results over a specific time period or to compare results against competitors.

Payback method

The payback method is a simple technique that determines the number of years before the cash flow from an investment or project "pays back" the initial investment. Obviously, the shorter the length of time it takes a project to pay off the initial investment, the more lucrative the project.

Example. A company wants to add a new product line and is evaluating two types of manufacturing equipment to produce this new line. The first equipment type costs $700,000 and can produce enough product to result in an estimated $225,000 in additional after-tax cash flow per year.

The second type of equipment is more expensive at \$900,000, but has a higher output that can bring an estimated \$420,000 per year in after-tax cash flows to the company. Which type of equipment should the company buy?

Solution:

Equipment Type 1
Payback Period = \$700,000/\$225,000 per year
= 3.11 years
Equipment Type 2
Payback Period = \$900,000/\$420,000 per year
= 2.14 years

Based upon this payback analysis, equipment type 2 should be purchased.

The advantage of the payback method is its simplicity, as shown in the example. However, there are several disadvantages that need to be mentioned. First, the payback method does not evaluate any cash flows that occur after the payback date, and therefore ignores long-term results and salvage values. Secondly, the payback method does not consider the time value of money (TVM) but assumes that each dollar of cash flow received in the second year and beyond is equal to that received the first year. Thirdly, the payback method has no way to take into account and evaluate any risk factors involved with the project.

Accounting rate of return (ROR) method

The *accounting rate of return (ROR)* for an existing company or project can be easily calculated directly from the company's or project's quarterly or annual financial accounting statements. If a potential (rather than existing) project is to be evaluated, financial results will have to be projected. Since ROR is an accounting measure, it is calculated for the same time period as the company's or project's income statement. The ROR on an investment is simply the net income generated during this time period, divided by the book value of the investment at the start of the time period.

$$\text{ROR} = \frac{\text{Net Income}}{\text{Net Investment (book value)}}$$

Net income is an accounting term defined as "revenues minus expenses plus gains minus losses" (i.e., the "bottom line" profit or loss on the income statement). Net investment is the "book value" of the investment, which is the original investment amount (or purchase price) less any depreciation in value of the investment.

Example. Company XYZ invested \$500,000 in plant and equipment 4 years ago. At the end of year 3, the company's financial statements show that the plant and equipment's value had depreciated \$100,000 and had a listed book value of \$400,000. If company XYZ has a net income of \$60,000 during its fourth year of operation, what is its ROR for this time period?

Solution:

$$\text{ROR} = \frac{60{,}000}{400{,}000} = 0.15\ (15\%)$$

Like the payback method, the advantage of the accounting ROR is that it is a simple calculation. It is especially simple for an existing company or project when the accounting statements are already completed and don't have to be projected. One disadvantage of ROR is that it doesn't consider TVM when it is used for more than one time period. It also uses accounting incomes rather than cash flows, which are not the same (see Accounting Fundamentals in this chapter). Another disadvantage is that ROR is a very subjective method, because the ROR is likely to vary considerably each time period due to revenue and expense fluctuations, accumulating depreciation, and declining book values. (Some textbooks recommend averaging ROR over a number of time periods to compensate for fluctuations and declining book values, but an "averaged" ROR still does not consider TVM.)

In general, accounting measurements such as ROR are best used for evaluating the performance of an *existing* company over a specific annual or quarterly time period or for comparing one company's performance against another. Net present value (NPV) and internal rate of return (IRR) are superior methods for evaluating *potential* investments and financial projects and in making the final decision whether to go ahead with them.

Internal rate of return (IRR) method

Internal rate of return (*IRR*) is an investment profitability measure that is closely related to net present value (NPV). The IRR of an investment is that rate of return which, when used to discount an investment's future cash flows, makes the NPV of an investment equal zero. In other words, when the future cash flows of an investment or project are discounted using the IRR, their PV will exactly equal the initial investment amount. Therefore, the IRR is an extremely useful quantity to know when you are evaluating a potential investment project. The IRR tells you the exact rate of return that will be earned on the original investment (or overall project with any additional investments) if the projected future cash flows occur. Similarly, when analyzing past investments, the IRR tells you the exact rate of return that was earned on the overall investment. The IRR decision rule for whether or not to go ahead with any potential investment or project being considered is simple: *If the IRR exceeds your opportunity cost of capital (rate of return that can be earned elsewhere), you should accept the project.*

The following example will make it clear how the IRR of an investment is calculated. It is highly recommended that a cash flow diagram be drawn first before trying to calculate the IRR. Particular attention should be paid to make sure cash inflows are drawn as positive (from the investor's point of view) and cash outflows are drawn as negative. It is also recommended that a financial calculator (or computer) be used to save time; otherwise, an iteration procedure will have to be used to solve for the IRR.

Example. Acme Tool and Die Company is considering a project that will require the purchase of special machinery to produce precision molds for a major customer, Gigantic Motors. The project will last only 4 years, after which Gigantic Motors plans to manufacture its own precision molds. The initial investment for the special machinery is \$450,000, and the machinery will also require a partial overhaul in 3 years at an estimated cost of \$70,000. Projected after-tax cash flows resulting from the project at the end of each of the 4 years are \$140,000, \$165,000, \$185,000, and \$150,000, respectively. These projections are based upon a purchase agreement that Gigantic is prepared to sign, which outlines the quantities of molds it will buy. Expected salvage value from the machinery at the end of the 4-year period is \$80,000. Should Acme proceed with this project, if its opportunity cost of capital is 23%?

Step 1: Cash flow diagram:

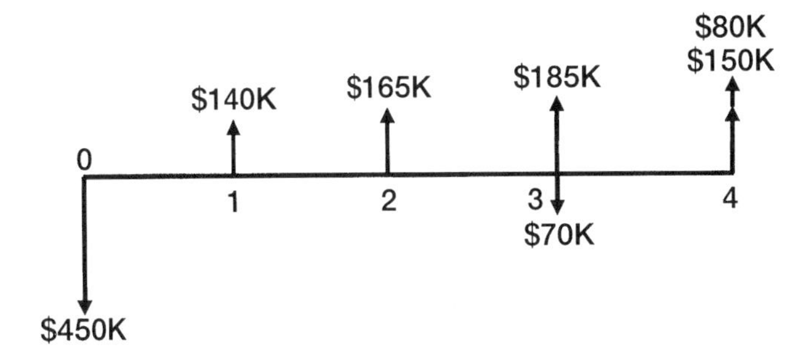

Step 2: Write an equation that sets the original investment equal to the PV of future cash flows, using the discounted cash flow formula. Solve for r, which is the IRR. (This is a simple process with a financial calculator or computer but otherwise requires iteration with any other calculator.)

$$450{,}000 = \frac{140{,}000}{(1+r)} + \frac{165{,}000}{(1+r)^2} + \frac{(185{,}000 - 70{,}000)}{(1+r)^3} + \frac{(150{,}000 + 80{,}000)}{(1+r)^4}$$

Solving for r (IRR):

IRR = 0.1537 (15.37%)

Step 3: Apply decision rule to results: Since the IRR for this project is 15.37%, and Acme's required rate of return (opportunity cost of capital) is 23%, Acme should not proceed with this project.

IRR has a number of distinct advantages (as does NPV) over both the payback and ROR methods as a decision-making tool for evaluating potential investments. The first advantage is that the IRR calculation takes into account TVM. Secondly, IRR considers all cash flows throughout the life of the project (including salvage values), rather than looking at only one time period's results like accounting ROR does, or the years until the initial investment is paid off like the payback method. Therefore, IRR is an objective criterion, rather than a subjective criterion, for making decisions about investment projects. Thirdly, IRR uses actual cash flows rather than accounting incomes like the ROR method. Finally, since IRR is calculated for the entire life of the project, it provides a basis for evaluating whether the overall risk of the project is justified by the IRR, since the IRR may be compared with the IRR of projects of similar risk and the opportunity cost of capital.

The IRR method has several disadvantages compared to the NPV method, though only one disadvantage is mentioned here for purposes of brevity. Further information about potential problems with the IRR method (compared to NPV) may be obtained from most finance textbooks. One major problem with IRR is the possibility of obtaining multiple rates of return (multiple "roots") when solving for the IRR of an investment. This can occur in the unusual case where cash flows change erratically from positive to negative (in large quantities or for sustained time periods) *more than once* during the life of the investment. The graph in the following example illustrates how multiple IRR roots can occur for investments with these types of cash flows. The second graph is for comparison purposes and typifies the IRR calculation and graph for most investments. (Normally NPV declines with increasing discount rate, thus giving only one IRR "root".) When multiple "roots" are obtained using the IRR method, it is best to switch to the NPV method and calculate NPV by discounting the cash flows with the opportunity cost of capital.

Example. A project under consideration requires an initial investment/cash flow of $3,000 (negative) and then has cash flows of $26,000 (positive) and $30,000 (negative) at the end of the following 2 years. A graph of NPV versus discount rate shows that two different IRR "roots" exist. (IRR is the point on the curve that intersects the horizontal axis at NPV = 0.) The second graph is from the previous example, showing that most "normal" investments have only one IRR root.

The IRR for this example is both 37.1% and 629.6%, as NPV equals zero at these two discount rates.

Illustration of Multiple I.R.R.s
(This Example)

NPV ($)
3000
2500
2000
1500
1000
500
0
-500
-1000
-1500
-2000
-2500
0
200
400
600
800
Discount Rate (%)

Illustration of Single I.R.R. Root
(Previous Example - Acme Tool & Die)

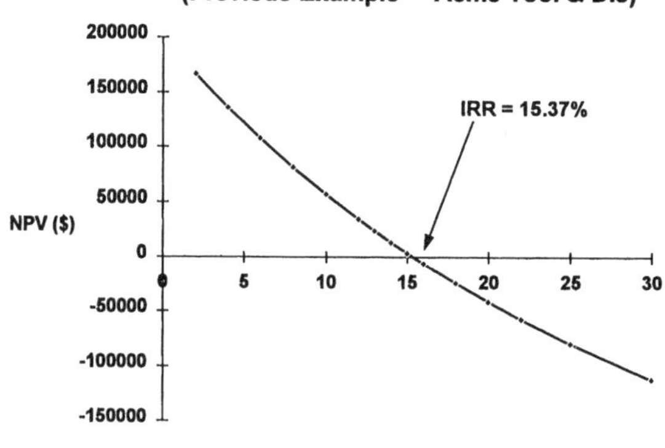

Discount Rate (%)

Net present value (NPV) method

Net present value (*NPV*), as its name suggests, calculates the net amount that the discounted cash flows of an investment exceed the initial investment. Using the discounted cash flow (DCF) formula, the future cash flows are discounted by the rate of return offered by comparable investment alternatives (i.e., the opportunity cost of capital) and then summed and added to the initial investment amount (which is usually negative).

$$NPV = C_0 + \sum_{i=1}^{n} \frac{C_i}{(1+r)^i}$$

where:

C_0 = Initial cash flow (negative for a cash "outflow")
C_i = Cash flow in time period i
n = Number of time periods
r = Opportunity cost of capital/discount rate

Though similar to the IRR method, NPV does not calculate an investment's exact rate of return but instead calculates the exact dollar amount that an investment exceeds, or fails to meet, the *expected* rate of return. In other words, if an investment provides a rate of return exactly equal to the opportunity cost of capital, then the NPV of this investment will be zero, because the discounted future cash flows equal the initial investment. Thus, NPV provides an excellent decision criterion for investments. An investment with a positive NPV should be accepted, since it provides a rate of return above the opportunity cost of capital. By the same reasoning, an investment with a negative NPV should be rejected. NPV also does *not* suffer from any of the drawbacks of the payback, accounting ROR, or IRR methods. Because of this, NPV is the method most recommended by financial experts for making investment decisions. IRR is still useful to determine the exact rate of return for an investment, but NPV has none of the problems that IRR may have with unusual investments (such as the "multiple root" problem illustrated previously).

The following example shows how NPV is used to decide between investment alternatives.

Project	C_0	C_1	C_2	C_3	C_4
A	– $125,000	$31,000	$43,000	$48,000	$51,000
B	– $210,000	$71,000	$74,000	$76,000	$79,000

Example. A company must choose between two alternative manufacturing projects, which require different amounts of capital investment and have different cash flow patterns. The company's opportunity cost of capital for projects of similar risk is 15%. Which project should it choose?

Solving with the NPV formula:

NPV(A) = – $4,809
NPV(B) = $2,833

Conclusion: The company should choose project B. Despite a higher initial cost than A, Project B earns $2,833 more than its required rate of return of 15%. Incidentally, IRR also could have been used, since future cash flows are all positive, and therefore no multiple roots exist. Project A's IRR is 13.23%; Project B's IRR is 15.65%.

SENSITIVITY ANALYSIS

The previous sections of this chapter show how to analyze investments and financial projects. These analysis methods can help quantify whether or not an investment is worthwhile and also help choose between alternative investments. However, no matter what method is used to analyze a financial project, that method and the decision about the project will ultimately rely upon the *projections* of the project's future cash flows. It is, therefore, extremely important that the future cash flows of any financial project or investment under consideration be forecast as accurately and realistically as possible. Unfortunately, no matter how precise and realistic you try to be with your projections of cash flows, the future is never certain, and actual results may vary from what is projected. Thus, it is often useful to perform a *sensitivity analysis* on the investment or financial project.

Sensitivity analysis is a procedure used to describe analytically the effects of uncertainty on one or more of the parameters involved in the analysis of a financial project. The objective of a sensitivity analysis is to provide the decision maker with quantitative information about what financial effects will be caused by variations from what was projected. A sensitivity analysis, once performed, may influence the decision about the financial project, or at least show the decision maker which parameter is the most critical to the financial success of the project. The NPV method lends itself nicely to sensitivity analysis, since the discount rate is already fixed at the opportunity cost of capital. One parameter at a time can be varied in steps (while the other parameters are held constant) to see how much effect this variance has on NPV. A financial calculator or computer spreadsheet program will greatly expedite the multiple, repetitive calculations involved in this analysis. Plotting the results graphically will help show the sensitivity of NPV to changes in each variable, as illustrated in the following example.

Example. A-1 Engineering Co. is contemplating a new project to manufacture sheet metal parts for the aircraft industry. Analysis of this project shows that a $100,000 investment will be required up front to purchase additional machinery. The project is expected to run for 4 years, resulting in estimated after-tax cash flows of $30,000 per year. Salvage value of the machinery at the end of the 4 years is estimated at $32,000. A-1's opportunity cost of capital is 15%. Perform a sensitivity analysis to determine how variations in these estimates will affect the NPV of the project.

Step 1: Cash flow diagram:

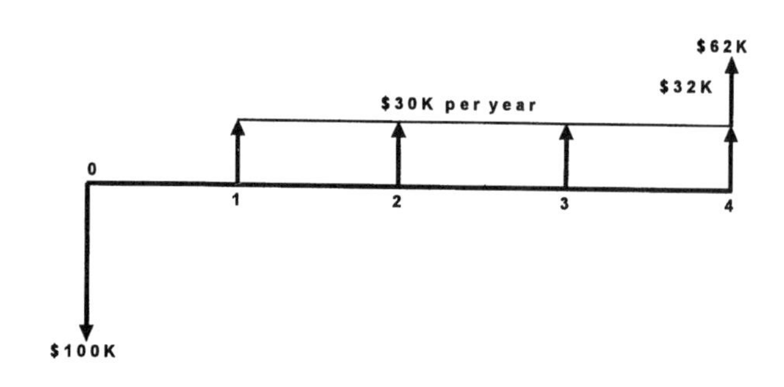

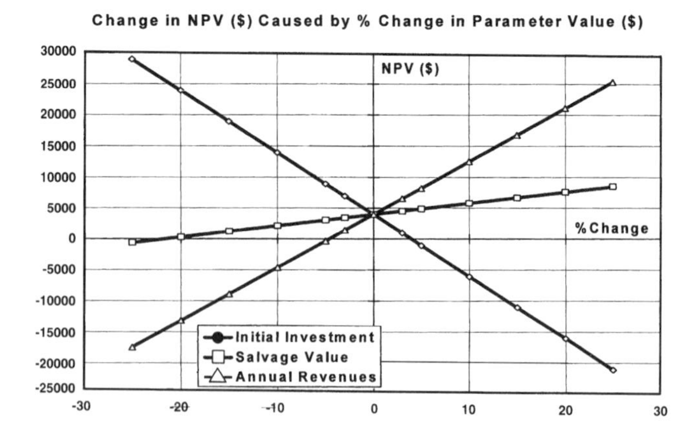

Step 2: Calculate baseline NPV:

$$NPV = -100{,}000 + \frac{30{,}000}{(1+.15)^1} + \frac{30{,}000}{(1+.15)^2} + \frac{30{,}000}{(1+.15)^3} + \frac{62{,}000}{(1+.15)^4} = \$3{,}945.45$$

Step 3: Change salvage value, annual cash flow revenues, and initial investment amount in steps. Calculate NPV for each step change.

Step 4: Plot values to illustrate the NPV sensitivity to changes in each variable.

Conclusion: The graph shows that this project's NPV is equally sensitive to changes in the required initial investment amount (inversely related) and the annual revenues (after-tax cash flow) from the project. This is easily seen from the slope of the lines in the graph. The graph also shows that NPV is not affected nearly as much by changes in salvage value, primarily because the revenues from salvage value occur at the end of the fourth year and are substantially discounted at the 15% cost of capital. In conclusion, this sensitivity analysis shows A-1 Machinery that it cannot accept anything more than a 5% increase in the initial cost of the machinery or a 5% decrease in the annual cash flows. If these parameters change more than this amount, NPV will be less than zero and the project will not be profitable. Salvage value, however, is not nearly as critical, as it will take more than a 20% decrease from its estimated value before the project is unprofitable.

DECISION TREE ANALYSIS OF INVESTMENTS AND FINANCIAL PROJECTS

Financial managers often use *decision trees* to help with the analysis of large projects involving sequential decisions and variable outcomes over time. Decision trees are useful to managers because they graphically portray a large, complicated problem in terms of a series of smaller problems and decision branches. Decision trees reduce abstract thinking about a project by producing a logical diagram that shows the decision options available and the corresponding results of choosing each option. Once all the possible outcomes of a project are diagrammed in a decision tree, objective analysis and decisions about the project can be made.

Decision trees also allow *probability theory* to be used, in conjunction with NPV, to analyze financial projects and investments. It is often possible to estimate the probability of success or failure for a new venture or project, based upon either historical data or business experience. If the new venture is a success, then the projected cash flows will be considerably higher than what they will be if the project is not successful. The projected cash flows for both of these possible outcomes (success or failure) can then be multiplied by their probability estimates and added to determine an *expected* outcome for the project. This expected outcome is the *average* payoff that can be expected (for multiple projects of the same type) based upon these probability estimates. Of course, the *actual* payoff for a single financial project will either be one or the other (success or failure), and not this average payoff. However, this average expected payoff is a useful number for decision making, since it incorporates the probability of both outcomes.

If there is a substantial time lapse between the time the decision is made and the time of the payoff, then the time value of money must also be considered. The NPV method can be used to discount the expected payoff to its present value and to see if the payoff exceeds the initial investment amount. The standard NPV decision rule applies: *Accept only projects that have positive NPVs*.

The following example illustrates the use of probability estimates (and NPV) in a simple decision tree:

Example. A ticket scalper is considering an investment of \$10,000 to purchase tickets to the finals of a major outdoor tennis tournament. He must order the tickets 1 year in advance to get choice seats at list price. He plans to resell the tickets at the gate on the day of the finals, which is on a Sunday. Past experience tells him that he can sell the tickets and double his money, but only if the weather on the day of the event is good. If the weather is bad, he knows from past experience that he will only be able to sell the tickets at an average of 70% of his purchase price, even if the event is canceled and rescheduled for the following day (Monday). Historical data from the *Farmer's Almanac* shows that there is a 20% probability of rain at this time of year. His opportunity cost of capital is 15%. Should the ticket scalper accept this financial project and purchase the tickets?

Step 1: Construct the decision tree (very simple in this example).

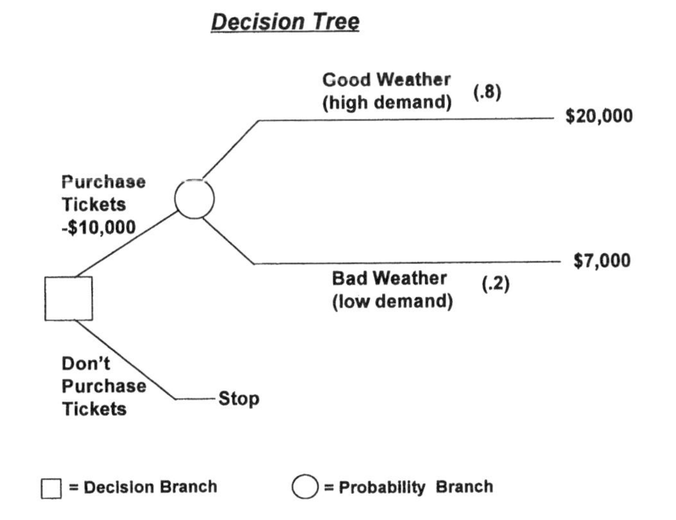

Step 2: Calculate the expected payoff from purchasing the tickets:

Expected payoff = (probability of high demand) × (payoff with high demand) + (probability of low demand) × (payoff with low demand)
= (.8 × 20,000) + (.2 × 7,000) = \$17,400

Step 3: Calculate NPV of expected payoff:

$$\text{NPV} = -\ 10{,}000 + \frac{\$17{,}400}{(1+1.5)^1} = \$5{,}130$$

Remember that the cash flow from selling the tickets occurs 1 year after the purchase. NPV is used to discount the cash flow to its present value and see if it exceeds the initial investment.

Conclusion: The ticket scalper should buy the tickets, because this project has a positive NPV.

The preceding example was relatively simple. Most financial projects in engineering and manufacturing applications are considerably more complex than this example; therefore, it is useful to have a procedure for setting up a decision tree and analyzing a complex project.

Here is a procedure that can be used:

1. Identify the decisions to be made and alternatives available throughout the expected life of the project.
2. Draw the decision tree, with branches drawn first for each alternative at every decision point in the project. Secondly, draw "probability branches" for each possible outcome from each decision alternative.
3. Estimate the probability of each possible outcome. (Probabilities at each "probability branch" should add up to 100%.)
4. Estimate the cash flow (payoff) for each possible outcome.
5. Analyze the alternatives, starting with the most distant decision point and working backwards. Remember the time value of money and, if needed, use the NPV method to discount the expected cash flows to their present values. Determine if the PV of each alternative's expected payoff exceeds the initial investment (i.e., positive NPV). Choose the alternative with the highest NPV.

The following example illustrates the use of this procedure.

Example. Space-Age Products believes there is considerable demand in the marketplace for an electric version of its barbecue grill, which is the company's main product. Currently, all major brands of grills produced by Space-Age and its competitors are either propane gas grills or charcoal grills. Recent innovations by Space-Age have allowed the development of an electric grill that can cook and sear meat as well as any gas or charcoal grill, but without the problems of an open flame or the hassle of purchasing propane gas bottles or bags of charcoal.

Based upon test marketing, Space-Age believes the probability of the demand for its new grill being high in its first year of production will be 60%, with a 40% probability of low demand. If demand is high the first year, Space-Age estimates the probability of high demand in subsequent years to be 80%. If the demand the first year is low, it estimates the probability of high demand in subsequent years at only 30%.

Space-Age cannot decide whether to set up a large production line to produce 50,000 grills a month or to take a more conservative approach with a small production line that can produce 25,000 grills a month. If demand is high, Space-Age estimates it should be able to sell all the grills the large production line can make. However, the large production line puts considerably more investment capital at risk, since it costs $350,000 versus only $200,000 for the small line. The smaller production line can be expanded with a second production line after the first year (if demand is high) to double the production rate, at a cost of an additional $200,000.

Should Space-Age initially invest in a large or small production line or none at all? The opportunity cost of capital for Space-Age is 15%, and cash flows it projects for each outcome are shown in the decision tree. Cash flows shown at the end of year 2 are the PV of the cash flows of that and all subsequent years.

Steps 1–4 of procedure: (Construct decision tree and label with probabilities and cash flow projections).

Step 5: Analysis of alternatives:

(a) Start the analysis by working backwards from the R/H side of the decision tree. The only decision that Space-Age needs to make at the start of year 2 is deciding

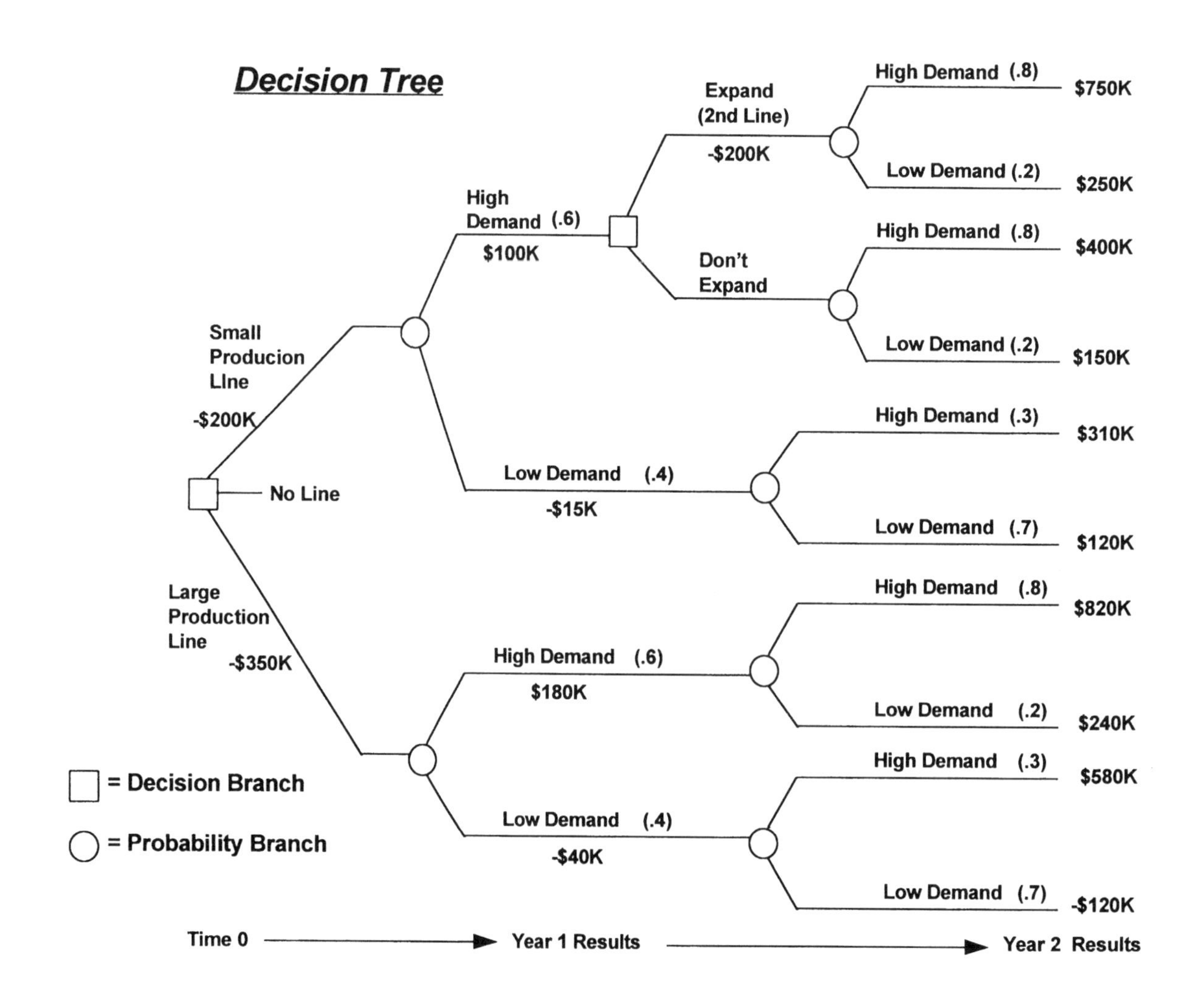

whether or not to expand and install a second production line if demand is high the first year (if the initial decision was for a small production line).

Expected payoff of expanding with a second small production line (at the end of year 2):

$$= (.8 \times \$750K) + (.2 \times \$250K) = \$650{,}000$$

Expected payoff without the expansion (at the end of year 2):

$$= (.8 \times \$400K) + (.2 \times \$150K) = \$350{,}000$$

To decide if expansion is worthwhile, the NPVs of each alternative (at end of year 1) must be compared:

$$\text{NPV } (\textit{with 2nd line}) = -\$200K + \frac{\$650K}{(1+.15)^1} = \$365{,}217$$

$$\text{NPV } (\textit{without 2nd line}) = \$0 + \frac{\$350K}{(1+.15)^1} = \$304{,}348$$

These NPV amounts show that Space-Age *should clearly expand with a second small production line* (if its initial decision at time 0 was to open a small production line and the demand was high the first year). Now that this decision is made, the $365,217 NPV of this expansion decision will be used to work backwards in time and calculate the overall NPV of choosing the small production line.

(b) The expected payoff (at the end of year 2) if the demand is low the first year with the small production line:

$$= (.3 \times \$310K) + (.7 \times \$120K) = \$177{,}000$$

This expected payoff must be discounted so that it can be added to the projected cash flow of year 1:

$$\text{PV} = \frac{\$177{,}000}{(1+.15)} = \$153{,}913$$

(c) The expected payoff (at the end of year 1) for the small production line can now be determined:

$$= [.6 \times (\$100{,}000 + \$365{,}217)] + [.4 \times (-\$15{,}000 + \$153{,}913)] = \$334{,}695$$

Notice that the cash flow for year 1 is the sum of the first year's projected cash flow plus the discounted expected payoff from year 2.

(d) Calculate the NPV (at time 0) of the small production line option:

$$\text{NPV} = \$200{,}000 + \frac{\$334{,}695}{(1+.15)} = \$91{,}039$$

(e) Calculate the NPV (at time 0) of the large production line option. (Since there are no decisions at the end of year 1, there is no need to determine intermediate NPVs, and the probabilities from each branch of each year can be multiplied through in the NPV calculation as follows:

$$\text{NPV} = -\$350 + \frac{(.6)(\$180K) + (.4)(\$40K)}{(1+.15)^1} + \frac{(.6)[(.8)(\$820K) + (.2)(\$240K)] \quad (.4)[(.3)(\$580K) + (.7)(-\$120K)]}{(1+.15)^2}$$

Solving: NPV = $76,616

(f) *Conclusion*: Both the NPVs of the large and small production lines are positive, and thus both production lines would be profitable based upon the expected payoffs. However, the NPV of the small production line is higher than the NPV of the large production line ($91,039 vs. $76,616). Therefore, Space-Age's decision should be to set up a small production line, and then expand with a second line if the demand for electric grills is high in the first year of production.

ACCOUNTING FUNDAMENTALS

Regardless of whether a person is managing the finances of a small engineering project or running a large corporation, it is important to have a basic understanding of accounting. Likewise, it is also impossible to study engineering economics without overlapping into the field of accounting. This section, therefore, provides a brief overview of accounting fundamentals and terminology.

Accounting is the process of measuring, recording, and communicating information about the economic activities of accounting entities (financial projects, partnerships, companies, corporations, etc.). Public corporations, and most other types of companies, issue financial statements from their accounting data on a periodic basis (usually quarterly and yearly). These financial statements are necessary because they provide information to outside investors and creditors and assist a company's own management with decision making. Four primary types of financial statements are normally issued:

- Balance sheet
- Income statement
- Statement of changes in retained earnings
- Statement of cash flows

The *balance sheet* provides information (at one particular instant in time) about financial resources of a company and claims against those resources by creditors and owners. The balance sheet's format reflects the *fundamental accounting equation*:

Assets = Liabilities + Owner's Equity

Assets are items of monetary value that a company possesses, such as plant, land, equipment, cash, accounts receivable, and inventory. *Liabilities* are what the company owes its creditors, or conversely, the claims the creditors have against the company's assets if the company were to go bankrupt. Typical liabilities include outstanding loan balances, accounts payable, and income tax payable. *Owner's equity* is the monetary amount of the owner's claim against the company's assets. (Owner's equity is also referred to as stockholder's equity for corporations that issue and sell stock to raise capital.) Owner's equity comes from two sources: (1) the original cash investment the owner used to start the business and (2) the amount of profits made by the company that are not distributed (via *dividends*) back to the owners. These profits left in the company are called *retained earnings*.

There are numerous asset, liability, and owner's equity *accounts* that are used by accountants to keep track of a company's economic resources. These accounts are shown in the following "balance sheet" illustration, which is a typical balance sheet for a small manufacturing corporation.

Notice in this illustration that the balance sheet's total assets "balance" with the sum of liabilities and owner's equity, as required by the fundamental accounting equation. There are also a number of additional *revenue* and *expense* accounts that are maintained but not shown in the balance sheet. The balances of these revenue and expense accounts are closed out (zeroed) when the financial statements are prepared, and their amounts are carried over into the income statement. The *income statement* is the financial statement that reports the profit (or loss) of a company over a specific period of time (quarterly or annually). The income statement reflects another basic and fundamental accounting relationship:

Revenues − Expenses = Profit (or Loss)

A sample "income statement" for our typical small corporation is illustrated.

The income statement simply subtracts the company's expenses from its revenues to arrive at its *net income* for the period. Large companies and corporations usually separate their income statements into several sections, such as income from operations, income or loss from financial activities and discontinuing operations, and income or loss from extraordinary items and changes in accounting principles.

Companies frequently show an expense for depreciation on their income statements. *Depreciation* is an accounting method used to allocate the cost of asset "consumption" over the life of the asset. The value of all assets (building, equipment, patents, etc.) depreciate with time and use, and depreciation provides a method to recover investment capital by expensing the portion of each asset "consumed" during each income statement's time period. Depreciation is a *noncash expense*, since no cash flow actually occurs to pay for this expense. The *book value* of an asset is shown in the balance sheet and is simply the purchase price of the asset minus its accumulated depreciation. There are several methods used to calculate depreciation, but the easiest and most frequently used method is *straight-line depreciation*, which is calculated with the following formula:

$$\text{Depreciation Expense} = \frac{\text{Cost} - \text{Salvage Value}}{\text{Expected Life of Asset}}$$

Straight-line depreciation allocates the same amount of depreciation expense each year throughout the life of the asset. *Salvage value* (residual value) is the amount that a company can sell (or trade in) an asset for at the end of its useful life. "Accelerated" depreciation methods (such as sum-of-the-years'-digits and declining balance methods) are available and may sometimes be used to expense the cost of

XYZ Company
Balance Sheet
As of March 31, 1994

ASSETS			
Current Assets:			
Cash		$25,000.00	
Accounts Receivable		$268,000.00	
Inventory		$163,000.00	
Total Current Assets			$456,000.00
Property, Plant, & Equipment:			
Building	$600,000.00		
less Accumulated Depreciation	($250,000.00)	$350,000.00	
Equipment	$300,000.00		
less Accumulated Depreciation	($125,000.00)	$175,000.00	
Land		$100,000.00	
Total Property, Plant, & Equipment			$625,000.00
Total Assets			***$1,081,000.00***
Liabilities			
Current Liabilties:			
Accounts Payable	$275,000.00		
Salaries Payable	$195,000.00		
Interest Payable	$13,000.00		
Income Tax Payable	$36,000.00		
Total Current Liabilities		$519,000.00	
Long-Term Liabilities:			
Notes Payable		$260,000.00	
Total Liabilities			$779,000.00
Owner's (Stockholder's) Equity			
Capital Stock	$200,000.00		
Retained Earnings	$102,000.00		
Total Owner's Equity			$302,000.00
Total Liabilities & Owner's Equity			***$1,081,000.00***

an asset faster than straight-line depreciation. However, the use of accelerated depreciation methods is governed by the tax laws applicable to various types of assets. A recent accounting textbook and tax laws can be consulted for further information about the use of accelerated depreciation methods.

The profit (or loss) for the time period of the income statement is carried over as retained earnings and added to (or subtracted from) owner's equity, in the "statement of changes in retained earnings" illustration.

If the company paid out any dividends to its stockholders during this time period, that amount would be shown as a deduction to retained earnings in this financial statement. This new retained earnings amount is also reflected as the retained earnings account balance in the current balance sheet.

The last financial statement is the *statement of cash flows*, which summarizes the cash inflows and outflows of the company (or other accounting entity) for the same time period as the income statement. This statement's purpose is to show where the company has acquired cash (inflows) and to what activities its cash has been utilized (outflows) during this time period. The statement divides and classifies cash flows into three categories: (1) cash provided by or used by operating activities; (2) cash provided by or used by investing activities; and (3) cash provided by or used by financing

XYZ Company

Income Statement

For the Quarter Ended March 31, 1994

Sales Revenues		$2,450,000.00
Less: Cost of Goods sold		($1,375,000.00)
Gross Margin		$1,075,000.00
Less: Operating Expenses		
Selling expenses	$295,000.00	
Salaries expense	$526,000.00	
Insurance expense	$32,000.00	
Property Taxes	$27,000.00	
Depreciation, Equipment	$35,000.00	
Depreciation, Building	$70,000.00	($985,000.00)
Income from Operations		$90,000.00
Other Income		
Sale of Assets	$16,000.00	
Interest Revenue	$2,000.00	$18,000.00
		$108,000.00
Less: Interest Expense		($22,000.00)
Income before Taxes		$86,000.00
Less: Income Tax Expense		($34,400.00)
Net Income		*$51,600.00*

XYZ Company

Statement of Changes in Retained Earnings

For the Quarter Ended March 31, 1994

Retained Earnings, Dec. 31, 1993	$50,400.00
Net Income, Quarter 1, 1994	$51,600.00
Retained Earnings, March 31, 1994	$102,000.00

activities. Cash flow from operations is the cash generated (or lost) from the normal day-to-day operations of a company, such as producing and selling goods or services. Cash flow from investing activities includes selling or purchasing long-term assets and sales or purchases of investments. Cash flow from financing activities covers the issuance of long-term debt or stock to acquire capital, payment of dividends to stockholders, and repayment of principal on long-term debt.

The statement of cash flows complements the income statement, because although income and cash flow are related, their amounts are seldom equal. Cash flow is also a better measure of a company's performance than net income, because net income relies upon arbitrary expense allocations (such as depreciation expense). Net income can also be manipulated by a company (e.g., by reducing R&D spending, reducing inventory levels, or changing accounting methods to make its "bottom line" look better in the short term. Therefore, analysis of a company's cash flows and cash flow trends is frequently the best method to evaluate a company's performance. Additionally, cash flow rather than net income should be used to determine the rate of return earned on the initial investment or to determine the value of a company or financial project. (Value is determined by discounting cash flows at the required rate of return.)

The statement of cash flows can be prepared directly by reporting a company's gross cash inflows and outflows over a time period, or indirectly by adjusting a company's net income to net cash flow for the same time period. Adjustments to net income are required for (1) changes in assets and liability account balances, and (2) the effects of noncash revenues and expenses in the income statement.

For example, in comparing XYZ Company's current balance sheet with the previous quarter, the following information is obtained about current assets and liabilities:

- Cash increased by $10,000
- Accounts receivable decreased by $5,000
- Inventory increased by $15,000
- Accounts payable increased by $25,000
- Salaries payable increased by $5,000

Additionally, the balance sheets show that the company has paid off $166,600 of long-term debt this quarter. This information from the balance sheets, along with the noncash expenses (depreciation) from the current income statement, is used to prepare the "statement of cash flows." The example demonstrates the value of the statement of cash flows, because it shows precisely the sources and uses of XYZ Company's cash. Additionally, it reconciles the cash account balance from the previous quarter to the amount on the current quarter's balance sheet.

XYZ Company

Statement of Cash Flows

For the Quarter Ended March 31, 1994

Net Income		$51,600.00	
Adjustments to reconcile net income to cash flow from op. activities:			
Decrease in accounts receivable	$5,000.00		
Increase in inventory	($15,000.00)		
Increase in accounts payable	$25,000.00		
Decrease in salaries payable	$5,000.00		
Depreciation expense, equipment	$35,000.00		
Depreciation expense, building	$70,000.00		
Total Adjustments:		$125,000.00	
Net cash flow from operating activities:			$176,600.00
Cash flows from financing activities:			
Retirement of long-term debt (notes payable)	($166,600.00)		
Net cash flow from financing activities:			($166,600.00)
Net Increase in cash:			$10,000.00
Cash account balance, Dec. 31, 1993			$15,000.00
Cash account balance, March 31, 1994			$25,000.00

REFERENCES AND RECOMMENDED READING

Engineering economics

1. Newinan, D. G., *Engineering Economic Analysis*. San Jose, CA: Engineering Press, 1976.
2. Canada, J. R. and White, J. A., *Capital Investment Decision Analysis for Management and Engineering*. Englewood Cliffs, NJ: Prentice-Hall, 1980.
3. Park, W. R., *Cost Engineering Analysis: A Guide to Economic Evaluation of Engineering Projects*, 2nd ed. New York: Wiley, 1984.
4. Taylor, G. A., *Managerial and Engineering Economy: Economic Decision-Making*, 3rd ed. New York: Van Nostrand, 1980.
5. Riggs, J. L., *Engineering Economics*. New York: McGraw-Hill, 1977.
6. Barish, N., *Economic Analysis for Engineering and Managerial Decision Making*. New York: McGraw-Hill, 1962.

Finance

1. Brealey, R. A. and Myers, S. C., *Principles of Corporate Finance*, 3rd ed. New York: McGraw-Hill, 1988.
2. Kroeger, H. E., *Using Discounted Cash Flow Effectively*. Homewood, IL: Dow Jones-Irwin, 1984.

Accounting

1. Chasteen, L. G., Flaherty, R. E., and O'Conner, M. C., *Intermediate Accounting*, 3rd ed. New York: McGraw-Hill, 1989.
2. Eskew, R. K. and Jensen, D. L., *Financial Accounting*, 3rd ed. New York: Random House, 1989.

Additional references

Owner's Manual from any Hewlett-Packard, Texas Instruments, or other make business/financial calculator.

Note: A business or financial calculator is an *absolute must* for those serious about analyzing investments and financial projects on anything more than an occasional basis. In addition to performing simple PV, FV, and mortgage or loan payment calculations, modern financial calculators can instantly calculate NPVs and IRRs (including multiple roots) for the most complex cash flow problems. Many calculators can also tabulate loan amortization schedules and depreciation schedules and perform statistical analysis on data. The owner's manuals from these calculators are excellent resources and give numerous examples of how to solve and analyze various investment problems.

Estimate the cost of a pipeline in the United States (based on 1994 data)

The following cost data are based on average costs for pipelines constructed on and offshore in the United States. Cost data are based on FERC permit filings. These costs are average and consideration will need to be given to the location, terrain, season of the year, length of line, environmental issues, and other variables before using these data.

Example. To find the cost of 100 miles of 36-in. diameter pipeline (onshore construction), 27,277 × 100 × 36 = $98,197,200.

Line size inches	Onshore ($/dia.in./mi.)	Offshore ($/dia.in./mi.)
8	25,321	
10	81,515	
12	21,160	
16	22,382	
20	54,688	
24	29,243	
26	29,923	
30	40,022	49,664
36	27,277	47,037
42	42,193	
48	26,305	

How to compare the cost of operating an engine on diesel and natural gas

Rule. For a rough estimate of the cost of diesel fuel per hour of operation, multiply: (.05) (engine hp) (cost per gallon of diesel fuel). To find the cost of natural gas per hour's operation, multiply: (.008) (engine hp) (cost of natural gas per Mcf).

Example. What would it cost per hour to operate a 500 hp engine on diesel fuel? On natural gas? Diesel fuel costs 10 cents per gallon, and natural gas costs 20 cents per Mcf.

Solution.

$$\text{Cost/hour for diesel fuel operation} = (.05)(500)(.10) = \$2.50$$

(A more accurate answer can be obtained by substituting .0533 for .05. This would make the cost of diesel fuel operation \$2.67 per hour.)

$$\text{Cost/hour for natural gas} = (.008)(500)(.20) = \$0.80$$

Example. Compare the cost of operating the engine on straight diesel fuel with dual fuel consisting of 90% natural gas and 10% diesel pilot fuel, if the conditions are the same as those just described.

Solution.

$$\text{Cost per hour} = (.10)(2.50) + (.90)(.80) = \$0.97$$

Generally speaking, 1,000 cu. ft of natural gas has a Btu content of 1,000,000 and a gallon of diesel fuel has a Btu content of 150,000. Thus 1 Mcf of natural gas is equivalent to $6\frac{2}{3}$ gallons of diesel fuel. To arrive at the figures above, the engine was assumed to require an average of 8,000 Btu per brake horsepower hour. This is an arbitrary figure, which is about the average for engines in good mechanical condition operating under high percentage of load. Some pipeline companies are equipped with engines that consume about 7,000 Btu/bhp; others run as high as 11,000 Btu/bhp. Condition of the engine and the loading account for most of the difference.

If an engine used an average of 7,000 Btu/bhp in developing its full horsepower of say, 1,000, naturally it would use 7,000,000 Btu per hour. This would require 7 Mcf of natural gas or about 46.7 gallons of diesel fuel.

How to estimate energy costs for different pipeline throughputs

The rule of thumb given below will apply to liquid pipelines where the flow is in the turbulent region only. It will serve to estimate energy requirements for pumping through pipelines between increments of +15% and decrements of − 15%.

Rule. Multiply the normal throughput in barrels times the cost of energy per barrel. Then multiply the increment or decrement (provided it is within a range of +15% or − 15%) times the cost of energy per barrel at normal throughput times 3. Then add or subtract the latter amount from the cost of normal throughput.

Or, restated in simpler form:
Cost of energy per barrel at normal throughput = Y
Cost of energy per barrel of increment or decrement within a range of ±15% = 3Y

Example. Assume energy requirements for pumping 2,000 barrels per hour through a 12-in. pipeline are \$.01 per barrel. What are energy requirements for pumping 2,200 barrels per hour?

Energy cost for 2,000b/h:

$$2{,}000 \times .01 = \$20 \text{ per hour}$$

Energy cost for additional 200b/h:

$$200 \times .01(3) = \$6 \text{ per hour}$$

Energy cost for 2,200 b/h = \$26 per hour

Comparing fuel costs for diesel and electric prime movers

Here's a formula that will give you a quick comparison of fuel costs of a pump station—for diesel engine or electric motor.

Just find out which of the expressions below, C_e or C_d, is the smaller:

$$C_e = \frac{0.7457}{n} \times E$$

$$C_d = 0.065\ D + 0.04$$

where:

C_e = Power consumption cost of electric motors, cents/bhp-hr
C_d = Fuel and lube oil cost of diesel engines, cents/bhp-hr
E = Electric power rate, cents/kW-hr
D = Diesel fuel cost, cents/gal
n = Efficiency of electric motor

Example. Let's see which is more economical: running a 1,000 bhp unit on diesel or electric driver.

The data are:
$n = 95\%$
$E = 1¢/kW\text{-}hr$
$D = 10¢/gal$

So, we have:

$$C_e = \frac{0.7457}{0.95} \times 1 = 0.785$$

$$C_d = 0.065 \times 10 + 0.04 = 0.690$$

Then, for a 1,000 hp unit the estimated operating cost per hour would be:

motor: $7.85/hr

engine: $6.90/hr

Keep in mind that this rule deals only with costs of power and fuel consumption. It does not take into consideration such expenses as operating personnel, repairs and renewals, depreciation, overhead, and so on.

Nomograph for calculating scale-up of equipment or plant costs

In correlating the cost of equipment or complete plants as a function of size, it has been shown that very often the cost varies as a fractional power of the ratio of capacities (or equipment sizes), as follows:

$$\left(\frac{C_2}{C_1}\right)^n = \frac{I_2}{I_1}$$

where C_1 and C_2 = The capacities (or sizes) of the equipment or plants in question
I_1 and I_2 = The cost or fixed investment of the known unit and the new unit, respectively
n = The exponent that applies to the particular case at hand

The nomograph enables one to calculate the capacity ratio raised to the fractional power n and the equipment or plant cost if the size for one capacity level is known.

Example. A rotary drum filter costs $20,000 when its filtering area is 50 ft^2. Assuming the cost varies as the 0.6 power of the capacity, how much will a 100-ft^2 filter cost?

The capacity ratio is:

$$C_2/C_1 = 100/50 = 2$$

Enter 2 on the capacity ratio scale (A) and align with 0.6 to the intersection point on the middle scale with $20,000 on the (D) scale and read I2 = $30,000 on the (E) scale.

Source

Kuong, J. F., *Brit. Chem. Eng.*, 7, No. 7, 546 (1962).

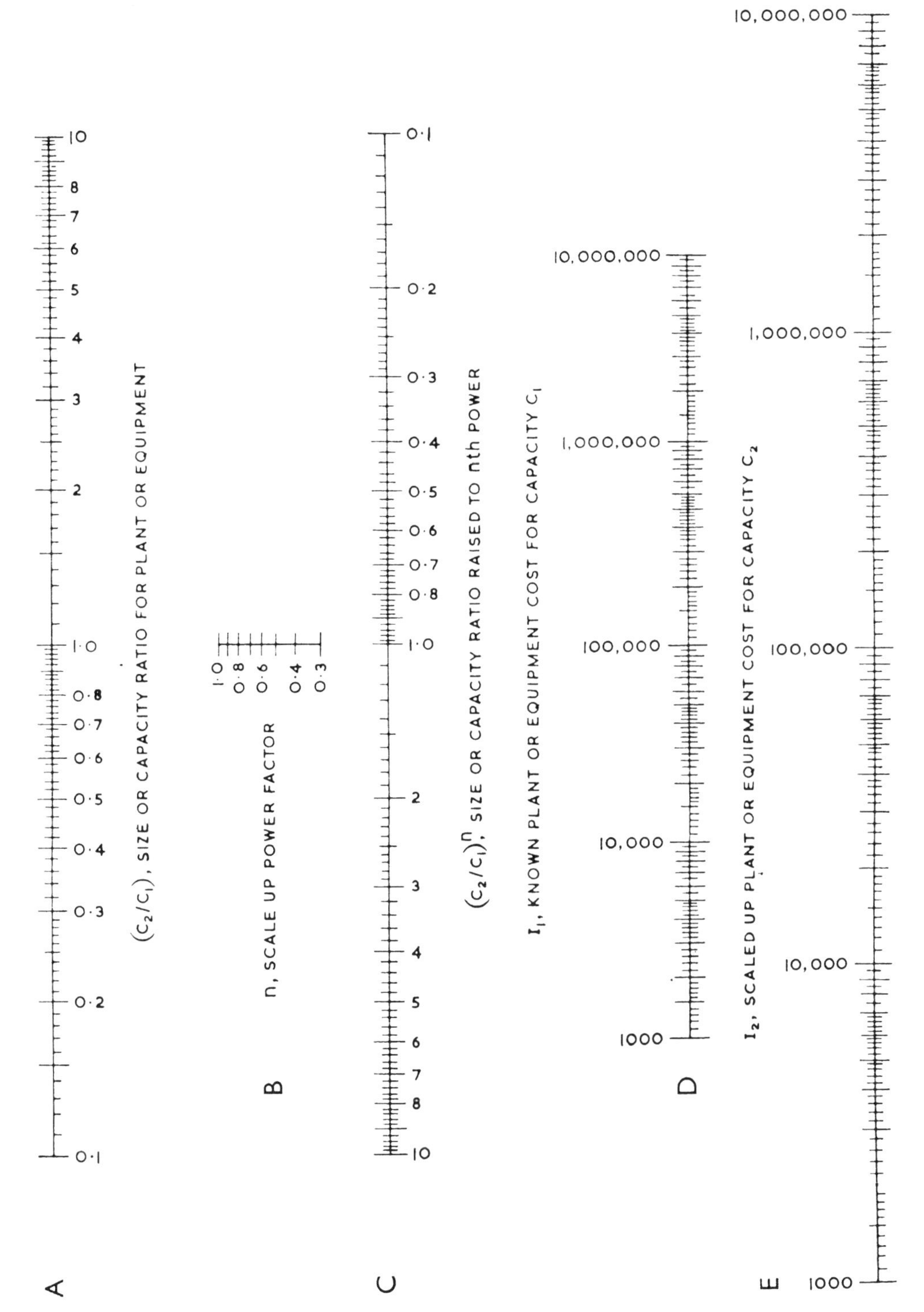

Figure 1. Scale-up of equipment and plant costs.

Nomograph for calculating scale-up of tank costs

The nomograph enables one to estimate the cost of a steel container of given volume when one knows the cost and volume of another or standard container.

The chart is based on the equation[1]:

$$C = cV_s^{0.3}V^{0.7}$$

where:

C = Cost of container in question, dollars
c = Cost of standard container, dollars per unit volume
V_s = Volume of standard container
V = Volume of container in question

V_s and V are expressed in the same units: gallons cubic ft, barrels.

Based on Dickson's suggestions, the chart can be used for drums of the same diameter and internal pressure and for spheres of the same internal pressure.

Example. If a 30,000-barrel tank costs $36,000, what is the cost of a 15,000-barrel tank? For the standard tank, c = 36,000/30,000 or $1.2 per barrel. Follow the key; connect 30,000 on the V_s scale with 15,000 on the V-scale and note the intersection with the α-axis. Connect this point with 1.2 on the c-scale and read the desired cost of approximately $22,000 on the C-scale.

Source

Davis, D. S., *Petrol Refiner*, 37, No. 4200 (1958).

Reference

1. Dickson, R. A., *Chem. Eng.*, 54, (8) 122 (1947).

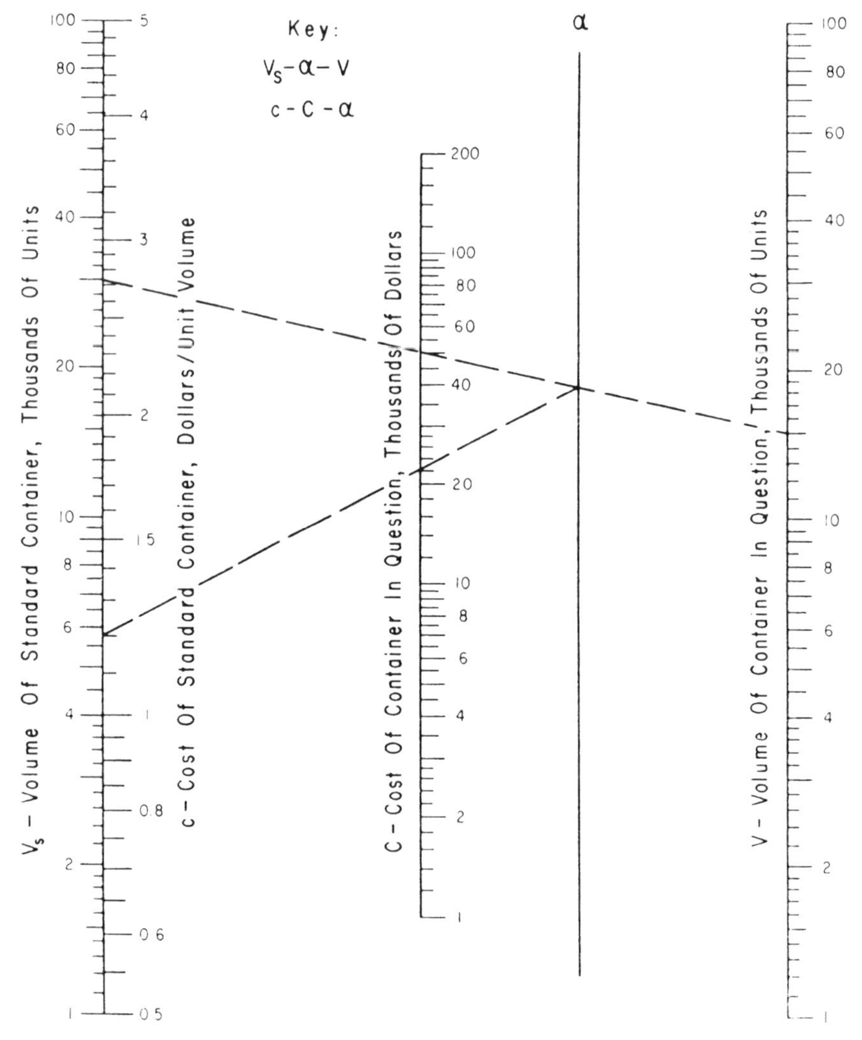

Figure 1. Scale-up of tank costs.

Nomograph for determining sum-of-years depreciation

In estimating annual depreciation of fixed capital assets, the sum-of-the-years-digits depreciation method is commonly used. This method allows for the effect of accelerated depreciation in the earlier years of a project. It is based on the number of years of service remaining each year and on the sum of the arithmetic series of the years from 1 to N. The depreciation factor for each year is calculated by dividing the number of remaining service years by the sum of the digits corresponding to the total service years. The depreciation charge can then be obtained by multiplying the depreciation factor each year by the total fixed asset cost.

The nomograph was constructed from the following equation:

$$D = \frac{N_D}{\sum_{1}^{N} N_i} \times V = \frac{N_D}{1+2+3+\ldots .N} \times V$$

where:

D = Annual depreciation, dollars
N = Number of years of service of the depreciable item
N_D = Number of years of service remaining each year ($N-0$, $N-1$, ... $N-(N-1)$ for the first, second and Nth year of depreciation)
N_i = Digit corresponding to the ith year

Example. The value of a piece of equipment is $5,200. Its salvage value at the end of 10 years is $200. What is the annual depreciation for the first, second, and third years?

Solution. The depreciable amount is 5,200 − 200 = $5,000. The first, second, and third years correspond to 10, 9, and 8 years of remaining service life, respectively.

Connect	With	Mark or Read
55 for l0yrs	$N_D = 10$ (1st yr)	R
R	V = 5,000	D = $910

Repeat the procedure for $N_D = 9$ and 8 and obtain D = 820 and 720 for the second and third years of depreciation, respectively.

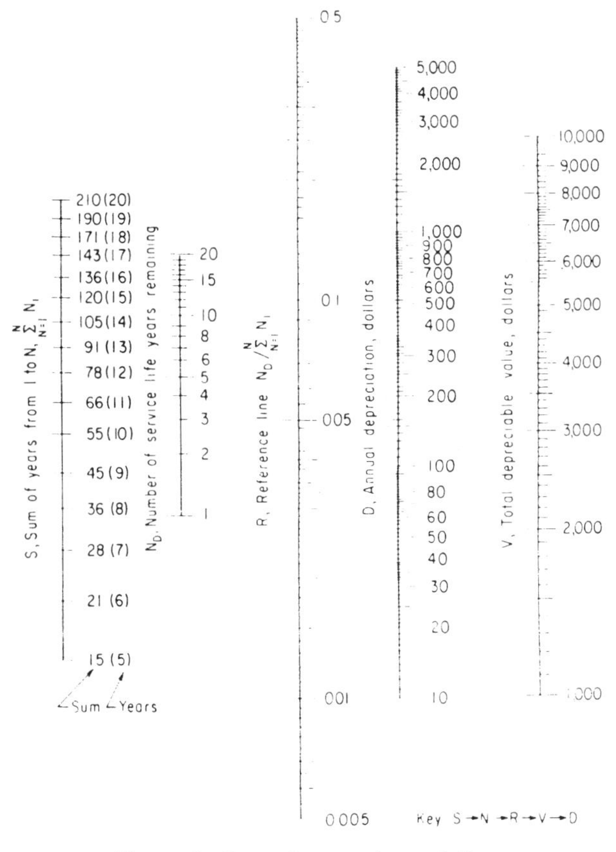

Figure 1. Sum-of-years depreciation.

If the total depreciable value V is larger than the range of the scale, divide by a factor that brings it within the range of the scale, then read the D value, but remember that the result must be multiplied by the same factor.

Source

Kuong, J. F., *Chem. Eng.*, 67, No. 23, 237 (1960).

Nomograph for estimating interest rate of return on investment ("profitability index")

Several methods are used for evaluating the profitability of a given project. A method for estimating profitability that has received considerable emphasis in the past few years is the "interest rate of return," (Figure 1), also known as "profitability index" or "discount cash flow" method. This profitability method takes into account the time

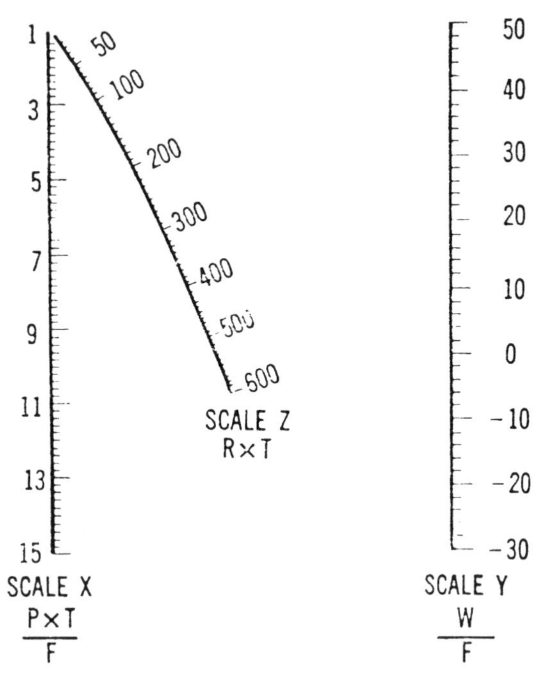

Figure 1. Interest rate of return.

value of money. Mathematically, interest rate of return may be defined as that rate at which the company's outstanding investment is repaid by proceeds from the project.[1]

The nomograph was developed for the case in which the yearly profits (P) (or the difference between income and costs) are equal through the life of a project. The nomograph is based on the following assumptions:

1. Depreciation method is the sum of the years-digits.
2. Project economic life is equal to depreciable life (T).
3. Construction time is zero, i.e., total investment occurs instantaneously at time zero.
4. There is no salvage value at the end of the project life.
5. There is no start-up expense, (S = 0).

Where these conditions apply, the nomograph gives the interest rate of return R, as a function of P and T, fixed investment (F), and working capital (W).

Example 1. *Basic Profitability*. For the preliminary evaluation of a project the following estimates are available:

Plant capacity and sales	10,000,000 lb/yr
Average price	\$0.20/lb
Unit cost (at capacity)	\$0.10/lb
Overhead 15% of price	\$0.03/lb
Project life 10 years (T)	
Depreciation is sum of the years-digits	
Plant investment (F)	\$1,000,000
Working capital (W)	\$200,000

It may be assumed that no start-up expense is incurred.

Estimate the interest rate of return if the project is undertaken. The reference or base case is not to invest.

First compute P, the yearly before-tax and depreciation profit.

Gross income: 10,000,000 × 0.20 = \$2,000,000
Less overhead @ 15%: 10,000,000 × 0.03 = \$300,000
Less operating cost: 10,000,000 × 0.10 = \$1,000,000
P, profit before tax and depreciation = \$700,000

Compute:

$$\frac{P \times T}{F} = \frac{700{,}000 \times 10}{1{,}000{,}000} = 7$$

Compute:

$$\frac{W}{F} = \frac{200{,}000}{1{,}000{,}000} = 0.2$$

Connect	**With**	**Read**
$\frac{P \times T}{F} = 7$	$\frac{W}{F} = 0.2$	$R \times T = 330$

Since T = 10, the interest rate of return for the project is 33%/yr.

This is equivalent to saying that the outlay of capital of \$1,000,000 in plant facilities plus 200,000 in working capital when invested in the project would generate enough cash to pay back the total investment plus interest at the rate of 33%/yr.

Example 2. *Comparison of alternatives*. Consider the following alternatives for replacing a pump:

- *Alternative A*. Replace existing pump with one of the same type at a cost of \$3,000, working capital of \$50, and maintenance expense of \$300/yr.
- *Alternative B*. Install a new-type pump costing \$3,500 with working capital of \$100 and maintenance costs of \$100/yr. Economic and depreciable life for both pumps is 10 years.

The savings per year for using pump B versus pump A are:

$300 – $100 = $200/yr
to justify an increment of investment of $500 ($3,500 – $3,000). What is the profitability index or interest rate of return on the differential investments?

Calculate:

$$\frac{P \times T}{F} = \frac{200 \times 10}{500} = 4$$

$$\frac{100(W)}{F} = \frac{50}{500} = 0.10$$

Connect	With	Read
$\frac{P \times T}{F} = 4$	0.1 on $\frac{W}{F} = 0.1$ scale	$R \times T = 210$

Since T = 10, R = 21%/yr = profitability index for project.

If 21% is considered an adequate rate of return, after consideration of risk, then pump B would be the preferred one.

Source

Greist, W. H., Bates, A. G., and Weaver, J. B., *Ind Eng. Chem.*, 53, 52A (1961).

References

1. Dean, J., *Harvard Bus.* Rev., 32, 120–130 Jan., Feb. (1954).
2. Martin, J. C., in *Chemical Process Economics in Practice*, Hur, J. J., Ed., Reinhold Publishing Corp., New York, 1956.
3. Weaver, J. B., and Reilly, R. J., *Chem. Eng. Progr.*, 52, 405 (1956).

Nomograph for determining break-even point

The nomograph gives the break-even point for a plant as a percent of design capacity when the unit fixed cost, Cf, the sales price, S, and the unit variable cost, Cv are known. It is based on the following equation:

BE = 100.Cf/(S – Cv)

The nomograph can also be used to determine the sensitivity of the break-even point to any combination of fixed variable costs or sales price or to determine the maximum fixed cost that can be absorbed for a given break-even point.

Example. Determine the break-even point when:

Design output	10,000 lb/yr
Unit fixed cost	$0.045/lb
Unit variable cost	$0.075/lb
Sales price	$0.200/lb

Calculate Δ = 0.200 – 0.075 = 0.125 or 12.5¢/lb

Align Cf = 4.5¢ with Δ = 12.5 and obtain BE = 36% of plant capacity.

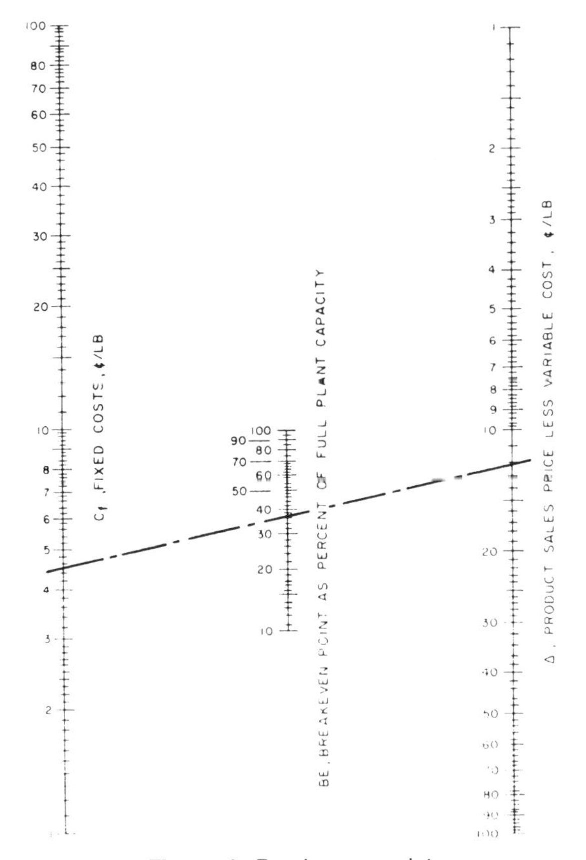

Figure 1. Break-even point.

Source

Kuong, J. F., *Petro/Chem. Engineer*, 35, No. 1, 44 (1963).

Chart gives unit cost per brake horsepower of reciprocating compressors with various types of prime movers

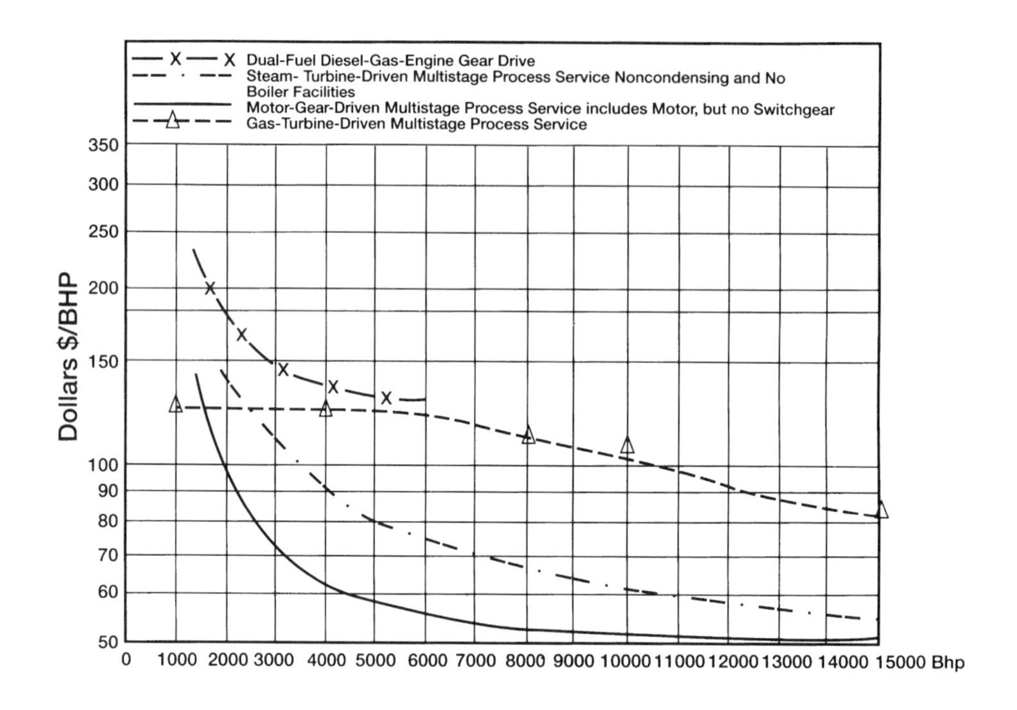

Chart shows influence on unit cost of numbers of reciprocating compressor units installed in one station

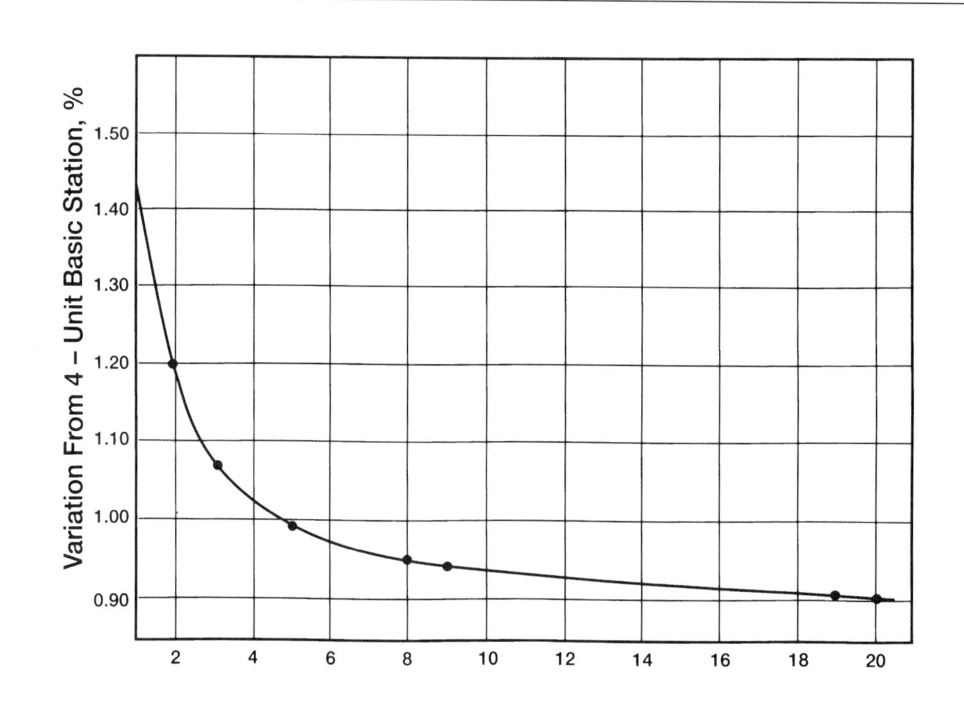

Chart gives unit cost per brake horsepower of centrifugal compressors with various types of prime movers

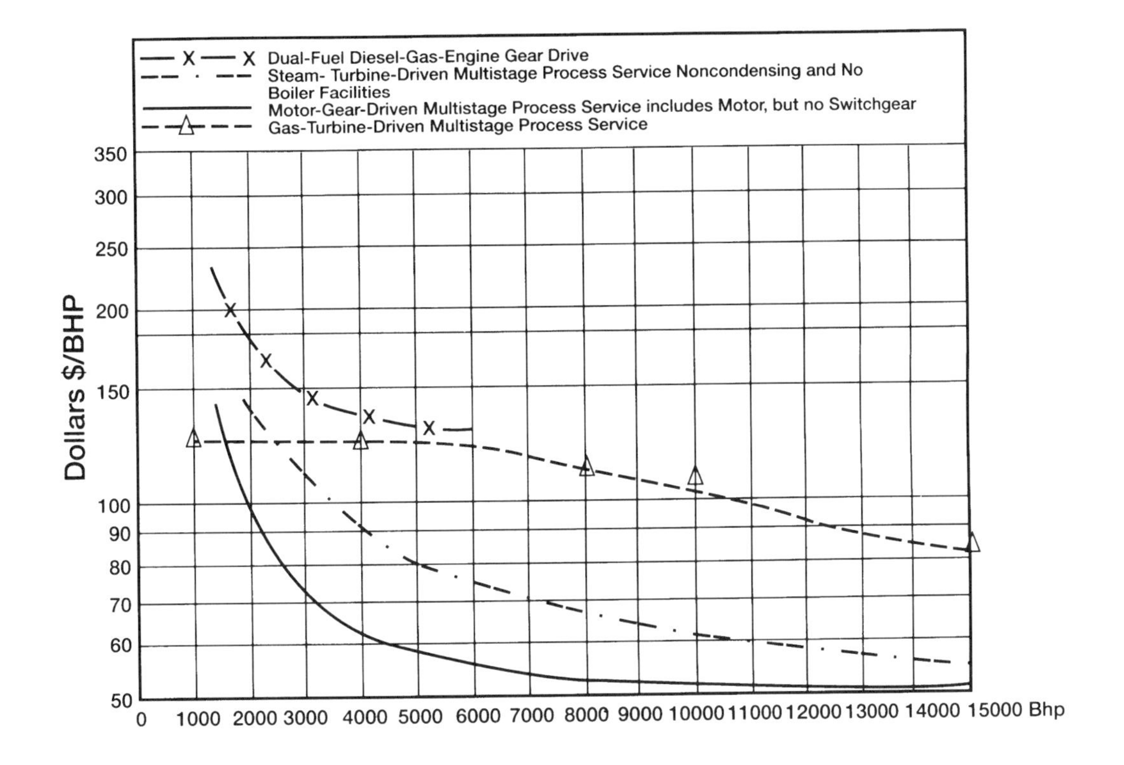

21: Rehabilitation–Risk Evaluation

When does a pipeline need revalidation? The influence of defect growth rates and inspection criteria on an operator's maintenance program

John F. Kiefner and Patrick H. Vieth, Kiefner and Associates, Inc. Worthington, Ohio

Introduction

Buried, high-pressure pipelines are subjected to loads and environmental effects that may cause them to become degraded with the passage of time. All pipeline operators are well aware of this, and the prudent operators have active programs to mitigate deterioration and to remediate defective pipe. The efficiency of an operator's revalidation program can be optimized if the necessary responses can be made when needed but not until then. The authors have assisted several pipeline operators in establishing timely intervention programs to assure continuing pipeline serviceability. A variety of techniques are used depending on the nature of the pipeline and the perceived problems. Some of the basic techniques are described in this document.

Background

Pipelines may suffer degradation from a variety of causes, including corrosion, mechanical damage, fatigue, and stress-corrosion cracking. The appropriate remedies for these problems are well known but other than the routine patrolling of the rights-of-way and monitoring of cathodic-protection potentials and rectifier currents, such remedies are too expensive to be applied on a regular or routine basis. By that we mean that most operators cannot afford to routinely and periodically utilize inline inspection and/or hydrostatic testing to revalidate their pipelines. Usually, these techniques are invoked when some special circumstances exist. The special circumstances may be the existence of excessive amounts of low pipe-to-soil potential readings, the occurrence of leaks or ruptures, or just an intuitive feeling that it is time to check the condition of a pipeline. Alternatively, as we are finding, more and more operators are coming to depend on more sophisticated models to determine when intervention is needed. The types of models we have used are described.

General concepts

Failure pressure versus defect size. Pipeline integrity is usually defined in terms of pressure-carrying capacity. For a given diameter, wall thickness, and grade of material, one can expect that a sound piece of line pipe will be able to sustain an internal pressure level of at least 100% of its specified minimum yield strength (SMYS). And, that pressure level can readily be calculated by means of the Barlow formula. For example, if one has a 16-in. O.D. by 0.250-inch wall thickness X52 pipeline, the pressure level corresponding to 100% of SMYS is 1,625 psig. For comparison, the 72% of SMYS maximum operating pressure level is 1,170 psig. Thus, one has the expectation that a new piece of this particular pipe has a safety margin against failure of 1,625/1,170 or 1.39. In fact, if it is free of defects, it will have a failure pressure of at least 1,938 psig. The latter is based upon research conducted by the pipeline research Committee of the American Gas Association.[1]

Realistically, not all pipe is defect free; otherwise there would never be pre-service hydrostatic test breaks. But, after a preservice test to a pressure level of 90%-100% of SMYS, one can expect a pipeline to perform satisfactorily at an operating stress level up to 72% of SMYS, that is, until or unless it becomes degraded in service by corrosion, mechanical damage, fatigue, or stress-corrosion cracking.

Thanks to extensive research by the previously mentioned Pipeline Research Committee of A.G.A.,[1] an experimentally validated model exists for calculating the effects of a longitudinally oriented part-through-the-wall defect on the pressure-carrying capacity of the pipe. For any given piece of pipe, the model can be used to generate the relationships between failure pressure and flaw size. For our example, the 16-in. O.D. by 0.250-in. wall thickness X52 material, the relationships for selected depths of flaws are shown in Figure 1. Each of the nine parallel curves represents the failure-pressure-versus-flaw-length relationship for a

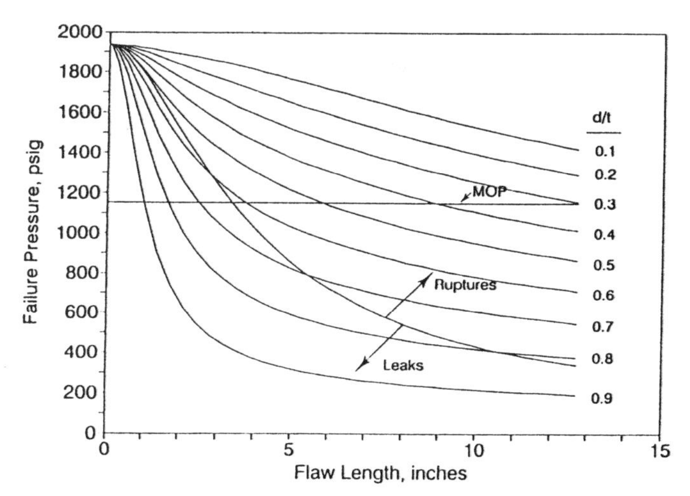

Figure 1. Failure pressure versus flaw size for a material of normal toughness (16 in. by 0.250 in., X52, normal toughness).

particular depth (d) to wall thickness (t) ratio. The curve that cuts across the others is the dividing line between leaks and ruptures. Defects with failure at pressures above and to the right of this curve will fail as ruptures, whereas those that have failure pressures below and to the left of this curve will fail as leaks.

A horizontal line is drawn on Figure 1 to represent the maximum operating pressure (MOP) of this pipe material, 1,170 psig, corresponding to a hoop stress level of 72% of SMYS. All flaws with length and d/t combinations that lie below this line will fail at pressure levels below the MOP. All flaws that lie above this line could exist in the pipeline in service. For example, the curve representing a d/t of 0.5 crosses the line at a flaw length of 5.8 inches. Flaws of that length which are more than halfway through the wall will fail at the MOP; flaws of that length which are less than halfway through will not.

The family of curves shown in Figure 1 is quite useful in the kinds of models we use to determine the time-dependent degradation of a pipeline. Two features of the curves should be kept in mind as the discussion proceeds. First, these curves represent failure pressures that are achievable with steadily increasing pressure over a relatively short period of time (minutes). Because of the phenomenon of time-dependent growth, it is possible to observe failures at pressure levels of 5%-10% lower than these curves predict if pressure levels 5%-10% below the predicted levels are held long enough.

Secondly, these curves are affected by the ductile-fracture toughness of the material. The curves shown in Figure 1 are characteristic of a garden-variety X52 line-pipe material having a Charpy V-notch upper-shelf energy (full-size specimens) of 25 ft-lb. The set of curves shown in Figure 2 was generated for a hypothetical material of the same pipe geometry and grade but with a Charpy energy of 250 ft-lb (a material of optimum toughness). Note that for a given defect size, the material of optimum toughness has a higher failure pressure. To illustrate, it is recalled from Figure 1 that a 5.8-in.-long flaw halfway through the wall fails at 1,170 psig. In contrast, the same 5.8-in.-long flaw with a d/t of 0.5 in the very tough material has a failure pressure of 1,210 psig. It is noted that blunt flaws such as corrosion-caused metal loss tend to behave as illustrated in Figure 2 rather than Figure 1 regardless of the toughness of the material. This is because the material in a corrosion pit is strained over a large "gauge length" unlike the situation of a crack-like flaw. For the latter, toughness becomes a very important parameter.

Representing the other extreme, Figure 3 presents the failure-pressure-versus-flaw-size relationship for a very low toughness material such as the ERW bond line in an older, low-frequency welded material. Even though the diameter, thickness, and grade are the same as that represented in Figures 1 and 2, the critical flaw sizes are much smaller.

Benefits of hydrostatic testing and in-line inspection

Another useful concept in evaluating pipelines for revalidation arises from the unique failure-pressure-versus-flaw-size relationships of the type described in Figures 1, 2, and 3. That is the manner in which those relationships illustrate the benefits of hydrostatic testing and in-line inspection. For illustrative purposes, let us consider the 25 ft-lb material for which Figure 1 gives the failure-pressure-versus-flaw-size relationship. This relationship is replotted in Figure 4, but Figure 4 contains additional information. It contains horizontal lines at 1,170 psig (maximum operating pressure), at 1,463 psig (90% of SMYS), and at

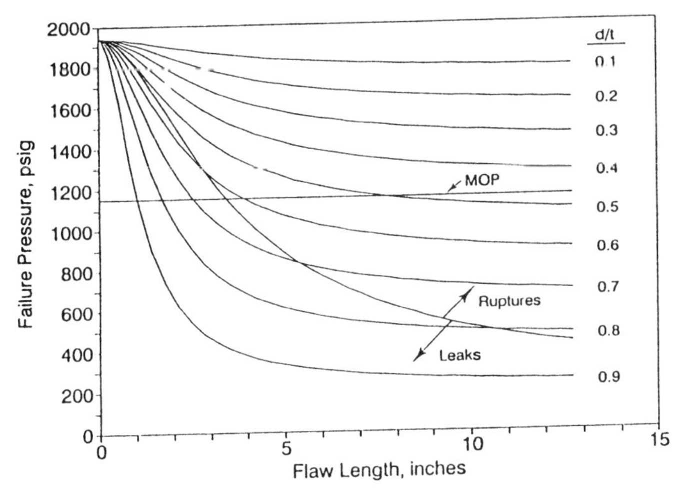

Figure 2. Failure pressure versus flaw size for a material of optimum toughness or for blunt defects (16 in. by 0.250 in., x52, optimum toughness or blunt defects).

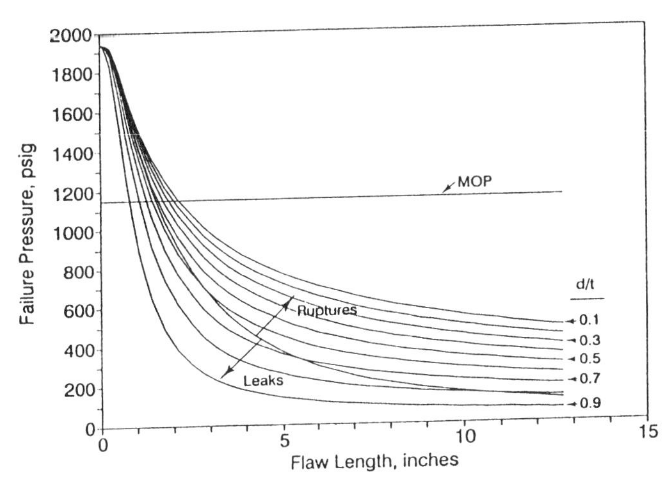

Figure 3. Failure pressure versus flaw size for a material of low toughness (16 in. by 0.250 in., x52, low toughness).

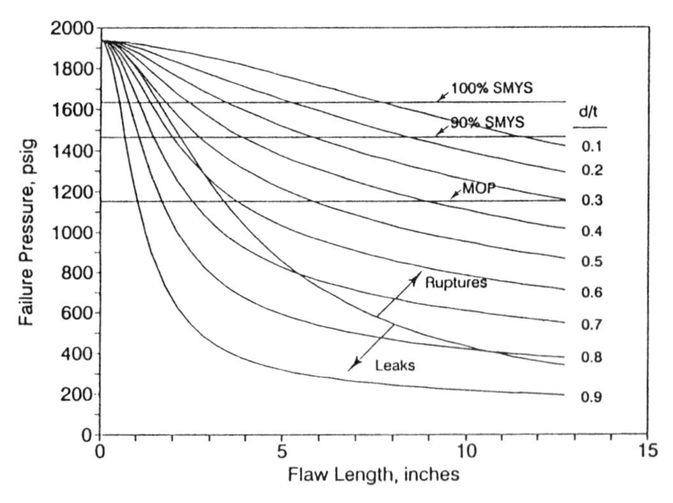

Figure 4. Sizes of flaws located by hydrostatic testing (all types except gouge and dent combinations) (16 in. by 0.250 in., x52, normal toughness).

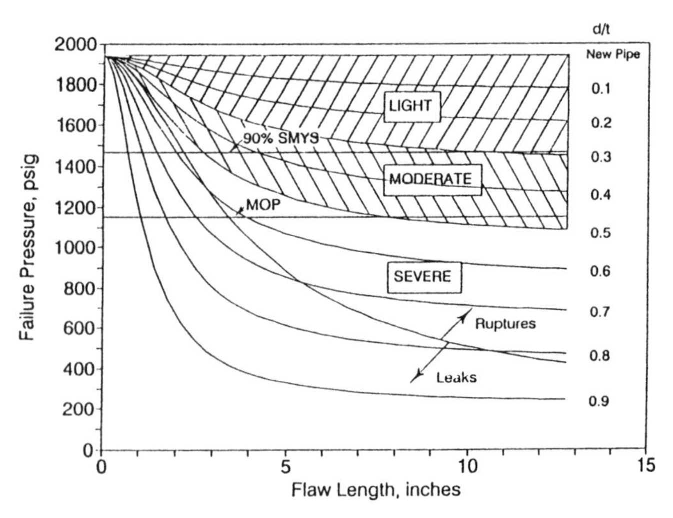

Figure 5. Sizes of flaws located by in-line inspection (corrosion) (16 in. by 0.250 in., x52, blunt defects).

1,625 psig (100% of SMYS). As illustrated previously, no flaws having failure pressures below 1,170 psig can exist if the pipeline is operating at 72% of SMYS; otherwise, they would fail (i.e., their particular lengths and d/t values would lie below the horizontal line at 1,170 psig). Similarly, if the pressure in the pipeline is raised to 1,463 psig (90% of SMYS), any flaws with length and d/t combinations lying above the line at 1,170 psig but below that at 1,463 psig will fail and be eliminated. If no failures occur, the absence of flaws in this size range is proven. And, if the pressure in the pipeline is raised to 1,625 psig (100% of SMYS), the absence of additional still smaller flaws will be proven. Hence, the benefit of a hydrostatic test is expressible as the ratio of test pressure to operating pressure. The higher the ratio, the more beneficial the test, because the flaws that can remain will be smaller.

With respect to in-line inspection, most pipeliners are aware that the conventional magnetic-flux leakage tools are reasonably capable of sorting metal-loss anomalies into categories by depth of light, moderate, and severe. Lights generally are those with depths of less than 30% of the wall thickness ($d/t < 0.3$). Moderate are those with depths of 30%-50% through the wall thickness ($d/t = 0.3$ to $d/t = 0.5$). Severes have depths greater than 50% of the wall thickness ($d/t > 0.5$). Recalling that the behavior of blunt metal-loss flaws is best represented by the optimum toughness relationships of Figure 2, one can see in Figure 5 (based on Figure 2) how in-line inspection data can be interpreted in terms of the effects on pressure-carrying capacity. The top cross-hatched area corresponds to light; the lower cross-hatched area corresponds to moderate, and the unshaded area corresponds to severe. Note that for long flaws, the failure pressures corresponding to moderates become low enough to intersect the MOP level. This illustrates the need to excavate and examine moderates in a program of conventional in-line inspection. With the use of the more expensive, high-resolution in-line inspection tools, it is possible to define flaw length and, hence, to avoid many excavations.

Figure 5 provides an important conceptual comparison between in-line inspection and hydrostatic testing. With the aid of this figure, one can see where each has its advantages and disadvantages. The hydrostatic test represented by the 90% SMYS line provides a demonstration of the immediate pressure-carrying capacity up to a pressure level of 1,463 psig. The use of conventional magnetic-flux tools cannot provide this kind of assurance unless all severe and moderate anomalies are excavated, leaving only the lights. However, the hydrostatic test does not locate any of the anomalies that do not fail. Hence, very deep pits can survive the test and develop leaks shortly thereafter (as illustrated by the amount of unshaded area—severe anomalies—lying above the horizontal line at 1,463 psig). At least with in-line inspection, one can locate and remove or repair the anomalies (especially the deep ones). After having removed or repaired all moderates and severes, one can have a high degree of confidence that the pipeline will not fail or leak for a long time as the result of corrosion-caused metal loss. The same cannot be said after a hydrostatic test. Very short but deep pits could have survived the test and may become leaks (not ruptures) within a short time after the test.

On the other hand, because none of the existing in-line inspection tools is capable of finding crack-like flaws, hydrostatic testing is the only means to assure the integrity of a pipeline that may have sustained cracking such as fatigue-crack growth or stress-corrosion cracking.

Determining when to revalidate

Having established the concepts illustrated previously, we can now proceed to the determination of when to revalidate a pipeline. This is done by different means depending on the type of anomaly of concern. First, let us consider corrosion-caused metal loss.

Corrosion-caused metal loss. A key parameter for scheduling the revalidation of an externally or internally corroded pipeline is the worst-case corrosion rate. A pipeline operator's corrosion control personnel may be able to estimate such a rate. Alternatively, one can obtain a reasonable estimate if prior corrosion leaks have occurred. If leaks have occurred, the decision to revalidate may be obvious without the use of a model. Nevertheless, having some idea of the corrosion rate is useful for scheduling the "next" revalidation. The rate can be estimated by dividing the nominal wall thickness of the pipeline by the number of years between the time of the original installation and the time of the first leak. This involves assuming a constant rate of corrosion over the life of the pipeline. For any pit that is not a leak, the corrosion rate will be less; it will be proportional to the d/t ratio of the pit. For our example pipeline (0.250-in. wall thickness), if we assume that it first developed a leak after 25 years of service, its worst-case corrosion rate is 0.25/25 or 0.010 in. per year (10 mils per year). If we postulate that there is a pit on this pipeline that is 80% through the wall, the corrosion rate for that pit is $0.8 \times 0.01 = 0.008$ in. per year (8 mils per year). With this kind of a corrosion-rate rationale, we can utilize a figure like Figure 2 to plan the revalidation of a corroded pipeline.

The family of curves represented in Figure 2 is reproduced in Figure 6 for the purpose of examining the effects of time-dependent metal loss. In Figure 6, the pipeline when it was installed is represented by the horizontal line at 1,938 psig ($d/t = 0$). If this pipeline is losing metal to corrosion, the effect of the metal loss is to lower the failure pressure.* In terms of Figure 6, one can see that as metal loss proceeds, the horizontal line originally at 1,938 psig would move downward as indicated by the arrows until it would eventually reach the 1,170 psig operating level, and a failure would occur. It is important to note that this occurs at a d/t of 0.75 for a 2-in.-long flaw (Arrow 1) and at a d/t of 0.47 for a 10-in.-long flaw (Arrow 2). If the

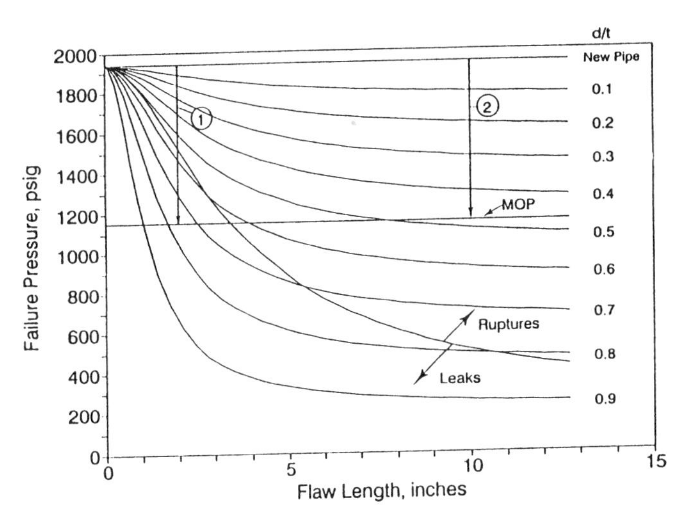

Figure 6. Determination of need for revalidation of a pipeline affected by corrosion (16 in. by 0.250 in., x52, blunt flaws).

company's corrosion specialist has determined that the worst-case corrosion rate is 8 mils per year, then 75% of the wall thickness (250 mils) will be penetrated in less than 23 years, and 47% of the wall thickness will be penetrated in less than 14 years. If the type of metal loss that is occurring is long (10 in. or more along the axis of the pipe), this pipeline will require revalidation in less than 14 years. On the other hand, if experience shows that the corrosion is occurring as isolated pits (less than 2 in. in axial extent), the time for revalidation would be less than 23 years.

Perhaps a more practical case is that corresponding to a pipeline which has either been hydrostatically tested or has been repaired following in-line inspection. Now the question becomes: When will revalidation be needed again? The question can be answered in terms of Figure 7. Let us consider two alternatives simultaneously. In one case, let us assume that our 16-in. O.D. by 0.250-in. wall thickness pipeline has been inspected by a conventional magnetic-flux leakage tool and that all anomalies except the lights have been repaired (see the cross-hatched area in Figure 7). In the other case, let us assume that the pipeline has been hydrostatically tested to 1,463 psig (90% of SMYS).

With respect to a 2-in.-long isolated pit (Arrow 1), the deepest pit remaining after the in-line inspection is represented by Point A ($d/t = 0.3$, pit depth = 75 mils). From the standpoint of the hydrostatic test, the deepest pit that can exist with a 2-in. length is represented by Point B ($d/t = 0.6$, pit depth = 150 mils). To cause a leak in service (and it will be a leak rather than a rupture if the length remains less than about 3 inches), the pit must grow to Point C ($d/t = 0.76$, pit depth = 190 mils). If the corrosion rate remains at 8 mils per year, the time to grow from A to C is $(190 - 75)/8 = 14.4$ years, and the time to grow from B to C is

*The ASME B31G method for evaluating corroded pipe is based on the same empirical equation that is used to derive the curves of Figure 6. The only difference is that the flaws represented in Figure 6 have been idealized as being rectangular in profile.

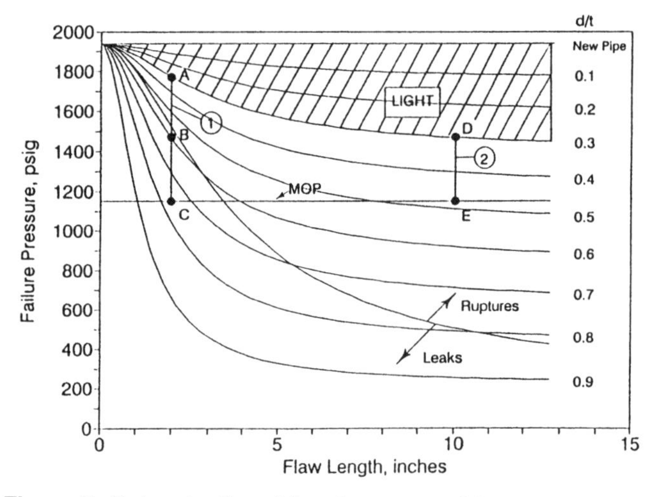

Figure 7. Determination of time to next revalidation after in-line inspection or hydrostatic test (corrosion) (16 in. by 0.250 in., x52, blunt flaws).

(190 – 150)/8 = 5 years. This shows again the advantage of in-line inspection over hydrostatic testing with respect to dealing with pitting corrosion.

With respect to a long corrosion anomaly (Arrow 2), the deepest possible remaining anomaly is the same whether the present revalidation was by in-line inspection or by hydrostatic test. The anomaly must grow from Point D to Point E to fail in service (and it will fail as a rupture rather than a leak, because it is well above the leak/rupture dividing line). To do so, d/t must change from 0.3 to 0.47. This is a change in wall thickness of 42.5 mils and will take place in 5.3 years at a corrosion rate of 8 mils per year. From this example it appears that if the corrosion is occurring over long regions as opposed to isolated pits, the advantage of in-line inspection over hydrostatic testing is eroded. This is not quite correct for two reasons. First, the high-resolution tools give reasonably accurate information on the length of the anomaly as well as the depth. Hence, one can focus the repairs on the basis of remaining pressure-carrying capacity. Second, even with the use of conventional tools, the advantage of locating the corrosion and having the option to examine and repair it provides benefits that do not accrue with hydrostatic testing.

Returning to our analysis, we see that for our example, a revalidation interval of less than 5 years is suggested. An appropriate factor of safety should be applied to this interval, so if this were a real case, the pipeline operator might arbitrarily decide to take action after 2 or 3 years. This probably is an excessively conservative example. Hopefully, with adequate corrosion control, a lower corrosion rate than 8 mils per year can be assured. If so, the revalidation interval would be longer than shown in this example. However, the principles are the same regardless of the corrosion rate.

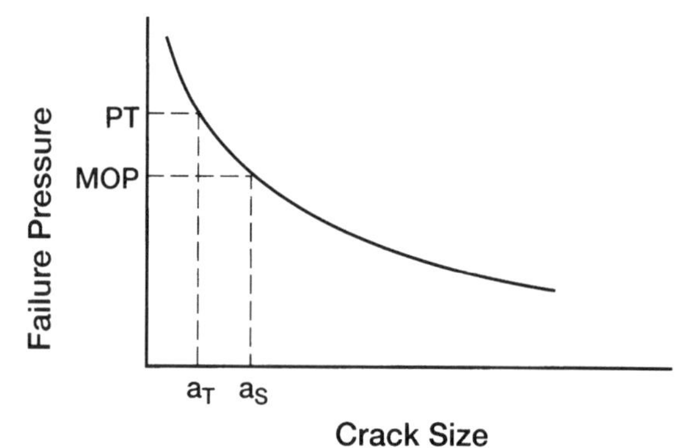

a. Failure pressure as a function of crack size.

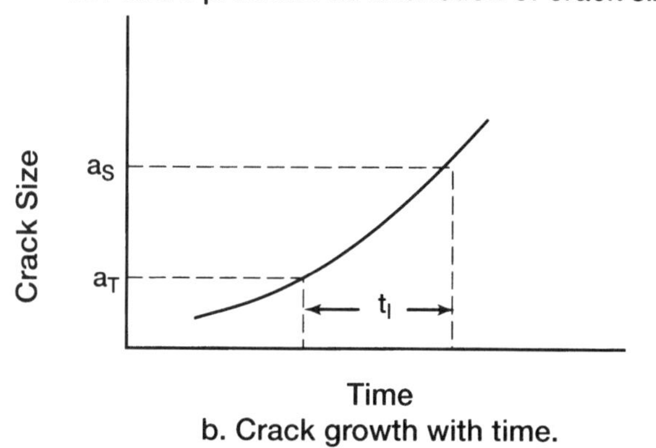

b. Crack growth with time.

Figure 8. Effect of crack growth on pipeline integrity.

Fatigue crack growth

Model for crack growth. As we have seen in terms of Figure 4, if a pipeline is hydrostatically tested to a given pressure, defects of a certain size that are present at the time of the test will fail. Smaller defects would be expected to survive the test but would not be expected to grow further at a static pressure level much below the test pressure level. However, defects of size a_T limited by pressure level PT in Figure 8 could grow during cyclic-pressure service until they reach the size a_s. At that point, failure will occur at the maximum operating pressure (MOP). If a defect as large as at remains after a hydrostatic test and if the pipeline is subjected to sufficient pressure fluctuations, the defect will fail after time t_1.

It should be obvious from Figure 8 that flaw enlargement by fatigue-crack growth will erode the effective safety margin provided by a hydrostatic test. This situation may be addressed by analyzing the effects of the service-pressure cycles in terms of the size of the margin between test pressure and operating pressure and assessing when it may be necessary to conduct a subsequent hydrostatic test in the event that flaws which survive the first test become enlarged.

For this purpose, one may use a crack-growth model developed explicitly for pipelines.[2] The model is based upon linear-elastic fracture mechanics. It assumes that crack growth takes place under fluctuating stress levels according to the relationship:

$$da/dN = C\Delta K^n \quad (1)$$

where:

da/dN is the increment of crack advance per cycle of loading.

ΔK is the range of stress intensity factor, a function of the change in stress and the initial crack size.

C and n are material properties determined through crack-growth-rate tests. In the use of the model, the initial crack size is fixed by the hydrostatic test. A defect that just barely survives the test is assumed to be the initial crack. The stress range is determined by the maximum and minimum pressure levels. The stress-intensity factor that provides the cracking-driving force is calculated via a Newman-Raju algorithm.[3] The model is capable of handling irregular-size pressure variations. For example, we have analyzed pressure spectra as complex as that shown in Figure 9. Equation (1) is numerically summed over a number of cycles until the crack becomes large enough to fail in service. The failure criterion for flaw size at failure is that used to generate Figures 1, 2, and 3.

To demonstrate the importance of a high test-pressure-to-operating-pressure ratio in a situation where service pressure cycles can cause flaw growth, one can consider an example based on an actual cyclic pressure test. The results of a full-scale pipe experiment involving crack growth are described in Reference 4 (Figure 47 of Reference 4) and are shown in Figure 10. In this experiment, a 2-in.-long part-through (0.095-in.-deep) flaw was placed in the fusion line of the ERW seam of a 12/-3/4-in. O.D. by 0.250-in. wall thickness X52 material. It was then subjected to constant pressure cycles from 0 to 1,020 psig until it grew through the wall and failed (after about 8,000 cycles). The actual failure pressure levels are not important, so for purposes of illustration, we can consider the implications of Figure 10 for a material having an ERW seam toughness equivalent to 2 ft-lb of Charpy V-notch impact energy. If this were a 2-ft-lb

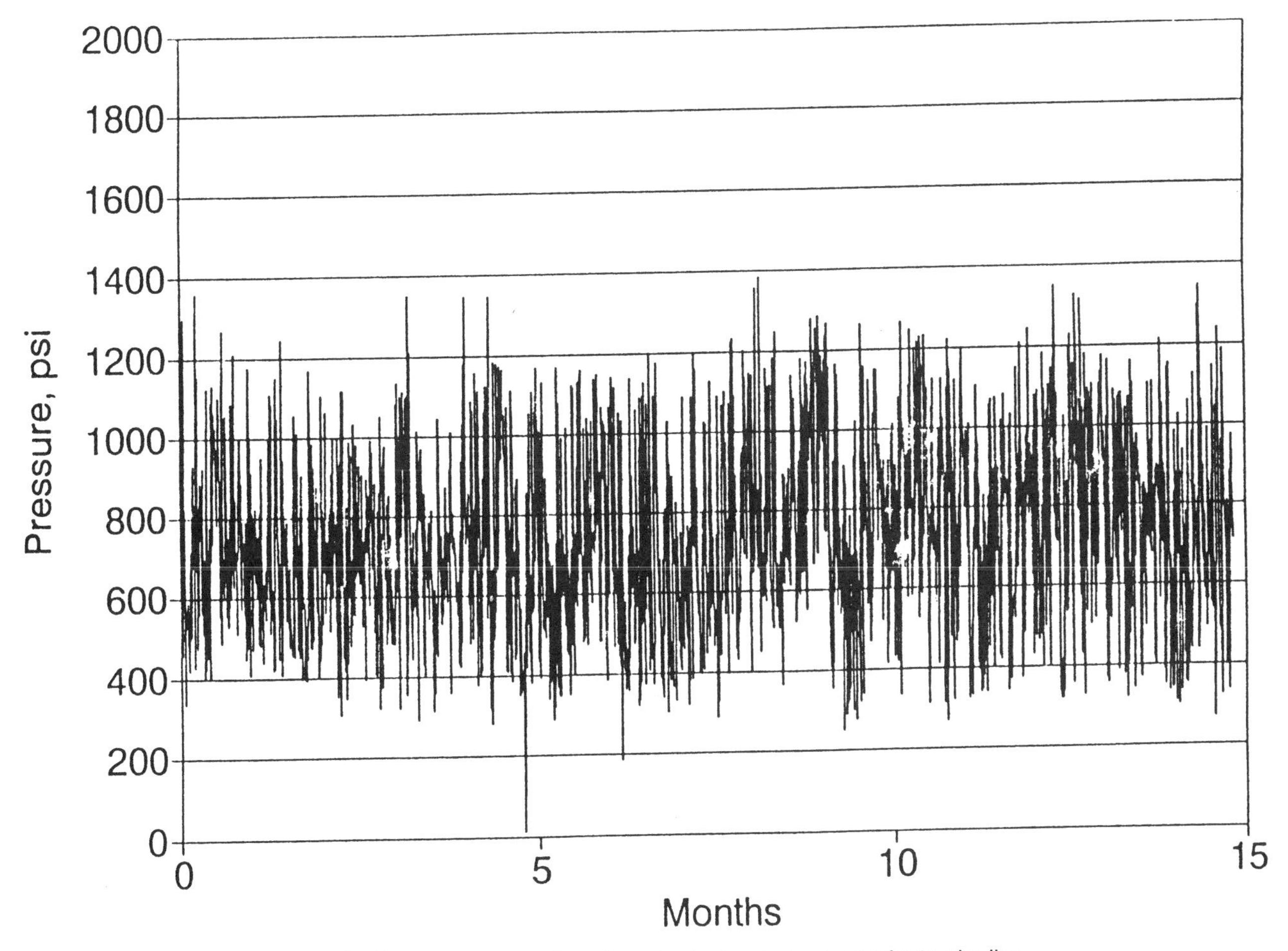

Figure 9. Pressure versus time for a typical petroleum products pipeline.

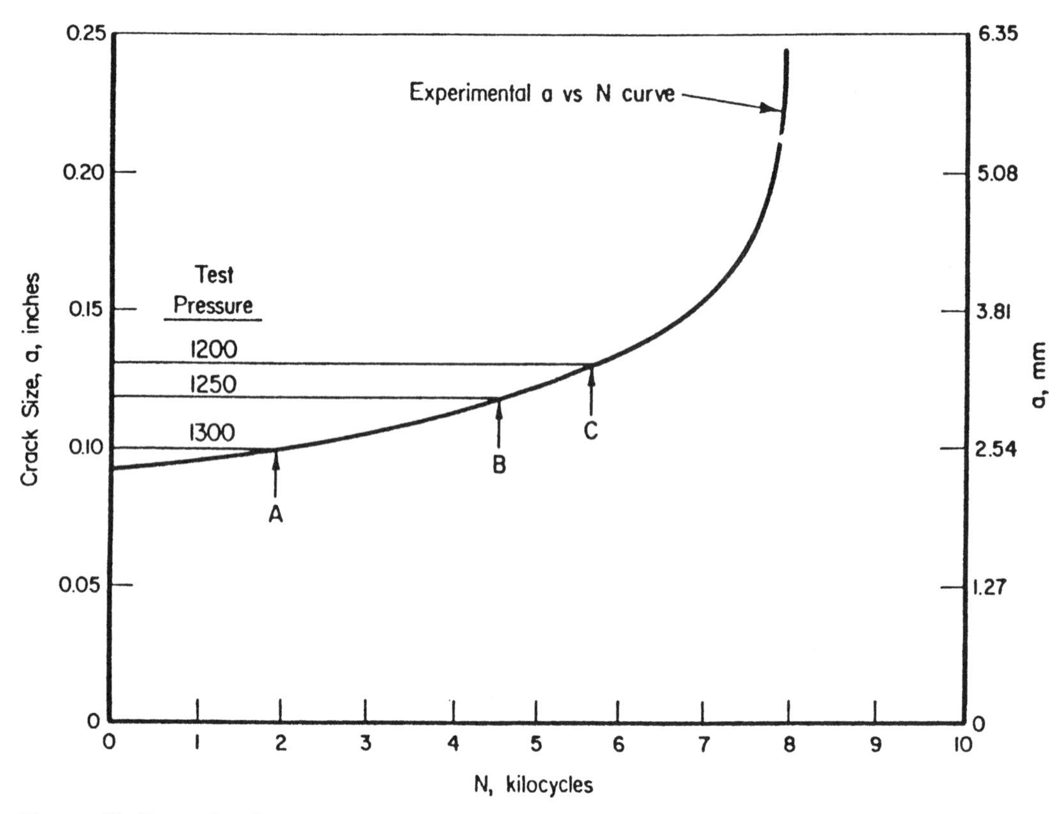

Figure 10. Example of an actual a versus N relationship used to compare effects of test pressure.

Charpy-energy material, the flaw would have had a failure pressure level of 1,300 psig after growing to a depth of 0.1 in. as indicated by the 1,300 psi "test-pressure" line. Similarly, it would have had failure pressure levels of 1,250 psig and 1,200 psig after growing to depths of 0.12 in. and 0.13 in., respectively. These values were calculated on the basis of the failure-pressure flaw-size relationship of the types shown in Figures 1–3 but for this particular pipe geometry.

Consider the following scenarios based on Figure 10. First, as service cycles are applied and the flaw has grown from 0.095 in. to 0.1 in. (N = 2,300 cycles corresponding to Point A), the remaining flaw is exactly the size that would just survive a test to 1,300 psig. Since the number of cycles to failure is 8,000, the flaw has a remaining life of 5,700 cycles. Second, after the flaw has grown to 0.12 in. (N = 4,800 cycles corresponding to Point B), the remaining flaw is exactly the size that would just survive a test to 1,250 psig. It has a remaining life of 3,200 cycles. Third, after the flaw has grown to 0.13 in. (N = 6,000 cycles corresponding to Point C), the remaining flaw is exactly the size that would just survive a test to 1,200 psig. It has a remaining life of 2,000 cycles. What this shows is the effectiveness of maximizing the test-pressure-to-operating pressure ratio. In terms of test pressure versus cycles to failure, the results would be as follows:

Test Pressure psig	Cycles to Failure After the Test	Relative Time to Failure
1,200	2,000	1.0
1,250	3,200	1.6
1,300	5,700	2.9

This exercise illustrates the valuable gain that can be made by small increases in test pressure. For a 4% increase over the base case (1,200 psig), the gain in time to failure is 60%. For an 8% increase, the gain is nearly 300%. The above-demonstrated effect of test-pressure-to-operating-pressure ratio on service life provides one of the strongest incentives for high-pressure testing.

Calculating a retest interval. In the example shown above, let us assume that the pipeline sees 200 cycles of pressure from 0 to 1,020 psig per year. Let us also assume that it was tested to 1,200 psig. Such a test probably does not give an adequate margin of safety and, in fact, it does not meet the D.O.T.'s minimum requirement of 1.25 × MOP for a liquid pipeline (Part 195). But it serves as a useful example. As the example shows, the worst-case flaw that might survive the test could be expected to fail in 2,000 cycles or 10 years at 200 cycles per year. Prudence dictates that one use a factor of safety as was suggested in our previous discussion on corrosion. The recommended factor of safety is 2, and it is based upon an examination of the "a" versus "N" relationship of Figure 10.

Shown in Figure 11 is a replot of the "a" versus "N" relationship of Figure 10. Early in the life of a crack, the growth is slow. It accelerates, however, as the crack grows deeper, because the stress intensity factor driving the crack is a function of both applied stress and crack size. The implications of this behavior are illustrated as follows. Consider the crack represented by the curve in Figure 11. This crack is initially 0.095 in. in depth and will grow to failure in 8,000 cycles of the service pressure.

The hydrostatic test and retest level of 1,200 psig is represented by the horizontal line at "a" = 0.13. This line is arbitrarily divided into four increments represented by Points 1, 2, 3, 4, and 5. The horizontal line intersects the a versus N relationship at about 5,500 cycles. Any time the pipeline is tested after 5,500 cycles, the flaw will fail in the test. Suppose one chooses a retest interval equal to the distance from Point 1 to Point 3 (5,000 cycles). If the pipeline is tested at Point 3, the flaw will not fail. But it will also not survive until the next test at Point 5 (10,000 cycles). Such a retest interval is too long even though no test failure occurred in the first test.

Now consider what happens if the test interval is cut in half. The retests are then conducted at Points 2, 3, and 4, and the flaw fails during the test at Point 4. Obviously, this is a somewhat contrived example. However, the point is that if the model is reasonably accurate, the times to failure it predicts should be adjusted by a substantial factor of safety to establish an appropriate retest interval. Because of this, we normally recommend a factor of safety of 2. In our example, this would correspond to a retest interval of only 5 years. For a higher test pressure, say 1,300 psig, the interval could be 4,700/(2 × 200) = 14 years where the 5,700 cycles come from our previous analysis based on Figure 10.

Mechanical damage

Although mechanical damage is frequently inflicted on pipelines and can cause severe degradation of their pressure-carrying capacities, we have not applied any revalidation model to this type of defect for three reasons. First, the state of analysis of mechanical-damage flaws is not as advanced as that for other types of flaws, although both Maxey[5] and British Gas[6] have made attempts to derive algorithms for dents and gouges. Second, the time-dependent nature of mechanical damage is nearly impossible to model. Since its occurrence is often concealed, the "zero" time is seldom known. The rate at which it degrades may depend on creep-like behavior and/or environmental effects. Third, we feel that the best action with respect to mechanical damage is preventative action, patrolling the rights-of-way, participating in one-call systems, conducting public education programs, and providing on-site locating of pipelines at excavation sites. In fact, our primary activity in this respect has been the development of ranking algorithms to calculate relative risk numbers for damage-prevention activities.

It should be mentioned that both in-line inspection and hydrostatic testing are viable remedies for mechanical damage. The problem is that since the existence of concealed damage is difficult to detect short of doing either of these, it

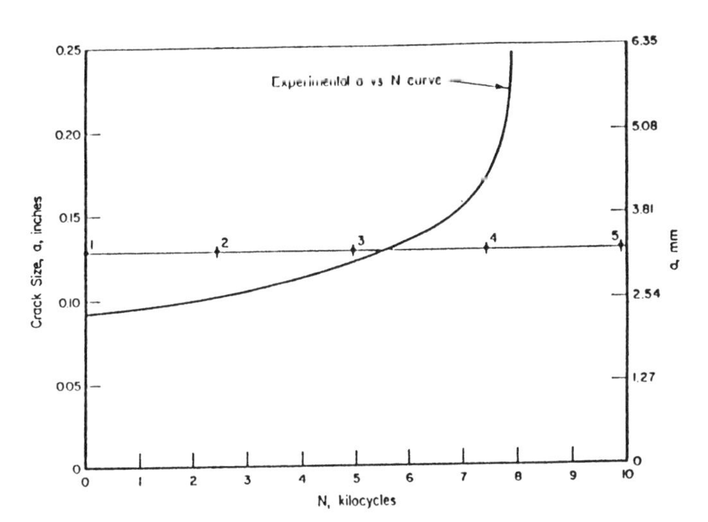

Figure 11. Example of a versus N relationship used to compare effects of retest interval.

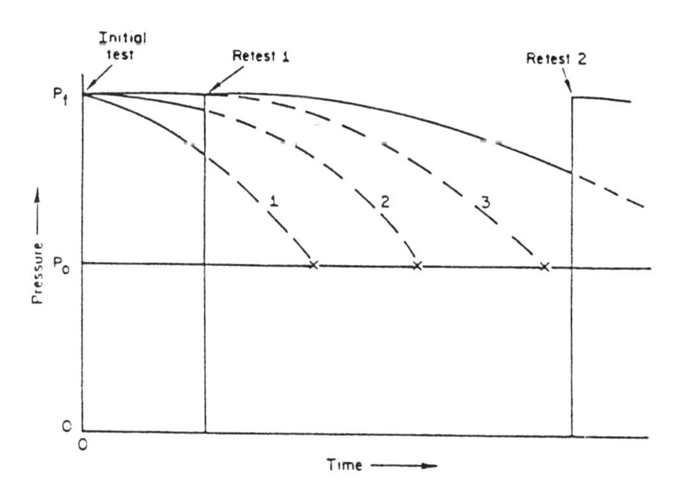

Figure 12. Concept of retesting when defect growth rate is unknown.

is difficult to know when or even if such remedies are needed.

Stress-corrosion cracking

Perhaps the hardest type of defect to deal with is stress-corrosion cracking. With this type of cracking, the main problem is that no reliable model exists to define the crack growth rate. Furthermore, it is suspected that the rate is highly variable with changes in environmental conditions. However, unlike the case for mechanical damage, it is possible to establish a "zero" time based on the date of installation. And, if the cracks exist, it is almost certain that they will grow.

A scheme used by several pipeline operators to control stress-corrosion cracking is periodic hydrostatic retesting. In the absence of a crack-growth-rate model, the retest interval has to be set arbitrarily at first and corrected as required by the results of the test. A generalized scheme for setting retest intervals to deal with defect growth at unknown rates is illustrated in Figure 12. First, an initial test is made in which defects may or may not be discovered. But, in either case, the maximum size flaw remaining is fixed by the test pressure, P_f in Figure 12, and the margin of effectiveness is proportional to P_f/P_o where P_o is the MOP. It is assumed that the defects will grow with the passage of time. Retest 1 is made after some arbitrary period of time short enough to assure that none of the remaining flaws can grow to a size that will fail in service. If this is done correctly, the growing flaws, 1, 2, and 3 in Figure 12, will fail in the test instead of in service at the times suggested by the dashed curves. If no service failures have occurred before Retest 1, then the first interval was adequate (based on the assumption that the crack-growth rate is fairly constant on the average). Depending on the outcome of Retest 1, the pipeline operator may decide to increase the interval between retests. If no test failures occur or if they occur only at levels at or near the test pressure, then it makes sense to consider a longer interval. If Retest 2 is then conducted before any flaws grow to failure in service, the new interval is apparently valid and can be lengthened depending again on the failure pressure levels of flaws in Retest 2. In this manner, the pipeline operator is able to control the phenomenon by preventing service failures, and in effect, determine the average flaw-growth rates.

It should be emphasized that the scheme portrayed in Figure 12 is valid only if the crack-growth rate is reasonably constant. For example, such a scheme would not be appropriate for fatigue cracks because of the shape of the "a" versus "N" curve as shown in Figures 10 and 11. Where the growth of the flaw accelerates rapidly, it is necessary to determine the actual growth rate. The setting of retests based upon arbitrarily assumed intervals could easily be defeated by the crack being in a rapid stage of growth, as was illustrated in the discussion of Figure 11.

Summary

We have presented concepts that are frequently used by us and others to attempt to determine when it is time to be concerned about the integrity of a pipeline. The models we use rely on well-established principles and require the use of data that is usually readily available. The models we have discussed can be used to assess revalidation needs for pipelines affected by corrosion, fatigue-crack growth, and stress-corrosion cracking. By means we alluded to but did not discuss, the risks associated with mechanical-damage defects can also be evaluated. While there is never an absolute certainty about the correctness of these types of models, we recommend them to our clients to provide guidance for establishing revalidation policies. We also recommend that they be used in conjunction with sound engineering judgement and generous factors of safety.

References

1. Maxey, W. A., Kiefner, J. E, Eiber, R. J., and Duffy, A. R., "Ductile Fracture Initiation, Propagation, and Arrest in Cylindrical Vessels," Fracture Toughness Proceedings of 1971 National Symposium on Fracture Mechanics, Part II, ASTM STP 514, American Society for Testing and Materials, pp 70–81 (1972).
2. Kiefner, J. F., and Forte, T P., "Model Predicts Hydrostatic Retest Intervals," Oil and Gas Journal (January 7 and January 14, 1985).
3. Newman, J. C., and Raju, I. S., "An Empirical Stress Intensity Factor Equation for Surface Crack," Engineering Fracture Mechanics, Vol. 15, No. 1-2 (1981).
4. Mayfield, M. E., and Maxey, W. A., ERW Weld Zone Characteristics, NG-18 Report No. 130 to American Gas Association, Catalog No. L51427 (June 18, 1982).
5. Maxey, W. A., "Outside Force Defect Behavior," NG-18 Report No. 162, Line Pipe Research Supervisory Committee, Pipeline Research Committee, American Gas Association (August 15, 1986).
6. Hopkins, P., Jones, D. G., and Clyne, A. J., "The Significance of Dents and Defects in Transmission Pipelines," Conference on Pipework Engineering and Operations, I Mech E., London, Paper C376/049 (February 1989).

Reprinted with permission of Systems Integrity and Specialty Pipeline Inspection & Engineering, Inc., organizers of the International Pipeline Rehabilitation Seminar Paper presented at the 1992 International Pipeline Rehabilitation Seminar, Houston, Jan. 28–31, 1992.

Modeling for pipeline risk assessment

W. Kent Muhlbauer, PE, WKM Consultancy, LLC

Pipeline risk assessment

The safety of pipelines relative to other modes of transportation is generally not in question: By most measures, moving products by pipeline is safer than by alternate means. Yet even as pipelines claim a superior safety record, catastrophic failures are a real possibility. With an aging infrastructure, ever-encroaching population areas, and increasing economic pressures, the perceived, if not actual, risk of pipeline failure is increasing. Pipeline safety probably can and should be improved.

The ability to predict pipeline failures—where and when the next is to occur—would obviously be a great advantage in reducing risk. Unfortunately, this is not possible except in extreme cases. So, modern risk assessment methodologies provide a surrogate for such predictions. Our risk efforts are NOT attempts to predict how many failures will occur or where the next failure will occur. Rather, the efforts are designed to systematically and objectively capture everything we know and use this information to make better decisions.

Formal versus informal risk management

Although formal pipeline risk management is growing in popularity among pipeline operators, it is important to note that risk management has always been practiced by these pipeline operators. Every time a decision is made to spend resources in a certain way, a risk management decision has been made. This informal approach to risk management has served us well, as evidenced by the very good safety record of pipelines versus other modes of transportation. An informal approach to risk management can have the further advantages of being simple, being easy to comprehend and to communicate, and being the product of expert engineering consensus built upon solid experience.

However, an informal approach to risk management does not hold up well to close scrutiny, because the process is often poorly documented and not structured to ensure objectivity and consistency of decision making. Expanding public concerns over human safety and environmental protection have contributed significantly to raising the visibility of risk management. Although the pipeline safety record is good, the violent intensity and dramatic consequences of some accidents, an aging pipeline infrastructure, and the continued urbanization of formerly rural areas has increased perceived if not actual risks.

Historical (informal) risk management therefore has these pluses and minuses:

Advantages:

- Simple/intuitive
- Consensus is often sought
- Utilizes experience and engineering judgment
- Somewhat successful, based upon pipeline safety record

Reasons to change:

- More at stake from mistakes
- Inefficiencies/subjectivities inherent in informal systems
- Lack of consistency and continuity in a changing workforce unless decision-support systems are in place
- Need to better consider complicated risk factors and their interactions

The beginning of the twenty-first century has witnessed the introduction of more formalized risk programs and studies, as well as an increasing number of private pipeline risk management ventures. Although the general conclusion is typically a recommendation to move the industry toward such formal risk management programs, there is considerably less consensus as to the type and extent of assessment programs to adopt.

Beginning the risk modeling process

Successful risk assessment modeling involves a balance among various issues:

- Identifying an exhaustive list of contributing factors versus choosing the critical few to incorporate in a model (complex versus simple)
- "Hard" data and engineering judgment (how to incorporate widely held beliefs that do not have supporting statistical data)
- Uncertainty versus statistics (how much reliance to place on the predictive power of limited data)
- Flexibility versus situation-specific model (ability to use same model for a variety of products, geographical locations, facility types, etc.)

It is important that all risk variables be considered, even if only to conclude that certain variables will not be included in the model. In fact, many variables will not be included when such variables do not add significant value but reduce the usability of the model. These decisions are done carefully and with full understanding of the role of the variables in the risk picture.

Note that many simplifying assumptions are often made, especially in complex phenomena like dispersion modeling, fire and explosion potentials, etc., in order to make the risk model easy to use and still relatively robust.

Both probability variables and consequence variables are examined in most formal risk models. This is consistent with the most widely accepted definition of risk:

$$(\text{event risk}) = (\text{event probability}) \times (\text{event consequence})$$

Several questions to the pipeline operator may better focus a risk management effort and direct the choice of a formal risk assessment technique:

- What data do you have?
- What is your confidence in the predictive value of those data?
- What are the resource demands (and availability) in terms of costs, man-hours, and time to set up and maintain a risk model?
- What benefits do you expect to accrue, in terms of cost savings, reduced regulatory burdens, improved public support, and operational efficiency?

Choices in Risk Assessment Techniques

The goal of any risk assessment technique is to quantify the risks in either a relative or an absolute sense. The risk assessment phase is the critical first step in practicing risk management. It is also the most difficult phase. Although we understand engineering concepts about corrosion and fluids flow, predicting failures beyond the laboratory in a complex "real" environment can prove impossible. No one can definitively state where or when an accidental pipeline failure will occur. However, the more likely failure mechanisms, locations, and frequencies can be estimated in order to focus risk efforts.

There is no universally accepted way to assess risks from a pipeline. It is apparent, however, that simply using historical accident counts is not enough. Because of the rare occurrence of pipeline failures, there is a high level of uncertainty associated with pipeline failure prediction and risk assessment. Although not always apparent, all risk models, from the most simple to the most complex, make use of probability theory and statistics. In a very simple application, these tools manifest themselves in hard-to-quantify experience factors and engineering judgments, that is, they are the underlying basis of sound judgments. In the more mathematically rigorous models, historical data, statistical theory, and/or mechanistic models (hard data) primarily drive the calculations.

Three general categories of more formal pipeline risk assessment models can be found in use today:

Simple decision support: Matrix models. One of the simplest risk assessment approaches is the matrix decision-analysis system. It ranks pipeline risks according to the likelihood and potential consequences of an event by a simple scale, such as high, medium, and low. Each threat to the pipeline is assigned to a cell of the matrix, the intersection of the perceived likelihood and perceived consequence. Events with both a high likelihood and a high consequence appear higher on the resulting prioritized list. This approach may be as simple as using an expert's opinion or as complicated as using quantitative data to rank risks.

The rigorous approach: Probabilistic/mechanistic models. The more rigorous and complex risk assessment is often called a Probabilistic Risk Assessment (PRA). It usually refers to a technique employed in the nuclear and aerospace industries that generally uses event trees and fault trees to model every aspect of a system. Initiating events are flowcharted forward to all possible concluding events, with probabilities being assigned to each branch along the way. Failures are backward flowcharted to all possible initiating events, again with probabilities assigned to all branches. All possible paths can then be quantified based upon the branch probabilities along the way. This technique usually relies heavily on historical failure rate data. It yields absolute risk assessments for all possible failure events. These more elaborate models are more costly than simpler risk assessments. They are technologically more demanding to develop and integrate via computer applications, require trained operators, and benefit from extensive databases.

The hybrid approach: Indexing models. Perhaps the most popular pipeline risk assessment technique in current use is the "index model." In this approach, a relative weight is assigned to every important condition and activity on the pipeline. This includes both risk-reducing and risk-increasing items. This relative weight reflects the importance of the item in the risk assessment and is based upon statistics where available and upon engineering judgment where data are not available. The risk component scores are subsequently summed for each pipeline section to obtain a relative risk ranking of all pipe sections. The various pipe segments may then be ranked according to their relative risk scores in order to prioritize repairs and inspections. Among pipeline operators today, this technique is widely used and ranges from a very simple 5–20 factor model (where only factors such as leak history and population density are considered) to models with hundreds of factors considering virtually every item that impacts risk. When an indexing model is created from a probabilistic approach—using scenarios, event/fault trees, and all available historical data—this approach is often a solution with the best cost–benefit ratio.

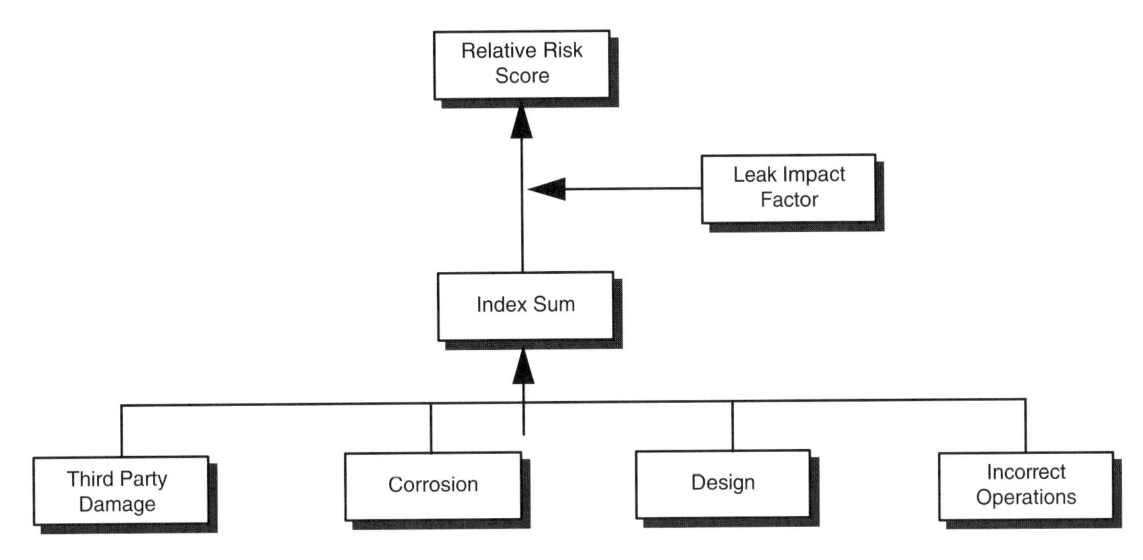

Typical Pipeline Risk Model Structure

Other issues in risk modeling

In comparing risk assessment approaches, some issues arise that often lead to confusion. The following section discusses some of those issues:

Absolute versus relative risk. Absolute risk implies a universal scale, usually in terms of a measurable consequence such as fatalities or costs per time period by which all risks can be compared. An example would be a risk assessment that produces a value such as 1.25 E-08 fatalities per thousand mile-years of pipeline. Relative risk is designed to prioritize risks by saying that pipeline section B is riskier than section A by an amount equal to X. The former is a predictive measure yielding the probability and consequence of failure, whereas the latter is a comparative tool yielding a ranking of current risks. A criticism of the relative scale is its inability to compare risks of several dissimilar systems (e.g., pipelines versus trucks) and its inability to directly provide failure predictions. The absolute scale fails in relying heavily upon point estimates, particularly for rare events that are extremely difficult to quantify, and in the unwieldy numbers that often generate a negative reaction from the public.

For purposes of communications and in practical applications, this is not really an important issue. Either scale can be readily converted to the other scale, if circumstances so warrant. A relative risk scale can be converted to an absolute scale by correlating the relative risk scores with known failure rates and ensuring that the relative scales appropriately reflect real-world processes. The absolute scale is converted to more manageable and understandable (nontechnical) relative scales by simple mathematical relationships.

A danger underlying this issue is the possible assumption that a complex number expressed in scientific notation is more accurate than a simple number. In reality, either method should use the same available data pool and be forced to make the same number of assumptions when data are not available.

Quantitative versus qualitative models. These terms are often used to distinguish the amount of historical failure-related data analyzed in the model and the amount of mathematical calculations employed in arriving at a risk answer. A model that exclusively uses hard data is sometimes referred to as quantitative, whereas a model employing relative scales, even if later assigned numbers, is referred to as qualitative or semiquantitative. The danger in such labeling is that they imply a level of accuracy. In reality, the labels often tell more about the level of modeling effort, cost, and data sources. All complete risk models must use both quantitative and qualitative information.

Subjectivity versus objectivity. In theory, a purely objective model will strictly adhere to statistical information and will have no opinion data. A purely subjective model implies complete reliance upon expert opinion. In practice, no pipeline risk model fully adheres to either. Objectivity cannot be purely maintained when dealing with the real-world situation of missing data and variables that are highly confounded. On the other hand, subjective models certainly use "hard" data to form or support judgments.

The use of unquantifiable evidence. In a difficult-to-quantify item, some would argue that nonstatistical analyses are potentially damaging. Although there is always this danger of misunderstanding the role of a factor in a risk assessment, there is similarly the more immediate danger of an incomplete analysis by omission of a factor. For example, public education is seen by most pipeline professionals to be a very important aspect of reducing the number of third-party damages, leak reporting, and emergency response. Quantifying this level of importance and correlating with the many varied approaches to public education is quite difficult. A concerted effort to study these data is needed to best assess the effect on risk, but in the absence of such a study, most would agree that a company that practices public education will achieve some level of risk reduction over a company that does not.

Uncertainty. False precision often is assigned to analysis yielding a specific number, especially when such a number appears to be the result of extensive calculations. In reality, uncertainty always exists. Uncertainty can be thought of as a property of the situation. We add to that uncertainty when we measure the situation. The measurement process itself adds uncertainty as measurement errors and inaccuracies accumulate. Separating the noise—the random variations caused in part by measurement uncertainties—from the signal—the true differences in risk—is an arduous task. At best, we hope for a high signal-to-noise ratio. The higher the ratio, the higher our confidence in identifying true signals. All risk-assessment techniques must address uncertainty, and no method can reliably eliminate it or even reduce it significantly.

Choosing a risk-assessment technique

Any or all of the previously described techniques might have a place in risk assessment/management. The choice may well be dependent on the situation. Understanding the strengths and weaknesses of the different risk assessment methodologies gives the user the basis for choosing one. A case can be made for using each in certain situations. For example, a simple matrix approach crystallizes thinking and is a step above informal risk assessment. However, if the need is to evaluate specific events at any point in time, a narrow focus assessment with well-known event probabilities, then probabilistic risk analysis might be the tool of choice. If the need is to weigh immediate risk trade-offs or perform inexpensive overall assessments, indexing models might be the best choice. The following shows some trade-offs associated with each choice:

Pros and cons I (PRA)

+ Rigorous, scientific approach
+ Accepted in other industries
+ Uses all available information
– Costly
– Difficult to do resource allocation
– Difficult to do overall risk management
– Might create the "illusion of knowledge"
– Intimidating to nontechnical audience

Pros and cons II (indexing)

+ Uses all available information
+ Intuitive
+ Flexible
– Possibly more subjective
– Must be well documented
– Often requires a subsequent linkage to absolute values

Pros and cons III (matrix)

+ Improvement over informal techniques
+ Forces more logical examination of situation
+ Inexpensive approach
– Limited number of risk factors (not comprehensive)
– Subjective

In addition to the three general categories of risk models, supporting tools such as fault trees, event trees, HAZOPS, what-if scenarios, and others play a role in performing risk assessment. These tools are almost always the foundation and building blocks of the complete risk assessment model.

The following are some ways a particular technique can support a specific need:

Application choices I (PRA, scenarios, event/fault trees)

- Study specific events and/or create risk models
- Perform postincident investigations
- Compare risks of specific failures
- Calculate specific event probabilities

Application choices II (indexing model built from scenarios, event/fault trees, PRA)

- Obtain an inexpensive overall risk model
- Create a resource allocation model
- Model the interaction of many potential failure mechanisms
- Study or create an operating discipline

Application choices III (matrix)

- Better quantify a belief
- Create a simple decision support tool
- Combine several beliefs into a single solution
- Document choices in resource allocation

The ideal risk assessment model

Regardless of the technique used, some properties of the ideal risk assessment tool can be identified. A partial list of the characteristics of the "ideal model" includes the following:

Costs. The value or benefits derived from the risk assessment process should clearly outweigh the costs of setting up, implementing, and maintaining the program.

Learning ability. Because risk is not constant over the length of a pipeline or over a period of time, the model must be able to "learn" as information changes. This means that new data should be easy to incorporate into the model.

Signal-to-noise ratio. Because the model is in effect a measurement tool, it must have a suitable signal-to-noise ratio. This means that the "noise," the amount of uncertainty in the measurement (due to numerous causes) must be low enough that the "signal," the risk value of interest, can be read. This is similar to the "accuracy" of the model but involves additional considerations that surround the high level of uncertainty associated with risk management.

In examining a proposed risk assessment effort, it may be wise to evaluate the risk assessment model to ensure the following:

- All failure modes are considered
- The model appropriately treats both time-dependent and independent failure mechanisms
- All risk elements are considered and the most critical ones are included
- Failure modes are considered, independently as well as in aggregate
- All available information is being appropriately utilized
- Real-world processes are appropriately represented
- Relative risk values are anchored by real failure rates and consequence potential
- Provisions exist for regular updates of information, including new types of data
- Consequence factors are separable from probability factors
- Weightings, or other methods to recognize relative importance of factors, are established
- The rationale behind weightings is well documented and consistent
- A sensitivity analysis has been performed
- The model reacts appropriately to failures of any type
- Risk elements are combined appropriately ("and" versus "or" combinations)
- Steps are taken to ensure consistency of evaluation
- Risk assessment results form a reasonable statistical distribution (outliers?)
- There is adequate discrimination in the measured results (signal-to-noise ratio)
- Comparisons can be made against fixed or floating "standards" or benchmarks

Finally, a view to the next step, risk management, should be taken. A good risk-assessment technique will allow a smooth transition into the management of the observed risks. This means that provisions for resource allocation modeling and the evolution of the overall risk model must be made.

Managing risks: The cost connection

In some sense, we have near-complete control of the risk. We can spend nothing on preventing accidents, or we can spend enormous sums of money overdesigning facilities, employing an army of inspectors, and routinely shutting down lines for preventive maintenance and replacement. Pragmatically, operators spending too little on preventing accidents will be out of business from regulatory intervention or from the cost of accidents. On the other hand, if an operator spends too much on accident prevention, he can be driven out of business by the competition, even if, perhaps, that competition has more accidents.

Risk management, to a large extent, revolves around the central process of making choices in the design and day-to-day operations of a pipeline system. Many of these choices are mandated by regulations, whereas others are economically (budget) constrained. By assigning a cost to pipeline accidents (a sometimes difficult and controversial thing to do) and including this in the cost of operations, the optimum balance point is the lowest cost of operations.

No operator will ever have all of the relevant information needed to guarantee safe operations. There will always be an element of the unknown. Managers must control the "right" risks with limited resources, as there will always be limits on the amount of time, manpower, or money to apply to a risk situation. Managers must weigh their decisions carefully in light of what is known and unknown. It is, perhaps, this assimilation of complex data and the subsequent integration of competing risk reduction–profit goals that lies at the heart of the debate about how best to manage pipeline risks.

Conclusion

The move to more formal risk techniques is intended to increase operational consistency and credibility, especially when such techniques are offered for public viewing. In a concerted effort to demonstrate its dedication to public and environmental safety while simultaneously seeking to reduce potential liabilities and regulatory requirements, the federal government and the pipeline industry (both gas and liquid) are embracing risk management techniques through integrity management programs.

Even without such broad-based support, risk management has long been recognized as a valuable effort in pipeline operations. Risk assessment approaches are available to serve many needs, although no single approach is the best in addressing all needs. It is important that those practicing (or planning to practice) risk management understand the strengths and limitations of various approaches and set their expectations accordingly.

References

1. W. Kent Muhlbauer, Pipeline Risk Management Manual, 3rd ed., Gulf Publishing Co., Houston (2004).

Sample risk variable list

Relative Risk = [Index Sum]/[Leak Impact Factor]
Index Sum = [Third Party] + [Corrosion] + [Design] + [Incorrect Operation]

Third party damage potential

A	Minimum depth cover	20%	0–20 pts
B	Activity level	20%	0–20 pts
C	Above-ground facilities	10%	0–10 pts
D	One-call system	15%	0–15 pts
E	Public education	15%	0–15 pts
F	Right-of-way condition	5%	0–5 pts
G	Patrol	15%	0–15 pts
		100%	100 pts

Corrosion potential

A. Atmospheric corrosion 0–10 pts

A1	Atmospheric corrosion	0–5 pts
A2	Atmospheric type	0–2 pts
A3	Atmospheric coating	0–3 pts

B. Internal corrosion 0–20 pts

B1	Product corrosivity	0–10 pts
B2	Internal protection	0–10 pts

C. Subsurface corrosion 0–70 pts

C1	Subsurface environment		0–20 pts
	Soil corrosivity	0–15 pts	
	Mechanical corrosion	0–5 pts	
C2	Cathodic protection		0–25 pts
	Effectiveness	0–15 pts	
	Interference potential	0–10 pts	
C3	Coating		0–25 pts
	Fitness	0–10 pts	
	Condition	0–15 pts	
			70 pts

Design risk

A	Safety factor	35%	0–35 pts
B	Fatigue	15%	0–15 pts
C	Surge potential	10%	0–10 pts
D	Integrity verification	25%	0–25 pts
E	Land movements	15%	0–15 pts
		100%	0–100 pts

Incorrect operations

A	Design	30%	0–30 pts
	Hazard identification		0–4 pts
	MAOP potential		0–12 pts
	Safety systems		0–10 pts
	Material selection		0–2 pts
	Checks		0–2 pts
B	Construction	20%	0–20 pts
	Inspection		0–10 pts
	Materials		0–2 pts
	Joining		0–2 pts
	Backfill		0–2 pts
	Handling		0–2 pts
	Coating		0–2 pts
C	Operation	35%	0–35 pts
	Procedures		0–7 pts
	SCADA/ communications		0–3 pts
	Drug testing		0–2 pts
	Safety programs		0–2 pts
	Surveys/maps/ records		0–5 pts
	Training		0–10 pts
	Mechanical error preventers		0–6 pts
D	Maintenance	15%	0–15 pts
	Documentation		0–2 pts
	Schedule		0–3 pts
	Procedures		0–10 pts
		100%	0–100 pts

Leak Impact Factor
= (Product Hazard) × (Receptors) × (Spill Size) × (Dispersion)

A Product hazard

1.	(Acute + chronic hazards)	
	Acute hazards	
	a. Nf	0–4
	b. Nr	0–4
	c. Nh	0–4
Total (Nh + Nr + Nf)		0–12
2.	*Chronic hazard, RQ*	0–1
B	Receptors	0–1
	Population	
	Environment	
	High value areas	
C	Spill size	0–1
D	Dispersion	0–1

22: Conversion Factors

Units of measurement convert from one system to another

To Convert	To	Multiply by
	A	
abcoulomb	statcoulombs	2.998 E+10
acre	square chain (gunters)	10
acre	rods	160
acre	square links (gunters)	1 E+05
acre	hectare or square hectometer	0.4047
acres	square feet	43,560
acres	square meters	4,047
acres	square miles	1.562 E–03
acres	square yards	4,840
acre-feet	cubic feet	43,560
acre-feet	gallons	3.259 E+05
acre-feet	cubic meters	1.233489 E+03
amperes/square centimeter	amperes/square inch	6.452
amperes/square centimeter	amperes/square meter	1 E–04
amperes/square inch	amperes/square centimeter	0.155
amperes/square inch	amperes/square meter	1,550
amperes/square meter	amperes/square centimeter	E–04
amperes/square meter	amperes/square inch	6.452 E–04
ampere-hours	coulombs	3,600
ampere-hours	faradays	0.03731
ampere-turns	gilberts	1.257
ampere-turns/centimeter	ampere-turns/inch	2.54
ampere-turns/centimeter	ampere-turns/meter	100
ampere-turns/centimeter	gilberts/centimeter	1.257
ampere-turns/inch	ampere-turns/centimeter	0.3937
ampere-turns/inch	ampere-turns/meter	39.37
ampere-turns/inch	gilberts/centimeter	0.495
ampere-turns/meter	ampere-turns/centimeter	0.01
ampere-turns/meter	ampere-turns/inch	0.0254
ampere-turns/meter	gilberts/centimeter	0.01257
angstrom unit	inch	3,937 E–09
angstrom unit	meter	1 E–10
angstrom unit	micron or Mu	1 E–04
are	acre (U.S.)	0.02471
ares	square yards	119.6
ares	acres	0.02471
ares	square meters	100
astronomical unit	kilometers	1.495 E+08
atmospheres	ton/square inch	0.007348
atmospheres	bar	1.01325
atmospheres	centimeters of mercury	76
atmospheres	feet of water (at 4°C)	33.9
atmospheres	inches of mercury (at 0°C)	29.92
atmospheres	kilograms/square centimeter	1.0333
atmospheres	kilograms/square meter	10,332
atmospheres	kiloPascal	101.33
atmospheres	pounds/square inch	14.7
atmospheres	tons/square foot	1.058
	B	
barrels (U.S., dry)	cubic inches	7,056
barrels (U.S., dry)	quarts (dry)	105
barrels (U.S., liquid)	gallons	31.5
barrels (oil)	gallons (oil)	42
barrels	cubic inches	9,702
barrels	cubic meters	0.15899
barrels/hour	cubic meters/hour	1.589873 E–01
barrels/day	cubic meters/day	1.589873 E–01
barrels/day	cubic meters/hour	6.624471 E–03
barrels/day	cubic meters/year	5.803036 E+01
barrels/day	gallons/minute	0.02917
barrels/foot	cubic meters/meter	5.216119 E–01
barrels/mile	cubic meters/km	9.879013 E–02
barrels/U.S. ton	cubic meters/ton	0.17525
bars	atmospheres	0.9869
bars	dynes/square centimeter	1 E+06
bars	kilograms/square meter	1.020 E+04
bars	kiloPascal	100
bars	pounds/square foot	2,089
bars	pounds/square inch	14.5
baryl	dyne/square centimeter	1
board feet	cubic inches	144
board feet	cubic meters	0.00236
bolt (U.S. cloth)	meters	36.576
btu	liter-atmosphere	10.409
btu	ergs	1.0550 E+10
btu	foot-pounds	778.3
btu	gram-calories	252
btu	horsepower-hours	3.931 E–04
btu	joules	1,054.80
btu	kilograin-calories	0.252
btu	kilograin-meters	107.5
btu	kilowatt-hours	2.928 E–04
btu/cubic foot	megajoules/cubic meter	3.725895 E–02
btu/hour	foot-pounds/second	0.2162
btu/hour	gram-calories/second	0.07
btu/hour	horsepower-hours	3.929 E–04

To Convert	To	Multiply by
btu/hour	watts	0.2931
btu/hour-foot 2°F/foot	watt/meter-Kelvin	1.73074
btu/pound	joules/gram	2.326
btu/pound	watt hour/kilogram	0.64611
btu/pound mol	joules/mol	2.326
btu/pound	megajoules/kilogram	2.326000 E–03
btu/minute	foot-pounds/second	12.96
btu/minute	horsepower	0.02356
btu/minute	kilowatts	0.01757
btu/minute	watts	17.57
btu/square foot/minute	watts/square inch	0.1221
btu/U.K. gallon	megajoules/cubic meter	2.320800 E–01
btu/U.S. gallon	megajoules/cubic meter	2.787163 E+04
bucket (Br. dry)	cubic centimeters	1.818 E+04
bushels	cubic feet	1.2445
bushels	cubic inches	2,150.40
bushels	cubic meters	0.03524
bushels	laters	35.24
bushels	pecks	4
bushels	pints (dry)	64
bushels	quarts (dry)	32
	C	
caliber	inches	0.01
calories	joules	4.184
calories/pound	joules/kilogram	9.22414
calories, gram (mean)	btu (mean)	3.9685 E–03
candle/square centimeter	lamberts	3,142
candle/square inch	lamberts	0.487
centares (centiares)	square meters	1
centigrade	Fahrenheit	(°C × 9/5) + 32
centigrams	grams	0.01
centiliter	ounce fluid (U.S.)	0.3382
centiliter	cubic inches	0.6103
centiliter	drams	2,705
centiliters	liters	0.01
centimeters	feet	3.281 E–02
centimeters	inches	0.3937
centimeters	kilometers	1 E–05
centimeters	meters	0.01
centimeters	miles	6.214 E–06
centimeters	millimeters	10
centimeters	mils	393.7
centimeters	yards	1.094 E–02
centimeter-dynes	centimeter-grams	1.020 E–03
centimeter-dynes	meter-kilograms	1.020 E–08
centimeter-dynes	pound-feet	7.376 E–08

To Convert	To	Multiply by
centimeter-grams	centimeter-dynes	980.7
centimeter-grams	meter-kilograms	1 E–05
centimeter-grams	pound-feet	7.233 E–05
centimeters of mercury	atmospheres	0.01316
centimeters of mercury	feet of water	0.4461
centimeters of mercury	kilograms/square meter	136
centimeters of mercury	pascal	1,333.2239
centimeters of mercury	pounds/square foot	27.85
centimeters of mercury	pounds/square inch	0.1934
centimeters/second	feet/minute	1.1969
centimeters/second	feet/second	0.03281
centimeters/second	kilometers/hour	0.036
centimeters/second	knots	0.1943
centimeters/second	meters/minute	0.6
centimeters/second	miles/hour	0.02237
centimeters/second	miles/minute	3.728 E–04
centimeters/second/second	feet/second/second	0.03281
centimeters/second/second	kilometers/hour/second	0.036
centimeters/second/second	meters/second/second	0.01
centimeters/second/second	miles/hour/second	0.02237
centipoise	newton-second/square meter	0.001
centipoise	pascal-second	0.001
centistoke	square millimeter/second	1
chain	inches	792
chain	meters	20.12
chains (surveyors' or gunter's)	yards	22
circular mils	square centimeters	5.067 E–06
circular mils	square mils	0.7854
circumference	radians	6.283
circular mils	sqinches	7.854 E–07
cords	cord feet	8
cord feet	cubic feet	16
coulomb	statcoulombs	2.998 E+09
coulombs	faradays	1.036 E–05
coulombs/square centimeter	coulombs/square inch	64.52
coulombs/square centimeter	coulombs/square meter	1 E+04
coulombs/square inch	coulombs/square centimeter	0.155
coulombs/square inch	coulombs/square meter	1,550
coulombs/square meter	coulombs/square centimeter	1 E–04
coulombs/square meter	coulombs/square inch	6.452 E–04
cubic centimeters	cubic feet	3.531 E–05
cubic centimeters	cubic inches	0.06102
cubic centimeters	cubic meters	1 E–06
cubic centimeters	cubic yards	1.308 E–06
cubic centimeters	gallons (U.S. liquid)	2.642 E–04
cubic centimeters	liters	0.001

(continued on next page)

To Convert	To	Multiply by
cubic centimeters	pints (U.S. liquid)	2.113 E–03
cubic centimeters	quarts (U.S. liquid)	1.057 E–03
cubic feet	bushels (dry)	0.8036
cubic feet	cubic centimeters	28,320
cubic feet	cubic inches	1,728
cubic feet	cubic meters	0.02832
cubic feet	cubic yards	0.03704
cubic feet	gallons (U.S. liquid)	7.48052
cubic feet	liters	28.32
cubic feet	pints (U.S. liquid)	59.84
cubic feet	quarts (U.S. liquid)	29.92
cubic feet/minute	cubic centimeters/ second	472
cubic feet/minute	gallons/second	0.1247
cubic feet/minute	liters/second	0.4719474
cubic feet/minute	pounds of water/minute	62.43
cubic feet/second	million gallons/day	0.646317
cubic feet/second	gallons/minute	448.831
cubic inches	cubic centimeters	16.39
cubic inches	cubic feet	5.787 E–04
cubic inches	cubic meters	1.639 E–05
cubic inches	cubic yards	2.143 E–05
cubic inches	gallons	4.329 E–03
cubic inches	liters	0.01639
cubic inches	mil-feet	1.061 E+05
cubic inches	pints (U.S. liquid)	0.03463
cubic inches	quarts (U.S. liquid)	0.01732
cubic meters	bushels (dry)	28.38
cubic meters	cubic centimeters	1 E+06
cubic meters	cubic feet	35.31
cubic meters	cubic inches	61,023
cubic meters	cubic yards	1.308
cubic meters	gallons (U.S. liquid)	264.2
cubic meters	liters	1,000
cubic meters	pints (U.S. liquid)	2,113
cubic meters	quarts (U.S. liquid)	1,057
cubic yards	cubic centimeters	7.646 E+05
cubic yards	cubic feet	27
cubic yards	cubic inches	46,656
cubic yards	cubic meters	0.7646
cubic yards	gallons (U.S. liquid)	202
cubic yards	liters	764.6
cubic yards	pints (U.S. liquid)	1,615.90
cubic yards	quarts (U.S. liquid)	807.9
cubic yards/minute	cubic feet/second	0.45
cubic yards/minute	gallons/second	3.367
cubic yards/minute	liters/second	12.74

To Convert	To	Multiply by
	D	
dalton	gram	1.650 E–24
days	seconds	86,400
decigrams	grams	0.1
deciliters	liters	0.1
decimeters	meters	0.1
degrees (angle)	quadrants	0.01111
degrees (angle)	radians	0.01745
degrees (angle)	seconds	3,600
degrees F	$(F - 32) \times 0.55555$	degrees C
degrees C	$(C + 17.78) \times 1.8$	degrees F
degrees/second	radians/second	0.01745
degrees/second	revolutions/minute	0.1667
degrees/second	revolutions/second	2.778 E–3
dekagrams	grams	10
dekaliters	liters	10
dekameters	meters	10
drams (apothecaries' or troy)	ounces (avoidupois)	0.1371429
drams (apothecaries' or troy)	ounces (troy)	0.125
drams (U.S., fluid or apothecaries')	cubic centimeters	3.6967
drams	grams	1.7718
drams	grains	27.3437
drams	ounces	0.0625
dyne/centimeter	erg/square millimeter	0.01
dyne/square centimeter	atmospheres	9.869 E–07
dyne/square centimeter	inch of mercury at 0°C	2.953 E–05
dyne/square centimeter	inch of water at 4°C	4.015 E–04
dynes	grams	1.020 E–03
dynes	joules/centimeter	1 E–07
dynes	joules/meter (newtons)	1 E–05
dynes	kilograms	1.020 E–06
dynes	poundals	7.233 E–05
dynes	pounds	2.248 E–06
dynes/square centimeters	bars	1 E–06
	E	
ell	centimeters	114.3
ell	inches	45
em, pica	inch	0.167
em, pica	centimeters	0.4233
erg/second	dyne–centimeter/second	1
ergs	btu	9.480 E–11
ergs	dyne-centimeters	1
ergs	foot-pounds	7.367 E–08
ergs	gram-calories	0.2389 E–07
ergs	gram-centimeters	1.020 E–03

To Convert	To	Multiply by
ergs	horsepower-hours	3.7250 E–14
ergs	joules	1 E–07
ergs	kilogram-calories	2.389 E–11
ergs	kilogram-meters	1.020 E–08
ergs	kilowatt-hours	0.2778 E–13
ergs	watt-hours	0.2778 E–10
ergs/second	btu/minute	5,688 E–09
ergs/second	foot-pounds/minute	4.427 E–06
ergs/second	foot-pounds/second	7.3756 E–08
ergs/second	horsepower	1.341 E–10
ergs/second	kilogram-calories/minute	1.433 E–09
ergs/second	kilowatts	1 E–10
	F	
farads	microfarads	1 E+06
faraday/second	ampere (absolute)	9.6500 E+04
faradays	ampere-hours	26.8
faradays	coulombs	9.649 E+04
fathom	meter	1.828804
fathoms	feet	6
feet	centimeters	30.48
feet	kilometers	3.048 E–04
feet	meters	0.3048
feet	miles (nautical)	1.645 E–04
feet	miles (statute)	1.894 E–04
feet	millimeters	304.8
feet	mils	1.2 E+04
feet of water	atmospheres	0.0295
feet of water	inches of mercury	0.8826
feet of water	kilograms/square centimeter	0.03048
feet of water	kilograms/square meter	304.8
feet of water	pounds/square feet	62.43
feet of water	pounds/square inches	0.4335
feet/minute	centimeters/second	0.508
feet/minute	feet/second	0.01667
feet/minute	kilometers/hour	0.01829
feet/minute	meters/minute	0.3048
feet/minute	miles/hour	0.01136
feet/second	centimeters/second	30.48
feet/second	kilometers/hour	1.097
feet/second	knots	0.5921
feet/second	meters/minute	18.29
feet/second	miles/hour	0.6818
feet/second	miles/minute	0.01136
feet/second/second	centimeters/second/second	30.48
feet/second/second	kilometers/hour/second	1.097
feet/second/second	meters/second/second	0.3048

To Convert	To	Multiply by
feet/second/second	miles/hour/second	0.6818
feet/100 feet	percent grade	1
foot-candle	lumen/square meter	10.764
foot-pounds	btu	1.286 E–03
foot-pounds	ergs	1.356 E+07
foot-pounds	grain-calories	0.3238
foot-pounds	horsepower-hours	5.050 E–07
foot-pounds	joules	1.356
foot-pounds	kilogram-calories	3.24 E–04
foot-pounds	kilogram-meters	0.1383
foot-pounds	kilowatt-hours	3.766 E–07
foot-pounds/minute	btu/minute	1.286 E–03
foot-pounds/minute	foot-pounds/second	0.01667
foot-pounds/minute	horsepower	3.030 E–05
foot-pounds/minute	kilogram-calories/minute	3.24 E–04
foot-pounds/minute	kilowatts	2.260 E–05
foot-pounds/second	btu/hour	4.6263
foot-pounds/second	btu/minute	0.07717
foot-pounds/second	horsepower	1.818 E–03
foot-pounds/second	kilogram-calories/minute	0.01945
foot-pounds/second	kilowatts	1.356 E–03
furlongs	miles (U.S.)	0.125
furlongs	rods	40
furlongs	feet	660
	G	
gallons	cubic centimeters	3,785
gallons	cubic feet	0.1337
gallons	pints (liquid)	8
gallons	quarts	3
gallons	cubic inches	231
gallons (U.S.)	cubic meters	3.785412 E–03
gallons (U.K.)	cubic meters	4.546092 E–03
gallons	cubic yards	4.951 E–03
gallons	liters	3.785
gallons (liquid British imperial)	gallons (U.S. liquid)	1.20095
gallons (U.S.)	gallons (imperial)	0.83267
gallons of water	pounds of water	8.3453
gallons/minute	cubic feet/second	2.228 E–03
gallons/minute	liters/second	0.06308
gallons/minute	cubic feet/hour	8.0208
gausses	lines/square inch	6.452
gausses	webers/square centimeter	1 E–08
gausses	webers/square inch	6.452 E–08
gausses	webers/square meter	1 E–04
gilberts	ampere-turns	0.7958
gilberts/centimeter	ampere-turns/centimeter	0.7958

(continued on next page)

To Convert	To	Multiply by
gilberts/centimeter	ampere-turns/inch	2.021
gilberts/centimeter	ampere-turns/meter	79.58
gills (British)	cubic centimeters	142.07
gills	liters	0.1183
gills	pints (liquid)	0.25
grade	radian	0.01571
grains	drams (avoirdupois)	0.03657143
grains (troy)	grains (avoirdupois)	1
grains (troy)	grams	0.0648
grains (troy)	ounces (avoirdupois)	2.0833 E–03
grains (troy)	pennyweight (troy)	0.04167
grains/U.S. gallon	parts/million	17.118
grains/U.S. gallon	pounds/million gallon	142.86
grains/imperial gallon	parts/million	14.286
grams	dynes	980.7
grams	grains	15.43
grams	joules/centimeter	9.807 E–05
grams	joules/meter (newtons)	9.807 E–03
grams	kilograms	0.001
grams	milligrams	1,000
grams	ounces (avoirdupois)	0.03527
grams	ounces (troy)	0.03215
grams	poundals	0.07093
grams	pounds	2.205 E–03
grams/centimeter	pounds/inch	5.600 E–03
grams/cubic centimeter	pounds/cubic foot	62.43
grams/cubic centimeter	pounds/cuin	0.03613
grams/cubic centimeter	pounds/mil-foot	3.405 E–07
grams/liter	grains/gallon	58.417
grams/liter	pounds/1,000 gallons	8.345
grams/liter	pounds/cubic foot	0.062427
grams/liter	parts/million	1,000
grams/square centimeter	pounds/square foot	2.0481
gram-calories	btu	3.9683 E–03
gram-calories	ergs	4.1868 E+07
gram-calories	foot-pounds	3.088
gram-calories	horsepower-hours	1.5596 E–06
gram-calories	kilowatt-hours	1.1630 E–06
gram-calories	watt-hours	1.1630 E–06
gram-calories/second	btu/hour	14.286
gram-centimeters	btu	9.297 E–08
gram-centimeters	ergs	980.7
gram-centimeters	joules	9.807 E–05
gram-centimeters	kilogram-calories	2.343 E–08
gram-centimeters	kilogram-meters	1 E–05
hands	centimeters	10.16

To Convert	To	Multiply by
	H	
hectares	acres	2.471
hectares	square feet	1.076 E+05
hectograms	grams	100
hectoliters	liters	100
hectometers	meters	100
hectowatts	watts	100
henries	millihenries	1,000
hogsheads (British)	cubic feet	10.114
hogsheads (U.S.)	cubic feet	8.42184
hogsheads (U.S.)	gallons (U.S.)	63
horsepower	btu/minute	42.44
horsepower	foot-pounds/minute	33,000
horsepower	foot-pounds/second	550
horsepower (metric) (542.5 foot pounds/second)	horsepower (550 foot pounds/second)	0.9863
horsepower (550 foot pounds/second)	horsepower (metric) (542.5 foot pounds/second)	1.014
horsepower	kilogram-calories/minute	10.68
horsepower	kilowatts	0.7457
horsepower	watts	745.7
horsepower (boiler)	btu/hour	33479
horsepower (boiler)	kilowatts	9.810554
horsepower-hours	btu	2,547
horsepower-hours	pegs	2.6845 E+13
horsepower-hours	foot-pounds	1.98 E+06
horsepower-hours	gram-calories	641,190
horsepower-hours	joules	2.684 E+06
horsepower-hours	kilogram-calories	641.1
horsepower-hours	kilogram-meters	2.737 E+05
horsepower-hours	kilowatt-hours	0.7457
hours	days	4.167 E–02
hours	weeks	5.952 E–03
hundredweights (long)	pounds	112
hundredweights (long)	tons (long)	0.05
hundredweights (short)	ounces (avoirdupois)	1600
hundredweights (short)	pounds	100
hundredweights (short)	tons (metric)	0.0453592
Hundredweights (short)	tons (long)	0.0446429
hydraulic horsepower	kilowatts	7.460430 E–01
	I	
inches	centimeters	2.54
inches	meters	2.540 E–02
inches	miles	1.578 E–05
inches	millimeters	25.4
inches	mils	1,000

To Convert	To	Multiply by
inches	yards	2.778 E–02
inches of mercury	atmospheres	0.03342
inches of mercury	feet of water	1.133
inches of mercury	kilograms/square centimeter	0.03453
inches of mercury	kilograms/square meter	345.3
inches of mercury	pounds/square foot	70.73
inches of mercury	pounds/square inch	0.4912
inches of mercury @ 60°F	kiloPascal	3.376850 E+00
inches of mercury @ 32°F	kiloPascal	3.386380 E+00
inches of water @ 4°C	atmospheres	2.458 E–03
inches of water @ 4°C	inches of mercury	0.07355
inches of water @ 4°C	kilograms/square centimeter	2.540 E–03
inches of water @ 4°C	ounces/square inch	0.5781
inches of water @ 4°C	pounds/square foot	5,204
inches of water @ 4°C	pounds/square inch	0.03613
inches of water @ 39.2°F	kiloPascal	2.490820 E–01
inches of water @ 60°F	kiloPascal	2.488400 E–01
international ampere	ampere (absolute)	0.9998
international volt	volts (absolute)	1.0003
international volt	joules (absolute)	1.593 E–19
international volt	joules	9.654 E+04
J		
joules	btu	9.480 E–04
joules	ergs	1 E+07
joules	foot-pounds	0.7376
joules	kilogram-calories	2.389 E–04
joules	kilogram-meters	0.102
joules	watt-hours	2.778 E–04
joules/centimeter	grams	1.020 E+04
joules/centimeter	dynes	1 E+07
joules/centimeter	joules/meter (newtons)	100
joules/centimeter	poundals	723.3
joules/centimeter	pounds	22.48
K		
kilograms	dynes	980,665
kilograms	grams	1,000
kilograms	joules/centimeter	0.09807
kilograms	joules/meter (newtons)	9.807
kilograms	poundals	70.93
kilograms	pounds	2,205
kilograms	tons (long)	9.842 E–04
kilograms	tons (short)	1.102 E–03
kilograms/cubic meter	grams/cubic centimeter	0.001
kilograms/cubic meter	pounds/cubic foot	0.06243
kilograms/cubic meter	pounds/cuinches	3.613 E–05

To Convert	To	Multiply by
kilograms/cubic meter	pounds/mil-foot	3.405 E–10
kilograms calorie-second	btu/second	3.968
kilograms calorie-second	foot-pounds/second	3.086
kilograms calorie-second	HP	5.6145
kilograms calorie-second	watts	4186.7
kilograms calorie-minute	foot-pounds/minute	3085.9
kilograms calorie-minute	HP	0.9351
kilograms calorie-minute	watts	69.733
kilograms/meter	pounds/foot	0.672
kilogram/square centimeter	dynes	980,665
kilograms/square centimeter	atmospheres	0.9678
kilograms/square centimeter	feet of water	32.81
kilograms/square centimeter	inches of mercury	28.96
kilograms/square centimeter	pounds/square foot	2,048
kilograms /square centimeter	pounds/square inch	14.22
kilograms/square meter	atmospheres	9.678 E–05
kilograms/square meter	bars	98.0 E–06
kilograms/square meter	feet of water	3.281 E–03
kilograms/square meter	inches of mercury	2.896 E–03
kilograms/square meter	pounds/square foot	0.2048
kilograms/square meter	pounds/sqinches	1.422 E–03
kilograms/square millimeter	kilograms/square meter	1 E+06
kilograin-calories	btu	3.968
kilograin-calories	foot-pounds	3,088
kilograin-calories	horsepower-hours	1.560 E–03
kilograin-calories	joules	4.186
kilograin-calories	kilogram-meters	426.9
kilograin-calories	kilojoules	4,186
kilograin-calories	kilowatt-hours	1.163 E–03
kilogram meters	btu	9.294 E–03
kilogram meters	ergs	9.804 E+07
kilogram meters	foot-pounds	7.233
kilogram meters	joules	9.804
kilogram meters	kilogram-calories	2.342 E–03
kilogram meters	kilowatt-hours	2.723 E–06
kilolines	maxwells	1,000
kiloliters	liters	1,000
kilometers	centimeters	1 E+05
kilometers	feet	3,281
kilometers	inches	3.937 E+04
kilometers	meters	1,000
kilometers	miles	0.6214
kilometers	millimeters	1 E+06
kilometers	yards	1,094
kilometers/hour	centimeters/second	27.78
kilometers/hour	feet/minute	54.68
kilometers/hour	feet/second	0.9113

(continued on next page)

To Convert	To	Multiply by
kilometers/hour	knots	0.5396
kilometers/hour	meters/minute	16.67
kilometers/hour	miles/hour	0.6214
kilometers/hour/second	centimeters/second/second	27.78
kilometers/hour/second	feet/second/second	0.9113
kilometers/hour/second	meters/second/second	0.2778
kilometers/hour/second	miles/hour/second	0.6214
kilowatts	btu/minute	56.92
kilowatts	foot-pounds/minute	4.426 E+04
kilowatts	foot-pounds/second	737.6
kilowatts	horsepower	1.341
kilowatts	kilogram-calories/minute	14.34
kilowatts	watts	1,000
kilowatt-hours	btu	3,413
kilowatt-hours	ergs	3.600 E+13
kilowatt-hours	foot-pounds	2.655 E+06
kilowatt-hours	gram-calories	859,850
kilowatt-hours	horsepower-hours	1.341
kilowatt-hours	joules	3.6 E+06
kilowatt-hours	kilogram-calories	860.5
kilowatt-hours	kilogram-meters	3.671 E+05
kilowatt-hours	pounds of water evaporated at 212° F	3.53
kilowatt-hours	pounds of water raised from 62° to 212° F	22.75
knots	feet/hour	6,080
knots	kilometers/hour	1.8532
knots	nautical miles/hour	1
knots	statute miles/hour	1.151
knots	yards/hour	2,027
knots	feet/second	1.689
	L	
league	miles (approx.)	3
light year	miles	5.9 E+12
light year	kilometers	9.46 E+12
lines/square centimeters	gausses	1
lines/square inch	gausses	0.155
lines/square inch	webers/square centimeter	1.550 E–09
lines/square inch	webers/square inch	1 E–08
lines/square inch	webers/square meter	1.550 E–05
links (engineer's)	inches	12
links (surveyor's)	inches	7.92
liters	bushels (U.S. dry)	0.02838
liters	cubic centimeters	1,000
liters	cubic feet	0.03531
liters	cubic inches	61.02

To Convert	To	Multiply by
liters	cubic meters	0.001
liters	cubic yards	1.308 E–03
liters	gallons (U.S. liquid)	0.2642
liters	pints (U.S. liquid)	2.113
liters	quarts (U.S. liquid)	1.057
liters/minute	cubic feet/second	5.886 E–04
liters/minute	gallons/second	4.403 E–03
lumens/square foot	foot-candles	1
lumen	spherical candle power	0.07958
lumen	watt	0.001496
lumen/square foot	lumen/square meter	10.76
lux	foot-candles	0.0929
	M	
maxwells	kilolines	0.001
maxwells	webers	1 E–08
megalines	maxwells	1 E+06
megohms	microhms	1 E+12
megohms	ohms	1 E+06
meters	centimeters	100
meters	feet	3.281
meters	inches	39.37
meters	kilometers	0.001
meters	miles (nautical)	5.396 E–04
meters	miles (statute)	6.214 E–04
meters	millimeters	1,000
meters	yards	1,094
meters	varas	1,179
meters/minute	centimeters/second	1,667
meters/minute	feet/minute	3.281
meters/minute	feet/second	0.05468
meters/minute	kilometers/hour	0.06
meters/minute	knots	0.03238
meters/minute	miles/hour	0.03728
meters/second	feet/minute	196.8
meters/second	feet/second	3,281
meters/second	kilometers/hour	3.6
meters/second	kilometers/minute	0.06
meters/second	miles/hour	2.237
meters/second	miles/minute	0.03728
meters/second/second	centimeters/second/second	100
meters/second/second	feet/second/second	3.281
meters/second/second	kilometers/hour/second	3.6
meters/second/second	miles/hour/second	2.237
meter-kilograms	centimeter-dynes	9.807 E+07

To Convert	To	Multiply by
meter-kilograms	centimeter-grams	1 E+05
meter-kilograms	pound-feet	7.233
microfarad	farads	1 E–06
micrograms	grams	1 E–06
microhms	megohms	1 E–12
microhms	ohms	1 E–06
microliters	liters	1 E–06
microns	meters	1 E–06
miles (nautical)	feet	6,080.27
miles (nautical)	kilometers	1.853
miles (nautical)	meters	1,853
miles (nautical)	miles (statute)	1.1516
miles (nautical)	yards	2,027
miles (statute)	centimeters	1.609 E+05
miles (statute)	feet	5,280
miles (statute)	inches	6.336 E+04
miles (statute)	kilometers	1.609
miles (statute)	meters	1,609
miles (statute)	miles (nautical)	0.8684
miles (statute)	yards	1,760
miles/hour	centimeters/second	44.7
miles/hour	feet/minute	88
miles/hour	feet/second	1.467
miles/hour	kilometers/hour	1.609
miles/hour	kilometers/minute	0.02682
miles/hour	knots	0.8684
miles/hour	meters/minute	26.82
miles/hour	miles/minute	0.1667
miles/hour/second	centimeters/second/second	44.7
miles/hour/second	feet/second/second	1.467
miles/hour/second	kilometers/hour/second	1.609
miles/hour/second	meters/second/second	0.447
miles/minute	centimeters/second	2,682
miles/minute	feet/second	88
miles/minute	kilometers/minute	1.609
miles/minute	knots/minute	0.8684
miles/minute	miles/hour	60
mil-feet	cuinches	9.425 E–06
milliers	kilograms	1,000
millimicrons	meters	1 E–09
milligrams	grains	0.01543236
milligrams	grams	0.001
milligrams/liter	parts/million	1
millihenries	henries	0.001
milliliters	liters	0.001
millimeters	centimeters	0.1
millimeters	feet	3.281 E–03

To Convert	To	Multiply by
millimeters	inches	0.03937
millimeters	kilometers	1 E–06
millimeters	meters	0.001
millimeters	miles	6.214 E–07
millimeters	mils	39.37
millimeters	yards	1.094 E–03
million gallons/day	cubic feet/second	1.54723
mils	centimeters	2.540 E–03
mils	feet	8.333 E–05
mils	inches	0.001
mils	kilometers	2.540 E–08
mils	yards	2.778 E–05
miner's inches	cubic feet/minute	1.5
minims (British)	cubic centimeters	0.059192
minims (U.S., fluid)	cubic centimeters	0.061612
minutes (angles)	degrees	0.01667
minutes (angles)	quadrants	1.852 E–04
minutes (angles)	radians	2.909 E–04
minutes (angles)	seconds	60
myriagrams	kilograms	10
myriameters	kilometers	10
myriawatts	kilowatts	10
	N	
nautical miles (international)	meters	1,852
nepers	decibels	8.686
newton	dynes	1 E+05
	O	
ohm (international)	ohm (absolute)	1.0005
ohms	megohms	1 E–06
ohms	microhms	1 E+06
ounces	drams	16
ounces	grains	437.5
ounces	grams	28.349527
ounces	pounds	0.0625
ounces	ounces (troy)	0.9115
ounces	tons (long)	2.790 E–05
ounces	tons (metric)	2.835 E–05
ounces (fluid)	cubic inches	1.805
ounces (fluid)	liters	0.02957
ounces (troy)	grains	480
ounces (troy)	grams	31.103481
ounces (troy)	ounces (avoirdupois)	1.09714
ounces (troy)	pennyweights (troy)	20
ounces (troy)	pounds (troy)	0.08333

(continued on next page)

To Convert	To	Multiply by
ounces/square inch	dynes/square centimeters	4,309
ounces/square inch	pounds/square inch	0.0625
	P	
parsecs	miles	19 E+12
parsecs	kilometers	3.084 E+13
parts/million	grains/U.S. gallon	0.0584
parts/million	grains/imperial gallon	0.07016
parts/million	pounds/million gallons	8.345
pecks (British)	cubic inches	554.6
pecks (British)	liters	9.091901
pecks (U.S.)	bushels	0.25
pecks (U.S.)	cubic inches	537.605
pecks (U.S.)	liters	8.809582
pecks (U.S.)	quarts (dry)	8
pennyweights (troy)	grains	24
pennyweights (troy)	ounces (troy)	0.05
pennyweights (troy)	grams	1.55517
pennyweights (troy)	pounds (troy)	4.1667 E–03
pints (dry)	cubic inches	33.6
pints (liquid)	cubic centimeters	473.2
pints (liquid)	cubic feet	0.01671
pints (liquid)	cubic inches	28.87
pints (liquid)	cubic meters	4.732 E–04
pints (liquid)	cubic yards	6.189 E–04
pints (liquid)	gallons	0.125
pints (liquid)	liters	0.4732
pints (liquid)	quarts (liquid)	0.5
Planck's quantum	erg-second	6.624 E–27
poise	gram/centimeters second	1
pounds (avoirdupois)	ounces (troy)	14.5833
poundals	dynes	13,826
poundals	grams	14.1
poundals	joules/centimeter	1.383 E–03
poundals	joules/meter (newtons)	0.1383
poundals	kilograms	0.0141
poundals	pounds	0.03108
pounds	drams	256
pounds	dynes	44.4823 E+04
pounds	grains	7,000
pounds	grams	453.5924
pounds	joules/centimeter	0.04448
pounds	joules/meter (newtons)	4.448
pounds	kilograms	0.4536
pounds	ounces	16
pounds	ounces (troy)	14.5833
pounds	poundals	32.17

To Convert	To	Multiply by
pounds	pounds (troy)	1.21528
pounds	tons (short)	0.0005
pounds/square inch/foot	kiloPascal/meter	2.262059 E+01
pounds/square inch/mile	kiloPascal/kilometer	4.284203 E+00
pounds (troy)	grains	5,760
pounds (troy)	grams	373.24177
pounds (troy)	ounces (avoirdupois)	13.1657
pounds (troy)	ounces (troy)	12
pounds (troy)	pennyweights (troy)	240
pounds (troy)	pounds (avoirdupois.)	0.822857
pounds (troy)	tons (long)	3.6735 E–04
pounds (troy)	tons (metric)	3.7324 E–04
pounds (troy)	tons (short)	4.1143 E–04
pounds/U.S. gallon (gas)	kilograms/cubic meter	1.601846 E+01
pounds/U.S. gallon (liquid)	kilograms/cubic meter	1.198264 E+02
pounds/U.K. gallon (liquid)	kilograms/cubic meter	9.977633 E+01
pounds/cubic foot	kilograms/cubic meter	1.601846 E+01
pounds/barrel	kilograms/cubic meter	2.853010 E+00
pounds of water	cubic feet	0.01602
pounds of water	cubic inches	27.68
pounds of water	gallons	0.1198
pounds of water/minute	cubic feet/second	2.670 E–04
pound-feet	centimeter-dynes	1.356 E+07
pound-feet	centimeter-grams	13,825
pound-feet	meter-kilograms	0.1383
pounds/cubic foot	grams/cubic centimeter	0.01602
pounds/cubic foot	kilograms/cubic meter	16.02
pounds/cubic foot	pounds/cubic inches	5.787 E–04
pounds/cubic foot	pounds/mil-foot	5.456 E–09
pounds/cubic inch	grams/cubic centimeter	27.68
pounds/cubic inch	kilograms/cubic meter	2.768 E+04
pounds/cubic inch	pounds/cubic foot	1,728
pounds/cubic inch	pounds/mil-foot	9.425 E–06
pounds/foot	kilograms/meter	1.488
pounds/inch	grams/centimeter	178.6
pounds/mil-foot	grams/cubic centimeter	2.306 E+06
pounds/square foot	atmospheres	4.725 E–04
pounds/square foot	feet of water	0.01602
pounds/square foot	inches of mercury	0.01414
pounds/square foot	kilograms/square meter	4.882
pounds/square foot	pounds/square inches	6.944 E–03
pounds/square inch	atmospheres	0.06804
pounds/square inch	feet of water	2.307
pounds/square inch	inches of mercury	2.036
pounds/square inch	kilograms/square meter	703.1
pounds/square inch	pounds/square foot	144

To Convert	To	Multiply by
	Q	
quadrants (angle)	degrees	90
quadrants (angle)	minutes	5,400
quadrants (angle)	radians	1.571
quadrants (angle)	seconds	3.24 E+05
quarts (dry)	cubic inches	67.2
quarts (liquid)	cubic centimeters	946.4
quarts (liquid)	cubic feet	0.03342
quarts (liquid)	cubic inches	57.75
quarts (liquid)	cubic meters	9.464 E–04
quarts (liquid)	cubic yards	1.238 E–03
quarts (liquid)	gallons	0.25
quarts (liquid)	liters	0.9463
	R	
radians	degrees	57.3
radians	minutes	3,438
radians	quadrants	0.6366
radians	seconds	2.063 E+05
radians/second	degrees/second	57.3
radians/second	revolutions/minute	9.549
radians/second	revolutions/second	0.1592
radians/second/second	revolutions/minute/minute	573
radians/second/second	revolutions/minute/second	9.549
radians/second/second	revolutions/second/second	0.1592
revolutions	degrees	360
revolutions	quadrants	4
revolutions	radians	6.283
revolutions/minute	degrees/second	6
revolutions/minute	radians/second	0.1047
revolutions/minute	revolutions/second	0.01667
revolutions/minute/minute	radians/second/second	1.745 E–03
revolutions/minute/minute	revolutions/minute/second	0.01667
revolutions/minute/minute	revolutions/second/second	2.778 E–04
revolutions/second	degrees/second	360
revolutions/second	radians/second	6.283
revolutions/second	revolutions/minute	60
revolutions/second/second	radians/second/second	6.283
revolutions/second/second	revolutions/minute/minute	3,600
revolutions/second/second	revolutions/minute/second	60
rod	chain (gunters)	0.25
rod	meters	5.029
rods (surveyors' measure)	yards	5.5
rods	feet	16.5

To Convert	To	Multiply by
	S	
scruples	grains	20
seconds (angle)	degrees	2.778 E–04
seconds (angle)	minutes	0.01667
seconds (angle)	quadrants	3.087 E–06
seconds (angle)	radians	4.848 E–06
slug	kilogram	14.59
slug	pounds	32.17
sphere	steradians	12.57
square centimeters	circular mils	1.973 E+05
square centimeters	square feet	1.076 E–03
square centimeters	square inches	0.155
square centimeters	square meters	0.0001
square centimeters	square miles	3.861 E–11
square centimeters	square millimeters	100
square centimeters	square yards	1.196 E–04
square feet	acres	2.296 E–05
square feet	circular mils	1.833 E+08
square feet	square centimeters	929
square feet	square inches	144
square feet	square meters	0.0929
square feet	square miles	3.587 E–08
square feet	square millimeters	9.290 E+04
square feet	square yards	0.1111
square inches	circular mils	1.273 E+06
square inches	square centimeters	6.452
square inches	square feet	6.944 E–03
square inches	square millimeters	645.2
square inches	square mils	1 E+06
square inches	square yards	7.716 E–04
square kilometers	acres	247.1
square kilometers	square centimeters	1 E+10
square kilometers	square feet	10.76 E+06
square kilometers	square inches	1.550 E+09
square kilometers	square meters	1 E+06
square kilometers	square miles	0.3861
square kilometers	square yards	1.196 E+06
square meters	acres	2.471 E–04
square meters	square centimeters	1 E+04
square meters	square feet	10.76
square meters	square inches	1,550
square meters	square miles	3.861 E–07
square meters	square millimeters	1 E+06
square meters	square yards	1.196
square miles	acres	640
square miles	square feet	27.88 E+06
square miles	square kilometers	2.59

(continued on next page)

To Convert	To	Multiply by
square miles	square meters	2.590 E+06
square miles	square yards	3.098 E+06
square millimeters	circular mils	1,973
square millimeters	square centimeters	0.01
square millimeters	square feet	1.076 E–05
square millimeters	square inches	1.550 E–03
square mils	circular mils	1.273
square mils	square centimeters	6.452 E–06
square mils	square inches	1 E–06
square yards	acres	2.066 E–04
square yards	square centimeters	8,361
square yards	square feet	9
square yards	square inches	1,296
square yards	square meters	0.8361
square yards	square miles	3.228 E–07
square yards	square millimeters	8.361 E+05
	T	
temperature	absolute temperature (°C)	1 (°C) + 273
temperature	temperature (°F)	1.8 (°C) + 17.78
temperature	absolute temperature (°F)	1 (°F) + 459.67
temperature (°F)	temperature (°C)	(°F – 32) × 0.5556
temperature (°R)	temperature (°K)	°R/1.8
therm/U.S. gallon	megajoules/cubic meter	2.787163 E+04
therm/U.S.	joule	1.05 E+08
tons (long)	kilograms	1,016
tons (long)	pounds	2,240
tons (long)	tons (short)	1.12
tons (metric)	kilograms	1,000
tons (metric)	pounds	2,205
tons (short)	kilograms	907.1848
tons (short)	ounces	32,000
tons (short)	ounces (troy)	29,166.66
tons (short)	pounds	2,000
tons (short)	pounds (troy)	2,430.56
tons (short)	tons (long)	0.89287
tons (short)	tons (metric)	0.9078
tons (U.K.)	kilograms/second	2.822353 E–01
tons (U.S.)	kilograms/second	2.519958 E–01
tons (short)/square foot	kilograms/square meter	9,765
tons (short)/square foot	pounds/sqinches	2,000
tons of water/24 hours	pounds of water/hour	83.333
tons of water/24 hours	gallons/minute	0.16643
tons of water/24 hours	cubic feet/hour	1.3349
tons of refigeration	btu/hour	12,000
tons of refigeration	watts	3517.2

To Convert	To	Multiply by
	V	
volt/inch	volt/centimeters	0 .39370
volt (absolute)	statvolts	0.003336
	W	
watts	btu/hour	3.4129
watts	btu/minute	0.05688
watts	ergs/second	107
watts	foot-pounds/minute	44.27
watts	foot-pounds/second	0.7378
watts	horsepower	1.341 E–03
watts	horsepower (metric)	1.360 E–03
watts	kilogram-calories/minute	0.01433
watts	kilowatts	0.001
watts (absolute)	btu (mean)/minute	0.056884
watts (absolute)	joules/second	1
watt-hours	btu	3.413
watt-hours	ergs	3.60 E+10
watt-hours	foot-pounds	2,656
watt-hours	gram-calories	859.85
watt-hours	horsepower-hours	1.341 E–03
watt-hours	kilogram-calories	0.8605
watt-hours	kilogram-meters	367.2
watt-hours	kilowatt-hours	0,001
watt (international)	watt (absolute)	1.0002
webers	maxwells	1 E+08
webers	kilolines	1 E+05
webers/square inch	gausses	1.550 E+07
webers/square inch	lines/square inch	1 E+08
webers/square inch	webers/square centimeter	0.155
webers/square inch	webers/square meter	1,550
webers/square meter	gausses	1 E+04
webers/square meter	lines/square inch	6.452 E+04
webers/square meter	webers/square centimeter	1 E–04
webers/square meter	webers/square inch	6.452 E–04
	Y	
yards	centimeters	91.44
yards	kilometers	9.144 E–04
yards	meters	0.9144
yards	miles (nautical)	4.934 E–04
yards	miles (statute)	5.682 E–04
yards	millimeters	914.4
year (common)	days	365
year (common)	hours	8,760
year (leap)	days	366
year (leap)	hours	8,784

Viscosity—equivalents of absolute viscosity

Absolute or Dynamic Viscosity		Centipoise (μ)	Poise, $\frac{Gram}{Cm\ Sec}$, $\frac{Dyne\ Sec}{Cm^2}$ ($100\ \mu$)	$\frac{Slugs}{Ft\ Sec}$, $\frac{*Pound_f\ Sec}{Ft^2}$ (μ'_e)	$\frac{†Pound_m}{Ft\ Sec}$, $\frac{Poundal\ Sec}{Ft^2}$ (μ_e)
Centipoise	(μ)	1	0.01	2.09 (10^{-5})	6.72 (10^{-4})
Poise, $\frac{Gram}{Cm\ Sec}$, $\frac{Dyne\ Sec}{Cm^2}$	($100\ \mu$)	100	1	2.09 (10^{-3})	0.0672
$\frac{Slugs}{Ft\ Sec}$, $\frac{*Pound_f\ Sec}{Ft^2}$	(μ'_e)	47 900	479	1	g or 32.2
$\frac{†Pound_m}{Ft\ Sec}$, $\frac{Poundal\ Sec}{Ft^2}$	(μ_e)	1487	14.87	$\frac{1}{g}$ or .0311	1

$*Pound_f$ = Pound of Force $†Pound_m$ = Pound of Mass

To convert absolute or dynamic viscosity from one set of units to another, locate the given set of units in the left-hand column and multiply the numerical value by the factor shown horizontally to the right under the set of units desired.

Courtesy Crane Co. (Technical Paper #410).

As an example, suppose a given absolute viscosity of 2 poise is to be converted to slugs/foot second. By referring to the table, we find the conversion factor to be 2.09 (10^{-3}). Then, 2 (poise) times 2.09 (10^{-3}) = 4.18 (10^{-3}) = 0.00418 slugs/foot second.

General liquid density nomograph

X and Y Values for Density Nomograph

Compound	X	Y	Compound	X	Y	Compound	X	Y
Acetic acid	40.6	93.5	Ethyl chloride	42.7	62.4	Methyl sulfide	31.9	57.4
Acetone	26.1	47.8	Ethylene	17.0	3.5	n-Nonane	16.2	36.5
Acetonitrile	21.8	44.9	Ethyl ether	22.6	35.8	n-Octadecane	16.2	46.5
Acetylene	20.8	10.1	Ethyl formate	37.6	68.4	n-Octane	12.7	32.5
Ammonia	22.4	24.6	Ethyl propionate	32.1	63.9	n-Pentadecane	15.8	44.2
Isoamyl alcohol	20.5	52.0	Ethyl propyl ether	20.0	37.0	n-Pentane	12.6	22.6
Ammine	33.5	92.5	Ethyl auindo	25.7	55.3	n-Nonadecane	14.9	47.0
Benzene	32.7	63.0	Fluorobenzene	41.9	87.6	Isopentane	13.5	22.5
n-Butyric acid	31.3	78.7	n-Heptadecane	15.6	45.7	Phenol	36.7	103.8
Isobutane	13.7	16.5	n-Heptane	12.6	29.8	Phosphine	28.0	22.1
Isobutyric acid	31.5	75.9	n-Hexadecane	15.8	45.0	Propane	14.2	12.2
Carbon dioxide	78.6	45.4	n-Hexane	13.5	27.0	Propionic acid	35.0	83.5
Chlorobenzene	41.7	105.0	Methanethiol	37.3	59.5	Piperidene	27.5	60.0
Cyclohexane	19.6	44.0	Methyl acetate	40.1	70.3	Propionitrile	20.1	44.6
n-Decane	16.0	38.2	Methyl alcohol	25.8	49.1	Propyl acetate	33.0	65.5
n-Dedecane	14.3	41.4	Methyl n-butyrate	31.5	65.5	Propyl alcohol	23.8	50.8
Diethylamine	17.8	33.5	Methyl isobutyrate	33.0	64.1	Propyl formate	33.8	66.7
n-Elconane	14.8	47.5	Methyl chloride	52.3	62.9	n-Tetradecane	15.8	43.3
Ethane	10.8	4.4	Methyl ether	27.2	30.1	n-Tridecane	15.3	42.4
Ethanethiol	32.0	55.5	Methyl ethyl ether	25.0	34.4	Triethylamine	17.9	37.0
Ethyl acetate	35.0	95.0	Methyl formate	46.4	74.6	n-Undecane	14.4	39.2
Ethyl alcohol	24.2	48.6	Methyl propionate	36.5	68.3			z

Reprinted by permission of Tubular Exchanger Manufacturers Association (TEMA).

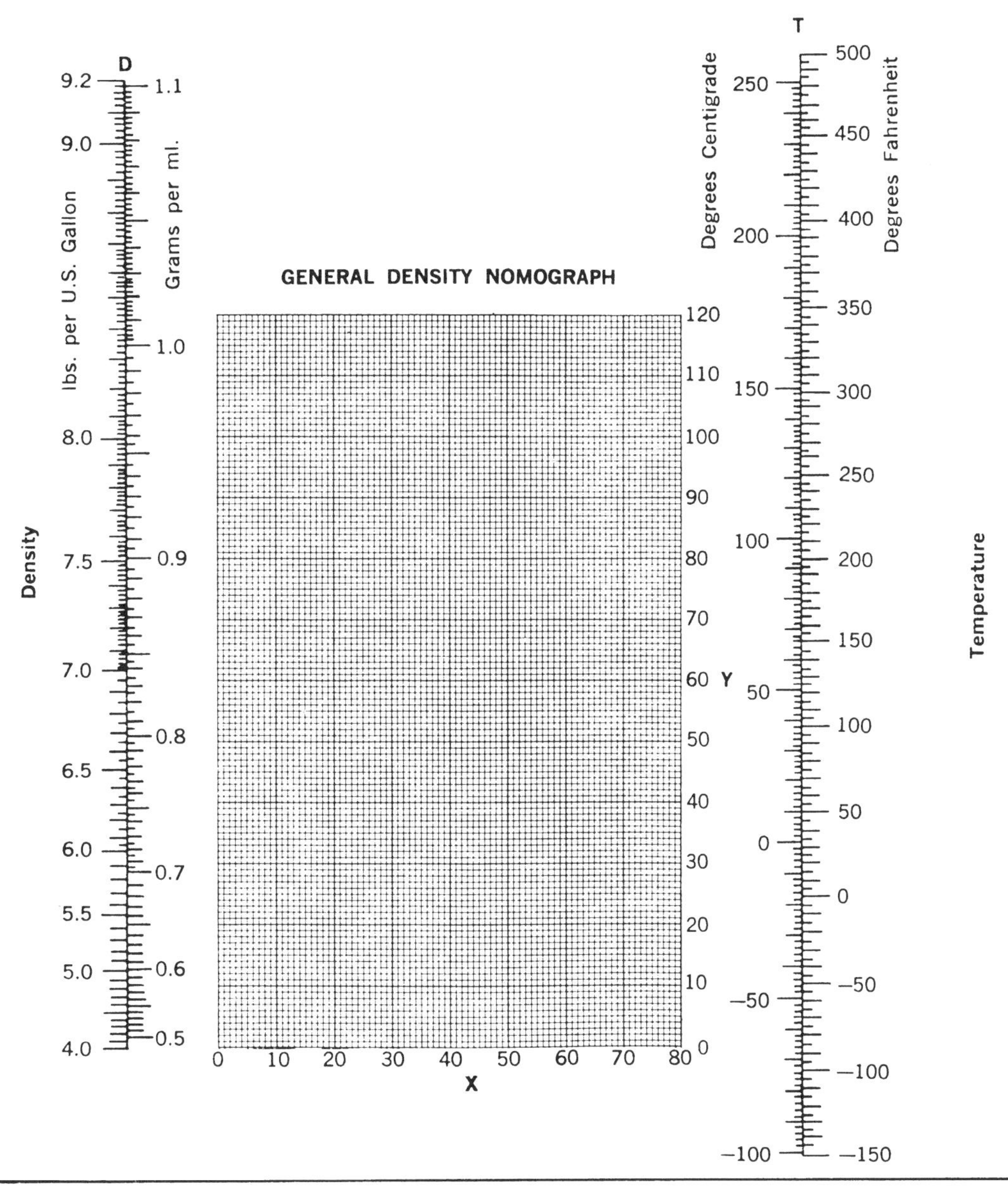

Density Conversions

To convert the numerical value of a property expressed in one of the units in the left-hand column of the table to the numerical value expressed in one of the units in the top row of the table, multiply the former value by the factor in the block common to both units.

Units→ ↓	g/cu cm	g/ml	lb/cu in.	lb/cu ft	lb/gal (U.S.)
g/cu cm	1	1.000028	0.036128	62.428	8.3455
g/ml	0.99997	1	0.036126	62.427	8.3452
lb/cu in.	27.680	27.681	1	1,728	231
lb/cu ft	0.016018	0.016019	5.7870×10^{-4}	1	0.13368
lb/gal (U.S.)	0.11983	0.11983	4.3290×10^{-3}	7.4805	1

Chart gives specific gravity/temperature relationship for petroleum oils

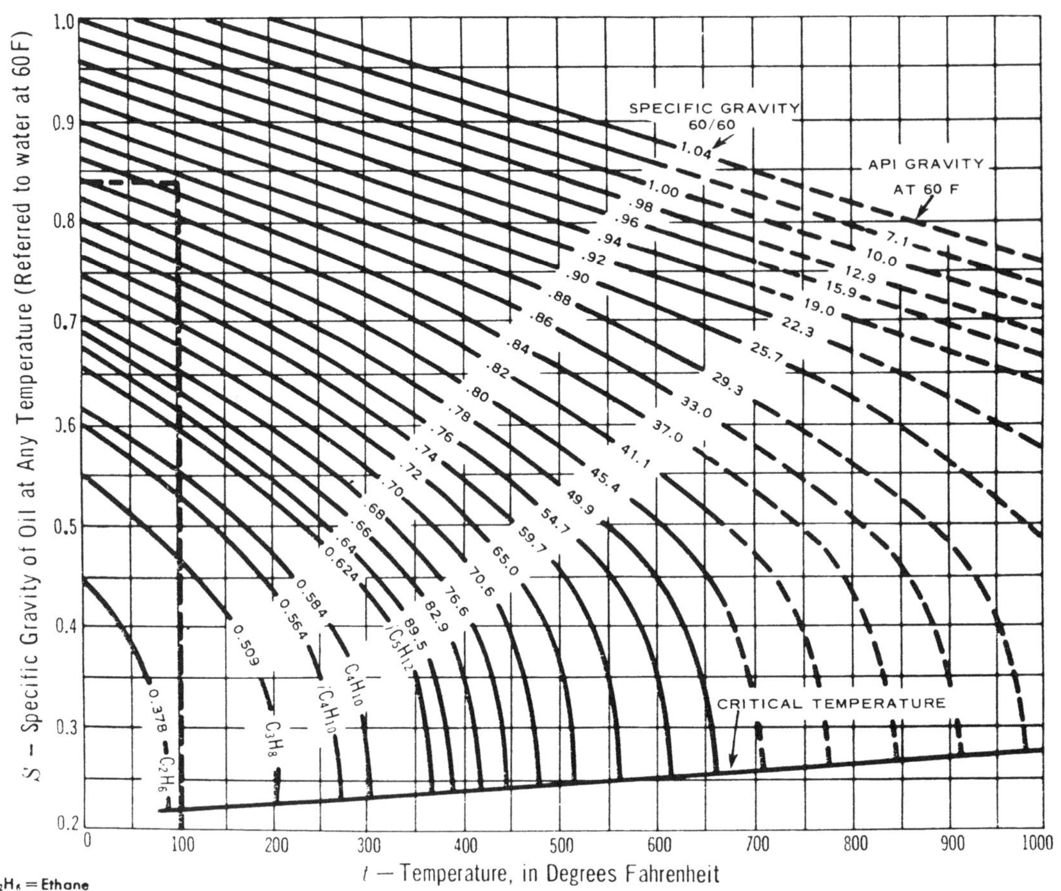

C_2H_6 = Ethane
C_3H_8 = Propane iC_4H_{10} = Isobutane
C_4H_{10} = Butane iC_5H_{12} = Isopentane

Example: The specific gravity of an oil at 60 F is 0.85. The specific gravity at 100 F = 0.83.

Courtesy Crane Co. (Technical Paper #410).

To find the weight density of a petroleum oil at its flowing temperature when the specific gravity at 60 F/60 F is known, multiply the specific gravity of the oil at flowing temperature (see chart above) by 62.4, the density of water at 60 F.

Weight density and specific gravity of various liquids

Liquid	Temp. t Deg. Fahr.	Weight Density ρ Lbs. per Cu. Ft.	Specific Gravity S	Liquid	Temp. t Deg. Fahr.	Weight Density ρ Lbs. per Cu. Ft.	Specific Gravity S
Acetone	60	49.4	0.792	Mercury	20	849.74	...
Ammonia, Saturated	10	40.9	...	Mercury	40	848.03	...
Benzene	32	56.1	...	Mercury	60	846.32	13.570
Brine, 10% Ca Cl	32	68.05	...	Mercury	80	844.62	...
Brine, 10% Na Cl	32	67.24	...	Mercury	100	842.93	...
Bunkers C Fuel Max.	60	63.25	1.014	Milk	...	†	...
Carbon Disulphide	32	80.6	...	Olive Oil	59	57.3	0.919
Distillate	60	52.99	0.850	Pentane	59	38.9	0.624
Fuel 3 Max.	60	56.02	0.898	SAE 10 Lube‡	60	54.64	0.876
Fuel 5 Min.	60	60.23	0.966	SAE 30 Lube‡	60	56.02	0.898
Fuel 5 Max.	60	61.92	0.993	SAE 70 Lube‡	60	57.12	0.916
Fuel 6 Min.	60	61.92	0.993	Salt Creek Crude	60	52.56	0.843
Gasoline	60	46.81	0.751	32.6° API Crude	60	53.77	0.862
Gasoline, Natural	60	42.42	0.680	35.6° API Crude	60	52.81	0.847
Kerosene	60	50.85	0.815	40° API Crude	60	51.45	0.825
M. C. Residuum	60	58.32	0.935	48° API Crude	60	49.16	0.788

*Liquid at 60 F referred to water at 60 F.

†Milk has a weight density of 64.2 to 64.6.

‡100 Viscosity Index.

Courtesy Crane Co. (Technical Paper #410).

True vapor pressure of crude oil stocks with a Reid vapor pressure of 2 to 15 psi

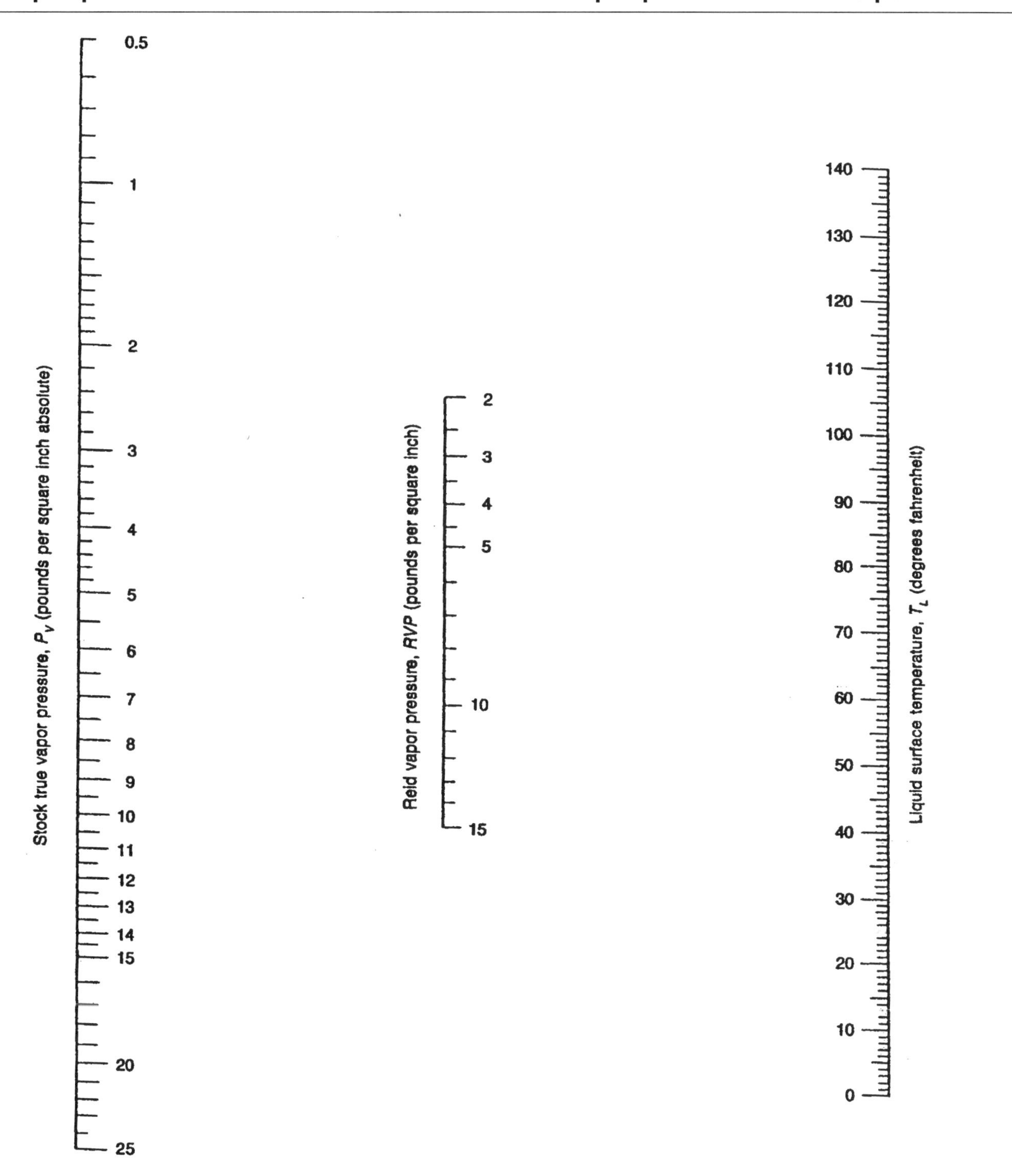

True vapor pressue (P_V) of crude oil stocks with a Reid vapor pressure of 2 to 15 psi. Reprinted courtesy of the American Petroleum Institute.

Low temperature vapor pressure of light hydrocarbons

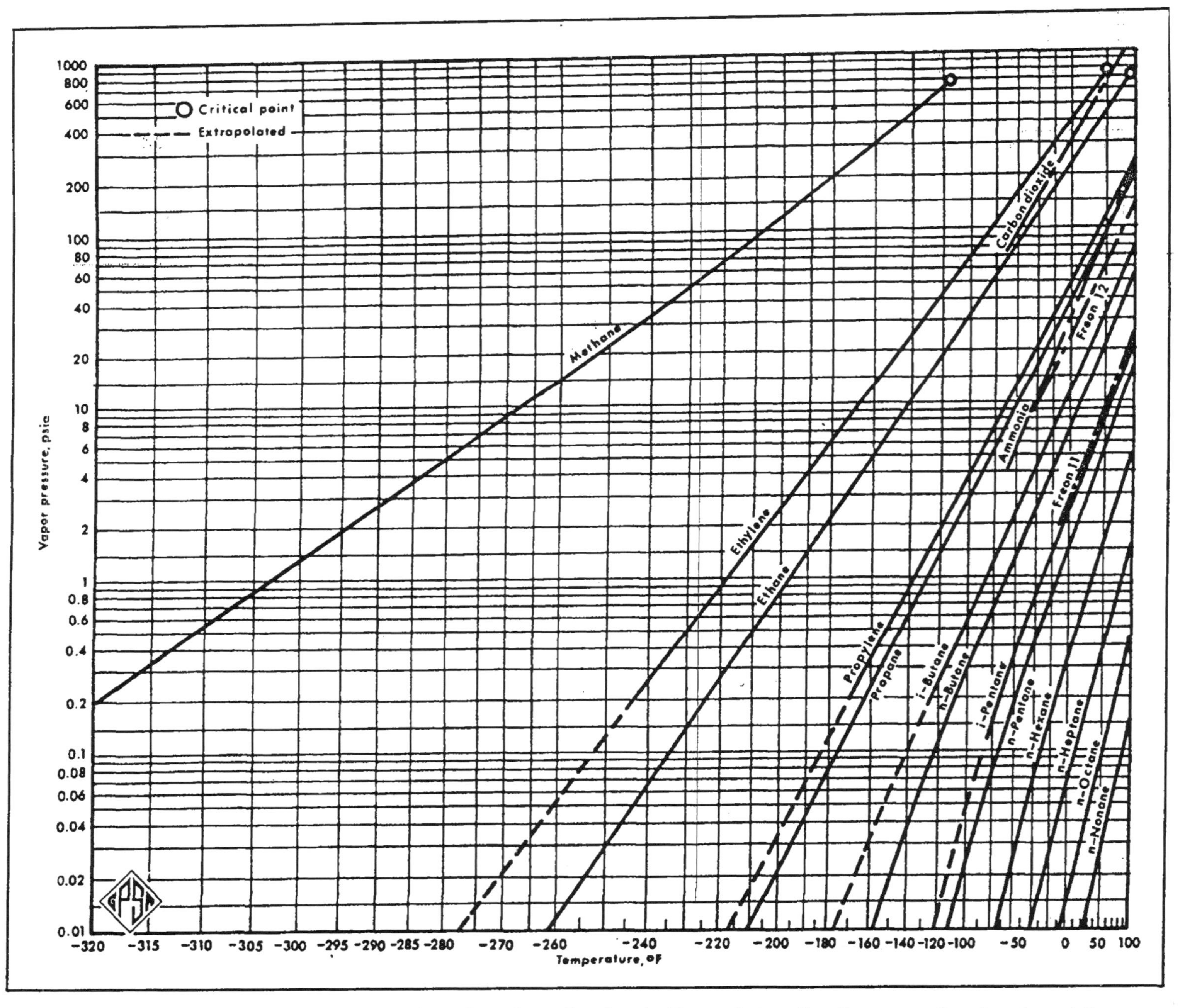

Low-temperature vapour pressures for light hydrocarbons. Reprinted with permission–Gas Processors Suppliers Association.

High temperature vapor pressure of light hydrocarbons

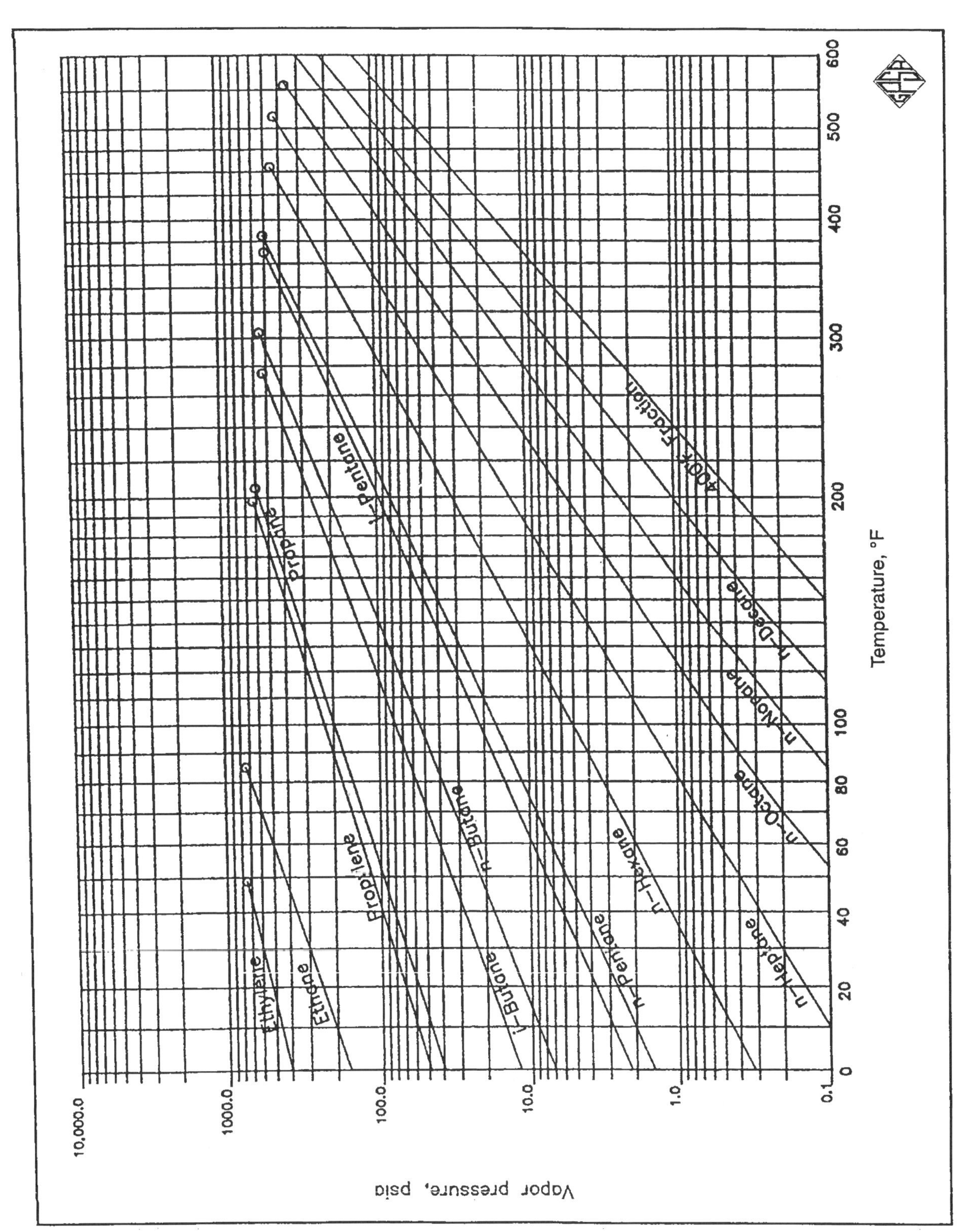

High-temperature vapor pressures for light hydrocarbons. Reprinted with permission—Gas Processors Suppliers Association.

Hydrocarbon gas viscosity

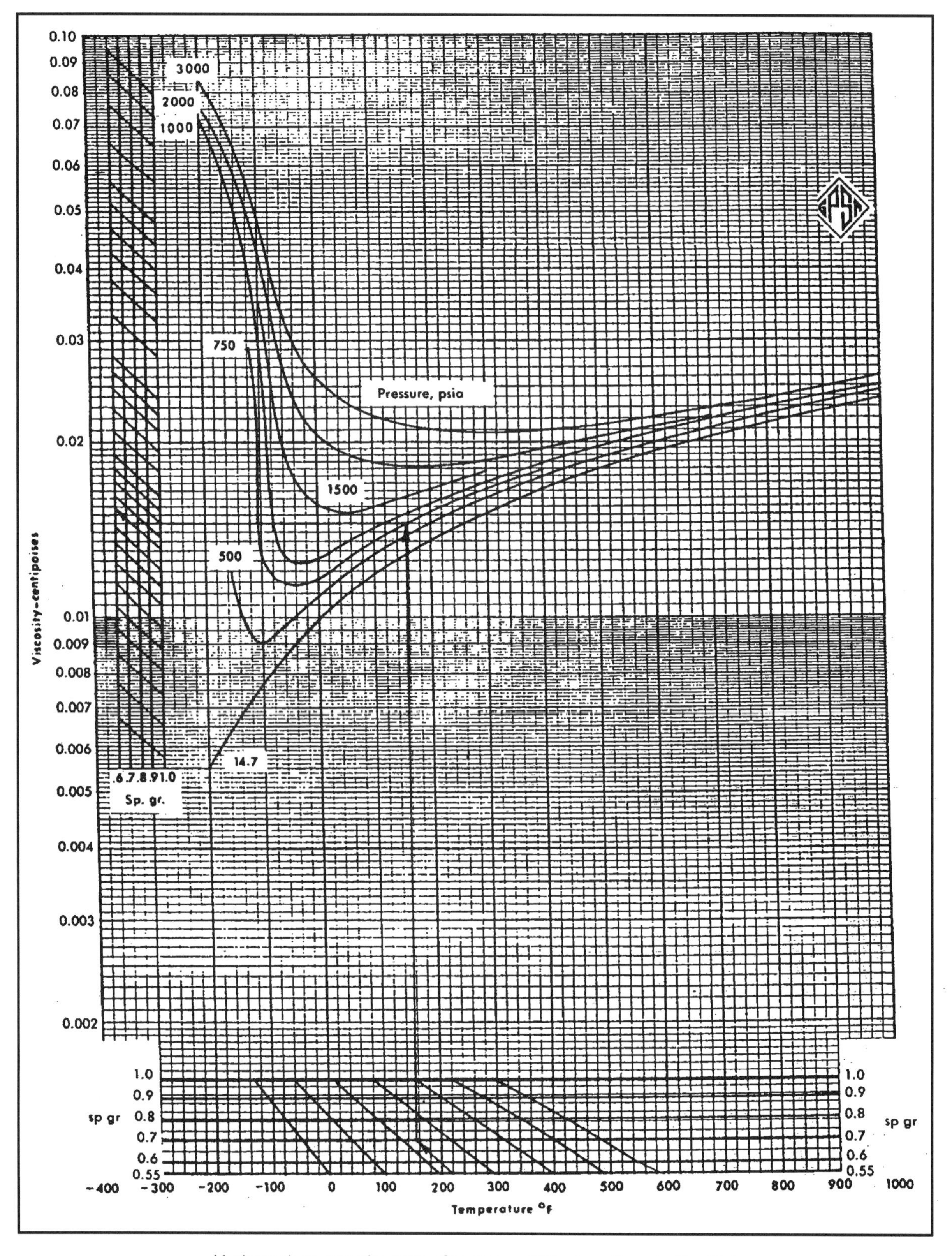

Hydrocarbon gas viscosity. Courtesy of Western Supply Co. Tulsa.

Metric conversions—metric to English, English to metric

To convert from	To	Multiply by	To convert from	To	Multiply by
Length			**Length**		
mm	inches	.03937	inches	mm	25.40
cm	inches	.3937	inches	cm	2.540
meters	inches	39.37	inches	meters	.0254
meters	feet	3.281	feet	meters	.3048
meters	yards	1.0936	feet	km	.0003048
km	feet	3280.8	yards	meters	.9144
km	yards	1093.6	yards	km	.0009144
km	miles	.6214	miles	km	1.609
Area			**Area**		
sq mm	sq inches	.00155	sq inches	sq mm	645.2
sq cm	sq inches	.155	sq inches	sq cm	6.452
sq meters	sq feet	10.764	sq feet	sq meters	.09290
sq meters	sq yards	1.196	sq yards	sq meters	.8361
sq km	sq miles	.3861	sq miles	sq km	2.590
hectares	acres	2.471	acres	hectares	.4047
Volume			**Volume**		
cu cm	cu inches	.06102	cu inches	cu cm	16.387
cu cm	fl ounces	.03381	cu inches	liters	.01639
cu meters	cu feet	35.314	cu feet	cu meters	.02832
cu meters	cu yards	1.308	cu feet	liters	28.317
cu meters	US gal	264.2	cu yards	cu meters	.7646
liters	cu inches	61.023	fl ounces	cu cm	29.57
liters	cu feet	.03531	US gal	cu meters	.003785
liters	US gal	.2642	US gal	liters	3.785
Weight			**Weight**		
grams	grains	15.432	grains	grams	.0648
grams	ounces†	.0353	ounces†	grams	28.350
kg	ounces†	35.27	ounces†	kg	.02835
kg	pounds†	2.2046	pounds†	kg	.4536
kg	US tons	.001102	pounds†	tonnes	.000454
kg	long tons	.000984	US tons	kg	907.2
tonnes	pounds†	2204.6	US tons	tonnes	.9072
tonnes	US tons	1.1023	long tons	kg	1016
tonnes	long tons	.9842	long tons	tonnes	1.0160

†avoirdupois pounds and ounces.

Courtesy Cameron Hydraulic Data Book, *Ingersoll-Rand Company.*

To convert from	To	Multiply by	To convert from	To	Multiply by
Unit weight			**Unit weight**		
gr/sq cm	lb/sq in	.01422	lb/ft	kg/m	1.4881
gr/cu cm	lb/cu in	.0361	lb/sq in	gr/sq cm	70.31
kg/sq cm	lb/sq in	14.22	lb/sq in	kg/sq cm	.07031
kg/cu m	lb/cu ft	.0624	lb/cu in	gr/cu cm	27.68
kg/m	lb/ft	.6720	lb/cu ft	kg/cu m	16.018
Unit volume			**Unit volume**		
liters/min	US gpm	.2642	US gpm	liters/min	3.785
liters/min	cfm	.03531	US gpm	liters/hr	227.1
liters/hr	US gpm	.0044	US gpm	cu m/hr	.2271
cu m/min	cfm	35.314	cfm	liters/min	28.317
cu m/hr	cfm	.5886	cfm	cu m/min	.02832
cu m/hr	US gpm	4.4028	cfm	cu m/hr	1.6992
Power			**Power**		
watts	ft-lb/sec	.7376	ft-lb/sec	watts	1.356
watts	hp	.00134	hp	watts	745.7
kw	hp	1.3410	hp	kw	.7457
cheval-vap	hp	.9863	hp	cheval-vap	1.0139
Heat			**Heat**		
gr-cal	Btu	.003969	Btu	gr-cal	252
kg-cal	Btu	3.9693	Btu	kg-cal	.252
kg-cal/kg	Btu/lb	1.800	Btu/lb	kg-cal/kg	.5556
gr-cal/sq cm	Btu/sq ft	3.687	Btu/sq ft	gr-cal/sq cm	.2713
kg-cal/cu m	Btu/cu ft	.1124	Btu/cu ft	kg-cal/cu m	8.899
Work or energy			**Work or energy**		
joule	ft-lb	.7376	ft-lb	joule	1.356
meter-kg	ft-lb	7.2330	ft-lb	meter-kg	.1383
gr-cal	ft-lb	3.087	ft-lb	gr-cal	.3239
kg-cal	ft-lb	3087	ft-lb	kg-cal	.000324
hp-hr	ft-lb	1,980,000	ft-lb	hp-hr	5.05×10^{-7}
kwhr	ft-lb	2,655,000	ft-lb	kwhr	3.766×10^{-7}
Btu	ft-lb	778.0	ft-lb	Btu	.001285

Temperature conversion—centigrade to Fahrenheit or Fahrenheit to centigrade

NOTE: The numbers in boldface refer to the temperature in degrees, either Centigrade or Fahrenheit, which it is desired to convert into the other scale. If converting from Fahrenheit to Centigrade degrees, the equivalent temperature will be found in the left column; while if converting from degrees Centigrade to degrees Fahrenheit, the answer will be found in the column on the right.

Centigrade		*Fahrenheit*	*Centigrade*		*Fahrenheit*	*Centigrade*		*Fahrenheit*	*Centigrade*		*Fahrenheit*
-73.3	**-100**	-148.0	2.8	**37**	98.6	33.3	**92**	197.6	293	**560**	1040
-67.8	**-90**	-130.0	3.3	**38**	100.4	33.9	**93**	199.4	299	**570**	1058
-62.2	**-80**	-112.0	3.9	**39**	102.2	34.4	**94**	201.2	304	**580**	1076
-59.4	**-75**	-103.0	4.4	**40**	104.0	35.0	**95**	203.0	310	**590**	1094
-56.7	**-70**	-94.0	5.0	**41**	105.8	35.6	**96**	204.8	316	**600**	1112
-53.9	**-65**	-85.0	5.6	**42**	107.6	36.1	**97**	206.6	321	**610**	1130
-51.1	**-60**	-76.0	6.1	**43**	109.4	36.7	**98**	208.4	327	**620**	1148
-48.3	**-55**	-67.0	6.7	**44**	111.2	37.2	**99**	210.2	332	**630**	1166
-45.6	**-50**	-58.0	7.2	**45**	113.0	37.8	**100**	212.0	338	**640**	1184
-42.8	**-45**	-49.0	7.8	**46**	114.8	43	**110**	230	343	**650**	1202
-40.0	**-40**	-40.0	8.3	**47**	116.6	49	**120**	248	349	**660**	1220
-37.2	**-35**	-31.0	8.9	**48**	118.4	54	**130**	266	354	**670**	1238
-34.4	**-30**	-22.0	9.4	**49**	120.2	60	**140**	284	360	**680**	1256
-31.7	**-25**	-13.0	10.0	**50**	122.0	66	**150**	302	366	**690**	1274
-28.9	**-20**	-4.0	10.6	**51**	123.8	71	**160**	320	371	**700**	1292
-26.1	**-15**	5.0	11.1	**52**	125.6	77	**170**	338	377	**710**	1310
-23.3	**-10**	14.0	11.7	**53**	127.4	82	**180**	356	382	**720**	1328
-20.6	**-5**	23.0	12.2	**54**	129.2	88	**190**	374	388	**730**	1346
-17.8	**0**	32.0	12.8	**55**	131.0	93	**200**	392	393	**740**	1364
-17.2	**1**	33.8	13.3	**56**	132.8	99	**210**	410	399	**750**	1382
-16.7	**2**	35.6	13.9	**57**	134.6	100	**212**	414	404	**760**	1400
-16.1	**3**	37.4	14.4	**58**	136.4	104	**220**	428	410	**770**	1418
-15.6	**4**	39.2	15.0	**59**	138.2	110	**230**	446	416	**780**	1436
-15.0	**5**	41.0	15.6	**60**	140.0	116	**240**	464	421	**790**	1454
-14.4	**6**	42.8	16.1	**61**	141.8	121	**250**	482	427	**800**	1472
-13.9	**7**	44.6	16.7	**62**	143.6	127	**260**	500	432	**810**	1490
-13.3	**8**	46.4	17.2	**63**	145.4	132	**270**	518	438	**820**	1508
-12.8	**9**	48.2	17.8	**64**	147.2	138	**280**	536	443	**830**	1526
-12.2	**10**	50.0	18.3	**65**	149.0	143	**290**	554	449	**840**	1544
-11.7	**11**	51.8	18.9	**66**	150.8	149	**300**	572	454	**850**	1562
-11.1	**12**	53.6	19.4	**67**	152.6	154	**310**	590	460	**860**	1580
-10.6	**13**	55.4	20.0	**68**	154.4	160	**320**	608	466	**870**	1598
-10.0	**14**	57.2	20.6	**69**	156.2	166	**330**	626	471	**880**	1616
-9.4	**15**	59.0	21.1	**70**	158.0	171	**340**	644	477	**890**	1634
-8.9	**16**	60.8	21.7	**71**	159.8	177	**350**	662	482	**900**	1652
-8.3	**17**	62.6	22.2	**72**	161.6	182	**360**	680	488	**910**	1670
-7.8	**18**	64.4	22.8	**73**	163.4	188	**370**	698	493	**920**	1688
-7.2	**19**	66.2	23.3	**74**	165.2	193	**380**	716	499	**930**	1706
-6.7	**20**	68.0	23.9	**75**	167.0	199	**390**	734	504	**940**	1724
-6.1	**21**	69.8	24.4	**76**	168.8	204	**400**	752	510	**950**	1742
-5.6	**22**	71.6	25.0	**77**	170.6	210	**410**	770	516	**960**	1760
-5.0	**23**	73.4	25.6	**78**	172.4	216	**420**	788	521	**970**	1778
-4.4	**24**	75.2	26.1	**79**	174.2	221	**430**	806	527	**980**	1796
-3.9	**25**	77.0	26.7	**80**	176.0	227	**440**	824	532	**990**	1814
-3.3	**26**	78.8	27.2	**81**	177.8	232	**450**	842	538	**1000**	1832
-2.8	**27**	80.6	27.8	**82**	179.6	238	**460**	860	566	**1050**	1922
-2.2	**28**	82.4	28.3	**83**	181.4	243	**470**	878	593	**1100**	2012
-1.7	**29**	84.2	28.9	**84**	183.2	249	**480**	896	621	**1150**	2102
-1.1	**30**	86.0	29.4	**85**	185.0	254	**490**	914	649	**1200**	2192
-0.6	**31**	87.8	30.0	**86**	186.8	260	**500**	932	677	**1250**	2282
0.0	**32**	89.6	30.6	**87**	188.6	266	**510**	950	704	**1300**	2372
0.6	**33**	91.4	31.1	**88**	190.4	271	**520**	968	732	**1350**	2462
1.1	**34**	93.2	31.7	**89**	192.2	277	**530**	986	760	**1400**	2552
1.7	**35**	95.0	32.2	**90**	194.0	282	**540**	1004	788	**1450**	2642
2.2	**36**	96.8	32.8	**91**	195.8	288	**550**	1022	816	**1500**	2732

The formulas at the right may also be used for converting Centigrade or Fahrenheit degrees into the other scale.

Degrees Cent., $C° = \frac{5}{9}(F° + 40) - 40$

Degrees Fahr., $F° = \frac{9}{5}(C° + 40) - 40$

Courtesy Natural Gas Processors Suppliers Association.

Viscosity—equivalents of kinematic viscosity

Kinematic Viscosity		Centistokes (ν)	Stokes $\frac{Cm^2}{Sec}$ (100 ν)	$\frac{Ft^2}{Sec}$ (ν')
Centistokes	(ν)	1	0.01	1.076 (10^{-5})
Stokes $\frac{Cm^2}{Sec}$	(100 ν)	100	1	1.076 (10^{-3})
$\frac{Ft^2}{Sec}$	(ν')	92,900	929	1

To convert kinematic viscosity from one set of units to another, locate the given set of units in the left-hand column and multiply the numerical value by the factor shown horizontally to the right, under the set of units desired.

As an example, suppose a given kinematic viscosity of 0.5 sq ft/sec is to be converted to centistokes. By referring to the table, we find the conversion factor to be 92,900. Then, 0.5 (sq ft/sec) times 92,900 = 46,450 centistokes.

Courtesy Crane Co. and American Society for Testing and Materials.

Viscosity—equivalents of kinematic and Saybolt Universal Viscosity

Kinematics Viscosity, Centistokes ν	Equivalent Saybolt Universal Viscosity, Sec	
	At 100°F Basic Values	At 210°F
1.83	32.01	32.23
2.0	32.62	32.85
4.0	39.14	39.41
6.0	45.56	45.88
8.0	52.09	52.45
10.0	58.91	59.32
15.0	77.39	77.93
20.0	97.77	98.45
25.0	119.3	120.1
30.0	141.3	142.3
35.0	163.7	164.9
40.0	186.3	187.6
45.0	209.1	210.5
50.0	232.1	233.8
55.0	255.2	257.0
60.0	278.3	280.2
65.0	301.4	303.5
70.0	324.4	326.7
75.0	347.6	350.0
80.0	370.8	373.4
85.0	393.9	396.7
90.0	417.1	420.0
95.0	440.3	443.4
100.0	463.5	466.7
120.0	556.2	560.1
140.0	648.9	653.4
160.0	741.6	
180.0	834.2	Saybolt Seconds equal Centistokes times 4.6673
200.0	926.9	
220.0	1019.6	
240.0	1112.3	
260.0	1205.0	
280.0	1297.7	
300.0	1390.4	
320.0	1483.1	
340.0	1575.8	
360.0	1668.5	
380.0	1761.2	
400.0	1853.9	
420.0	1946.6	
440.0	2039.3	
460.0	2132.0	
480.0	2224.7	
500.0	2317.4	
Over 500	Saybolt Seconds equal Centistokes times 4.6347	

Note: To obtain the Saybolt Universal viscosity equivalent to a kinematic viscosity determined at t, multiply the equivalent Saybolt Universal viscosity at 100°F by 1 + (t − 100) 0.000064. For example, 10ν at 210°F are equivalent to 58.91 multiplied by 1.0070 or 59.32 see Saybolt Universal at 210°F.

Courtesy Crane Co. (Technical Paper #410).

Viscosity—equivalents of kinematic and Saybolt Furol Viscosity at 122°F

Kinematics Viscosity, Centistokes v	Equivalent Saybolt Universal Viscosity, Sec	
	At 122°F Basic Values	At 210°F
48	25.3	
50	26.1	25.2
60	30.6	29.8
70	35.1	34.4
80	39.6	39.0
90	44.1	43.7
100	48.6	48.3
125	60.1	60.1
150	71.7	71.8
175	83.8	83.7
200	95.0	95.6
225	106.7	107.5
250	118.4	119.4
275	130.1	131.4
300	141.8	143.5
325	153.6	155.5
350	165.3	167.6
375	177.0	179.7
400	188.8	191.8
425	200.6	204.0
450	212.4	216.1
475	224.1	228.3
500	235.9	240.5
525	247.7	252.8
550	259.5	265.0
575	271.3	277.2
600	283.1	289.5
625	294.9	301.8

Kinematics Viscosity, Centistokes v	Equivalent Saybolt Universal Viscosity, Sec	
	At 122°F Basic Values	At 210°F
650	306.7	314.1
675	318.4	326.4
700	330.2	338.7
725	342.0	351.0
750	353.8	363.4
775	365.5	375.7
800	377.4	388.1
825	389.2	400.5
850	400.9	412.9
875	412.7	425.3
900	424.5	437.7
925	436.3	450.1
950	448.1	462.5
975	459.9	474.9
1000	471.7	487.4
1025	483.5	499.8
1050	495.2	512.3
1075	507.0	524.8
1100	518.8	537.2
1125	530.6	549.7
1150	542.4	562.2
1175	554.2	574.7
1200	566.0	587.2
1225	577.8	599.7
1250	589.5	612.2
1275	601.3	624.8
1300	613.1	637.3
Over 1300	*	†

*Over 1300 Centistokes at 122°F:
Saybolt Fluid Sec = Centistokes × 0.4717
†Over 1300 Centistokes at 210°F
Log (Saybolt Furol Sec − 2.87) = 1.0276 [Log (Centistokes)] − 0.3975

Courtesy Crane Co. and American Society for Testing and Materials.

Viscosity—general conversions

Approximate Conversion Table

SECONDS SAYBOLT UNIVERSAL	SECONDS SAYBOLT FUROL	ENGLER DEGREES	ENGLER TIME	BARBEY	REDWOOD STANDARD	REDWOOD ADMIRALTY	KINEMATIC VISCOSITY STOKES
35		1.18	60	2,800	32		0.026
50		1.60	82	880	44		0.074
75		2.30	102	460	65		0.141
100	15	3.00	153	320	88		0.202
150	19	4.40	230	205	128		0.318
200	23	5.90	305	148	170	18	0.431
250	28	7.60	375	118	212	23	0.543
300	33	8.90	450	98	254	27	0.651
400	42	11.80	550	72	338	36	0.876
500	52	14.50	750	59	423	45	1.10
600	61	17.50	900	48	518	53	1.32
700	71	20.6	1,050	41	592	62	1.54
800	81	23	1,200	36.5	677	71	1.76
900	91	27	1,300	32.0	762	78	1.98
1,000	100	29	1,500	29.5	846	87	2.20
1,500	150	42	2,300	19.5	1,270	135	3.30
2,000	200	59	3,000	14.5	1,695	175	4.40
2,500	250	73	3,750	11.5	2,120	230	5.50
3,000	300	87	4,500	9.6	2,540	260	6.60
4,000	400	117	6,000	7.4	3,380	350	8.80
5,000	500	145	7,500	6.0	4,230	435	11.00
6,000	600	175	9,000	5.2	5,080	530	13.20
7,000	700	205	10,500	4.1	5,925	610	15.40
8,000	800	230	12,000	3.7	6,770	700	17.60
9,000	900	260	13,500	3.2	7,620	780	19.8
10,000	1,000	290	15,000	2.9	8,460	870	22
20,000	2,000	590	30,000	1.4	16,920	1,760	44
40,000	4,000	1,170	60,000		33,850	3,600	88
60,000	6,000	1,750	90,000		50,800	5,300	132
80,000	8,000	2,300	120,000		67,700	7,000	176
100,000	10,000	2,900	150,000		84,600	8,700	220
200,000	20,000	5,900	300,000		169,200	17,600	440
400,000	40,000	11,700	600,000		338,500	36,000	880
600,000	60,000	17,500	900,000		508,000	53,000	1,320
800,000	80,000	23,000	1,200,000		677,000	70,000	1,760
1,000,000	100,000	29,000	1,500,000		846,000	87,000	2,200

ABSOLUTE VISCOSITY is the force required to move a unit plane surface over another plane surface at a unit velocity when surfaces are separated by a layer of fluid of unit thickness.

KINEMATIC VISCOSITY is the absolute viscosity divided by the density of the fluid.

SAE VISCOSITY NUMBERS are arbitrary numbers that have been assigned to classify lubricating oils and do not have any functional relationship with other systems.

VISCOSITY INDEX is an arbitrary system relating the effect of temperature change to viscosity. It is the slope of a plot of viscosity versus temperature for a given fluid.

1 Stoke = 1 Poise/Density in Grams
1 Centistoke = 1,076 × 10^{-5} Ft^2/Sec
1 Poise = 1Gram/(Cm)(Sec)
1 Centipoise = 1/100 Poise

1 Centistoke = 1/100 Stoke
1 Centipoise = 0.000672 Lbs/(Ft)(Sec)
1 Centipoise = 0.000672 (Poundals)(Sec)/Ft^2
1 Centipoise = 0.0000209(Lbs)(Sec)/Ft^2

1 Centipoise = 0.0000209 Slugs/(Ft)(Sec)
1 Centipoise = 2.42 Lbs/(Ft)(Hr)
1 Slug/(Ft)(Sec) = 1/0.002089 Poise
1 Ft^2/Sec = 92,903 Centistokes

A.S.T.M. standard viscosity temperature chart

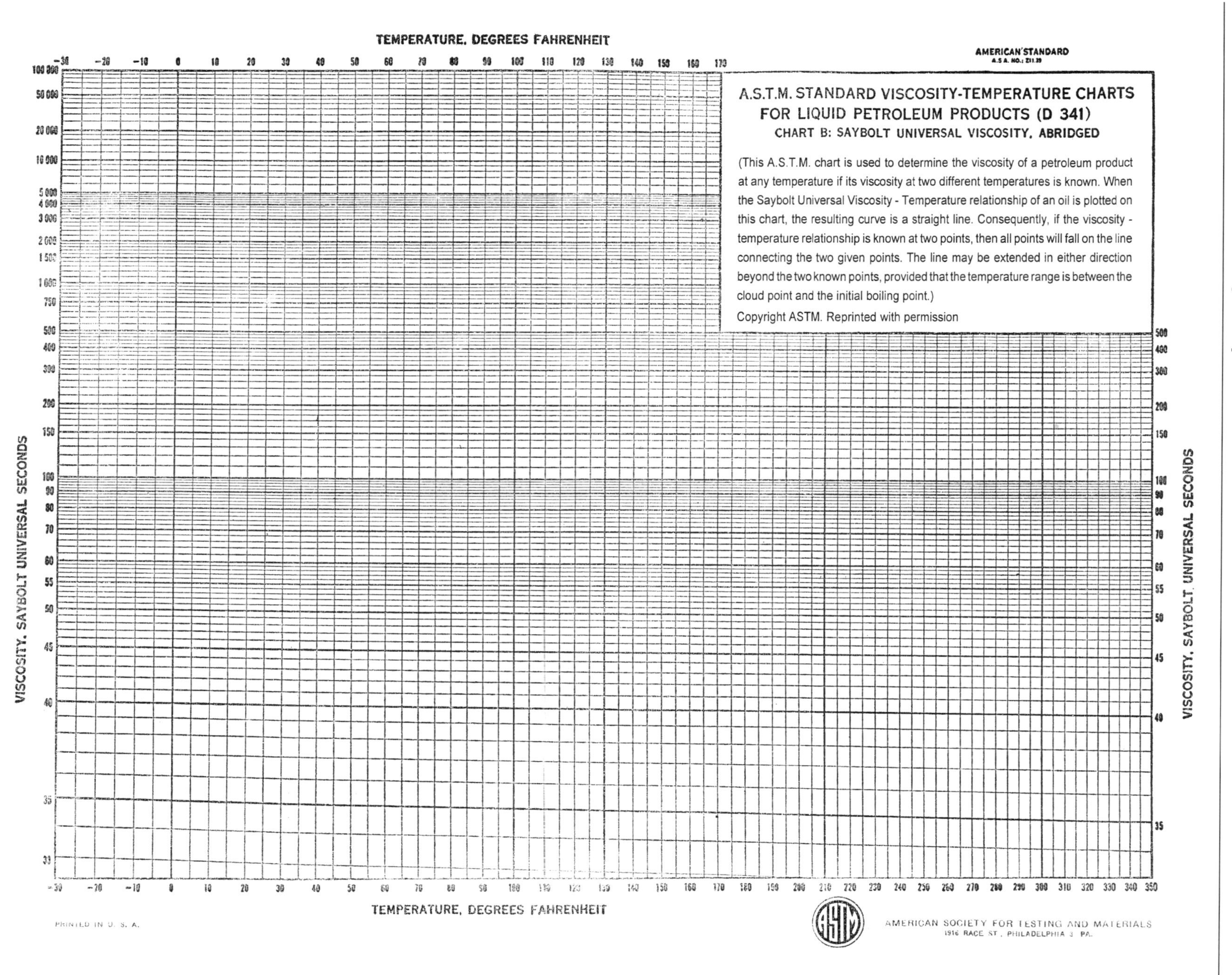

Pressure conversion chart

This line chart provides an easy method for converting units of pressure. Draw a line perpendicular to the scale lines through a known value of flow and read the equivalent value on any of the other scales. For values off the scale, known values can be read in multiples of 10 or 100.

Example. Known value—120 psi is equivalent to:

278 ft of water
1,920 oz. per in.^{2}
17,400 lb/ft^2
3,300 in. of water
242 in. of mercury
6,180 mm of mercury

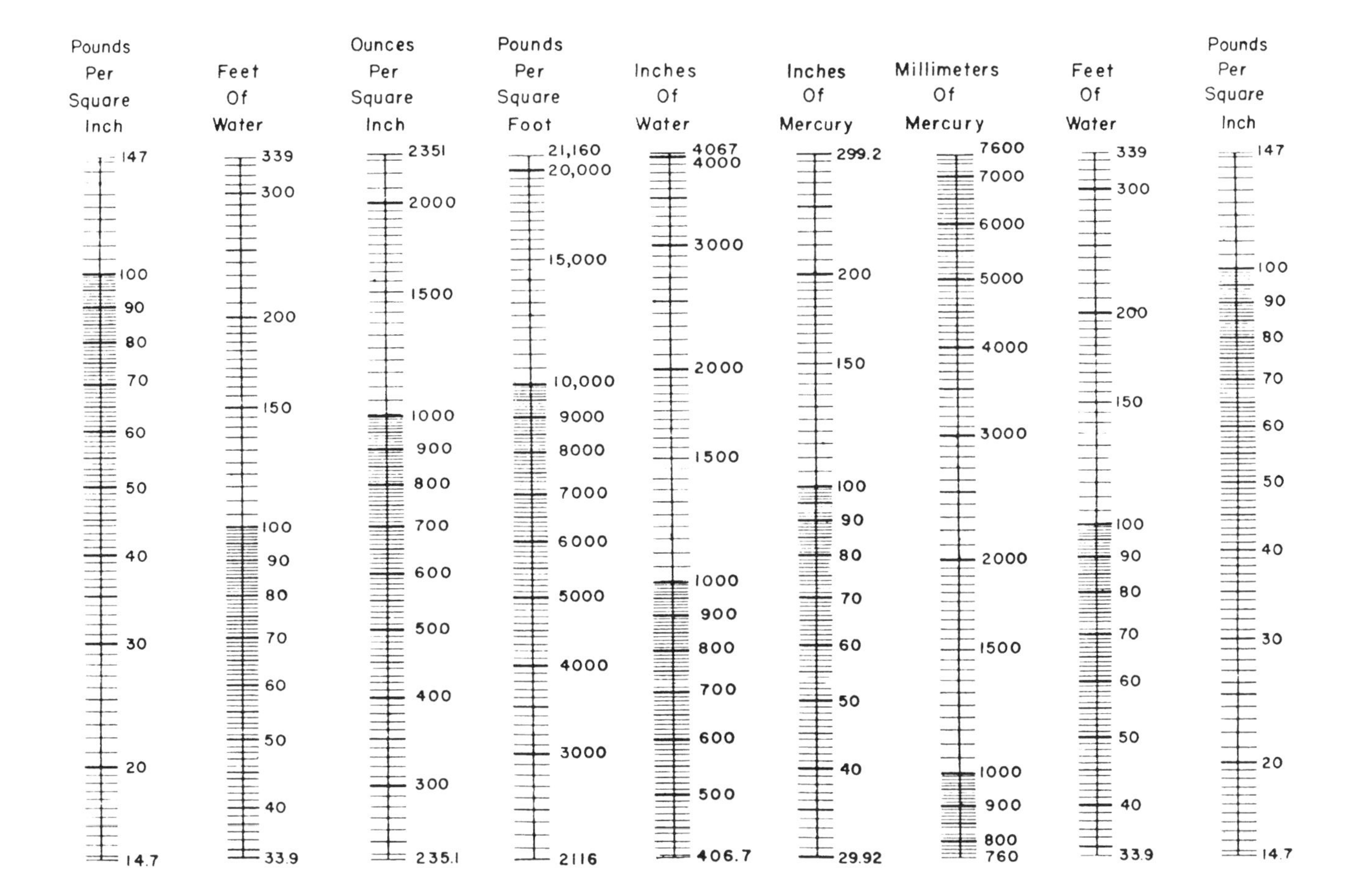

A simple method to determine square root

Problem. Determine the square root of the number 6,280.

Solution.

1. *Guess the square root to an even number.* For the example, use 80 since $(80)^2$ equals 6,400.
2. *Divide the number used into the number of which the square is desired* and carry the quotient to only two figures. Thus:

   $6{,}280 \div 80 = 78$

3. *Average this quotient with the divisor.*

80 + 78 = 158, and 158 ÷ 2 = 79

4. *Divide this figure into the number* and carry quotient to four or five figures, if desired.

6,280 ÷ 79 = 79.5

5. *Average this quotient with* #3 for the square root.

79 + 79.5 = 158.5, and 158.5 ÷ 2 = 79.25

6. *To check answer,* multiply answer by itself.

79.25 × 79.25 = 6,280.5

SI data*

Examples of SI-Derived Units

Quantity	SI Unit Name	SI Unit Symbol
Area	square metre	m^2
Volume	cubic metre	m^3
Speed, velocity	metre per second	m/s
Acceleration	metre per second squared	m/s^2
Density	kilogram per cubic metre	kg/m^3
Concentration (of amount of substance)	mole per cubic metre	mol/m^3
Specific volume	cubic metre per kilogram	m^3/kg
Luminance	candela per square metre	cd/m^2

SI Prefixes

Factor	Prefix	Symbol
10^{18}	exa	E
10^{15}	peta	P
10^{12}	tera	T
10^{9}	giga	G
10^{6}	mega	M
10^{3}	kilo	k
10^{2}	hecto	h
10^{1}	deka	da
10^{-1}	deci	d
10^{-2}	centi	c
10^{-3}	milli	m
10^{-6}	micro	μ
10^{-9}	nano	n
10^{-12}	Pico	p
10^{-15}	femto	f
10^{-18}	atto	a

Examples of SI-Derived Units with Special Names

Quantity	SI Unit Name	Symbol	Expression in Terms of Other Units	Expression in Terms of SI Unit
Frequency	hertz	Hz		s^{-1}
Force	newton	N		$m{\cdot}kg{\cdot}s^{-2}$
Pressure	Pascal	Pa	N/m2	$m^{-1}{\cdot}kg{\cdot}s^{-2}$
Energy, work, quantity of heat	joule	J	N·m	$m^2{\cdot}kg{\cdot}s^{-3}$
Power, radiant flux	watt.	W	J/s	$m^2{\cdot}kg{\cdot}s^{-3}$
Electric potential, potential differences, electromotive force	volt	V	W/A	$m^2{\cdot}kg{\cdot}s^{-3}{\cdot}A^{-1}$
Electric resistance	ohm	Ω	V/A	$m^2{\cdot}kg{\cdot}s^{-3}{\cdot}A^{-2}$
Conductance	siemens	S	AV	$m^{-2}{\cdot}kg{\cdot}s^{-1}{\cdot}A^2$

*Reprinted from API *Manual of Petroleum Measurement Standards*, Chapter 15, "Guidelines for the Use of the International System of Units (SI) in the Petroleum and Allied Industries," with permission of American Petroleum Institute.

Energy conversion chart

This nomograph provides an easy method for determining the fuel requirements for various types of fuel. Since the Btu content per unit volume of natural gas and crude oil vary over a limited but significant range of values, nomographs for the energy value of several gases and oils are included rather than using an average value.

Example. 30 cu. ft of a gas with 1,100 Btu per cu. ft yields 33,000 Btu. Conversion of 33,000 Btu to the other units of energy is obtained by drawing a straight line connecting the values 33,000 on two end Btu scales. Using the right-hand nomograph, an oil with 11,000 Btu per gallon (about 4.6 million Btu per barrel) will produce 33,000 Btu from 0.3 gallon (about 0.0072 barrel).

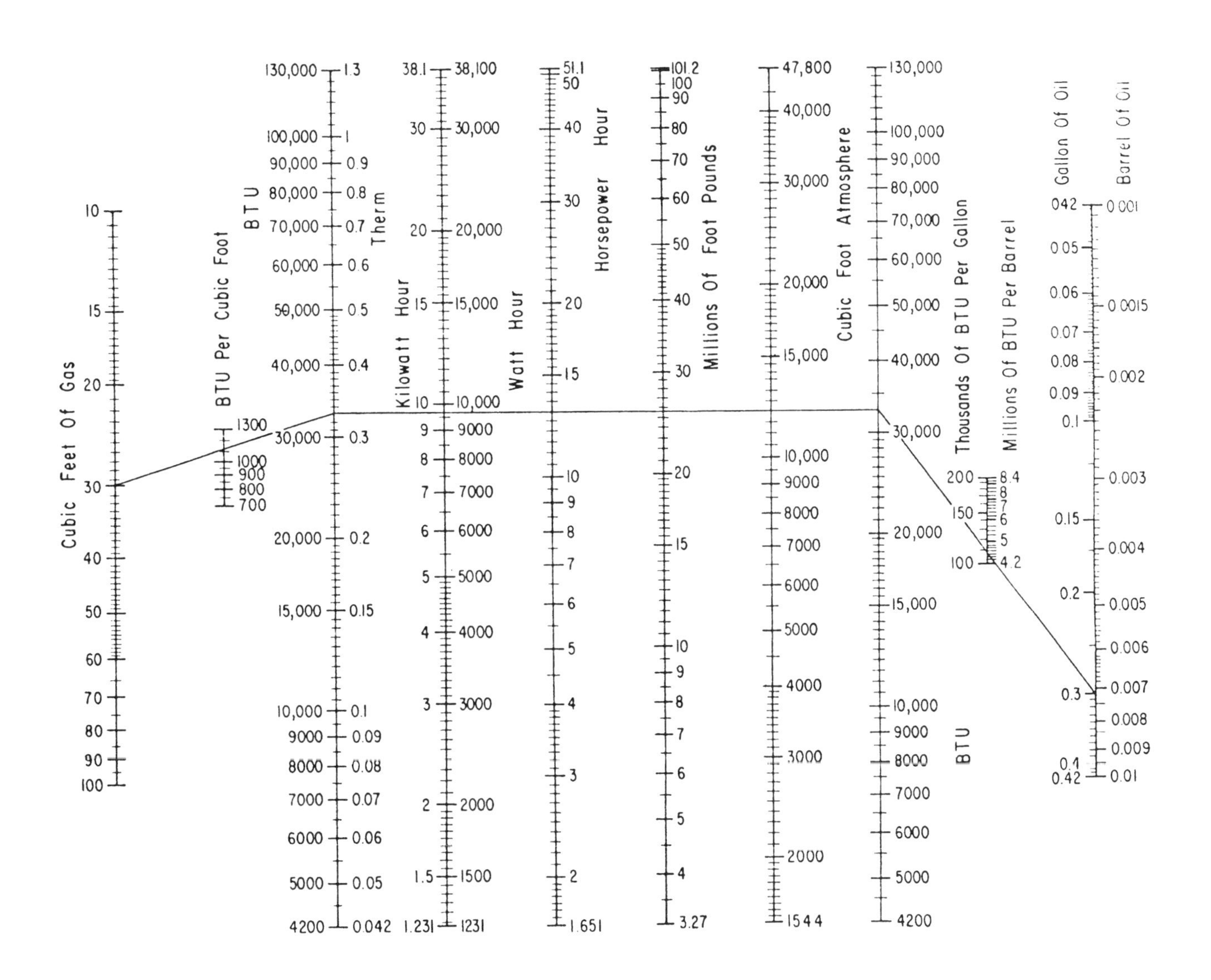

Flow conversion chart

This line chart provides an easy method for converting units of volume flow. Simply draw a line perpendicular to the scale lines through a known value of flow and read the equivalent value on any of the other scales.

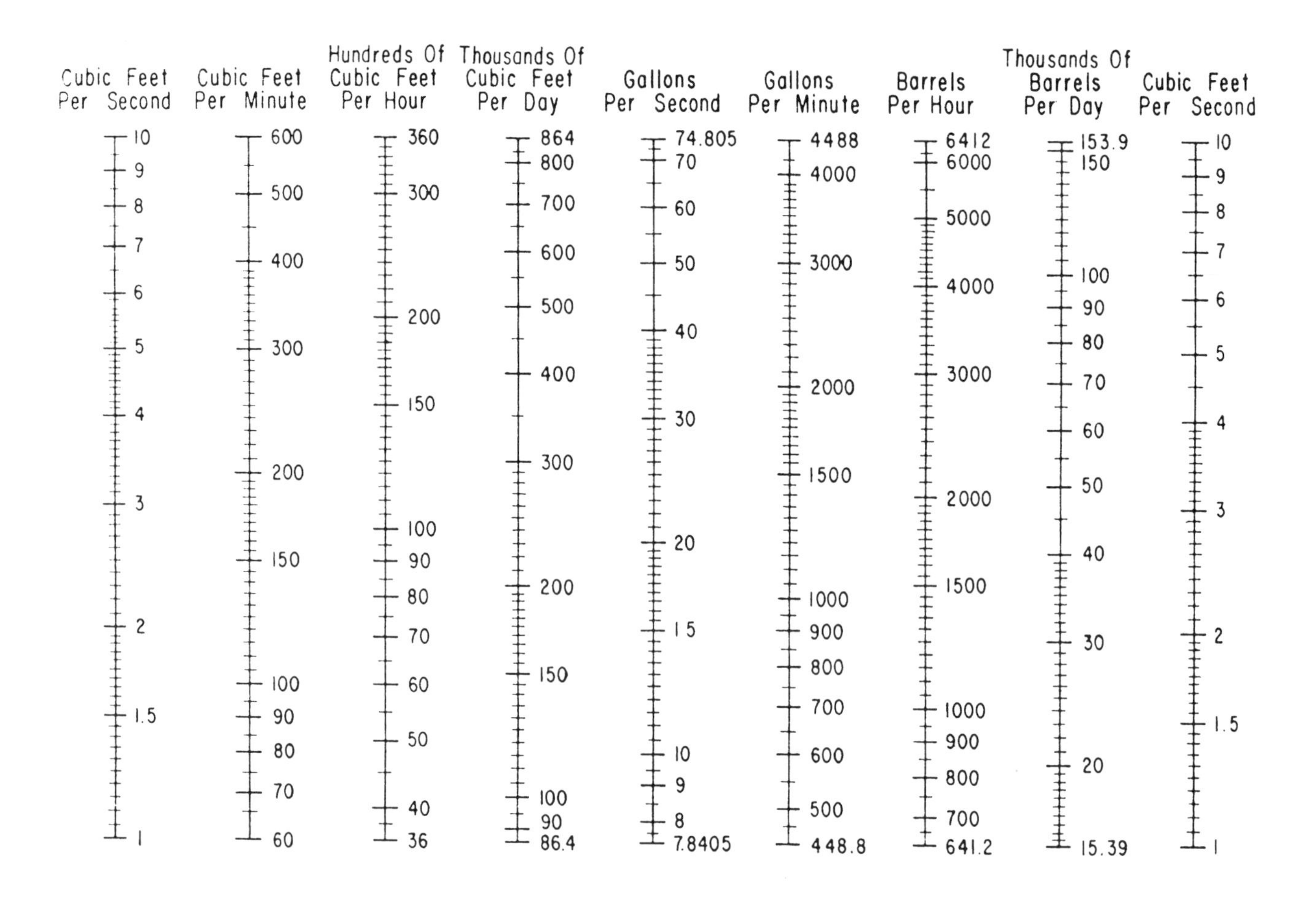

Conversions involving different types of fuel

Natural gas

The calorific value of natural gas varies according to its composition, and virtually no two fields produce gas of identical quality; the composition of gas from any one field can vary over its producing life. Thus, there is no such thing as a "standard" natural gas, nor a corresponding "standard" calorific value. However, when referring to natural gas in hypothetical cases it is quite common to indicate an "average" gross calorific value of 1,000 Btu/ft^3 (the equivalent of which is taken, for purposes here, to be 9,400 kcal/Nm^3 gross; see note on Table 3). Although in the United States most natural gas transmitted is at or slightly above this heat value, this does not generally apply in other countries. As an important example, Groningen (Netherlands) gas, because of a high nitrogen content, has a gross calorific value no greater than 898 Btu/ft^3 measured at 60°F, 30 in Hg, dry (7,980 kcal/m^3_s or 8,420 kcal/Nm^3).

When precise values are required, reference should be made to the proprietors of the gas field concerned or to published data.

Manufactured gases

These are derived from primary energy sources by processes involving chemical reaction, e.g., gases produced from coal and naphtha, also propane, butane, etc. Gas as piped to consumers can comprise both manufactured gases and natural gas used for enrichment. Calorific values of the component gases can thus vary widely depending on origin, but the resulting blend of what is known as town *gas* has a fairly constant heating value that, depending on locality, is usually about half that of natural gas. Exact figures can only come from the distribution company concerned.

Solid fuels

The calorific value of solid fuels varies widely between countries, grades, qualities, and even from seam to seam. It is particularly important, therefore, to use local data wherever possible. Measurements in Btu/lb or kcal/kg will normally include moisture and ash content as received.

N.B. Btu/lb × 0.5556 = kcal/kg

kcal/kg × 1.8 = Btu/lb

Where actual values are not known, an average gross calorific value of 12,400 Btu/lb (6,890 kcal/kg) is considered reasonable for hard coal grades of good quality. This also serves as the basis of a common unit, "coal equivalent," to which the many grades of solid fuel and other forms of

Table 1
Petroleum fuels volume/weight/calorific value factors

Hydrocarbons	Specific Gravity	Weight/Volume	Gross Calorific Value						
		bbl/ton	Btu/Imp.gal	Btu/lb	kcal/kg	10^6 Btu/bbl	10^6 kcal/bbl	b.o.e./bbl
Propane	0.501	12.790	......	21,500	11,943	3.795	0.956	0.654
Butane	0.579	11.060	......	21,140	11,742	4.315	1.088	0.744
Motor Gasoline	0.733	8.740	148,000	20,160	11,200	5.176	1.304	0.892
Kerosene	0.780	8.213	156,000	20,000	11,110	5.456	1.377	0.941
Vaporizing Oil	0.823	7.783	163,000	19,800	11,000	5.701	1.442	0.983
Gas Oil	0.840	7.625	164,000	19,500	10,835	5.736	1.445	0.988
Fuel Oils	0.850	7.536	165,660	19,490	10,830	5.794	1.460	0.999
	0.860	7.448	166,840	19,400	10,780	5.835	1.470	1.006
	0.870	7.362	168,080	19,320	10,730	5.878	1.481	1.013
	0.880	7.278	169,310	19,240	10,690	5.921	1.492	1.021
	0.890	7.196	170,440	19,150	10,640	5.961	1.502	1.028
	0.900	7.116	171,640	19,070	10,590	6.002	1.513	1.035
	0.910	7.038	172,730	18,980	10,540	6.041	1.522	1.042
	0.920	6.961	173,800	18,890	10,490	6.078	1.532	1.048
	0.930	6.887	174,860	18,800	10,440	6.115	1.541	1.054
	0.940	6.813	175,890	18,710	10,390	6.151	1.550	1.061
	0.950	6.741	176,910	18,620	10,340	6.187	1.559	1.067
	0.960	6.671	177,910	18,530	10,290	6.222	1.568	1.073
	0.970	6.602	178,890	18,440	10,240	6.256	1.577	1.079
	0.980	6.535	179,860	18,350	10,190	6.290	1.585	1.084
	0.990	6.469	180,800	18,260	10,140	6.323	1.593	1.090
	1.000	6.404	181,730	18,170	10.090	6.356	1.602	1.096

Sources:
W.P.C. Publication "Technical Data on Fuel" (Editor H. M. Spiers) 6th Edition 1961; all products except fuel oil. Latter from H. Hyams, "Petroleum Cargoes, Measurement and Sampling, 1966."

Notes:
1. Btu/gal × 34.9726 = Btu/bbl
2. Btu/bbl × 0.252 = kcal/bbl
3. b.o.e./bbl = $\frac{10^6 \text{Btu/bbl}}{5.8}$
4. kcal/kg × 1.8 = Btu/lb
5. To convert 10^6 Btu/bbl to ft^3 natural gas (1,000 Btu), multiply by 10^3
6. To convert 10^6 kcal/bbl to m^3 natural gas (9,400 kcal), multiply by 106.4

Conversion factors for calorific values of gases under different conditions of measurement

Table 2

A \ B	Btu/ft³ @ 60°F 30 InHg dry	Btu/ft³ @ 60°F 30 inHg saturated	Btu/ft³ @ 60°F 760 mmHg (60°F) dry	kcal/m³ @ 0°C 760 mmHg dry	kcal/m³ @ 0°C 760 mmHg saturated	kcal/m³ @ 15°C 760 MmHg dry	kcal/m³ @ 15°C 760 mmHg saturated
Btu/ft³ @ 60°F 30 inHg dry	1	0.9826	0.9946	9.381	9.325	8.893	8.743
Btu/ft³ @ 60°F 30 inHg saturated	1.0177	1	1.0122	9.547	9.490	9.050	8.898
Btu/ft³ @ 60°F 760 mmHg (60°F) dry	1.0054	0.9879	1	9.431	9.375	8.940	8.790
kcal/m³ @ 0°C 760 mmHg dry	0.10659	0.10474	0.10602	1	0.9940	0.9480	0.9320
kcal/m³ @ 0°C 760 mmHg saturated	0.10724	0.10537	0.10666	1.0060	1	0.9537	0.9376
kcal/m³ @ 15°C 760 mmHg dry	0.11245	0.11049	0.11184	1.0548	1.0486	1	0.9832
kcal/m³ @ 15°C 760 mmHg saturated	0.11437	0.11238	0.11375	1.0729	1.0665	1.0171	1

Notes

1. For equal conditions of temperature, pressure and water vapor content, the following factors apply:

   $1 Btu/ft^3 = 8.8992 kcal/m^3$

   $1 kcal/m^3 = 0.11237 Btu/ft^3$

2. To convert one unit of A to its equivalent heat value in terms of alternative units/conditions B, multiply by the appropriate factor.

energy may be reduced for the purposes of aggregation. Standard fuel for the purposes of European Economic Community statistics is considered to have a heat value of 7,000 kcal/kg.

Electricity

The basic unit of electric power is the *watt* (W), with the *kilowatt-hour* (kWh) as the derivative of most importance.

1kilowatt – hour(kWh) = 3,412.14 Btu
= 859.845 kcal
= 3,600 kJ

See note to Table 4.

Liquid fuels

In the case of liquid fuels, the actual calorific value of the particular fuel under consideration should be used where possible for conversions to other units, for example when calculating the natural gas equivalent of some specific application. For more general conversions, refer to Table 1.

Standard reference fuels

To convert different types and grades of fuel to a common basis for the purposes of presentation or comparison, it is the practice to express them in terms of some standard unit of a defined calorific value, for example:

Barrel of Oil Equivalent (b.o.e.), with a heat content of 5.8×10^6 Btu gross.
Tons of Oil Equivalent (t.o.e.), simply a hypothetical "ton of oil" with an average heating value of 43×10^6 Btu gross.
Oil equivalents may, if required, also be expressed in terms of standard fuel oil equivalent with a calorific value of 41.4×10^6 Btu per ton gross, i.e., 18,500 Btu/lb × 2,240.
Metric Tonne Coal Equivalent (t.c.e.), with a heat content of 27.337×10^6 Btu gross, i.e., 12,400 Btu/lb × 2,204.62.

The heat values of these different units and conversions between them are given in Table 3, together with natural gas equivalents.

Heat value conversions and natural gas equivalents of various fuel units

Table 3

	10^6 Btu gross	10^6 kcal gross	bbl oil equiv.	tons oil equiv.	metric t.c.e.	ft^3 natural gas	m^3 natural gas	10^6 kWh
1 barrel oil equivalent	5.8	1.462	1	0.1349	0.21217	5,800	155.50	0.0017
1 metric tonne coal equivalent	27.337	6.888	4.713	0.6357	1	27,337	732.90	0.0080
1 ton oil equivalent	43	10.836	7.414	1	1.5730	43,000	1,152.82	0.0126
1 ton fuel oil equivalent	41.4	10.433	7.138	0.9628	1.5144	41,400	1,109.92	0.0121

Notes
The natural gas equivalents above and in Tables 1 (notes) and 4 are based on gas with a gross heat content of 1,000 Btu/ft^3 measured dry at 60°F and 30 inHg, which within 0.2% is equal to 9,400 kcal/m^3 gross measured dry at 0°C and 760 mmHg. On this basis, 37.3 cubic feet = 1 cubic meter.

The abbreviation m.t.c.e. is often used for metric tonnes of coal equivalent or million tons coal equivalent. Neither practice is sound, as small m is the symbol for meter and not for metric or million. Instead, the ton unit used should be stated in full and the quantity expressed by the use of exponents; e.g., 36×10^6 metric t.c.e.

Conversions above and in Table 4 are based on gross calorific values in line with normal statistical practice both inside and outside the gas industry. Net calorific values may, however, be more appropriate in situations where it is desired to compare or convert fuels in terms of useful heat content, due allowance having been made for changes in end-use efficiency.

Conversion for daily/annual rates of energy consumption (gross heat basis)

Table 4

	1×10^6 metric t.c.e. per year	**1×10^6 ft^3 nat. gas per day**	**1×10^9 m^3 nat. gas per year**	**bbl daily oil equiv.**	**1×10^{12} (US trillion) ft^3 nat. gas per year**	**1×10^6 kWh per year**	**1×10^6 thermies per year**
1×10^6 metric t.c.e. per year =	1	74.896	0.7329	12,913	0.027337	8,012	6,891
100×10^6 ft^3 nat. gas per day =	1.3352	100	0.9785	17,241	0.0365	10,697	9,201
1×10^9 m^3 nat. gas per year =	1.3645	102.2	1	17,620	0.0373	10,932	9,403
1×10^4 bbl daily oil equiv. =	0.7744	58	0.5675	1×10^4	0.0212	6,204	5,336
1×10^{12} (US trillion) ft^3 nat. gas per year =	36,580	2,740	26,808	472,367	1	293,071	252,074
1×10^6 kWh per year =	0.1248×10^{-3}	0.9348×10^{-2}	0.9147×10^4	1.6118	0.341×10^{-5}	1	0.8601
1×10^6 thermies per year =	0.1451×10^{-3}	0.1087×10^{-1}	0.1064×10^{-3}	1.8739	0.397×10^{-5}	1.1626	1

Notes

The table is good for natural gas conversions on the basis of a gas of 1,000 Btu/ft^3 gross (9,400 kcal/Nm^3). See note to Table 3 for conditions of measurement. Corrections may be applied as follows:

Gases with other calorific values: apply a correction based on the ratio of the calorific values; e.g., to use for Groningen gas, multiply the horizontal figures shown against $1 \times 10^9 m^3$ per year by 8,420 ÷ 9,400, i.e., 0.896 and the corresponding column by 1.116.

Gases measured at other conditions of temperature and pressure: apply a correction based on the appropriate factor shown in Table 2; e.g., for gas measured saturated at 60°F and 30_6 inHg, multiply the horizontal figures shown against $100 \times 10^6 ft^3$ per day by 0.9826 and the corresponding column by 1.0177.

Calorific value and conditions of measurement different: apply a combined correction; e.g., to use for Gronigen gas measured dry at 15°C and 760 mmHg (m^3_s), multiply the horizontal figures shown against $1 \times 10^9 m^3$ by 0.948 × 8,420 = 9,400, i.e., 0.849 and the corresponding column by the reciprocal of this number, i.e., 1.178.

To use the table for 1×10^6 *therms per year*, multiply U.S. trillion row by 10^{-4} and the corresponding column by 10^4.

To use the table for 1×10^6 *tons oil equivalent*, multiply the horizontal figures shown against 1×10^6 metric t.c.e. per year by 1.573 and the corresponding column by 0.6357.

The factors relating to kilowatt-hours are based on the heat content of the electricity when used and not to the corresponding power station intakes of coal, oil, or natural gas, which would need to reflect station-generating efficiencies and load factor.

Weight of water per cubic foot at various temperatures

Temperature °F.	Weight per Cubic Foot, Lbs.	Temperature °F.	Weight per Cubic Foot, Lbs.	Temperature °F.	Weight per Cubic Foot, Lbs.	Temperature °F	Weight per Cubic Foot, Lbs.	Temperature °F	Weight per Cubic Foot, Lbs.
32	62.42	150	61.18	260	58.55	380	54.36	500	48.7
40	62.42	160	60.98	270	58.26	390	53.94	510	48.1
50	62.41	170	60.77	280	57.96	400	53.5	520	47.6
60	62.37	180	60.55	290	57.65	410	53.0	530	47.0
70	62.31	190	60.32	300	57.33	420	52.6	540	46.3
80	62.23	200	60.12	310	57.00	430	52.2	550	45.6
90	62.13	210	59.88	320	56.66	440	51.7	560	44.9
100	62.02	212	59.83	330	56.30	450	51.2	570	44.1
110	61.89	220	59.63	340	55.94	460	50.7	580	43.3
120	61.74	230	59.37	350	55.57	470	50.2	590	42.6
130	61.56	240	59.11	360	55.18	480	49.7	600	41.8
140	61.37	250	58.83	370	54.78	490	49.2	...	...

From "National Pipe Standards," The National Tube Co.

Engineering constants

Atmosphere:

Theoret. air to burn 1 lb. carbon = 11.521bs.
Vol. of 1 lb air at 32 deg. F and 14.7 lbs. per sq. in. = 12.387 cu. ft.
Vol. of 1 lb. air at 62 deg. F and 14.7 lb. per sq. in. = 13.144 cu. ft.
14.223 lb. per sq. in. = 1 Kg. per sq. cm. = 1 metric atm.
28.8 = equivalent mol. wgt. of air
30 in. Hg. at 62 deg. = atm. press
0.07608 lb. = wgt. 1 cu. ft. air at 62 deg. F and 14.7 lb. per sq. in.
0.08073 lb. = wgt. 1 cu. ft. air at 32 deg. F and 14.7 lb. per sq. in.

Heat:

8 lb. oxygen required to burn 1 lb. hydrogen (H)
777.5 ft. lb. = 1 B.T.U.
62,000 B.T.U. = cal. val. hydrogen (H)
14,600 B.T.U. per lb. cal. val. of carbon (C)
1.8 F dog. = 1 C deg.
2,547 B.T.U. per hr. = 1 h.p.
−270 deg. C = absolute zero
3,145 B.T.U. = 1 kw.hr.
34.56 lb. = wt. air to burn 1 lb. hydrogen (H)
3.9683 B.T.U. = 1 Kg. calorie
4,050 B.T.U. = cal. val. of sulphur (S)
4.32 lb. = wgt. air to burn 1 lb. sulphur (S)

Water:

Freezing point = 0 deg.C, 32 deg. F
Boiling point, atm. press= 100 deg. C, 212 deg. F
Water power available from 1 cu.ft.-sec. falling 1 ft. = 0.1134 h.p.
1.134 ft. at 62 deg. F = 1 in. Hg. at 62 deg. F
2.309 ft. at 62 deg. F = 1 lb. per sq. in.
27.71 in. at 62 deg. F = 1 lb. per sq.in.
33.947 ft. at 62 deg. F = atm. press.
5.196 lb. per sq.ft. = 1 in. at 62 deg. F
59.76 lb. = wgt. 1 cu. ft. at 212 deg. F
62.355 lb. = wgt. 1 cu. ft. at 62 deg. F
1 in. at 62 deg. F = 0.0735 in. Hg. at 62 deg. F
8.3356 lb. = wgt. 1 gal. at 62 deg. F
970.4 B.T.U. = latent heat of evap. at 212 deg. F

Steam:

Total heat sat. at atm. press = 1,150.4 B.T.U.
0.47 B.TU. per lb. deg. F = approx. specific heat of superheated steam at atm. press.

Surfaces:

1,273,239 circular mils = 1 sq. in.
1.4142 = square root of 2
144 square ins. = 1 sq. ft.
1,728 cu. in. = 1 cu. ft.
1.7321 = square root of 3
2.54 cm = 1 in.
57.296 deg. = 1 radian (angle)
0.0624281 b. per cu. ft. = 1 Kg. per cu. meter
area of circle = 0.7854 (= 3.1416 ÷ 4) × diameter squared

Miscellaneous:

2,116.3 lb. per sq.ft. = atm. press
2.3026 × log 10a log ea
2.7183 = c = base hyperbolic logs
288,000 B.T.U. per 24 hrs. = 1 ton of refrigeration
3.1416 = π = ratio of circumference of circle to diameter = ratio area of circle to square of radius
550ft.-lb. per sec. = 1 h.p.
58.349 grains per gal. = 1 gram per liter
61.023 cu.in. = 1 liter
745.7 watts = 1 h.p.
8,760 hr. = 1 year of 365 days
88 ft. per sec. (min.) = 1 mile per min. (hr.)
2.666 lb. = wgt. oxygen required to burn 1 lb. carbon

Mensuration units

Diameter of a circle × 3.1416 = Circumference
Radius of a circle × 6.283185 = Circumference

Square of the radius of a circle × 3.1416 = Area
Square of the diameter of a circle × 0.7854 = Area
Square of the circumference of a circle × 0.07985 = Area
Half the circumference of a circle × half its diameter = Area

Circumference of a circle × 0.159155 = Radius
Square root of the area of a circle × 0.56419 = Radius

Circumference of a circle × 0.31831 = Diameter
Square root of the area of a circle × 1.12838 = Diameter

Diameter of a circle × 0.86 = Side of an inscribed equilateral triangle
Diameter of a circle × 0.7071 = Side of an inscribed square
Circumference of a circle × 0.225 = Side of an inscribed square
Circumference of a circle × 0.282 = Side of an equal square
Diameter of a circle × 0.8862 = Side of an equal square

Base of a triangle × ½ the altitude = Area
Multiplying both diameter and .7854 together = Area of an ellipse
Surface of sphere × ⅙ of its diameter = Solidity

Circumference of a sphere × its diameter = Surface
Square of the diameter of a sphere × 3.1416 = Surface
Square of the circumference of a sphere × 0.3183 = Surface

Cube of the diameter of a sphere × 0.5236 = Solidity
Cube of the radius of a sphere × 4.1888 = Solidity
Cube of the circumference of a sphere × 0.016887 = Solidity

Square root of the surface of a sphere × 0.56419 = Diameter
Square root of the surface of a sphere × 1.772454 = Circumference

Cube root of the solidity of a sphere × 1.2407 = Diameter
Cube root of the solidity of a sphere × 3.8978 = Circumference

Radius of a sphere × 1.1547 = Side of inscribed cube
Square root of (⅓ of the square of) the diameter of a sphere = Side of inscribed cube

Area of its base × ⅓ of its altitude = Solidity of a cone or pyramid, whether round, square, or triangular

Area of one of its sides × 6 = Surface of cube
Altitude of trapezoid × ½ the sum of its parallel sides = Area

Minutes to decimal hours conversion table

Minutes	Hours	Minutes	Hours
1	.017	31	.517
2	.034	32	.534
3	.050	33	.550
4	.067	34	.567
5	.084	35	.584
6	.100	36	.600
7	.117	37	.617
8	.135	38	.634
9	.150	39	.650
10	.167	40	.667
11	.184	41	.684
12	.200	42	.700
13	.217	43	.717
14	.232	44	.734
15	.250	45	.750
16	.267	46	.767
17	.284	47	.784
18	.300	48	.800
19	.317	49	.817
20	.334	50	.834
21	.350	51	.850
22	.368	52	.867
23	.384	53	.884
24	.400	54	.900
25	.417	55	.917
26	.434	56	.934
27	.450	57	.950
28	.467	58	.967
29	.484	59	.984
30	.500	60	1.000

How to compare costs of gas and alternate fuels

Growing shortage of natural gas prompts consideration of practical replacements, including coal, heavy fuel oil, middle distillates, and LPG

Conversion from natural gas to other fuels has progressed from possibility to reality for many companies and will become necessary for many others in months and years ahead.

Fuels considered practical replacements for gas include coal, heavy fuel oils, middle distillates (such as kerosine-type turbo fuel and burner fuel oils) and liquefied petroleum gas (LPG).

To compare relative costs of fuels, the quoted prices normally are converted to the price per million British thermal units (Btu). With HHV being the higher (gross) heating value of the fuel, the following equations may be used.

Coal:

$$\text{Cents/million Btu} = \frac{\text{dollars/ton} \times 100 \times 10^6}{\text{lb/ton} \times \text{HHV}}$$

Heavy fuel oils:

$$\text{Cents/million Btu} = \frac{\text{dollars/barrel} \times 100 \times 10^6}{\text{gallons/barrel} \times \text{HHV}}$$

Middle distillates and LPG:

$$\text{Cents/million Btu} = \frac{\text{cents/gallon} \times 10^6}{\text{HHV}}$$

Natural gas:

$$\text{Cents/million Btu} = \frac{\text{cents/Mcf} \times 10^6}{\text{HHV} \times 1{,}000}$$

The approximate heat contents (HHV) of the fuels are as follows:

Coal:

Anthracite	13,900 Btu/lb
Bituminous	14,000 Btu/lb
Sub-bituminous	12,600 Btu/lb
Lignite	11,000 Btu/lb

Heavy fuel oils and middle distillates:

Kerosine	134,000 Btu/gal
No. 2 burner fuel oil	140,000 Btu/gal
No. 4 heavy fuel oil	144,000 Btu/gal
No. 5 heavy fuel oil	150,000 Btu/gal
No. 6 heavy fuel oil, 2.7% sulfur	152,000 Btu/gal
No. 6 heavy fuel oil, 0.3% sulfur	143,800 Btu/gal

Gas:

Natural	1,000 Btu/cubic ft
Liquefied butane	103,300 Btu/gal
Liquefied propane	91,600 Btu/gal

Typical characteristics of fuel oils

	Kerosine	No. 2 burner fuel
Gravity, °API	41.4	36.3
Pounds per gallon	6.814	7.022
Flash point, °F	133	165
Btu/gal.	134,000	140,000
Viscosity, SSU at 100°F	31.5	35.0
Pour point, °F	−30	0
Sulfur, wt.%	0.04	0.18
Carbon residue, wt. %	0.01	0.05
BS&W, %	–	1.30[2]

[1]Saybolt Furol Viscosity at 122°F.
[2]mg/100 ml.

LPG

Vapor pressure, psig at 100°F	188
Distillation, 95% point, °F	−40
Residue, vol. %	0.05
Oil-ring test	none

Index